THE VOCABULARY AND CONCEPTS OF ORGANIC CHEMISTRY

THE VOCABULARY AND CONCEPTS OF ORGANIC CHEMISTRY

SECOND EDITION

Milton Orchin
Roger S. Macomber
Allan R. Pinhas
R. Marshall Wilson

WILEY-INTERSCIENCE
A John Wiley & Sons, Inc., Publication

Published by John Wiley & Sons, Inc., Hoboken, New Jersey.
Published simultaneously in Canada.

Library of Congress Cataloging-in-Publication Data is available.

ISBN 0-471-68028-1

Printed in the United States of America
10 9 8 7 6 5 4 3 2 1

CONTENTS

1 Atomic Orbital Theory 1

2 Bonds Between Adjacent Atoms: Localized Bonding, Molecular Orbital Theory 25

3 Delocalized (Multicenter) Bonding 54

4 Symmetry Operations, Symmetry Elements, and Applications 83

5 Classes of Hydrocarbons 110

6 Functional Groups: Classes of Organic Compounds 139

7 Molecular Structure Isomers, Stereochemistry, and Conformational Analysis 221

8 Synthetic Polymers 291

9 Organometallic Chemistry 343

10 Separation Techniques and Physical Properties 387

11 Fossil Fuels and Their Chemical Utilization 419

12 Thermodynamics, Acids and Bases, and Kinetics 450

13 Reactive Intermediates (Ions, Radicals, Radical Ions, Electron-Deficient Species, Arynes) 505

14 Types of Organic Reaction Mechanisms 535

15 Nuclear Magnetic Resonance Spectroscopy 591

16 Vibrational and Rotational Spectroscopy: Infrared, Microwave, and Raman Spectra 657

17 Mass Spectrometry 703

18 Electronic Spectroscopy and Photochemistry 725

Name Index 833

Compound Index 837

General Index 849

PREFACE

It has been almost a quarter of a century since the first edition of our book *The Vocabulary of Organic Chemistry* was published. Like the vocabulary of every living language, old words remain, but new ones emerge. In addition to the new vocabulary, other important changes have been incorporated into this second edition. One of the most obvious of these is in the title, which has been expanded to *The Vocabulary and Concepts of Organic Chemistry* in recognition of the fact that in addressing the language of a science, we found it frequently necessary to define and explain the concepts that have led to the vocabulary. The second change from the first edition is authorship. Three of the original authors of the first edition have participated in this new version; the two lost collaborators were sorely missed. Professor Hans Zimmer died on June 13, 2001. His contribution to the first edition elevated its scholarship. He had an enormous grasp of the literature of organic chemistry and his profound knowledge of foreign languages improved our literary grasp. Professor Fred Kaplan also made invaluable contributions to our first edition. His attention to small detail, his organizational expertise, and his patient examination of the limits of definitions, both inclusive and exclusive, were some of the many advantages of his co-authorship. We regret that his other interests prevented his participation in the present effort. However, these unfortunate losses were more than compensated by the addition of a new author, Professor Allan Pinhas, whose knowledge, enthusiasm, and matchless energy lubricated the entire process of getting this edition to the publisher.

Having addressed the changes in title and authorship, we need to describe the changes in content. Two major chapters that appeared in the first edition no longer appear here: "Named Organic Reactions" and "Natural Products." Since 1980, several excellent books on named organic reactions and their mechanisms have appeared, and some of us felt our treatment would be redundant. The second deletion, dealing with natural products, we decided would better be treated in an anticipated second volume to this edition that will address not only this topic, but also the entire new emerging interest in biological molecules. These deletions made it possible to include other areas of organic chemistry not covered in our first edition, namely the powerful spectroscopic tools so important in structure determination, infrared spectroscopy, NMR, and mass spectroscopy, as well as ultraviolet spectroscopy and photochemistry. In addition to the new material, we have updated material covered in the first edition with the rearrangement of some chapters, and of course, we have taken advantage of reviews and comments on the earlier edition to revise the discussion where necessary.

The final item that warrants examination is perhaps one that should take precedence over others. Who should find this book useful? To answer this important question, we turn to the objective of the book, which is to identify the fundamental vocabulary and concepts of organic chemistry and present concise, accurate descriptions of them with examples when appropriate. It is not intended to be a dictionary, but is organized into a sequence of chapters that reflect the way the subject is taught. Related terms appear in close proximity to each other, and hence, fine distinctions become understandable. Students and instructors may appreciate the concentration of subject matter into the essential aspects of the various topics covered. In addition, we hope the book will appeal to, and prove useful to, many others in the chemical community who either in the recent past, or even remote past, were familiar with the topics defined, but whose precise knowledge of them has faded with time.

In the course of writing this book, we drew generously from published books and articles, and we are grateful to the many authors who unknowingly contributed their expertise. We have also taken advantage of the special knowledge of some of our colleagues in the Department of Chemistry and we acknowledge them in appropriate chapters.

MILTON ORCHIN
ROGER S. MACOMBER
ALLAN R. PINHAS
R. MARSHALL WILSON

1 Atomic Orbital Theory

1.1	Photon (Quantum)	3
1.2	Bohr or Planck–Einstein Equation	3
1.3	Planck's Constant h	3
1.4	Heisenberg Uncertainty Principle	3
1.5	Wave (Quantum) Mechanics	4
1.6	Standing (or Stationary) Waves	4
1.7	Nodal Points (Planes)	5
1.8	Wavelength λ	5
1.9	Frequency ν	5
1.10	Fundamental Wave (or First Harmonic)	6
1.11	First Overtone (or Second Harmonic)	6
1.12	Momentum (**P**)	6
1.13	Duality of Electron Behavior	7
1.14	de Broglie Relationship	7
1.15	Orbital (Atomic Orbital)	7
1.16	Wave Function	8
1.17	Wave Equation in One Dimension	9
1.18	Wave Equation in Three Dimensions	9
1.19	Laplacian Operator	9
1.20	Probability Interpretation of the Wave Function	9
1.21	Schrödinger Equation	10
1.22	Eigenfunction	10
1.23	Eigenvalues	11
1.24	The Schrödinger Equation for the Hydrogen Atom	11
1.25	Principal Quantum Number n	11
1.26	Azimuthal (Angular Momentum) Quantum Number l	11
1.27	Magnetic Quantum Number m_l	12
1.28	Degenerate Orbitals	12
1.29	Electron Spin Quantum Number m_s	12
1.30	*s* Orbitals	12
1.31	1*s* Orbital	12
1.32	2*s* Orbital	13
1.33	*p* Orbitals	14
1.34	Nodal Plane or Surface	14
1.35	2*p* Orbitals	15
1.36	*d* Orbitals	16
1.37	*f* Orbitals	16
1.38	Atomic Orbitals for Many-Electron Atoms	17

The Vocabulary and Concepts of Organic Chemistry, Second Edition, by Milton Orchin, Roger S. Macomber, Allan Pinhas, and R. Marshall Wilson

1.39 Pauli Exclusion Principle 17
1.40 Hund's Rule 17
1.41 Aufbau (*Ger.* Building Up) Principle 17
1.42 Electronic Configuration 18
1.43 Shell Designation 18
1.44 The Periodic Table 19
1.45 Valence Orbitals 21
1.46 Atomic Core (or Kernel) 22
1.47 Hybridization of Atomic Orbitals 22
1.48 Hybridization Index 23
1.49 Equivalent Hybrid Atomic Orbitals 23
1.50 Nonequivalent Hybrid Atomic Orbitals 23

The detailed study of the structure of atoms (as distinguished from molecules) is largely the domain of the physicist. With respect to atomic structure, the interest of the chemist is usually confined to the behavior and properties of the three fundamental particles of atoms, namely the electron, the proton, and the neutron. In the model of the atom postulated by Niels Bohr (1885–1962), electrons surrounding the nucleus are placed in circular orbits. The electrons move in these orbits much as planets orbit the sun. In rationalizing atomic emission spectra of the hydrogen atom, Bohr assumed that the energy of the electron in different orbits was quantized, that is, the energy did not increase in a continuous manner as the orbits grew larger, but instead had discrete values for each orbit. Bohr's use of classical mechanics to describe the behavior of small particles such as electrons proved unsatisfactory, particularly because this model did not take into account the uncertainty principle. When it was demonstrated that the motion of electrons had properties of waves as well as of particles, the so-called dual nature of electronic behavior, the classical mechanical approach was replaced by the newer theory of quantum mechanics.

According to quantum mechanical theory the behavior of electrons is described by wave functions, commonly denoted by the Greek letter ψ. The physical significance of ψ resides in the fact that its square multiplied by the size of a volume element, $\psi^2 d\tau$, gives the probability of finding the electron in a particular element of space surrounding the nucleus of the atom. Thus, the Bohr model of the atom, which placed the electron in a fixed orbit around the nucleus, was replaced by the quantum mechanical model that defines a region in space surrounding the nucleus (an atomic orbital rather than an orbit) where the probability of finding the electron is high. It is, of course, the electrons in these orbitals that usually determine the chemical behavior of the atoms and so knowledge of the positions and energies of the electrons is of great importance. The correlation of the properties of atoms with their atomic structure expressed in the periodic law and the Periodic Table was a milestone in the development of chemical science.

Although most of organic chemistry deals with molecular orbitals rather than with isolated atomic orbitals, it is prudent to understand the concepts involved in atomic orbital theory and the electronic structure of atoms before moving on to consider the behavior of electrons shared between atoms and the concepts of molecular orbital theory.

1.1 PHOTON (QUANTUM)

The most elemental unit or particle of electromagnetic radiation. Associated with each photon is a discrete quantity or quantum of energy.

1.2 BOHR OR PLANCK–EINSTEIN EQUATION

$$E = h\nu = hc/\lambda \tag{1.2}$$

This fundamental equation relates the energy of a photon E to its frequency ν (see Sect. 1.9) or wavelength λ (see Sect. 1.8). Bohr's model of the atom postulated that the electrons of an atom moved about its nucleus in circular orbits, or as later suggested by Arnold Summerfeld (1868–1951), in elliptical orbits, each with a certain "allowed" energy. When subjected to appropriate electromagnetic radiation, the electron may absorb energy, resulting in its promotion (excitation) from one orbit to a higher (energy) orbit. The frequency of the photon absorbed must correspond to the energy difference between the orbits, that is, $\Delta E = h\nu$. Because Bohr's postulates were based in part on the work of Max Planck (1858–1947) and Albert Einstein (1879–1955), the Bohr equation is alternately called the Planck–Einstein equation.

1.3 PLANCK'S CONSTANT *h*

The proportionality constant $h = 6.6256 \times 10^{-27}$ erg seconds (6.6256×10^{-34} J s), which relates the energy of a photon E to its frequency ν (see Sect. 1.9) in the Bohr or Planck–Einstein equation. In order to simplify some equations involving Planck's constant h, a modified constant called $\hbar$, where $\hbar = h/2\pi$, is frequently used.

1.4 HEISENBERG UNCERTAINTY PRINCIPLE

This principle as formulated by Werner Heisenberg (1901–1976), states that the properties of small particles (electrons, protons, etc.) cannot be known precisely at any particular instant of time. Thus, for example, both the exact momentum p and the exact position x of an electron cannot both be measured simultaneously. The product of the uncertainties of these two properties of a particle must be on the order of Planck's constant: $\Delta p \cdot \Delta x = h/2\pi$, where Δp is the uncertainty in the momentum, Δx the uncertainty in the position, and h Planck's constant.

A corollary to the uncertainty principle is its application to very short periods of time. Thus, $\Delta E \cdot \Delta t = h/2\pi$, where ΔE is the uncertainty in the energy of the electron and Δt the uncertainty in the time that the electron spends in a particular energy state. Accordingly, if Δt is very small, the electron may have a wide range of energies. The uncertainty principle addresses the fact that the very act of performing a measurement of the properties of small particles perturbs the system. The uncertainty principle is at the heart of quantum mechanics; it tells us that the position of an electron is best expressed in terms of the *probability* of finding it in a particular region in space, and thus, eliminates the concept of a well-defined trajectory or orbit for the electron.

1.5 WAVE (QUANTUM) MECHANICS

The mathematical description of very small particles such as electrons in terms of their wave functions (see Sect. 1.15). The use of wave mechanics for the description of electrons follows from the experimental observation that electrons have both wave as well as particle properties. The wave character results in a probability interpretation of electronic behavior (see Sect. 1.20).

1.6 STANDING (OR STATIONARY) WAVES

The type of wave generated, for example, by plucking a string or wire stretched between two fixed points. If the string is oriented horizontally, say, along the x-axis, the waves moving toward the right fixed point will encounter the reflected waves moving in the opposite direction. If the forward wave and the reflected wave have the same amplitude at each point along the string, there will be a number of points along the string that will have no motion. These points, in addition to the fixed anchors at the ends, correspond to nodes where the amplitude is zero. Half-way between the nodes there will be points where the amplitude of the wave will be maximum. The variations of amplitude are thus a function of the distance along x. After the plucking, the resultant vibrating string will appear to be oscillating up and down between the fixed nodes, but there will be no motion along the length of the string—hence, the name standing or stationary wave.

Example. See Fig. 1.6.

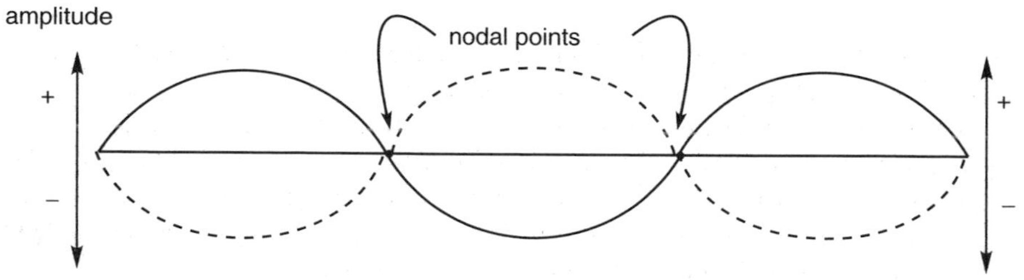

Figure 1.6. A standing wave; the two curves represent the time-dependent motion of a string vibrating in the third harmonic or second overtone with four nodes.

1.7 NODAL POINTS (PLANES)

The positions or points on a standing wave where the amplitude of the wave is zero (Fig. 1.6). In the description of orbitals, the node represent a point or plane where a change of sign occurs.

1.8 WAVELENGTH λ

The minimum distance between nearest-neighbor peaks, troughs, nodes or equivalent points of the wave.

Example. The values of λ, as shown in Fig. 1.8.

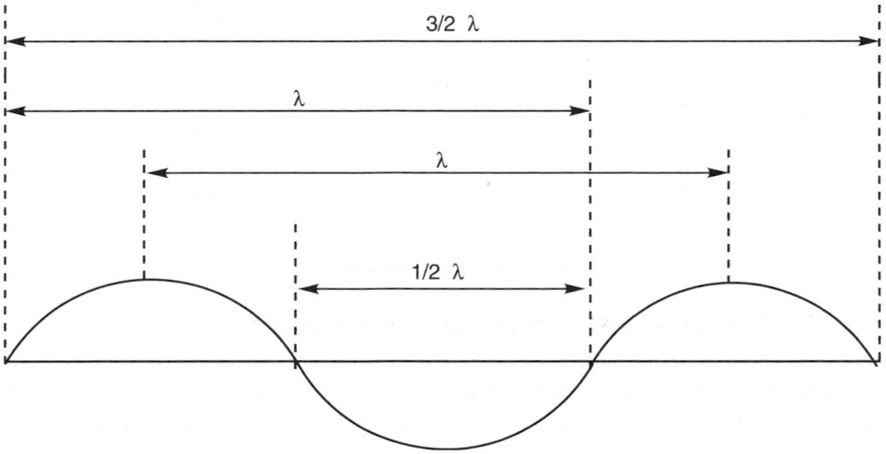

Figure 1.8. Determination of the wavelength λ of a wave.

1.9 FREQUENCY ν

The number of wavelengths (or cycles) in a light wave that pass a particular point per unit time. Time is usually measured in seconds; hence, the frequency is expressed in s^{-1}. The unit of frequency, equal to cycles per second, is called the Hertz (Hz). Frequency is inversely proportional to wavelength; the proportionality factor is the speed of light c (3×10^{10} cm s^{-1}). Hence, $\nu = c/\lambda$.

Example. For light with λ equal to 300 nm (300×10^{-7} cm), the frequency $\nu = (3 \times 10^{10}\ \text{cm s}^{-1})/(300 \times 10^{-7}\ \text{cm}) = 1 \times 10^{15}\ \text{s}^{-1}$.

1.10 FUNDAMENTAL WAVE (OR FIRST HARMONIC)

The stationary wave with no nodal point other than the fixed ends. It is the wave from which the frequency ν' of all other waves in a set is generated by multiplying the fundamental frequency ν by an integer n:

$$\nu' = n\nu \tag{1.10}$$

Example. In the fundamental wave, $\lambda/2$ in Fig. 1.10, the amplitude may be considered to be oriented upward and to continuously increase from either fixed end, reaching a maximum at the midpoint. In this "well-behaved" wave, the amplitude is zero at each end and a maximum at the center.

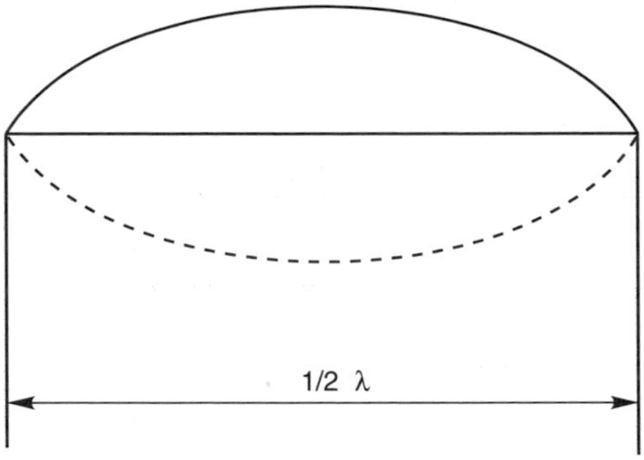

Figure 1.10. The fundamental wave.

1.11 FIRST OVERTONE (OR SECOND HARMONIC)

The stationary wave with one nodal point located at the midpoint ($n = 2$ in the equation given in Sect. 1.10). It has half the wavelength and twice the frequency of the first harmonic.

Example. In the first overtone (Fig. 1.11), the nodes are located at the ends and at the point half-way between the ends, at which point the amplitude changes direction. The two equal segments of the wave are portions of a single wave; they are not independent. The two maximum amplitudes come at exactly equal distances from the ends but are of opposite signs.

1.12 MOMENTUM (P)

This is the vectorial property (i.e., having both magnitude and direction) of a moving particle; it is equal to the mass m of the particle times its velocity $\mathbf{v}$:

$$\mathbf{p} = m\mathbf{v} \tag{1.12}$$

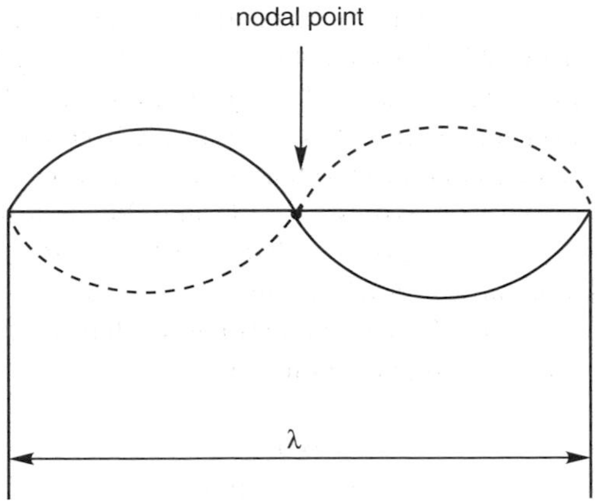

Figure 1.11. The first overtone (or second harmonic) of the fundamental wave.

1.13 DUALITY OF ELECTRONIC BEHAVIOR

Particles of small mass such as electrons may exhibit properties of either particles (they have momentum) or waves (they can be defracted like light waves). A single experiment may demonstrate either particle properties or wave properties of electrons, but not both simultaneously.

1.14 DE BROGLIE RELATIONSHIP

The wavelength of a particle (an electron) is determined by the equation formulated by Louis de Broglie (1892–1960):

$$\lambda = h/p = h/mv \tag{1.14}$$

where h is Planck's constant, m the mass of the particle, and v its velocity. This relationship makes it possible to relate the momentum p of the electron, a particle property, with its wavelength λ, a wave property.

1.15 ORBITAL (ATOMIC ORBITAL)

A wave description of the size, shape, and orientation of the region in space available to an electron; each orbital has a specific energy. The position (actually the probability amplitude) of the electron is defined by its coordinates in space, which in Cartesian coordinates is indicated by $\psi(x, y, z)$. ψ cannot be measured directly; it is a mathematical tool. In terms of spherical coordinates, frequently used in calculations, the wave function is indicated by $\psi(r, \theta, \varphi)$, where r (Fig. 1.15) is the radial distance of a point from the origin, θ is the angle between the radial line and the

z-axis, and φ is the angle between the x-axis and the projection of the radial line on the xy-plane. The relationship between the two coordinate systems is shown in Fig. 1.15. An orbital centered on a single atom (an *atomic orbital*) is frequently denoted as ϕ (phi) rather than ψ (psi) to distinguish it from an orbital centered on more than one atom (a molecular orbital) that is almost always designated ψ.

The projection of r on the z-axis is $z = OB$, and OBA is a right angle. Hence, $\cos\theta = z/r$, and thus, $z = r\cos\theta$. Cos $\varphi = x/OC$, but $OC = AB = r\sin\theta$. Hence, $x = r\sin\theta\cos\varphi$. Similarly, $\sin\varphi = y/AB$; therefore, $y = AB\ \sin\ \varphi = r\ \sin\theta\sin\varphi$. Accordingly, a point (x, y, z) in Cartesian coordinates is transformed to the spherical coordinate system by the following relationships:

$$z = r\cos\theta$$
$$y = r\sin\theta\sin\varphi$$
$$x = r\sin\theta\cos\varphi$$

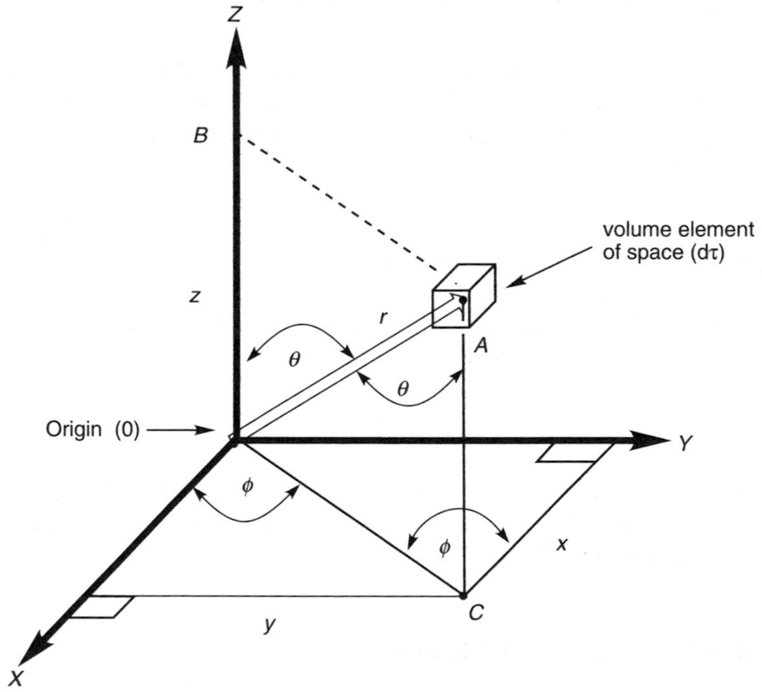

Figure 1.15. The relationship between Cartesian and polar coordinate systems.

1.16 WAVE FUNCTION

In quantum mechanics, the wave function is synonymous with an orbital.

1.17 WAVE EQUATION IN ONE DIMENSION

The mathematical description of an orbital involving the amplitude behavior of a wave. In the case of a one-dimensional standing wave, this is a second-order differential equation with respect to the amplitude:

$$d^2f(x)/dx^2 + (4\pi^2/\lambda^2)\,f(x) = 0 \tag{1.17}$$

where λ is the wavelength and the amplitude function is $f(x)$.

1.18 WAVE EQUATION IN THREE DIMENSIONS

The function $f(x, y, z)$ for the wave equation in three dimensions, analogous to $f(x)$, which describes the amplitude behavior of the one-dimensional wave. Thus, $f(x, y, z)$ satisfies the equation

$$\partial^2f(x)/\partial x^2 + \partial^2f(y)/\partial y^2 + \partial^2f(z)/\partial z^2 + (4\pi^2/\lambda^2)\,f(x, y, z) = 0 \tag{1.18}$$

In the expression $\partial^2f(x)/\partial x^2$, the portion $\partial^2/\partial x^2$ is an operator that says "partially differentiate twice with respect to x that which follows."

1.19 LAPLACIAN OPERATOR

The sum of the second-order differential operators with respect to the three Cartesian coordinates in Eq. 1.18 is called the Laplacian operator (after Pierre S. Laplace, 1749–1827), and it is denoted as ∇^2 (del squared):

$$\nabla^2 = \partial^2/\partial x^2 + \partial^2/\partial y^2 + \partial^2/\partial z^2 \tag{1.19a}$$

which then simplifies Eq. 1.18 to

$$\nabla^2f(x, y, z) + (4\pi^2/\lambda^2)\,f(x, y, z) = 0 \tag{1.19b}$$

1.20 PROBABILITY INTERPRETATION OF THE WAVE FUNCTION

The wave function (or orbital) $\psi(r)$, because it is related to the amplitude of a wave that determines the location of the electron, can have either negative or positive values. However, a *probability*, by definition, must always be positive, and in the present case this can be achieved by squaring the amplitude. Accordingly, the probability of finding an electron in a specific volume element of space $d\tau$ at a distance r from the nucleus is $\psi(r)^2d\tau$. Although ψ, the orbital, has mathematical significance (in that it can have negative and positive values), ψ^2 has physical significance and is always positive.

1.21 SCHRÖDINGER EQUATION

This is a differential equation, formulated by Erwin Schrödinger (1887–1961), whose solution is the wave function for the system under consideration. This equation takes the same form as an equation for a standing wave. It is from this form of the equation that the term *wave mechanics* is derived. The similarity of the Schrödinger equation to a wave equation (Sect. 1.18) is demonstrated by first substituting the de Broglie equation (1.14) into Eq. 1.19b and replacing f by ϕ:

$$\nabla^2\phi + (4\pi^2 m^2 v^2/h^2)\phi = 0 \tag{1.21a}$$

To incorporate the total energy E of an electron into this equation, use is made of the fact that the total energy is the sum of the potential energy V, plus the kinetic energy, $1/2\ mv^2$, or

$$v^2 = 2(E - V)/m \tag{1.21b}$$

Substituting Eq. 1.21b into Eq. 1.21a gives Eq. 1.21c:

$$\nabla^2\phi + (8\pi^2 m/h^2)(E - V)\phi = 0 \tag{1.21c}$$

which is the Schrödinger equation.

1.22 EIGENFUNCTION

This is a hybrid German-English word that in English might be translated as "characteristic function"; it is an acceptable solution of the wave equation, which can be an orbital. There are certain conditions that must be fulfilled to obtain "acceptable" solutions of the wave equation, Eq. 1.17 [e.g., $f(x)$ must be zero at each end, as in the case of the vibrating string fixed at both ends; this is the so-called boundary condition]. In general, whenever some mathematical operation is done on a function and the same function is regenerated multiplied by a constant, the function is an eigenfunction, and the constant is an eigenvalue. Thus, wave Eq. 1.17 may be written as

$$d^2f(x)/dx^2 = -(4\pi^2/\lambda^2)\,f(x) \tag{1.22}$$

This equation is an eigenvalue equation of the form:

(Operator) (eigenfunction) = (eigenvalue) (eigenfunction)

where the operator is (d^2/dx^2), the eigenfunction is $f(x)$, and the eigenvalue is $(4\pi^2/\lambda^2)$. Generally, it is implied that wave functions, hence orbitals, are eigenfunctions.

1.23 EIGENVALUES

The values of λ calculated from the wave equation, Eq. 1.17. If the eigenfunction is an orbital, then the eigenvalue is related to the orbital energy.

1.24 THE SCHRÖDINGER EQUATION FOR THE HYDROGEN ATOM

An (eigenvalue) equation, the solutions of which in spherical coordinates are

$$\phi(r, \theta, \varphi) = R(r)\, \Theta(\theta)\, \Phi(\varphi) \tag{1.24}$$

The eigenfunctions ϕ, also called orbitals, are functions of the three variables shown, where r is the distance of a point from the origin, and θ and φ are the two angles required to locate the point (see Fig. 1.15). For some purposes, the *spatial* or radial part and the *angular* part of the Schrödinger equation are separated and treated independently. Associated with each eigenfunction (orbital) is an eigenvalue (orbital energy). An exact solution of the Schrödinger equation is possible only for the hydrogen atom, or any one-electron system. In many-electron systems wave functions are generally approximated as products of modified one-electron functions (orbitals). Each solution of the Schrödinger equation may be distinguished by a set of three quantum numbers, n, l, and m, that arise from the boundary conditions.

1.25 PRINCIPAL QUANTUM NUMBER *n*

An integer 1, 2, 3, . . . , that governs the size of the orbital (wave function) and determines the energy of the orbital. The value of n corresponds to the number of the shell in the Bohr atomic theory and the larger the n, the higher the energy of the orbital and the farther it extends from the nucleus.

1.26 AZIMUTHAL (ANGULAR MOMENTUM) QUANTUM NUMBER *l*

The quantum number with values of $l = 0, 1, 2, \ldots, (n - 1)$ that determines the shape of the orbital. The value of l implies particular angular momenta of the electron resulting from the shape of the orbital. Orbitals with the azimuthal quantum numbers $l = 0$, 1, 2, and 3 are called *s*, *p*, *d*, and *f* orbitals, respectively. These orbital designations are taken from atomic spectroscopy where the words "sharp", "principal", "diffuse", and "fundamental" describe lines in atomic spectra. This quantum number does not enter into the expression for the energy of an orbital. However, when electrons are placed in orbitals, the energy of the orbitals (and hence the energy of the electrons in them) is affected so that orbitals with the same principal quantum number n may vary in energy.

Example. An electron in an orbital with a principal quantum number of $n = 2$ can take on l values of 0 and 1, corresponding to 2s and 2p orbitals, respectively. Although these orbitals have the same principal quantum number and, therefore, the same energy when calculated for the single electron hydrogen atom, for the many-electron atoms, where electron–electron interactions become important, the 2p orbitals are higher in energy than the 2s orbitals.

1.27 MAGNETIC QUANTUM NUMBER m_l

This is the quantum number having values of the azimuthal quantum number from $+l$ to $-l$ that determines the orientation in space of the orbital angular momentum; it is represented by m_l.

Example. When $n = 2$ and $l = 1$ (the p orbitals), m_l may thus have values of $+1$, 0, -1, corresponding to three 2p orbitals (see Sect. 1.35). When $n = 3$ and $l = 2$, m_l has the values of $+2$, $+1$, 0, -1, -2 that describe the five 3d orbitals (see Sect. 1.36).

1.28 DEGENERATE ORBITALS

Orbitals having equal energies, for example, the three 2p orbitals.

1.29 ELECTRON SPIN QUANTUM NUMBER m_s

This is a measure of the intrinsic angular momentum of the electron due to the fact that the electron itself is spinning; it is usually designated by m_s and may only have the value of 1/2 or $-1/2$.

1.30 s ORBITALS

Spherically symmetrical orbitals; that is, ϕ is a function of $R(r)$ only. For s orbitals, $l = 0$ and, therefore, electrons in such orbitals have an orbital magnetic quantum number m_l equal to zero.

1.31 1s ORBITAL

The lowest-energy orbital of any atom, characterized by $n = 1$, $l = m_l = 0$. It corresponds to the fundamental wave and is characterized by spherical symmetry and no nodes. It is represented by a projection of a sphere (a circle) surrounding the nucleus, within which there is a specified probability of finding the electron.

Example. The numerical probability of finding the hydrogen electron within spheres of various radii from the nucleus is shown in Fig. 1.31*a*. The circles represent contours of probability on a plane that bisects the sphere. If the contour circle of 0.95 probability is chosen, the electron is 19 times as likely to be inside the corresponding sphere with a radius of 1.7 Å as it is to be outside that sphere. The circle that is usually drawn, Fig. 1.31*b*, to represent the 1*s* orbital is meant to imply that there is a high, but unspecified, probability of finding the electron in a sphere, of which the circle is a cross-sectional cut or projection.

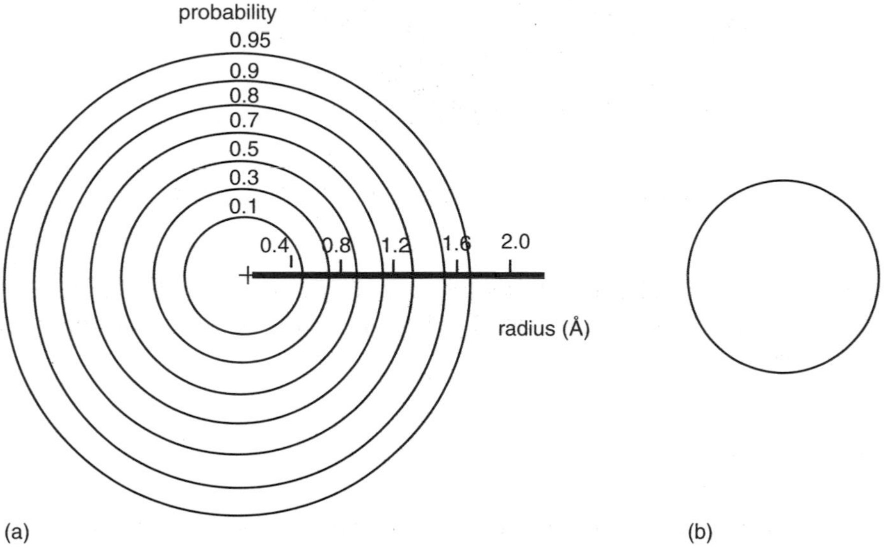

Figure 1.31. (*a*) The probability contours and radii for the hydrogen atom, the probability at the nucleus is zero. (*b*) Representation of the 1*s* orbital.

1.32 2*s* ORBITAL

The spherically symmetrical orbital having one spherical nodal surface, that is, a surface on which the probability of finding an electron is zero. Electrons in this orbital have the principal quantum number $n = 2$, but have no angular momentum, that is, $l = 0$, $m_l = 0$.

Example. Figure 1.32 shows the probability distribution of the 2*s* electron as a cross section of the spherical 2*s* orbital. The 2*s* orbital is usually drawn as a simple circle of arbitrary diameter, and in the absence of a drawing for the 1*s* orbital for comparison,

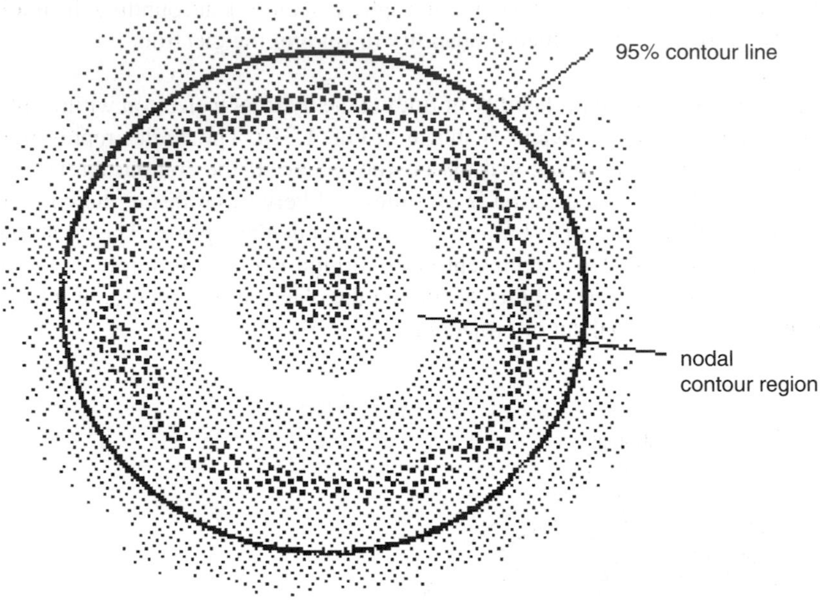

Figure 1.32. Probability distribution ψ^2 for the 2*s* orbital.

the two would be indistinguishable despite the larger size of the 2*s orbital* and the fact that there is a nodal surface within the 2*s* sphere that is not shown in the simple circular representation.

1.33 *p* ORBITALS

These are orbitals with an angular momentum *l* equal to 1; for each value of the principal quantum number *n* (except for $n = 1$), there will be three *p* orbitals corresponding to $m_l = +1, 0, -1$. In a useful convention, these three orbitals, which are mutually perpendicular to each other, are oriented along the three Cartesian coordinate axes and are therefore designated as p_x, p_y, and p_z. They are characterized by having one nodal plane.

1.34 NODAL PLANE OR SURFACE

A plane or surface associated with an orbital that defines the locus of points for which the probability of finding an electron is zero. It has the same meaning in three dimensions that the nodal point has in the two-dimensional standing wave (see Sect. 1.7) and is associated with a change in sign of the wave function.

1.35 2p ORBITALS

The set of three degenerate (equal energy) atomic orbitals having the principal quantum number (n) of 2, an azimuthal quantum number (l) of 1, and magnetic quantum numbers (m_l) of $+1$, 0, or -1. Each of these orbitals has a nodal plane.

Example. The 2*p* orbitals are usually depicted so as to emphasize their angular dependence, that is, $R(r)$ is assumed constant, and hence are drawn for convenience as a planar cross section through a three-dimensional representation of $\Theta(\theta)\Phi(\varphi)$. The planar cross section of the $2p_z$ orbital, $\varphi = 0$, then becomes a pair of circles touching at the origin (Fig. 1.35*a*). In this figure the wave function (without proof) is $\phi = \Theta(\theta) = (\sqrt{6}/2)\cos\theta$. Since $\cos\theta$, in the region $90° < \theta < 270°$, is negative, the top circle is positive and the bottom circle negative. However, the physically significant property of an orbital ϕ is its square, ϕ^2; the plot of $\phi^2 = \Theta^2(\theta) = 3/2\cos^2\theta$ for the p_z orbital is shown in Fig. 1.35*b*, which represents the volume of space in which there is a high probability of finding the electron associated with the p_z orbital. The shape of this orbital is the familiar elongated dumbbell with both lobes having a positive sign. In most common drawings of the *p* orbitals, the shape of ϕ^2, the physically significant function, is retained, but the plus and minus signs are placed in the lobes to emphasize the nodal property, (Fig. 1.35*c*). If the function $R(r)$ is included, the oval-shaped contour representation that results is shown in Fig. 1.35*d*, where $\phi^2(p_z)$ is shown as a cut in the *yz*-plane.

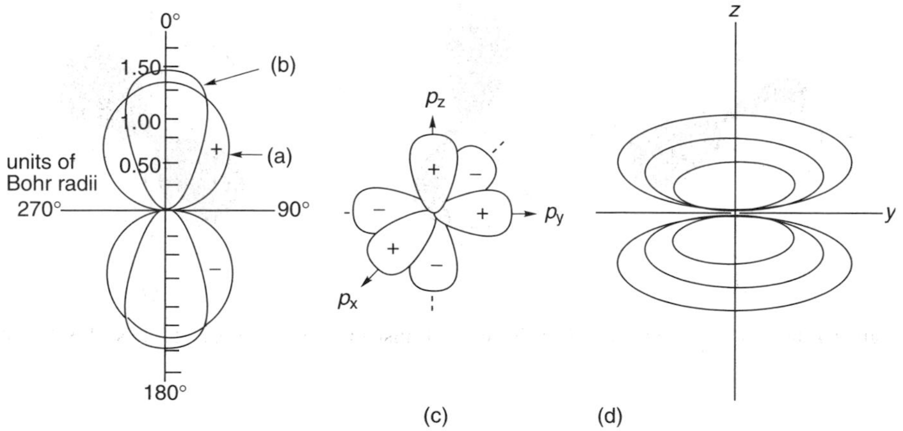

Figure 1.35. (*a*) The angular dependence of the p_z orbital; (*b*) the square of (*a*); (*c*) the common depiction of the three 2*p* orbitals; and (*d*) contour diagram including the radial dependence of ϕ.

1.36 *d* ORBITALS

Orbitals having an angular momentum l equal to 2 and, therefore, magnetic quantum numbers, (m_l) of $+2$, $+1$, 0, -1, -2. These five magnetic quantum numbers describe the five degenerate d orbitals. In the Cartesian coordinate system, these orbitals are designated as d_{z^2}, $d_{x^2-y^2}$, d_{xy}, d_{xz}, and d_{yz}; the last four of these d orbitals are characterized by two nodal planes, while the d_{z^2} has surfaces of revolution.

Example. The five d orbitals are depicted in Fig. 1.36. The d_{z^2} orbital that by convention is the sum of $d_{z^2-x^2}$ and $d_{z^2-y^2}$ and, hence, really $d_{2z^2-x^2-y^2}$ is strongly directed along the z-axis with a negative "doughnut" in the xy-plane. The $d_{x^2-y^2}$ orbital has lobes pointed along the x- and y-axes, while the d_{xy}, d_{xz}, and d_{yz} orbitals have lobes that are pointed half-way between the axes and in the planes designated by the subscripts.

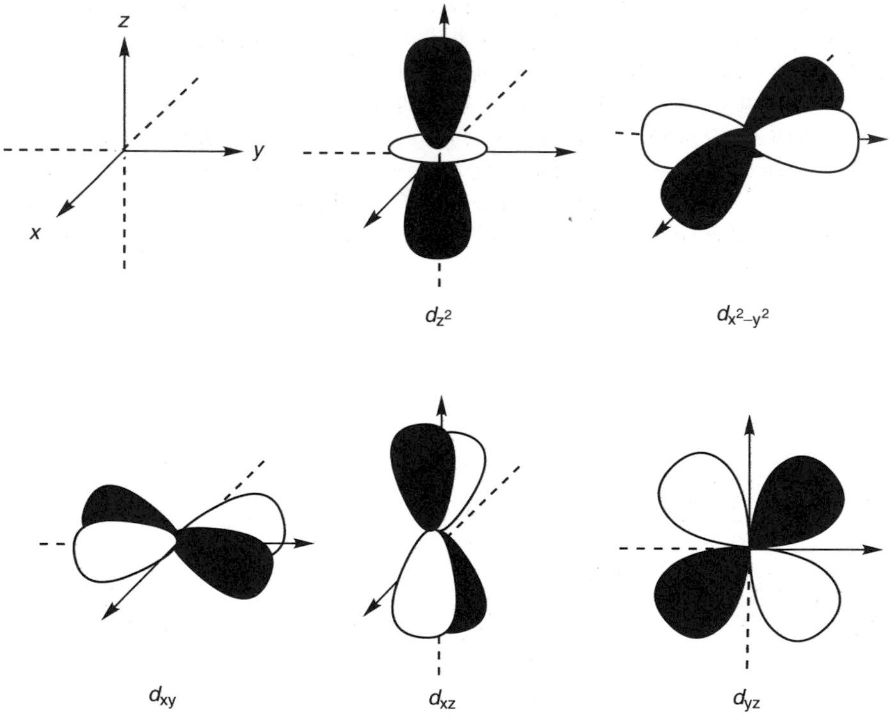

Figure 1.36. The five d orbitals. The shaded and unshaded areas represent lobes of different signs.

1.37 *f* ORBITALS

Orbitals having an angular momentum l equal to 3 and, therefore, magnetic quantum numbers, m_l of $+3$, $+2$, $+1$, 0, -1, -2, -3. These seven magnetic quantum numbers describe the seven degenerate *f* orbitals. The *f* orbitals are characterized by three nodal planes. They become important in the chemistry of inner transition metals (Sect. 1.44).

1.38 ATOMIC ORBITALS FOR MANY-ELECTRON ATOMS

Modified hydrogenlike orbitals that are used to describe the electron distribution in many-electron atoms. The names of the orbitals, *s*, *p*, and so on, are taken from the corresponding hydrogen orbitals. The presence of more than one electron in a many-electron atom can break the degeneracy of orbitals with the same *n* value. Thus, the 2*p* orbitals are higher in energy than the 2*s* orbitals when electrons are present in them. For a given *n*, the orbital energies increase in the order $s < p < d < f < \ldots$.

1.39 PAULI EXCLUSION PRINCIPLE

According to this principle, as formulated by Wolfgang Pauli (1900–1958), a maximum of two electrons can occupy an orbital, and then, only if the spins of the electrons are opposite (paired), that is, if one electron has $m_s = +1/2$, the other must have $m_s = -1/2$. Stated alternatively, no two electrons in the same atom can have the same values of n, l, m_l, and m_s.

1.40 HUND'S RULE

According to this rule, as formulated by Friedrich Hund (1896–1997), a single electron is placed in all orbitals of equal energy (degenerate orbitals) before a second electron is placed in any one of the degenerate set. Furthermore, each of these electrons in the degenerate orbitals has the same (unpaired) spin. This arrangement means that these electrons repel each other as little as possible because any particular electron is prohibited from entering the orbital space of any other electron in the degenerate set.

1.41 AUFBAU (*GER.* BUILDING UP) PRINCIPLE

The building up of the electronic structure of the atoms in the Periodic Table. Orbitals are indicated in order of increasing energy and the electrons of the atom in question are placed in the unfilled orbital of lowest energy, filling this orbital before proceeding to place electrons in the next higher-energy orbital. The sequential placement of electrons must also be consistent with the Pauli exclusion principle and Hund's rule.

Example. The placement of electrons in the orbitals of the nitrogen atom (atomic number of 7) is shown in Fig. 1.41. Note that the 2*p* orbitals are higher in energy than the 2*s* orbital and that each *p* orbital in the degenerate 2*p* set has a single electron of the same spin as the others in this set.

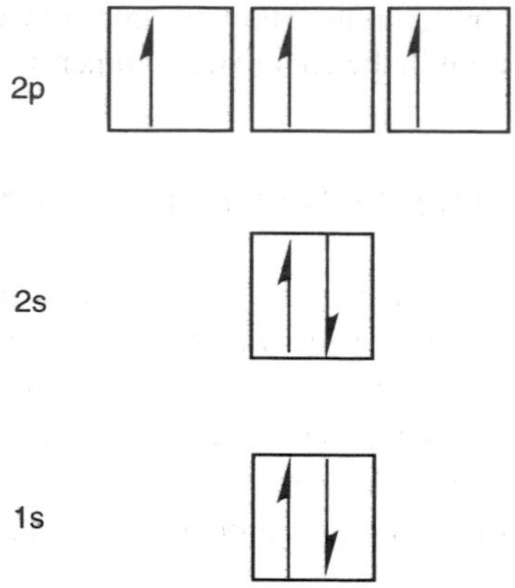

Figure 1.41. The placement of electrons in the orbitals of the nitrogen atom.

1.42 ELECTRONIC CONFIGURATION

The orbital occupation of the electrons of an atom written in a notation that consists of listing the principal quantum number, followed by the azimuthal quantum number designation (*s*, *p*, *d*, f), followed in each case by a superscript indicating the number of electrons in the particular orbitals. The listing is given in the order of increasing energy of the orbitals.

Example. The total number of electrons to be placed in orbitals is equal to the *atomic number* of the atom, which is also equal to the number of protons in the nucleus of the atom. The electronic configuration of the nitrogen atom, atomic number 7 (Fig. 1.41), is $1s^2\ 2s^2\ 2p^3$; for Ne, atomic number 10, it is $1s^22s^22p^6$; for Ar, atomic number 18, it is $1s^22s^22p^63s^23p^6$; and for Sc, atomic number 21, it is [Ar]$4s^23d^1$,where [Ar] represents the rare gas, 18-electron electronic configuration of Ar in which all *s* and *p* orbitals with $n = 1$ to 3, are filled with electrons. The energies of orbitals are approximately as follows: $1s < 2s < 2p < 3s < 3p < 4s \approx 3d < 4p < 5s \approx 4d$.

1.43 SHELL DESIGNATION

The letters K, L, M, N, and O are used to designate the principal quantum number *n*.

Example. The 1*s* orbital which has the lowest principal quantum number, $n = 1$, is designated the K shell; the shell when $n = 2$ is the L shell, made up of the 2*s*, $2p_x$, $2p_y$, and $2p_z$ orbitals; and the shell when $n = 3$ is the M shell consisting of the 3*s*, the three 3*p* orbitals, and the five 3*d* orbitals. Although the origin of the use of the letters K, L, M, and so on, for shell designation is not clearly documented, it has been suggested that these letters were abstracted from the name of physicist Charles Barkla (1877–1944, who received the Nobel Prize, in 1917). He along with collaborators had noted that two rays were characteristically emitted from the inner shells of an element after

X-ray bombardment and these were designated K and L. He chose these mid-alphabet letters from his name because he anticipated the discovery of other rays, and wished to leave alphabetical space on either side for future labeling of these rays.

1.44 THE PERIODIC TABLE

An arrangement in tabular form of all the known elements in rows and columns in sequentially increasing order of their atomic numbers. The Periodic Table is an expression of the periodic law that states many of the properties of the elements (ionization energies, electron affinities, electronegativities, etc.) are a periodic function of their atomic numbers. By some estimates there may be as many as 700 different versions of the Periodic Table. A common display of this table, Fig. 1.44*a*, consists of boxes placed in rows and columns. Each box shown in the table contains the symbol of the element, its atomic number, and a number at the bottom that is the average atomic weight of the element determined from the natural abundance of its various isotopes. There are seven rows of the elements corresponding to the increasing values of the principal quantum number *n*, from 1 to 7. Each of these rows begins with an element having one electron in the *ns* orbital and terminates with an element having the number of electrons corresponding to the completely filled K, L, M, N, and O shell containing 2, 8, 18, 32, and 32 electrons, respectively. Row 1 consists of the elements H and He only; row 2 runs from Li to Ne; row 3 from Na to Ar, and so on. It is in the numbering of the columns, often called groups or families, where there is substantial disagreement among interested chemists and historians.

The table shown in Fig. 1.44*a* is a popular version (sometimes denoted as the American ABA scheme) of the Periodic Table. In the ABA version the elements in a column are classified as belonging to a group, numbered with Roman numerals I through VIII. The elements are further classified as belonging to either an A group or a B group. The A group elements are called *representative* or *main group elements.* The last column is sometimes designated as Group 0 or Group VIIIA. These are the rare gases; they are characterized by having completely filled outer shells; they occur in monoatomic form; and they are relatively chemically inert. The B group elements are the *transition metal elements*; these are the elements with electrons in partially filled $(n-1)d$ or $(n-2)f$ orbitals. The 4th and 5th row transition metals are called *outer transition metals*, and the elements shown in the 6th and 7th row at the bottom of Fig. 1.44*a* are the *inner transition metals*.

Although there is no precise chemical definition of *metals*, they are classified as such if they possess the following group characteristics: high electrical conductivity that decreases with increasing temperature; high thermal conductivity; high *ductility* (easily stretched, not brittle); and *malleability* (can be hammered and formed without breaking). Those elements in Fig. 1.44*a* that are considered metals are shaded either lightly (A group) or more darkly (B group); those that are not shaded are *nonmetals*; those having properties intermediate between metals and nonmetals are cross-hatched. The members of this last group are sometimes called *metalloids* or *semimetals*; these include boron, silicon, germanium, arsenic antimony, and tellurium. The elements in the A group have one to eight electrons in their outermost

Period	I A	II A	III B	IV B	V B	VI B	VII B	VIII (VII B)	VIII (VII B)	VIII (VII B)	I B	II B	III A	IV A	V A	VI A	VII A	0 (VIII A)
1	1 H 1.0080																1 H 1.0080	2 He 4.00260
2	3 Li 6.941	4 Be 9.01218											5 B 10.81	6 C 12.0111	7 N 14.0067	8 O 15.9994	9 F 18.9984	10 Ne 20.179
3	11 Na 22.9898	12 Mg 24.305											13 Al 26.9815	14 Si 28.086	15 P 30.9738	16 S 32.06	17 Cl 35.453	18 Ar 39.948
4	19 K 39.102	20 Ca 40.08	21 Sc 44.9559	22 Ti 47.90	23 V 50.9414	24 Cr 51.996	25 Mn 54.9380	26 Fe 55.847	27 Co 58.9332	28 Ni 58.71	29 Cu 63.546	30 Zn 65.37	31 Ga 69.72	32 Ge 72.59	33 As 74.9216	34 Se 78.96	35 Br 79.904	36 Kr 83.80
5	37 Rb 85.4678	38 Sr 87.62	39 Y 88.9059	40 Zr 91.22	41 Nb 92.9064	42 Mo 95.94	43 Tc 98.9062	44 Ru 101.07	45 Rh 102.9055	46 Pd 106.4	47 Ag 107.868	48 Cd 112.40	49 In 114.82	50 Sn 118.69	51 Sb 121.75	52 Te 127.60	53 I 126.9045	54 Xe 131.30
6	55 Cs 132.9055	56 Ba 137.34	*	72 Hf 178.49	73 Ta 180.9479	74 W 183.85	75 Re 186.2	76 Os 190.2	77 Ir 192.22	78 Pt 195.09	79 Au 196.9665	80 Hg 200.59	81 Tl 204.37	82 Pb 207.2	83 Bi 208.9806	84 Po (210)	85 At (210)	86 Rn (222)
7	87 Fr (223)	88 Ra 226.0254	†	104 Rf (260)	105 Ha	106												

Representative elements: I A–II A; III A–0 (VIII A). Transition elements: III B–II B.

Inner transition elements

*Lanthanides	57 La 138.9055	58 Ce 140.12	59 Pr 140.9077	60 Nd 144.24	61 Pm (145)	62 Sm 150.4	63 Eu 151.96	64 Gd 157.25	65 Tb 158.9254	66 Dy 162.50	67 Ho 164.9303	68 Er 167.26	69 Tm 168.9342	70 Yb 173.04	71 Lu 174.97
†Actinides	89 Ac (227)	90 Th 232.0381	91 Pa 231.0359	92 U 238.029	93 Np 237.0482	94 Pu (242)	95 Am (243)	96 Cm (247)	97 Bk (247)	98 Cf (251)	99 Es (254)	100 Fm (257)	101 Md (256)	102 No (256)	103 Lr (257)

Figure 1.44. (*a*) A Periodic Table of the elements.

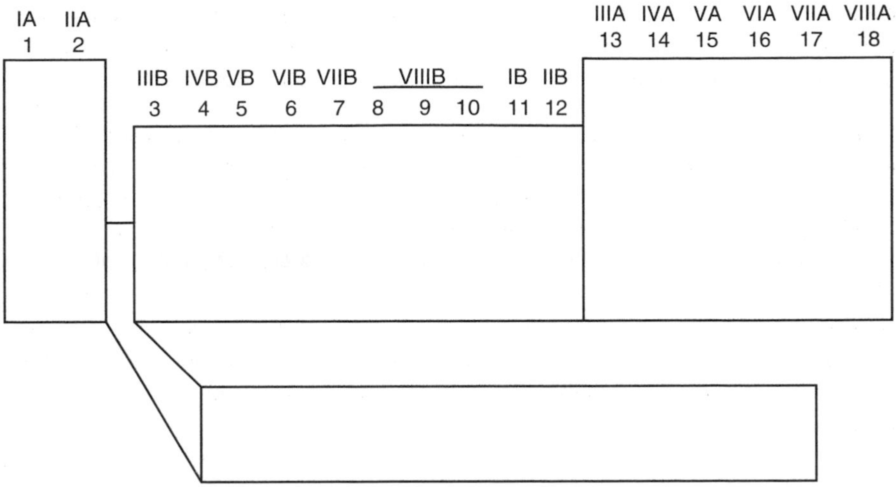

(b)

Figure 1.44. (*b*) A block outline showing the Roman numeral American ABA designation and the corresponding Arabic numeral IUPAC designation for families of elements in the Periodic Table.

shell and their group Roman number corresponds to the number of electrons in this shell, for example, Ca(IIA), Al(IIIA), C(IVA), and so on. Elements in Group IA are called *alkali metals* and those in Group IIA are called *alkaline earth metals*.

Recently, the International Union of Pure and Applied Chemistry (IUPAC) recommended a version of the Periodic Table in which the A and B designations are eliminated, the Roman numerals of the columns are replaced with Arabic numerals, and the columns are numbered from 1 to 18. These column numbers make it possible to assign each of the outer transition metals to a separate group number, thus, for example, the *triads of Group VIIIB transition metals*: Fe, Co, Ni; Ru, Rh, Pd; and Os, Ir, Pt in Fig. 1.44*a* become, respectively, members of Groups 8, 9, and 10 in the IUPAC version. This version has many advantages; for example, it eliminates the ambiguity of the definition of transition metals as well as the group assignments of H and He. It does not, however, indicate a group number assignment to any of the two rows of inner transition metals consisting of 14 elements each (which would require 32 instead of 18 groups), nor does it provide the chemical information, for example, the number of valence electrons in each group, that is provided by the older labels. Thus, the valuable advantage of correlating the B group with the same number A group inherent in the ABA system is lost, for instance, the fact that there are five valence electrons in the structure of both nitrogen (Group VA) and vanadium (Group VB). Nevertheless the IUPAC version is gaining increasing acceptance.

1.45 VALENCE ORBITALS

The orbitals of an atom that may be involved in bonding to other atoms. For the main group or representative elements, these are the *ns* or *ns* + *np* orbitals, where *n* is the quantum number of the highest occupied orbital; for the outer transition metals, these are the $(n-1)d + ns$ orbitals; and for the inner transition metals, these are the $(n-2)f + ns$ orbitals. Electrons in these orbitals are *valence electrons.*

Example. The valence orbitals occupied by the four valence electrons of the carbon atom are the $2s + 2p$ orbitals. For a 3rd row element such as Si (atomic number 14) with the electronic configuration $[1s^2 2s^2 2p^6]3s^2 3p^2$, shortened to $[\text{Ne}]3s^2 3p^2$, the $3s$ and $3p$ orbitals are the valence orbitals. For a 4th row ($n = 4$) element such as Sc (atomic number 21) with the electronic configuration $[\text{Ar}]\ 3d^1 4s^2$, the valence orbitals are $3d$ and $4s$, and these are occupied by the three valence electrons. In the formation of coordination complexes, use is made of lowest-energy vacant orbitals, and because these are involved in bond formation, they may be considered vacant valence orbitals. Coordination complexes are common in transition metals chemistry.

1.46 ATOMIC CORE (OR KERNEL)

The electronic structure of an atom after the removal of its valence electrons.

Example. The atomic core structure consists of the electrons making up the noble gas or *pseudo-noble gas* structure immediately preceding the atom in the Periodic Table. A pseudo-noble gas configuration is one having all the electrons of the noble gas, plus, for the outer transition metals, the 10 electrons in completely filled $(n-1)d$ orbitals; and for the inner transition metals, the noble gas configuration plus the $(n-2)f^{14}$, or the noble gas plus $(n-1)d^{10}(n-2)f^{14}$. Electrons in these orbitals are not considered valence electrons. The core structure of Sc, atomic number 21, is that corresponding to the preceding rare gas, which in this case is the Ar core. For Ga, atomic number 31, with valence electrons $4s^2 4p^1$, the core structure consists of the pseudo-rare gas structure $\{[\text{Ar}]3d^{10}\}$.

1.47 HYBRIDIZATION OF ATOMIC ORBITALS

The mathematical mixing of two or more different orbitals on a given atom to give the same number of new orbitals, each of which has some of the character of the original component orbitals. Hybridization requires that the atomic orbitals to be mixed are similar in energy. The resulting hybrid orbitals have directional character, and when used to bond with atomic orbitals of other atoms, they help to determine the shape of the molecule formed.

Example. In much of organic (carbon) chemistry, the $2s$ orbital of carbon is mixed with: (a) one p orbital to give two hybrid sp orbitals (digonal linear); (b) two p orbitals to give three sp^2 orbitals (trigonal planar); or (c) three p orbitals to give four sp^3 orbitals (tetrahedral). The mixing of the $2s$ orbital of carbon with its $2p_y$ to give two carbon sp orbitals is shown pictorially in Fig. 1.47. These two hybrid atomic orbitals have the form $\phi_1 = (s + p_y)$ and $\phi_2 = (s - p_y)$.

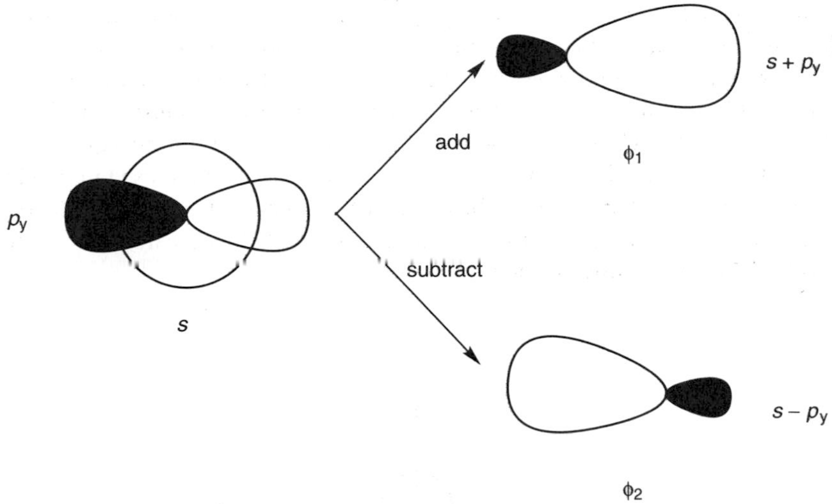

Figure 1.47. The two hybrid *sp* atomic orbitals, ϕ_1 and ϕ_2. The shaded and unshaded areas represent lobes of different mathematical signs.

1.48 HYBRIDIZATION INDEX

This is the superscript x on the p in an sp^x hybrid orbital; such an orbital possesses $[x/(1+x)]$ (100) percent p character and $[1/(1+x)]$ (100) percent s character.

Example. The hybridization index of an sp^3 orbital is 3 (75% p-character); for an $sp^{0.894}$ orbital, it is 0.894 (47.2% p-character).

1.49 EQUIVALENT HYBRID ATOMIC ORBITALS

A set of hybridized orbitals, each member of which possesses precisely the same value for its hybridization index.

Example. If the atomic orbitals 2s and 2p_z are distributed equally in two hybrid orbitals, each resulting orbital will have an equal amount of s and p character; that is, each orbital will be sp ($s^{1.00}p^{1.00}$) (Fig. 1.47). If the 2s and two of the 2p orbitals are distributed equally among three hybrid orbitals, each of the three equivalent orbitals will be sp^2 ($s^{1.00}p^{2.00}$) (Fig. 1.49). Combining a 2s orbital equally with three 2p orbitals gives four equivalent hybrid orbitals, $s^{1.00}p^{3.00}$ (sp^3); that is, each of the four sp^3 orbitals has an equal amount of s character, $[1/(1+3)] \times 100\% = 25\%$, and an equal amount of p character, $[3/(1+3)] \times 100\% = 75\%$.

1.50 NONEQUIVALENT HYBRID ATOMIC ORBITALS

The hybridized orbitals that result when the constituent atomic orbitals are not equally distributed among a set of hybrid orbitals.

Example. In hybridizing a 2*s* with a 2*p* orbital to form two hybrids, it is possible to put more *p* character and less *s* character into one hybrid and less *p* and more *s* into the other. Thus, in hybridizing an *s* and a p_z orbital, it is possible to generate one hybrid that has 52.8% *p* ($sp^{1.11}$) character. The second hybrid must be 47.2% *p* and is therefore $sp^{0.89}$ ($[x/(1+x)] \times 100\% = 47.2\%$; $x = 0.89$). Such nonequivalent carbon orbitals are found in CO, where the *sp* carbon hybrid orbital used in bonding to oxygen has more *p* character than the other carbon *sp* hybrid orbital, which contains a lone pair of electrons. If dissimilar atoms are bonded to a carbon atom, the *sp* hybrid orbitals will always be nonequivalent.

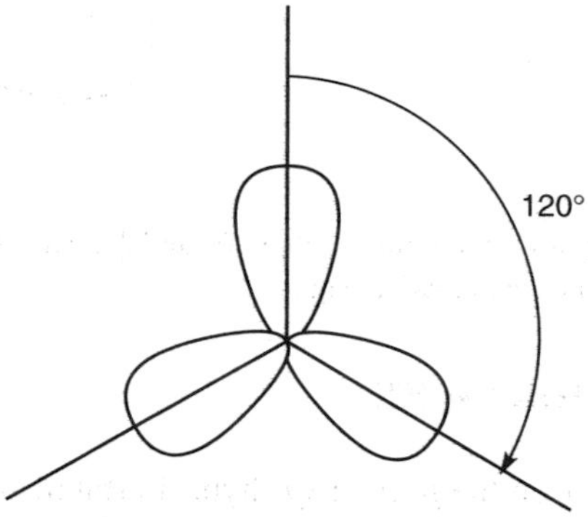

Figure 1.49. The three hybrid sp^2 atomic orbitals (all in the same plane).

Acknowledgment. The authors thank Prof. Thomas Beck and Prof. William Jensen for helpful comments.

SUGGESTED READING

See, for example,

The chemistry section of **Educypedia** (The Educational Encyclopedia) http://users.telenet.be/educypedia/education/chemistrymol.htm.

Atkins, P. W. *Molecular Quantum Mechanics*, 2nd ed. Oxford University Press: London, 1983.

Coulson, C. A. *Valence*. Oxford University Press: London, 1952.

Douglas, B.; McDaniel, D. H.; and Alexander, J. J. *Concepts and Models of Inorganic Chemistry*, 3rd ed. John Wiley & Sons: New York, 1994.

Gamow, G. and Cleveland, J. M. *Physics*. Prentice-Hall: Englewood Cliffs, NJ, 1960.

Jensen, W. B. *Computers Maths. Appl., 12B*, 487 (1986); *J. Chem. Ed. 59*, 634 (1982).

Pauling, L. *Nature of the Chemical Bond*, 3rd ed. Cornell University Press: Ithaca, NY, 1960.

For a description of the *f* orbitals, see:

Kikuchi, O. and Suzuki, K. *J. Chem. Ed. 62*, 206 (1985).

2 Bonds Between Adjacent Atoms: Localized Bonding, Molecular Orbital Theory

2.1	Chemical Bond	27
2.2	Covalent Bond	28
2.3	Localized Two-Center, Two-Electron (2c-2e) Bond; Electron Pair Bond	28
2.4	Valence Bond (VB) Theory	28
2.5	Lone Pair Electrons	28
2.6	Lewis Electron (Dot) Structures	28
2.7	Octet Rule	29
2.8	Electronegativity	29
2.9	Valence, Ionic Valence, Covalence	30
2.10	Oxidation Number (Oxidation State)	31
2.11	Formal Charge	32
2.12	Nonpolar Covalent Bond	32
2.13	Dipole Moment	33
2.14	Dipole Moments of Polyatomic Molecules; Vectorial Addition of Dipole Moments	33
2.15	Polar Covalent Bond; Partially Ionic Bond	34
2.16	Ionic Bond	35
2.17	Single, Double, and Triple Bonds	35
2.18	Morse Curve	35
2.19	Bond Length d_0	36
2.20	Bond Dissociation	36
2.21	Bond Dissociation Energy D_0	37
2.22	Bond Angle	37
2.23	Atomic Radius r_0	37
2.24	Ionic Radius r^+ and r^-	38
2.25	van der Waals Radius	39
2.26	Coordinate Covalent Bond (Dative Bond)	40
2.27	Hydrogen Bond	40
2.28	Valence Shell Electron Pair Repulsion (VSEPR)	41
2.29	Molecular Orbitals	42
2.30	Molecular Orbital (MO) Theory	43
2.31	Bonding Molecular Orbitals	43
2.32	Antibonding Molecular Orbitals	43
2.33	Linear Combination of Atomic Orbitals (LCAO)	43
2.34	Basis Set of Orbitals	44

The Vocabulary and Concepts of Organic Chemistry, Second Edition, by Milton Orchin, Roger S. Macomber, Allan Pinhas, and R. Marshall Wilson

2.35 σ Bonding Molecular Orbital (σ Orbital); σ Bond 44
2.36 σ Antibonding Molecular Orbital (σ* Orbital) 45
2.37 *pπ* Atomic Orbital 45
2.38 π Bonding Molecular Orbital (π Orbital) 45
2.39 Localized π Bond 46
2.40 π Antibonding Molecular Orbital (π* Orbital) 46
2.41 σ Skeleton 47
2.42 Molecular Orbital Energy Diagram (MOED) 47
2.43 Electronic Configuration of Molecules 48
2.44 MOED for 2nd Row Homodiatomic Molecules 48
2.45 MOED for the 2nd Row Heteroatomic Molecule; Carbon Monoxide 50
2.46 Coefficients of Atomic Orbitals c_{ij} 50
2.47 Normalized Orbital 51
2.48 Normalization 52
2.49 Orthogonal Orbitals 52
2.50 Orthonormal Orbitals 52
2.51 Wave Functions in Valence Bond (VB) Theory 52

Electrons are the cement that binds together atoms in molecules. Knowledge concerning the forces acting on these electrons, the energy of the electrons, and their location in space with respect to the nuclei they hold together are fundamental to the understanding of all chemistry. The nature of the bonding of atoms to one another is usually described by either of two major theories: valence bond (VB) theory and molecular orbital (MO) theory. The starting point for the development of VB theory was a 1927 paper by Walter Heitler and Fritz London that appeared in *Z. Physik* dealing with the calculation of the energy of the hydrogen molecule. Several years later J. Slater and then Linus Pauling (1901–1994) extended the VB approach to organic molecules, and VB theory became known as HLSP theory from the first letters of the surnames of the men who contributed so much to the theory. The popularity of VB theory owes much to the brilliant work of Pauling and his success in explaining the nature of the chemical bond using resonance concepts.

According to VB theory, a molecule cannot be represented solely by one valence bond structure. Thus, CO_2 in valence bond notation is written as :Ö=C=Ö:, which shows that eight electrons surround each oxygen atom as well as the carbon atom. This one structure adequately describes the bonding in CO_2, and one does not ordinarily consider all the other less important but relevant resonance structures, such as $^{-}$:Ö̤—C≡O:$^{+}$ and $^{+}$:O≡C—Ö̤:$^{-}$. Because resonance theory is such a powerful tool for understanding delocalized bonding, the subject of the next chapter, we will defer further discussion of it here. Despite the merits of the VB approach with its emphasis on the electron pair bond, the theory has several drawbacks even for the description of the bonding in some simple molecules such as dioxygen, O_2. In VB notation one is tempted to write the structure of this molecule as :Ö=Ö:, but this representation implies that all the electrons of oxygen are paired and hence the molecule should be diamagnetic, which it is not. On the other hand, according to the MO description, the two highest occupied molecular orbitals of O_2 are degenerate and antibonding and each contains one electron with identical spin, thus accounting for the observed paramagnetism, the most unusual property of dioxygen.

In MO theory the behavior of each electron in a molecule is described by a wave function. But calculations of wave functions for many electron atoms become very complicated. Fortunately, considerable simplification is achieved by use of the linear combination of atomic orbitals, (LCAO) method first described by Robert S. Mulliken (1895–1986). In this approach it is assumed that when one electron is near one nucleus, the wave function resembles the atomic orbital of that atom, and when the electron is in the neighborhood of the other atom, the wave function resembles that of the neighboring atom. Since the complete wave function has characteristics separately possessed by the two atomic orbitals, it is approximated by the linear combination of the atomic orbitals.

To further illustrate the difference in the two theories, consider the bonding in methane, CH_4. According to VB theory, the four C–H bonds are regarded as though each bond were a separate localized two-center, two-electron bond formed by the overlap of a carbon sp^3 orbital and a hydrogen $1s$ orbital. Each bond is a result of the pairing of two electrons, one from each of the bonded atoms, and the electron density of the shared pair is at a maximum between the bonded atoms. In the molecular orbital treatment, the four $1s$ hydrogen orbitals are combined into four so-called group (or symmetry-adapted) orbitals, each of which belongs to a symmetry species in the T_d point group to which tetrahedral methane belongs. These four hydrogen group orbitals are then combined by the LCAO method with the $2s$ and three $2p$ orbitals of the carbon atom of similar symmetry to generate the four bonding and four antibonding molecular orbitals, necessary for the MO description. The eight valence electrons are then placed in the four bonding molecular orbitals, each of which is delocalized over the five atoms. For the treatment of the bonding in methane, the valence bond approach is simpler and usually adequate. However, for insight into some areas of chemical importance such as, for example, molecular spectroscopy, the molecular orbital approach is more satisfactory.

This chapter deals with bonds between atoms in molecules in which adjacent atoms share a pair of electrons, giving rise to what is called two-center, two-electron bonding. Both VB theory and MO theory are used with more emphasis on the latter.

2.1 CHEMICAL BOND

A general term describing the result of the attraction between two adjacent atoms such that the atoms are held in at relatively fixed distances with respect to each other. The bond may be said to occur at the distance between the two atoms that corresponds to the minimum in the potential energy of the system as the two atoms are brought into proximity to one another (see Morse curve, Fig. 2.18).

2.2 COVALENT BOND

A chemical bond resulting from the sharing of electrons between adjacent atoms. If the sharing is approximately equal, the bond is designated as *nonpolar covalent* (Sect. 2.12), and if substantially unequal, the bond is *polar covalent* (Sect. 2.15). Only in the case where the bond between two atoms coincides with a center of symmetry of a molecule is the sharing of electrons between the two atoms exactly equal.

2.3 LOCALIZED TWO-CENTER, TWO-ELECTRON (2c-2e) BOND; ELECTRON PAIR BOND

The covalent bond between two adjacent atoms involving two electrons. Such bonds may be treated theoretically by either molecular orbital (MO) theory or valence bond (VB) theory (see introductory material).

2.4 VALENCE BOND (VB) THEORY

This theory postulates that bond formation occurs as two initially distant atomic orbitals, each containing one valence electron of opposite spin, are brought into proximity to each other. As the overlap of the atomic orbitals increases, each electron is attracted to the opposite nucleus eventually to form a localized two-center, two-electron bond at a distance between the atoms corresponding to a minimum in the potential energy of the system (see Morse curve, Sect. 2.18).

2.5 LONE PAIR ELECTRONS

A pair of electrons in the valence shell of an atom that is not involved in bonding to other atoms in the molecule.

2.6 LEWIS ELECTRON (DOT) STRUCTURES

Gilbert N. Lewis (1875–1946) devised the use of dots to represent the valence electrons (usually an octet) surrounding an atom in molecules or ions. For convenience, most authors now use a dash to represent a single two-electron bond shared between adjacent atoms and a pair of dots on a single atom to symbolize a lone pair of electrons.

Example. Water, ammonia, hydrogen cyanide, in Figs. 2.6*a*, *b*, and *c*. Some authors also indicate the lone pair electrons as a dash or bar, as shown in Fig. 2.6*d*.

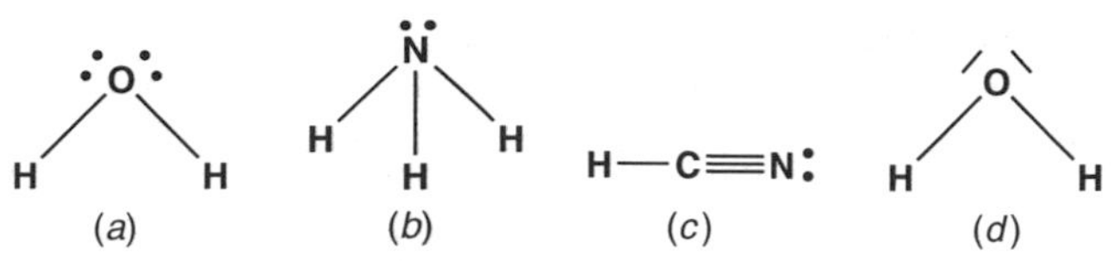

Figure 2.6. Lewis electron structures for *(a)* H_2O, *(b)* NH_3, *(c)* HCN, and *(d)* the use of bars to represent lone pair electrons in H_2O.

2.7 OCTET RULE

The tendency of the main group elements (Sect. 1.44) to surround themselves with a total of eight valence electrons, the number of valence electrons characteristic of the noble gases (with the exception of He, which has a closed shell of only two electrons). The octet rule rationalizes the bonding arrangement in most Lewis structures, but there are exceptions involving both fewer and more than eight valence electrons.

Example. The oxygen, nitrogen, and carbon atoms of the molecules shown in Fig. 2.6. Compounds involving the elements boron (e.g., BF_3) and aluminum (e.g., $AlBr_3$), each of which has six rather than eight electrons in their valence shells, not unexpectedly react with a partner molecule having a lone pair of electrons available for bonding. The formation of the addition complex (Fig. 2.7*a*) involving the lone pair on nitrogen (see Sect. 2.11) and the dimer of $AlBr_3$ (Fig. 2.7*b*) involving a lone pair on each of the bridging bromine atoms are examples of the operation of the octet rule (for the explanation of the charges on the atoms, see Sect. 2.11). Exceptions involving more than eight valence electrons involve the 3rd row elements phosphorus and sulfur in compounds, for example, such as PCl_5 and SF_6 where vacant 3*d* valence orbitals are presumably utilized.

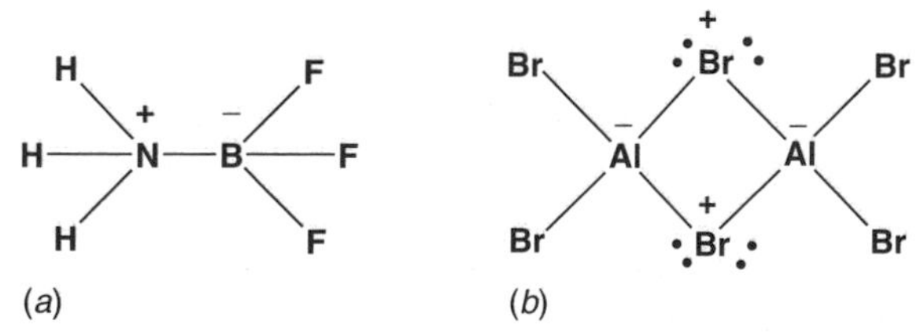

Figure 2.7. Complexes of boron and aluminum that obey the octet rule.

2.8 ELECTRONEGATIVITY

The relative attraction by an atom for the valence electrons on or near that atom. Pauling who originated the concept of electronegativity recognized that the experimental bond energy of the bond A–B was greater than the average bond energies of A–A and B–B.

The additional bond strength is due to the ionic resonance energy arising from the contributions of ionic resonance structures A^+B^-, and if A is more electronegative than B, A^-B^+. The ionic resonance energy Δ can be calculated from the equation:

$$\Delta = E(A - B) - [E(A - A) \times E(B - B)]^{1/2} \tag{2.8a}$$

where E is the energy of the bond between the atoms shown in parentheses. Pauling set the square root of Δ equal to the electronegativity difference between A and B. Then if an electronegativity value of 2.20 is *arbitrarily* assigned to the element hydrogen, the electronegativities of most other atoms may be calculated.

Several other scales for rating the electronegativity of atoms have been proposed. One of the most useful is the one suggested by Mulliken. He proposed that the electronegativity of an atom is the average of its *ionization energy* or *IE* (the energy required for the removal of an electron from an atom in the gas phase) and its *electron affinity* or *EA* (the energy released by adding an electron to the atom in the gas phase):

$$\text{Electronegativity} = (\text{IE} + \text{EA})/2 \tag{2.8b}$$

Nearly all methods of calculating electronegativities lead to approximately the same values, which are almost always expressed as dimensionless numbers.

Example. Typically, Pauling electronegativity values, for example, that of the F atom, are obtained as follows: The experimentally observed bond energy of H–F is 5.82 electron volts or eV (1 eV = 96.49 kJ mol^{-1} or 23.06 kcal mol^{-1}). The ionic resonance energy Δ of H–F calculated from Eq. 2.8a is 3.15 and therefore $\Delta^{1/2}$ is 1.77. If the electronegativity of H is 2.20, then the electronegativity of F is (2.20 + 1.77) = 3.98. The Pauling electronegativities of the 2nd row elements in the Periodic Table are Li, 0.98; Be, 1.57; B, 2.04; C, 2.55; N, 3.04; O, 3.44; F, 3.98. The extremes in the scale of electronegativities are Cs, 0.79, and F, 3.98. From these values it is clear that the electrons in, for example, a C–F bond, will reside much closer to the F atom than to the C atom. The unequal sharing of the electrons in a bond gives rise to a partial negative charge on the more electronegative atom and a partial positive charge on the less electronegative atom. This fact is sometimes incorporated into the structure of the molecule by placing partial negative and partial positive signs above the atoms as in $^{\delta+}C–F^{\delta-}$. It is not always possible to assign fixed values for the electronegativity of a particular atom because its electronegativity may vary, depending on the number and kind of other atoms attached to it. Thus, the electronegativity of an *sp* hybridized carbon (50% *s* character) is 0.6 higher than that of an sp^3 hybridized carbon (25% *s* character).

2.9 VALENCE, IONIC VALENCE, COVALENCE

Terms used to describe the capacity of an element to form chemical bonds with other elements. In the case of a covalent compound, the valence, more precisely called covalence, corresponds to the number of bonds attached to the atom in question. In the case of ions, the valence, more precisely called ionic valence, is the absolute charge on a monoatomic ion.

Example. The valence of Mg^{++} is 2. The covalency of carbon in carbon monoxide, written as :C=Ö: is 2, but in carbon dioxide, :Ö=C=Ö:, it is 4. The ionic valence of both Ca and O in CaO is 2. The word "valence" standing alone is rather ambiguous and the more precise terms such as ionic valence, covalence, valence orbitals, valence electrons, oxidation number, and formal charge are preferred.

2.10 OXIDATION NUMBER (OXIDATION STATE)

A whole number assigned to an atom in a molecule representative of its formal ownership of the valence electrons around it. It is calculated by first assuming that all the electrons involved in bonding to the atom in question in the Lewis structure are assigned to either that atom or to its partner, if its partner is more electronegative. The number of valence electrons remaining on the atom is then determined if the atom is bonded to the same element, as in a C–C bond where the bonding electrons are divided equally, and this number is then subtracted from the number of valence electrons associated with the atom in its elemental form. The difference between the two numbers is the oxidation number of the atom in question.

Example. In the structures shown in Fig. 2.10, the bonding electrons are removed with the more electronegative atoms as shown and the oxidation numbers for carbon (which can range from +4 to −4) and sulfur are displayed below the structure. The oxidation number for oxygen in all these compounds is −2 and for the hydrogen

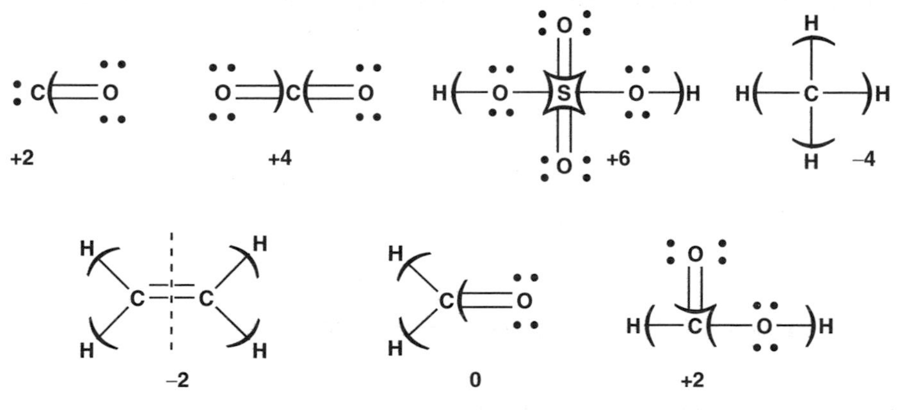

Figure 2.10. The oxidation numbers of carbon and sulfur in various compounds. The atoms inside the curves between atoms are in each case the more electronegative atoms, and the bonding electrons are, therefore, associated with those atoms in determining the oxidation numbers.

atoms it is +1. For monoatomic ions, the oxidation number is the same as the charge on the ion. For a neutral compound (all the compounds shown in Fig. 2.10 are neutral, i.e., they have no net charge), the sum of the oxidation numbers of all atoms must equal zero.

2.11 FORMAL CHARGE

This is the positive or negative charge of an atom in a molecule indicating that the atom has a fewer or greater number of valence electrons associated with it than it would have as an isolated atom in its elemental form. To determine the magnitude and sign of the formal charge, the atom is assigned all its lone pair electrons plus half of those electrons involved in the bonds with neighboring atoms; this number is then subtracted from the number of valence electrons in the isolated atom.

Example. In the neutral complex $H_3\overset{+}{N}-\overset{-}{B}F_3$ (Fig. 2.11*b*), the nitrogen atom is surrounded by four electron-paired bonds. The isolated nitrogen atom has five valence electrons and hence the formal charge is $5 - 8/2 = +1$. The formal charge on the boron atom is $3 - 8/2 = -1$, leaving a net charge of zero on the complex. The formal charge on ions is calculated in the same way. In the negatively charged hydoxide ion $[OH]^-$, the oxygen atom is surrounded with three lone pairs of electrons plus the pair it shares with the hydrogen. Thus, the formal charge on oxygen is $6 - (6 + 2/2) = -1$. For nitrogen in $[NO_3]^-$, (Fig. 2.11*a*), the formal charge on nitrogen is $5 - 8/2 = +1$. The formal charge on each of the two singly bonded oxygen atoms is $6 - (6 + 2/2) = -1$, and on the doubly bonded oxygen it is $6 - (4 + 4/2) = 0$, leaving a net formal charge of $2(-1) + (+1) + 0 = -1$ on the ion. Formal charges should not be confused with oxidation numbers, which for the N atom in $H_3\overset{+}{N}-\overset{-}{B}F_3$ is $5 - 8 = -3$, and for the B atom is $3 - 0 = +3$. For the N in NH_3 and the C in CH_4, both with formal charges of zero, the oxidation numbers are -3 and -4, respectively.

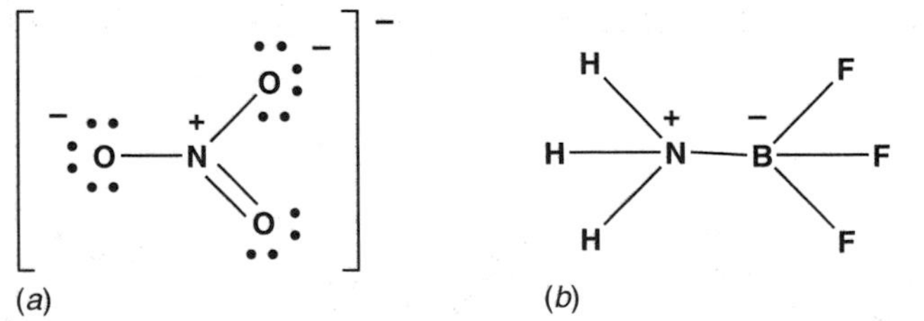

Figure 2.11. Formal charges on atoms in (*a*) $[NO_3]^{-1}$ and (*b*) $H_3\overset{+}{N}–\overset{-}{B}F_3$.

2.12 NONPOLAR COVALENT BOND

A bond between atoms involving equal or almost equal sharing of the bonding electrons. As a rule of thumb or rough approximation, and quite arbitrarily, the difference in the electronegativities of the bonded atoms should be less than 0.5 (Pauling scale) for the covalent bond to be classified as nonpolar.

Example. The C–H and the C–P bonds; the electronegativity difference in these bonds in each case is 0.4.

2.13 DIPOLE MOMENT

A vectorial property of individual bonds or entire molecules that characterizes their polarity. A diatomic molecule in which the electrons are not shared equally gives rise to a dipole moment vector with a negative end and a positive end along the bond connecting the two atoms. Therefore, such a molecule acts as a dipole and tends to become aligned in an electrical field. The (electric) dipole moment μ (see also Sect. 4.31) is obtained by multiplying the charge at either atom (pole) q (in electrostatic units or esu) by the distance d (in centimeters) between the atoms (poles): $q \times d = \mu$ (in esu-cm). Dipole moments are usually expressed in Debye units (named after Peter Debye, 1884–1966), abbreviated D, equal to 10^{-18} esu-cm.

Example. Typical dipole moments (in Debye units) of some C–Z bonds are shown in Fig. 2.13. The dipole moment of the C–Cl bond is greater than that of C–F because the C–Cl distance is larger than the C–F distance even though fluorine is more electronegative than chlorine. The direction of the dipole of a bond is frequently indicated by a crossed arrow over the bond in question with the crossed tail at the positive end and the head of the arrow over the negative end, as shown in the examples.

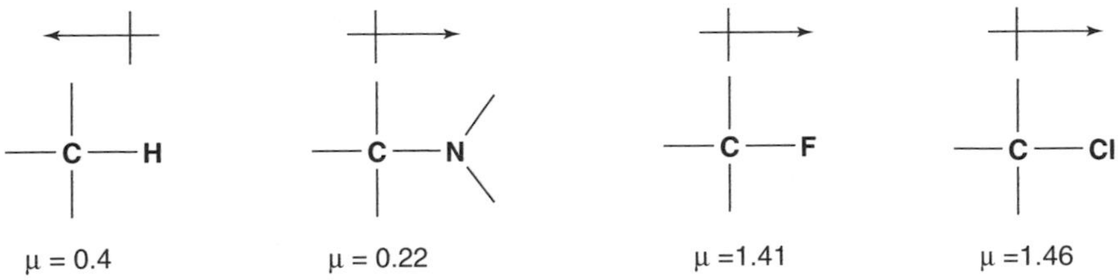

Figure 2.13. Dipole moments for the C–H, C–N, C–F, and C–Cl bonds.

2.14 DIPOLE MOMENTS OF POLYATOMIC MOLECULES; VECTORIAL ADDITION OF DIPOLE MOMENTS

The dipole moment of a molecule may be calculated from the vectorial sum of the individual bond dipole moments. Each bond between atoms in a molecule has an associated directed dipole moment that is approximately independent of the nature of the groups in the rest of the molecule. The resultant of the vectorial addition of all bond moments yields the overall dipole moment of the molecule.

Example. Chlorobenzene has a measured dipole moment of 1.70 D; see Fig. 2.14*a* (the point O represents the center of the hexagon). The dipole moment of 1,2-dichlorobenzene can be approximated from the dipole moment of chlorobenzene by vector addition, as shown in Fig. 2.14*b*. The component vector **Ob** is equal to 1.70 cos 30° = 1.47, and the resultant dipole moment (**OB**, Fig. 2.14*b*) is the sum of the two component vectors, 2.94 D. However, the experimental value for the dipole moment of 1,2-dichlorobenzene is 2.25 D. If one substitutes this experimental value into the rearranged vector sum equation and calculates the individual component vectors, one gets

$$\mathbf{OCl} = \mathbf{OB}/2 \cos 30° = 2.25/2 \cos 30° = 1.30 \text{ D}$$

These vector components are considerably less that the single vector (1.70 D) in monochlorobenzene, indicating that the dipole vectors in dichlorobenzene interact with each other, resulting in a vector sum less than that calculated on the basis of no interaction.

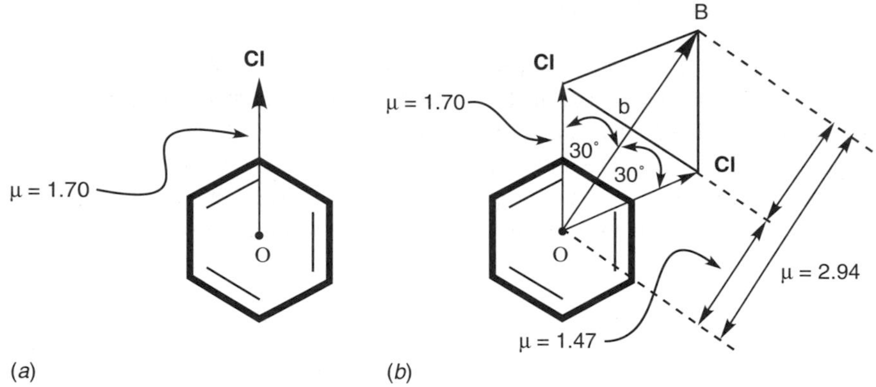

Figure 2.14. Vectorial addition of bond moments.

2.15 POLAR COVALENT BOND; PARTIALLY IONIC BOND

This is a covalent bond with appreciable ionic character, that is, a bond between atoms in which the shared electrons reside much closer to the atom of greater electronegativity. The distinction between a nonpolar and a polar covalent bond is arbitrary; if the difference in electronegativities of the bonded atoms is greater than 0.5 (Pauling scale), the bond has an appreciable ionic character and may be considered a polar covalent bond.

Example. The percent ionic character of a bond can be approximated by the magnitude of the bond's dipole moment, which in turn is dependent on the relative electronegativities of the two atoms comprising the bond. Thus, if we assume, for purposes of calculation, that in the molecule H–F, the entire charge of one electron

(4.8×10^{-10} esu) is on the F atom, we would have the completely ionic form, H^+F^-. The H-F distance determined experimentally is 0.917Å $= 0.917 \times 10^{-8}$ cm, leading for the completely ionized species to a dipole moment of $\mu = q \times d = [(4.8 \times 10^{-10}) \times (0.917 \times 10^{-8})]/10^{-18} = 4.4$ D. However, since the experimental value is 1.98 D, it may be concluded that the bond has (1.98/4.4)100 = 45% ionic character. When one desires to indicate the partial ionic character of a bond, the superscripts δ^+ and δ^- are placed above the electropositive and electronegative element, for example, $H^{\delta+}$–$F^{\delta-}$. A useful guide for correlating electronegativity differences with percent ionic character (in parentheses) is 1.0 (20%), 1.5 (40%), 2.0 (60%), and 2.5 (80%).

2.16 IONIC BOND

The result of electrostatic attraction between oppositely charged ions. Such bonds can be viewed as theoretically resulting from the complete transfer of an electron from an electropositive atom to an electronegative atom and not as a result of any unequal sharing of electrons between the atoms.

Example. Sodium chloride, Na^+Cl^-, is an ionic compound. However, in the solid state, both ions are present in a lattice network in which each sodium ion is surrounded by six chloride ions and each chloride ion is surrounded by six sodium ions. There is no such species as a diatomic Na^+Cl^- molecule except in the vapor phase.

2.17 SINGLE, DOUBLE, AND TRIPLE BONDS

The covalent bonding between adjacent atoms involving two electrons (single bond), four electrons (double bond), and six electrons (triple bond). Conventionally, such bonding is indicated by one dash (single), two dashes (double), and three dashes (triple) between the bonded atoms.

Example. H_3C–CH_3, $H_2C{=}CH_2$, $HC{\equiv}CH$.

2.18 MORSE CURVE

This is a plot, named after Phillip M. Morse (1903–1985), showing the relationship between the potential energy E_p of a chemical bond between two atoms as a function of the distance between them.

Example. The Morse curve for the hydrogen molecule is shown in Fig. 2.18.

The minimum in the curve occurs at the equilibrium interatomic distance d_0. At distances smaller than d_0, the two nuclei repel each other and the potential energy E_p rises sharply; at distances larger than d_0, the E_p increases because of reduced orbital

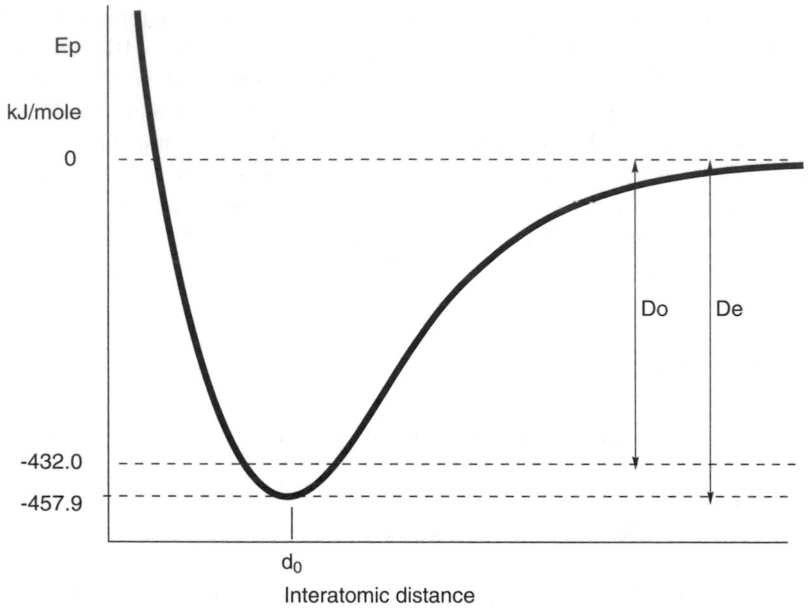

Figure 2.18. The Morse curve for the hydrogen molecule; d_0, the bond length, is 0.74 Å.

overlap, and at very large distances, E_p approaches zero as the atoms become essentially separated or dissociated.

2.19 BOND LENGTH d_0

The interatomic distance d_0 corresponding to the minimum in the Morse curve for the bonded atoms of interest.

Example. The bond length for the H–H molecule, d_0 in Fig. 2.18, is 0.74 Å. For the molecule H–F, $d_0 = 0.92$ Å.

2.20 BOND DISSOCIATION

The cleavage of a covalent bond into two fragments. The dissociation may be *homolytic*, in which case the two fragments each possess one electron of the original covalent bond, or the cleavage may be *heterolytic*, in which case one fragment possesses both electrons of the orginal covalent bond.

Example. Homolytic cleavage of the H–H bond gives the fragments H· and H·, while heterolytic cleavage gives the fragments $H^+ + :H^-$.

2.21 BOND DISSOCIATION ENERGY D_0

The energy (heat or light) that must be supplied to a covalent bond in order to break the bond by homolytic cleavage.

Example. For H_2, the energy represented by D_0 in Fig. 2.18 is the dissociation energy $\{0 - (-432.0) = +432.0\,\text{kJ mol}^{-1}\}$; the plus sign signifies an endothermic process, that is, it requires energy. Note that the level marked -432.0 in Fig. 2.18 represents the lowest vibrational state of the molecule. The lower level, $D_e = -457.9$ in Fig. 2.18, represents the potential energy at the minimum of the Morse curve, which is 25.9 kJ mol^{-1} below the energy of the lowest vibrational level. This energy difference $(D_e - D_0)$ is called the *zero point energy* of the bond.

2.22 BOND ANGLE

The angle formed by two bonds joined to a common atom. In measuring the angle, all three atoms are treated as points.

Example. The H–O–H angle (104.5°) in H_2O (Fig 2.22*a*); the six H–C–H angles in CH_4 (Fig. 2.22*b*), all of which are 109.5°.

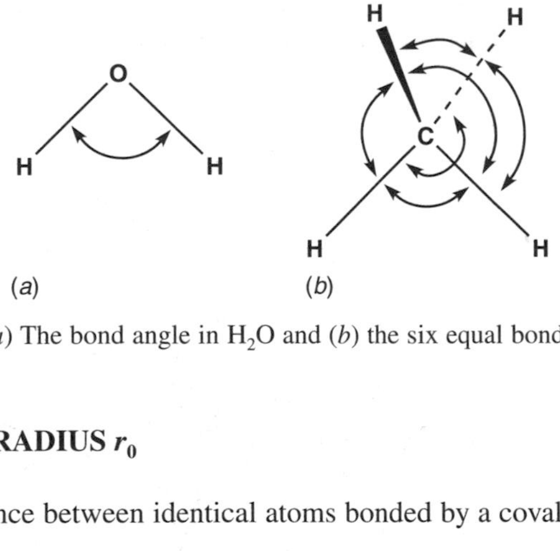

Figure 2.22. (*a*) The bond angle in H_2O and (*b*) the six equal bond angles in CH_4.

2.23 ATOMIC RADIUS r_0

One-half the distance between identical atoms bonded by a covalent bond.

Example. For a particular covalent bond between two nonidentical atoms, the individual atomic radii of each of the two atoms are added to calculate the bond length. Thus for the C–Cl bond, the covalent radius of the chlorine atom is taken as one-half the Cl–Cl distance of 1.998 Å in $Cl_2 \approx 1.00$ Å, and the carbon radius is taken as one-half the C–C bond distance in diamond of 1.54 Å = 0.77 Å. The C–Cl distance calculated from the sum of the above atomic radii is then 1.00 Å + 0.77 Å = 1.77 Å.

This is approximately equal to the C–Cl distance of 1.761 Å found in CCl_4. If atoms are bonded by double and triple bonds, the atomic radii will, of course, be considerably less than that of the comparable singly bonded atoms. Thus, the atomic radius of carbon in $CH_2{=}CH_2$ is one-half the C=C distance of 1.35 Å or 0.68 Å compared to the 0.77 Å atomic radius of single-bonded carbon in diamond. The atomic radius of carbon in HC≡CH is 0.65 Å.

2.24 IONIC RADIUS r^+ AND r^-

Because the outer edge of an ion is not well defined, the radius (r^+ for cations and r^- for anions) is evaluated from the measured distances between centers of nearest-neighbor ions.

Example. The crystal lattice of sodium fluoride shows that the shortest Na^+ F^- distance is 2.31 Å, and if it is assumed that these ions are in contact, then this distance is the sum of the ionic radii: $r^+(Na^+) + r^-(F^-) = 2.31$ Å. However, this result does not tell us what the atomic radius of each of the ions actually is, and as a matter of fact, the solution to this problem remains unresolved. Many approaches, both empirical and theoretical, have been used to determine ionic radii.

To give one example: In an ion pair in which the anion is much larger than the cation, for example, LiI, it may be assumed that the anions touch each other and hence the ionic radius of the iodide ion would be half the measured distance between adjacent iodides: $r^- = 4.28$ Å$/2 = 2.14$ Å (Fig. 2.24). The experimental Li–I distance is 3.02 Å, and if it is assumed that the cations and anions are in contact, then the ionic

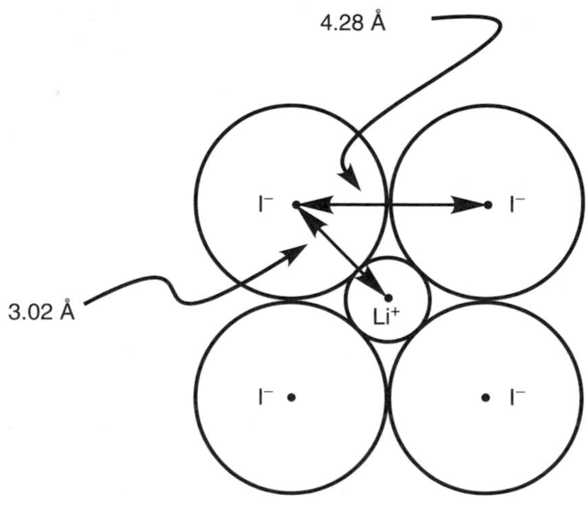

Figure 2.24. The measured I–I and Li–I distances in crystals of LiI from which the ionic radii of the iodide ion ($r^- = 2.14$ Å) and the lithium ion ($r^+ = 3.02$ Å $- 2.14$ Å $= 0.88$ Å) are calculated.

radius for Li^+ must be $r^+ = 3.02$ Å − 2.14 Å = 0.88 Å. Now knowing the ionic radius of Li^+ and measuring the distance between Li^+ and other halides in LiX crystals, we may determine the ionic radius of the other halides, X^-.

However, the values obtained this way usually do not correspond exactly to values obtained by other methods. Nevertheless, more or less similar values for ionic radii appear in many tables. The ionic radius will depend on whether the ion is univalent or multivalent and on the number of ions having different charges that are in contact with ions of opposite charge. For different ions having the same number of electrons, the ion with larger nuclear charge has the smaller radius. Thus, the ionic radius of Cl^- is 1.70 Å, but that of isoelectronic Ca^{2+} is 1.18 Å. Cations have smaller radii than their parent element [r_0(Na) = 1.57 Å; $r^+(Na^+)$ = 0.98 Å] and anions have larger radii than their parent element [r_0(F) = 0.64 Å; $r^-(F^-)$ = 1.33 Å].

2.25 VAN DER WAALS RADIUS

One-half the smallest distance between nonbonded atoms on adjacent molecules. This distance corresponds to the minimum in the potential energy of the two atoms as they approach each other. The minimum corresponds to a balance between the *van der Waals attraction* of the nuclei of one of the atoms for the valence electrons of the neighboring atom (and vice versa) and the *van der Waals repulsion* resulting from the inner penetration of the electron clouds of the near-neighbor atoms (both named after Johannes D. van der Waals, 1837–1923). The van der Waals radius is a measure of the effective size of an atom that is, the volume it occupies in space and into which no other atom can intrude. In a sense, the sum of the two van der Waal radii reflects the distance between atoms when they just barely "touch" each other. The interaction between nonbonded atoms on adjacent molecules applies as well to nonbonded atoms within the same molecule that touch each other, and thus, van der Waals forces influence the shape of a molecule.

Example. The general situation in which the potential energy E_p changes as a function of the distance between nonbonded atoms in the same or adjacent molecules can be represented by a Morse curve, analogous to that shown for the hydrogen molecule in Fig. 2.18. Just as in the case of the hydrogen molecule, when two nonbonded atoms in different molecules approach each other from a large distance, there is at first a decrease in E_p as the nucleus of one partner attracts electrons of the other partner. This van der Waals attraction is followed by a very sharp increase in E_p, van der Waals repulsion, as the neighbors come close and the electrons associated with one partner start to encounter the repulsive effect of the electrons of its neighbor. For van der Waal interactions, the minimum in the Morse curve represents a distance corresponding to the sum of the van der Waals radii of the two atoms involved. In CCl_4, for example, the shortest distance between Cl atoms on nearest-neighbor molecules is 3.6 Å and the van der Waals radius of the chlorine atom is therefore taken to be 1.8 Å.

2.26 COORDINATE COVALENT BOND (DATIVE BOND)

A covalent bond that may be represented as arising from the donation of an unshared pair of valence electrons on one atom (the donor) to an empty valence orbital on another atom (the acceptor). After the bond is formed, there is no way to distinguish this bond from any other covalent bond.

Example. The coordinate covalent bond is also called a dative (from the Latin for "giving") bond and frequently represented by an arrow from the donor atom to the acceptor atom, for example, $H_3N\!:\rightarrow BF_3$. The ammonium ion may be represented as being formed by the donation of the unshared pair of electrons on nitrogen into the empty orbital of the proton: $H_3N\!: + H^+ \rightarrow [NH_4]^+$.

2.27 HYDROGEN BOND

The bond that results when a hydrogen atom serves as a bridge between two electronegative elements. In such bonding one hydrogen atom is bonded to an electronegative atom A by a conventional covalent bond, and is loosely bonded to the second electronegative element B by electrostatic (dipole–dipole) forces, A–H$\cdots$B. The electronegative atom A withdraws electrons from the covalently bonded hydrogen atom attached to it, and the resulting partial positive character of the hydrogen atom attracts electrons from the donating partner.

Example. Molecules involved in strongly hydrogen bonded interactions are typically those containing O, N, and F atoms. Hydrogen atoms on any of these elements have partial positive character, whereas lone pair electrons on any of the same atoms on neighboring molecules act as donor sites to form intermolecular hydrogen bonds. Homonuclear intermolecular hydrogen bonding (B=A) can be very important in determining the properties of molecules. The relatively high boiling point of water is due to the strong intermolecular hydrogen bonding in H_2O (Fig. 2.27*a*), which is much stronger than that in liquid NH_3 (Fig. 2.27*b*), as evidenced by the fact that H_2O boils at 100°C while liquid NH_3 boils at −33°C. The hydrogen bonding between NH_3 and H_2O (Fig. 2.27*c*) is even stronger than that between H_2O molecules and is particularly effective because the hydrogen on the more electronegative oxygen of water is a better electron pair acceptor (more acidic) than the hydrogen on the less electronegative nitrogen. In addition, the lone pair on nitrogen in NH_3 is a better donor (more basic) than the lone pair on oxygen in H_2O. Indeed, ammonia gas is almost explosively soluble in water. Hydrogen bonds usually have a bond energy of about 15 to 20 kJ (3–6 kcal) mol^{-1} compared to the typical covalent bond energy that usually ranges from 200 to 400 kJ (50–100 kcal) mol^{-1}. Hydrogen bonding may have a profound effect on some of the physical as well as chemical properties of molecules. It is very important in many biochemical systems; the base pairs in DNA, for example, are held together by

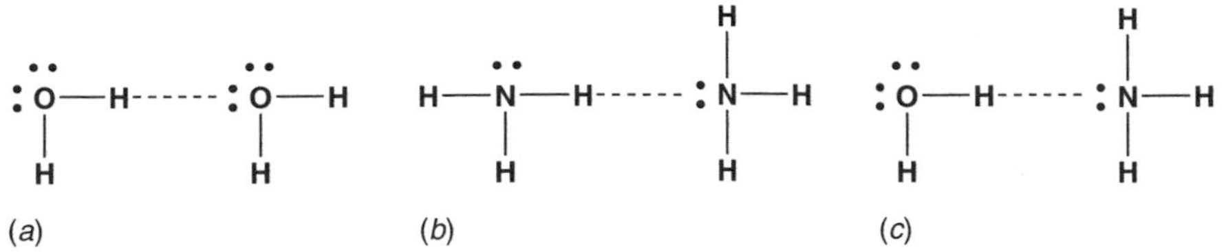

Figure 2.27. Hydrogen bonding in (*a*) H_2O, (*b*) NH_3, and (*c*) a solution of NH_3 in H_2O.

hydrogen bonds. Both cations and anions may be involved in hydrogen bonding; indeed, the hydrogen bond in $F^- \cdots H–F$ has a bonding energy of 150 to 159 kJ mol^{-1} (36–38 kcal mol^{-1}).

2.28 VALENCE SHELL ELECTRON PAIR REPULSION (VSEPR)

Pairs of electrons in the valence shell of an atom in a molecule, whether present as covalent bonds or as electron lone pairs, tend to repel each other and move as far apart as possible. Such repulsion determines, in the first approximation, the shape of the molecule.

Example. When the 2nd row elements C, N, O, and F are present in a molecule, they tend to be surrounded by an octet of electrons in their valence shells (isoelectronic with Ne). This means that carbon will share its four valence electrons with four valence electrons from other atoms, thereby generating four covalent bonds. Nitrogen having five valence electrons will require three additional valence electrons for its octet, which can be obtained by covalent bonding with three atoms; oxygen having six valence electrons will bond with two atoms; and fluorine, with seven valence electrons, will bond with one atom. In forming covalent bonds with other atoms, the resulting molecule will assume a preferred geometry that minimizes electron repulsion between pairs of surrounding electrons. A carbon atom bonded to four other atoms leads to a tetrahedral disposition of the four atoms so that, for example, in methane all six H–C–H angles are 109.47° (Fig. 2.22*b*). In compounds involving the N atom covalently bonded to three other atoms, there will be one unshared pair of electrons on N. The volume requirements of an unshared pair of electrons is greater than that of a shared pair, and thus, the three H–N–H bond angles in NH_3, for example, (Fig. 2.28*a*) are 107° and, thus, somewhat smaller than the tetrahedral angles in CH_4. In compounds of O bonded to two other atoms, there will be two unshared pairs on the O, and since these have relatively larger volume requirements than is the case with NH_3 with only one unshared pair, the result is that the H–O–H angle in H_2O (Fig. 2.22*a*) is 104.5°.

Molecular geometry, therefore, depends on the number of shared and unshared pairs of electrons that surround the atom and can usually be qualitatively predicted

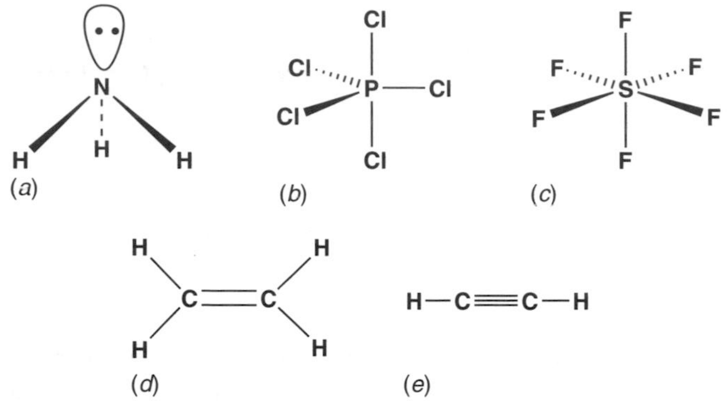

Figure 2.28. Geometry of some simple molecules: (*a*) trigonal pyramidal NH_3, (*b*) trigonal bipyramidal PCl_5; (*c*) octahedral SF_6; (*d*) trigonal planar carbons in C_2H_4; (*e*) linear carbons in C_2H_2.

on the basis of this VSEPR model. In the case of a compound such as PCl_5 with five electron pairs surrounding the P atom, the model suggests a trigonal bipyramidal geometry (Fig. 2.28*b*). In the compound SF_6 with six electron pairs surrounding the S atom, an octahedral geometry (Fig. 2.28*c*) is preferred. In every case the preferred geometry is that in which the electron pairs are arranged as far from each other as possible. In the VSEPR model all multiply bonded atoms are treated as though they were singly bonded atoms. Thus, in considering the H–C–H angle in ethylene (Fig. 2.28*d*), the VSEPR model suggests that each carbon atom is surrounded by three (not four) atoms. Optimum bond angles for three atoms disposed around a central atom are 120°. In acetylene (Fig 2.28*e*), the model requires two atoms around each carbon atom, suggesting H–C–C angles of 180°, and indeed acetylene is a linear molecule. Thus, the shape is determined (to a first approximation) by the number of atoms plus lone pairs of electrons around the central atom.

The description of the shape of the NH_3 molecule as *trigonal pyramidal* warrants comment. This is a term that describes the geometric arrangement of the three hydrogen atoms bonded to the nitrogen atom. The VESPR model focuses on electron pair geometry around the nitrogen atom. The nitrogen atom is surrounded by four electron pairs and the H–N–H angles in NH_3 are therefore approximately tetrahedral.

2.29 MOLECULAR ORBITALS

A wave description of the size, shape, and orientation of the region in space available to an electron, which extends over two or more atoms and is generated by combining two or more localized atomic orbitals (see Sect. 1.15).

2.30 MOLECULAR ORBITAL (MO) THEORY

This theory postulates that individual atomic orbitals (already at equilibrium bonding distance) are combined into molecular orbitals. In combining atomic orbitals, the core orbitals of the atoms remain unperturbed, but the valence orbitals are stripped of electrons; the electrons are then pooled and placed one by one into the polycentric molecular orbitals according to the Aufbau principle (see Sect. 1.41) and the Pauli exclusion principle (see Sect. 1.39).

2.31 BONDING MOLECULAR ORBITALS

Molecular orbitals that are generated by positive (attractive) overlap of the atomic orbitals involved in their formation. Bonding molecular orbitals are lower in energy than the isolated atomic orbitals that are combined to make them (see MOEDs, Sect. 2.42). *Nonbonding molecular orbitals* have approximately the same energy as the atomic orbitals from which they are generated. Electrons in bonding molecular orbitals are more likely to be found somewhere between the interacting atoms and they serve to bond together these atoms.

2.32 ANTIBONDING MOLECULAR ORBITALS

Molecular orbitals that are generated by negative (repulsive) overlap of the atomic orbitals involved in their formation. Antibonding molecular orbitals are higher in energy than the isolated atomic orbitals from which they are generated. Electrons in these orbitals are not found midway between the interacting atoms because of the nodal plane, but tend to maximize in the vicinity of the atoms involved in the bond. Electrons in antibonding orbitals serve to repel these atoms.

2.33 LINEAR COMBINATION OF ATOMIC ORBITALS (LCAO)

This is a method of generating molecular orbitals by adding and subtracting atomic orbitals; the number of molecular orbitals so generated must be equal to the number of atomic orbitals that are combined (the so-called basis set; see Sect. 2.34).

Example. When two *s* atomic orbitals, ϕ_A and ϕ_B, are combined, the linear combinations that result are $(\phi_A + \phi_B)$ and $(\phi_A - \phi_B)$. The combinations are shown pictorially in Fig. 2.33*a*. The combination made by addition, ψ_b, is a bonding molecular orbital, and that made by subtraction, ψ_a, is an antibonding molecular orbital. In these depictions black and white areas represent lobes or orbitals of different signs; thus, a white area adjacent to a black area represents a change in sign or a node (see Sect. 1.7). Figure 2.33*b* shows the end-on combinations of *p* orbitals. The potential energy of a bonding molecular orbital is lower, and that of an antibonding molecular orbital

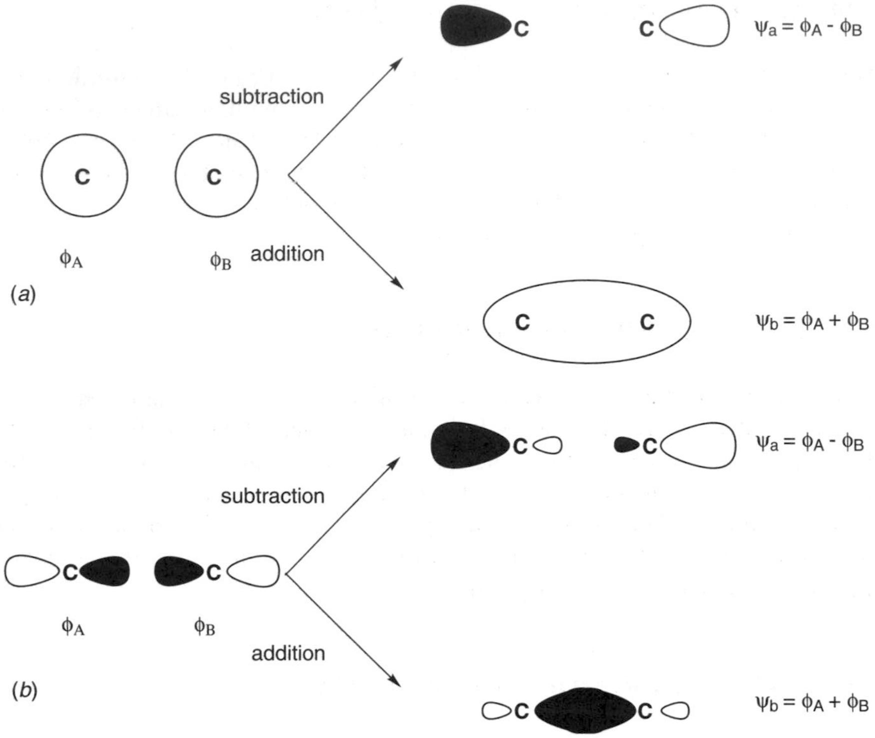

Figure 2.33. (*a*) The linear combination of *s* orbitals; (*b*) the linear combination of end-on *p* orbitals. The subscripts *a* and *b* refer to antibonding and bonding orbitals, respectively.

higher, than the potential energy of the atomic orbitals from which the molecular orbitals are generated.

2.34 BASIS SET OF ORBITALS

The combination or set of atomic orbitals that is used in generating the molecular orbitals of interest in a molecule.

Example. Two molecular orbitals (Fig. 2.33*a*) are generated from the basis set of the two 1*s* atomic orbitals.

2.35 σ BONDING MOLECULAR ORBITAL (σ ORBITAL); σ BOND

The molecular orbital produced by the in-phase (same sign) overlap of atomic orbitals on adjacent atoms and aligned along the internuclear axis. Such orbitals are symmetric with respect to rotation around the internuclear axis, and they are lower in energy than the corresponding atomic orbitals from which they are generated. The bond formed by electron occupation of this molecular orbital is called a *σ bond* because this molecular orbital is symmetric with respect to rotation around the bond axis.

Example. ψ_b in Fig. 2.33*a*; also the orbital ψ_b produced by end-on overlap of atomic *p* orbitals on adjacent atoms (Fig. 2.33*b*). This latter orbital has cylindrical symmetry analogous to ψ_b in Fig. 2.33*a*.

2.36 σ ANTIBONDING MOLECULAR ORBITAL (σ* ORBITAL)

The antibonding molecular orbital encompassing adjacent atoms generated by the out-of-phase (opposite signs) overlap of atomic orbitals that are aligned along the internuclear axis. Such orbitals are symmetric with respect to rotation around the internuclear axis but have a nodal plane perpendicular to the internuclear axis, and are higher in energy than the corresponding atomic orbitals from which they were generated.

Example. ψ_a in Figs. 2.33*a* and *b* are σ antibonding orbitals. To distinguish these from σ bonding orbitals, they are written with an asterisk superscript (σ*) and called *sigma star* orbitals. Note that these orbitals have the cylindrical symmetry of a σ orbital.

2.37 *pπ* ATOMIC ORBITAL

This is a *p* orbital that takes part in π-type bonding.

2.38 π BONDING MOLECULAR ORBITAL (π ORBITAL)

The molecular orbital encompassing adjacent atoms produced by the in-phase overlap (addition) of parallel *p* atomic orbitals perpendicular to the internuclear axis. Such orbitals have a nodal plane and are antisymmetric with respect to rotation around the internuclear axis.

Example. In constructing the LCAO of *pπ* atomic orbitals of ethylene, two possible combinations are shown in Fig. 2.38. The π bonding molecular orbital ψ_b is shown in Fig. 2.38*b*, the π antibonding molecular orbital ψ_a in Fig. 2.38*a*, and the wave character corresponding to molecular orbitals ψ_a and ψ_b in Fig. 2.38*c*.

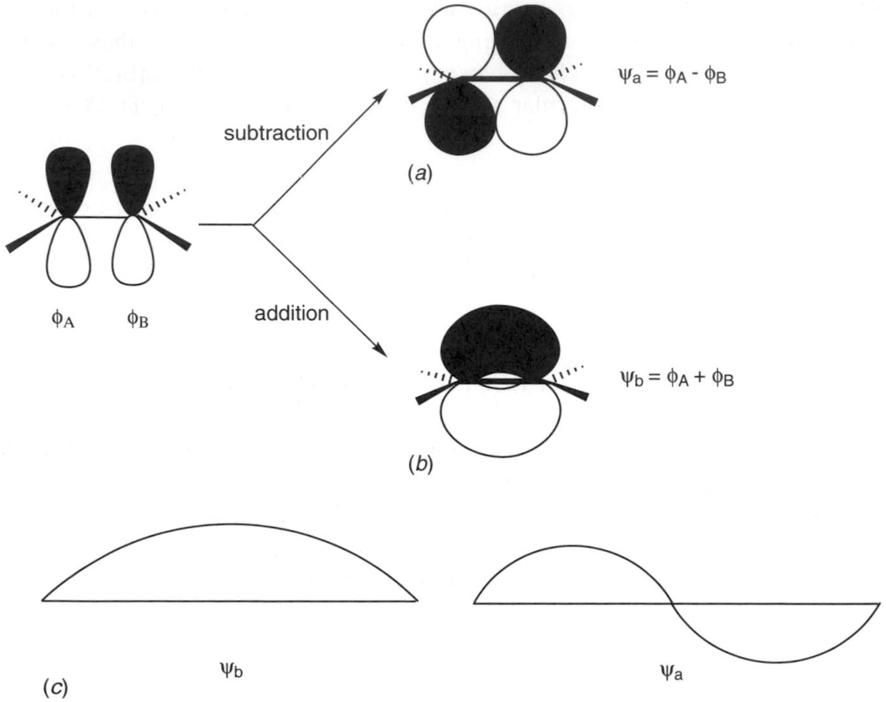

Figure 2.38. The linear combination of p orbitals to form π orbitals: (a) ψ_a antibonding (out-of-phase overlap); (b) ψ_b bonding (in-phase overlap); and (c) the wave character of ψ_a and ψ_b. The subscripts a and b refer to antibonding and bonding orbitals, respectively.

2.39 LOCALIZED π BOND

The bond that is formed by electron occupation of the π bonding molecular orbital. Like the p atomic orbitals from which it is formed, the π bond has a nodal plane.

Example. The π bond of ethylene (Fig. 2.38b). This orbital is antisymmetric with respect to rotation around the internuclear axis and has ungerade (antisymmetric with respect to the center of the molecule) character, hence π_u. (See Chapter 4 for an explanation of symmetry terms.)

2.40 π ANTIBONDING MOLECULAR ORBITAL (π* ORBITAL)

The molecular orbital generated by the out-of-phase overlap (subtraction) of parallel p orbitals on adjacent atoms. Such orbitals are antisymmetric with respect to 180° rotation around the internuclear axis and have a nodal plane perpendicular to that bond axis. This orbital has gerade (symmetric with respect to the center of the molecule) character, hence π_g^*.

Example. See Fig. 2.38*a*.

2.41 σ SKELETON

A framework of all the atoms in a molecule attached to each other by σ bonds and neglecting all π bonds.

Example. The σ bond skeleton for ethylene and acetylene, Fig. 2.41. The complete C–C bonding in ethylene is composed of five σ bonds and one π bond whereas in acetylene in addition to three σ bonds, there are two π bonds, one in the plane of the paper and the other perpendicular to the plane of the paper.

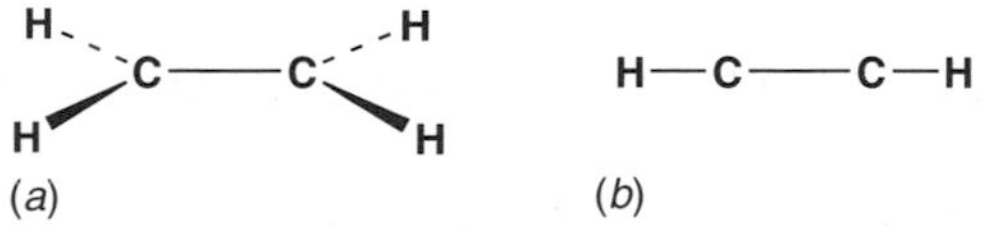

Figure 2.41. The σ skeleton of (*a*) ethylene and (*b*) acetylene.

2.42 MOLECULAR ORBITAL ENERGY DIAGRAM (MOED)

This diagram displays the relative energies of the atomic orbitals to be combined (the basis set) in the LCAO treatment and the energies of the resulting molecular orbitals formed by the combination.

Example. The MOED for H_2 and that for the *pπ* combinations in ethylene (C_2H_4) are shown in Figs. 2.42*a* and *b*, respectively. The crosses on the lines represent electrons, placed in the lowest-energy available orbitals.

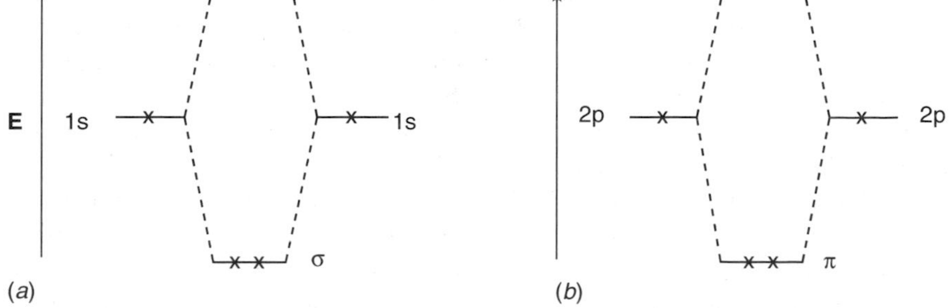

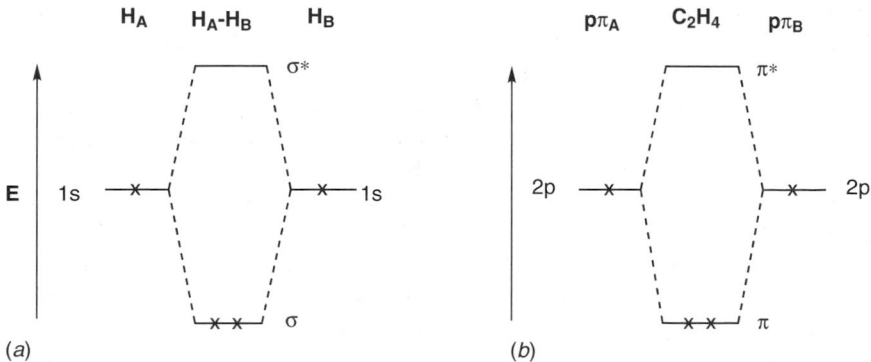

Figure 2.42. Molecular orbital energy diagrams: (*a*) H_2 and (*b*) ethylene π system.

2.43 ELECTRONIC CONFIGURATION OF MOLECULES

The enumeration of the occupied molecular orbitals, in order of energy, consisting of the orbital designation, followed by an appropriate superscript indicating the number of electrons in the orbital. If the molecule has a center of symmetry then the *gerade* or *ungerade* character of the orbital is added as the subscript *g* or *u* as appropriate. Usually for single-electron occupancy, the superscript is omitted. In discussions of molecules with π electron systems where the emphasis is on the π bonds, the σ bond orbitals are omitted. The enumeration of electronic configurations of molecules is analogous to that used for atoms (see Sect. 1.42).

Example. The electronic configuration of the hydrogen molecule, H_2 (Fig. 2.42*a*), is σ_g^2 and that of ethylene (Fig. 2.42*b*) is π_u^2. If one of the electrons in the doubly occupied π orbital of the π system in ethylene were promoted to the π* orbital, the resulting electronic configuration would be denoted as $\pi_u\pi_g^*$.

2.44 MOED FOR 2ND ROW HOMODIATOMIC MOLECULES

The molecular orbital energy diagrams for dioxygen, O_2, constructed from the combination of the valence orbitals of the two atoms are shown in Fig. 2.44*a*. This figure neglects the hybridization of the atomic orbitals. The combination of the two 2*s* atomic orbitals generates the two molecular orbitals $1\sigma_g$ and $1\sigma_u^*$. The $2p_z$ orbitals on each atom point to each other along the internuclear axis, and these two orbitals combine to generate the two molecular orbitals $2\sigma_g$ and $2\sigma_u^*$. The degenerate $2p_x$ and $2p_y$ atomic orbitals on each atom combine to generate the degenerate π_u molecular orbitals and the degenerate π_g^* orbitals. To complete the MOED, it is now only necessary to place the 12 valence electrons (six from each oxygen atom) in the molecular orbitals according to the Aufbau principle and Hund's rule. The resulting MOED, (Fig. 2.44*a*) shows one electron in each of the two antibonding π_g^* orbitals. This accounts for one of the most unusual features of O_2, namely its biradical character and, hence, its paramagnetic properties. For the MOED of F_2, two valence electrons more than for dioxygen are required and these completely fill the π_g^* orbitals. The four electrons in π_g^* antibonding orbitals cancel the bonding effect of four electrons in π_u bonding orbitals, and the two electrons in $1\sigma_u^*$ cancel the bonding effect of the two electrons in $1\sigma_g$. These 12 electrons, six pairs total or three pairs on each F atom, may be considered as three unshared pairs on each F since they contribute nothing to the bonding. As a result, the two F atoms are held together by the σ sigma bonding ($2\sigma_g$) molecular orbital. There is thus considerable justification for writing the Lewis dot structure for F_2 as :F̤̈–F̤̈:.

The MOED for dinitrogen, N_2 (Fig. 2.44*b*), is somewhat different from that of O_2 and F_2. The $1\sigma_g$ and $2\sigma_g$ molecular orbitals although generated from *s* and *p* orbitals, respectively, are both σ orbitals and thus possess the same symmetry. Accordingly, they may mix and split as shown in Fig. 2.44*b* to give two new σ orbitals, one lower in energy than the orginal $1\sigma_g$ and one higher in energy than the original $2\sigma_g$. Such

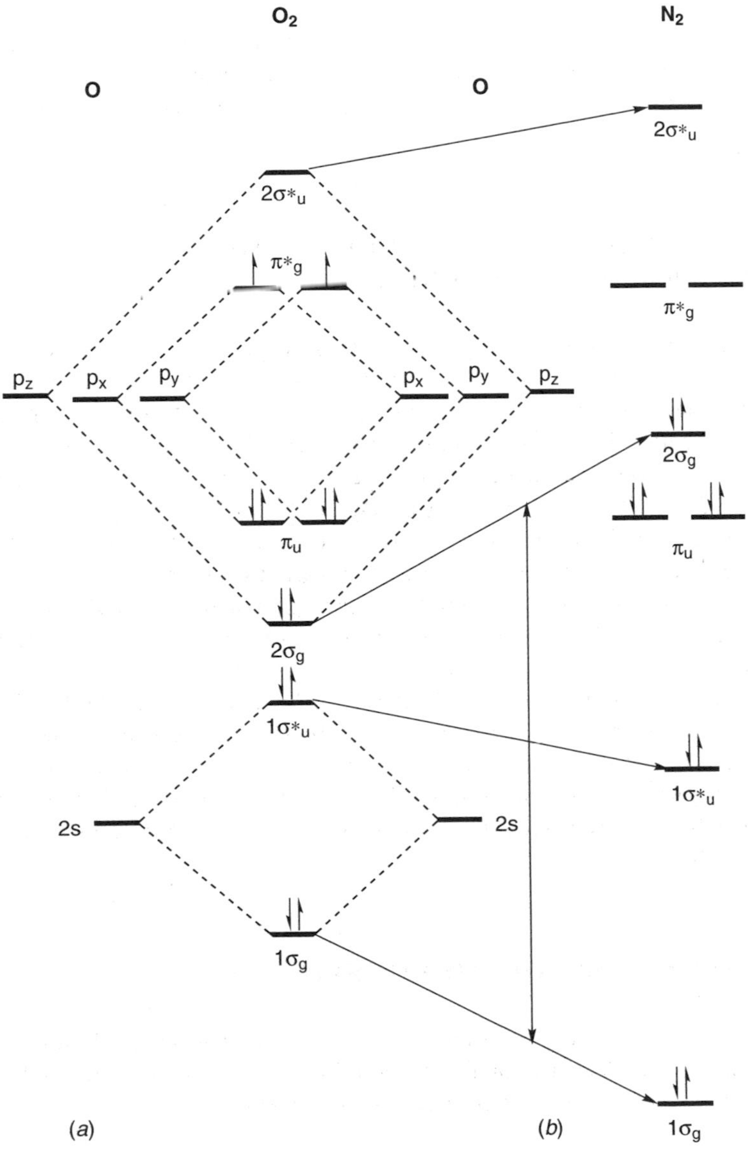

Figure 2.44. The MOED for (*a*) dioxygen, O_2, and (*b*) for dinitrogen, N_2, with the large mixing between $1\sigma_g$ and $2\sigma_g$ and the smaller mixing of $1\sigma_u^*$ and $2\sigma_u^*$ emphasized. The bonding and antibonding molecular orbitals are given numbers in order of their increasing energies.

mixing of molecular orbitals is analogous to the hybridization of atomic orbitals (see Sect. 1.47). In the case of dioxygen and difluorine, this mixing is not very large, but in the case of dinitrogen the mixing is quite large (partly because of the lower nuclear charge on nitrogen atoms as compared to oxygen and fluorine), which makes these two orbitals closer in energy than is the case with oxygen and fluorine. The result of the splitting is that the $2\sigma_g$ orbital of N_2 becomes higher in energy than the π_u orbitals, a fact that is unusual because π bonding molecular orbitals in organic molecules are usually higher in energy than σ bonding molecular orbitals.

2.45 MOED FOR THE 2ND ROW HETEROATOMIC MOLECULE; CARBON MONOXIDE

The molecular orbital energy diagram for CO is shown in Fig. 2.45. Only the valence orbitals on each atom are considered. In this case, it is convenient to use the artificial technique of hybridize first, by combining the s and p_z orbitals on each atom. The hybridization in each case raises the energy of the s orbitals and lowers the energy of the p_z orbitals. However, since the higher-energy sp hybrid orbital on oxygen, hybrid (2), which has more p than s character, is a good energy match with the lower energy of hybrid (1) on carbon, which has more s than p character, these two hybrid orbitals combine to form σ and σ^* orbitals as shown. The p_x and p_y orbitals on each atom combine to form degenerate pairs of π and π^* orbitals. Now with all the molecular orbitals in place, it remains to distribute the 10 valence electrons in accordance with the Aufbau and Pauli principles with the result shown in Fig. 2.45. The electrons in the higher-energy hybrid (2) orbital on carbon, which has larger p character than hybrid (1), are not involved in bonding to oxygen and are nonbonding electrons, and the orbital that they occupy is called a *nonbonding molecular orbital* designated as n. Analogously, the electrons in the lower hybrid orbital on oxygen, hybrid(1), are not involved in bonding to carbon and are thus also in a nonbonding orbital n.

2.46 COEFFICIENTS OF ATOMIC ORBITALS c_{ij}

The number c_{ij} $(-1 < c_{ij} < +1)$ indicates the degree to which each atomic orbital participates in a molecular orbital. This definition and the ones immediately following are those used in simple π electron (Hückel) theory.

Example. Figures 2.33*b* and 2.38 show how two atomic orbitals combine to generate one bonding and one antibonding molecular orbital. The correct wave functions require that the coefficient on each atomic orbital in all of the molecular orbitals be $\pm 1/\sqrt{2}$ (see below); thus, $\psi_b = (1/\sqrt{2})\phi_A + (1/\sqrt{2})\phi_B$ and $\psi_a = (1/\sqrt{2})\phi_A - (1/\sqrt{2})\phi_B$. When coefficients need to be designated, especially for the purposes of calculation, they are indicated by the letter c followed by two subscripts ij, where j refers to the jth atomic orbital in the ith molecular orbital. Thus, in the molecular orbital ψ_a (Fig. 2.38), $c_{21} = 1/\sqrt{2}$ and $c_{22} = -1/\sqrt{2}$.

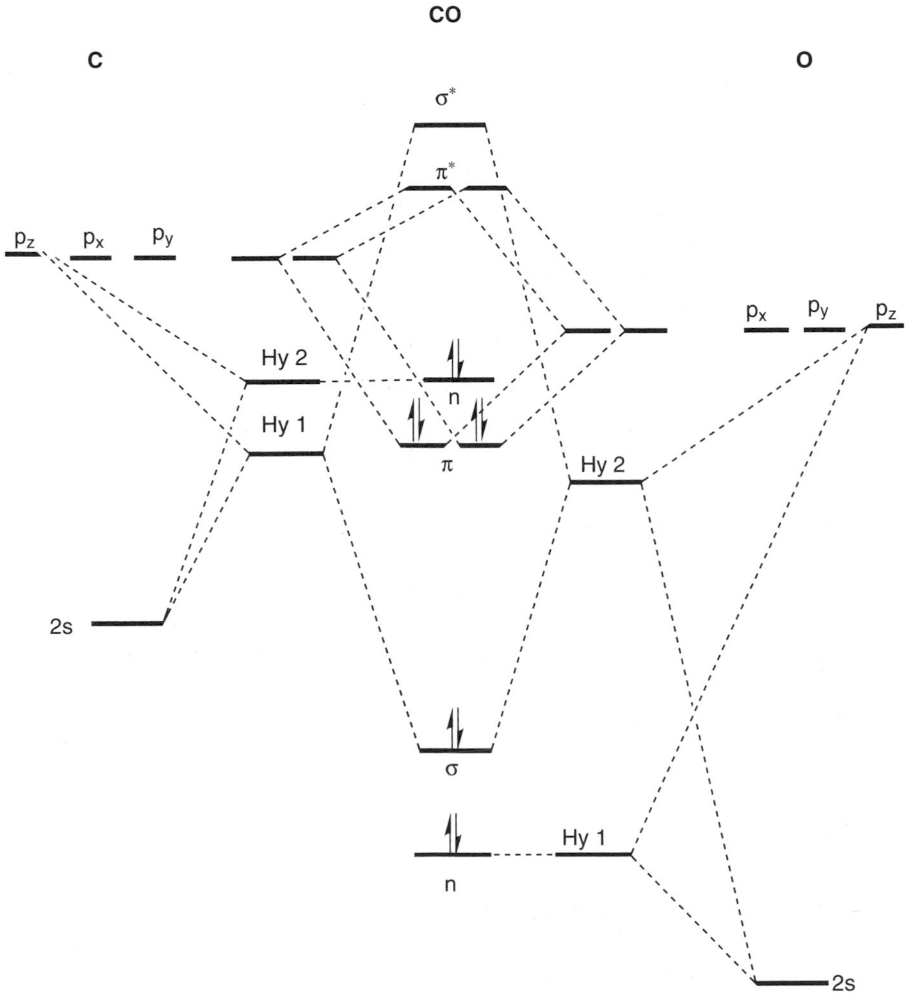

Figure 2.45. The MOED for carbon monoxide.

2.47 NORMALIZED ORBITAL

This is an orbital meeting the requirement that the integration of the square of its wave function over all space is equal to unity; that is, $\int \psi^2 d\tau = 1$, that is, the probability of finding an electron within the bounds of space defined by the orbital is 1 or 100%. An alternate statement of this requirement is that the sum of the squares of the coefficients of all atomic orbitals in a molecular orbital is equal to unity; that is, $\sum_j c_{ij}^2 = 1$. Atomic orbitals ϕ are always assumed to include the normalization factor and hence $\int \phi^2 d\tau = 1$.

2.48 NORMALIZATION

The process of calculating a normalized orbital ψ, which is accomplished by multiplying it by a number N, such that $\int(N\psi)^2 d\tau = 1$.

Example. Neither the orbital ψ_b nor ψ_a (Figs. 2.38*a* and *b*), are normalized. To normalize, for example, the wave function $\psi_a = (\phi_A - \phi_B)$: $\int N^2(\phi_A - \phi_B)^2\, d\tau = \int N^2(\phi_A^2 - 2\phi_A\phi_B + \phi_B^2)\, d\tau = N^2\int\phi_A^2\, d\tau - 2N^2\int\phi_A\phi_B\, d\tau + N^2\int\phi_B^2\, d\tau = 1$. In Sect. 2.47 it was pointed out that $\int\phi_A^2 d\tau = \int\phi_B^2 d\tau = 1$. The integral $\int\phi_A\phi_B\, d\tau$, called the overlap integral, is for most simple approximations assumed to be equal to zero. Hence, the above equation simplifies to $N^2[1 + 0 + 1] = 1$, and $N = \frac{1}{\sqrt{2}}$ and $\psi_a = \frac{1}{\sqrt{2}}(\phi_A - \phi_B) = \frac{1}{\sqrt{2}}\phi_A - \frac{1}{\sqrt{2}}\phi_B$. That this orbital is indeed normalized is apparent from the requirement $c_{21}^2 + c_{22}^2 = 1$, since $\left(\frac{1}{\sqrt{2}}\right)^2 + \left(-\frac{1}{\sqrt{2}}\right)^2 = 1$

2.49 ORTHOGONAL ORBITALS

These are any two molecular orbitals ψ_r and ψ_s meeting the requirement that their product integrated over all space is equal to zero; that is, $\int\psi_r\psi_s\, d\tau = 0$ any two molecular orbitals in a set must be completely independent of each other.

Example. The orbitals $\psi_b = \frac{1}{\sqrt{2}}\phi_A + \frac{1}{\sqrt{2}}\phi_B$ and $\psi_a = \frac{1}{\sqrt{2}}\phi_A - \frac{1}{\sqrt{2}}\phi_B$ are seen to be orthogonal, that is,

$$\int\left(\frac{1}{\sqrt{2}}\phi_A + \frac{1}{\sqrt{2}}\phi_B\right)\left(\frac{1}{\sqrt{2}}\phi_A - \frac{1}{\sqrt{2}}\phi_B\right) d\tau = \frac{1}{2}\int\phi_A^2\, d\tau - \frac{1}{2}\int\phi_B^2\, d\tau = \frac{1}{2} - \frac{1}{2} = 0$$

2.50 ORTHONORMAL ORBITALS

This is a set of orbitals that are both orthogonal and normalized.

2.51 WAVE FUNCTIONS IN VALENCE BOND (VB) THEORY

The orbital description of the theory that considers the bond between adjacent atoms to be formed by bringing together two distinct valence atomic orbitals on these atoms, each containing one electron of the opposite spin.

Example. As two hydrogen atoms [$H_A(1)$ and $H_B(2)$] approach each other, the 1*s* orbital overlap increases and the electron in each orbital feels the influence of the opposite nucleus and is increasingly associated with it. The molecule is stabilized at the appropriate H–H bond distance by the exchange between two equivalent structures: $H_A(1)\ H_B(2)$ and $H_A(2)\ H_B(1)$, where A and B refer to the hydrogen atoms and (1) and (2) refer to electrons of opposite spin. The wave function for the occupied orbital of the hydrogen molecule may then be written as $\psi = [(\phi_A(1)\phi_B(2) + \phi_A(2)\phi_B(1)]$, where ϕ_A and ϕ_B are the atomic 1*s* orbitals on hydrogens A and B. This wave function indicates that the two electrons are never associated together at one atom.

SUGGESTED READING

Bowser, J. R. *Inorganic Chemistry*. Brooks/Cole: Pacific Grove, CA, 1993.

Coulson, C. A. *Valence*. Oxford University Press: London, 1952.

Gray, H. B. *Electrons and Chemical Bonding*. W.A. Benjamin: New York, 1964.

Harris, D. C., and Bertolucci, M. D. *Symmetry and Spectroscopy*. Oxford University Press: New York, 1978.

Jaffé, H. H. and Orchin, M. “Hybridization in Carbon Monoxide.” *Tetrahedron 10*, 212 (1960).

Jean, Y., Volatron, F., and Burdett, J. *An Introduction to Molecular Orbitals*. Oxford University Press: New York, 1993.

Orchin, M., and Jaffé, H. H. *Symmetry, Orbitals and Spectra*. Wiley-Interscience: New York, 1971.

Streitwieser, A. *Molecular Orbital Theory for Organic Chemists*. John Wiley & Sons: New York, 1961.

3 Delocalized (Multicenter) Bonding

3.1	Delocalized Bond	56
3.2	Delocalized Bonding	56
3.3	Delocalized π Molecular Orbitals	56
3.4	Resonance Structures	58
3.5	Curved Arrow (Electron Pushing) Notation	59
3.6	Conjugated Systems	59
3.7	Essential Single Bond	60
3.8	Essential Double Bonds	61
3.9	Hyperconjugation	61
3.10	Resonance Energy	62
3.11	The Free Electron Method (FEM)	62
3.12	Electron Density q_r	64
3.13	Charge Density ζ (Charge Distribution)	64
3.14	π Bond Order ρ_{rs}	64
3.15	Free Valence Index F_r	65
3.16	Hamiltonian Operator H	66
3.17	Coulomb Integral α_i	66
3.18	Overlap Integral S_{ij}	66
3.19	Resonance Integral β_{ij}	67
3.20	Hückel Molecular Orbital (HMO) Method	67
3.21	Secular Determinant	68
3.22	Determination of the Coefficients of Hückel Molecular Orbitals	69
3.23	Hückel's Rule	70
3.24	π Molecular Orbital Energy	70
3.25	Hückel ($4n + 2$) Molecular Orbital Energy Levels	70
3.26	Delocalization Energy	71
3.27	Antiaromatic Compounds	72
3.28	Localization Energy	72
3.29	Alternant Hydrocarbons	73
3.30	Even Alternant Hydrocarbons	73
3.31	Odd Alternant Hydrocarbon Systems	73
3.32	π Nonbonding Molecular Orbital	74
3.33	Aromatic Hydrocarbons (Arenes)	75
3.34	Kekulé Structures	75
3.35	Nonalternant Hydrocarbons	76
3.36	Annulenes	77
3.37	Heteroaromatic Compounds	77
3.38	π-Electron-Excessive Compounds	78

The Vocabulary and Concepts of Organic Chemistry, Second Edition, by Milton Orchin, Roger S. Macomber, Allan Pinhas, and R. Marshall Wilson

3.39 π-Electron-Deficient Compounds 78
3.40 Highest Occupied Molecular Orbital (HOMO) 79
3.41 Lowest Unoccupied Molecular Orbital (LUMO) 79
3.42 Frontier Orbitals 79
3.43 Multicenter σ Bonding 79
3.44 Three-Center, Two-Electron Bonding (3c-2e) 80
3.45 Open Three-Center, Two-Electron Bonding 80
3.46 Closed Three-Center, Two-Electron Bonding 80
3.47 Extended Hückel Molecular Orbital Theory 81

The preceding chapter was devoted to the topic of localized bonding in which a pair of electrons is shared by nearest-neighbor atoms, giving rise to the two-center, two-electron bond. In this chapter we will discuss the concepts and define the terms that are used to describe multicenter bonding in which the pair of electrons is involved in bonding three or more adjacent atoms. The multicenter bonding of most interest to organic chemists involves delocalized *p*π electrons. As is the case with localized bonding, either the valence bond approach or the molecular orbital approach can be used to describe the wave functions of the system. The principal tool of the valence bond approach to multicenter π bonding is resonance theory, and its principal architect was Linus Pauling. The power of resonance theory resides in the facts that it can rationalize the properties and reactions of countless organic compounds having conjugated π electron systems and its application requires little mathematical background. Despite its enduring popularity, there are many situations in which resonance theory proves less than adequate. Although it is possible to make valence bond calculations for the wave functions of aromatic compounds such as benzene, naphthalene, and more complex polynuclear structures, these calculations are incomparably more complicated than corresponding molecular orbital calculations. Thus, for the entire field of excited state chemistry in which knowledge of the wave functions of the highest occupied molecular orbital (HOMO) and the lowest unoccupied molecular orbital (LUMO) of a molecule is of critical importance, molecular orbital theory is of inestimable value while resonance theory is completely inadequate. The HOMO-LUMO relationship is important not only in understanding reactions such as intramolecular cyclization, but also it is of increasing utility in rationalizing the pathways of intermolecular reactions.

The molecular orbital treatment is particularly simple and fruitful in determining the wave functions of linear conjugated polyenes. In such molecules, the electrons are not restricted to specific atomic nuclei, but are distributed over the entire length of the conjugated system. In these systems the wave functions may be calculated by an approximate quantum mechanical method called the free electron method (FEM). According to this method, the wave function of each molecular orbital can be evaluated from a simple sine curve. The calculation of wave functions of other conjugated π systems are vastly simplified through the procedures developed by Eric Hückel in a paper devoted to the structure of benzene [*Z. Physik 70*, 204 (1930)].

The Hückel molecular orbital treatment (HMO) treatment allows the calculation of the energies and other important properties of compounds with delocalized π bonding. The Hückel approach resulted in the now famous $(4n + 2)$ π electron rule, where n is an integer including zero. Molecules having this number of π electrons show the remarkable stability associated with "aromatic" character, while species having $4n$ π electrons show the instability associated with "antiaromatic" character. The LCAO methodology (Sect. 2.33) of molecular orbital theory has also been a powerful tool in explaining multicenter bonding in molecules such as the boron hydrides, for example, B_2H_6, where eight σ bonds are generated from a total of only 12 valence electrons.

3.1 DELOCALIZED BOND

A bond that is formed when a pair of electrons is shared between three or more atoms, as distinguished from the bond between two adjacent atoms in a localized two-center, two-electron bond (Sect. 2.3).

3.2 DELOCALIZED BONDING

This kind of bonding occurs when a single pair of electrons contributes to the bonding of more than two atoms, and a delocalized bond is present. Such bonding may be treated either by the valence bond approach, which makes extensive use of resonance concepts, or by the molecular orbital approach. Organic chemists generally use both approaches interchangeably, with the choice dictated by which is most effectively and simply applied to the phenomenon under study.

3.3 DELOCALIZED π MOLECULAR ORBITALS

The molecular orbitals that result from the combination of three or more parallel p atomic orbitals on adjacent atoms; usually, they are just called π molecular orbitals. There are always the same number of molecular orbitals generated as the number of $p\pi$ atomic orbitals combined. Each of these molecular orbitals is delocalized that is, they encompass all the atoms involved and they may be either bonding, nonbonding, or antibonding.

Example. The $n\pi$ molecular orbitals are generated from the n parallel atomic p orbitals by the LCAO procedure (Sect. 2.33), and like the component p atomic orbitals, are perpendicular to the molecular plane. When n is even, the number of bonding π MOs equals the number of antibonding MOs. The four MOs of 1,3-butadiene are shown in Fig. 3.3. The energies of molecular orbitals always increase as the number of nodes increases. ψ_1, the lowest-energy MO of butadiene has no nodes,

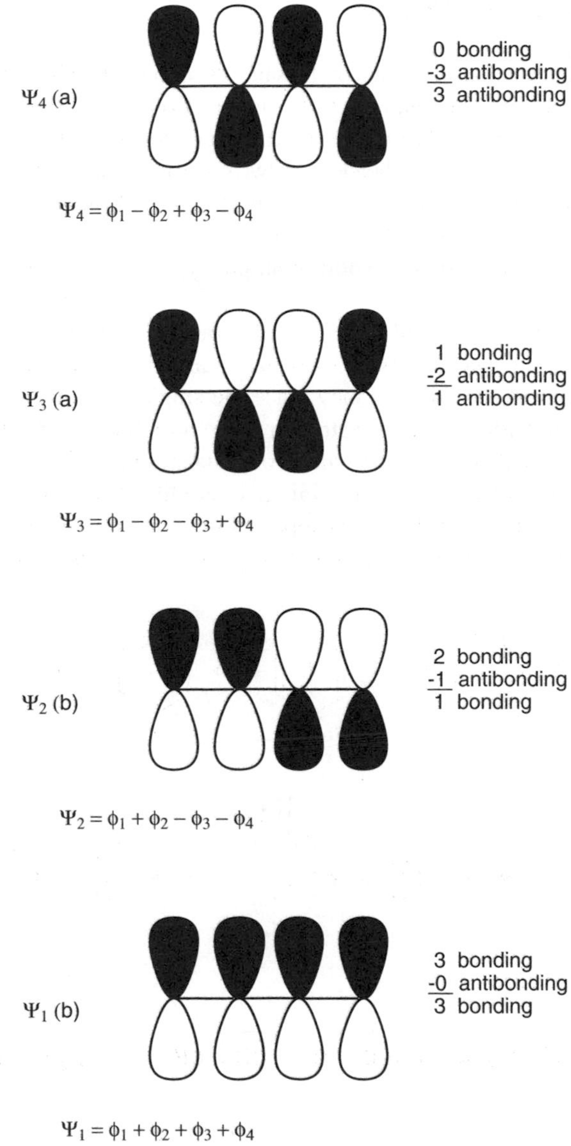

Figure 3.3. The π molecular orbitals of 1,3-butadiene, $CH_2{=}CH{-}CH{=}CH_2$.

and ψ_2, ψ_3, and ψ_4 have one, two, and three nodes, respectively (the node in the horizontal plane of the molecule is neglected since all $p\pi$ orbitals have such a nodal plane). Figure 3.3 is drawn to emphasize bonding and antibonding interactions and does not reflect the magnitude of the coefficients (see Fig. 3.11) of the individual atomic orbitals.

3.4 RESONANCE STRUCTURES

Two or more valence bond structures of the same compound that have identical nuclear geometries, possess the same number of paired electrons, but differ in the spatial distribution of these electrons. Resonance structures are conventionally shown as related to each other by a single double-headed arrow (↔) to emphasize that such structures are not different molecules in equilibrium, that is, they do not have separate, independent existence but are different representations of the same molecule. The actual molecule has a structure that is a hybrid of all possible resonance structures.

Example. Benzene is commonly used to illustrate resonance structures. The two most important resonance structures of benzene are shown in Fig. 3.4*a*; these two structures are equivalent, although they place the single and double bonds in different positions. Many other resonance structures are possible including structures with long bonds, the so-called *Dewar structures* of benzene shown in Fig. 3.4*b*. The three resonance structures of CO_3^{-2} shown in Fig. 3.4*c* are likewise equivalent. The importance of resonance structures becomes apparent in explaining the fact that each C–C bond in every molecule of benzene is of equal length and is intermediate between a

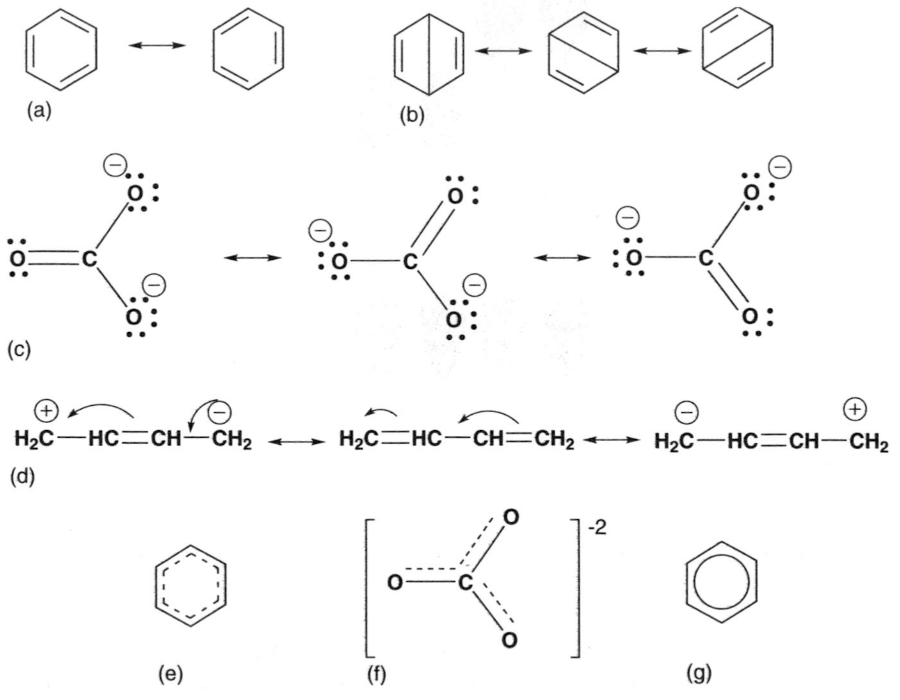

Figure 3.4. Resonance structures: (*a*) benzene, (*b*) Dewar benzene, (*c*) carbonate anion, (*d*) 1,3-butadiene; the single structure representations for (*e*) benzene and (*f*) carbonate anion; and (*g*) the circle inside the hexagon that represents delocalization of the six π electrons of benzene.

single and double bond, as is also the case with each C–O bond in CO_3^{2-}. Resonance structures need not be equivalent; three of the resonance structures of 1,3-butadiene are shown in Fig. 3.2*d*. The middle structure is a lower-energy (more stable) structure than the other two, which are equivalent charge-separated species. The middle structure is said to make a larger contribution to the ground-state structure than either of the charge-separated structures, and hence it is the preferred single representation. However, the double bond distance between atoms 1 and 2 (as well as between 3 and 4) in the real molecule is longer than the carbon–carbon double bond distance in ethylene, and the distance between atoms 2 and 3 is shorter than the carbon–carbon single bond distance in ethane. These distances indicate that the actual molecule is a hybrid involving the three resonance structures (as well as others not shown).

The problem of selecting one best structure to depict the location of the $p\pi$ electrons in compounds, ions, or radicals that are hybrids of resonance structures is frequently solved by the use of broken lines connecting the atoms over which the $p\pi$ electrons are delocalized. Thus, the single structures shown for benzene (Fig. 3.4*e*) and CO_3^{-2} (Fig. 3.4*f*) are representations showing $p\pi$ electron distribution. In the case of $p\pi$ electrons delocalized around a polygon, a solid circle inside the polygon, as shown for benzene, (Fig. 3.4*g*), is sometimes used.

3.5 CURVED ARROW (ELECTRON PUSHING) NOTATION

The use of curved arrows, as formulated by Robert Robinson (1886–1975), to depict the movement of electrons pairs (lone pairs, σ, or π electrons) from one position to another. The arrow starts at the pair of electrons undergoing the migration and terminates with the head of the arrow at the destination of the pair following migration. The migration may be intramolecular (Figs. 3.5*a, b* and *c*) or intermolecular (Figs. 3.5*d* and *e*). Many authors also use a curved arrow to represent the migration of a single electron, but in this case a curved arrow resembling a "fish hook" (one-half the arrowhead) is used (Fig. 3.5).

Example. The relationship between resonance structures of benzene (Fig. 3.5*a*), and carbonate anion (Fig. 3.5*b*) by shifting of π electrons; the shifting of σ electrons in the Cope rearrangement (Fig. 3.5*c*); intermolecular movement of an electron pair in a displacement reaction (Fig. 3.5*d*); the use of fish hook arrows to show single electron movements in a hydrogen abstraction reaction (Fig. 3.5*e*).

3.6 CONJUGATED SYSTEMS

An arrangement of atoms connected by alternating single and double bonds, Fig. 3.6. If all the connected atoms are carbon atoms to which hydrogen atoms are attached, the hydrocarbon belongs to a class of compounds called *polyenes*. In conjugated systems, the $p\pi$ orbitals on adjacent carbon atoms must be parallel for overlap to occur and to generate delocalized molecular orbitals.

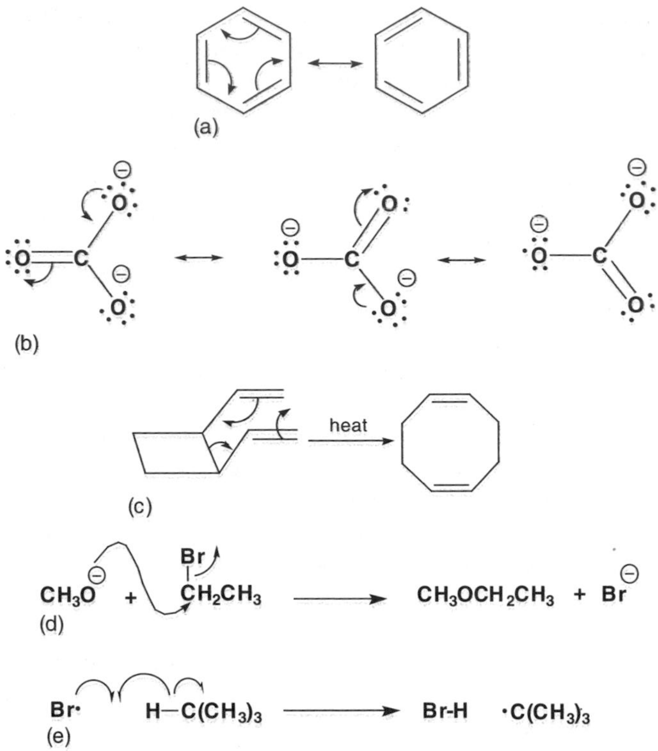

Figure 3.5. Curved arrows showing the interconversion of resonance structures of (*a*) benzene and (*b*) the carbonate anion; (*c*) the Cope rearrangement of *cis*-1,2-divinylcyclobutane to 1,5-cyclooctadiene; (*d*) the displacement of bromide by methoxide in the preparation of methyl ethyl ether; (*e*) the single electron "fish hook" notation in the reaction involving the hydrogen atom abstraction by a bromine atom.

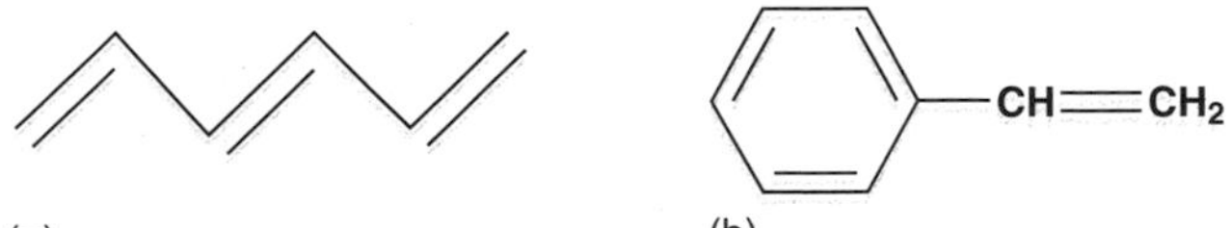

(a) (b)

Figure 3.6. Conjugated hydrocarbons: (*a*) 1,3,5-hexatriene and (*b*) styrene.

3.7 ESSENTIAL SINGLE BOND

A single bond in a conjugated system that can be written only as a single bond in all resonance structures which do not involve charge-separated species.

Example. The single bond between atoms 2 and 3 in the uncharged 1,3-butadiene structure (Fig. 3.4*d*).

3.8 ESSENTIAL DOUBLE BONDS

Double bonds in a conjugated system that can be written only as double bonds in all resonance structures not involving charge-separated species.

Example. The double bonds between atoms 1 and 2, and 3 and 4 in 1,3-butadiene (Fig. 3.4*d*). Neither the single nor double carbon–carbon bonds in benzene (Fig. 3.4*a*) are *essential*.

3.9 HYPERCONJUGATION

Overlap between neighboring σ and π (or *p*) orbitals giving rise to delocalization of the σ electrons.

Example. A common example of hyperconjugation shown in Fig. 3.9*a* helps explain the stabilization of suitably substituted carbocations. In cyclopentadiene (Fig. 3.9*b*), the C–H σ bonds of the methylene group can overlap with the π bonds of the diene portion of the molecule. This overlap is more readily visualized by combining the two 1*s* orbitals of methylene hydrogens into two molecular orbitals (called *group orbitals*, Section 4.32) that are linear combinations of atomic orbitals (Fig. 2.33*a*). These group orbitals have the form [H_1 (*s*) + H_2 (*s*)] (Fig. 3.9*c*) and [H_1 (*s*) − H_2 (*s*)] (Fig. 3.9*d*); this latter orbital with a node in the plane of the molecule has the same symmetry as the π system, hence can overlap with it (Fig. 3.9*e*).

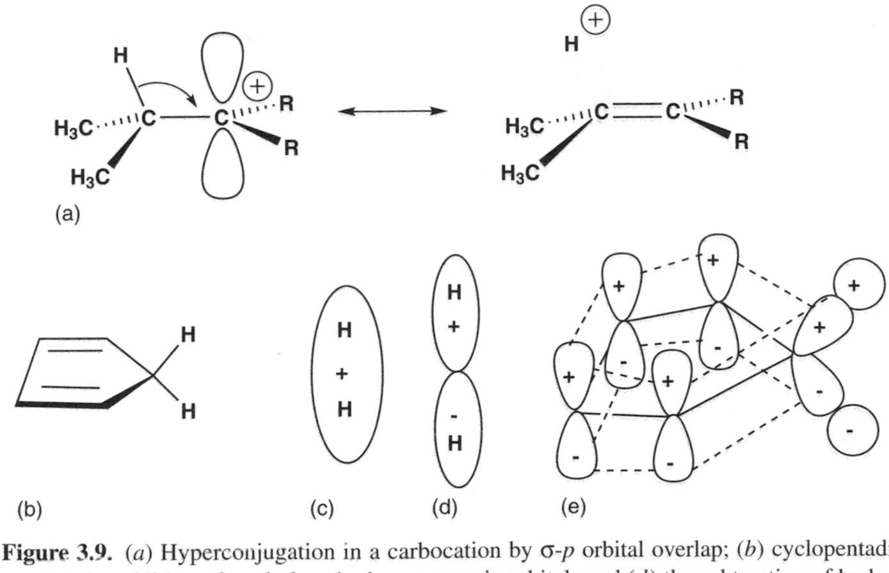

Figure 3.9. (*a*) Hyperconjugation in a carbocation by σ-*p* orbital overlap; (*b*) cyclopentadiene; (*c*) the addition of methylene hydrogen atomic orbitals and (*d*) the subtraction of hydrogen atomic orbitals; (*e*) the overlap of (*d*) with the *p* orbitals of cyclopentadiene.

3.10 RESONANCE ENERGY

Additional stabilization (lowering of energy due to electron delocalization) over that calculated for a single resonance structure. The resonance stabilization arises as a result of the delocalization of the electrons over a conjugated system and is essentially equivalent to the delocalization energy (Sect. 3.26).

Example. Benzene is considerably more stable than would be expected if all three double bonds were independent of each other. Experimentally, the additional stabilization can be demonstrated by measuring the heat liberated on hydrogenation of benzene and comparing it with the heat of hydrogenation of three moles of cyclohexene. Thus, the conversion of one mole of benzene to cyclohexane liberates 208.4 kJ (49.8 kcal), while the conversion of three moles of cyclohexene to cyclohexane liberates 361.5 kJ (86.4 kcal). The difference between these two numbers, 153.1 kJ (36.6 kcal), corresponds to the increased stabilization of benzene with its three interacting double bonds over that of three isolated noninteracting double bonds in a six-membered ring. Thus, the resonance energy of benzene is 153 kJ (36.6 kcal) mol^{-1}.

3.11 THE FREE ELECTRON METHOD (FEM) [FOR CALCULATING THE WAVE FUNCTIONS OF CONJUGATED COMPOUNDS]

An approximate quantum mechanical method of calculating the wave functions of π electron systems of linear conjugated compounds. It is assumed that the π electrons in the system are free to travel in a box of length *a,* assumed to be one bond length beyond each end of the conjugated system. It is further assumed that the potential energy of the π electrons inside the box is constant and that outside the box, it is infinite. In this situation the Schrödinger equation (Sect. 1.21) implies a wave function that involves a single coordinate *x,* which is the length of the box:

$$(-h^2/8\pi^2 m)d^2\psi/dx^2 = E\psi \tag{3.11a}$$

Functions that are solutions of this differential equation are

$$\psi = \sqrt{\frac{2}{a}}\sin n\pi x/a \tag{3.11b}$$

where $n = 1,2,3, \ldots$, and $\sqrt{\frac{2}{a}}$ is the normalization constant. The energies E_n of these orbitals will be

$$E_n = n^2h^2/8ma^2 \tag{3.11c}$$

Example. The wave functions for the four MOs of 1,3-butadiene are all of the form $\psi_n = c_{n1}\phi_1 + c_{n2}\phi_2 + c_{n3}\phi_3 + c_{n4}\phi_4$, where the ϕ's are the atomic orbitals of each numbered carbon atom. If we use Eq. 3.11 with a (the length of the box) equal to five bond lengths, for the lowest-energy MO where $n = 1$, the value of c_{11}, the coefficient of ϕ_1 that represents the amplitude of the wave at carbon 1 is $\psi = \sqrt{\frac{2}{5}} \sin 1\pi \cdot 1/5 = \sqrt{\frac{2}{5}} \cdot \sin 36° = 0.632 \cdot 0.589 = 0.372$. For c_{12}, $\psi = \sqrt{\frac{2}{5}} \cdot \sin 1\pi \cdot 2/5 = \sqrt{\frac{2}{5}} \cdot \sin 72° = 0.602$. Since by symmetry, carbons 1 and 4 and carbons 2 and 3 are equivalent, the complete wave function is $\psi_1 = 0.372\phi_1 + 0.602\phi_2 + 0.602\phi_3 + 0.372\phi_4$. For $\psi_2 = \sqrt{\frac{2}{5}} \cdot \sin 2\pi \cdot x/a$, when $x/a = 1/2$, $\psi_2 = \sqrt{\frac{2}{5}} \cdot \sin \pi = 0$; hence, ψ_2 has a node at 5/2 bond lengths. The calculations for the coefficients are as follows: at C_1, $\psi = \sqrt{\frac{2}{5}} \cdot \sin 2\pi \cdot 1/5 = \sqrt{\frac{2}{5}} \cdot \sin 72° = 0.602$; at C_2, $\psi = \sqrt{\frac{2}{5}} \cdot \sin 2\pi \cdot 2/5 = 0.632 \sin 144° = 0.372$. By symmetry, $c_{21} = -c_{24}$ and $c_{22} = -c_{23}$, and the complete wave function is $\psi_2 = 0.602\phi_1 + 0.372\phi_2 - 0.372\phi_3 - 0.602\phi_4$. When we use calculations analogous to those given above, the wave functions for the other molecular orbitals of butadiene may be determined, and the equations for all four wave functions are shown in Fig. 3.11. The vertical lines in the figure represent the relative magnitude (coefficients) at each of the carbon atoms in the four wave functions. Note that the amplitudes of the FEM orbital at each atom are equal to the coefficient of the AO of this atom in the corresponding MO, as evaluated by the HMO method (Sect. 3.20).

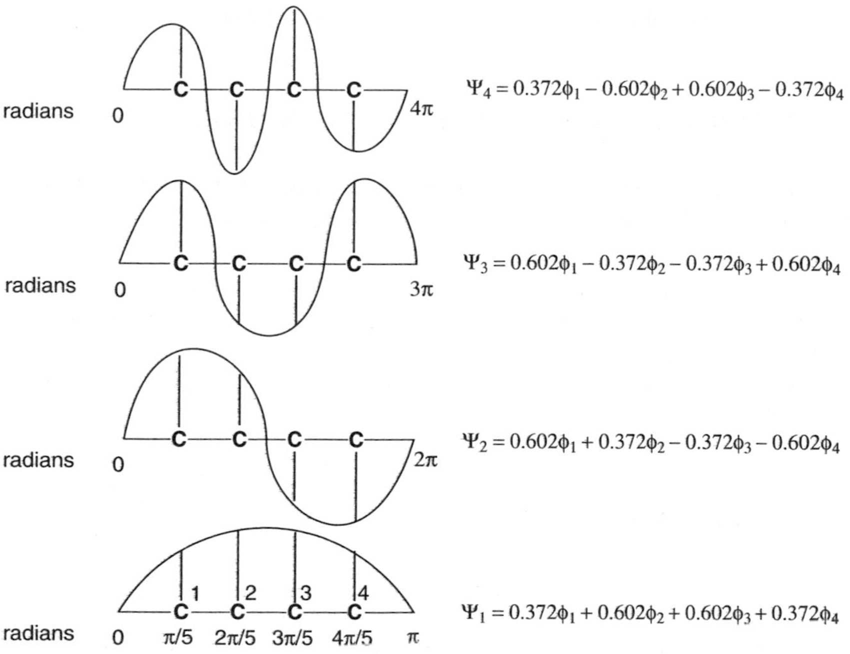

Figure 3.11. The four MOs of butadiene with vertical lines corresponding to the coefficients at each atom in each MO.

3.12 ELECTRON DENSITY q_r

$$q_r = \sum n_j c_{jr}^2 \tag{3.12}$$

where c_{jr} is the coefficient of atom r in the jth MO, which is occupied by n_j electrons, and the summation is over all occupied MOs. In the LCAO approximation (Sect. 2.33), the probability that an electron in a molecular orbital is associated with a particular atomic orbital in that molecular orbital is the square of that atomic orbital coefficient c_{jr}^2. The total electron density at the atom r, q_r, is the sum of the electron densities at that atom contributed by each electron (i.e., by each occupied orbital).

Example. In 1,3-butadiene ψ_1 and ψ_2 are each occupied by two electrons, whereas both ψ_3 and ψ_4 are unoccupied in the ground state, that is, $n = 0$. The coefficients for the MOs have been determined in Sect. 3.11; hence, $q_1 = 2c_{11}^2 + 2c_{21}^2 = 2(0.372)^2 + 2(0.602)^2 = 1$, and $q_2 = 2c_{12}^2 + 2c_{22}^2 = 2(0.602)^2 + 2(0.372)^2 = 1$. By symmetry, atoms 4 and 3 give the same answers, respectively, and thus the electron density is unity at each atom. The sum of the π electron densities over all the atoms must be equal to the total number of π electrons; thus, $q_1 + q_2 + q_3 + q_4 = 4$. The electron density at each carbon atom is unity, not only in butadiene but also in all *alternant hydrocarbons* (Sect. 3.29), that is, compounds that contain only conjugated carbon atoms and even-numbered rings, with all Coulomb integrals being equal (see Section 3.17) and having as many π electrons as atoms in the conjugated system. According to the last criterion, the electron density is not 1 at the three carbon atoms of the allyl anion or cation (see Sect. 3.31).

3.13 CHARGE DENSITY ζ (CHARGE DISTRIBUTION)

The difference between the electron density at an isolated atom (normally 1 in a π system) and its electron density when part of a conjugated system: $\zeta_r = 1 - q_r$.

Example. Since the electron density at each of the carbon atoms in 1,3-butadiene is unity (Sect. 3.12), the charge density (the Greek zeta or ζ) at each carbon atom is $1 - 1 = 0$. The charge densities at both carbon atoms 1 and 3 in the allyl system (Sect. 3.31) are $1 - 2(1/2)^2 = +1/2$ for the cation, and for the anion, $1 - \{2(1/2)^2 + 2(1/\sqrt{2})^2\} = -1/2$.

3.14 π BOND ORDER ρ_{rs}

$$\rho_{rs} = \sum n_j c_{jr} c_{js} \tag{3.14}$$

where n_j is the number of electrons in the jth molecular orbital, c_{jr} and c_{js} are the coefficients of bonded atoms r and s in the jth molecular orbital, and ρ_{rs} is the bond order between atoms r and s. The π bond order has meaning only for adjacent atoms and is a measure of the amount of π bond character between the two adjacent atoms;

it is related to the coefficients of the atomic orbitals on the atoms involved in the bond.

Example. For 1,3-butadiene in the ground state:

$$\rho_{12} = 2c_{11}c_{12} + 2c_{21}c_{22} = 2(0.372 \times 0.602) + 2(0.602 \times 0.372) = 0.896$$

$$\rho_{23} = 2c_{12}c_{13} + 2c_{22}c_{23} = 2(0.602 \times 0.602) + 2(0.372 \times (-0372)) = 0.448$$

By symmetry, $\rho_{34} = \rho_{12}$. The essential double bonds between carbons 1 and 2 and between 3 and 4 have a large measure of double bond character but because of delocalization, the bond between carbons 2 and 3 also has substantial double bond character, approximately half that between carbons 1 and 2 and between 3 and 4.

3.15 FREE VALENCE INDEX F_r

$$F_r = F_{\text{max}} - \sum \rho_{rs} \qquad (3.15)$$

where ρ_{rs} is the π bond order between atom r and an atom to which it is bonded, atom s, and the summation extends over all atoms bonded to r. The free valence index measures the extent to which the maximum valence of that atom in the conjugated system is *not* satisfied by the delocalized bonds. F_{max} is $\sqrt{3} = 1.732$, the ρ value for the central atom in trimethylenemethane whose sigma skeleton is shown below.

Example. The free valence at carbon 2 in 1,3-butadiene is calculated as follows: $F_2 = 1.732 - (\rho_{12} + \rho_{23}) = 1.732 - (0.896 + 0.448) = 0.388$. Information about bond orders and free valence numbers is summarized in molecular diagrams, in which the ρ_{rs} is written on each bond and the value of F_r is placed at the end of a short arrow starting at an atom of each type. For 1,3-butadiene this is

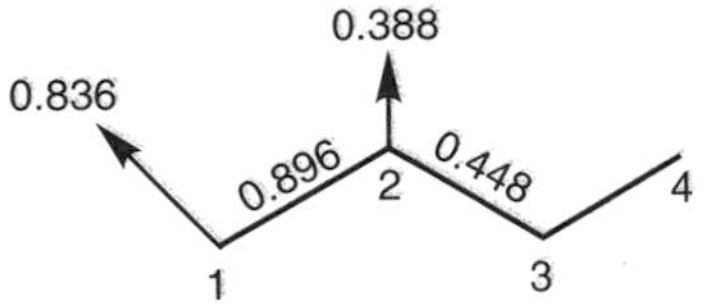

The F_{max} of 1.732 represents the sum of the ρ_{rs} for the central carbon atom (C_4) in trimethylenemethane. In this molecule, each methylene group can focus all its bonding power onto the central carbon atom, which, because it is involved in $p\pi$ bonding with three other $p\pi$ orbitals, has a very small residual free valence, which is assumed to be zero:

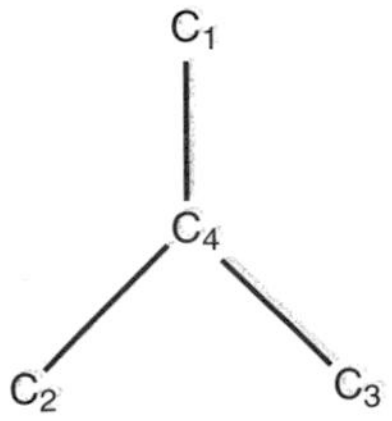

The bond order for each of the three equivalent bonds is $\rho_{14} = \rho_{24} = \rho_{34} = 1/\sqrt{3}$. The sum of the bond order around C_4, the central carbon atom, is thus $3 \times 1/\sqrt{3} = \sqrt{3} = 1.732$, and this is taken as F_{max}.

3.16 HAMILTONIAN OPERATOR *H*

A mathematical expression associated with the sum of the potential and kinetic energies of a system.

The Schrödinger equation (Eq. 1.21c) may be written in operator form as $H\psi = E\psi$, where H is the Hamiltonian, ψ the wave function (or eigenfunction, Sect. 1.22), and E the energy (eigenvalue) of the system. The total energy of a molecule is frequently given the symbol H in honor of Lord Hamilton, an English physicist, who made important contributions to the mathematics of classical mechanics. In wave mechanics the fundamental equation involving the energy of a molecule is the Schrödinger equation (Sect. 1.21), where H is now an energy operator and $H\psi$ is the result of the operation of H on ψ.

3.17 COULOMB INTEGRAL α_i

An integral that corresponds in the crudest (zero-order) approximation to the energy of an electron in an orbital ϕ_i in the field of the atom i, that is, $\int \phi_i H \phi_i d\tau$, where H is the Hamiltonian energy operator. This integral is denoted by α in the Hückel π electron approximation.

Example. The value of the Coulomb integral α_i is assumed to be independent of any other neighboring atom j. It is roughly equivalent to the energy required to remove an electron from the atomic orbital in which it resides (the ionization potential); for an electron in a carbon $2p$ orbital, this energy is 11.16 eV.

3.18 OVERLAP INTEGRAL S_{ij}

A measure of the extent of overlap of the orbitals on neighboring atoms i and j, that is, the value of $\int \phi_i \phi_j d\tau$. In the Hückel π electron approximation, this integral, frequently denoted by S_{ij}, is taken to be zero. The extent of overlap of an atomic orbital with itself is unity, that is, $S_{ii} = S_{jj} = 1$.

Example. When i and j are nearest-neighbor atoms with $p\pi$ orbitals such as those on the carbons in ethylene, the actual value of S_{ij} *is* about 0.25, which would appear to be too large to neglect. Although the overlap integral is responsible for π bonding between atoms, the consequences of neglecting S_{ij} are not very serious in simple Hückel calculations (Sect. 3.20).

3.19 RESONANCE INTEGRAL β_{ij}

An integral that determines the energy of an electron in the fields of two atoms i and j involving the orbitals ϕ_i and ϕ_j, that is, $\int \phi_i H \phi_j d\tau$, where H is the Hamiltonian energy operator. This integral is denoted by β_{ij} in the Hückel approximation.

Example. When i and j are not adjacent atoms, β_{ij} is neglected and set to zero. The resonance integral is approximately proportional to the actual value of the overlap integral S_{ij}.

3.20 HÜCKEL MOLECULAR ORBITAL (HMO) METHOD

A procedure, devised by Eric Hückel (1896–1980), for calculating the energy of the π electrons in conjugated systems. The procedure includes a number of approximations; for example, all σ electrons and their interaction with π electrons are neglected, all overlap intergals S_{ij} are neglected, the Coulomb integrals α are all set equal, all resonance integrals β_{ij} are set equal. However, it should be noted that all α's and β's are equal only if there are no heteroatoms.

Example. The wave functions ψ of the two MOs of ethylene may be written as

$$\psi_1 = c_{11}\phi_1 + c_{12}\phi_2$$
$$\psi_2 = c_{21}\phi_1 + c_{22}\phi_2$$

In order to determine the energies E of these MOs, use is made of approximate solutions of the Schrödinger equation (and other approximations) to obtain a set of two homogeneous linear equations with two unknowns, called *secular equations*:

$$c_{11}(H_{11} - ES_{11}) + c_{12}(H_{12} - ES_{12}) = 0$$
$$c_{21}(H_{21} - ES_{21}) + c_{22}(H_{22} - ES_{22}) = 0$$

Solutions of these equations can be found by solving the determinant (called the *secular determinant,* Sect. 3.21) of the factors by which the AO coefficients are multiplied (Sect. 2.46) in the secular equations. The calculations are simplified by assuming the following: $S_{12} = S_{21} = 0$; setting $H_{11} = H_{22}$ and replacing them by α, the Coulomb integral; setting $H_{12} = H_{21}$ and replacing them by β, the resonance integral. Then since $S_{11} = S_{22} = 1$, the secular determinant becomes

$$\begin{vmatrix} \alpha - E & \beta \\ \beta & \alpha - E \end{vmatrix} = 0$$

Dividing through by β gives

$$\begin{vmatrix} \dfrac{\alpha - E}{\beta} & 1 \\ 1 & \dfrac{\alpha - E}{\beta} \end{vmatrix} = 0$$

Now if we let $x = (\alpha - E)/\beta$, we obtain

$$\begin{vmatrix} x & 1 \\ 1 & x \end{vmatrix} = 0$$

Hence, $x^2 = 1$ and $x = \pm 1$. For $x = 1$, $E = \alpha - \beta$ (for the antibonding MO) and for $x = -1$, $E = \alpha + \beta$ (for the bonding MO).

3.21 SECULAR DETERMINANT

The general form of a determinant is

$$\begin{vmatrix} a_{11} & a_{12} & a_{13} & a_{1n} \\ a_{21} & a_{22} & a_{23} & a_{2n} \\ a_{31} & a_{32} & a_{33} & a_{3n} \\ \vdots & \vdots & \vdots & \vdots \\ a_{n1} & a_{n2} & a_{n3} & a_{nn} \end{vmatrix} = 0$$

Each a is called an element, and the subscripts refer to the position, that is, the row and column, respectively, of the element. Thus, element a_{32} is the element in the third row and second column. In HMO theory such a determinant is used to solve the sets of secular equations (Sect. 3.20) in order to obtain the energies of the wave functions and the value of the coefficients of the atomic orbitals making up each wave function.

Example. The secular determinant for calculating the Hückel molecular orbitals of butadiene, $H_2C{=}CH{-}CH{=}CH_2$, is set up as follows: The carbon atoms are numbered consecutively starting from the left. The first element a_{11} represents carbon 1 and is given the value $x = (\alpha - E)/\beta$. The next element in the first row, a_{12}, represents carbon atom 1 bonded to carbon 2 and is given the value 1. Resonance integrals between nonbonded atoms are neglected, and since in butadiene carbon 1 is not directly bonded to carbons 3 and 4, elements a_{13} and a_{14} are assigned values of zero. Carbon atom 2 is bonded to 1 and 3 but not to 4, and hence elements $a_{21} = a_{23} = 1$ and $a_{24} = 0$; the element a_{22} (a diagonal element) is again x. Each element of the diagonal (a_{11}, a_{22}, a_{33}, . . .) in the secular determinant is designated as x, the elements with the subscripts of nearest-neighbor numbers (a_{12}, a_{21}, a_{23}, . . .) are given the number 1 (if all bonds are assumed equivalent, i.e., all β equal), and other elements are zero. Thus, for 1,3-butadiene the secular determinant is

$$\begin{vmatrix} x & 1 & 0 & 0 \\ 1 & x & 1 & 0 \\ 0 & 1 & x & 1 \\ 0 & 0 & 1 & x \end{vmatrix} = 0$$

An evaluation of this determinant (see the first reference at the end of this chapter for a procedure for doing this) gives $x^4 - 3x^2 + 1 = 0$, and the solution of this equation gives $x = \pm\frac{1 \pm \sqrt{5}}{2}$. Since $x = (\alpha - E)/\beta$, the energies of the four wave functions are in increasing order: $E_1 = \alpha + 1.618\beta$; $E_2 = \alpha + 0.618\beta$; $E_3 = \alpha - 0.618\beta$; $E_4 = \alpha - 1.618\beta$.

3.22 DETERMINATION OF THE COEFFICIENTS OF HÜCKEL MOLECULAR ORBITALS

The calculation involves determining the ratio of coefficients of the various AOs in any particular MO and then taking advantage of the normalization condition. The coefficients are obtained directly from the secular equations.

Example. The four linear simultaneous equations corresponding to the secular determinant for 1,3-butadiene are

$$\begin{array}{lllllll} \text{(a)} & c_1x & + c_2 & & & = 0 \\ \text{(b)} & c_1 & + c_2x & + c_3 & & = 0 \\ \text{(c)} & & c_2 & + c_3x & + c_4 & = 0 \\ \text{(d)} & & & c_3 & + c_4x & = 0 \end{array}$$

For the determination of the coefficients of the AOs in ψ_4, for example, where $x_4 = 1.618\beta$, first determine all the coefficients in terms of c_1:

From (a), $c_2 = -c_1x = -1.618c_1$.
From (b), $c_3 = -c_2x - c_1 = c_1(x^2 - 1) = 1.618c_1$.
From (c), $c_4 = -c_3x - c_2 = -c_1[x(x^2 - 1) - x] = -c_1[x(x^2 - 2)] = -c_1$.

Then from the normalization condition: $c_1^2 + c_2^2 + c_3^2 + c_4^2 = 2(c_1^2 + 1.618^2 c_1^2) = 7.24c_1^2 = 1$ ∴ $c_1 = 0.372$, $c_2 = -0.602$, $c_3 = 0.602$, $c_4 = -0.372$ and the wave function $\psi_4 = 0.372\phi_1 - 0.602\phi_2 + 0.602\phi_3 - 0.372\phi_4$. Note that the wave functions for 1,3-butadiene calculated by the HMO method are identical to those obtained by the FEM (Sect. 3.11).

3.23 HÜCKEL'S RULE

These are planar monocyclic systems possessing $(4n + 2)$ $p\pi$ electrons delocalized over the ring, where n is any integer, 0, 1, 2, 3, . . . , and possessing unusual thermodynamic stability; that is, they will be aromatic (*vide infra*).

Example. Benzene, cyclopentadienyl anion, cyclopropenyl cation, cycloheptatrienyl cation (Figs. 3.23 *a–d*, respectively) are compounds and ions that obey this rule ($n = 1$, 1, 0, 1, respectively); hence, they are characterized by large resonance energies. Although this rule is strictly applicable only to monocyclic carbocycles, it has been applied to polynuclear hydrocarbons as well.

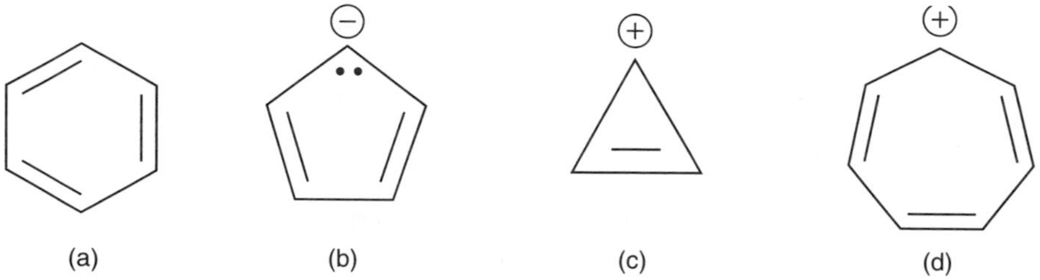

Figure 3.23. Some $4n + 2$ π electron (Hückel) species: (*a*) benzene; (*b*) cyclopentadienyl anion; (*c*) cyclopropenyl cation; (*d*) cycloheptatrienyl cation.

3.24 π MOLECULAR ORBITAL ENERGY

The energy of a π molecular orbital. In Hückel π electron theory, this is usually expressed in terms of the Coulomb integral α and the resonance integral β.

Example. The energy of the bonding π molecular orbital in ethylene (ψ_1 in Sect. 3.20) is $(\alpha + \beta)$. In the ground state both $p\pi$ electrons are in ψ_1; since α is the energy of a single electron in a $p\pi$ atomic orbital, and β for each electron is 1, the total π molecular orbital energy for ethylene is $(2\alpha + 2\beta)$.

3.25 HÜCKEL (4n + 2) MOLECULAR ORBITAL ENERGY LEVELS

These energy levels consist of one low-lying energy level accommodating two electrons and then n pairs of degenerate levels, each pair accommodating four electrons, hence, $(4n + 2)$ π electrons. Actually, n is a quantum number and is equal to the number of nodal planes bisecting the ring.

Example. A useful technique for ordering the energy levels in the various $(4n + 2)$ ring systems involves inscribing the polygon representing the carbocycle in a circle

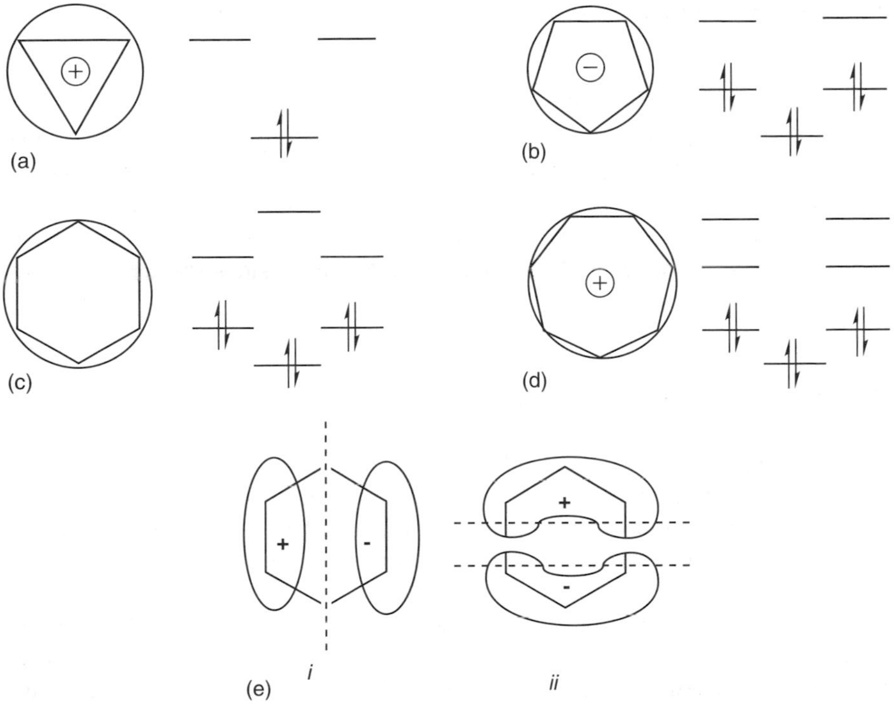

Figure 3.25. Molecular orbital levels of some $(4n + 2)$ ring systems: (*a*) cyclopropenyl cation; (*b*) cyclopentadienyl anion; (*c*) benzene (an even alternant hydrocarbon); (*d*) cycloheptatrienyl cation, and (*e*) a sketch of the degenerate occupied MOs of benzene showing the nodal plane bisecting the ring through a set of opposite atoms e_i and a set of opposite sides e_{ii}.

with one vertex at the 6 o'clock position. The level at each vertex corresponds to the relative energy level of a π molecular orbital and the center of the circle corresponds to the zero level (Fig. 3.25). The MOs below the zero level are bonding and the MOs above this level are antibonding. The stability of the $(4n + 2)$ systems stems from the fact that all bonding MOs are completely filled, giving a "closed-shell" configuration. The MOs of even alternant hydrocarbons such as benzene have a mirror image relationship between bonding and antibonding MOs, (Fig. 3.25*c*), which is not the case with the odd alternant ions shown in Figs. 3.25 *a, b,* and *d.* The degenerate pair of molecular orbitals of benzene with quantum number $n = 1$, that is, with one node, is shown in Fig. 3.25*e*.

3.26 DELOCALIZATION ENERGY

This is the stabilization energy associated with electron delocalization; it is the difference in calculated electronic energy between a localized electron system and a delocalized system. In HMO theory this energy is usually expressed in terms of β (the resonance integral) and corresponds to the resonance energy in valence bond theory (Sect. 3.4)

Example. For 1,3-butadiene the π energies of the two lowest-energy molecular orbitals (Fig. 3.3) are $\psi_1 = \alpha + 1.618\beta$ and $\psi_2 = \alpha + 0.618\beta$ (Sect. 3.21). Since two electrons are in each of these orbitals in the ground state, the total π energy is $4\alpha + 4.472\beta$. The energy for ethylene in the ground state is $2\alpha + 2\beta$ and for two isolated ethylenes it is $4\alpha + 4\beta$. The delocalization energy, the difference in energy between resonance-stabilized 1,3-butadiene and two isolated ethylenes, is thus 0.472β.

3.27 ANTIAROMATIC COMPOUNDS

These are conjugated cyclic hydrocarbons in which if π electron delocalization were to occur, it would result in an *increase* in energy (destabilization); they are also monocyclic systems with $4n$ π electrons, hence called "Hückel antiaromatic" in contrast to Hückel aromatic systems that possess $(4n + 2)$ π electrons.

Example. Cyclobutadiene and cyclopropenyl anion (Figs. 3.27*a* and *b*, respectively). The delocalization energy of 1,3-butadiene is 0.472β, while the π electronic energy of cyclobutadiene is $4\alpha + 4\beta$, corresponding to a delocalization energy equal to zero β. The zero delocalization energy is the same as that for two isolated ethylenes and implies that there is no interaction between the two double bonds. More refined calculations which include overlap (Sect. 3.47) show, however, that cyclobutadiene is less stable than two isolated ethylenes, hence its antiaromatic character. Using the technique described in Fig. 3.25 shows that cyclobutadiene in the ground state is a triplet biradical.

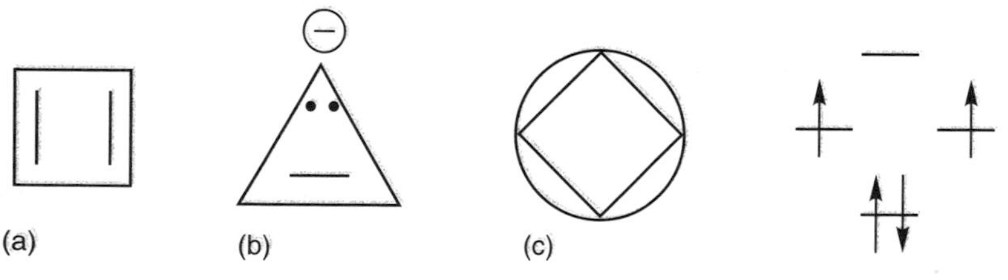

Figure 3.27. Antiaromatic species: (*a*) cyclobutadiene; (*b*) cyclopropenyl anion; (*c*) the molecular orbital energy level diagram of cyclobutadiene.

3.28 LOCALIZATION ENERGY

The decrease in delocalization energy involved in the removal of one (or more) orbitals from a delocalized π system of which it was a part. This theoretical process localizes π electrons, which were part of a conjugated system, onto a reduced set of carbon atoms in the system.

Example. For benzene and its derivatives (and other aromatics), the localization energy is a measure of the π energy change between the parent aromatic and the intermediate σ complex in aromatic substitution reactions. Thus when benzene undergoes electrophilic substitution, one of the π electron pairs is converted into a σ bond at a carbon atom with the formation of a cation, which has its charge delocalized over the remaining five carbon atoms:

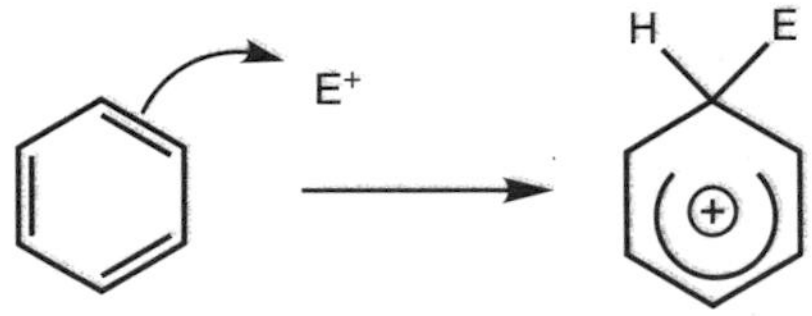

The localization energy concept has been useful in evaluating reactivates, for example, the effect of substitutents on aromatic substitution reactions.

3.29 ALTERNANT HYDROCARBONS

Any conjugated hydrocarbon that does *not* include an odd number of carbon atoms in a ring system. This classification is useful in MO theory because such hydrocarbons possess complementary bonding and antibonding orbitals; that is, the bonding and antibonding levels have a mirror image relationship. The alternant hydrocarbons can be further classified into even or odd alternants.

3.30 EVEN ALTERNANT HYDROCARBONS

If the alternant hydrocarbon contains an even number of carbon atoms, it is called an *even alternant hydrocarbon* (nonalternant hydrocarbons are discussed in Sect. 3.35).

Example. 1,3-Butadiene, 1,3,5-hexatriene, benzene, and naphthalene are even alternant hydrocarbons.

3.31 ODD ALTERNANT HYDROCARBON SYSTEMS

A system consisting of an odd number of linked carbon atoms with overlapping *p*π orbitals. Such a system is theoretically generated by removing an atom, with or without its electrons, from a covalently saturated carbon center linked to a doubly bonded carbon atom; the resulting delocalized (or conjugated) cation, radical, or anion is called an odd alternant hydrocarbon system.

$$\psi_3 = \frac{1}{2}\phi_1 - \frac{1}{\sqrt{2}}\phi_2 + \frac{1}{2}\phi_3 \qquad \alpha - \sqrt{2}\,\beta$$

$$\psi_2 = \frac{1}{\sqrt{2}}\phi_1 - \frac{1}{\sqrt{2}}\phi_3 \qquad \alpha$$

$$\psi_1 = \frac{1}{2}\phi_1 + \frac{1}{\sqrt{2}}\phi_2 + \frac{1}{2}\phi_3 \qquad \alpha + \sqrt{2}\,\beta$$

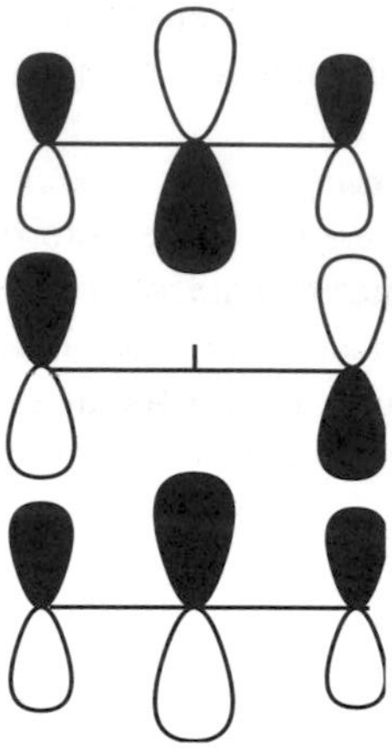

Figure 3.31. The three π molecular orbitals of the allyl system and their π molecular orbital energy levels.

Example. The allyl system, $[CH_2{=}CH{-}CH_2]$, may have two $p\pi$ electrons (cation), three $p\pi$ electrons (radical), or four $p\pi$ electrons (anion), with electronic configurations of ψ_1^2, $\psi_1^2\psi_2$, or $\psi_1^2\psi_2^2$, respectively. The energies and sketches of the three π allyl MOs generated from the three AOs are shown in Fig. 3.31. The delocalization energy of the cation is 0.83β, which is also equal to that of the radical and the anion as well. This equality is a consequence of the fact that the additional one and two electrons required in going from the cation to the radical and anion, respectively, occupy the nonbonding π orbital (see below) that contributes nothing to the *bonding* energy. Thus, although the *total* energies of the three π electron (radical) and four π electron (anion) systems are greater than that of the two electron allyl cation (since more electrons are present), all three systems would appear to be equally stable because of their equal delocalization energy as calculated from simple Hückel theory. In fact, because the additional electrons lead to greater electron–electron repulsions, the order of stability should be cation > radical > anion.

3.32 π NONBONDING MOLECULAR ORBITAL

A π molecular orbital, which when occupied by electrons, contributes nothing to the bonding energy of the molecule.

Example. ψ_2 of the allyl system. In this orbital the coefficient of ϕ_2 is zero; there is a node through this central carbon atom (Fig. 3.31), indicating no interaction between nearest-neighbor $p\pi$ atomic orbitals. The benzyl system $[C_6H_5CH_2]$ Fig. 3.35*f*, which is also an odd alternant hydrocarbon, possesses a nonbonding molecular orbital. The energy of the nonbonding molecular orbital is α, the same as the energy of an isolated sp^2 atomic orbital.

3.33 AROMATIC HYDROCARBONS (ARENES)

A large class of hydrocarbons, of which benzene is the first member, consisting of cyclic assemblages of conjugated carbon atoms and characterized by large resonance energies. There are several subclasses of aromatic hydrocarbons. (See Section 5.40.)

Example. Some polynuclear aromatic hydrocarbons in which there are two, three, and four rings: naphthalene, phenanthrene, anthracene, and benz[*a*]anthracene (Fig. 3.33).

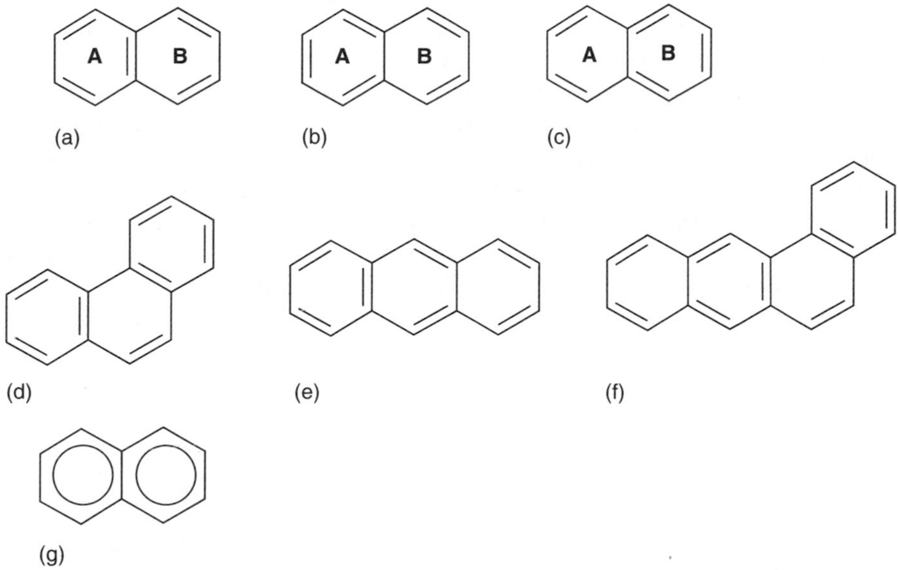

Figure 3.33. Resonance structures of naphthalene: (*a*) with two Kekulé rings; (*b, c*) with one Kekulé ring; (*d*) phenanthrene; (*e*) anthracene; (*f*) benz[*a*]anthracene; (*g*) a misleading representation of naphthalene. The use of two (or more) circles inside fused six-membered rings should be avoided because this circle represents six delocalized $p\pi$ electrons and the use of two circles for naphthalene, as in (*g*), implies the presence of 12 $p\pi$ electrons, whereas in fact only 10 are present.

3.34 KEKULÉ STRUCTURES

Six-membered rings in which three conjugated double bonds are placed.

Example. The three most important resonance structures of naphthalene are shown in Figs. 3.33*a* through *c*. In Fig. 3.33*a* rings A and B are both Kekulé rings, but in structure 3.33*b* only ring A and in Fig. 3.33*c* only ring B are Kekulé. Since, as a matter of convenience, most polynuclear aromatic compounds are represented by only one structure, the structure chosen is the one that can be written with the maximum number of Kekulé rings; in the case of naphthalene it is structure *a*. Structures *b* and *c* each contain one "quinoidal" ring (labeled B and A, respectively).

3.35 NONALTERNANT HYDROCARBONS

Conjugated hydrocarbons containing at least one odd numbered ring.

Example. Azulene, fulvene, and fluoranthene (Figs. 3.35*a*, *b*, and *c*, respectively). A more fundamental distinction between alternant and nonalternant hydrocarbons can be illustrated by means of a "starring" procedure. In this procedure all carbon atoms in a molecule are divided into two classes or sets, one starred and the other unstarred. In an alternant molecule no two atoms of the same set can be bonded together. This procedure is shown for the even alternant hydrocarbon naphthalene (Fig. 3.35*e*) and for the odd alternant benzyl ion (Fig. 3.35*f*). In the starring procedure for nonalternant hydrocarbons, either one pair of adjacent bonded atoms will end up being starred (Figs. 3.35*a* and *c*) or one pair of adjacent bonded atoms will end up being unstarred (Fig. 3.35*d*) no matter where in the system the starring procedure begins. Thus in a nonalternant hydrocarbon, a complete *alternating* starring pattern of bonded atoms is not possible. This difference between alternant and nonalternant hydrocarbons is reflected in some of their properties, for example, ultraviolet spectra.

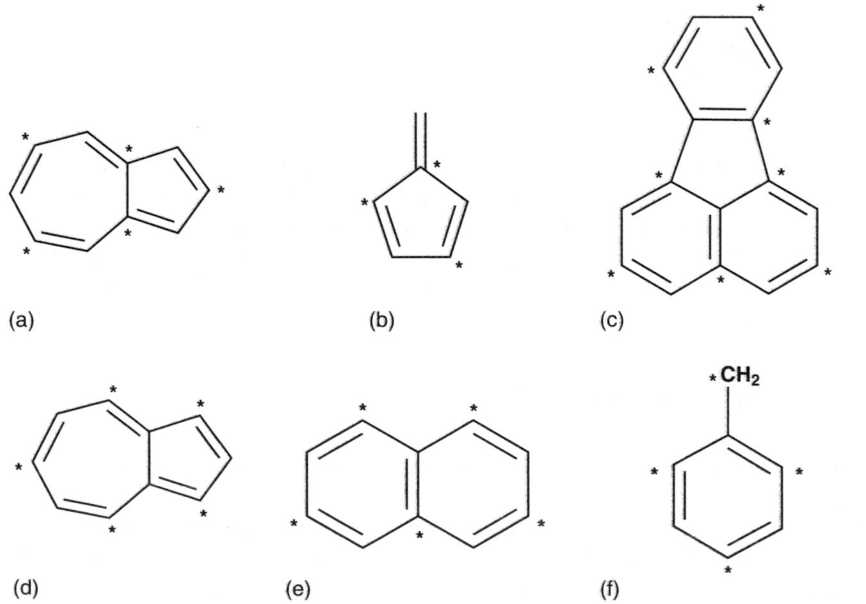

Figure 3.35. Nonalternant hydrocarbons: (*a* and *d*) azulene, (*b*) fulvene, and (*c*) fluoranthene; alternant hydrocarbons: (*e*) naphthalene and (*f*) benzyl.

3.36 ANNULENES

A family of monocyclic compounds having the empirical formula $(CH)_n$. If the annulene is essentially planar and has $(4n + 2)$ delocalized *p* electrons, it has aromatic character.

Example. Cyclooctadecanonaene (Fig. 3.36) or [18] annulene. The number in brackets indicates the number of carbon atoms in the cycle. This 18π electron hydrocarbon is a relatively stable brown-red compound whose X-ray structure suggests that all the C–C bonds as well as the internal C–H bonds are of equal length. The 12 hydrogen atoms on the "outside" of the ring are all equivalent, but differ from the six equivalent hydrogens on the "inside" of the ring.

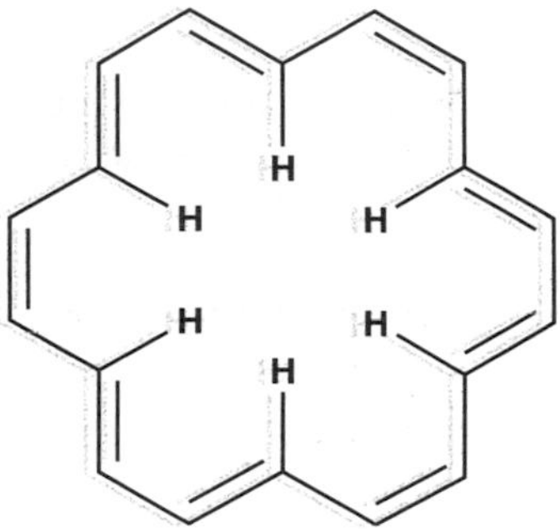

Figure 3.36. [18] Annulene.

3.37 HETEROAROMATIC COMPOUNDS

Aromatic compounds in which one or more –CH= groups of an aromatic hydrocarbon are replaced by a hetero atom, most frequently by nitrogen, oxygen, or sulfur. Oxygen and sulfur are usually part of a five-membered ring, but –N= commonly replaces –CH= in most ring systems.

Example. The five-membered ring systems pyrrole, furan, and thiophene (Figs. 3.37*a–c*, respectively). One of the lone pair of electrons on the heteroatom in each of these compounds is involved in conjugation with the ring π electrons, and hence they are all π-isoelectronic with the cyclopentadienide ion (Fig. 3.23*b*), and all possess six *p*π electrons. The six-membered ring system containing one nitrogen is pyridine (Fig. 3.37*d*). There are three possible six-membered rings containing two nitrogens: 1,2-, 1,3-, and 1,4-diazabenzene: pyridazine, pyrimidine, and pyrazine, respectively (Figs. 3.37*e–g*). Figure 3.37*h*, phenanthroline, is an example of a polynuclear heteraromatic. When the nitrogen atom is part of a six-membered ring, the lone pair of electrons is not involved in conjugation with the ring π system and such compounds are, therefore, basic.

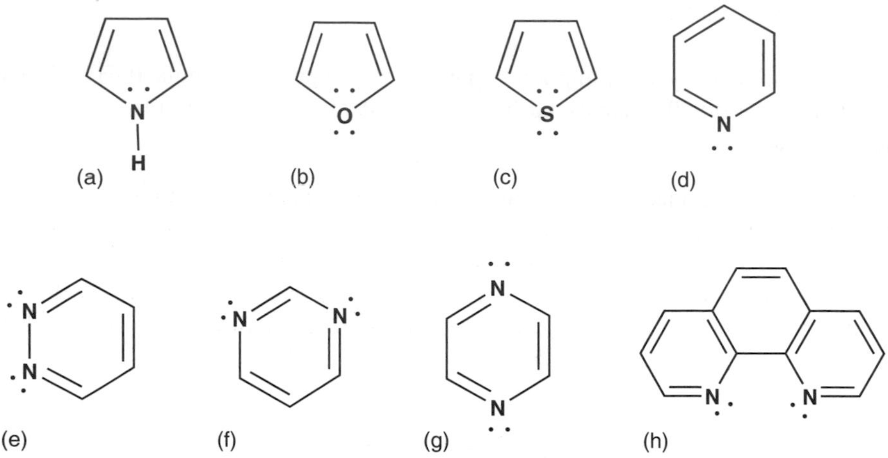

Figure 3.37. Heteroaromatic compounds: (*a*) pyrrole; (*b*) furan; (*c*) thiophene; (*d*) pyridine; (*e*) pyridazine; (*f*) pyrimidine; (*g*) pyrazine; (*h*) phenanthroline.

3.38 π–ELECTRON-EXCESSIVE COMPOUNDS

Compounds with more delocalized π electrons than atoms over which they are delocalized.

Example. Pyrrole and furan (Figs. 3.37*a* and *b*, respectively). Each possesses six *p*π electrons delocalized over five atoms and thus the ring carbons are more electron-rich than those in benzene.

3.39 π-ELECTRON-DEFICIENT COMPOUNDS

Heteroaromatics in which the π electron density on the ring carbons is less than that in the analogous aromatic hydrocarbon.

Example. Pyridine (Fig. 3.39); its resonance structures show the relative positive charge of the ring carbons. The substitution of the more electronegative nitrogen for the isoelectronic CH group of benzene results in less available electron density in

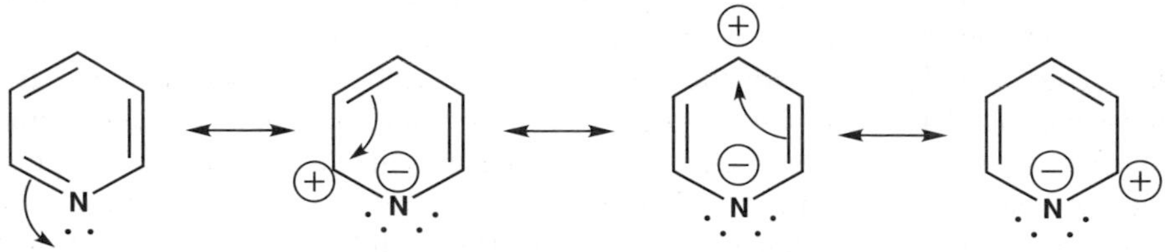

Figure 3.39. Resonance structures of pyridine.

the heterocycle than in benzene, hence the compound's relative inertness toward electron-seeking reagents.

3.40 HIGHEST OCCUPIED MOLECULAR ORBITAL (HOMO)

The occupied molecular orbital of highest energy. Electrons in this orbital are those usually involved in reactions pertaining to the donation of electrons into empty orbitals of other atoms or molecules. On exposure to light (electromagnetic radiation), electrons in these orbitals can be promoted to unoccupied molecular orbitals of higher energy in the same molecule.

Example. In 1,3-butadiene in the normal or ground state, ψ_2 is the highest occupied molecular orbital (HOMO) (Fig. 3.3). However, if irradiated at the proper wavelength (approximately 210 nm), an electron in ψ_2 is promoted to ψ_3 and this excited state then becomes the HOMO.

3.41 LOWEST UNOCCUPIED MOLECULAR ORBITAL (LUMO)

The molecular orbital of lowest energy that is unoccupied. This is the orbital that accepts electrons from other atoms or molecules in many reactions. It is also the orbital that usually becomes populated on exposure of the molecule to light (electromagnetic radiation).

Example. In 1,3-butadiene (Fig. 3.3), ψ_3 is the LUMO in the ground state and it is the orbital that accepts an electron when butadiene is irradiated to give an excited state with a π electronic configuration of $\psi_1^2\ \psi_2\ \psi_3$.

3.42 FRONTIER ORBITALS

The combination of HOMO and LUMO either in the same molecule or in reacting partner molecules. These are the orbitals of interest and importance in predicting whether certain bimolecular reactions will proceed thermally or photochemically by donation of electrons from the HOMO of one partner into the LUMO of the other partner.

3.43 MULTICENTER σ BONDING

In its broadest sense, synonymous with delocalized bonding (Sect. 3.2). However, the term is commonly restricted to the bonding in stable species that results when a series of atoms are linked together with fewer valence electrons than required for separate localized bonds.

3.44 THREE-CENTER, TWO-ELECTRON BONDING (3c-2e)

The multicenter bonding between three atoms by two electrons.

3.45 OPEN THREE-CENTER, TWO-ELECTRON BONDING

Three-center, two-electron bonding in which the two outside atoms (or groups) use orbitals pointed toward the middle atom (or group); the two outside atoms are not directly bonded to each other.

Example. The bridging B–H–B bonding in boron hydrides such as B_2H_6 (Fig. 3.45*a*). In 3c-2e bonds such as the B–H–B bonds, only two, not four, electrons are present to bond the three atoms. This "deficiency" of electrons is apparent, since in B_2H_6 there are 2(3) + 6(1) = 12 valence electrons; eight electrons are involved in conventional H–B bonds, leaving two electrons for each of the two three-center B–H–B bonds. The atomic orbitals used in this type of bonding are illustrated in Fig. 3.45*b*. The LCAO treatment in such situations gives

$$\psi_1 = 1/2\varphi_1 + 1/\sqrt{2}\varphi_2 + 1/2\varphi_3 \quad \text{(strongly bonding)}$$
$$\psi_2 = 1/\sqrt{2}\varphi_1 - 1/\sqrt{2}\varphi_3 \quad \text{(nonbonding)}$$
$$\psi_3 = 1/2\varphi_1 - 1/\sqrt{2}\varphi_2 + 1/2\varphi_3 \quad \text{(antibonding)}$$

The two available electrons occupy ψ_1.

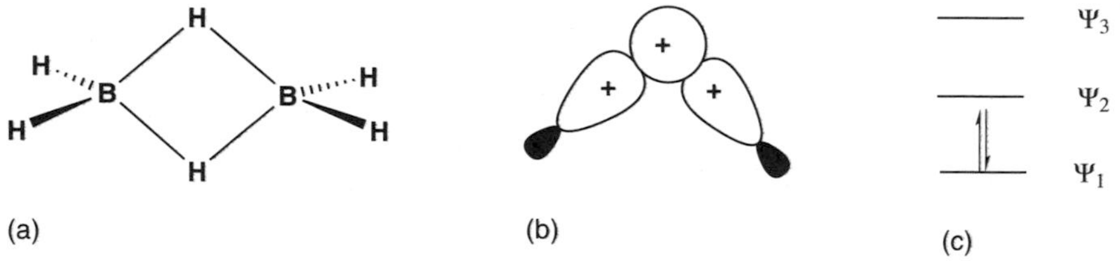

Figure 3.45. (*a*) The structure of B_2H_6; (*b*) the atomic orbitals involved in open three-center bonding, ψ_1; (*c*) the molecular orbital energy diagram of the σ orbitals involved in the 3c-2e bond.

3.46 CLOSED THREE-CENTER, TWO-ELECTRON BONDING

Three-center, two-electron bonding in which the three atoms (or groups) use orbitals directed mutually at each other; all three atoms are directly bonded to each other.

Example. The boron bonding shown in Fig. 3.46*a*. The atomic orbitals used in this bonding to generate the molecular orbitals are shown in Fig. 3.46*b*; the two electrons

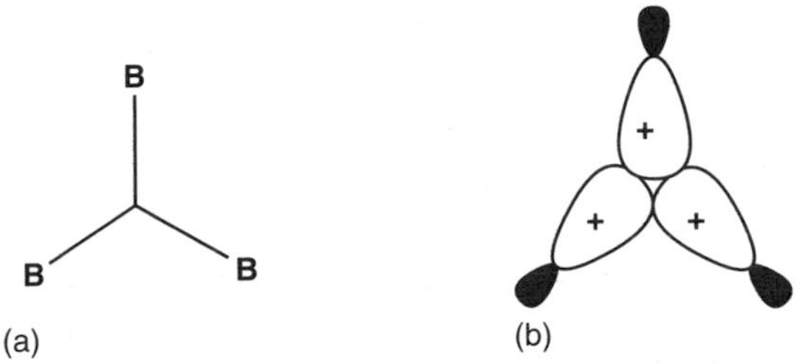

Figure 3.46. (*a*) The arrangement of atoms in closed three-center bonding and (*b*) the σ atomic orbitals involved in three-center bonding.

occupy the MO $\psi_1 = (1/3)\,\phi_1 + (1/3)\,\phi_2 + (1/3)\,\phi_3$. In some of the complicated boron hydrides, both open and closed 3c-2e bonds may be present.

3.47 EXTENDED HÜCKEL MOLECULAR ORBITAL THEORY

A method for performing molecular orbital calculations that, unlike simple Hückel theory, incorporates all valence σ and π orbitals. In addition, unlike the simple Hückel method, extended Hückel theory includes overlap between all atomic orbitals and a resonance integral between all atomic orbitals, not just nearest neighbors.

The secular determinant (Sect. 3.21) is solved in extended Hückel theory by making the following approximations:

1. S_{ii} are set equal to 1 because all atomic orbitals are first normalized.
2. H_{ii} are numerical parameters that must be supplied by the user.
3. S_{ij} are explicitly evaluated, based on some rather complicated formulae, which take into account the atomic orbitals involved and the internuclear distances.
4. H_{ij} are calculated from the formula $[k\, S_{ij}\, (H_{ii} + H_{jj})/2]$, where k is a constant, usually taken to be equal to 1.75.

Example. When the energies of the molecular orbitals for a simple diatomic like He_2 are calculated using the approximations of simple Hückel theory, the bonding and antibonding orbitals are symmetrically displaced from the energy of the 1*s* atomic orbitals. However, according to extended Hückel calculations, which includes the overlap of atomic orbitals, the energy of the antibonding level goes up more than the bonding level goes down: "Antibonding is more antibonding than bonding is bonding." This is illustrated in the molecular orbital energy diagram for the (theoretical) molecule, dihelium (Fig. 3.47), from which it can be correctly concluded that helium prefers to be monoatomic rather than diatomic. Note that extended Hückel theory, being a one-electron method, does not take into account electron–electron repulsion and correlations.

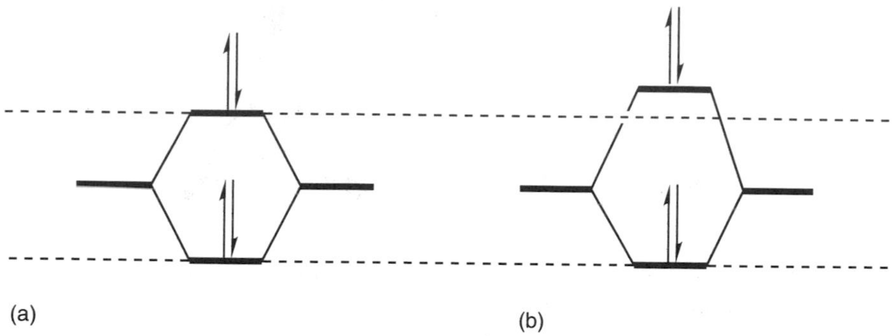

Figure 3.47. The molecular orbital energy level diagrams for He_2 (*a*) using the simple Hückel approximation of no overlap and (*b*) the same orbital energy diagram taking into account the overlap between the He 1*s* orbitals.

SUGGESTED READING

Liberles, A. *Introduction to Molecular Orbital Theory*. Holt, Reinhart and Winston: New York, 1966.

Orchin, M. and Jaffé, H. H. *Symmetry, Orbitals, and Spectra*. Wiley-Interscience: New York, 1971.

Roberts, J. D. *Molecular Orbital Calculations*. W.A. Benjamin: New York, 1961.

Streitwieser, A., Heathcock, C. H., and Kosower, E. M. *Introduction to Organic Chemistry*, 4th ed. Macmillan: New York, 1992.

Jean, Y., Volatron, F., and Burdett, J. *An Introduction to Molecular Orbitals*. Oxford University Press: New York, 1993.

4 Symmetry Operations, Symmetry Elements and Applications

4.1	Symmetry Operation	84
4.2	Symmetry Element	84
4.3	Axis of Symmetry C_n (Rotation)	85
4.4	Fold of a Rotational Axis, the n in C_n	85
4.5	Plane of Symmetry σ (Reflection)	86
4.6	Center of Symmetry i (Inversion)	87
4.7	Alternating or Improper Axis of Symmetry S_n (Rotation-Reflection)	88
4.8	Newman Projection	89
4.9	Equivalent Orientation	90
4.10	Identity Operation E	91
4.11	Equivalent Symmetry Operations	91
4.12	Point Groups	91
4.13	Classes of Symmetry Elements; Order of Classes	92
4.14	Inverse Operation	93
4.15	Nondegenerate and Degenerate Point Groups	94
4.16	Symmetric and Antisymmetric Behavior	94
4.17	Gerade g and Ungerade u Behavior	95
4.18	Dissymmetric	96
4.19	Asymmetric	97
4.20	Local Symmetry	98
4.21	Spherical Symmetry	98
4.22	Cylindrical Symmetry	98
4.23	Conical Symmetry	99
4.24	Symmetry Number	99
4.25	Symmetry-Equivalent Atoms or Group of Atoms	100
4.26	Character Tables of Point Groups	101
4.27	Symmetry Species or Irreducible Representations	102
4.28	Transformation Matrices	104
4.29	Trace or Character of a Transformation Matrix	105
4.30	Character Tables of Degenerate Point Groups	106
4.31	Symmetry and Dipole Moments	107
4.32	Group (or Symmetry) Orbitals	107

The Vocabulary and Concepts of Organic Chemistry, *Second Edition*, by Milton Orchin, Roger S. Macomber, Allan Pinhas, and R. Marshall Wilson

Symmetry is one of the most pervasive and unifying concepts in the universe. Its manifestations, as D'Arcy Thompson pointed out, are ubiquitous; it unites diverse disciplines such as anatomy, architecture, astronomy, botany, chemistry, engineering, geology, mathematics, physics, and zoology as well as the creative arts such as painting, poetry, literature, music, graphic arts, and sculpture. Symmetry is everywhere in nature; it shapes our sense of form, pattern, and beauty. Although the power of symmetry arguments has been recognized and employed for decades in mathematics (H. Weyl), physics (E. Wigner), and spectroscopy (G. Herzberg), its application to organic chemistry, apart from questions of optical activity, has been relatively recent. Applications of symmetry considerations to molecular orbital theory, group theory, and control of reaction pathways (R. Woodward, R. Hoffmann, K. Fukui, R. Pearson, F.A. Cotton) are now well understood and extensively utilized in organic chemistry.

In this chapter we discuss simple symmetry elements and operations at a level that requires little chemical knowledge, although we will almost always use molecules rather than objects to illustrate symmetry properties.

4.1 SYMMETRY OPERATION

An imaginary manipulation performed on a molecule (e.g., rotation, reflection, inversion, rotation-reflection) that results in a new orientation which is superimposable on, and hence, indistinguishable from the original orientation. The test for indistinguishability is superimposability, and hence, these two words will be used interchangeably.

Example. A 180° rotation of the HOH molecule around the axis shown in Fig. 4.1*a* transforms the molecule into the orientation *b*, which is superimposable on *a*.

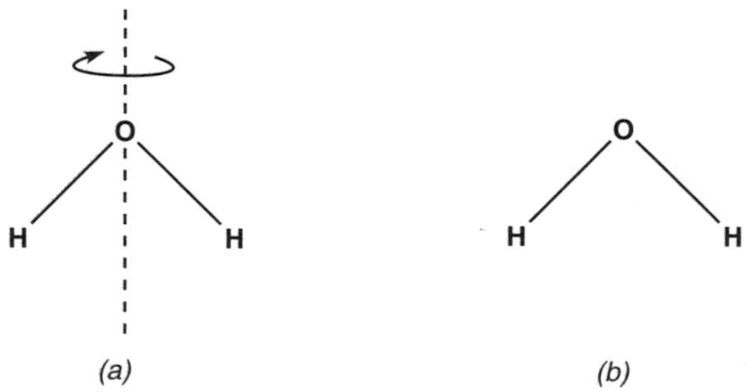

Figure 4.1. The symmetry operation that converts orientation *(a)* to *(b)*.

4.2 SYMMETRY ELEMENT

The axis, plane, or point in a molecule about which a symmetry operation is performed. There is a subtle distinction between a symmetry element and a symmetry operation (in mathematical terms this is the distinction between an operand and an operator, respectively). In the discussions in this chapter, the two terms frequently will be used interchangeably. For individual molecules, the possible symmetry elements (operations) are a rotational axis of symmetry (rotation); a plane of symmetry (reflection); a center of symmetry (inversion); and an alternating axis of symmetry (rotation-reflection).

4.3 AXIS OF SYMMETRY C_n (ROTATION)

An imagined axis passing through the center of mass of a molecule, around which the molecule can be rotated by some angle less than or equal to 360° such that the resulting orientation is superimposable on the original. This symmetry element, denoted by C (for cyclic), is also called a proper axis and the symmetry operation is rotation.

Example. The axis shown by the broken line in Fig. 4.1*a* is a rotational axis; rotation around it by 180° gives the orientation shown in *b*, which is superimposable on the original.

4.4 FOLD OF A ROTATIONAL AXIS, THE n IN C_n

This is an integer equal to 360°/θ, where θ is the smallest angle of rotation leading to the indistinguishable orientation; it also is equal to the number of times that orientations superimposable on the original are obtained when the molecule is rotated 360° around the axis.

Example. Rotation around the axis shown in Fig. 4.1*a* by 180° gives the orientation shown in *b* and a second rotation of 180° in the same direction gives Fig. 4.1*a* again. Thus in a 360° rotation, two indistinguishable orientations occur, and hence the axis is a two-fold axis (360°/180°) or C_2. The molecular framework of the benzene molecule may be represented by a planar regular hexagon (Fig. 4.4), where it is assumed that each numbered vertex represents a carbon atom to which a hydrogen atom is attached. Such a hexagon possesses a six-fold rotational axis perpendicular to the plane of the hexagon, as shown in Fig. 4.4. A rotation of 60° about this axis takes vertex 1 into 2, 2 into 3, and so on; hence, this axis is a (360°/60° = 6) C_6 axis. There is also a C_3 as well as a C_2 axis coincident with this C_6 axis.

There are also rotational axes *in* the plane of the benzene molecule but these will be discussed later. Every rotational axis is also a C_1 axis (360° rotation) but this is a trivial symmetry element; any of the infinite number of axes passing through the center of mass of a molecule (or for that matter anywhere through the molecule) is a C_1 axis.

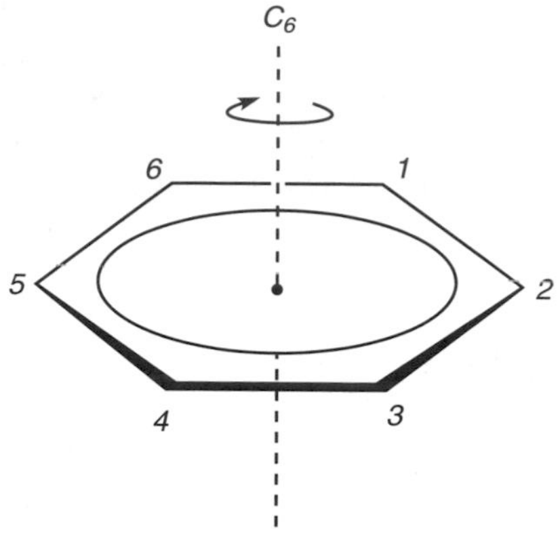

Figure 4.4. The C_6 axis of benzene.

4.5 PLANE OF SYMMETRY σ (REFLECTION)

This refers to a plane passing through the center of mass of a molecule that bisects it such that each atom on one side of the plane of the molecule when translated normal to and through the plane encounters an indistinguishable atom an equal distance from the other side of the plane. The plane of symmetry is the symmetry element; the symmetry operation is translation through this plane (reflection). The symmetry plane is

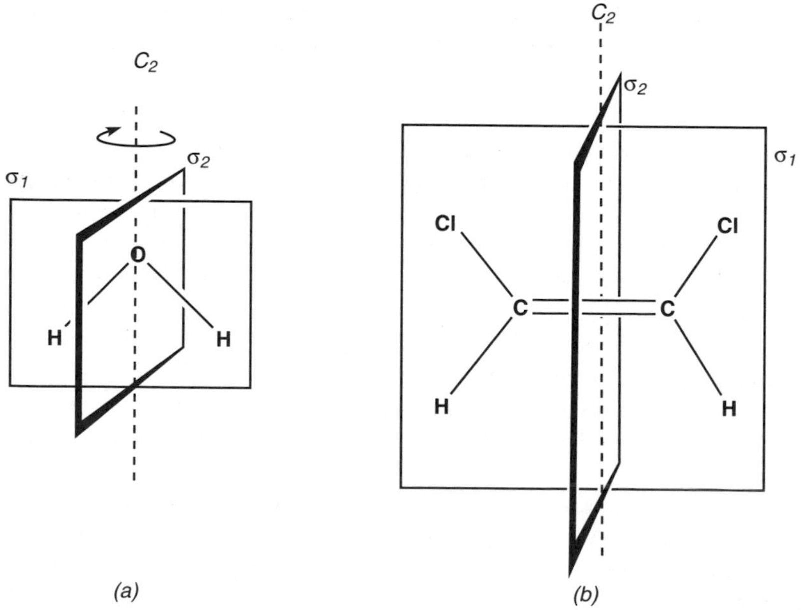

Figure 4.5. The two planes of symmetry in *(a)* H_2O and *(b) cis*-dichloroethylene.

also called a mirror plane. The designation σ comes from *spiegel* (the German for "mirror").

Example. Every planar molecule necessarily has a plane of symmetry, namely the molecular plane. Thus in HOH (Fig. 4.5*a*), the plane of the paper is one of the planes of symmetry, σ_1. The plane perpendicular to the molecular plane, σ_2 (Fig. 4.5*a*), is also a plane of symmetry since translation through it takes the hydrogen atoms into each other and the oxygen atom into itself. The intersection of the two planes is a C_2 axis (dotted line). The two analogous planes of symmetry in *cis*-1,2-dichloroethylene are shown in Fig. 4.5*b*.

4.6 CENTER OF SYMMETRY *i* (INVERSION)

A point at the center of mass of a molecule such that every atom in the molecule can be translated through this point and encounter an indistinguishable atom an equal distance on the other side of the point. To ascertain whether this symmetry element is present, the imagined operation is performed by drawing a straight line from each atom through the center of mass of the molecule and continuing this line an equal distance from this center, whereupon the line must encounter an equivalent atom. The letter *i*, denoting a center of symmetry, stands for inversion that is the operation associated with a center of symmetry. Any point (x, y, z) in a Cartesian coordinate system is transformed into the point $(-x, -y, -z)$ upon inversion through the origin.

Example. The inversion operation can be carried out on ethane when it is in the orientation shown in Fig. 4.6*a* (called the *staggered conformation*; see Sect. 4.8). If one draws a line from H_1 through the center of the carbon–carbon bond, the center of symmetry, and continues an equal distance, one will encounter H_6. Similarly, H_2 would transform into H_5, H_3 into H_4, and the two carbon atoms would transform into each other. In $[PtCl_4]^{2-}$ (Fig 4.6*b*), which has the shape of a square plane, the Pt atom

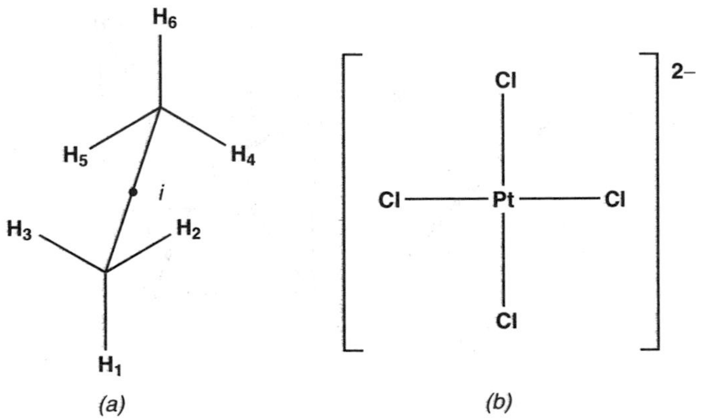

Figure 4.6. Molecules with a center of symmetry: *(a)* staggered ethane and *(b)* $[PtCl_4]^{2-}$.

lies at the center of symmetry. Inversion transforms each Cl atom into the one opposite it, and the Pt atom into itself.

4.7 ALTERNATING OR IMPROPER AXIS OF SYMMETRY S_n (ROTATION-REFLECTION)

An imagined axis passing through the center of mass of a molecule, rotation around which by an angle $\Theta \leq 360°$, followed by reflection through a plane perpendicular to this axis, results in an orientation indistinguishable from the original. The alternating axis is designated as *S* and subscripted with an integer that indicates its fold (see Sect. 4.4).

Example. The molecule of ethane in the orientation shown in Fig. 4.6*a* possesses an S_6 axis coincident with the C–C bond axis; clockwise rotation of the molecule by 60° around this axis followed by reflection through a plane perpendicular to the C–C bond axis takes H_1 into the position originally occupied by H_5, H_2 into H_4, and H_3 into H_6 and, of course, the two carbon atoms into one another, a 60° counterclockwise rotation followed by reflection takes 1 into 4, 2 into 6, and 3 into 5. Under the S_n operation the equivalent atoms are carried from one side of the reflection plane to the other in an alternating sequence, hence the name alternating axis. The molecule of methane, CH_4, possessing tetrahedral geometry, has three S_4 axes. This is readily visualized by inscribing the CH_4 molecule (Fig. 4.7*a*) in a cube (Fig. 4.7*b*). An axis passing through the center of opposite faces is an S_4 axis, one of which is shown in Fig. 4.7*b*. It passes through the center of the molecule and bisects opposite HCH angles. Clockwise rotation of the molecule by 90° around this axis followed by reflection in a horizontal plane (perpendicular to the axis and bisecting the cube)

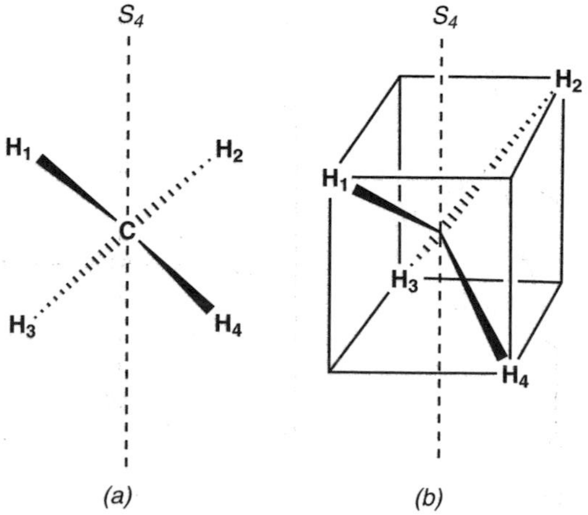

Figure 4.7. *(a)* An S_4 axis in methane and *(b)* the same axis with CH_4 in a cube.

restores the molecule to an orientation superimposable on the original. The S_4 axis shown bisects opposite sides of the cube, and since there are six sides to a cube, three S_4 axes exist.

The methane molecule also possesses other symmetry elements, for example, the axis going through any H atom and the C atom is a C_3 axis, and since there are four H atoms, four C_3 axes exist; in the cube, these are axes that go through diagonally opposite vertices. It is important to understand that while the C_3 axes interchange three hydrogen atoms at a time, the only *single* symmetry operation that results in transforming all four hydrogen atoms into each other is the S_4 operation. The importance of the alternating axis is reflected in its position in the hierarchy of symmetry operations, which is usually considered to be $C_n < \sigma < i < S_n$.

4.8 NEWMAN PROJECTION

This is a representation, as formulated by Melvin S. Newman (1908–1993), showing the spatial relationship of the atoms in a molecule as viewed down one particular carbon–carbon bond axis and drawing a projection of the carbons and all the attached groups onto the plane of the paper. In such a drawing, the bonds from the atoms attached to the front carbon are represented by solid lines from these atoms and meeting at a central point that represents the front carbon atom. The bonds from the atoms attached to the rear carbon atom, which is eclipsed by the front carbon, are represented by short lines from these atoms to the periphery of a circle drawn around the central point. This type of projection drawing in which the eight atoms involved are projected onto the plane of the paper is not only used to show spatial relationships around a particular carbon–carbon bond of interest, but for other systems as well.

Example. Ethane in the staggered conformation shown in Fig. 4.6*a* is represented in the Newman projection as illustrated in Fig. 4.8. This representation is particularly useful for illustrating the high degree of symmetry in staggered ethane. Thus, the six-fold rotation-reflection operation S_6, which is the only single symmetry operation that carries all six hydrogen atoms into each other, is more readily visualized in Fig. 4.8

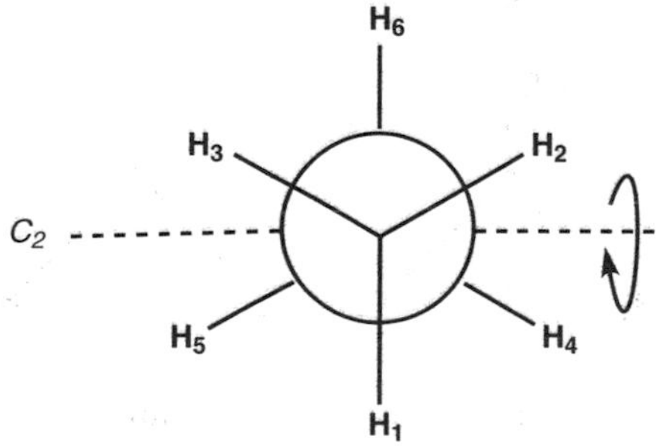

Figure 4.8. Newman projection of staggered ethane.

than in Fig. 4.6*a*. The S_6 axis is coincident with the carbon-carbon bond axis, which is perpendicular to the plane of the paper and goes through the point where the three front and the three rear C–H bonds converge. Clockwise rotation of 60° about this axis followed by reflection through the plane of the paper exchanges H_1 with H_5, H_2 with H_4, and H_3 with H_6, and the C atoms forming the axis transform into each other. A C_3 axis coincident with the S_6 axis is also present. One of the three C_2 axes, all of which are perpendicular to the S_6, is also shown in the figure; these are very difficult to visualize in the three-dimensional model of staggered ethane shown in Fig. 4.6*a*.

4.9 EQUIVALENT ORIENTATION

The orientation of a molecule resulting from a symmetry operation, as distinct from the (original) orientation of the molecule before anything has been done to it. Of course, by definition the orientation resulting from a symmetry operation is indistinguishable from the original. Accordingly, one cannot distinguish between any equivalent orientation and the original orientation simply by looking at the molecule without some sort of imaginary labels that provide information as to the original orientation.

Example. A rotation of 180° of the HOH molecule (Fig. 4.9*a*) around the C_2 axis transforms the molecule into the equivalent orientation *b*. Since the two hydrogen atoms, although labeled 1 and 2 for the purposes of illustration, are in fact indistinguishable, the new orientation is equivalent to the original. However, to obtain the *original* orientation, it would be necessary to conduct a second 180° rotation in either direction. The fictional labels are used here only to document what has been done. The same results, transforming (*a*) into (*b*) as obtained by the C_2 rotation, can also be achieved by reflection in the vertical plane perpendicular to the plane of the paper.

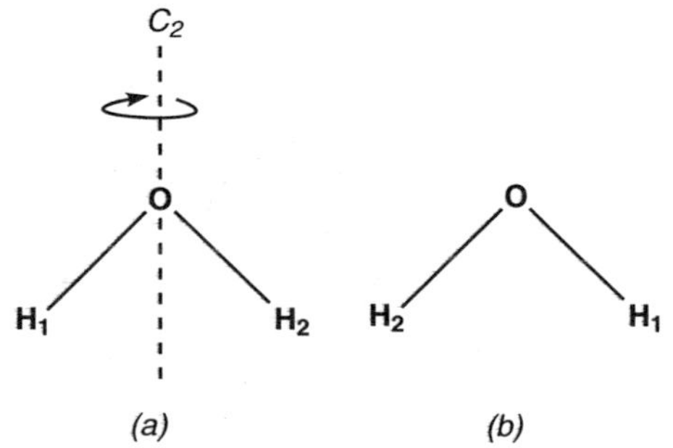

Figure 4.9. Equivalent orientations of H_2O.

4.10 IDENTITY OPERATION *E*

An instruction to do nothing to a molecule. It is included in the list of symmetry operations because it is a mathematical requirement for a group, for example, a point group (see Sect. 4.12). There is an operational distinction between actually doing nothing (the identity operation) and performing a symmetry operation since by definition a symmetry operation leads to a result that *appears* as though nothing has been done.

Example. Consider again Fig. 4.9*a* in which we imagine that the hydrogen atoms are labeled. The single symmetry operation C_2 leads to the equivalent orientation Fig. 4.9*b*. However in order to restore the molecule to (*a*) [i.e., the result of doing nothing to (*a*), which is the identity operation], we must perform a second C_2 on (*b*). In simple mathematical terms $C_2^2 = E$. Similarly, if we return to the vertical axis of benzene shown in Fig 4.4, we can see that $C_6^6 = C_3^3 = C_2^2 = E$.

4.11 EQUIVALENT SYMMETRY OPERATIONS

Two different symmetry operations, each performed separately, that lead to the same result. Specifying either one of them thus implies the other.

Example. The operations *i* and S_2 are equivalent. Consider again staggered ethane (Fig. 4.8*a*). The operation *i* transforms H_1 into H_6, H_3 into H_4, and H_2 into H_5. For the equivalent S_2 operation, the improper axis is in the plane of the C_1–H_1 and C_2–H_6 bonds and bisects the C–C bond (this is not a C_2 axis!). A 180° rotation around this axis followed by reflection through a plane perpendicular to this axis (this is not a plane of symmetry!) results in the exchange of atoms (e.g., H_1 into H_6, etc.), identical to the result achieved by *i*.

Another important equivalence is that between S_1 and σ. For example, consider the molecule *cis*-1,2-dichloroethylene (Fig. 4.5*b*). To perform the S_1 operation, rotate the molecule around the horizontal axis coincident with the double bond axis by 360°, followed by reflection through the vertical plane of symmetry (σ_2 in Fig. 4.5*b*) perpendicular to this axis. This combined rotation-reflection operation, S_1, results in an exchange of atoms identical to that obtained by performing the single reflection operation through the plane σ_2. Demonstrating the equivalence of S_1 to σ, and S_2 to *i* are not trivial exercises since these equivalencies allow one to eliminate a plane of symmetry and a center of inversion as fundamental independent operations. Thus, all point symmetry operations can be reduced to just two types of operations, proper (C_n) and improper (S_n) axes (where $n \geq 1$).

4.12 POINT GROUPS

A classification of molecules based on the number and kind of symmetry operations, for example, C_n, *i*, σ, S_n, *E*, that can be performed on them. All the symmetry operations described thus far are *point symmetry operations* because they all leave one point, the center of mass of the molecule, unmoved. Molecules that possess all the same symmetry elements and only these elements belong to the same point group.

Example. The completely dissimilar molecules of water, methylene chloride, formaldehyde, and phenanthrene (Figs. 4.12*a*, *b*, *c*, and *d*, respectively) all belong to the same point group C_{2v}. Molecules in this point group possess a C_2 axis and two vertical planes (σ_v) of symmetry perpendicular to each other as their only symmetry elements. The planes of symmetry are considered to be vertical (as distinct from horizontal σ_h) because they include the rotation axes.

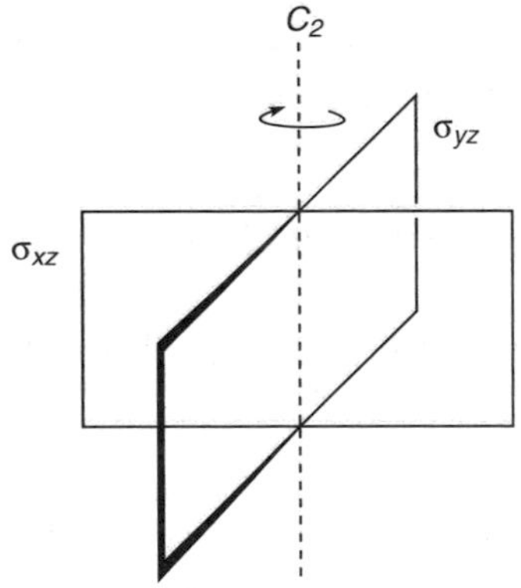

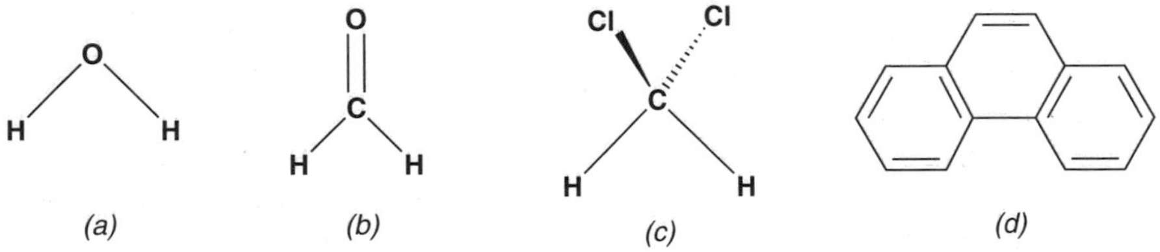

Figure 4.12. Molecules belonging to point group C_{2v}: (*a*) H_2O, (*b*) CH_2O, (*c*) CH_2Cl_2, and (*d*) phenanthrene.

4.13 CLASSES OF SYMMETRY ELEMENTS; ORDER OF CLASSES

These are separate symmetry elements that can be transformed into each other by a symmetry operation appropriate to the point group; the number of these elements belonging to the same class is the order of that class.

Example. Many molecules have more than one C_n, S_n, or σ. Consider the regular hexagon representing benzene oriented in the plane of the paper, as shown in Fig. 4.13. All the C_2 axes shown are also in the plane of the paper. Now consider the C_2 axis going through opposite vertices 1 and 4. On rotation around the perpendicular C_6 axis by 60°,

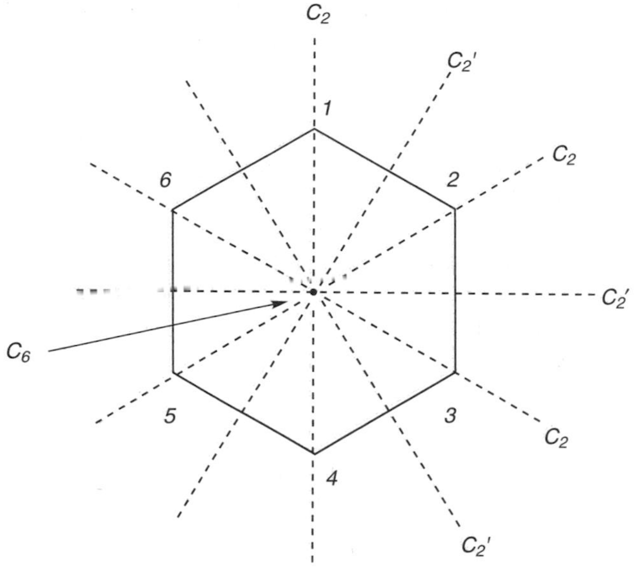

Figure 4.13. Two classes of C_2 axes in benzene.

a symmetry operation, this C_2 axis is transformed into the equivalent C_2 axis bisecting opposite vertices 2 and 5, and every C_2 axis passing through vertices is transformed into an equivalent one by the C_6 operation. Hence, the three axes bisecting opposite vertices are said to belong to the same class of C_2 with order 3 or $3C_2$. The same can be said of the three C_2 axes going through opposite sides, denoted as C_2' in the figure. If one wishes to transform C_2 into C_2', a rotation of 30° around the vertical axis would be required, but since a rotation of 30° is not a symmetry operation, the two sets of C_2 belong to different symmetry classes. They are frequently designated separately by priming one set as we have done in this case. Each of the two planes of symmetry in molecules having C_{2v} symmetry (Fig. 4.12) belong to a separate class understood to be of order 1.

There is another important point to be made about the order of a class. Consider again the regular hexagon in Fig. 4.13; the clockwise rotation of 60° around the C_6 axis that is perpendicular to the plane of the paper takes position 1 into 2, 2 into 3, and so on. However, counterclockwise rotation takes 1 into 6, 6 into 5, and so on, an equivalent but nonidentical orientation. Hence, the order of the class C_6 is 2. Similarly, there will be $2C_3$. In tetrahedral CH_4 (Fig. 4.7), where each C–H bond is a three-fold axis, the order for the C_3 is 8, that is, $8C_3$.

4.14 INVERSE OPERATION

The inverse or reciprocal of each symmetry operation is indicated by placing the superscript -1 over the operation. The inverse operation is a mathematical requirement of a group; it tells us that for each operation A, there must be another operation in the group, A^{-1}, which is its inverse, such that $AA^{-1} = E$. This means that if the operations A and A^{-1} are performed in sequence, the product of the operations is equal to the identity E.

Example. Consider tetrahedral CH_4 shown in Fig. 4.14. If one performs a C_3 (counterclockwise) operation around the axis shown, H_1 goes to H_2, H_3 to H_1, and H_2 to H_3. If this operation is followed by C_3^{-1}, the molecule is restored to its original orientation. For any C_2 axis, for example, that in the HOH molecule (Fig. 4.9), the C_2 operation is its own inverse, for example, 180° rotation around the C_2 axis of the HOH molecule, followed by 180° rotation in either direction, returns the molecule to its original orientation. The same is true of a reflection through a plane of symmetry; thus in point group C_{2v}, each symmetry element is its own inverse.

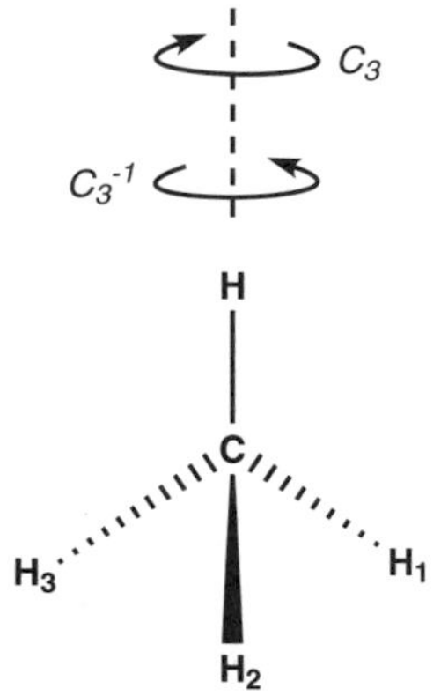

Figure 4.14. The inverse relationship of C_3 and C_3^{-1} axes in CH_4.

4.15 NONDEGENERATE AND DEGENERATE POINT GROUPS

Nondegenerate point groups describe molecules possessing no C_n or S_n axes greater than two-fold. Degenerate point groups describe molecules having either C_n or S_n with n greater than 2.

Example. All the molecules shown in Fig. 4.12 belong to the nondegenerate point group C_{2v}, as does *cis*-dichloroethylene (Fig. 4.5*b*). The tetrahedral molecule methane that, among other symmetry elements, has S_4 and C_3 axes (Figs. 4.7 and 4.14) belongs to the degenerate point group called T_d. (T stands for tetrahedral, d for dihedral.)

4.16 SYMMETRIC AND ANTISYMMETRIC BEHAVIOR

If the sign of a wave function or direction of a vector associated with a molecule is unchanged upon application of a symmetry operation, the property is said to be *symmetric* with respect to that symmetry operation (or the symmetry element associated with it). In contrast, if the sign of a wave function or direction of a vector associated with a molecule is reversed, but the magnitude is unchanged, the property is said to be *antisymmetric* with respect to the symmetry operation. Common nonstationary properties are translational, rotational, vibrational, and electron (orbital) motions of a molecule; all these can be characterized in terms of the symmetry operations appropriate to the point group of the molecule.

Example. The two possible stretching vibrations of the water molecule are illustrated in Figs. 4.16*a* and *b*. The motion of the O–H bonds is indicated by the direction and magnitude of the arrows. One of the symmetry operations that can be performed on the vibrational motion of this C_{2v} molecule is reflection through the plane bisecting the H–O–H angle. If we perform this operation on the vibration shown in Fig. 4.16*a*, the arrow on the right exchanges with the arrow on the left without a reversal of direction and vice versa. This stretching vibration is thus symmetric with respect to reflection through the plane. Now if we perform the same operation on the vibration shown in Fig. 4.16*b*, which shows that one O–H bond is being stretched while the other is being compressed, the arrows are exchanged but their directions are reversed. This stretching vibration is thus antisymmetric with respect to reflection through the plane. The magnitude of the vectors (arrows) associated with each bond must be equal; otherwise, the molecule will move and we would be dealing with a translation or rotation rather than with a pure vibration.

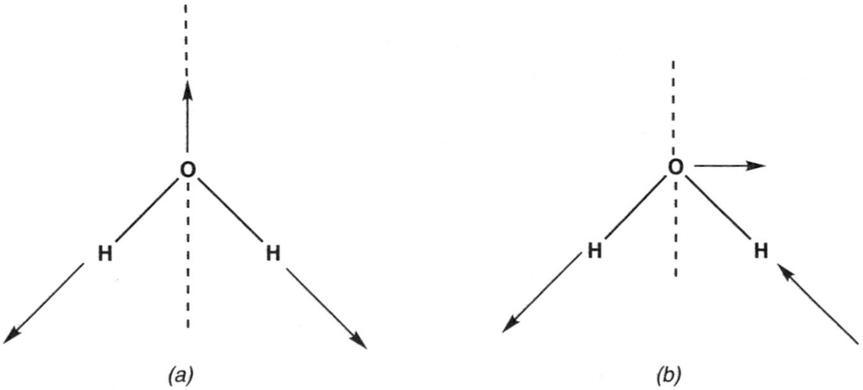

Figure 4.16. The vibrational stretching modes of H_2O: *(a)* symmetric; *(b)* antisymmetric with respect to reflection across the perpendicular plane or with respect to rotation about the twofold axis.

4.17 GERADE g AND UNGERADE u BEHAVIOR

These refer to symmetric and antisymmetric behavior with respect to molecules possessing a center of symmetry; symmetric behavior is denoted as g (gerade) and antisymmetric behavior as u (ungerade). *Gerade* and *ungerade* are the German words for even and odd, respectively.

Example. The CO_2 molecule is linear and has a center of symmetry. When this molecule undergoes a vibration in which the two oxygen atoms are moving in the same direction (Fig. 4.17*a*), such motion is antisymmetric with respect to inversion at the center of symmetry and is denoted as u; the symmetric stretch g is shown in *b*.

Figure 4.17. The two stretching vibrations of CO_2: *(a)* antisymmetric u and *(b)* symmetric g.

4.18 DISSYMMETRIC

The symmetry characterization of molecules possessing as their *only* symmetry element one or more C_n axes where $n \geq 2$. Such molecules are not superimposable on their mirror image.

Example. trans-1,2-Dichlorocyclopropane (Fig. 4.18*a*) is dissymmetric; it has a C_2 axis (in the plane of the ring and bisecting the front C–C bond) as the only symmetry element; its mirror image (Fig. 4.18*b*) is not superimposable on *a*. The twisted or nonplanar ethylene (Fig. 4.18*c*) has a C_2 axis coincident with the C=C bond. Perpendicular to this axis are two other C_2 axes at right angles to each other, shown as broken lines in the Newman projection of the twisted ethylene (Fig. 4.18*d*).

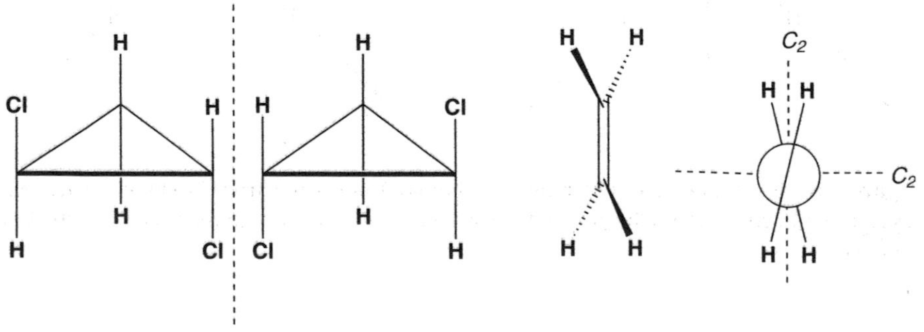

Figure 4.18. *(a)* The dissymmetric molecule *trans*-1,2-dichlorocyclopropane; *(b)* its nonsuperimposable mirror image; *(c)* dissymmetric twisted nonplanar ethylene; and *(d)* a Newman projection of *(c)*.

4.19 ASYMMETRIC

The absence of any symmetry element (other than E and the trivial C_1). As with a dissymmetric molecule, the mirror image of an asymmetric molecule is not superimposable on the original. A molecule whose mirror image is not superimposable on the original is characterized as *chiral*; all other molecules are *achiral*. Dissymmetric and asymmetric molecules represent the two types of chiral molecules. All molecules possessing an S-axis of any fold (a plane of symmetry S_1, a center of symmetry S_2, or any other $S_n > 2$ axis) are achiral.

Example. A molecule that contains at least one carbon atom to which four different groups are attached (Fig. 4.19*a*). Such a carbon atom is called an asymmetric or chiral carbon atom. The cyclic molecule, 1,1,2-trichlorocyclopropane (Fig. 4.19*b*), is asymmetric; it has four different groups attached to carbon atom 2 and hence the molecule is asymmetric. A molecule may contain several individual asymmetric carbon

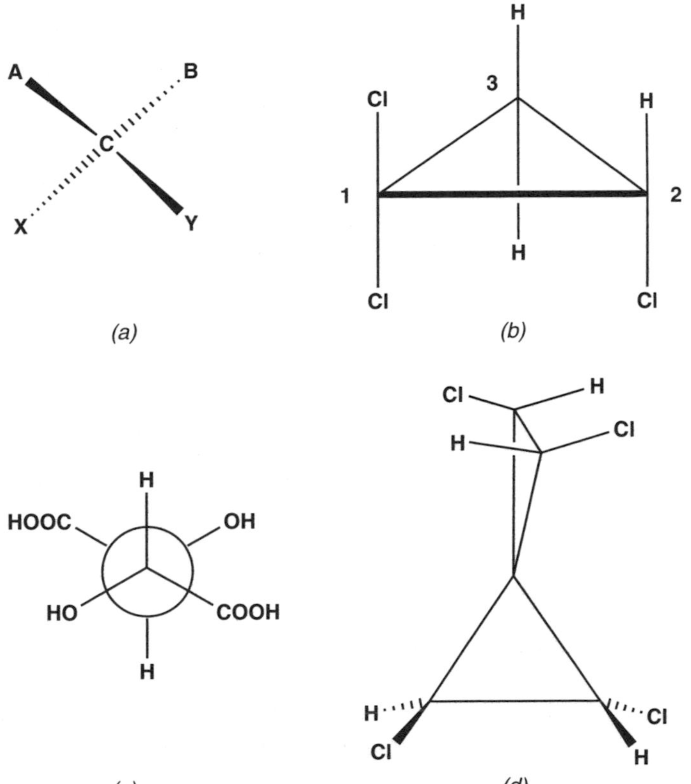

Figure 4.19. Two examples of asymmetric or chiral molecules: *(a)* a molecule with four different groups attached to carbon and *(b)* 1,1,2-trichlorocyclopropane. Two examples of achiral molecules: *(c)* *meso*-tartaric acid and *(d)* a spiro compound.

atoms, but if the groups attached to these atoms are arranged such that the entire molecule has a center or plane of symmetry, the molecule is achiral, for example, tartaric acid shown in Fig. 4.19*c*. The spiro compound shown in Fig. 4.19*d* has a C_2 axis, but it is not chiral because this axis is also an S_4 axis.

4.20 LOCAL SYMMETRY

The symmetry of a small group of atoms within a molecule without reference to the symmetry of the entire molecule. Such local symmetry usually is not characteristic of the entire molecule. One frequent application of local symmetry involves the methyl group, CH_3. Although a methyl group may be a small portion of a molecule, for local symmetry purposes the molecule can be assumed to have the structure CH_3–Z, where Z represents the collection of all atoms attached to the methyl group. Such an assumption is reasonable because rapid rotation around the C–Z bond makes the three hydrogens equivalent and this bond defines a local C_3 axis.

Example. Methanol, CH_3OH, shown in Fig. 4.20*a* does not have a C_3 axis in any orientation. However, the methyl group has local C_3 symmetry (Fig. 4.20*b*), because for local symmetry purposes, the OH group may be considered a single atom Z.

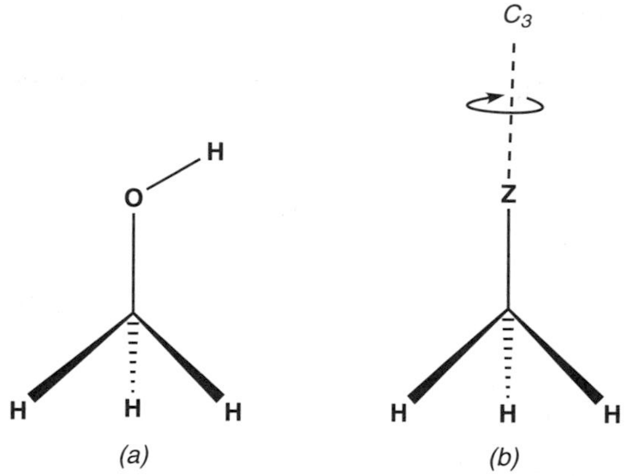

Figure 4.20. *(a)* Methanol (no rotational axis) and *(b)* the local C_3 axis.

4.21 SPHERICAL SYMMETRY

This is symmetry characteristic of a sphere; it is sometimes referred to as perfect symmetry since the sphere has an infinite number (except *E* and *i*) of all symmetry elements.

4.22 CYLINDRICAL SYMMETRY

Symmetry characteristic of a cylinder. All linear molecules possess such symmetry characterized by a C_∞ axis coincident with the center of the long axis of the cylinder, an infinite number of planes of symmetry that include this axis, as well as a plane of symmetry perpendicular to the axis.

Example. Linear molecules with a center of symmetry have cylindrical symmetry, for example, O=C=O (carbon dioxide), HC≡CH (acetylene). Common objects with cylindrical symmetry are barrels, bagels, and footballs.

4.23 CONICAL SYMMETRY

This is symmetry characteristic of a cone or funnel; molecules possessing it have a C_∞ axis, an infinite number of planes of symmetry intersecting this axis, but do not have a plane of symmetry perpendicular to the C_∞ axis or a center of symmetry.

Example. Linear molecules without a center of symmetry have conical symmetry, for example, H–F (hydrogen fluoride), H–C≡N (hydrogen cyanide).

4.24 SYMMETRY NUMBER

The number of indistinguishable orientations that a molecule can assume by rotation around its symmetry axes.

Example. The symmetry numbers for H_2O and NH_3 are 2 and 3, respectively. The four indistinguishable orientations of the planar ethylene molecule C_2H_4 are shown in Fig. 4.24*a*; these are obtained by rotating the original orientation around the three

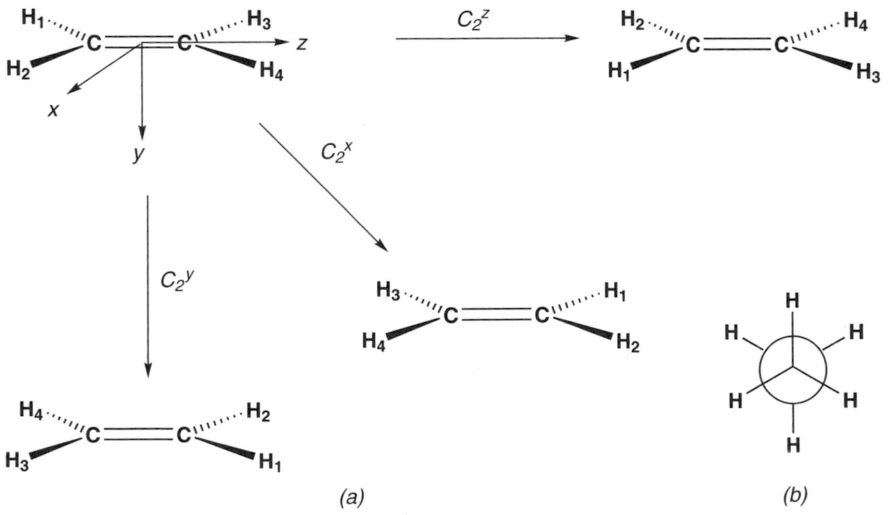

Figure 4.24. *(a)* The four orientations of ethylene made possible by rotation around the three C_2 axes; *(b)* staggered ethane.

C_2 axes noted by the x-, y-, and z-coordinates; hence, the symmetry number is 4. Ethane, shown in the staggered conformation (Fig. 4.24*b*) has a symmetry number of 6: three orientations obtained by rotation around the C_3 axis plus each of two orientations obtained by rotation around the three C_2 axes.

4.25 SYMMETRY-EQUIVALENT ATOMS OR GROUP OF ATOMS

Atoms or groups of atoms in a molecule that can be interchanged by any symmetry operation or combination of symmetry operations. In organic chemistry, the symmetry equivalence of the hydrogen atoms in a molecule is most frequently of interest and they will be used for illustrative purposes. The determination of sets of hydrogen and carbon atoms that can be exchanged by any symmetry operation permits ready determination of the various isomers that can be derived from a particular compound by the substitution of hydrogen with another atom.

Example. In cyclopropane (Fig. 4.25*a*), the hydrogen atoms above the plane of the ring are exchangeable by a C_3 operation that also exchanges the hydrogen atoms below the plane. The top and bottom hydrogens are exchanged by rotation about any of the C_2 axes that bisect a C–C bond and the opposite C atom, or by reflection through the plane defined by the three carbon atoms. Because every H atom can be transformed into any other H atom by one or more symmetry operations, the six hydrogens of the molecule

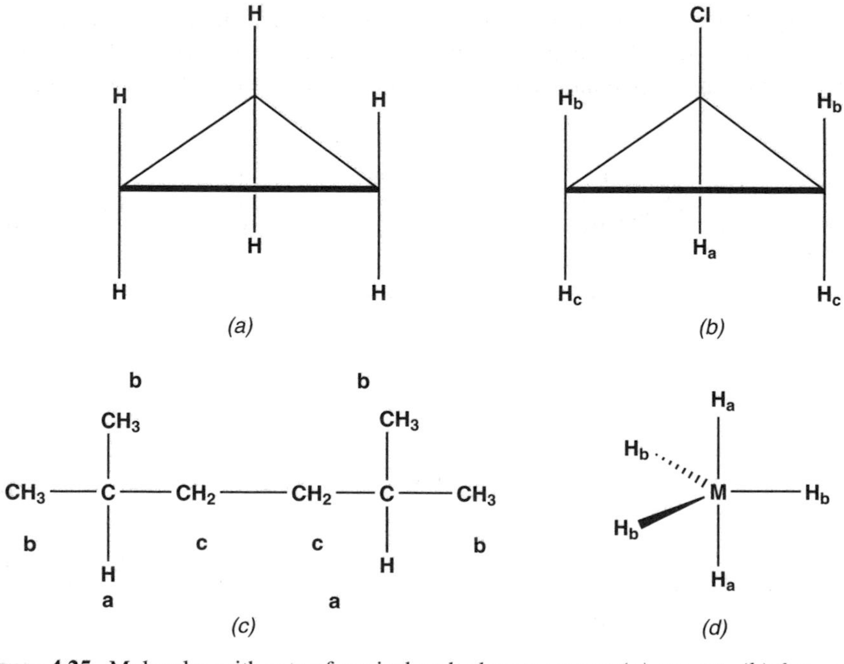

Figure 4.25. Molecules with sets of equivalent hydrogen atoms: *(a)* one set, *(b)* three sets, *(c)* three sets, and *(d)* two sets.

belong to one set, and therefore, there is only one possible mono-substituted cyclopropane. In monochlorocyclopropane (Fig. 4.25*b*), there are three sets of symmetry-equivalent H atoms, hence three possible dichlorocyclopropanes.

The three hydrogen atoms of a methyl group are symmetry-equivalent by local symmetry. Thus, in 2,5-dimethylhexane (Fig. 4.25*c*), there are three sets of symmetry-equivalent hydrogen atoms: one set of 12 hydrogens (the hydrogens of the four methyl groups), one set of two hydrogens (the methine hydrogens), and one set of four hydrogens (the hydrogens of the two methylene groups). There will, thus, be only three different carbon positions for a monochloro derivative.

In the trigonal bipyramidal metal hydride, MH_5 (for which there is, as yet, no real example), the three equatorial hydrogens (one set) cannot be transformed into either of the two apical H atoms (the second set) by any symmetry operation appropriate to MH_5; hence, there are two distinct sets of hydrogen atoms.

4.26 CHARACTER TABLES OF POINT GROUPS

Tables in which all the various symmetry elements that characterize the point group are typically listed along the top row. In columns, under each element, are listed the behavior patterns (character) that various molecular orientations display when subjected to the symmetry operation at the top of the column. *Character tables of nondegenerate point groups* can be developed rather simply because all motions of a molecule [e.g., the vibrations of HOH (Fig. 4.16) as well as its translations] belonging to a nondegenerate point group must be either symmetric or antisymmetric with respect to the symmetry operations of the point group. Understanding character tables of degenerate point groups is more complicated and requires some knowledge of matrices; these will be discussed in Sect. 4.30.

Example. In order to construct the character table of a point group, we must first point out another property of a group, namely that the product of any two operations in the group must be equal to another operation in the group. In point group C_{2v} to which HOH belongs (Fig 4.26), there are four operations: E, the identity operation that is trivial; and the three nontrivial operations C_2^z, σ_{xz}, and σ_{yz}. If we characterize symmetric behavior as $+1$ and antisymmetric behavior as -1 under each of these four operations, we can readily verify that the product of any two of the nontrivial operations will be equal to the third, for example, $C_2^z \times \sigma_{xz} = \sigma_{yz}$. Thus, only two independent operations are sufficient to construct the character table. Since each of the two operations can have the character of ± 1, there will be a total of four possible combinations. If, in this case we arbitrarily consider C_2^z and σ_{xz} as the two independent operations, both can be $+1$; or they can be $+1$, -1; -1, $+1$; or finally both can be -1.

Now we can place this information in a table (Table 4.26) and develop what is called a character table. It shows that under the identity operation E, every property is symmetric ($+1$) since this operation does nothing. The four possible combinations of C_2^z and σ_{xz} are shown in the second and third columns, and in each case the character of σ_{yz}, shown in the fourth column, is the product of $C_2^z \times \sigma_{xz}$.

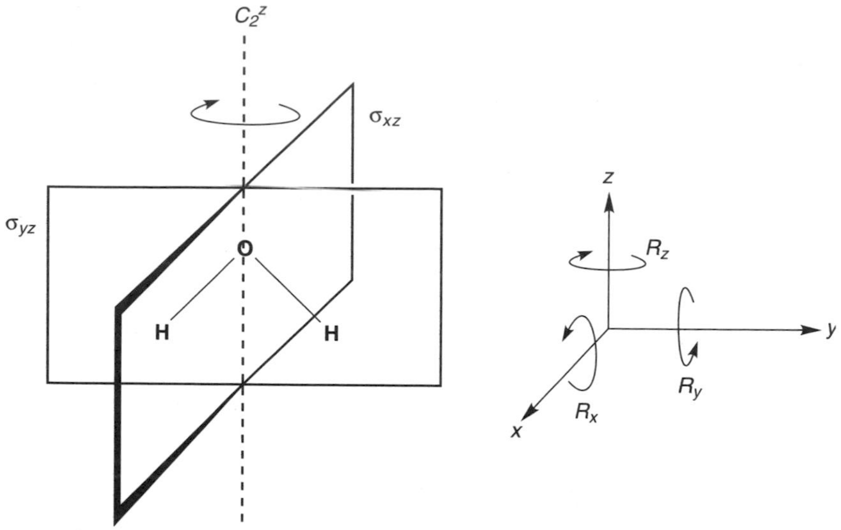

Figure 4.26. H_2O and its symmetry elements.

TABLE 4.26. The Character Table for Point Group C_{2v}

C_{2v}	E	C_2^z	σ_{xz}	σ_{yz}	
A_1	+1	+1	+1	+1	z, α_{xx}, α_{yy}, α_{zz}
A_2	+1	+1	−1	−1	R_z, α_{xy}
B_1	+1	−1	+1	−1	x, R_y, α_{xz}
B_2	+1	−1	−1	+1	y, R_x, α_{yz}

4.27 SYMMETRY SPECIES OR IRREDUCIBLE REPRESENTATIONS

Each of the four rows in Table 4.26 represents a separate behavior pattern, and each is called a symmetry species (or an irreducible representation). For convenience and by convention, the symmetry species are designated by certain symbols, displayed in the first column of the character table. All symmetry species that are symmetric with respect to the rotational axis (the highest-order axis if more than one) are designated by *A* and those that are antisymmetric by *B*. When more than one symmetry species is symmetric with respect to the axis (as in C_{2v}), they are distinguished by subscripts. The totally symmetric species (+1 with respect to all operations) is A_1. Subscripting the *B* species is somewhat arbitrary; in the case of C_{2v}, the *B* species that is symmetric to σ_{xz} is denoted as B_1.

Example. Let us analyze the behavior of a property of the HOH molecule, specifically its motion in the *y* direction (Fig. 4.26). Rotation around the C_2 axis changes the direction of motion; hence, this behavior is antisymmetric or -1. Reflection on σ_{xz} gives a change in sign, hence also -1, while reflection in σ_{yz} gives itself, hence $+1$. The behavior pattern of motion along the y-axis is therefore $+1$, -1, -1, $+1$ and such motion is thus B_2. The symmetry species to which motion (a vector) in the x, y, and z directions belongs, is of special interest, because such symmetry species correspond to those of the p orbitals. Consequently, in the character table the letters x, y, and z that appear in the last column indicate the symmetry species of the three p orbitals, as well as the symmetry species of motion (translation) in those three directions. The letter R in the last column refers to rotation about the subscripted axes. Thus, the position of R_z in the table indicates that rotation around the z-axis belongs to species A_2. The α's in the table are associated with the d orbitals and represent polarizability tensors with six distinct components.

In point groups where there is an operation i, indicating a center of symmetry, the subscripts for all symmetry species in the character table are given the letter u or g, depending on whether antisymmetric or symmetric behavior is exhibited with respect to i. Table 4.27 gives the character table for C_{2h}. *trans*-Dichloroethylene (Fig. 4.27) belongs to this point group. It has a center of symmetry i, a rotational axis C_2^z, and a horizontal plane of symmetry σ_h perpendicular to the C_2^z axis. Note from the table that motions (vectors) in any of the x, y, or z directions (and the corresponding p orbitals) all behave as a u symmetry species.

TABLE 4.27. Character table for C_{2h}

C_{2h}	E	C_2^z	σ_h	i	
A_g	$+1$	$+1$	$+1$	$+1$	R_z, α_{xx}, α_{yy}, α_{zz}, α_{xy}
A_u	$+1$	$+1$	-1	-1	z
B_g	$+1$	-1	-1	$+1$	R_x, R_y, α_{xz}, α_{yz}
B_u	$+1$	-1	$+1$	-1	x, y

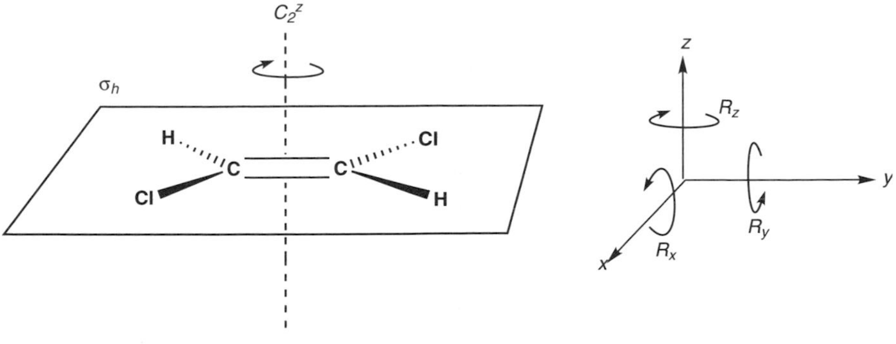

Figure 4.27. *trans*-Dichloroethylene, a molecule with C_{2h} symmetry.

4.28 TRANSFORMATION MATRICES

The matrix that transforms one set of properties (vectors) into another set by some symmetry operation is the transformation matrix. A matrix is a rectangular array of numbers set in parentheses or brackets; in the general case the matrix Z with elements a_{ij} may be represented as shown below. The first subscript of the letter a indicates the row number and the second digit of the subscript indicates the column number. Thus, the element a_{ij} is the element in the ith row and jth column:

$$Z = \begin{pmatrix} a_{11} & a_{12} & a_{13} & \cdots & a_{1n} \\ a_{21} & a_{22} & a_{23} & \cdots & a_{2n} \\ a_{31} & a_{32} & a_{33} & \cdots & a_{3n} \end{pmatrix}$$

We will be interested in multiplying this matrix by a column of numbers representing a vector **b** to give the transformation matrix, which converts the one vector into another.

Example. Consider a molecule with a C_4 axis and a center of symmetry, for example, $[PtCl_4]^{2-}$ (Fig. 4.6*b*), with the coordinate axes shown in Fig. 4.28. The coordinate axes may be considered as vectors (which might correspond to any vectorial property such as a vibration, translation, orbital) having both magnitude and direction.

We will perform a C_4^z operation, a clockwise rotation by 90° on these vectors. This operation transforms x to y, and y to $-x$. The new x that we will label x' has no (zero) contribution from the old x, but is equal to the old y. We can write this mathematically as

$$x' = 0x + 1y$$

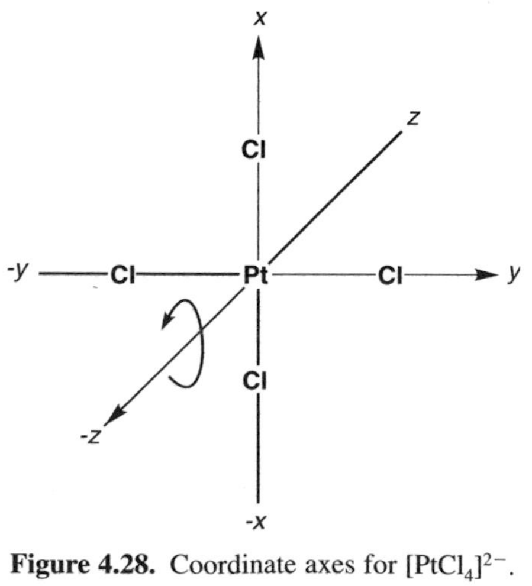

Figure 4.28. Coordinate axes for $[PtCl_4]^{2-}$.

Simultaneously on the 90° rotation, the old y is transformed into the new y, labeled y', and it is equal to the product of -1 times the old x plus zero times the old y, that is,

$$y' = -1x + 0y$$

In matrix notation these two equations may be written as

$$\begin{pmatrix} x' \\ y' \end{pmatrix} = \begin{pmatrix} 0 & +1 \\ -1 & 0 \end{pmatrix} \begin{pmatrix} x \\ y \end{pmatrix}$$

The set of numbers in the middle parentheses is the transformation matrix; it transforms the old vectors x and y through a 90° rotation into the new set of vectors, x' and y'. The two vectors transform as a set. Multiplication of the 2×2 matrix:

$$\begin{pmatrix} 0 & +1 \\ -1 & 0 \end{pmatrix}$$

by the 2×1 column vector:

$$\begin{pmatrix} x \\ y \end{pmatrix}$$

according to the rules of matrix multiplication gives the original equations:

$$x' = 0x + 1y$$

$$y' = -1x + 0y$$

4.29 TRACE OR CHARACTER OF A TRANSFORMATION MATRIX

The sum of the numbers appearing in the diagonal from upper left to lower right of the transformation matrix is called the trace (in German the *spur*), and the actual number is termed the character of the transformation matrix. This is the number that appears in the character tables of the point groups under the various symmetry operations.

Example. The transformation matrix given for C_4 in Sec. 4.28 was

$$\begin{pmatrix} 0 & +1 \\ -1 & 0 \end{pmatrix}$$

The character, the sum of the diagonal elements, is zero. The numbers 0 and ± 1 which occur in this matrix describing behavior on rotation correspond to the trigonometric values of the sin and cos of 90°. Accordingly and in general, the transformation matrices for a symmetry rotation by $\theta°$ may be written as

$$\begin{pmatrix} \cos\theta & \sin\theta \\ -\sin\theta & \cos\theta \end{pmatrix}$$

and the character is 2 cos θ. Thus, for C_4^z, where $\theta = 90°$, the character is $2 \cos 90° = 0$. In this example x and y are a degenerate set of two; under the operation, they transform together in the manner described.

Let us now consider the transformation matrix for the operation i. Under this operation, x and y in Fig. 4.28 transform together into x' and y' and we thus have

$$x' = -1x + 0y$$

$$y' = 0x - 1y$$

for which the transformation matrix is

$$\begin{pmatrix} -1 & 0 \\ 0 & -1 \end{pmatrix}$$

with a character of -2.

4.30 CHARACTER TABLES OF DEGENERATE POINT GROUPS

Character tables of point groups involving a symmetry axis C_n or S_n with $n \geq 2$. Such character tables contain degenerate symmetry species. In the case of double degeneracy, these species are denoted in character tables with the letter *e* (or *E*) (not to be confused with the similarly labeled identity operation *E*). In the case of triple degeneracy, the symmetry species is denoted as *T*.

Example. The ion $[PtCl_4]^{2-}$ (Fig. 4.6*b*) belongs to point group D_{4h}, whose character table is given in Table 4.30. The table indicates that there are $2C_4^z$ axes. Note that the symmetry species E_u under the operation C_4 has the character of zero, and under the operation i, E_u has the character -2. These were the characters we derived in Sec. 4.29 from the transformation matrices that describe the behavior of the doubly degenerate x- and y-axes with respect to the C_4 and i operations. Indeed, in the last column of the character table we see the letters x and y showing that the p_x and p_y orbitals of the central Pt atom in the D_{4h} ion $[PtCl_4]^{2-}$ belong jointly to species E_u. The vector z in Fig. 4.28 cannot be transformed into x or y by any operation in D_{4h}, and hence, the p_z orbital belongs to a different symmetry species. Reference to the character table for D_{4h} shows that the p_z orbital behaves as species A_{2u}. Recall that in CH_4, a molecule with tetrahedral symmetry belongs to point group T_d, with each C–H bond being a three-fold axis; see Figs. 4.7 and 4.14. All three p orbitals on the central carbon atom, therefore, transform as a set on rotation around these C_3 axes, and hence, the three p orbitals are triply degenerate and belong to a T species.

TABLE 4.30. Character Table for D_{4h}

D_{4h}	E	$2C_4$	C_4^2	$2C_2$	$2C_2'$	σ_h	$2\sigma_v$	$2\sigma_d$	$2S_4$	i	
A_{1g}	+1	+1	+1	+1	+1	+1	+1	+1	+1	+1	α_{xx+yy}, α_{zz}
A_{1u}	+1	+1	+1	+1	+1	−1	−1	−1	−1	−1	
A_{2g}	+1	+1	+1	−1	−1	+1	−1	−1	+1	+1	R_z
A_{2u}	+1	+1	+1	−1	−1	−1	+1	+1	−1	−1	z
B_{1g}	+1	−1	+1	+1	−1	+1	+1	−1	−1	+1	α_{xx-yy}
B_{1u}	+1	−1	+1	+1	−1	−1	−1	+1	+1	−1	
B_{2g}	+1	−1	+1	−1	+1	+1	−1	+1	−1	+1	α_{xy}
B_{2u}	+1	−1	+1	−1	+1	−1	+1	−1	+1	−1	
E_g	+2	0	−2	0	0	−2	0	0	0	+2	(R_x, R_y) α_{xz}, α_{yz}
E_u	+2	0	−2	0	0	+2	0	0	0	−2	(x, y)

4.31 SYMMETRY AND DIPOLE MOMENTS

Although the dipole moment is a vectorial property, it is a stationary property. This implies that the dipole moment vector of a molecule must lie in each of the symmetry elements of the molecule. For example, in H_2O (or any molecule belonging to point group C_{2v}) (see Sect. 4.12), the dipole moment vector lies along the C_2 axis, which also is coincident with the line of intersection of the two planes of symmetry. Thus, the symmetry of a molecule determines the direction of the dipole moment vector. However, symmetry says nothing about the magnitude of the dipole or which end of the vector is positive or which end negative. Molecules having a center of symmetry *i* have no net dipole moment since a vector will always invert when transformed by the inversion operation *i*. Molecules with more than one noncoincident rotational axis cannot have a dipole moment since a vector cannot be coincident with two noncoincident symmetry axes. Nonzero dipole moments are limited to the following: asymmetric molecules (point group C_1); molecules whose only symmetry consists of one plane of symmetry (point group C_s); one rotational axis (point group C_n, $n > 1$); or finally, those molecules with an n-fold axis and n-planes of symmetry intersecting this axis, that is, point group C_{nv} [(C_{2v}), e.g., H_2O; (C_{3v}), e.g., CH_3Cl; *etc.*]. Molecules belonging to all other point groups will not have a dipole moment for the reasons given above. Dipole moments of some common molecules are CH_3Cl, 1.87; CH_2Cl_2, 1.60; $CHCl_3$, 1.01; CCl_4, 0.0; *trans*-1,2-dichloroethylene, 0.0; H_2O, 1.85. The dipole moment is a stationary property of a molecule and, therefore, must remain unchanged under every symmetry operation appropriate to the molecule.

4.32 GROUP (OR SYMMETRY) ORBITALS

Because orbitals describe the distribution of electrons, they must be either symmetric or antisymmetric with respect to the operations of the point group to which the molecule belongs. If the atomic orbitals of the individual atoms of a molecule do not conform to the symmetry properties of the entire molecule, then linear combinations of them are constructed to make them conform. Such combinations of atomic orbitals are called group or symmetry orbitals.

Example. Consider the water molecule which has C_{2v} symmetry, placed in the coordinate system shown in Fig. 4.32. The character table for C_{2v} is shown in Table 4.26. The valence atomic orbitals to be used in constructing the molecular orbitals are the two 1*s* orbitals of the hydrogen atoms, and the 2*s*, $2p_x$, $2p_y$, and $2p_z$ orbitals of the oxygen atom; these six are the so-called basis set of orbitals. The 1*s* orbital of H_a, Fig. 4.32*c*, considered by itself does not have the symmetry properties of C_{2v}; e.g., if we were to perform the C_2 operation, rotation of 180° around the *z* axis, it would not transform into

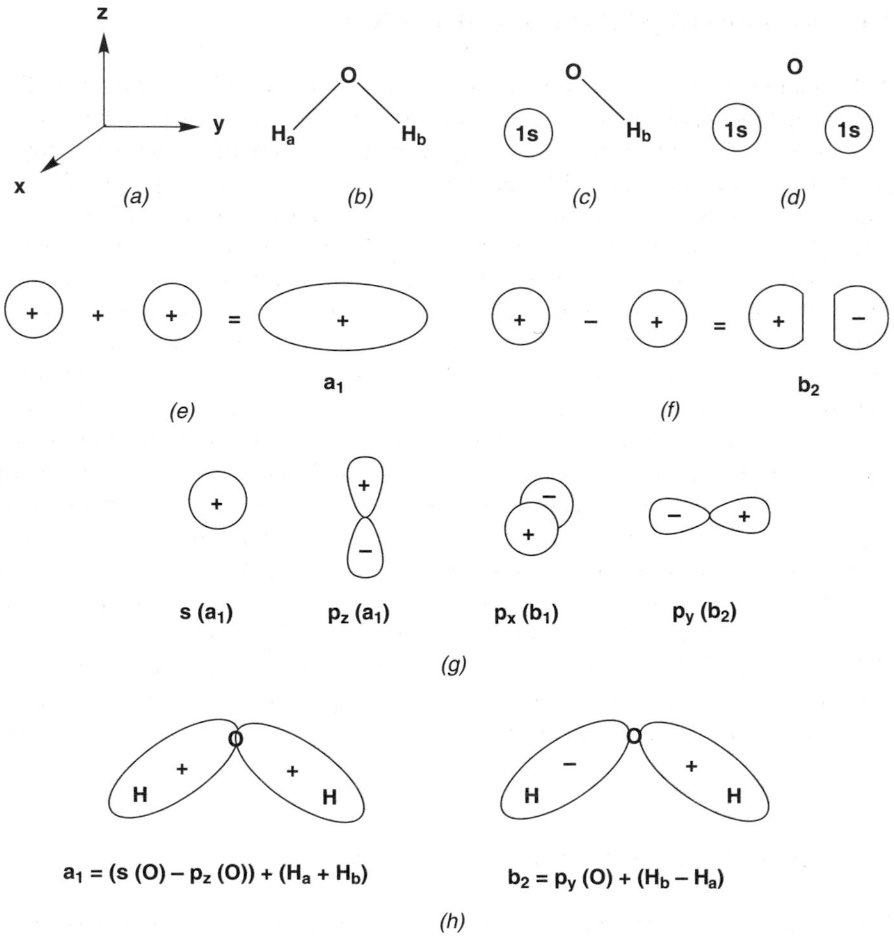

Figure 4.32. The orbitals of the water molecule.

anything; the rotation is neither symmetric nor antisymmetric. However, if we make linear combinations (addition and subtraction) of the 1*s* orbitals of both hydrogen atoms as shown in Fig. 4.32*e* and *f*, the two resulting group orbitals now possess C_{2v} symmetry and belong, respectively, to a_1 and b_2 symmetry species. We now can proceed to construct the molecular orbitals of the water molecule by combining the group orbitals of the hydrogen atoms with the appropriate matching oxygen atomic orbitals, which are shown in Fig. 4.32*g*. Before doing this we note that the 2*s* (O) and $2p_z$ (O) orbitals belong to the a_1 species and we can, therefore, construct two hybrid *sp* orbitals: [(2*s* (O)+ $\lambda 2p_z$ (O)] and [$\lambda 2s$ (O) $-$ $2p_z$ (O)]. The ($\lambda s - p$) hybrid has the large positive lobe pointing in the $-z$ direction toward the hydrogen atoms and the other hybrid orbital ($s + \lambda p$) pointing in the $+z$ direction. The Greek λ is a weighting factor, which results in putting more than 50% *p* character in the *sp* hybrid pointing toward the hydrogen orbitals, and hence, yielding better overlap. We now combine the group orbitals of the hydrogens with the appropriate (same symmetry species) oxygen orbitals giving without normalization $\psi_1 = [1s\ (H_a + 1sH_b)] + [\lambda 2s\ (O) - 2p_z\ (O)]$ (bonding) (Fig. 4.32*h*) and $\psi_6 = [1s\ (H_a + H_b)] - [\lambda 2s\ (O) - 2p_z\ (O)]$ (antibonding). Combining the other group orbital of the hydrogens (Fig. 4.32*f*) with the $2p_y$ of oxygen, both having b_2 symmetry, gives the combination of molecular orbitals $\psi_2 = [1s\ (H_a - H_b) + 2p_y\ (O)]$ (bonding) (Fig 4.32*h*) and $\psi_5 = [1s\ (H_a - H_b) - 2p_y\ (O)]$ (antibonding). Next we have $\psi_3 = 2p_x$ (O) (nonbonding) and finally the nonbonding hybrid orbital pointing in the $+z$ direction $\psi_4 = [(2s\ (O) + \lambda 2p_z\ (O)]$. The total of eight valence electrons then are placed in each of the two bonding MOs, ψ_1 and ψ_2, and in each of the two nonbonding atomic oxygen orbitals, ψ_3 and ψ_4. The numbers which appear as subscripts to ψ indicate the approximate order of the energy of the orbitals with the two lowest bonding orbitals, the two highest antibonding orbitals, and the two nonbonding atomic orbitals on oxygen of intermediate energies.

Acknowledgment. The late Hans H. Jaffe left a legacy of profound insights that were a constant inspiration in the construction of this chapter.

SUGGESTED READING

Cotton, F. A. *Chemical Applications of Group Theory.* John Wiley & Sons, Inc. New York, 1990, 3rd ed. New York, 1963.

Ferraro, J. R. and Ziomek, J. S. *Introductory Group Theory.* Plenum Press: New York, 1969.

Hall, L. H. *Group Theory and Symmetry in Chemistry.* McGraw Hill: New York, 1969.

Hargittai, I. and Hargittai, M. *Symmetry Through the Eyes of a Chemist.* VCH Publishers: New York, 1987.

Harris, D. C. and Bertolucci, M. D. *Symmetry and Spectroscopy.* Oxford University Press: London, 1978.

Jaffé, H. H. and Orchin, M. *Symmetry in Chemistry.* John Wiley & Sons: New York, 1965. Reprinted by Dover Publishing Company, 2003.

Mislow, K. *Introduction to Stereochemistry.* W. A. Benjamin: New York, 1966.

Wells, A. F. *The Third Dimension in Chemistry.* Oxford University Press: London, 1962.

5 Classes of Hydrocarbons

5.1	Saturated Hydrocarbons	112
5.2	Unsaturated Hydrocarbons	112
5.3	Acyclic Hydrocarbons	113
5.4	Alkanes	113
5.5	Aliphatic Hydrocarbons	114
5.6	Paraffins	114
5.7	Straight-Chain Alkanes	114
5.8	Branched-Chain Alkanes	115
5.9	Primary, Secondary, Tertiary, and Quaternary Carbon Atoms	115
5.10	Primary, Secondary, and Tertiary Hydrogen Atoms	116
5.11	The IUPAC System of Nomenclature	116
5.12	Homologs	116
5.13	Alkyl Group	116
5.14	Normal Alkyl Group	117
5.15	Isoalkanes	117
5.16	Isoalkyl Group	118
5.17	Methylene Group ($-CH_2-$)	118
5.18	Methine Group $—\overset{\vert}{C}H—$	119
5.19	Cycloalkanes	119
5.20	Alicyclic Hydrocarbons	119
5.21	Alkenes	120
5.22	Olefins	120
5.23	Vinyl Group, $-CH{=}CH_2$	120
5.24	Cycloalkenes	120
5.25	Bridged Hydrocarbons	121
5.26	Bridgehead Carbon Atoms	121
5.27	Bridged Bicyclic Hydrocarbons	122
5.28	Bredt's Rule	123
5.29	Angular Groups	123
5.30	Tricyclic and Polycyclic Compounds	124
5.31	Spirans	125
5.32	Conjugated Polyenes	125
5.33	Cross-Conjugated Hydrocarbons	126
5.34	Exocyclic and Endocyclic Double Bonds	126
5.35	Cumulenes	127
5.36	Alkynes	127
5.37	Benzenoid Hydrocarbons	127

The Vocabulary and Concepts of Organic Chemistry, Second Edition, by Milton Orchin, Roger S. Macomber, Allan Pinhas, and R. Marshall Wilson,

5.38 Aryl Group 128
5.39 *ortho*, *meta*, and *para* 129
5.40 Polynuclear Aromatic Hydrocarbons (PNA) 129
5.41 Cata-Condensed Polynuclear Aromatic Hydrocarbons 130
5.42 Acenes 130
5.43 Phenes 131
5.44 Peri-Condensed Polynuclear Aromatic Hydrocarbons 131
5.45 Meso Carbon Atoms 131
5.46 The Bay, K, and L Regions in Polycyclic Aromatic Hydrocarbons 132
5.47 Pseudo-Aromatic Compounds 132
5.48 Cyclophanes 133
5.49 Index of Unsaturation (Hydrogen Deficiency) *i* 133
5.50 Crystalline Forms of Elemental Carbon 135
5.51 Graphite 135
5.52 Diamond 135
5.53 Fullerenes and Related Compounds 135
5.54 Carbon Nanotubes 137
5.55 Carbon Fibers 137

The first four chapters of this book have dealt with topics that are fundamental to all of chemistry. The concepts of atomic orbital theory, localized and delocalized bonding principles, molecular orbital theory, and symmetry are all essential concepts and are useful in the many subdisciplines of chemistry. Starting with this chapter, the focus is almost exclusively on the description and classification of compounds containing carbon atoms, commonly known as organic chemistry. The first use of the term *organic chemistry* has been ascribed to Jons J. Berzelius (1779–1840) who wrote a textbook in 1827 in which he treated organic chemistry as a separate independent entity. Organic chemistry developed much more slowly than inorganic or mineral chemistry. Most of the organic compounds that were known and used in the early nineteenth century were natural products of rather great complexity. There was the common belief that compounds produced by living organisms, organic compounds, required a *vital force* that directed the biological processes leading to them. But in 1828 Fríedrich Woehler (1800–1882), working at the University of Goettingen, prepared urea, $H_2NC(O)NH_2$, an organic compound, from its structural isomer, ammonium cyanate or NH_4CNO, an inorganic compound. Thereafter, the vitalistic theory was gradually abandoned and organic chemistry underwent explosive growth during the last half of the nineteenth century. The spectroscopic tools for determining structure and synthetic stategies grew much more slowly. Thus, for example, it took more than 150 years from the time cholesterol was first isolated to its proof of structure in 1932 and 23 more years elapsed before the first synthesis of cholesterol by Robert B. Woodword (1917–1979, who received the Nobel Prize in 1967), then a spectacular achievement considering the necessity of constructing the correct stereochemistry at its eight stereogenic centers.

Practically all organic compounds contain hydrogen as well as carbon, and this chapter is almost exclusively devoted to *hydrocarbons, compounds that contain only carbon and hydrogen*. Compounds with the highest ratio of hydrogen atoms to carbon atoms are the alkanes. As hydrogen atoms are removed, unsaturation increases, the H/C ratios decrease, leading finally to highly condensed polynuclear aromatic compounds and culminating in the recently discovered fullerenes and carbon nanotubes, which, like diamonds and graphite, contain carbon atoms only. Although structures consisting of multiple carbon atoms only may be considered by some chemists as polymers of carbon, we have chosen to discuss these in this chapter partly on the basis that the consecutive loss of hydrogen from hydrocarbons finally leads to carbon-only structures. Thus, readers will find the topics of fullerenes, carbon nanotubes, and carbon fibers discussed at the end of this chapter.

The first class of compounds discussed in practically all textbooks of organic chemistry is the alkanes, followed later by alkenes and alkynes, and after these aliphatic compounds are considered, a discussion of aromatic compounds ensues. However, because earlier chapters in this book have been devoted to all types of bonding, it is logical to include aromatic compounds, as well as aliphatic compounds in this chapter, even though some of the polynuclear aromatics have been described in Chapter 3, in connection with the topics of resonance and delocalized bonding.

5.1 SATURATED HYDROCARBONS

These are hydrocarbons in which no multiple bonds are present; every valence orbital on carbon, which is not involved in carbon–carbon bonding, is used in bonding to hydrogen atoms.

Example. Propane (Fig. 5.1*a*) and cyclohexane (Fig. 5.1*b*).

(*a*) (*b*)

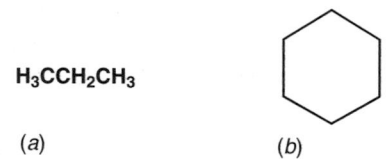

Figure 5.1. (*a*) Propane and (*b*) cyclohexane.

5.2 UNSATURATED HYDROCARBONS

These are hydrocarbons with one or more carbon-carbon multiple bonds; unsaturated hydrocarbons can usually be readily converted to saturated hydrocarbons by the catalytic addition of dihydrogen (catalytic hydrogenation). The reverse reaction, the removal of dihydrogen, is an oxidation.

Example. Propyne (Fig. 5.2*a*) and the hydrogenation of cyclohexene to cyclohexane and its reverse (Fig. 5.2*b*).

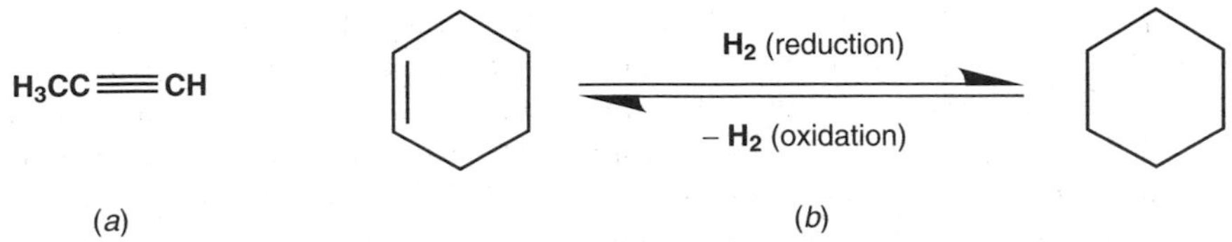

Figure 5.2. (*a*) Propyne and (*b*) hydrogenation of cyclohexene to cyclohexane and its reverse.

5.3 ACYCLIC HYDROCARBONS

These are hydrocarbons in which no rings are present; they are synonymous with aliphatic hydrocarbons (see Sect. 5.5).

5.4 ALKANES

These are saturated hydrocarbons having the molecular formula C_nH_{2n+2} (n = an integer); they are characterized by their relative inertness toward chemical reagents.

Example. The simplest alkane is methane, CH_4 ($n = 1$). The tetrahedral arrangement of methane, inscribed in a cube and showing one of its S_4 axes (Sect. 4.7), has been illustrated in Fig. 4.7*b*. In order to emphasize its geometry, methane is frequently represented as a solid tetrahedron. In such a representation (Fig. 5.4*a*), it is understood that the four hydrogen atoms are located at the corners *a*, *b*, *c*, and *d* and the carbon atom is imagined to be exactly in the center of the tetrahedron. Each of the four faces of the tetrahedron is an equilateral triangle, characterized by three-fold symmetry. Since there are four faces, methane possesses four C_3 axes, each one of which coincides with a C–H bond and penetrates the carbon atom and the center of

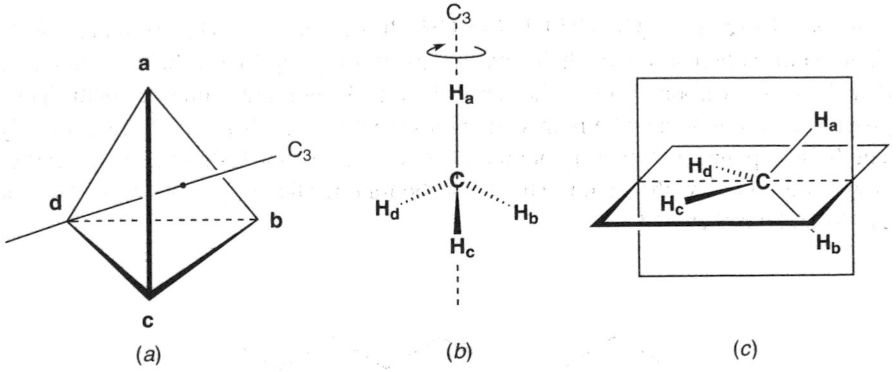

Figure 5.4. (*a*) The tetrahedron; (*b*) perspective drawing of methane emphasizing a three-fold axis; (*c*) perspective drawing showing the perpendicular relationship of the two sets of hydrogen atoms.

the opposite face. Thus, for example, in Fig. 5.4*a* one of the C_3 axes coincides with the C–H_d bond (Fig. 5.4*b*); this axis passes through the center of the face *abc*. Another C_3 axis, shown in Fig. 5.4*b*, coincides with the C–H_a bond and passes through the center of face *bcd*. Another feature of the tetrahedral structure of methane is shown in the perspective drawing of it given in Fig. 5.4*c*. This drawing illustrates the fact that when any two hydrogen atoms reside in a plane of symmetry (e.g., H_a and H_b in the vertical plane), the other two hydrogen atoms must reside in a plane of symmetry perpendicular (orthogonal) to the first (e.g., H_c and H_d in the horizontal plane).

5.5 ALIPHATIC HYDROCARBONS

Aliphatic (from the Greek word *aleiphatos* for "fat") hydrocarbons (alkanes, alkenes, alkynes) do not possess rings of any kind (open-chain hydrocarbons). Hydrocarbons are frequently divided into two categories: aliphatic (fatlike) and carbocyclic; most of the latter are aromatic (sweet-smelling) hydrocarbons.

5.6 PARAFFINS

Paraffins (from the Latin word *parum affinis* for "low affinity") are synonymous with alkanes, but are usually limited to those alkanes that are solid at room temperature.

5.7 STRAIGHT-CHAIN ALKANES

Alkanes in which all carbon atoms except the terminal ones are bonded only to two other carbon atoms, thus forming a chain of carbon atoms with no branches.

Example. Hexane, $CH_3CH_2CH_2CH_2CH_2CH_3$. It is common practice to represent the skeleton of carbon atoms in chains by means of a zig-zag line, it being understood that there is a carbon atom at the termini of each individual line segment. Thus, propane and hexane may be represented as shown in Figs. 5.7*a* and *b*, respectively. The boiling points and melting points increase in a regular fashion as more carbon atoms are added to the chain. The 15-carbon-atom alkane, pentadecane, boils at 270°C and melts at 10°C.

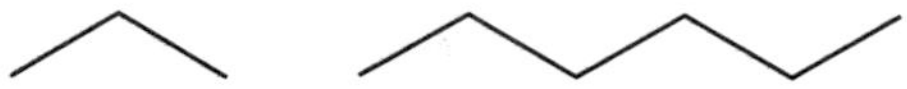

(*a*) (*b*)

Figure 5.7. Representations of straight-chain hydrocarbons: (*a*) propane and (*b*) hexane.

5.8 BRANCHED-CHAIN ALKANES

Acyclic saturated hydrocarbons in which at least one of the carbon atoms is attached to more than two other carbon atoms.

Example. Isobutane (methylpropane) (Fig. 5.8*a*) and 3,3,4-trimethylhexane (Fig. 5.8*b*); these compounds may be represented in skeletal form as shown in Figs. 5.8*c* and *d*, respectively. Branching has important influence on the properties of the alkanes, for example, branched-chain hydrocarbons always boil lower than their straight-chain alkane counterparts. Thus, octane with eight carbon atoms boils at 126°C and melts at −57°C, while its branched-chain isomer 2,2,3,3-tetramethylbutane boils at 106°C and melts at 101°C.

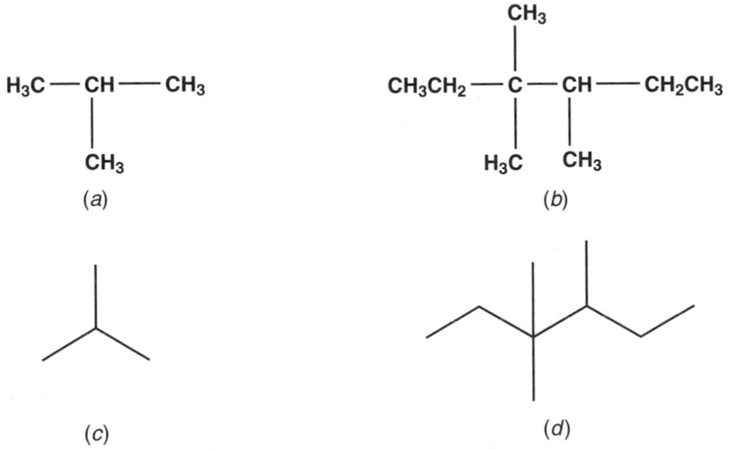

Figure 5.8. (*a*) Isobutane (methylpropane); (*b*) 3,3,4-trimethylhexane; and (*c*) and (*d*) their carbon skeleton representations, respectively.

5.9 PRIMARY, SECONDARY, TERTIARY, AND QUATERNARY CARBON ATOMS

These terms refer to carbon atoms singly bonded to only one other carbon atom (*primary*), those singly bonded to two other carbons (*secondary*), those singly bonded to three other carbons (*tertiary*), and those singly bonded to four other carbon atoms (*quaternary*).

Example. In 3,3,4-trimethylhexane (Fig. 5.8*b*), there are five primary, two secondary, one tertiary, and one quaternary carbon atoms. In neopentane (dimethylpropane) (Fig. 5.9), there are four primary carbons and one quaternary carbon atom; the four primary carbon atoms in this molecule are equivalent [by virtue of the S_4 axis (Sect. 4.7)], as are all 12 hydrogen atoms (by virtue of the S_4 axis and the local symmetry of the CH_3 group).

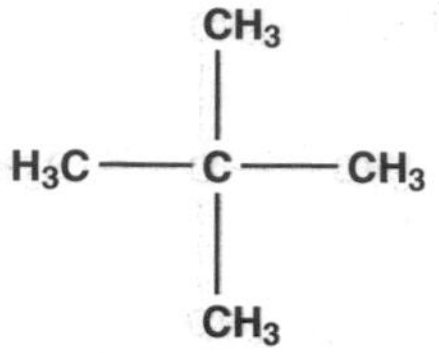

Figure 5.9. Neopentane (dimethylpropane).

5.10 PRIMARY, SECONDARY, AND TERTIARY HYDROGEN ATOMS

The hydrogen atoms bonded to primary, secondary, and tertiary carbon atoms, respectively.

Example. 2-Methylbutane (isopentane) (Fig. 5.10) has nine primary, two secondary, and one tertiary hydrogen atoms.

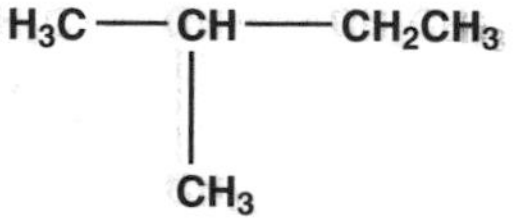

Figure 5.10. Isopentane (2-Methylbutane).

5.11 THE IUPAC SYSTEM OF NOMENCLATURE

The system of nomenclature first proposed in 1892 by the organization called the International Union of Pure and Applied Chemistry (IUPAC). Committees of this organization work continuously on systematic schemes for naming all known (now more than 7 million) organic compounds, as well as all yet to be discovered organic compounds such that each different compound has a unique name. Versions of this system can be found in every elementary textbook on organic chemistry.

5.12 HOMOLOGS

Two compounds that only differ by a $-CH_2-$ group.

Example. Propane, $CH_3CH_2CH_3$, and butane, $CH_3CH_2CH_2CH_3$, are alkane homologs; methanol, CH_3OH, and ethanol, CH_3CH_2OH, are homologous alcohols.

5.13 ALKYL GROUP

The group of atoms remaining after one hydrogen atom is removed from a carbon atom in the parent hydrocarbon. There will be as many different alkyl groups derived from the parent hydrocarbon as there are nonequivalent hydrogen atoms in the corresponding parent. Cycloalkyls are derived from cycloalkanes (Sect. 5.19) by the removal of one hydrogen atom.

Example. All the hydrogen atoms in methane and ethane are equivalent; therefore, only one alkyl group can be derived from each of these parent hydrocarbons. The name of an alkyl group derived from a particular alkane is obtained by replacing the suffix *-ane* of the alk*ane* by the letters *yl*. Thus, methane and ethane lead, respectively, to *methyl* (CH_3–) and *ethyl* (CH_3CH_2–) groups. Propane has two different sets of hydrogen atoms, giving rise to two different alkyl groups. Three different alkyl groups can be derived from the five-carbon, straight-chain, saturated hydrocarbon pentane (Fig. 5.13). When a hydrogen atom is removed from an alkyl group to give a doubly deficient carbon atom, the name of the alkyl group is given the ending *-idene*; thus, $CH_3CH=$ is called *ethylidene* and $CH_2=$ is methylidene.

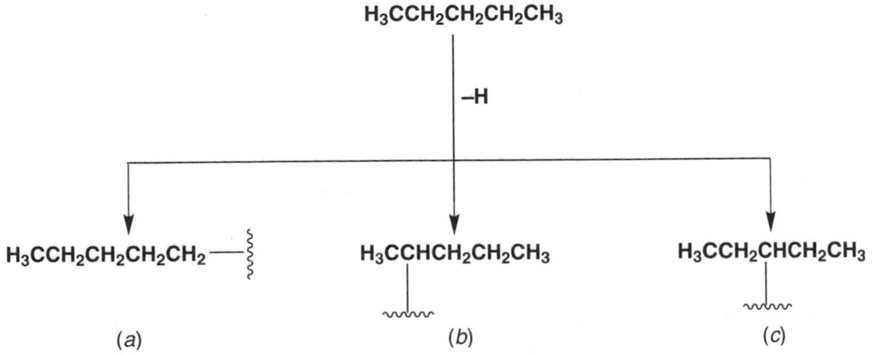

Figure 5.13. The three alkyl groups derived from pentane.

5.14 NORMAL ALKYL GROUP

The alkyl group, abbreviated *n*, derived by removing a hydrogen atom from the terminal methyl group of a straight-chain alkane.

Example. *n*-propyl, $CH_3CH_2CH_2$–.

5.15 ISOALKANES

Branched-chain alkanes having a single methyl branch on the next to the last (penultimate) carbon atom.

Example. Isopentane (2-methylbutane) (Fig. 5.10).

5.16 ISOALKYL GROUP

The group remaining after the removal of one hydrogen atom from the methyl group at the nonbranched end of an isoalkane.

Example. The structures shown in Fig. 5.16.

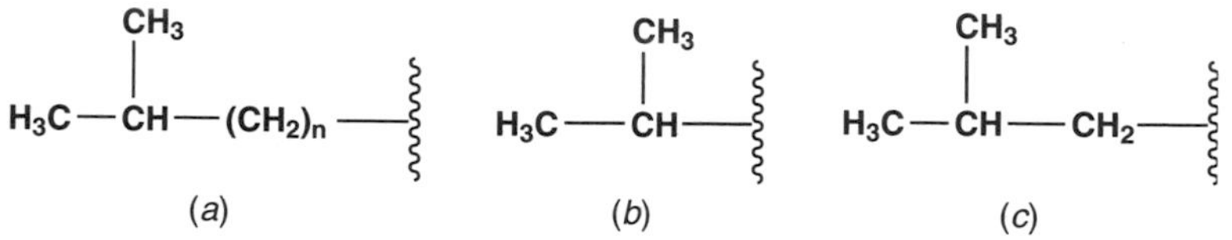

Figure 5.16. (*a*) The general skeletal structure for an isoalkyl group, $n = 0, 1, 2, \ldots$ (*b*) isopropyl; (*c*) isobutyl.

5.17 METHYLENE GROUP ($-CH_2-$)

The divalent triatomic grouping, CH_2 (consisting of one carbon atom and its two attached hydrogens). As the first part of the name implies, the methylene group may be considered as a derivative of the methyl group, CH_3-, by removal of a hydrogen atom to give a doubly hydrogen-deficient carbon atom, hence the suffix *-ene*.

Example. Various kinds of methylene groups are shown in Fig. 5.17. The methylene group can be bonded to two other atoms as in Figs. 5.17*a* and *b*, or to a single other atom as in Fig. 5.17*c*. A third kind of methylene group is a stand-alone CH_2 group (Fig. 5.17*d*) and is called carbene. The two unshared electrons may be either paired or unpaired. The form with the paired electrons has an empty orbital centered on carbon, and it and some of its derivatives such as Cl_2C: are important intermediates in several reactions.

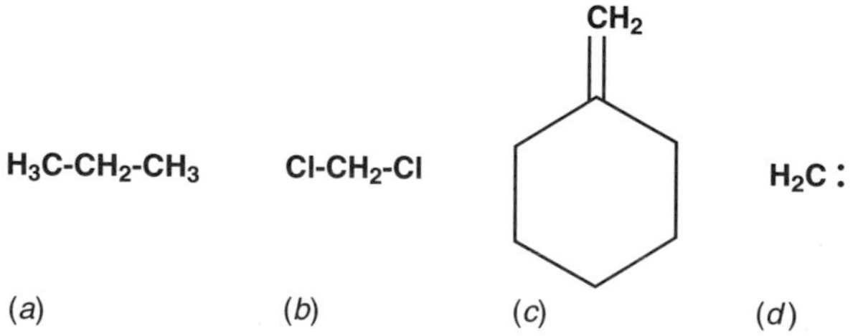

Figure 5.17. Methylene groups in (*a*) propane; (*b*) methylene chloride; (*c*) methylenecyclohexane or methylidene cyclohexane; (*d*) methylene, the parent reactive intermediate known as carbene.

5.18 METHINE GROUP (—CH— with a bond up)

The trivalent diatomic grouping consisting of one carbon atom and one attached hydrogen, that is, —CH— (with a bond up). It may be considered as a derivative of the methylene group by removal of a hydrogen atom to give a triply hydrogen-deficient carbon atom.

Example. The —CH— (with a bond up) group in isopentane (Fig. 5.10). The carbon atom of the methine group is called a methine carbon and the hydrogen atom a methine hydrogen.

5.19 CYCLOALKANES

These are saturated hydrocarbons containing one or more rings; the monocyclic compounds have the general formula C_nH_{2n}; n is an integer corresponding to the number of carbon atoms in the ring.

Example. The compounds shown in Fig. 5.19. Most chemists regard cycloalkanes as limited to monocyclic saturated hydrocarbons. The cycloalkanes have higher boiling points than their counterpart straight-chain compounds, for example, octane has a boiling point of 126°C, cyclooctane a boiling point of 150°C.

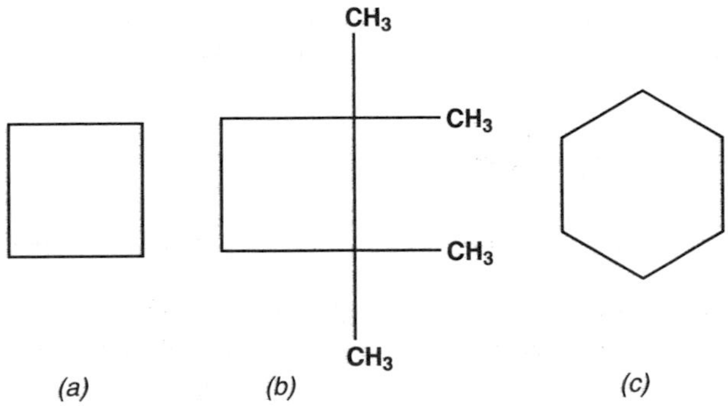

Figure 5.19. (*a*) Cyclobutane; (*b*) 1,1,2,2-tetramethylcyclobutane; (*c*) cyclohexane (planar representation; for the correct geometry, see Section 7.119).

5.20 ALICYCLIC HYDROCARBONS

Hydrocarbons having the same general properties as *ali*phatic hydrocarbons (e.g., insoluble in polar solvents, unreactive) but that are *cyclic*, hence the name alicyclic.

5.21 ALKENES

This is a class of hydrocarbons characterized by the presence of a carbon–carbon double bond; just as the suffix *-ane* in alkane denotes saturation, the suffix *-ene* in alkene denotes unsaturation. A compound containing one double bond is a monoene; two, a diene; three, a triene; and so on. A compound with many double bonds is a polyene. Monoenes have the molecular formula C_nH_{2n}; dienes, C_nH_{2n-2}; and so on.

Example. 2,3-Dimethyl-2-butene (tetramethylethylene) (Fig. 5.21). It is sometimes useful to think of C=C as a two-membered ring.

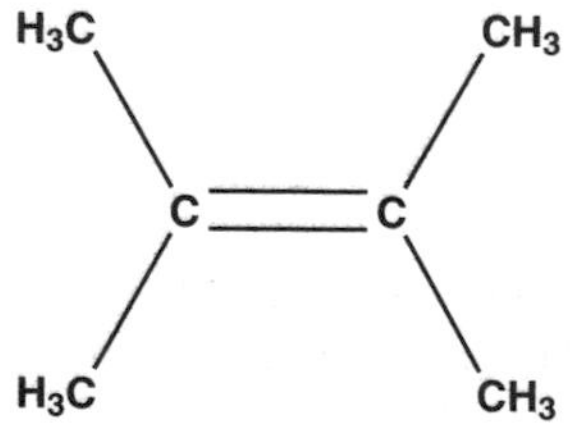

Figure 5.21. 2,3-Dimethyl-2-butene (tetramethylethylene).

5.22 OLEFINS

Synonymous with alkenes. The name olefin had its origins in *olefiant* ("oil-making"), the ancient name for ethylene, because the addition of chlorine gas to ethylene gas yielded an oil (1,2-dichloroethane). Olefiant was then shortened to olefin to denote a general class of hydrocarbons containing a double bond.

5.23 VINYL GROUP, $-CH=CH_2$

The monovalent group remaining after one hydrogen is removed from ethylene.

Example. Vinyl chloride (Fig. 5.23*a*); 4-vinylcyclohexene (Fig. 5.23*b*). Here the vinyl group is considered a substituent in the 4 position of cyclohexene. The position of the ring double bond is given priority over the position of the substituent, and the 1 and 2 numbered positions are assigned so as to include the carbons of the ring double bond. Finally, the ring is numbered so as to give the substituent as low a number as possible.

5.24 CYCLOALKENES

Cyclic hydrocarbons containing a double bond.

Example. 3,3-Dimethylcyclohexene (Fig. 5.24*a*). The configuration around the double bond in rings of seven or fewer carbon atoms is necessarily *cis*, but with rings of

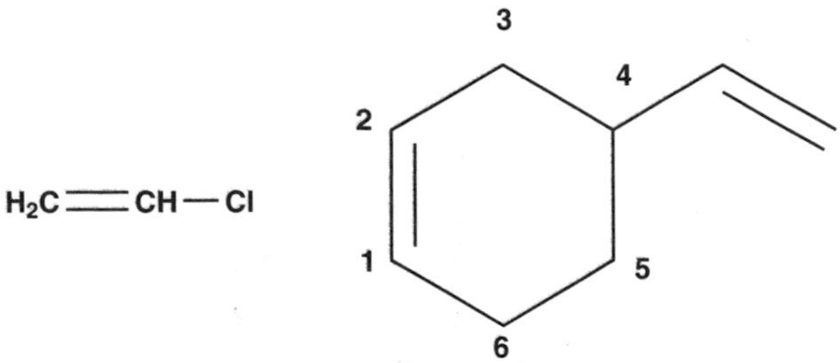

(a) (b)

Figure 5.23. (*a*) Vinyl chloride and (*b*) 4-vinylcyclohexene.

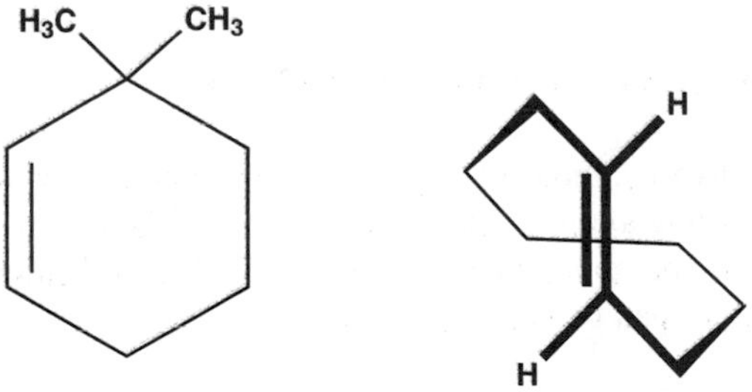

(a) (b)

Figure 5.24. (*a*) 3,3-Dimethylcyclohexene and (*b*) *trans*-cyclooctene.

eight or more atoms, a *trans* configuration is possible, as in *trans*-cyclooctene (Fig. 5.24*b*). (For definitions of *cis* and *trans*, see Sect. 7.25.)

5.25 BRIDGED HYDROCARBONS

The class of cyclic hydrocarbons containing one (or more) pairs of carbon atoms common to two (or more) rings.

Example. Bicyclo[2.2.2]octane (Fig. 5.25).

5.26 BRIDGEHEAD CARBON ATOMS

Carbon atoms common to two or more rings.

Example. The carbon atoms numbered 1 and 4 (Fig 5.25) are bridgehead carbons; in the naming of such compounds, one of the bridgehead carbon atoms is always numbered 1.

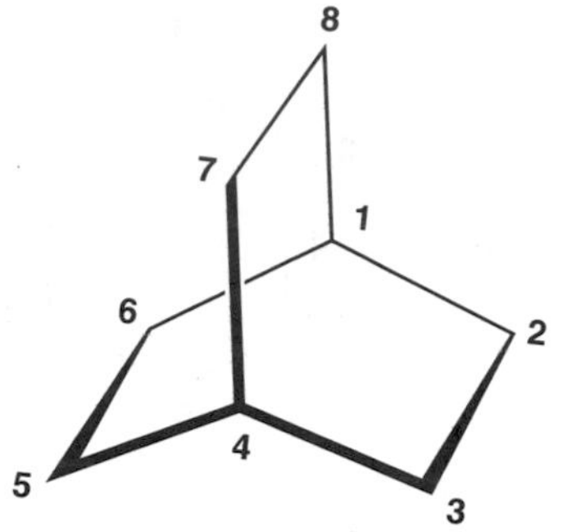

Figure 5.25. Bicyclo[2.2.2]octane.

5.27 BRIDGED BICYCLIC HYDROCARBONS

A class of bicyclic hydrocarbons having two bridgehead carbon atoms and that possess at *least one carbon atom in each of the three bridges between the bridgehead carbons*. If a bicyclic compound has one bridge with zero bridging atoms, it is not considered to be a bridged bicyclic compound.

Example. Bicyclo[2.2.2]octane (Fig. 5.25); the general structure (Fig. 5.27*a*); bicyclo[2.2.1]hept-2-ene (Fig. 5.27*b*). Bicyclo[1.1.1]pentane (Fig. 5.27*c*) is the first member of a class of strained bicyclic compounds called the *propellanes* because of their resemblance to a propeller. The prefix *bicyclo-* is part of the *von Baeyer system* of naming bridged alicyclic hydrocarbons and refers to a system of *two* rings. In general, the number of rings in a polycyclic hydrocarbon system is equal to the minimum number of bond scissions or cuts required to convert the bridged ring system into an acyclic hydrocarbon having the same number of carbon atoms; for bicyclics two cuts or scissions are necessary to open all the rings. If three cuts or scissions are required, then the compound is tricyclic, and so on. [The number of rings can also

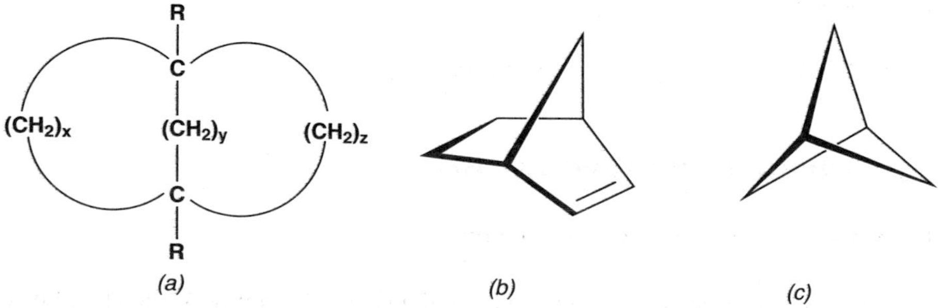

Figure 5.27. (*a*) General structure for a bridged bicyclo compound (*x*, *y*, *z*>0); (*b*) bicyclo[2.2.1]hept-2-ene (norbornene); (*c*) bicyclo[1.1.1] pentane (a propellane).

be determined using the equation for the *index of unsaturation* and the empirical formula for a hydrocarbon (Sect. 5.49).] The total number of carbon atoms in the bridging system is reflected in the suffix name. The numbers in brackets correspond, in descending order, to the number of atoms in each bridge between bridgehead atoms. The names for the compounds shown in Figs. 5.25, and 5.27*b* and *c*, are derived from these rules.

5.28 BREDT'S RULE

A bridged bicyclic hydrocarbon cannot have a double bond involving a bridgehead carbon unless one of the rings contains at least eight carbons atoms. Bicyclo[2.2.2]octane (Fig. 5.25) has three two-carbon atom bridges and so cannot have a double bond at the bridgehead carbons. In this situation, the *p* orbitals required for the formation of the double bond would be rotated so that they could not effectively overlap with each other to form a π bond. However, *trans*-bicyclo[4.4.0]decane (also called *trans*-decahydronaphthalene or *trans*-decalin; Fig. 5.28*a*) may be considered to have three bridges and is a bicyclic hydrocarbon. But since one of the bridges (C_1–C_6) has zero atoms, it does not meet the criteria for a *bridged* bicyclic, and is therefore outside the scope of Bredt's rule. A *p* orbital at the zero bridging positions 1 and 6 can overlap with a *p* orbital on an adjacent carbon atom, and it is therefore possible in this situation to have a π bond at the bridgehead carbon, as in bicyclo[4.4.0]dec-1-ene (Fig. 5.28*b*).

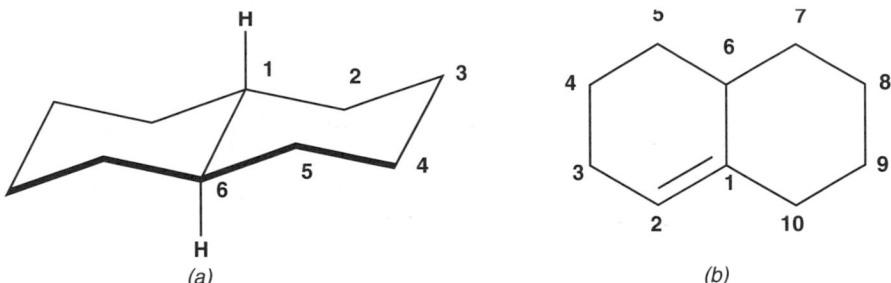

Figure 5.28. (*a*) *trans*-Bicyclo[4.4.0]decane (the prefix *trans*- refers to the geometry of the fusion of the two cyclohexane rings); (*b*) bicyclo[4.4.0]dec-1-ene. Although these two hydrocarbons are bicyclic, they are not considered *bridged* bicyclic compounds.

5.29 ANGULAR GROUPS

Atoms or groups attached to carbons atoms that are common to two rings.

Example. The hydrogen atoms at positions 1 and 6 in Fig. 5.28. Methyl groups in the angular positions (Fig. 5.29) are common in many naturally occurring compounds such as the steroids.

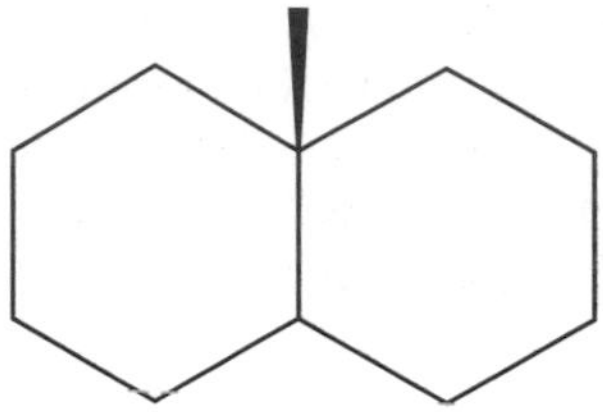

Figure 5.29. 1-Methyl[4.4.0]decane.

5.30 TRICYCLIC AND POLYCYCLIC COMPOUNDS

Tricyclic compounds possess four bridgehead carbons, each common to two rings.

Example. Adamantane (Fig. 5.30*a*) is tricyclo[3.3.1.1^{3,7}]decane. This unusually stable hydrocarbon has tetrahedral symmetry; its four bridgehead carbon atoms are numbered 1, 3, 5, and 7. It is tricyclic because three cuts would be required to generate an acyclic analog by opening all the rings. The superscript numbers indicate the positions of the two termini of the bridge that is superscripted, C-10, Fig. 5.30*a* The compound shown in Fig. 5.30*b* is tricyclo[8.4.0.0^{3,8}]tetradecane; the name indicates a ring of 14 carbon atoms with two bridges, a bridge with zero atoms between

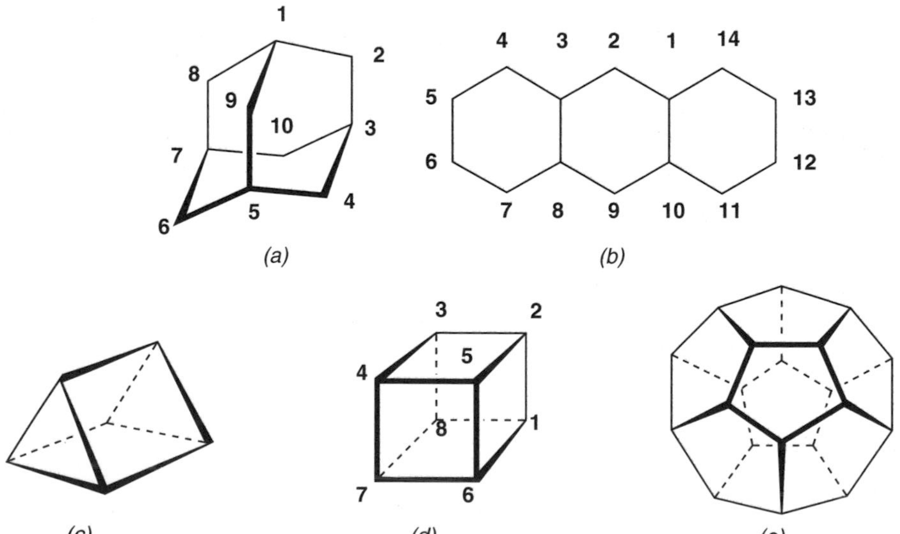

Figure 5.30. Tricyclic and polycyclic hydrocarbons: (*a*) adamantane (tricyclo[3.3.1.1^{3,7}]-tetradecane); (*b*) tetradecahydroanthracene (tricyclo[8.4.0.0^{3,8}]tetradecane); (*c*) prismane; (*d*) cubane; (*e*) dodecahedrane.

bridgehead carbons 1 and 10 and a bridge with zero atoms between bridgehead carbons 3 and 8. For simplicity, compounds such as those in Fig. 5.30*b* are usually named after a parent aromatic hydrocarbon whose double bonds have been saturated with hydrogen, in this case tetradecahydroanthracene.

Organic chemists have long been challenged to synthesize unusual polycyclic hydrocarbons with solid geometry counterparts and to give them common names that reflect their geometry. Some of the more important of these exotic compounds are *prismane* (Fig. 5.30*c*), *cubane* (Fig. 5.30*d*), and *dodecahedrane* (Fig. 5.30*e*). The reason such simple imaginative names have become popular is readily appreciated when one considers, for example, the alternative systematic name of the eight-carbon cube, $(CH)_8$ or cubane, which is pentacyclo[4.2.0.$0^{2,5}$.$0^{3,8}$.$0^{4,7}$]octane.

5.31 SPIRANS

Compounds containing only a single carbon atom common to two rings.

Example. The hydrocarbon shown in Fig. 5.31 is spiro[3.3]heptane; the two rings joined by the common tetrahedral carbon atom are at right angles to each other. The numbers in brackets refer to the number of carbon atoms in the bridges spanning the spiro atom; the last part of the name refers to the saturated hydrocarbon with the same number of carbon atoms as the total number of carbon atoms in the two rings, including the common spiro atom.

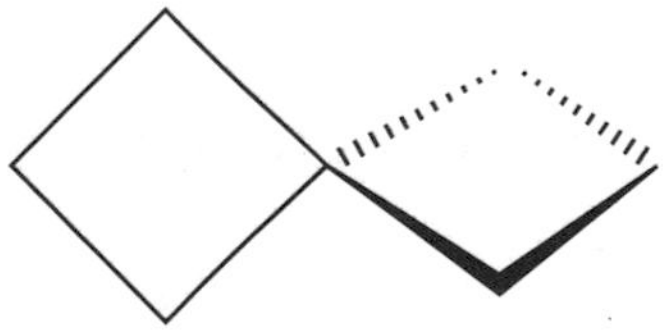

Figure 5.31. Spiro[3.3]heptane.

5.32 CONJUGATED POLYENES

Hydrocarbons possessing a system of alternating carbon–carbon double and carbon–carbon single bonds. The carbon skeleton may be cyclic or acyclic, or a combination of rings and chains, but aromatic hydrocarbons are not considered conjugated polyenes.

Example. 1,3,5-hexatriene (Fig. 5.32*a*); *fulvene*, the parent of a class of related conjugated hydrocarbons based on cyclopentadiene (Fig. 5.32*b*); *fulvalene* (Fig. 5.32*c*).

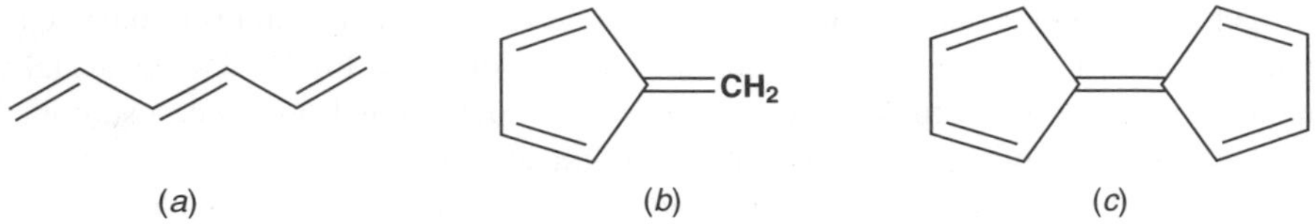

Figure 5.32. (*a*) 1,3,5-hexatirene; (*b*) fulvene; (*c*) fulvalene.

5.33 CROSS-CONJUGATED HYDROCARBONS

These hydrocarbons possess a grouping of at least three carbon–carbon double bonds, with two of them each separately conjugated to a third, but the two not conjugated to each other.

Example. 3-Methylene-1,4-pentadiene (Fig. 5.33). The central double bond at position 3 is conjugated to each of the other double bonds at positions 1 and 4, but these two main chain double bonds are not conjugated to each other. The classical resonance structures would show only the interaction of the conjugated double bonds. Molecular orbital theory postulates delocalization of the six $p\pi$ electrons over the entire six-carbon atom system.

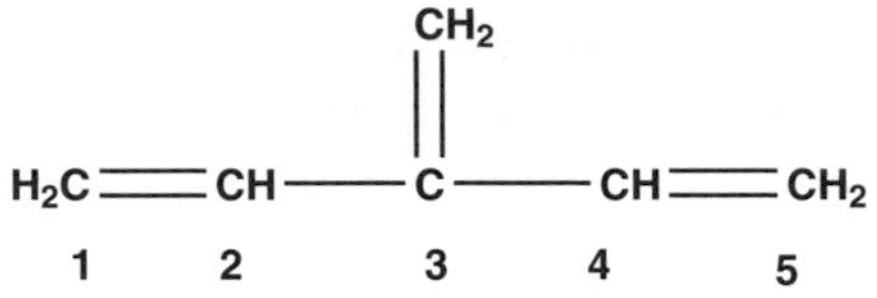

Figure 5.33. 3-Methylene-1,4-pentadiene, a cross-conjugated polyene.

5.34 EXOCYCLIC AND ENDOCYCLIC DOUBLE BONDS

Double bonds with only a single carbon atom as part of a ring are exocyclic (external) to that ring, and double bonds with both carbon atoms as part of a ring are endocyclic (internal).

Example. Methylenecyclohexane (Fig. 5.17*c*) possesses an exocyclic double bond. The molecule shown in Fig. 5.34 has two double bonds, both of which are exocyclic to ring A and endocyclic in ring B.

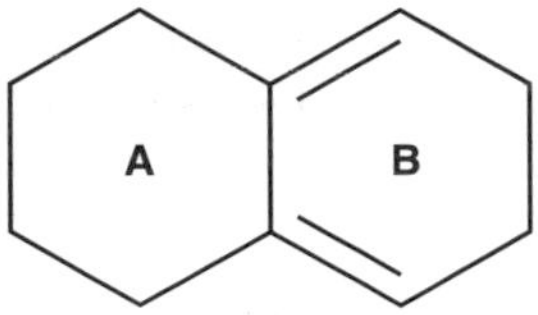

Figure 5.34. Exo- and endocyclic double bonds in bicyclo[4.4.0]deca-1,5-diene.

5.35 CUMULENES

Compounds containing two or more consecutive (cumulated) double bonds.

Example. The simplest member in this series of hydrocarbons is *allene* (1,2-propadiene) (Fig. 5.35*a*), and the next member of the series is 1,2,3-butatriene (Fig. 5.35*b*). Although commonly not so regarded, the internal *sp* hybridized carbon atoms can be thought of as spiro carbon atoms. The triene (Fig. 5.35*b*) may be thought of as having three two-membered rings, each ring at right angles to the ring either preceding or following it. In allene or any other cumulene with an even number of double bonds, the terminal sets of hydrogen atoms are in planes perpendicular to each other, whereas in the triene or any other cumulene with an odd number of double bonds, they are in the same plane.

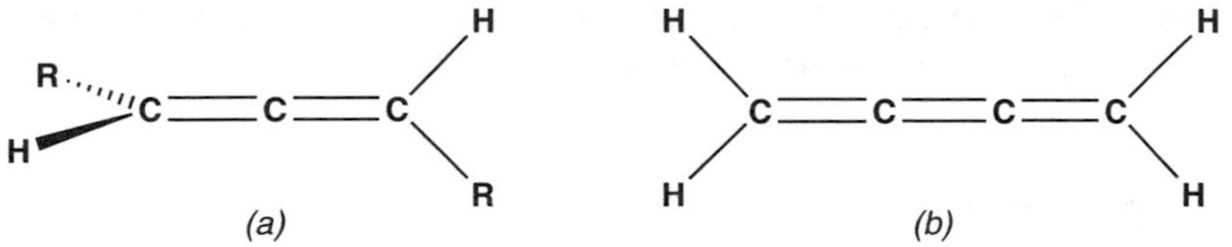

Figure 5.35. (*a*) Allene (R = H, 1,2-propadiene) and (*b*) 1,2,3-butatriene. When, for example, R = CH_3 (2,3-pentadiene), the diene exists in enantiomeric forms as a consequence of the lack of an S_n axis (see Sect. 4.7).

5.36 ALKYNES

Hydrocarbons containing a carbon–carbon triple bond.

Example. 5-Methyl-2-hexyne (Fig. 5.36); in the IUPAC numbering system the π bond is always given priority over the alkyl substituent.

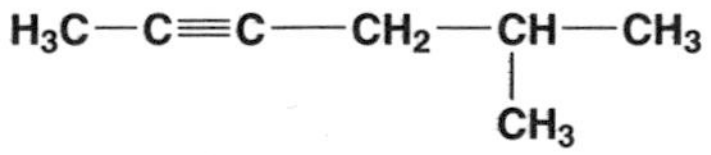

Figure 5.36. 5-Methyl-2-hexyne.

5.37 BENZENOID HYDROCARBONS

Hydrocarbons containing a substituted benzene ring. Benzene (Fig. 5.37*a* with R = H) was first isolated by Michael Faraday (1791–1867).

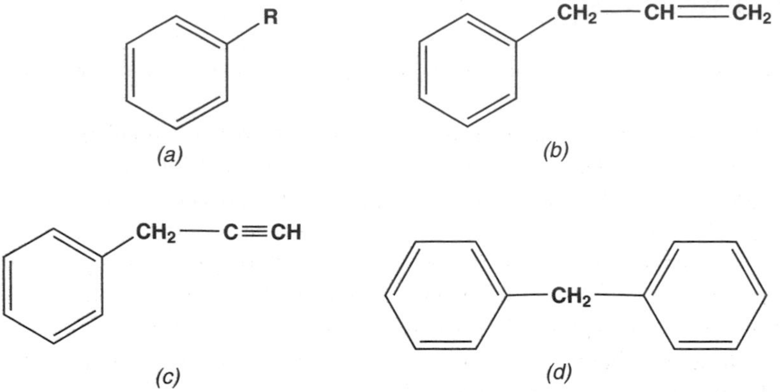

Figure 5.37. (*a*) Alkylbenzene; (*b*) allylbenzene; (*c*) propargylbenzene; (*d*) diphenylmethane.

Example. Alkylbenzenes (Fig.5.37*a*); allylbenzene (Fig. 5.37*b*); propargylbenzene (Fig. 5.37*c*); and diphenylmethane (Fig. 5.37*d*).

5.38 ARYL GROUP

The class or generic name for the aromatic grouping that remains after the imaginary removal of a hydrogen from the ring position of a parent arene (aromatic) nucleus (Sect. 3.33). Like alkyl groups, the names of the aryl groups are characterized by the suffix *-yl.*

Example. Phenyl (Fig. 5.38*a*; the name phenyl is derived from the Greek word *phainein* for "to shine" for what came to be known as benzene); 2-naphthyl (Fig. 5.38*b*); 9-phenanthryl (Fig. 5.38*c*; note the unusual but customary numbering system of phenanthrene). Phenyl groups are frequently abbreviated as Ph and are sometime represented by the Greek letter φ (phi).

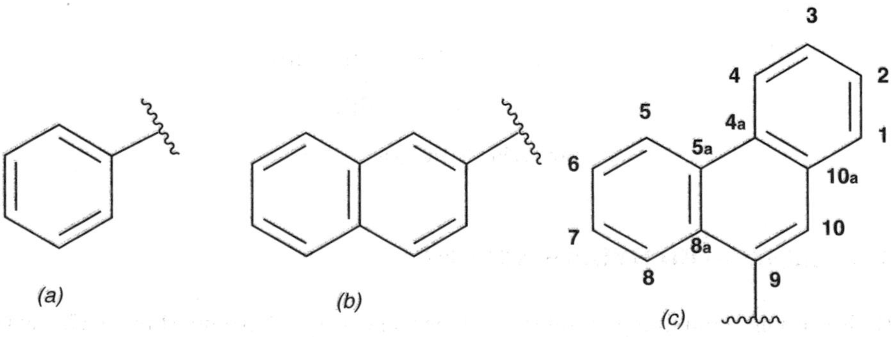

Figure 5.38. Aryl groups: (*a*) Phenyl; (*b*) 2-naphthyl; (*c*) 9-Phenanthryl.

5.39 ORTHO, META, AND PARA

These prefixes are used to distinguish the three possible disubstitution patterns of benzene: *ortho-* meaning substituents on adjacent carbons, (1, 2−); *meta-* with one intervening open position (1, 3−); and *para-* with substituents opposite each other (1, 4−); they are abbreviated as *o-*, *m-*, and *p-*, respectively.

Example. *o*-Xylene, *m*-xylene, and *p*-xylene (Figs. 5.39*a–c*, respectively).

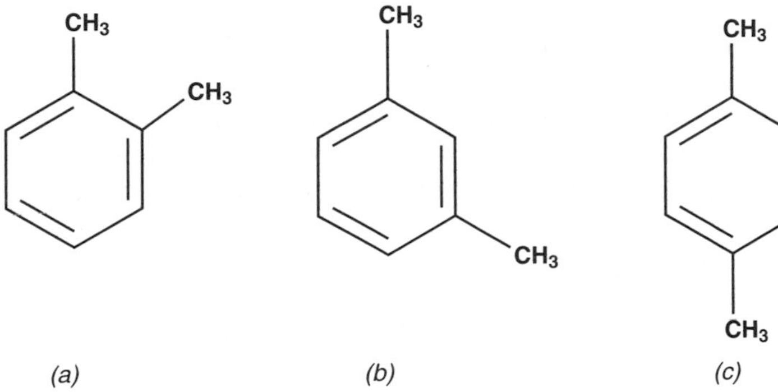

Figure 5.39. The three xylenes: (*a*) *ortho*; (*b*) *meta*; (*c*) *para.*

5.40 POLYNUCLEAR AROMATIC HYDROCARBONS (PNA)

The class of hydrocarbons consisting of cyclic assemblages of conjugated carbon atoms and characterized by large resonance energies (see also Sect. 3.33). The PNAs have been implicated as possible carcinogens in humans.

Example. 1,2-Benzpyrene (Fig. 5.40) is also called benzo[*a*]pyrene to indicate that a benzene ring is fused to side [*a*] of the parent pyrene. Side [*a*] is the side between

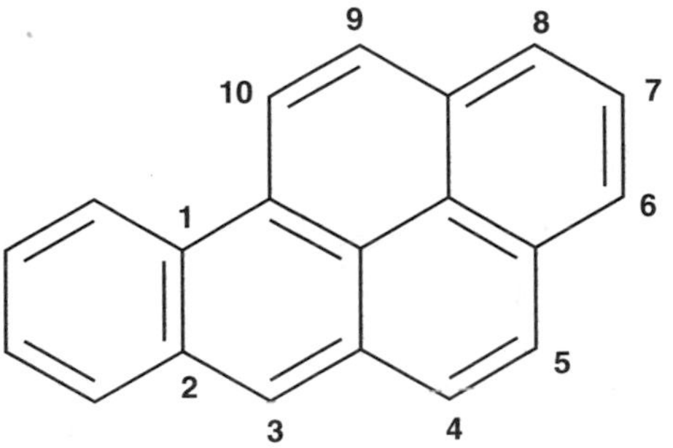

Figure 5.40. Benzo[*a*]pyrene (1,2-benzpyrene).

carbons numbered 1 and 2 in the numbering system of the parent PNA. Nonalternant polycyclic hydrocarbons, for example, fluoranthene (Fig. 3.35*c*) are also considered PNAs. Benzo[*a*]pyrene was isolated from coal tar in 1933 and shown to be a potent carcinogen in mice and rabbits. Because a high incidence of cancer of the scrotum was known to occur among chimney sweeps in England, it was suggested that exposure to benzo[*a*]pyrene was the causative agent and, thus, began the long crusade against environmental hazards.

For further examples of PNAs, see Fig. 3.33.

5.41 CATA-CONDENSED POLYNUCLEAR AROMATIC HYDROCARBONS

These are polynuclear aromatic compounds (PNAs) in which no carbon atoms belong to more than two rings, implying that all the carbon atoms are on the periphery of the polynuclear system and that there are no internal carbon atoms.

Example. All the PNAs shown in Fig. 3.33. These compounds consist of six-membered rings only, with no more than two carbon atoms common to two rings, and therefore, the number of $p\pi$ electrons obeys the Hückel $4n + 2$ rule (Sect. 3.23).

5.42 ACENES

Cata-condensed PNAs in which all the rings are fused in a linear array.

Example. Naphthacene (also called tetracene) (Fig. 5.42). It is interesting to note that all carbon atoms of this compound are on the periphery, and if the double bonds are distributed in the resonance structure as shown, this tetracyclic compound may be considered to be an 18-carbon cyclopolyene cross-linked at three positions; it is not unexpected, therefore, that naphthacene is a colored (deep orange) compound. The color deepens as the number of fused rings increases; pentacene is blue, hexacene is green, and heptacene is almost black. Although the four-ring isomeric benz[*a*]anthracene (Fig. 3.33*f*) has one resonance structure that can be written analogously, it is not a major resonance structure and the compound is colorless.

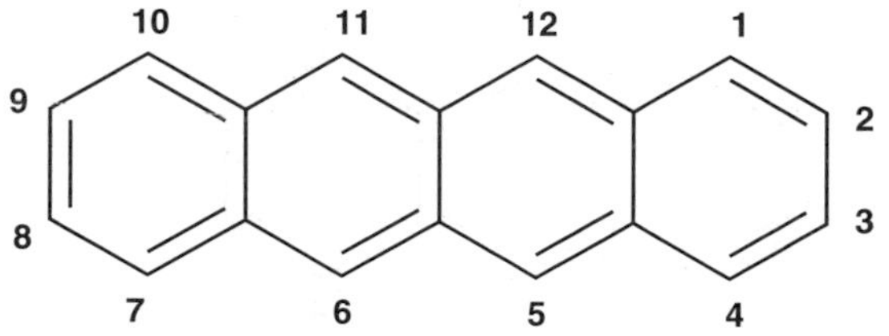

Figure 5.42. Naphthacene or tetracene.

5.43 PHENES

Cata-condensed PNAs containing an angular or bent system of rings.

Example. Phenanthrene (Fig. 3.33*d*) and benz[*a*]anthracene (Fig. 3.33*f*).

5.44 PERI-CONDENSED POLYNUCLEAR AROMATIC HYDROCARBONS

PNAs possessing some carbon atoms that are common to three rings.

Example. Pyrene (Fig. 5.44*a*); perylene (Fig. 5.44*b*); anthanthrene (Fig. 5.44*c*); such compounds need not obey the $4n + 2$ rule, for example, pyrene has 16 $p\pi$ electrons. The central carbon atoms common to three rings are indicated by solid dots. Note that in pyrene, which has D_{2h} symmetry, the four sides labeled [*a*] are all equivalent, and thus, if a benzene ring were fused at any [*a*] position, it would produce an identical compound, although the orientation would be different.

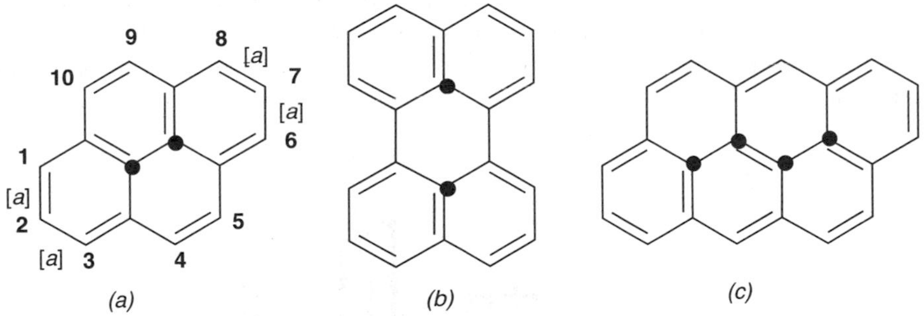

Figure 5.44. (*a*) Pyrene (the four sides labeled [*a*] are all equivalent); (*b*) perylene; (*c*) anthanthrene (the carbon atoms indicated by heavy dots are common to three rings).

5.45 MESO CARBON ATOMS

In polycyclic aromatics, the carbon atoms of the CH groups on the middle ring of an anthracene framework. (For orther examples of **meso** see sect 7.35 and 10.84)

Example. The 9,10-carbon atoms of anthracene (Fig. 3.33*e*) and the 5,6,11,12-carbon atoms of tetracene (Fig. 5.42). Anthracene undergoes a Diels–Alder reaction with the meso carbon atoms functioning as the sites of the 1,4-addition. Thus, with maleic anhydride, a 1,4-adduct is formed in quantitative yield (Fig. 5.45).

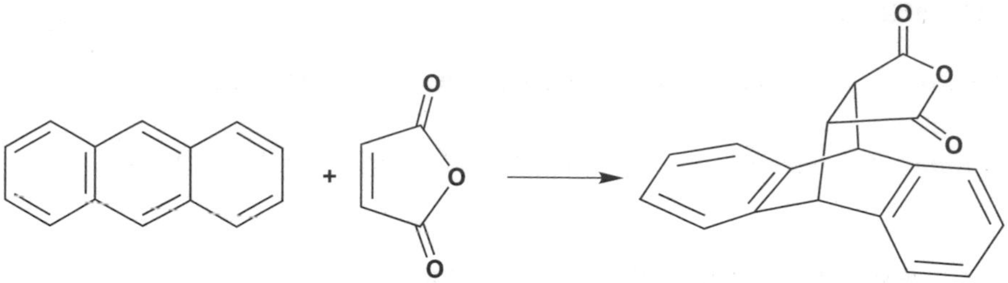

Figure 5.45. Diels–Alder reaction between anthracene and maleic anhydride.

5.46 THE BAY, K, AND L REGIONS IN POLYCYCLIC AROMATIC HYDROCARBONS

These positions in polycyclic compounds, as shown in Fig. 5.46, have special significance in correlating the structures of polycyclic aromatic compounds with their carcinogenic properties. The *L* region corresponds to the meso positions in anthracene, and the *K* region to the 9,10-position in phenanthrene (Figs. 3.33*d* and 5.38*c*). Rather recently, a different molecular area of polycyclic hydrocarbons, the *bay region*, has been implicated in carcinogenic-structure activity relationships. The bay region is defined as a concave exterior region of a polycyclic aromatic hydrocarbon bordered by three phenyl rings, at least one of which is a terminal ring. Such a region in benz[*a*]anthracene (1,2-benzanthracene) is shown in Fig. 5.46, along with the K and L regions in this molecule.

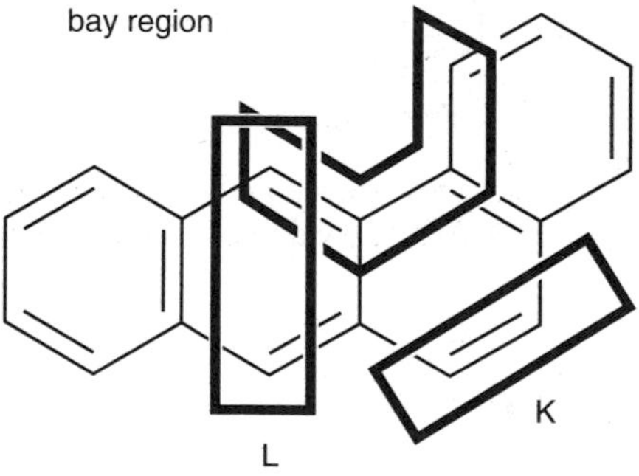

Figure 5.46. The K, L, and bay regions in PNAs.

5.47 PSEUDO-AROMATIC COMPOUNDS

Unstable cyclic conjugated hydrocarbons possessing two or more rings. This nomenclature is not widely used in the recent literature.

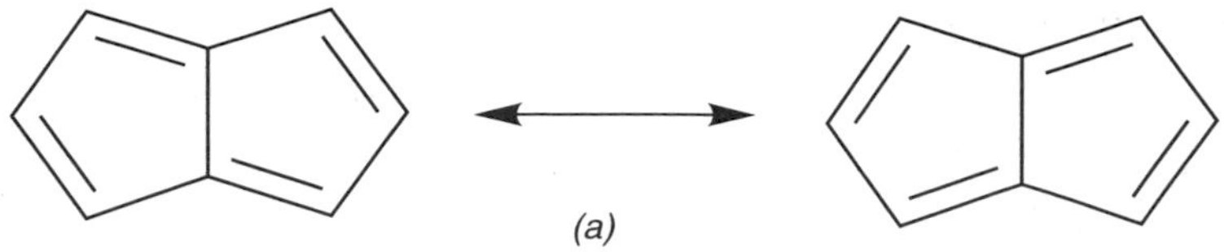

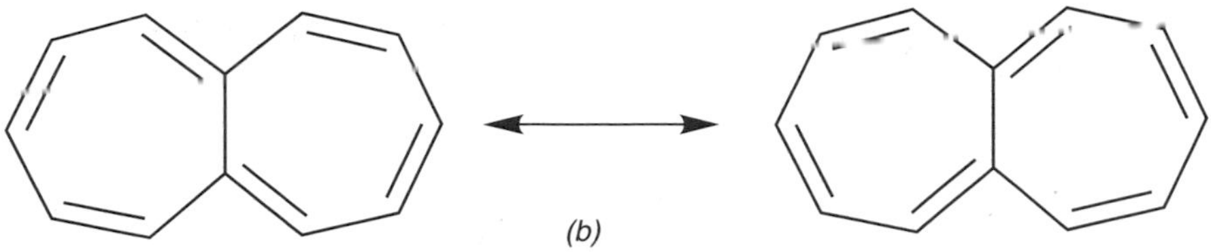

Figure 5.47. Pseudo-aromatic compounds: (*a*) pentalene and (*b*) heptalene.

Example. Pentalene (Fig. 5.47*a*) and heptalene (Fig. 5.47*b*). Although both of these ring systems can be written in two equivalent resonance structures, neither resonance structure can be written with a central double bond and both possess $4n$ π electrons. These compounds thus resemble a cross-linked cyclic conjugated polyene, and presumably, this accounts for their instability. Such molecules do not possess a totally symmetric ground-state wave function in the valence bond quantum mechanical treatment.

5.48 CYCLOPHANES

Compounds having a benzenoid ring that is *meta*- or *para*-disubstituted by a closed chain of carbon atoms, usually methylene groups. The carbon chain may or may not be interrupted by other benzenoid rings.

Example. [3.4]Paracyclophane (Fig. 5.48*a*). The methylene groups in this compound connect *para* positions in both benzenes. The numbers in brackets refer to the number of methylene groups in the two connecting chains. Other cyclophanes are [2.3]metacyclophane (Fig. 5.48*b*) and [10]paracyclophane (Fig. 5.48*c*).

5.49 INDEX OF UNSATURATION (HYDROGEN DEFICIENCY) *i*

A number whose value indicates the total number of rings and/or π bonds (a triple bond is counted as two π bonds) present in a molecule of known molecular formula. For a hydrocarbon it may be calculated from the equation

$$i = (2\text{C} + 2 - \text{H})/2 \tag{5.49}$$

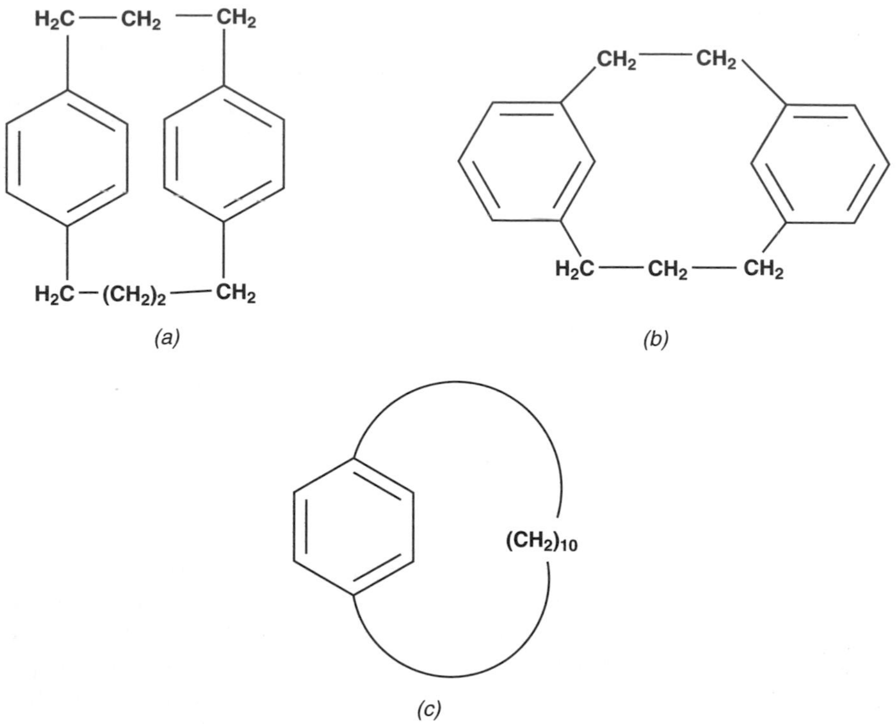

Figure 5.48. Cyclophanes: (*a*) [3.4] paracyclophane; (*b*) [2.3] metacyclophane; (*c*) [10] paracyclophane.

where C and H represent the *number* of carbon and hydrogen atoms, respectively, in the molecule.

Example. A compound with the molecular formula C_4H_6, according to the formula given above, has $i = (2 \times 4 + 2 - 6)/2 = 2$. The structure of the compound may thus possess one of the following structural features: two π bonds (two double bonds or one triple bond); a ring plus a π bond; or two rings. Some of the choices for C_4H_6 are thus 1,2-butadiene, 1,3-butadiene, cyclobutene, vinylcyclopropane, 2-butyne, or bicyclo[1,1,0]butane. A hydrocarbon with the molecular formula $C_{10}H_8$ must have $i = (2 \times 10 + 2 - 8)/2 = 7$ and, thus, some combination of rings or π bonds totaling seven must be present. One reasonable structure consistent with this requirement is naphthalene (Fig. 3.33*c*), with two rings and five double bonds.

For a generalized molecular formula $A_I B_{II} C_{III} D_{IV}$, where I, II, III, and IV equal the total number of all monovalent atoms A (H, F, Cl, etc.), divalent atoms B (O, S, etc.), trivalent atoms C (N, P, etc.), and tetravalent atoms D (C, Si, etc.), respectively, the index of hydrogen deficiency is $i = \mathrm{IV} - 1/2\,(\mathrm{I}) + 0\,(\mathrm{II}) + 1/2\,(\mathrm{III}) + 1$. Thus for a compound $C_{17}H_{19}O_3N$, $i = 17 - 1/2(19) + 1/2(1) + 1 = 9$ and any combination of nine rings and/or π bonds is possible. A compound consistent with these requirements is morphine, which has five double bonds and four rings.

5.50 CRYSTALLINE FORMS OF ELEMENTAL CARBON

Two forms of elemental carbon, graphite and diamond, have been known for thousands of years. It was therefore a shocking surprise to learn that there is a third and quite common form of crystalline carbon (discovered in 1985) which is present in some forms of soot, but which had escaped characterization for as long as civilization existed. The three forms in which elemental carbon exists are described below.

5.51 GRAPHITE

This form of elemental carbon consists of a two-dimensional network of hexagonal rings and, hence, resembles a huge sheet of carbon atoms arranged in a "chicken wire" pattern. These sheets of hexagons are stacked in layers with an interlayer spacing of about 3.35 Å, which is about the sum of two carbon van der Waals radii (1.70 Å). The in-plane carbon–carbon bond distance is about 1.42 Å, which corresponds to a bond order of approximately 1.33; each carbon atom is bonded to three other carbon atoms by one double bond and two single bonds. The weak interlayer bonding is responsible for the greasy feel of graphite and accounts for its effectiveness as a lubricant because two surfaces covered with a graphite coating, and in contact with each other, will tend to slip easily past one another.

5.52 DIAMOND

This form of crystalline carbon consists of a network of fused cyclohexane rings, all in the chair conformation, with each carbon atom tetrahedrally bonded to four other carbon atoms by single bonds. The carbon–carbon bond distance is 1.54 Å. The simplest hydrocarbon compound exhibiting a similar tetrahedral array of carbon atoms is adamantane (Fig. 5.30a); extending the adamantane structure in three dimensions would give the diamond structure. Diamond is one of the hardest substances known and therefore is used to cut other hard materials. Its hardness is a result of the fact that the diamond crystal is one very large molecule consisting of millions of strong carbon–carbon σ bonds.

5.53 FULLERENES AND RELATED COMPOUNDS

This is a class of pure crystalline all-carbon compounds, each having the shape of a hollow geodesic spheroid or dome; a structure of this geometry is thought to be the most efficient shape known for enclosing space. The first in the series and the first to be discovered (1985) is the C_{60} cluster, prepared and characterized by Harold W. Kroto (Sussex University) and Richard E. Smalley (Rice University) who shared the 1996 Nobel Prize in Chemistry with Robert F. Curl, Jr. (Rice University) for their work on fullerenes. The surface of the C_{60} carbon cluster cage (Fig. 5.53) consists of 60 equivalent carbon atoms (^{13}C NMR) arranged in 12 pentagons and 20 hexagons. Such a geometric arrangement is called a *truncated* icosahedron, because it results when each of the 12 vertices of the (20-faced) icosahedron is theoretically cut off (truncated), thereby generating a five-membered ring at each of the 12 sites. Each of the 12 resulting pentagons (which allows for the curvature) is completely surrounded by five fused hexagons. The resulting figure still has icosohedral symmetry and 59 rotational axes ($12C_5$, $12C_5^2$, $20C_3$, and $15C_2$). The magic number of 60 is the largest number of carbon atoms that can be arranged on the surface of a sphere such that every carbon atom is exchanged with every other carbon atom by rotation around a symmetry axis; there is no more symmetrical molecule possible.

In tribute to the architect Richard Buckminster Fuller (1895–1983) who invented the geodesic dome (patented in the United States in 1954), Kroto and Smalley gave the compound the eponymic 20-letter name buckminsterfullerene (one letter per face of the icosahedron). The geometric shape of buckminsterfullerene resembles that of a soccer ball, hence the colloquial name "bucky ball" for this C_{60} species. The general or generic name for the series of these cluster cage, domelike compounds is fullerene, and buckminsterfullerene, the first in the series, is [60]Fullerene. The fullerenes are even-numbered carbon cluster compounds C_n, with n ranging from 60 to 90 and larger. They all contain 12 five-membered rings and $(n - 20)/2$ hexagonal rings. The diameter of [60] fullerene is 7.1 Å, and allowing for the van der Waals radii of carbon, this leaves a hollow space whose diameter is about 3.5 Å. It is thus possible to accommodate the atom of almost any element inside the cage. When potassium is trapped in the cage, the internally "doped" fullerene is a semiconductor. Such internally doped fullerenes are called *endohedral complexes* and a symbolism denoting such complexes has been developed in which the enclosed atom is indicated by the @ symbol. Thus, the potassium C_{60} complex is written as $K@C_{60}$.

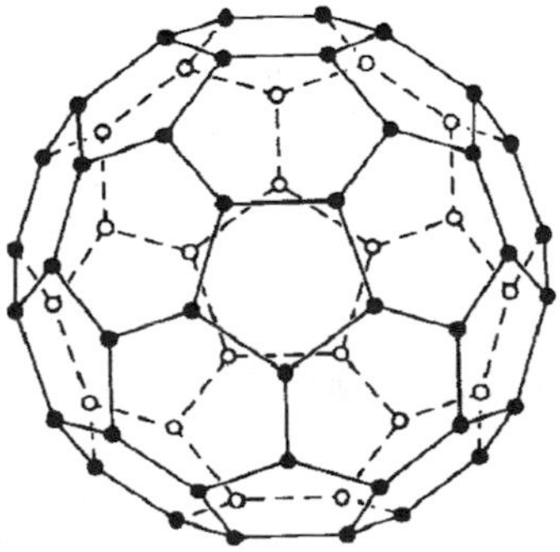

Figure 5.53. [60] Fullerene (buckminsterfullerene or bucky ball).

The fullerenes can be produced from a simple carbon arc generated from carbon electrodes in a helium atmosphere and are found on the cathode surface. The positive electrode is consumed in the arc. The plasma produced from the carbon has a temperature of about 3700°C. The C_{60} species can be sublimed above 350°C and is a brownish powder soluble in aromatic solvents. It is easily separated from the other fullerenes (the major contaminant is the C_{70} fullerene) by chromatography and promises to be readily available at very low cost.

NOTE. C_{60} fullerene was discovered as a peak in the mass spectrum of quenched carbon vapor independently by two groups of workers: Rohlfing, E. A., Cox, D. M., and Kaldor, A. *J. Chem. Phys. 81*, 3322 (1984); Kroto, H. W., Heath, J. R., O'Brien, S. C., and Smalley, R. E. *Nature 318*, 162 (1985).

5.54 CARBON NANOTUBES

These hollow microtubes of graphitic carbon have diameters in the nanometer range, hence nanotubes. They were first produced on the cathode surfaces of a carbon arc-discharge apparatus similar to that used for preparing fullerenes. The carbon structures (soot, whiskers, fullerenes) produced in this apparatus can be varied by controlling the type and pressure of gas surrounding the arc and by the configuration of the cooling system. Nanotubes can be viewed as single sheets of graphite (graphenes) rolled onto the surface of a cylinder in a helical (chiral) arrangement of carbon hexagons. Successive sheets can be rolled into concentric shells, resulting in a tube consisting of 2 to 50 individual shells, with a hollow core of about 2 to 25 nm separated by an annulus of approximately 0.34 nm between concentric shells. Single shell tubes with diameters on the order of 1 nm have also been prepared. Most nanotubes, unlike carbon fibers, are closed at both ends. Such capping is caused by defects and consists of half a fullerene molecule. Just as in the case of the fullerenes, the presence of pentagonal rings allows sufficient curvature of the hexagonal network to form closed capped structures. A nonotube may be considered as derived from fullerene by stretching the fullerene into a cylinder.

NOTE. The identification of multiple-shell graphitic nanotubes was first reported by Sumio Iijima, Nature *354*, 56 (1991). The nanotubes have unique mechanical and electrical properties and many practical applications are anticipated. An ordered array of single wall nontubes is called a *rope*.

5.55 CARBON FIBERS

Fibers resulting from the filamentary growth of graphitic carbon. They differ from carbon nanotubes in their much higher length-to-diameter ratio (the *aspect ratio*, whereby some carbon fibers can be spun into yarn) and their much larger (about 100–1,000 times) diameters. Most carbon fibers are generated from precursor well-defined polymers or they are vapor-grown. Polymer-based carbon fibers are most commonly prepared from poly(acrylonitrile) that after being spun out as a fiber is then carbonized first at about 1000°C for devolatilization and next at about 2000°C in the graphitization step. These fibers are very strong because of the complete sp^2 carbon–carbon bonding. Vapor-grown fibers are produced from the thermal (700–2000°C) decomposition of hydrocarbons on an inert surface. The diameter of these fibers can be as little as 100 nm. At these small diameters, the fibers approach the size range of nanotubes and have properties that start to resemble those of nanotubes.

NOTE. The first carbon fiber was prepared by Thomas Edison to serve as a filament in his electric lightbulb. He prepared these filaments from carbon fibers starting with Japanese bamboo filaments wound into a coil and pyrolyzed. When the aerospace and aircraft industries were looking for strong, lightweight materials that could be used in composites, they turned their attention to carbon fibers and the technology for producing them accelerated.

SUGGESTED READING

Bredt, J., Thouet, H., and Schmitz, J. *Ann. 47*, 219 (1924). This article contains the formal statement of what has become known as Bredt's rule, but Bredt's first article outlining the chemistry leading to the rule was published in 1902.

Curl, R. F. and Smalley, R. E. "Probing C_{60}." *Science 242*, 1017 (1988).

Ebbesen, T. W., ed. *Carbon Nanotubes*. CRC Press: Boca Raton, FL, 1997.

Fletcher, J. H., Dermer, O. C., and Fox, R. B., eds. *Nomenclature of Organic Compounds*. Advances in Chemistry Series 126; American Chemical Society: Washington, DC, 1974.

Noller, C. R. *Chemistry of Organic Compounds*, 3rd ed. W.B. Saunders Co.: Philadelphia, PA, 1965.

Patterson, A. M., Capell, L. T., and Walker D. F. *The Ring Index*. American Chemical Society: Washington, DC, 1960. This book contains a systematic listing of the structures and names of practically all known polycyclic ring systems.

Smalley, R. E. "The Third Form of Carbon." *Naval Research Reviews XLIII*, 3 (1991).

Solomons, T. W. G. *Organic Chemistry*, 5th ed. John Wiley & Sons: New York, 1992.

6 Functional Groups: Classes of Organic Compounds

6.1	Hetero Atom	144
6.2	Functional Group	144
6.3	Alkyl Halides, RX	145
6.4	Aryl Halides, ArX	145
6.5	Chlorocarbons	146
6.6	Fluorocarbons and Chlorofluorocarbons (CFCs)	146
6.7	Radicofunctional Names	147
6.8	Alcohols, ROH	147
6.9	Alkoxyl Group, RO–	148
6.10	Phenols, ArOH	149
6.11	Polyhydric Alcohols (Polyols)	149
6.12	Substitutive Nomenclature	149
6.13	Symmetrical Polyfunctional Compounds	152
6.14	Ethers, ROR	152
6.15	Ether Alcohols	153
6.16	Oligoethers and Podands	153
6.17	Supramolecules	154
6.18	Ligands	155
6.19	Crown Ethers or Coronands	155
6.20	Cryptands	156
6.21	Spherands	157
6.22	Calixarenes	157
6.23	Cryptophanes	158
6.24	Anticrowns	159
6.25	Catenanes and Rotaxanes	159
6.26	Carbonyl Group, –C(O)–	160
6.27	Aldehydes, RCH=O	161
6.28	Hemiacetals, RCH(OH)(OR)	162
6.29	Acetals, $RCH(OR)_2$	162
6.30	Ketones, RC(O)R	162
6.31	Hemiketals, $R_2C(OH)(OR')$, and Hemiaminals, $R_2C(OH)NR'_2$	163
6.32	Ketals, $R_2C(OR)_2$	163
6.33	Cyanohydrins, RCH(OH)CN or $R_2C(OH)CN$	164
6.34	Enols, $-\overset{\mid}{C}=\underset{\mid}{C}-OH$	164
6.35	Keto-Enol Isomers	164

The Vocabulary and Concepts of Organic Chemistry, *Second Edition*, by Milton Orchin, Roger S. Macomber, Allan Pinhas, and R. Marshall Wilson

6.36 Enediols 165

6.37 Enol Ethers, $-\overset{|}{C}=\underset{|}{C}-OR$ 166

6.38 Aldehyde and Ketone Derivatives 166

6.39 Carboxylic Acids, RC(O)OH 166

6.40 Carboxylate Anion, RCO_2^- 168

6.41 Carboxylic Acid Derivatives, RC(O)Z 168

6.42 Acyl and Aroyl Halides, RC(O)X 168

6.43 Esters, RC(O)OR, and Ortho Esters, $RC(OR)_3$ 169

6.44 Carboxylic Acid Anhydrides, $[RC(O)]_2O$ 169

6.45 Amides, $RC(O)NH_2$ 170

6.46 Substituted Amides, RC(O)NHR and $RC(O)NR_2$ 171

6.47 Secondary and Tertiary Amides, $[RC(O)]_2NH$ and $[RC(O)]_3N$ 171

6.48 Substituted Acidimides, $-\overset{\overset{X}{|}}{C}=N-$ 172

6.49 Imides, $O=C-NH-C=O$ (ring closed by $(CH_2)_x$) 172

6.50 α-Diketones, RC(O)C(O)R, and 1,2,3-Triketones, RC(O)C(O)C(O)R 172

6.51 β-Dicarbonyl Compounds, $RC(O)CR_2C(O)R$ 174

6.52 Keto Acids, $RC(O)(CH_2)_nC(O)OH$ 175

6.53 Di- and Polycarboxylic Acids 175

6.54 Conjunctive Names 175

6.55 Ketenes, $R_2C{=}C{=}O$ 177

6.56 Unsaturated Carbonyl Compounds, $RCH{=}CH(CH_2)_nC(O)Z$ 177

6.57 α,β-Unsaturated Carbonyl Compounds, RCH=CHC(O)Z, RC≡CC(O)Z 178

6.58 Hydroxy Carbonyl Compounds, $R_2C(OH)(CH2)_nC(O)R$ 178

6.59 Amino Carbonyl Compounds, $R_2C(NR_2)(CH_2)_nC(O)R$ 179

6.60 Amines, RNH_2, R_2NH, and R_3N 180

6.61 Imines, RCH=NH, $R_2C{=}NH$, and $R_2C{=}NR$ 180

6.62 Hydroxylamines, R_2NOH 181

6.63 Amine Oxides, $R_3N^+{-}O^-$ $(R_3N \rightarrow O)$ 181

6.64 Nitrones, $RCH{=}\overset{\overset{O^-}{|}}{N^+}{-}R$ and $R_2C{=}\overset{\overset{O^-}{|}}{N^+}{-}R$ 182

6.65 Ammonium Salts, $R_4N^+X^-$ 182

6.66 Nitriles, RCN 183

6.67 Isocyanides, $R{-}\overset{+}{N}{\equiv}\overset{-}{C}{:}$ 183

6.68 Nitroalkanes, RNO_2 184

6.69 Aci Form of the Nitro Group (Nitronic Acid or Azinic Acid) 184

6.70 Nitroso Alkanes, RN=O 185

6.71 Nitrate Esters, $R{-}ONO_2$ 185

6.72 Nitrite Esters, R–ON=O 186

6.73 Nitramines, $R_2N{-}NO_2$ 186

6.74 Nitrosamines, $R_2N{-}N{=}O$, and Nitrosamides, RRC(O)N–N=O 187

6.75 Diazo Alkanes, $R_2C{=}N^+{=}N^-$ 187

6.76 α-Diazo Ketones, $RC(O)CR'(N_2)$ 188

6.77 Azides, $RN{=}N^+{=}N^-$ 188

6.78 Acid (or Acyl) Azides, $RC(O)N{=}N^+{=}N^-$ 188

6.79 Azo Compounds, RN=NR 189

6.80	Azoxy Compounds, $R{-}N{=}\overset{+}{N}(O^-){-}R$	190
6.81	Aromatic Diazonium Salts, $ArN{\equiv}N^+\ X^-$	190
6.82	Cyanates, ROCN	190
6.83	Fulminates, $RON^+{\equiv}C^-$	191
6.84	Isocyanates, RN=C=O	191
6.85	Carbodiimides, RN=C=NR	192
6.86	Heterocyclic Systems	192
6.87	Oxa-Aza or Replacement Nomenclature	193
6.88	Hantzsch–Widman Heterocyclic Nomenclature	193
6.89	Lactones, $CH_2{-}(CH_2)_x{-}C{=}O$ (ring closed through O)	195
6.90	Lactams, $R{-}CH{-}(CH_2)_x{-}C{=}O$ (ring closed through NH)	195
6.91	Boranes	196
6.92	Organoboranes	196
6.93	Organoborates, $(RO)_3B$, $(RO)_2B(OH)$, and $ROB(OH)_2$	196
6.94	Heterocyclic Boranes	196
6.95	Carboranes	197
6.96	Hydroperoxides, ROOH	197
6.97	Peracids, RC(O)OOH	198
6.98	Dialkyl Peroxides, ROOR, and Diacyl or Diaroyl Peroxides, RC(O)OO(O)CR	198
6.99	Dioxetanes	199
6.100	Endoperoxides	199
6.101	Ozonides	200
6.102	Thiols (Thioalcohols, Mercaptans), RSH	200
6.103	Thiophenols, ArSH	201
6.104	Sulfides, RSR	201
6.105	Disulfides, RSSR	201
6.106	Sulfoxides, $R_2S(O)$ ($R_2S^+{-}O^-$)	202
6.107	Sulfones, $R_2S(O)_2$	202
6.108	Sulfonic Acids, RSO_2OH	203
6.109	Sultones, $CH_2{-}(CH_2)_x{-}SO_2$ (ring closed through O)	203
6.110	Sulfonates, $RS(O)_2OR$	204
6.111	Sulfates, $(RO)_2S(O)_2$	204
6.112	Bisulfates, $ROS(O)_2OH$	204
6.113	Sulfinates, RS(O)OH	205
6.114	Sulfinic Esters, RS(O)OR	205
6.115	Sulfenic Acids, RSOH	205
6.116	Sulfenates, RSOR	206
6.117	Thiocarboxylic Acids, RC(S)OH or RC(O)SH	206
6.118	Thioic *S*-Acids, RC(O)SH	206
6.119	Thioic *S*-Esters, RC(O)SR	206
6.120	Thioic *O*-Acids, RC(S)OH	207
6.121	Thioic *O*-Esters, RC(S)OR	207

6.122 Dithioic Acids, RC(S)SH 207
6.123 Dithioic Esters, RC(S)SR 208
6.124 Thioaldehydes, RCH=S 208
6.125 Thioketones, RC(S)R 208
6.126 Thiocyanates, RSCN 209
6.127 Isothiocyanates, RN=C=S 209
6.128 Sulfonamides, $RS(O)_2NH_2$ 209
6.129 Sulfinamides, $RS(O)NH_2$ 210
6.130 Sulfenamides, $RSNH_2$ 210
6.131 Thioyl Halides, RC(S)X 210
6.132 Sulfur Heterocyclic Compounds 210
6.133 Thiohemiacetals, RCH(OH)SR 211
6.134 Dithiohemiacetals, RCH(SH)SR 211
6.135 Thioacetals, $RCH(SR)_2$ 211
6.136 Sulfuranes, R_4S 212
6.137 Organophosphorus Compounds 212
6.138 Phosphines, RPH_2, R_2PH, R_3P, and RP=O 212
6.139 Phosphoranes, R_5P 213
6.140 Phosphine Oxides, $RPH_2(O)$, $R_2PH(O)$, and R_3PO 213
6.141 Phosphine Imides, R_3P=NH 214
6.142 Organophosphorus Acids 214
6.143 Phosphates, $(RO)_3P$=O 214
6.144 Phosphinic Acid, $H_2P(O)OH$ 214
6.145 Hypophosphorous Acid, $H_2P(O)OH$ 215
6.146 Phosphonous Acid, $HP(OH)_2$ 215
6.147 Alkyl or Aryl Phosphonous Acids, $RP(OH)_2$ 215
6.148 Alkyl or Aryl Phosphinic Acids, RPH(O)OH 215
6.149 Dialkyl or Diaryl Phosphinic Acids, $R_2P(O)OH$ 216
6.150 Phosphinous Acid, H_2POH 216
6.151 Dialkyl or Diaryl Phosphinous Acids, R_2POH 216
6.152 Phosphonic Acid, $HP(O)(OH)_2$ 216
6.153 Alkyl or Aryl Phosphonic Acids, $RP(O)(OH)_2$ 217
6.154 Phosphorus Acid, $P(OH)_3$ 217
6.155 Phosphites, $(RO)_3P$ 217
6.156 Heterocyclic Phosphorus Compounds 218
6.157 Halophosphorus Compounds 218
6.158 Organophosphorus Acid Amides 219
6.159 Organophosphorus Thioacids 219
6.160 Phosphonium Salts, $R_4P^+ X^-$ 219
6.161 Acid Halides of Inorganic Acids 220

One of the most useful concepts in the organization of organic chemistry is the grouping of molecules into families on the basis of the functional groups present within the molecule. A functional group is that portion of a molecule where reaction is most likely to occur. Because the only atoms attached to carbon atoms in saturated hydrocarbons are hydrogen atoms, there is generally little selectivity with respect to where a hydrocarbon molecule will undergo reaction. Of course, there are differences in the reactivity of primary, secondary, and tertiary hydrogen atoms, but these differences are rather small even though they may become important in certain hydrogen abstraction reactions. When unsaturation is introduced, the molecule will react selectively at this site of high electron density. Hence, a carbon–carbon multiple bond may be considered a functional group. Most functional groups, however, consist of a heteroatom or a group of atoms including a heteroatom that replaces a hydrogen atom on the parent hydrocarbon molecule. Such molecules may be represented by the general formula R–Z, where R is an alkyl group, the parent hydrocarbon minus one hydrogen atom, and Z is the functional group. The representation of the alkyl group by the letter R probably arose from the use (actually misuse) of the word "radical" (from the Latin *radic-* or *radix*, meaning "root") to designate the R group. However, the modern definition of radical is a species with an unpaired electron and the use of R in R–Z has no such implications. Although misleading, the use of the symbol R to represent the alkyl group or radical is so common as to be immune to change.

The nature of the Z group in a molecule R–Z not only determines the chemical behavior of R–Z, but also strongly influences its other properties. When R is small, the influence of Z is greatest. Thus, when R consists of less than four carbons, the alcohols ROH resemble HOH rather than the parent hydrocarbon RH, and hence, these alcohols are soluble in water and boil at a considerably higher temperature than hydrocarbons of comparable molecular weight, because, like water, they are polar and have strong hydrogen bonding interactions. Other low molecular weight oxygen (and nitrogen) compounds, for example, acetone, acetic acid, and triethylamine, are also polar and water-soluble (hydrophilic). As the size of R increases, RZ increasingly reflects the properties of the parent hydrocarbon RH, with an increasing hydrophobic and nonpolar character.

Of course, most molecules contain more than a single functional group. Thus, a molecule may have only a single type of functional group that is repeated one or more times. These types of molecules are called polyfunctional and tend to be characterized by molecules assembled from repeating units such as the man-made polymers polyvinyl alcohol or polyvinyl chloride (PVC). Molecules that contain more than a single type of functional group are called compounds of mixed function. Such molecules are typically encountered among natural products, in which a moderately large molecule usually will contain a number of different functional groups. Obviously, the possibilities are limitless. Therefore, in this chapter, the discussion has been confined to the more common types of simple polyfunctional and mixed functional molecules. Deliberately omitted in this volume is one of the most important areas of mixed functional molecules, the primary metabolites such as amino acids, proteins, sugars, and carbohydrates. Also omitted are nucleic acids, RNA, DNA, and polynucleotides. However, we do include the concept of supramolecules. These are groups of associated molecules that are held together by multiple, weak, noncovalent interactions, which when active in combination bind together the molecular constituents so strongly that the ensemble behaves as a discrete molecule.

Many of the functional groups discussed in this chapter contain the carbonyl group, >C=O, or thiocarbonyl group, >C=S. In order to distinguish this doubly bonded oxygen or sulfur from the singly bonded oxygen or sulfur, >C–O– or >C–S–, respectively, in a line formula, it is helpful to enclose the double bonded oxygen or sulfur atom in parentheses, for example, amides, $RC(O)NH_2$; thioketones, RC(S)R; and so on, and this is the usual practice in this chapter. However, such a formalism applies only to single oxygen or sulfur atoms. Other multielement groups may also be enclosed in parentheses for purposes of clarity. In these cases, the group of atoms in parentheses is generally attached to the rest of the molecule via a single bond. In addition, lone pair electrons are generally shown in the functional group structures because they are frequently important in the understanding of reaction mechanisms and resonance structures. Formal charges are usually circled for the same reasons. In subsequent chapters this emphasis may not always be followed.

No consideration by organic chemists of the many functional groups and the chemical and physical properties they impart to molecules can fail to refer to the monumental work of Saul Patai and Zvi Rappoport of the Hebrew University in Jerusalem who have edited more than 100 volumes of the well-known series *Chemistry of Functional Groups*. Readers are referred to appropriate volumes of this series if they are interested in learning more about the characteristic chemistry of particular functional groups.

6.1 HETEROATOM

Any atom, other than carbon or hydrogen, incorporated into a hydrocarbon backbone. The most common heteroatoms in organic chemistry are nitrogen, oxygen, and sulfur (see Sect. 6.86). Strictly speaking, hetero atoms are those that differ from the major atomic constituents of the molecule. Thus, in various classes of inorganic molecules, such as the carboranes (Sect. 6. 95), carbon atoms might be considered hetero atoms.

6.2 FUNCTIONAL GROUP

A heteroatom or group of atoms containing a heteroatom that replaces a hydrogen atom, or is inserted between two carbon atoms, in a hydrocarbon framework. Carbon–carbon double and triple bonds are frequently regarded as functional groups even though they contain no heteroatom. Therefore, some authors prefer to define a functional group as that portion of a molecule where reaction is most likely to occur. It is useful to represent by R, the alkyl, or by Ar, the aryl group that remains after the removal of a hydrogen atom from the parent hydrocarbon, and then to attach the functional group to the R- or Ar-, for example, $R{-}NO_2$, R–Cl, Ar–C(O)OH. The presence of a functional group confers characteristic physical and chemical properties on the molecule in which it occurs.

The coverage in this chapter is restricted to those functional groups that contain the heteroatoms B, N, O, S, P, or the halogens, either alone or in combination.

6.3 ALKYL HALIDES, RX

Derivatives of alkanes in which one of the hydrogen atoms of an alkane is replaced by halogen.

Example. 2-Chloropropane (isopropyl chloride) (Fig. 6.3*a*). As indicated by the alternate name in parentheses, halogenated compounds may be named as alkyl salts of the halogen acids (radicofunctional names, Sect. 6.7); other examples are *tert*-butyl iodide, benzyl fluoride, isopropyl bromide, and cyclohexyl chloride (Fig. 6.3*b*–*e*, respectively). The naming of the compound shown in Fig. 6.3*f* as an alkyl salt poses a special problem. A bivalent group, derived from an acyclic saturated hydrocarbon in which both free bonds are on the same carbon atom, is named by replacing the suffix *-ane* of the hydrocarbon with *-ylidene*. Thus, $-CH_2-$ and $CH_2=$ are methylidenes, although the former is almost always called methylene, and $CH_3CH=$ is ethylidene. In the case of Cl_2CH_2 (Fig 6.3*f*), the use of the name methylene chloride is so common that the correct systematic radical functional name (Sect. 6.7) for it, methylidene chloride, is seldom used.

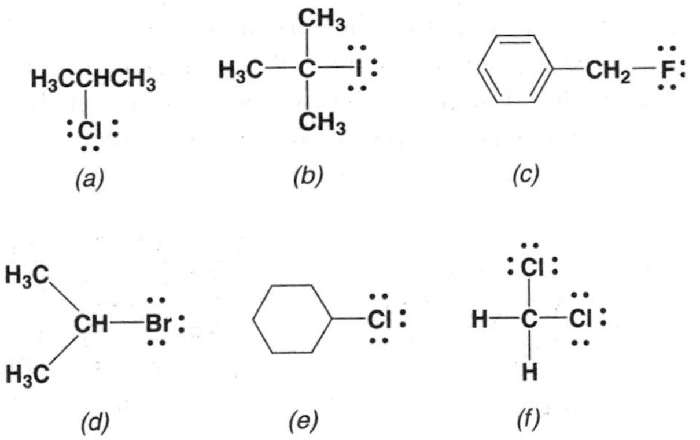

Figure 6.3. Alkyl halides: (*a*) Isopropyl chloride (2-chloropropane); (*b*) *tert*-butyl iodide (2-iodo-2-methylpropane); (*c*) benzyl fluoride; (*d*) isopropyl bromide (2-bromopropane); (*e*) cyclohexyl chloride (chlorocyclohexane); (*f*) methylene chloride (dichloromethane).

6.4 ARYL HALIDES, ArX

Derivatives of aromatic hydrocarbons in which one of the ring hydrogens has been replaced by halogen.

Example. 1-Chloronaphthalene and bromobenzene (Figs. 6.4*a* and *b*, respectively).

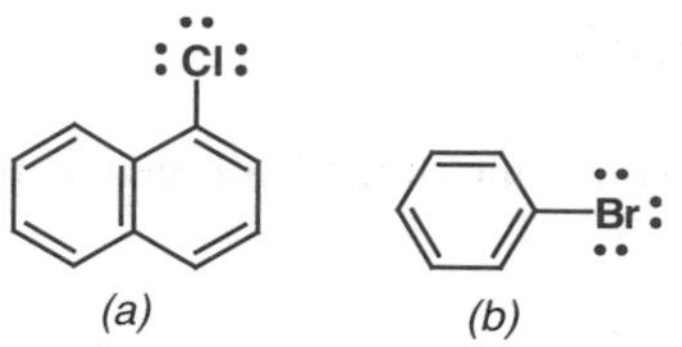

Figure 6.4. (*a*) 1-Chloronaphthalene and (*b*) bromobenzene.

6.5 CHLOROCARBONS

Molecules in which the hydrogens of the parent hydrocarbon have been largely or completely replaced by chlorine atoms.

Example. Chlorocarbons are the subject of great environmental concern. Since these materials tend to be very stable, they are not readily degraded in the environment and accumulate to the extent that they can become significant ecological and health hazards. A notorious example is the insecticide DDT (4,4′-dichlorodiphenyl-1,1,1-trichloroethane, Fig. 6.5*a*) that at one time accumulated to the extent of about 16 kg/acre in the sediment at the bottom of Long Island Sound. PCBs (polychlorobiphenyls, Fig. 6.5*b*), once widely used in electrical equipment such as transformers, as hydraulic fluids, lubricants, and plasticizers, and *dioxin* (Fig. 6.5*c*), which is a byproduct formed in the manufacture of agricultural chemicals such as weed killers and defoliants, and produced in the combustion of chlorine-containing compounds, are both well-known compounds of concern in *environmental chemistry*.

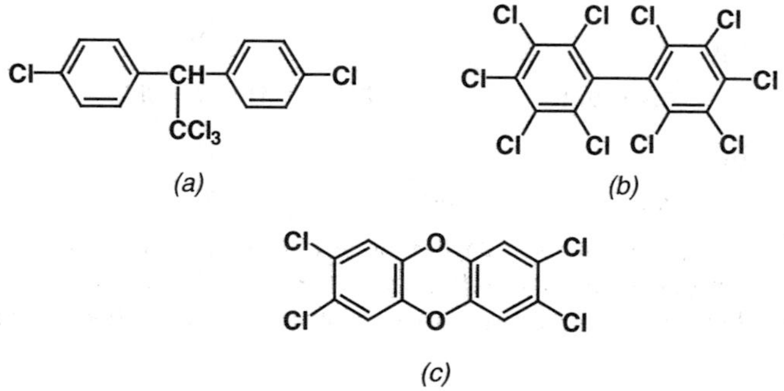

Figure 6.5. Chlorocarbons: (*a*) DDT (4,4′-dichlorodiphenyl-1,1,1-trichloroethane); (*b*) PCB (polychorobiphenyl); (*c*) dioxin.

6.6 FLUOROCARBONS AND CHLOROFLUOROCARBONS (CFCS)

Molecules in which the hydrogen atoms of the parent hydrocarbon have been largely or completely replaced by fluorine atoms. Many environmentally important fluorocarbons contain chlorine atoms as well and would be more accurately classified as chlorofluorocarbons (CFCs).

CCl_3F CCl_2F_2 CF_4

(a) (b) (c)

Figure 6.6. (*a*) Freon 11 (trichlorofluoromethane); (*b*) Freon 12 (dichlorodifluoromethane); (*c*) Freon 14 (tetrafluoromethane).

Example. The chlorofluorocarbons Freon 11, Freon 12 and tetrafluoromethane, Freon 14 (Figs. 6.6 *a*, *b*, and *c*, respectively) have been widely used as refrigerants, cooling media for air-conditioning systems, and propellants for aerosols. They are being replaced by other materials, and restricted by international law, since recent evidence indicates that these substances interfere with ozone accumulation in the upper atmosphere. Since ozone absorbs the shorter wavelengths of ultraviolet light, the "holes" produced in the ozone layer by the massive use of these materials is leading to increasing amounts of biologically damaging short-wavelength ultraviolet light reaching the surface of the earth (the 1995 Nobel Prize in Chemistry was awarded to F. Sherwood Rowland, Marid Molina, and Paul Crutzen for their research on ozone).

6.7 RADICOFUNCTIONAL NAMES

A system of naming compounds consisting of two major components. The first component corresponds to the name(s) of the alkyl or aryl group(s) involved (loosely called radicals), and the second component consists of the functional group name.

Example. The names of the alkyl halides written as salts of the halogen acids are radicofunctional names, for example, isopropyl bromide and methyl iodide. The class name *alkyl halides* is itself a generic radicofunctional name. Other examples are ethyl alcohol, dimethyl ether, and ethyl methyl ketone.

6.8 ALCOHOLS, ROH

Compounds in which a hydrogen atom of a parent hydrocarbon has been replaced by the hydroxyl group, –OH. Alcohols are classified as *primary, secondary*, or *tertiary* depending on whether the –OH group is attached to a primary, secondary, or tertiary carbon atom; the three classes have the general structures RCH_2OH, R_2CHOH, and R_3COH, respectively. Alcohols are most commonly named according to the IUPAC system in which the letter *-e* at the end of the name of the parent hydrocarbon is changed to *-ol*, a suffix denoting an alcohol.

Example. Ethanol, 2-propanol, and 2-methyl-2-propanol are primary, secondary, and tertiary alcohols, respectively. Such compounds may also be named according to the radicofunctional name that also has IUPAC approval, that is, ethyl alcohol, isopropyl alcohol, and *tert*-butyl alcohol, respectively. In a third system of nomenclature, now rarely used, the alcohols may be named as though the alcohol in question was *derived* from the first member of the series, methanol, CH_3OH. For this purpose, methanol is

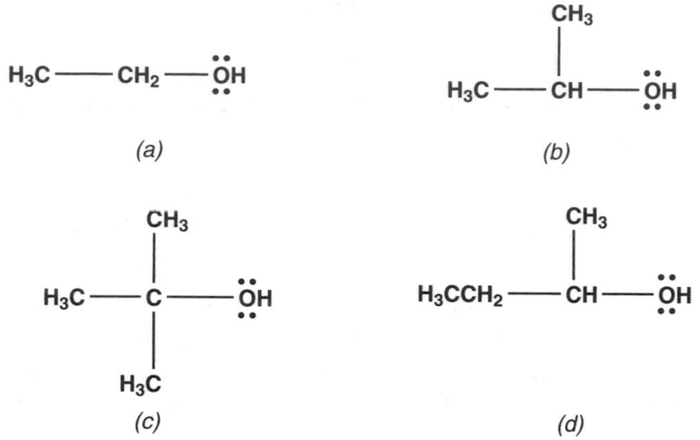

Figure 6.8. Classes of alcohols: (*a*) ethanol (primary); (*b*) 2-propanol (secondary); (*c*) 2-methyl-2-propanol (tertiary); (*d*) 2-butanol (secondary).

called carbinol, and other alcohols are considered as derivatives of carbinol in which one or more hydrogen atoms of the methyl group are replaced by alkyl groups. According to this system, the compounds shown in Figs. 6.8*a* through *c* may be called methylcarbinol, dimethylcarbinol, and trimethylcarbinol, respectively. [The (apocryphal) story is told that the carbinol nomenclature was invented to fool tipplers into believing that a bottle containing ethyl alcohol and (correctly) labeled methylcarbinol was unfit for drinking, since the use of methyl in the name would be associated with highly toxic methyl alcohol.] The compound shown in Fig. 6.8*d* may be called by either of three names: *sec*-butyl alcohol (radicofunctional), 2-butanol (the preferred IUPAC name), or ethylmethylcarbinol (derived).

6.9 ALKOXYL GROUP, RO–

The group remaining after the hydrogen atom attached to the oxygen of an alcohol is theoretically removed. When the RO– is attached to a parent compound, the letter *l* is dropped from the group name.

Example. Methoxyl anion (Fig. 6.9*a*), isopropoxyl anion (Fig. 6.9*b*), and 1,3-dimethoxybenzene (Fig. 6.9*c*).

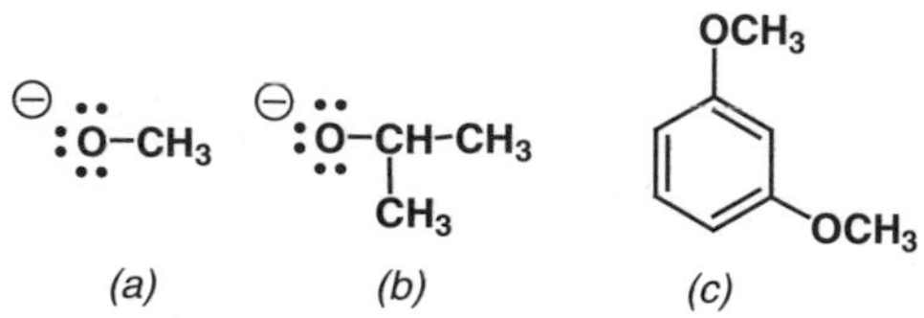

Figure 6.9. (*a*) Methoxyl anion; (*b*) isopropoxyl anion; (c) 1,3-dimethoxybenzene.

6.10 PHENOLS, ArOH

This is a class of compounds possessing a hydroxyl group (OH) attached directly to an aromatic ring; it is named after the first member of the series PhOH, phenol (Fig. 6.10*a*). The abbreviation Ph is frequently used to denote the *phenyl group*, C_6H_5–.

Example. Phenols: (*a*) phenol, (*b*) 1-naphthol, (*c*) 9-phenanthrol (9-hydoxyphenanthrene) in Fig. 6.10.

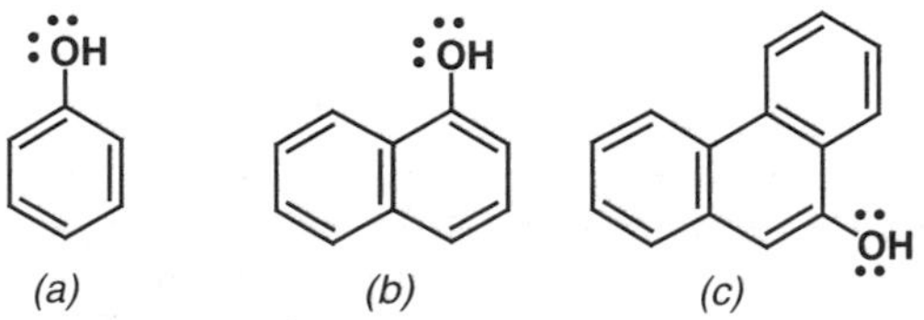

Figure 6.10. (*a*) Phenol; (*b*) 1-naphthol; (*c*) 9-phenanthrol.

6.11 POLYHYDRIC ALCOHOLS (POLYOLS)

Compounds possessing two or more hydroxyl groups. *Glycols* are a class of compounds containing two hydroxyl groups on adjacent carbon atoms. Sugars and carbohydrates are extremely important polyols, but will not be discussed in this volume.

Example. Ethylene glycol (Fig. 6.11*a*). It is used as a heat transfer medium, for example, in antifreeze; glycerol (1,2,3-propanetriol) (Fig 6.11*b*) is obtained from the hydrolysis of fats; and pyrogallol or hydroquinol (1,2,3-benzenetriol or 1,2,3-trihydroxybenzene) (Fig. 6.11*c*).

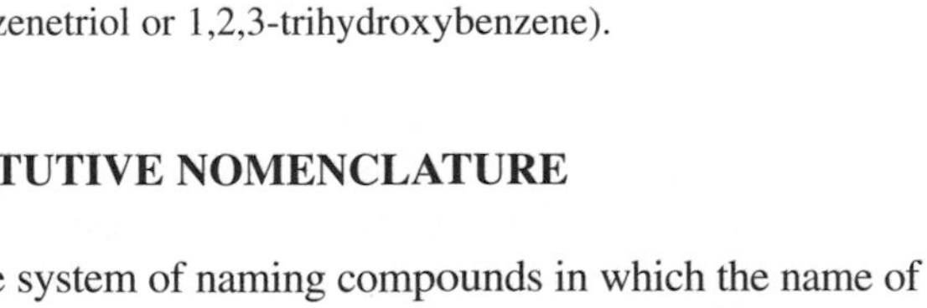

Figure 6.11. (*a*) Ethylene glycol; (*b*) glycerol (1,2,3-propanetriol); (*c*) pyrogallol or hydroquinol (1,2,3-benzenetriol or 1,2,3-trihydroxybenzene).

6.12 SUBSTITUTIVE NOMENCLATURE

This refers to the system of naming compounds in which the name of the functional or replacement group, called the *substituent*, provides the suffix after the name of the parent hydrocarbon from which the substituted compound is derived by the formal replacement of a hydrogen atom (Table 6.12*a*). Certain functional groups, for example, halogens, nitro, nitroso, and hydroperoxy (–OOH), are always cited as prefixes, in

TABLE 6.12*a*. Partial List of Common Functional Suffixes in Descending Order of Priority

Structure	Name	Structure	Name
Cations	*-onium*	$-C(O)NH_2$	*-amide*[b]
Anions	*-ate, -ide*	$-CH{=}O$	*-al*
$-C(O)OH$	*-oic* acid[a]	$>C{=}O$	*-one*
$-S(O)_2OH$	*-sulfonic* acid	$>C{=}S$	*-thione*
$-S(O)OH$	*-sulfinic* acid	$-OH$	*-ol*
$-SOH$	*-sulfenic* acid	$-SH$	*-thiol*
$-C(O)X$	*-oyl* halide	$-NH_2$	*-amine*
$-C(O)N_3$	*-oyl* azide		

[a] When this suffix is used, the carbon on the carboxyl group is included in the root name, for example, propanoic acid ($CH_3CH_2C(O)OH$), as contrasted with the suffix carboxylic acid, as in cyclohexanecarboxylic acid ($C_6H_{11}C(O)OH$), where the carboxyl carbon is named separately.
[b] For example, $CH_3CH_2C(O)NH_2$ is propanamide, or less commonly, ethanecarboxamide, analogous to suffix names *-oic* acid and carboxylic acid (see "Conjunctive Names," Sect. 6.54). The convenience of the suffix name carboxamide is apparent, for example, in 1,1-butanedicarboxamide rather than *n*-propylmalondiamide (generic), $CH_3CH_2CH_2CH[C(O)NH_2]_2$.

TABLE 6.12*b*. Partial List of Aromatic Compounds with Approved Common Names

Structure	Name	Structure	Name
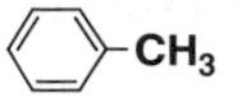	toluene[a]	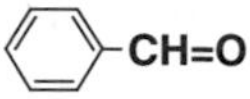	benzaldehyde
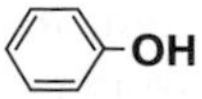	phenol		benzoic acid
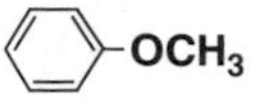	anisole		benzamide
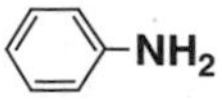	aniline	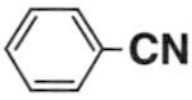	benzonitrile
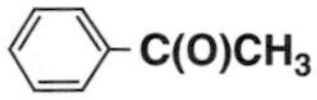	acetophenone	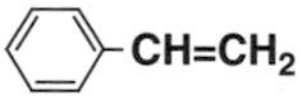	styrene
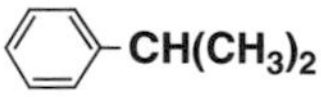	cumene	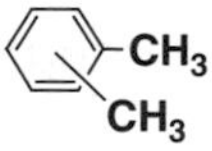	*o, m, p*-xylene[b]

[a] When a common name is used, the group that determines this name is always assigned position 1 (see examples).
[b] Equivalent groups must be treated equally. Thus, the xylenes should not be named as methyltoluenes.

which case the parent compound is the corresponding hydrocarbon. If two or more functional groups have suffix names, for example, carboxylic acid, acyl halide, amide, alcohol, and amine, then the functional group that stands highest in a table of piority (the principal group, IUPAC rules) is selected and used in deriving the name of the parent compound. However, for aromatic compounds, more than for any other class of organic substances, a considerable number of nonsystematic (generic) names are in vogue. Although the use of such names is discouraged, IUPAC rules permit some of the more widely used ones to be retained (Table 6.12*b*). Benzene derivatives with three or more substituents are named by numbering the position of each substituent on the ring. The numbering is ordered such that the lowest possible set of numbers is used. The substituents are arranged alphabetically. If an alkyl substituent on a benzene ring has more than six carbon atoms, then the benzene ring becomes a substituent of the alkane and is named a phenyl group.

Example. 2-Chloropropane (Fig. 6.3*a*). The compounds shown in Figs. 6.10*b* and *c* may be called l-hydroxynaphthalene and 9-hydroxyphenanthrene, respectively (from the parent hydrocarbons naphthalene and phenanthrene). Nonaromatic compounds containing only the –OH function are not usually named by the substitutive name. Thus, the alcohol in Fig. 6.8*b* is usually called by its radicofunctional name, isopropyl alcohol, or frequently by its IUPAC name, 2-propanol, but it should not be called isopropanol, which is a hybrid name that mixes the two systems of nomenclature and, hence, is inappropriate. The functional group priority that determines the root endings is illustrated in the following examples: Alcohols have higher priority than amines, thus, 3-amino-1-propanol (Fig 6.12*a*). Esters, which are named as anions, have among the highest priorities, thus, methyl 2-methoxy-6-methyl-3-cyclohexene-1-carboxylate (Fig 6.12*b*). Although the IUPAC name is preferred, 1-fluoro-4-methylbenzene, the substitutive name *p*-fluorotoluene is often used (Fig 6.12*c*). Low priority groups leave the root name unaffected, thus, 2-bromo-1-chloro-4-nitrosobenzene (Fig 6.12*d*). Alternatively, high-priority groups determine the root name, thus, 2-ethyl-4-nitrobenzaldehyde (Fig 6.12*e*). When the "substituent" is larger (> C_6) and more complex than the aromatic ring, the aromatic ring becomes the substituent, thus, 3-chloro-2-(4-methoxyphenyl)-5-methylheptane (Fig 6.12*f*).

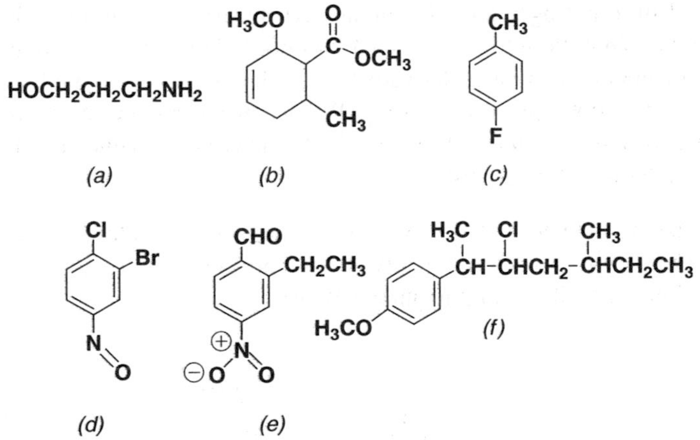

Figure 6.12. (*a*) 3-Amino-1-propanol; (*b*) methyl 2-methoxy-6-methyl-3-cyclohexene-1-carboxylate; (*c*) *p*-fluorotoluene; (*d*) 2-bromo-1-chloro-4-nitrosobenzene; (*e*) 2-ethyl-4-nitrobenzaldehyde; (*f*) 3-chloro-2-(4-methoxyphenyl)-5-methylheptane.

6.13 SYMMETRICAL POLYFUNCTIONAL COMPOUNDS

Compounds containing symmetry-equivalent functional groups.

Example. If two symmetry-equivalent functional groups are present within a molecule, such as the chloroethyl groups in 1,4-bis(2-chloroethyl)benzene (Fig. 6.13*a*), then the prefix *bis* replaces the usual prefix *di*-; if three such groups are present, as in tris(2,3-dibromopropyl)phosphate, formerly used as a flame retardant and commonly known as tris (Fig. 6.13*b*), *tris*-replaces the usual prefix *tri*-; if four such symmetry-equivalent groups are present, as in tetrakis(hydroxymethyl)methane (pentaerythritol) (Fig 6.13*c*), *tetrakis*-replaces the usual prefix *tetra*-.

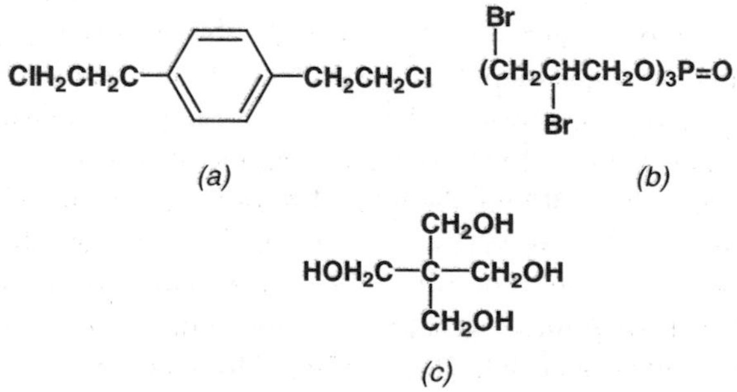

Figure 6.13. (*a*) 1,4-Bis(2-chloroethyl)benzene; (*b*) tris(2,3-tribromopropyl)-phosphate; (*c*) tetrakis(hydroxymethyl)methane or pentaerythritol).

6.14 ETHERS, ROR

Compounds that may be considered as derived from hydrocarbons either by replacement of a hydrogen atom by an alkoxyl (or aryloxyl) group or by the insertion of an oxygen atom into a carbon–carbon bond. However, it is frequently more useful to consider an ether as a derivative of an alcohol in which the hydrogen atom on the oxygen is replaced by an R group. Whenever more than one R group appears in a generic formula, it is understood that the R's may be the same or different unless it is specifically noted otherwise.

Example. Diethyl ether, $C_2H_5OC_2H_5$, is a common anesthetic. The compound shown in Fig. 6.14*a* is isopropyl methyl ether, and in Fig. 6.14*b* it is methyl phenyl ether, also known by its common name, anisole.

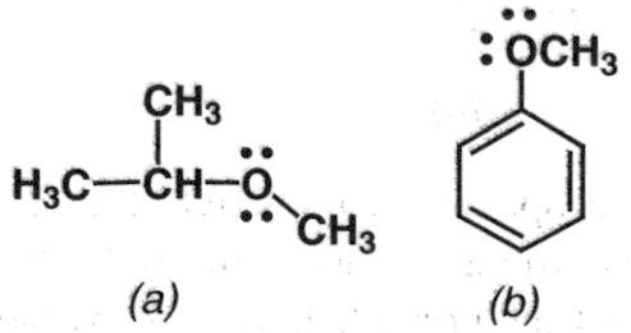

Figure 6.14. (*a*) Isopropyl methyl ether and (*b*) methyl phenyl ether (anisole).

6.15 ETHER ALCOHOLS

Compounds possessing both an ether and a hydroxyl function.

Example. 3-(2-Methoxyphenoxy)-1,2-propanediol is an expectorant (Fig. 6.15*a*); diethylene glycol (3-oxapentane-1,5-diol; see Sect. 6.87 for oxa-aza nomenclature) (Fig 6.15*b*).

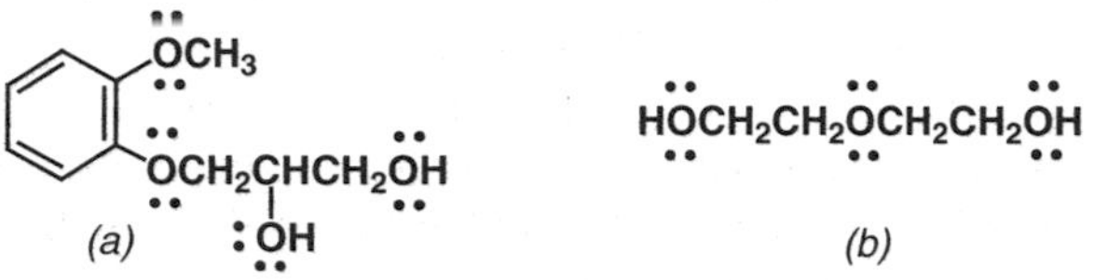

Figure 6.15. (*a*) 3-(2-Methoxyphenoxy)-1,2-propanediol and (*b*) diethylene glycol (3-oxapentane-1,5-diol).

6.16 OLIGOETHERS AND PODANDS

These terms designate molecules containing several ether linkages. The widely used prefix *oligo-* refers to several but less than many (*poly-*), and, while not prescisely defined, is usually taken to mean from about 4 to 20 of the units in question. Acyclic ethers with at least four oxygen atoms are also referred to as *open crowns.* Podands (from the Greek word *poda* for "foot" and the suffix *-nd* from ligand; see Sect. 6.18) are a class of oligoethers with two or more ether-containing chains extending from a central structural unit. Nitrogen and sulfur atoms may replace the ether oxygen atoms in these molecules, many of which form stable complexes with metal cations.

Example. Glyme, an acronym for ethylene glycol dimethyl ether (Fig 6.16*a*); *diglyme*, an acronym for diethylene glycol dimethyl ether (Fig 6.16*b*); the podand tris(3,6-dioxaheptyl)amine (Fig 6.16*c*).

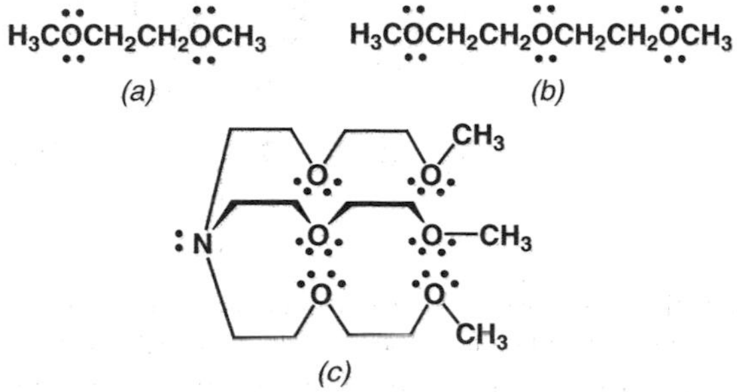

Figure 6.16. (*a*) Ethylene glycol dimethyl ether (glyme); (*b*) diethylene glycol dimethyl ether (diglyme); (*c*) tris(3,6-dioxaheptyl)amine.

6.17 SUPRAMOLECULES

Groups of two or more polyfunctional molecules "that are held together and organized by means of intermolecular (non-covalent) binding interactions" [J.-M. Lehn, *Science 260*, 1762 (1993)]. A supramolecule (from the Latin *supra*, meaning "above or beyond," as contrasted with the Latin *super*, meaning "over") behaves as a distinct molecule with properties differing from those of its molecular components.

Example. A variety of different types of bonding interactions can lead to supramolecule formation: (1) Simple van der Waals interactions can lead to organized forms of matter such as *liquid crystals* (Sect. 10.83), *micelles* (Sect. 10.72), *vesicles*, and catenanes and rotaxanes (Sect. 6.25). (2) *Hydrogen bonding* is the primary source of binding in the classical supramolecule, double-stranded DNA. Strong association between the two strands of DNA requires that the two strands have the proper sequence of complementary bonding sites. If the binding sites on the two chains

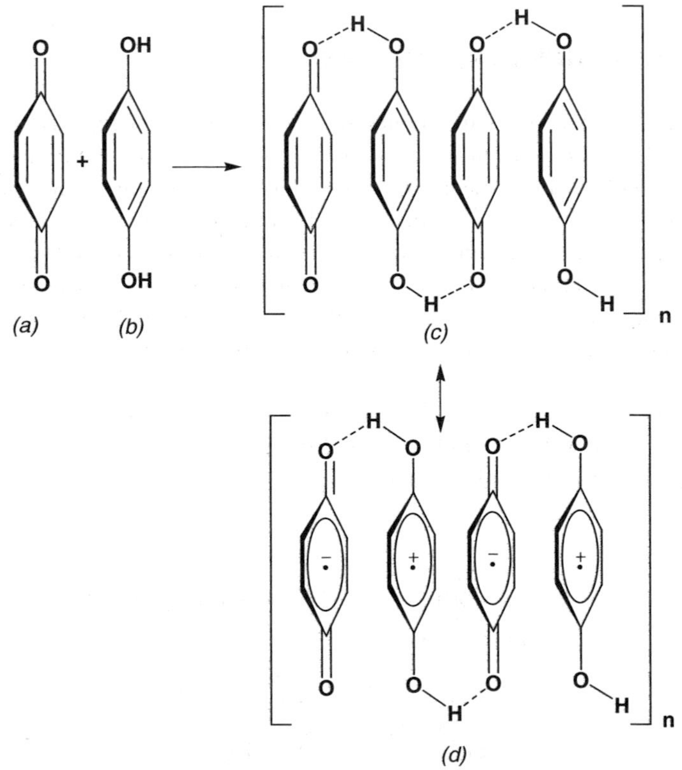

Figure 6.17. Association of *p*-benzoquinone (*a*) and hydroquinone (*b*) to form the supramolecule quinhydrone (*c*); (*d*) the alternating radical cation-radical anion resonance structure of quinhydrone, a charge transfer complex formed by electron transfer from the hydroquinone donor to benzoquinone, the acceptor.

are not properly sequenced, the two DNA chains will not "recognize" and bond to each other. For this reason, *supramolecule chemistry* involving complementary binding sites is frequently called *molecular recognition chemistry.* (3) Electrostatic attraction between electron-rich (*donor*) and electron-poor (*acceptor*) molecules leads to the formation of *charge-transfer complexes* (*donor-acceptor pairs*). And finally, (4) weak noncovalent ligand-metal bonding (Sects. 6.19 and 6.24).

Supramolecule formation may involve any combination of the above binding types. For example, *p*-benzoquinone (Fig. 6.17*c–d*) is a bright yellow-orange solid and hydroquinone (Fig. 6.17*b*) is a colorless material. When these two substances are mixed, a very insoluble black solid known as quinhydrone (Fig. 6.17*c*) is formed. Quinhydrone consists of stacks of alternating *p*-benzoquinone and hydroquinone molecules held together by a combination of hydrogen bonding (type 2 above) and electrostatic attraction (type 3 above). The electrostatic attraction is due to partial *electron transfer* from the hydroquinone (donor) to the *p*-benzoquinone (acceptor) that forms an extended charge transfer complex as shown in Fig. 6.17*d*, where it is represented by a resonance structure consisting of stacked and alternating *radical anions* and *radical cations*. Because of the greater mobility of electrons up and down the quinhydrone supramolecular stacks, quinhydrone has very different electrical conducting properties than either of its component molecules.

6.18 LIGANDS

Although the term ligand is most commonly associated with organometal chemistry (see Sect. 9.4), the use of the word has been extended to the complex ligands described in the following sections. The names of these special ligands are usually based on Latin verbs of action. Indeed, in the case of the word "ligand" itself, the name is based on the infinitive form of the Latin verb meaning "to bind": *ligase*. The suffix *-nd* is the Latin ending for gerunds, a noun derived from a verb and expressing generalized action. Thus, the noun ligand, something that acts to bind, might translate efficiently, although somewhat awkwardly, to the noun "binder," and the special ligands coronand (Sect. 6.19; from the Latin for "to crown") to "crowner" and cryptand (Sect. 6.20; from the Latin for "to hide") to "hider."

6.19 CROWN ETHERS OR CORONANDS

Cyclic oligoethers that form complexes with metal cations. Crown ethers are also called coronands (from the Latin word for "*coronare*, to crown;" see also Sect. 6.18), because when complexed with a metal ion (Fig. 6.19), they resemble a crown on the metal ion. Crown ethers themselves do not constitute supramolecules (see Sect. 6.17), but their complexes with metal cations do.

Example. Crown ethers are highly selective in their formation of complexes with metal cations. In general, the complexation will be most favorable when the diameter

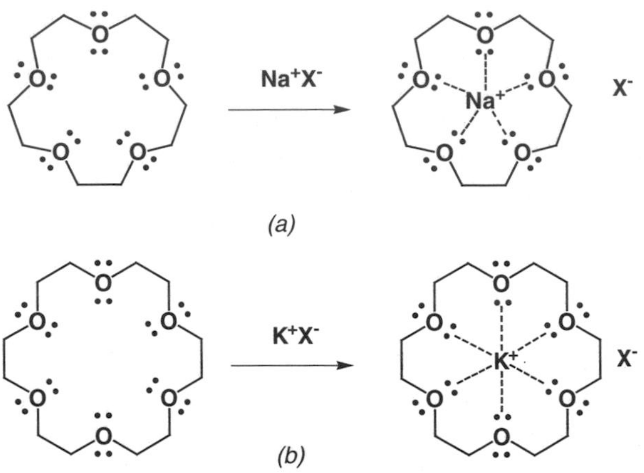

Figure 6.19. (*a*) 15-Crown-5 and its sodium ion complex; (*b*) 18-crown-6 and its potassium ion complex, where X^- represents any anion.

of the metal cation exactly matches the dimensions of the intracrown cavity encircled by the lone pair electrons on the ether oxygens. Thus, 15-crown-5 (15-membered ring incorporating five symmetrically arranged oxygen atoms) complexes most strongly with Na^+ ions (Fig. 6.19*a*); while 18-crown-6, with its larger cavity, forms more stable complexes with the larger K^+ ions (Fig. 6.19*b*). As a result of this selective ion bonding, either of these crowns may be used to separate Na^+ from K^+. The basis for such separations is the fact that metal cations and their associated anions become much more soluble in nonpolar organic solvents when they are encased in the nonpolar shell of the crown ether.

Thus, metal ions can be transported or carried from a polar aqueous medium into a nonpolar organic medium by crown ethers and related substances. For this reason, such substances are frequently called *ionophores* (ion carriers, from the Greek -*phoros* or *pherein*, meaning "carrying" or "to carry"). One very important characteristic of the complexation of the cation is that the associated anion is brought into the organic phase as a "naked" anion, free of its usual associated water molecules. This results in enhancing the nucleophillic character of the anion, and many synthetic procedures depend on this phenomenon. Finally, chemistry in which a small molecule or ion (the guest) becomes strongly associated with a larger molecule (the host) is often called *host-guest chemistry*.

6.20 CRYPTANDS

From the Latin word *cryptare*, meaning "to hide," these bridged crown ethers provide a three-dimensional cavity for metal ion complex formation. The metal ion complexes are supramolecules called cryptates.

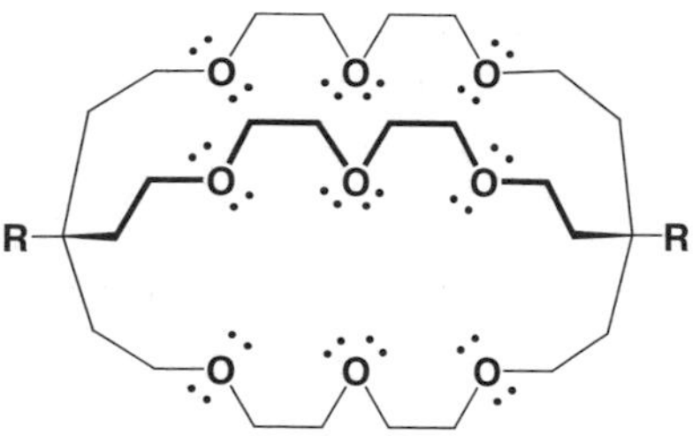

Figure 6.20. A [3,3,3]cryptand.

Example. A [3,3,3]cryptand has three oxygen atoms in each of three bridges (Fig. 6.20).

6.21 SPHERANDS

A crown ether variation in which the oxygen atoms form a spherical cavity into which metal cations can be incorporated through bonding with the lone pairs on the oxygen atoms.

Example. In the spherand shown in Fig. 6.21, the ether oxygens are alternately situated above and below the plane of the macrocyclic ring to form a spherical cavity that can accommodate metal ions.

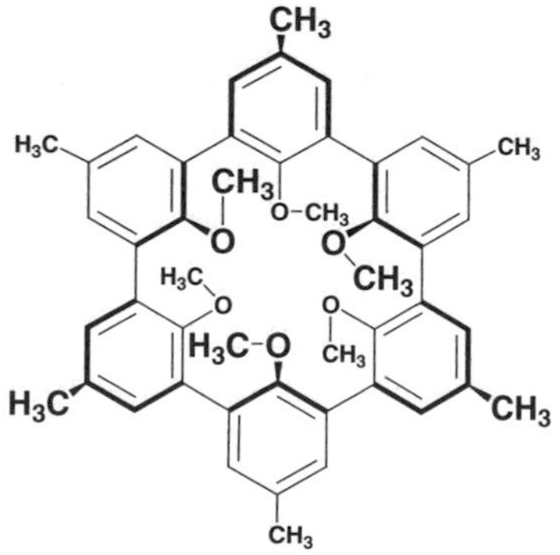

Figure 6.21. A spherand with oxygen atoms alternating above and below the plane of the macrocyclic ring to form a spherical cavity in the center of the molecule.

6.22 CALIXARENES

From the Greek word *calix*, meaning "vase," this is a crown ether variation in which the oxygen atoms encircle the smaller diameter end of a conical cavity, thus providing a vaselike cavity for the complexation of metal cations. Because aromatic rings always provide the rigid framework for molecules of this type, the suffix *-arene* is used.

Example. A calixarene (Fig. 6.22*a*), in which a vaselike cavity is provided as shown in Fig. 6.22*b*.

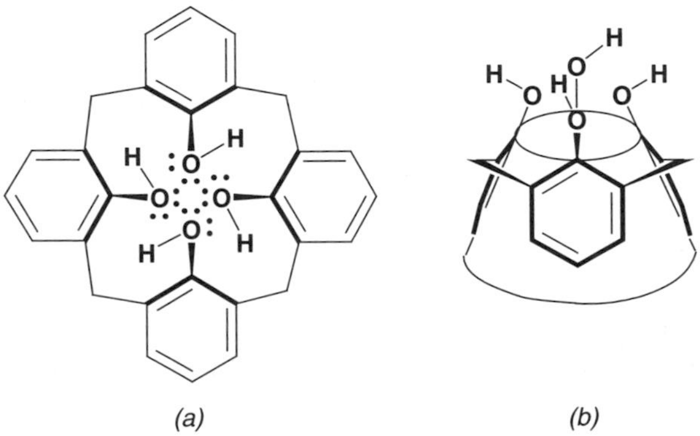

Figure 6.22. (*a*) A calixarene and (*b*) the conical, vaselike cavity formed by this molecule.

6.23 CRYPTOPHANES

Oligoethers formed by joining two calixarenes edge to edge to produce a cylindrical cavity of ether oxygen atoms.

Example. Cryptophane (Fig. 6.23).

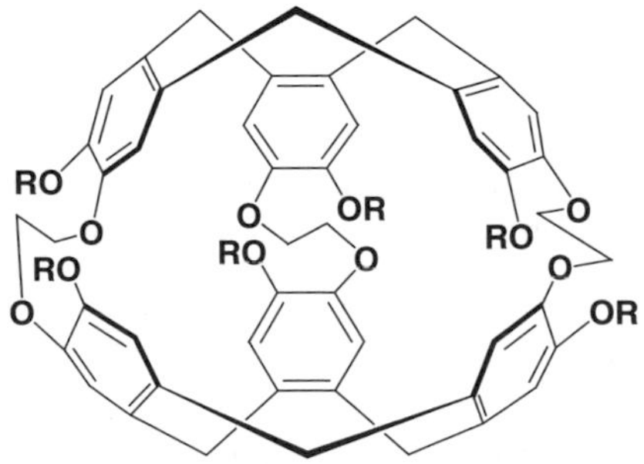

Figure 6.23. Cryptophane.

6.24 ANTICROWNS

Lewis acid hosts that provide cavities capable of binding anions in contrast to the crowns, which provide cavities into which cations are held.

Example. Anticrowns are not as well known as crowns. The macrocyclic carborane 12-mercurocarborand-4 (see Sect. 6.95) provides an anion-binding cavity (Fig. 6.24).

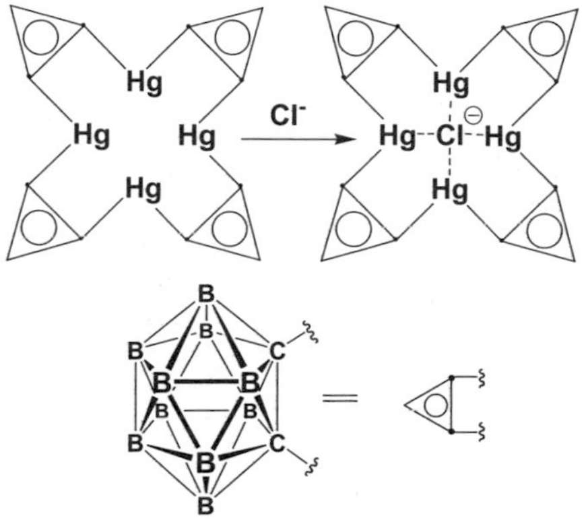

Figure 6.24. Binding of chloride ion by the anticrown 12-mercurocarborand-4 where each boron atom is bonded to a single hydrogen atom (BH), not shown in the figure for clarity.

6.25 CATENANES AND ROTAXANES

Catenanes, from the Latin word *catena*, meaning "chain," are molecules in which two or more macrocyclic rings are interlinked like the links of a chain. If *n* rings are interlinked, the molecule is called an [*n*]catenane (Figs. 6.25*a* and *b*). Rotaxanes are molecules in which a carbon chain is threaded through a ring and large groups are attached to both ends of the chain so that it cannot be withdrawn from the ring (Fig. 6.25*c*). Although molecules of these types need not contain any functional groups, since they can be constructed from simple hydrocarbon chains and rings, they are included in this chapter, since they are supramolecules. They are unique among molecules in that they are not held together by any form of bonding, but rather by the mutual repulsion of the interlinked groups (supramolecules of type 1, Sect. 6.17).

Example. Catenanes may be synthesized by the random interlinking of two large rings during the cyclization process. More rational approaches involve the prior assembly of a supramolecule with a geometry suitable for catenane formation. This strategy is illustrated in Fig. 6.25*d*, where the two components are prearranged as a copper complex in which the copper atom serves as a template. This utilization of the *template effect* produces relatively high yields of the catenane (Fig. 6.25*e*) following construction of the second macrocycle and removal of the copper ion. This type of catenane formed by two interlinked crown ethers is called a *caterand*, since the oligoether ring segment can bind metal ions.

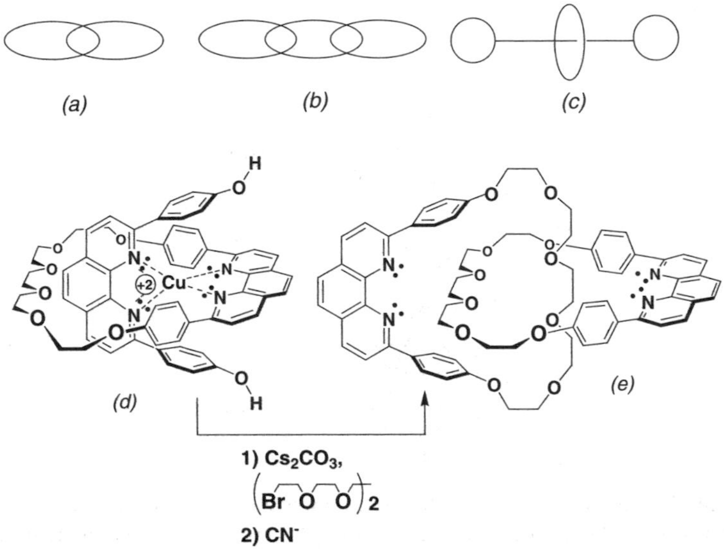

Figure 6.25. (*a*) [2]catenane; (*b*) [3]catenane; (*c*) rotaxane; copper complex precursor (*d*) to catenane (*e*) with lone pair electrons shown on nitrogen only.

6.26 CARBONYL GROUP, –C(O)–

An oxygen atom doubly bonded to a carbon atom.

Example. When writing structures as a sequence of atoms on a line, ambiguity can be minimized by enclosing in parentheses the atom that is joined by a double bond to the preceding atom; thus, the line formula for acetone (propanone) is $CH_3C(O)CH_3$.

The carbonyl function is one of the most important functional groups in organic chemistry and is present in aldehydes, ketones, esters, and so on (*vide infra*). The oxygen atom of the carbonyl group possesses two lone pairs of electrons. It is usually assumed that one electron lone pair is in an *sp* orbital which points in the direction opposite to the carbon–oxygen σ bond (n_1 in Fig. 6.26) and that the other electron lone pair is in the p_y orbital, n_2, which is perpendicular (orthogonal) to both the σ and π bonds. The π bond of the carbonyl is formed from overlap of the *p*π atomic orbitals on carbon and oxygen (Fig. 6.26*a*). The molecular orbital energy

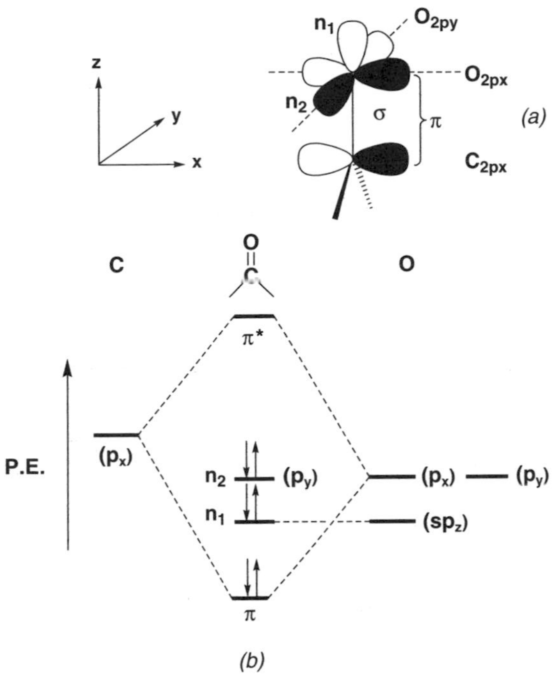

Figure 6.26. (*a*) The π bond of the carbonyl group; (*b*) the MOED for the carbonyl group showing the orbital distribution of the electrons in the carbonyl group.

diagram (MOED) for the carbonyl group involving the *p*π electrons is shown in Fig. 6.26*b*.

6.27 ALDEHYDES, RCH=O

Compounds in which a carbonyl group is bonded to both a hydrogen atom, and an alkyl group, aryl group, or a second hydrogen atom.

Example. The first member of the family of aldehydes is $H_2C{=}O$, formaldehyde (common name). The second member of the aldehyde series is CH_3CHO, acetaldehyde (common name). These common names are derived from those of the corresponding carboxylic acids to which they are related, namely formic and acetic acid, respectively. The IUPAC name for an aldehyde is derived from the parent hydrocarbon by changing the terminal *-e* in the name of the hydrocarbon to the functional group suffix, *-al*. Thus, formaldehyde is methanal from methane, and acetaldehyde is ethanal from ethane. The simplest aromatic aldehyde is benzaldehyde (Fig. 6.27).

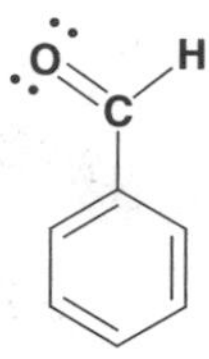

Figure 6.27. Benzaldehyde.

6.28 HEMIACETALS, RCH(OH)(OR)

The compounds that result from the addition of one molecule of an alcohol, ROH, to the carbonyl group of an aldehyde, RCHO.

Example. A hemiacetal of ethanal, 1-isopropoxyethanol (Fig. 6.28).

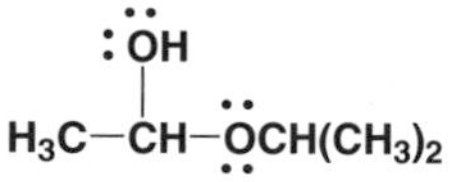

Figure 6.28. A hemiacetal of ethanal, 1-isopropoxyethanol.

6.29 ACETALS, $RCH(OR)_2$

These are compounds in which the double-bonded carbonyl oxygen of an aldehyde is replaced by two alkoxyl groups; such compounds result from the (acid-catalyzed) addition of two molecules of alcohol to the aldehyde with the associated loss of a molecule of water.

Example. 1,1-Diethoxypropane (Fig. 6.29).

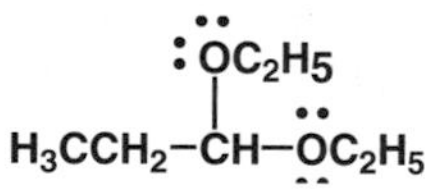

Figure 6.29. 1,1-Diethoxypropane.

6.30 KETONES, RC(O)R

Compounds having the bivalent carbonyl group bonded to two carbon substituents.

Example. The simplest ketone is $CH_3C(O)CH_3$, acetone, which is its common name. The IUPAC name is propanone, a name derived from the parent three-carbon

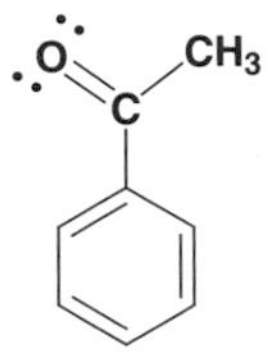

Figure 6.30. Acetophenone.

hydrocarbon, propane, by replacing the final *-e* with the suffix *-one*, which denotes the ketone function. The radicofunctional name (Sect. 6.7) for acetone is dimethyl ketone, but this name is seldom used. The carbonyl carbon may also be part of a ring, as in cyclopentanone. The two R groups of a ketone may be the same or different; for example, $CH_3C(O)CH(CH_3)_2$ is isopropyl methyl ketone (radicofunctional name) or 3-methyl-2-butanone (IUPAC name), and $PhC(O)CH_3$ (Fig. 6.30) is methyl phenyl ketone (radicofunctional name) or acetophenone (common name).

6.31 HEMIKETALS, $R_2C(OH)(OR')$, AND HEMIAMINALS, $R_2C(OH)NR'_2$

Hemiketals are compounds that result from the addition of one molecule of an alcohol, R′OH, to the carbonyl group of a ketone, $R_2C(O)$, whereas hemiaminals result from the addition of either ammonia (R′ = H), or a primary or secondary amine, R′ = alkyl or aryl, to the carbonyl groups of a ketone.

6.32 KETALS, $R_2C(OR)_2$

Compounds in which the carbonyl oxygen of a ketone is replaced by two alkoxyl groups. A separate name for this class of compound is no longer recommended; ketals are included in the category of acetals (Sect. 6.29).

Example. The acetal (ketal), 1,1-dimethoxy-*n*-butylbenzene (Fig. 6.32*a*), and the cyclic acetal derived from acetone and ethylene glycol (Fig. 6.32*b*).

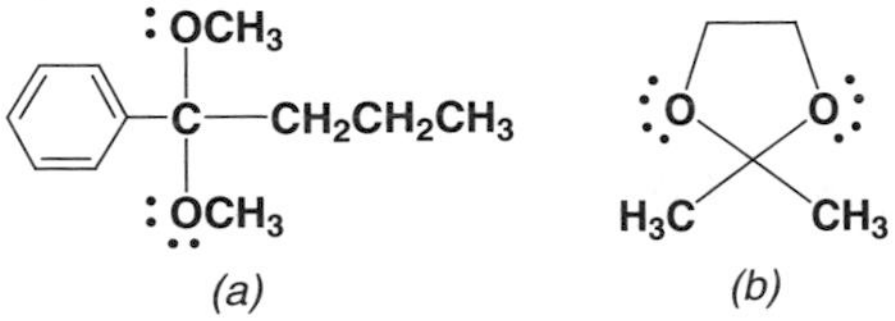

Figure 6.32. (*a*) 1,1-Dimethoxy-*n*-butylbenzene and (*b*) the ethylene glycol acetal (ketal) of acetone (2,2-dimethyl-1,3-dioxolane, see Table 6.88*b*).

6.33 CYANOHYDRINS, RCH(OH)CN OR $R_2C(OH)CN$

The cyanohydrins are compounds that result from the addition of HCN (hydrogen cyanide) to the carbonyl group of either an aldehyde, RCHO, or (less frequently) a ketone (see also "Nitriles," Sect. 6.66).

Example. The cyanohydrin derived from acetone and HCN (Fig. 6.33).

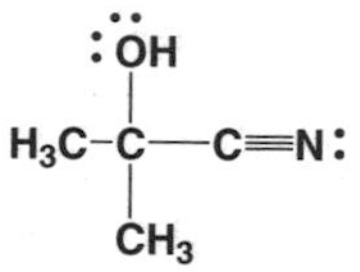

Figure 6.33. 2-Hydroxy-2-methylpropanenitrile or acetone cyanohydrin.

6.34 ENOLS, $-\overset{|}{C}=\underset{|}{C}-OH$

These are compounds in which one of the hydrogen atoms attached to a carbon–carbon double bond is replaced by a hydroxyl group; the enol (enol form) is usually in equilibrium with the corresponding carbonyl compound (keto form).

Example. Propen-2-ol (or 2-propenol) (Fig. 6.34*a*) is the enol of acetone. This enol is present to a very small extent (estimated to be about 0.00025%) in pure acetone. Cyclohexen-1-ol (Fig. 6.34*b*) is the enol of cyclohexanone (Fig. 6.34*c*). It should be noted that phenol (Fig. 6.10*a*) possesses the enolic grouping. The molecule is so stable that it exists completely in the enol form because conversion to the corresponding ketone (keto form) would involve the loss of aromaticity.

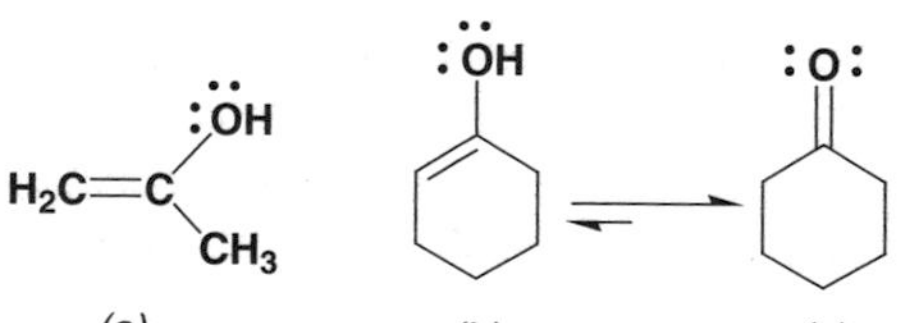

(*a*) (*b*) (*c*)

Figure 6.34. (*a*) The enol of acetone; (*b*) cyclohexen-1-ol; (*c*) cyclohexanone.

6.35 KETO-ENOL ISOMERS

Isomers that result from *reversible* hydrogen migration from a carbon atom, adjacent to a carbonyl group, to the oxygen of the carbonyl group, or the reverse process. The hydrogen atom bonded to the oxygen is called the enolic hydrogen.

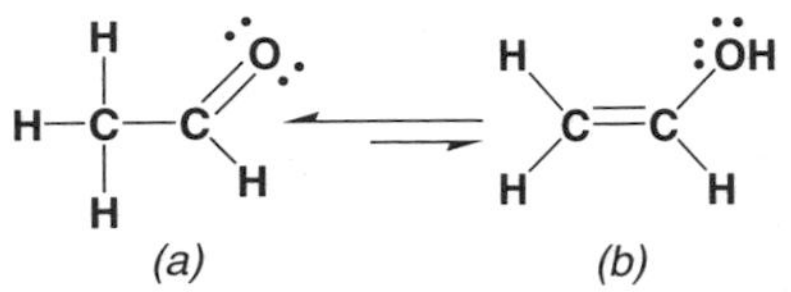

Figure 6.35. (*a*) Acetaldehyde and (*b*) its enol (vinyl alcohol).

Example. The keto-enol equilibrium between isomers (also called *proton tautomerism* or *prototropy*) is characteristic of all carbonyl-containing compounds possessing a hydrogen atom on the carbon adjacent to the carbonyl group (α-H atom); for example, acetaldehyde and its enol (Figs. 6.35*a* and *b*, respectively) or cyclohexanone and its enol (Figs. 6.34*c* and *b*, respectively).

6.36 ENEDIOLS

Compounds containing a double bond to which two hydroxyl groups are attached to the carbon atoms of a carbon–carbon double bond. This term might refer either to a 1,1-enediol (Fig. 6.36*a*) or a 1,2-enediol (Fig. 6.36*b*). The enolic form of a carboxylic acid (see Sect. 6.39) would constitute a 1,1-enediol, whereas the enolic form

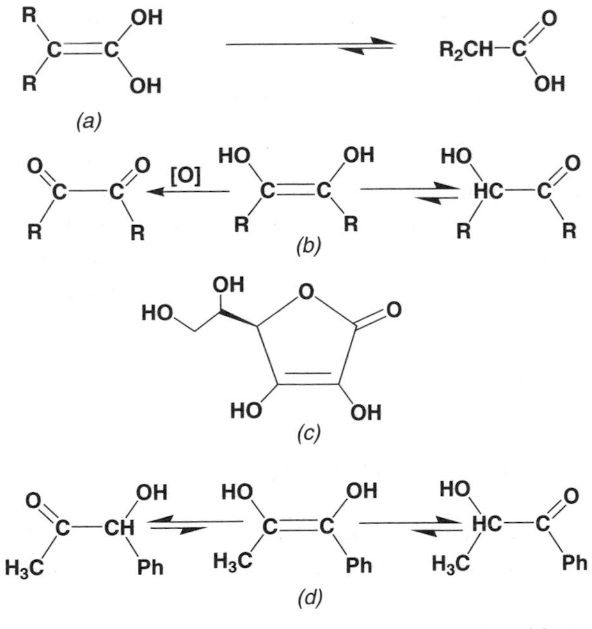

Figure 6.36. (*a*) 1,1-enediol and its equilibrium with a carboxylic acid; (*b*) 1,2-enediol and its equilibrium with an α-ketol and oxidation to an α-diketone; (*c*) vitamin C; (*d*) enediol intermediate in the interconversion of α-ketol isomers.

of an α-hydroxyketone (α-ketol, see Sect. 6.58) would constitute a 1,2-enediol. When the term enediol is used without designation of the positions of the hydroxyl groups, it is usually taken to mean the 1,2-enediol form (Fig. 6.36*b*); such structures are commonly encountered in carbohydrate chemistry.

Example. Vitamin C (Fig. 6.36*c*) is the best-known example of a stable enediol. Since enediol are excellent reducing agents, the parent α-ketols are sometimes referred to as *reductones* (Fig. 6.36*b*). Finally, enediols are intermediates in the facile interconversion of α-ketol isomers (Fig. 6.36*d*).

6.37 ENOL ETHERS, —C=C—OR

Compounds in which the enolic hydrogen atom of an enol is replaced by an alkyl group.

Example. Methyl vinyl ether (Fig. 6.37*a*) and trimethylsilyl vinyl ethers (Fig. 6.37*b*) are important in many synthetic procedures.

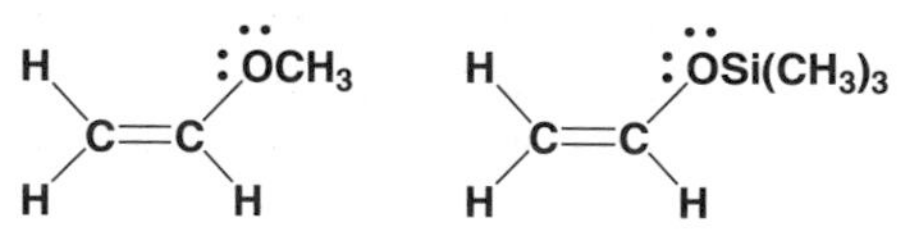

(*a*) (*b*)

Figure 6.37. (*a*) Methyl vinyl ether and (*b*) trimethylsilyl vinyl ether.

6.38 ALDEHYDE AND KETONE DERIVATIVES

Derivatives formerly used to characterize carbonyl compounds; these derivatives may be regarded in a formal way as resulting from a splitting out of water by loss of the oxygen of the carbonyl group of the aldehyde or ketone and two hydrogen atoms from the derivatizing agent where these hydrogens are usually on a nitrogen atom. Some typical derivatizing reagents and the products of their reactions with carbonyl compounds are shown in Table 6.38. Such "wet" chemistry reactions for the characterization of aldehydes and ketones have largely been discarded in favor of spectroscopic methods of identification.

6.39 CARBOXYLIC ACIDS, RC(O)OH

A class of compounds containing a functional group, called the carboxyl group, consisting of a hydroxyl group attached to a carbonyl group, –C(O)OH or $-CO_2H$.

Example. The group names for the –OH and –C(O)OH functions when used as substituents in a parent compound are written as hydroxy and carboxy and not hydroxyl and carboxyl. The first member of the carboxylic acid series (R=H) is HCO_2H, called

TABLE 6.38. Some Derivatives of Carbonyl Compounds Used for this Characterization:
$R_2C{=}O + H_2N{-}Z \longrightarrow R_2C{=}N{-}Z + H_2O$

Reagents		Derivative	
Name	Structure	Generic Name	Structure
Hydroxylamine	$H_2N{-}OH$	Oxime	$R_2C{=}N{-}OH$[a]
Hydrazine	$H_2N{-}NH_2$	Hydrazone	$R_2C{=}N{-}NH_2$
Phenylhydrazine	$H_2N{-}NHPh$	Phenylhydrazone	$R_2C{=}N{-}NHPh$
Tosylhydrazine	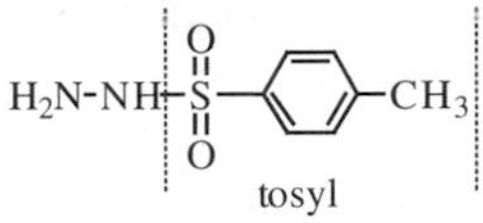	Tosylhydrazone	$R_2C{=}N{-}NHTs$
2,4-Dinitro-phenylhydrazine	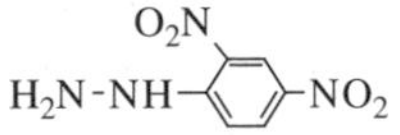	2,4-Dinitro-phenylhydrazine	$R_2C{=}N{-}NHC_6H_3(NO_2)_2$
Semicarbazide	$H_2N{-}NHCONH_2$	Semicarbazone	$R_2C{=}N{-}NHCONH_2$
Primary amine	$H_2N{-}R$	Imine or Schiff base	$R_2C{=}N{-}R$
Sodium bisulfite[b]	$NaHSO_3$	Sodium bisulfite addition compound	$R{-}CH(OH){-}SO_3Na$

[a] See Sect. 6.70.
[b] Reacts only with aldehydes, cyclic ketones, and methyl ketones to give the addition product without loss of water.

formic acid (common name) or less commonly, methanoic acid (its IUPAC name). The second member of the series (Fig. 6.39*a*) is CH_3CO_2H, acetic acid (common name) or ethanoic acid (IUPAC). The three-carbon acid is $CH_3CH_2CO_2H$, propionic acid (common name) or propanoic acid (IUPAC); as with methanoic and ethanoic acid, the IUPAC name is derived by replacing the *-e*, in the name of the parent hydrocarbon possessing the same number of carbon atoms, with the suffix *-oic*, denoting the acid function. When a carboxyl group is attached to a ring, the compound is named by adding "carboxylic acid" to the name of the ring system to which the carboxyl group is attached (substitutive nomenclature, Sect. 6.12). Thus, the compounds shown in Figs. 6.39*b* and *c* are, respectively, cyclohexanecarboxylic acid and 1-phenanthrenecarboxylic acid.

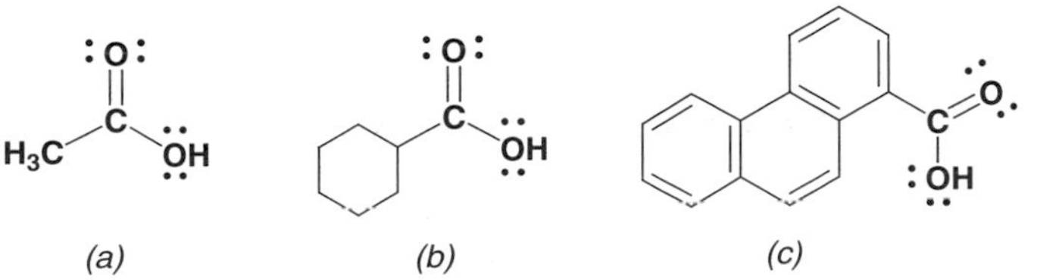

Figure 6.39. (*a*) Acetic acid; (*b*) cyclohexanecarboxylic acid; (*c*) 1-phenanthrenecarboxylic acid.

6.40 CARBOXYLATE ANION, RCO_2^-

The anion formed by the loss of a proton from the hydroxyl group of a carboxylic acid.

Example. The acetate anion. The ion can be represented by either of the equivalent resonance structures (Fig. 6.40*a*), or by the structure in which a broken line represents delocalization of the negative charge over the three-atom system (Fig. 6.40*b*). The carbon atom to which the carboxylate anion is attached as well as the three atoms of the carboxylate anion all lie in the same plane. In this anion there are four $p\pi$ electrons; the two electrons associated with the carbonyl group (one from carbon, one from oxygen) and a pair of $p\pi$ electrons on the second oxygen atom. There are three atoms involved in the O–C–O π system, and thus, there are three π-type orbitals. One of these orbitals is very strongly bonding, one approximately nonbonding or slightly antibonding, and the third, strongly antibonding. The four electrons occupy the two lower-energy π orbitals. Note that the CO_2^- group is isoelectronic with the NO_2 group, and hence, the molecular orbital descriptions are essentially identical (Fig. 6.68).

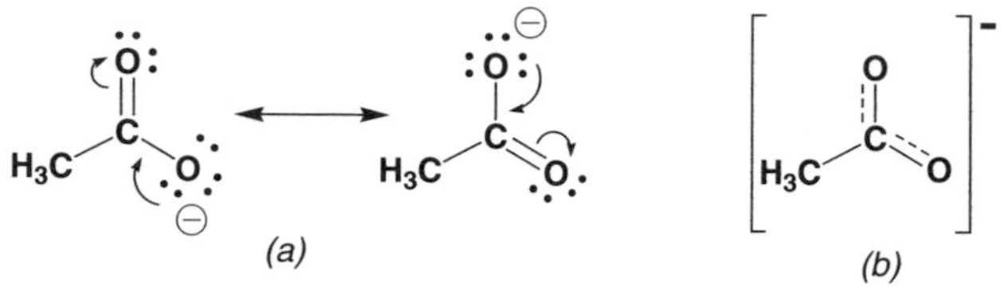

Figure 6.40. (*a*) Resonance structures of the acetate anion and (*b*) structure showing delocalization of the $p\pi$ electrons of the carboxyl anion.

6.41 CARBOXYLIC ACID DERIVATIVES, RC(O)Z

Compounds of the structure RC(O)Z, where Z represents a group bonded to the carbonyl via a hetero atom such as –X, –OR, $-NR_2$, –SR, and so on. The resulting classes of compounds are frequently considered to be derived from parent carboxylic acids and are named accordingly. The RC(O)- unit is referred to as an *acyl group* if R is derived from an aliphatic hydrocarbon, or an *aroyl group* if R is derived from an aromatic hydrocarbon.

6.42 ACYL AND AROYL HALIDES, RC(O)X

Compounds in which the hydroxyl group of a carboxylic acid has been replaced by halogen (carboxylic acid derivatives {RC(O)Z} with Z = halogen). When R is aliphatic, these compounds are called acyl halides, and when R- is aromatic, they are called aroyl halides.

Example. Acetyl fluoride and benzoyl bromide (Figs. 6.42*a* and *b*, respectively). These compounds are named as though they were salts of HF and HBr; the organic portion of the name is derived from the name of the corresponding acids, acetic and benzoic acids, by replacing the suffix *-ic* in the acid name with *-yl*.

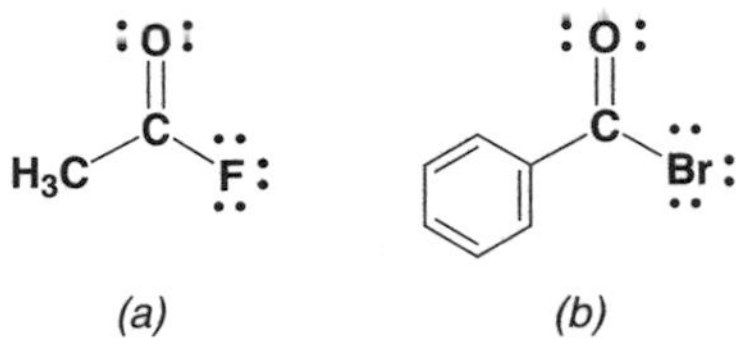

Figure 6.42. (*a*) Acetyl fluoride (ethanoyl fluoride) and (*b*) benzoyl bromide.

6.43 ESTERS, RC(O)OR, AND ORTHO ESTERS, $RC(OR)_3$

Simple esters are compounds in which the hydroxyl group of a carboxylic acid is replaced by an -OR group (carboxylic acid derivatives {RC(O)Z} with Z = OR). Esters incorporated into rings are referred to as *lactones* (see Sect. 6.89). Ortho esters are derivatives of carboxylic acids in which three OR groups are attached to the carboxy carbon atom.

Example. Methyl 2-methylbutanoate (Fig. 6.43*a*); isopropyl benzoate (Fig. 6.43*b*); and the ortho ester, trimethyl orthoformate (Fig. 6.43*c*).

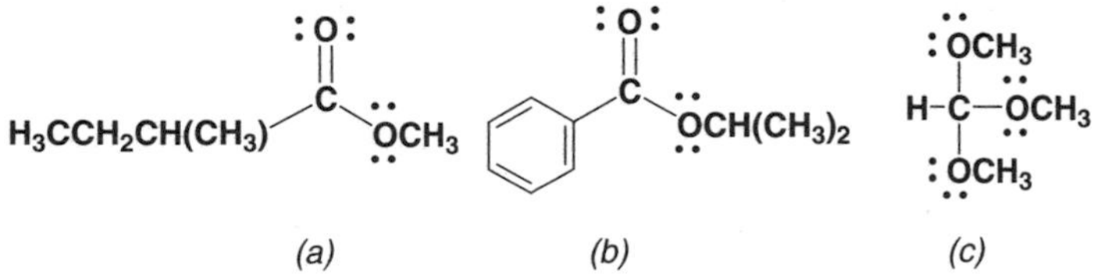

Figure 6.43. (*a*) Methyl 2-methylbutanoate; (*b*) isopropyl benzoate; and (*c*) trimethyl orthoformate.

6.44 CARBOXYLIC ACID ANHYDRIDES, $[RC(O)]_2O$

Compounds in which the hydroxyl group of an carboxylic acid has been replaced by a –OC(O)R group. As the name implies, these compounds are derived from two molecules of a carboxylic acid by the intermolecular loss of a molecule of water (Fig. 6.44*a*).

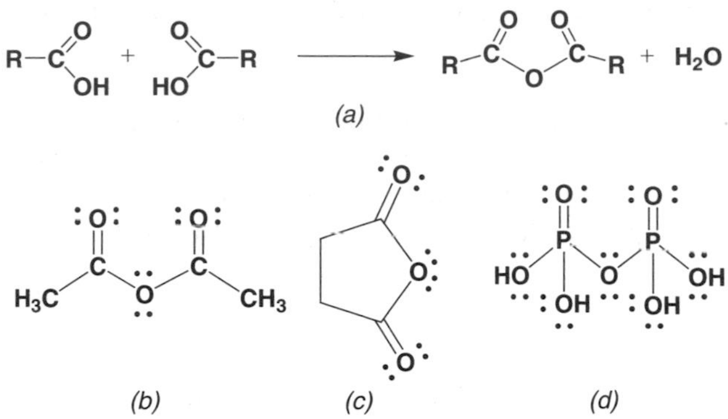

Figure 6.44. (*a*) Formation of carboxylic acid anhydride by condensation with the loss of a molecule of water; (*b*) acetic anhydride; (*c*) succinic anhydride; (*d*) the inorganic acid anhydyride, phosphoric acid anhydride (diphosphoric acid).

Example. Acetic anhydride (Fig 6.44*b*); succinic anhydride (Fig. 6.44*c*) is formed by intramolecular loss of water from the dibasic acid, succinic acid. Acid anhydrides may also be prepared from two different carboxylic acids to form mixed anhydrides, Acids other than carboxylic acids can also form anhydrides. Anhydrides of phosphoric acid are very important components of many biological molecules (Fig. 6.44*d*).

6.45 AMIDES, $RC(O)NH_2$

Compounds in which the hydroxyl group of a carboxylic acid has been replaced by $-NH_2$ (carboxylic acid derivatives $\{RC(O)Z\}$ with $Z = NH_2$).

Example. 2-Methylbutanamide (Fig. 6.45*a*), or since the common name of the C-4 acid is butyric acid, an alternate name is α-methylbutyramide. Amides involving

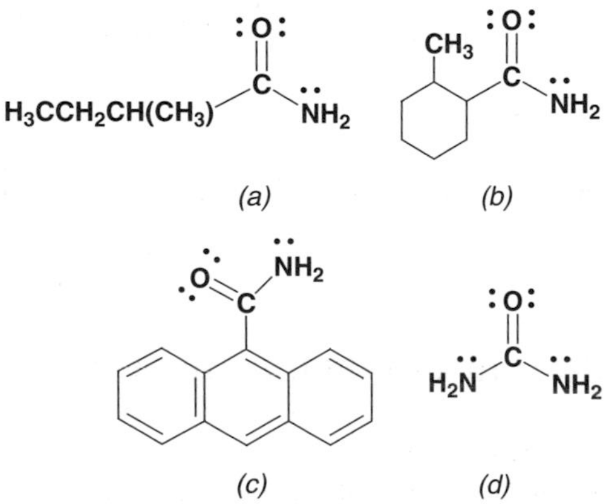

Figure 6.45. (*a*) 2-Methylbutanamide (α-methylbutyramide); (*b*) 2-methylcyclohexanecarboxamide; (*c*) 9-anthracenecarboxamide; (*d*) urea.

attachment of the $-CONH_2$ to a ring are named substitutively; the compounds shown in Figs. 6.45*b* and *c* are 2-methylcyclohexanecarboxamide and 9-anthracenecarboxamide, respectively. The compound in which the R group in $RC(O)NH_2$ is replaced by an amino group leads to urea (Fig. 6.45*d*), the parent of the substituted ureas, RNHC(O)NHR and $R_2NC(O)NR_2$

6.46 SUBSTITUTED AMIDES, RC(O)NHR AND $RC(O)NR_2$

Compounds in which one (or both) of the nitrogen-bonded hydrogen atoms of the amide function is replaced by an R or Ar group. Substituted amides incorporated into rings are referred to as *lactams* (see Sect. 6.90).

Example. *N*-Methylpentanamide (Fig. 6.46*a*); *N,N*-di-*n*-propylcyclopentane-carboxamide (Fig. 6.46*b*). The *hydroxamic acids* (Fig. 6.46*c*) may be considered as substituted amides in which a hydrogen on the nitrogen atom has been replaced by a hydroxy group; they are important in peptide chemistry.

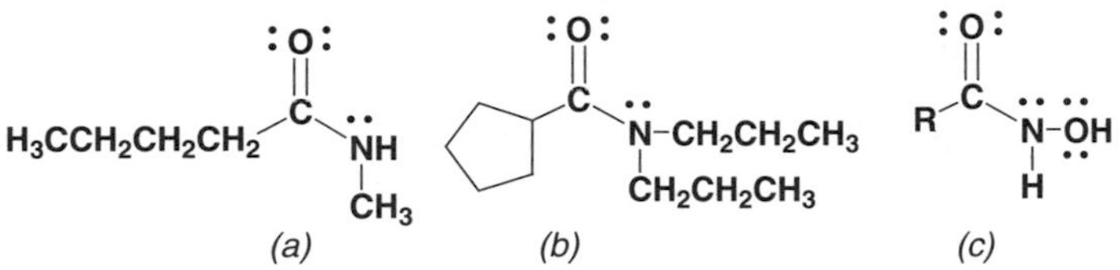

Figure 6.46. (*a*) *N*-Methylpentanamide; (*b*) *N,N*-di-*n*-propylcyclopentane-carboxamide; (*c*) an hydroxamic acid.

6.47 SECONDARY AND TERTIARY AMIDES, $[RC(O)]_2NH$ AND $[RC(O)]_3N$

Amides in which one hydrogen (secondary) or both hydrogens (tertiary) bonded to the nitrogen of an amide are replaced by acyl or aroyl groups, RC(O)–.

Example. Diacetamide and triacetamide (Figs. 6.47*a* and *b*, respectively), not to be confused with alkyl- or aryl-substituted amides (Sect. 6.46).

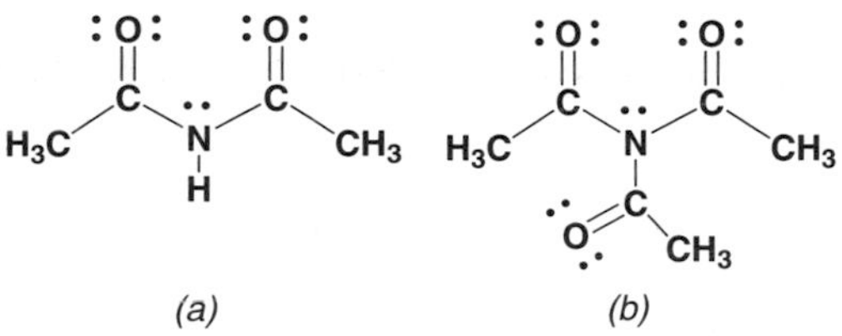

Figure 6.47. (*a*) Diacetamide and (*b*) triacetamide.

6.48 SUBSTITUTED ACIDIMIDES, $-\overset{X}{\overset{|}{C}}{=}N-$

Compounds in which the carbonyl oxygen atom of a carboxylic acid derivative is replaced by an NR group.

Example. *Acidimide chlorides*, a class of compounds resulting from the reaction of nitriles with hydrochloric acid (Fig. 6.48*a*); *acidimido esters* or *acidimido ethers*, a class of compounds obtained from the alcoholysis of acidimido chlorides (Fig. 6.48*b*); and *amidines*, a class of compounds formed by the aminolysis of orthocarboxylic esters (Sect. 6.43) or acidimide chlorides (Fig. 6.48*c*). These compounds are important starting materials in the synthesis of many heterocyclic systems.

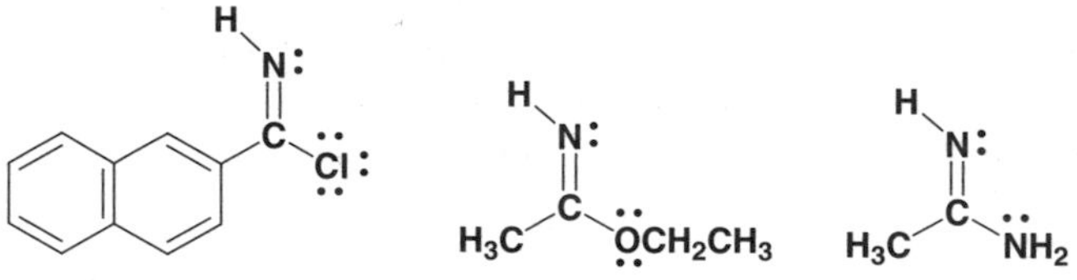

(*a*) (*b*) (*c*)

Figure 6.48. (*a*) Naphthalene-2-carboxamide chloride; (*b*) ethyl acetimidoacetate; (*c*) acetamidine.

6.49 IMIDES, $O{=}\overset{\ulcorner\;(CH_2)_x\;\urcorner}{C{-}NH{-}C}{=}O$

Cyclic secondary amides.

Example. Phthalimide (Fig. 6.49*a*); glutarimide (Fig. 6.49*b*); and the linkage of these two imides to give thalidomide (Fig. 6.49*c*), the sedative that when taken by pregnant women caused a disastrous rash of severe birth defects (*teratogenic activity*) in the early 1960s. Imide drugs such as thalidomide resemble the pyrimidine bases, thymine and uracil, of nucleic acids. From a strict nomenclature point of view, compounds having the acyclic structure RCONHCOR are secondary amides and not imides.

6.50 α-DIKETONES, RC(O)C(O)R, AND 1,2,3-TRIKETONES, RC(O)C(O)C(O)R

Di- and triketones in which the carbonyl groups are directly bonded to each other.

Example. The diketone benzil (Fig. 6.50*a*) is not capable of enolization, since it possesses no α-hydrogen atoms. However, when α-hydrogens are available, the

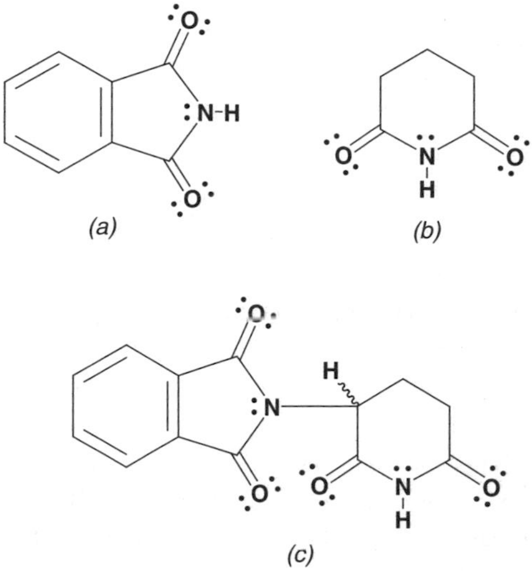

Figure 6.49. (*a*) Phthalimide; (*b*) glutarimide; (*c*) *N*-(2,6-dioxo-3-piperidyl)phthalimide (thalidomide).

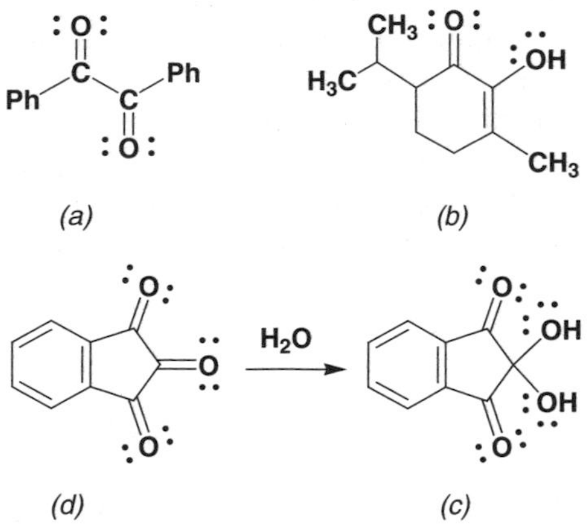

Figure 6.50. (*a*) Benzil, an α-diketone; (*b*) enolized α-diketone, diosphenol; (*c*) hydrated triketone, ninhydrin; (*d*) 1,2,3-triketone.

enolic form often predominates as in the case of diosphenol, 2-hydroxy-3-methyl-6-(1-methylethyl)-2-cyclohexene-1-one (Fig. 6.50*b*). The stability of the enolic form in such molecules is due to conjugation between the enol and carbonyl double bonds. Ninhydrin (Fig. 6.50*c*) is a *hydrated carbonyl* derived from the addition of water to the extremely electron-deficient central carbonyl group of the 1,2,3-triketone shown in Fig. 6.50*d*. This is typical behavior for carbonyl groups bonded to strongly electron-withdrawing groups, for example, chloral-hydrate, $Cl_3CCHC(OH)_2$, the so-called knockout drops. Ninhydrin is the classical reagent for the detection of α-amino acids. Its reaction with these compounds produces a deep blue color.

6.51 β-DICARBONYL COMPOUNDS, $RC(O)CR_2C(O)R$

Dicarbonyl compounds in which the two carbonyl groups are separated by a single sp^3 hybridized carbon atom. The two carbonyl groups may be aldehydes, ketones, carboxylic acids, their derivatives, or any combination of these carbonyl classes.

Example. β-Dicarbonyl compounds are starting materials for the synthesis of many organic compounds, because they undergo extremely facile condensation reactions. These reactions are greatly facilitated by the relative high acidity of the α-protons on the carbon atom flanked by two carbonyl groups. In addition, β-dicarbonyl systems involving aldehydes and ketones with at least one α-hydrogen exist in the enolic form to a significant extent (Figs. 6.51*a* and *b*). In acyclic β-dicarbonyls, the enolic form is stabilized both by the conjugation between the enol and carbonyl double bonds and the internal hydrogen bonding between the enol hydrogen and the carbonyl oxygen atom (Fig. 6.51*b*). Hydrogen bonding of this type is not possible in cyclic β-dicarbonyls (Fig. 6.51*c*); nevertheless, the enolic form is present in these molecules to a significant extent as well, due to the stabilizing influence of the conjugated double bonds. Finally, there is a distinctly greater propensity for aldehyde and ketone carbonyls to undergo enolization in β-dicarbonyl molecules than there is for carboxylic acids and their derivatives. The influence of these various factors on the enol content of β-dicarbonyl compounds can be seen from the data listed in Table 6.51.

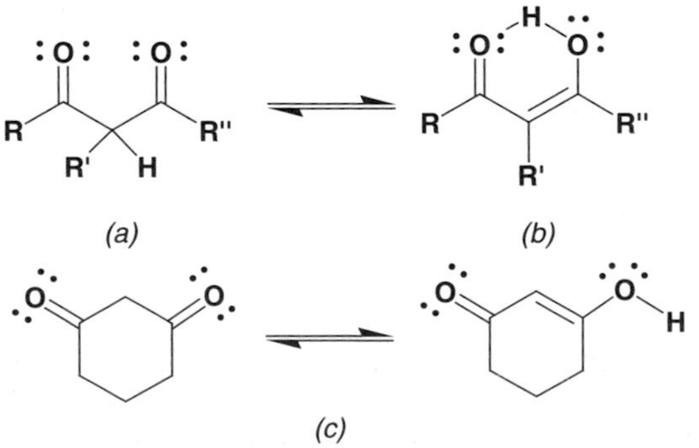

Figure 6.51. Enolizable acyclic β-dicarbonyl molecules: (*a*) the keto form and (*b*) the enol form; (*c*) enolization of the cyclic β-diketone, 1,3-cyclohexadione.

TABLE 6.51. Enol Content of β-Dicarbonyl Molecules

β-Dicarbonyl Compound	% Enol
$EtO_2CCH_2CO_2Et$	7.7×10^{-3}
$CH_3C(O)CH_2CO_2Et$	8.4[a]
$CH_3C(O)CH_2C(O)CH_3$	80
$PhC(O)CH_2C(O)CH_3$	84.2[b]

[a] Ketone carbonyl group enolized.
[b] Aryl ketone carbonyl group enolized.

6.52 KETO ACIDS, $RC(O)(CH_2)_nC(O)OH$

These are compounds containing both ketone and carboxylic acid functional groups; the position of the ketone (keto) group relative to the carboxylic acid group is usually indicated by the Greek symbol α, β, γ, and so on. An α indicates that the ketone is immediately adjacent to the carboxyl group ($n = 0$), a β that it is once removed ($n = 1$) from the carboxyl group, and a γ twice ($n = 2$) removed.

Example. γ-Ketovaleric acid (common name) or 4-oxopentanoic acid (its IUPAC name) (Fig. 6.52).

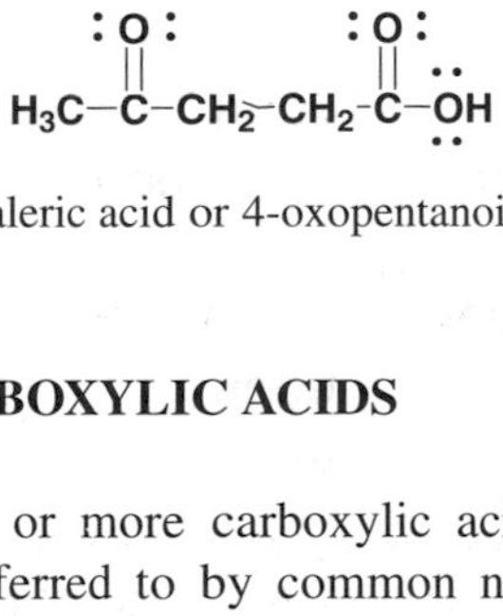

Figure 6.52. γ-Ketovaleric acid or 4-oxopentanoic acid (levulinic acid).

6.53 DI- AND POLYCARBOXYLIC ACIDS

Compounds containing two or more carboxylic acid groups. A number of dicarboxylic acids are still referred to by common names, although this practice is discouraged by IUPAC rules. Some common dicarboxylic acids are listed in Table 6.53.

Example. 1,2-Benzenedicarboxylic acid (Fig. 6.53*a*), also known as phthalic acid; 1,6-hexanedioic acid (Fig. 6.53*b*), also known as adipic acid; and 1,2,3,4,5,6-benzenehexacarboxylic acid (Fig. 6.53*c*), also known as mellitic acid.

6.54 CONJUNCTIVE NAMES

A type of approved nomenclature sometimes employed when the principal functional group occurs in a side chain attached to a ring. In such nomenclature, the

TABLE 6.53. Common Names for Some Dicarboxylic Acids

Structure	Name
HOOC–$(CH_2)_n$–COOH	Oxalic acid ($n = 0$) Malonic acid ($n = 1$) Succinic acid ($n = 2$) Glutaric acid ($n = 3$) Adipic acid ($n = 4$) Pimilic acid ($n = 5$) Suberic acid ($n = 6$)
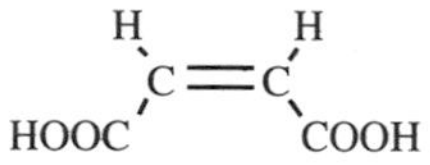	Maleic acid
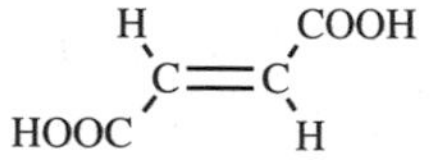	Fumaric acid

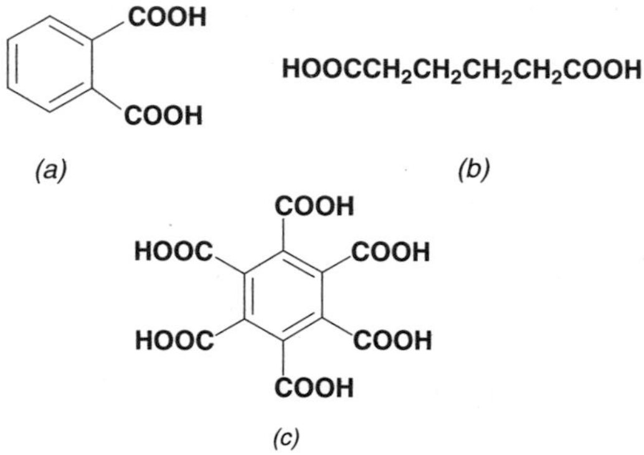

Figure 6.53. (*a*) 1,2-Benzenedicarboxylic acid (phthalic acid); (*b*) 1,6-hexanedioic acid (adipic acid); (*c*) 1,2,3,4,5,6-benzenehexacarboxylic acid (mellitic acid).

acyclic component is considered the parent to which the full name of the cyclic component is attached. According to IUPAC, in conjuctive nomenclature, the name of the molecule is formed by juxtaposition of the names of the two component molecules linked to each other by carbon–carbon bonds. This nomenclature is particularly useful when a side chain is attached to a cyclic compound, in which case, the cyclic molecule is named first.

Example. 2-Naphthaleneacetic acid (Fig. 6.54*a*) and β-chloro-α-methyl-3β-pyridineethanol (Fig. 6.54*b*).

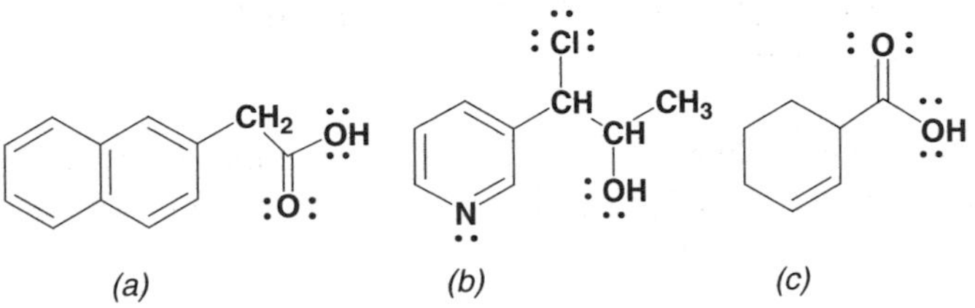

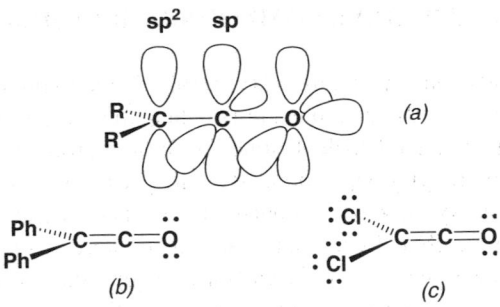

Figure 6.54. (*a*) 2-Naphthaleneacetic acid; (*b*) β-chloro-α-methyl-3β-pyridineethanol; (*c*) 3-cyclohexenecarboxylic acid.

6.55 KETENES, $R_2C{=}C{=}O$

A class of carbonyl compounds in which the carbonyl double bond and a carbon–carbon double bond form a cumulative double-bond system (Fig. 6.55*a*). The central carbon atom of the ketene group is *sp* hybridized and the terminal carbon atom has the usual sp^2 hybridization.

Example. Ketenes are perhaps the most reactive class of carbonyl functional groups. Many of them might be best categorized as reactive intermediates. However, a number of ketenes are sufficiently stable to be isolated and stored for prolonged periods. These include diphenylketene (Fig. 6.55*b*) and dichloroketene (Fig. 6.55*c*).

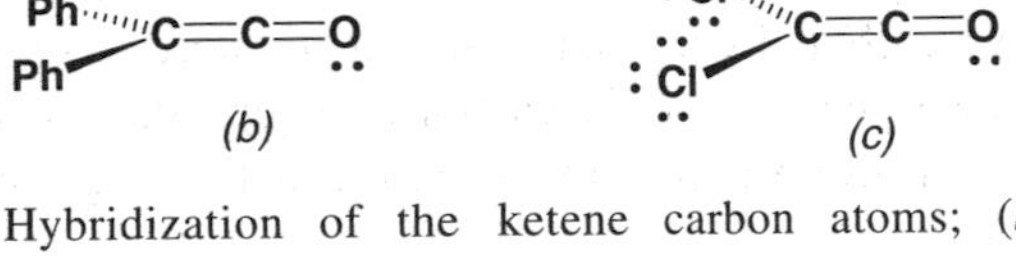

Figure 6.55. (*a*) Hybridization of the ketene carbon atoms; (*b*) diphenylketene; (*c*) dichloroketene.

6.56 UNSATURATED CARBONYL COMPOUNDS, $RCH{=}CH(CH_2)_nC(O)Z$

Carbonyl compounds that contain olefinic linkages (Z = H, alkyl, aryl, OH, OR, Hal, NR_2; $n = 0, 1, 2, \ldots$). Carbonyl compounds containing acetylenic linkages and aryl groups are included in this category as well.

6.57 α,β-UNSATURATED CARBONYL COMPOUNDS, RCH=CHC(O)Z, RC≡CC(O)Z

Carbonyl compounds having a conjugated unsaturated linkage between the carbon atoms α and β to the carbonyl group. The unsaturated unit might be a double bond, a triple bond, or an aryl group.

Example. Acrylic acid (common name) (Fig 6.57*a*) or propenoic acid (its IUPAC designation), *trans*-cinnamaldehyde (Fig 6.57*b*), propynoic acid (its IUPAC name) (Fig. 6.57*c*), or propiolic acid (common name) and *p*-benzoquinone (Fig. 6.57*d*).

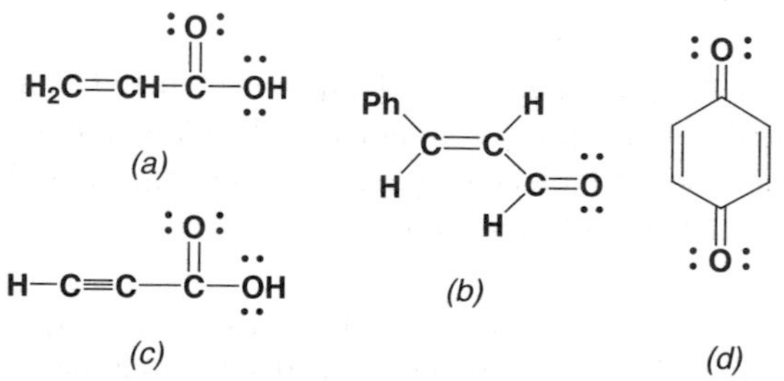

Figure 6.57. (*a*) Propenoic acid (acrylic acid); (*b*) *trans*-cinnamaldehyde; (*c*) 3-propynoic acid (propiolic acid); (*d*) *p*-benzoquinone.

6.58 HYDROXY CARBONYL COMPOUNDS, $R_2C(OH)(CH_2)_nC(O)R$

Carbonyl compounds containing a hydroxyl group. If the hydroxyl group is bonded to an α-carbon atom ($n = 0$), the molecule is called an α-hydroxy carbonyl, and the carbonyl group may be an aldehyde, ketone (α-ketol), carboxylic acid, or any of its derivatives. If the hydroxyl group is bonded to the β-carbon atom ($n = 1$), the molecule is called a β-hydroxy carbonyl. Of course, many other variations exist where $n > 1$. In addition, hydroxy carbonyl compounds are encountered in α- and β-dicarbonyl systems where enolization of one of the carbonyl groups leads to a hydroxy carbonyl in which the hydroxyl group is part of an enol functional group (see Sects. 6.35, 6.36, 6.50, and 6.51).

Example. The α-hydroxyketone (α-ketol) benzoin (Fig. 6.58*a*, see Sect. 6.36) is a member of a class of molecules sometimes called *acyloins*. β-Hydroxyaldehydes and ketones are commonly encountered molecules that are formed in *aldol* condensations (Fig. 6.58), such as that of acetone to form 4-hydroxy-4-methyl-2-pentanone (diacetone alcohol) (Fig. 6.58*b*). β-Hydroxy carbonyls are synthetically important substances since they undergo a very facile loss of water to form α,β-unsaturated carbonyl compounds; for example, the loss of water from diacetone alcohol leads to the formation of mesityl oxide (4-methyl-3-penten-2-one) (Fig. 6.58*c*). Finally,

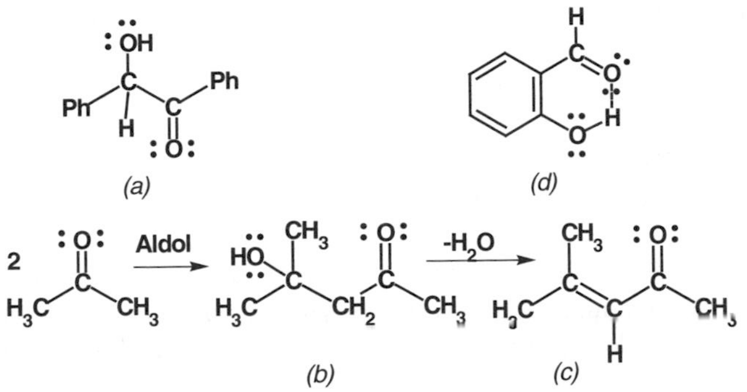

Figure 6.58. (*a*) The α-hydroxyketone (α-ketol or acyloin) benzoin; (*b*) the aldol reaction leading to β-hydroxyketone diacetone alcohol or 4-hydroxy-4-methyl-2-pentanone; (*c*) the α,β-unsaturated ketone called mesityl oxide (common name) or 4-methyl-3-penten-2-one (IUPAC name); (*d*) internal hydrogen bonding in salicylaldehyde.

salicylaldehyde (Fig. 6.58*d*) is a hydroxyaldehyde in which the hydroxyl group is part of a phenol or enol moiety (Sects. 6.10 and 6.34).

6.59 AMINO CARBONYL COMPOUNDS, $R_2C(NR_2)(CH_2)_nC(O)R$

Carbonyl compounds containing an amine group. If the amine group is bonded to an α-carbon atom ($n = 0$), the molecule is called an α-amino carbonyl and the carbonyl group may be part of an aldehyde, ketone, carboxylic acid, or any of its derivatives. If the amine group is bonded to the β-carbon atom ($n = 1$), they are called β-amino carbonyls. Of course, many other variations exist where $n > 1$.

Example. α-Aminocarboxylic acids constitute an important functional unit, because they are the building blocks of proteins. β-Aminoaldehydes and ketones are formed in the nitrogen equivalent of the aldol condensation, the *Mannich reaction* (Fig. 6.59*a*). For this reason, β-amino carbonyls are commonly referred to as *Mannich bases* (Fig. 6.59*b*), the salts of which undergo extremely facile conversion to α,β-unsaturated carbonyls as shown for the formation of methyl vinyl ketone (MVK, Fig. 6.59*c*).

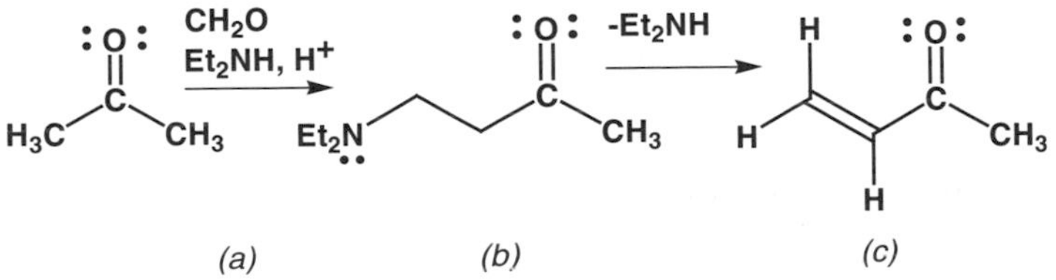

Figure 6.59. (*a*) Mannich reaction; (*b*) β-aminoketone or Mannich base; (*c*) α,β-unsaturated ketone, methyl vinyl ketone (MVK).

6.60 AMINES, RNH_2, R_2NH, AND R_3N

Compounds in which one or more of the hydrogen atoms of ammonia is replaced by an R group. Replacement of one hydrogen in NH_3 by an R group gives a *primary amine*, RNH_2; replacement of two hydrogens gives a secondary amine, R_2NH, and replacement of all three hydrogens gives a *tertiary amine*, R_3N.

Example. n-Propylamine (Fig. 6.60*a*), a primary amine; diisopropylamine (Fig. 6.60*b*), a secondary amine; and ethyldimethylamine (Fig. 6.60*c*), a tertiary amine (the name amine is derived from ammonia). Replacement of a hydrogen atom in a primary amine by a carboxy group gives a *carbamic acid*, RNHC(O)OH. Such acids decarboxylate spontaneously to the corresponding amine, but the ester derivatives, RNHC(O)OR, called *carbamates* or *urethanes*, are stable. Accordingly, the –C(O)OR group is often used as a protecting group for the amine group of amino acids in peptide synthesis (Fig. 6.60*d*).

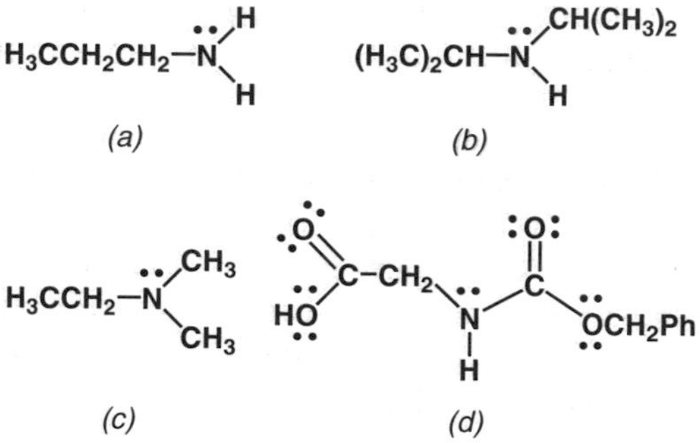

Figure 6.60. (*a*) *n*-Propylamine; (*b*) diisopropylamine; (*c*) ethyldimethylamine; (*d*) *N*-benzyloxycarbonyl glycine, which on hydrolysis gives glycine.

6.61 IMINES, RCH=NH, R_2C=NH, AND R_2C=NR

Compounds in which two of the hydrogens of ammonia are replaced by a double bond to a carbon atom (see also Table 6.38). The remaining hydrogen atom on nitrogen can also be replaced by an R group to give *N*-substituted imines or *Schiff bases*, R_2C=NR. Such *N*-substituted imines are in tautomeric equilibrium with their corresponding

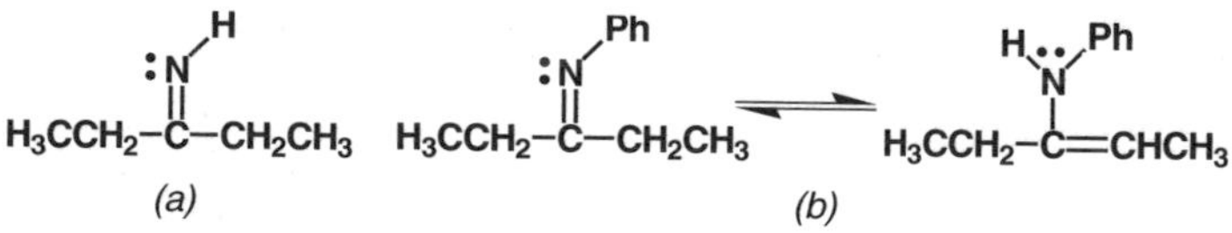

Figure 6.61. (*a*) 3-Pentanimine and (*b*) the tautomeric equilibrium involving *N*-phenyl-3-pentanimine.

enamines. This tautomerization is analogous to the keto-enol tautomerization (see Sect. 6.35).

Example. 3-Pentanimine (Fig. 6.61*a*); the tautomeric equilibrium between an *N*-substituted imine and its enamine (Fig. 6.61*b*).

6.62 HYDROXYLAMINES, R_2NOH

Compounds in which one of the hydrogen atoms on the nitrogen of an amine is replaced by a hydroxyl group. This class of compounds derives its generic name from the parent compound hydroxylamine, NH_2OH.

Example. (*a*) *N*-Phenylhydroxylamine (Fig. 6.62*a*); (*b*) *O*-methylhydroxylamine (Fig. 6.62*b*) is the hydroxylamine analog of an ether.

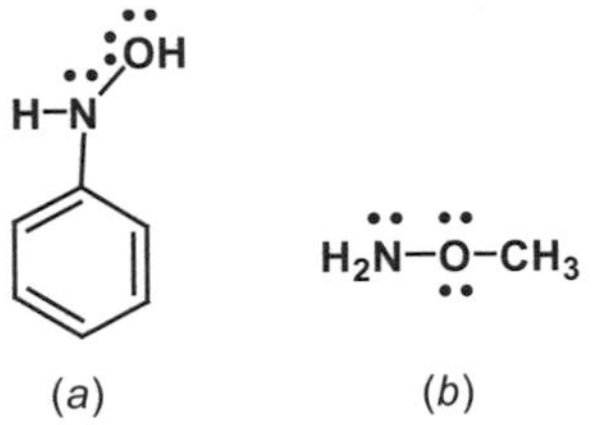

Figure 6.62. (*a*) *N*-Phenylhydroxylamine and (*b*) *O*-methylhydroxylamine.

6.63 AMINE OXIDES, $R_3N^+–O^-$ ($R_3N \rightarrow O$)

Compounds in which an oxygen atom is bonded to the nitrogen of a tertiary amine.

Example. Trimethylamine oxide (Fig. 6.63*a*); 4-methylpyridine *N*-oxide, (Fig. 6.63*b*). The notation $N \rightarrow O$ implies a coordinate covalent bond involving the donation of the lone pair on nitrogen into an empty orbital on the oxygen atom. The alternative notation, $N^+–O^-$, requires a formal plus charge on nitrogen and a formal negative charge on

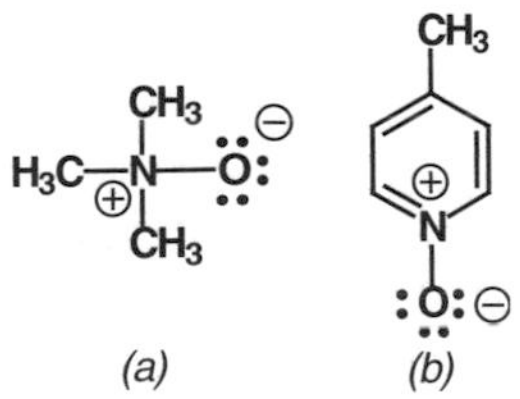

Figure 6.63. (*a*) Trimethylamine oxide and (*b*) 4-methylpyridine *N*-oxide.

oxygen. One should choose between the two notations: either an arrow with no charges, or charges without an arrow, but not charges with an arrow.

6.64 NITRONES, $\mathrm{RCH{=}\overset{\overset{O^-}{|}}{N}{}^{+}{-}R}$ AND $\mathrm{R_2C{=}\overset{\overset{O^-}{|}}{N}{}^{+}{-}R}$

These are compounds in which an oxygen atom is bonded to the nitrogen atom of an imine (RCH=NR or $R_2C{=}NR$); see Table 6.38.

Example. Ethylidenemethylazane oxide, the preferred name (Fig. 6.64), but commonly called *N*-methyl-α-methylnitrone. Because of the double bond, E–Z (*cis-trans*) isomers of nitrones are possible.

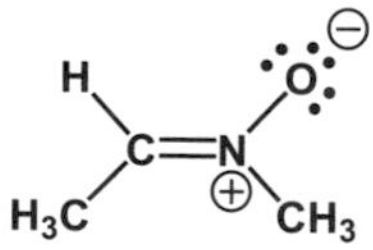

Figure 6.64. (*E*) Ethylidenemethylazane oxide (*N*-methyl-α-methylnitrone).

6.65 AMMONIUM SALTS, $R_4N^+X^-$

Ammonium salts are amine derivatives in which the nitrogen atom has a formal charge of +1 and has four bonds to other atoms or groups, $R_4N^+X^-$, R = alkyl, aryl, *or* H. *Quarternary ammonium salts* are a class of organic ammonium salts usually formed by the reaction between R_3N, a tertiary amine, and RX, $R_4N^+X^-$ R = alkyl or aryl, *but not* H.

Example. Ammonium chloride (Fig. 6.65*a*) and trimethylammonium chloride (Fig. 6.65*b*) are ammonium salts in which the nitrogen is bonded to four other groups or atoms. Pyridinium hydrochloride (Fig. 6.65*c*), is an ammonium salt in which the nitrogen is bonded to only three groups or atoms, but in which there is

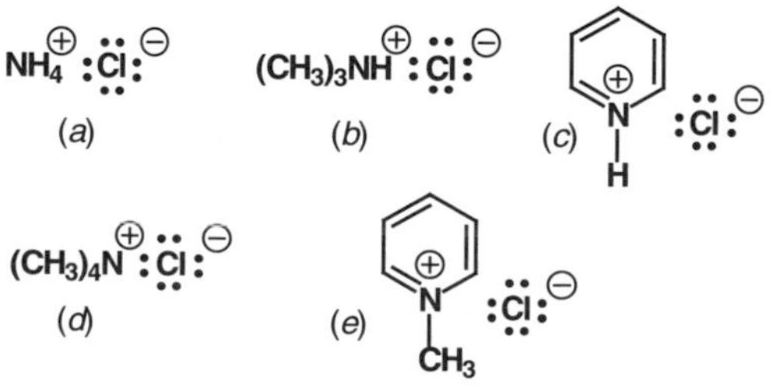

Figure 6.65. (*a*) Ammonium chloride; (*b*) trimethylammonium chloride; (*c*) pyridinium hydrochloride; (*d*) tetramethylammonium chloride; (*e*) *N*-methylpyridinium chloride.

a double bond to one of these groups. Tetramethylammonium chloride (Fig. 6.65*d*) and methylpyridinium chloride (Fig. 6.65*e*) are quaternary ammonium salts; they do not possess N–H bonds.

6.66 NITRILES, RCN

Compounds having a terminal nitrogen atom triply bonded to carbon.

Example. Acetonitrile, $CH_3C{\equiv}N$, a name derived from the name of acetic acid to which it is chemically related. Acrylonitrile, $CH_2{=}CHCN$, is named after acrylic acid, $CH_2{=}CHCO_2H$. The functional group name of the –CN group is nitrile. According to IUPAC rules, this name is added as a suffix to the alkane corresponding to the longest carbon chain, including the nitrile carbon *e.g.*, hexanenitrile, Fig 6.66*a*. Nitriles may also be named using the suffix carbonitrile as part of a conjunctive name by adding it to the name of the ring system. Cyclohexanenitrile (Fig. 6.66*b*) is analogous to the correcponding name cyclohexanecarboxylic acid. In some cases with multiple functions, where radicofunctional names are used, –CN is called the cyano group, for example, cyanoacetic acid (Fig. 6.66*c*).

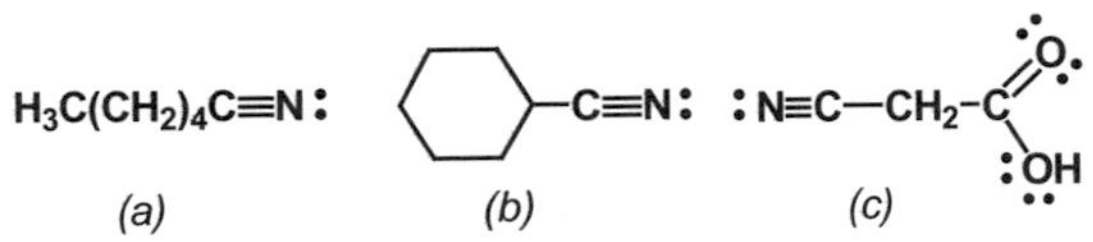

Figure 6.66. (*a*) Hexanenitrile (IUPAC); (*b*) cyclohexanecarbonitrile; (*c*) cyanoacetic acid (common name).

6.67 ISOCYANIDES, $R{-}\overset{+}{N}{\equiv}\overset{-}{C}{:}$

These are compounds in which a hydrogen atom of a hydrocarbon has been replaced by the –NC group; any particular RNC is isomeric with the corresponding nitrile, RCN. Hence, isocyanides are sometimes (inappropriately) called *isonitriles*.

Example. Phenylisocyanide (Fig 6.67). The substitutive name for phenyl isocyanide is carbylaminobenzene, and it is preferred by some nomenclaturists to the radicofunctional name. Isocyanides have two important resonance structures, shown in

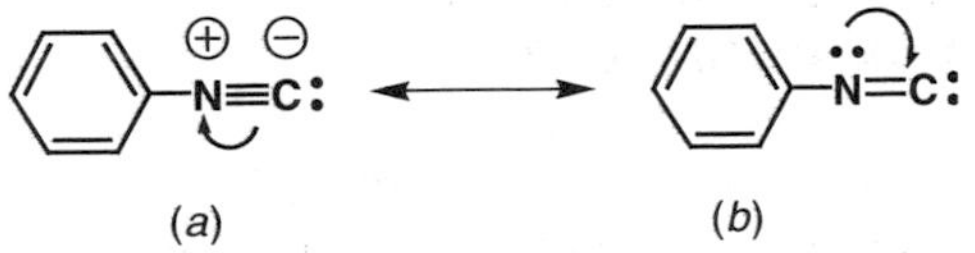

Figure 6.67. Phenyl isocyanide (carbylaminobenzene): (*a*) polar resonance structure and (*b*) nonpolar resonance structure.

Figs. 6.67*a* and *b*. These indicate that isocyanides may behave either as nucleophiles (Fig. 6.67*a*) or as carbenes (Fig. 6.67*b*). In fact, both types of behavior are known.

6.68 NITROALKANES, RNO_2

Compounds in which a hydrogen atom of an alkane is replaced by a nitro group $-NO_2$.

Example. Nitromethane, CH_3NO_2. Figures 6.68*a* and *b* show the two important resonance structures for this compound, and Fig. 6.68*c* is a composite structure illustrate electron delocalization. The structure of the nitro group can be used to highlight some useful concepts in MO theory. The σ bond skeleton of the nitro group involves the use of three sp^2-like orbitals on nitrogen, leaving a pure *p* orbital that can overlap with the *p*π orbitals on each of the oxygen atoms. Combining the three *p*π atomic orbitals gives three molecular orbitals: one strongly bonding MO (Fig. 6.68*d*), one nonbonding MO (Fig. 6.68*e*), and one antibonding MO (Fig. 6.68*f*). These $-NO_2$ orbitals are analogous to those describing the carboxylate anion, $-CO_2$, with which it is isoelectronic (Sect. 6.40). The four *p*π electrons occupy the bonding and nonbonding molecular orbitals.

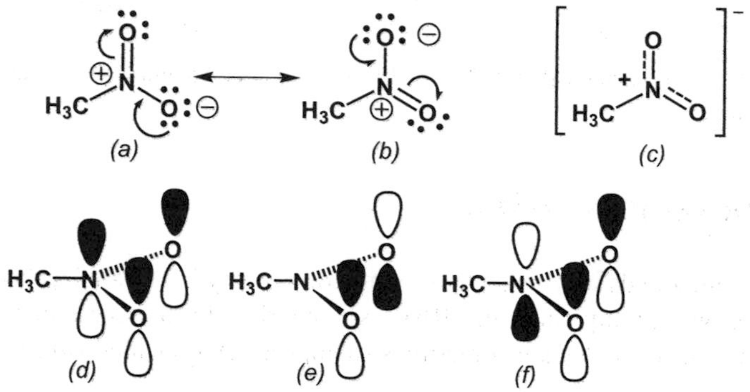

Figure 6.68. (*a*) and (*b*) Resonance structures of CH_3NO_2; (*c*) the composite structure; the three π molecular orbitals of the nitro group: (*d*) bonding, (*e*) nonbonding, and (*f*) antibonding.

6.69 ACI FORM OF THE NITRO GROUP (NITRONIC ACID OR AZINIC ACID)

The tautomer of aliphatic nitro compounds with hydrogen atoms on the carbon α to the nitro group. In this tautomer, one of the α-hydrogens has migrated to one of the nitro oxygen atoms with the formation of a double bond between the α-carbon and nitro nitrogen atom. Aci forms are sometimes called *nitronic acids* or *azinic acids*.

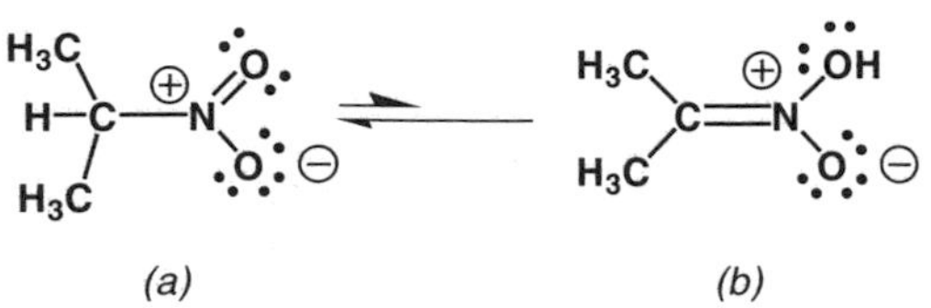

Figure 6.69. (*a*) Nitro form of an aliphatic nitro compound and (*b*) aci form.

Example. The equilibrium between the nitro form (Fig. 6.69*a*) and its aci form (Fig. 6.69*b*) is analogous to the equilibrium between a ketone and its enol (Sect. 6.35). In the case of nitro groups, the nitro form predominates over the aci form. At the other extreme is the equilibrium between aliphatic nitroso groups and their tautomeric oxime forms (Fig. 6.70*b*). In this case, the oxime is so stable that it is formed irreversibly.

6.70 NITROSO ALKANES, RN=O

Compounds in which a hydrogen atom of a hydrocarbon is replaced by the –NO group.

Example. Nitrosobenzene (Fig. 6.70*a*). The only stable nitroso compounds with α-carbons are those with R = aryl (ArN=O), and tertiary alkyl (R_3CN=O). If a hydrogen atom is present on the α-carbon atom, the tautomeric oxime (see Sect. 6.38) is formed irreversibly (Fig. 6.70*b*).

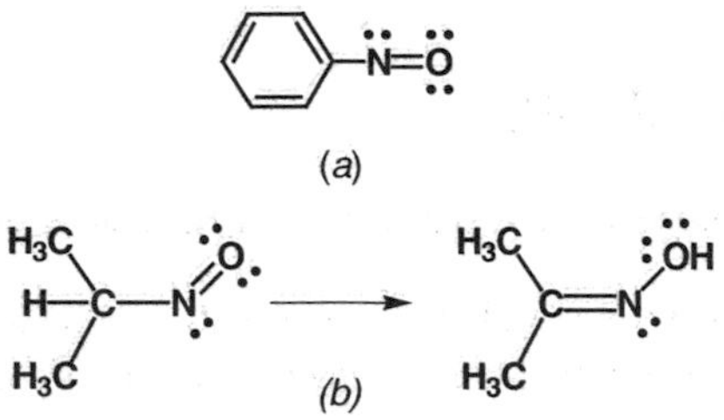

Figure 6.70. (*a*) Nitrosobenzene and (*b*) transformation of nitroso group to oxime.

6.71 NITRATE ESTERS, R–ONO_2

Esters formed between alcohols, R–OH, and nitric acid, HO–NO_2.

Example. The trinitrate of glycerol, nitroglycerin (Fig. 6.71) is an extremely shock-sensitive high explosive. In 1866, Alfred Nobel (1833–1896) discovered that if nitroglycerin is absorbed on kieselguhr, a diatomaceous earth, a dry powder results that is much less shock-sensitive, but can still be detonated with a small detonator cap. This

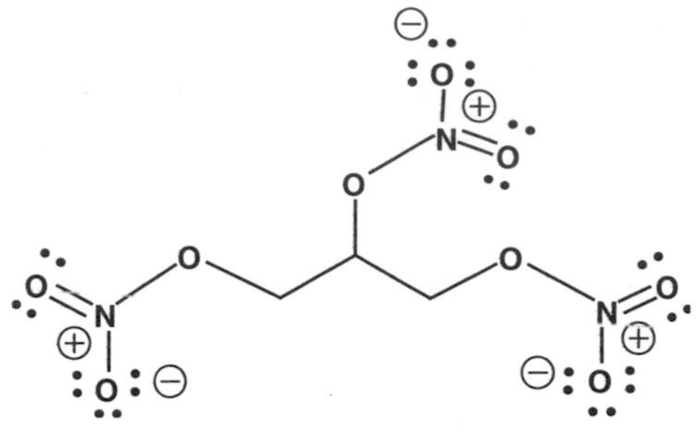

Figure 6.71. Nitroglycerin.

formulation is known as dynamite. The wealth accumulated by Nobel as a result of his discovery of dynamite was used in large part to establish the Nobel Prizes.

6.72 NITRITE ESTERS, R–ON=O

Esters formed between alcohols, R–OH, and nitrous acid, HO–NO.

Example. Ethyl nitrite (Fig. 6.72) is a commonly used reagent for the introduction of a nitroso group.

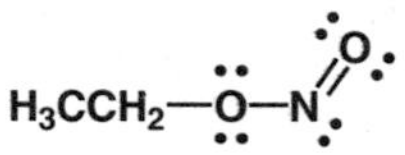

Figure 6.72. Ethyl nitrite.

6.73 NITRAMINES, $R_2N–NO_2$

Amides formed between secondary amines, $R_2N–H$, and nitric acid, $HO–NO_2$.

Example. Cyclonite (RDX) is a high explosive with a very high shattering power, or *brisänce,* that contains three nitramines incorporated into a single six-membered ring (Fig. 6.73).

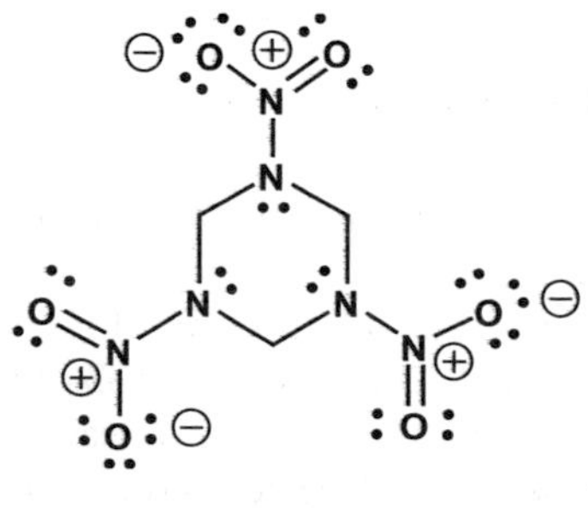

Figure 6.73. The nitramine cyclonite (RDX).

6.74 NITROSAMINES, $R_2N{-}N{=}O$, AND NITROSAMIDES, RRC(O)N–N=O

Nitrosamines are amides formed from the reaction between a secondary amine, $R_2N{-}H$, and nitrous acid, HO–NO, and nitrosamides are imides formed from the reaction between amides, RC(O)NHR, and nitrous acid, HO–NO.

Example. *N,N*-Dimethylnitrosamine (Fig. 6.74*a*) is a powerful cancer-causing (*carcinogenic*) agent. *N*-methyl-*N*-nitrosourea (Fig. 6.74*b*) is a common precursor for diazomethane (Sect. 6.75), and because the nitroso group is attached to an amide or amidelike nitrogen, it is a nitrosamide.

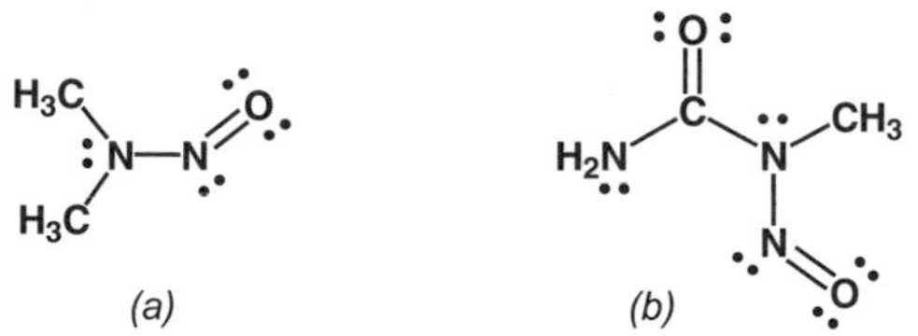

Figure 6.74. (*a*) *N,N*-Dimethylnitrosamine and (b) *N*-methyl-*N*-nitrosourea.

6.75 DIAZO ALKANES, $R_2C{=}N^+{=}N^-$

Alkanes in which two hydrogen atoms attached to the same carbon are replaced by two nitrogens, $>CN_2$.

Example. Diazomethane (Fig. 6.75*a*) and diphenyldiazomethane (Fig. 6.75*b*). The diazo group is isoelectronic with the azido group (see Fig. 6.77*b*). Diazomethane is used to insert the CH_2 group between an H–O bond in carboxylic acids (and phenols) to give methyl esters. The reaction probably involves a proton transfer from the acid to the diazomethane in an acid–base reaction, followed by an S_N2 displacement by the carboxylate anion on $H_3C{-}N^+{\equiv}N$ in which dinitrogen is one of the best leaving groups.

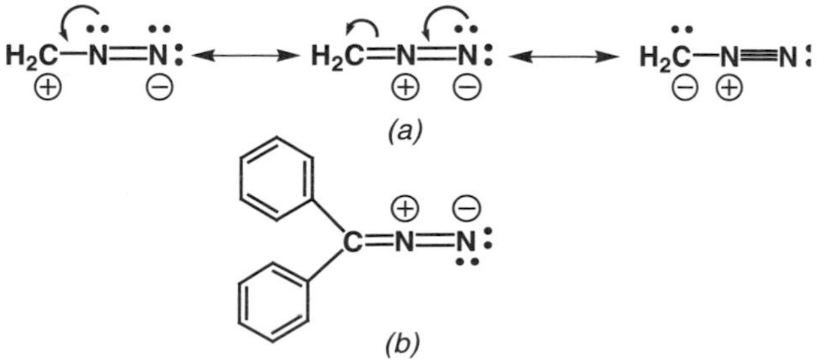

Figure 6.75. (*a*) The three major resonance structures of diazomethane and (*b*) diphenyldiazomethane.

6.76 α-DIAZO KETONES, $RC(O)CR'(N_2)$

Compounds in which two hydrogens on a carbon atom adjacent to a carbonyl group are replaced by two nitrogens, $>CN_2$; R′ may be either alkyl, aryl, or H.

Example. α-Diazoacetophenone (Fig 6.76). α-Diazoketones are isoelectronic with acyl azides (see Fig. 6.78).

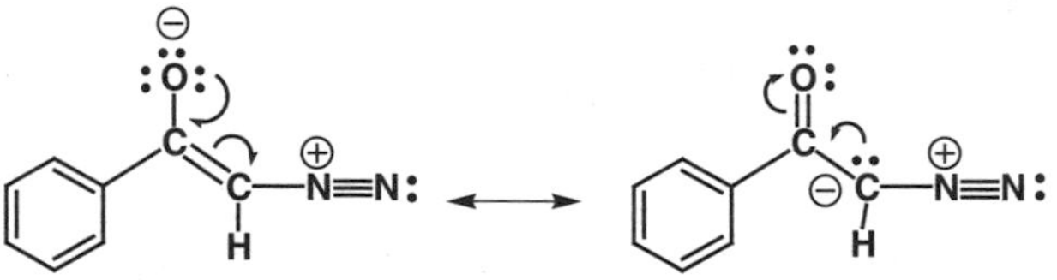

Figure 6.76. Resonance structures of α-diazoactophenone.

6.77 AZIDES, $RN{=}N^+{=}N^-$

Compounds in which a hydrogen atom of a hydrocarbon is replaced by the azido, $-N_3$, group, or alternatively, alkyl or aryl derivative of hydrazoic acid, HN_3.

Example. Phenyl azide (Fig. 6.77*a*). The azido group is isoelectronic with the diazo group (see Fig. 6.77*b*).

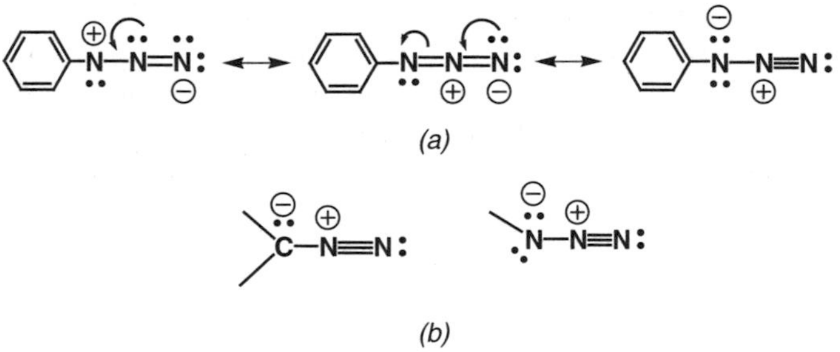

Figure 6.77. (*a*) Resonance structures of phenyl azide and (*b*) structures showing that the diazo group and azido group are isoelectronic.

6.78 ACID (OR ACYL) AZIDES, $RC(O)N{=}N^+{=}N^-$

These are compounds in which the hydroxyl group of a carboxylic acid is replaced by the azido group, $-N_3$; such compounds may be regarded as acyl or aroyl derivatives of hydrazoic acid, HN_3.

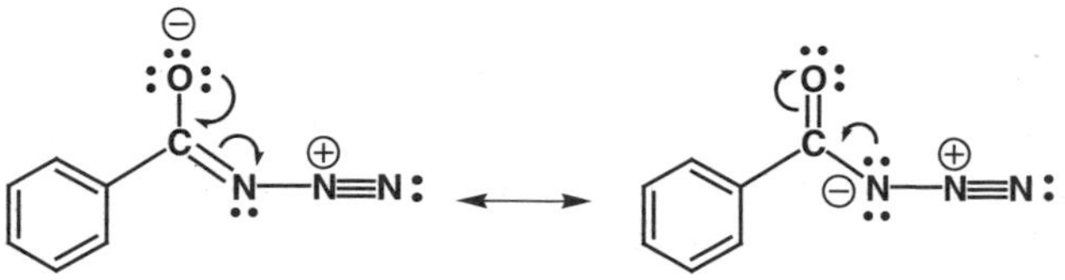

Figure 6.78. Resonance structures of benzoyl azide.

Example. Benzoyl azide (Fig. 6.78). Note that acyl azides are isoelectronic with α-diazoketones (see Fig. 6.76). As with α-diazoketones, acyl azides are represented by two major resonance structures.

6.79 AZO COMPOUNDS, RN=NR

Compounds in which two R groups are connected by a −N=N− (azo) group. Azo compounds (RN=NR) should not be confused with diazo compounds ($R_2C{=}N_2$). The two classes of compounds have very different chemistry. Azo compounds can occur as either (E) or (Z) isomers. The R groups may either be the same or different; when they are identical, they are named by adding the prefix *azo-* to the name of the unsubstituted R−H compound.

Example. Azomethane or 1,2-dimethyldiazene (Fig. 6.79*a*). The latter name is derived from the parent *diazene*, HN=NH (its common name is *diimine*). The systematic name of the corresponding parent unsubstituted saturated analog, $H_2N{-}NH_2$, is *diazane* (its common name is *hydrazine*). The suffixes *-ane* and *-ene* as usual denote saturated and unsaturated species, as in alk*ane* and alk*ene*. Hydrazines have both laboratory and commercial significance; *1,1-dimethylhydrazine* is an important component of rocket fuels. Figure 6.79*b* shows diphenyldiazene (*azobenzene*). Because of the double bond, both *cis* (Z) and *trans* (E) isomers of azo compounds are possible. For example, the azo group in conjugation with aromatic rings is an important *chromophore* (from the Greek words *chroma-*, for "color," and *pherein* or *-phoros*, for "to carry") found in

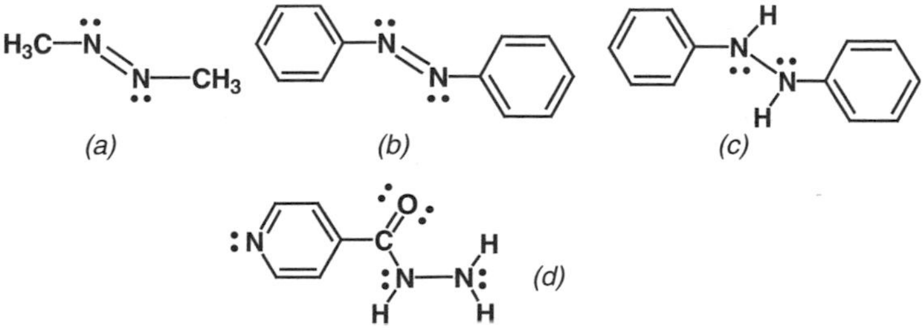

Figure 6.79. (*a*) *trans*-Dimethyldiazene (azomethane); (*b*) *trans*-diphenyldiazene (azobenzene); (*c*) diphenyldiazane (hydrazobenzene or 1,2-diphenylhydrazine); (*d*) isonicotinic acid hydrazide.

many dyes. Both stereoisomers of azobenzene are known and absorb visible light: *trans*-azobenzene (yellow-orange, $\lambda_{max} = 350\,nm$) and *cis*-azobenzene (red-orange, $\lambda_{max} = 433\,nm$). The hydrogenation of the double bond of azo compounds gives hydrazo compounds; Fig. 6.79*c* shows diphenyldiazane (*hydrazobenzene*). Acylated derivatives of hydrazines are called *hydrazides* (Fig. 6.79*d*).

6.80 AZOXY COMPOUNDS, $R{-}N{=}\overset{+}{N}(O^-){-}R$

Compounds in which an oxygen atom is bonded to one of the nitrogen atoms of an azo compound. Azoxy compounds can occur as either (E) or (Z) isomers.

Example. Azoxybenzene (Fig. 6.80).

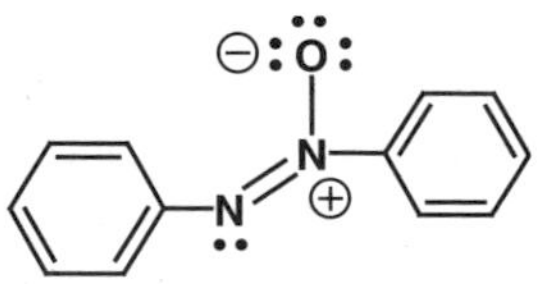

Figure 6.80. *trans*-Azoxybenzene.

6.81 AROMATIC DIAZONIUM SALTS, $ARN{\equiv}N^+ X^-$

Salts in which a $-N_2^+$ group is bonded directly to an aryl group. Note that alkyl diazonium salts, in most cases, constitute protonated diazo compounds and are highly unstable species.

Example. Benzenediazonium chloride (Fig. 6.81*a*). The protonation of diazomethane (Fig. 6.75*a*) on carbon leads to the methyl diazonium ion (Fig. 6.81*b*).

(*a*) (*b*)

Figure 6.81. (*a*) Benzenediazonium chloride; and (*b*) methyl diazonium ion.

6.82 CYANATES, ROCN

Compounds in which a hydrogen of a hydrocarbon is replaced by $-O{-}C{\equiv}N$, the cyanato group.

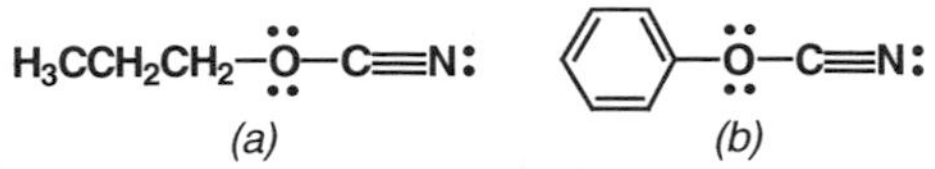

Figure 6.82. (*a*) *n*-Propyl cyanate and (*b*) phenyl cyanate.

Example. *n*-Propyl cyanate (Fig. 6.82*a*); this compound is named as though it were an ester of cyanic acid despite the fact that cyanic acid has the structure H–N=C=O. Phenyl cyanate (Fig. 6.82*b*).

6.83 FULMINATES, RON$^+$≡C$^-$

Compounds in which a hydrogen of a hydrocarbon R–H is replaced by –O–N$^+$≡C:$^-$, the fulminate group. The fulminates are isomers of *nitrile oxides*, RC≡N→O.

Example. *n*-Propyl fulminate (Fig. 6.83). This compound is named as though it were an ester of fulminic acid, which at one time was thought to have the structure HONC but is now known to have the structure HC≡N→O, and is not really an acid. Heavy metal salts of fulminic acids are extremely shock-sensitive and are used as detonators (primers) to initiate the explosion of other less sensitive materials.

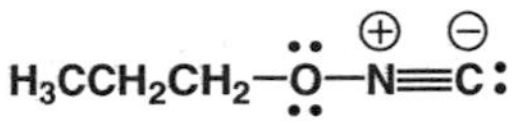

Figure. 6.83. *n*-Propyl fulminate.

6.84 ISOCYANATES, RN=C=O

Compounds in which a hydrogen atom of a hydrocarbon is replaced by –N=C=O, the isocyanato group.

Example. Phenyl isocyanate (Fig. 6.84). This compound is not the salt of an acid; hence, although it is commonly referred to as phenyl isocyanate, it should be called correctly by its substitutive name, carbonylaminobenzene. It is isoelectronic with the ketenes, RCH=C=O, and similarly reacts rapidly with water. Four isomers having a different linear (noncyclic) sequence of carbon, oxygen, and nitrogen atoms bonded to an R group have been discussed above: R–OCN (cyanates); R–ONC (fulminates); R–CNO (nitrile oxides); and R–NCO (isocyanates). The two other possible linear permutations of these three atoms, namely R–NOC and R–CON, are not known and are unlikely to exist.

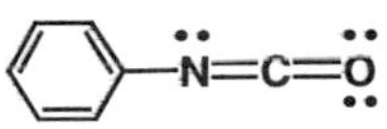

Figure 6.84. Carbonylaminobenzene (phenyl isocyanate).

6.85 CARBODIIMIDES, RN=C=NR

Compounds formally derived from the parent carbodiimide, HN=C=NH, by replacement of hydrogens by R groups. The stable form of the parent molecule is the tautomer, H_2N–C≡N (*cyanamide*).

Example. Dicyclohexylcarbodiimide (Fig. 6.85). The systematic name for this compound is *dicyclohexylmethanediimine* and for the parent, HN=C=NH, it is methanediimine.

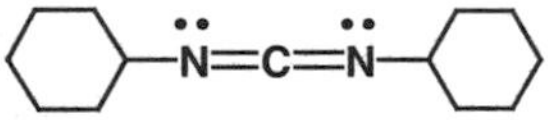

Figure 6.85. Dicyclohexylmethanediimine [dicyclohexylcarbodiimide (DCC)].

6.86 HETEROCYCLIC SYSTEMS

Cyclic compounds containing atoms of at least two different elements as ring members. Because organic chemistry is concerned principally with carbon atom chemistry, any element in a ring other than carbon is called a *heteroatom* (Sect. 6.1).

Example. There are hundreds of thousands of known heterocyclic compounds. Many of the more common systems contain nitrogen, oxygen, and/or sulfur as the heteroatom. Heteroaromatic compounds have already been discussed in Sect. 3.37. It is far beyond the scope of this book to treat all heterocyclic systems. Heterocyclic systems containing boron (Sect. 6.94), sulfur (Sect. 6.132), and phosphorus (Sect. 6.156) are mentioned separately. Suffice it here to give a few examples of non-aromatic oxygen and nitrogen heterocyclic compounds. Figures 6.86*a* and *b* are examples of oxygen heterocycles, and Figs. 6.86*c, d* and *e* are examples of nitrogen heterocycles, whereas Fig. 6.86*f* is an example of a heterocycle containing both oxygen and nitrogen.

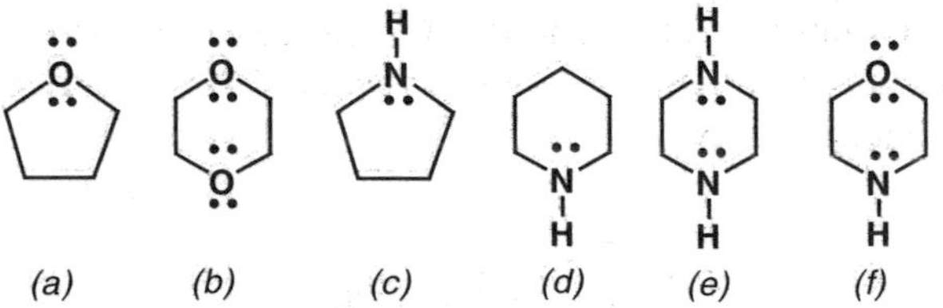

Figure 6.86. (*a*) Tetrahydrofuran; (*b*) 1,4-dioxane; (*c*) pyrrolidine; (*d*) piperidine; (*e*) piperazine; (*f*) morpholine.

6.87 OXA-AZA OR REPLACEMENT NOMENCLATURE

The system of naming heterocyclic or heteroacyclic compounds by relating them to the corresponding parent carbon system. The hetero element is denoted by prefixes ending in *a*, for example, *oxa-*, *aza-*, *thia-*, *phospha*, and so on. The prefix *oxa-*, which refers to oxygen in an ether, is not to be confused with *oxo-*, which refers to a carbonyl group.

Example. 1,4-Diazabicyclo[2.2.2]octane (DABCO) (Fig. 6.87*a*). This compound is used commercially to catalyze the formation of polyurethanes. 2,5-Dioxaheptane (Fig. 6.87*b*).

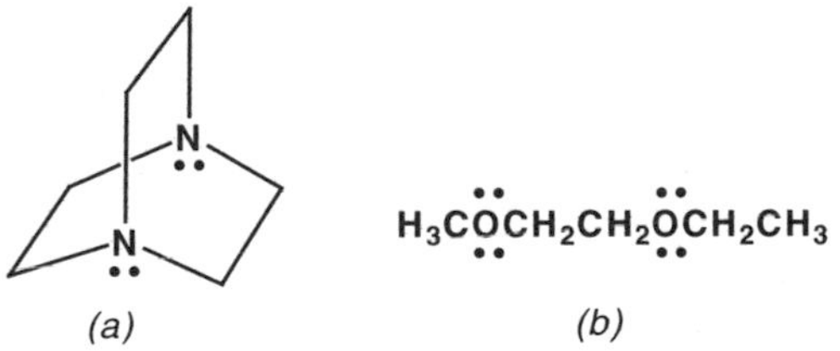

Figure 6.87. (*a*) 1,4-Diazobicyclo[2.2.2]octane (DABCO) and (*b*) 2,5-dioxaheptane.

6.88 HANTZSCH–WIDMAN HETEROCYCLIC NOMENCLATURE

This is a system of naming heterocyclics, proposed by the named chemists, which consists of adding identifying suffixes to the prefix used in replacement nomenclature; see Tables 6.88*a* and 6.88*b*. The H.-W. system was designed originally to

TABLE 6.88*a*. Examples of simple heterocyclic systems

Number of atoms in saturated ring (heteroatom)	Replacement name	H.-W.[a] name	Common name
3(N)	Azacyclopropane	Aziridine	Ethylenimine
3(O)	Oxacyclopropane	Oxirane	Ethylene oxide[b]
3(S)	Thiacyclopropane	Thiirane	Episulfide
4(N)	Azacyclobutane	Azetidine	Trimethyleneimine
4(O)	Oxacyclobutane	Oxetane	Trimethylene oxide
4(S)	Thiacyclobutane	Thietane	Trimethylene sulfide
5(N)	Azacyclopentane	Azolidine	Pyrrolidine
5(O)	Oxacyclopentane	Oxolane	Tetrahydrofuran
5(S)	Thiacyclopentane	Thiolane	Tetramethylene sulfide

[a] Hantzsch-Widman (A. Hantzsch and J. H. Weber *Berichte* **1887**, *20*, 319; O. Widman *J. Prakt. Chem.* **1885**, *38*, 185).

[b] The cyclic ethers having three or four atoms in the ring are sometimes called 1,2- and 1,3-epoxides, respectively.

TABLE 6.88*b*. Extension of the H.-W. Names by Adding the Indicated Suffixes to Designate Heterocycle Ring Size

	Nitrogen heterocycles			Other heterocycles		
Ring Size	No Double Bonds	One or More Double Bond	Maximum Number of Double Bonds	No Double Bonds	One or More Double Bonds	Maximum Number of Double Bonds
3	*-iridine*	—	*-irine*	*-irane*	—	*-irene*
4	*-etidine*	*-etine*	*-ete*	*-etane*	*-etene*	*-ete*
5	*-olidine*	*-oline*	*-ole*	*-olane*	*-olene*	*-ole*
6	[a]	[b]	*-ine*	*-ane*	[b]	*-in*
7	[a]	[b]	*-epine*	*-epane*	[b]	*-epine*

[a] This is expressed by using the prefix *perhydro-*.
[b] This is expressed by using the prefix *dihydro-* or *tetrahydro-*.

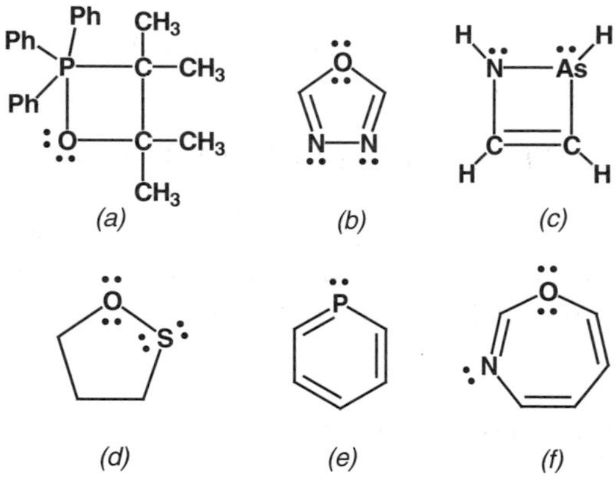

Figure 6.88. (*a*) 3,3,4,4-Tetramethyl-2,2,2-triphenyl-1-oxa-2-phosphoxetane; (*b*) 1-oxa-3,4-diazole; (*c*) 1,2-azarset-3-ene; (*d*) 1,2-oxathiolane; (*e*) phosphorine; (*f*) 1-oxa-3-azepine.

name five- and six-membered heterocyclic systems by adding the suffixes *-ole* (five-membered) and *-ine* (six-membered) to a prefix denoting the heteroatom.

For heterocycles with more than one type of heteroatom, the priority order for the prefixes in the name of the compound is *oxa-*, *thia-*, *selena-*, *tellura-*, *aza-*, *phospha-*,* *arsa-*,* *stiba-*,* *bisma-*, *sila-*, *germa-*, *stanna-*, *plumba-*, *bora-*, and *mercura-*. (*In cases where these prefixes would immediately be followed by *-in* or by *-ine*, they are replaced by *phosphor-*, *arsen-*, and *antimon-*, respectively.)

Example. 3,3,4,4-Tetramethyl-2,2,2-triphenyl-1-oxa-2-phosphoxetane (Fig. 6.88*a*); 1-oxa-3,4-diazole (Fig. 6.88*b*); 1,2-azarset-3-ene (Fig. 6.88*c*); 1,2-oxathiolane (Fig. 6.88*d*); phosphorine (Fig. 6.88*e*); and 1-oxa-3-azepine (Fig. 6.88*f*).

6.89 LACTONES, $\overline{CH_2-(CH_2)_x-C}{=}O$ (ring closed through —O—)

Intramolecular (cyclic) esters, compounds formed by the elimination of water from a hydroxyl and carboxyl group in the same molecule.

Example. The ease with which hydroxycarboxylic acids are converted to lactones depends on the number of atoms separating the two functional groups (x); five- and six-membered lactones, $x = 2$ and 3, respectively, are the most easily formed, and hence, the most common members of this family. The Greek letters in β-propiolactone (Fig. 6.89*a*) and γ-butyrolactone (Fig. 6.89*b*), where $x = 1$ and 2, respectively, refer to the position of the saturated ring carbon atom to which the ring oxygen atom is attached relative to the carbonyl group. Any hydrogen atom in the lactone may be replaced by an R group.

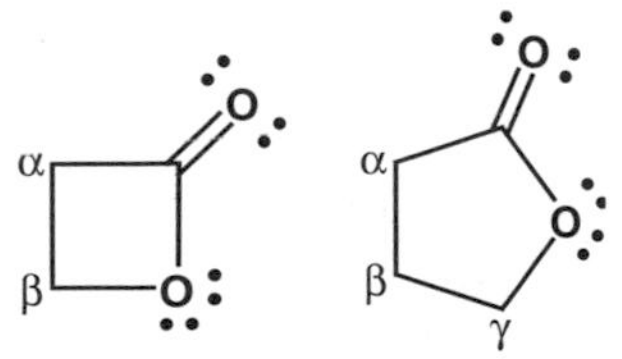

(*a*) (*b*)

Figure 6.89. (*a*) β-Propiolactone and (*b*) γ-butyrolactone.

6.90 LACTAMS, $R-\overline{CH-(CH_2)_x-C}{=}O$ (ring closed through —NH—)

Intramolecular (cyclic) amides, compounds formed by the elimination of water from an amino and carboxyl group in the same molecule.

Example. The ease of lactam formation from aminocarboxylic acids depends on the number of atoms separating the two functional groups (x); five- and six-membered lactams, $x = 2$ and 3, respectively, are the most easily formed and hence the most common members of this family. Caprolactam (Fig. 6.90*a*) is used in nylon manufacture. Many compounds of biological interest (*e.g., penicillin*) are derivatives of β-propiolactam (Fig. 6.90*b*).

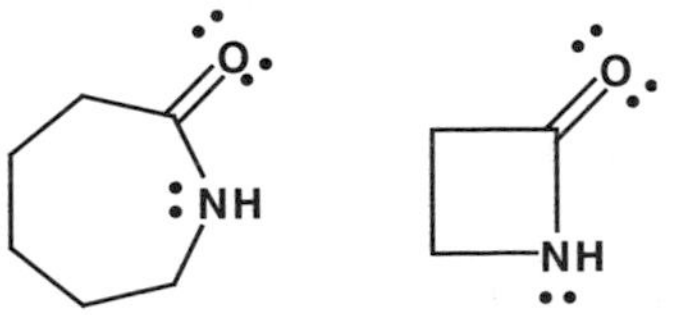

(*a*) (*b*)

Figure 6.90. (*a*) Caprolactam and (*b*) β-propiolactam.

6.91 BORANES

Derivatives of boron hydrides, B_xH_y. BH_3 and B_2H_6 are borane and diborane, respectively (Sect. 3.45). Hydrides containing more than one boron atom are named by using the familiar multiplying prefixes and adding an Arabic numeral in parentheses as a suffix to denote the number of hydrogens present in the parent; for example, B_5H_9 is pentaborane (9).

6.92 ORGANOBORANES

Compounds in which one or more R groups replace the hydrogens in a boron hydride.

Example. Trans-1,2-dimethyldiborane(6) (Fig. 6.92*a*) and 9-borabicyclo[3.3.1]-nonane or 9-BBN (Fig. 6.92*b*).

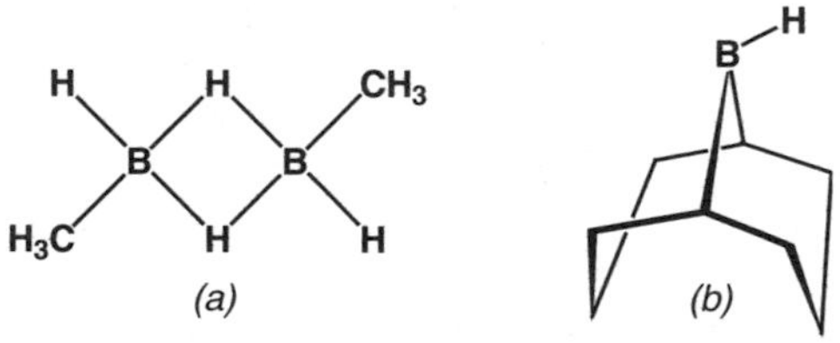

Figure 6.92. (*a*) *trans*-1,2-Dimethyldiborane(6) and (*b*) 9-borabicyclo[3.3.1]nonane or 9-BBN.

6.93 ORGANOBORATES, $(RO)_3B$, $(RO)_2B(OH)$, AND $ROB(OH)_2$

Esters of boric acid, $B(OH)_3$, in which one or more of the hydrogen atoms is replaced by −R.

Example. Triethyl borate (Fig. 6.93*a*), diethyl hydrogen borate (Fig. 6.93*b*), and ethyl dihydrogen borate (Fig. 6.93*c*).

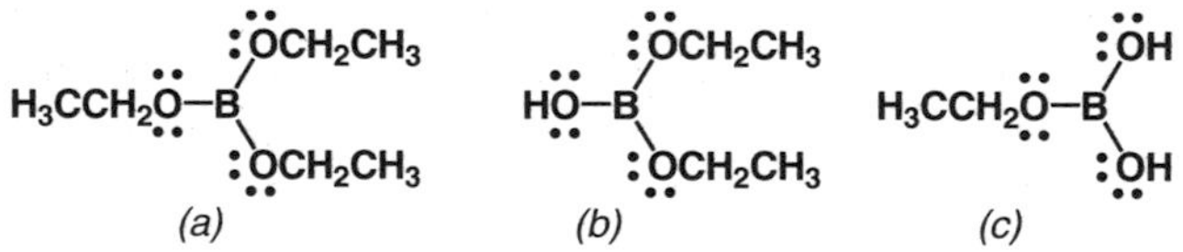

Figure 6.93. (*a*) Triethyl borate; (*b*) diethyl hydrogen borate; (*c*) ethyl dihydrogen borate.

6.94 HETEROCYCLIC BORANES

Organic ring systems containing one or more boron atoms.

Example. The replacement (oxa-aza) nomenclature is commonly used, for example, 1,4-diboracyclohexane (Fig. 6.94).

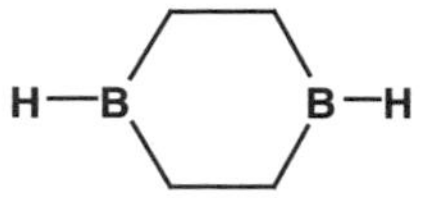

Figure 6.94. 1,4-Diboracyclohexane.

6.95 CARBORANES

This name is a shortened form of the word "carbaborane," denoting a class of compounds in which one or more carbon atoms replace boron in polyboron hydrides.

Example. 1,2-Dicarbadodecaborane(12), $B_{10}C_2H_{12}$ (Fig. 6.95, see Sect. 6.24); there are three isomers of this compound; the 1,2-isomer (the numbers indicate the location of the carbon atoms) is shown in Fig. 6.95. Such triangulated polyhedral structures have the formula $B_{n-2}C_2H_n$.

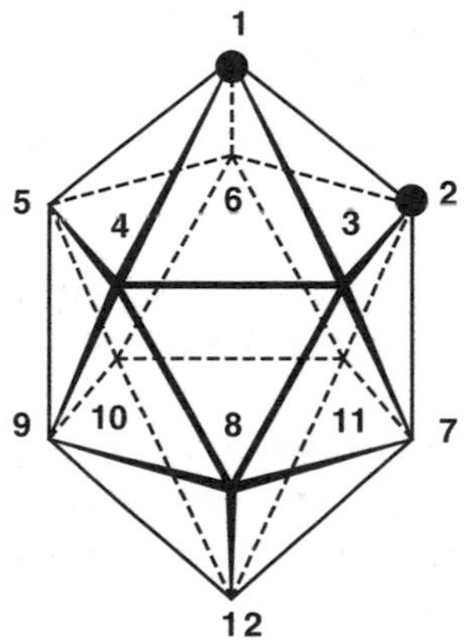

Figure 6.95. 1,2-Dicarbadodecaborane(12); the solid dots represent carbon atoms at positions 1 and 2.

6.96 HYDROPEROXIDES, ROOH

Compounds in which a hydrogen of a hydrocarbon is replaced by an –OOH group.

Example. tert-Butyl hydroperoxide (Fig. 6.96). The –OOH functional group is called the hydroperoxy group; hence, compounds containing it may be named by the substitutive nomenclature. Thus, Fig. 6.96 is also called 2-hydroperoxy-2-methylpropane.

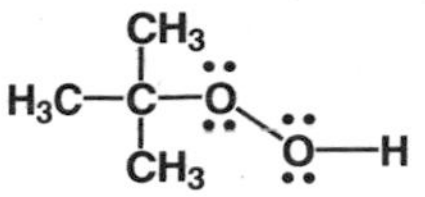

Figure 6.96. *tert*-Butyl hydroperoxide (2-hydroperoxy-2-methylpropane).

6.97 PERACIDS, RC(O)OOH

Carboxylic acids in which an oxygen atom has been inserted between the carbonyl carbon and the oxygen of the hydroxyl group. The R group may be either acyl or aroyl.

Example. Perbenzoic acid (Fig. 6.97); the systematic name for this compound is peroxybenzoic acid. Although named as an acid, this compound has the very low acidity of an alcohol owing to the lack of resonance stabilization of the corresponding anion. Furthermore, this compound is relatively nonpolar due to the internal hydrogen bonding shown. Esters of peracids are peresters or *peroxy esters*.

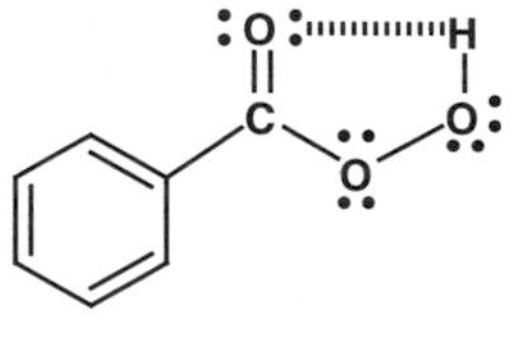

Figure 6.97. Perbenzoic acid (peroxybenzoic acid).

6.98 DIALKYL PEROXIDES, ROOR, AND DIACYL OR DIAROYL PEROXIDES, RC(O)OO(O)CR

Compounds in which the hydrogen atoms of hydrogen peroxide, HOOH, are replaced by alkyl groups. Aryl peroxides, where either or both Rs = aryl, are unknown substances. Diacyl and diaroyl peroxides are the corresponding peroxides, where R = alkylC(O)- and arylC(O)-, respectively.

Example. Di-*tert*-butyl peroxide, (Fig. 6.98*a*). When the two R groups are different, peroxides are preferably named substitutively as dioxy derivatives of a parent hydrocarbon. Thus, Fig. 6.98*b* is methyldioxycyclohexane. The diesters of hydrogen per-

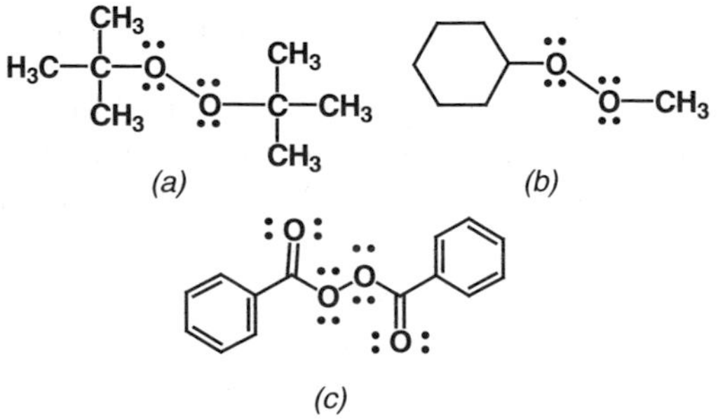

Figure 6.98. (*a*) Di-*tert*-butyl peroxide; (*b*) methyldioxycyclohexane; (*c*) dibenzoyl peroxide.

oxide, such as dibenzoyl peroxide (Fig. 6.98*c*), are readily available stable peroxides.

6.99 DIOXETANES

Cyclic peroxides incorporated into a four-membered ring, 1,2-dioxacyclobutanes.

Example. A few examples of stable dioxetanes are known. These molecules are often the products of the reactions of singlet oxygen with olefins, and tend to fragment thermally to form two carbonyl molecules and a photon of light. Thus, dioxetanes are the most common species that give rise to *chemiluminescence* (see Sect. 18.151). The dioxetane of adamantylidene adamantane (Fig. 6.99*a*) is stable under ambient conditions, whereas 1,2-dioxacyclobutadione (Fig. 6.99*b*) is the reactive intermediate involved in chemiluminescent "light sticks."

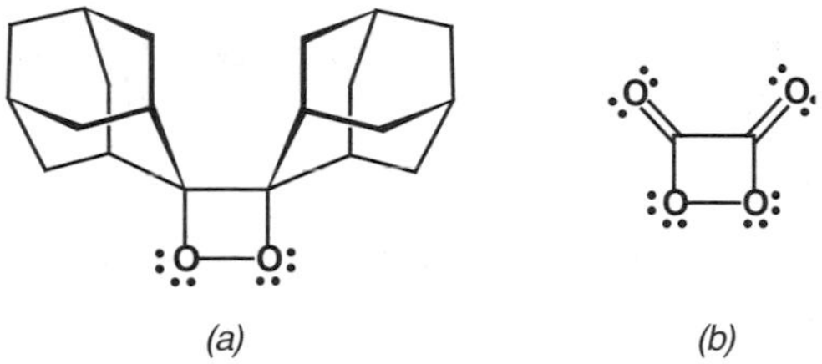

Figure 6.99. (*a*) The dioxetane of adamantylidene adamantane and (*b*) 1,2-dioxacyclobutadione.

6.100 ENDOPEROXIDES

Peroxides in which the peroxide linkage is incorporated into a ring.

Example. Endoperoxides typically are formed in the reactions of singlet oxygen with cyclic dienes, as, for example, with cyclopentadiene to form 2,3-dioxabicyclo-[2.2.1]hept-5-ene (Fig. 6.100). The prefix *endo-* should not be confused with the stereochemical term endo (and exo); see Sect. 7.28. As used here, *endo-* simply means that the peroxide group is incorporated into the ring system.

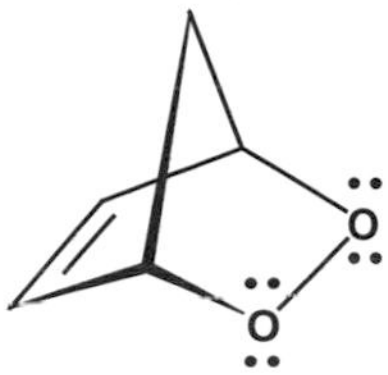

Figure 6.100. The endoperoxide 2,3-dioxabicyclo[2.2.1]hept-5-ene.

6.101 OZONIDES

Cyclic peroxides incorporated into five-membered rings, 1,2,4-trioxacyclopentanes.

Example. Ozonides are formed in the reaction of ozone, O_3, with olefins. The initial adduct formed in these reactions contains the highly unstable functional group consisting of three oxygen atoms bonded in a lincar array (Fig. 6.101*a*). It is called a *molozonide*, 1,2,3-trioxacyclopentane, and rapidly rearranges to the relatively stable ozonide (Fig. 6.101*b*) under ambient conditions. The *looped arrow* in Fig. 6.101 symbolizes a rearrangement or reorganization of the skeletal atoms. Ozonides of low molecular weight olefins are very unstable and tend to explode when solutions containing these materials are concentrated.

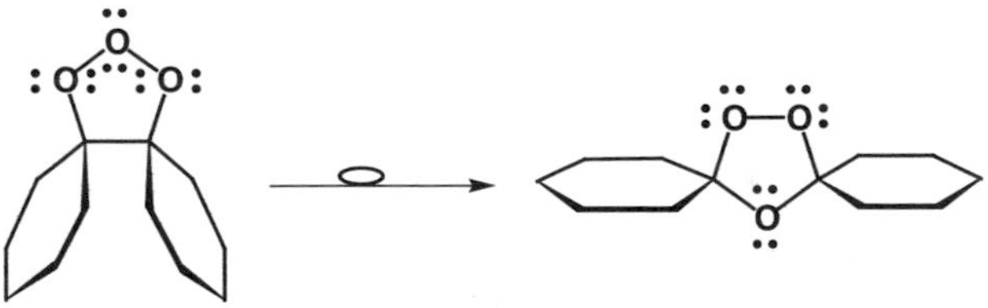

(a) (b)

Figure 6.101. Rearrangement of (*a*) molozonide, 1,2,3-trioxacyclopentane, to (*b*) ozonide, 1,2,4-trioxacyclopentane.

6.102 THIOLS (THIOALCOHOLS, MERCAPTANS), RSH

Compounds in which a hydrogen atom of a parent hydrocarbon is replaced by a mercapto (−SH) group. Compounds of this class are frequently called *mercaptans*. They have a distinctively unpleasant odor. One part of ethanethiol in 50 billion parts of air can be detected by the average human nose.

Example. The thiols are named in the IUPAC system like the alcohols using the suffix *-thiol* in place of *-ol*, and retaining the letter *e* in the parent name. Ethanethiol (Fig. 6.102*a*), phenylmethanethiol (Fig. 6.102*b*), and cyclohexanethiol (Fig. 6.102*c*); these names are substitutive names. The first two of these compounds are usually called ethyl mercaptan and benzyl mercaptan, respectively, but such radicofunctional names are not sanctioned by IUPAC.

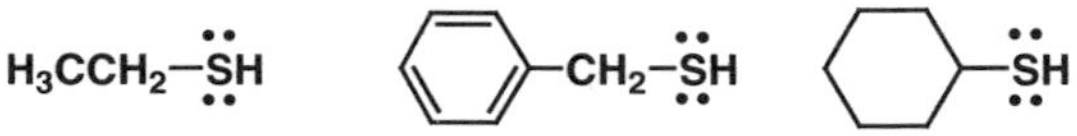

(a) (b) (c)

Figure 6.102. (*a*) Ethanethiol (ethyl mercaptan); (*b*) phenylmethanethiol (benzyl mercaptan); (*c*) cyclohexanethiol.

6.103 THIOPHENOLS, ArSH

Compounds in which a ring hydrogen atom of an aromatic hydrocarbon is replaced by a mercapto (-SH) group. The generic name thiols actually includes both thioalcohols and thiophenols.

Example. Benzenethiol (phenyl mercaptan or thiophenol) (Fig. 6.103*a*); 2-naphthalenethiol (2-naphthyl mercaptan or 2-thionaphthol) (Fig. 6.103*b*).

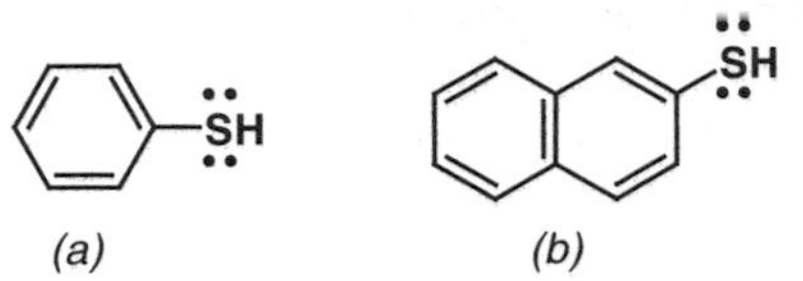

Figure 6.103. (*a*) Benzenethiol (phenyl mercaptan, thiophenol) and (*b*) 2-naphthalenethiol (2-naphthyl mercaptan, 2-thionaphthol).

6.104 SULFIDES, RSR

Compounds in which a hydrogen atom of a hydrocarbon is replaced by an −SR group, or alternatively, hydrocarbons in which a sulfur atom is inserted into a carbon–carbon bond.

Example. Ethyl methyl sulfide, (Fig. 6.104*a*) and cyclopentyl phenyl sulfide (Fig. 6.104*b*), also called (cyclopentylthio)benzene.

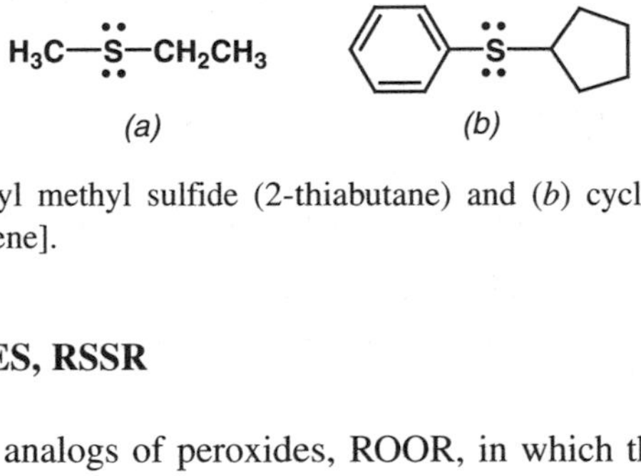

Figure 6.104. (*a*) Ethyl methyl sulfide (2-thiabutane) and (*b*) cyclopentyl phenyl sulfide [(cyclopentylthio)benzene].

6.105 DISULFIDES, RSSR

These are the sulfur analogs of peroxides, ROOR, in which the oxygen atoms are replaced by sulfur atoms; they are the oxidation products of thiols, RSH, and are generally formed from the dimerization of RS• radicals.

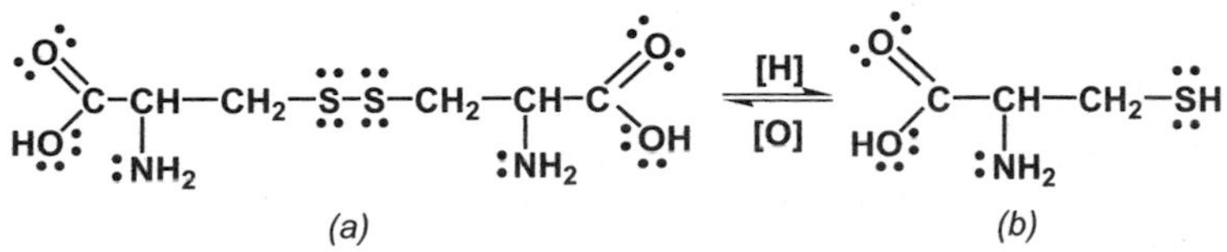

Figure 6.105. (*a*) Cystine and (*b*) cysteine.

Example. Cystine (Fig. 6.105*a*), obtained from the corresponding amino acid, cysteine (Fig. 6.105*b*). The disulfide linkage is very important in the chemistry of amino acids and proteins.

6.106 SULFOXIDES, $R_2S(O)$ ($R_2S^+{-}O^-$)

Compounds in which an oxygen atom is bonded to the sulfur atom of a sulfide.

Example. Ethyl methyl sulfoxide, also called (methylsulfinyl)ethane, two resonance structures of which are shown in Figs. 6.106*a* and *b*. The resonance structure Fig. 6.1060*a* (and others related to it) is frequently written with a coordinate covalent bond, as in Fig. 6.106*c*. Sulfoxides are chiral (Sect. 7.21), indicating that the sulfur atom is sp^3 hybridized; thus, the S-O bond of sulfoxides has very little double bond character. The prefixes *alkylthio-* and *alkylsulfinyl-* represent RS– and $R{-}S^+{-}O^-$ groups, respectively. In the older literature, the –S(O)– group was referred to as thionyl, and $SOCl_2$ is commonly called *thionyl chloride* (Sect. 6.161).

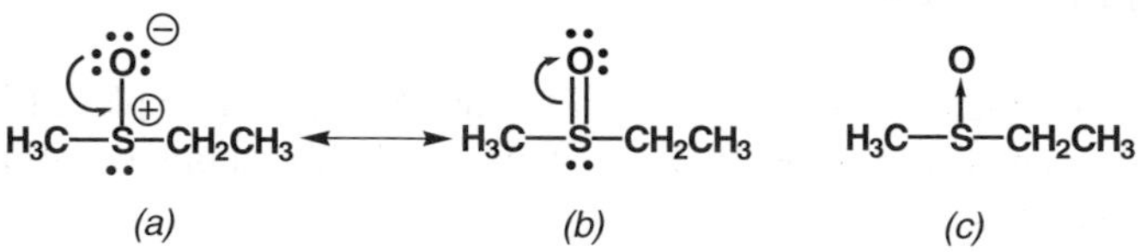

Figure 6.106. (*a*) and (*b*), Resonance structures of ethyl methyl sulfoxide [(methylsulfinyl)-ethane] (see Fig. 6.107); (*c*) the notation showing a coordinate $S \rightarrow O$ bond. Lone pair electrons are not shown in structure (*c*) because this representation is a hybrid of representations (*a*) and (*b*); therefore, the number of lone pair electrons on oxygen and sulfur are undefined.

6.107 SULFONES, $R_2S(O)_2$

Compounds in which the sulfur of a sulfide is bonded to two oxygen atoms.

Example. Isopropyl methyl sulfone (Fig. 6.107), also called 2-methylsulfonylpropane (its IUPAC name). The three major resonance structures for this sulfone group are

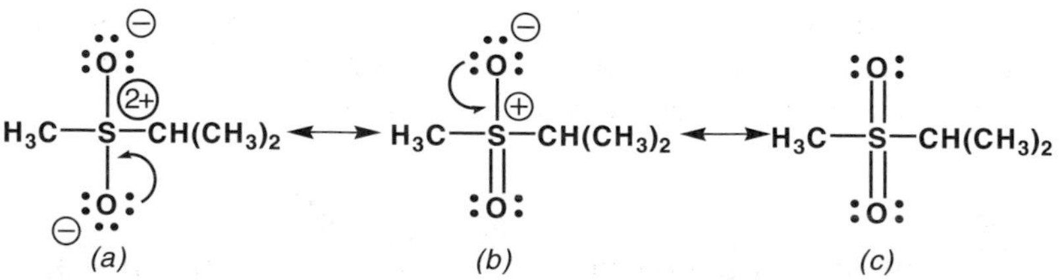

Figure 6.107. Resonance structures of isopropyl methyl sulfone (2-methylsulfonylpropane). Note that structures *b* and *c* require *d*- orbitals and expansion of the sulfur valence shell to 10 and 12 electrons, respectively.

shown in Fig. 6.107. In the older literature the $-SO_2-$ group was called sulfuryl and SO_2Cl_2 is referred to as *sulfuryl chloride* (Sect. 6.161).

6.108 SULFONIC ACIDS, RSO_2OH

Compounds in which a hydrogen atom of a hydrocarbon is replaced by a sulfo ($-SO_2OH$) group.

Example. Benzenesulfonic acid (Fig. 6.108). Because of the positive charge on sulfur, this is a vey strong protic acid. Replacement of the OH group by halogen gives a *sulfonyl halide*, $RS(O)_2X$.

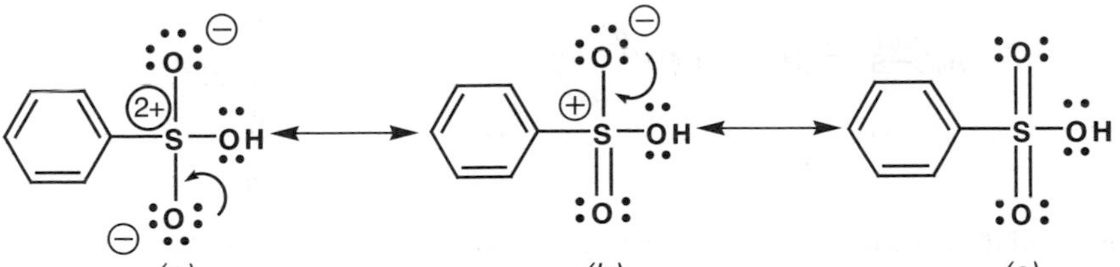

Figure 6.108. Three resonance structures of benzenesulfonic acid. For convenience in discussing the structure of this acid and its derivatives, only one resonance structure will usually be shown.

6.109 SULTONES, $CH_2-(CH_2)_x-SO_2$ (ring closed through O)

This is the intramolecular dehydration product of an hydroxysulfonic acid; it is analogous to lactones, which are dehydration products of hydroxycarboxylic acids.

Example. γ-Hydroxypropane sulfonic acid sultone, or 1,3-propanesultone (Fig. 6.109).

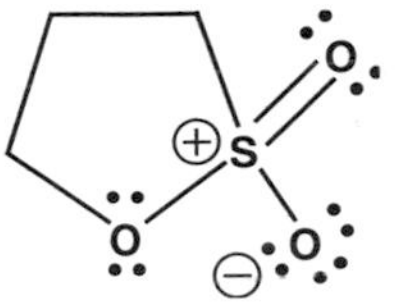

Figure 6.109. γ-Hydroxypropane sulfonic acid sultone (1,3-propanesultone).

6.110 SULFONATES, $RS(O)_2OR$

Esters of sulfonic acids. The sulfonates are particularly useful in synthetic chemistry since the anions RSO_3^- are excellent "leaving groups."

Example. Methyl methanesulfonate (Fig. 6.110*a*); methyl *p*-toluenesulfonate (Fig. 6.110*b*); and methyl trifluormethanesulfonate (Fig. 6.110*c*). Because these sulfonate esters are used so frequently, their sulfonic acid moieties have been given abbreviated common names. Thus, Fig. 6.110*a* is called a mesylate ester; Fig. 6.110*b* is a tosylate ester; and Fig. 6.110*c* is a triflate ester. The anions that are the corresponding leaving groups are written in shorthand as MsO^-, TsO^-, and TfO^-, respectively.

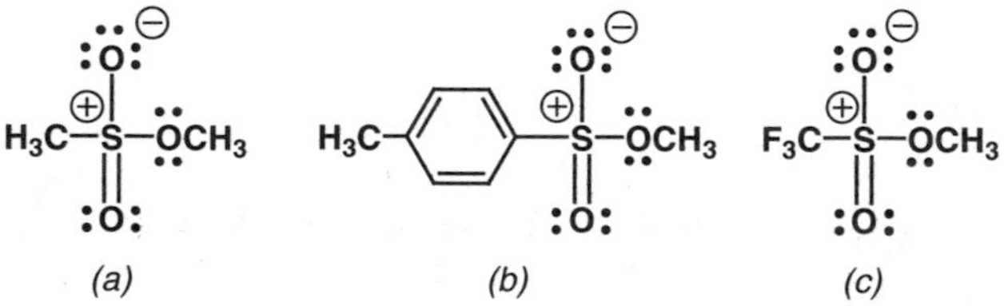

Figure 6.110. (*a*) Methyl methanesulfonate (methyl mesylate, Ms-OCH_3); (*b*) methyl *p*-toluenesulfonate (methyl tosylate, Ts-OCH_3); (*c*) methyl trifluromethanesulfonate (methyl triflate, Tf-OCH_3).

6.111 SULFATES, $(RO)_2S(O)_2$

Diesters of sulfuric acid.

Example. Dimethyl sulfate, (Fig. 6.111).

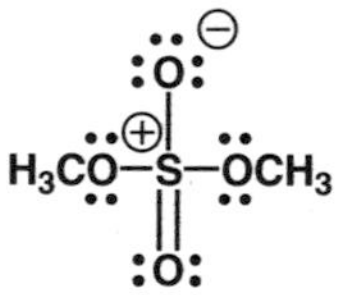

Figure 6.111. Dimethyl sulfate.

6.112 BISULFATES, $ROS(O)_2OH$

Mono esters of sulfuric acid.

Example. Methyl hydrogen sulfate (Fig. 6.112).

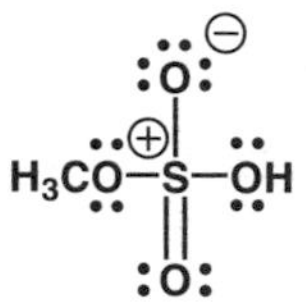

Figure 6.112. Methyl hydrogen sulfate.

6.113 SULFINATES, RS(O)OH

Compounds in which a hydrogen atom of a hydrocarbon is replaced by a sulfino ($-SO_2H$) group.

Example. Ethanesulfinic acid, (Fig. 6.113); replacement of the OH group by a halogen give a *sulfinyl halide*, RS(O)X.

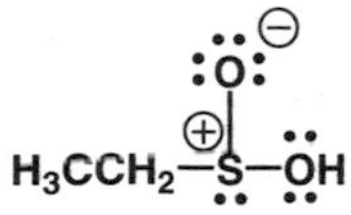

Figure 6.113. Ethanesulfinic acid.

6.114 SULFINIC ESTERS, RS(O)OR

Esters of sulfinic acid.

Example. Ethyl methanesulfinate (Fig. 6.114).

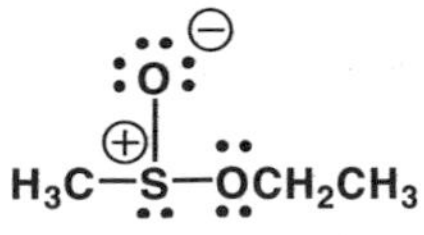

Figure 6.114. Ethyl methanesulfinate.

6.115 SULFENIC ACIDS, RSOH

Compounds in which a hydrogen atom of a hydrocarbon is replaced by a sulfeno (−SOH) group.

Example. Butanesulfenic acid (Fig. 6.115); replacement of the OH group by halogen gives a *sulfenyl halide*, RSX.

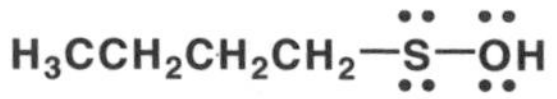

Figure 6.115. Butanesulfinic acid.

6.116 SULFENATES, RSOR

Esters of sulfenic acid.

Example. Methyl propanesulfenate (Fig. 6.116).

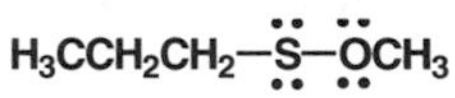

Figure 6.116. Methyl propanesulfenate.

6.117 THIOCARBOXYLIC ACIDS, RC(S)OH OR RC(O)SH

A generic name for carboxylic acids in which either of the oxygen atoms of the carboxyl group has been replaced by sulfur.

6.118 THIOIC *S*-ACIDS, RC(O)SH

Carboxylic acids in which the oxygen of the hydroxyl group has been replaced by sulfur.

Example. Propanethioic *S*-acid (Fig. 6.118).

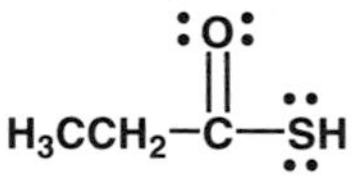

Figure 6.118. Propanethioic *S*-acid.

6.119 THIOIC *S*-ESTERS, RC(O)SR

Esters of thioic *S*-acids.

Example. *S*-Methyl ethanethioate (Fig. 6.119).

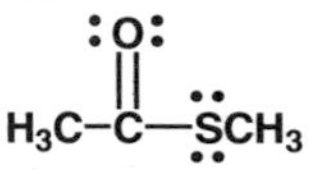

Figure 6.119. *S*-Methyl ethanethioate.

6.120 THIOIC *O*-ACIDS, RC(S)OH

Compounds in which the carbonyl oxygen of the carboxyl group is replaced by sulfur.

Example. Propanethioic *O*-acid (Fig. 6.120*a*) and benzenecarbothioic *O*-acid (Fig. 6.120*b*).

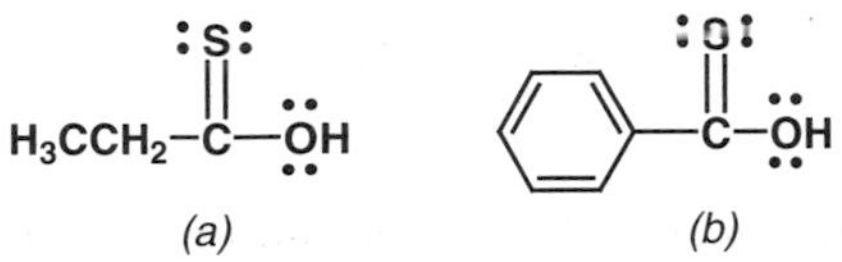

Figure 6.120. *(a)* Propanethioic *O*-acid and *(b)* benzenecarbothioic *O*-acid.

6.121 THIOIC *O*-ESTERS, RC(S)OR

Esters of thioic *O*-acids.

Example. *O*-Ethyl propanethioate (Fig. 6.121).

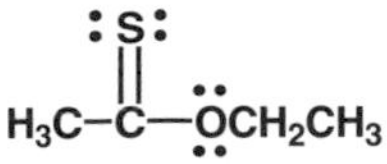

Figure 6.121. *O*-Ethyl propanethioate.

6.122 DITHIOIC ACIDS, RC(S)SH

Compounds in which both of the oxygen atoms of the carboxyl group are replaced by sulfur.

Example. Butanedithioic acid (Fig. 6.122*a*) and 1-naphthalenecarbodithioic acid (Fig. 6.122*b*).

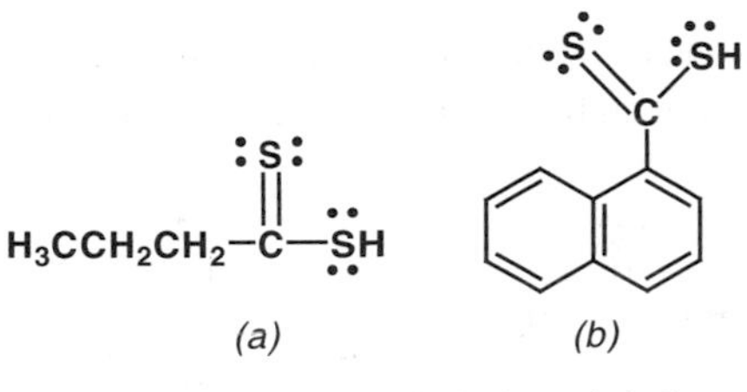

Figure 6.122. (*a*) Butanedithioic acid and (*b*) 1-naphthalenecarbodithioic acid.

6.123 DITHIOIC ESTERS, RC(S)SR

Esters of dithioic acids.

Example. Methyl 1-naphthalenecarbodithioate (Fig. 6.123).

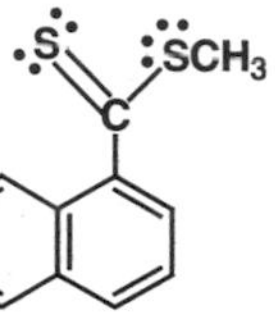

Figure 6.123. Methyl 1-naphthalenecarbodithioate.

6.124 THIOALDEHYDES, RCH=S

These are aldehydes in which the carbonyl oxygen is replaced by sulfur; always dimeric or trimeric, they are analogous to dimers and trimers of formaldehyde.

Example. Ethanethial (thioacetaldehyde) (Fig. 6.124*a*); benzenecarbothialdehyde (Fig. 6.124*b*) 1,3,5-trisulfa-2,4,6-trialkylcylcohexane (Fig. 6.124*c*).

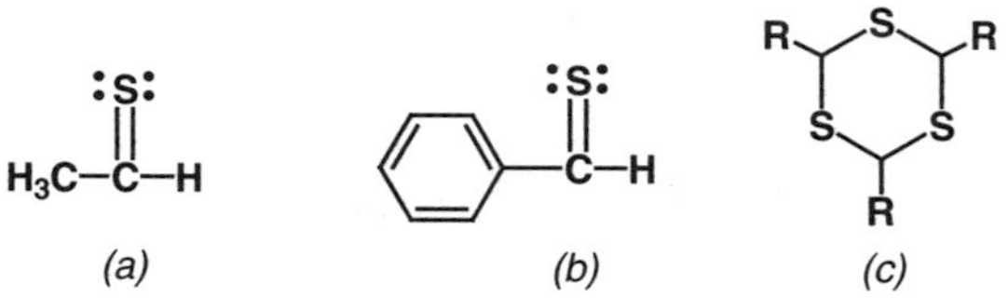

Figure 6.124. (*a*) Ethanethial; (*b*) benzenecarbothialdehyde; (*c*) 1,3,5-trisulfa-2,4,6-trialkylcyclohexane.

6.125 THIOKETONES, RC(S)R

These are ketones in which the carbonyl oxygen is replaced by a sulfur atom; such compounds often occur as trimers or dimers. Diaryl thioketones are relatively stable as the monomers.

Example. 2-Propanethione (dimethyl thioketone) (Fig. 6.125*a*); the name thioacetone is frequently used, but is not an IUPAC name. Thiobenzophenone (Fig. 6.125*b*) is blue.

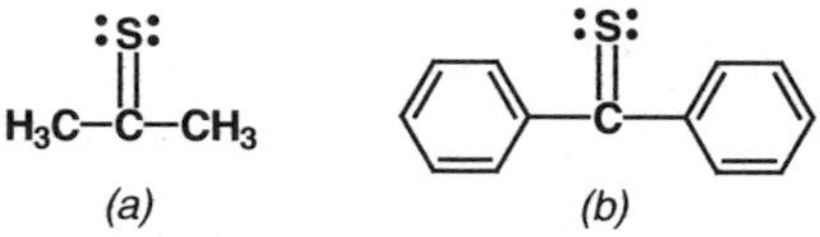

Figure 6.125. (*a*) 2-Propanethione (dimethyl thioketone) and (*b*) thiobenzophenone.

6.126 THIOCYANATES, RSCN

Compounds in which a hydrogen atom of a hydrocarbon is replaced by the thiocyanato –S–C≡N group.

Example. 2-Thiocyanatopropane (Fig. 6.126).

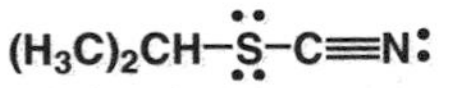

Figure 6.126. 2-Thiocyanatopropane (isopropylthiocyanate).

6.127 ISOTHIOCYANATES, RN=C=S

Compounds in which a hydrogen atom of a hydrocarbon is replaced by an isothiocyanato group –N=C=S.

Example. Isothiocyanatobenzene (Fig. 6.127*a*). The isothiocyanato compound shown in Fig 6.127*b* is of particular interest because it is believed to be the anticancer compound in broccoli. This natural product is optically active because of the chiral center at the sp^3 sulfoxide sulfur. Its full name is (−)-1-isothiocyanato-(4R)-(methylsulfinyl)-butane, and it has been given the common name sulforaphane. (The minus sign indicates levorotatory and the R refers to the configuration at the chiral center; see Sect. 7.32).

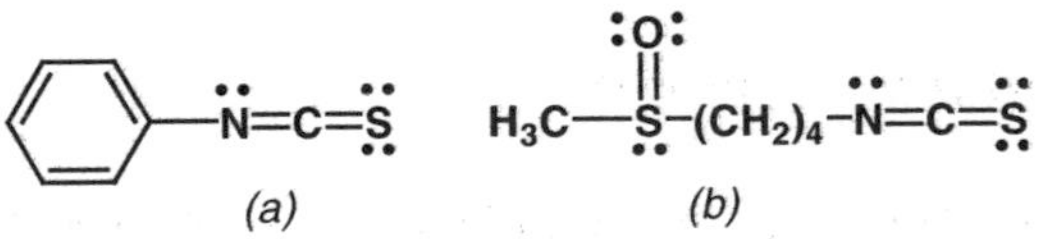

Figure 6.127. (*a*) Isothiocyanatobenzene and (*b*) sulforaphane.

6.128 SULFONAMIDES, $RS(O)_2NH_2$

Amides of sulfonic acids.

Example. The first antibacterial sulfa drug 4-aminophenylsulfonamide (sulfanilamide) (Fig. 6.128*a*); *N*-ethylmethanesulfonamide (Fig. 6.128*b*).

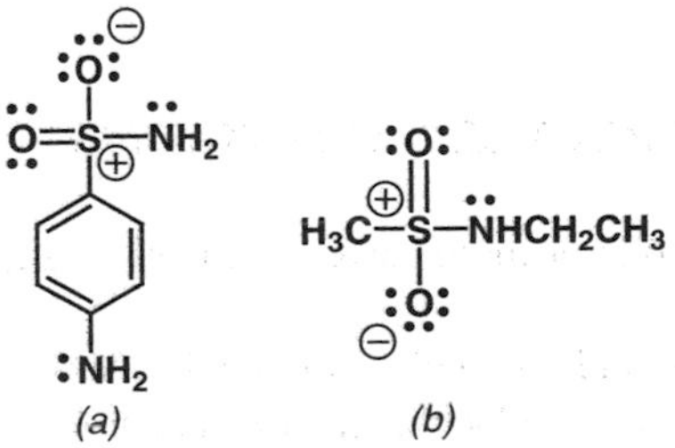

Figure 6.128. (*a*) 4-Aminophenylsulfonamide (sulfanilamide) and (*b*) *N*-ethylmethanesulfonamide.

6.129 SULFINAMIDES, $RS(O)NH_2$

Amides of sulfinic acids (Sect. 6.113).

Example. 2-Propanesulfinamide (Fig. 6.129).

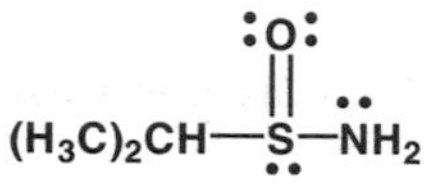

Figure 6.129. 2-Propanesulfinamide.

6.130 SULFENAMIDES, $RSNH_2$

Amides of sulfenic acid (Sect. 6.115).

Example. Ethanesulfenamide (Fig. 6.130).

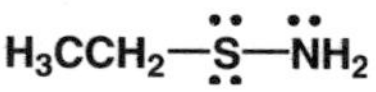

Figure 6.130. Ethanesulfenamide.

6.131 THIOYL HALIDES, RC(S)X

Compounds in which the oxygen of the carbonyl group of an acyl halide, RC(O)Cl, is replaced by sulfur.

Example. Ethanethioyl chloride (Fig. 6.131).

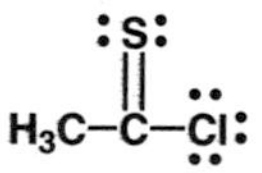

Figure 6.131. Ethanethioyl chloride.

6.132 SULFUR HETEROCYCLIC COMPOUNDS

Compounds that contain one or more sulfur atoms in a ring system.

Example. There are thousands of such compounds. They frequently have common names or are named by the replacement nomenclature system; thiacyclopentane (Fig. 6.132*a*) and 1-thiacyclopenta[*b*]naphthalene (Fig. 6.132*b*). (The letter *b* in

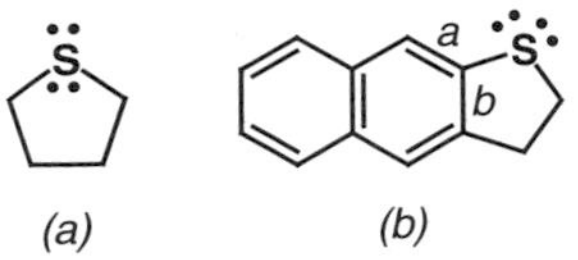

Figure 6.132. (*a*) Thiacyclopentane and (*b*) 1-thiacyclopenta[*b*]naphthalene.

brackets denotes the side between carbons 2 and 3 of the naphthalene nucleus to which the thiacyclopentane is fused.; see Sect. 5.40.)

6.133 THIOHEMIACETALS, RCH(OH)SR

These are compounds resulting from the addition of a molecule of a thiol to the carbonyl group of an aldehyde; they are analogous to hemiacetal (Sect. 6.28).

Example. 1-(Ethylthio)-1-propanol (Fig. 6.133).

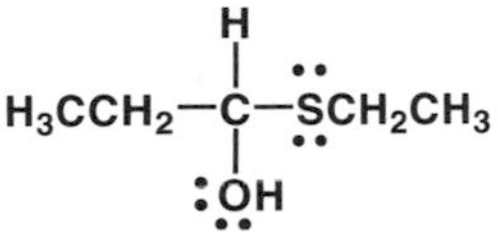

Figure 6.133. 1-(Ethylthio)-1-propanol.

6.134 DITHIOHEMIACETALS, RCH(SH)SR

These are compounds that may be regarded as arising from the addition of a molecule of thiol to a thioaldehyde, in which the SH and SR groups are singly bonded to what was the aldehyde carbon atom; such compounds are sometimes also called thiohemiacetals. Replacement of the H on the carbon of RCH(SH)SR by a second R group affords $R_2C(SH)(SR)$, a *dithiohemiketal.*

Example. 1-(Methylthio)-1-butanethiol (Fig. 6.134).

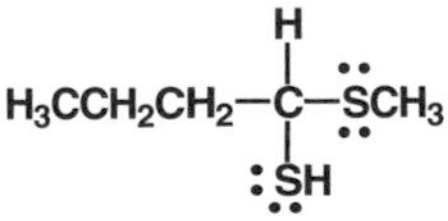

Figure. 6.134. 1-(Methylthio)-1-butanethiol.

6.135 THIOACETALS, $RCH(SR)_2$

Compounds in which thc oxygen atom of an aldehyde is replaced by two alkylthio (or arylthio) groups. When the H on carbon is replaced by a second R group to afford $R_2C(SR)_2$, the compound also is called a thioacetal, formerly known as a thioketal.

Example. 1,1-Bis(ethylthio)propane (Fig. 6.135*a*). The cyclic thioacetals (Fig. 6.135*b*) are used in some synthetic procedures to protect aldehydes and ketones.

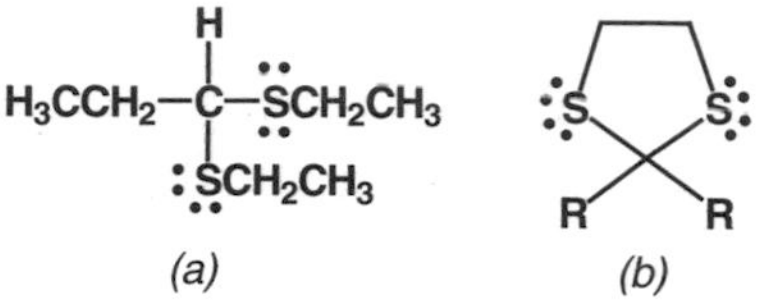

Figure 6.135. (*a*) 1,1-Bis(ethylthio)propane and (*b*) a cyclic thioacetal.

6.136 SULFURANES, R_4S

Neutral, four-coordinate organosulfur compounds. Such materials are not stable and decompose to form diphenyl sulfide and biphenyl.

Example. Tetraphenylsulfurane (Fig. 6.136).

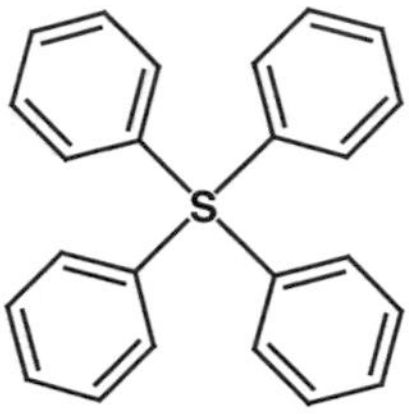

Figure 6.136. Tetraphenylsulfurane.

6.137 ORGANOPHOSPHORUS COMPOUNDS

Compounds containing at least one phosphorus atom linked to at least one carbon atom. Thousands of such compounds are known, and there are multivolume series of books devoted exclusively to them (see "Suggested Reading, p. 220").

6.138 PHOSPHINES, RPH_2, R_2PH, R_3P, AND RP=O

Compounds in which one or more of the hydrogen atoms of phosphine, PH_3, are replaced by R groups, or compounds in which the two hydrogen atoms of RPH_2 are replaced by an oxygen atom. (Some chemists prefer the name *phosphane* in place of phosphine, and thus, the derivatives are called phosphanes.)

Example. Dimethylphosphine (Fig. 6.138*a*) and ethyloxophosphine (Fig. 6.138*b*). When the group R_2P-, R= H, alkyl, or aryl is treated as a substituent, it is given the

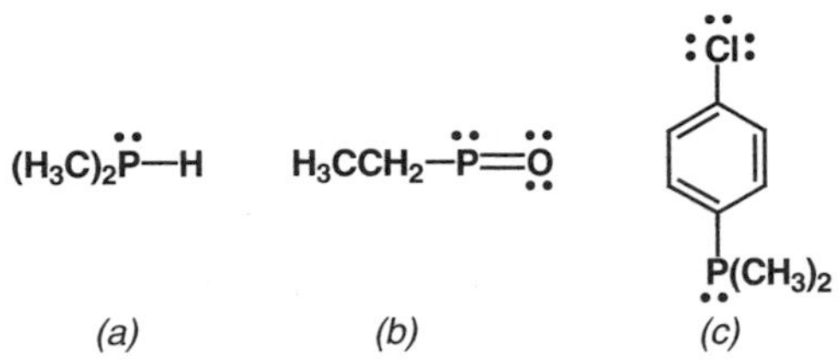

Figure 6.138. (*a*) Dimethylphosphine; (*b*) ethyloxophosphine; (*c*) 1-chloro-4-dimethylphosphinobenzene.

generic prefix name *phosphino-*, for example, 1-chloro-4-dimethylphosphinobenzene (Fig. 6.138*c*). In all these compounds, phosphorus has a valence of 3.

6.139 PHOSPHORANES, R_5P

Compounds in which one or more of the hydrogen atoms of phosphorane, PH_5, are replaced by R groups.

Example. Triethylphosphorane (Fig. 6.139*a*). When the group $-PR_4$, R = H, alkyl, or aryl is treated as a substituent, it is given the generic prefix name *phosphoranyl-*, for example, 2,3-dimethyl-1-ethylphosphoranylbenzene (Fig. 6.139*b*).

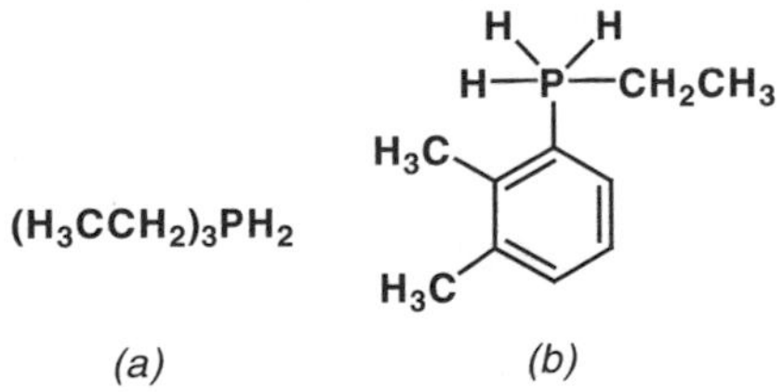

Figure 6.139. (*a*) Triethylphosphorane and (*b*) 2,3-dimethyl-1-ethylphosphoranylbenzene.

6.140 PHOSPHINE OXIDES, $RPH_2(O)$, $R_2PH(O)$, AND R_3PO

Phosphoranes in which two of the R groups are replaced by a doubly bonded oxygen.

Example. Triphenylphosphine oxide (Fig. 6.140*a*); trimethylenebis(phosphine oxide) (Fig. 6.140*b*). Most phosphines are oxidized in air to give the corresponding phosphine oxides.

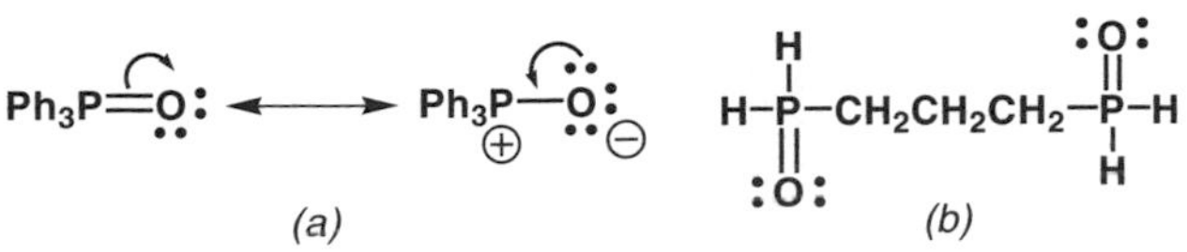

Figure 6.140. (*a*) Triphenylphosphine oxide and (*b*) trimethylenebis(phosphine oxide).

6.141 PHOSPHINE IMIDES, $R_3P{=}NH$

Analogous to phosphine oxides (Sect. 6.140), but with an imino (=NH) group rather than an isoelectronic oxygen atom attached to phosphorus.

Example. *P,P,P*-Triphenylphosphine imide (Fig. 6.141).

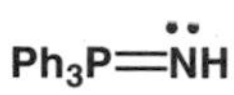

Figure 6.141. *P,P,P*-Triphenylphosphine imide.

6.142 ORGANOPHOSPHORUS ACIDS

Organophosphorus compounds containing one or more hydroxyl groups bonded to phosphorus.

6.143 PHOSPHATES, $(RO)_3P{=}O$

Esters of phosphoric acid, $P(O)(OH)_3$ (also called orthophosphoric acid), in which one (or more) of the hydrogens is replaced by R groups.

Example. Trimethyl phosphate (Fig. 6.143*a*); di-*n*-propyl hydrogen phosphate (Fig. 6.143*b*); ethyl dihydrogen phosphate (Fig. 6.143*c*).

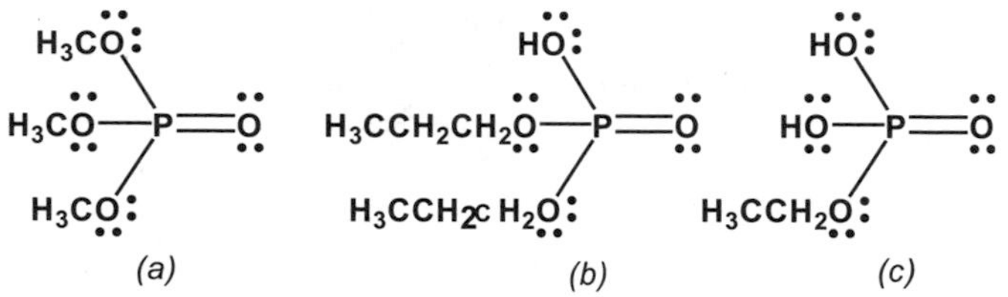

Figure 6.143. (*a*) Trimethyl phosphate; (*b*) di-*n*-propyl hydrogen phosphate; (*c*) ethyl dihydrogen phosphate.

6.144 PHOSPHINIC ACID, $H_2P(O)OH$

A four coordinate phosphorus acid. This acid is in equilibrium with a three coordinate species called *phosphonous acid* (Sect. 6.146). These acids, by most definitions, are

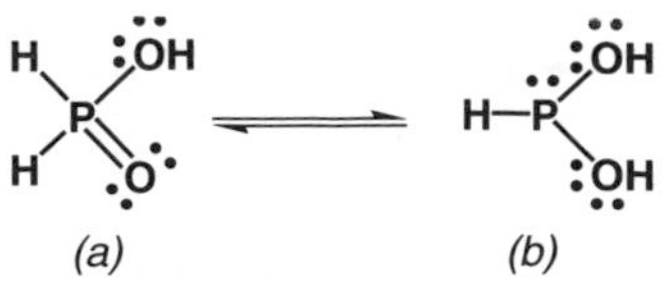

Figure 6.144. (*a*) phosphinic acid; (*b*) phosphonous acid

inorganic acids, but because the distinction between organic and inorganic compounds of phosphorus is often difficult to determine, these and several other acids of phosphorus are included in this chapter.

Example. The equilibrium between phosphinic acid and phosphonous acid, Figure 6.144.

6.145 HYPOPHOSPHOROUS ACID, $H_2P(O)OH$

Synonymous with phosphinic acid (Sect. 6.144), which is the preferred name.

6.146 PHOSPHONOUS ACID, $HP(OH)_2$

A three coordinate phosphorus acid in equilibrium with phosphinic acid (Sect. 6.144).

6.147 ALKYL OR ARYL PHOSPHONOUS ACIDS, $RP(OH)_2$

Compounds in which the hydrogen atom attached to phosphorus in phosphonous acid is replaced by R.

Example. Ethylphosphonous acid, Fig. 6.147.

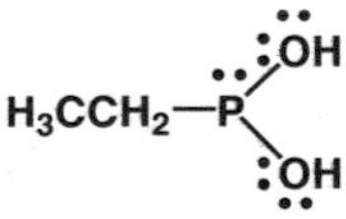

Figure 6.147. Ethylphosphonous acid.

6.148 ALKYL OR ARYL PHOSPHINIC ACIDS, RPH(O)OH

Compounds in which one of the hydrogen atoms attached to phosphorus in phosphinic acid is replaced by R– or Ar–.

Example. Phenylphosphinic acid (Fig. 6.148).

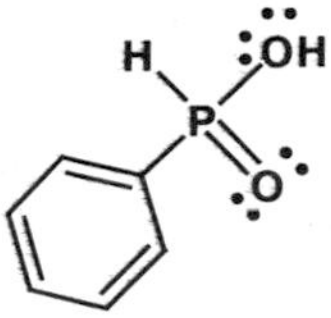

Figure 6.148. Phenylphosphinic acid.

6.149 DIALKYL OR DIARYL PHOSPHINIC ACIDS, $R_2P(O)OH$

Compounds in which both hydrogen atoms attached to phosphorus in phosphinic acid are replaced by R groups.

Example. Ethylmethylphosphinic acid (Fig. 6.149).

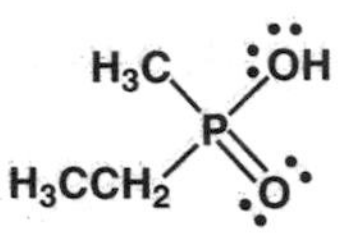

Figure 6.149. Ethylmethylphosphinic acid.

6.150 PHOSPHINOUS ACID, H_2POH

A three-coordinate, monobasic phosphorus acid.

6.151 DIALKYL OR DIARYL PHOSPHINOUS ACIDS, R_2POH

Compounds in which the hydrogen atoms of phosphinous acid are replaced by alkyl or aryl groups.

Example. Dimethylphosphinous acid (Fig. 6.151).

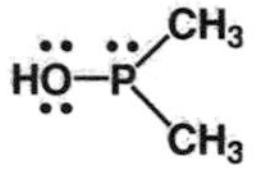

Figure 6.151. Dimethylphosphinous acid.

6.152 PHOSPHONIC ACID, $HP(O)(OH)_2$

A four-coordinate phosphorus acid. This acid is in equilibrium with the three-coordinate species called phosphorous acid.

Example. The equilibrium between phosphonic acid and phosphorous acid (Sect. 6.154) (Fig. 6.152).

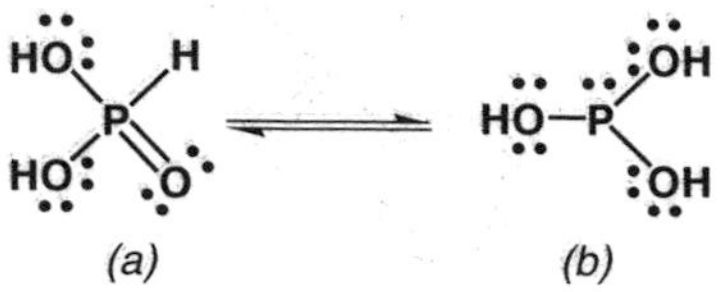

Figure 6.152. (*a*) Phosphonic acid and (*b*) phosphorous acid.

6.153 ALKYL OR ARYL PHOSPHONIC ACIDS, $RP(O)(OH)_2$

Compounds in which the H atom bonded to phosphorus in a phosphonic acid is replaced by an R group.

Example. Ethylphosphonic acid (Fig. 6.153*a*). One or both of the hydrogens on the oxygen atom may be replaced to give phosphonates. Dimethyl methyl phosphonate (Fig. 6.153*b*) is the starting material for the synthesis of highly toxic sarin, (Fig. 6.153*c*) the substance used in a 1995 terror attack in the Japanese subway system.

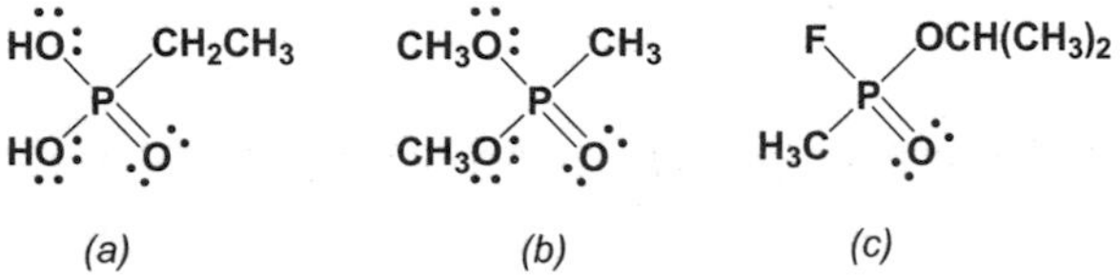

Figure 6.153. (*a*) Ethylphosphonic acid; (*b*) dimethyl methyl phosphonate; (*c*) sarin.

6.154 PHOSPHOROUS ACID, $P(OH)_3$

A three-coordinate phosphorus acid containing three hydroxyl groups attached to phosphorus. This acid is in equilibrium with the four-coordinate species, phosphonic acid (Sect. 6.152).

6.155 PHOSPHITES, $(RO)_3P$

Derivatives of phosphorous acid. The trialkyl or triaryl phosphites are compounds in which the three hydrogen atoms in $P(OH)_3$ are replaced by R groups. The dialkyl phosphites are better formulated as four-coordinate species, $HP(O)(OR)_2$, while the monoalkyl phosphites are monobasic, four-coordinate acids (phosphonic acid structure) HP(O)(OR)(OH).

Example. Trimethyl phosphite (Fig. 6.155*a*). If the dialkyl phosphite were generically related to phosphorus acid, it would have structure in Fig. 6.155*b* and be called

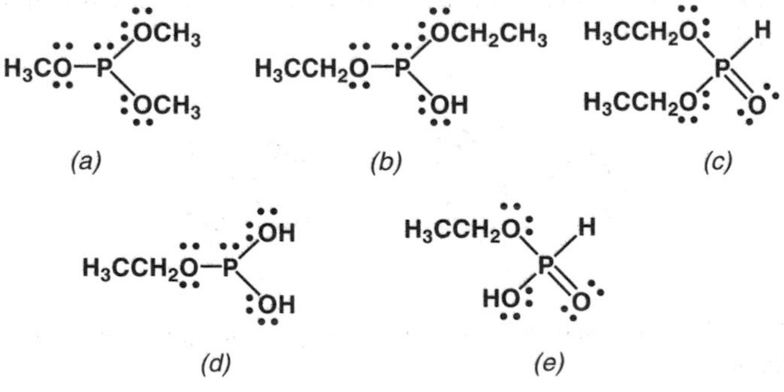

Figure 6.155. (*a*) Trimethyl phosphite; (*b*) diethyl hydrogen phosphite; (*c*) diethyl phosphonate; (*d*) ethyl dihydrogen phosphite; (*e*) monoethyl phosphonate.

diethyl hydrogen phosphite. Actually, diethyl phosphite has the structure of Fig. 6.155*c* and is the diester of phosphonic acid. The monoalkyl phosphite, if it were related to phosphorus acid, would have the structure $ROP(OH)_2$, for example, ethyl dihydrogen phosphite (Fig. 6.155*d*). In fact, the compound has the structure in Fig. 6.155*e* and is a monoester of phosphonic acid.

6.156 HETEROCYCLIC PHOSPHORUS COMPOUNDS

Organic compounds in which a phosphorus atom has replaced a carbon atom in a carbocyclic system.

Example. 2-Phosphanaphthalene (Fig. 6.156*a*) and 1-phosphacyclopentane-1-oxide (Fig. 6.156*b*).

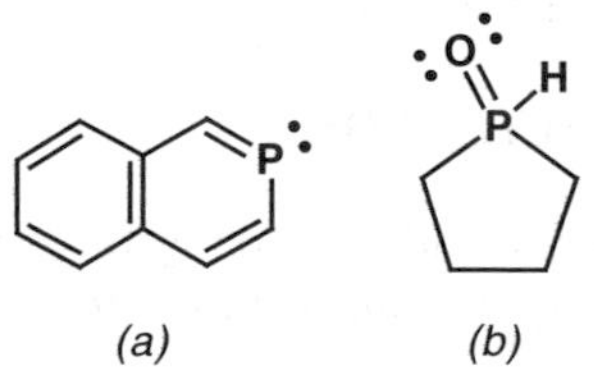

Figure 6.156. (*a*) 2-Phosphanaphthalene and (*b*) 1-phosphacyclopentane-1-oxide.

6.157 HALOPHOSPHORUS COMPOUNDS

Organophosphorus compounds that possess halogens attached to phosphorus.

Example. Phenylphosphonous dichloride (Fig. 6.157*a*). This compound is more commonly called phenyldichlorophosphine, but according to recommended nomenclature, when one or more halogens are attached to a phosphorus atom carrying no OH or OR groups, the compound should be named as an acid halide; in this example the corresponding acid is phenylphosphonous acid, $PhP(OH)_2$. Dimethylphosphinic chloride (Fig. 6.157*b*).

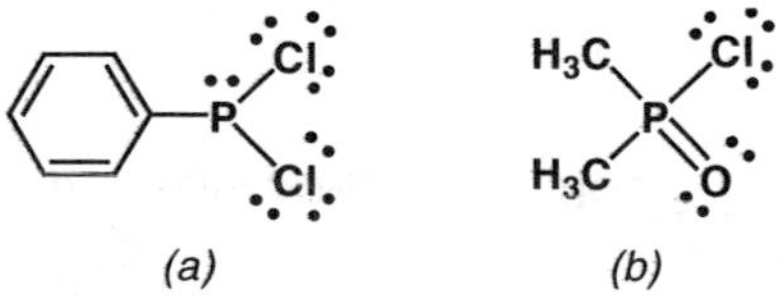

Figure 6.157. (*a*) Phenylphosphonous dichloride (phenyldichlorophosphine) and (*b*) dimethylphosphinic chloride.

6.158 ORGANOPHOSPHORUS ACID AMIDES

Organophosphorus acids having one or more NH_2 groups attached to a phosphorus atom.

Example. P,P-Diphenylphosphinous amide (Fig. 6.158*a*), also called aminodiphenylphosphine but preferably named after diphenylphosphinous acid, $(Ph)_2P\text{-}OH$. Compounds carrying no OH or OR groups attached to phosphorus are preferably called amides. *P*-Methylphosphonic diamide (Fig. 6.158*b*).

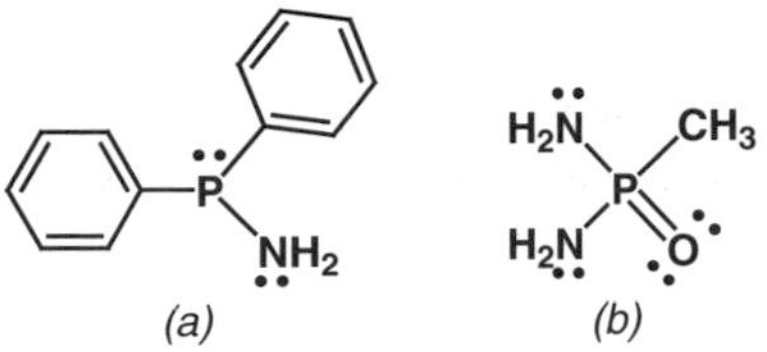

Figure 6.158. (a) *P,P*-Diphenylphosphinous amide (aminodiphenylphosphine) and (*b*) *P*-methylphosphonic diamide.

6.159 ORGANOPHOSPHORUS THIOACIDS

Organophosphorus acids and esters in which one or more sulfur atoms are attached to phosphorus.

Example. Dimethylphosphinothioic *O*-acid (Fig. 6.159*a*) and methylphosphonomonothioic *S*-acid, (Fig. 6.159*b*).

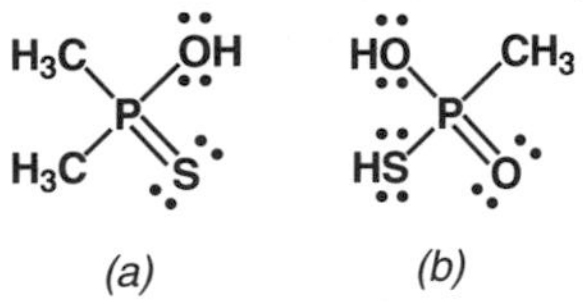

Figure 6.159. (*a*) Dimethylphosphinothioic *O*-acid and (*b*) methylphosphonomonothioic *S*-acid.

6.160 PHOSPHONIUM SALTS, $R_4P^+ X^-$

Tetracoordinated phosphorus cations.

Example. Ethyltrimethylphosphonium bromide (Fig. 6.160).

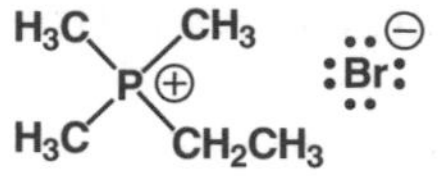

Figure 6.160. Ethyltrimethylphosphonium bromide.

6.161 ACID HALIDES OF INORGANIC ACIDS

A class of inorganic compounds, some of which are used extensively in the preparation of organic compounds. The inorganic acid halides of sulfur and phosphorus are particularly useful for this purpose.

Example. The acid halides of phosphoric and phosphorous acid (Figs. 6.161*a* and *b*, respectively) and those of sulfurous and sulfuric acids (Figs. 6.161*c* and *d*, respectively).

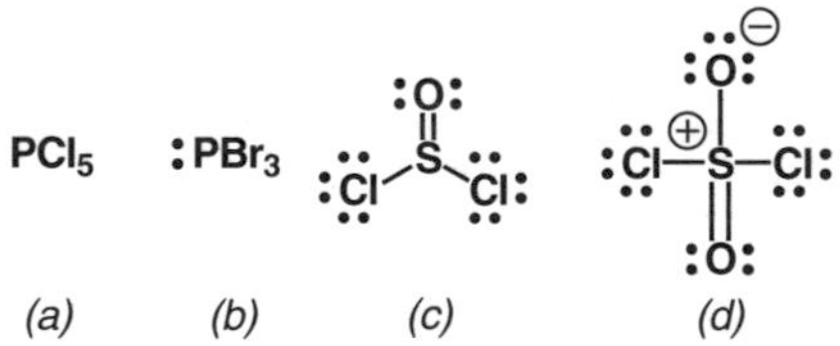

Figure 6.161. (*a*) Phosphorus pentachloride; (*b*) phosphorus tribromide; (*c*) thionyl chloride; (*d*) sulfuryl chloride.

SUGGESTED READING

Fletcher, J. H., Dermer, O. C., and Fox, R.B., eds. *Nomenclature of Organic Compounds.* Advances in Chemistry Series 126, American Chemical Society: Washington, DC, 1974.

Katritsky, A. R., ed. *Advances in Heterocyclic Chemistry*. Academic Press: New York, The most recent of the 26 volumes appeared in 1994.

Organophosphorus Chemistry, published by the Royal Society of Chemistry, Cambridge, UK. There are 24 volumes in the this series; the most recent was published in 1993.

Patai, S. and Rappoport, Z. *The Chemistry of Functional Groups*. Wiley Interscience: New York, 1994. This series was first published in 1964; 30 years later approximately 81 volumes have appeared. The most recent volume is "The Chemistry of Enamines, Part 2."

Quin. L. D. *The Heterocyclic Chemistry of Phosphorus*. Wiley-Interscience: New York, 1981.

Sandler, S. R. and Karo, W. *Organic Functional Group Preparations*. Vols. I, II, and III. Academic Press: New York, The 2nd ed. of Vol. III appeared in 1989.

Weissberger, A. and Taylor, E. C., eds. *The Chemistry of Heterocyclic Compounds*: John Wiley and Sons: New York. This series appears to have ended with Vol. 26 in 1972.

7 Molecular Structure Isomers, Stereochemistry, and Conformational Analysis

7.1	Stereochemistry	225
7.2	Connectivity (Constitution)	225
7.3	Configuration	226
7.4	Molecular Structure	226
7.5	Representations of Connectivity in Molecular Structure	226
7.6	Representations of Configuration in Molecular Structure	227
7.7	Isomers	228
7.8	Constitutional Isomers (Structural Isomers)	229
7.9	Functional Group Isomers	229
7.10	Positional Isomers	229
7.11	Skeletal Isomers	229
7.12	Stereoisomers (Stereochemical Isomer, Configurational Isomers)	230
7.13	Stereogenic	230
7.14	Stereogenic Center	230
7.15	Enantiomers (Enantiomorphs)	230
7.16	Diastereomers (Diastereoisomers)	230
7.17	Enantiogenic	231
7.18	Enantiogenic Center (Enantio Center, Chiral Center, Asymmetric Center)	231
7.19	Diastereogenic	231
7.20	Diastereogenic Center (Diastereocenter)	231
7.21	Chiral	232
7.22	Achiral	232
7.23	Descriptor	232
7.24	Geometrical Isomers	232
7.25	*cis* and *trans* Configurations	232
7.26	*r* (Reference)	234
7.27	*syn* and *anti* Configurations	234
7.28	*exo* and *endo* Configurations	234
7.29	α and β Configurations	235
7.30	CIP (Cahn–Ingold–Prelog) Priority	236
7.31	*Z* and *E* Configurations	237
7.32	(*R*) and (*S*) Configurations	237

The Vocabulary and Concepts of Organic Chemistry, Second Edition, by Milton Orchin, Roger S. Macomber, Allan Pinhas, and R. Marshall Wilson,

7.33	Absolute Configuration	238
7.34	Relative Configuration	238
7.35	*meso* Configuration	239
7.36	*erythro* and *threo* Configurations	240
7.37	Electromagnetic Radiation (Light)	240
7.38	Plane-Polarized Light (PPL)	240
7.39	Unpolarized Light	241
7.40	Generation of Plane-Polarized Light	242
7.41	Circularly-Polarized Light (CPL)	243
7.42	Relationships Between Plane- and Circularly-Polarized Light	244
7.43	Elliptically-Polarized Light	245
7.44	Birefringence	245
7.45	Optical Activity	247
7.46	Optical Isomers (Optical Antipodes)	248
7.47	Optical Rotation α (Observed Rotation)	248
7.48	Specific Rotation $[\alpha]$	248
7.49	Polarimeter	249
7.50	(+)-Enantiomer and (−)-Enantiomer	249
7.51	*d*-(Dextrorotatory) and *l*-(Levorotatory) Enantiomers	249
7.52	D-Enantiomer and L-Enantiomer	250
7.53	Inversion (of Configuration)	250
7.54	Retention (of Configuration)	251
7.55	Racemic Modification (*d,l* Pair)	251
7.56	Conglomerate (Racemic Mixture)	251
7.57	Racemate	252
7.58	Racemize	252
7.59	Resolution	252
7.60	Rules for Correlation of Configuration with Optical Rotation	252
7.61	Enantiomeric Purity	254
7.62	Enantiomeric Excess (ee, Optical Purity)	254
7.63	Scalemic Mixture	254
7.64	Optical Yield	254
7.65	Optical Stability	255
7.66	Homotopic Atoms or Groups	255
7.67	Symmetry Equivalent Atoms or Groups	255
7.68	Chemically Equivalent Atoms or Groups	256
7.69	Equivalent Faces	256
7.70	Enantiotopic Atoms or Groups	257
7.71	Prochiral Atoms or Groups	257
7.72	Enantiotopic (Prochiral) Faces	258
7.73	*re* and *si* Faces	259
7.74	Diastereotopic Atoms or Groups	259
7.75	Diastereotopic Faces	260
7.76	Symmetry Classification of Atoms, Groups, and Faces (Summary)	261
7.77	Ogston's Principle	261
7.78	Like *l* and Unlike *u*	262
7.79	Epimers	262
7.80	Epimerize	263
7.81	Anomers	263

7.82	Mutarotation	263
7.83	Pseudo-Asymmetric Center	264
7.84	Stereospecific Reaction	264
7.85	Stereoselective Reaction	265
7.86	Enantioselective Reaction	266
7.87	Diastereoselective Reaction	266
7.88	Regioselective Reaction	266
7.89	Regiospecific Reaction	267
7.90	Asymmetric Induction	267
7.91	Asymmetric Synthesis	267
7.92	Chiral Auxiliary	268
7.93	Chiral Pool	269
7.94	Chiral Solvents	269
7.95	Cram's Rule	269
7.96	Chemzymes	270
7.97	Strain	270
7.98	Bond-Stretching Strain	271
7.99	Angle Strain (Baeyer Strain, I Strain)	271
7.100	Steric Strain (Prelog Strain, F Strain)	272
7.101	Torsional Strain (Eclipsing Strain, Pitzer Strain)	272
7.102	Ring Strain	272
7.103	Torsional Angle (Dihedral Angle)	272
7.104	Conformation (Conformer, Rotomers)	273
7.105	Conformational Isomer	273
7.106	Invertomers (Inversion Isomers)	274
7.107	gem (Geminal)	274
7.108	Vic (Vicinal)	275
7.109	Eclipsed Conformation	275
7.110	Staggered Conformation	275
7.111	*cis* Conformation (Syn-Periplanar)	276
7.112	*gauche* Conformation (Skew Conformation, Synclinal)	276
7.113	Anticlinal Eclipsed Conformation	277
7.114	*anti* Conformation (*trans* Conformation, Antiperiplanar)	277
7.115	*s-cis*	277
7.116	*s-trans*	277
7.117	W Conformation and Sickle Conformation	277
7.118	Ring Puckering	277
7.119	Chair Conformation	279
7.120	Axial Positions	279
7.121	Equatorial Positions	279
7.122	Ring Flipping	279
7.123	Configurations of Disubstituted Chair Cyclohexanes	279
7.124	Boat Conformation	280
7.125	Pseudo-Equatorial	280
7.126	Pseudo-Axial	280
7.127	Flagpole	281
7.128	Bowsprit	281
7.129	Twist Boat (Skew Boat)	281
7.130	A Values	281

7.131 Folded, Envelope, Tub, and Crown Conformations 282
7.132 Anancomeric Ring 282
7.133 Atropisomers 282
7.134 In(side), Out(side) Isomerism 283
7.135 Homeomorphic Isomers 283
7.136 Isotopomers 283
7.137 Apical and Equatorial Positions of a Trigonal Bipyramid (TBP) 284
7.138 Apicophilicity 284
7.139 Berry Pseudo-Rotation 284
7.140 Walden Inversion 284
7.141 Octahedral Configuration 285
7.142 Helicity (*M*, *P*) 286
7.143 Allylic Isomers 286
7.144 Tautomers 286
7.145 Valence Isomers (Valence Tautomers) 287
7.146 Fluxional Molecules 287
7.147 Stereoelectronic Effects 288
7.148 Regiochemistry and Stereochemistry of Ring Closure Reactions 288

The notion that all molecules of a pure compound have the same well-defined structure is at the core of organic chemistry, indeed of all chemistry. However, exactly what constitutes molecular structure is not without ambiguity. If a compound is a crystalline solid and its X-ray structure is determined, then its molecular structure is viewed as the highly ordered arrangement of its constituent atoms with respect to measured bond distances and bond angles. From this information not only can the bonding sequence of the molecule (*i.e.*, the connectivity) be derived, but also its configuration, that is, the three-dimensional relationship of all the atoms. By contrast, in the gas or liquid phase the atoms constituting a molecule are in constant relative motion, and distances and angles are constantly changing, though time-averaged, as well as "instantaneous," and values for these parameters can be determined. As a result of this motion, the exact structure of a molecule depends on the time scale of the observation, which may range from hours to picoseconds (10^{-12} s). Thus, a molecule may have an infinite number of instantaneous arrangements of its atoms (called conformations), while still having only one connectivity and one configuration. Since the time-dependent conformations do not affect the time-independent connectivity and configuration of the molecule, conformations are considered distinct from isomers in this chapter. Nonetheless, there are molecules that can be isomeric under one set of conditions, and conformations under another.

One of the most important developments in the concept of molecular structure resulted from the revolutionary work of Louis Pasteur (1822–1895). In 1848, Pasteur crystallized the racemic form of sodium ammonium tartrate and noted that two crystalline forms of the salt resulted that were related as nonsuperimposable mirror images. Using tweezers to separate the two forms, he found that each form rotated the plane of polarized light by an equal amount but in opposite directions. He suggested that since the crystals of the two forms exhibited chirality, the molecules themselves might also be chiral (*i.e.*, enantiomers). Subsequently, the independent proposals of Jacobus H. van't Hoff (1852–1911) and Joseph A. LeBel (1847–1930) regarding the tetrahedral nature of the carbon atom rationalized the stereochemical implications of Pasteur's suggestion. Subsequent developments in cyclohexane stereochemistry and conformational analysis, in which D. H. R. Barton (1918–1998; he shared the 1969 Nobel Prize with O. Hassel) played a major role, have helped in the understanding of the chemical behavior of many complex ring systems such as the steroids.

In a more recent development, A. Ogston (1911–1996) has described a theory of how three-dimensional chirality can be projected onto a two-dimensional molecular surface. In today's sophisticated pharmaceutical industry, it is recognized that enantiomers present in racemic mixtures of biologically active drugs can no longer be regarded as having comparable or inconsequential activity. The tragedy of thalidomide where one enantiomer exhibits useful medicinal properties but the other enantiomer is a potent teratogen has led to the search for methodologies that produce only the desired enantiomer even when the other enantiomer can be established to be totally benign. This chapter is devoted to the terms and concepts that are essential to the understanding of chemical structure, including its three-dimensional aspects.

7.1 STEREOCHEMISTRY

The subdiscipline of chemistry dealing with the spatial (i.e., three-dimensional) relationships between the atoms in a molecule. See also Sect. 7.3.

7.2 CONNECTIVITY (CONSTITUTION)

A specification of the bonding sequence of atoms within a molecule, that is, the sequence by which atoms are interconnected and the type of bonds involved in these interconnections. Connectivity alone does not convey three-dimensional information.

Example. Both ethanol and dimethyl ether (Fig. 7.2) have the same molecular formula (C_2H_6O), but they possess different connectivity.

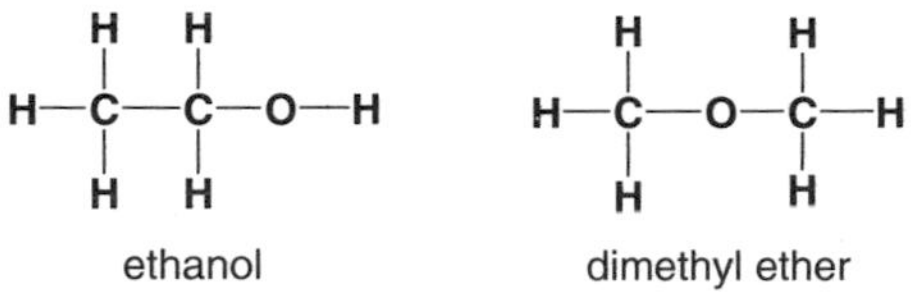

Figure 7.2. Different connectivites of ethanol and dimethyl ether.

7.3 CONFIGURATION

A specification of the fixed three-dimensional relationship between the atoms in a molecule. The term *stereochemistry* (Sect. 7.1) is often used as equivalent to the configuration.

7.4 MOLECULAR STRUCTURE

A specification of the connectivity and configuration of the atoms in a molecule. (Several representations of molecular structure are described in Sects. 7.5 and 7.6.)

7.5 REPRESENTATIONS OF CONNECTIVITY IN MOLECULAR STRUCTURE

These are depictions of molecular structure showing only connectivity between atoms; no three-dimensional information is implied. For purposes of comparison, the same molecule, 2,3,4,5-tetrahydroxypentanoic acid, is used in Figs. 7.5*a* through *c* to illustrate all three representations.

A *Kekulé (line-bond)* structure (Fig. 7.5*a*) is a planar representation in which all atoms are shown by the appropriate atomic symbol and each bond is represented by a dash. In a *condensed* structure (Fig. 7.5*b*), all the atoms and functional groups are written in a single line of text, with bonds not explicitly shown. In a *skeletal* structure, the carbon backbone is depicted by connected lines denoting carbon–carbon bonds, and chemical symbols for all heteroatoms (atoms other than C and H), and the hydrogens attached to them, are shown in Fig. 7.5*c*. There is understood to be a carbon at each end of any line where no heteroatom appears, and each carbon is

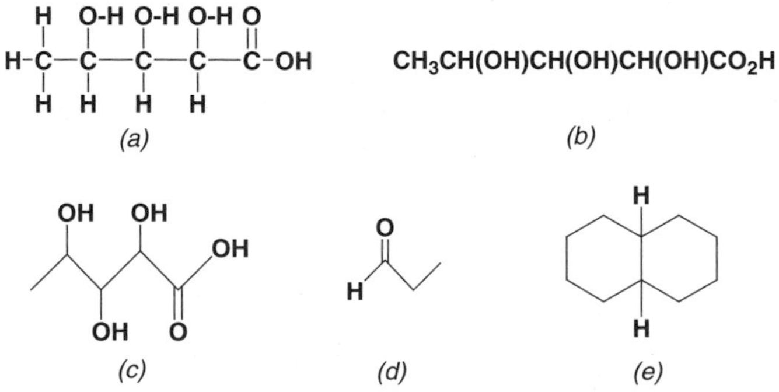

Figure 7.5. Representations of the structure of 2,3,4-trihydroxypentanoic acid: (*a*) Kekulé structure; (*b*) condensed structure; (*c*) skeletal structure; skeletal structures of (*d*) propionaldehyde and (*e*) decalin (decahydronaphthalene).

understood to be connected to the number of hydrogens required to fulfill its valence of 4. Occasionally, certain hydrogen atoms bonded to carbon atoms are shown explicitly [*e.g.*, in the carbonyl group of aldehydes (Fig. 7.5*d*) and at the ring junctions of carbocycles (Fig. 7.5*e*)] in order to provide emphasis.

7.6 REPRESENTATIONS OF CONFIGURATION IN MOLECULAR STRUCTURE

There are several ways to convey three-dimensional perspective in a two-dimensional drawing of molecular structure:

A. The *zigzag structure* (Fig. 7.6*a*) is a variant of the skeletal representation (Sect. 7.5), where each carbon atom along the carbon backbone is understood to be *in* the plane of the paper. The three-dimensional position of atoms and groups connected to the chain is conveyed by solid wedges for those located *above* the plane and dashed lines (or dashed wedges) for groups located *below* the plane.

B. The *sawhorse (andiron) structure* (Fig. 7.6*b*) focuses attention on the relative positions of the three substituents attached to the tetrahedral carbon atoms 3 and 4. This is done by showing these two carbon atoms as the ends of a sawhorse. Figure 7.6*d* shows the same sawhorse structure in which the groups around carbon 3 and carbon 4 are rotated to a conformation that minimizes steric interactions between the two largest groups on the adjacent carbon atoms. Such rotations are conformational changes and do not alter the configuration. Due to an optical illusion with this sawhorse representation, there can be ambiguity as to which set of groups [H, OH, CH_3 vs. H, OH, $CH(OH)CO_2H$] is "in front" and which set is "behind." For this reason, solid and dashed wedges are sometimes used in conjunction with the sawhorse to give a structure resembling the zigzag structure (Fig. 7.6*c*).

C. The *Newman projection* (see Sect. 4.8) is a representation in which the viewer is looking directly down the carbon–carbon bond of interest. Figure 7.6*e* is a Newman projection of Fig. 7.6*d*.

D. Substituted *cyclic compounds* can be depicted by thickening ring bonds that are closer to the viewer (Fig. 7.6*f*), or by drawing the carbocycle itself as a regular polygon and using wedges and dashes to indicate the positions of attached groups (Fig. 7.6*g*). The latter representation unfortunately fails to convey the fact that most carbocyclic rings are not planar. Hydrogen atoms located *above* the rings at a ring fusion can be implied by a solid dot on the carbon (Fig. 7.6*h*); the hydrogen *below* the rings is not depicted.

E. In the *Fischer projection* [devised by Emil Fischer (1852–1919)], the carbon backbone is arranged vertically (Fig. 7.6*i*). The intersection of the backbone with each horizontal line represents a tetrahedral backbone carbon, with the horizontal line depicting bonds to the substituents of that carbon. By convention,

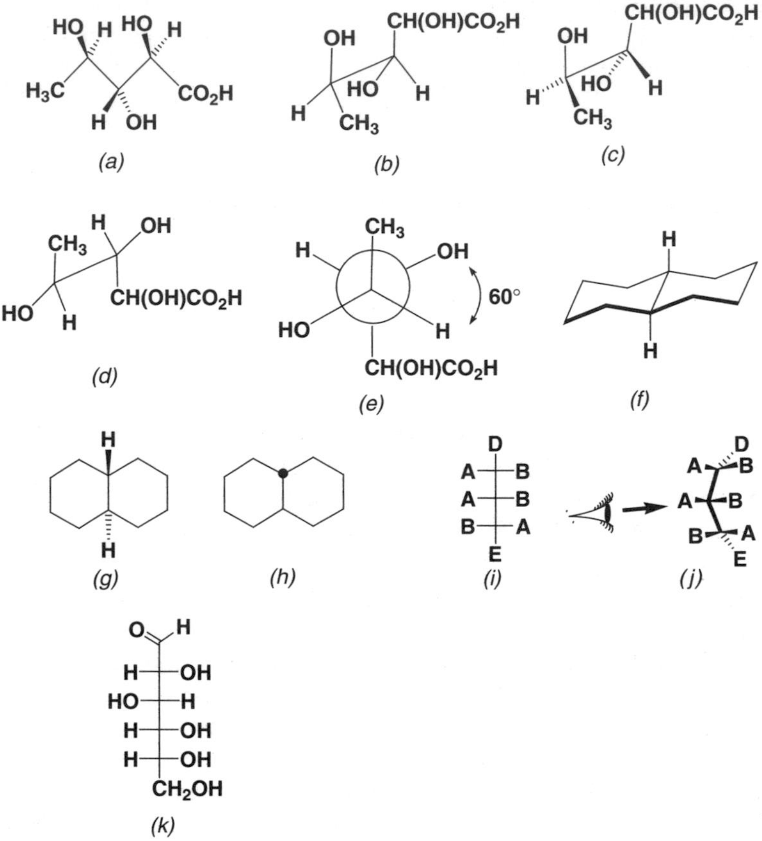

Figure 7.6. (*a*) Zigzag structure; (*b*) sawhorse structure; (*c*) modified sawhorse structure; (*d*) rotated sawhorse structure; (*e*) Newman projection of *d*; (*f, g, h*) depictions of cyclic molecules; (*i*) Fischer projection of structure (*j*); (*k*) D-glucose.

the horizontal substituents are understood to be above the page, while the backbone carbons above and below the horizontal line are located behind the page (i.e., Fig. 7.6*i* is the Fischer projection of the perspective drawing, Fig. 7.6*j*). The Fischer projection is commonly used in carbohydrate chemistry, as illustrated by the structure of the acyclic form of D-glucose (Fig. 7.6*k*).

7.7 ISOMERS

Different molecular structures that have the same molecular formula (i.e., the same number and kinds of atoms). Isomers are subdivided into two major types: constitutional (or structural) isomers and stereoisomers.

7.8 CONSTITUTIONAL ISOMERS (STRUCTURAL ISOMERS)

Isomers whose structures differ only in the connectivity of their atoms. Examples include functional group (Sect. 7.9), positional (Sect. 7.10), and skeletal isomers (Sect. 7.11).

7.9 FUNCTIONAL GROUP ISOMERS

Constitutional isomers with different functional groups.

Example. 1-Butanol, $CH_3(CH_2)_3OH$, and diethyl ether, $CH_3CH_2OCH_2CH_3$, are functional group isomers (Fig. 7.11*a*).

7.10 POSITIONAL ISOMERS

Constitutional isomers that differ only with respect to the point of attachment of a substituent.

Example. 1-Chloropropane ($ClCH_2CH_2CH_3$) and 2-chloropropane ($CH_3CHClCH_3$) (Fig. 7.11*b*).

7.11 SKELETAL ISOMERS

Constitutional isomers that differ with respect to branching of the carbon skeleton.

Example. Octane and 2,5-dimethylhexane (Fig. 7.11*c*).

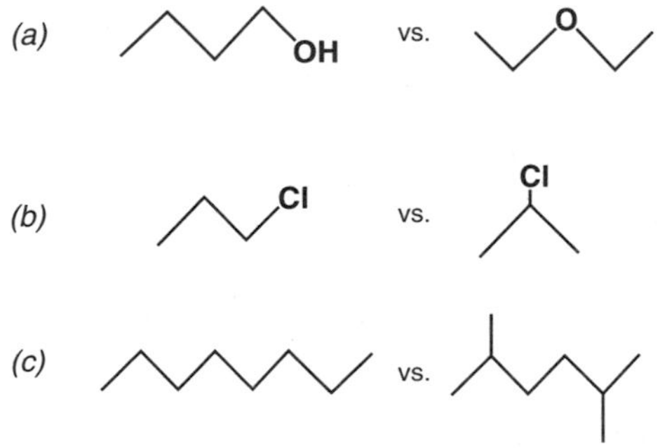

Figure 7.11. (*a*) Functional group isomers; (*b*) positional isomers; (*c*) skeletal isomers.

7.12 STEREOISOMERS (STEREOCHEMICAL ISOMER, CONFIGURATIONAL ISOMERS)

Isomers with the same connectivity but differing in their configurations. Stereoisomers are further subdivided into enantiomers (see Sect. 7.15) and diastereomers (see Sect. 7.16).

7.13 STEREOGENIC

Capable of giving rise to stereoisomers.

7.14 STEREOGENIC CENTER

An atom (or grouping of atoms) within a molecule at which interchange of two substituents gives rise to a stereoisomer of the original molecule.

Example. The two olefinic carbons in a suitably substituted alkene constitute stereogenic centers. Exchange of the H and CH_3 groups on the same olefinic carbon of 2-butene (Fig. 7.14) gives rise to a stereoisomer. Stereogenic centers can be subdivided as enantiogenic (see Sect. 7.18) and diastereogenic (see Sect. 7.20) centers.

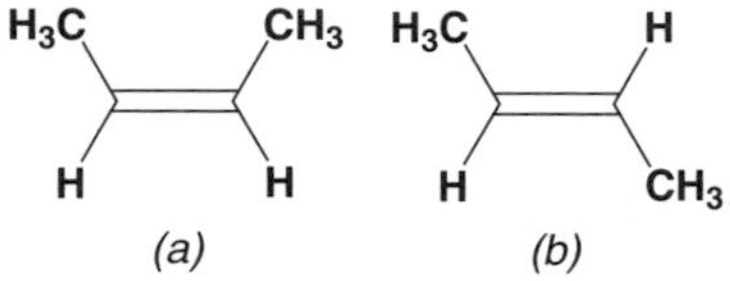

Figure 7.14. (*a*) *cis*-2-butene and (*b*) *trans*-2-butene.

7.15 ENANTIOMERS (ENANTIOMORPHS)

Molecules that are related as an object and its nonsuperimposable mirror image. This is the same as the relationship between the left and the right hand. As with hands, enantiomers arise in pairs. Enantiomers have identical physical and chemical properties under normal (achiral, Sect. 7.22) conditions. See also Sect. 7.45.

Example. bromochlorofluoroiodomethane (Fig. 7.15*a*); 1,3-dimethylallene (Fig. 7.15*b*); and hexahelicene (Fig. 7.15*c*).

7.16 DIASTEREOMERS (DIASTEREOISOMERS)

Stereoisomers that are not enantiomers. Unlike enantiomers, each diastereomer has a unique set of physical and chemical properties. Diastereomers are usually encountered in molecules with multiple stereogenic centers.

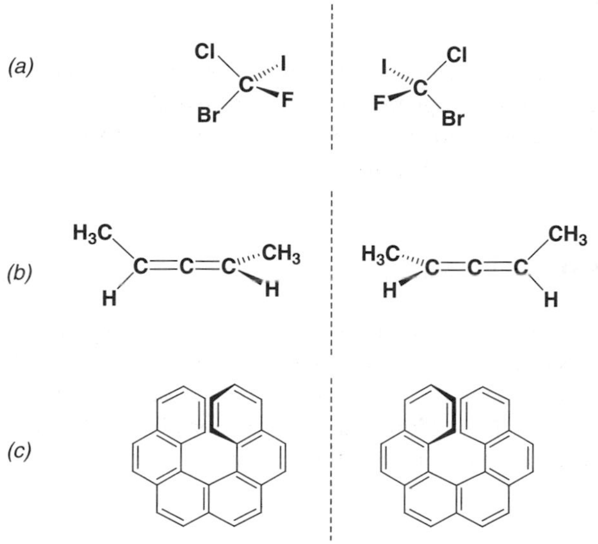

Figure 7.15. Enantiomeric pairs: (*a*) bromochlorofluoroiodomethane; (*b*) 1,3-dimethylallene; (*c*) hexahelicene.

7.17 ENANTIOGENIC

Capable of giving rise to enantiomers.

7.18 ENANTIOGENIC CENTER (ENANTIO CENTER, CHIRAL CENTER, ASYMMETRIC CENTER)

An atom (or grouping of atoms) within a molecule where interchange of any two substituents gives rise to the enantiomer of the original molecule. Enantiogenic center is essentially synonymous with the older terms *chiral center* and *asymmetric center*.

7.19 DIASTEREOGENIC

Capable of giving rise to diastereomers.

7.20 DIASTEREOGENIC CENTER (DIASTEREOCENTER)

An atom (or group of atoms) within a molecule where interchange of two substituents gives rise to a diastereomer of the original molecule. Tetrahedral diastereogenic centers are often designated with an asterisk (*).

7.21 CHIRAL

A molecule (or other object) that possesses the property of handedness or, equivalently, possesses a mirror image that is not superimposable on the original object. In conformationally flexible molecules, every accessible conformation must be chiral for the molecule to be chiral. See also Sect. 7.17.

7.22 ACHIRAL

A molecule (or other object) that lacks handedness or, equivalently, that possesses a mirror image superimposable on the original object. Any object with an alternating axis of any fold [S_1 (plane of symmetry), S_2 (center of symmetry), or any higher-order S_n (see Sect. 4.7)] is achiral. If a conformationally flexible molecule has even one accessible achiral conformation, it is achiral.

7.23 DESCRIPTOR

A label specifying the configuration of atoms about a stereogenic center.

Examples. *cis* and *trans*; *syn* and *anti*; *exo* and *endo*; α and β; *meso*, *erythro*, and *threo*; *R* and *S*; *E* and *Z*; *D* and *L*; *re* and *si*; *P* and *M*; and *like l* and *unlike u* are all descriptors that are defined below. (Descriptors are always italicized when they are part of the name of a molecule.)

7.24 GEOMETRICAL ISOMERS

The class of diastereomers differing in configuration about double bonds or rings. The configurations of geometrical isomers are distinguished by the descriptors *cis* and *trans*, *syn* and *anti*, or *E* and *Z*.

7.25 *CIS* AND *TRANS* CONFIGURATIONS

Two groups possess the *cis* (*c*) configuration when they are on the same side of the plane containing the π bond, or the plane of a ring. Two groups are *trans* (*t*) when they are on opposite sides of the π bond plane or the plane of a ring.

Example. The diastereomers in Figs. 7.25*a*, *b*, and *c* are *cis/trans* pairs. The *cis/trans* descriptors for both cyclic alkanes and acyclic olefins are assigned by the relationship of the two groups that constitute the main chain. If the two groups that constitute the main chain (groups A and B in Fig. 7.25*d*) lie on the same side

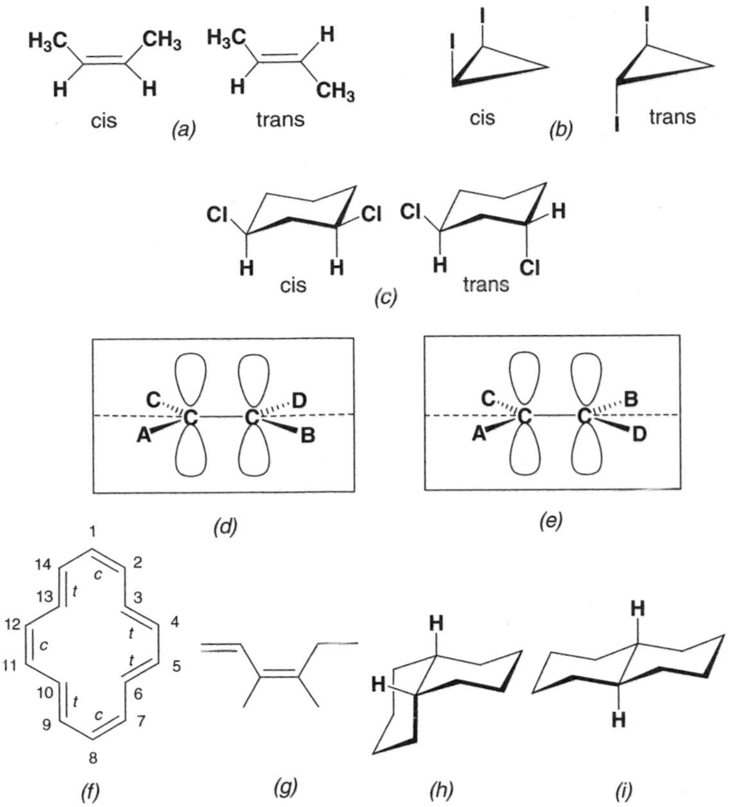

Figure 7.25. Three sets of *cis/trans* isomers: (*a*) 2-butene, (*b*) 1,2-diiodocyclopropane, (*c*) 1,3-dichlorocyclohexane; (*d*) general depiction of a *cis* double bond in relation to the π bond plane; (*e*) general depiction of a *trans* double bond in relation to the π bond plane where (*a*) and (*b*) are parts of the main chain; (*f*) [14]annulene stereochemistry; (*g*) *cis*-3,4-dimethyl-1,3-hexadiene; (*h*) *cis*-decalin; (*i*) *trans*- decalin.

of the π bond plane, the configuration is *cis* at that double bond, whereas if they lie on opposite sides, the configuration is *trans* at the double bond (Fig. 7.25*e*). Thus, the [14]annulene illustrated in Fig. 7.25*f* is *cis, trans, trans, cis, trans, cis, trans*-1,3,5,7,9,11,13-cyclotetradecaheptaene, and Fig. 7.25*g* shows *cis*-3,4-dimethyl-1,3-hexadiene. The terms *cis* and *trans* are also used to designate configuration at a ring junction. It is important to note the relationship of the hydrogens (or other substituents) about the ring junction. Substituents that are on the same side of a ring plane are *cis* (Fig. 7.25*h*) or, on opposite sides, *trans* (Fig. 7.25*i*).

When there is ambiguity as to which groups constitute the main chain, the *E*/*Z* (Sect. 7.31) descriptors are used.

7.26 *r* (REFERENCE)

When there are more than two substituent-bearing atoms in a ring, the symbol *r* designates the substituent to which the others are referenced when assigning *cis* and *trans* stereochemistry.

Example. *trans*-4-Ethyl-*cis*-3-methyl-*r*-1-cyclohexanecarboxylic acid (Fig. 7.26*a*) and methyl 2-cyano-*trans*-3-phenyl-*r*-2-oxiranecarboxylate (Fig. 7.26*b*).

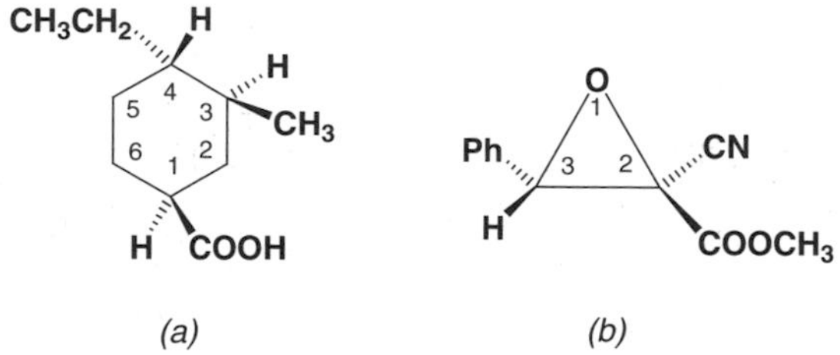

Figure 7.26. (*a*) *trans*-4-Ethyl-*cis*-3-methyl-*r*-1-cyclohexanecarboxylic acid and (*b*) methyl 2-cyano-*trans*-3-phenyl-*r*-2-oxiranecarboxylate.

7.27 *SYN* AND *ANTI* CONFIGURATIONS

Syn, a term similar in meaning to *cis*, and *anti*, a term similar in meaning to *trans*, are used to name configurations in such structures as oximes, azo compounds, and multiple rings.

Example. Figure 7.27*a*. For special cases, as in dioximes (Fig. 7.27b), the relationship of the oxime substituents requires a third descriptor. Thus, dioximes are described as *anti*, *syn*, or *amphi* (both of the hydroxyl groups pointing in a direction parallel to one another), respectively.

7.28 *EXO* AND *ENDO* CONFIGURATIONS

Exo refers to the position in a bicyclic molecule nearer the reference bridge. The reference bridge is determined by applying the priorities (with the highest first): (1) bridge with hetero atoms; (2) bridge with fewest members; (3) saturated bridge; (4) bridge with fewest substituents; (5) bridge with the lowest-priority substituents. The *exo* substituent is considered to be on the “outside” or less hindered position. *Endo* represents the position in a bicyclic molecule opposite the reference bridge. The *endo* substituent is considered to be on the “inside” or more hindered position.

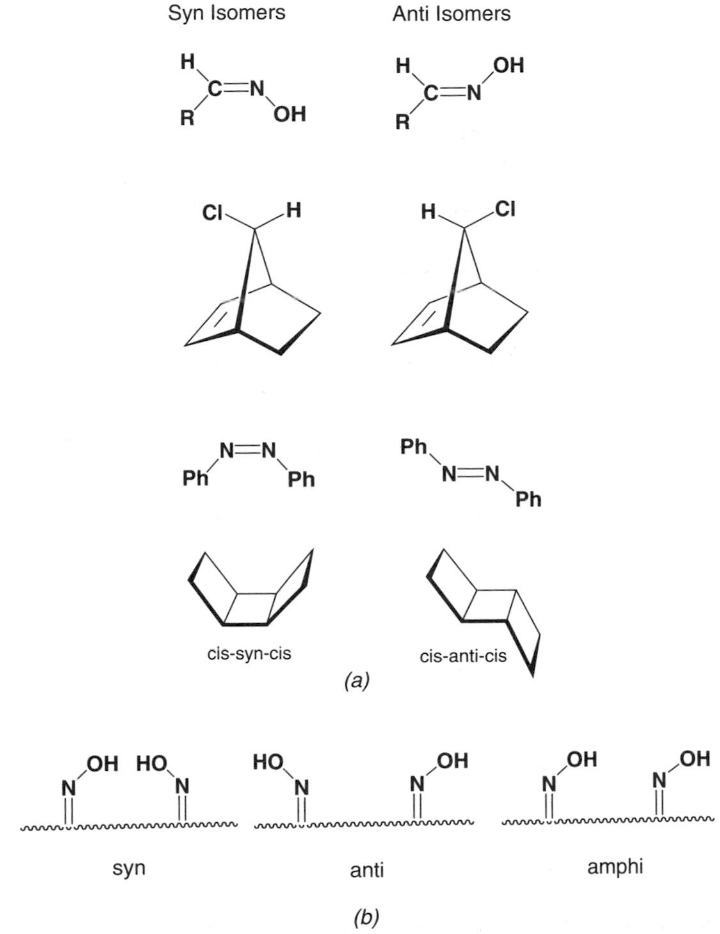

Figure 7.27. (*a*) *Syn* and *anti* isomers and (*b*) dioxime stereoisomers.

Example. The *exo* and *endo* positions in a bicyclo[2.2.1]heptane and generalized bicyclic system are given in Fig. 7.28.

7.29 α AND β CONFIGURATIONS

The term α configuration is used primarily with steroid-type molecules to denote a downward-directed substituent when the molecule is viewed as shown in Fig. 7.29*a*, or a backward-projecting substituent when viewed as in Fig. 7.29*b*. The term β configuration is used to denote the upward-directed substituent when the molecule is

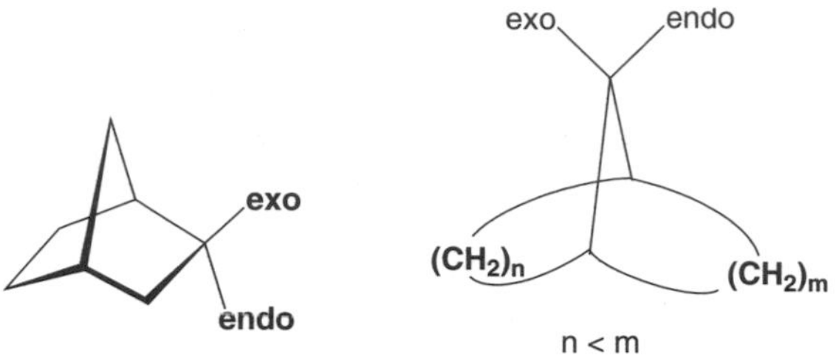

Figure 7.28. *Exo* and *endo* isomers.

viewed as shown in Fig. 7.29*a*, or the forward-projecting substituent when viewed as in Fig. 7.29*b*.

NOTE. It may be a source of confusion that α, β, γ, and so on, are also used to denote the position of a substituent relative to a functional group. Thus, CH_3CHCl–CH_2CO_2H is sometimes called β-chlorobutyric acid, although 3-chlorobutanoic acid is preferred.

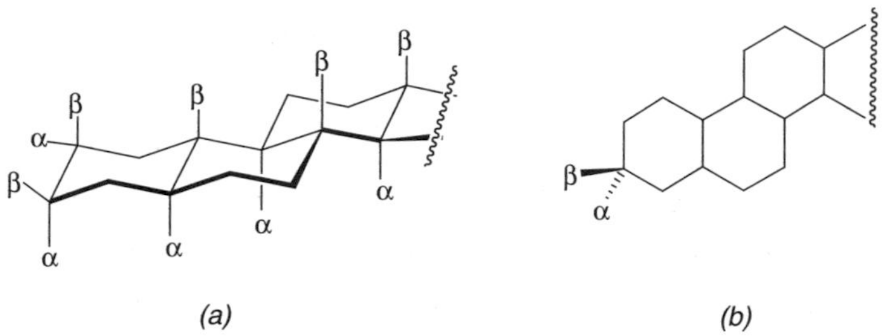

Figure 7.29. α Configuration and β configuration: (*a*) perspective and (*b*) planar.

7.30 CIP (CAHN–INGOLD–PRELOG) PRIORITY

A system of prioritizing groups for the purpose of labeling configuration, named after Robert S. Cahn (1899–1981), Christopher Ingold (1893–1970), and Vladimir Prelog (1906–1998) who received the Nobel Prize in 1975. Priority is based on atomic number (with the heaviest isotope first) of the first atom of the group: The higher the atomic number, the higher the priority. If the first atoms in two groups are identical, the second atoms are considered. A partial list (with the highest priority first) has I > Br > Cl > SO_2R > SOR > SR > SH > F > OCOR > OR > OH > NO_2 > NO > NHCO > NR_2 > NHR > NH_2 > CX_3 (X = halogen) > COX > CO_2R > CO_2H > $CONH_2$ > COR > CHO > CR_2OH > CH(OH)R > CH_2OH > C≡CR > C(R)=CR_2 > C_6H_5 > C≡CH > CR_3 > $CH(C_2H_5)_2$ > CH=CH_2 > $CH(CH_3)_2$ > C H_2R > CH_3 > D > H > electron pair.

7.31 *Z* AND *E* CONFIGURATIONS

The term *Z* (from the German word *zusammen*, meaning "together") designates the *cis*-like relationship of the higher CIP priority groups connected to two different atoms of a double bond or ring. *E* (from the German word *entgegen*, meaning "opposite") designates the *trans*-like configuration.

Example. The *E* and *Z* isomers of various compounds are shown in Figs. 7.31*a* through *d*.

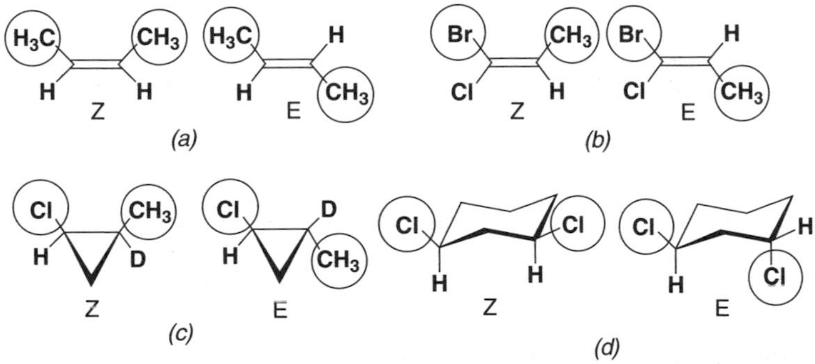

Figure 7.31. Examples of *E*, *Z* isomers; the higher-priority groups are circled. (*a*) *Z*- and *E*-2-butene; (*b*) *Z*- and *E*-1-bromo-1-chloropropene; (*c*) *Z*- and *E*-1-chloro-2-deuterio-2-methylcyclopropane; (*d*) *Z*- and *E*-1,3-dichlorocyclohexane.

7.32 (*R*) AND (*S*) CONFIGURATIONS

Descriptors for the configuration at chiral (enantiogenic and tetrahedral diastereogenic) centers. The enantiogenic or diastereogenic center of interest is oriented so that the group of lowest priority is pointing away from the observer and therefore hidden from view (as shown schematically in Figs. 7.32*a* and *b*). *R* (from the Latin word *rectus*, meaning "right") is that configuration about the center with a clockwise pattern of CIP priorities (highest to third highest, Fig. 7.32*a*), while *S* (from the Latin word *sinister*, meaning "left") is that configuration with a counterclockwise pattern (Fig. 7.32*b*).

Many chiral molecules that do not possess individual enantiogenic atoms instead have one or more multiatom enantiogenic centers. For such molecules (e.g., allenes, spiro compounds, hindered unsymmetrical biphenyls, and unsymmetrical paracyclophanes), the (*R*) or (*S*) configuration is determined by assigning priority 1 to the higher-priority group in front of the molecule (toward the viewer), 2 to the lower-priority group in front, 3 to the higher-priority group in back, and so on, then examining the helical path $1 > 2 > 3 > 4$. For example, the left-hand enantiomer of 1,3-dimethylallene in Fig. 7.15*b* has the *R* configuration, as shown in Fig. 7.32*c*.

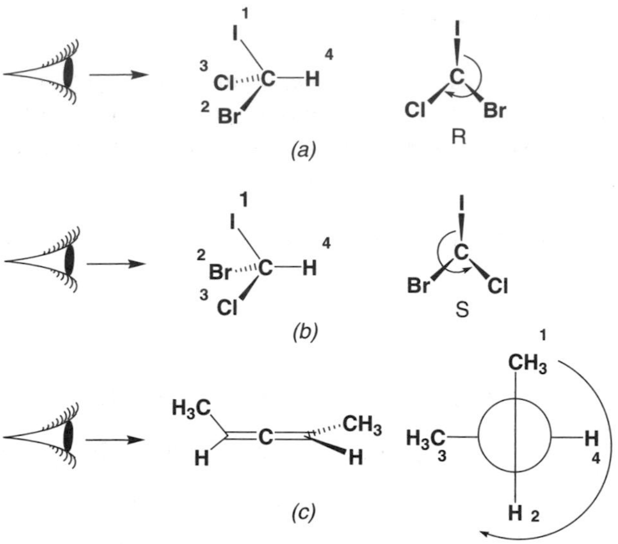

Figure 7.32. (*a*) (*R*)-bromochloroiodomethane; (*b*) (*S*)-bromochloroiodomethane; (*c*) (*R*)-1,3-dimethylallene.

7.33 ABSOLUTE CONFIGURATION

The specification of (*R*) or (*S*) at each enantiogenic or diastereogenic center in a molecule.

Example. Figure 7.32.

7.34 RELATIVE CONFIGURATION

A similarity between the configurations of two chiral molecules. When two such molecules can be interconverted (at least in principle) without breaking any bonds to the stereogenic center, they have the same relative configuration regardless of the (*R*) or (*S*) descriptors. Relative configurations were used to interrelate structures before absolute configurations were known.

Example. Oxidation of (−)-2-methyl-1-butanol to (+)-2-methybutanoic acid (Fig. 7.34) does not involve any changes in the bonds to the enantiogenic center. Thus,

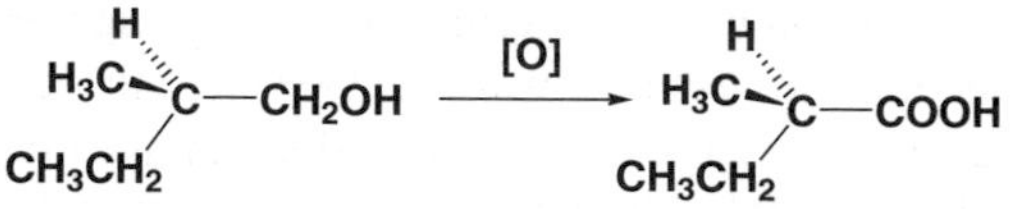

Figure 7.34. Oxidation of (−)-2-methyl-1-butanol to (+)-2-methylbutanoic acid.

two molecules have the same relative configuration, even though their absolute configurations may differ.

7.35 *MESO* CONFIGURATION

A molecule with multiple tetrahedral diastereogenic centers that is achiral by virtue of a plane or center of symmetry in at least one of its accessible conformations (Sect. 7.104). A *meso* (from the Greek word, meaning "between") form has different physical and chemical properties than any of its diastereomers.

Example. Figure 7.35*a* shows *meso*-2,3-dibromobutane, whereas Fig. 7.35*b* shows a pair of enantiomers of 2,3-dibromobutane. The *meso* compound is shown in a conformation with a plane of symmetry; another conformation has a center of symmetry. Based on *R*/*S* descriptors (see Sect. 7.32), the *meso* compound in Fig. 7.35*a* has stereocenters of configuration *S* and *R*, while the enantiomers in Fig. 7.35*b* are *R*,*R* and *S*,*S*, respectively. In principle, to convert the *R*,*R* enantiomer into the *S*,*S* enantiomer by changing the configuration of one stereocenter at a time, the molecule must pass through the *meso* diasteromer. Thus, the *meso* isomer is encountered *between* the *R*,*R* and *S*,*S* isomers (Fig. 7.35*c*). There are two additional common uses of the term *meso* in organic chemistry that are unrelated to stereochemistry: (1) the ring positions of a porphyrin that are in between the constituent pyrrole rings (**m** in Fig. 7.35*d*) and

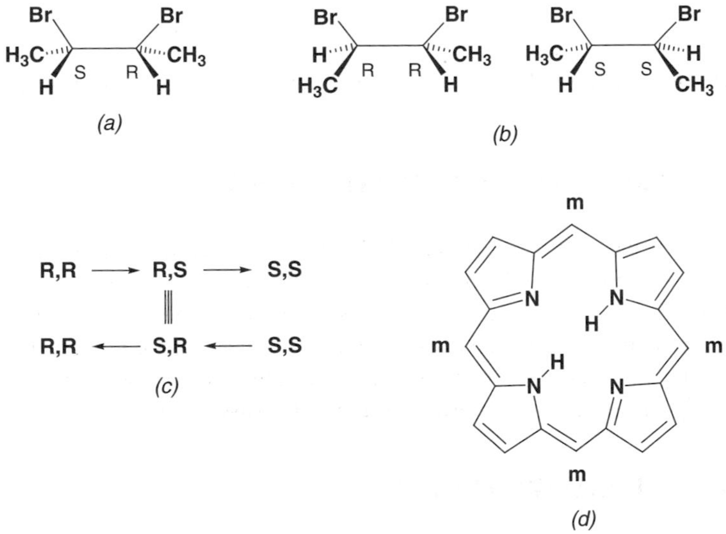

Figure 7.35. (*a*) (2*S*,3*R*) (or *meso*) 2,3-dibromobutane; (*b*) the (2*R*,3*R*) and (2*S*,3*S*) enantiomers of 2,3-dibromobutane; (*c*) stepwise conversion of *R*,*R* to *S*,*S* via *S*,*R*; (*d*) the *meso* position (*m*) in porphyrins.

(2) a liquid crystal is a mesophase (see Sect. 10.84) or a phase between the solid and liquid phase.

7.36 *ERYTHRO* AND *THREO* CONFIGURATIONS

The prefix *erythro-* is given to the set of enantiomers with a configuration similar to the two adjacent stereogenic centers of erythrose (Fig. 7.36*a*, where only one enantiomer is shown). The *erythro* form is often described as *meso*-like because in the orientation shown, there would be a plane of symmetry if instead of the two dissimilar groups ($HOCH_2$ and CHO in this case), the groups were identical. Thus, *erythro*-3-bromo-2-butanol is the configuration shown in Fig. 7.36*b*. The prefix *threo-* is given to the set of enantiomers with a configuration similar to the two adjacent stereogenic centers of threose (Fig. 7.36*c*, where only one enantiomer is shown). The *threo* isomer of 3-bromo-2-butanol is shown in Fig. 7.36*d*.

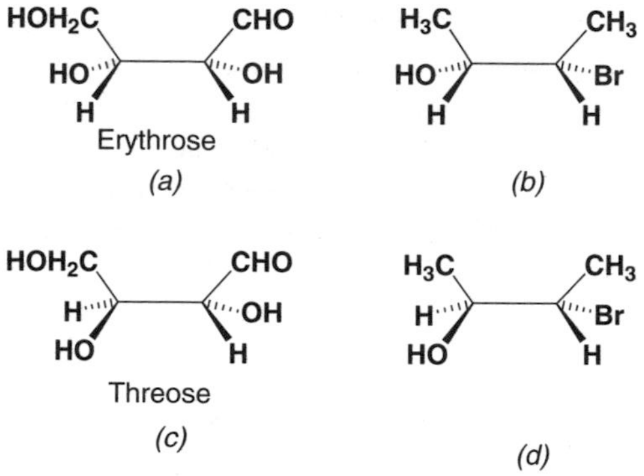

Figure 7.36. (*a*) Erythrose; (*b*) *erythro*-3-bromo-2-butanol; (*c*) threose; (*d*) *threo*-3-bromo-2-butanol.

7.37 ELECTROMAGNETIC RADIATION (LIGHT)

Time-dependent electric and magnetic fields (waves) oscillating as a unit in phase, in perpendicular planes and normal to the direction of light wave propagation (Fig. 7.37). While all electromagnetic waves are fundamentally identical except for their frequency, the term *light* is reserved for those wavelengths between 10 and 10,000 nm (nanometers, 10^{-9} meters). For most chemical purposes, it is sufficient to consider only the electric field portion of the light wave. (see Sect. 18.)

7.38 PLANE-POLARIZED LIGHT (PPL)

A light wave in which the electric field oscillates in a single plane. The light wave shown in Fig. 7.37 is plane-polarized, with the electric vector oscillating in the *xz*-plane. (see Sect. 18.3)

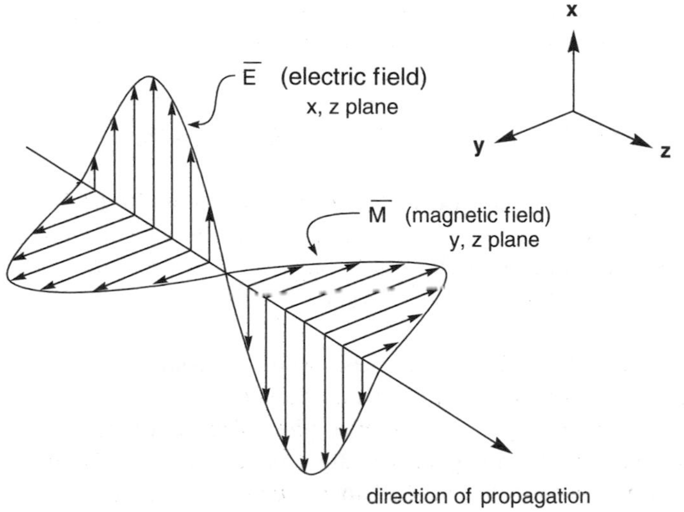

Figure 7.37. Relationship of the oscillating electric and magnetic fields of light.

7.39 UNPOLARIZED LIGHT

A light beam in which the component electric field elements are oriented at all possible angles about the direction of propagation (Fig. 7.39*a*).

Each of these electric field elements may be resolved into two orthogonal components directed along the *x*- and *y*-axes (Fig. 7.39*b*). The sum of all *x*- and *y*-axes electric field components constitutes two plane-polarized light waves oscillating at right angles to one another. Thus, unpolarized light can be regarded as being composed of

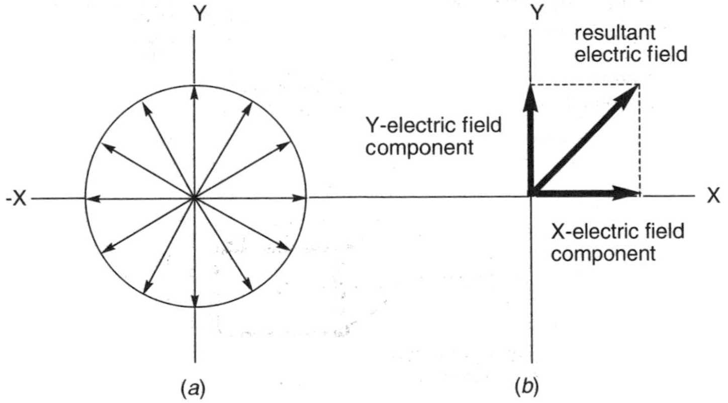

Figure 7.39. Unpolarized light. (*a*) Randomly oriented plane-polarized electric field elements of a light wave propagating normal to the page surface and (*b*) *x*- and *y*-electric field components of one electric field element.

equal contributions from two plane-polarized waves oscillating at right angles to each other: one oscillating in the *xz*-plane and the other in the *yz*-plane.

7.40 GENERATION OF PLANE-POLARIZED LIGHT

The two orthogonal plane-polarized components of unpolarized light can be separated from each other by either of two processes: (1) the selective absorption of one of the components or (2) the refraction of the two components into two waves traveling in different directions. (see Sect. 18.3)

Example. (1) The first of these processes is achieved using a *polarizer*. Two of the various types of polarizers are represented schematically in Figs. 7.40*a* and *b*. In Fig. 7.40*a*, a parallel array of microscopic wires placed very close to one another serves as a polarizer. When unpolarized light passes through this array, the plane-polarized component oriented parallel to the wires will generate a current in the wires and, thus, be absorbed. The plane-polarized component oriented perpendicular to the wires is unattenuated by the wires and affords a plane-polarized (horizontally polarized in this example) light wave one-half the intensity of the original wave. In order to obtain plane-polarized light oriented in the vertical direction, the polarizer must be rotated 90°.

Contemporary polarizers are more commonly constructed from parallel arrays of "molecular wires." These molecular wires consist of oriented polymer molecules on to which have been attached heavy, polarizable atoms such as iodine (Fig. 7.40*b*). The orientation process is achieved by stretching the polymer film

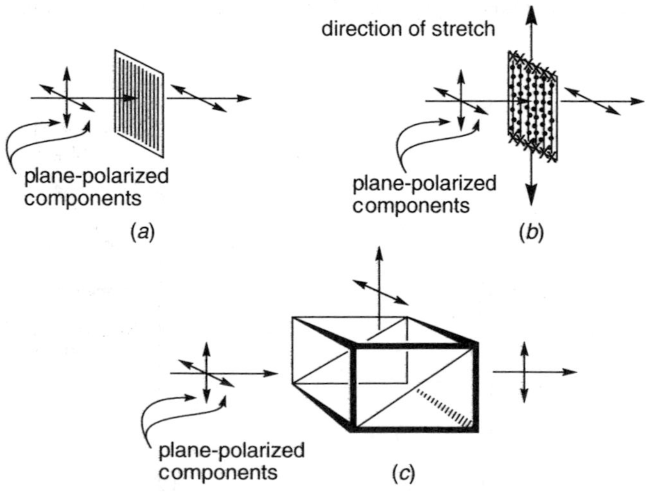

Figure 7.40. Devices for the generation of plane-polarized light. (*a*) wire array polarizer; (*b*) stretched, iodinated polymer film polarizer; (*c*) calcite prism polarizer.

that tends to align the polymer molecules in a parallel array. The iodine atom provide a source of polarizable electrons analogous to that present in the metal wires of the previous example. This stretched, iodinated polymer film functions to produce plane-polarized light in the same way as the wire array polarizer described above.

(2) The second type of polarizer is a prism constructed from a crystal of a highly refracting material such as calcite (Fig. 7.40*c*). When unpolarized light passes through such an oriented crystal, the plane-polarized component aligned parallel to the diagonally cut crystal face is refracted 90° and passes out the top of the prism. The orthogonal plane-polarized component passes directly through the prism and exits the rear of the prism. This type of polarizer affords both plane-polarized light forms in a single operation. See also "Birefringence" (Sect. 7.44).

7.41 CIRCULARLY-POLARIZED LIGHT (CPL)

A light wave in which the electric field strength remains constant, but rotates around the axis of propagation (Fig. 7.41).

Example. CPL has associated with it a chiral electric field (and magnetic field), and consequently, CPL light waves can exist in either of two forms that bear nonsuperimposable mirror images relationships to one another. These are left circularly-polarized light (LCPL) and right circularly-polarized light (RCPL), as shown in Figs. 7.41 and *b*, respectively. If these circularly-polarized waves are viewed down the axis of propagation as shown in Fig. 7.41, the electric vector of each form remain constant in magnitude, but in LCPL, rotates in a counterclockwise direction, and in RCPL, rotates in a clockwise direction. (See Sect. 18.4)

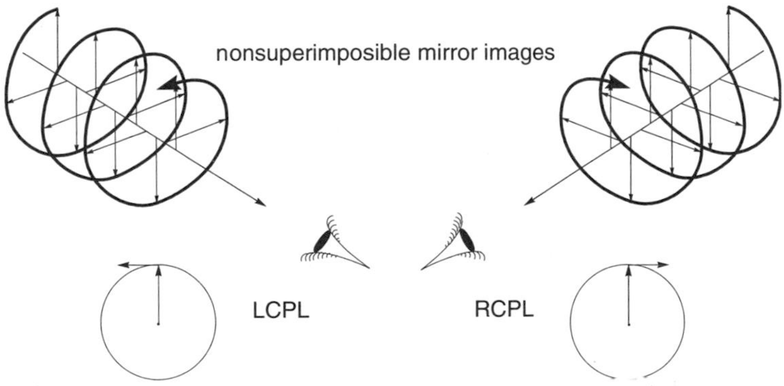

Figure 7.41. Electric fields of left and right circularly-polarized light.

7.42 RELATIONSHIPS BETWEEN PLANE- AND CIRCULARLY-POLARIZED LIGHT

There are two important phase relationships between plane- and circularly-polarized light:

1. Any circularly-polarized light wave can be described with two equally intense, but out-of-phase plane-polarized components (Figs. 7.42*a* and *b*).

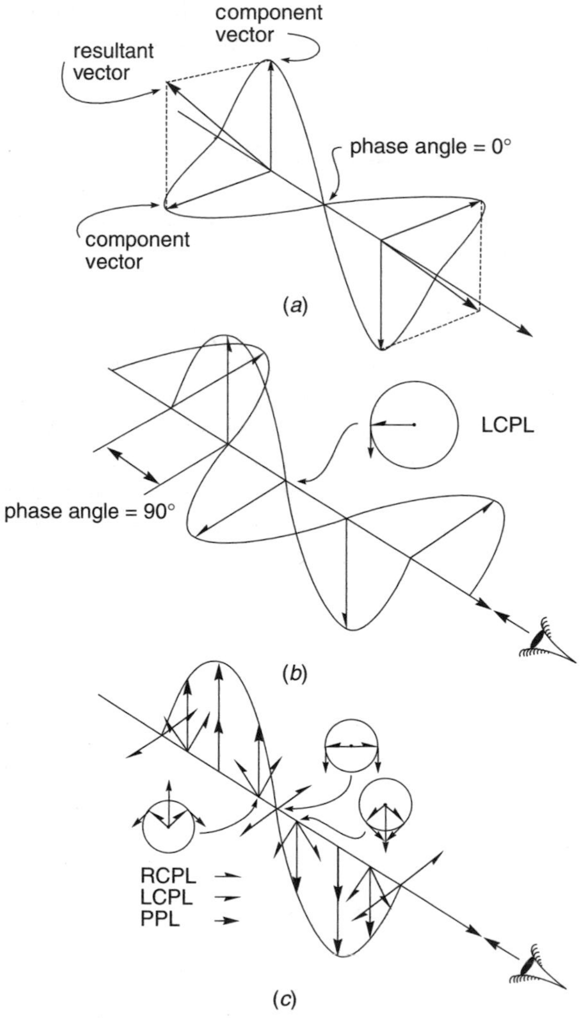

Figure 7.42. (*a*) Plane-polarized components of a plane-polarized light wave oriented at 45° to the horizontal; (*b*) plane-polarized components of a left circularly-polarized light wave; (*c*) circularly-polarized components of a plane-polarized light wave.

2. Any plane-polarized light wave can be described by two equally intense, right- and left-hand circularly-polarized components (Fig. 7.42*c*).

Example. (1) Any plane-polarized light wave can be resolved into two equally intense, orthogonal plane-polarized components, as shown in Fig. 7.42*a*. These two-component waves must be "in phase" (i.e., have a difference in phase angle of 0°). If one of the plane-polarized components is shifted 90° in advance of the other component, the resultant wave becomes left circularly-polarized, as shown in Fig. 7.42*b*. In this case, the two-component plane-polarized waves are said to have a phase angle of 90°. Alternatively, if the same plane-polarized component is retarded relative to the other component by 90°, the resultant wave becomes right circular-polarized (not shown) and the two-component waves have a phase angle of −90°.

(2) The circularly-polarized components of a plane-polarized light wave are shown in Fig. 7.42*c*. When viewed down the axis of propagation, as shown in Fig. 7.41, two oppositely polarized circularly-polarized component waves (LCPL and RCPL) will combine to form a resultant plane-polarized wave in which the electric field vector is oscillating in a plane.

7.43 ELLIPTICALLY-POLARIZED LIGHT

Any form of light intermediate between plane- and circularly-polarized forms. That is any form of light with a phase angle of $>-90°$ to $<0°$ and $>0°$ to $<90°$ between the plane-polarized components shown in Fig. 7.42*b*. The electric field of an elliptically-polarized light wave varies in magnitude as it rotates about the axis of light propagation (Fig. 7.43) (See Sect. 18.5).

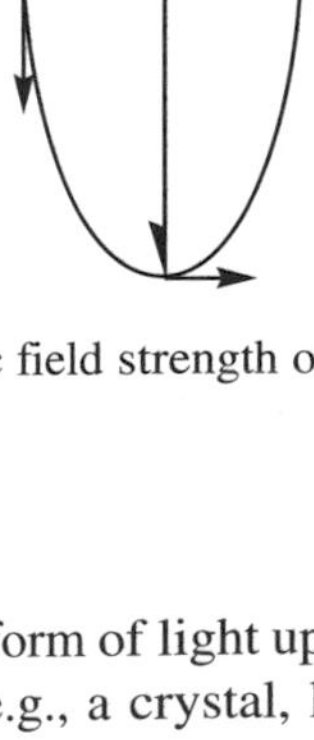

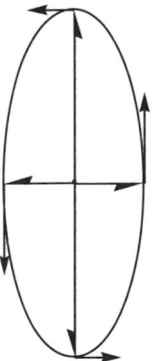

Figure 7.43. Variation in electric field strength of an elliptical-polarized light wave.

7.44 BIREFRINGENCE

The alteration of the polarization form of light upon passing through a sample of oriented, nonabsorbing molecules (e.g., a crystal, liquid crystal, or stretched polymer film). Birefringence arises from the different speeds (phase velocities) with which different orientations of plane-polarized light pass through samples that are highly ordered at the molecular level. It is a fundamental property of all forms of organized matter and manifests itself in the brilliant rainbow-colored patterns observed when a crystal, for example, is placed between two crossed polarizers (polarizers with their transmission axes rotated at 90° to each other). In fact, this is the simplest and most often used test for crystallinity. A sample may appear to be crystalline to the unaided eye, but if it does not exhibit birefringence, it is *amorphous* (matter without form or organization, a *glass*). (See Sect. 18.26)

Example. Any sample of oriented molecules will have two optical axes. One of these, when aligned with the plane of plane-polarized light, will afford the maximum velocity of light transmission, while the other axis affords the minimum velocity of light transmission. Quartz crystals are such birefringent materials. Quartz crystals cut so that these two optical axes are at 90° to each other are used to alter the polarization form of light as shown in Fig. 7.44. In this example, the plane of polarization of the entering light is oriented at 45° to either of the optical axes. If the quartz plate is cut to a thickness that will retard one of the plane-polarized components by a phase angle of 90° relative to the other, the resulting light will be circularly-polarized. Such a device is called a *1/4-wave plate* since it retards one plane-polarized component by one-quarter of a wavelength relative to the other. The degree of retardation is dependent on the wavelength of the light; a 1/4-wave plate for one wavelength will not be a 1/4-wave plate for a different wavelength. Furthermore, for a given wavelength of light, different plate thicknesses will afford different degrees of retardation. For example, a thicker plate might provide a 180° retardation that would produce a rotation of the plane of polarization by 90°. This would be a 1/2-wave plate. Devices constructed from two sliding wedges of quartz can provide any thickness desired

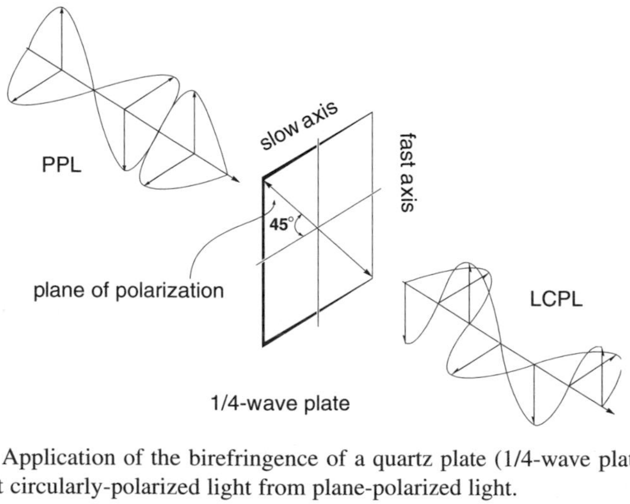

Figure 7.44. Application of the birefringence of a quartz plate (1/4-wave plate) in the production of left circularly-polarized light from plane-polarized light.

and, thus, they can be adjusted as a 1/4-, 1/2-, or 3/4-wave plate for any wavelength of light desired. Such a device is called a *compensator*.

7.45 OPTICAL ACTIVITY

The property of a collection of molecules containing an excess of one enantiomer over the other to rotate the polarization plane of plane-polarized light. Such molecules are said to be optically active. The terms *optically active* and *chiral* are frequently used interchangeably.

The rotation of plane-polarized light by chiral molecules arises from the interaction of these chiral molecules with the chiral circularly-polarized components of the plane-polarized light (Fig. 7.42*c*). Thus, the velocity of the two circularly-polarized components (LCPL and RCPL) through a chiral medium will not be equal. One circularly-polarized component will be retarded or advanced relative to the other, as shown in Fig. 7.45. In this example, either the LCPL has been retarded or the RCPL has been advanced so that the resultant plane-polarized light undergoes a rotation of $\alpha°$ from its original vertically polarized orientation.

Thus, optical activity constitutes a special form of birefringence associated with chiral molecules. Furthermore, this rotation of plane-polarized light by chiral molecules constitutes a manifestation of one of the fundamental relationships in nature: Enantiomers will behave in an identical fashion in all interactions with achiral agents such as with achiral reagents, solvents, and fields (see Sect. 7.15). Only under the

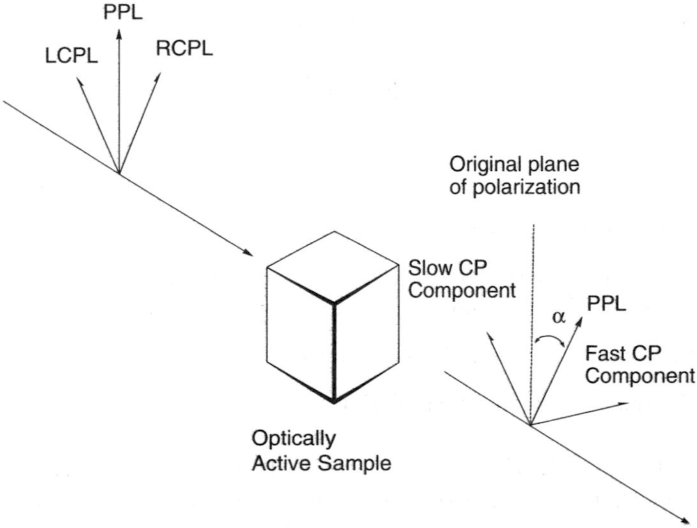

Figure 7.45. Role of the relative velocities of the circularly-polarized components of plane-polarized light in the rotation of the polarization plane.

influence of chiral agents such as chiral reagents, solvents, and circularly-polarized light fields will enantiomers exhibit different chemical and physical properties. Thus, the axiom, "chirality begets chirality." In this example, the *racemic mixture* (see Sect. 7.55) of left and right circularly-polarized light is affected differentially by the chiral sample molecules. Conversely, a racemic mixture of molecules might undergo a selective reaction of one enantiomer if irradiated with circularly-polarized light, but not with plane-polarized light.

7.46 OPTICAL ISOMERS (OPTICAL ANTIPODES)

Synonymous with enantiomers.

7.47 OPTICAL ROTATION α (OBSERVED ROTATION)

The angle by which the polarization plane is rotated as plane-polarized light passes through a sample of optically active molecules. A quantitative measure of optical activity (Sect. 7.45).

Example. Figure 7.49; the angle, α, is $-40°$ in this example.

7.48 SPECIFIC ROTATION $[\alpha]$

A characteristic physical property of a pure optically active compound. It is related to the observed rotation (α) by the following equation:

$$[\alpha]_{\lambda}^{T} = \alpha/(l \cdot c)$$

where

T = the temperature (°C)

λ = the wavelength (nm) of the plane-polarized light; often, the D line emission (589–590 nm) of a sodium vapor lamp is used

l = the sample cell length in decimeters

c = the sample concentration in g/mL

The sign and magnitude of $[\alpha]$ are functions of these variables, as well as the nature of the solvent. Under a given set of conditions, the specific rotation of one enantiomer has the same magnitude but opposite sign as that of its mirror image. The magnitude of optical rotation is strongly wavelength-dependent, increasing in magnitude as the wavelength decreases until reaching a maximum rotation, then decreasing to zero at the wavelength of an electronic transition. At still shorter wavelengths, the sign of the optical rotation reverses. This wavelength-dependence of rotation is known as the *optical rotatory dispersion (ord) curve*; see also Sect. 7.60 and 18.21. For convenience and reproducibility, monochromic light sources (e.g., a sodium vapor lamp with "D line" emission maxima at 589–590 nm) are usually employed when making measurements of optical rotation.

7.49 POLARIMETER

The instrument that measures optical rotation (Fig. 7.49). Monochromatic light, plane-polarized by a polarizer, is passed through a solution of the sample. The plane of polarization (initially vertical in the figure) emerges from the sample rotated by an angle α (counterclockwise in the figure, a negative observed rotation). The exact magnitude of rotation is determined using a second polarizer called an analyzer.

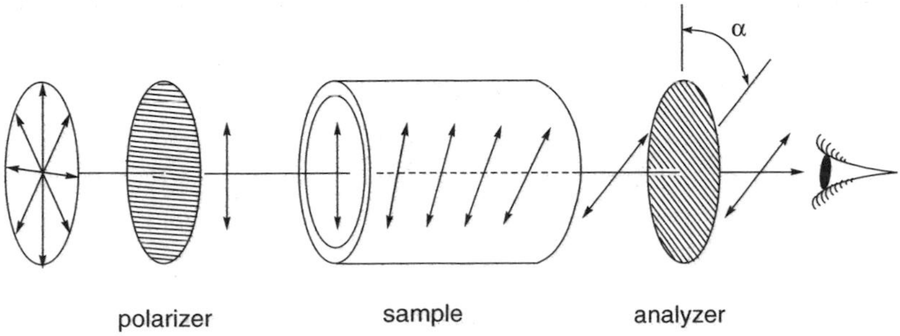

Figure 7.49. Polarimeter.

7.50 (+)-ENANTIOMER AND (−)-ENANTIOMER

An indicator of the sign of the optical rotation of an enantiomer. The (+)-enantiomer rotates the polarization plane in the clockwise direction, when viewed as shown in Fig. 7.49. The (−)-enantiomer rotates it in the counterclockwise direction. Note that the sign of optical rotation for an enantiomer cannot normally be predicted from its absolute configuration alone; see Sect. 7.60.

7.51 *d*-(DEXTROROTATORY) AND *l*-(LEVOROTATORY) ENANTIOMERS

Synonymous with the (+)-enantiomer and the (−)-enantiomer, respectively. Note the use of lowercase letters.

7.52 D-ENANTIOMER AND L-ENANTIOMER

Having the same relative configuration (Sect. 7.34) as (+)-glyceraldehyde or (−)-glyceraldehyde, respectively. Note the use of smaller capital letters (small caps).

Example. (+)-Glyceraldehyde is shown in Fig. 7.52*a*. Its conversion to glyceric acid (Fig. 7.52*b*) does not involve breaking any bonds to the chiral center. Thus, this acid has the D configuration. (−)-Glyceraldehyde is shown in Fig. 7.52*c*.

Before absolute configurations (Sect. 7.33) were known, glyceraldehyde was the reference compound for the stereochemical elaboration of the sugars. The simple sugars can be synthesized by starting with either (+)- or (−)-glyceraldehyde. Thus, of the 16 possible aldohexoses, eight may be considered as being derived from D-(+)-glyceraldehyde and the other eight from L-(−)-glyceraldehyde. The D and L in the names of these sugars specify only the configuration at the starred chiral carbon (Figs. 7.52*d* and *e*). Thus, the name D-(+)-glucose (Fig. 7.52*f*) indicates that dextrorotatory glucose has the same configuration at the number 4 chiral carbon as D-glyceraldehyde. In the Fischer projection (Sect. 7.6*E*), the OH group is placed on the right-hand side of the structure to indicate the D configuration on the carbon related to D-glyceraldehyde.

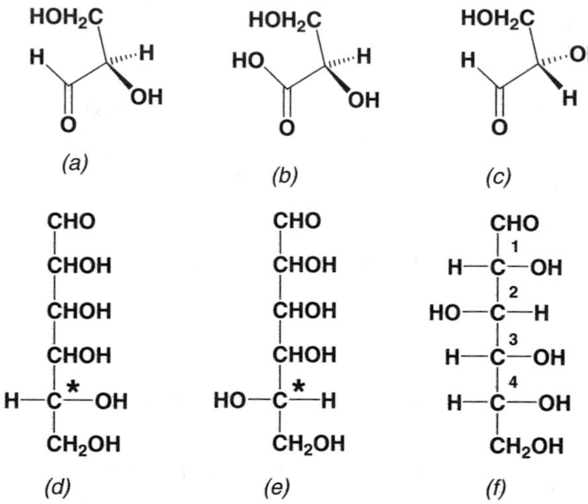

Figure 7.52. Assignments of D and L configurations: (*a*) D-glyceraldehyde; (*b*) D-glyceric acid; (*c*) L-glyceraldehyde; (*d*) D-Aldohexose; (*e*) L-Aldohexose; (*f*) D-glucose.

7.53 INVERSION (OF CONFIGURATION)

The conversion of one molecule into another that has the opposite relative configuration at each and every enantiocenter or tetrahedral diastereocenter.

Example. Two reactions that proceed with inversion are shown in Figs. 7.53*a* and *b*. Inversion does not necessarily require a change in designated absolute configuration,

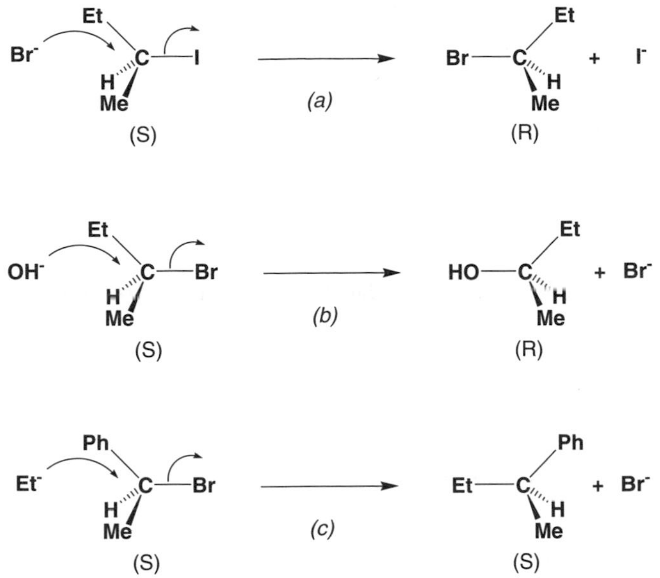

Figure 7.53. Inversion of relative configuration: (*a, b*) with change in absolute configuration; (*c*) without change in absolute configuration.

nor does a change in designated absolute configuration require inversion. The process illustrated in Fig. 7.53*c* proceeds with inversion, although (*S*)-reactant gives (*S*)-product! This is because the incoming and departing groups have different relative CIP priorities.

7.54 RETENTION (OF CONFIGURATION)

The conversion of one molecule into another with the same relative configuration at each stereocenter; the opposite of inversion of configuration.

7.55 RACEMIC MODIFICATION (*d,l* PAIR)

The generic term for any 50:50 combination of a pair of enantiomers, with the mixture having zero optical rotation. See also Sects. 7.56 and 7.57.

7.56 CONGLOMERATE (RACEMIC MIXTURE)

An exactly 50:50 mixture of crystals of pure (+)-enantiomer with crystals of the pure (−)-enantiomer. Each individual crystal is composed of a single enantiomer, but the mixture, when well mixed, has a lower melting point than either pure enantiomer and shows zero optical rotation. A small amount of a pure enantiomer added to a conglomerate always raises the melting point of the mixture.

7.57 RACEMATE

A compound, individual crystals of which contain equal numbers of *d* and *l* molecules. A racemate has a sharp melting point (characteristic of a pure compound) that usually differs from that of the pure enantiomers. Other physical properties of the racemate differ from those of the pure enantiomers, but it has zero optical rotation. A small amount of pure enantiomer added to a racemate always lowers the melting point of the mixture.

7.58 RACEMIZE

To generate a racemic modification.

7.59 RESOLUTION

The separation of a racemic modification into its pure enantiomers.

7.60 RULES FOR CORRELATION OF CONFIGURATION WITH OPTICAL ROTATION

Brewster's Rules. A series of semiempirical rules that allow a summation of the contributions to the sign and magnitude of optical rotation that are introduced by substituents about a stereogenic center. Configurational and conformational factors, group polarizabilities, the medium's refractive index, and hydrogen bonding effects are all significant variables. (J. H. Brewster, *Topics in Stereochemistry*, N. L. Allinger and E. L. Eliel, eds., John Wiley and Sons: New York, 1967, Vol. 2, pp. 33–39.)

Octant Rule. An empirical method of interpreting the relationship between optical rotatory dispersion (ord) or circular dichroism (cd) curves and the three-dimensional structure of certain classes of molecules. (P. Laszlo and P. J. Stang, *Organic Spectroscopy, Principles and Applications*, Harper and Row: New York, 1971, pp. 93–98.) (See Sect. 18.23)

Example. There is a correlation between the appearance of the ord curve (see Sect. 7.48) and the absolute configuration of five-, six-, and seven-membered cyclic ketones. When the ord curve reaches a maximum before it reaches a minimum as the wavelength decreases (Fig. 7.60*a*), the compound is said to exhibit a *positive cotton effect (CE)*. If the minimum precedes the maximum, it exhibits a negative CE (Fig. 7.60*b*). When the carbon of a carbonyl chromophore in a dissymmetric molecular environment is drawn at the origin of a Cartesian coordinate system (Fig. 7.60*c*), the presence of substituents

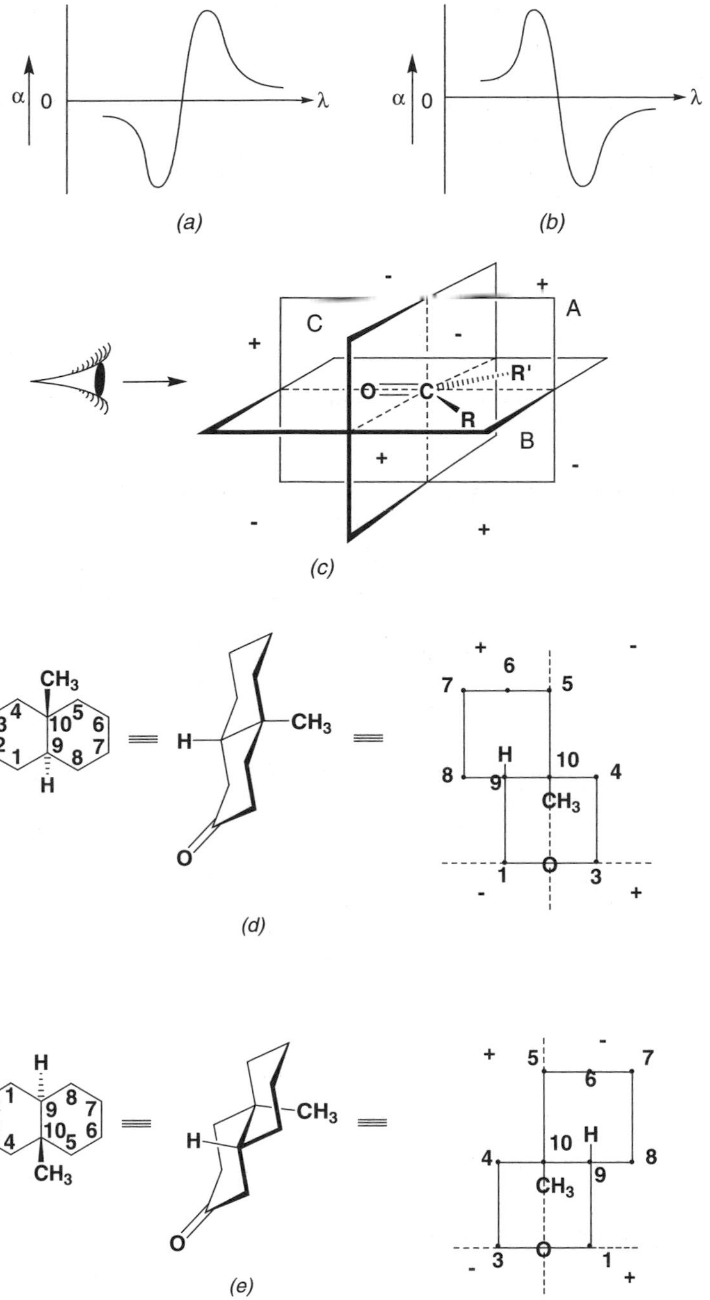

Figure 7.60. (*a*) An ord curve with a positive Cotton effect (CE); (*b*) an ord curve with a negative CE; (*c*) specification of the octants and their influence on the sign of the CE; (*d*) prediction of the sign of the CE for (+)-*trans*-10-methyl-2-decalone (only the signs of the four rear octants are shown); (*e*) prediction of the sign of the CE for (−)-*trans*-10-methyl-2-decalone.

within each of the surrounding octants provides positive or negative contributions to the compound's CE, as indicated by the sign of each quadrant. The octant rule says that if excess mass exists in the positive quadrants (compared to that in the negative quadrants), the compound will exhibit a positive CE, and vice versa. This is illustrated for the two enantiomers of *trans*-10-methyl-2-decalone (Figs. 7.60*d* and *e*). For the configuration shown in Fig. 7.60*d*, excess mass in the upper-left rear octant correlates with a positive CE, while the enantiomer in Fig. 7.60*e* has excess mass in the negative octant leading to a negative CE.

7.61 ENANTIOMERIC PURITY

The fraction or percentage of the predominant enantiomer in a mixture of both.

7.62 ENANTIOMERIC EXCESS (ee, OPTICAL PURITY)

The excess of one enantiomer over the other in a mixture of the two. The enantiomeric excess (ee) is related to the enantiomeric purity by the formula:

$$\text{ee} = 2\ (\%\ \text{enantiomeric purity}) - 100\% \quad \text{or}$$

$$\text{ee} = \%\ \text{major enantiomer} - \%\ \text{minor enantiomer}$$

Example. The specific rotation of any *d,l* mixture is equal to the specific rotation of the pure enantiomer times the ee (expressed as a fraction). These relationships are illustrated in the following table for a compound whose maximum specific rotation is 150°.

Enantiomeric Purity	Enantiomeric Excess	[α] (°)
100% *d*	100%	+150
75% *d* (25% *l*)	50%	+75
50% *d* (50% *l*)	0	0
75% *l* (25% *d*)	50%	−75
100% *l*	100%	−150

7.63 SCALEMIC MIXTURE

An unequal mixture of two enantiomers, that is, with $0\% < \text{ee} < 100\%$.

7.64 OPTICAL YIELD

The enantiomeric excess of the product(s) from the reaction starting with enantiomerically pure reactant or reagent, regardless of the chemical yield.

Example. If a reaction of enantiomerically pure A gives a 100% yield of racemic B, the optical yield is 0%. If enantiomerically pure A gives a 66% chemical yield of enantiomerically pure B, the optical yield is 100%. If enantiomerically pure A gives a 50% chemical yield of 36% ee B, the optical yield is 36%. This definition is valid regardless of whether the process involves retention or inversion.

7.65 OPTICAL STABILITY

A measure of how resistant a pure enantiomer is toward racemization under a given set of conditions.

7.66 HOMOTOPIC ATOMS OR GROUPS

A set of atoms (or groups of atoms) that can be interchanged by a rotational axis of symmetry C_n ($n > 1$, Sect. 4.3). The homotopic relationship can also be ascertained by the atom replacement test: If any atom in the set under consideration is replaced by a different atom, and this leads to a compound indistinguishable from one obtained by like replacement of any other atom in the set, all atoms in the set are homotopically related.

Example. The following are sets of homotopic hydrogens: the three H's of NH_3 or CH_3Cl; the four H's of $CH_2{=}CH_2$ or CH_4; the six H's of benzene or CH_3CH_3. The three H's of the methyl group in CH_3OH are homotopic (e.g., only one CH_2DOH is possible) because of the local symmetry (Sect. 4.20) of the CH_3 group (local C_3 axis) and free rotation of the C–O bond. However, the two H's of CH_2ClF (Fig. 7.66*a*) are not homotopic because of the absence of a C_n, $n > 1$. The hydrogens on C–1 and C–2 of *trans*-1,2-dichlorocyclopropane (Fig. 7.66*b*) are homotopic by virtue of a C_2 axis.

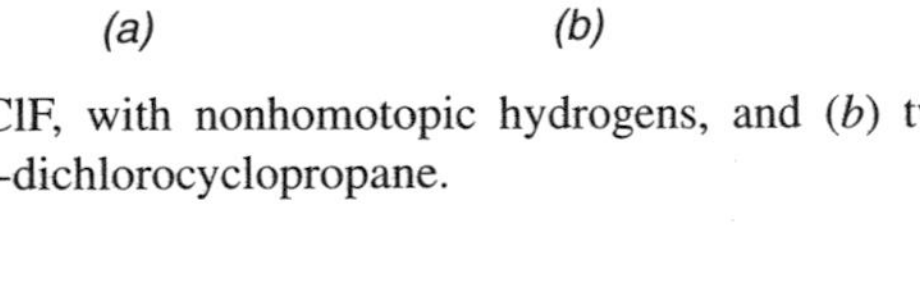

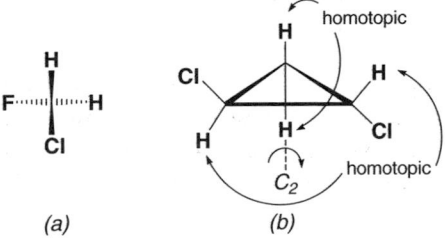

Figure 7.66. (*a*) CHClF, with nonhomotopic hydrogens, and (*b*) two sets of homotopic hydrogens in *trans*-1,2-dichlorocyclopropane.

7.67 SYMMETRY EQUIVALENT ATOMS OR GROUPS

Synonymous with homotopic atoms or groups. All homotopic (equivalent) atoms are necessarily symmetry-equivalent (Sect. 4.25), but not all symmetry-equivalent atoms are homotopic (see Sects. 7.68 and 7.70).

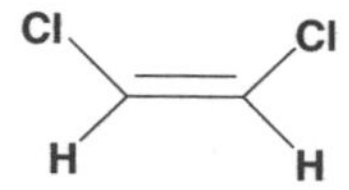

Figure 7.67. (*Z*)-1,2-Dichloroethylene; all like atoms are homotopic.

Example. cis-Dichloroethylene, shown in Fig. 7.67, has the following sets of equivalent atoms: two H's, two Cl's, and two C's.

7.68 CHEMICALLY EQUIVALENT ATOMS OR GROUPS

Synonymous with symmetry-equivalent atoms or groups (Sect. 4.25). In the *absence* of a chiral environment, symmetry-equivalent atoms are chemically indistinguishable; for example, they undergo reactions at the same rate. In the *presence* of a chiral environment, homotopic atoms are still chemically indistinguishable, but enantiotopic atoms (see Sect. 7.70) become chemically distinguishable. Hence, in such an environment enantiotopic atoms are not chemically equivalent.

7.69 EQUIVALENT FACES

Opposite faces of a molecule that can be exchanged by an axis of rotation C_n ($n > 1$, Sect. 4.3). Chemical reaction at either equivalent face leads to the same product. In the context of stereochemistry, the *faces* of a molecule (or a part of a molecule) are most commonly defined as two surfaces of the molecule that are related by a plane between them. Thus, a molecule with a double bond presents two faces, located on either side of the trigonal plane (the plane containing the two sp^2 hybridized atoms and the atoms connected directly to them).

Example. Both cyclopropanone (Fig. 7.69*a*) and *trans*-2,3-dichlorocyclopropanone (Fig. 7.69*b*) have equivalent faces, although the former molecule is achiral and the latter is dissymmetric and chiral. The opposite faces of (*Z*)-2-butene are equivalent, while those of (*E*)-2-butene (Sect. 7.31) are enantiotopic (Sect. 7.72).

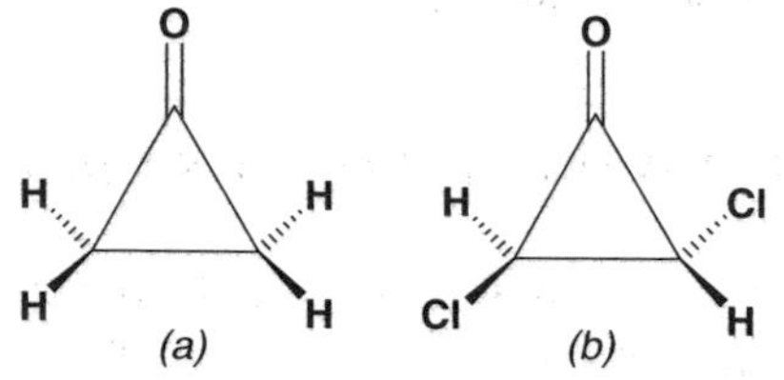

Figure 7.69. Molecules with equivalent faces: (*a*) cyclopropanone and (*b*) (*E*)-2,3-dichlorocyclopropanone.

7.70 ENANTIOTOPIC ATOMS OR GROUPS

A pair of atoms (or groups of atoms) in a molecule that are interchangeable only by an S_n axis, usually an S_1 (a plane of symmetry) or S_2 (a center of inversion). Substitution of one enantiotopic atom by another atom or group leads to a dissymmetric molecule and like substitution of the other atom yields the enantiomer.

Example. Most commonly, enantiotopic hydrogens are bonded to the same carbon atom and are symmetry-equivalent by virtue of a plane of symmetry. Figure 7.70*a* shows the pair of enantiotopic H's of $ClCH_2F$ (the plane of symmetry lies in the plane of the paper) and the enantiomers formed by substitution. Enantiotopic hydrogens are pervasive in organic chemistry, since all compounds having the general structure shown in Fig. 7.70*b* possess them. It is not a necessary requirement of the enantiotopic relationship that the two symmetry-equivalent hydrogen atoms be on the same carbon atom. Thus, the hydrogen atoms on C–1 and C–2 in *cis*-1,2-dichlorocyclopropane (Fig. 7.70*c*), and on 3-chlorocyclopropene (Fig. 7.71), are enantiotopic and exchangeable by an S_1. Enantiotopic hydrogens (exchangeable only by an S_2) are shown in Fig. 7.70*d*.

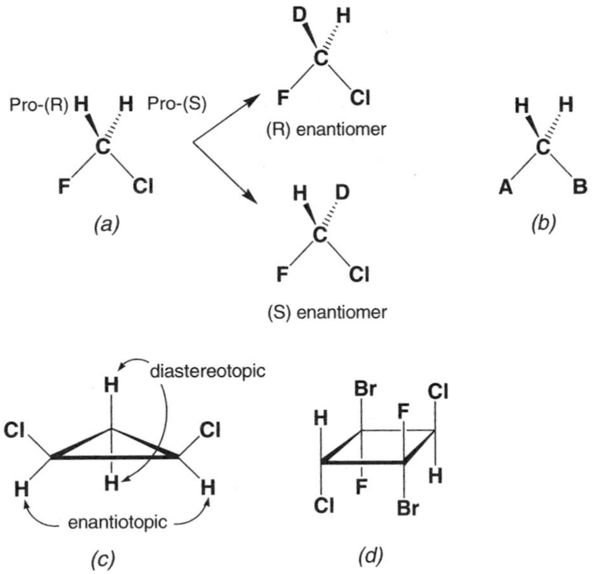

Figure 7.70. Enantiotopic hydrogens: (*a*) in CH_2ClF; (*b*) in a general case; (*c*) in *cis*-1,2-dichlorocyclopropane; (*d*) in a molecule with a center of inversion (S_2) only.

7.71 PROCHIRAL ATOMS OR GROUPS

Synonymous with enantiotopic atoms or groups.

Example. The enantiotopic hydrogens in Fig. 7.70*a* are prochiral atoms. They may be separately indicated as pro-(*R*) and pro-(*S*). The hydrogen under consideration is

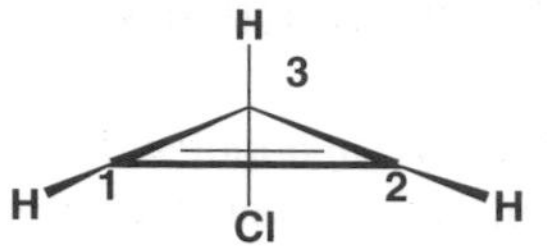

Figure 7.71. 3-Chlorocyclopropene, in which the olefinic hydrogens are prochiral.

pro-(*R*) if in the orientation that places this hydrogen in front and the other hydrogen in the rear, one must move in the clockwise direction in going from the highest-priority to the lower-priority grouping. Note that in the example shown in Fig. 7.70*a*, the replacement of the pro-(*R*) hydrogen with deuterium, leads to an (*R*) configuration. The configuration of the product depends on the nature of the group that replaces the pro-(*R*) hydrogen. In 3-chlorocyclopropene (Fig. 7.71), the two hydrogen atoms on the 1 and 2 positions are prochiral (enantiotopic); the replacement of either one of them by another atom leads to chirality that is centered on the methine carbon and not on either of the olefinic carbons.

7.72 ENANTIOTOPIC (PROCHIRAL) FACES

Opposite faces (Sect. 7.69) of a molecule that can be exchanged *only* by an alternating axis (S_n). Attack of a reagent on one side of the π bond leads to one enantiomer, whereas attack on the opposite side leads to the other enantiomer.

Example. A carbonyl compound of the general structure shown in Fig. 7.72*a* has enantiotopic faces, exchangeable by an S_1. Reaction of an achiral reagent (e.g., H_2)

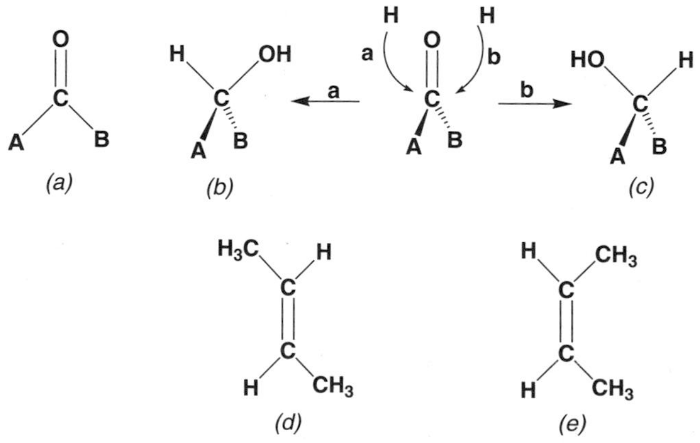

Figure 7.72. Enantiotopic faces: Addition of H_2 to ketone (*a*) gives enantiomers (*b*) and (*c*); (*d*) *E*-2-butene (enantiotopic faces); (*e*) *Z*-2-butene (homotopic faces).

at one face gives rise to a molecule that is the enantiomer of the molecule obtained by reaction at the other face (Figs.7.72*b* and *c*). The two faces of (*E*)-2-butene (Fig. 7.72*d*) are enantiotopic (S_1 and S_2, equivalent to a plane of symmetry and a center of symmetry, respectively), in contrast to those of (*Z*)-2-butene in Fig.7.72*e* (S_1 but C_2 also, hence homotopic). The difference in the faces of these isomers can be visualized by considering the structures of the corresponding epoxides (or analogous derivatives). The epoxide from (*E*)-2-butene is chiral (C_2 only), whereas the one from (*Z*)-2-butene (plane of symmetry) is achiral; hence, the faces of the butenes from which they were formed are enantiotopic and homo topic, respectively.

7.73 *RE* AND *SI* FACES

Stereochemical descriptors that specify the configuration of molecular faces above and below a trigonal plane (e.g., that surrounding an sp^2 hybridized atom).

Example. [Begin by assigning CIP priorities to the three groups attached to the trigonal center. Then determine the pattern of rotation from highest priority group (1) to lowest priority (3), as viewed from the face of interest. When the rotation is clockwise, the face is *re* (*R*-like; see Sect. 7.32). When the rotation is counterclockwise, the face is *si* (*S*-like). Even the facial rotation about the two carbon ends of a double bond can be assigned as *si* or *re* as shown in Fig. 7.73.]

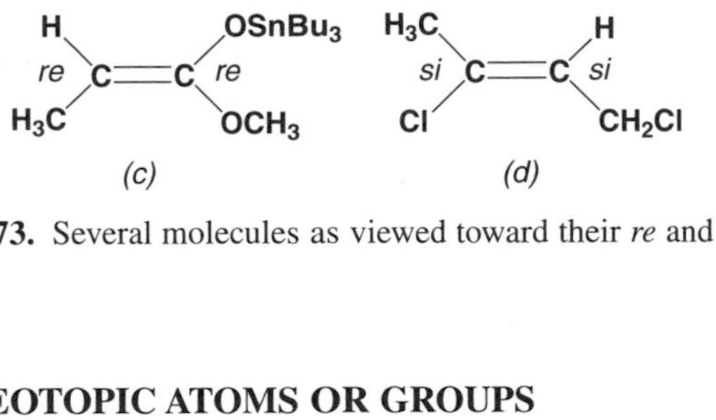

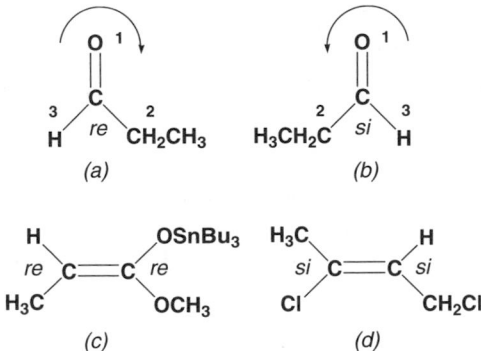

Figure 7.73. Several molecules as viewed toward their *re* and *si* faces.

7.74 DIASTEREOTOPIC ATOMS OR GROUPS

A pair of like atoms (or groups of atoms) that are *not* symmetry-equivalent, where substitution of one by a different group leads to a diastereomer of the molecule obtained by like substitution of the other atom.

Example. The two H's on the same carbon of chloroethylene (Fig. 7.74*a*) constitute a set of diastereotopic hydrogens in this achiral molecule. The relationship between either of the diastereotopic hydrogens and the hydrogen on the chlorine-containing carbon is constitutional (positional) and not stereochemical; replacement of this hydrogen by another atom gives a positional isomer of the compound obtained by replacement of either diastereotopic hydrogen by the same atom. The methylene hydrogens of the chiral molecule 1,2-fluorochlorocyclopropane, shown in Fig. 7.74*b*, are diastereotopic, as are those of *cis*-1,2-dichlorocyclopropane (Fig. 7.74*c*), an achiral molecule. The methylene hydrogens of the chiral molecule *trans*-1,2-dichlorocyclopropane (Fig. 7.66*b*) are equivalent (homotopic), not diastereotopic. Substitution of the diastereotopic methylene hydrogens α to the carbonyl group in the compound shown in Fig.7.74*d* by OH leads to the diastereomers, erythrose and threose.

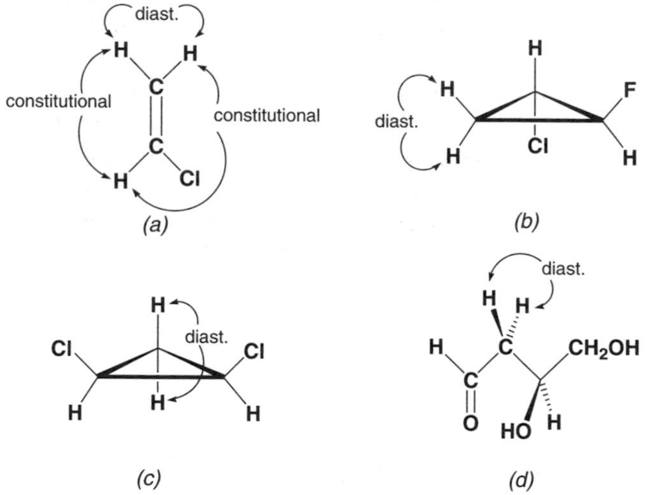

Figure 7.74. (*a*) Relationships between hydrogen atoms in chloroethylene (diast. = diastereotopic); diastereotopic methylene hydrogens in (*b*) *trans*-1-chloro-2-fluorocyclopropane, (*c*) *cis*-1,2-dichlorocyclopropane, and (*d*) 3,4-dihydroxybutanal.

7.75 DIASTEREOTOPIC FACES

Opposite faces (Sect. 7.69) of a molecule that cannot be interchanged by any symmetry operation.

Example. 5-Bromocyclopentadiene (Fig. 7.75*a*). The chiral molecule 3-chloro-2-butanone (Fig. 7.75*b*) has diastereotopic faces; reduction of the carbonyl (e.g., by $LiAlH_4$) gives rise to two diastereomers. The achiral molecule shown in Fig. 7.75*c* possesses diastereotopic faces (this is a *meso* compound, Sect. 7.35).

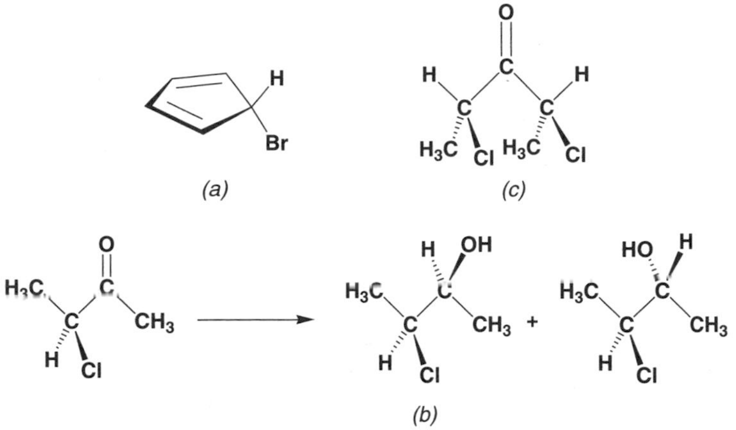

Figure 7.75. Diastereotopic faces: (*a*) in 5-bromocyclopentadiene; (*b*) in chiral 3-chloro-2-butanone; (*c*) in the achiral molecule *meso*-2,4-dichloropentanone.

7.76 SYMMETRY CLASSIFICATION OF ATOMS, GROUPS, AND FACES (SUMMARY)

Like atoms (or groups of atoms) in a molecule may be classified according to their symmetry relationships as follows: (1) *Symmetry-equivalent*: interchangeable by any symmetry operation (also called *chemically equivalent*). (2) *Homotopic*: symmetry-equivalent by virtue of a rotation (proper) axis (also called equivalent). (3) *Enantiotopic* (or *prochiral*): symmetry-equivalent by virtue of an alternating (improper) axis (S_n) *only.* (4) *Diastereotopic*: not symmetry-equivalent, that is, not related by any symmetry operation (but excluding, e.g., atoms whose substitution by the same atom would lead to positional isomers rather than diastereomers). Analogously, the faces (Sect. 7.69) of a molecule are equivalent, enantiotopic, or diastereotopic depending on whether they can be exchanged by C_n, S_n (only), or are nonexchangeable, respectively.

7.77 OGSTON'S PRINCIPLE

A theory developed by Alexander Ogston describing how the three-dimensional asymmetry of a molecule is projected onto a two-dimensional molecular surface. When an enantiogenic center, for example, makes facial (three-point) contact with a "two-dimensional" surface of another molecule, the chirality of the original molecule is fully described in terms of the three substituents in contact with the surface (see Fig. 7.77). Thus, when a molecule such as citrate (Fig. 7.77*a*), in which the two $-CH_2CO_2-$ groups are prochiral, binds to (is projected onto) an enzyme surface, only three groups need be involved in the binding. If the enzyme contains specific binding sites for these groups, *X*, *Y*, and *Z*, as shown in Figs. 7.77*b* and *c*, only one of the two possible binding configurations will bring the three groups of the substrate molecule into contact with the appropriate binding sites. Upon binding to the enzyme surface,

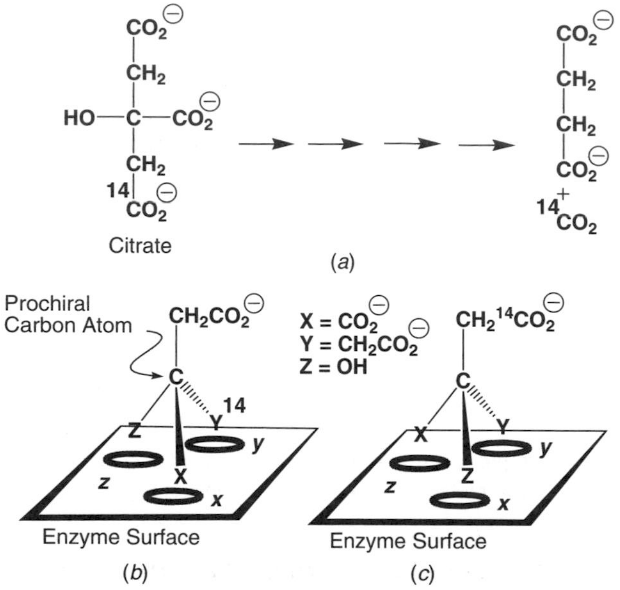

Figure 7.77. (*a*) Fate of specifically ^{14}C-labeled citrate in the citric acid cycle; binding of a prochiral carbon atom to a specific enzyme receptor site: (*b*) substrate matched to binding site and (*c*) substrate mismatched to binding site.

the two $-CH_2CO_2-$ groups at the prochiral carbon atom become differentiated, and only one undergoes reaction. Thus, the concept of facial three-point binding to an enzyme surface explains why enzymatic reactions almost always produce only one of the two possible enantiomers of the product.

7.78 LIKE *l* AND UNLIKE *u*

A stereochemical descriptor developed to describe the relative configuration of molecules that have two tetrahedral stereocenters: *u* (unlike) for (*R*,*S*) or (*S*,*R*); *l* (like) for (*R*,*R*) or (*S*,*S*). The terminology is also applicable for describing the faces (Sect. 7.73) that are approaching each other during a reaction: a *re*-to-*re* or *si*-to-*si* approach is *l*, while a *re*-to-*si* or *si*-to-*re* approach would be classified as *u*.

Example. Figure 7.78.

7.79 EPIMERS

Diastereomers differing in configuration at only one (of possibly several) tetrahedral diastereogenic centers.

Example. Figure 7.79.

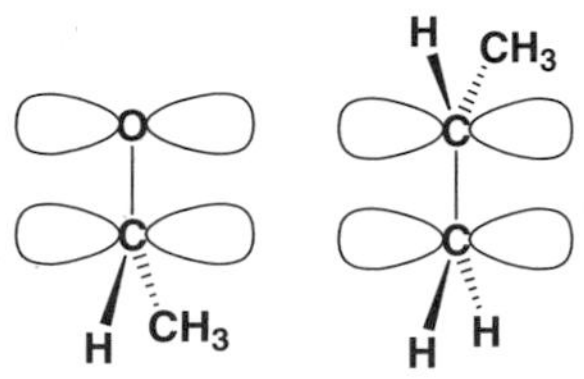

Figure 7.78. The *l* approach of the *re* face of ethanal to the *re* face of propene.

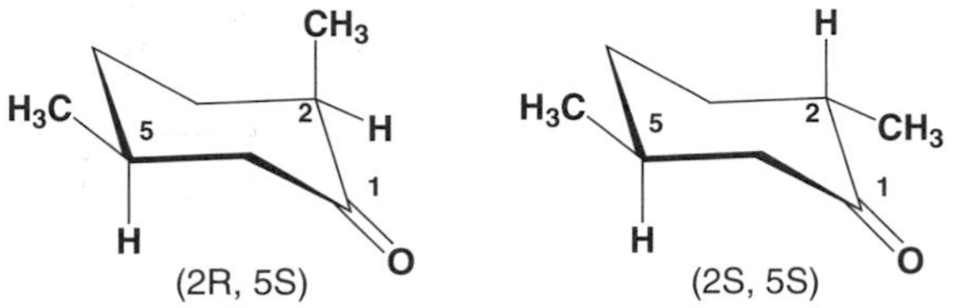

Figure 7.79. Epimers of 2,5-dimethylcyclohexanone that differ only in the configuration at carbon 2.

7.80 EPIMERIZE

To invert configuration at only one (of several) tetrahedral diastereogenic centers of a molecule.

Example. Any reaction in which (2*R*,5*S*)-2,5-dimethylcyclohexanone (Fig. 7.79) is converted to its (2*S*,5*S*) isomer involves epimerization at the 2 position.

7.81 ANOMERS

Used primarily in carbohydrate chemistry to indicate the two epimers (diastereomers) at the hemiacetal or acetal carbon of a sugar in its cyclic form.

Example. Figure 7.81 shows the two anomers of the aldohexose sugar D-(+)-glucose; their interconversion at the anomeric carbon (*anomerization*) occurs through the open form shown in Fig. 7.81*b*.

7.82 MUTAROTATION

The change in optical rotation of a (solution of a) pure stereoisomer, caused by reversible epimerization to form a mixture of two epimers, each with a nonzero optical rotation. Thus, the equilibrium value of the optical rotation of the mixture is generally not zero.

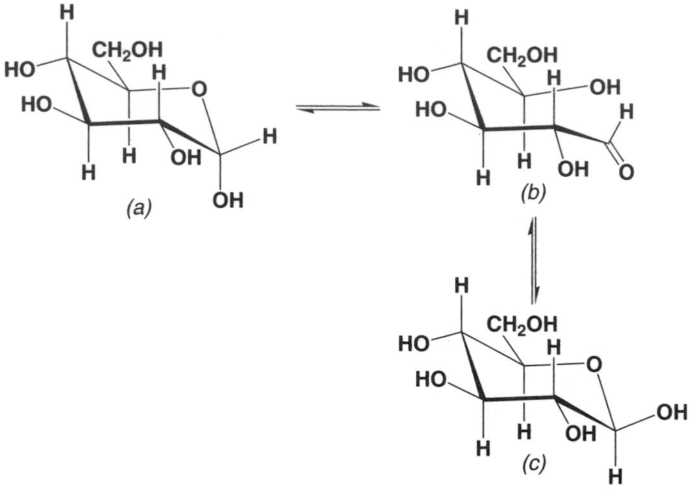

Figure 7.81. (*a*) The α anomer of glucose; (*b*) the open form of glucose; (*c*) the β anomer of glucose.

Example. When the pure crystalline α anomer (α-D-glucose) shown in Fig. 7.81*a* ($[\alpha]_D = +111°$) is dissolved in water, the specific rotation gradually decreases to a value of +52.5°, corresponding to a mixture of 38% α and 62% β (Fig. 7.81*c*, $[\alpha]_D = +19°$). The intermediate is the open-chain aldehyde (Fig. 7.81*b*).

7.83 PSEUDO-ASYMMETRIC CENTER

A tetrahedral diastereogenic center lying in the mirror plane of a *meso* form (Sect. 7.35). Substitution at this center gives different *meso* diastereomers. Molecules with pseudo-asymmetric centers are achiral.

Example. Figure 7.83.

7.84 STEREOSPECIFIC REACTION

A reaction in which one specific stereoisomer of the reactant gives one specific stereoisomer (or *d,l* pair) of product. Thus, two different stereoisomers of the reactant must be separately convertible into two different stereoisomers of the product.

Example. The addition of Br_2 to a π bond is generally stereospecifically *trans,* as shown in Fig. 7.84*a*. A direct nucleophilic displacement generally proceeds with stereospecific inversion, as shown in Fig. 7.84*b*.

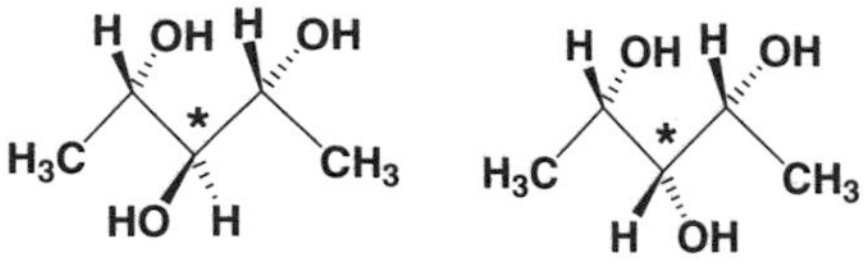

Figure 7.83. Pseudo-asymmetric centers (*) in the *meso* diastereomers of 2,3,4-pentanetriol.

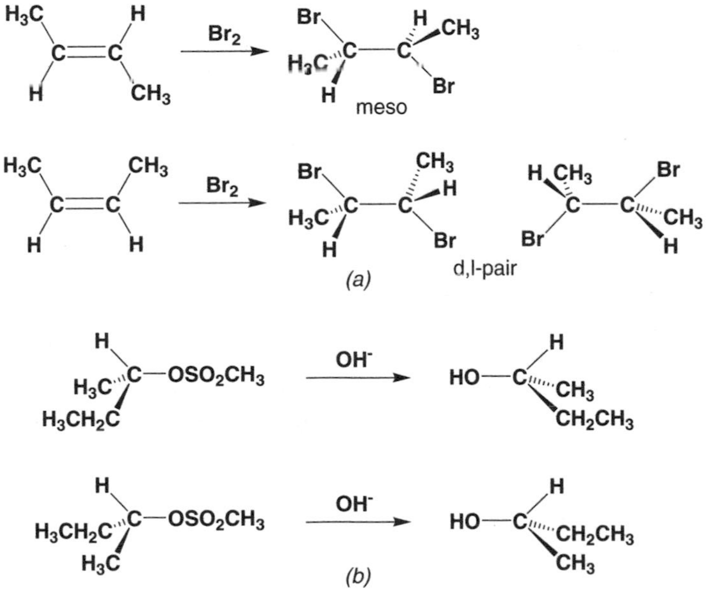

Figure 7.84. (*a*) Stereospecific *trans* addition of bromine to *trans*- and *cis*-2-butene and (*b*) stereospecific inversion during backside nucleophilic substitution.

7.85 STEREOSELECTIVE REACTION

A reaction in which one stereoisomer in a mixture of substrate stereoisomers is destroyed (or one stereoisomer in a mixture of product stereoisomers is formed) more rapidly than another, resulting in the predominance of one stereoisomer in the mixture of products.

Example. The epimers shown in Fig. 7.85 react at vastly different rates. The reactions shown in Fig. 7.84*a* are also stereoselective. *All* stereospecific reactions are necessarily stereoselective. However, a stereoselective reaction need not *be* stereospecific. Thus, if conditions existed under which both (*Z*) and (*E*)-2-butene individually gave the same reaction mixture richer in one stereoisomer, both reactions would be stereoselective, but neither would be stereospecific.

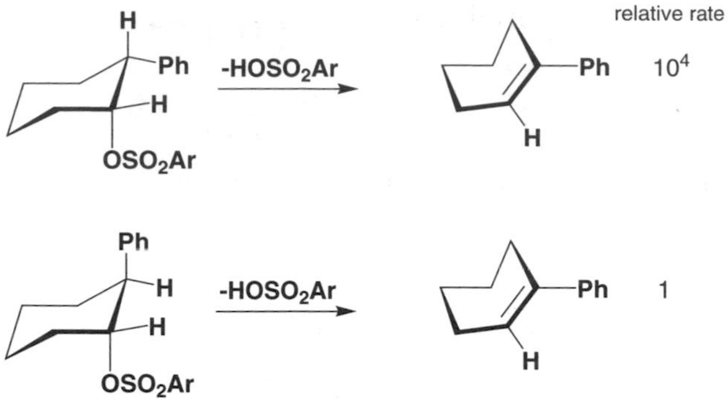

Figure 7.85. A stereoselective elimination of the stereoisomers of 2-phenylcyclohexyl sulfonate to form 1-phenylcyclohexene.

7.86 ENANTIOSELECTIVE REACTION

One class of stereoselective reactions. A reaction in which one enantiomer in a mixture of substrate enantiomers is destroyed (or one enantiomer in a mixture of product enantiomers is formed) more rapidly than another, resulting in the predominance of one enantiomer in the mixture of products.

Example. See Sect. 7.77.

7.87 DIASTEREOSELECTIVE REACTION

One class of stereoselective reactions. A reaction in which one diastereomer in a mixture of substrate diastereomers is destroyed (or one diastereomer in a mixture of product diastereomers is formed) more rapidly than another, resulting in the predominance of one diastereomer in the mixture of products.

Example. The reaction in Fig. 7.85 is diastereoselective.

7.88 REGIOSELECTIVE REACTION

A reaction in which one reaction site in the substrate molecule is more reactive than any other site, leading to the preferential formation of one positional isomer of the product.

Example. The fact that neither of the elimination reactions shown in Fig. 7.85 gives an appreciable amount of the nonconjugated olefin (Fig. 7.88*a*) suggests that this elimination involves regioselective elimination of the methine hydrogen. The ionic

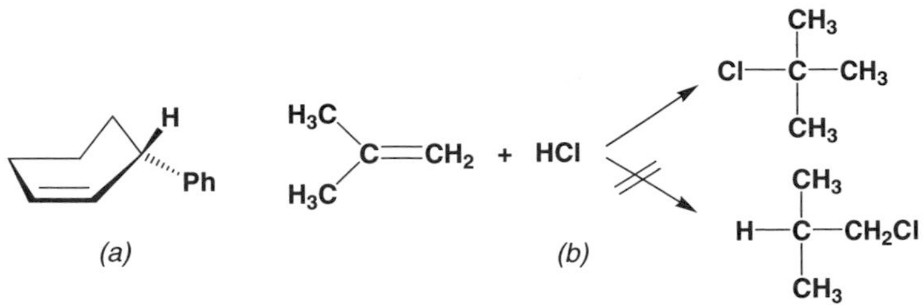

Figure 7.88. (*a*) The nonconjugated 3-phenylcyclohexene *not* formed during the regioselective eliminations in Fig. 7.85 and (*b*) the 100% regioselective addition of HCl to isobutene.

addition of HCl to substituted olefins proceeds regioselectively to place the Cl on the more substituted carbon (Fig. 7.88*b*).

7.89 REGIOSPECIFIC REACTION

A reaction in which one specific positional isomer of the product is formed exclusively when other isomers are possible. A regiospecific reaction is 100% regioselective.

Example. The reaction shown in Fig. 7.88*b* actually gives only the upper product, so the reaction is regiospecific.

7.90 ASYMMETRIC INDUCTION

Control of the stereoselectivity of a reaction exerted by an existing chiral center during the formation of a new chiral center. The new chiral center can be created in the same molecule that contains the existing center (intramolecular) or in a different molecule (intermolecular). Asymmetric induction makes a reaction either diastereoselective (Sect. 7.87) or enantioselective (Sect. 7.86).

Example. Figure 7.90*a* shows a new chiral center being generated near an existing chiral center in 2-phenylpropanal during the diastereoselective addition of methylmagnesium iodide (intramolecular asymmetric induction). (*b*) The prochiral carbonyl carbon in 3,3-dimethyl-2-butanone is converted to a chiral carbon in the alkoxide by the enantioselective reaction with a chiral Grignard reagent (intermolecular asymmetric induction).

7.91 ASYMMETRIC SYNTHESIS

A synthesis that includes one or more reactions involving asymmetric induction.

Example. See Fig. 7.92.

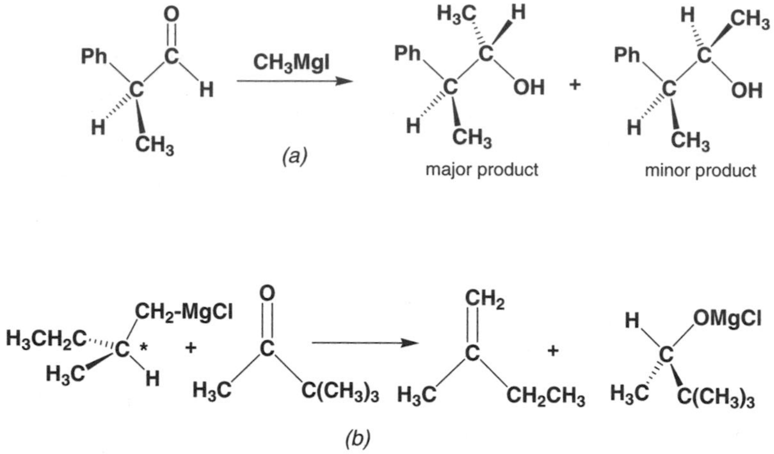

Figure 7.90. (*a*) Intramolecular and (*b*) intermolecular asymmetric induction.

7.92 CHIRAL AUXILIARY

A chiral group, covalently bonded to the reagent or substrate, that causes stereoselectivity during subsequent formation of a new stereogenic center. Such groups render the molecular environment around the reaction site chiral. Because in some ways they resemble protecting groups, chiral auxiliaries must be readily attachable, operate as efficient stereochemical directing groups, and then be readily detachable after they have performed their function.

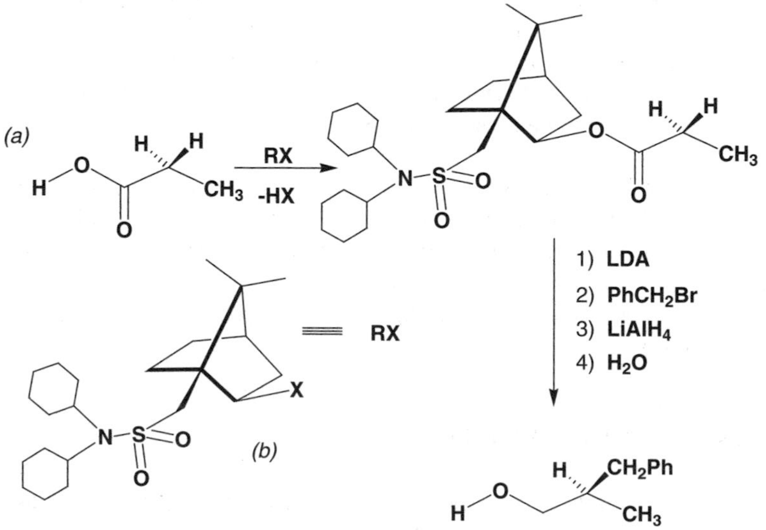

Figure 7.92. (*a*) An asymmetric synthesis involving the chiral auxiliary (*b*).

Example. The methylene hydrogens of propionic acid are enantiotopic. Reaction with RX in Fig. 7.92*b* converts propanoic acid to a chiral ester by attaching camphor-based chiral auxiliary R. This renders the methylene hydrogens diastereotopic. Base (lithium diisopropylamide, LDA)-promoted alkylation, followed by removal of the auxiliary, leads to the indicated enantiomer of 3-phenyl-1-butanol in 98% enantiomeric excess.

7.93 CHIRAL POOL

Readily available, inexpensive, enantiomerically pure compounds that can serve as the starting materials for introducing chirality into synthetically produced compounds. These substrates can be converted into enantiomerically pure synthetic intermediates by making use of conventional stereospecific organic transformations.

Example. Many natural products (e.g., amino acids, terpenes, carbohydrates, lactic and tartaric acid derivatives) have been used as substrates from the chiral pool.

7.94 CHIRAL SOLVENTS

An enantiomerically pure solvent. Such solvents can be used to provide the chiral environment necessary for discriminating between the enantiomers of a solute.

Example. Optically pure 1-phenyl-2,2,2-trifluoroethanol has been used to differentiate the enantiomers of a chiral solute during examination of the mixture by nuclear magnetic resonance spectroscopy (Sect. 15.81).

7.95 CRAM'S RULE

Donald Cram (1919–2001), who shared the 1987 Nobel Prize with Charles Pederson and Jean-Marie Lehn, formulated this semiempirical rule for predicting the diastereoselectivity of a reaction involving nucleophilic addition to the carbonyl group of a chiral ketone or aldehyde.

Example. The reaction in Fig. 7.95*a* involves the nucleophilic addition of organometallic reagent RM to a chiral aldehyde, where the three R substituents attached to the enantiogenic center of the aldehyde are labeled S (small), M (medium), and L (large) to reflect their steric bulk. The presence of chirality in the substrate causes the two faces of the carbonyl to be diastereotopic. As a result, the amount of product 1 (from attack nearer the S group) should be different than the amount of product 2 (from attack nearer the M group).

Initially, it was believed that the asymmetric induction seen in such reactions could be predicted solely on the basis of steric differences between the groups attached to the asymmetric center. A Newman projection of the ketone in its reactive conformation (Fig. 7.95*b*) shows that the sterically least crowded path for reagent

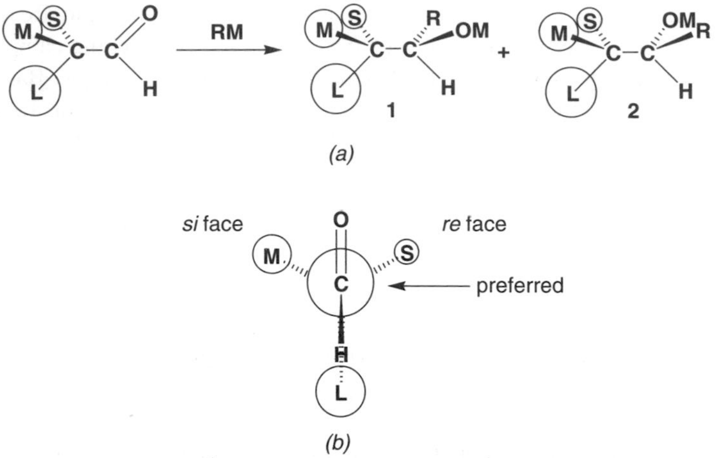

Figure 7.95. (*a*) Addition of reagent RM to a chiral ketone, leading to two diastereomeric products, and (*b*) Newman projection of the chiral ketone.

attack is nearer the S group, the smallest substituent. Therefore, product 1 is predicted to be formed preferentially.

Subsequent work has demonstrated that other factors, such as the coordination of Lewis acids at the carbonyl oxygen, can also have a significant influence on the stereochemical outcome.

7.96 CHEMZYMES

A class of relatively small molecules conceptualized by Elias J. Corey (1928– , who received the Nobel Prize in 1990), capable of catalyzing reactions to give levels of enantio- and diastereoselectivity approaching that induced by naturally occurring enzymes.

Example. The chiral dipeptide in Fig. 7.96 catalyzes the addition of HCN to benzaldehyde, yielding only the (*R*)-enantiomer of the cyanohydrin product.

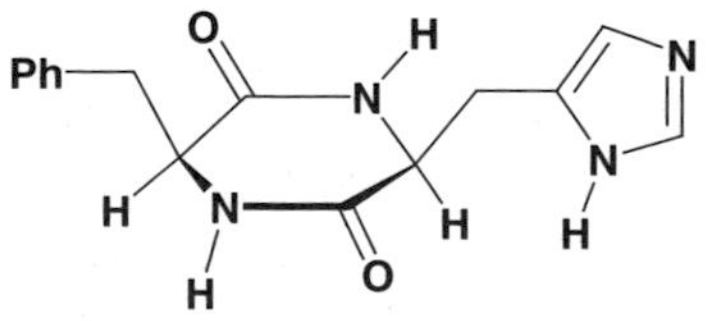

Figure 7.96. An example of a chemzyme.

7.97 STRAIN

A deformation in the structure of a molecule (or molecular complex) that raises its energy compared to a structure (or hypothetical structure) that is not deformed. The principal types of strain are due to deformations in bond lengths, bond angles, torsional angles, and steric contact. These stains have been described, perhaps facetiously, as FBI strain for front, back, and internal.

7.98 BOND-STRETCHING STRAIN

A structurally induced lengthening or shortening of a bond relative to the length of that bond in its lowest-energy geometry.

Example. The preferred C_{sp^3}–C_{sp^3} bond length in alkanes is 1.54 Å. The C1-C4 bond in Fig. 7.98 is constrained by the molecular geometry to be 1.75 Å, giving rise to a total strain energy (including all forms of strain) of 104 kcal/mol. As a result, this exceptionally long bond is unusually reactive.

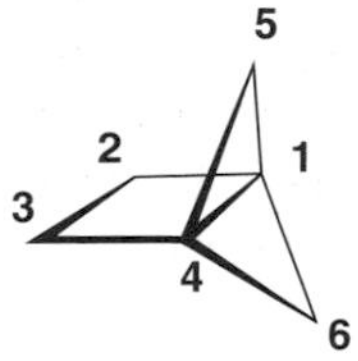

Figure 7.98. Bond-stretching strain in tricyclo[2.1.1.0^{1,4}]hexane.

7.99 ANGLE STRAIN (BAEYER STRAIN, I STRAIN)

A structurally induced opening or closing of a bond angle relative to the same bond angle in its preferred lowest-energy geometry.

Example. Small rings (three- to five-membered, Figs. 7.99*a–c*) have internuclear C–C–C bond angles smaller than the preferred tetrahedral angle of 109.5°. The interior angles of a regular polygon with *n* sides are 180° − (360°/*n*). Cyclopropane (Fig. 7.99*a*), with 28 kcal/mol of total strain, manifests this strain in the unusually high reactivity of its C–C bonds.

		internuclear C-C-C- angle	total strain energy per CH_2 kcal/mol
(a)		60°	9.2
(b)	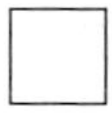	90°	6.6
(c)	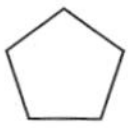	108°	1.3

Figure 7.99. Internuclear C–C–C bond angles and total strain energy per CH_2 in (*a*) cyclopropane, (*b*) cyclobutane, and (*c*) cyclopentane. The latter two rings are assumed to be planar (flat).

7.100 STERIC STRAIN (PRELOG STRAIN, F STRAIN)

An increase in energy caused by the intra- or intermolecular approach of two or more atoms not directly bonded to each other within a distance where their electron clouds begin to repel each other. Such strain occurs when the internuclear distance between the atoms is less than the sum of their van der Waals radii (Sect. 2.25).

7.101 TORSIONAL STRAIN (ECLIPSING STRAIN, PITZER STRAIN)

An increase in the energy of a molecule arises when the orbitals on neighboring atoms are brought into alignment by rotation around the bond that connects them. It is known as torsional strain, eclipsing strain, and Pitzer strain, after Kenneth Pitzer (1913–1996). The reason such alignment (eclipsing, Sect. 7.109) of orbitals increases the energy of a molecule is not only a result of electrostatic repulsion between the electrons in these orbitals. Torsional strain also appears to be partially caused by a weak antibonding interaction (Sect. 2.32) between the orbitals.

Example. Figure 7.101.

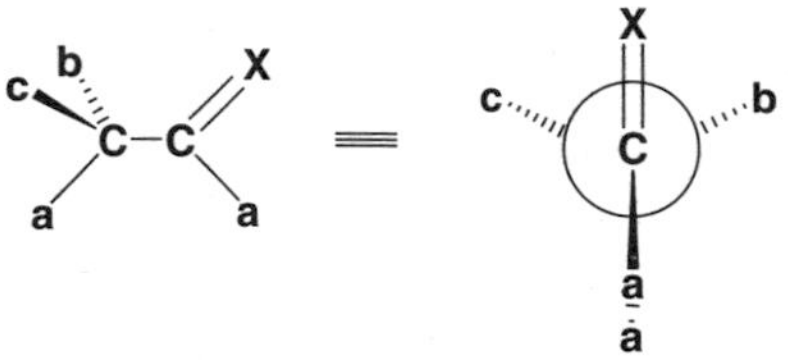

Figure 7.101. The two C–a bonds are aligned (eclipsed), giving rise to torsional strain.

7.102 RING STRAIN

The total strain in a cyclic or polycyclic molecule, representing the sum of all angle strain, torsional strain, steric strain, and bond-stretching strain.

7.103 TORSIONAL ANGLE (DIHEDRAL ANGLE)

The angle around the B–C bond between the A–B bond and the C–D bond in the nonlinear bond sequence A–B–C–D. Alternatively, it is the angle between the ABC plane and the BCD plane. *Dihedral angle* (literally, the angle between two planes) is the more general term, extending to the situation where the two sets of atoms defining the planes are not connected to each other.

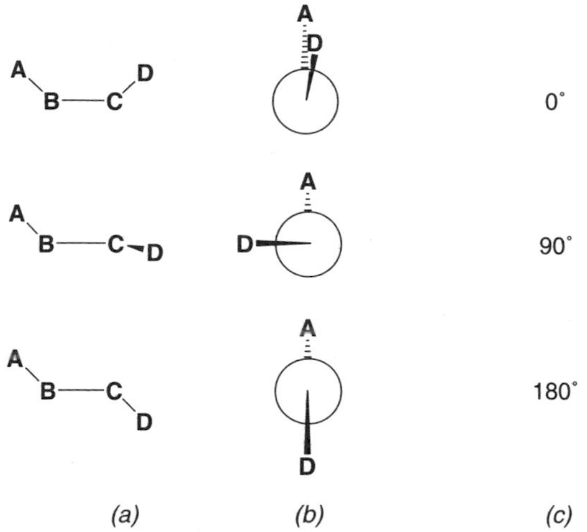

Figure 7.103. Torsional (dihedral) angles around the B–C bond.

Example. Figure 7.103 shows arrangements of atoms with various torsional angles.

7.104 CONFORMATION (CONFORMER, ROTOMERS)

A particular instantaneous orientation of the atoms in a molecule, differing from other possible orientations by rotation(s) around single bonds (or in special cases by inversion, Sect. 7.106). Unless it is held rigid by small rings, double bonds, or other structural units with large barriers to rotation, a molecule can have an infinite number of conformations, while retaining one configuration.

Example. Whether two stereoisomeric structures constitute different configurations or merely different conformations depends on how readily they interconvert. For example, the various *conformations* of ethane derivatives (Sect. 7.109) interconvert trillions of times per second at 25°C, while the *Z* and *E configurations* of ethylene derivatives (Sect. 7.31) do not interconvert spontaneously even at 200°C. This is true because normally there is relatively free (unrestricted) rotation around single bonds, while rotation around a double bond has an activation barrier of ca. 65 kcal/mol. Since, in principle, it should be possible to slow or even stop conformational interconversion by sufficient cooling, the operational distinction between configurations and conformations is based on whether the structures interconvert rapidly under the conditions of the experiment.

7.105 CONFORMATIONAL ISOMER

A term usually reserved for those conformations that represent shallow local minima on the conformational energy surface. Because these energy minima are shallow, there is rapid interconversion between conformational isomers. However, see "Atropisomers" (Sect. 7.133).

7.106 INVERTOMERS (INVERSION ISOMERS)

Two conformers that interconvert by (pyramidal) inversion at an atom possessing a nonbonding pair of electrons.

Example. The classic example of pyramidal inversion involves the ammonia molecule (Fig. 7.106*a*), where the inversion of the electron pair and the three hydrogen atoms occurs at ca. 10^6 Hz, and serves as the basis of the so-called ammonia clock. Note that the transition state for the inversion is trigonal planar (sp^2 hybridized) at nitrogen. Thus, although individual molecules of appropriately substituted acyclic amines are chiral (Fig. 7.106*b*), inversion is too rapid to permit resolution (Sect. 7.59) of either enantiomeric invertomer. Various ploys, such as using larger central atoms (e.g., P, As), inclusion of the heteroatom in a small ring (e.g., certain aziridines), or inclusion of additional heteroatoms in the small ring, slow down the rate of inversion and have led to resolution of the invertomers.

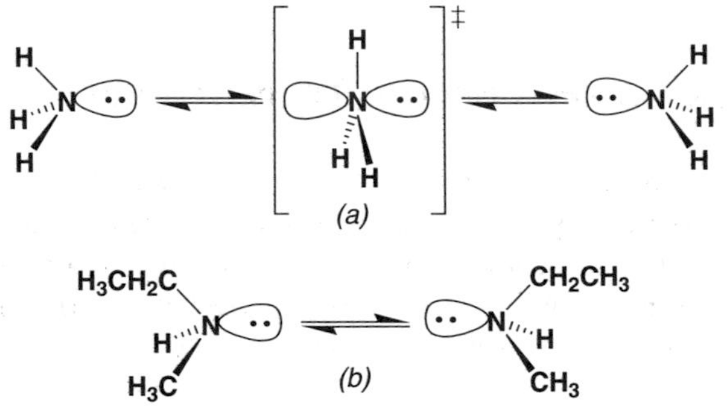

Figure 7.106. (*a*) The (equivalent) invertomers of ammonia, separated by a planar transition state; (*b*) the enantiomeric invertomers of ethylmethylamine, which also occurs via a planar transition state (not shown).

7.107 GEM (GEMINAL)

Two identical atoms or groups separated by two bonds and attached to the same atom in a molecule.

Example. Figure 7.107 shows geminal methyls.

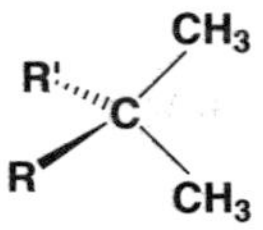

Figure 7.107. Geminal methyl groups.

7.108 VIC (VICINAL)

Two identical atoms or groups separated by three bonds and attached to directly bonded atoms.

Example. Figure 7.108 shows vicinal chlorines.

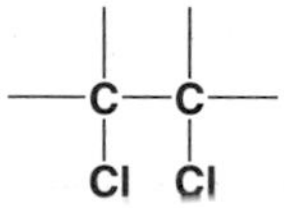

Figure 7.108. Vicinal chlorines.

7.109 ECLIPSED CONFORMATION

A conformation with a 0° dihedral angle (Sect. 7.103). In an eclipsed conformation one or more pairs of vicinal bonds (orbitals) will be aligned.

Example. Ethane (Fig. 7.109) has an infinite number of conformations, each with a different torsional angle between vicinal hydrogens. Because both carbons are tetrahedral, whenever one set of vicinal C–H bonds is eclipsed (aligned), the other two sets must also be eclipsed (Fig. 7.109*a*); this is the eclipsed conformation, and it suffers the maximum torsional strain.

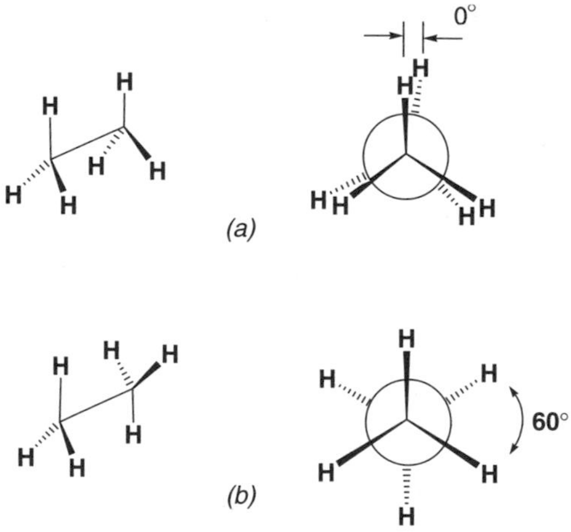

Figure 7.109. (*a*) Eclipsed conformation of ethane and (*b*) staggered conformation of ethane (see Sect. 4.8).

7.110 STAGGERED CONFORMATION

A conformation with a 60° dihedral angle. In this conformation vicinal bonds are perfectly spaced for minimal torsional strain.

Example. Figure 7.109*b* shows the staggered conformation of ethane. In ethane the staggered conformation is the most stable, and the eclipsed conformation is the least stable, 2.9 kcal/mol above the staggered (i.e., ca. 1.0 kcal/mol of torsional strain per set of eclipsed hydrogens). Note that all conformations of ethane other than staggered and eclipsed are chiral.

7.111 *cis* CONFORMATION (SYN-PERIPLANAR)

In molecules of the type X–CH_2–CH_2–Y, the fully eclipsed conformation, with a X–C–C–Y torsional angle of 0° (Fig. 7.111*a*). See also Sect. 7.115.

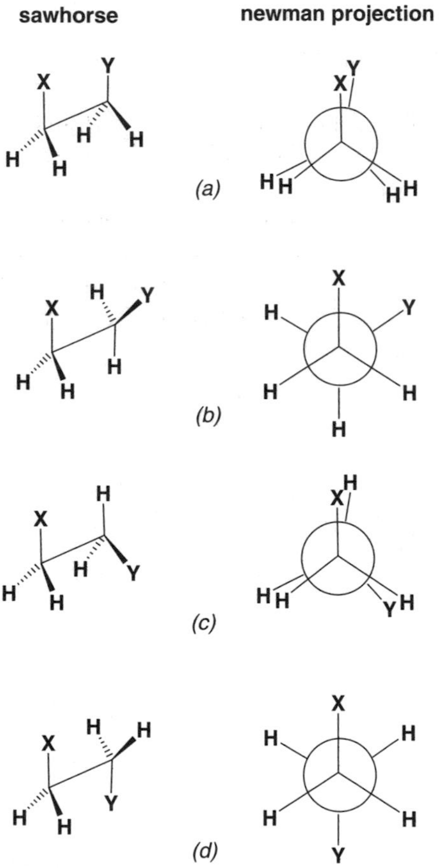

Figure 7.111. Conformations of the molecule X–CH_2–CH_2–Y: (*a*) *cis* (or syn-periplanar); (*b*) gauche; (*c*) anticlinal eclipsed; (*d*) *trans* (or antiperiplanar).

7.112 *gauche* CONFORMATION (SKEW CONFORMATION, SYNCLINAL)

The staggered conformation of X–CH_2–CH_2–Y with a X–C–C–Y torsional angle of 60° (Fig. 7.111*b*).

7.113 ANTICLINAL ECLIPSED CONFORMATION

The eclipsed conformation of X–CH_2–CH_2–Y with a X–C–C–Y torsional angle of 120° (Fig. 7.111*c*).

7.114 *anti* CONFORMATION (*trans* CONFORMATION, ANTIPERIPLANAR)

Staggered conformation of X–CH_2–CH_2–Y with a X–C–C–Y torsional angle of 180° (Fig. 7.111*d*). The relative stability of the above four conformations in Fig. 7.111 is $d > b > c > a$. See also Sect. 7.116.

7.115 *s-cis*

In molecules of the type A=C–C=B (or A=C–B–C), the eclipsed conformation with a torsional angle of 0° between the A=C and C=B (or B–C) bonds. See also Sect. 7.111.

Example. *s-cis*-1,3-Butadiene and *s-trans*-1,3-butadiene (Figs. 7.115*a* and *b*, respectively) and *s-cis*- and *s-trans*-*N*-methylformamide (Figs. 7.115*c* and *d*, respectively). The descriptors *s*-cis and *s*-trans refer to *cis* and *trans* relationships with respect to the formally single ($s = \sigma$) bond, which possesses some double bond character (due to resonance) that provides a modest torsional barrier to rotation.

7.116 *s-TRANS*

In molecules of the type A=C–C=B (or A=C–B–C), the conformation with a torsional angle of 180° between the A=C and C=B (or B–C) bonds. See also Sect. 7.114.

Example. See Fig. 7.115.

7.117 W CONFORMATION AND SICKLE CONFORMATION

Conformational arrangements of X–C–C–C–Y that resemble a W or sickle, respectively.

Example. See Fig. 7.117.

7.118 RING PUCKERING

With the exception of three-membered rings, saturated carbocyclic rings are considerably more strained when forced to be planar, than in other nonplanar (puckered)

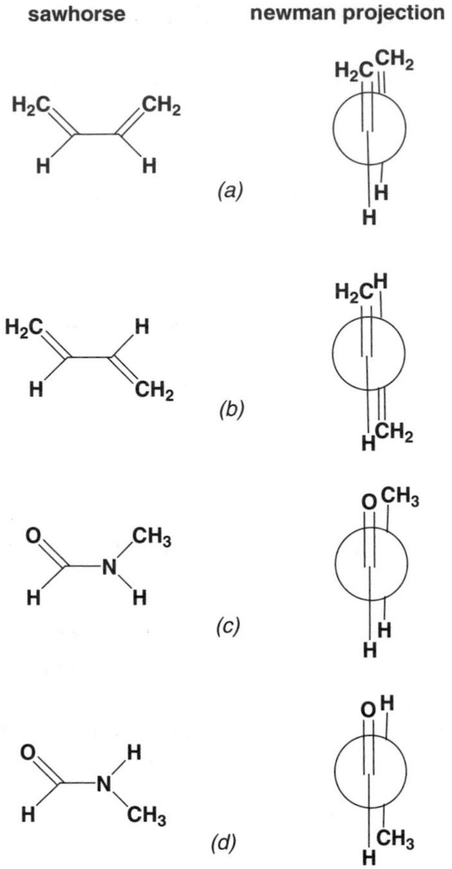

Figure 7.115. (*a*) *s-cis*-1,3-butadiene; (*b*) *s-trans*-1,3-butadiene; (*c*) *s-cis-N*-methylformamide; (*d*) *s-trans-N*-methylformamide.

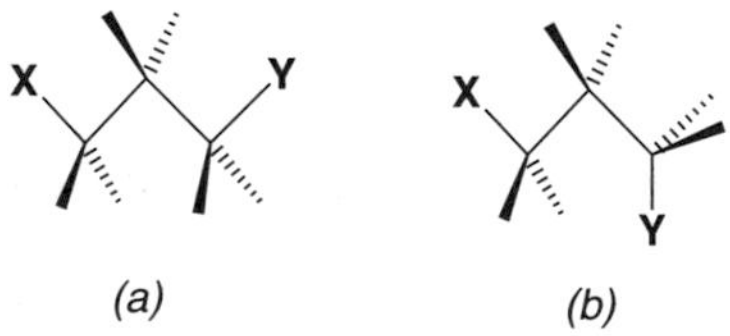

Figure 7.117. (*a*) The W conformation and (*b*) sickle conformation.

conformations. This is because the planar conformations give rise to both angle and torsional strain that can be relieved when the molecule relaxes to a less strained nonplanar conformation. See Sects. 7.119, 7.129, and 7.131.

7.119 CHAIR CONFORMATION

The all-staggered conformation of a saturated six-membered ring. The term is less commonly applied to certain conformations of seven- and eight-membered rings.

Example. Chair cyclohexane (Fig. 7.119), where **a** = an axial substituent and **e** = an equatorial substitutent.

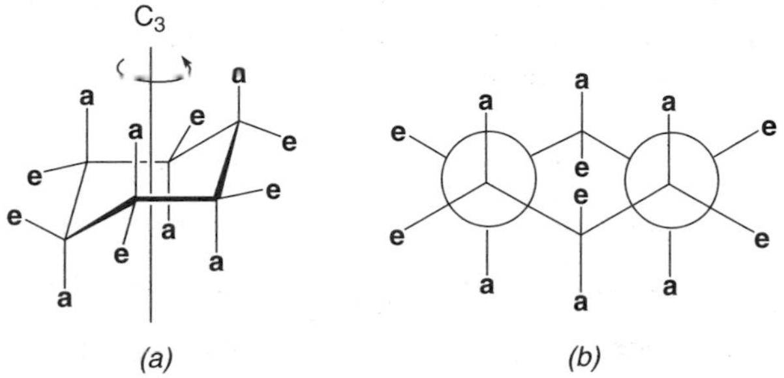

Figure 7.119. Chair conformation of cyclohexane: (*a*) in side view and (*b*) Newman projection.

7.120 AXIAL POSITIONS

The six bond positions (three up and three down) labeled **a** in Fig. 7.119, which are parallel to the molecular C_3 axis. Any axial position is *gauche* to both vicinal equatorial positions and *anti* to both vicinal axial positions.

7.121 EQUATORIAL POSITIONS

The bond positions around the imagined "equator" of the molecule, labeled **e** in Fig. 7.119. Any equatorial position is *gauche* to both vicinal axial and equatorial positions.

7.122 RING FLIPPING

By a series of correlated rotations, one chair form (Sect. 7.119) can be interconverted with another chair form in which each originally axial position has become equatorial and *vice versa* (Fig. 7.122).

7.123 CONFIGURATIONS OF DISUBSTITUTED CHAIR CYCLOHEXANES

The stereochemical relationship between two groups on different carbons of a chair cyclohexane ring is described as *cis* (*Z*) or *trans* (*E*) according to Fig. 7.123.

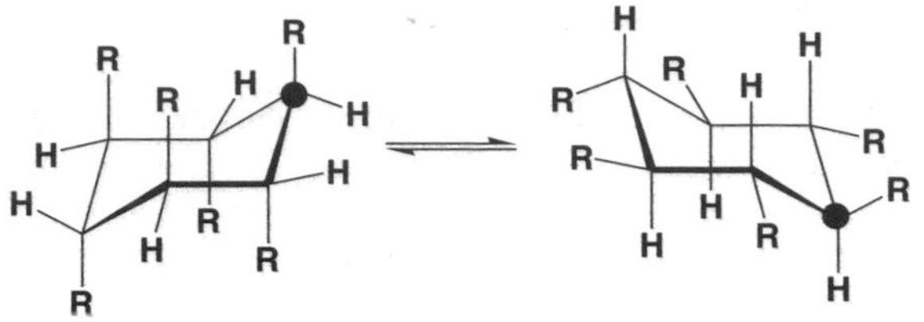

Figure 7.122. Axial/equatorial interchange during chair–chair ring flip. The **R** groups, originally axial, become equatorial, while the originally equatorial hydrogens become axial. The dotted carbon is for visual reference only.

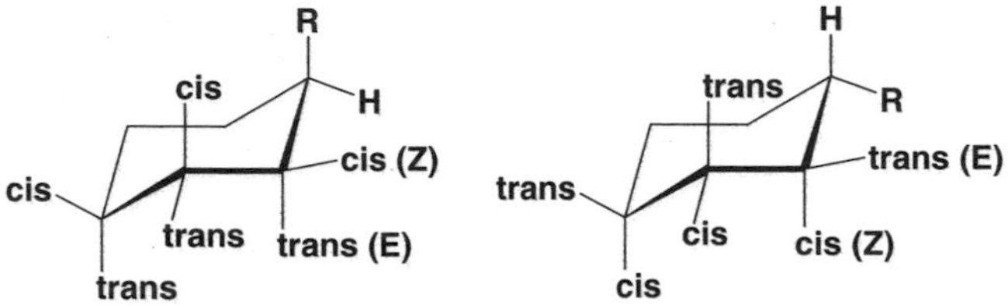

Figure 7.123. Configurations of chair cyclohexane substituents with respect to the **R** group.

7.124 BOAT CONFORMATION

The least stable conformation of a six-membered ring, resembling a boat, as shown in Fig. 7.124. The boat conformation is rendered unstable by many eclipsing interactions, as well as flagpole–flagpole (bow–stern) repulsion (Sect. 7.127). In cyclohexane itself, the boat conformation constitutes a transition state between the somewhat more stable twist boat conformations (Sect. 7.129).

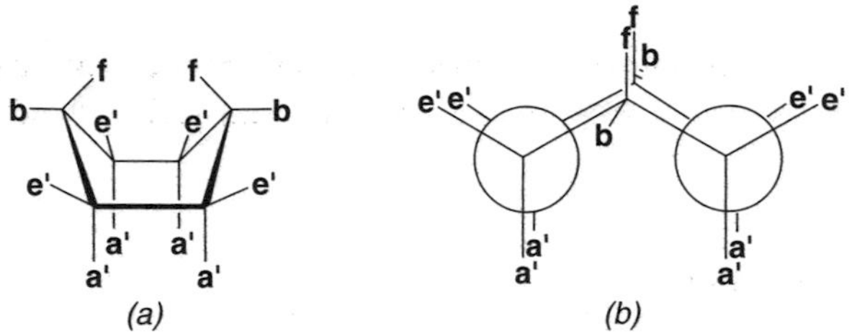

Figure 7.124. Boat conformation of cyclohexane: (*a*) side view and (*b*) Newman projection.

7.125 PSEUDO-EQUATORIAL

Positions labeled **e′** (Fig. 7.124).

7.126 PSEUDO-AXIAL

Positions labeled **a′** (Fig. 7.124).

7.127 FLAGPOLE

Positions labeled **f** (Fig. 7.124).

7.128 BOWSPRIT

Positions labeled **b** (Fig. 7.124).

7.129 TWIST BOAT (SKEW BOAT)

A partially staggered conformation of a six-membered ring (Fig. 7.129). The order of stability of six-membered ring conformations is boat < twist boat < chair.

Figure 7.129. Twist boat conformations of cyclohexane.

7.130 A VALUES

The experimentally determined differences in free energy ($\Delta G°$) for equatorial versus axial substituents on a chair cyclohexane. These energy differences, which are equal to twice the 1,3-diaxial interaction between the group and a hydrogen atom, are useful for estimating the effective steric size of substituents, and are assumed to be additive. Some selected A values are given below. The [equatorial]/[axial] equilibrium constant K (Sect. 12.49) is given by exp(A/RT). Thus, at 25°C (298 K) the [equatorial]/[axial] ratio for methylcyclohexane is $\exp(1.74/0.00199 \times 298) = 17.8$, or 95% equatorial, 5% axial.

Group	$A = -\Delta G$ (kcal/mole)[a]
OH	0.7[b]
OCH_3	0.7
F	0.25
Cl	0.4
Br	0.5
I	0.4
$CO_2C_2H_5$	1.1
CH_3	1.74
C_2H_5	1.8
$CH(CH_3)_2$	2.1
$C(CH_3)_3$	5
C_6H_5 (phenyl)	3.1

[a] Values from Eliel, E. L., Allinger, N. L., Angyal, S. J., and Morrison, G. A., *Conformational Analysis*, John Wiley & Sons: New York, 1965.
[b] Value increases in protic solvents.

7.131 FOLDED, ENVELOPE, TUB, AND CROWN CONFORMATIONS

Descriptive terms applied to puckered conformations of four-, five-, eight-, and larger-membered rings, respectively.

Example. Figure 7.131.

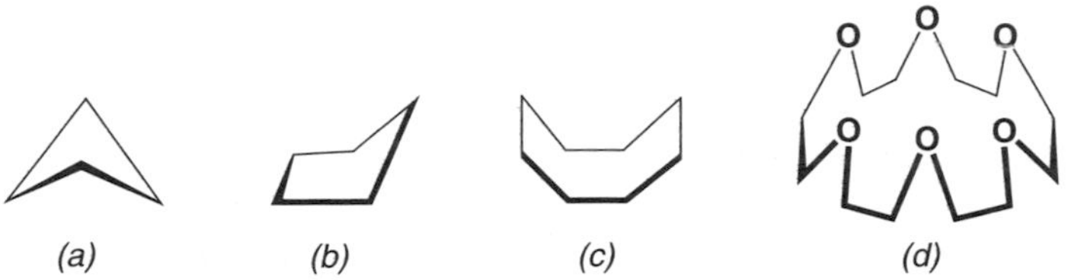

Figure 7.131. (*a*) Folded conformation of cyclobutane; (*b*) envelope conformation of cyclopentane; (*c*) tub conformation of cyclooctane; (*d*) crown conformation of 18-crown-6.

7.132 ANANCOMERIC RING

A ring that is "locked" into a specific conformation by virtue of a large substituent.

Example. The bulky *tert*-butyl group attached to a cyclohexane ring has a 4.9 kcal/mol preference for the equatorial position over the axial position (Sect. 7.130). This means that at any instant, >99.9% (virtually all) of the molecules have an equatorial substituent. Although ring-flipping (Sect. 7.122) is still occurring, fewer than 0.1% of the rings have the substituent in the axial position.

7.133 ATROPISOMERS

Two conformations of a molecule whose interconversion is sufficiently slow under ambient conditions to allow separation and isolation, but which are capable of interconverting at higher temperatures. Atropisomers constitute the "grey area" between conformations and stereoisomers.

Example. The two enantiomeric conformations of the ortho tetrasubstituted biphenyl (Fig. 7.133) exhibit hindered rotation about the bond joining the two rings. If the *ortho* groups are large enough, they completely inhibit rotation (and therefore interconversion), enabling the two forms to be resolved into enantiomers.

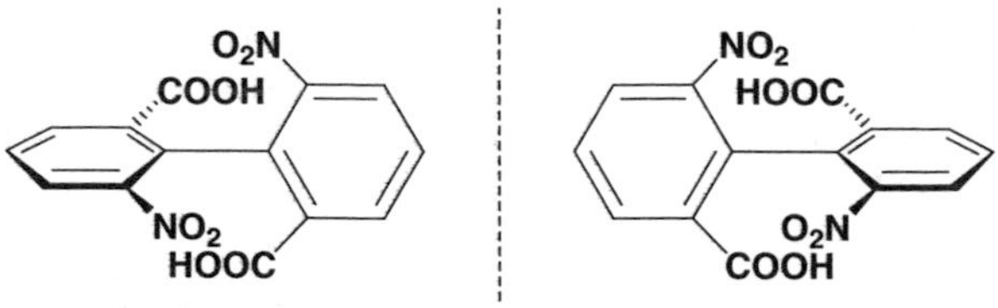

Figure 7.133. Two enantiomeric atropisomers of 2,2′-dinitro-6,6′-biphenyldicarboxylic acid.

7.134 IN(SIDE), OUT(SIDE) ISOMERISM

Terms used to describe stereoisomers of large ring bicyclic molecules where the bridgehead substituent (usually hydrogen) can be either "inside" or "outside" the ring system.

Example. The three possible stereoisomers of bicyclo[5.5.5]heptadecane (Fig. 7.134).

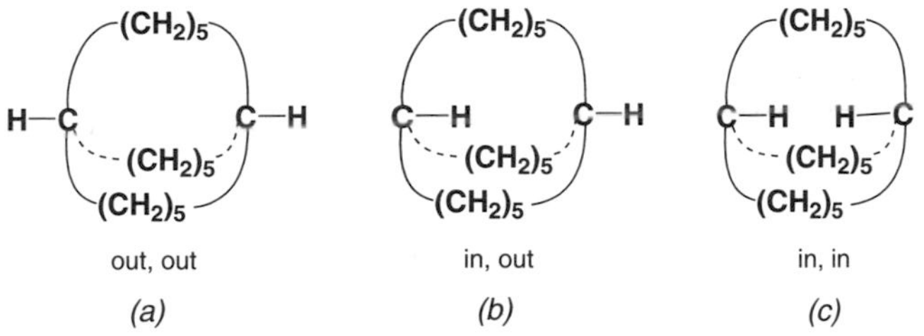

Figure 7.134. The three stereoisomers of bicyclo[5.5.5]heptadecane: (*a*) out(side), out(side); (*b*) in(side), out(side); (*c*) in(side), in(side).

7.135 HOMEOMORPHIC ISOMERS

Objects that are topologically equivalent (same sequence of atoms and bonds) even though they have a different shape. Homeomorphic isomerism is a special type of stereoisomerism that involves molecules which have identical connectivity, but are differentiable because there is a significant energy barrier for their interconversion by simple motion about bonds.

Example. Atropisomers (Sect. 7.133) and in-out isomers (Sect. 7.134).

7.136 ISOTOPOMERS

Stereoisomers in which the stereochemical differences are due to isotopic substitution at one of a pair of prochiral atoms (Sect. 7.71).

Example. Enantiotopic hydrogens on a prochiral methylene group may be differentially substituted to generate enantiomeric isotopomers, as in Figs. 7.136*a* and *b*. As another example, the three hydrogens on a methyl group have been stereospecifically replaced by H, D, and T to afford a chiral methyl group. When such a substrate is subjected to enzymatic incubation, the isotopomers will react at different rates. Such experiments have been of critical importance in elucidating various enzyme reaction mechanisms.

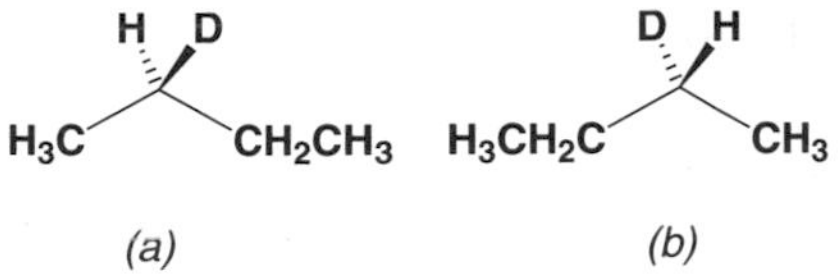

Figure 7.136. Isotopomers of 2-deuteriobutane: (*a*) the (*R*) isomer and (*b*) the (*S*) isomer.

7.137 APICAL AND EQUATORIAL POSITIONS OF A TRIGONAL BIPYRAMID (TBP)

The positions labeled **a** and **e** in Fig. 7.137*a* are the apical and equatorial positions, respectively. The apical bonds are generally slightly longer than equatorial bonds. Trigonal bipyramidal systems are often associated with the chemistry of pentavalent phosphorus compounds.

Example. PF_5 has three equivalent equatorial fluorines and two equivalent apical fluorines (Fig. 7.137*b*).

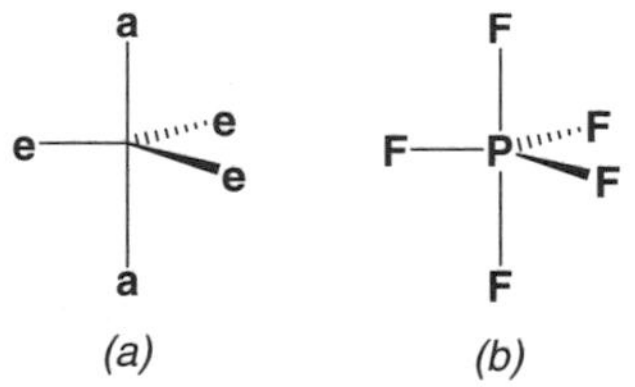

Figure 7.137. (*a*) The equatorial **e** and apical **a** positions of a trigonal bipyramid (TBP) and (*b*) The TBP structure of PF_5.

7.138 APICOPHILICITY

A measure of the preference of certain electronegative ligands attached to pentacoordinate phosphorus to occupy the apical positions of the trigonal bipyramid unit.

7.139 BERRY PSEUDO-ROTATION

A process by which the five groups around a TBP center undergo stereoisomerization, the two apical substituents exchanging positions with two of the equatorial ones. (See Sect. 9.49)

Example. Any one of the three equatorial substituents of a TBP can serve as the pivot. The 120° bond angle between the other two equatorial substituents increases to 180° (making them apical), while at the same time the 180° angle between the two apical substituents decreases to 120° (making them equatorial). See Fig. 7.139. The term pseudo-rotation is also used to describe the series of correlated rotations that certain conformations of cyclic molecules undergo during interconversion, for example, the interconversion of cyclopentane envelope conformations.

7.140 WALDEN INVERSION

The inversion of relative configuration (Sect. 7.53) that accompanies a concerted backside nucleophilic displacement at a chiral tetrahedral center.

Example. See Fig. 7.140, as well as Fig. 7.84*b*.

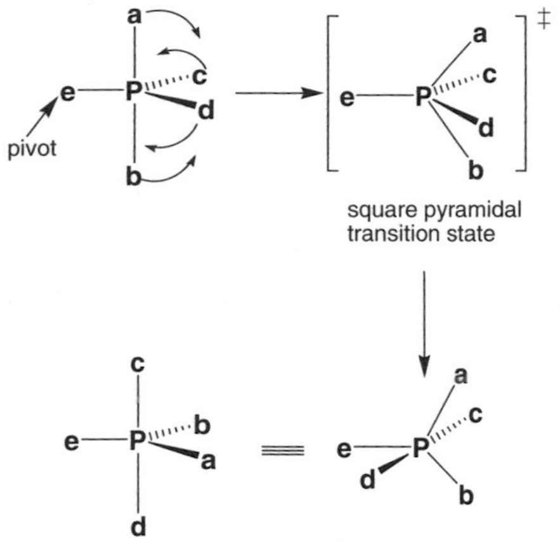

Figure 7.139. Berry Pseudo-rotation of a TBP around one equatorial pivot.

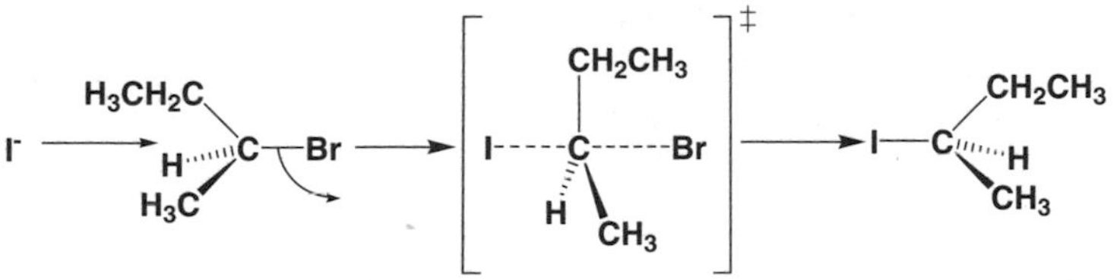

Figure 7.140. The Walden inversion accompanying nucleophilic attack by iodide ion on (*S*)-2-bromobutane to give (*R*)-2-iodobutane.

7.141 OCTAHEDRAL CONFIGURATION

A molecular shape characteristic of many hexacoordinate molecules and transition metal complexes (Fig. 7.141). The term *octahedral* comes from the eight-faced polyhedron defined by drawing lines between all pairs of the six substituent atoms. Unlike a TBP, all six positions of an octahedral molecule are equivalent.

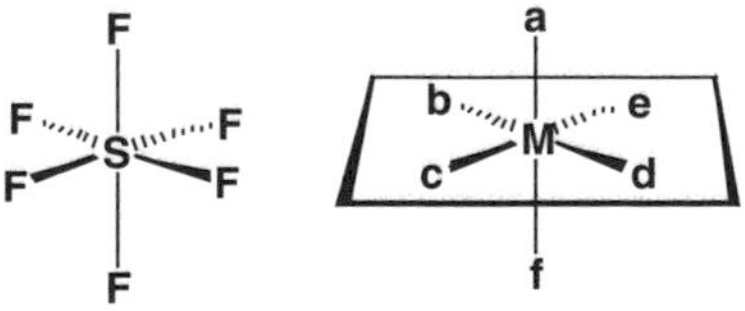

(a) *(b)*

Figure 7.141. (*a*) Sulfur hexafluoride and (*b*) an octahedral structure with ligands **b**, **c**, **d**, and **e** defining a plane, with ligands **a** and **f** defining a line perpendicular to the plane. Alternatively, the plane can be defined by **a**, **e**, **f**, **c** (or **a**, **b**, **f**, **d**), with the remaining two groups above and below the plane.

7.142 HELICITY (*M*, *P*)

A descriptor for extended chiral structures that resemble a spiral staircase, the threads on a screw, or the three-dimensional twist of some seashells. The descriptors for helicity assume the helix is approaching the viewer; a clockwise helix is designated as *M* and a counterclockwise helix as *P*.

Example. Refer to Fig. 7.41. The helix on the left is *P*, while the one on the right is *M*.

7.143 ALLYLIC ISOMERS

These are the constitutional isomers that result from rearrangements involving the allyl (propenyl) group; they are sufficiently common in organic chemistry as to warrant specific recognition.

Example. Figures 7.143*a* and *c* are allylic isomers. Such rearrangements usually occur via delocalized ion or radical intermediates, which are shown by the resonance structures in Fig. 7.143*b*. The loss of X at one end and its return at the opposite end of the allyl group result in the overall rearrangement. The interconversion of allylic isomers is sometimes described as a 1,3-rearrangement.

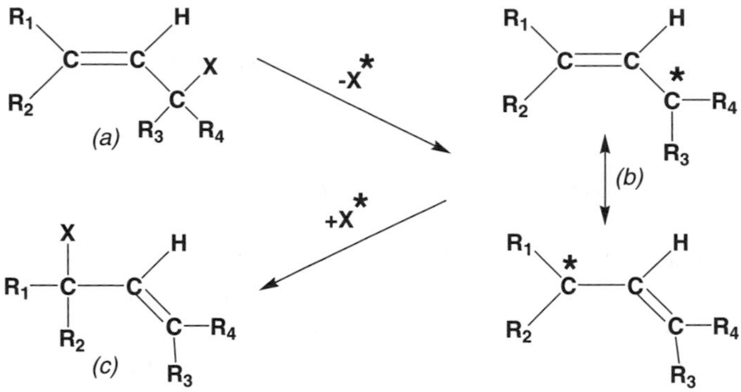

Figure 7.143. (*a*) and (*c*) are allylic isomers interconverting via intermediate (*b*), where the * represents an empty, half-filled, or filled orbital (i.e., cation, radical, or anion, respectively).

7.144 TAUTOMERS

Constitutional isomers that interconvert rapidly via migration of a small atom or group. The rapid interconversion is a consequence of a low activation barrier.

Example. Proton tautomers (prototropic tautomers) involve interconversion via proton (H^+) migration. Thus, the proton tautomer of acetone (the *keto form*, Fig. 7.144*a*) is its *enol form* (Fig. 7.144*b*).

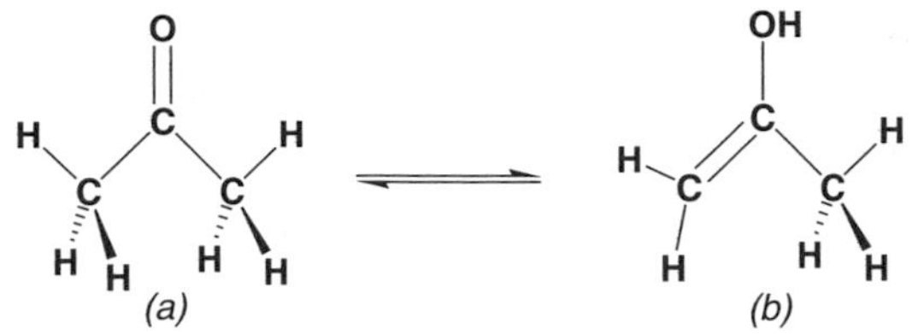

Figure 7.144. (*a*) Acetone and (*b*) its proton tautomer, the enol.

7.145 VALENCE ISOMERS (VALENCE TAUTOMERS)

Constitutional isomers or equivalent structures that are related by conversion of one network of contiguous σ and π bonds into another, usually by a pericyclic reaction (Sect. 14.88). These interconversions are accompanied by atom movement and rehybridization. Valence isomers can have different structures or, if they have equivalent structures (degenerate species), the fate of the individual atoms can, in principle, be determined (e.g., by isotope labeling). Valence isomers should not be confused with resonance structures (Sect. 3.4) that represent different electron distributions of the same molecule with no movement of any atoms.

Example. The interconversion of cyclooctatetraene and its bicyclic valence isomer (Fig. 7.145).

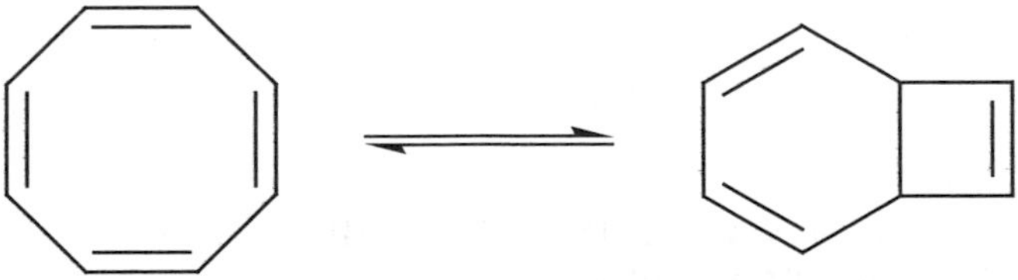

Figure 7.145. Two valence tautomers of C_8H_8.

7.146 FLUXIONAL MOLECULES

These are molecules that undergo rapid degenerate rearrangements, that is, rearrangements into indistinguishable molecules; the rearrangements may involve either bond reorganization or atom (group) migration. (See Sigmatropic Shifts, Sect. 14.102)

Example. Tricyclo[3.3.2.0^{4,6}]deca-2,7,9-triene (Fig. 7.146*a*), called bullvalene because of the disbelief with which its proposed structure was greeted, is fluxional at room temperature. It has been calculated that, if each of the 10 carbon atoms of bullvalene were individually labeled, 1,209,600 structures of the labeled compound would be possible! Bullvalene is an example of degenerate valence isomerism (Sect. 7.145).

The fluxionality of the σ-bonded metal cyclopentadienide (Fig. 7.146*b*) involves atom migration, but because the structures are degenerate, this is not considered to be a case of simple tautomerism.

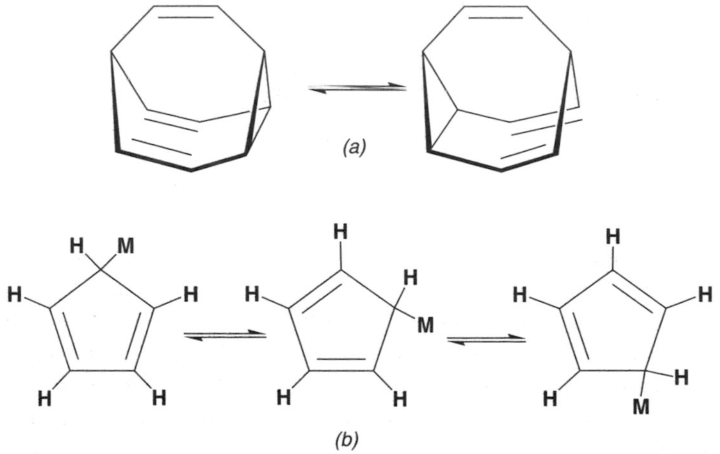

Figure 7.146. Examples of fluxional molecules: (*a*) bullvalene and (*b*) σ-bonded metal cyclopentadienide.

7.147 STEREOELECTRONIC EFFECTS

Factors related to the three-dimensional proximity of interacting orbitals that control the stability of a structure or the speed of a reaction.

7.148 REGIOCHEMISTRY AND STEREOCHEMISTRY OF RING CLOSURE REACTIONS

Several sets of largely empirical rules for predicting the regiochemistry and stereochemistry of ring closure reactions.

Baldwin's Rules. Relate to cyclization reactions involving intramolecular attack by a nucleophilic (e.g., carbanionic) center on an electrophilic (often unsaturated) center. The predictions have a theoretical basis in stereoelectronic factors. Jack Baldwin (1938–) devised a set of rules to emphasize that certain types of cyclization are favorable and others are less so. Baldwin's rules focus on the so-called exo- and endo-cyclizations. For cyclizations between a nucleophile and an electrophilic center, there are two possible regiochemical pathways: exo or endo (Fig. 7.148*a*). Besides the total number of atoms in the product ring *n*, as well as whether the cyclization is exo or endo, another important factor is the hybridization of the original electrophilic center, which can be either tetrahedral (Tet), trigonal (Trig), or digonal (Dig).

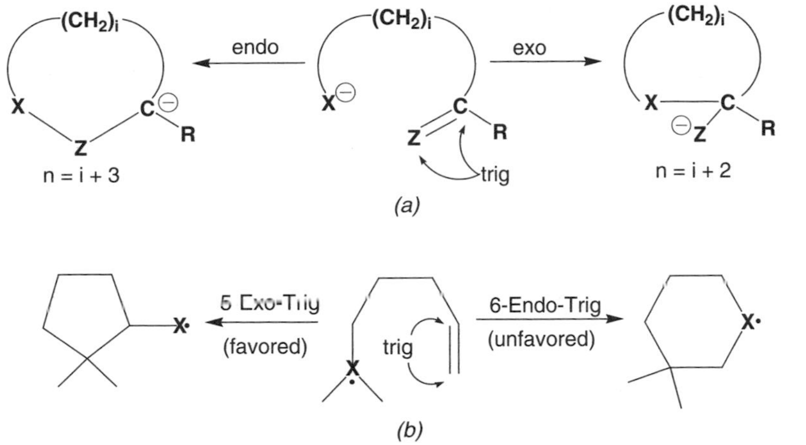

Figure 7.148. (*a*) Endo and exo Baldwin cyclization at a Trig center and (*b*) preferred 5-exo and 6-endo-Trig Beckwith cyclization of a radical.

The rules can be summarized as follows:

n	Class	Ease of Cyclization
3–7	Exo-Tet	Favorable
5–6	Endo-Tet	Unfavorable
3–7	Exo-Trig	Favorable
3–5	Endo-Trig	Unfavorable
6–7	Endo-Trig	Favorable
3–7	Endo-Dig	Favorable
3–4	Exo-Dig	Unfavorable
5–7	Exo-Dig	Favorable

Beckwith's Rules. For radical cyclizations, a related set of rules based on the interplay of entropic and stereoelectronic factors has been formulated by Beckwith and others. For cyclizations involving a 5,6 double bond (Fig. 7.148*b*), the five-membered ring product from exo-Trig cyclization is favored kinetically over the six-membered ring resulting from endo-Trig cyclization. Similarly, from 3,4 to 7,8 double bonds, exo-Trig cyclization is preferred over endo-Trig. However, if there is a 5-alkyl substituent on a 5,6 double bond, then endo-Trig cyclization generally occurs.

SUGGESTED READING

Bassindale, A. *The Third Dimension in Organic Chemistry*. John Wiley & Sons: Chichester, 1984.

Collins, A. N., Sheldrake, G. N., and Crosby, J., eds. *Chirality in Industry, The Commercial Manufacture and Applications of Optically Active Compounds*. John Wiley & Sons: Chichester, UK, 1992.

Eliel, E. L., Allinger, N. L., Angyal, S. J., and Morrison, G. A. *Conformational Analysis*. John Wiley & Sons: New York, 1965.

Eliel, E. L. and Wilen, S. H. *Stereochemistry of Organic Compounds*. John Wiley & Sons: New York, 1994.

Juaristi, E. *Introduction to Stereochemistry & Conformational Analysis.* Wiley-Interscience: New York, 1991.

Newman, M.S. ed. Sterie Effects in Organic Chemistry. John Wiley & Sons: New York, 1956.

Nogradi, M. *Stereochemistry, Basic Concepts and Applications.* Pergamon Press: Oxford, UK, 1981.

Nogradi, M. *Stereoselective Synthesis.* VCH: Weinheim, Germany, 1986.

Potapov, V. M. *Stereochemistry.* MIR Publishers: Moscow, 1979.

Seco, J. M., Quinoa, E., and Riguera, R. *Chem. Rev. 104*, 17 (2004).

8 Synthetic Polymers

8.1	Macromolecule	294
8.2	Polymer	295
8.3	Oligomer	295
8.4	Monomeric Unit	295
8.5	Constitutional Repeating Unit (CRU)	295
8.6	Chemical Formula of Constitutional Repeating Units (CRUs)	296
8.7	Polymerization	297
8.8	Homopolymer	297
8.9	Nomenclature of Homopolymers	297
8.10	Chain-Growth Polymerization (Addition Polymerization)	298
8.11	Radical Polymerization	299
8.12	Linear (Single-Strand) Polymer	299
8.13	Chain Transfer	301
8.14	Branched Polymer	301
8.15	Head-to-Tail (Tail-to-Head) Orientation (of Vinyl Polymers)	302
8.16	Head-to-Head (Tail-to-Tail) Orientation (of Vinyl Polymers)	303
8.17	Cationic Polymerization	303
8.18	Anionic Polymerization (Monoanions)	303
8.19	Living Polymers (Dianionic Polymerization)	304
8.20	Copolymers	305
8.21	Block Polymer	306
8.22	Graft Polymer	307
8.23	Cross-Linking	307
8.24	Ladder Polymer	308
8.25	Polymer Architectures	310
8.26	Ring-Opening Chain-Growth Polymerization of Heterocycles	310
8.27	Ring-Opening Metathesis Polymerization (ROMP)	311
8.28	Emulsion Polymerization	311
8.29	Low-Density Polyethylene (LDPE)	312
8.30	High-Density Polyethylene (HDPE)	312
8.31	Plastics	313
8.32	Plasticizer	313
8.33	Thermoplastic Polymers	313
8.34	Thermosetting Polymers	313
8.35	Molding	314
8.36	Compression Molding	314
8.37	Injection Molding	315

The Vocabulary and Concepts of Organic Chemistry, Second Edition, by Milton Orchin, Roger S. Macomber, Allan Pinhas, and R. Marshall Wilson

8.38	Melt Transition Temperature T_m	315
8.39	Glass Transition Temperature T_g	315
8.40	Configurational Base Unit	316
8.41	Tactic Polymer	316
8.42	Isotactic Polymer	316
8.43	Syndiotactic Polymer	318
8.44	Stereoregular Polymer	318
8.45	Atactic Polymer	319
8.46	Ziegler–Natta Catalysis (Coordination Polymerization)	319
8.47	Metallocene Catalysts	321
8.48	Molecular Weight of Polymers	322
8.49	Number-Average Molecular Weight M_n	322
8.50	Weight-Average Molecular Weight M_w	323
8.51	Polydispersity Index $\bar{M}_w/\bar{M}_n$	324
8.52	Determination of Polymer Molecular Weight	324
8.53	Step-Growth Polymerization (Condensation Polymerization)	325
8.54	Foamed Polyurethanes	328
8.55	Nylon	329
8.56	Dendrimers	329
8.57	Epoxy Resins	331
8.58	Silicones	331
8.59	Fibers	332
8.60	Spinning	332
8.61	Synthetic Fibers	332
8.62	Denier	334
8.63	Elastomers	334
8.64	Rubber	334
8.65	Natural Rubber	335
8.66	Gutta-percha	335
8.67	Synthetic Rubbers	335
8.68	Foamed Rubber	336
8.69	Vulcanization	336
8.70	Elastomeric Fibers	337
8.71	Animal Fibers	337
8.72	Wool	337
8.73	Silk	337
8.74	Leather	338
8.75	Vegetable Fibers	338
8.76	Cotton	338
8.77	Rayon	338
8.78	Composites	339
8.79	Poly(phosphazenes), $\text{-(N=PX}_2\text{--)}_n$	340
8.80	Biodegradable Polymers	340
8.81	Recycle of Polymers (Plastics)	341

In today's world, polymers are so ubiquitous that one literally cannot take a step without encountering some form of synthetic polymer, since the soles of most shoes, and many items of clothing as well, are no longer made exclusively from naturally occurring polymers such as leather, cotton, wool, and silk. Beverage bottles, rugs, rubber tires, food wraps, glues, paints, toys, films, cleaners, packing materials, lubricants, pharmaceuticals, artificial hips, breast implants, contact lenses, and many common household goods are manufactured from synthetic polymers that are constructed to impart the special characteristics required in each particular application. More than 30,000 polymers have been patented in the United States. Polymers, some in the form of composites, are also increasingly used for protective and structural materials, replacing wood and metals. The primary metabolites of life—DNA, RNA, proteins, and carbohydrates—are natural polymers (biopolymers). These biopolymers will not be discussed, but other naturally occurring polymers useful in consumer products, such as wool, cotton, silk, leather, and rubber, are discussed in this chapter. Most textbooks of organic chemistry place a discussion of synthetic polymers toward the end of the book, as though it were a necessary afterthought, included to make the coverage of organic chemistry more complete. Perhaps this practice arose because most textbooks are written by academicians, whereas commercial interests propelled polymer development.

The great Swedish chemist Jöns J. Berzelius introduced the word "polymeric" in 1832 to describe the relation between compounds having the same empirical formula but different molecular weight ("equality of composition" according to Berzelius), for example, ethylene (C_2H_4) and butene (C_4H_8). Although the word remains, and polymers produced by chain-growth (addition) polymerization do indeed have the same composition as the monomers from which they are generated, the term polymeric describes substances of very high molecular weight made up of repeating small units. Chemists made use of polymers long before their structures were elucidated. By the mid-nineteenth century, industries already had been established to produce artificial polymeric substances. For example, the first *plastic* (a common term for polymers capable of being shaped or molded and thus possessing plastic properties) was celluloid, invented in 1856. It consisted of the polymer, nitrocellulose, mixed with camphor. However, its widespread use was discontinued after several disastrous fires and it was replaced by cellulose acetate, which is used extensively for the manufacture of photographic film. Viscose rayon and phenol-formaldehyde resins (Bakelite) were articles of commerce in the late nineteenth century, marking the beginning of the polymer industry. By 1979 the annual volume of plastics produced in the United States exceeded that of steel. The period of polymer development before 1914 (World War I) also was characterized by speculations regarding the structure of natural polymers (such as rubber, starch, cotton, and cellulose) and proteins (including silk and wool).

Hermann Staudinger (1881–1965, who received the Nobel Prize in 1953) was probably the first chemist to recognize, in 1920, that the structure of a polymer consisted of repeating units of a monomer. Herman Mark (1895–1992) was another important contributor to the development of polymer chemistry in the period between World Wars I and II. Although he was trained as an organic chemist, Mark recognized that X-ray crystallography could have a profound impact on advancing the structural knowledge of polymers, and he proceeded to become an authority on the use of this tool. After his forced immigration from Austria to Canada and from there to the United States, he introduced, at the Polytechnic Institute of Brooklyn, the first academic teaching and research program in polymer chemistry in the United States.

Polymer development in the period between World Wars I and II was marked by substantiation of Staudinger's views of addition polymerization, followed by the discovery of the synthetic fiber nylon (whose patent was issued in 1938), a product of condensation polymerization and the fruit of the E. I. du Pont Company's venture into fundamental chemistry. The introduction of nylon had a major impact on the wearing apparel industry (Betty Grable aided a World War II bond drive by auctioning off her nylon hose, then a novelty much in demand, for $40,000). The spectacular financial success of nylon, the brain-child of a research team headed by Wallace Carothers (1896–1937), prodded other chemical corporations to invest in long-range research efforts, creating career opportunities for many Ph.D. organic chemists. The requirements for a variety of materials to successfully fight World War II provided the incentive for developments in the chemistry of synthetic rubbers and other elastomers. In the early post-World War II period, the brilliant work of Paul Flory (1910–1985, who received the Nobel Prize in 1974) on chain-growth polymers and the statistical analysis of their conformational mobility helped advance polymer science to its chemical maturity. Determining the exact structure of a long-chain macromolecule such as polyethylene is a daunting problem considering the number of possible conformations. The preferred conformation around each carbon–carbon single bond in a polymer chain is the staggered conformation, and there will be three of these around each single bond (one lowest-energy *trans* and two *gauche*). If there are n bonds in a chain, 3^n preferred conformations will exist. Thus for a modestly long chain having 10,000 bonds, there will be $3^{10,000} = 10^{4,771}$ distinct comformations, a number that is virtually impossible to comprehend.

Presently, a wide variety of specialty polymers and composite materials having unusual properties for a variety of applications are being developed. The recent discovery of new metallocene catalysts has led to polymers with improved performance. Polymer chemistry has a rich past, an exciting present, and an even more promising future.

8.1 MACROMOLECULE

A relatively high molecular weight molecule with a defined molecular weight, composed almost entirely of a large number of repeating units of identical or closely related structure called monomeric units. The number of repeat units is sufficiently large so that the addition or removal of one or several units does not change the properties of the molecule. Macromolecules composed of as many as 10^6 units are not uncommon. The term *macromolecules* was coined by Staudinger in 1922.

8.2 POLYMER

A polymer (from the Greek words *polys* for "many" and *meros* for "part") is a substance composed of a collection of individual macromolecules of varying molecular weight. A single macromolecule may be called a polymer molecule; however, a polymer in the generic sense is a *substance* having an average molecular weight. Each individual polymer molecule (macromolecule) has one end group consisting of the initiating group (corresponding to an ion or radical) and a terminating group consisting of either a hydrogen atom or other small group (but not if two radical chains dimerize). In terms of the entire molecule, both end groups are so small that they are neglected in representing the formula of the macromolecule. Most authors (including the present ones) generally do not distinguish between a single macromolecule and the collection of macromolecules that constitute the polymer. The words "macromolecule" and "polymer" are invariably used interchangeably, because the process (polymerization) that generates macromolecules from monomeric units results in polymers.

8.3 OLIGOMER

An oligomer (from the Greek words *oligos* for "few" and *meros* for "part") is similar to a polymer except that the individual molecules comprising an oligomer consist of about 5 to 10 units of monomer, and hence, the addition or subtraction of one or more units may influence the properties of the substance. There is no specific number of linked monomeric units that defines the resulting substance as either a polymer or an oligomer, and hence, no sharp delineation exists between relatively high molecular weight oligomers and relatively low molecular weight polymers.

8.4 MONOMERIC UNIT

The molecule whose skeletal structure is repeated in the polymer; it is the precursor compound to the constitutional repeating unit (see CRU below).

Example. Styrene, $PhCH{=}CH_2$, is the monomeric unit of polystyrene, whose skeletal or constitutional repeating unit is $\left(CH_2-\underset{\displaystyle Ph}{\underset{|}{CH}}\right)_n$.

8.5 CONSTITUTIONAL REPEATING UNIT (CRU)

The smallest repeating structural unit of which the polymer is a multiple. The unit (for single-stranded polymers) is a bivalent group, that is, a group having two open valences, one at each end, conveniently viewed as being generated from a precursor multiple bond.

Example. Figure 8.5 shows the CRU for the general case of a substituted vinyl polymer (for polystyrene, R= Ph). In writing this structure for a single-strand polymer, the CRU is enclosed in parentheses in the manner shown. This representation implies that the CRU can be repeated in either direction, with the result that the R group(s) will appear on alternate carbon atoms in the chain. The subscript n in the figure is an integer denoting the *average degree of polymerization*, that is, the average number of repeat units in the collection of individual macromolecules constituting the polymer. Of course, there must be some atom or group at either end of the chain of the macromolecule to satisfy the two free valences shown as dashes, but these are omitted from the representative structure. The parentheses are placed to bisect the bonds representing the terminal free valencies to indicate the connection between units at either end of the enclosed unit:

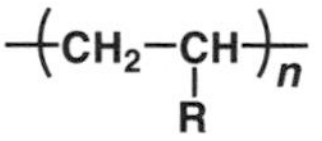

Figure 8.5. The constitutional repeating unit (CRU) of a polymer prepared from a substituted vinyl monomer

8.6 CHEMICAL FORMULA OF CONSTITUTIONAL REPEATING UNITS (CRUs)

In the IUPAC system, the polymer is considered as an essentially endless chain and the manner of its construction is of no importance. Accordingly, there is really no distinction between equivalent units A and B in the figure representing a portion of the polystyrene chain (Fig. 8.6*a*), and the CRU is correctly expressed by either of the equivalent formula, with Figs. 8.6*b* or *c* corresponding to units A and B, respectively. However, in considering the chemistry of chain-growth or addition polymerization, both the initiation and propagation steps (Sect. 14.63) proceed via the most stable radical (or ion) and this is the most, rather than the least, substituted radical (or ion). Hence, if importance is attached to the mechanism of the growth of a vinyl polymer in the left to right direction, the choice in writing the CRU goes to unit A, with the CRU formula shown in Fig. 8.6*c*. In naming this CRU, one uses the right to left

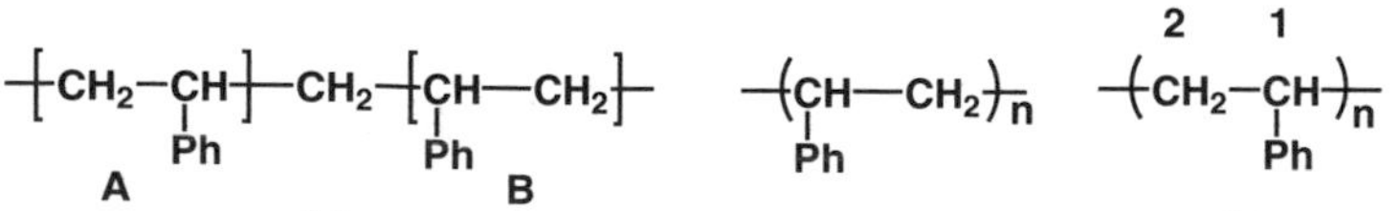

(*a*) (*b*) (*c*)

Figure 8.6. CRU structures for a vinyl polymer: (*a*) the two equivalent units A and B (shown in brackets) of the polystyrene chain; (*b*) the CRU structure corresponding to B; (*c*) the (preferred) CRU structure corresponding to A.

numbering system in order to give the most substituted carbon atom the number 1 position in a chain regardless of the chain's orientation.

8.7 POLYMERIZATION

The process that converts monomers to polymers. There are two broad types of polymerization: (1) chain-growth polymerization and (2) step-growth polymerization. Chain-growth polymers (also called addition polymers) have the same empirical formula as the monomer from which they are formed (except for the initiator). However, step-growth polymers (also called condensation polymers) usually have a different empirical formula from that corresponding to the monomers because step-growth polymerization commonly proceeds by the intermolecular loss of small molecules such as water, alcohol, or ammonia.

Example. Polymerization of ethylene and substituted ethylenes, for example, propylene and various acrylates, are instances of chain-growth polymerization, while, for example, nylon, poly(hexamethyleneadipamide), results from a step-growth polymerization.

8.8 HOMOPOLYMER

A polymer made from a single monomeric precursor.

Example. Polystyrene, poly(vinyl chloride), polyethylene.

8.9 NOMENCLATURE OF HOMOPOLYMERS

The names of single-strand homopolymers may be either source-based or structure-based. Source-based names consist of the prefix *poly-* followed by the *common* name of the monomeric unit. If the monomeric unit has a complex name (more than one word or a compound name), then the name is enclosed in parentheses. Thus, polystyrene, poly(vinyl alcohol), poly(acrylonitrile), and so on, are source-based names for the polymers derived from the indicated monomers. Structure-based names for polymers also start with the prefix *poly-* followed by a set of parentheses enclosing the IUPAC *systematic* name of the monomer. The structure-based names for the polymers mentioned above are poly(l-phenylethylene), poly(1-hydroxyethylene) and poly(l-cyanoethylene), respectively. Possible ambiguity can arise in the source name system. Consider the problem of naming the polymer derived from 1,3-butadiene. There are three different structural isomers of the polymer derived from this monomer (Fig. 8.9), and thus the source name, polybutadiene, is not very precise. The IUPAC name for $CH_2{=}CH_2$ is ethene, but the common name ethylene is so entrenched in the literature that it is usually used. The structure-based name of the polymer from unsubstituted ethylene should be written as poly(ethylene), while the source-based and preferable name is written without the parentheses.

Example. An interesting anomaly involves the names for polymers such as polyethylene and poly(tctrafluoroethylene), whose CRU structures are written as $-(-CH_2$-$CH_2-)_n-$ and $-(-CF_2$-$CF_2-)_n-$, respectively. Strictly speaking, the structure-based names should be poly(methylene) and poly(difluoromethylene), respectively, corresponding to the structures $-(CH_2-)_n-$ and $-(-CF_2-)_n-$, because the single carbon species actually is the smallest CRU in both cases. However, the source name (and corresponding structure) are the accepted names (and structures). As a matter of fact, $-(-CF_2$-$CF_2-)_n-$ is best known by its trade name, *Teflon*. The polymer generated from diazomethane, CH_2N_2, would be correctly source-named polymethylene, although it has the same CRU as polyethylene. The above discussion of names has been limited to single-strand homopolymers. The names of other types of polymers will be discussed in the sections where they are defined.

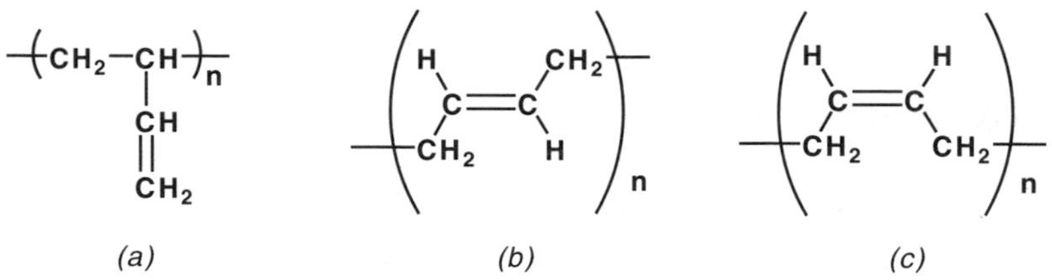

Figure 8.9. Three possible polymers of 1,3-butadiene: (*a*) 1,2-additon; (*b*) 1,4-*trans*-addition; (*c*) 1,4-*cis*-addition.

8.10 CHAIN-GROWTH POLYMERIZATION (ADDITION POLYMERIZATION)

The consecutive addition of monomeric units to form chains (macromolecules), each chain consisting of a large but variable number of monomeric units consecutively linked to each other. Every propagation step involves a monomer molecule. The resulting polymer has essentially the same empirical formula as the monomer. The polymerization proceeds by the addition reaction of a monomer with the growing end of the chain, and ideally at any stage of the polymerization, there should be present only a monomer plus a high molecular weight polymer because monomers do not react with each other. Such polymerization processes may be initiated by radicals (generated thermally or photochemically), cations (or Lewis acids), anions (or Lewis bases), or transition metal coordination compounds.

8.11 RADICAL POLYMERIZATION

A chain-growth polymerization of monomers via a chain reaction initiated by a free radical. Like all chain reactions (see Sect. 14.60), radical polymerization involves three steps: (1) initiation, (2) propagation, and (3) termination.

Example. The polymerization of vinyl chloride can be initiated by thermal decomposition of an initiator such as AIBN (Fig. 8.11). In the propagation step, the addition of the radical initiator R• to vinyl chloride can theoretically result in the formation of either R–CH_2–CH(Cl)• (a secondary radical) or R–CHCl–CH_2• (a primary radical). Of the two possibilities, the more stable secondary radical is formed, and in the propagation (steps 2 and 3) each succeeding monomer is added regiospecifically in the same fashion. The polymerization in which chain growth proceeds by successive additions at one preferred site of the monomer is called *regiospecific polymerization.*

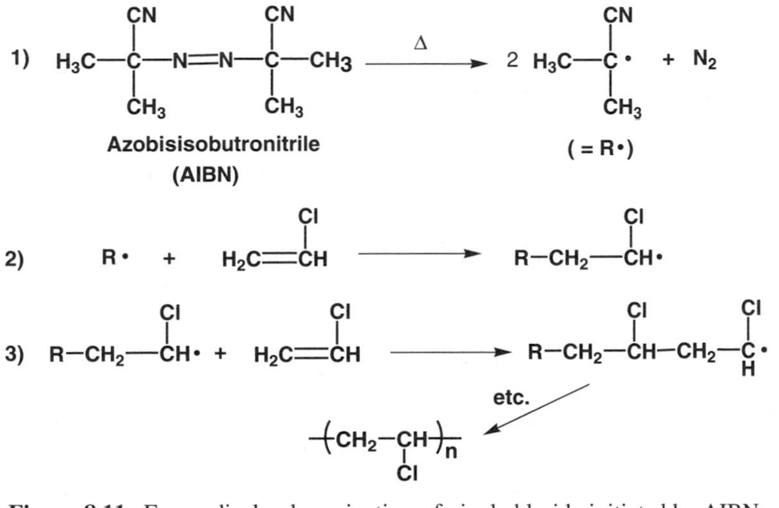

Figure 8.11. Free radical polymerization of vinyl chloride initiated by AIBN.

8.12 LINEAR (SINGLE-STRAND) POLYMER

A polymer in which the chain backbone is formed by consecutive additions of monomeric units, whereby the units become arranged in a linear chain. The smallest repeating unit in the single-strand polymer is a bivalent group.

Example. Polyethylene, poly(vinyl chloride), and all other polymers discussed above. The bivalent repeating unit in vinyl polymers is $-(CH_2\text{-}CH(R)-)_n-$, where R may be H, CH_3, Ph, CO_2CH_3, Cl, $OCOCH_3$, or OH. Table 8.12 lists a variety of commercial vinyl polymers.

TABLE 8.12. Some Common Vinyl Polymers

Monomer	CRU	Source Name	Commercial Designation
Ethylene	$-(CH_2-CH_2)_n-$	Poly(ethylene)	LDPE, HDPE
Propylene	$-(CH_2-CH(CH_3))_n-$	Poly(propylene)	PP
Isobutylene	$-(CH_2-C(CH_3)_2)_n-$	Poly(isobutylene)	PIB
Styrene	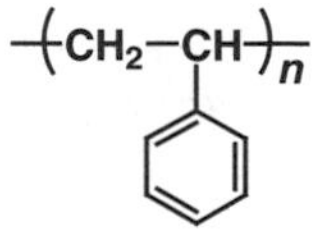	Poly(styrene)	PS
Acrylonitrile	$-(CH_2-CH(CN))_n-$	Poly(acrylonitrile)	PAN ORLON
Vinyl alcohol	$-(CH_2-CH(OH))_n-$	Poly(vinyl alcohol)	PVAL
Vinyl acetate	$-(CH_2-CH(OAc))_n-$	Poly(vinyl acetate)	PVAC
Vinyl chloride	$-(CH_2-CH(Cl))_n-$	Poly(vinyl chloride)	PVC
Methyl methacrylate	$-(CH_2-C(CH_3)(COOCH_3))_n-$	Poly(methyl methacrylate)	PMMA, Lucite
Tetrafluoro-ethylene	$-(CF_2-CF_2)_n-$	Poly(tetrafluoro-ethylene)	Teflon
Vinyl fluoride	$-(CH_2-CH(F))_n-$	Poly(vinyl fluoride)	Tedlar
1,1-Difluoroethylene	$-(CH_2-CF_2)_n-$	Poly(vinylidene fluoride)	PVDF
1-Chloro-1,2,2-trifluoroethylene	$-(CF_2-CF(Cl))_n-$	Poly(1-chloro-1,2,2-trifluoro-ethylene)	KEL-F
1,1-Dichloroethylene + vinyl chloride	Random copolymer		Saran
1,1-Dichloroethylene + perfluoroethylene	Random copolymer		Vitron

8.13 CHAIN TRANSFER

The termination of a growing polymer chain by atom abstraction from a foreign molecule with the simultaneous initiation of a new chain by the atom-depleted foreign radical. Chain transfer is a technique that is frequently used to generate polymers of shorter chain length than would otherwise be obtained in the absence of a chain transfer agent. Intramolecular chain transfer frequently involves *backbiting.* (See Sect. 8.14)

Example. When styrene is polymerized in the presence of carbon tetrachloride (the foreign molecule), the resulting polystyrene, initiated by CCl_3 and terminated by Cl, has a relatively low molecular weight owing to the reaction sequence shown in Fig. 8.13. The overall effect is to add CCl_4 to the oligomeric styrene. Oligomers generated by such chain transfer reagents are called *telomers.*

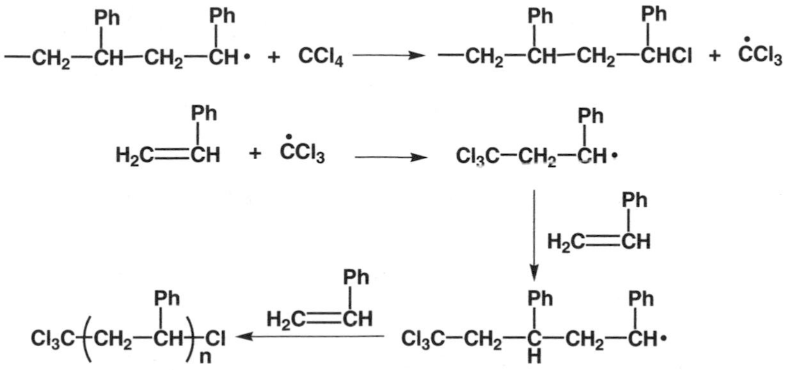

Figure 8.13. The use of CCl_4 as a chain transfer reagent in the telemerization of styrene.

8.14 BRANCHED POLYMER

A homopolymer in which occasional chains are attached to the polymer backbone, that is, the polymer occasionally grows branches instead of growing as a continuous linear chain.

Example. Branching may occur as a result of chain transfer between a growing but rather short polymer chain with another and longer polymer chain, as shown in Fig. 8.14*a*. Branching may also occur if the radical end of a growing chain abstracts a hydrogen atom from a carbon atom four or five carbons removed from the end, as shown in Fig. 8.14*b*. This phenomenon is called backbiting. In the special case of polyethylene, methyl branching presumably occurs because of hydrogen migration driven by the somewhat greater thermodynamic stability of secondary compared to primary radicals (see Sect. 11.7, Fig. 8.14*c*). Branching may be deliberately introduced by copolymerization (Sect. 8.20). Thus, a mixture of ethylene and 1-hexene

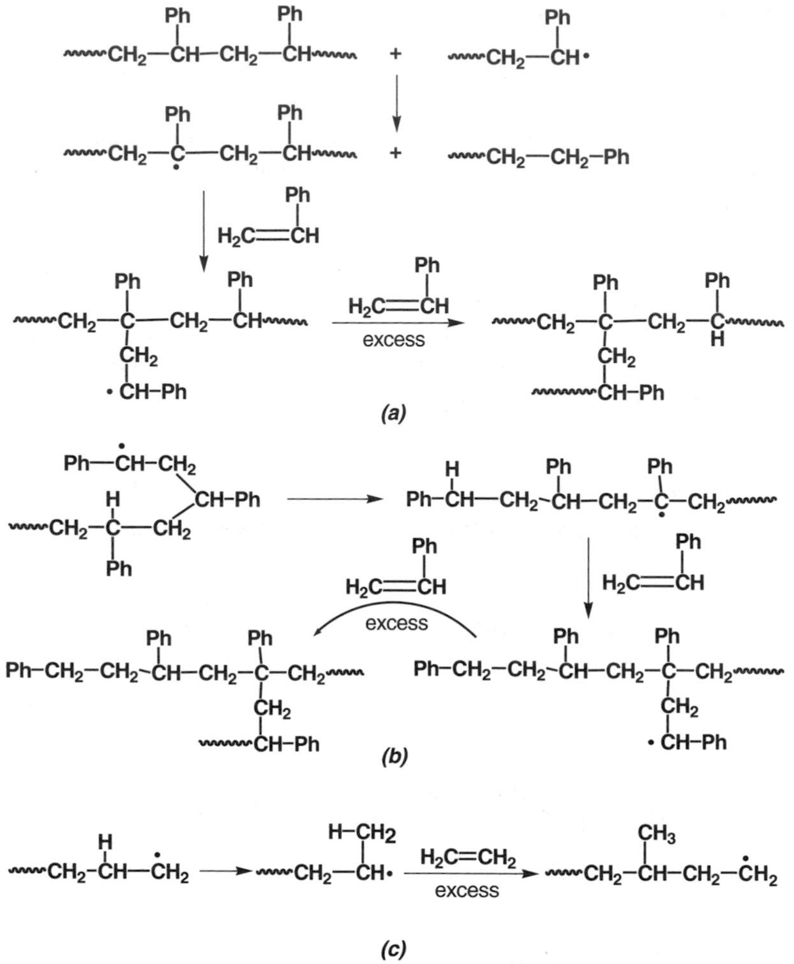

Figure 8.14. Branched-chain polymer formation: (*a*) by intermolecular chain transfer; (*b*) by intramolecular backbiting; (*c*) methyl branching in polyethylene resulting from hydrogen migration.

leads to branching *n*-butyl groups, spaced approximately on the polyethylene chain, in proportion to the concentration of 1-hexene.

8.15 HEAD-TO-TAIL (TAIL-TO-HEAD) ORIENTATION (OF VINYL POLYMERS)

The polymer structure that results from the sequential addition of the more substituted carbon atom (the head) of the vinyl group to the least substituted carbon atom (the tail) of the adjacent vinyl group (or *vice versa*). As a result of such sequencing, the substituent(s) on the vinyl groups appear on alternate carbon atoms in the resulting polymer.

Example. The free radical polymerization of vinyl chloride (Fig. 8.11). Head-to-tail polymerization is the preferred sequencing of all vinyl polymers.

8.16 HEAD-TO-HEAD (TAIL-TO-TAIL) ORIENTATION (OF VINYL POLYMERS)

The polymer structure that results from the sequential addition of the head carbon atom of the vinyl group to the head carbon atom of the adjacent vinyl group (head-to-head) followed by tail-to-tail addition (or *vice versa*). Head-to-head sequencing results in the substituent appearing on adjacent carbon atoms in the resulting polymer. It is possible that a growing polymer can have an occasional sequencing of this nature to give an irregular, random arrangement, but the probability of obtaining a regular alternating chain structure resulting from head-to-tail and head-to-head addition is remote.

Example. Figure 8.16. A portion of a vinyl polymer resulting from a head-to-head, tail-to-tail juncture.

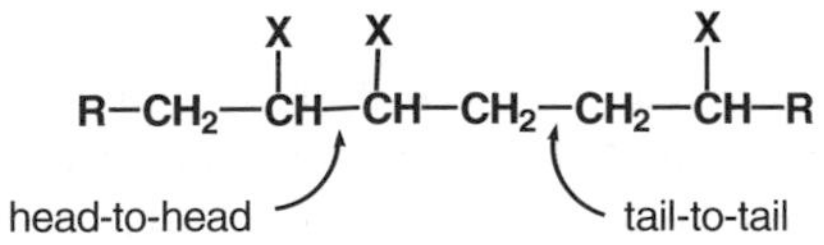

Figure 8.16. Alternating head-to-head and tail-to-tail sequencing in a vinyl polymer

8.17 CATIONIC POLYMERIZATION

The polymerization of monomers, initiated by proton donors or Lewis acids.

Example. The common mineral acids (particularly H_3PO_4) are frequently used as initiators, but the resulting polymers generally possess quite low molecular weights. The reaction of Lewis acids with water is also used to provide an initiating proton source.

8.18 ANIONIC POLYMERIZATION (MONOANIONS)

The polymerization of monomers initiated by anions, usually carbanions, or less frequently by the amide ion, NH_2^-. Anionic polymerization is generally a slow reaction and is effective when the substrate monomer is either a diene such as butadiene or a vinyl compound with a strong electron-withdrawing group attached to the double bond. Simple olefins such as propylene and homologs are not satisfactory substrates because initiating anions may abstract the allylic protons of such substrates to form stable unreactive allylic anions.

Example. The anionic polymerization of methyl methacrylate with *n*-butyllithium (Fig. 8.18).

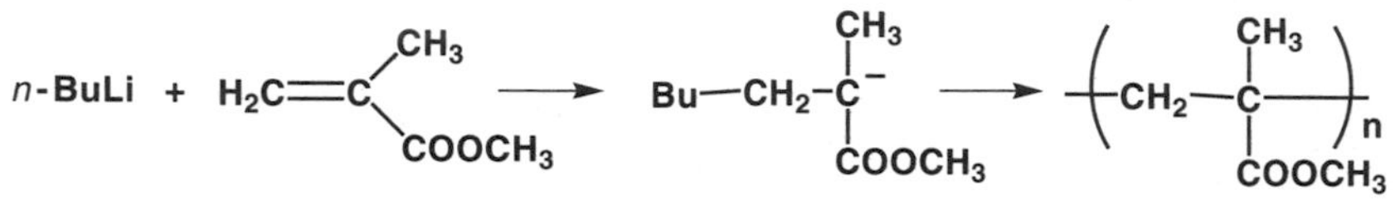

Figure 8.18. *n*-Butyllithium-initiated anionic-polymerization of methyl methacrylate.

8.19 LIVING POLYMERS (DIANIONIC POLYMERIZATION)

A polymer formed under conditions such that there are no termination steps, resulting in a chain that continues to grow until all of the monomer is consumed. The polymer will continue to grow upon the addition of more polymer or the addition of another monomer of a different structure. This is the technique used to produce important block copolymers (Sect. 8.21).

Example. The dianion-initiated polymerization of styrene shown in Fig. 8.19. The anionic polymerization is initiated by a reactive metal donating an electron to a polynuclear aromatic compound in a one-electron reduction to form a radical anion.

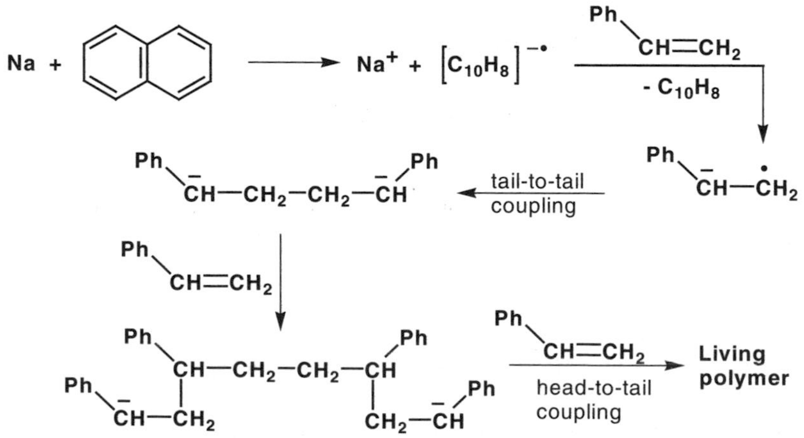

Figure 8.19. The Polymerization of styrene initiated by sodium naphthalide to give a living polymer.

This radical anion acts as a catalyst to transfer an electron to a vinyl monomer substrate forming a new radical anion that dimerizes to form a dianion. In the process, the aromatic compound is regenerated and, thus, needs to be added only in catalytic quantities. The dianion then undergoes anionic polymerization at both ends to generate a "living" polymer. Under ideal conditions, chains of nearly identical length can be generated, the degree of polymerization being determined by the ratio of monomer to initiator molecules. The living polymer is "killed" when desired, by the addition of an impurity, such as water or other protic compounds that react with carbanions. Alternately, the polymer can be capped purposefully, for example, by the addition of CO_2 to give a dicarboxylate of the polymer. Living polymers, first described in detail by Michael Szwarc (1909–2000), are usually prepared under carefully controlled anionic polymerization conditions, where no impurities (e.g., water) are present to unintentionally terminate the chain.

8.20 COPOLYMERS

A polymer made from a mixture of two or more structurally different monomers. In chain-growth polymerization (Sect. 8.10), the two or more different monomeric units may be distributed either in a random sequence (*random copolymer*), or in alternating sequence (*alternating copolymer*). Using the living polymer technique, the copolymers may be produced in blocks of one monomer followed by blocks of the other (*block polymers,* Sect. 8.21). In chain-growth polymerization, the composition of the final copolymer is determined in large measure by the rates of the various possible propagations steps.

Example. In a free radical copolymerization involving two different monomers A and B, four separate propagation steps are possible:

1. Reaction of a propagating radical chain terminating in A• with monomer A:

   --A• + A → ---AA•   (rate constant k_{aa})

2. Reaction of a propagating radical chain terminating in A• with monomer B:

   --A• + B → ---AB•   (rate constant k_{ab})

3. Reaction of a propagating radical chain terminating in B• with monomer B:

   --B• + B → ---BB•   (rate constant k_{bb})

4. Reaction of a propagating radical chain terminating in B• with monomer A:

   --B• + A → ---BA•   (rate constant k_{ba})

The *reactivity ratios* r for monomers A and B are defined, respectively, as

$$r_A = k_{aa}/k_{ab}, \qquad r_B = k_{bb}/k_{ba} \tag{8.20}$$

The relative reactivity ratio for monomer A (r_A) is thus the ratio of the propagation rate constants for the addition of A (homopolymerization) and for the addition of B (copolymerization) to the propagating radical chain terminating in A• while r_B is analogously defined.

From these definitions, it follows that if $r_A = r_B = 1$, the resulting copolymer will be completely random, whereas if $r_A = r_B = 0$ ($k_{aa} = k_{bb} = 0$), the sequencing will be perfectly alternating. It is, of course, unlikely that either situation will prevail exactly, but the two extreme situations may be approached. If both reactivity ratios are close to unity, as with styrene and 4-chlorostyrene ($r_S = 0.816$ and $r_{CIS} = 1.062$), the copolymer will be mostly random, while if the reactivity ratios are small but not exactly zero, the copolymer with be highly alternating with some randomness, for example, styrene and acrylonitrile ($r_S = 0.290$ and $r_A = 0.020$). Of course, the sequencing of the copolymer is also influenced by the relative concentrations of the two monomers, as well as their individual rate constants. Knowledge of monomer concentration and determination of the copolymer composition can be used to obtain experimental values of the reactivity ratios.

One of the most common polymeric elastomers (Sect. 8.63) is made from a mixture of 75% butadiene (B) and 25% styrene (S). A portion of the resulting random copolymer may be represented by ···BBSBBBBBSBSB···, which differs from a block polymer (Sect. 8.21). On the other hand, styrene and methyl methacrylate form an alternating copolymer (Fig. 8.20).

The copolymerization of a nonpolar olefin, like styrene, with 15% or less of an ionic vinyl monomer such as methacrylic acid sodium salt, leads to a polymer whose bulk properties are governed by ionic interactions in discrete regions (ionic aggregates) of the copolymer. Such copolymers are called *ionomers*.

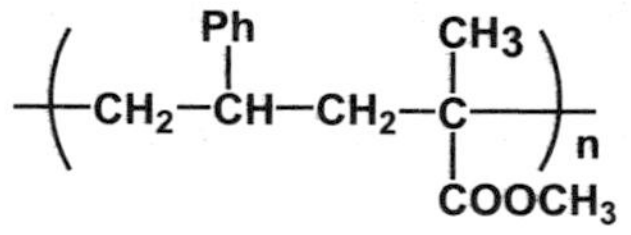

Figure 8.20. An alternating copolymer of styrene and methyl methacrylate.

8.21 BLOCK POLYMER

A copolymer whose structure consists of a homopolymer chain of one monomer, A, attached to a homopolymer chain of a different monomer, B. Such a polymer is prepared by first polymerizing monomer A by the *living polymer* technique until all of it is consumed, and then adding monomer B to the same flask to continue the polymerization. The resulting copolymer, A_nB_m, is called a *diblock copolymer*.

Example. After the living polymer from styrene, shown in Fig. 8.19, has formed and all the monomeric styrene has been utilized in forming a homopolymeric block, the monomer butadiene is then added, whereupon the addition occurs at both ends of the living polymeric polystyrene dianion to form a diblock of living polymer of styrene and butadiene (Fig. 8.21*a*). At this stage, a third monomer may be added if desired to form a *triblock copolymer*. When one of the blocks in a diblock polymer is water-soluble, such as the polymer from an *N*-alkylvinylpyridinium halide, the diblock is called an *amphiphilic copolymer* (Fig. 8.21*b*). Triblock copolymers with structures A–B–A or A–B–C are also known, and a wide variety of both diblock and triblock copolymers are available commercially. What may be considered a variation of the block polymer is the construction of a polymer having terminal reactive functional groups enabling the formation of a larger macromolecule. Such polymers are called *telechelic polymers*, from the Greek words *tele*, meaning "far-off," and *chele*, meaning "claw."

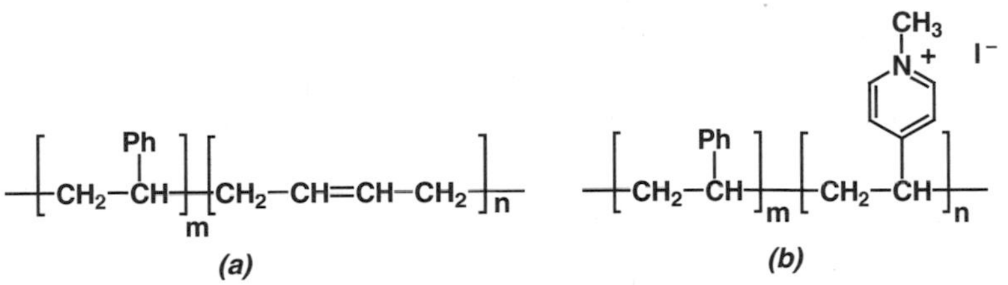

Figure 8.21. Diblock copolymers: (*a*) poly(styrene-b-butadiene) (1,4-addition) and (*b*) the amphiphilic diblock copolymer, poly(styrene-b-4-vinylpyridinium iodide). The letter b in the source name of the diblock copolymers stands for the word "block."

8.22 GRAFT POLYMER

A copolymer in which the chain backbone is made up of one kind of monomer and the branches are composed of another kind of monomer.

Example. Polystyrene may be grafted onto a poly(butadiene) polymer by treating the latter with styrene in the presence of a radical initiator. The reactions are shown in Fig. 8.22*a*. The schematic formula for a graft polymer made from a single-strand polymer A and a grafting monomer B is shown in Fig. 8.22*b*.

8.23 CROSS-LINKING

The linking of two or more independent polymer chains by groups that span or link the two chains.

Example. Cross-linked polystyrene. In practice, cross-linking of linear polystyrene is achieved by adding some divinylbenzene to the styrene before polymerization. When

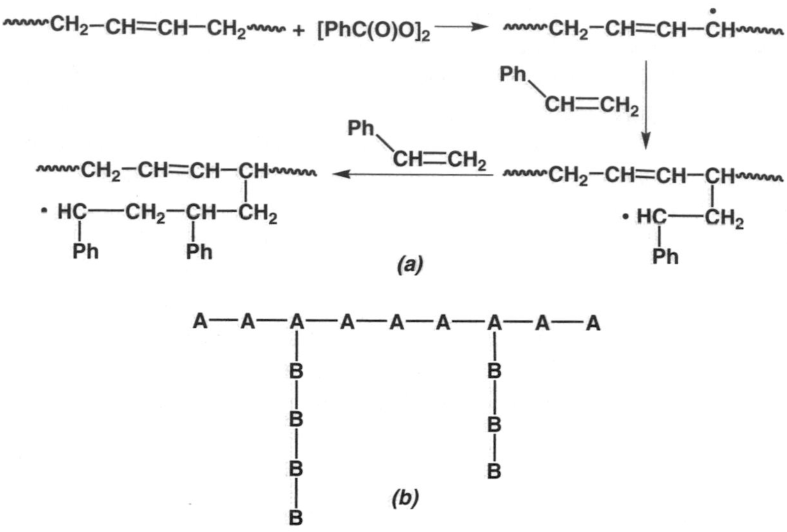

Figure 8.22. Graft polymers: (*a*) reactions leading to a graft polymer and (*b*) schematic representation of a graft polymer having a main chain of monomer A units and branches of monomer B units.

each of the two vinyl groups of divinylbenzene on the same ring participate in chain propagation steps, a cross-link is formed (Fig. 8.23*a*). A randomly spaced cross-linked network is shown schematically in Fig. 8.23*b*. Cross-linking can also be achieved by exposing the polymer to *ionizing radiation* from neutrons or γ rays, which strips hydrogen atoms from polymer chains, leaving radicals that then combine to form cross-links.

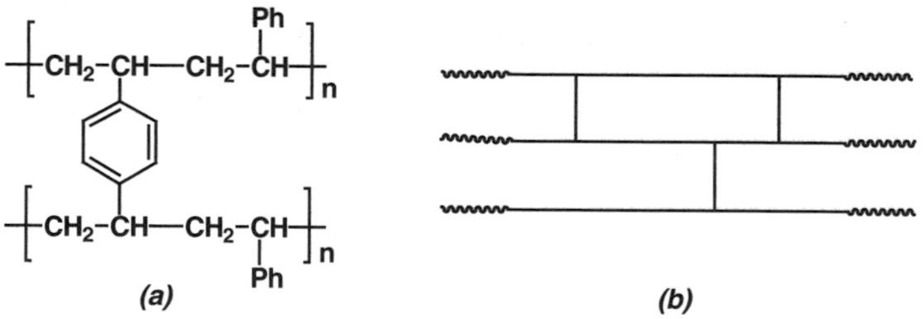

Figure 8.23. (*a*) A cross-link in polystyrene and (*b*) a cross-linked polymer network.

8.24 LADDER POLYMER

The polymers described thus far possess bivalent (two free valences) constitutional repeating units. When the constitutional repeating units have four free valences, a polymer consisting of two backbone chains regularly cross-linked at short intervals (a double-stranded polymer) results; this is called a ladder polymer.

Example. The simplest possible constitutional repeating unit of a ladder polymer is shown in Fig. 8.24*a*. A ladder polymer can be constructed so that a single-strand polymer is first formed, followed by the construction of the second strand to give the double-strand structure of the ladder polymer. In one such case, vinyl isocyanate (Fig. 8.24*b*) is treated at −55° with NaCN to give the linear polymer (Fig. 8.24*c*), which on subsequent treatment with a radical initiator gives the ladder polymer shown in Fig. 8.24*d*.

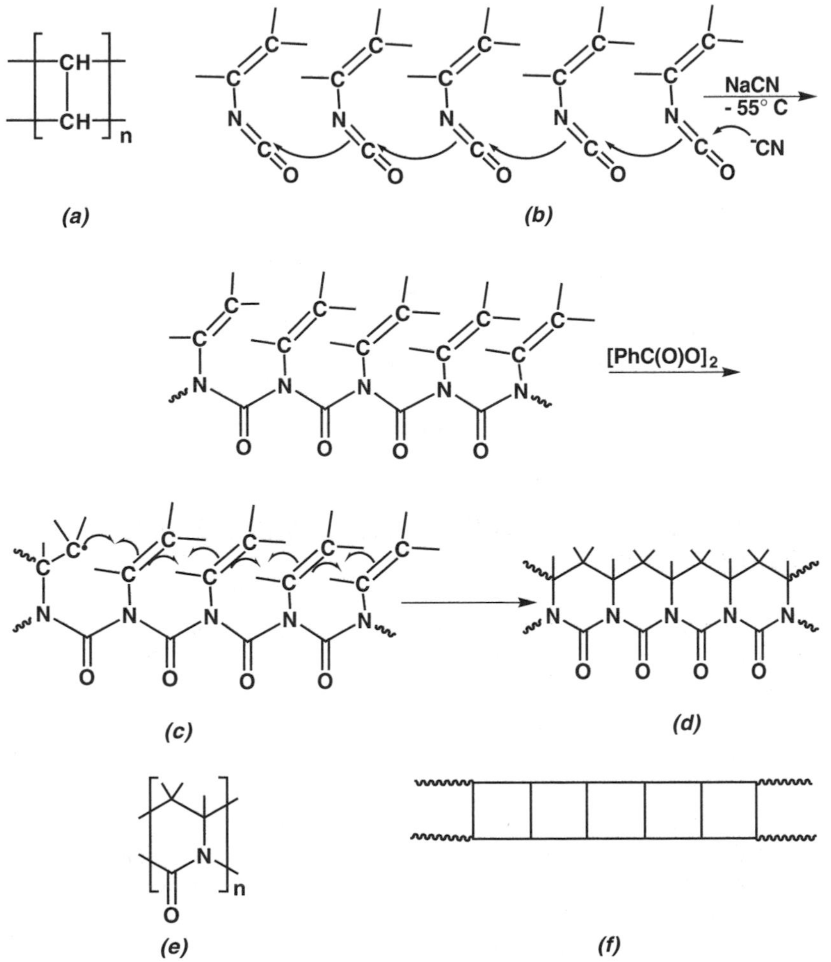

Figure 8.24. Ladder polymers: (*a*) the simplest repeating unit; (*b*) vinyl isocyanate polymerization; (*c*) the single-strand polymer polymerizing further to give (*d*) the ladder polymer; (*e*) the repeating tetravalent constitutional unit; (*f*) schematic representation of a ladder polymer.

The tetravalent constitutional repeating unit is illustrated in Fig. 8.24*e*, and the general schematic structure of a ladder polymer is shown in Fig. 8.24*f*.

8.25 POLYMER ARCHITECTURES

The variety of forms or shapes that chain growth polymers may assume as a result of the polymerization process. These are quite diverse and the various shapes affect the properties of the polymer.

Example. Some shapes of various kinds of polymers are shown in Fig. 8.25.

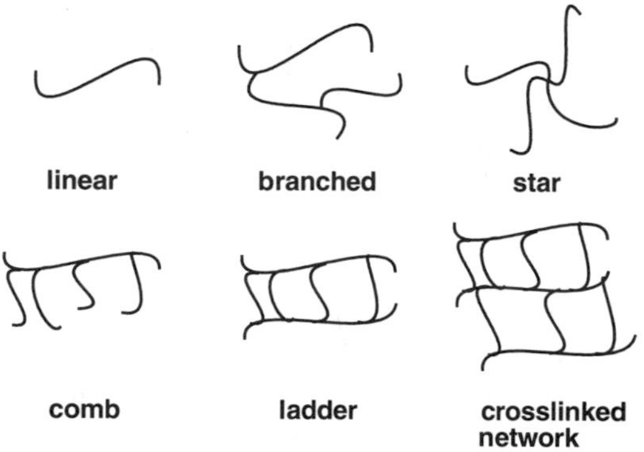

Figure 8.25. Some possible polymer architectures.

8.26 RING-OPENING CHAIN-GROWTH POLYMERIZATION OF HETEROCYCLES

A chain-growth polymerization of reactive heterocyclic monomers. The monomers may be lactones, lactams, or cyclic ethers (including epoxides). Polymerization of these reactive cyclic monomers may be initiated by radicals, anions, or protons and cations, in the same fashion as other chain-growth polymerizations, such as vinyl monomer polymerization.

Example. The ring-opening polymerization of γ-butyrolactone (Fig. 8.26*a*). The polymerization of caprolactam to 6-nylon (Fig. 8.26*b*) can be achieved either by using a strong base in an anion-catalyzed chain-growth polymerizaton or by heating the lactam with water, which converts the lactam to the open-chain amino acid. The amino acid then proceeds to polymerize by a step-growth process (Sect. 8.53). Polyethers are prepared from epoxides; ethylene oxide is widely used to prepare poly(ethylene glycol), PEG (Fig. 8.26*c*), a polymer that is compatible with water and is therefore used in cosmetic creams and similar products.

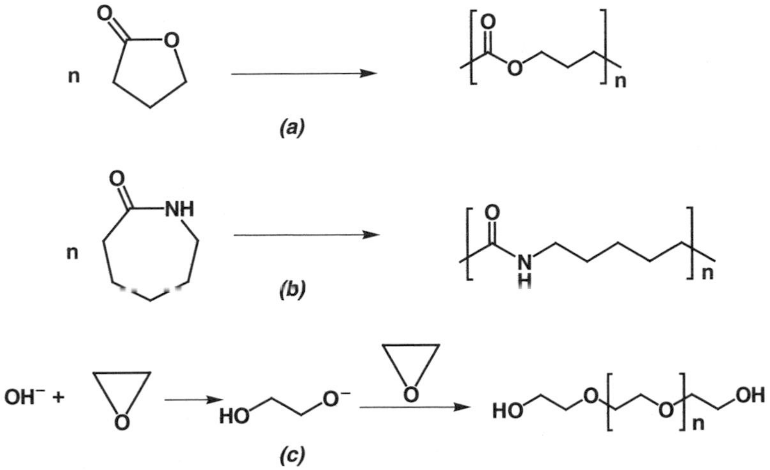

Figure 8.26. Ring-opening chain-growth polymerization of (*a*) γ-butyrolactone, (*b*) caprolactam, and (*c*) ethylene oxide.

8.27 RING-OPENING METATHESIS POLYMERIZATION (ROMP)

This is the transition-metal catalyzed polymerization of certain cycloalkenes to give acyclic polyolefin polymers; the greater the strain in the cycloalkene monomer, the greater the reactivity. The polymer has the same number of double bonds as the monomer and this fact alone indicates that the mechanism for ROMP polymerization is fundamentally different than the conventional chain-growth acyclic olefin polymerization. The presence of the double bond in the polymer provides a site for further functionalization.

Example. The ring-opening polymerization is assumed to proceed via metal-carbene intermediates (see "Olefin Metathesis," Sect. 9.40) and is catalyzed by, for example, an appropriate combination of tungsten hexachloride, diethylaluminum chloride, and ethanol. The polymerization of cyclopentene to give an elastomer (Sect. 8.63), known commercially as polypentamer, is shown in Fig. 8.27*a*. (M represents the transition metal catalyst with its associated ligands.) 1-Methylcyclobutene is polymerized by a preformed carbene complex, $(C_6H_5)_2C{=}W(CO)_5$, to give a polymer having essentially the same structure as cis-1,4-poly(isoprene), an isotactic polymer (Fig. 8.27*b*).

8.28 EMULSION POLYMERIZATION

A polymerization process (usually free radical) in which finely divided water-insoluble monomers are allowed to react with each other in an aqueous medium

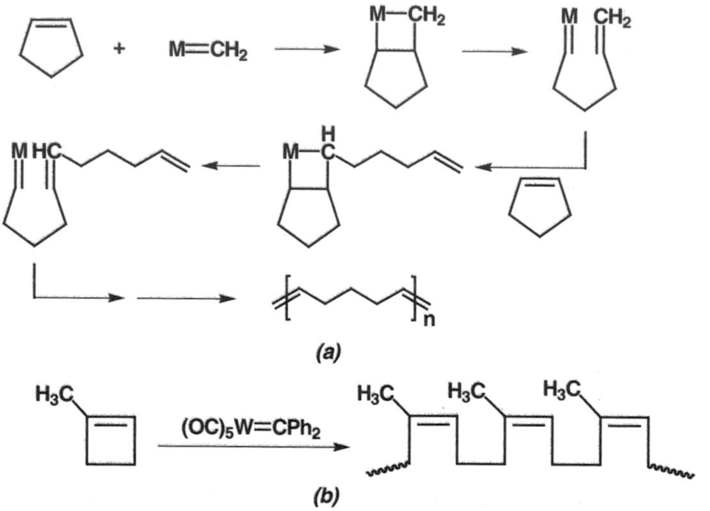

Figure 8.27. Ring-opening polymerization: (*a*) polypentamer from cyclopentene and (*b*) *cis*-1,4-poly(isoprene) from 1-methylcyclobutene.

containing an emulsifying agent. The resultant milky emulsion is called a latex, a generic term for such emulsions of polymers. The polymeric material can be separated as a coagulated latex and is processed further.

8.29 LOW-DENSITY POLYETHYLENE (LDPE)

A branched polyethylene polymer, partially crystalline, that melts at about 115°C and has a density of 0.91 to 0.94 $g \cdot cm^{-3}$. The branches can be as long as the main chain or as short as one to four carbon atoms; most of the short branches consist of ethyl and *n*-butyl groups. LDPE is prepared from ethylene at very high pressure with traces of oxygen to initiate the polymerization.

8.30 HIGH-DENSITY POLYETHYLENE (HDPE)

Essentially, this is a linear polymer; it is highly crystalline, melts at about 135°C, and has a density of 0.95 to 0.97 $g \cdot cm^{-3}$. Linear polyethylene has greater tensile strength and hardness than the branched low-density variety. It is frequently prepared via Ziegler–Natta catalysis (Sect. 8.46) and more recently by the use of metallocene catalysts (Sect. 8.47). The tensile strength of polyethylene polymers can be enhanced by stretching the long chains in order to obtain *oriented polymers* having chains parallel to each other instead of disordered. This difference has been described as the difference between cooked spaghetti (high degree of disorder, high entropy) and raw spaghetti (oriented, highly ordered, low entropy).

8.31 PLASTICS

Finished articles that are made from polymeric materials by molding techniques. The polymeric material, destined for plastic production, either pure or with appropriate additives, is called a *resin*. Conversion of resins to plastics is usually achieved by subjecting the resin to heat and/or pressure, whereupon the polymer softens, is then shaped or molded, and subsequently allowed to harden to become the article of commerce called a plastic.

8.32 PLASTICIZER

A relatively nonvolatile inert compound mechanically incorporated into a polymer to increase its flexibility and workability.

Example. Phthalic acid esters such as dioctyl phthalate. When incorporated into poly(vinyl chloride), which is an extremely hard resin, it softens and adds flexibility to the polymer, making it more suitable for common use, for example, as a shower curtain or Tygon tubing.

8.33 THERMOPLASTIC POLYMERS

Polymers that *reversibly* soften on heating and harden on cooling and are therefore molded in their softened state to form desirable solid objects upon cooling. This type of polymer usually is a single-strand or linear polymer with few if any cross-links, and such polymers either swell appreciably in organic solvents, or are even soluble in them.

Example. Many common products, such as bottles for soft drinks and bottled water, are made from thermoplastic polymers. The sulfur-containing polymer poly(phenylene sulfide), 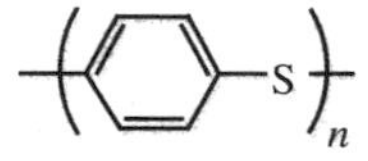, is important in materials known as *engineering thermoplastics*.

8.34 THERMOSETTING POLYMERS

Polymers that are *permanently* hardened on heating, that is, do not reversibly soften. This property is a result of the formation, on heating, of a cross-linked polymer network.

Plastic articles generated from such polymers (*thermosets*) are stable to heat and insoluble in organic solvents. The market for thermosets is about a tenth that of thermoplastics.

Example. It has been suggested that in concept a thermoset polymer is similar to a raw egg because on heating it undergoes irreversible change much like a hard-boiled egg. Phenol-formaldehyde plastics were the first synthetic thermosetting materials; they were invented in 1907 and are called *Bakelite* after their inventor, Leo Baekeland (1863–1944). Bakelite manufacturers call it the material of a thousand uses. The structure is shown in Fig. 8.34*a*.

The reaction of formaldehyde with the triamine, melamine, the cyclic trimer of cyanamide (Fig. 8.34*b*), gives rise to another commercially important thermoset. The polymerization reaction involves the intermolecular loss of water: $-NH_2 + CH_2O + H_2N- \rightarrow -NHCH_2NH-$ at each of the three amino groups and results in the highly cross-linked, colorless polymer called Melmac.

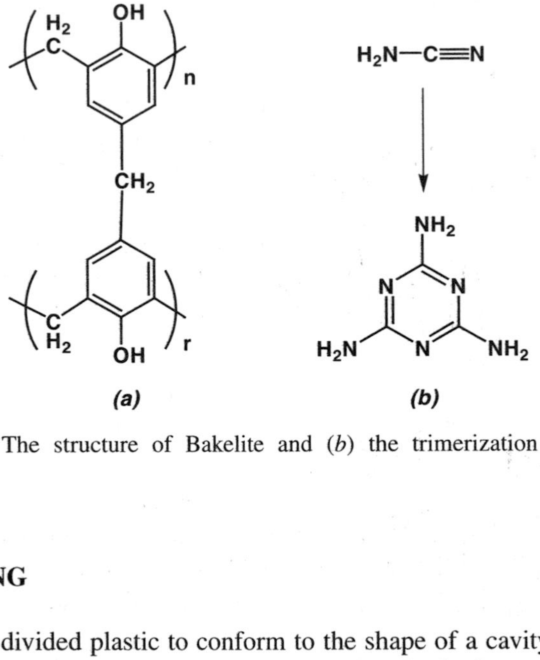

Figure 8.34. (*a*) The structure of Bakelite and (*b*) the trimerization of cyanamide to melamine.

8.35 MOLDING

Forcing a finely divided plastic to conform to the shape of a cavity, form, or mold, usually by applying heat and pressure to make the plastic flow to conform to the mold.

8.36 COMPRESSION MOLDING

Forming a molded plastic by placing the polymer in a stationary molding cavity and forcing the matching mold into the polymer while it is being heated. Both thermoplastic and thermosetting polymers may be compression-molded, but if the former is used, the mold must be cooled before the pressure is released.

8.37 INJECTION MOLDING

Forming a molded plastic by heating the polymer, usually in a cylinder carrying a plunger, until it can flow and then forcing it, hydraulically, through a nozzle into a relatively cold, closed mold cavity. Most thermoplastic materials are molded by this technique. The most common configuration of an injection molding machine is a reciprocating screw machine. The polymer feed is first softened by heating and then delivered to the mold by a screw arrangement rotating in a cylinder.

8.38 MELT TRANSITION TEMPERATURE T_m

Most polymers in a solid state are composed of both ordered or crystalline domains (*crystallites*) and amorphous or glassy domains. The temperature at which the crystalline domain of the polymer undergoes a melt transition is called the T_m of the polymer. It is analogous to, but not usually as sharp as, the melting point of a crystalline compound. Both thermodynamic properties and physical properties change with crystalline content. Usually, the more crystalline a polymer, the more opaque it becomes owing to the increased light scattering by the crystalline portion. Also, increasing crystallinity usually results in increased strength and rigidity and higher T_m. Amorphous phases do not undergo melting; on heating they are transformed from a hard glass to a rubbery state.

Example. Silicon rubber, poly(dimethylsiloxane), has a T_m of −40°C, while for natural rubber the T_m is 28°C; the value for nylon 6,6 (Sect. 8.55) is 270°C. Nylon polymers are thus hard and durable and are not affected by commonly encountered temperature changes.

8.39 GLASS TRANSITION TEMPERATURE T_g

This is the temperature at which an amorphous polymer (or the amorphous domain portion) is transformed from a hard glass into a mobile rubbery state; or starting from a high-temperature melt, which is slowly cooled, it is the temperature at which the liquid polymer becomes hard and glassy. At the transition temperature a remarkable change in the *modulus of elasticity* (a measure of the extensibility of a polymer, see Sect. 8.63) occurs. Amorphous polymers are rather transparent and are weak and flexible. Completely amorphous polymers exhibit only a glass transition temperature, but since most polymers are composed of both amorphous and crystalline domains, they show both T_m and T_g. If an *elastomeric polymer* (Sect. 8.63) is cooled below its

T_g value, it loses its elasticity and turns into a rigid glassy solid that is hard, stiff, and transparent.

Example. The elastomeric O-rings used to seal the solid booster rockets on the spacecraft *Challenger* had a T_g value of about 0°C, but unfortunately at launch time the ambient temperature was below this value and the O-rings lost their elastomeric properties and became glassy. This sealing failure was the most probable cause of the horrible disaster that ensued. For some polymers, there may be as much as a 1,000-fold change in the modulus of elasticity over a 10 to 20° range.

8.40 CONFIGURATIONAL BASE UNIT

The constitutional repeating unit in a polymer possessing at least one site whose chirality is specified.

Example. In polypropylene, the constitutional repeating unit is $-CH_2-CH(CH_3)-$ and the two possible configurational base units are shown in Fig. 8.40. These two configurational base units are enantiomeric (nonsuperimpossable mirror images).

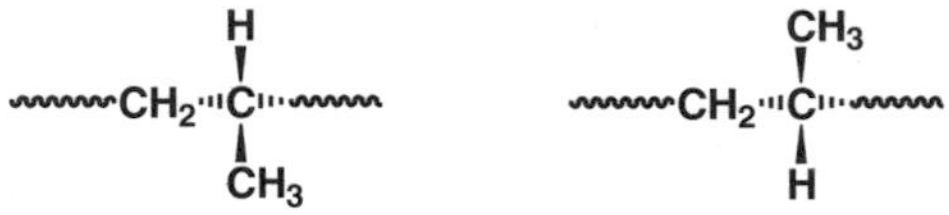

Figure 8.40. Representation of enantiomeric configurational base units of polypropylene.

8.41 TACTIC POLYMER

A polymer, whose name derives from the Greek word *tactikos* for "order," in which the configuration is specified around one or more stereoisomeric sites in the configurational repeating unit (the smallest set of successive configurational base units). If there are multiple chiral centers in the configurational repeating unit, at least one of them must be configurationally defined for the polymer to be a tactic polymer.

8.42 ISOTACTIC POLYMER

A tactic polymer having only one configurational base unit in the repeating unit. If a vinyl polymer, $-[-CH_2\text{-}CH(R)-]_n-$, is written in the form of a chain with the backbone carbon atoms in the plane of the paper, this requires that all the similar substituents on the chiral atoms in the polymer chain appear on the same side of the chain; all must be either in front of, or behind, the plane of the paper. In isotactic polymers, the configurational repeating unit is identical to the configurational base unit.

Example. A portion of the polymer chain of isotactic polypropylene is shown in Fig. 8.42*a*. In order to make an assignment of absolute configuration at every chiral center, it would be necessary to consider the end groups. If the polymer chain were terminated at the left end of the chain by an atom or group with a higher priority (Sect. 7.30) than the terminal group on the right end, then the configuration at each chiral center of this isotactic polypropylene would be (*S*) and this polymer chain would theoretically be optically active. However, the end groups of the polymer chain represent such a small fraction of the molecule that their presence can be neglected. If the end groups are neglected (or if they are equivalent), then the chain, whatever its length, has a plane of symmetry and is optically inactive. The polymer has multiple chiral centers but, because it has a plane of symmetry, it is a *meso* compound (Sect. 7.35). If each chiral center on the left side of the middle of the chain has an (*R*) configuration, then every chiral center on the right side of the middle of the chain has the (*S*) configuration! Because the chain has a plane of symmetry, the mirror image of the chain shown in Fig. 8.42*a* (which would have all the methyl groups on the far side of the chain) is identical and not enantiomeric with the chain of Fig. 8.42*a*. The same type of argument with respect to optical activity can be made with respect to syndiotactic (Sect. 8.43) vinyl polymers, and in fact no optically active vinyl polymers having chiral centers in the backbone chain are known.

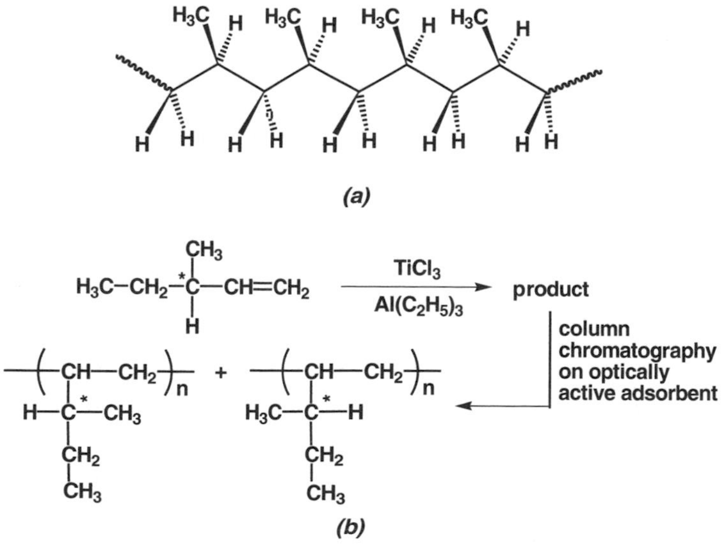

Figure 8.42. Isotactic polymers: (*a*) polypropylene and (*b*) an optically active polymer.

However, if the olefinic monomer possesses a chiral center in the substituent group, optically active isotactic polymers can be prepared (using Ziegler–Natta catalysis, Sect. 8.46). When a racemic mixture of the monomer is polymerized, resolution can lead to an optically active polymer (Fig. 8.42*b*).

8.43 SYNDIOTACTIC POLYMER

A tactic polymer, whose name derives from the Greek word *syndyo* for "two together," and whose configurational repeating unit contains successive configurational base units that are enantiomeric.

Example. Syndiotactic polypropylene (Fig. 8.43*a*); the chiral centers have opposite configurations. In the polymer chain, successive methyl groups would appear alternately on one side and then on the other side of the chain. If, in addition to the enantiomeric configurational sites, the repeating unit contains another, but undefined, chiral center, it is still considered to be syndiotactic (Fig. 8.43*b*), but such a polymer is not stereoregular.

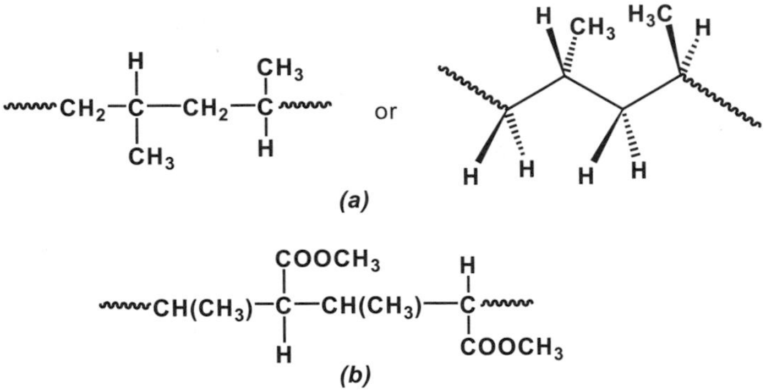

Figure 8.43. Syndiotactic polymers: (*a*) syndiotactic polypropylene and (*b*) a nonstereoregular syndiotactic polymer.

8.44 STEREOREGULAR POLYMER

A polymer in which every chiral center has a defined configuration.

Example. Isotactic and syndiotactic polypropylene. According to IUPAC rules, in a polymer such as $-[-CH(CO_2CH_3)CH(CH_3)-]_n-$, if only one of the main chain chiral sites of each constitutional repeating unit is defined (Fig. 8.44*a*), then the polymer (Fig. 8.44*b*) is isotactic but not stereoregular. A stereoregular polymer is always tactic but a tactic polymer need not be stereoregular.

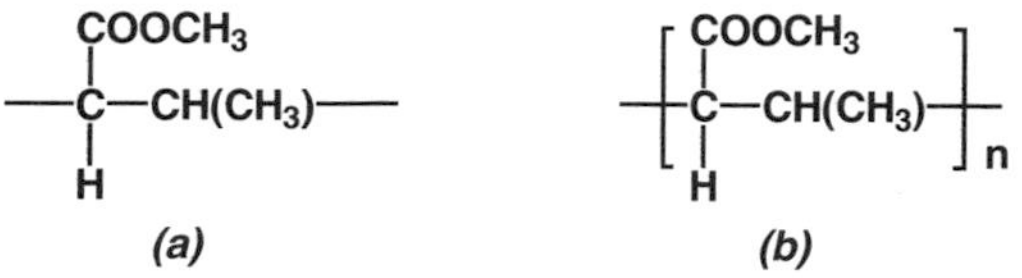

Figure 8.44. (*a*) A configurational base unit with one defined and one undefined chiral center and (*b*) the isotactic polymer that it forms.

8.45 ATACTIC POLYMER

A polymer possessing a random distribution of the possible configurational base units.

Example. Atactic polypropylene (Fig. 8.45*a*). According to the IUPAC definition, an atactic polymer is one having a random distribution of equal numbers of the possible configurational base units. Most polymer chemists call a regular polymer that does not possess a configurationally specified repeating unit an atactic polymer, even though an equal number of the configurational base units may not be present. It should be pointed out that a polymer conceivably can be tactic without being either iso- or syndiotactic, for example, Fig. 8.45*b*. This polymer contains three configurational base units in the configurational repeating unit.

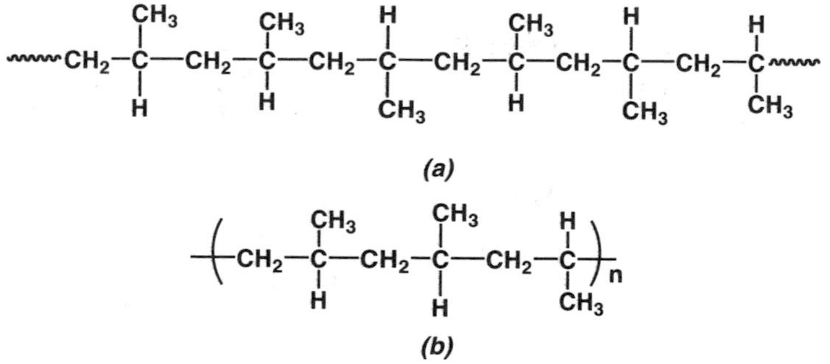

Figure 8.45. (*a*) Atactic polypropylene and (*b*) tactic polypropylene showing a configurational repeating unit that is neither isotactic nor syndiotactic.

8.46 ZIEGLER–NATTA CATALYSIS (COORDINATION POLYMERIZATION)

The catalytic system discovered by Karl Ziegler (1898–1973) and developed by Guido Natta (1903–1979) (co-Nobel Laureates in 1963), consisting of a bimetallic coordination complex, for example, $Et_3Al{:}TiCl_4$, on a magnesium chloride support. With this catalytic system, the polymerization of ethylene can be carried out at ambient conditions and leads to a stereoregular isotactic polymer of high density polyethylene (HDPE) and high crystallinity (Sect. 8.30).

Example. The exact structure of the active catalytic bimetallic coordination complex is not known. Presumably, the $TiCl_4$ is reduced by the $AlEt_3$. It is very likely that the olefin to be polymerized is coordinated to the titanium ion, although aluminum alkyls by themselvcs are known to be oligomerization catalysts. The active catalyst is probably a reduced, insoluble $TiCl_3$ with the $AlEt_3$ chemisorbed on the $TiCl_3$. The chain-growth mechanism has, at least until very recently, been assumed to be that shown in Fig. 8.46*a*. The key step in the mechanism is assumed to be olefin insertion into a metal-alkyl bond. It has been suggested that Ziegler–Natta catalysis may

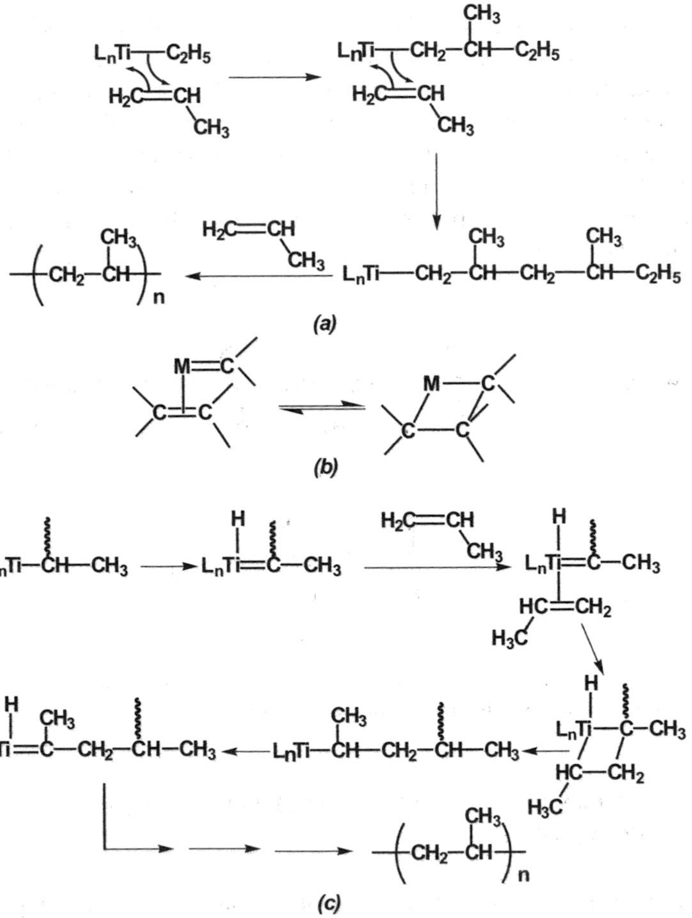

Figure 8.46. Zeigler–Natta catalysis: (*a*) olefin insertion mechanism; (*b*) metallocyclobutane intermediate in metathesis; (*c*) metallocyclobutane intermediate in polypropylene polymerization.

involve metal-carbene formation, followed by carbene addition to coordinated olefin to give a metallocyclobutane, the same type of reaction involved in the metathesis of olefins (Sect. 8.27, Fig. 8.46*b*). In order for the metal-carbene intermediate to form in the polymerization reaction, a 1,2-hydrogen shift from the α carbon to the metal is required, followed by metallocyclobutane formation and then the reverse 1,2-hydrogen shift (Fig. 8.46*c*).

The Ziegler–Natta catalyst system has been used to polymerize acetylene to give polyacetylene, a polymer with conjugated double bonds, *trans* $-(-CH{=}CH-)_n-$. The long chain of alternating single and double bonds gives rise to a large number of very closely spaced occupied orbitals called the valence band. This band is separated by a small energy gap from a corresponding band consisting of higher-energy empty orbitals. Polyactelylene is therefore a *semiconductor*, a material that has conductivity intermediate between a metal and an insulator. When properly doped, that is, adding a small quantity of impurity, polyacetylene has the electrical conducting properties of a metal. Such polymers belong to a special class of polymers called *conducting polymers.* Alan MacDiarmid (1927–) shared (with Alan Heegar and Hidelei Shirakawa) the 2000 Nobel Prize in Chemistry for his work in this field.

8.47 METALLOCENE CATALYSTS

A class of soluble coordination polymerization catalysts based on the metallocene structure (Sect. 9.22) $M(\eta^5\text{-}C_5H_5)_2$, where M is a metal, generally from group 4 (Ti, Zr, Hf). The metallocene catalyst is used along with a cocatalyst prepared by reacting trimethylaluminum with a small amount of water to give a complex mixture of methylaluminum oxide oligomers, called *methaluminoxane* MAO, $-(-O\text{-}Al(CH_3)-)_n$. The third component of the metallocene system is a solid support such as silica gel. The two cyclopentadiene rings of the metallocene can be bridged with another grouping, such as $(H_3C)_2C<$ or $(H_3C)_2Si<$ and otherwise modified to affect their steric and electronic character.

Example. Ethylene polymerization using a metallocene and MAO cocatalyst. Presumably, the MAO activates the metal by abstracting a methyl anion forming a cationic species that is the active polmerization catalyst (Fig. 8.47). Many modifications of metallocene catalysts have been developed. Particularly active catalysts have been prepared in which the two cyclopentadienyl rings are linked together by a silicon or carbon atom, called the *ansa bridge*, to give a constrained metal site. Monocyclopentadienyl complexes as well as complexes with nitrogen ligands bonded to iron and cobalt are active catalysts. The catalytic metallocene systems are regarded as *single-site catalysts (SSC)*, in contrast to the bimetallic Ziegler–Natta catalysts, and are highly reactive for the polymerization of ethylene; they can polymerize up to 20,000 ethylene monomers per second. They also catalyze the polymerization of other terminal olefins and presently are the commercial catalysts of choice.

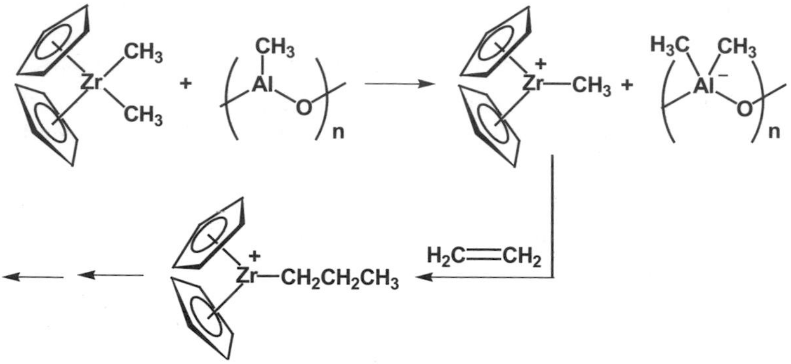

Figure 8.47. Initiation of ethylene polymerization by bis(cyclopentadienyl)dimethylzirconium and methaluminoxane.

8.48 MOLECULAR WEIGHT OF POLYMERS

Any random sample of polymer contains a mixture of chains of different molecular weights. The experimental measurement of molecular weight, thus, gives only an average molecular weight; there will be many chains with molecular weights greater and many chains with molecular weights smaller than this average. However, several different kinds of averages are possible.

8.49 NUMBER-AVERAGE MOLECULAR WEIGHT M_n

Experimental procedures that reflect the colligative properties of a sample, that is, the number of molecules in a known weight of sample, lead to the number-average molecular weight M_n. The total number of moles in a polymer sample is the sum, over all molecular species, of the number of moles N_i of each species present:

$$\sum_{i=1}^{\infty} N_i$$

The total weight w of the sample is the sum of the weights of each particular molecular species, which for each species i is equal to N_i times its molecular weight M_i:

$$w = \sum_{i=1}^{\infty} w_i = \sum_{i=1}^{\infty} N_i M_i \tag{8.49a}$$

The number-average molecular weight is the weight of sample per mole:

$$\overline{M}_n = \frac{w}{\sum_{i=1}^{\infty} N_i} = \frac{\sum_{i=1}^{\infty} N_i M_i}{\sum_{i=1}^{\infty} N_i} \tag{8.49b}$$

Example. Suppose a sample of oliomeric polyethylene prepared from pure ethylene (C_2H_4, mol. wt. 28) contains only two molecular weight species in the following amounts: four moles of a pentamer (mol. wt. $= 5 \times 28 = 140$) and nine moles of a decamer (mol. wt. $= 10 \times 28 = 280$):

$$\overline{M}_n = \frac{4(5\times 28)+9(10\times 28)}{13} = 237 \quad (8.49c)$$

M_n is very sensitive to changes in the percentage of low molecular weight species in a polymer and relatively insensitive to changes in the percent of high molecular weight species. Some common molecular weight measurements that depend on colligative properties are vapor pressure lowering, boiling point elevation, and osmotic pressure. Most molecular weights of polymer solutions are determined by various kinds of osmometry.

8.50 WEIGHT-AVERAGE MOLECULAR WEIGHT M_w

This measure of molecular weight is obtained experimentally by light scattering techniques. The measurement depends on the fact that the intensity of scattering is proportional to the square of the particle mass. $\overline{M}_w$ is calculated as follows:

$$\overline{M}_w = \frac{\sum_{i=1}^{\infty} N_i M_i^2}{\sum_{i=1}^{\infty} N_i M_i} \quad (8.50a)$$

where the quantities are defined as in the preceding section.

Example. Assuming the same sample as in the preceding section, we get

$$\overline{M}_w = \frac{4[(5\times 28)^2]+9[(10\times 28)^2]}{4(5\times 28)+9(10\times 28)} = 255 \quad (8.50b)$$

Note that $\overline{M}_w$ is always equal to or greater than $\overline{M}_n$. $\overline{M}_w$ is very sensitive to changes in the fraction of high molecular weight species. The difference between $\overline{M}_n$ and $\overline{M}_w$ can also be illustrated by comparing their respective values in a mixture of two polymers with the same CRU but different chain length, that is, different molecular weights, say, one of 40,000 and the other of 100,000. If equal *molar* concentrations are present in the mixture, the calculated values are shown in the second column of Table 8.50, but if equal *weights* of the same two species are present, quite different values are obtained.

As can be seen from the table, the number-average molecular weight of an equal molar mixture is equal to the weight-average molecular weight of an equal weight mixture. The statistical distribution of molecular weights in a typical polymer is shown in Fig. 8.50, where the percentage of polymer per unit interval of molecular weight is plotted against the molecular weight.

TABLE 8.50. Combinations of Species A (mol. wt. = 40,000) and Species B (mol. wt. = 100,000)

Mol. Wt.	Equal Molar	Equal Weight
$\bar{M}_n$	70,000	57,143
$\bar{M}_w$	82,857	70,000

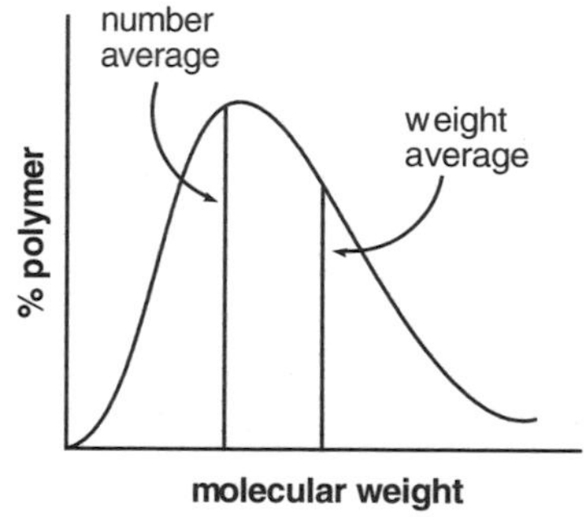

Figure 8.50. The distribution of molecular weights in a polymer.

8.51 POLYDISPERSITY INDEX $\bar{M}_w/\bar{M}_n$

This index is a measure of the breadth of molecular weight distribution; it is the ratio of weight-average molecular weight to number-average molecular weight. Ratios close to unity indicate a narrow molecular-weight distribution that is frequently an advantage for obtaining desirable physical properties. Polymers prepared by ring-opening metathesis polymerization (ROMP) or certain polymers produced by anionic polymerization can sometimes achieve an index as low as 1.02, whereas some commercial polyethylenes can have an index as high as 20.

8.52 DETERMINATION OF POLYMER MOLECULAR WEIGHT

Molecular weights of polymers may be obtained by a variety of techniques. Developments in instrumentation have made it possible to use the mass spectrometer (Chapter 17) for the determination of molecular weights of polymers. Other methods of determining molecular weights are (1) size-exclusion chromatography, (2) light scattering, (3) measurements of viscosity, and (4) measurements of osmotic pressure (Chapter 10).

Size-exclusion Chromatography (SEC). Is a chromatographic technique that separates molecules according to their size. The powder in a chromatographic column usually consists of a silica gel of specified size and surface area. When solutions of the polymer are allowed to flow through the column, the smaller molecular weight species diffuse more completely into the pores of the absorbent than do the larger molecules and therefore take more time to be eluted. The columns are calibrated with polymer standards of known molecular weight. It has been found that there is a linear relationship between the log of the molecular weight of a polymer and the volume of the eluant required for the weight-average molecular weight, the number-average molecular weight, and the polydispersity index.

Light Scattering. Is an optical method for the determination of the weight-average molecular weight of polymers. The scattering of light occurs whenever a beam of light encounters matter. When a solution of the polymer at a known concentration is exposed to a laser beam of light, the polymer particles scatter the incident light in all directions with the same wavelength as the incident beam. This scattered light is monitored by a number of detectors situated at different angles to the incident light. The amplitude of the scattered light is proportional to the mass of the scattering particle. The polymer solution is diluted and the process repeated several times in succession. Appropriate plots and calculations allow the determination of the weight average molecular weight with more accuracy than that obtained by SEC.

Viscosity Measurements. These measurements of polymers are proportional to their molecular weights. Solution viscosities can be measured by comparing the time required for a polymer solution to flow through a measured length of glass capillary tubing with that for the same volume of the pure solvent to flow through the same tube. The time difference is related to the molecular weight of the polymer.

Osmotic Pressure. This pressure is required to prevent the differential flow of a pure solvent through a semipermeable membrane into a solution of the polymer in question in the same solvent; it is related to the molar concentration and, hence, the molecular weight of the polymer. Osmotic pressure is a colligative (related to the number of particles present) property of a solution similar to freezing-point lowering or boiling-point elevation.

8.53 STEP-GROWTH POLYMERIZATION (CONDENSATION POLYMERIZATION)

The polymerization that results when bifunctional monomers react with other bifunctional molecules, usually accompanied by the intermolecular loss of small molecules such as water or alcohols. This type of polymerization has also been called (less desirably) *condensation polymerization.* When the two functional groups, A and B, that condense with each other are on different interacting bifunctional monomers, an A-A/B-B type polymer is formed, whereas if A and B are on the same monomer, the polymerization leads to an A-B type. The polymerization occurs in a stepwise manner beginning with two different bifunctional monomers reacting with each other to give dimers that in turn gives trimers, then tetramers, and so on. Unlike the case of chain-growth polymerization where monomer concentration decreases steadily, in step-growth polymerization, the monomer disappears rapidly in the early stages of polymerization. Naturally, the process is complicated because fragments of all sizes can react with each other or with the monomer, but the growth is characterized by functional group A reacting exclusively with functional group B and B only with A. At the beginning of the reaction, small chains are most likely to react with other small chains; unlike chain-growth polymerization, high molecular weight polymers are not formed until the reaction begins to approach completion.

Examples.

POLYAMIDES. An A/B type: poly(6-aminohexanoic acid) (Fig. 8.53*a*); A-A/B-B types: poly(hexamethylene adipate) (Fig. 8.53*b*). Either or both components of the polymer may have aromatic groups, in which case the polymer is called an *aramid.* In order to be classed as an aramid, the polymer must contain 85% or more of amide bonds attached to aromatic rings. An example is *Kevlar* (Fig. 8.53*c*), an unusual polymer in that it has a tensile strength greater than steel and is used to fabricate bulletproof vests.

POLYIMIDES. The polymer called *Kapton* is prepared from a dianhydride and diamine (Fig. 8.53*d*); It is used as an insulator in the electronics industry.

POLYESTERS. Poly(ethylene terphthalate) or *PET* (Fig. 8.53*e*) is a polymer used for the fabrication of containers for bottled water and soft drinks. A related polyester made from a naphthalenedicarboxylic acid and ethylene glycol, abbreviated *PEN*, is said to have some advantages over PET. These polymers, like the majority of commercial polyester polymers, are thermoplastic. There are some thermoset polyesters as well; they are prepared from unsaturated dibasic acids such as maleic anhydride and they are highly cross-linked.

POLYCARBONATES. Poly(4,4′-carbonato-2,2′-diphenylpropane), called *Lexan* (Fig. 8.53*f*).

POLYURETHANES. The step-growth intermolecular polymerization reaction between difunctional monomers leading to polyurethanes proceeds without loss of small molecules, and accordingly, the polymer has essentially the same empirical formula as the sum of the empirical formulae of the constituent monomers. A urethane is the generic name for the compounds formed from the reaction between an isocyanate and an alcohol: $RN{=}C{=}O + R'OH \rightarrow RNHC(O)OR'$. The polymer resulting from the reaction of 1,4-diisocyanatobenzene and ethylene glycol is shown in Fig. 8.53*g*. Because the polyurethanes are polymers formed by step-growth polymerization without the loss of small molecules, they are sometimes classed as *step addition* to distinguish them from the more common step-growth polymers. Most of these polymers can be pulled or drawn to generate fibers with special characteristics (Sect. 8.59).

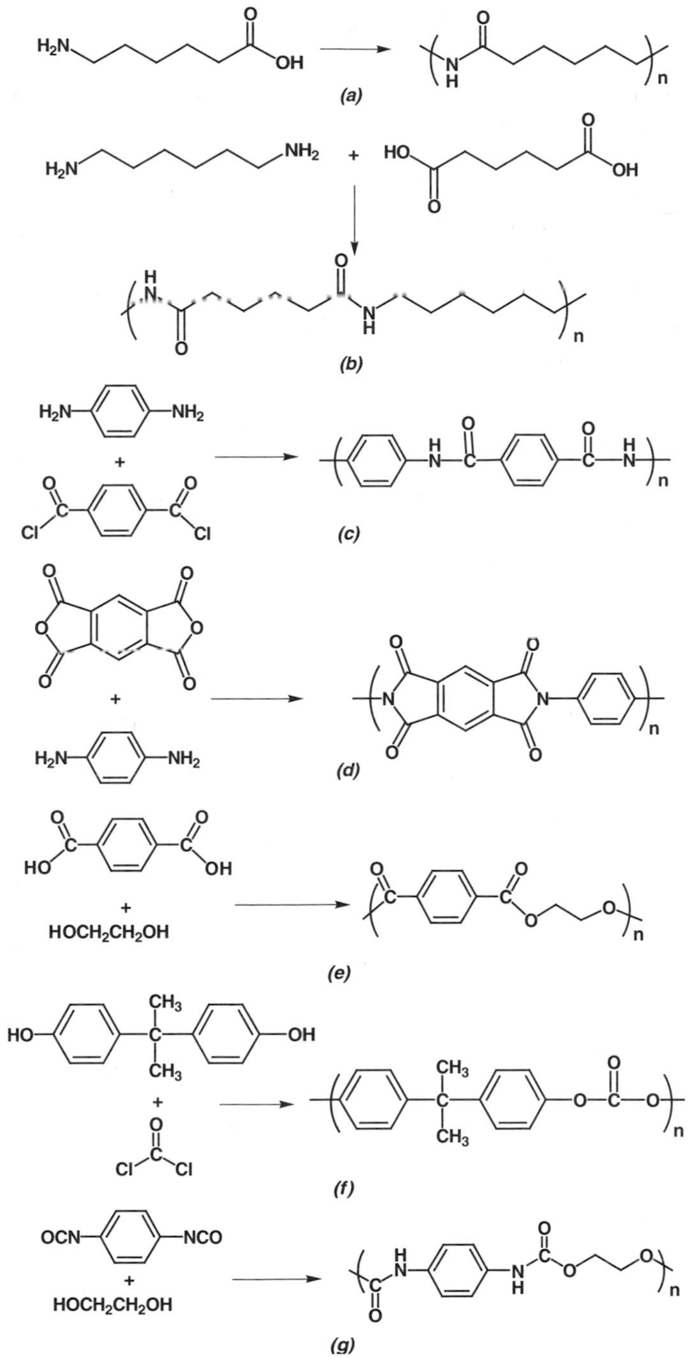

Figure 8.53. Step-growth polymers: polyamides (*a*) nylon 6, an A-B type, and (*b*) nylon 6,6, an A-A/B-B type; (*c*) Kevlar, an aromatic polyamide of the A-A/B-B type; (*d*) Kapton, a polyimide; (*e*) PET, a polyester type; (*f*) Lexan, a polycarbonate type; and (*g*) a polyurethane polymer.

8.54 FOAMED POLYURETHANES

The formation of a polyurethane accompanied by the liberation of carbon dioxide formed according to the typical reaction:

$$R{-}N{=}C{=}O + H_2O \rightarrow RNHCO_2H \rightarrow RNH_2 + CO_2$$

The CO_2 liberated during the polymerization causes the polymer to foam or froth, generating a finished product that is cellular in structure, similar to a sponge.

Example. In the simplest model, a prepolymer prepared from a polymeric diol and excess diisocyanate is treated with water (or carboxylic acids) in the presence of a tertiary amine catalyst with low steric requiremens such as *DABCO* (1,4-diaza[2.2.2]bicyclo octane) to give the reactions shown in Fig. 8.54. The formation of the amino group occurs while free isocyanate groups are still present, and the amine-isocyanate reaction leads to disubstituted urea formation and chain lengthening. The protons on the nitrogen atoms of the urethane and urea linkages can also react with

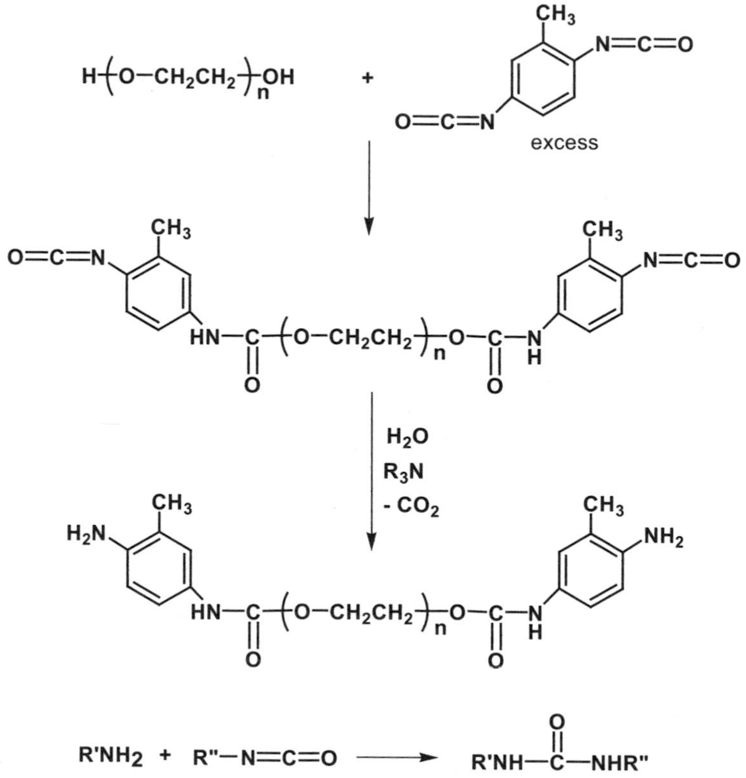

Figure 8.54. Reactions leading to foamed polyurethanes.

free isocyanate groups to generate cross-linkages. Urethane foams can be made in either flexible or rigid form, depending on the extent of cross-linking, the latter being highly cross-linked, very rigid, and very strong. Almost 5 billion pounds of polyurethanes are consumed in the United States annually in the manufacture of flexible and rigid foams.

8.55 NYLON

A generic term for any long-chain synthetic polyamide polymer. The original nylon was first produced in 1936 by a team of du Pont research chemists led by Wallace Carothers. The conversion of the polyamide to a fiber is achieved by melting the polymer and then forcing the melt through a spinneret. The emerging thin filaments are then stretched to 400 to 600% of their original length (cold drawing) to bring about orientation in the direction of the axis of the filament. Such orientation is driven by the intermolecular hydrogen bonding between C=O----HN bonds, whereby the crystallinity and, hence, the tensile strength of the filament are increased. The filaments can then be wound into fibers suitable for conversion into fabrics. In April 1937, samples of nylon were woven for the first time into "silk" stockings.

Example. The structure of the popular nylon prepared from hexamethylenediamine and adipic acid, [poly(hexamethyleneadipamide)], is shown in Fig. 8.53*b*. Equimolar amounts of the diamine and the diacid are mixed to give the amine salt that is then heated under vacuum to eliminate water. On a laboratory scale, instead of the diacid, the diacid chloride may be used and sodium carbonate is added to take up the HCl. Commercial nylons of the A-A/B-B type are frequently designated by a two-digit number. The first digit indicates the number of carbon atoms in the diamine component and the second digit the number of carbon atoms in the diacid component. The A/B type nylons are given a one-digit designation. Thus, the nylon prepared either from ε-aminocaproic acid or from its lactam is designated nylon 6 (Fig. 8.53*a*). Systematic IUPAC names for polymers, although awkward, are based on the CRU (Sect. 8.5) and these are the names used by Chemical Abstracts for indexing polymers. In the IUPAC scheme, polyamides are named as derivatives of repeating divalent, nitrogen-substituted (designated as imino) carbon backbones. The carbonyl function is designated as an oxo substituent on the carbon backbone. In the hierarchy of functional group designation, amines rank higher than carbonyls. Thus, the IUPAC name for nylon 6 (Fig. 8.53*a*) is poly[imino-(1-oxo-1,6-hexanediyl)] and nylon 6,6 (Fig. 8.53*b*) is poly[imino(1,6-dioxo-1,6-hexanediyl)imino-hexanediyl].

8.56 DENDRIMERS

Highly branched treelike macromolecules formed by successive reactions of polyfunctional (usually bifunctional) monomers around a core monomer.

Example. A treelike schematic replica of a two-generation dendimer is shown in Fig. 8.56*a*. The structure of a polyether dendrimer grown from 3,5-dihydroxybenzyl ether core by the intermolecular loss of water is shown in Fig. 8.56*b*. Two (2) generations of the growing structure are shown. In the first generation there are four phenolic hydroxyl groups on the periphery, and this number increases geometrically with each additional generation. Thus, in the second generation there are 8 OH as shown, in the third generation there would be 16 OH groups, and in the fourth there would be 32. The dendrimer may be capped with a monofunctional molecule; in the above example the capping may be accomplished with benzyl alcohol (not shown). Dendrimers of various structures can be purchased commercially. Their molecular weights can be accurately controlled as can their level of branching and the structure of the end groups. Many uses of these unusual polymers have been developed.

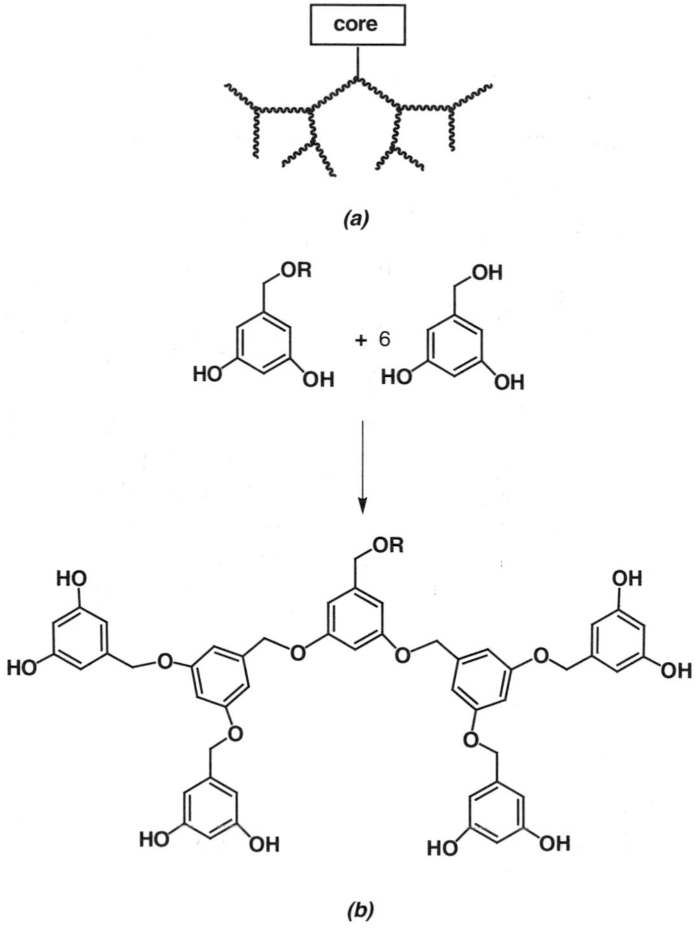

Figure 8.56. (*a*) Schematic of a typical dedrimer and (*b*) the second generation dendrimer formed from 3,5-dihydroxybenzyl ether.

8.57 EPOXY RESINS

Relatively low molecular weight polymers formed by condensing an excess of epichlorhydrin with a dihydroxy compound, most frequently 2,2,-bis(*p*-hydroxyphenyl)propane, commonly called *bisphenol A*.

Example. The condensation shown in Fig. 8.57 leaves epoxy end groups that are then reacted in a separate step (cured) with nucleophilic compounds (alcohols, acids, or amines). For use as a household adhesive, the epoxy resin and the curing resin, usually a polyamine, are packaged separately and mixed together immediately before use.

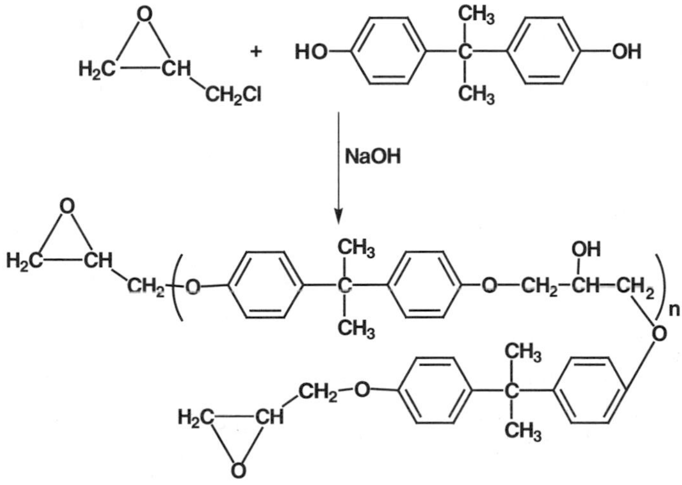

Figure 8.57. The formation of an epoxy resin.

8.58 SILICONES

Low molecular weight polymers containing Si-O-Si linkages that are produced by intermolecular condensation of silanols. The silanols, in turn, are generated by hydrolysis of halosilanes.

Example. Silicone oils are typically made by the reactions represented in Fig. 8.58. The silicones are not wet by water; hence, they are used for waterproofing. They also have excellent low-temperature properties (i.e., they do not become viscous or crystalline) and they have remarkable stability at high temperatures.

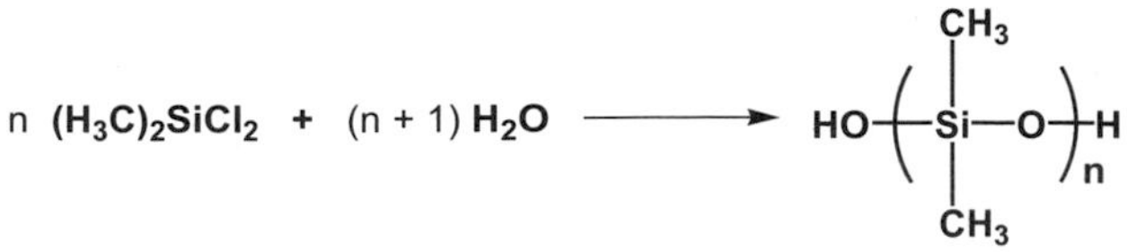

Figure 8.58. Silicone formation.

8.59 FIBERS

Slender, threadlike crystalline polymers, natural or synthetic. They are characterized by great tensile strength in the longitudinal direction of the fiber. Bulk synthetic polymer powder is converted to a fiber by the process called spinning (Sect. 8.60). The classification of a substance as a fiber is rather arbitrary; usually, a fiber is a matcrial that has a length at least 100 times its diameter. Of course, artificial or synthetic fibers can be made into any desired ratio of length to diameter. Although this chapter is devoted primarily to synthetic polymers, the natural polymers that are used for fiber production will be discussed below.

Example. Cotton, wool, silk, and flax fibers have lengths 1,000 to 3,000 times their diameter; coarser fibers such as jute and hemp have lengths 100 to 1,000 times their diameter. The polymer molecules of a fiber must be at approximately 100 nm long (when extended), hence have a molecular weight of at least 10,000.

8.60 SPINNING

The conversion of bulk polymer to fiber form. Usually, the polymer is dissolved and then metered through orifices into a vertical, heated chamber where the volatile solvent is removed in a stream of air and the remaining threadlike polymer is pulled out of the chamber as a yarn. This is called *dry spinning*. In the *wet spinning* process, the polymer is coagulated or precipitated from solution and the polymer removed as yarn. Polymers that can be melted without decomposition (nylons, dacron, polyproplene) are converted to fibers by *melt spinning*. In this process, polymer chips are melted and the melt is pumped through an extrusion nozzle and into a cooling chamber. The resolidified polymer is removed as yarn. The extrusion nozzle is called a *spinneret* and may consist of a 3 in. steel disk about 4 in. thick, into which 50 to 60 holes, 0.010 in. or less in diameter, are drilled. Spinning develops molecular orientation due to the intermolecular interaction between polar groups on neighboring chains. As the filaments become solid, they are brought together and wound on spools.

8.61 SYNTHETIC FIBERS

Fiber-forming polymers of crystalline polyamides, polyesters, poly(acrylonitrile), polyolefins, polyurethanes. Practically all polyvinyl-type polymers can be spun into fibers. Graft polymers can also be spun.

Example. The structures and names of some commercial synthetic fibers are listed in Table 8.61. Synthetic fiber-forming polymers are characterized by high tensile strength and a high modulus of stiffness. The strongest fabric, known as Dyneema, consists of a polymer of oriented polyethylene that has a molecular weight 100 times that of high-density polyethylene. A rope made from it is said to able to lift

119,000 lb, while a steel rope of similar size fails with a load of 13,000 lb. Other fibers possess these properties, as well as resistance to heat to an unusually high degree. This is particularly true of some of the aromatic polyamides that have been given the generic name aramids. One of these, Kevlar (Table 8.61), can be fabricated

TABLE 8.61. Some Commercial Synthetic Fibers

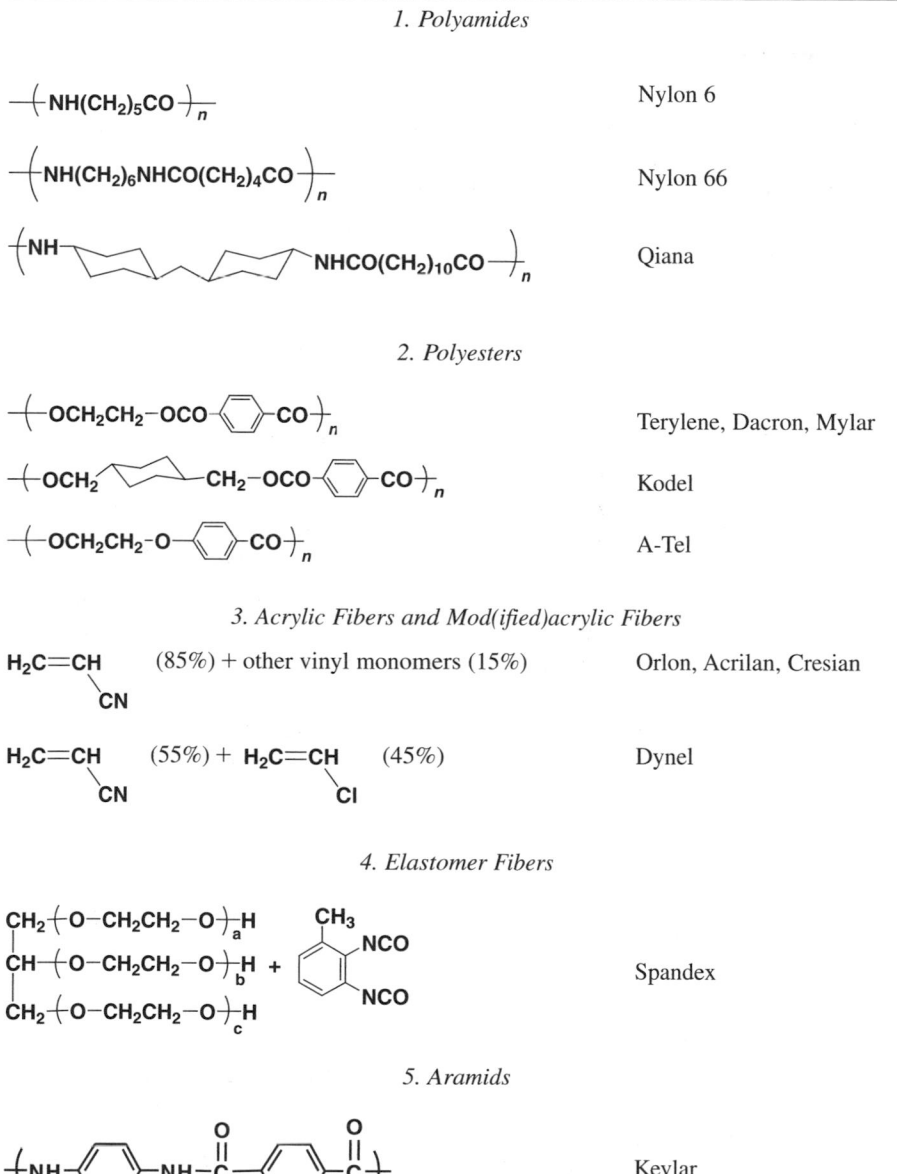

1. Polyamides	
$\left[\!-NH(CH_2)_5CO-\right]_n$	Nylon 6
$\left[\!-NH(CH_2)_6NHCO(CH_2)_4CO-\right]_n$	Nylon 66
	Qiana
2. Polyesters	
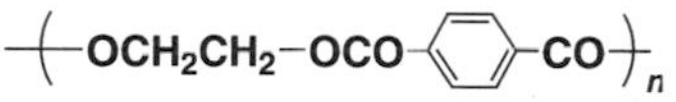	Terylene, Dacron, Mylar
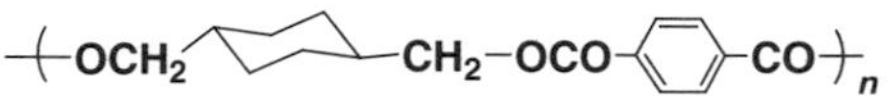	Kodel
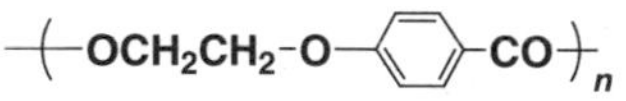	A-Tel
3. Acrylic Fibers and Mod(ified)acrylic Fibers	
$H_2C{=}CH(CN)$ (85%) + other vinyl monomers (15%)	Orlon, Acrilan, Cresian
$H_2C{=}CH(CN)$ (55%) + $H_2C{=}CH(Cl)$ (45%)	Dynel
4. Elastomer Fibers	
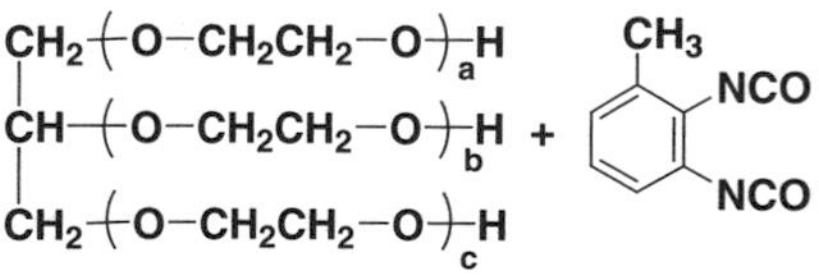	Spandex
5. Aramids	
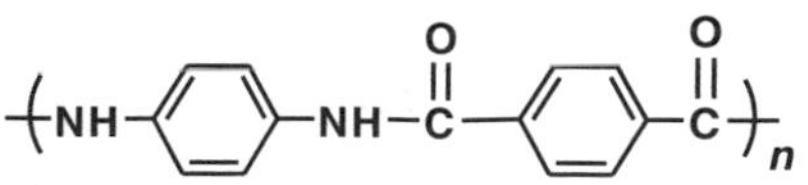	Kevlar

into cables and ropes that possess a breaking stress higher than steel wire of identical diameter but are only approximately one-sixth the weight of the steel wire. Recently, items such as tennis rackets have been fabricated from carbon fibers. Most *carbon fibers* are produced by the controlled thermal treatment of acrylic fibers. The acrylic is heated in air at 200 to 300°C while held under tension to give orientation. The oxidized fibers are then carbonized in an inert atmosphere by gradual heating to 1500°C. Finally, the carbon is graphitized by heating to 2500 to 3000°C.

8.62 DENIER

The weight in grams of 9,000 m of a spun fiber. Because this weight will depend among other things on the diameter of the thread, the denier is frequently used as a measure of the thickness of a thread. The filaments that emerge from a spinneret may be 2 or 3 deniers, and the size of the thread that is made by bringing together such filaments may vary from 15 to 500 deniers. The smaller the denier, the finer the thread.

8.63 ELASTOMERS

A generic name for polymers that exhibit rubberlike elasticity, that is, they can be stretched several hundred percent under tension and, when the stretching force is removed, they retract rapidly and recover their original dimensions. Such deformations are called elastic to differentiate them from the behavior of plastics, which after removal of an applied force remain in the deformed state, having undergone what is termed *plastic flow*, an *inelastic deformation*. The restorative force in the case of elastomers is entropy-based. In the stretched state the polymer is in a more ordered higher-energy (low-entropy) state, while in the relaxed form where the polymer has a more coiled or disordered conformation, the polymer is in a lower-energy, more random (high-entropy) form. Elasticity is unique to long-chain molecules. One of the most useful techniques for characterizing an elastomer is determining its stress-strain curve, which is a measure of the force developed as the polymer is elongated at a constant rate of extension. The slope of the resulting curve is called the *modulus of elasticity.*

8.64 RUBBER

The name given to the solid polymeric material isolated from the white fluid or latex found in a variety of plants and trees. It was called rubber by Joseph Priestley (1733–1804), who used it to erase or rub out pencil marks. It now has a generic meaning and applies to all those substances, regardless of origin, natural or synthetic, that have elastomeric properties. The terms rubber and elastomer are now used synonymously; the latter not only sounds more elegant, but also combines the notion of elasticity and polymer character.

8.65 NATURAL RUBBER

The rubber isolated from natural sources such as trees and plants. The principal commercial source is from the rubber tree, *Hevea brasiliensis*. It is a 1,4-addition polymer of isoprene, $-(CH_2-CH(CH_3)=CH-CH_2)_n-$, and the configuration around each double bond is *cis* (Z) (Fig. 8.65).

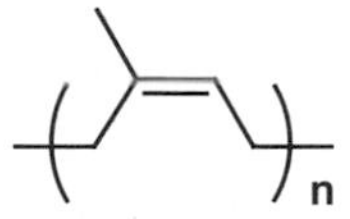

Figure 8.65. Natural rubber polymer.

8.66 GUTTA-PERCHA

A natural nearly white rubber obtained from the milky latex of certain Malaysian trees of the sapondilla family. It is also a polymer of isoprene, but its configuration around each double bond, in contrast to the all-*cis* diastereoisomer of natural rubber (Sect. 8.65), is *trans* or *E* (Fig. 8.66). As a result, it is hard and tough rather than soft and deformable.

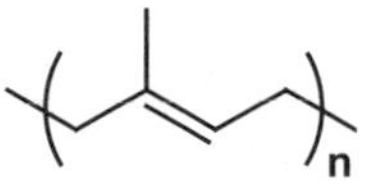

Figure 8.66. Gutta-percha polymer.

8.67 SYNTHETIC RUBBERS

The polymers with rubberlike characteristics prepared synthetically from dienes or olefins. Other polymers such as polyurethanes, fluorinated hydrocarbons, and polyacrylates can be synthesized to generate rubbers with special properties.

Example. Styrene-butadiene rubber (SBR) is prepared from the free radical copolymerization of one part by weight of styrene and three parts of butadiene. The butadiene is incorporated by both 1,4- (80%) and 1,2-addition (20%), and the configuration around the double bond of the 1,4-adduct is about 80% *trans*. It is a random copolymer and a typical portion of the polymer is shown in Fig. 8.67; several types of commercial synthetic rubbers are listed in Table 8.67.

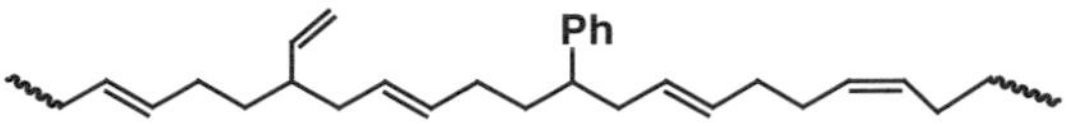

Figure 8.67. A portion of the butadiene-styrene copolymer chain.

TABLE 8.67. Some Commercial Synthetic Rubbers

Name	Monomer(s)
GRS, Buna S, SBR	Styrene and 1,3-butadiene
Neoprene	Chloroprene (2-chloro-1,3-butadiene)
cis-Polybutadiene, BR	1,3-Butadiene, mostly 1,4-addition; *Z* configuration
cis-Polyisoprene, IR	Isoprene, only 1,4-addition; *Z* configuration
Butyl rubber, IIr, GRI	Isobutylene, isoprene (acid-catalyzed)
Nitrile rubber, NBR, GRN, Buna N	1 part Acrylonitrile to 2 parts 1,3-butadiene
Ethylene-propylene rubber, EPR	3 Ethylene to 2 propylene
Polysulfide rubbers, Thiokol ST	1,2-Dichloroethane + Na_2S_4 $HS(CH_2CH_2SSSS)_nCH_2CH_2SH$
Silicone rubbers	$(H_3C)_2SiCl_2 \xrightarrow[\text{steps}]{\text{Several}} \left(Si(CH_3)_2 - O \right)_n$
Tygon	Poly (vinyl chloride) + special plasticizers

8.68 FOAMED RUBBER

Rubber that has been mixed with a special compound so that on heating, the softened mixture liberates a gas such as nitrogen, forming a low-density, cellular rubber, much like a sponge. The most commonly used nitrogen generator is the thermally unstable *N,N′*-dinitrosopentamethylenetetramine (Fig. 8.68). When the nitrogen is liberated, the resulting product acts as a cross-linking agent.

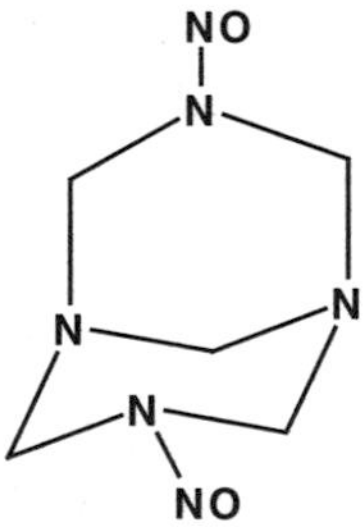

Figure 8.68. Nitrogen precursor (*N,N′*-dinitrosopentamethylenetetramine or 3,7-dinitroso-1,3,5,7-tetraazabicyclo[3.3.1]nonane) used for foaming rubber.

8.69 VULCANIZATION

Originally, the reaction of rubber hydrocarbons with elemental sulfur at about 150 to 200°C to produce sulfur cross-links between the chains of the rubber polymer to impart elastic properties. This process, formulated by Charles Goodyear (1800–1860), makes the rubber flexible, insoluble, and elastic. Specific properties can be introduced by regulating the amount of sulfur; for example, rubber vulcanized with 1 to 3% sulfur results in the soft rubber used for rubber bands. When finely divided fillers, called reinforcing agents (e.g., carbon black), are added to rubber vulcanized with 3 to 10% sulfur, the result is to give the modified rubber abrasion and tear resistance, making it suitable for use in tire manufacture. In general, any agent that causes cross-linking of the linear rubber polymer, for example, certain peroxides, is a vulcanization agent.

8.70 ELASTOMERIC FIBERS

Fibers that possess elastic properties (Sect. 8.63). One of the most common groups of polymers used for elastomeric fiber production are the polyurethanes (Fig. 8.53*g*), which are prepared from polymeric polyols (e.g., Fig. 8.70), and diisocyanates.

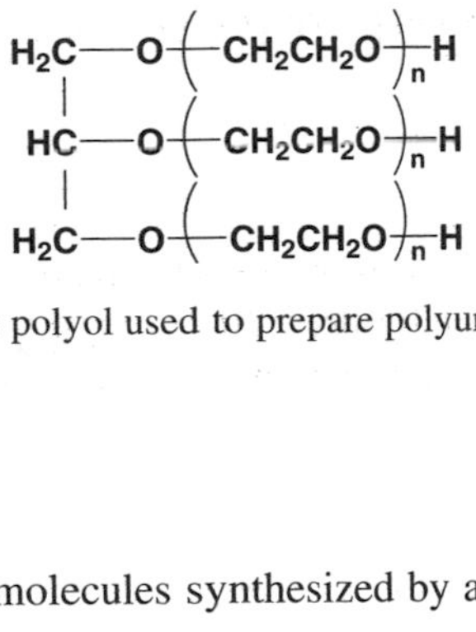

Figure 8.70. Polymeric polyol used to prepare polyurethane elastomeric fibers.

8.71 ANIMAL FIBERS

Fibers made up of protein molecules synthesized by animals or insects; principally wool and silk.

8.72 WOOL

The hair of certain animals, notably sheep. The hair consists of proteins made from about 20 different amino acids. The hair strands are composed principally of interacting protein chains, highly cross-linked through sulfur–sulfur bonds and hydrogen bonding, or saltlike ($RCO_2^- \, .. \, {}^+NH_3R'$) ionic interactions or combinations of these. The sulfur-containing amino acid cysteine is present in significant quantities in wool.

8.73 SILK

Fibrous proteins produced by spiders, butterflies, and moths. The cocoon spun by the larva of the moth *Bombyx mori* provides most of the silk used in fabrics. The threads of *B. mori* silk consist of a highly oriented fibrous protein, called fibroin. This protein has a high content (86%) of the small amino acids glycine, alanine, and serine (see "Rayon," Sect. 8.77).

8.74 LEATHER

Most of the skin, bone, and muscle of animals consists of the complex fibrous protein called collagen. *Collagen* has a high content of the amino acids proline and hydroxyproline and resembles a block copolymer. The skin of animals is converted into leather by the introduction of stable cross-links between the collagen strands in a process called tanning. Formaldehyde is one of the most frequently used tanning agents.

8.75 VEGETABLE FIBERS

Vegetable fibers are obtained from the stems, leaves, or seed hairs of plants, the chief fibrous constituent of which is the polymer cellulose.

8.76 COTTON

A pure form of cellulose obtained from the seed hairs of the plant *Gossypium* spp. Its structure consists of a partially crystalline polymer containing four cellobiose units in the unit cell. Hydrolysis of cotton with hot, dilute, mineral acid ultimately yields D-glucose.

8.77 RAYON

The first known synthetic fiber, discovered accidentally in France in attempts to replace silk. It was originally prepared from nitrocellulose and displayed at the Paris Exposition in 1891. The manufacture of rayon fibers began about 1893 in England and was introduced in the United States in 1910. It is now prepared by solublizing cellulose and then regenerating it by some form of precipitation. Rayon is thus a regenerated cellulose. It was called rayon because its sheen appeared to give off rays of light. *Viscose rayon* is obtained by the reaction of carbon disulfide with the hydroxy groups of cellulose in the presence of alkali to give xanthates. When this solution is spun into an acid medium, the reaction is reversed and the cellulose is coagulated.

Example. Cellulose in the form of cotton linters (the fibers adhering to the cotton seeds) is treated with carbon disulfide (approximately one mole per C_6 unit), and the mixture is dissolved in 3% NaOH to yield a viscous solution called viscose, in

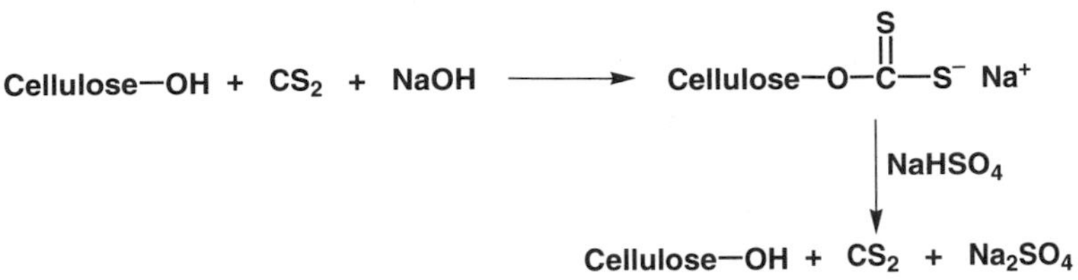

Figure 8.77. Reactions for preparing viscose rayon.

which the average chain length is 400 to 500 C_6 units. After a certain period of aging, the solution is put through a spinneret and the filaments passed through a bath containing $NaHSO_4$ and other additives to give a regenerated cellulose. The reactions involved are shown in Fig. 8.77. Solubilization of the cellulose can also be achieved by treatment with cupric ammonium hydroxide, prepared from copper sulfate and ammonia. Cuprammonium rayon is obtained by spinning the solution into an acid-coagulating bath.

8.78 COMPOSITES

Two or more materials in *macroscopic* combination. This definition is designed to specifically exclude homogeneous materials such as alloys that involve the *microscopic* combination of metals. However, like alloys, composites produce performance properties not displayed by the separate constituents. Most composites consist of fibers embedded in a polymer matrix. For high-temperature performance, ceramics are used as the matrix material. The preparation of high-quality ceramics based on TiO_2 requires high-purity powders, and this is achieved through a process called the *sol-gel method*. In this method, $Ti(OEt)_4$ in an organic solvent is hydrolyzed with water and the resulting $Ti(OH)_4$ forms what is called a *sol*, that is, extremely small particles suspended in a liquid as a colloidial dispersion. Intermolecular loss of water from the hydroxide causes these particles of the sol to link together through Ti–O–Ti bonds with the formation of a rigid, three-dimensional, gelatinlike material called a *gel*. Heating this gel produces TiO_2 in finely divided pure form, suitable for conversion to a high-quality ceramic.

Example. Perhaps the earliest example of the use of composites were the building bricks manufactured by the biblical Egyptians or their Hebrew slaves. These consisted of a clay matrix to which was added straw fibers. Some common fibers presently used in composites are Kevlar, glass, carbon, aramides, boron, and silicon carbide. The fibers are embedded in a polymer matrix of polyesters, polyimides, epoxy, or phenolic resins. The performance of monolithic ceramics in high-temperature applications is enhanced by the use of a second phase or reinforcing material to produce ceramic matrix composites (CMC). Mother of pearl, or nache, is a naturally occurring composite found, for example, in the shells of abalone. It consists of tiny sheets of calcium carbonate held together in a protein matrix and is amazingly resistant to fracture.

8.79 POLY(PHOSPHAZENES), -(N=PX_2--)$_n$

Although the early polymers of this structure were completely inorganic (X = Cl), the halogens are replaceable by organic groups and a host of such polymers containing organic groupings is now known (e.g., X = CH_3, OR, OCH_2CF_3, $NHCH_2$ $CO_2C_2H_5$).

Example. The poly(phosphazenes) can be prepared from the reaction between NH_4Cl and PCl_5 (Fig. 8.79). All the chlorines can be replaced by alkoxyl or aryloxyl groups to give poly[bis(alkoxy or aryloxy)]phosphorazenes. The polymers have unusual properties, especially high flexibility.

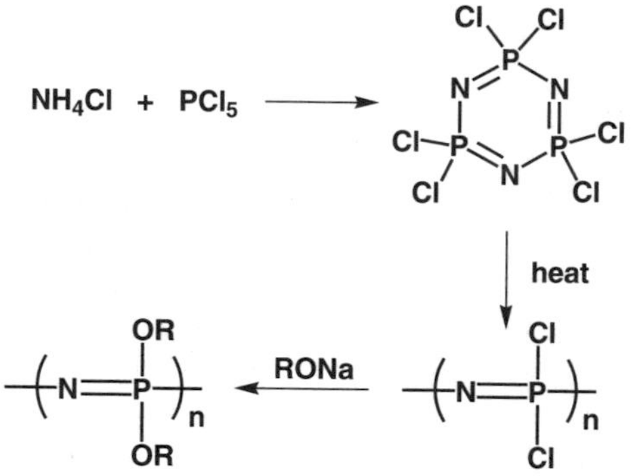

Figure 8.79. The preparation of poly[bis(alkoxy or aryloxy)]phosphazenes.

8.80 BIODEGRADABLE POLYMERS

Synthetic polymers that incorporate various functional groups which are susceptible to attack, either by naturally occurring enzymes or ultraviolet light. The degradation is achieved by hydrolysis of photochemical cleavage at these vulnerable sites.

Example. Biodegradable structures; these were first fabricated from poly(glycolic acid) –(–O-CH_2-C(O)–)$_n$– and subsequently its properties were improved by copolymerization with lactic acid to give poly(lactic acid-co-glycolic acid) (Fig. 8.80).

Biodegradable polymers are also particularly useful in the fabrication of various medical implants in humans where they slowly decompose and disappear, a property that is also important in the manufacture of time-release medications.

$$-[O\text{-}CH(CH_3)\text{-}C(O)]_m[O\text{-}CH_2\text{-}C(O)]_n-$$

Figure 8.80. Poly(lactic acid-co-glycolic acid); a biodegradable polymer.

8.81 RECYCLE OF POLYMERS (PLASTICS)

The collection of used plastic materials and their conversion to either the original polymer form or to alternate plastic materials. The enormous volume of plastic materials used in packaging, wrapping, and various gadgets places an enormous burden on their disposal after use, accounting for more than 20% by volume of all solid waste. Currently, procedures exist for the recycling of about six polymers, the most common ones being polyethylene terphthalate (PET) and high-density polyethylene (HDPE). The former is used extensively for the fabrication of soft drink bottles and the latter for milk containers and water jugs. The used bottles are shredded into small chips, and after removal of extraneous material by air cycloning, the PET and HDPE, which have different densities, are separated by floatation or hydrocycloning and then dried and sold to manufacturers for reuse. The mixed plastics can also be used without separation for the fabrication of construction materials. Plastic containers are now labeled on their bottoms with a triangle consisting of three angled arrows with a code number from 1 to 6 in the center of the triangle and an abbreviation beneath the triangle, which identifies the polymer corresponding to the code.

Example. The label on the bottom of a bottle made of high-density polyethylene (Fig. 8.81).

Figure 8.81. The label on the bottom fabricated from high-density polyethylene.

SUGGESTED READING

Billmeyer, Jr., F. W. *Textbook of Polymer Science*, 2nd ed. Wiley-Interscience: New York, 1971.

Bruice, P. Y. *Organic Chemistry*, 2nd ed. Prentice Hall: Englewood Cliffs, NJ, 1998.

Chemical and Engineering News, issues dated June 3, 1996, and September 22, 1997. Contain extensive discussion of dendrimers.

Fried, J. *Polymer Science and Technology*. Prentice Hall: Englewood Cliffs, NJ, 1995.

Kirk Othmer, *Encyclopedia of Chemical Technology*, 4th ed., Vol. 7, Composite materials; Vol. 19, Polymers Wiley-Interscience, NY (Vol. 7, 1993); (Vol.19, 1996).

Mandelkern, L. *An Introduction to Macromolecules*: Heidelberg Science Library, Springer-Verlag: New York, 1983.

Morawetz, H. *Polymers. The Origins and Growth of a Science.* John Wiley & Sons: New York, 1985.

Morton, J. and Cantwell, W. J. In *Encyclopedia of Chemical Technology*, 4th ed., Vol. 7.

Novak, B. M. In *Organic Polymer Chemistry: A Primer, Supplement to Organic Chemistry*, W. H. Brown, ed. Harcourt Brace: 1995.

Szwarc, M. and van Beylen, M. *Ionic Polymerization and Living Polymers.* Chapman and Hall: New York, 1993.

9 Organometallic Chemistry

9.1	Organometallic Compounds	347
9.2	Main Group Organometallic Compounds	348
9.3	Transition Metal Organometallic Compounds	349
9.4	Ligands	349
9.5	Coordination Compounds	350
9.6	Chelating Ligand	350
9.7	Bridging Ligands	351
9.8	π Complexes	352
9.9	Metal-to-Ligand Back-Bonding	352
9.10	Hapto	353
9.11	Oxidation Number or Oxidation State	354
9.12	Coordination Number (CN)	355
9.13	Effective Atomic Number (EAN)	355
9.14	Rule of 18	355
9.15	d^n Configuration	356
9.16	Coordinative Unsaturation	356
9.17	Oxidative Addition	357
9.18	Reductive Elimination	357
9.19	Intramolecular Insertion Reaction (Migratory Insertion)	357
9.20	Intramolecular β-Hydride Transfer (β-Hydrogen Elimination)	358
9.21	Agostic Hydrogen	360
9.22	α-Hydrogen Abstraction	360
9.23	Ligand Substitution	360
9.24	*trans* Effect	361
9.25	Cone Angle	362
9.26	Catalytic Activation	362
9.27	Metallocenes	363
9.28	Ambidentate Ligands	364
9.29	Dioxygen Complexes	364
9.30	Dinitrogen Complexes	364
9.31	Nomenclature of Metal Complexes	365
9.32	Names of Bridging Groups	366
9.33	Metal Cluster Compounds	366
9.34	Edge-Bridging Ligand	367
9.35	Face-Bridging Ligands	367
9.36	*Closo, Nido,* and *Arachno* Clusters	368
9.37	Metallacycles	368
9.38	Crystal Field Theory (CFT)	369

The Vocabulary and Concepts of Organic Chemistry, Second Edition, by Milton Orchin, Roger S. Macomber, Allan Pinhas, and R. Marshall Wilson

9.39 Crystal Field Splitting (of *d* Orbitals) 370
9.40 Crystal Field Stabilization Energy (CFSE) 371
9.41 Pairing Energy 372
9.42 Low-Spin and High-Spin Complexes 372
9.43 Ligand Field Theory (LFT) 373
9.44 Organometallic Carbene Complexes 373
9.45 Olefin Metathesis (*trans*-Alkylidenation or Dismutation Reaction) 375
9.46 Carbyne (Alkylidyne) Complexes, M≡C–R 375
9.47 Transmethylation 376
9.48 Stereochemically Nonrigid Molecules 377
9.49 Pseudo-Rotation 377
9.50 Isolobal Analogies 378
9.51 Chemical Vapor Deposition (CVD) 380
9.52 Organometal Compounds as Catalysts 380
9.53 The Wacker Reaction 381
9.54 Acetic Acid Synthesis 382
9.55 The Hydroformylation (*oxo*) Reaction 382
9.56 The Heck Reaction 384
9.57 Catalytic Coupling Reaction 384
9.58 Suzuki Reaction 385
9.59 The Stille Reaction 385

The adjective *organometallic* when used in conjunction with the word "chemistry" suggests that this particular kind of chemistry involves compounds consisting of two components, an organic moiety, plus a metal atom, linked by some type of bond. However, this constitutional description leads to some ambiguities. One ambiguity involves the definition of a metal. No precise chemical definition of a metal exists. One suggested definition is that a metal is an element, which yields positively charged ions when its salts are dissolved in water. Perhaps a more acceptable definition is one that classifies elements in terms of the properties that characterize what we designate as metals, namely metals are those elements that possess high thermal conductivity, high electrical conductivity (which decreases with temperature), high ductility (easily stretched and not brittle), and malleability (easily hammered and formed without breaking). This still leaves open the question of whether the elements B, Si, Ge, As, Sb, and Te, the so-called semimetals, should also be included. For the purposes of this book, such compounds will not be considered as organometallic compounds. The second ambiguity involves the question of what constitutes an organic moiety. For example, is trimethylphosphine, PPh_3, an *organic* compound? Because it contains carbon atoms, the answer is yes. However, when triphenylphosphine acts as a *ligand* bonded to a metal, the bonding to the metal does not occur through a carbon atom but rather through the phosphorus atom, as in $Ni(PPh_3)_4$, tetrakis(triphenylphosphine)nickel(0). Since the accepted definition of an organometallic compound requires that the compound in question have a metal atom directly bonded to one or more carbon atoms, $Ni(PPh_3)_4$ is not considered to be an organometallic compound. On the other hand, the related compound, $Ni(CO)_4$, is considered to be an organometal compound. It is thus very difficult to develop a meaningful definition that is simultaneously inclusive and exclusive.

When a metal atom is surrounded by two or more neutral or anionic ligands, each separately bonded to the metal, the entire assembly is called a *coordination compound.* The ligands that surround the central atom almost always provide the pair of bonding electrons and, thus, are coordinated to it. The study of coordination chemistry is usually dated back to the 1798 discovery of the compound corresponding to the composition $CoCl_3 \cdot 6NH_3$. Chemists at the time were puzzled as to how NH_3 and $CoCl_3$, two independent, neutral, and stable compounds, each with an apparent saturated valence structure, could be combined into a new compound that exhibited properties different from those of its constituent moieties. Almost a hundred years later, the compound was shown to have the structure $[Co(NH_3)_6]Cl_3$.

This correct structural formulation resulted from the insights of Alfred Werner (1867–1919, who received the Nobel Prize in 1913), who, at age 26, proposed what is commonly referred to as Werner's coordination theory. The major concepts of this theory were as follows: (1) The central metal atom of such coordination compounds exhibits a *primary* valence or oxidation number; (2) the central metal atom possesses a second or *secondary* type of valence called its coordination number; and (3) the ligands surrounding the central atom are directed to fixed positions (coordinated) in space, that is, coordination complexes have a preferred stereochemical orientation. Werner, thus, correctly deduced that the structure of $[Co(NH_3)_6]Cl_3$ involved Co in the 3^+ oxidation state, that the *coordination number* was 6 (the number of ammonia ligands coordinated to the cobalt to form the cation), that the oxidation number and the coordination number satisfied both the primary and secondary valences of cobalt, and finally, that the ion possessed octahedral geometry. In accordance with the Werner proposal, the total number of ions is four, three of which are chloride ions that precipitate as AgCl when treated with Ag^+. The term *complex ion* was used to distinguish a cation complexed (bonded) to two or more ligands such as $[Co(NH_3)_6]^{3+}$ or $[Ag(NH_3)_2]^+$ from the simple ions such as Co^{3+} or Ag^+ (later work showed that such "simple" ions are probably coordinated to water and are, in fact, aqua complexes).

Gradually, the term "complex" was given a broader meaning. Although the definition of coordination compounds was originally limited to compounds having a central metal atom surrounded by neutral or anionic inorganic ligands such as ammonia, water, and chloride, the name has gradually been extended to include metals coordinated to organic ligands. Some authors make an arbitrary distinction between coordination compounds and organometallic compounds. Thus, any species where the number of metal–carbon bonds is less than half the coordination number is classified by such authors as a coordination compound (the definition used, e.g., to qualify for inclusion in the 25,000 entries in the well-known seven-volume compendium *Comprehensive Coordination Chemistry* edited by G. Wilkinson). However, in what follows no such arbitrary distinction is made. A compound such as $K_3[Fe(CN)_6]$, with its six Fe–C bonds, may be called either a coordination compound, or an organometallic compound, or, even more ambiguously, a complex. Many early Werner compounds were ionic complexes having anions such as CN^- rather than neutral ligands coordinated to a transition metal. Presently, a great deal of organometallic chemistry involves transition metal complexes with neutral ligands.

The *main group* metals form many organometallic compounds and generally the formulas for these are relatively simple despite the fact that their structures are, in many cases, quite complex; examples are methyllithium (CH_3Li), ethylmagnesium bromide (EtMgBr), tetraethyllead ($PbEt_4$), and so on. Transition metals, on the other hand, form many thousands of organometallic compounds of an almost infinite and bewildering variety. Their properties can be amazingly different from their individual constituent parts. Consider, for example, iron pentacarbonyl, $Fe(CO)_5$, one of the metal carbonyls discovered in the 1890s. It consists of an iron atom coordinated to five carbon monoxide molecules. The element iron is a dense, solid metal with a boiling point of about 3000°C, while carbon monoxide is a gas (extremely toxic), with a boiling point of about −190°C, almost 3200°C lower. These two totally unlike substances when combined give iron pentacarbonyl with properties remarkably different from those of either of its component parts; it is a stable, light yellow liquid, which boils without decomposition at about 100°C.

The first organotransition metal complex possessing a coordinated neutral organic ligand (ethylene) capable of independent existence, in 1827, was Zeise's salt, $K[(C_2H_4)PtCl_3] \cdot H_2O$ [named after its discoverer, William C. Zeise, a Danish pharmacist; an account of his work, reported by J. J. Berzelius, appeared in *Ann. Phys. Chem. 9*, 632 (1827)]. This is a water-soluble salt whose monovalent anion is considered to be a four-coordinate complex in which the ethylene is directly bonded to platinum. Because the anionic complex has an over-all charge of −1 and the three chloride ligands each has a charge of −1, the platinum must be in the 2^+ oxidation state. None of the three chlorine atoms can be precipitated with silver ion. It was not until 1951 through 1953, well over 100 years after its discovery, that the nature of the olefin-metal bonding in such complexes was elucidated by Michael J. S. Dewar, using Ag^+-olefin complexes as a model [*Bull. Soc. Chim. Fr. 18*, C79 (1951)], and Joseph Chatt and Leonard A. Duncanson [*J. Chem. Soc.* 2939 (1953)], who focused specifically on the molecular description of Zeize's salt. The latter authors proposed that the bonding involves a *σ bond* generated by donation of the localized π electrons of ethylene into an empty *d* orbital of the 2^+ platinum atom accompanied by the back-donation of a pair of electrons in a filled *d* orbital on platinum into the π* orbital of the olefin to form a π-type bond. The availability of partially filled *d* orbitals (one definition of a transition metal) allows the metal to act as both an acceptor and a donor of electrons and, if the energy of these *d* orbitals is a good match with the energies of the π and π* orbitals of the ligand (as they are in olefin-metal complexes), strong metal-ligand bonding occurs.

The discovery of ferrocene, $Fe(C_5H_5)_2$, first reported in 1951 by T. J. Kealy and P. J. Pauson [*Nature 168*, 1039 (1951)] was a giant leap forward in the development of organometallic chemistry. This remarkably stable compound involves the Fe atom sandwiched between two cyclopentadiene moieties. The metal-ligand bonding in this complex is perhaps best represented by two cyclopentadienyl anions, each donating six delocalized *p*π electrons into three empty acceptor orbitals of appropriate symmetry centered on the Fe^{2+} ion [Myron Rosenblum, Mark C. Whiting, and R. B. Woodward, *J. Am. Chem. Soc. 74*, 6148 (1952)]. In the 2^+ oxidation state, Fe has a d^6 (18 electron core and 6 valence electrons) configuration; on accepting 12 additional electrons, it attains the closed shell of 18 valence electrons (36 total electrons), isoelectronic with Kr, the stable rare gas. The element Cr in the zero oxidation state is also a d^6 (24-electron) species; hence, it is not surprising that Cr(0) is able to complex with two benzene molecules in the same sandwich-type geometry since each benzene molecule acts as a 6 $p\pi$ electron donor to form again the stable 18 valence electron (36 total) configuration surrounding the central Cr(0) atom. Many of these sandwich-type organometallic complexes are known, and they have been given the generic name of metallocenes. The oxidation state of the metal in organotransition metal complexes is assigned by means of formalisms (see Sect. 9.11) and may not reflect the actual positive or negative charge on the metal.

Since the 1950s, organotransition metal chemistry has grown at an unprecedented pace. A bewildering array of organometallic compounds have been prepared and their sometimes astonishing structures have been confirmed, usually by X-ray diffraction. Thus, a mononuclear tungsten complex is known in which one ligand is bonded to the metal through a triple carbon–metal bond (carbyne); a second through a carbon–metal double bond (carbene); and a third through a carbon–metal single bond. Organometallic complexes containing multiple metal–metal single, double, triple, and even quadruple bonds are known. Multimetal complexes with metal–metal bonding or with various groups such as carbon monoxide, halogens, or hydrogen bridging two or more metals are well established. But organometallic compounds are no longer simple laboratory curiosities. They are often important intermediates in synthetic procedures both as catalysts and as stoichiometric reagents, and in the last few years the metallocenes have become important as catalysts for olefin polymerization. The development of this vast new area of chemistry also has led to some new concepts and new vocabulary to describe the structure, properties, and reactions of these fascinating complexes.

9.1 ORGANOMETALLIC COMPOUNDS

Compounds consisting of one or more carbon atoms directly bonded to a metal atom. Organometallic compounds may be divided into two broad classes based on the nature of the metal bonded to the carbon atom: main group organometallic compounds and transition metal organometallic compounds. The latter compounds often are classified further according to either the nature of the ligands bonded to the metal or the nature of the bonding between metal and ligand (σ, π, or delocalized π).

Example. Common main group organometallics are Grignard reagents and lithium alkyls (Sect. 9.2). The compound called Prussian blue, $Fe_4[Fe(CN)_6]_3$, ferric ferrocyanide, because it contains a Fe–C bond, may be considered an organometallic compound, even though most chemists would regard it as an inorganic complex.

Compounds such as aluminum triisopropoxide, $Al[OCH(CH_3)_2]_3$; tris(ethylenediamine)cobalt(III) chloride, $[Co(en)_3]Cl_3$; and tetrakis (triphenyl phosphine) platinum(0), $[Pt(PPh_3)_4]$, are not considered organometallic compounds because in each of these complexes, even though they possess carbon atoms, the metal is not directly attached to a carbon atom. The transition metal carbonyls such as $Co_2(CO)_8$ and $Fe_2(CO)_9$ are organometallic compounds, as are the metallocenes such as ferrocene: $Fe(C_5H_5)_2$.

9.2 MAIN GROUP ORGANOMETALLIC COMPOUNDS

Compounds containing a bond between carbon and a main group element.

Example. Organoalkali metal compounds such as methyllithium, CH_3Li; the alkaline earth metal compounds such as diethylmagnesium, $(C_2H_5)_2Mg$; compounds of Sn and Pb, the heavier of the Group 14 elements (Group IVA in what is called the American ABA nomenclature, where the A elements in a group are the main group elements and the B elements in the same group are the transition metal elements), such as tetrabutyltin, $(C_4H_9)_4Sn$, and tetraethyllead, $(C_2H_5)_4Pb$, are all main group organometallic compounds (see Sect. 1.44). The organometallic compounds of magnesium are particularly useful in many syntheses and merit special discussion. These compounds, called *Grignard reagents* [named after Victor Grignard (1871–1935), who received the Nobel Prize in 1912], are formulated as RMgX, but they are best represented by the so-called Schlenk equilibrium named after W. J. Scehlenk (1879–1943):

$$RMgX \rightleftharpoons R_2Mg + MgX_2$$

The intermediate in the interconversion has not been established, but the most reasonable structure is one involving a bridging R group and a bridging halide:

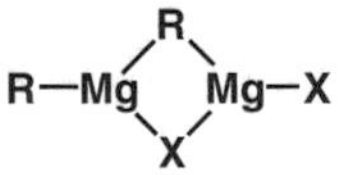

The starting materials for their preparation are magnesium metal turnings; the appropriate halide, RX; and a coordinating solvent, most commonly diethyl ether. The first step in the reaction involves a single electron transfer (SET) from surface magnesium, $Mg_{(s)}$, to an adjacent RX, to form a radical anion–radical cation pair, which rearranges before going into solution as RMgX:

$$RX + Mg_{(s)} \xrightarrow{e^{\bullet -}} R\text{-}\overset{\bullet -}{X}\cdots\cdots\overset{+\bullet}{Mg}_{(s)} \longrightarrow R\cdots\cdots MgX_{(s)} \longrightarrow RMgX_{(soln)}$$

There is substantial evidence that before RMgX goes into solution, the R group diffuses into the solvent as the radical R•. The Schlenk equilibrium occurs after the formation of RMgX.

9.3 TRANSITION METAL ORGANOMETALLIC COMPOUNDS

Compounds possessing a chemical bond directly linking carbon and a transition metal. (The terms transition metal and transition element are used synonymously.) The transition elements, as the word "transition" suggests, are those elements that appear in the middle of the Periodic Table, breaching the transition from the alkali and alkaline earth metals on the left side, to the nonmetals on the right side of this table. Many organometal chemists prefer to define transition metals as those elements (either as elements in their zero oxidation state or in their common oxidation states) that have partially filled $(n-1)d$ or $(n-2)f$ orbitals, where n is the principal quantum number, an integer with the values 1, 2, 3, and so on, corresponding to the number of the shell or row in the Periodic Table. The transition metals are also called the B group elements.

Example. The transition metals begin with the 4th row element, Sc, atomic number 21 with a $4s^23d^1$ valence electron configuration. The 4th row transition elements are usually considered to end with Cu. However, the free element Cu(0) has a outer shell electronic configuration of $4s^13d^{10}$, that is, the $(n-1)d$ orbitals are completely filled, and thus on the basis of the above definition, Cu(0) (as well as the two other so-called *coinage metals*, Ag and Au, below it in the Periodic Table) would not qualify as transition metals. Nevertheless, the Cu^{2+} ion (with two fewer electrons) has a d^9 configuration and thus qualifies, whereas Cu(I) has a d^{10} configuration and therefore would not qualify. (See "d^n Configuration," Sect. 9.15.) This example demonstrates the ambiguity of the definition and indeed almost any rational definition will lead to ambiguities. The 4th and 5th row transition metals are called outer transition metals and the transition metals in the 6th and 7th row are called inner transition metals.

9.4 LIGANDS

The various ions, atoms, or neutral molecules that surround and are directly bonded to a central atom, usually a metal atom. Ligands (from the Latin word *ligare*, meaning "to bond") form bonds to the central metal atom by functioning as electron donors, either by virtue of lone pair electrons on one of their atoms (carbon in :C=O, N, O, P, S, X) or by virtue of π electrons, either localized (olefin) or delocalized (conjugated polyenes, cyclopentadienyl, aromatics). In complexes involving transition metals, ligands may serve both as electron donors and acceptors.

Example. The neutral ligands H_2O and NH_3, when bonded to metal ions in inorganic complexes such as $[Fe(H_2O)_6]^{3+}$ and $[Ag(NH_3)_2]^+$, are ubiquitous in inorganic chemistry. Anions such as SCN^- may also be ligands as in the mixed complex, $[Fe(H_2O)_5(SCN)]^{2+}$. One of the common neutral ligand in organometal complexes is carbon monoxide, as in iron pentacarbonyl $Fe(CO)_5$. Complexes in which all the ligands bonded to the metal are identical are called *homoleptic complexes*. The term ligand was introduced by Alfred Stock (1876–1946) in 1916. However, it did not come into common usage in the English language until the early 1950s.

9.5 COORDINATION COMPOUNDS

The generic name for compounds consisting of a central metal atom coordinated, that is, bonded, to a specific number of ligands present as ions, atoms, or neutral molecules. The metal and associated ligands act as a unit and tend to retain their identity in solution, although, depending on the solvent, partial dissociation may occur. In coordination compounds, ligands are said to complex, that is, bind to the metal, forming complexes. The two terms, coordination compounds and complexes, presently are used interchangeably. The central metal and the ligands bonded to it constitute the *coordination sphere*.

9.6 CHELATING LIGAND

From the Greek word *chela*, meaning "claw," a ligand possessing two or more bonding sites (usually atoms with lone pair electrons) capable of bonding to a common metal atom. A ligand with two bonding sites is called a *bidentate* (common usage) or *didentate* (IUPAC) ligand; with three such bonding sites, a *tridentate*; with four,

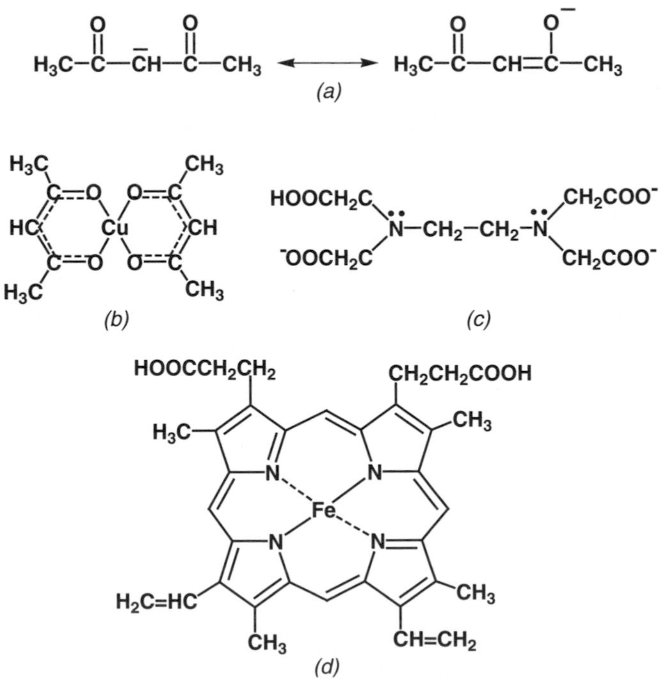

Figure 9.6. (*a*) Acetylacetonato anion (acac); (*b*) copper acetylacetonate; (*c*) ethylenediaminetetraacetic acid (EDTA) trianion, a pentadentate ligand; (*d*) heme.

a *tetradentate* ligand; and so on (*dent*, from the Latin for "tooth"). In the development of transition metal chemistry the *monodentate* ligand triphenylphosphine, PPh_3, has played a particularly prominent role. The angle formed between the ligand–metal–ligand bonds of a polydentate ligand is called the *bite angle*. Many chelating ligands are given (approximate) acronymic abbreviations *in lieu of* writing out their complete line structure. Thus, the ligand $Ph_2PCH_2CH_2PPh_2$, 1,2-bis(diphenylphosphino)ethane, is abbreviated dppe as in the six-coordinate octahedral complex $[Mn(CO)_3(dppe)F]$. When there are two (or more) different bonding sites on a ligand, the "kappa" convention is used. The ligating atoms are designated by the lowercase Greek letter κ (kappa) preceeding the italicized atom. Thus, the ligand $(CH_3)_2NCH_2CH_2PPh_2$, abbreviated (pn), when chelated to for example, a Mn atom, is written as Mn(pn-κ^2*P,N*). The superscript indicates that the two italicized atoms are both bonded to the Mn. If the complex involved bonds solely to the phosphorus end of the ligand, it is written as Mn(pn-κ^1*P*).

Example. Acetylacetonato anion (acac) (Fig. 9.6*a*) functions as a didentate ligand (Fig. 9.6*b*); ethylenediaminetetraacetic acid (EDTA) could function as a hexadentate ligand but usually behaves as a pentadentate ligand with a free carboxyl group (Fig. 9.6*c*). The porphine nucleus consisting of four pyrrole units linked to form a planar 16-membered outer ring is a tetradendate ligand; it is present in the important heme proteins, hemoglobin, myoglobin, and cytochromes. A substituted porphine, the iron porphyrin complex called heme, is shown in Fig. 9.6*d*. Polyethers in the form of polydentate crown ethers are very useful in the separation of main group metal ions.

9.7 BRIDGING LIGANDS

Ligands, consisting of a single atom, or a group of atoms, that are bonded simultaneously to at least two metal atoms. Bridging ligands are denoted by the lowercase Greek letter, mu (μ), preceding the name of the bridging group.

Example. $Fe_2(CO)_9$ (Fig. 9.7*a*) has three bridging carbonyl groups. The structure consists of two octahedra with a common face. The three bridging or μ-COs, each bonded to the two Fe atoms, share the corners of the face. Such bridging group ligands are referred to as *face-sharing ligands*. The presence of bridging carbonyl groups can be detected from the infrared spectrum associated with the carbonyl stretching frequencies, which occur in the 1,750 to 1,850 cm^{-1} range compared to the 1,850 to 2,140 cm^{-1} range for terminal carbonyls. Zeise's dimer (Fig. 9.7*b*) has two bridging chlorine atoms. The structure consists of two square planes with a common edge; the two bridging ligands are called *edge-sharing ligands*. A single bridging atom leads to a shared corner; in the case of $[(RuCl_5)_2O]^{4-}$ (Fig. 9.7*c*), the two octahedra are joined at a corner by the oxygen atom. Molecular nitrogen can act as a bridging ligand and alkyl groups can act similarly.

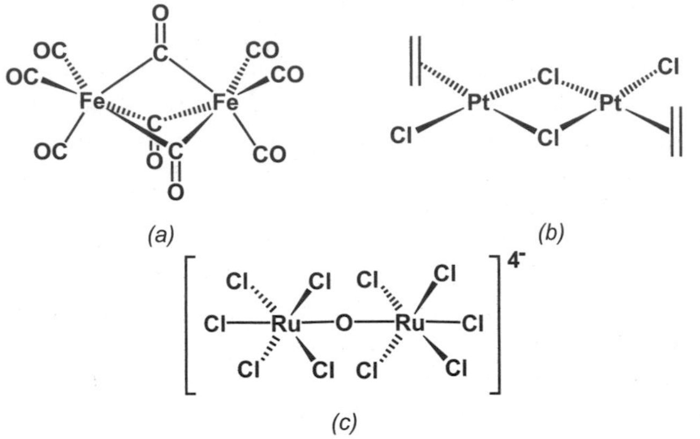

Figure 9.7. (*a*) tri-μ-carbonylhexacarbonyldiiron or diironenneacarbonyl, $Fe_2(CO)_9$; (*b*) di-μ-dichloro-*trans*-dichloro-*trans*-diethylenediplatinium(0) or Zeise's dimer, $[PtCl_2(C_2H_4)]_2$; (*c*) μ-oxodecachlorodiruthenate(4^-), $[(RuCl_5)_2O]^{4-}$.

9.8 π COMPLEXES

Organometallic complexes having at least one organic ligand containing a π electron system that overlaps an orbital of the central metal atom.

Example. The prototype is Zeise's salt, $K[(C_2H_4)PtCl_3 \bullet H_2O]$, devised by William C. Zeise (1789–1847), the first π complex to be prepared, in 1828. The bonding of the unsaturated organic moiety in such complexes is assumed to consist of two component parts: a σ bond formed by overlap of a filled π orbital on the olefin with an empty metal orbital (Fig. 9.8*a*) and a second component part in which an appropriate filled *d* orbital of the metal overlaps an antibonding orbital of the ligand (Fig. 9.8*b*). The two interactions are sometimes referred to as the σ donor–π acceptor concept.

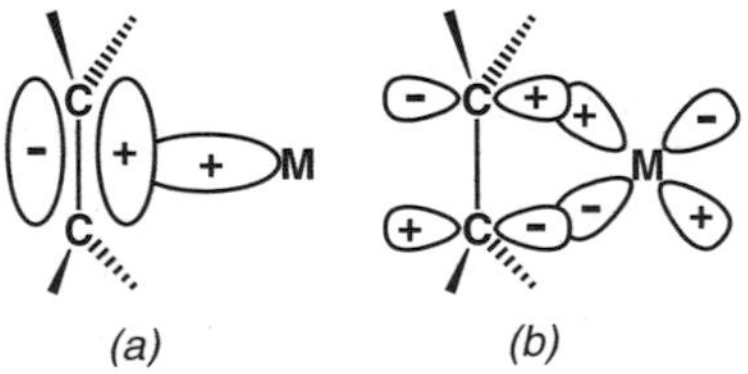

Figure 9.8. (*a*) σ Bonding in a π complex and (*b*) *d*-π* (back-bonding) bonding in a π complex.

9.9 METAL-TO-LIGAND BACK-BONDING

The contribution to the overall ligand-metal bonding in π complexes that involves the overlap of a filled atomic orbital on the metal with a vacant π* antibonding orbital on the ligand.

Example. The component of the bonding in Zeise's salt, shown in Fig. 9.8*b*. Because electrons occupy the antibonding orbital of the ethylene, the carbon–carbon π bonding in the ethylene is weakened and attack on the ethylene is facilitated, that is, the ethylene is *activated* by its coordination to platinum.

9.10 HAPTO

A generic prefix, from the Greek word *hapto*, meaning "to fasten," affixed to the name of a ligand, indicating that the ligand is capable of bonding to the central metal atom by more than one atom of the ligand. The prefix *hapto-* is followed by a superscripted number that specifies the number of atoms of the ligand, which are bonded to the central metal atom. If two adjacent atoms are bonded simultaneously, as, for example, the ethylene in Zeise's salt (Fig. 9.8), the prefix is *dihapto-*; if an allyl group is σ-bonded, it is *monohapto-*, but if it is π-bonded through its delocalized three-carbon system, the prefix is *trihapto-*, and so on. In writing a hapto structure in abbreviated line form, the ligand name is prefixed with the lowercase Greek letter eta η (italicized). An Arabic number is placed as a superscript on η to indicate the number of carbon atoms in the π system of the ligand that are bonded to the metal.

Example. The hapto nomenclature was first suggested by F. Albert Cotton (1930–) in describing (1,2,3,4-tetrahaptocyclooctatetraene)tricarbonyl iron. Thus, Cotton's tetrahapto compound may be written as η^4-cyclooctatetraenetricarbonyl iron, Fig. 9.10*a*. A few authors use the English lowercase letter *h* for the prefix *hapto-*. Other examples are: ferrocene, di-(η^5-cyclopentadienyl)iron(II) (Fig. 9.10*b*) and di-(η^5-cyclopentadienyl)-di-(η^1-cyclopentadienyl)titanium(IV), $Ti(\eta^5\text{-}C_5H_5)_2(\eta^1\text{-}C_5H_5)_2$ (Fig. 9.10*c*).

In order to explain certain associative substitution reactions of stable 18-electron (Sect. 9.14) η^5 complexes, it was suggested that the oncoming ligand is able to coordinate to the metal by forming an intermediate in which the η^5 ligand undergoes a *ligand slippage* or "haptotropic" conversion to form an η^3 ligand. The η^3 intermediate proceeds to eject one of the ligands and regenerate the η^5 system as shown for the indenyl ligand (Fig. 9.10*d*).

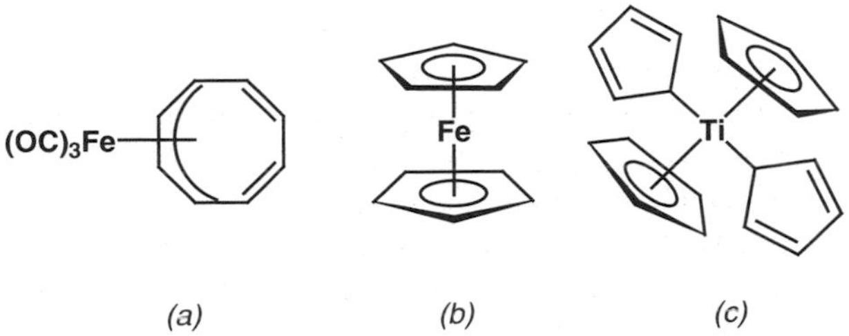

Figure 9.10. (*a*) η^4-cyclooctatetraene tricarbonyl iron; (*b*) ferrocene $Fe(\eta^5\text{-}C_5H_5)_2$; (*c*) $Ti(\eta^5\text{-}C_5H_5)_2(\eta^1\text{-}C_5H_5)_2$.

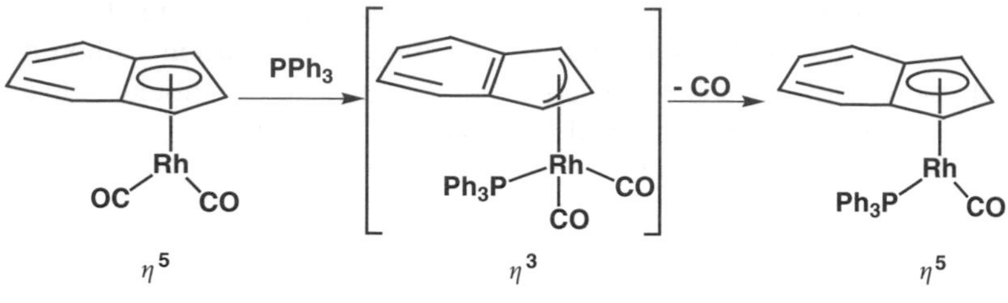

Figure 9.10*d*. An associative ligand–substitution reaction with an intermediate involving ligand slippage.

9.11 OXIDATION NUMBER OR OXIDATION STATE

A number equal to the charge remaining on the central metal atom of a complex after all the ligands have been (theoretically) removed in their closed-shell electronic configurations. The determination of the oxidation number by this agreed upon formalism facilitates consistency in communication between organometal chemists.

Example. In dealing with complexes possessing ligands that have a separate existence as a neutral molecule, or with ligands that form well-characterized anions, there is generally very little trouble or ambiguity in assigning an oxidation number to the central metal. Thus in $Cr(\eta^6\text{-}C_6H_6)_2$, chromium is zero. In $K[PtCl_3(C_2H_4)]$, ethylene is removed with its π electrons and the three chlorines are removed as anions. Inasmuch as the complex ion $[PtCl_3(C_2H_4)]^-$ has a 1^- charge (evident from the presence of the counterion K^+), the net oxidation number of Pt is 2^+. In $Fe_4[Fe(CN)_6]_3$, each of the six CN groups is -1, and since each $Fe(CN)_6$ anion has a net charge of 4^-, each Fe in the complex anion must have an oxidation number of $6 - 4 = +2$. Each Fe cation is 3^+ (to give a total of 12 net plus charges to balance the 12 net negative charges contributed by the three anions), hence the accepted name of ferric hexacyanoferrate for the complex.

In dealing with the organic ligands that involve carbon σ-bonded to the metal, the organic moiety R is removed from the metal with the pair of electrons on that carbon atom, that is, as R^-. Thus, Pt in $Pt(CH_3)_4$ is 4^+. According to this formalism, a hydrogen atom attached to metals is arbitrarily removed with two electrons as hydride, H^-, regardless of whether or not such complexes possess hydridic character. For example, $HCo(CO)_4$ is a very strong acid in aqueous solution, but Co is still considered to be 1^+ in the undissociated complex. An even more arbitrary situation arises with the organic ligands that are bonded to the metal by a delocalized π system, such as in the cyclopentadienyl complexes. The closed-shell configuration of such ligands involves the cyclopentadienyl as an anion, with 6 *p*π electrons (aromatic); as a consequence, Fe is regarded as 2^+ in $Fe(\eta^5\text{-}C_5H_5)_2$ and the overall complex is neutral. The cycloheptatrienyl cation is also a 6 *p*π-electron donor, but it carries an actual charge of $+1$, and if it and other neutral ligands are bonded to a transition metal, the resulting complex has a 1^+ charge, as in $[(\eta^7\text{-}C_7H_7)Mn(CO)_3]^{2+}$ with Mn in the +1 oxidation state.

9.12 COORDINATION NUMBER (CN)

The number of pairs of electrons contributed by ligand atoms in bonding to the central metal atom of a coordination compound.

Example. The coordination number of Fe in $Fe(CO)_5$ is 5; in $[Fe(CN)_6]^{4-}$, it is 6; and in bis(acetylacetonato)copper(II) (Fig. 9.6*b*), it is 4. Complexed alkenes contribute 1, alkynes 2, to the coordination number of the metal to which they are attached, while π-bonded cyclopentadienyl ligands contribute 3 to the coordination number of the central metal. Thus in ferrocene (Fig. 9.10*b*), each C_5H_5 ligand is an η^5 ligand but contributes 3 to the total coordination number of 6.

9.13 EFFECTIVE ATOMIC NUMBER (EAN)

In a coordination complex, the total number of electrons surrounding the central metal atom. It is equal to the atomic number of the metal minus its oxidation number plus the formal number of bonding electrons donated to the metal by the ligands.

Examples.

Metal Complex	Metal Atomic No.	Metal Oxidation No.	Electrons from Ligand	EAN
$Cr(\eta^6\text{-}C_6H_6)_2$	24	0	12	36
$[PtCl_3(C_2H_4)]^-$	78	+2	8	84
$[Fe(CN)_6]^{4-}$	26	+2	12	36
$Co(H)(CO)_4$	27	+1	10	36
$Fe(\eta^5\text{-}C_5H_5)_2$	26	+2	12	36
$[\eta^5\text{-}C_7H_7)Mn(CO)_3]^{2+}$	25	+1	12	36

9.14 RULE OF 18

The rule that a transition metal in a complex tends to acquire the 18 electrons required to completely fill its *valence dsp* orbitals. Chemical behavior involves the outer metal *dsp* orbitals and these are completely filled when 18 valence electrons are present. Transition metal complexes with EANs of 18, 36, 54, and 86 are isoelectronic with the rare gases Ar, Kr, Xe, and Rn, respectively, and often exhibit unusual kinetic and thermodynamic stability.

9.15 d^n CONFIGURATION

The number of electrons in the valence orbitals of the metal in a transition metal complex, based on the assignment of the metal's formal oxidation number. All the electrons beyond the closed-shell configurations are assumed to be *d* orbital electrons even though they may be in *s* or *p* orbitals. The electron configuration is designated as d^n, where *n* can be any integer from zero to 10.

Example. The first transition metal, Sc(0), whose actual configuration is $4s^1 3d^2$, is said by the d^n convention to have a d^3 configuration. Cu(II) is d^9 and Pt(0) is d^{10}. Because Pt in $K[PtCl_3(C_2H_4)]$ is Pt^{2+}, it has a d^8 configuration. Other examples:

Complex ML_x	Oxidation State of M	d^n	$(d^n)^a$	CN^b
$IrCl(CO)(PPh_3)_2$	+1	d^8	(d^9)	4
$CoH_3(PPh_3)_3$	+3	d^6	(d^9)	6
$Ni(CO)_4$	0	d^{10}	(d^{10})	4
$[Co(CN)_5]^{3-}$	+2	d^7	(d^9)	5

[a]Electron configuration of zero valent metal.
[b]Coordination number.

9.16 COORDINATIVE UNSATURATION

When the central metal atom in a transition metal complex has an effective atomic number (Sect. 9.13) short of the number of electrons characteristic of a rare gas structure, it is considered to be coordinatively unsaturated.

Example. The number of electrons surrounding the central metal in a complex is readily ascertained by multiplying the coordination number by 2 (the number of electrons contributed by each ligand) and adding the product to the number *n* of the d^n configuration; if the resulting number is smaller than 18 (the rule of 18), the metal is coordinatively unsaturated. The number of valence electrons surrounding Co in the complex $CoH_3(PPh_3)_3$ (where Co has a formal oxidation number of 3^+ and is therefore d^6) is equal to $(6 \times 2) + 6 = 18$; hence, Co is coordinatively saturated. However, the number of valence electrons around Co in the complex anion $[Co(CN)_5]^{3-}$ (where Co has a formal oxidation number of 2^+ and is therefore d^7) is equal to $(5 \times 2) + 7 = 17$, and because Co in this complex is short one electron, it is coordinatively unsaturated. Therefore, it is not surprising that it reacts with many substrates by homolytic cleavage, leading to coordinative saturation at Co and to radical intermediates from the remaining substrate fragment. The concept of coordinative unsaturation is useful in predicting (albeit sometimes poorly) when a complex might function as a catalyst. Catalytic activity (in solution) requires that, in order to react, a substrate must be coordinated to the metal site; this is usually only possible if the metal is initially coordinatively unsaturated with or becomes so under conditions of the reaction.

9.17 OXIDATIVE ADDITION

The addition of both atoms associated with a single bond in a reactant molecule to the central metal atom of a complex accompanied by the simultaneous increase in the metal's oxidation state. In the cleavage-addition process, new bonds are formed between the metal and the entering fragments. Such bond formation requires *formal* electron donation from the metal to the fragments, resulting in an increase in both the coordination number of the metal as well as its oxidation number.

Example. Oxidative addition of an alkyl halide RX to a square planar Rh(I) d^8 complex leads to an octahedral Rh(III) d^6 complex. Thus,

$$Rh^{I}Cl(CO)(PPh_3)_2 + CH_3Br \longrightarrow Rh^{III}Cl(Br)(CH_3)(CO)(PPh_3)_2$$

An important homogeneous hydrogenation cobalt catalyst is prepared by dihydrogen cleavage and oxidative addition to a d^7 five-coordinate trigonal bipyramidal complex to give two moles of an octahedral d^6 complex:

$$2\,[Co^{II}(CN)_5]^{3-} + H_2 \longrightarrow 2\,[Co^{III}(CN)_5H]^{3-}$$

9.18 REDUCTIVE ELIMINATION

This is the reverse of oxidative addition; in the usual case, it involves the loss of two σ-bonded ligands A and B from the central metal atom M of a complex A–M–B to form the molecule A–B. Because two (two-electron) bonds in the complex are broken and only one (two-electron) bond is formed in the process, the reaction constitutes a formal two-electron reduction of the metal, as well as a decrease of 2 in its coordination number.

Example.

$$Rh^{III}(C_2H_5)(Cl)(H)(PPh_3)_3 \longrightarrow Rh^{I}(Cl)(PPh_3)_3 + C_2H_6$$

9.19 INTRAMOLECULAR INSERTION REACTION (MIGRATORY INSERTION)

The rearrangement of a complex consisting of the intramolecular migration of one ligand onto another ligand. An oversimplified version of the rearrangement can be represented by

$$\text{A-M-L} \longrightarrow \text{L-A-M}$$

in which M is the transition metal atom, A is an unsaturated ligand (carbon monoxide, olefin, isocyanide), and L is a σ-bonded ligand (alkyl or hydride). It appears that A has been inserted between M and L, but actually L has undergone an intramolecular migration onto A and the M–A bond remains essentially intact during the rearrangement. The site vacated on M by the migrating ligand L can be filled simultaneously with, or subsequently to, the insertion by an incoming ligand.

Example. Olefin insertion into an M–H bond (Figs. 9.19*a* and *b*). In the carbon monoxide insertion in Fig. 9.19*c*, it has been demonstrated unequivocally that the reaction involves a migration of CH_3 to the position *trans* to the triphenylphosphine (with retention of configuration if the methyl group is replaced by a chiral group) onto an intact coordinated CO to give an acetyl group. Although the CH_3 was *cis* to Ph_3P in the original, the acetyl group is *trans* to the Ph_3P in the product. Hence, although this reaction is called a migratory insertion because of the apparent insertion of CO between Mn and CH_3, it actually involves the migration of CH_3.

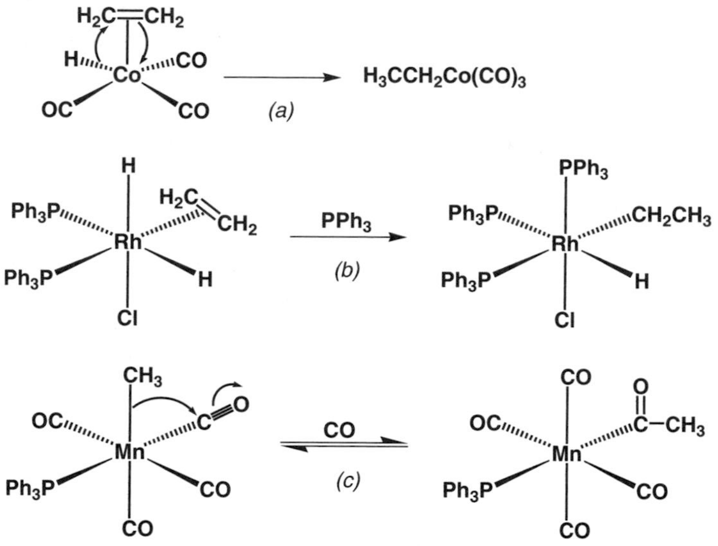

Figure 9.19. (*a, b*) Olefin insertion; (*c*) carbon monoxide insertion.

9.20 INTRAMOLECULAR β-HYDRIDE TRANSFER (β-HYDROGEN ELIMINATION)

The transfer of a hydrogen atom (formally with its pair of electrons) from the β-position on a ligand to the transition metal bonded to that ligand (Fig. 9.20*a*).

The metal hydride complex may be stable or it may dissociate into its component parts depending on a variety of factors; in either case, the net result is a intramolecular hydrogen atom transfer from the β position of a ligand to the metal. The hydride transfer requires that there be an empty coordination site at the *cis* position on the transition metal in order for the intermediate olefin complex to form. If such a site is not present, it may become available by prior dissociation of a labile ligand at this

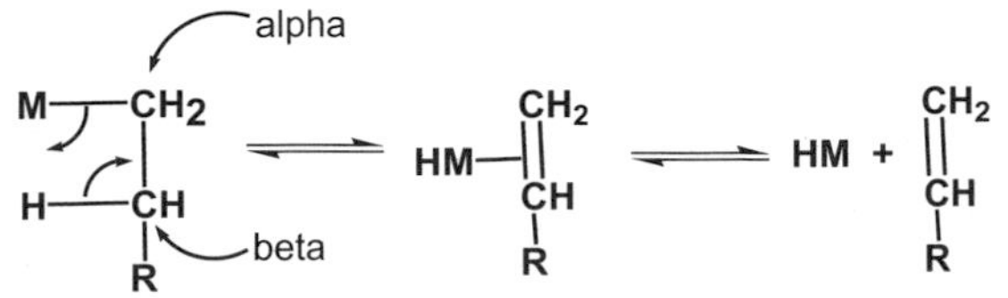

Figure 9.20*a*. β-Hydrogen elimination from a transition metal-alkyl ligand.

position. The elimination is frequently spontaneous. β Hydride transfer is the reverse of olefin insertion into a metal–hydrogen bond. Actually, the hydrogen to be transferred as hydride from the β carbon to the transition metal can be from the β position of any group σ-bonded to the metal, provided the necessary four-centered transition state can be realized.

Example. The reverse of the formation of the alkyl metal bonds shown in Figs. 9.19*a* and *b*. The stereochemistry for both the olefin insertion and the β-hydrogen transfer (β-hydride elimination) involves a four-center transition state (Fig. 9.20*b*). This transition-state requirement explains why the catalytic hydrogenation of olefins is always *cis*. It also explains the kinetic instability of transition metal alkyl and alkoxyl compounds; the latter are prone to decompose to aldehydes, for example,

$$L_xM\text{-}OCH_3 \longrightarrow L_xM\text{-}H + CH_2O$$

Although metallacyclic compounds (Sect. 9.37) may be regarded as transition metal dialkyl compounds, they are relatively stable compared to their acyclic analogs because of steric constraints to the formation of the four-centered transition state with its characteristic small angles (about 90°):

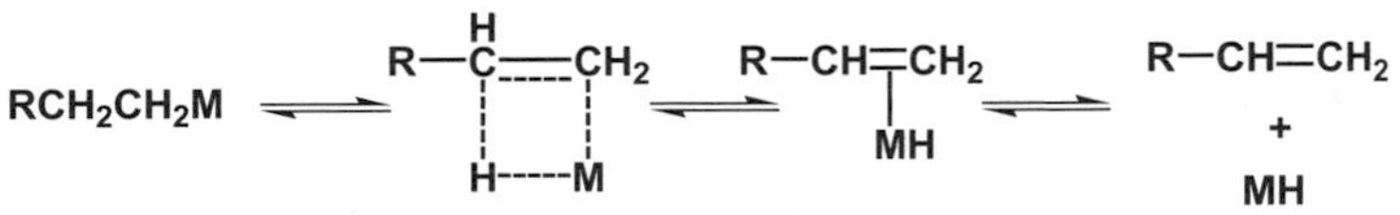

Figure 9.20*b*. The four-center transition state for β-hydrogen elimination (as well as the reverse metal hydride addition to an olefin); its geometry is responsible for stereospecific *cis*-elimination (and *cis*-addition) of HM to a complexed olefin.

The *hydrozirconation* reaction involves the insertion of an internal olefin into the zirconium-hydrogen bond of Cp_2ZrHCl (the *Schwartz reagent*), to give a straight-chain compound (Fig. 9.20*c*) via a series of olefin insertions and β-hydride eliminations. 1-, 2-, and 3-hexene all give the same straight-chain product:

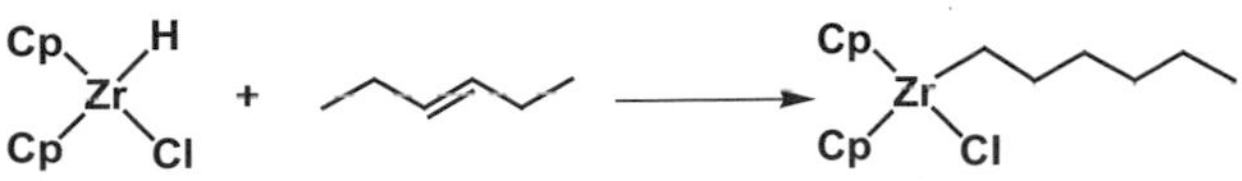

Figure 9.20*c*. Hydrozirconation reaction.

9.21 AGOSTIC HYDROGEN

Derived from the Greek word *agostic*, meaning "to hold onto oneself," the interaction (partial transfer) between the hydrogen atom on a carbon attached to a metal (α-hydrogen) and the metal to form a three-center, two-electron bond. This phenomenon has been observed with transition metal carbenes [See M. Brookhart, M. L. H. Green, *J. Organomet. Chem. 250*, 395 (1983)]. The C–H stretching frequency of such complexes is unusually low ($\nu = 2{,}600\,cm^{-1}$ or less), the C–H bond is unusually long (greater than 1.10 Å) and the coupling constant, ^{13}C-^{1}H, unusually low (100 Hz or less).

Example. The H atom on the carbene carbon in the tantalum complex, $Cl_3[(CH_3)_3P]Ta{=}C\underline{H}C(CH_3)_3$.

9.22 α-HYDROGEN ABSTRACTION

The removal of hydrogen atom from the α position on a ligand attached to a transition metal to form an alkylidene carbene, (Sect. 9.44) complex. This is generally an intramolecular process, requiring a strong base.

Example. The preparation of the alkylidene Ta neopentyl complex, the first reported transition metal alkylidene (carbene) compound [R = neopentyl = $CH_2C(CH_3)_3$]:

$$R_3TaCl_2 + 2\,LiR \longrightarrow R_3Ta{=}CHC(CH_3)_3 + 2\,LiCl + C(CH_3)_4$$

The expected product R_5Ta is not formed probably because of the large steric requirements of the neopentyl groups.

9.23 LIGAND SUBSTITUTION

The substitution of a ligand L coordinated to a metal center M by another ligand L′:

$$ML + L' \longrightarrow ML' + L$$

There are two limiting mechanisms by which such substitution is thought to occur: *dissociative ligand substitution* and *associative ligand substitution.*

Example. Mechanistically, the dissociative and associative ligand substitution reactions in organometallic chemistry resemble the S_N1 and S_N2 substitution reactions of conventional organic chemistry (see Sect. 14.15 and 14.19). In both cases, the pure (or limiting) S_N1-type mechanism involves dissociative bond-breaking prior to bond-making with the formation of an intermediate of *decreased* coordination number. In contrast, the pure (or limiting) S_N2-type mechanism, in both organometallic chemistry and conventional organic chemistry, involves associative bond-making to give a transition state (or intermediate) with an *increased* coordination number. The organometallic ligand substitution process is somewhat more complicated because of transition metal *d* orbital participation, as compared with the carbon-based system where only *s* and *p* orbitals are involved. Hence, an expanded vocabulary is used by some authors to describe the kinetic behavior of ligand substitution reactions in organometallic chemistry. The term *intimate mechanism* is used to address the issue of whether the main factor controlling the activation energy is bond-making (S_N2) associative activation or bond-breaking (S_N1) dissociative activation. The term *stoichiometric mechanism* addresses the sequence of elementary steps, especially the molecularity involved in each step, in going from reactants to products. Some of the most extensively investigated associative substitution reactions are those of the coordinatively unsaturated 16-electron square-planar complexes of Pt(II) and Pd(II), for example, the replacement of a halide *trans* to the ethylene of Zeise's salt by an amine:

$$[(C_2H_4)PtCl_3]^- + RNH_2 \longrightarrow \textit{trans}\text{-}[(C_2H_4)PtCl_2(NH_2R)] + Cl^-$$

Such reactions proceed via a five-coordinate 18-electron intermediate and the rate depends on the nature of the nucleophile and the nature of the leaving group. Associative substitution can also occur with five- and six-coordinate 18-electron complexes. Such substitution is usually slower than with 16-electron complexes. In complexes with η^3 or η^5 ligands, the high-energy 20-electron transition state required for the S_N2-type substitution is avoided by so-called *bond slippage*, in which bonding to lower hapticity occurs, for example, from five to three in the indenyl ligand, (Fig. 9.10*d*). A dissociative mechanism (S_N1) is involved in the replacement of a CO by PPh_3 in $HCo(CO)_4$, where the rate is zero-order in PPh_3. All the above examples involve a two-electron substitution process where the nucleophile with its pair of nonbonding electrons attacks the metal center and the departing group leaves with two electrons. One-electron ligand substitution radical reactions are also known and are becoming of increased importance; these generally involve 17-electron species such as $\bullet V(CO)_6$ and most frequently such substitutions proceed by an associative mechanism.

9.24 *trans* EFFECT

The relative ability of a ligand, in a substitution reaction in square-planar complexes, to labilize the group *trans* to the ligand in preference to a group *cis* to it.

Example. The synthesis of *cis*- and *trans*-$[PtCl_2(NO_2)(NH_3)]$ (Fig. 9.24) illustrates that the *trans*-directing ability of NO_2^- is greater than that of Cl^- is greater than that of NH_3. The magnitude of the *trans* effect of some common ligands in decreasing order is as follows: $CN^- \sim C_2H_4 \sim CO > PR_3 > NO_2^- > I^- \sim > Br^- \sim > Cl^- \sim > C_5H_5^- > RNH_2 > NH_3 > HO^- > H_2O$. The labilizing effect is not necessarily related to the weakening of the metal–ligand bond opposite the *trans*-directing ligand, a thermodynamic effect often distinguished by calling it the trans *influence*, but is more likely a consequence of the reaction mechanism of displacement. The labilization

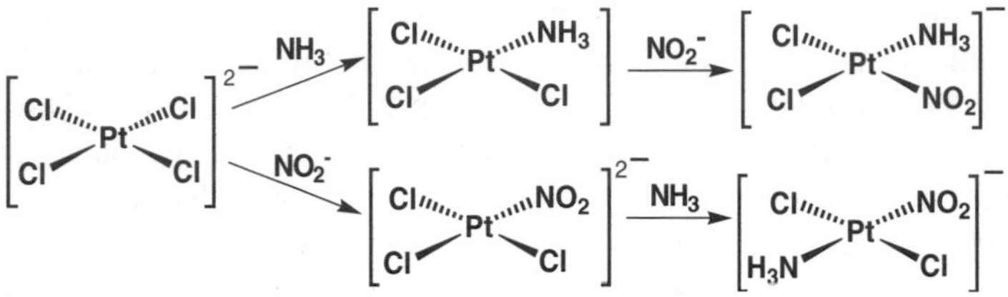

Figure 9.24. The preparation of *trans* and *cis* isomers using the *trans* effect principle.

accordingly should be more accurately called the *kinetic* trans *effect*. One factor influencing the *trans* effect in complexes with phoshpine ligands is the size of the phosphine.

9.25 CONE ANGLE

The cone angle, a measure of the steric bulk of phosphines, is the angle shown in Fig. 9.25. It should be noted that this angle is measured with the metal and not the phosphorus atom as the cone angle vertex. It is the angle swept out by rotation of the ligand around the metal–ligand bond axis. It varies from 87° in PH_3, 118° in $P(CH_3)_3$, 145° in PPh_3, 182° in $P(t\text{-}Bu)_3$, and even higher when substituents are placed on the *ortho* positions of the phenyl group, as in $P(mesityl)_3$ with a cone angle of 212°, which implies that almost 60% of the coordination sphere of the metal complex is protected from attack by incoming reagents.

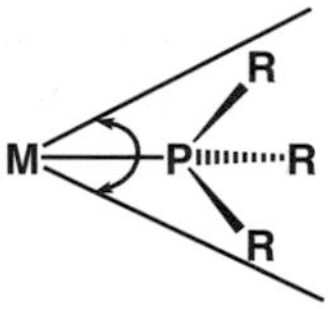

Figure 9.25. The cone angle in phosphines.

9.26 CATALYTIC ACTIVATION

A term used loosely to describe the conversion of a relatively inactive form of a chemical species to a more reactive form, usually by adsorption on a metal surface or by actual coordination of the chemical species (compound) to the metal.

Example. In hydrogenation reactions catalyzed by soluble transition metal complexes, the dihydrogen in solution is assumed to be coordinated by the overlap of its σ bond with an empty orbital on the metal. Back-donation of metal *d* electrons into the antibonding σ* orbital of dihydrogen is presumed to weaken the H–H bond further and, thus, *activate* the hydrogen prior to its oxidative addition to generate a dihydrido species. Analogous reactions are assumed to occur when hydrogen is activated by heterogeneous catalysts.

9.27 METALLOCENES

An unsystematic generic name for the class of organometal compounds having the structure $M(\eta^5\text{-}C_5H_5)_2$. Ferrocene, in which M = Fe (Fig. 9.10*b*) was the first such compound discovered, and it was given that common name because the *-ene* suffix emphasized the *arene* character of the complex, namely the unusual stability associated with aromaticity. The term metallocene has been extended by some chemists to include complexes in which the transition metal is "sandwiched" between other aromatic π electron systems, for example, cyclooctatetraene uranocene, $U(\eta^8\text{-}C_8H_8)_2$. Metallocenes, especially those of Group IV elements (Ti, Zr, Hf), have become part of important catalytic systems for the polymerization of ethylene and propylene on a commercial scale. Such metallocenes are used along with a cocatalyst prepared by reacting trimethylaluminum with a small quantity of water to give a complex mixture of methylaluminum oxide oligmers, called methylaluminoxane (MAO), $\text{-(-OAl(CH}_3\text{)-)}_n\text{-}$. The third component of the metallocene systsem is a solid support such as silica gel. The two cyclopentadiene rings of the metallocenes can be bridged with another grouping such as $(H_3C)_2C<$ or $(H_3C)_2Si<$ and otherwise modified to affect the steric and electronic character of the metallocene. These bridging groups are called *ansa ligands* (from the Latin word *ansa*, meaning "handle"). The catalysis of the polymerization of ethylene in one such system is depicted in Fig. 9.27. The function of the MAO is to activate the Zr by abstracting a methyl anion, forming a cationic species that is the active polymerization catalyst. The metallocenes are called *single-site catalysts* (SCC) in contrast to the Zeigler–Natta-type catalysts,

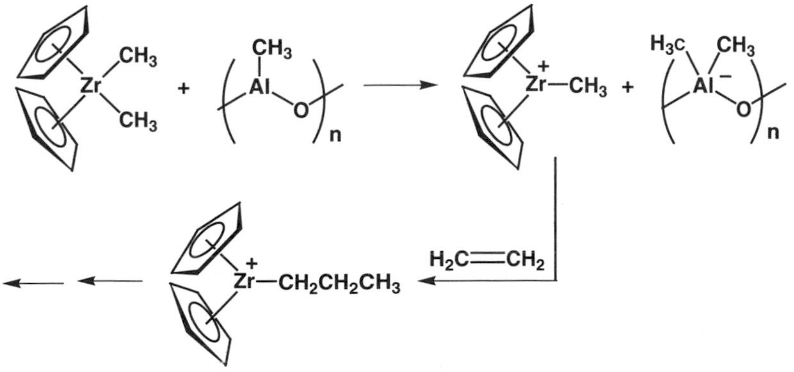

Figure 9.27. Initiation of ethylene polymerization by bis(cyclopentadienyldimethyl zirconium and methaluminoxane.

where both Zr and Ti sites are required for such activity. The use of metallocenes in the catalytic system leads to highly reactive catalysts; they can polymerize up to 20.000 ethylene monomers per second.

9.28 AMBIDENTATE LIGANDS

Anionic ligands possessing two or more different donor atoms, each having the potential of complexing with a metal.

Example. The SCN^- ligand may be bonded as M–SCN (thiocyanato) or M–NCS (isothiocyanato); the NO_2 group as M–NO_2 (nitro) or M–O–N=O (nitrito).

9.29 DIOXYGEN COMPLEXES

Complexes in which molecular oxygen (dioxygen) is coordinated to a metal as a ligating group. In such complexes, the oxygen–oxygen bond is retained.

Example. Most dioxygen complexes involve η^2 symmetrical bonding of the oxygen atoms to the metal in a triangular arrangement (Fig. 9.29). There are some η^1 oxygen complexes, notably of transition-metal porphyrins, with end-on bonding of the oxygen molecule to the metal.

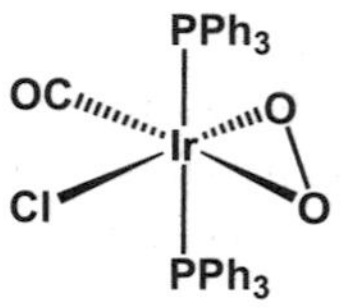

Figure 9.29. A dioxygen complex.

9.30 DINITROGEN COMPLEXES

Complexes in which molecular nitrogen is coordinated to a metal as a ligating group. In such complexes, the nitrogen molecule retains the nitrogen-nitrogen triple-bond linkage.

Example. In contrast to dioxygen complexes, most dinitrogen complexes involve end-on bonding with only one nitrogen atom bonded to a metal. Dinitrogen complexes in which N_2 is a bridging ligand, for example, $[Ru(NH_3)_5–N\equiv N–Ru(NH_3)_5]^{4+}$ (Fig. 9.30), are also known. Because CO and N_2 are isoelectronic, it was perhaps not unexpected that the latter should be able to function as a ligand, but it is generally accepted that N_2 is both a poorer donor and a poorer electron acceptor than CO and hence many more complexes of CO are known than those of dinitrogen. The "fixation" of dinitrogen by certain plants probably involves a dinitrogen complex of Mo.

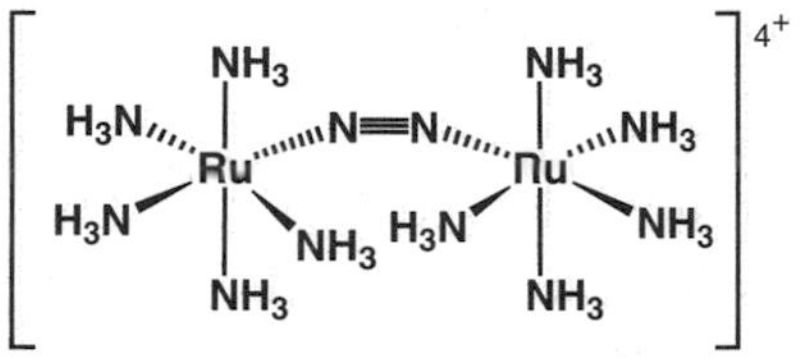

Figure 9.30. A dinitrogen complex of ruthenium.

9.31 NOMENCLATURE OF METAL COMPLEXES

Preferred systematic nomenclature is embodied in the approved IUPAC rules for the nomenclature of inorganic chemistry and organometal complexes. The rules deal with the naming of the ligands (uncharged or ionic) and the specific atom of attachment (if not obvious); the oxidation number of the metal; the charged or uncharged nature of the complex; and the stereochemistry, if relevant. The naming procedure is as follows:

1. The name of the central metal atom is placed last and, to avoid ambiguity, its oxidation number is given immediately thereafter in Roman numerals placed in parentheses.
2. Ligands. The descending order of citation of ligands attached to the metal is as follows: anionic, neutral, cationic. In each category alphabetical ordering is desirable. The names of all anionic ligands end in the letter *o*, and usually the names of all neutral and cationic ligands are used without change, except that coordinated water and ammonia are called aqua and ammine, respectively. The neutral ligands NO and CO are also exceptions for historical reasons; they are called nitrosyl and carbonyl, respectively. Alkyl and aryl groups that are σ-bonded to the metal are given their conventional radicofunctional names (methyl, allyl, phenyl, etc.). The usual multiplying prefixes, *di-*, *tri-*, *tetra-*, *penta-*, and so on, are used for simple ligands such as chloro. The prefixes *bis-*, *tris-*, *tetrakis-*, *pentakis-*, and so on, are used for multiword ligands or for ligands that already contain a multiplying prefix, such as triphenylphosphine.
3. Anionic complex ions are given the suffix *-ate* appended to the English name (or a shortened version) of the metal atom. In a few cases (iron, copper, lead, silver, gold, tin), Latin stems are used (ferrate, cuprate, plumbate, argentate, aurate, stannate).

Example. Frequently, but always desirably, the name of a metal complex is accompanied by its formula written in line form. Although practice varies considerably, a systematic procedure for writing a line formula is the following: The symbol for the central atom element is placed first, and the ligand structures are given in the same order (anionic, neutral, cationic, and alphabetical within each category) as in the naming procedure. If present, the formula of the entire complex ion is placed in brackets. Some examples of the names of metal complexes and their formulas written in line form are as follows:

Calcium hexacyanoferrate(II): $Ca_2[Fe(CN)_6]$
Potassium trichloro(ethylene)platinate (II): (Zeise's salt) $K[PtCl_3(C_2H_4)]$
Hexakis(phenylisocyanide)chromium(0): $Cr(C_6H_5NC)_6$
Decacarbonyldimanganese(0): $Mn_2(CO)_{10}$
Potassium pentacyanocobaltate(II): $K_3[Co(CN)_5]$
Bis(acetylacetonato)copper(II): $Cu(CH_3COCHCOCH_3)_2$
Tricarbonyl(cyclooctatetraene)iron(0): $Fe(CO)_3(C_8H_8)$
Hexamminecobalt(III) chloride: $[Co(NH_3)_6]Cl_3$
(Acetyl)pentacarbonylmanganese(I): $Mn(COCH_3)(CO)_5$
Bis(η^5-cyclopentadienyl)hydridorhenium(III): $Re(C_5H_5)_2H$

9.32 NAMES OF BRIDGING GROUPS

A bridging group is indicated by the Greek letter mu (μ) immediately before its name. Multiple bridging groups of the same kind are indicated by the usual multiplying prefix.

Example. $[Ru(NH_3)_5{-}N{\equiv}N{-}Ru(NH_3)_5]Cl_4$ is μ-Dinitrogenbis[pentammineruthenium(II)] tetrachloride (Fig. 9.30); $Fe_2(CO)_9$ is tri-μ-carbonylhexacarbonyldiiron(0), and its common name is diiron enneacarbonyl (Fig. 9.7*a*); di-μ-chloro-1,3-dichloro-2,4-bis(triphenylphosphine)diplatinum(II) (Fig. 9.32).

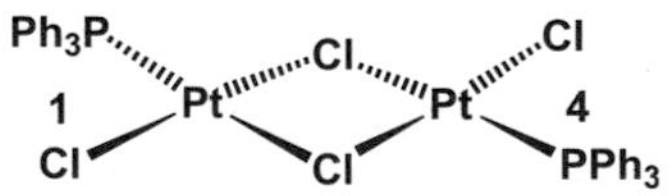

Figure 9.32. Di-μ-chloro-1,3-dichloro-2,4-bis(triphenylphosphine)diplatinum(II).

9.33 METAL CLUSTER COMPOUNDS

A limited number of metal atoms (as distinguished from an infinite number as in bulk metal) that are held together by direct metal–metal bonding; each metal atom is bonded to at least two other metal atoms. The metal atoms frequently define a complete or nearly complete triangulated polyhedron. Some nonmetal atoms may be associated with the cluster.

Example. $Ir_4(CO)_{12}$ (four metal atoms at the vertices of a tetrahedron, Fig. 9.33*a*): and $Fe_5(CO)_{15}C$ (five metal atoms at the vertices of a square pyramid, all bonded to a single central carbon atom, Fig. 9.33*b*).

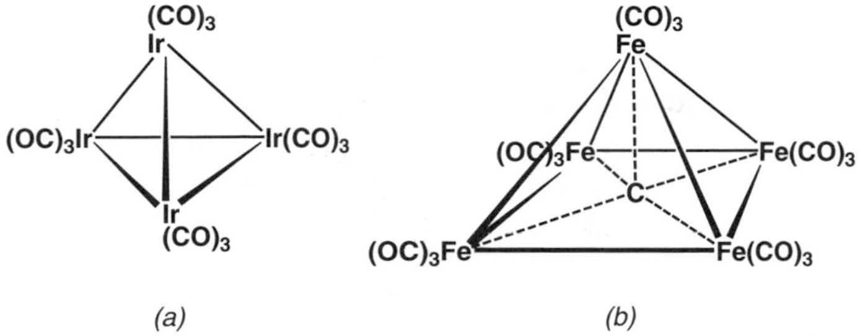

Figure 9.33. Metal cluster compounds: (*a*) tetrahedral and (*b*) square pyramidal.

9.34 EDGE-BRIDGING LIGAND

A ligand that bridges one edge of the polyhedron of a metal cluster.

Example. The octahedral metal cluster carbonyl (Fig. 9.34), which possesses one edge-bridging (μ, η^1)-carbonyl group.

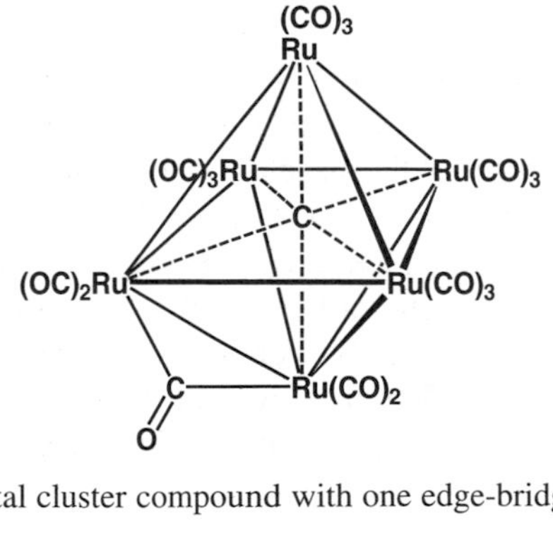

Figure 9.34. A metal cluster compound with one edge-bridging carbonyl group.

9.35 FACE-BRIDGING LIGANDS

A ligand that bridges one triangular face of a metal cluster polyhedron.

Example. The metal cluster compound $Rh_6(CO)_6$ has four face-bridging (μ^3, η^1)-carbonyl groups staggered on the eight faces of an octahedron; see Fig. 9.35 (top faces 1, 2, 5, and 3, 4, 5; bottom faces 1, 4, 6, and 2, 3, 6).

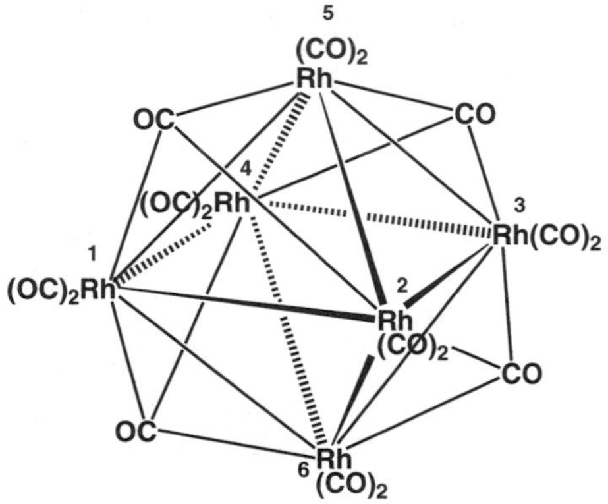

Figure 9.35. A metal cluster compound having four face-bridging carbonyl groups.

9.36 *CLOSO*, *NIDO*, AND *ARACHNO* CLUSTERS

Many cluster compounds including boron hydrides and carboranes form complete or nearly complete triangular-faced polyhedra, often called detahedra. These clusters form three structural types of polyhedra: *closo* (closed); *nido* (nestlike); and *arachno* (cob-web) structures. The *closo* structures have a metal (or boron or carbon) atom at each vertex of the polyhedron. The *nido* structures have incomplete polyhedral structures; they lack a skeletal atom at one of the vertices. The *arachno* structures lack skeletal atoms at two adjacent vertices of an approximate polyhedron.

Example. The complexes $Rh_6(CO)_{16}$ (Fig. 9.35) and $Co_4(EtC{\equiv}CEt)(CO)_{10}$ (Fig. 9.36*a*) are *closo* structures. In the latter, four cobalt atoms plus the two carbon atoms of the acetylene are at the six vertices of the skeletal octahedron. The metal carbonyl carbide cluster, $Fe_5(CO)_{15}C$ (Fig. 9.33*b*), is a *nido* structure. The five Fe atoms occur at the vertices of an incomplete octahedron, and thus, one skeletal metal atom at one vertex is absent. The square pyramidal borane, B_5H_9 (Fig. 9.36*b*), is also a *nido* structure; its planar representation is shown in Fig. 9.36*c*. For purposes of the present classification, the platinum π complex $Pt(C_2H_4)(PPh_3)_3$ may even be thought of as an *arachno* structure, with the missing two vertices of a trigonal bipyramid shown as circles in Fig. 9.36*d*.

9.37 METALLACYCLES

Compounds that consist of a cyclic array of atoms, one of which is a metal and the rest carbon atoms. The most frequently encountered metallacycles are those

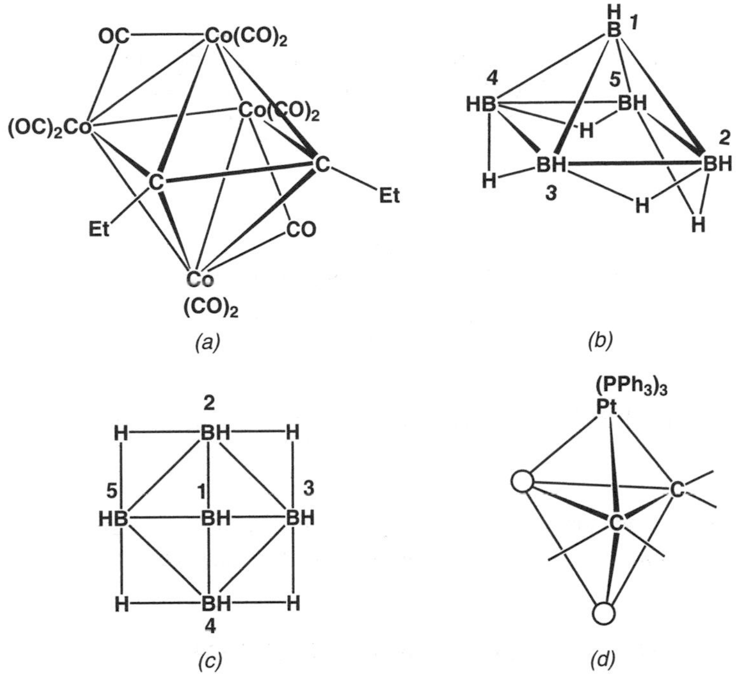

Figure 9.36. (*a*) A *closo* structure; (*b*) a *nido* structure, pentaborane (9); (*c*) planar projection of (*b*); (*d*) an *arachno* structure.

consisting of three or four carbon atoms plus a transition metal as part of the ring. Such compounds are usually much more stable than the acyclic metal alkyl analogs because the metal alkyls are prone to undergo β-hydrogen elimination, a process that is made difficult with the metallacycles because their cyclic structures are a barrier to the stereochemistry required for concerted hydrogen transfer.

Example. A platinacyclobutane (Fig. 9.37*a*) and 3,4-dimethyltungstacyclopentane (Fig. 9.37*b*) (usually the metal carries other ligands L as well). Figure 9.37*c* shows the structure of a titanium-aluminum metallacycle, $Cp_2Ti(\mu\text{-}Cl)(\mu\text{-}CH_2)AlMe_2$, called *Tebbe's reagent* [see F. N. Tebbe, G. W. Parshall, and G. S. Reddy, *J. Am. Chem. Soc. 100*, 3611 (1978)], which in the presence of a base such as pyridine, behaves as though it were the equivalent of the carbene $Ti{=}CH_2$ and undergoes the dismutation reaction with olefins. (Sect. 9.45).

9.38 CRYSTAL FIELD THEORY (CFT)

The theory of bonding in metal complexes that views the interaction between the ligands and the metal to which they are attached as a strictly ionic or ion-dipole interaction resulting from electrostatic attractions between the central metal and the

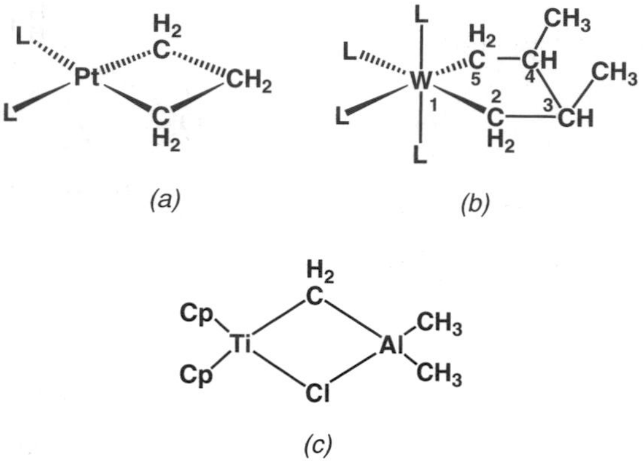

Figure 9.37. Metallacycles; (*a*) containing platinum; (*b*) containing tungsten; (*c*) Tebbe's reagent.

ligands. The ligands are regarded simply as point negative charges surrounding a central metal atom; covalent bonding is completely neglected.

9.39 CRYSTAL FIELD SPLITTING (OF *d* ORBITALS)

The splitting or separation of the energy levels of the five degenerate *d* orbitals of a transition metal in the gas phase when the metal is surrounded by ligands arranged in a particular geometry with respect to the metal center. If the ligands (or charges they represent) were arranged spherically around the metal, the energy of all the *d* orbitals would be raised equally relative to the energy of the *d* orbitals in the isolated gaseous metal atom, but all the *d* orbitals would remain degenerate. However, when the ligands assume particular geometries (e.g., square planar, tetrahedral, octahedral) the energies of the *d* orbitals are affected differently, resulting in the lifting of degeneracy which is called ***d* orbital splitting**.

Example. The octahedral complex anion $[CoF_6]^{3-}$ is depicted in Fig. 9.39*a*. Fig. 9.39*b* shows the splitting of the *d* orbitals in the octahedral field of the six fluoride ions generating two sets of orbitals: a lower energy, triply degenerate set given the designation t_{2g} and a higher energy, doubly degenerate set given the designation e_g (symmetry species in the O_h point group to which the molecule belongs). The difference in energy is the crystal field splitting energy Δ_O (where the subscript stands for octahedral). Δ_O is also designated as $10\,Dq$ (the 10 is a multiplier of convenience). The magnitude of $10\,Dq$ depends on the polarizability D of the metal (related to the charge and size of the metal ion) and particularly on q, the magnitude of the electronic and steric properties of the ligands. The energy corresponding to Δ_O or $10\,Dq$ can be determined experimentally (usually spectroscopically). Ligands are frequently classified on the basis of the value of q, which is called the ligand field splitting strength of the ligand. The decreasing order of q of some common ligands

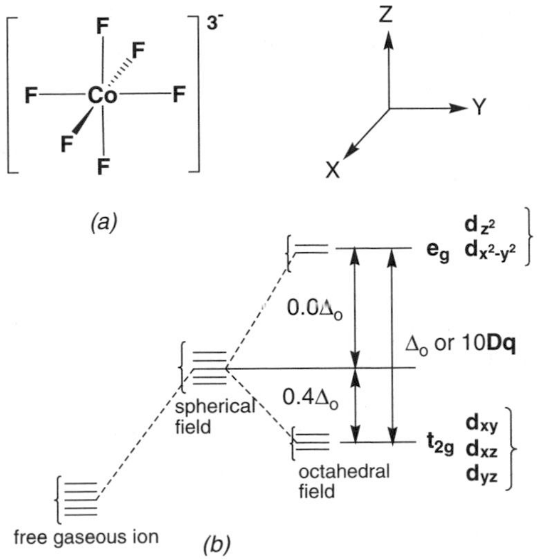

Figure 9.39. (*a*) The octahedral complex $[CoF_6]^{3-}$ and (*b*) the relative energies of the *d* orbitals in a gaseous ion in a spherical field and in an octahedral field.

TABLE 9.39. The Splitting of the Five *d* Orbitals (*and their symmetry species*) in Complexes of Various Geometries (Listed in Increasing Order of Energy)

Octahedral (O_h)	Tetrahedral (T_d)	Square Planar (D_{4h})	Trigonal Bipyramidal (D_{3h})
$(x^2 - y^2, z^2)\ e_g$	$(xy, xz, yz)\ t_2$	$(x^2 - y^2)\ a_{1g}$	$z^2\ a^{1\prime}$
$(xy, xz, yz)\ t_{2g}$	$(x^2 - y^2, z^2)\ e$	$xy\ b_{2g}$	$(x^2 - y^2, xy)\ e'$
		$z^2\ a_{1g}$	$(xz, yz)\ e''$
		$(xz, yz)\ e_g$	

is $CN^- > NH_3 > H_2O >$ $OH^- > F^- > Cl^- > Br^- >$ I. In an octahedral field, for example, electrons occupying the $d_{x^2-y^2}$ and d_{z^2} set (e_g) of orbitals are repelled most because these orbitals point directly at the ligands, whereas electrons in the d_{xz}, d_{yz}, d_{xy} set (t_{2g}) of the metal *d* orbitals are repelled least because such orbitals point between the axes on which the ligands are located. In other geometries, the splitting of the *d* orbitals is different. In a tetrahedral field, for example, the *d* orbital splitting is exactly opposite to that of the octahedral field, with the d_{xz}, d_{yz}, d_{xy} set being repelled more than the $d_{x^2-y^2}$ and d_{z^2} set. The splitting of the *d* orbitals in various geometries is summarized in Table 9.39.

9.40 CRYSTAL FIELD STABILIZATION ENERGY (CFSE)

The stabilization of a complex that results from preferential electron occupation of the lower-energy *d* orbitals split by the crystal field.

Example. The typical crystal field splitting resulting from an octahedral environment is displayed in Fig. 9.39*b*. Specifically, if Co^{3+} is the metal ion, it has the electronic configuration of d^6. Each electron that goes into a t_{2g} orbital is assigned a value of $0.4\,\Delta_o$ and each electron that goes into an e_g orbital is assigned a value of $-0.6\,\Delta_o$. The six *d* electrons go into the lower-energy t_{2g} set; hence, the crystal field stabilization energy for $(t_{2g})^6$ is

$$6(0.4\,\Delta_o) + 0(-0.6\,\Delta_o) = 1.8\,\Delta_o$$

If the five *d* orbitals had remained degenerate, there would be no CFSE, and likewise, if the metal had a d^{10} configuration, the CFSE would again be zero:

$$6(0.4\,\Delta_o) + 4(-0.6\,\Delta_o) = 0\,\Delta_o$$

9.41 PAIRING ENERGY

The electron–electron repulsion energy that must be overcome to enable two electrons of opposite spin to occupy the same orbital.

Example. The *d* electron configuration of $[Co(NH_3)]^{3+}$ is $(t_{2g})^6$ rather than $(t_{2g})^4(e_g)^2$. This arises because Δ_o in this case is greater [about 66 kcal mol^{-1} (276 kJ mol^{-1})] than the pairing energy [about 37 kcal mol^{-1} (155 kJ mol^{-1})]. This represents the low-spin case see Sect. 9.42. When the pairing energy is greater than Δ_o, electrons occupy some high-energy orbitals before all the low-energy orbitals are filled (the high-spin case).

9.42 LOW-SPIN AND HIGH-SPIN COMPLEXES

A classification of complexes based on the number of their unpaired electrons. After the ligand splitting of the *d* orbitals of the metal in a particular complex is ascertained (Sect. 9.39), the *n* electrons of the d^n configuration are placed in these orbitals in accordance with pairing energy considerations (Sect. 9.41): *Aufbau principle* (electrons are placed in orbitals in order of increasing energy) and *Hund's rule* (one electron is placed in every orbital of equal energy before placing a second in the same orbital). If the resulting electron distribution is such that only the low-energy orbitals are occupied, the complex is a low-spin complex. If some of the higher-energy *d* orbitals are occupied before all the lower-energy ones are completely filled, then the complex is a high-spin complex.

Example. In $[Co(NH_3)_6]^{3+}$, the electronic configuration is $(t_{2g})^6$, as explained in Sect. 9.41; hence, this is a low-spin complex. In the complex $[CoF_6]^{3-}$, Co is again 3^+ and has a d^6 configuration. However, the crystal field splitting of the F^- ligands is not nearly as great as that of the NH_3 ligands and, in fact, the splitting (Δ_o) is somewhat less than the pairing energy. Accordingly, in building up the electronic configuration of Co^{3+} in this complex, one electron is placed in each of the three degenerate t_{2g} orbitals, then the

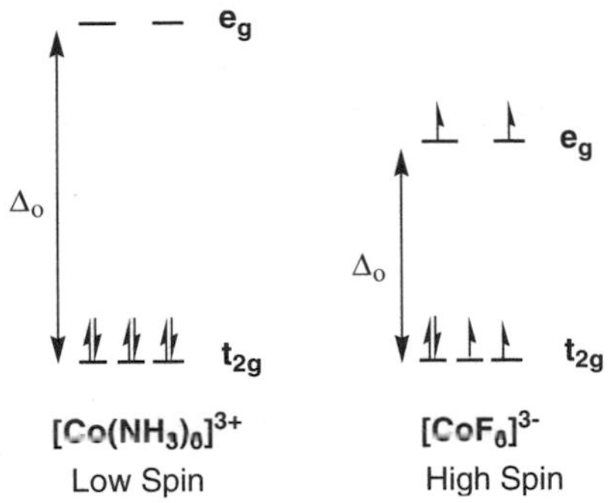

Figure 9.42. Octahedral low-spin and high-spin complexes.

fourth and fifth electrons are placed in each of the two degenerate e_g orbitals, since the pairing energy required to place two electrons in a t_{2g} orbital is greater than Δ_o. The sixth and final electron necessarily goes into a t_{2g} orbital, giving the high-spin complex with an electronic configuration of $(t_{2g})^4(e_g)^2$. Such a configuration involves four unpaired electrons and, indeed, $[CoF_6]^{3-}$ displays a paramagnetism corresponding to four unpaired electrons. Figure 9.42 summarizes this discussion.

9.43 LIGAND FIELD THEORY (LFT)

A slightly modified crystal field theory (CFT) that takes into account certain parameters as variables reflecting the existence of covalent contributions to M–L bonding, rather than assume these parameters to be equal to the values found for free ions as required by simple CFT. One of the most important parameters is the spin-orbit coupling constant, which affects the magnetic properties of many complex ions. For the reasons given above, LFT has also been called *adjusted crystal field theory.*

9.44 ORGANOMETALLIC CARBENE COMPLEXES

Complexes containing a divalent carbon doubly bonded to a metal atom. Stable uncomplexed carbenes are now well known and fully characterized. Thus, for example, stable imidazol-2-ylidines, shown below, are now readily available. However, much of carbene chemistry presently involves organometal carbenes, in which the carbene functions as a powerful ligand because it has both a lone pair of electrons and an empty orbital, two characteristics shared with the carbon of CO.

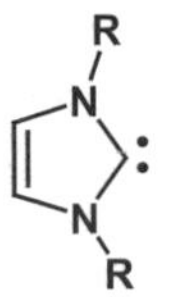

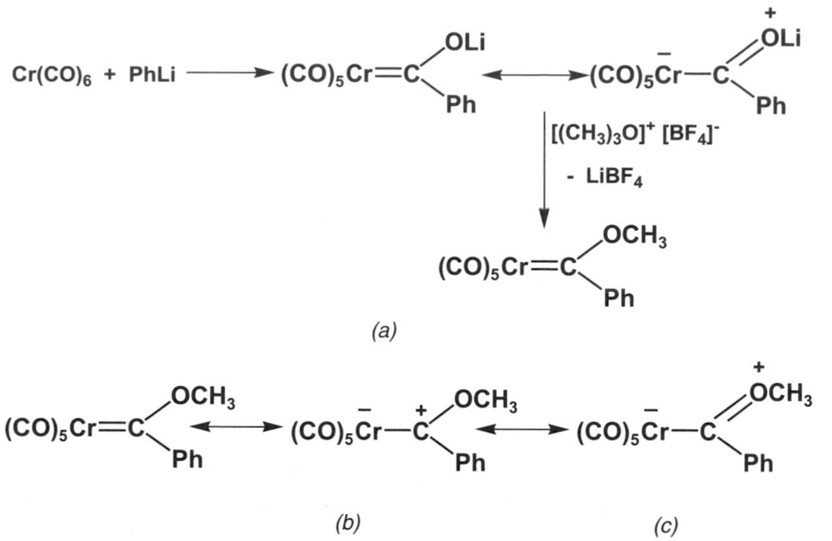

Figure 9.44. (*a*) The preparation of a Fischer carbene; (*b*) and (*c*) resonance structures of (*a*). No resonance structure comparable to (*c*) can be drawn for a Schrock carbene.

Example. The first organometallic carbene (Fig. 9.44*a*) was prepared in 1964 by Ernst O. Fischer (who received the Nobel Prize in 1973) [see Fischer, *Adv. Organomet. Chem. 14*, 1 (1976)]. A general method for the preparation of such carbenes consists of converting a metal carbonyl to an acyl complex and then alkylating with a strong acid as shown in Fig. 9.44*a*. A carbene complex in which a heteroatom is bonded to the carbene carbon may be regarded as a metal-stabilized carbenium ion because it may be represented by the resonance structures shown in Figs. 9.44*b* and *c*. Such carbenes are frequently called *Fischer carbenes*. These resonance-stabilized heteroatom-carbenes, characteristic of the late transition metals (d^6–d^{10}), are electrophilic; for example, they can undergo alcohol exchange reactions at the electrophilic carbon atom by nucleophilic substitution (such as replacement of the methoxide in Fig. 9.44 by ethoxide). The first organometallic carbene in which a heteroatom is not bonded to the carbene carbon, $[(CH_3)_3CCH_2]_3Ta{=}CHC(CH_3)_3$, was prepared by Richard Schrock in 1974. Because the complex has no β-hydrogens, it cannot undergo decomposition via β-hydrogen elimination. This and other related early transition metal (d^0–d^5) carbenes are frequently called *Schrock carbenes* [see R. R. Schrock, *Acc. Chem. Res. 12*, 98 (1979)]. In contrast to the Fischer carbenes, the carbene carbon atoms of the Schrock-type carbenes are nucleophilic. In the formalisms of electron counting, a Fischer-type carbene ligand may be regarded as a neutral, two-electron donor similar to carbon monoxide; hence, Cr in Fig. 9.44*a* is d^6 with an oxidation number of zero and an EAN of 36, isoelectronic with $Cr(CO)_6$ and in compliance with the rule of 18. However, in the Schrock alkylidene-type carbenes, for example, $Cp_2ClTa{=}CHC(CH_3)_3$, the carbene moiety is considered not as a neutral ligand but as a four-electron, dianionic donor ligand, and this leads to Ta in the above example with a formal oxidation number of 5^+ (d^0) and an EAN of (0 + 12 + 2 + 4 + Ar core) 36. It is therefore coordinatively saturated. Obviously, there is some arbitrariness to using the formalism for calculating oxidation numbers in the case of metal complexes with carbene (and carbyne) ligands.

9.45 OLEFIN METATHESIS (*trans*-ALKYLIDENATION OR DISMUTATION REACTION)

A reaction involving the exchange of alkylidene groups (dismutation) between two olefins:

$$\text{R-CH=CH-R} + \text{R'-CH=CH-R'} \longrightarrow 2\ \text{R-CH=CH-R'}$$

(R and/or R′ may also be H). The reaction is catalyzed by transition metal complexes and presumably involves metal carbene intermediates.

Example. The metathesis reaction employing propylene to give 2-butene and ethylene is used commercially. Metathesis reactions, carried out with stable carbenes as catalysts, suggest the intermediate formation of a metallacycle (Sect. 9.37); Fig. 9.45*a* shows such an intermediate formed in the reaction of a diphenyltungsten carbonyl carbene with a terminal olefin. An intermediate metallacycle rationalizes both the major and minor products of the reaction. A normal (ground-state) 2 + 2 addition is symmetry forbidden, but the availability of *d* orbitals on the metal allows the reaction to proceed. Metallacycles are also important intermediates in the *ring-opening metathesis polymerization (ROMP)* of cyclic olefins (see Sect. 8.27), for example, the formation of an acyclic polymer of cyclopentene (Fig. 9.45*b*), where the number of carbon atoms enclosed by the brackets is five. One of the most successful catalysts for this polymerization reaction is a di(tricyclohexyl)phosphine ruthenium carbene, $(Cy_3P)_2Cl_2Ru{=}CHPh$) [T. M. Trnka and R. H. Grubbs, *Acc. Chem. Res. 34*, 18 (2001)]. The reverse of the ROMP reaction in which an α, ω-diene can be cyclized may also be realized with the use of the same Ru carbene catalyst (Fig. 9.45*c*). Such reactions are called *ring closure metathesis (RCM)*.

$$2\ CH_3CH{=}CH_2 \longrightarrow CH_3CH{=}CHCH_3 + H_2C{=}CH_2$$

9.46 CARBYNE (ALKYLIDYNE) COMPLEXES, M≡C–R

Complexes containing a transition metal bonded to an organic moiety through a monocovalent triply bonded *sp* hybridized carbon atom.

Example. The carbyne, $Br(CO)_4W{\equiv}C{-}Ph$, is prepared from the precursor carbene, $(CO)_5W{=}C(OEt)Ph$, by treatment with BBr_3 in pentane at −50°C with the subsequent elimination of CO and Br_2BOEt from the intermediate. Several other procedures are known for preparing carbynes but usually a precursor carbene is the starting compound. In the formalism for counting electrons around the central metal atom, the carbyne

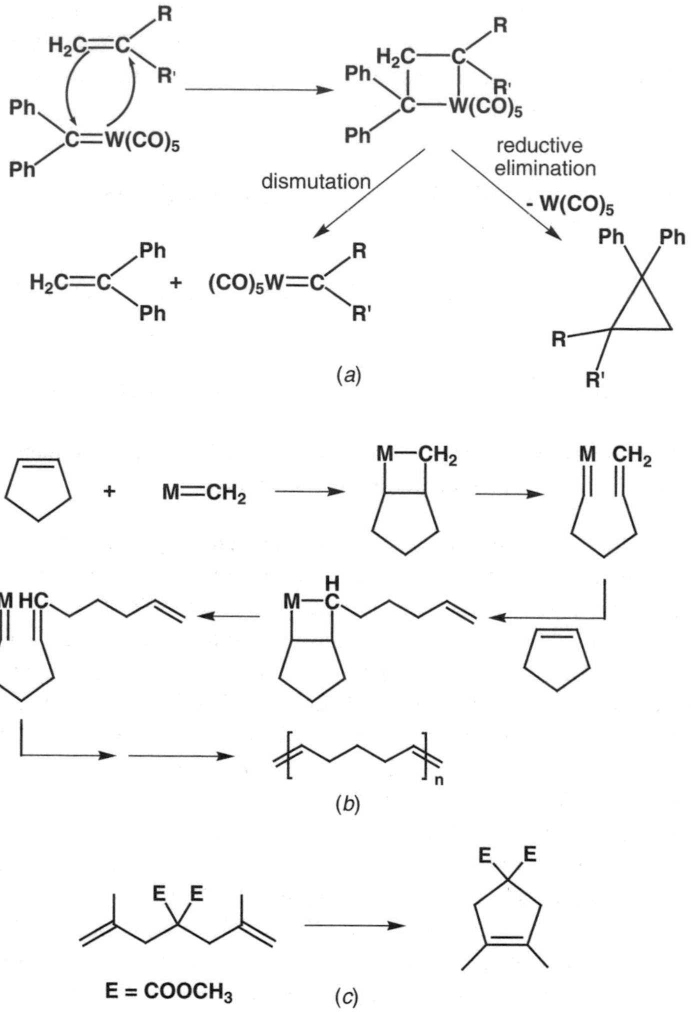

Figure 9.45. (*a*) A mechanism of the olefin metathesis (or trans alkylidene or dismutation) reaction involving a metallocycle intermediate. (*b*) The ROMP of cyclopentene and (*c*) RCM reaction of a substituted heptadiene to a cyclopentene derivative.

ligand is regarded as a trianionic species and a six-electron donor. E. O. Fischer discovered the first carbyne in 1973. Some carbyne (alkylidyne) complexes can catalyze acetylene dismutation reactions.

9.47 TRANSMETHYLATION

A reaction in which a methyl group attached to a transition metal is transferred to another acceptor molecule. This reaction is particularly important in vitamin B_{12}

chemistry where enzymatic transfer of a methyl group bonded to cobalt in a reductive elimination reaction is involved (Sect. 9.18).

Example. Methionine has been found to be formed from the reaction between an enzyme-bound methyl group (E–Co–CH_3) and homocysteine:

$$\text{E-Co}^{n+}\text{-CH}_3 + \text{HSCH}_2\text{CH}_2\text{CH(NH}_2\text{)COOH} \longrightarrow \text{E-Co}^{(n-1)+} + \text{CH}_3\text{SCH}_2\text{CH}_2\text{CH(NH}_2\text{)COOH} + \text{H}^+$$

9.48 STEREOCHEMICALLY NONRIGID MOLECULES

Molecules that undergo rapid intramolecular reorganization. In transition metal complex chemistry, such reorganization involve ground and transition states (or intermediates) viewed as idealized polygons and polyhedrons. The reorganization occur by ligand rotations around the ligand metal bonds and no bond-breaking is involved.

Example. Although not a transition metal complex, $(CH_3)_2NPF_4$ is an excellent example of a stereochemical nonrigid molecule that readily undergoes thermal reorganization. The original trigonal bipyramidal molecule goes through a square pyramidal intermediate in the reorganization, as shown in Fig. 9.48. The transformation can be demonstrated by ^{19}F nuclear magnetic resonance (NMR). At 25°C all the fluorine atoms are equivalent (on the NMR timescale); the ^{19}F spectrum consists of one doublet due to P-F spin–spin coupling. At low temperatures, however, the ^{19}F spectrum shows two signals, each of which is a doublet of triplets (two sets of nonequivalent F's, one apical and one equatorial, each set split by the P atom and each set split by the two F's of the other set).

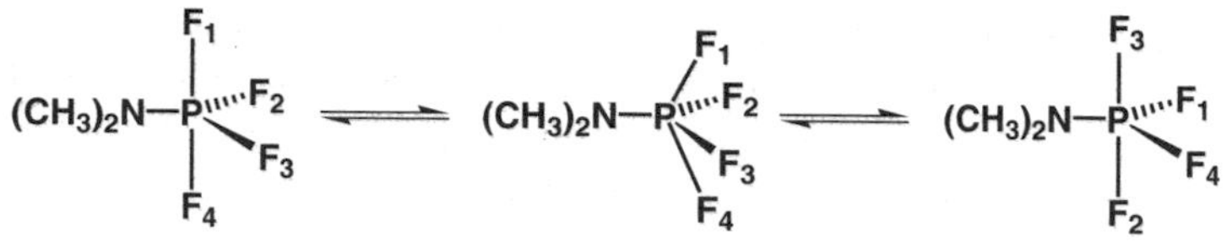

Figure 9.48. Pseudo-rotation in a trigonal bipyramidal complex.

9.49 PSEUDO-ROTATION

The permutation of nuclear positions, as in the trigonal bipyramidal complexes shown in Fig. 9.49. This particular type of reorganization is known as the *Berry mechanism of reorganization* (named after R. Stephen Berry). The Berry pseudo-rotation involves positional changes of four of the five ligands of a bipyramidal complex, with the two apical positions becoming equatorial as a result of the rotation, as in Fig. 9.49. This is the reason that $Fe(CO)_5$ exhibits only one resonance signal in the ^{13}C NMR. The

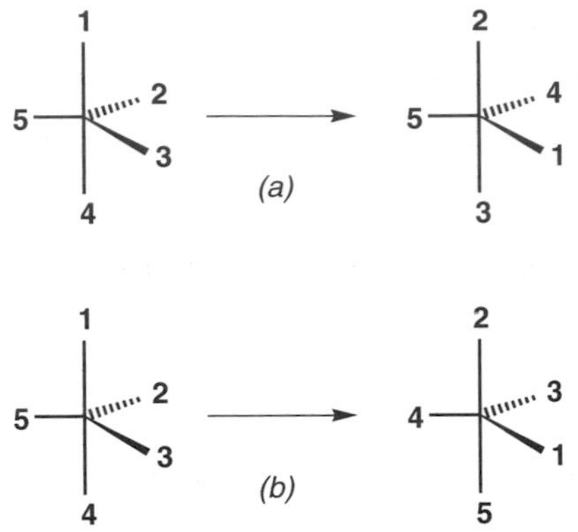

Figure 9.49. Rotational isomerism in trigonal bipyramidal complexes: (*a*) the general case for Berry pseudo-rotation and (*b*) the general case for the turnstyle rotation.

general case is shown in Fig. 9.49*a*. A stepwise, but complete, exchange of positions is involved in the *turnstyle* mechanism illustrated in Fig. 9.49*b*. In this rotation the two apical positions again become equatorial, but all positions are changed.

9.50 ISOLOBAL ANALOGIES

A concept introduced by R. Hoffmann (who received the Nobel Prize in 1981) to provide a bridge between inorganic and organic chemistry. This bridge connects fragments of organometallic complexes with analogous fragments of methane in terms of the symmetry, energy, and electronic configuration of the respective frontier orbitals (highest occupied and lowest unoccupied). One of the easiest ways to illustrate this relationship is to compare the isolobal fragment •$Mn(CO)_5$ (a d^7 ML_5 species) with •CH_3. Both fragments may be considered as having been formed by removal of a hydrogen atom from their stable parent compounds, octahedral $HMn(CO)_5$ and tetrahedral CH_4, respectively, and both therefore are short one electron of the closed-shell electron configuration of the parent. At Hoffmann's suggestion, the analogy is expressed by a two-headed arrow with half an orbital below the arrow [A. R. Pinhas, T. A. Albright, P. Hofmann, and R. Hoffmann, *Helv. Chim. Acta 63*, 29 (1980); R. Hoffmann, *Angew. Chem. Eng. Ed. 21*, 711 (1982)]:

Example. Many organometallic fragments such as $Mn(CO)_5$ and $Fe(CO)_4$ may be regarded as pieces of an octahedron, just as organic fragments like CH_3, CH_2, and CH may be regarded as pieces of a tetrahedron. Consider the isolobal analogy:

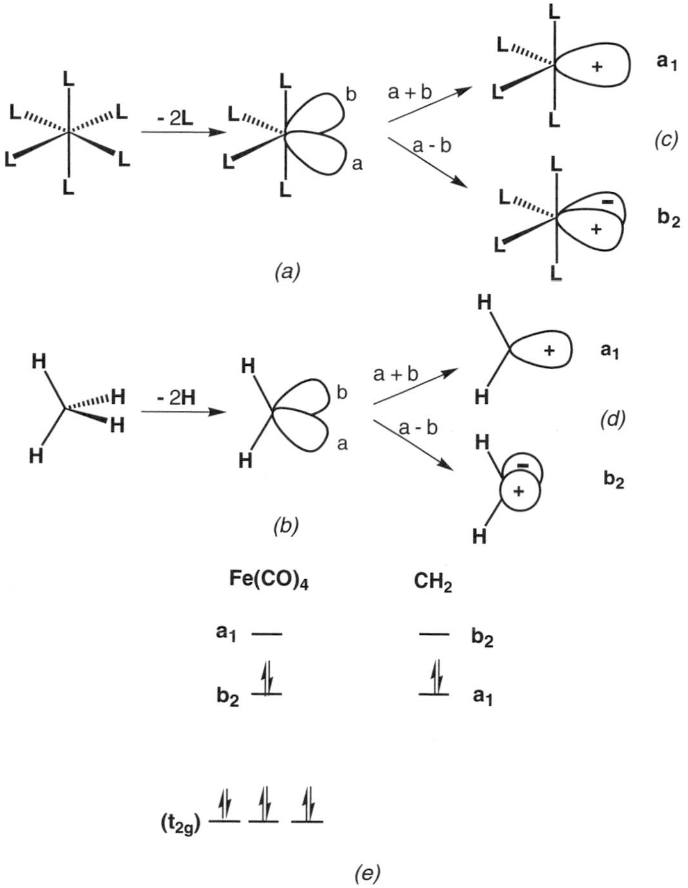

Figure 9.50. The isolobal relationship between the fragments $Fe(CO)_4$ and CH_2: (*a*) the two orbitals remaining after removing two ligands from the octahedron; (*b*) the two orbitals remaining on the carbene; (*c*) the combination of the two empty orbitals on the octahedron; (*d*) the combination of the two empty orbitals on the tetrahedron; (*e*) the orbital energy diagram.

The four-coordinate $Fe(CO)_4$ fragment, or, in general, d^8 ML_4, may be regarded as an octahedral complex minus two of the usual six ligands characteristic of an octahedral complex, leaving two vacant orbitals in the fragment (Fig. 9.50*a*). Analogously, the removal of two ligands (hydrogens) from the usual four of a tetrahedral complex (CH_4) gives two vacant orbitals on the CH_2 (carbene) fragment (Fig. 9.50*b*). The two orbitals on each separate fragment must be combined in a linear combination (addition and subtraction), to give in each case a_1 and b_2 molecular orbitals as depicted in Figs. 9.50*c* and *d*. The two orbitals on each fragment are the two frontier orbitals. As shown in the energy-level diagram (Fig. 9.50*e*), the energy-level order is reversed in each fragment. The b_2 orbital of the octahedral fragment has mostly *d* character and would overlap better with an attached ligand, while the a_1 orbital of the carbene fragment with mostly *p* character would better overlap with a ligand attached to it. However, the order is of little concern because their energies are similar, and hence these orbitals can be combined with the orbitals of similar symmetry of incoming groups. The orbital designations of a_1 and b_2 are symmetry species in point group C_{2v}, which is a subgroup of both the octahedral point group O_h and the tetrahedral point group T_d.

9.51 CHEMICAL VAPOR DEPOSITION (CVD)

The deposition of thin films of metals onto appropriate substrates by vaporizing volatile metal organics or organometallic precursors into a chamber where they are decomposed by heating at high temperature.

Example. Thin films of gallium arsenide, GaAs, for use as semiconductors are produced by copyrolyzing $(CH_3)_3Ga$ and AsH_3 inside a chamber containing these compounds at 600 to 700°C. Organometal compounds are also used as precursors, providing single-source precursors. The use of CVD has provided organometal chemists with another entry into the burgeoning field of material science. The thin films provided by CVD on inert surfaces reduce friction and wear.

9.52 ORGANOMETAL COMPOUNDS AS CATALYSTS

Probably, the majority of all organic chemical syntheses both in the laboratory and industry involve the use of catalysts at some stage or other. Many organic reactions are catalyzed by acids or bases in homogeneous systems and many of them by transition metals in their elementary state in heterogeneous systems. In the latter category, it is likely that even though the metal is present initially in the elementary or zero oxidation state and after reaction can be recovered unchanged, the chemistry during the reaction occurs at the surface and invariably involves either transient or intermediate organometal compounds. If such intermediates can be identified and prepared, they may serve as catalysts in homogeneous systems. In some systems, simple soluble transition metal salts can function as catalysts. One of the most important advantages of working in a homogenous system is that greater selectivity is usually accomplished and mass transfer factors are relatively unimportant. Such selectivity is particularly important in the synthesis of chiral compounds, especially for pharmaceutical purposes, where enantiomers may have different biological activity. One of the disadvantages of homogenous systems is the problem of product separation and, most important, the problem of catalyst separation and recycling. Presently, there are several large-scale industrial processes that involve organometal complexes as catalysts. Several special laboratory procedures also exist for forming carbon–carbon bonds that are catalyzed by organotransition metal complexes. A few of the reactions in both these categories are described in the sections that follow.

9.53 THE WACKER REACTION

The oxidation of ethylene to acetaldehyde with air or oxygen in aqueous solution in the presence of catalytic quantities of Pd^{2+} (as $PdCl_2$ or K_2PdCl_4) and $CuCl_2$ discovered by Alexander Wacker. The essential steps can be represented by three stoichiometric reactions:

$$\text{(1)}\quad C_2H_4 + PdCl_2 + H_2O \longrightarrow CH_3CHO + Pd(0) + 2\,I$$

$$\text{(2)}\quad Pd(0) + 2\,CuCl_2 \longrightarrow PdCl_2 + Cu_2Cl_2$$

$$\text{(3)}\quad Cu_2Cl_2 + 2\,HCl + 1/2\,O_2 \longrightarrow 2\,CuCl_2 + H_2O$$

$$C_2H_4 + 1/2\,O_2 \longrightarrow CH_3CHO$$

A one-stage process involves passing an ethylene-oxygen mixture into an aqueous solution of $PdCl_2$ and $CuCl_2$ at 125°C and 10 atmospheres pressure. The gases coming out of the reactor are fractionated, producing a yield of acetaldehye in excess of 90%. A possible catalytic cycle describing the reactions is outlined in Fig. 9.53. The first step involves the complexation of ethylene to give a water-soluble Pd^{2+} complex analogous to the well-characterized Pt complex, Zeise's salt. This is followed by nucleophilic attack of water on the carbon of the complexed ethylene. Subsequent additions and eliminations lead to acetaldehye as shown. The Pd is reoxidized by the $CuCl_2$ (reaction 2 above), and the reduced Cu is reoxidized by oxygen (reaction 3). Monosubstituted olefins also react by similar Wacker chemistry, and nucleophilic attack by water on the complexed olefin occurs at the more substituted carbon atom, leading to formation of a ketone. Propylene, for example, reacts very rapidly and the product is, almost exclusively, acetone:

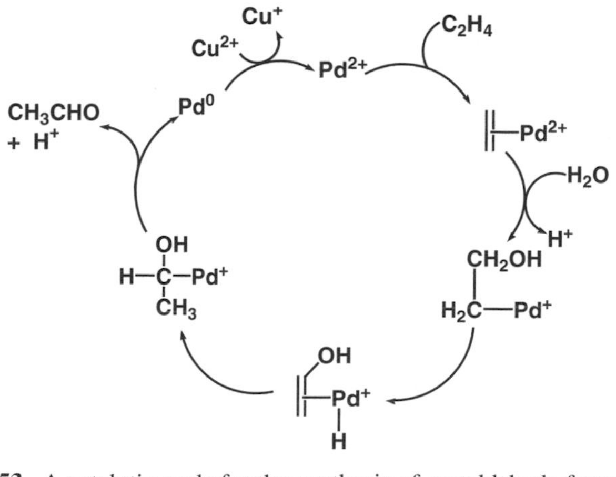

Figure 9.53. A catalytic cycle for the synthesis of acetaldehyde from ethylene.

9.54 ACETIC ACID SYNTHESIS

The synthesis of acetic acid by the reaction of methanol with carbon monoxide in the presence of methyl iodide and a soluble rhodium catalyst:

$$CH_3OH + CO \xrightarrow{Rh/CH_3I} CH_3COOH$$

The commercial process is sometimes called the Monsanto acetic acid process, after the company at which the reaction was developed. When the Rh is added as $RhCl_3$, Rh(III), its most common form, it is rapidly converted in the presence of iodide and CO to $[Rh(CO)_2I_2]^-$, where Rh is present as Rh(I). Oxidative addition of CH_3I results in $[CH_3Rh(CO)_2I_3]^-$, where octahedral Rh(III) is present. The catalytic cycle that focuses on the forms of Rh in the synthesis of acetic acid is shown in Fig. 9.54. Conditions maximizing the Rh(I)/Rh(III) ratio have been found to make for better yields and greater selectivity.

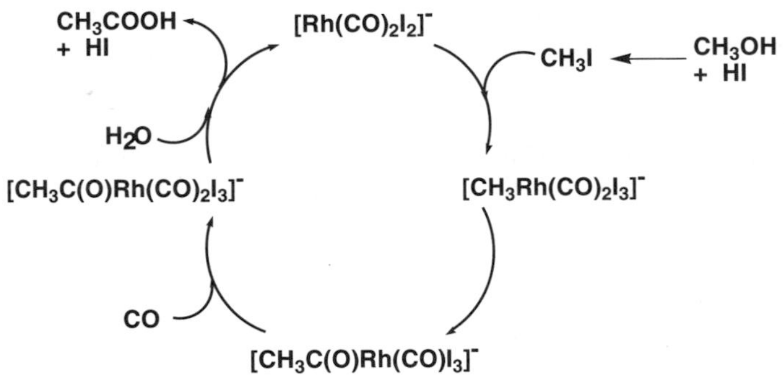

Figure 9.54. A catalytic cycle for the synthesis of acetic acid.

9.55 THE HYDROFORMYLATION (*oxo*) REACTION

The conversion of olefins to a mixture of aldehydes containing one carbon atom more that the starting olefin, by reacting the olefin with synthesis gas ($H_2 + CO$) in the presence of a cobalt or rhodium catalyst:

$$CH_3CH{=}CH_2 + CO + H_2 \xrightarrow{Co_2(CO)_8} CH_3CH_2CH_2CHO + CH_3CH(CH_3)CHO$$

Under the conditions of the reaction, almost all forms of the cobalt (or rhodium) are converted to the active forms of the catalyst, the carbonyl hydrides of cobalt ($HCo(CO)_4$) or rhodium ($HRh(CO)_4$). The reaction was discovered by Otto Roelen (1897–1993) in Germany while he was working on the *Fischer–Tropsch reaction*

(Sect. 11.75). When he added ethylene to synthesis gas and passed the gaseous mixture over a solid supported cobalt catalyst, he determined that the major products of the reaction were carbonyl (from the German *oxo*) compounds and hence used the term *oxo* to describe the reaction. Since then, the reaction has been commercialized and the catalysts modified to improve selectivity. Because the overall reaction appears to be formal addition of a hydrogen atom and a formyl group (H–CHO) across an olefinic bond, the reaction is also called the hydroformylation reaction.

Example. One application that is being successfully employed involves the conversion of propylene to *n*-butyraldehyde (plus a small quantity of isobutyraldehyde), in which a rhodium-triphenylphosphine catalyst is used with triphenylphosphine as the solvent. In this system, the Rh is present as a relatively stable phosphine complex and the product aldehydes can be separated by distillation, leaving the catalyst intact. Recently, a modification using a biphasic system has been employed involving a water-soluble Rh catalyst. In this continuous system, propylene is hydroformylated to give a water-insoluble layer consisting of the product butyraldehydes, which contain 96% of the straight-chain compound. The aqueous layer containing the Rh catalyst is essentially free of organic material, and is separated and recycled.

A catalytic cycle in which the cobalt catalyst is added as dicobalt octacarbonyl is shown in Fig. 9.55. One of the catalytic cobalt species in the cycle that has been identified and separately prepared is $HCo(CO)_4$. In the hydroformylation reaction, it readily dissociates to coordinatively unsaturated $HCo(CO)_3$, and after complexation with the olefinic substrate, it undergoes a concerted insertion reaction to give an alkylcobalt tricarbonyl species. All the tetracarbonyl species are in equilibrium with their tricarbonyl counterparts. Studies with pure $HCo(CO)_4$ have shown it to have quite remarkable properties; it can deliver its hydrogen to a substrate either as a proton, hydrogen atom, or hydride, depending on the structure of the substrate [M. Orchin, *Acct. Chem. Res. 14*, 259 (1981)]. Presently, the Rh-catalyzed biphasic hydroformylation is the preferred industrial process.

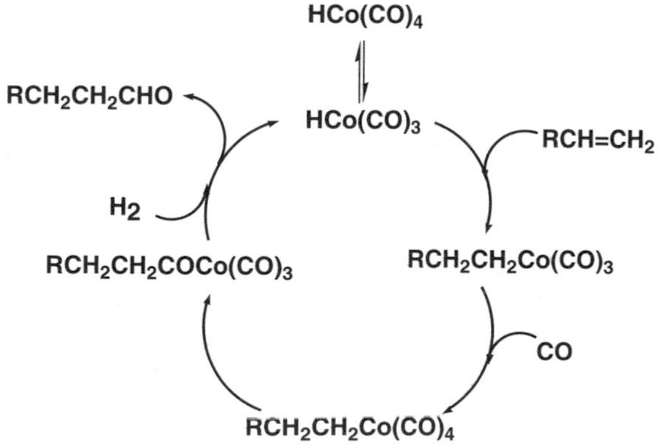

Figure 9.55. Simplified catalytic scheme for hydroformylation of olefins.

9.56 THE HECK REACTION

The Pd-catalyzed reaction (see R. F. Heck, *Org. React. 27*, 345 (1982)] of an aryl halide with an olefinic substrate in the presence of a base, resulting in a vinylic alkylation of the aryl group:

$$\text{Ar-X} + H_2C{=}CH(Z) \xrightarrow[\text{base}]{\text{Pd}} \text{Ar-CH}{=}CH(Z) + [\text{base-H}]X$$

The Z group may be alkyl, aryl, or one of a variety of functional groups, thus enhancing the reaction's versatility. The palladium may be added as the metal Pd(0) or as a Pd(II) salt such as $Pd(OAc)_2$ or Li_2PdCl_4, but because a Pd(0) species is required for the catalysis, an *in situ* reduction of the Pd(II) must occur during the reaction. This probably occurs by reductive elimination of a X-Pd-H species formed by β-hydride elimination from an intermediate Pd complex. A catalytic cycle for the reaction is shown in Fig. 9.56.

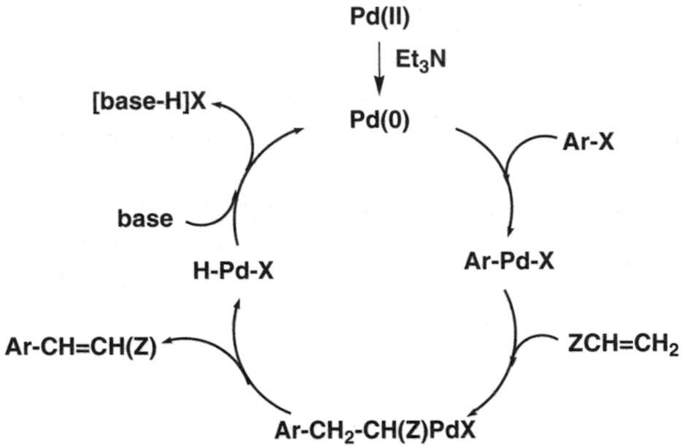

Figure 9.56. The catalytic cycle for the Heck arylation reaction.

9.57 CATALYTIC COUPLING REACTION

A general carbon–carbon coupling reaction involving three steps: (1) oxidative addition of RX to a low or zero valent transition metal catalyst (Pd, Ni) to produce a σ-alkyl or σ-aryl metal halide complex; (2) *trans*-metallation in which the halide is replaced by a desired R group attached to a main group atom (Mg, Zn, Sn, B, Al, Li) or an early transition metal atom (Zr) to generate a dialkyl, alkyl-aryl, or diaryl transition metal species; and (3) reductive elimination to give the R-R coupling product and regenerating the low or zero transition metal catalyst. The general catalytic system for the overall reaction is shown in Fig. 9.57.

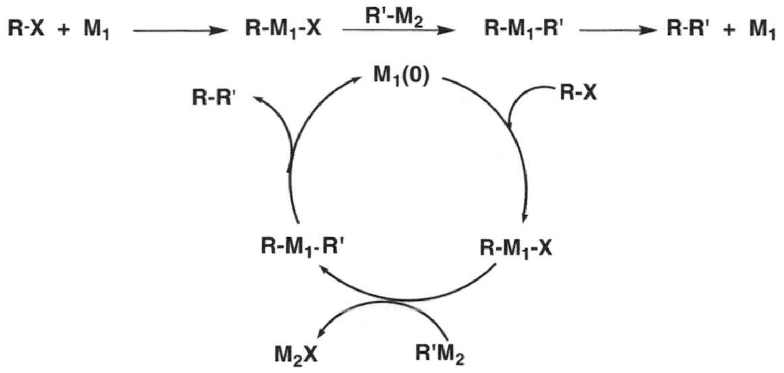

Figure 9.57. A general scheme for a catalytic coupling reaction.

9.58 SUZUKI REACTION

The palladium catalyzed aryl coupling reaction [see N. Miyaura, T. Yanagi, and A. Suzuki, *Synth. Commun. 11*, 513 (1981)] involves an aryl halide and aryl boronic acid in the presence of a base. A catalytic cycle for this reaction is shown in Fig. 9.58.

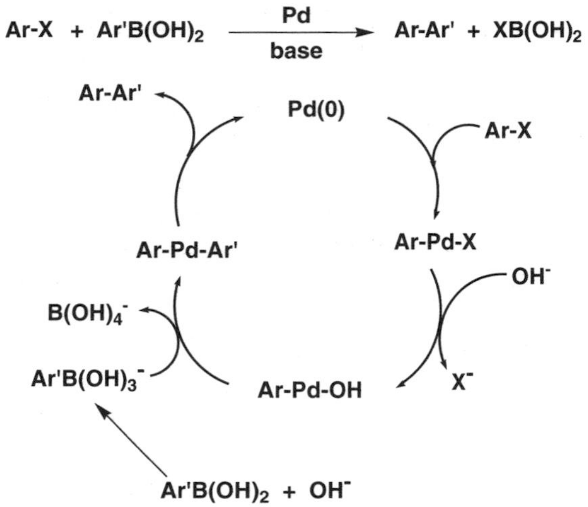

Figure 9.58. A catalytic cycle for the Suzuki reaction.

9.59 THE STILLE REACTION

The Stille reaction [see J. K. Stille, *Angew. Chem., Int. Ed. Engl. 25*, 508 (1968)] involves alkyltin complexes as the main group metal source. A catalytic cycle for the Stille reaction is shown in Fig. 9.59.

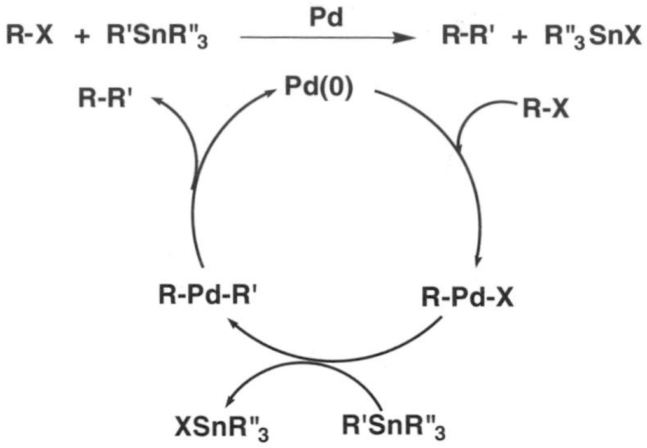

Figure 9.59. A catalytic cycle for the Stille palladium catalyzed coupling reaction.

SUGGESTED READING

Arduenco, A. J. *Accts. Chem. Res. 32*, 913 (1999).

Basolo, F. and Johnson, R. *Coordination Chemistry*. W. A. Benjamin: New York, 1964.

Chemical Engineering News, "Organozirconium Chemistry Arrives," April 19, 2004.

Collman, J. P., Hegedus, L. S., Norton, J. R., and Finke, R. G. *Principles and Applications of Organotransition Metal Chemistry*. University Science Books: Mill Valley, CA, 1987.

Crabtree, R. H. *The Organometallic Chemistry of the Transition Metals*, 2nd ed. John Wiley & Sons: New York, 1994.

Douglas, B., McDaniel, D. H., and Alexander, J. J. *Concepts and Models of Inorganic Chemistry,* 3rd ed. John Wiley & Sons: New York, 1994. Chapter 12 of this book is a particularly good resource.

Hoffmann, R. "Building Bridges Between Inorganic and Organic Chemistry." *Angew. Chemie, English ed. 21*, 711 (1982).

Jones, M. and Moss, R. A., eds. *Reactive Intermediates*, Vol. 2. John Wiley & Sons: New York, 1981.

Kodas, T. T. and Hampden-Smith, M., eds. *The Chemistry of Metal CVD*. VCH Press: Weinheim, Germany, 1944.

Lieber, S. and Brintzinger, H. H. *Macromolecules, 33*, 9192 (2000). This is one in a long series of articles by Brintzinger and coworkers on polymerization using ansa-organometallic complexes.

Negishi, E. *Acct. Chem. Res. 15*, 340 (1982).

Orchin, M. and Jaffe, H. H. *Symmetry, Orbitals and Spectra*. Wiley-Interscience: New York, 1971.

Parshall, G. W. *Homogeneous Catalysis*. John Wiley & Sons: New York, 1980.

Pines, H. *The Chemistry of Catalytic Hydrocarbon Conversions*. Academic Press: New York, 1981.

10 Separation Techniques and Physical Properties

10.1	Refluxing	390
10.2	Distillation	390
10.3	Raoult's Law	391
10.4	Ideal Solution	391
10.5	Relative Volatility α	391
10.6	Vapor–Liquid Equilibrium	391
10.7	Fractional Distillation	392
10.8	Theoretical Plate	393
10.9	Height Equivalent to a Theoretical Plate (HETP)	393
10.10	Reflux Ratio	394
10.11	Azeotrope	394
10.12	Minimum Boiling Azeotrope	394
10.13	Maximum Boiling Azeotrope	395
10.14	Molal Boiling Point Elevation Constant K_b	395
10.15	Ebullioscopic Constant	395
10.16	Extractive Distillation	395
10.17	Steam Distillation	395
10.18	Sublimation	396
10.19	Flash Point	396
10.20	Ignition Temperature (Spontaneous Ignition Temperature)	396
10.21	Explosive Limits	396
10.22	Critical Temperature and Critical Pressure	397
10.23	Chromatography	398
10.24	Column Chromatography	398
10.25	Partition Coefficient	398
10.26	Thin-Layer Chromatography (TLC)	399
10.27	Paper Chromatography	399
10.28	R_f Value	400
10.29	Electrophoresis	400
10.30	Gas–Liquid Partition Chromatography (GLPC, GLC, GC)	401
10.31	High-Performance Liquid Chromatography (HPLC)	402
10.32	Molecular Exclusion Chromatography	402
10.33	Reversed-Phase Chromatography	402
10.34	Cation Exchange Resin	402
10.35	Anion Exchange Resin	403
10.36	Ion Chromatography	403

The Vocabulary and Concepts of Organic Chemistry, Second Edition, by Milton Orchin, Roger S. Macomber, Allan Pinhas, and R. Marshall Wilson

10.37	Ion Exchange Chromatography	404
10.38	Chiral Chromatography	404
10.39	Affinity Chromatography	404
10.40	Thermal Conductivity Detector	405
10.41	Hydrogen Flame Detector	405
10.42	Flame Ionization Detector (FID)	405
10.43	Electron Capture Detector	406
10.44	Molal Freezing Point Depression Constant K_f	406
10.45	Cryoscopic Constant	406
10.46	Van't Hoff Factor i	406
10.47	Viscosity (Coefficient of Viscosity) η	407
10.48	Viscosity Index	407
10.49	Viscometer	407
10.50	Refractive Index n	407
10.51	Colligative Properties	408
10.52	Semipermeable Membrane	408
10.53	Osmosis	408
10.54	Osmotic Pressure π	408
10.55	Reverse Osmosis	409
10.56	Polarized Bond	410
10.57	London Dispersion Forces	410
10.58	Dipole–Dipole Interaction	410
10.59	Ion–Dipole Interaction	410
10.60	Polarizability	410
10.61	Dielectric Constant ε	410
10.62	Surface Tension	411
10.63	Adsorption	411
10.64	Physical Adsorption	411
10.65	Chemisorption (Chemical Adsorption)	411
10.66	Surface Area	412
10.67	BET Equation	412
10.68	Absorption	412
10.69	Interfacial Tension	412
10.70	Colloid	412
10.71	Emulsion	412
10.72	Micelle	413
10.73	Hydrophilic and Hydrophobic	413
10.74	Emulsifying Agent	414
10.75	Detergents	414
10.76	Wetting Agents	414
10.77	Surfactant (Surface Active Agent)	414
10.78	Cationic Surfactant	415
10.79	Invert Soap	415
10.80	Nonionic Surfactant	415
10.81	Aerosol	415
10.82	Phase-Transfer Catalyst	415
10.83	Liquid Crystals	415
10.84	Mesophases	416
10.85	Isotropic	416

10.86 Anisotropic 416
10.87 Thermotropic Liquid Crystals 416
10.88 Lyotropic Liquid Crystals 416
10.89 Smectic Mesophase 416
10.90 Nematic Mesophase 417
10.91 Cholesteric Mesophase 417
10.92 Ionic Liquids 418

The last half-century has been a time of enormous change in every aspect of everyday life. The awesome power of the computer has infiltrated and transformed the way of life in every advanced country. The teaching and practice of organic chemistry have not been immune to these changes. The science has broadened and deepened. Organic chemistry provides the skeletal framework for fantastic advances in molecular biology and material science. Although not widely emphasized or publicized, organic chemistry laboratory practice has changed as well. Most chemists would agree that despite advances in theory, experiment is still both desirable and necessary. In the 1940s and 50s the attribute of laboratory experimental skill, epitomized, for example, by the contributions of Louis Fieser and his talented collaborators at Harvard and the amazing synthetic expertise of Robert Woodward, was the model of excellence in carrying out the practice of organic chemistry. Teaching hands-on laboratory skills has been largely supplanted in the last half-century by teaching skills in the manipulation of instruments and interpretation of their output. The drawers in organic laboratories are now filled with pipettes and syringes rather than Erlenmeyer flasks and condensers. Two elements of scale have been dramatically affected in laboratory practice. The quantity of material used in most experiments has shrunk; experiments are now conducted with micrograms instead of grams, microliters instead of milliliters. The other element of change in scale is the time element; instead of calculating change in seconds, observations can now be routinely made in milliseconds and indeed events occurring in as little as a pico second (1×10^{-12} sec) can be observed.

In this chapter, we take our readers into the laboratory. When carrying out reactions, organic chemists want to end up with a pure product that they can place in a vial or other container, but first they must isolate a pure compound from the reaction mixture. This requires separation of the desired compound from the remaining starting material, if any, and from side products of the reaction, since it is rare to achieve 100% conversion to the desired product. If reactions are carried out on the gram scale or larger, and if the product has a reasonable boiling point, separations can be performed by fractional distillation. Although this procedure is an essential separation and purification technique in industry, particularly the petroleum industry, it is almost a lost art in the organic laboratory.

The most popular and powerful separations and purifications involve various types of chromatography. In 1910, the Italian-Russian botanist and chemist Mikail Tsvet, also spelled "Tswett" (1872–1919), performed the first chromatographic experiment when he passed an alcohol-ether solution of plant pigments over a column of calcium carbonate. He observed separate layers of different colors on the column, hence the term chromatography, meaning color writing. Variations of this basic procedure, the selective partitioning of mixtures of compounds based on their differential absorbance on stationary solid supports, followed his discovery. The 1952 Nobel Prize in Chemistry was awarded jointly to Archer S. P. Martin (1910–2002) and Richard L. M. Synge (1914–1994) for developing gas-liquid chromatography (GLC) and paper chromatography. Although the early chromatographic techniques were used primarily for analytical purposes, the development of high-performance liquid chromatography (HPLC) enhanced the use of this technique for preparative purposes as well. The pioneering work of Emanuel Gil-Av and Volker Schurig at the Weizmann Institute on the separation of enantiomers by chromatography and subsequent developments and refinements led to the widespread industrial use of chromatography, particularly in the pharmaceutical industry. If a drug under evaluation is optically active, the U.S. Food and Drug Administration (FDA), as well as most industrialized countries, requires the separate evaluation of enantiomers. The application of chromatography to problems in molecular biology has also led to the development of a technique called affinity chromatography. All these variations of chromatographic separations are briefly described in this chapter.

Before the advent of modern instrumental techniques for structure determination, chemists relied heavily on the physical properties of compounds for their characterization. The relationship between structure and physical and chemical properties is important not only in understanding organic chemistry, but also these relationships are very useful in a wide variety of practical applications where a choice between various compounds is critical; predictive power saves time and expense. Accordingly, discussions of physical properties and their measurements are also included in this chapter.

10.1 REFLUXING

The process of boiling a liquid contained in a flask, which is fitted with a condenser, such that the vapor is condensed and the resulting condensate continuously returned to the vessel from which it was vaporized. Refluxing is most commonly used when it is desired to conduct a reaction at approximately the temperature corresponding to the boiling point of the solvent chosen for the reaction.

10.2 DISTILLATION

The process of heating a flask containing a multicomponent liquid mixture to its boiling point and collecting the condensate in a second vessel, called the receiver. Thus, the condensate becomes richer in the more volatile component. A column, with an inert packing material to facilitate vapor–liquid equilibrium and enhance the separation of the various components, is inserted between the flask and the receiver, and a thermometer is usually placed at the top of the column. When changes in the vapor temperature are noted, the receiver is changed to collect various fractions (see Sect. 10.7).

10.3 RAOULT'S LAW

Named after Francois-Marie Raoult (1830–1901), the partial pressure P_A of a particular component A over a liquid solution of components is equal to the mole fraction X_A of A in the liquid times the vapor pressure P_A° of pure A at the temperature of interest:

$$P_A = X_A P^\circ_A \tag{10.3}$$

10.4 IDEAL SOLUTION

A solution that obeys Raoult's law. Such solutions have properties that are the weighted averages of the properties of the components, and the interactions between molecules of each individual component are unaffected by the presence of the other components.

10.5 RELATIVE VOLATILITY α

This is the ratio of vapor pressures at a particular temperature of two components of a solution; by convention the vapor pressure of the more volatile pure component P_A° is the numerator, that is, $\alpha > 1$:

$$\alpha = \frac{P^\circ_A}{P^\circ_B} = \frac{P_A / X_A}{P_B / X_B} = \frac{P_A \bullet X_B}{P_B \bullet X_A} \tag{10.5a}$$

Because the partial pressures P_A and P_B are directly proportional to the moles of A and B in the vapor phase, these terms can be replaced by the mole fractions of the components in the vapor phase, Y_A and Y_B. Hence,

$$\alpha = \frac{Y_A \bullet X_B}{Y_B \bullet X_A} \tag{10.5b}$$

10.6 VAPOR–LIQUID EQUILIBRIUM

In a mixture of liquids, the composition of the vapor phase relative to the composition of the liquid phase in equilibrium with it at any particular temperature.

Example. Figure 10.6 shows a typical vapor–liquid equilibrium diagram for an ideal two-component mixture of A and B.

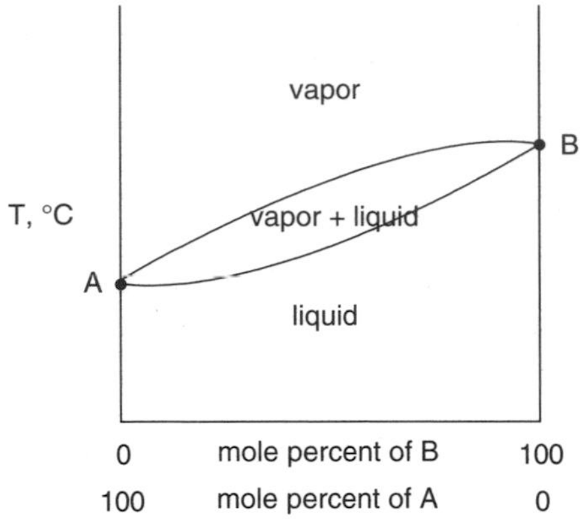

Figure 10.6. Vapor–liquid equilibrium diagram.

10.7 FRACTIONAL DISTILLATION

A distillation performed on a mixture of liquids such that the most volatile component is separated first, followed by those with successively higher boiling points.

Example. Consider the binary mixture shown in Fig. 10.6 and redrawn in Fig. 10.7, and suppose the mixture to be distilled has the composition represented by X_1. On bringing this mixture to its boiling point t_1, and then condensing the first small quantity of vapor in equilibrium with the liquid at t_1, the condensate would have the composition X_2 and boil at t_2. At the boiling point of liquid X_2, namely t_2, the vapor in equilibrium would have the composition X_3. These hypothetical successive distillations could be repeated until a very small quantity of pure A with boiling point t_A is obtained. To obtain good liquid–vapor equilibrium and simulate a succession of stages up the fractionating col-

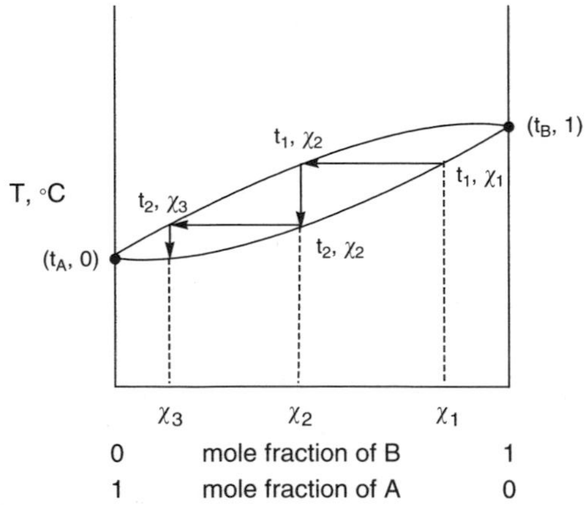

Figure 10.7. Temperature-composition curve showing the effect of two theoretical plates.

umn, an inert packing material is sometimes placed in a column above the distilling flask. Many expensive laboratory and commercial distilling columns are filled with perforated trays or plates (see Sect. 10.8) to enhance vapor–liquid equilibrium.

10.8 THEORETICAL PLATE

A plate in a distilling column on which complete liquid–vapor equilibrium is theoretically achieved. Since in practice it is impossible to obtain equilibrium, such a plate is a theoretical plate even though some distillation columns have actual plates. Under even the best of distilling conditions, several plates are required to obtain equilibrium. The closer the boiling points of components of a mixture are, the more theoretical plates are required to achieve separation.

Example. In Fig. 10.7, the dotted lines within the liquid–vapor curve represented by the step t_1X_1, t_1X_2, t_2X_2 show the change in liquid composition obtained by one equilibration with vapor, that is, one equilibrium stage; hence, this change is equivalent to that which can be achieved by one theoretical plate. The distillation column that is capable of achieving an enrichment of the more volatile component from the liquid composition represented by X_1 to that represented by X_3 in Fig. 10.7 is said to have an efficiency of two theoretical plates.

10.9 HEIGHT EQUIVALENT TO A THEORETICAL PLATE (HETP)

The height of a column packing, which can produce a separation equivalent to one theoretical plate.

Example. The HETP of a particular column is evaluated by determining the total number of theoretical plates divided by the height. To determine the number of theoretical plates the following are required: (1) a standard binary test mixture whose vapor–liquid diagram (Fig. 10.6), hence its relative volatility α, is known; (2) a method for determining the composition of this mixture in the distilling flask and the composition of the condensate under particular operating conditions. The distilling apparatus is operated under total reflux until everything connected with the column operation appears to be in a steady state; then a small amount of condensate is removed for analysis. When these data are obtained, the number of plates *n* may be determined from the formula

$$n = \frac{1}{\log\alpha} \log \frac{X_D{}^A / X_D{}^B}{X_S{}^A / X_S{}^B} \tag{10.9}$$

where α is the relative volatility, X_D^A and X_D^B are the mole fractions of *A* and *B*, respectively, in the distillate or overhead, and X_S^A and X_S^B are the mole fractions of *A* and *B*, respectively, in the still pot or distilling flask. Merrell Fenske at Pennsylvania State University developed this equation, which has become known as the Fenske equation.

10.10 REFLUX RATIO

The quantity of liquid being condensed in a distillation apparatus and returned to the column divided by the quantity collected in the receiver.

10.11 AZEOTROPE

A mixture of two (or more) liquids that distills at a constant temperature which is different from the temperature of the boiling point of the individual components. Such systems deviate from Raoult's law; that is, the total pressure is not the sum of the individual partial pressures at a particular temperature. The components of an azeotrope cannot be separated by fractional distillation.

10.12 MINIMUM BOILING AZEOTROPE

An azeotrope that boils at lower temperature than any of the pure components; such solutions show a positive deviation from Raoult's law.

Example. A typical liquid–vapor equilibrium diagram for a minimum boiling azeotrope is shown in Fig. 10.12. The azeotrope of water and alcohol containing 96.0% ethanol and 4.0% water by weight boils at 78.17°C, whereas pure ethanol boils at 78.3°C and water at 100.0°C. Water, benzene, and ethanol form a ternary azeotrope that

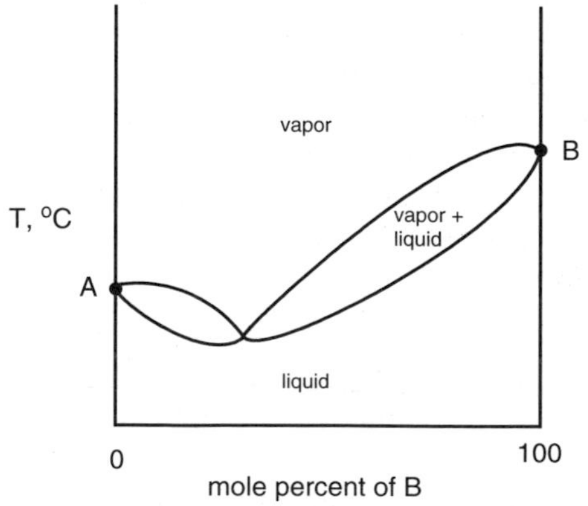

Figure 10.12. A typical minimum boiling azeotrope.

boils at 64.9°C (at 760 mm). This fact is taken advantage of for the removal of traces of water from ethanol. A small quantity of benzene is added to the 96% ethanol and a few milliliters of distillate containing the ternary azeotrope is collected. Distillation is continued until all the benzene is distilled, leaving behind anhydrous ethanol.

10.13 MAXIMUM BOILING AZEOTROPE

An azeotrope that boils at higher temperature than any of its components; such solutions show a negative deviation from Raoult's law.

Example. The azeotrope of HCl and water consists of 20.2% HCl and 79.8% water and boils at 108.6°C, whereas HCl gas has a boiling point of −80°C.

10.14 MOLAL BOILING POINT ELEVATION CONSTANT K_b

This refers to the increase in the boiling point of a pure liquid that is 1 molal (1 mole of solute per 1,000 g of solvent) in nonvolatile solute; the value is a constant, that is, independent of the nature of the solute.

Example. The boiling point of a 1 molal aqueous solution of ethylene glycol, $(CH_2OH)_2$, is 100.512°C; hence, the molal boiling point constant K_b for water is $0.512°C\,m^{-1}$.

10.15 EBULLIOSCOPIC CONSTANT

Synonymous with the molal boiling point elevation constant.

10.16 EXTRACTIVE DISTILLATION

A distillation process in which a solvent that has a higher boiling point than any of the components of the sample is added to the distillation column containing the vapor–liquid mixture of sample, in order to enhance the relative volatility of the components to be separated. The extractive solvent functions analogously to the stationary phase (*vide infra*) of a chromatographic column by interacting to a differing extent with the individual sample components and, thus, facilitating their separation.

10.17 STEAM DISTILLATION

The distillation of water mixed with a higher-boiling, immiscible liquid that enables the high-boiling liquid to be rapidly distilled at temperatures near that of boiling water. The process is made possible by the fact that in such a system the total vapor pressure at any temperature is the sum of the vapor pressures of water and the liquid at that temperature. The steam helps to transport the higher-boiling substance from the pot to the receiver.

Example. Bromobenzene boils at 157°C, but it steam-distills at 95°C because at this temperature the vapor pressure of bromobenzene is 120 mm Hg and that of water is 640 mm Hg, and thus, the total pressure is 1 atm (760 mm Hg). This is an example of Dalton's law, which states that the total pressure of a mixture of vapors is the sum of the vapor pressures of the individual components. The boiling point of a pure liquid is defined as the temperature at which the vapor pressure of the liquid corresponds to the atmospheric pressure (at sea level, 760 mm Hg).

10.18 SUBLIMATION

The process of converting a solid directly to its vapor and after a relatively short travel distance, condensing the vapor back to a solid, all without going through an intermediate liquid phase. It is a procedure used for the purification of solids with a relatively high vapor pressure.

10.19 FLASH POINT

This is the temperature to which a substance must be heated in air before its vapor can be ignited by a free flame; it is a measure of the flammability of a substance.

Example. The flash point of a typical gasoline is about −45°C and that of a typical lubricating oil is about 230°C.

10.20 IGNITION TEMPERATURE (SPONTANEOUS IGNITION TEMPERATURE)

The temperature at which a combustible mixture of vapor and air ignites in the absence of a flame.

Example. A mixture of pentane and air must be heated to about 300°C before it will ignite spontaneously. Carbon disulfide (CS_2) has an ignition temperature of about 100°C; heating it on a steam bath may ignite it without a flame source.

10.21 EXPLOSIVE LIMITS

The range of concentration of a vapor of a compound that, when mixed with air (or oxygen) and exposed to a spark, or other source of ignition, results in an explosion.

Example. Mixtures of pentane and air are explosive in the range of 1.5 to 7.5% by volume of pentane in air; if less or more pentane is present, an explosion will not occur. Hydrogen in air is explosive in the range of 4 to 74%, and of course, in the presence of pure oxygen, this range is considerably expanded so that mixtures of oxygen and hydrogen in any proportions are extremely hazardous.

10.22 CRITICAL TEMPERATURE AND CRITICAL PRESSURE

The critical temperature is the temperature above which the vapor of a compound cannot be condensed to a liquid regardless of pressure. The pressure at this temperature is the compound's critical pressure.

Any compound that is not a liquid at room temperature, that is, is a gas or a solid, can be converted to a liquid, where it exists in equilibrium with its vapor. If the compound is a solid, it can be converted to a liquid at its melting point. If the compound is a gas, it can be liquefied by subjecting it to pressure at sufficiently low temperature, because as the pressure is increased, the volume is decreased and the molecules are eventually brought so close together that the attractive force between them causes condensation to a liquid. As the liquid is heated, it becomes less dense; however, the gas becomes more dense as the pressure is increased. Eventually, the densities of the two phases become identical. The temperature at which this occurs is the critical temperature and the pressure of the compound at this temperature is the critical pressure. In a phase diagram describing this system, Fig. 10.22, this is called the critical point and is the point at which a further increase in either temperature or pressure produces the fourth phase, the super critical state. Compounds existing in the critical state are said to be supercritical, and in this state, they exhibit properties that can be quite different from those which they exhibit in either the liquid or gaseous state. This fact has led to whole new industries that use compounds such as carbon dioxide and water under supercritical conditions.

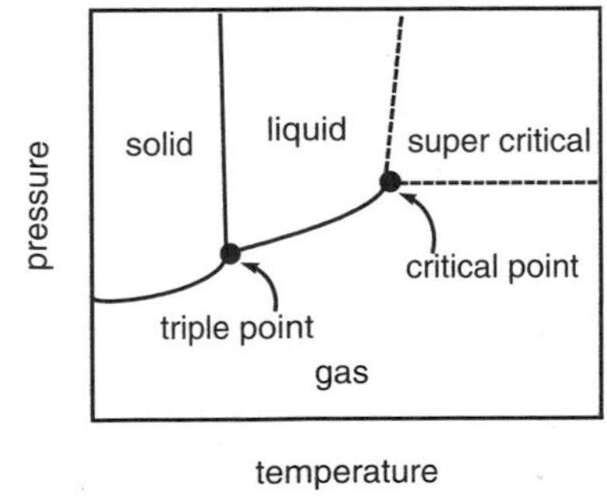

Figure 10.22. Phase diagram showing the critical temperature-critical pressure range. The triple point is the temperature-pressure point at which sold, liquid, and gas coexist.

Example. Liquid water has a dielectric constant of 78.5 at 20°C, a critical temperature of 374°C, and a critical pressure of 218 atm. Water is an excellent solvent for polar materials and a poor solvent for nonpolar materials. In contrast, supercritical water has a dielectric constant of 6.0 and in this state becomes an excellent solvent for nonpolar materials. Supercritical carbon dioxide is also widely used as a solvent. Its critical temperature is 31.3°C, and its critical pressure 73.8 atm. Supercritical CO_2 is used in the commercial extraction of caffeine from the coffee bean to produce decaffeinated coffee. Figure 10.22 shows the phase diagram indicating the critical temperature-critical pressure range.

10.23 CHROMATOGRAPHY

A general separation procedure that depends on the differential partitioning of the components of a mixture between a stationary phase, and a mobile liquid or gas phase. In the first recorded use of this technique (see the introductory section of this chapter), an extract of plant pigments in alcohol-ether solution (the mobile phase) was passed over a column of calcium carbonate (the stationary phase), and then a series of colored bands was observed on the solid phase corresponding to the different pigments in the sample. The observation of a series of colored bands inspired the name chromatography (color writing).

10.24 COLUMN CHROMATOGRAPHY

A type of chromatography in which a solid stationary phase, such as silica gel or alumina, is placed in a column and a solution of compounds to be separated is poured onto the head of the chromatographic bed. Various solvents of increasing polarity are subsequently passed through the column. The components in the mixture become distributed between the stationary phase and the solvent mobile phase by a combination of adsorption and desorption processes. The least strongly adsorbed and most easily desorbed compound, under the condition of operation, passes through the column more rapidly and is eluted from the column first, followed by the more strongly adsorbed components. This is the type of chromatography that was used in the separation of plant pigments described above.

10.25 PARTITION COEFFICIENT

The ratio of the quantities of a pure material distributed between immiscible phases. Thorough mixing of the phases is required to insure equilibrium distribution.

Example. The 1-octanol/water partition coefficient of a compound is the ratio of equilibrium concentrations of the compound in these two phases 1-octanol/water after mixing the two phases at a specified temperature. This parameter is used as a measure of the hydrophobicity/hydrophilicity of a compound, and is important in determining the efficacy of many drugs. In chromatography, the two phases of interest in the partitioning are the stationary and mobile phases.

10.26 THIN-LAYER CHROMATOGRAPHY (TLC)

A type of adsorption chromatography that utilizes a thin layer of solid adsorbent impregnated with an immobilized flourescent dye on a glass or plastic plate onto which a tiny amount of the solution of components (solutes) to be separated is placed. The bottom of the plate is then immersed in an appropriate solvent or mixture of solvents in a closed container. The solvent diffuses up the plate through the sample to be resolved. The various components move up the plate in accordance with their partition coefficients. After an appropriate time interval, the plate is removed and air-dried. Visualization of the spots on the plate can be accomplished through a variety of methods, for example, illumination by exposure to a hand-held ultraviolet lamp or exposure to I_2 vapor. Because of the low cost, speed, and convenience of this procedure, it is taught in most undergraduate chemistry laboratories, and it is used in many research and clinical laboratories.

Example. A typical arrangement is shown in Fig. 10.26.

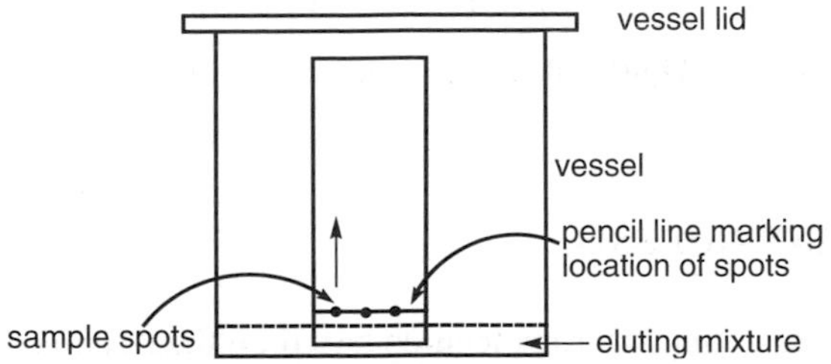

Figure 10.26. Apparatus for thin-layer chromatography.

10.27 PAPER CHROMATOGRAPHY

Liquid–liquid partition chromatography in which the components of a mixture of solutes are separated through their differing partition coefficients between a water phase adsorbed on the paper's surface and an eluting solvent that moves the components at differing rates up the paper.

Example. A typical setup is similar to that shown in Fig. 10.26, except that a strip of paper replaces the glass plate. In most applications water is the solvent and cellulose of the paper is the stationary phase.

10.28 R_f VALUE

In TLC or paper chromatography, the distance the solute moves divided by the distance traveled by the solvent front relative to the position of the original sample. The measurement is made when the solvent front has moved quite far from the origin.

Example. Typically, a mixture of compounds *A* and *B* is spotted on a TLC plate or paper very close to the bottom, as shown in Fig. 10.28, and the paper or plate is placed in a beaker with an eluting solvent. After an appropriate length of time the eluting solvent can be seen at the solvent front, as depicted in Fig. 10.28. The plate or paper is removed and the spots developed appropriately, and then the R_f values for *A* and *B* may be calculated and compared with the R_f values of authentic *A* and *B*, each determined as above.

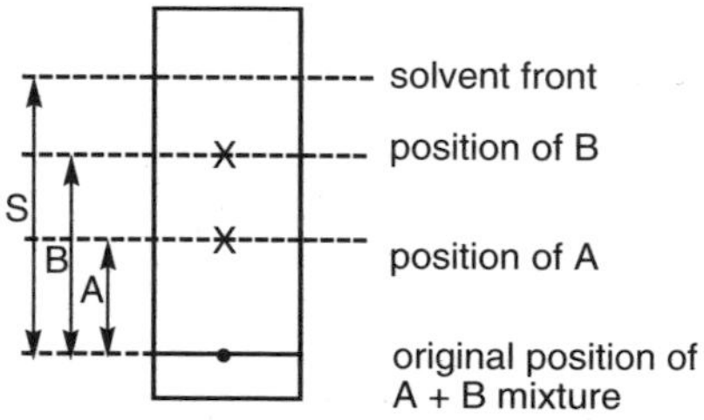

Figure 10.28. R_f values for *A* and *B*.

10.29 ELECTROPHORESIS

A procedure for separating macromolecules based on their charge to mass ratios. It involves the migration of charged particles (ions) under the influence of an electric field. The medium separating the electrodes is most frequently a gel (a solid colloid, see Sect. 10.70), hence the term gel electrophoresis (from the Greek word *phoresis*, meaning "to carry across"). In this technique, molecules are forced across a span of gel. Important biological molecules such as amino acids, peptides, nucleotides, and nucleic acids are all charged ions at appropriate pH and depending on net charge will migrate as cations (to the anode) or as anions (to the cathode). The rate of migration through the pores of the gel depends on the size and shape of the ions, as well as the strength of the electric field and the pH of the buffer solution. After staining, the separated macromolecules are seen as a series of bands spread between the ends of the gel. Classical electrophoretic methods using slab gels require post-column visualization similar to methods used in thin-layer chromatography. In gel electrophoresis, this is usually accomplished by staining. Instrumental-based electrophoretic separations, such as capillary electrophoresis, incorporate many of the same detection methods that are used in HPLC.

10.30 GAS–LIQUID PARTITION CHROMATOGRAPHY (GLPC, GLC, GC)

A type of chromatography in which the sample solution is first vaporized by passing it into a hot (250°C) injection port along with a stream of inert gas (usually He as the mobile phase) and the gaseous mixture is then passed though a column of adsorbent before going to a detector. Two general column configurations (with many variations) are used. One is a capillary column, Fig. 10.30, consisting of a flexible fused silica capillary typically 10 to 60 m long and about 0.3 mm in diameter, coated on the inside with a high-boiling, stable liquid (such as a silicone, the stationary phase). The other type of column is a packed column, shorter (usually less than 6 m) and wider (2–4 mm) than a capillary, and contains a finely divided, inert solid material like diatomaceous earth onto which the stationary phase is adsorbed. In both cases, when the stationary phase is polar, the relatively nonpolar compounds (or analytes as analytical chemists prefer to call individual compounds investigated with analytical instruments) in the mixture passed through the column more rapidly than the polar (analytes) compounds. It is also possible to use a nonpolar stationary phase (e.g., squalene), in which case polar compounds elute more rapidly than nonpolar compounds. This chromatographic technique, in which compounds in a mixture are differentially distributed between a gaseous mobile phase and a stationary phase, is frequently referred to as gas–liquid chromatography (GLC). Such chromatography is not only thousands of times more efficient than distillation techniques, but can be employed with thousands of times smaller quantities (10^{-3} mL). It is no wonder that such chromatography has revolutionized the way liquid mixtures are separated and purified in most chemical laboratories.

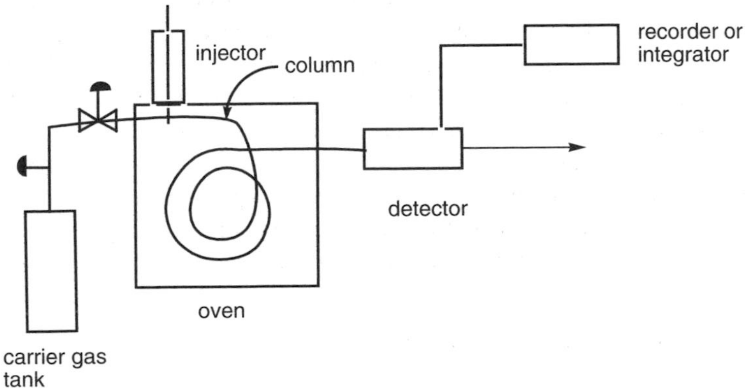

Figure 10.30. Schematic diagram of a gas–liquid chromatograph.

10.31 HIGH-PERFORMANCE LIQUID CHROMATOGRAPHY (HPLC)

A liquid chromatographic procedure carried out at high pressures to enhance the processing capacity of the chromatographic instrument by reducing the time required for the separation of components of a complex mixture, hence the term *high-performance*. All forms of column chromatography involving a liquid mobile phase can be configured to HPLC. Some HPLC equipment is designed to operate at pressures as high as 6,000 psi. Such high pressures are utilized in many types of chromatography (ion, reversed phase, chiral, affinity) and are now routinely carried out for effecting the separation of pure compounds for preparative purposes instead of for purely analytical purposes. The stainless steel columns used in HPLC laboratory equipment are very small bore (1–7 mm) and are packed with a stationary phase consisting of very small particles (5 μm diameter). The high-pressure modification has induced some chemists to refer to HPLC as *high-pressure liquid chromatography*. Two common detectors for HPLC are based on changes in the refractive index (see Sect. 10.50) and changes in the ultraviolet spectrum (see Sect. 18.39).

10.32 MOLECULAR EXCLUSION CHROMATOGRAPHY

A chromatography procedure using porous particles in which the separation depends on the ability of relatively small molecules to be trapped in the pores of the packing, while the large molecules are excluded and more rapidly pass through the medium. Packings with large pores are used to fractionate high molecular weight substances, whereas the denser and smaller packings are used for the separation of relatively low molecular weight substances. The variation of this type of chromatography used for the separation of biopolymers is referred to as gel permeation chromatography.

10.33 REVERSED-PHASE CHROMATOGRAPHY

A type of partition chromatography in which hydrocarbons, as well as polar compounds, are partitioned between a nonpolar stationary phase and a polar mobile phase. Under these conditions, the most polar substances elute most rapidly. The stationary phase often consists of hydrocarbon chains chemically bonded to silica surfaces and the eluting solvents are frequently aqueous methanol or aqueous acetonitrile. Liquid partition chromatography techniques were vastly improved by the introduction of small packing particles in small-diameter columns operating under high pressures with the stationary phase chemically bonded to the solid packing. Reversed-phase chromatography is one of the most widely used variations of HPLC.

10.34 CATION EXCHANGE RESIN

The exchange of cations in solution for protons provided by the cation exchange resin. The exchange resin is usually a cross-linked polystyrene, in the form of beads, that have been sulfonated (or carboxylated) and thus have -SO_3H or -C(O)OH functions on the aromatic rings of the polymeric resin.

Example. The exchange of metal cations, M^{n+}, with a strong acid cation exchange resin depends on the equilibrium

$$n\ [\text{P-SO}_3\text{H}] + M^{n+} \rightleftharpoons (\text{P-SO}_3)_n\text{M} + n\ H^+ \quad (10.34a)$$

With a weak carboxylic acid cation exchange resin, there is a similar equilibrium:

$$n\ [\text{P-CO}_2\text{H}] + M^{n+} \rightleftharpoons (\text{P-CO}_2)_n\text{M} + n\ H^+ \quad (10.34b)$$

where *P* represents the polystyrene framework to which the acids are attached. The weaker acid ion exchange resin has a greater affinity for protons and, therefore, requires a higher pH for a given ion to exchange.

10.35 ANION EXCHANGE RESIN

A resin that allows the exchange of anions in solution for OH^- anions provided by the anion exchange resin.

Example. The equilibrium with a strong base exchange resin (alkyl quaternary ammonium salt) is

$$n\ [\text{P-NR}_3\text{OH}] + A^{n-} \rightleftharpoons (\text{P-NR}_3)_n\text{A} + n\ OH^- \quad (10.35a)$$

and with a weaker base exchange resin (ammonium salt) a similar equilibrium occurs:

$$n\ [\text{P-NH}_3\text{OH}] + A^{n-} \rightleftharpoons (\text{P-NH}_3)_n\text{A} + n\ OH^- \quad (10.35b)$$

The weaker base ion exchange resin has greater affinity for hydroxide groups, and a lower pH is therefore required for a given ion to exchange.

10.36 ION CHROMATOGRAPHY

A chromatography technique useful for the separation of inorganic ions and amino acids, based on the exchange of ions in the sample, with ions in a stationary phase. The stationary phase is usually a polystyrene resin in which some of the aromatic hydrogens have been replaced by appropriate functional groups.

Example. The chloride and nitrate anions can be separated and their concentrations determined in an aqueous solution of the their sodium salts by using a chromatographic column packed with an anion exchange resin (Sect. 10.35). A small quantity of the solution of analytes is injected onto the column and the eluent consisting of an aqueous solution of sodium bicarbonate-carbonate is added to the column. The Cl^- and NO_3^- compete for the resin sites according to the equations in Sect. 10.35, and since the Cl^- is less strongly associated with the resin, it is eluted first as NaCl. Usually, the ion exchange column is followed by what is called a suppressor column that contains an ion exchange resin, which converts the salts into their corresponding acids, HCl and HNO_3. The acids are detected by a conductivity cell.

10.37 ION EXCHANGE CHROMATOGRAPHY

Ion exchange chromatography relies on basically the same types of resins as are used in ion chromatography (see Sect. 10.36), but with a higher ion exchange capacity and generally lower column efficiencies. Most analytical chemists prefer to use the term ion chromatography when they refer to the technique to separate and determine the concentration of simple cations (e.g., Na^+, K^+, NH_4^+) or anions (e.g., Cl^-, Br^-, NO_3^-). But since the principles of ion chromatography using ion exchange resins can be applied to the separation of mixtures of any species of charged molecule, its use for the separation of amino acids, peptides, proteins, and nucleotides has become a powerful tool for the biochemist who generally prefers to call the technique ion exchange chromatography. The terms are frequently used interchangeably.

10.38 CHIRAL CHROMATOGRAPHY

A chromatographic procedure that results in the separation of enantiomers by selective interaction of the enantiomers with a chiral stationary phase called the selector. Chiral stationary phases may consist of a variety of chiral selectors including proteins, macrocyclic antibiotics, cyclodextrins, polysaccharides, crown ethers, or peptide-derived selectors. In general, the differential retention of enantiomers is a result of selective interactions between the enantiomers of the solute and the chiral centers of the stationary phase. A large variety of stationary phases are available through commercial sources. Chiral chromatography has been incorporated into HPLC and is in widespread use in the pharmaceutical industry.

10.39 AFFINITY CHROMATOGRAPHY

A type of chromatography that depends on the affinity of immobilized ligand for a particular partner present in a mixture of substances that has complementary properties. The ligand is immobilized by bonding to an inert matrix material, such as polyacrylamide, cellulose, or agarose gel (a copolymer of D-galactose and 3,6-anhydro-L-galactose, which in bead form is called sepharose). The ligand is biospecific; for example, it may be an antibody, an enzyme, or a receptor protein, which has an affinity for the substance of interest in the mixture, for example, an antigen or a hormone. This type of chromatography is an implementation of the classical concept of lock and key specificity and is particularly useful in the study of biological materials. In some variations, a chain of carbon atoms called a spacer is inserted between the matrix and the ligand. After the sample mixture has entered the column and the compound of interest has been adsorbed, the noninteracting components of the mixture are washed out of the column, usually by a buffered aqueous solution, and finally the desired compound is eluted, usually by changing the pH of the buffer.

10.40 THERMAL CONDUCTIVITY DETECTOR

The effluent from a chromatographic separation must go to a detector of some type. Frequently, the column is linked to a mass spectrometer incorporating in one integrated piece of equipment GC/MS, a powerful tool for identification, as well as separation (see Sect. 17.23). We will consider here the stand-alone detectors and disregard at this time the mass spectrometer component.

The thermal conductivity detector is used in many GLPC instruments. Its operation depends on the fact that when a gas is passed over a heated thermister, the temperature, hence the resistance, of the thermister varies with the thermal conductivity of the gas. The difference in thermal conductivity, measured as a function of the resistance between pure carrier gas (usually, helium because of its low thermal conductivity) and a similar stream of carrier gas carrying a minute quantity of one of the components, is detected by means of a Wheatstone bridge circuit. The change in resistance occasioned by the presence of one of the components of the sample in the carrier gas, which is proportional to its concentration, is registered on a recorder.

10.41 HYDROGEN FLAME DETECTOR

A detector used in some GLPC instruments, This type of detector is based on the difference in luminosity between a flame produced by burning pure hydrogen and a flame in which the same quantity of hydrogen is diluted with a small amount of organic material present in the sample. The change in flame temperature can also be measured by a thermocouple.

10.42 FLAME IONIZATION DETECTOR (FID)

This is a detector that is almost 10^3 times more sensitive than the thermal conductivity detector; its functioning depends on the fact that most organic compounds form ions in a hydrogen flame. A pair of oppositely charged electrodes measures the ion current in the flame.

10.43 ELECTRON CAPTURE DETECTOR

Extremely sensitive for compounds that contain electronegative atoms. The cathode in the detector cell consists of a metal foil impregnated with an element (usually, tritium or ^{63}Ni) that emits β particles or high-energy electrons. On applying a potential to the cell, the β particles emitted at the cathode migrate to the anode, generating a current. If a compound with a high electron affinity is introduced into the cell, it captures electrons, decreasing the current; the resulting change is suitably recorded. Such a detector is very selective for compounds containing halogen, carbonyl, and nitro groups, but has overall low sensitivity for hydrocarbons other than aromatics.

10.44 MOLAL FREEZING POINT DEPRESSION CONSTANT K_f

This is the decrease in the freezing point of a pure liquid that is 1 molal (1 mole of solute per 1,000 g of solvent) in a nondissociating solute; the value is characteristic of each solvent and is a constant independent of the nature of the solute (provided that no dissociation of the solute occurs).

Example. Pure benzene has a freezing point of 5.5°C. When 2.40 g of biphenyl (mol. wt. 154) is dissolved in 75.0 g of benzene, the freezing point of the solution is lowered 1.06°C (ΔT). The molal freezing point depression constant K_f of benzene can then be calculated from the equation

$$k_f = \frac{\Delta T}{\text{molality}} = \frac{1.06}{0.208} = 5.1\ ^\circ\text{C}\ m^{-1} \tag{10.44}$$

Thus, a 1 molal solution of any organic solute in benzene would freeze at 0.4°C.

10.45 CRYOSCOPIC CONSTANT

Synonomous with molal freezing point depression constant.

10.46 VAN'T HOFF FACTOR *i*

Named after Jacobus H. van't Hoff (1852–1911), the ratio of the observed freezing point depression (ΔT) to that expected on the basis of no dissociation of the solute ($m \bullet K_f$):

$$i = \frac{\Delta T}{m\ K_f} \tag{10.46}$$

where m is the molality of the solution and K_f the cryoscopic constant of the solvent.

Example. 100% H_2SO_4 has a convenient freezing point (−12°C). The fact that a solution of o-benzoylbenzoic acid in this solvent gives a van't Hoff factor of 4 is evidence that the reaction shown in Fig. 10.46 occurs, giving rise to four ions.

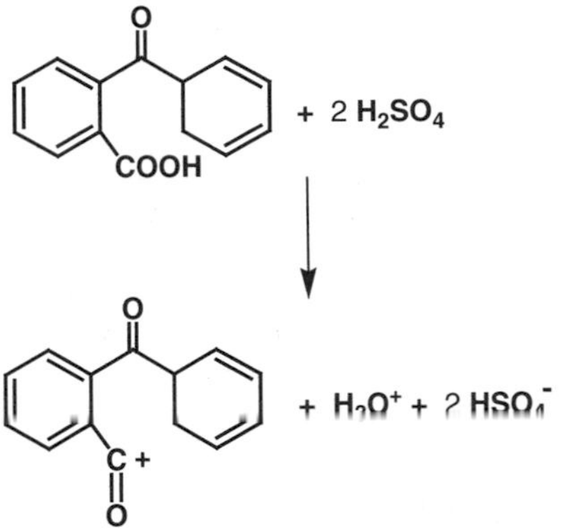

Figure 10.46. The ionization of *o*-benzoylbenzoic acid.

10.47 VISCOSITY (COEFFICIENT OF VISCOSITY) η

This is a measure of the resistance to flow of any fluid; the greater the resistance, the higher the viscosity. The coefficient of viscosity η, a term used interchangeably with viscosity, is given in units of poise (after Jean L. M. Poiseuille, 1799–1869) and has the dimension of 10^{-1} kg m^{-1} sec^{-1}. η is frequently determined from the rate of flow of a liquid through cylindrical tubes.

10.48 VISCOSITY INDEX

A measure of the change in viscosity as a function of temperature. Good lubricants have a low viscosity index.

10.49 VISCOMETER

An instrument for measuring viscosity. These are of various types. Three common methods of measurement are the determination of the rate of flow of the sample through a calibrated tube; the rate of settling for a ball in the liquid to be measured; and measurement of the force required to turn one of two concentric cylinders, filled with the liquid being investigated, at a certain angular velocity.

10.50 REFRACTIVE INDEX *n*

A measure of the apparent speed of light c' passing through matter relative to the speed of light c passing through a vacuum: $n = c/c'$. The change in the refractive index of a substance as a function of the wavelength of light is known as dispersion.

The density and refraction index of a liquid are useful in characterizing a pure liquid by means of its molecular refractivity [*R*], calculated from the equation

$$[R] = \frac{n^2 - 1}{n^2 + 2} \bullet \frac{M}{d} \tag{10.50}$$

where *M* is the molecular weight of the substance, *n* its refraction index, and *d* its density (determined at the same temperature as *n*). Refractive index measurements are commonly carried out using the sodium D line.

10.51 COLLIGATIVE PROPERTIES

Properties of solutions that depend principally on the concentration of dissolved particles and not the characteristics of these particles, for example, molal freezing point depression and boiling point elevation.

10.52 SEMIPERMEABLE MEMBRANE

A membrane that permits some types of molecules, but not others, to pass through it (see Fig. 10.54).

Example. When an aqueous solution of sucrose is placed in one arm of a U-tube, the bottom of which contains a semipermeable membrane separating the two arms (e.g., cellophane), and pure water is placed in the other arm, the water will diffuse into the solution of sucrose because the concentration of water on one side of the membrane is greater than that on the other, and only the water but not the sugar can diffuse through the membrane.

10.53 OSMOSIS

The process by which one type of molecule in a solution will preferentially pass through a semipermeable membrane, but another type will not.

10.54 OSMOTIC PRESSURE π

The hydrostatic pressure that develops at equilibrium on one side of a semipermeable membrane as a result of selective diffusion.

Example. The U-tube shown in Fig. 10.54 has unequal hydrostatic pressure on the two sides of the membrane. If the pure solvent in one arm were water and an aqueous solution of sucrose were in the other arm, the difference in the heights of the two

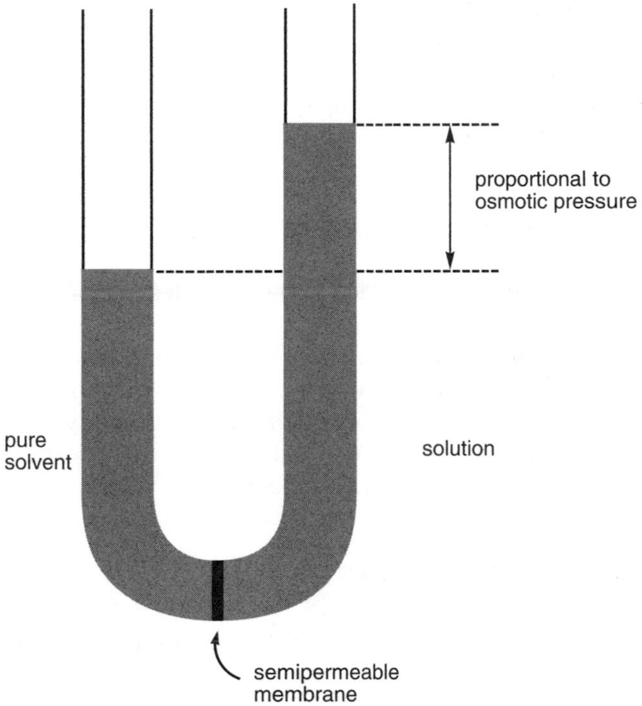

Figure 10.54. Measurement of osmotic pressure with a U-tube.

arms of the tubes would be a measure of the osmotic pressure. The osmotic pressure π is very similar to p (pressure) in the gas equation ($pV = nRT$):

$$\pi V = nRT \tag{10.54}$$

where π is the osmotic pressure (in atmospheres), n equals the number of moles of solute dissolved in volume V (in liters), R is the gas constant (0.08206 L atm K^{-1} mol^{-1}), and T is the absolute temperature.

10.55 REVERSE OSMOSIS

An osmosis process in which external pressure is applied to force a solvent to flow in the direction opposite to that normally expected.

Example. If, at equilibrium, the right arm of the U-tube shown in Fig. 10.54 containing the solution of sugar is subjected to pressure, water can be made to flow through the membrane to the left arm of the U-tube. This is one of the most important processes used in water desalination.

10.56 POLARIZED BOND

The bond that results when two different atoms are bonded and the electrons bonding them are unequally shared, for example, H–Cl. Other types of bonding, as well as dipole moments, are discussed in Chap. 2.

10.57 LONDON DISPERSION FORCES

Named after Fritz London (1900–1954), this is the interaction between uncharged molecules that arises because of the transient noncoincidence of the centers of negative and positive charge; these transient dipoles interact with similar transient dipoles in neighboring molecules. The attraction caused by London forces varies with the seventh power of the distance separating the molecules involved. Van der Waals forces and London forces are frequently used interchangeably, but the latter are really a specific type of van der Waals interaction.

10.58 DIPOLE–DIPOLE INTERACTION

An interaction between molecules having permanent electric dipole moments.

10.59 ION–DIPOLE INTERACTION

This is the interaction or solvation of ions by molecules with electrical dipoles; the solubility of most inorganic salts in water results from such interactions.

10.60 POLARIZABILITY

The susceptibility of an uncharged molecule to the development of an electric dipole under the influence of a transient or permanent dipole usually provided by a neighboring molecule. (For polarization induced by electromagnetic radiation, see Sect. 16.64.) The polarizability of a molecule increases as the number of electrons increases and as their distance from the nucleus increases.

10.61 DIELECTRIC CONSTANT ε

A constant that serves as a measure of the polarity of a solvent. It is determined by applying an electric field across a solvent positioned between two plates of a capacitor. When an electric field is applied, the solvent molecules are polarized and align themselves to reduce the strength of the field. The dielectric constant is measured relative to the effect the same applied field has on a vacuum, that is, ε is always greater than 1.

Example. Solvents with high dielectric constants are much better solvents for ions and polar organic compounds than solvents with low dielectric constants. Some organic solvents and their dielectric constants are given in Table 10.61.

TABLE 10.61. Dielectric Constants of Some Common Organic Solvents

Organic Solvent	Dielectric Constant ε (at 20°C)
CH_3Cl	12.6
CH_2Cl_2	9.1
$CHCl_3$	4.8
CCl_4	2.2
C_6H_6	2.3
C_6H_{14}	1.9
$H_3C\text{-}C(O)\text{-}CH_3$	21
CH_3CN	39
H_2O	80
$H\text{-}C(O)\text{-}NH_2$	109

10.62 SURFACE TENSION

A measure of the force acting on surface molecules by the molecules immediately below this surface. It is expressed in force per unit width, and hence, the units are dyn cm^{-1}. Surface tension causes liquids to maintain a minimum surface area, and the smallest surface-to-volume ratio generates a sphere, hence water droplets.

10.63 ADSORPTION

A process in which molecules (either gas or liquid) adhere to the surface of a solid. Molecules and atoms adhere in two ways as described below.

10.64 PHYSICAL ADSORPTION

An adsorption process that depends on the van der Waals force of interaction between molecules (gas or liquid) and the solid surface on which they are adsorbed. During this process, the chemical identity of the molecules remains intact. It is an exothermic process.

10.65 CHEMISORPTION (CHEMICAL ADSORPTION)

An adsorption process that depends on chemical interaction (weak chemical bonds) between molecules (gas or liquid) and a solid surface.

Example. The chemisorption of molecules on a solid catalyst is presumed to be the first step in many examples of heterogeneous catalysis.

10.66 SURFACE AREA

The area of an adsorbent surface that can be covered completely with a unimolecular layer of gas. It is most commonly calculated from the BET equation.

10.67 BET EQUATION

An equation named after Stephen Brunauer (1903–1986), Paul Emmett (1900–1985), and Edward Teller (1908–2003) [*J. Am. Chem. Soc.*, *60*, 309 (1938)], used in the experimental determination of surface areas of solids.

Example. At low temperature, N_2 forms a monolayer on the surface of a solid. Because the size of N_2 is known (15.8 Å^2), by measuring the quantity of N_2 adsorbed on the sample at 77 K, as a function of the pressure of N_2 over the sample, the surface area of the sample can be determined. A typical silica gel may have a surface area of $500\,m^2g^{-1}$.

10.68 ABSORPTION

The intimate mixing of the atoms or molecules of one phase with the atoms or molecules of a second phase.

10.69 INTERFACIAL TENSION

The surface tension at the interface between two liquids mutually saturated with each other, which tends to contract the interface area.

10.70 COLLOID

A two-phase system consisting of very fine solid particles (in the 10–10,000 Å range) dispersed in a second phase. The very large surface area of the dispersed phase is responsible for the unique properties of a colloidal system.

10.71 EMULSION

A system of two immiscible liquids, one of which is dispersed in small drops throughout the other. Not all emulsions are composed of micelles (see Sect. 10.72).

10.72 MICELLE

The stable, ordered aggregation of molecules that constitutes the dispersed phase of an emulsion.

Example. In water, soap molecules, for example, sodium oleate [$CH_3(CH_2)_7-CH{=}CH(CH_2)_7C(O)ONa$], orient themselves in the micelle form depicted in Fig. 10.72.

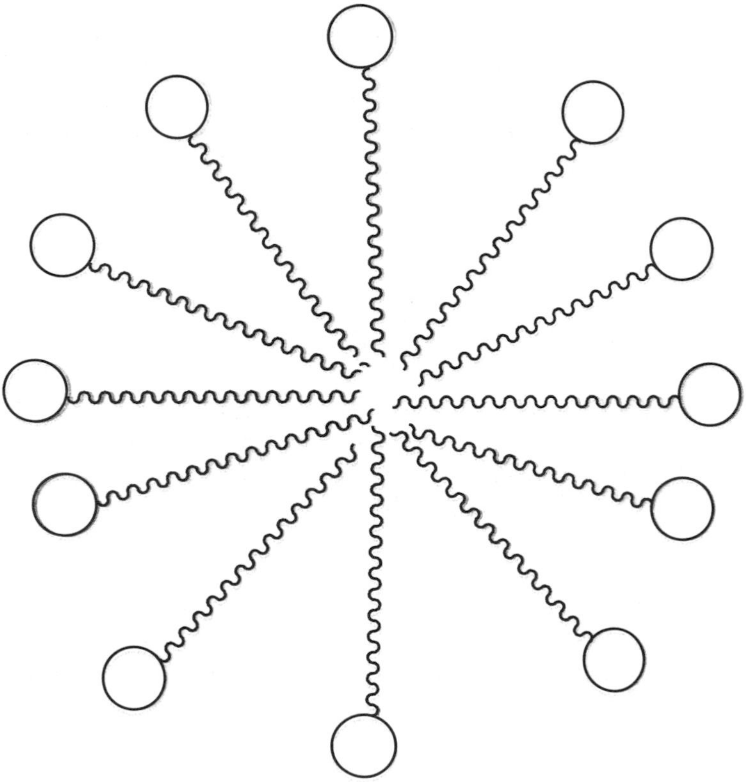

Figure 10.72. A micelle [the circles represent the polar end group ($C(O)O^-$] and the wavy lines represent the long hydrocarbon chain.

10.73 HYDROPHILIC AND HYDROPHOBIC

Water-attractive and water-repellant, respectively.

Example. Surface active agents or surfactants (soaps), such as sodium oleate, have a hydrophilic site [the $-C(O)O^-$] and a hydrophobic site (the hydrocarbon chain).

The self-aggregation properties of soaps in water allow for hydrocarbons to be drawn into the hydrophobic interior of the micelle, thus allowing the hydrocarbon to be dispersed in the water as an emulsion.

10.74 EMULSIFYING AGENT

A substance, which when added to an emulsion that tends to aggregate, keeps the system more or less permanently emulsified.

Example. The emulsifying action of soap depends on its ability to reduce the interfacial tension between a liquid hydrocarbon layer and water. The interfacial tension of benzene-water (35 dyn cm^{-1}) is reduced to 2 dyn cm^{-1} by the addition of sodium oleate. The emulsion that results may have as much as 100 parts of benzene, spread out as drops throughout only 1 part of water.

10.75 DETERGENTS

Compounds that act as emulsifying agents.

Example. Soaps (sodium salts of long-chain fatty acids) are detergents, but other types of molecules have been synthesized that are even better detergents than soaps. One group of commercially important synthetic detergents are the sodium salts of alkylbenzenesulfonic acids, $RC_6H_4SO_3^- Na^+$.

10.76 WETTING AGENTS

A substance adsorbed on the surface of a solid that interacts more strongly with water than the solid surface, thus permitting the otherwise inactive surface to become wet with water.

Example. The effectiveness of detergents in removing soil from fabrics depends on wetting and emulsification. The adsorption of the detergent at the solid surface allows the fabric to be wet by water and allows any oils to be emulsified and removed.

10.77 SURFACTANT (SURFACE ACTIVE AGENT)

This is a general term for a substance that reduces the surface tension between immiscible liquids; it is essentially synonymous with emulsifying agent (Sect. 10.74).

10.78 CATIONIC SURFACTANT

A surfactant whose polar group is a cation.

Example. Quaternary ammonium salts having as one of the R groups a long-chain alkyl group, for example, $C_{16}H_{33}N^{+}(CH_3)_3\ Cl^{-}$. Compounds such as $RC_6H_4SO_3^{-}Na^{+}$ are anionic surfactants.

10.79 INVERT SOAP

Synomymous with a cationic surfactant; so-called because the polar group is a cation rather than an anion, as in the more conventional soaps and detergents.

10.80 NONIONIC SURFACTANT

A surfactant that is an uncharged (neutral) species without cationic or anionic sites.

Example. Alkyl aryl ethers of poly(ethylene glycol) prepared from alkylphenols and ethylene oxide: $RC_6H_4O(CH_2CH_2O)_xCH_2CH_2OH$. This type of surfactant is used in formulating low-suds products for automatic washers.

10.81 AEROSOL

A colloidal system in which a solid or liquid is dispersed in a gas.

10.82 PHASE-TRANSFER CATALYST

A compound whose addition to a two-phase organic-water system helps transfer a water-soluble reactant across the interface into the organic phase, where a homogeneous reaction can occur, thus enhancing the rate of reaction between the reactants in the separate phases.

Example. A quaternary salt that is a phase-transfer catalyst function as shown in Fig. 10.82 for the reaction between 1-bromooctane and sodium thiophenoxide, in the presence of tetramethylammonium bromide.

10.83 LIQUID CRYSTALS

Generally, rodlike molecules that undergo transformations into stable, intermediate semifluid states before passing into the liquid state.

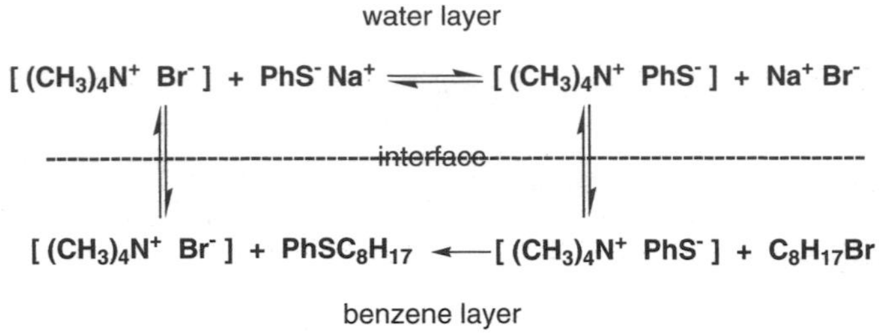

Figure 10.82. Phase-transfer catalysis for the reaction $PhS^- Na^+ + RBr \rightarrow PhSC_8H_{17} + Na^+Br^-$.

10.84 MESOPHASES

The stable phases exhibited by the liquid crystal molecules as they pass from the crystalline solid to the liquid state.

10.85 ISOTROPIC

Having identical physical properties in all directions. All liquids are isotropic because complete randomness occurs in the arrangements that the molecules may assume.

10.86 ANISOTROPIC

Having variations in physical properties along different axes of the substance, as contrasted to isotropic.

10.87 THERMOTROPIC LIQUID CRYSTALS

Liquid crystals that are formed by heating. One of the two main classes of liquid crystals.

10.88 LYOTROPIC LIQUID CRYSTALS

This is a second main class of liquid crystals; these are formed by mixing two components, one of which is highly polar such as water. Such liquid crystals are found in living systems.

10.89 SMECTIC MESOPHASE

The mesophase of liquid crystals that shows more disorder than the crystalline phase but still possesses a layered structure. In the smectic mesophase the layers of molecules have moved relative to each other, but the molecules retain their parallel orientation with respect to one another.

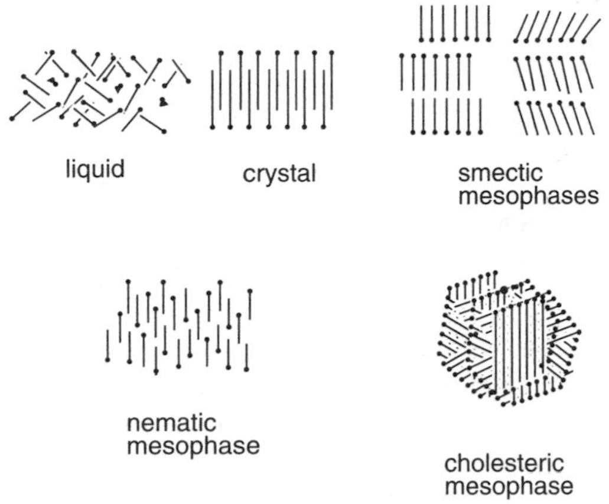

Figure 10.89. The various forms and phases of a liquid crystal.

10.90 NEMATIC MESOPHASE

In this phase, the liquid crystal molecules are still constrained to be parallel to each other, but they are no longer separated into the layers that characterize the smectic mesophase (see Fig. 10.89).

Example. Butyl *p*-(*p*-ethoxyphenoxycarbonyl)phenyl carbonate (Fig. 10.90) is a crystalline material that, on heating to −55°C, undergoes a transition to a nematic mesophase and at −87°C becomes an isotropic liquid.

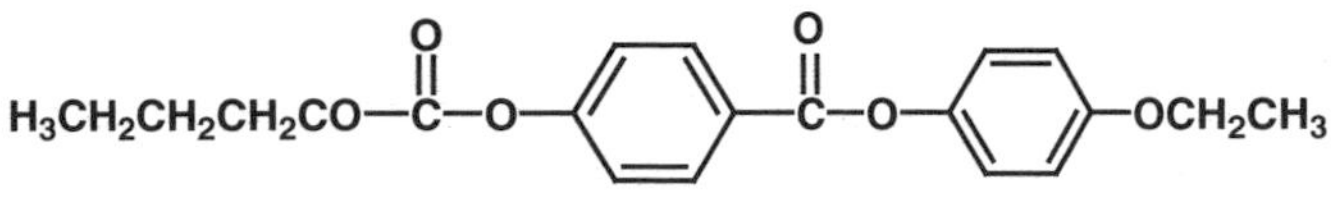

Figure 10.90. Butyl *p*-(*p*-ethoxyphenoxycarbonyl)phenyl carbonate.

10.91 CHOLESTERIC MESOPHASE

In this phase, the liquid crystal molecular arrangement resembles a twisted structure. The direction of the long axis of the molecules in the nematic phase is displaced from the long axis in an adjacent sheet of nematic phase molecules. The overall displacement results in a helical arrangement.

Example. See Fig. 10.89.

10.92 IONIC LIQUIDS

Stable, nonvolatile, low-melting (below 100°C) salts of rather large organic cations with various counter ions. Many of them are liquid even at room temperature or below. Because they are salts, they have virtually no vapor pressure. They can be used as solvents for many organic reactions, replacing traditional volatile chlorinated solvents, and are, thus, a major player in the development of green chemistry, chemistry practiced with an effort toward minimizing the use of, and exposure to, potentially hazardous materials. The present commercial availability of hundreds of different ionic liquids allows the chemist a choice of desired properties covering a range of melting points, hydrophilicity, and hydrophobicity.

Example. Some of the most popular ionic liquids are based on *N,N′*-dialkylimidazolium cations with a variety of counterions (Fig. 10.92). The first of the new ionic liquids (Fig. 10.92, $R = CH_3$, $R' = C_2H_5$, $Y = BF_4$) was prepared in 1992; it melts at 12°C and can be prepared from the corresponding iodide by treatment with $(NH_4^+)(BF_4^-)$ in acetone. It is miscible with many common organic solvents. The structure of the anions can have a strong influence on properties. Those with Y = halogens are water-soluble, whereas those with counterions such as PF_6 are hydrophobic; the size of the R groups usually has relatively little affect on properties.

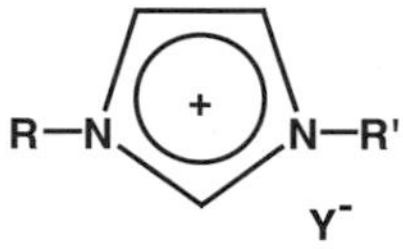

Figure 10.92. N,N′-dialkylimidazolium salts, $Y = X, BF_4, PF_6, R = CH_3, C_4H_9, C_{18}H_{37}$.

Acknowledgment. The authors thank Profesor Apryll Stalcup for helpful comments.

SUGGESTED READING

Adams, D. J., Dyson, P. J., and Tavener, S. J. *Chemistry in Alternative Reaction Media.* John Wiley & Sons: New York, 2004.

Denney, R. C. *Dictionary of Chromatography.* John Wiley & Sons: New York, 1982.

Gubitz, G. "Separation of Drug Enantiomers by HPLC Using Chiral Stationary Phases—A Selective Review." *Chromatographia 30*, 555 (1990).

Pirkle, W. H. and Pochapsky, T. C. In *Advances in Chromatography*, Giddings, J. C. E. Grushka, and P. R. Brown. Marcel Dekker: New York, 1987.

Small. H. *Ion Chromatography.* Plenum Press: New York, 1989.

11 Fossil Fuels and Their Chemical Utilization

11.1	Petroleum	422
11.2	Crude Oil	423
11.3	Sour Crude	423
11.4	API Gravity	423
11.5	Thermal Cracking	424
11.6	Steam Cracking	424
11.7	Catalytic Cracking	424
11.8	Petrochemicals	426
11.9	Zeolites	427
11.10	Molecular Sieves	429
11.11	Heterogeneous Catalysis	429
11.12	Homogeneous Catalysis	429
11.13	Dual Function (or Bifunctional) Catalyst	430
11.14	Fixed-Bed Operation	430
11.15	Fluidized-Bed Operation	430
11.16	Laminar and Turbulent Flow	431
11.17	Reynolds Number	431
11.18	Hydrocracking	431
11.19	Reforming (to High-Quality Gasoline)	431
11.20	Alkylate	432
11.21	Space Velocity	432
11.22	Contact Time	433
11.23	Percent Conversion	433
11.24	Selectivity	433
11.25	Yield	433
11.26	Space Time Yield	433
11.27	Turnover Number	434
11.28	Turnover Rate	434
11.29	Octane Number	434
11.30	Leaded Gasoline	435
11.31	Cetane Number	435
11.32	Diesel Fuel	435
11.33	Lubricating Oils	436
11.34	Greases	436
11.35	Coal	436

The Vocabulary and Concepts of Organic Chemistry, Second Edition, by Milton Orchin, Roger S. Macomber, Allan Pinhas, and R. Marshall Wilson

11.36 Rank of Coal 436
11.37 Proximate Analysis of Coal 436
11.38 Heating Value of Coal 437
11.39 Banded Coal 437
11.40 Vitrain 437
11.41 Fusain 438
11.42 Attrital Coal 438
11.43 Nonbanded Coal 438
11.44 Petrographic Composition, Coal Macerals 438
11.45 Mineral Matter in Coal 438
11.46 Ash 439
11.47 Fly Ash 439
11.48 Coke 439
11.49 Free Swelling Index (FSI) 439
11.50 Delayed Coking 439
11.51 MAF Coal 439
11.52 Pott–Broche Process 440
11.53 Bergius Process 440
11.54 Solvent-Refined Coal (SRC) 440
11.55 H-Coal Process 441
11.56 Exxon Donor Solvent Process 441
11.57 Hydroclone 441
11.58 Natural Gas 442
11.59 Liquefied Natural Gas (LNG) 442
11.60 Liquefied Petroleum Gas (LPG) 442
11.61 Coal Gas (Town Gas) 442
11.62 Coal Gasification 442
11.63 Producer Gas (Power Gas) 443
11.64 Lurgi Coal Gasification 443
11.65 Lurgi Slagging Gasifier 443
11.66 Koppers–Totzek Gasifier 443
11.67 Texaco Gasifier 443
11.68 CO_2-Acceptor Process 444
11.69 Synthesis Gas 444
11.70 Substitute Natural Gas (SNG) 444
11.71 Water–Gas Shift 444
11.72 Steam Reforming (to Produce Synthesis Gas) 445
11.73 Methanation 445
11.74 Gasohol 445
11.75 Fischer–Tropsch (FT) Process 445
11.76 Methanol Synthesis 446
11.77 Ammonia Synthesis 447
11.78 Oil Shale 447
11.79 Shale Oil 447
11.80 Kerogen 447
11.81 Fischer Assay 447
11.82 Tar Sand 448
11.83 Fuel Cells 448

Organic chemistry as a separate subdiscipline of chemistry developed rather late in the history of chemistry. In 1828, Friedrich Woehler (1800–1882) reported the synthesis of urea (an animal metabolic organic compound) from ammonium cyanate. This discovery presaged the demise of the so-called vital force theory that held compounds produced by living organisms (organic compounds) required the intervention of a vital force. In the following decades, it became apparent that the essential difference between inorganic and organic compounds was that the latter always contained carbon, and so the rather simplistic definition developed that organic chemistry is the chemistry of carbon and its compounds. Definitions of chemical terms that are precise and uniformly applicable are difficult to generate as our experience with this book testifies to. Hardly any chemist would consider $CaCO_3$, $Fe_2(CN)_6$ or NH_4CNO organic compounds although they contain one or more carbon atoms. Ionic liquids and compounds containing one or more metal atoms also raise questions of classification.

Despite the nit-picking, the foregoing definition of organic chemistry serves quite well. Practically all the known millions of compounds containing carbon also contain hydrogen, the most abundant element in the universe. When compounds contain only hydrogen atoms in addition to carbon, they are called hydrocarbons and much structural information can be gleaned simply from the H/C atomic ratio, which ranges from 4.0 (methane), to 2.0 (mono-olefins or monocyclic alkanes), to 1.4 (tricyclic aromatics) all the way to zero (fullerenes). The fact that this ratio in coal is about 0.8, whereas in petroleum it is about 1.2, suggests that the desired conversion of coal to liquids requires the addition of hydrogen for the transformation. Indeed, the liquefaction of coal, discussed in this chapter, is achieved by such hydrogenation. Organic compounds also contain oxygen, nitrogen, sulfur, phosphorous and other elements, and these elements, extraneous to hydrocarbons, are called heteroatoms. (Biological molecules, such as DNA, carbohydrates, and proteins, are examples of molecules that contain many heteroatoms.

In considering the millions of organic compounds that are known, one might ask what are the ultimate sources of carbon and hydrogen on which so much of organic chemistry is based. Fortunately, several of the most important requirements for human survival, fuel, food, oxygen, and water, also provide a source of the elements organic chemists manipulate. Fuels, as the name implies, are used primarily as fuels, a source of energy. They are rich in carbon and their combustion releases the energy needed for electricity generation, which in turn is used for lighting, manufacturing, transportation, heating, agriculture, and so on. The most important fossil fuels are coal, natural gas, and petroleum, and to a lesser extent, oil-shale and tar sands. But in addition to providing energy, these resources also provide, by appropriate conversion, the building blocks for a large proportion of compounds and materials synthesized by organic chemists.

In this chapter, we consider the properties and chemistry of these fuels and the many reactions by which they are transformed into a variety of organic compounds and their derivatives. The pyrolysis of coal generates a mixture of organic compounds, which can be readily separated into an acidic fraction (tar acids, consisting mostly of phenols), a basic fraction (tar bases, consisting mostly of pyridines), and a neutral fraction (consisting mostly of aromatics such as naphthalene). Pure compounds can be isolated from these fractions and they provide the starting materials for many organic reactions. However, the more intriguing chemical use of coal is its conversion by steam treatment (the water-gas reaction) into a mixture of hydrogen and carbon monoxide called synthesis (syn) gas. Synthesis gas, as its name implies, is a remarkably versatile starting point for the commercial synthesis of a wide range of hydrocarbons and oxygen-containing compounds produced in the million pounds per year category (commodity chemicals) as well as compounds in lesser quantities (specialty chemicals) that have special applications. Syn gas also provides (via the Fischer–Tropsch reaction) the indirect route from coal to transportation liquids.

Petroleum is the major provider of transportation fuels. It is also more important than coal as a source of organic chemicals (petrochemicals). Petroleum is produced and consumed (principally as a source of transportation fuel) in almost incomprehensibly large quantities (about 82 million barrels each day) so that the quantity diverted for chemical use is relatively minor, but nevertheless critical to the chemical industry. The remarkable technology of converting crude oil to gasoline in such large quantities is one of the industrial marvels of the past half-century, permitting the price (before taxes) of gasoline to be even lower than that of bottled water. Hydrogen, partner to the carbon of organic chemical compounds, which many view as the fuel of the future in the so-called hydrogen economy, can be extracted from both petroleum and coal, either directly or indirectly.

The great attraction for a hydrogen-based economy is that its combustion produces only water and can be used in a fuel cell to directly generate electricity. The most abundant potential source of hydrogen is, of course, that most remarkable compound, H_2O, which covers about 70% of the world's surface and constitutes about an equal percentage of the human body and whose unique properties sustain life on this planet. Making hydrogen economically available from water remains a major challenge.

In a discussion of fuels in general, one cannot neglect the constant source of energy supplied by the sun. Harnessing that energy is one of the most important tasks confronting chemists. Together with water and oxygen, the sun not only makes this planet comfortable, but the food supply of the world depends on the photosynthetic process of converting CO_2 and water to starch and cellulose. Perhaps environmentalists should be reminded that proper incineration of these and related materials in the form of waste products to CO_2, H_2O, and useful energy is the ultimate goal in recycle chemistry.

11.1 PETROLEUM

From the Latin words *petro* ("rock") and *oleum* ("oil"), an oily, colored (from amber to black, depending on origin) complex liquid mixture of hundreds of organic compounds, principally hydrocarbons, found beneath the earth's surface and generated from decaying organic material in a marine environment. Total world petroleum production is now about 82 million barrels per day. The United States consumes almost 22 million barrels per day, with about 60% of it imported.

11.2 CRUDE OIL

Petroleum prior to its conversion to commercially desirable components. When piped from its reservoirs, petroleum is usually accompanied by gas, water, salt, and minerals. Most of these impurities are removed before shipment to installations where the petroleum (called crude) is separated into various fractions (refined), suitable for use for specific purposes. The various fractions are usually treated further to improve their suitability for their intended use. The installations where this processing is carried out are called refineries. A typical refinery may process between 60,000 to 100,000 barrels per day. In the petroleum industry, one barrel is defined as containing 42 gal. Crude oil is fractionally distilled to give the various boiling fractions shown in Table 11.2.

TABLE 11.2. Distillate fractions of crude oil

Boiling Point Range (°C)	Name of Fraction	Specific Gravity
30–180	Gasoline[a]	0.75
150–260	Kerosine[b]	0.78
180–370	Gas oil[c]	0.84
>370	Fuel oil[d]	0.96

[a] The gasoline fraction obtained directly from crude oil before any processing is called straight-run gasoline or light naphtha.
[b] This fraction, also called heavy naphtha, is frequently used as a heating oil in space heaters.
[c] This fraction is used as a diesel fuel.
[d] This fraction is usually further fractionally distilled in vacuum to give distillate fractions used for central heating of homes (heating oils, numbers 1, 2, 3, and 4). The residual material, often called bunker oils (numbers 5 and 6), are heavy viscous oils that are used as boiler fuels for power production. The undistillable residue is called resid and is used as asphalt for paving streets and highways. It is also utilized as a feedstock for generating synthesis gas ($CO + H_2$) by reaction with steam.

11.3 SOUR CRUDE

Crude oil containing relatively large amounts of sulfur compounds, hence requiring special refining techniques. A typical Indonesian oil may contain less than 0.1% sulfur and is classed as a sweet crude, but the crude from Saudi Arabia may contain more than 1.5% sulfur and, hence, is considered sour crude.

11.4 API GRAVITY

An arbitrary scale adopted by the American Petroleum Institute (API) for comparing the density (hence, in part, the quality) of petroleum oils:

$$\text{API} = \frac{141.5}{\text{specific gravity (60°F)}} - 131.5 \qquad (11.4)$$

When we use this equation, a typical crude with a density of 0.860 g/mL has an API gravity of 33.0.

11.5 THERMAL CRACKING

The reduction in molecular weight of various fractions of crude oil through pyrolysis. Thermal cracking is mainly used to produce a mixture rich in ethylene and propylene. It has largely been replaced by steam cracking.

11.6 STEAM CRACKING

Cracking carried out in the presence of steam. Typically, a naphtha feedstock (with a boiling point or bp of 70–200°C) is passed, along with steam, through a coiled tube heated by a furnace. Process conditions are as follows: temperature of 750 to 900°C; very short contact or residence time at these temperatures (approximately 0.1–0.5 sec); and rapid quenching. The steam serves two purposes: It acts as a diluent and, thus, favors unimolecular reactions, which minimize radical chain termination steps, allowing cracking to continue; and it lowers the vapor pressure of the hydrocarbons (see Steam Distillation, Sect. 10.17), thereby reducing resid concentration and maximizing desired product formation.

Example. The mechanism that best explains the course of the conversion is the free radical chain mechanism consisting of the usual steps: initiation; propagation; termination (Fig. 11.6).

When steam cracking units are operated to produce acetylene rather than liquids, the temperature is raised to above 1200°C because, unlike other hydrocarbons, the free energy of formation of acetylene decreases with increasing temperature.

11.7 CATALYTIC CRACKING

The reduction in molecular weight of naphtha fractions of crude oil, achieved by passing them over a catalyst, usually at about 450 to 600°C. The catalyst is generally a crystalline zeolite (Sect. 11.9) that functions as a strong protic acid. The molecular weight reduction involves carbocation intermediates.

Example. The generation of the initiating carbocation can proceed either by hydride abstraction (by the catalyst) from an alkane, or by protonation of an olefin by surface protons on the catalyst. The small amount of olefin required for initiation can also be formed by a thermal reaction. The carbocation resulting from either of these initiating processes will undergo ß-scission to generate a primary carbocation. The primary carbocation then undergoes a 1,2-hydride migration to form a more stable secondary carbocation, and degradation proceeds by ß-cleavage (Fig. 11.7). In both

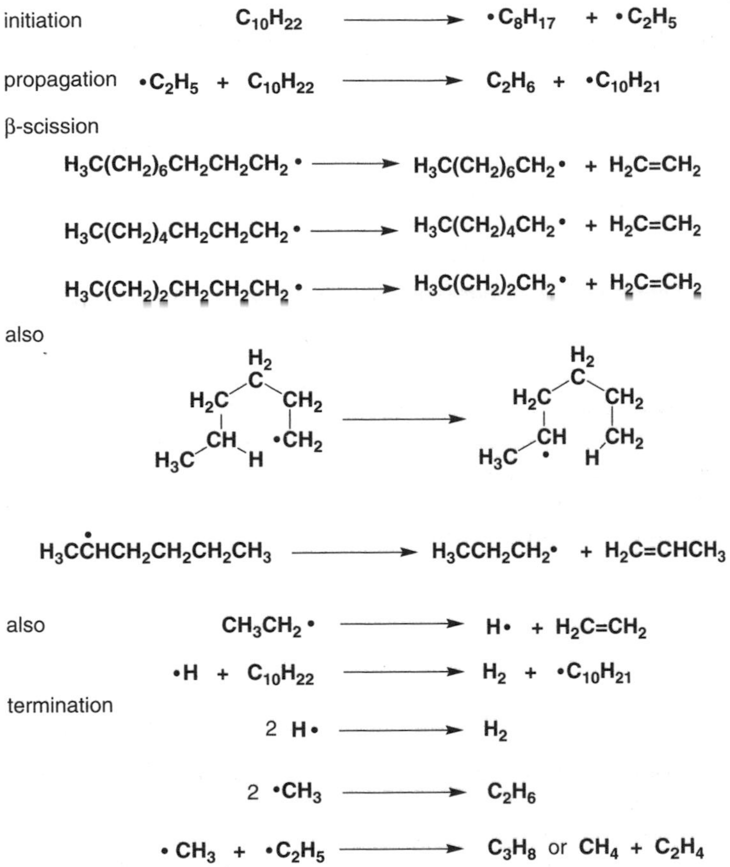

Figure 11.6. Examples of free radical reactions in steam cracking.

radical and carbocation chemistry, tertiary alkyl species are more stable than secondary, which are more stable than primary. However, the differences in energy between the three types of alkyl species in the case of radicals are relatively small, whereas the differences in the carbocation case are significantly larger, contributing to the facile rearrangement of primary to secondary, and where possible, secondary to tertiary alkyl carbocations. The relatively large differences in carbocation stability are ascribed to extensive hyperconjugation, particularly in tertiary carbocations (see Sect. 13.27). The mechanism shown in Fig. 11.7 rationalizes the preponderance of propylene in catalytic cracking compared to steam cracking. Because steam cracking involves radical reactions and radicals are not prone to undergo1,2-shifts, steam cracking leads to a preponderance of ethylene. [Almost 57 billion lb of ethylene are produced annually in the United States, making it by far the organic compound produced in the largest quantity in this country; propylene is a poor second (about 30 billion lb).]

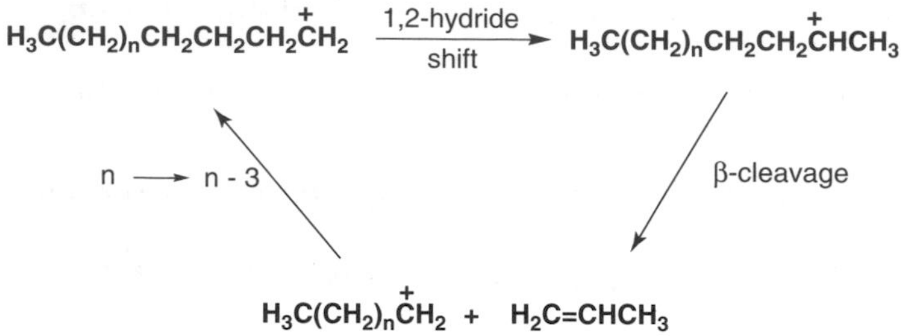

Figure 11.7. The carbocation mechanism for catalytic cracking.

Since the hydride shift (and an analogous methyl shift) is so important in petroleum carbocation chemistry, further discussion of this rearrangement is warranted. The mechanism for the shift involves a three-center, two-electron (3c-2e, Sect. 3.44) transition state, involving the carbocation carbon atom, the nearest-neighbor carbon atom, and a hydrogen atom attached to that carbon:

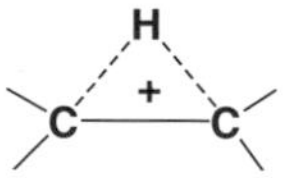

The molecular orbital (MO) description of such a transition state involves a basis set of three atomic orbitals: the *s* atomic orbital of the migrating hydrogen; the *p* orbital on the carbon attached to it; and the *p* orbital on the adjacent carbocation carbon. (These *p* orbitals have partial *s* character.) The linear combination of these three atomic orbitals leads to three MOs: One is a low-energy, strongly bonding orbital; one a high-energy, strongly antibonding orbital; and an intermediate-energy, slightly antibonding MO. The two electrons in the system occupy the bonding MO, with a resulting energy-favorable transition state for the hydrogen migration, yielding the more substited carbocation. If we now consider a radical of a similar structure as the cation, the same three molecular orbitals are generated, but in the radical case there are three electrons to be considered. Two of these electrons occupy the bonding MO as in the cation case, but the third electron requires occupation of the weakly antibonding MO, which would give rise to an unfavorable transition energy state. Accordingly, the hydride shift is a much more favorable process than the corresponding reorganization of radicals.

11.8 PETROCHEMICALS

Specific compounds (or materials) of industrial importance, which are formed from precursor compounds derived from crude oil. The precursor compounds are commonly generated from the steam cracking of selected fractions of the crude oil.

Example. Steam cracking of the 70 to 200°C (naphtha) fraction of crude oil typically produces a hydrocarbon mixture consisting of about 30% ethylene and 13% propylene. These olefins can be polymerized to give various types of polymers, such as the petrochemicals polyethylene and polypropylene, respectively (Sect. 8.7). The first petrochemical compound of any significance, discovered in 1920, was isopropyl alcohol, derived from the acid-catalyzed hydration of propylene. Ethylene is also used to provide the petrochemicals ethylene oxide, ethylene glycol, vinyl chloride, ethylbenzene, and others. Although only a small fraction of crude is diverted from use as a fuel to petrochemicals, this is an example of the concept of added value. Because ethylene and propylene as a fuel have value counted in cents per pound, their conversion for use as chemicals results in products worth dollars per pound. Although steam cracking of naphtha is usually the preferred route to ethylene, the dehydrogenation of ethane provides an alternate route.

11.9 ZEOLITES

This refers to a family of crystalline aluminosilicate minerals; many are naturally occurring but also now are available by synthesis. The structure consists of a three-dimensional network of tetrahedra linked by shared oxygen atoms. The tetrahedra consist of either a silicon atom or an aluminum atom attached to four oxygen atoms, but Al–O–Al bonding is forbidden. Four-coordinate aluminum requires a negative charge on each aluminum atom, which must be balanced by a countercation such as Na^+ or Ca^{2+}. The molecular formula may be expressed in terms of the oxides of aluminium and silicon by the molecular formula $M_{x/n}\,[(AlO_2)_x \cdot (SiO_2)_y] \cdot mH_2O$, where M is the cation of valence n, $x + y$ is the total number of tetrahedra in the crystallographic unit cell, and m is the number of water molecules. Note that the ratio of the number of oxygen atoms to the sum of the Al + Si atoms is 2:1. The stoichiometry of the zeolites is given in units of AlO_2 and SiO_2. The framework of tetrahedra gives a network of uniform channels. The size of these channels or pores is differentiated by the number of oxygen atoms in the rim around the pores; thus, the number 10 indicates that 10 oxygen atoms are linked around the rim of the pore, such as that shown in Fig. 11.9*a* for the synthetic zeolite mordenite ($y/x = 10$, 12 tetrahedra in a ring). These channels have room to accommodate neutral molecules of many kinds whose cross section does not exceed certain diameters (for mordenite, this is 8.0 Å).

Product selectivity occurs because only molecules of a certain size corresponding to the diameter of the pore can diffuse in and out of the channels. The cations are located close to the apertures, and changes in the cation result in substantial changes in pore size and, hence, adsorbent characteristics. The metal cations can readily be exchanged, frequently by protons, in which case the zeolite behaves like a strong protic or Brönsted acid. The zeolites are used as ion exchange resins, inclusion agents, drying agents, and most importantly as catalysts, where, after

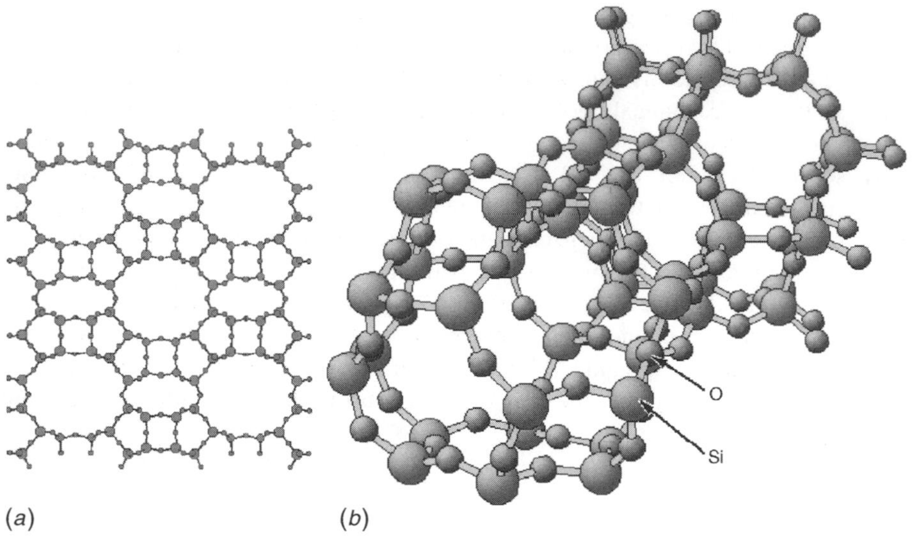

Figure 11.9. (*a*) The spatial arrangement of tetrahedra in the zeolite mordenite and (*b*) the faujasite elementary cell.

$[H^+]$ exchange, they function as proton donors. Some zeolites include as much as 265 mL of H_2O per g. This water can be driven out by heating to 350°C without disturbing the structure. As a matter of fact, the word zeolite is derived from the Greek word meaning "boiling stone." Because chemists during the Middle Ages, when analyzing some stone samples, used blowpipes to examine the effect of heat, and when certain silicates were treated in this manner, they extruded steam. Most zeolites in use today are prepared synthetically and designed for specific purposes.

Example. The naturally occurring faujasite zeolites having the formula $(Na_2Ca)_{32}$ $[AlO_2)_{64} \bullet (SiO2)_{128}] \bullet 216\ H_2O$ (Fig. 11.9*b*) have been studied extensively because of their catalytic properties. These are large-pore zeolites; erionite and chabazite are naturally occurring small-pore-size zeolites. One of the most useful synthetic zeolites is the high silica ZSM-5 zeolite (Z for Zeolite, SM for Socony-Mobil, the forerunner to the present Exxon-Mobil Co.); it has the formula $Na_3(AlO_2)_3 \bullet$ $(SiO_2)_{93} \bullet 16H_2O$. After the Na is exchanged for protons, the strongly acidic catalyst is used in the novel and highly successful methanol-to-gasoline (MTG) process. In this reaction, methanol is converted to a high-octane hydrocarbon gasoline in excellent conversion. The first step in the reaction is assumed to be the acid-catalyzed intermolecular dehydration of methanol to dimethyl ether. The pore size of the catalyst promotes the selective formation of only hydrocarbons in the gasoline boiling point range.

11.10 MOLECULAR SIEVES

Synonymous with zeolites. Because of the molecular-size selectivity of zeolites, they are also called the alternate name molecular sieves.

11.11 HETEROGENEOUS CATALYSIS

A catalytic system in which the catalyst constitutes a separate phase. The catalyst is a solid, and the reactants and products are either gases or liquids. The catalytic reaction occurs at the surface of the catalyst; hence, the rate of the reaction is sensitive to the surface area (Sect. 10.66) of the catalyst. One of the most important advantages of heterogeneous catalysis is that no separation of product from catalyst is required. There is also the concomitant engineering advantage of being able to process huge quantities of material continuously over a stationary catalyst.

Example. The most important practitioners of heterogeneous catalysis are the petroleum companies whose vast refineries convert petroleum to transportation fuels and heating oils. In the United States alone, each day about 22 million barrels (close to a billion gallons) of petroleum are processed, requiring large quantities of catalysts for cracking and reforming operations. Thus, it is not completely surprising that Sohio (Standard Oil Co. of Ohio, now part of British Petroleum Co.) developed one the most important and successful extensions of heterogeneous catalysis, designed to produce petrochemicals rather than fuels, namely the ammoxidation of propylene to acrylonitrile. The catalytic conversion of propylene to acrylonitrile by treatment with oxygen and NH_3 is a highly exothermic reaction:

$$C_3H_6 + NH_3 + 1.5\,O_2 \longrightarrow H_2C{=}CH{-}CN + 3\,H_2O$$

$$\Delta H = -515\,\text{kJ mol}^{-1} \quad (-123\,\text{kcal mol}^{-1}) \qquad (11.11)$$

The reaction is carried out in a fluidized bed reactor (Sect. 11.15) at about 450°C for 10 to 20 sec of contact time. In the early developments, either a bismuth molybdate or an antimony oxide promoted with a uranium salt were used as catalysts, but they have been supplanted by even more effective catalysts. Small quantities of the byproducts carbon dioxide (from combustion), HCN, and acetonitrile are also formed. An ammoxidation process based on propane rather than propylene is being developed, which will make acrylonitrile even more economical.

11.12 HOMOGENEOUS CATALYSIS

A catalytic system in which the catalyst and all the reactants are molecularly dispersed in one phase. The fact that the catalyst is soluble in the reaction medium facilitates the elucidation of molecular structure changes in the catalyst as the reaction proceeds. The sequence of reactions typically involves coordination of the reactant to an open site on the catalyst, intramolecular reaction of the newly coordinated ligand with another ligand on the catalyst, and finally an elimination step that leads to production and regeneration of the catalyst. One important advantage of homogeneous catalysis is that better heat transfer is achieved in the synthesis of molecules involving bond-forming reactions, because such reactions are exothermic. When carried out in large-scale reactors, heat buildup must be avoided and heat transfer is a major factor in design. The major disadvantage to homogeneous catalysis is the problem of catalyst separation and recycle.

Example. The hydroformylation reaction (Sect. 9.55) was one of the first commercial successes of homogenous catalysis. The biphasic modification of the hydroformylation of propylene, in which a water-soluble Rh catalyst is used and a water-insoluble product (C-4 aldehydes) is formed, avoids the problem of catalyst separation and recycle.

11.13 DUAL FUNCTION (OR BIFUNCTIONAL) CATALYST

This is a catalyst that possesses two different catalytic sites, hence, it can catalyze two different kinds of reactions.

Example. In the reforming reactions (Sect. 11.19) of hydrocarbons, a zeolite can be prepared that also has platinum on the surface, and therefore, reactions that are acid-catalyzed, as well as noble metal-catalyzed, may proceed.

11.14 FIXED-BED OPERATION

A heterogeneous catalytic processing operation in which the catalyst is in the form of solid particles held stationary in a reactor. The reacting gases or liquids are pumped through the catalyst bed. One disadvantage of this mode of operation is the problem of relatively poor heat transfer and resultant coke formation on the catalyst, requiring expensive regeneration of catalytic activity.

11.15 FLUIDIZED-BED OPERATION

A catalytic processing operation in which the catalyst in the form of small solid particles is maintained in suspension by the upward flow of a gas at an appropriate velocity, creating very thorough mixing, uniform temperature throughout the bed, and very rapid heat transfer. In this mode of operation, the bed of solid takes on the appearance of a vigorous boiling liquid having many properties of a fluid, such as exerting a hydrostatic head.

11.16 LAMINAR AND TURBULENT FLOW

Laminar flow occurs when the flow rate of a liquid through a pipe is relatively slow and steady. The flow velocity inside the pipe, owing to the friction or drag at the walls of the pipe, increases from a low velocity where the fluid contacts the inner surface or wall of the pipe to a higher velocity at the center of the pipe. The flowing fluid can be pictured as a series of parallel circular layers or lamina moving at different velocities, hence laminar flow. If the flow velocity is increased past a critical velocity, their flow characteristics change from laminar to turbulent flow, giving rise to flow eddies and vortices.

11.17 REYNOLDS NUMBER

Named after Osbourne Reynolds (1842–1912), a dimensionless number whose magnitude determines whether a flow is laminar or turbulent. When the number is below 2,300, the flow is laminar, and when higher than this number, the flow is turbulent. For flow through a pipe, it is calculated by taking the product of the density of the fluid ρ times the diameter of the pipe D, times the velocity V, and dividing the product of these three parameters by the viscosity of the fluid μ (the units are chosen so that they cancel in the quotient):

$$R = \rho DV/\mu \tag{11.17}$$

11.18 HYDROCRACKING

Catalytic cracking processes carried out in the presence of 150 to 200 atm of hydrogen at about 450°C.

11.19 REFORMING (TO HIGH-QUALITY GASOLINE)

A combination of dehydrogenation, isomerization, and hydrocracking reactions performed on distillates from crude oil by passing them over dual functional catalysts. The object is to prepare gasoline of high quality (high octane number, Sect. 11.29). Operating conditions are chosen to maximize aromatic content. Platinum metal (0.3–1.0 wt.%) supported on pure, high-surface-area alumina (Al_2O_3) was for many years the preferred catalyst; the platinum acts as a catalytic site for hydrogenation–dehydrogenation reactions, and the alumina as a Lewis acid site that catalyzes isomerization reactions. Presumably, both sites are capable of catalyzing hydrocracking reactions. Typical operational conditions are temperatures of 500 to 525°C and pressures of 10 to 40 atm. Feedstocks may have an octane number of 60, whereas the product (called reformate) can have an octane number of 100. In a rather recent development, a bimetallic catalyst consisting of platinum-rhenium replaced the single metal system. About half the gasoline used in the United States is produced by reforming reactions. In the older descriptions of fuel compositions, it was common to characterize fuels in terms of their percentages of paraffins (alkanes), naphthenes (cycloalkanes), and aromatics.

Example. Typical reforming reactions for C-6 hydrocarbons are shown in Table 11.19 along with equilibrium constants and heats of reaction at 500°C for each of the reactions.

TABLE 11.19. Reforming Reactions of C-6 Hydrocarbons at 500°C

Reaction	K^a	ΔH mol^{-1}	
		kJ	kcal
 + $3 H_2$	6×10^5	221	52.8
	8.6×10^{-2}	−15.9	−3.8
$CH_3(CH_2)_4CH_3$ ⇌  + $4 H_2$	7.8×10^4	266	63.6
$CH_3(CH_2)_4CH_3$ ⇌ $CH_3CH_2CH_2CH(CH_3)_2$	1.1	−5.9	−1.4
$CH_3(CH_2)_4CH_3$ ⇌ $CH_3CH_2CH(CH_3)CH_2CH_3$	7.6×10^{-1}	−4.6	−1.1
$CH_3(CH_2)_4CH_3$ ⇌ $CH_3CH_2CH_2CH_2CH{=}CH_2$ + H_2	3.7×10^{-2}	130	31.1

[a] Partial pressure in atmospheres.

11.20 ALKYLATE

In the petroleum industry, this refers to the product produced by the acid-catalyzed reaction between an olefin and alkane (alkylation). The reaction proceeds via carbocation intermediates and leads to a highly branched, low molecular weight product (alkylate) with a very high octane number.

Example. 2,2,4-Trimethylpentane (incorrectly called isooctane) is prepared by treating a mixture of isobutane (in excess) and isobutylene with a proton source. The catalytic cycle for this reaction is shown in Fig. 11.20.

11.21 SPACE VELOCITY

This is the volume of feed per unit volume of catalyst per unit of time at operating conditions; the usual unit is hr^{-1}.

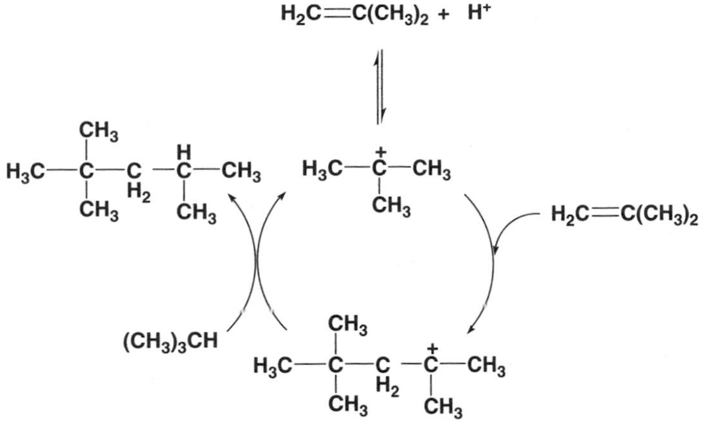

Figure 11.20. The preparation of 2,2,4-trimethylpentane (isooctane) by alkylation.

11.22 CONTACT TIME

This is the time the feed is in contact with the catalyst; it is calculated (in seconds) by dividing the space velocity (hr^{-1}) into 3,600 sec hr^{-1}.

11.23 PERCENT CONVERSION

The percent conversion is $[(F_0 - F_t)/F_0] \times 100$, where F_0 is the molar equivalents of feed before reaction and F_t is the molar equivalents of feed at time t.

11.24 SELECTIVITY

The ratio of the quantity of desired product over the quantity of feed converted. The percent selectivity is $[P_t/(F_0 - F_t)] \times 100$, where P_t is the molar equivalents of desired product at time t, and F_0 and F_t are as defined above.

11.25 YIELD

This is the fraction of feed converted at time t, multiplied by the selectivity; stated equivalently, it is the quantity of desired product over the quantity of feed. The percent yield is $[(F_0 - F_t)/F_0] \times P_t/(F_0 - F_t) \times 100 = (P_t/F_0) \times 100$, where all the terms are the same as defined above.

11.26 SPACE TIME YIELD

The number of moles of product P per liter of catalyst per hour. In heterogeneous catalytic systems, the bulk volume of the catalyst is measured, and in homogeneous catalytic systems, the total volume of the feed solution containing dissolved catalyst is measured.

11.27 TURNOVER NUMBER

In heterogeneous catalysis, the number of moles of feed or substrate converted per number of catalytic sites. In homogeneous catalysis, the number of moles of feed or substrate converted per mole of catalyst.

11.28 TURNOVER RATE

The turnover number per unit of time (usually, the second, but the unit must be specified).

11.29 OCTANE NUMBER

A number usually ranging from 60 to 100 assigned to a gasoline sample that reflects its relative performance as a fuel in the internal combustion engine. It is the vol. % of 2,2,4-trimethylpentane (100 octane) in a blend with heptane (0 octane) that gives the same intensity of engine knock (the sound produced by premature ignition of the fuels) under the same operating conditions as the fuel sample being tested. A single-cylinder engine with a variable compression ratio is used for the evaluation.

Octane numbers obtained in this test engine operating under mild engine severity and low speed (600 rpm, 52°C inlet temperature) are called research octane numbers. The research octane number (RON) for a typical leaded gasoline of regular grade was about 93 (in 1979). Octane numbers obtained with an engine operating under high engine severity and high speed (900 rpm, 150°C) are somewhat less than RONs and correlate well with actual road tests. Such octane numbers are called motor octane numbers (MON) and a gasoline with an RON of 93 might show a MON of 85. The octane number required by federal law to be posted at the gasoline pump is the average of the two octane numbers, (RON + MON)/2 (the antiknock index), which in the above example would be (85 + 93)/2 = 89. The blending octane number is the number assigned to an additive to gasoline that reflects the effect such an additive has on the octane number of the gasoline. The addition of 10 vol.% of ethanol to a gasoline having a RON of 87 results in a mixture (see "Gasohol," Sect. 11.74) having a RON of 91.8. The blending research octane number of ethanol can be calculated as 135:

$$(87 \times 0.9) + (y \times 0.1) = 91.8$$

$$\text{Blending RON, } y = 135 \tag{11.29}$$

The octane number is a measure of performance. But the performance of a gasoline ultimately depends on its composition and there are many studies relating composition to performance. Accordingly, it is rather simple, knowing the approximate composition, to make an educated guess as to the octane number without testing its performance in an engine.

11.30 LEADED GASOLINE

Gasoline to which small quantities of liquid tetraethyllead (TEL) have been added. The TEL used commercially is a mixture of tetraethyllead, $Pb(C_2H_5)_4$ (63%); 1,2-dibromoethane, $BrCH_2CH_2Br$ (26%); and 1,2-dichloroethane, $ClCH_2CH_2Cl$ (9%); plus a red dye (2%). The purpose of the bromo compound (and to a lesser extent, the chloro compound) is to "scavenge" the lead oxide formed in the automobile cylinder by reacting with it to form volatile lead compounds. The addition of TEL (1–3 mL gal^{-1}) increases the research octane number by about 7. Motor octane numbers over 100 can be secured by using 2,2,4-trimethylpentane, to which some TEL has been added. Owing to the toxicity of lead, leaded gasoline is no longer commercially available in the United States. An alternative antiknock agent is η^5-methylcyclopentadienyltricarbonylmanganese, $[Mn(\eta^5–(C_5H_4CH_3)(CO)_3]$ (in the trade called MMT). After the U.S. Congress passed an amendment to the Clean Air Act, which mandated that oxygenated compounds be added to gasoline, methyl tert-butyl ether (MTBE) became increasingly popular as an octane booster. It is readily prepared from methanol and isobutylene, and is reported to have an RON of about 125 and an MON of about 100. However, when MTBE was discovered to have infiltrated some water sources because of leakage from storage tanks, and concern arose as to its toxicity, its use was banned in some states.

11.31 CETANE NUMBER

A number given to a diesel fuel sample that is a measure of its performance in a compression engine. It is the vol.% of hexadecane (cetane) in a blend with 1-methylnaphthalene that gives the same performance in a standard compression engine as the fuel sample under the same operating conditions. Hexadecane is given an arbitrary cetane number of 100 and 1-methylnaphthalene a number of 0. The diesel engine charges air into a chamber and compresses it during intake. Liquid fuel is then injected into the compressed air, whereupon spontaneous combustion occurs, generating the energy required for locomotion. Fuels with cetane numbers greater than 45 are required for good performance in a compression engine.

11.32 DIESEL FUEL

This is a mixture of hydrocarbons boiling in the range of 180 to 240°C; hence, it is considerably less volatile than gasoline. In contrast to gasoline, good diesel fuels are low in aromatic content and high in straight-chain alkane content. Current interest in renewable sources of transportation fuels have focused on bio-diesel fuels prepared from plant materials such as olive and canola oils. The fatty acid glycerides in these oils are transesterified with methanol and their methyl esters make acceptable diesel fuels.

11.33 LUBRICATING OILS

The oils used to reduce friction between solid moving parts, usually obtained by vacuum distillation of a high boiling fraction (~375–475°C) of crude oil. Synthetic lubricating oils are becoming more common; typically, these are esters of sebacic (1,10-decadioic) acid.

11.34 GREASES

High boiling petroleum oils thickened with metallic soaps, such as lithium stearate, which are the salts of fatty acids.

11.35 COAL

Buried plant remains that have been chemically altered and consolidated through geologic time by heat and pressure. Coals exhibit a wide range of composition and properties because of the different kinds and quantities of plant material and intermixed mineral matter from which they were formed and the extent of physical and chemical alteration before, during, and after burial.

China is the world's largest producer of coal with the United States a close second. A little over a billion tons were produced in the United States in 2003, with 90% of it used for the generation of electricity. Of the fuels utilized in 2001 for the generation of the 3,532 billion kWh of electricity consumed in the United States, 52% was based on coal; nuclear provided 21%, natural gas 12%, and petroleum about 3%.

11.36 RANK OF COAL

A classification of coal based on the relative degree of alteration or metamorphosis of the plant material. The precursor to coal from the metamorphosis of wood and other plant material is peat, which is not considered coal. The least altered peat is the lowest-rank (lignite) coal and the most altered is the highest-rank (anthracite) coal; hence, the rank of the coal corresponds roughly to its relative age. The classification of coals is made on the basis of the "fixed carbon" for the high-rank coals and the calorific value for the lower-rank coals (Table 11.36). Geologic age is also reflected in the elemental composition of the coal; older, more altered coals show progressively lower oxygen content and lower hydrogen-to-carbon ratios than younger, less altered coals.

11.37 PROXIMATE ANALYSIS OF COAL

The direct analytical determination of three quantities (moisture, volatile matter, and ash) and the calculation of a fourth quantity called the fixed carbon. Moisture is obtained from loss of weight of a sample heated at 104 to 110°C; volatile matter as the percent loss in weight, other than moisture, when a separate sample is heated to 950°C;

TABLE 11.36. Classification of Coal by Rank

Classification	Group	Fixed Carbon (%)[a]	Volatile Matter (%)[a]	Caloric Value (Btu lb^{-1})[b]
Anthracite	Metaanthracite	>98	—	
	Anthracite	92–98	2–8	
	Semianthracite	86–92	8–14	
Bituminous	Low-volatile bituminous	78–86	14–22	
	Medium-volatile bituminous	69–78	22–31	
	High-volatile A bituminous	<69	>31	14,000
	High-volatile B bituminous			13,000–14,000
	High-volatile B bituminous			11,500–13,000
Subbituminous	Subbituminous A[c]			10,500–11,500
	Subbituminous A			10,500–11,500
	Subbituminous B			9,500–10,500
	Subbituminous B			8,300–9,500
Lignite	Lignite A			6,300–8,300
	Lignite B			<6,300

[a] Dry, mineral matter-free basis.
[b] Mineral matter-free basis.
[c] A subbituminous A coal that has agglomerating properties.

and ash as the percent residue remaining after combustion of a sample in a high-temperature (~1500°C) furnace. The fixed carbon is 100 minus the sum of the other three quantities. All the analyses are carried out under specified controlled conditions.

11.38 HEATING VALUE OF COAL

The amount of heat liberated per unit weight, obtained by combustion at constant volume in an oxygen bomb calorimeter. The usual units of measurement are Btu's per pound of coal [Btu = British thermal unit, the amount of energy necessary to raise the temperature of 1 lb of water by 1°F at 39.2°F (4°C); 1 Btu is equivalent to 1,055.6 J or 252.3 cal].

11.39 BANDED COAL

Coal that appears to the eye to consist of a series of distinguishable bands whose visual classification and characterization constitute the petrographic descriptions of coal. These bands or layers are of three kinds: vitrain, attrital coal, and fusain, distinguishable by their degree of luster or shininess.

11.40 VITRAIN

Shiny, lustrous black layers in banded coal.

11.41 FUSAIN

Dull black layers resembling charcoal in which the original plant tissue structure is preserved. It is very brittle.

11.42 ATTRITAL COAL

The layers of coal that vary in luster and are usually intermediate in this respect as compared to the other constituents because it is the matrix in which vitrain and fusain bands are embedded. It has a striated or rough texture. It is visibly heterogeneous, and accordingly, some petrographers further classify attrital coal into clairain (with a bright luster similar to that of vitrain) and durain (with a dull luster).

11.43 NONBANDED COAL

Fine, granular, dull-appearing coal in which bands are absent. The principal kinds of nonbanded coal are of two types: (1) cannels, which are rich in spores, and (2) bogheads, rich in algae. Both are differentiated on the basis of the constitution of the waxy component (exinite) that is present.

11.44 PETROGRAPHIC COMPOSITION, COAL MACERALS

The classification of coals based on microscopic examination (200+ magnification) and the observation of the distinctive kinds of organic matter (macerals) that are present in a cross-section sample. There are six macerals: vitrinite, exinite, resinite, semifusinite, micrinite, and fusinite. The first three are present in the vitrain band and are relatively reactive; they are the macerals that can be converted to chemicals by hydrogenation or oxidation. The last three are present in the fusain band and are relatively inert. The macerals are analogous to minerals in inorganic rocks, but do not have the stoichiometric integrity of composition characteristic of minerals. The kind of macerals present can be determined nondestructively from reflectance measurements, the magnitudes of which are mainly a measure of the increasing proportion and number of rings in polynuclear aromatic structures.

11.45 MINERAL MATTER IN COAL

The various minerals associated with coal. Low-temperature ashing using an oxygen plasma (radio-frequency-activated) at low pressure permits standard X-ray analysis of the original minerals associated with the coal sample.

11.46 ASH

The residue remaining from a coal sample after heating it in a muffle furnace (~1500°C). This treatment converts most of the mineral matter to metal oxides.

11.47 FLY ASH

The small solid particles (particulates) of noncombustible mineral residue carried out of a bed of coal by the gases formed during combustion. They are usually captured from the effluent gas stream before escape and now find many useful applications, particularly in the formulation of concretes.

11.48 COKE

The residue that remains when bituminous coal is heated rather rapidly at high temperature (~800°C) in the absence of air (or in the presence of a limited supply of air). The coal softens as it is heated and goes through a plastic stage in which the volatile gases are eliminated, causing the coke to develop a porous structure. Coke provides the source of carbon for the reduction of iron oxides in the blast-furnace manufacture of steel. In this application, the coke must be sufficiently strong to support the column of iron ore and limestone used in the process, while sufficiently porous for the air blast to pass up the vertical furnace. Good coking coals therefore are highly valued.

11.49 FREE SWELLING INDEX (FSI)

A measure of the caking characteristics of a coal. A standard sample is heated rapidly to 800°C and cooled. If the residue is a powder, it is given a value of 0; if it is coherent but does not support a 500-g weight, it is given a FSI of $\frac{1}{2}$. Coherent residues that do support the weight are graded from 1 to 9 in half-units, depending on its size comparison with a standard chart. This information is helpful in evaluating the coal as a source of coke.

11.50 DELAYED COKING

The process of forming coke by relatively long heating at a relatively low (~500°C) temperature.

11.51 MAF COAL

Moisture and ash-free coal. When the percentage of carbon, hydrogen, nitrogen, sulfur, and so on, is determined directly on a sample of coal, the results are frequently normalized to the MAF basis to reflect the percentage of these elements in terms of the organic content of the coal. In proximate analysis reports, the equivalent term dry, mineral matter-free (dmmf) is often used.

11.52 POTT–BROCHE PROCESS

The dissolution or dispersion of coal by heating at atmospheric pressure its suspension in an inert gas at 450°C in high boiling (350–450°C) aromatic solvents. The mixture can be filtered while hot to remove ash and inerts, and the solvent recovered by vacuum distillation. It is uncertain whether a true solution or colloidal suspension is obtained by this treatment. The product is an ash-free high-Btu fuel. The process is named after the two German investigators who were largely responsible for researching the process, A. Pott and H. Broche.

11.53 BERGIUS PROCESS

Named after Friedrich Bergius (1884–1949, who was Nobel Laureate in 1931), the high-temperature (~400°C), high-pressure (~200 atm) direct hydrogenation of coal (suspended in a recycle oil so that the slurry fed to the reaction consists of about 1.5 parts oil to 1 part coal) to convert it to a liquid product suitable for processing into transportation fuels. Germany utilized the process during World War II to provide more than half the fuel used in transportation and combat. Inexpensive catalysts such as iron oxides may be employed, but better results are obtained with tin and zinc salts, principally the chlorides. Much effort has been devoted to improving on the original process, but all modification of the direct coal liquefaction processes involves three fundamental steps: mixing powdered coal into a slurry with a solvent (usually one recycled from a portion of the product), solubilizing the coal slurry with high-pressure hydrogen at a high temperature, and separating the liquid product from ash and unconverted coal. Three of these processes are described below.

11.54 SOLVENT-REFINED COAL (SRC)

A process designed to convert high-sulfur (>1.0%), high-mineral matter (3.5–15%) coals to a product that is relatively low in sulfur and mineral matter and, therefore, environmentally acceptable as a boiler fuel. Pulverized coal is mixed with a coal-derived recycle solvent (~2:1 solvent/coal), and the slurry is treated with hydrogen at ~400°C and ~100 atm H_2 with no added catalyst. From the dissolver vessel the mixture passes to a separator, where gases are bled off and the liquid mixture is filtered or centrifuged. The organic material in the solids is gasified for recycle hydrogen and the liquid product is distilled to recover recycle solvent. The residue from the distillation, solvent-refined coal (SRC), softens at ~150°C and has a heating value of about 16,000 Btu lb^{-1}. It can be pulverized and used like coal. In a modification (SRC II), the filtration is avoided by distilling off the recycle oil, recycling a portion of the residue and gasifying the remainder.

11.55 H-COAL PROCESS

Named after Hydrocarbon Research, Inc., a catalytic hydroliquefaction process for converting high-sulfur coal to boiler fuels and synthetic oil (syn crude). Dried, crushed coal (minus 60 mesh, i.e., sufficiently fine to pass through a screen having 60 holes per linear inch) is slurried with recycle oil, pressured to 200 atm, and then compressed hydrogen is added. The preheated mixture is charged continuously to the bottom of a modified fluid bed (Sect. 11.15) reactor containing a cobalt molybdate catalyst. The upward passage of the reaction mixture maintains the catalyst in a fluidized state. The reactor is kept at ~360°C. After the gas and light liquids are removed, the mixture is fed to a flash separator to yield a liquid product as distillate and a residue that is centrifuged in a hydroclone (*vide infra*). The oil from this separator is recycled along with some of the lighter distillate fractions. The solid material can be treated in a variety of ways to obtain additional oil or char.

11.56 EXXON DONOR SOLVENT PROCESS

A hydroliquefaction process in which the principal conversion of coal to oil is achieved by hydrogen transfer from a donor solvent to the coal in the absence of added catalyst. After liquefaction, the converted coal and solvent are vacuum-distilled and a portion of the overhead sent to a catalytic hydrogenation unit to restore the hydrogen donating ability of the recycle solvent. It is estimated that over 3 barrels of oil per ton of dry coal can be made by this process developed by the Exxon Corporation, now Exxon-Mobil.

More recent developments were framed by the recognition that the conversion of coal to liquid products involves first its solubilization, and then its subsequent hydrocracking. This led to two-stage variations of the coal liquefaction process and the optimization of both stages.

11.57 HYDROCLONE

A liquid–solid separating device in which the liquid–solid suspension is passed through a tight conical vortex and the solids are separated by the centrifugal force that results from the passage. It is an example of a cyclone extractor and was developed in connection with the H-Coal process.

11.58 NATURAL GAS

A gaseous fossil fuel consisting of about 60 to 80% methane, 5 to 9% ethane, 3 to 18% propane, and 2 to 14% higher hydrocarbons. Much of the ethane and higher boiling hydrocarbons are removed from the gas before commercial utilization because they are more valuable when sold separately. Typical pipeline natural gas consists of about 75% methane, 18% nitrogen, 10% ethane, 5% hydrogen, 5% propane, and 3% carbon dioxide. t-Butyl thiol [$(CH_3)_3CSH$] is also present in trace quantities, in order to detect leaks. The energy content varies from about 900 to 1,200 Btu SCF^{-1} (standard cubic foot). Most of the natural gas (casinghead gas) is produced in association with crude oil, but there are gas wells with little or no accompanying liquid. Some commercial sources of natural gas are also found to be associated with coal (coal bed methane) in shallow formations.

11.59 LIQUEFIED NATURAL GAS (LNG)

Natural gas that has been cooled to about −160°C for shipment or storage as a liquid. It is shipped in cryogenic tankers. Such cooling results in reducing 600 ft^3 of the gas to 1 ft^3 of LNG.

11.60 LIQUEFIED PETROLEUM GAS (LPG)

A mixture of propane and butane (also called bottled gas) recovered either from natural gas or petroleum refineries and having an energy content of 2,000 to 3,500 Btu SCF^{-1}.

11.61 COAL GAS (TOWN GAS)

The gas produced by the pyrolysis of coal in a limited supply of air. It was the by-product gas generated, collected and transported for lighting and heating purposes in connection with the early manufacture of coke, and was also called “coke oven gas.” It contained hydrogen, carbon monoxide, carbon dioxide, nitrogen, some oxygen, ammonia, and a range of low boiling hydrocarbons with a Btu SCF^{-1} of about 450. The manufacture of coke no longer provides gas for household distribution and so-called town gas is largely of historical interest.

11.62 COAL GASIFICATION

The chemical transformation that converts coal to gas, principally carbon monoxide and hydrogen, along with methane and carbon dioxide and other gases depending on the particular gasification process, some of which will be described below. The first gasification plant in the United States designed to provide gas for heating and lighting was constructed in 1816 in Baltimore, Maryland.

11.63 PRODUCER GAS (POWER GAS)

A low-Btu gas (~150 Btu SCF^{-1}) produced in early gasifiers that operated by blowing the hot coal in a moving bed with air and steam. Because it is uneconomical to transport such low-quality gas, it was used only for on-site electrical generation.

11.64 LURGI COAL GASIFICATION

A proven commercial process for coal gasification to provide synthesis gas ($H_2 + CO$, Sect. 11.69) developed by the Lurgi Company in Germany. The gasifier unit operates at 28 atm (called low or medium pressure depending on who is describing the process) and 570 to 700°C. Crushed coal is fed into a vertical reactor from the top, and steam and oxygen are fed into the bottom. In the first stage of the reaction, the crushed coal is contacted by the rising synthesis gas and volatile matter is driven off in this stage. In the second stage, the dry char reacts with the steam and oxygen to produce the synthesis gas. The remaining ash is removed through a rotating grate at the bottom of the reactor.

11.65 LURGI SLAGGING GASIFIER

A Lurgi gasifier operated at higher temperatures so that the ash runs off as a liquid slag. The disadvantages of the Lurgi process are that it must operate with noncaking coals and the size of the units must be relatively small. When oxygen (rather than air) is used, the synthesis gas produced has a heating value of about 300 Btu SCF^{-1} and the efficiency (Btu in the outgas per Btu in feed and fuel) is about 60%.

11.66 KOPPERS–TOTZEK GASIFIER

A commercial gasifier that can utilize all types of coal. A mixture of pulverized, predried coal and oxygen is introduced through four coaxial burners into a horizontal reactor at 900°C and atmospheric pressure.

11.67 TEXACO GASIFIER

A gasifier designed to operate at 1200 to 1500°C and 20 to 30 atm. This gasifier is used not only to produce synthesis gas ($H_2/CO = 0.68$), but also the hot gases are used to generate steam to operate a turbine that produces electricity. The combination of gas and electricity production is called cogeneration and is the basis of what is now called "integrated gasification cycle" (IGCC) technology. The synthesis gas produced in the Texaco gasifier was designed to be used for chemical production. The concept of cogeneration is also utilized in the combustion of coal for power production in a gas turbine-steam turbine combined cycle.

11.68 CO_2-ACCEPTOR PROCESS

A gasification process that utilizes calcium oxide, along with coal, to react with the CO_2 produced to form calcium carbonate in an exothermic reaction.

11.69 SYNTHESIS GAS

A mixture of carbon monoxide and hydrogen. Synthesis gas can be prepared, theoretically, from any carbon source by treatment with steam at about 950°C:

$$\mathrm{C\,(s) + H_2O\,(g) \longrightarrow CO\,(g) + H_2\,(g)}$$

$$\Delta H = 162.8\,\mathrm{kJ\ mol^{-1}} \quad (38.9\,\mathrm{kcal\ mol^{-1}}) \tag{11.69}$$

Since the reaction is highly endothermic, oxygen is added to the steam in order to provide heat derived from the combustion of some of the carbon to CO_2.

11.70 SUBSTITUTE NATURAL GAS (SNG)

A gas prepared from coal (or naphtha) that has approximately the same energy as natural gas, roughly 1,000 Btu SCF^{-1}, and can be fed into the gas pipeline system to augment or replace natural gas. Pipeline gas is principally methane, and since coal has atomic $H/C = \sim 0.8$ and methane has $H/C = 4$, the preparation of SNG from coal requires the addition of hydrogen. Processes for producing SNG involve the preparation of a synthesis gas that, after clean-up, is sent to a water–gas shift converter (Sect. 11.71) to increase the proportion of hydrogen, since the stoichiometry for H_2/CO conversion to methane is 3/1. After an intense purification, the hydrogen-rich mixture is sent to a methanation unit (Sect. 11.73) to provide a gas that is essentially the equivalent ($\sim$1,000 Btu SCF^{-1}) of natural gas.

11.71 WATER–GAS SHIFT

The catalytic reaction:

$$\mathrm{CO + H_2O \longrightarrow H_2 + CO_2}$$

$$\Delta H = -41.0\,\mathrm{kJ\ mol^{-1}} \quad (-9.8\,\mathrm{kcal\ mol^{-1}}) \tag{11.71}$$

The reaction is usually used in conjunction with synthesis gas production to shift the concentration of hydrogen in the gas mixture from approximately 1:1 to a higher concentration (3:1) required for conversion to methane. The reaction temperature is about 325°C.

11.72 STEAM REFORMING (TO PRODUCE SYNTHESIS GAS)

The catalytic conversion of methane or other low molecular weight hydrocarbons to carbon monoxide, carbon dioxide, and hydrogen. The two principal reactions are

$$CH_4 + H_2O \longrightarrow CO + 3\,H_2$$

$$\Delta H = +206.3\,\text{kJ mol}^{-1} \quad (49.3\,\text{kcal mol}^{-1}) \qquad (11.72a)$$

$$CH_4 + 2\,H_2O \longrightarrow CO_2 + 4\,H_2$$

$$\Delta H = +163\,\text{kJ mol}' \quad (39\,\text{kcal mol}^{-1}) \qquad (11.72b)$$

11.73 METHANATION

The conversion of synthesis gas to methane (Eq. 11.73) at about 375°C:

$$CO + 3\,H_2 \longrightarrow CH_4 + H_2O$$

$$\Delta H = -206.3\,\text{kJ mol}^{-1} \quad (-49.3\,\text{kcal mol}^{-1}) \qquad (11.73)$$

This is essentially the reverse of the steam reforming reaction.

11.74 GASOHOL

An acronym for a mixture of gasoline and alcohol. It is usually a mixture of 10 vol.% ethanol and gasoline. Pure ethanol has a research octane number of 105 and a blending research octane number of 135 (Sect. 11.29). Because pure ethanol readily takes up water, the alcohol is usually added to the gasoline as late in the distribution as possible. The similar mixture with methanol is also called gasohol. However, water compatibility, corrosion, and toxicity problems associated with methanol make it a less popular choice.

11.75 FISCHER–TROPSCH (FT) PROCESS

The process developed by F. Fischer (1877–1948) and H. Tropsch (1889–1935) in Germany about 1925, which involves the catalytic conversion of synthesis gas to hydrocarbons and low molecular weight oxygenated compounds. The overall reaction is highly exothermic:

$$n\,CO + 2n\,H_2 \longrightarrow -\!(CH_2)_n\!- + n\,H_2O$$

$$\Delta H = -192\,\text{kJ mol}^{-1} \quad (-46\,\text{kcal mol}^{-1}) \qquad (11.75a)$$

Typical catalysts are supported (alumina or silica gel) cobalt-thoria or iron catalysts. The reaction temperature is about 250 to 300°C and pressures range from 1 to about 20 atm. The hydrocarbons produced are almost completely alkanes, which include high molecular weight waxes. The liquid product is a very satisfactory Diesel fuel but a relatively poor internal combustion engine fuel. The FT synthesis is thought to result from the polymerization of (CH_2) units and follows polymerization kinetics. The distribution of products can be described by the Anderson–Schultz–Flory equation, which in logarithmic form is

$$\log(W_n/n) = n \log \alpha + \log[(1-\alpha)^2/\alpha)] \qquad (11.75b)$$

where W_n is the weight fraction of products with carbon number n, α is the chain growth probability, $0 \leq \alpha \leq 1$ (called the Schultz–Flory alpha), whose value depends on the catalyst and operating conditions. When α equals 0, the product is pure methane, and when its value is 1, the product is high molecular weight wax. A plot of log (W_n/n) against n gives a straight line.

The South African company, Sasol, is a highly successful enterprise that has commercialized the FT process and is the world's largest synthetic fuels plant based on coal. It was developed to produce gasoline and Diesel fuels because South Africa is rich in coal, but has no indigenous source of petroleum. The plant can be modified to enhance the production of oxygenated compounds, which are designed for chemical uses.

11.76 METHANOL SYNTHESIS

The reaction of synthesis gas at 300 to 400°C and 200 to 300 atm over a zinc-oxide-based catalyst promoted with chromium oxide to produce methanol in high selectivity. Presently, a low-pressure process (50–100 atm) using a mixture of copper, zinc, and aluminum oxides is preferred. The reaction is

$$CO + 2\,H_2 \longrightarrow CH_3OH$$

$$\Delta H = -109\,\text{kJ mol}^{-1} \quad (-26\,\text{kcal mol}^{-1}) \qquad (11.76)$$

11.77 AMMONIA SYNTHESIS

Commercially, ammonia is usually prepared from nitrogen and hydrogen by the Haber process (named after Fritz Haber, 1868–1934):

$$N_2 + 3\,H_2 \longrightarrow 2\,NH_3$$

$$\Delta H = -92\,\text{kJ mol}^{-1} \quad (-22\ \text{kcal mol}^{-1}) \tag{11.77}$$

It is a high-temperature (400–550°C), high-pressure (100–1,000 atm) reaction that is carried out with an iron-based catalyst (Fe_3O_4) containing small amounts of K_2O and $A1_2O_3$. The impetus to build coal gasification plants in many areas of the world arose from the need for fertilizers whose production requires a source of hydrogen to combine with the nitrogen. The hydrogen required for the synthesis of NH_3 is generated via synthesis gas (usually by steam reforming from methane) followed by the water–gas shift, followed by CO_2 removal for conversion to ammonia. About 85% of the ammonia produced in the world is converted to fertilizers, although a small quantity is also used directly as liquid NH_3 for the same purpose. About 20% is directed to chemical syntheses such as acrylonitrile and caprolactam. It is also used for closed-circuit refrigeration because of its high latent heat. Many chemists have viewed liquid ammonia chemistry as being analogous to water chemistry.

11.78 OIL SHALE

A laminated sedimentary rock that contains 5 to 20% organic material.

11.79 SHALE OIL

The oil that can be obtained from shale oil by heating or retorting the oil shale. Commercial processes have been developed for this purpose, and yields of about 30 gal of shale oil per ton of oil shale have been obtained.

11.80 KEROGEN

The waxy organic material in shale that is the precursor of the oil obtained upon pyrolysis.

11.81 FISCHER ASSAY

A standardized analytical retorting procedure used to evaluate the quantity of shale oil that can be produced from a particular oil shale.

11.82 TAR SAND

A naturally occurring mixture of tar, water, and sand. The tar probably results from petroleum quite near the surface of the earth that has been partially devolatilized. The large tar sand deposit called the Athabasca deposit in Alberta, Canada, is presently being used for commercial extraction. The mined tar sand is treated with hot water and steam to separate the tar as a liquid. A typical sample ton of tar sand contains about 80 lb of recoverable tar.

11.83 FUEL CELLS

An electrochemical (voltaic) cell in which electricity is produced directly from a fuel. Common fuels are hydrogen, carbon monoxide, and methane.

Example. In one type of fuel cell, hydrogen and oxygen are bubbled through porous carbon electrodes that are impregnated with active catalysts such as platinum metal. The electrodes are immersed in an aqueous solution of sodium hydroxide. The reactions are

anode: $$2\,H_2(g) + 4\,OH^- \longrightarrow 4\,H_2O(l) + 4\,e^-$$

cathode: $$2\,H_2O(l) + O_2(g) + 4\,e^- \longrightarrow 4\,OH^-$$

cell reaction: $$2\,H_2(g) + O_2(g) \longrightarrow 2\,H_2O(l)$$

$$E^0 = 1.229\ \mathrm{V}$$

The gases are thus continuously consumed, and since the cell is maintained at an elevated temperature, the water evaporates as it is formed.

Perhaps the most promising fuel cell being developed for widespread applications is the Proton Exchange Membrane Fuel Cell (PEMFC or PEFC). The electrolyte in this cell is a solid polymer membrane, thus eliminating corrosion and safety problems, prolonging its lifetime and simplifying handling. The polymeric membrane is prepared from a fluorinated hydrocarbon backbone (Nafion, similar to teflon), containing strong protic acid, SO_3H, groups. The electrodes consist of graphite coated on the inside with platinum black, which functions as the catalyst for reactions at both the anode and cathode. The reaction at the anode is $H_2 \rightarrow 2H^+ + 2e^-$. The protons migrate through the membrane, where they are exchanged for the protons of the SO_3H groups and then migrate to the cathode. The catalyzed cathode exothermic reaction is $1/2\ O_2 + 2H^+ \rightarrow H_2O$. The membrane is impervious to electrons as they flow externally from anode to cathode, doing their work in the external circuit. The

PEMFC operates at about 70 to 80°C. The heat liberated can be utilized in a variety of ways but is not sufficient to generate steam for utility use.

SUGGESTED READING

Green, M. M. and Witcoff, H. A. *Organic Chemistry Principles and Industrial Practice.* Wiley-VCH: New York, 2003.

Kent, J. A., ed. *Riegel's Handbook of Industrial Chemistry*, 10th ed. Kluwer Academic/Plenum: New York, 2003.

Orchin, M. "Solvation and Depolymerization of Coal." U.S. Patent 2,476, 999 (1949).

Wender, I. "Reactions of Synthesis Gas." *Fuel Proc. Technol. 48*, no. 3, 189 (1996).

12 Thermodynamics, Acids and Bases, and Kinetics

12.1	Thermodynamics	453
12.2	Thermochemistry	453
12.3	System	453
12.4	Reversible Process	453
12.5	Irreversible Process	453
12.6	Adiabatic Process	454
12.7	Isothermal Process	454
12.8	State	454
12.9	Intensive/Extensive Properties	454
12.10	State Function (State Variable)	454
12.11	Standard State	455
12.12	Kinetic Energy, E_k	455
12.13	Potential Energy E_p (or V)	455
12.14	Heat, q	456
12.15	Internal Energy U (or E)	456
12.16	Enthalpy (Heat Content) H	456
12.17	Enthalpy of Reaction ΔH_r	457
12.18	Exothermic Reaction	457
12.19	Endothermic Reaction	457
12.20	Standard Enthalpy (or Standard Heat) of Reaction ΔH_r^0	457
12.21	Standard Enthalpy (or Standard Heat) of Formation ΔH_f^0	458
12.22	Hess's Law of Constant Heat Summation	458
12.23	Heat Capacity C and Specific Heat Capacity c (Specific Heat)	458
12.24	Heat of Hydrogenation ΔH_{hydro}	459
12.25	Heat of Combustion	460
12.26	Heat of Fusion (or Melting) ΔH_m	460
12.27	Heat of Vaporization ΔH_{vap}	460
12.28	Heat of Sublimation ΔH_s	461
12.29	Bond Dissociation Energy (Bond Strength, Bond Enthalpy) $D^0{}_{(\text{AB})}$	461
12.30	Average Bond Dissociation Energy D^a	461
12.31	Ionization Energy (Ionization Potential) (IE)	462
12.32	Electron Affinity (EA)	463
12.33	Heterolytic Bond Dissociation Energy, $D(A^+B^-)$	464
12.34	Enthropy S	465

The Vocabulary and Concepts of Organic Chemistry, Second Edition, by Milton Orchin, Roger S. Macomber, Allan Pinhas, and R. Marshall Wilson

12.35	Standard Enthropy S^0	465
12.36	First Law of Thermodynamics	466
12.37	Second Law of Thermodynamics	466
12.38	Thrid Law of Thermodynamics	466
12.39	Gibbs Free Energy G	467
12.40	Standard Free Energy Change ΔG^0	467
12.41	Exergonic (Exoergic) Process	467
12.42	Endergonic (Endoergic) Process	467
12.43	Standard Free Energy of Formation ΔG_f^0	468
12.44	Helmholtz Free Energy A	468
12.45	Activity a and activity coefficient γ	468
12.46	Chemical Potential	469
12.47	Mass Action Expression Q	469
12.48	Equilibrium	469
12.49	Thermodynamic Equilibrium Constant K	469
12.50	Brönsted Acid and Brönsted Base	470
12.51	Lewis Acid and Lewis Base	471
12.52	Stability Constant	472
12.53	Hard/Soft Acids and Bases (HSAB)	472
12.54	Superacid	473
12.55	Ionization Constant	473
12.56	Acid Dissociation Constant K_a and pK_a	474
12.57	Acidity Function H and Hammett Acidity Function H_0	474
12.58	Base Dissociation Constant K_b and pK_b	475
12.59	Proton Affinity (PA)	476
12.60	Hydride Affinity	476
12.61	Autoprotolysis	477
12.62	Aprotic Solvent	477
12.63	Leveling Effect of Solvent	477
12.64	Transition State	478
12.65	Elementary Reaction Steps, Molecularity	478
12.66	Concerted Reaction	478
12.67	Multistep Reactions	479
12.68	Mechanism of a Reaction	479
12.69	Reaction Coordinate (Profile) Diagram	479
12.70	Free Energy Diagram	480
12.71	Hammond's Postulate	482
12.72	Kinetics	483
12.73	Rate of Reaction ν	483
12.74	Differential Rate Equation (Rate Law), Rate Constant	483
12.75	Integrated Rate Expression	484
12.76	Rate-Determining (or Rate-Limiting) Step	484
12.77	Michaelis–Menton Kinetics	486
12.78	Pseudo-Rate Law	487
12.79	Half-Life $t_{1/2}$	487
12.80	Lifetime τ	488
12.81	Arrhenius Energy of Activation E_a	488
12.82	Transition-State Theory	488
12.83	Thermodynamic Stability	489

12.84	Kinetic Stability	490
12.85	Thermodynamic (or Equilibrium) Control of a Reaction	490
12.86	Kinetic Control	490
12.87	Curtin–Hammett Principle	491
12.88	Diffusion- (or Encounter-) Controlled Rate	492
12.89	Catalysis	492
12.90	General Acid or Base Catalysis	493
12.91	Specific Acid or Base Catalysis	494
12.92	Reactivity	494
12.93	Relative Rate	495
12.94	Partial Rate Factor f	495
12.95	Substituent Effect	496
12.96	Kinetic Isotope Effect	496
12.97	Double Labeling Experiment	498
12.98	Common Ion Effect	498
12.99	Ion Pair	499
12.100	Salt Effect	499
12.101	Migratory Aptitude	500
12.102	Linear Free Energy Relationships	500
12.103	Hammett Equation	500
12.104	Taft Equation	501
12.105	Brönsted Equation	502
12.106	Winstein–Grunwald Equation	502
12.107	Tunneling	502
12.108	Microscopic Reversibility	503
12.109	Isokinetic Temperature	503
12.110	Marcus Theory	503

One extremely important aspect of all chemical reactions involves the energy changes that accompany the chemical changes. Is energy liberated or absorbed? How much energy is involved? Is the reaction favorable, that is, will it proceed spontaneously? These questions are the province of chemical thermodynamics.

Many common terms are encountered in the context of thermodynamics—terms such as energy, heat, work, temperature, and force. These terms have very specific meanings in thermodynamics, meanings that are not always as simple as first imagined. For example, physicists define energy as "the capacity to do work." *Work* is the operation of a force through a distance. However, *force* is best defined as the derivative of potential energy with respect to distance. *Heat* is a form of energy that is equated somehow with the temperature and motion of particles. Yet most textbooks on thermodynamics delay the definition of temperature until they cover something called the Boltzman distribution function.

In the definitions below, the most concise and useful definitions of the common terms of thermodynamics, acids and bases, and kinetics are presented, without overburdening the reader with too much theoretical obfuscation.

12.1 THERMODYNAMICS

The scientific discipline that deals with the energy changes which accompany chemical and physical processes. The suffix-*dynamics* in the word thermodynamics is somewhat misleading because thermodynamics is concerned only with where the reaction starts and where it ends, never with how rapidly or slowly the reaction proceeds.

12.2 THERMOCHEMISTRY

The branch of chemistry dealing with the quantity of heat energy that is released or absorbed during chemical reactions and physical processes.

12.3 SYSTEM

The region of the universe under investigation. Boundaries, real or imaginary, delineate the system from the remainder of the universe, known as the *surroundings*. An *isolated system* is not affected by changes in the surroundings and does not exchange energy (heat or work) or mass with the surroundings. A *closed system* is one in which no transfer of mass occurs across the boundaries; energy may, however, be exchanged. A *microscopic system* is one viewed on the atomic or molecular scale. A *macroscopic system* is viewed on the bulk scale, for example, moles, grams, and so on. Thermodynamics, which predates our detailed understanding of molecular and atomic structure, is applied mainly to macroscopic systems.

12.4 REVERSIBLE PROCESS

In theory, a process (chemical or physical change) that the system undergoes at an infinitesimally slow rate such that, at any instant, all forces acting on the system are in balance. Thus, an infinitesimal change in any one of these forces can change the direction of the process from forward to backward, or *vice versa*. In principle, a reversible process is achievable, but in practice such is not the case because the forces cannot be readily balanced and the rates cannot be made slow enough. See also Sect. 12.48 (equilibrium).

In practice, chemists use the term *reversible reaction* to connote one that can be observed to occur in both directions under a given set of conditions.

12.5 IRREVERSIBLE PROCESS

In theory, a process (chemical or physical change) that a system undergoes at a finite rate because the forces acting on the system are not in balance. Thus, any chemical reaction that occurs spontaneously is occurring irreversibly.

In practice, chemists use the term *irreversible reaction* to connote one that can be observed to occur predominantly in one direction under a given set of conditions.

12.6 ADIABATIC PROCESS

A process during which the system neither gains nor loses heat. (See Sect. 12.31 for an alternate use of the term adiabatic when dealing with ionization potential.)

12.7 ISOTHERMAL PROCESS

This is a process during which the system remains at a constant temperature, and therefore, the particles making up the system remain at constant average kinetic energy (Sect. 12.12).

12.8 STATE

The specific set of conditions that completely describes the properties of the system. These include such factors as the quantity of each chemical species present, as well as physical variables such as temperature, pressure, electric and magnetic field strength.

Example. A system might be described as 1.00 mole of liquid acetic acid under 1 atm of external pressure at 25°C. The description of the state must enable one to reproduce the system exactly.

12.9 INTENSIVE/EXTENSIVE PROPERTIES

An extensive property is one that is proportional to the amount of material present in the system; an intensive property is one that is independent of the amount of material present in the system.

Example.

EXTENSIVE PROPERTIES. The quantity of energy produced by a reaction depends on the amount of material (grams, moles) present.

INTENSIVE PROPERTY. Temperature and applied pressure are independent of the amount of material present.

Often, an extensive property can be converted into an intensive one by creating a ratio of two extensive variable. Examples include density (grams/liter), heat capacity (calories/mole, Sect. 12.23), and various other ratios of energy/mole.

12.10 STATE FUNCTION (STATE VARIABLE)

A thermodynamic function that characterizes the system and is dependent only on the state of the system, not the path by which it is created. When a system undergoes a change from one state to another, the change in a state function is also dependent on only the initial and final states and not the path taken.

Example. Internal energy (Sect. 12.15), enthalpy (Sect. 12.16), entropy (Sect. 12.34), and free energy (Sect. 12.39) of a system. Each is a state function.

12.11 STANDARD STATE

An arbitrary reference state. The standard state for a solid or liquid is the pure substance in its most stable form under a pressure of 1 atm. The standard state for a gas is the (hypothetical) ideal gas at 1 atm pressure. Although the temperature of the standard state is arbitrary, the one most frequently used is 298.15 K (25.00°C). A superscript of zero is attached to a state function (e.g., enthalpy H^0) to indicate that it represents a system in a standard state. The temperature of the state sometimes is indicated by a subscript (H^0_{298K}).

12.12 KINETIC ENERGY, E_k

The energy that a system possesses by virtue of the motion of its constituent particles. Translational energy (arising from translations and consequent changes in the motion of the center of mass of a particle), rotational energy (arising from rotation of a particle about its center of mass), and vibrational energy (arising from motion that results in changes in bond angles and distances without displacing the center of mass)—all contribute to a particle's kinetic energy. The average kinetic energy of a system is directly proportional to the absolute *temperature* (Kelvin or K).

Example. The translational kinetic energy of 1 mole of an ideal gas is equal to 3/2*RT*, where R is the gas constant (1.987 cal mol^{-1} K^{-1} = 8.315 J mol^{-1} K^{-1}) and T is the absolute temperature (K). At 25°C, the translational kinetic energy is 3,715 J mol^{-1}. Thus, one molecule has an average kinetic energy of 6.17×10^{-21} J. The kinetic energy of a molecule is related to its mass m and velocity v by the equation

$$E_k = 1/2(mv^2) = 3/2(RT) \tag{12.12}$$

If we use this equation, the average speed of a gaseous water molecule at 25°C is 1,214 mph.

12.13 POTENTIAL ENERGY E_p (OR V)

The energy a particle possesses by virtue of its position relative to the forces acting on it. Potential energy is often a consequence of the attraction or repulsion of a particle or object by surrounding particles or objects. Kinetic energy may be interconverted with potential energy, but their sum must remain constant for an isolated system.

Example. Any airborne object has potential energy because of the force of gravity acting on it. When the object falls, gravitational potential energy is converted to kinetic energy. The higher an object is, the more potential energy it possesses, and thus, the more kinetic energy it acquires as it falls.

Figure 2.18 shows a Morse curve (also see Sect. 18.58), a plot of the potential energy of the H_2 molecule as a function of internuclear distance. The two nuclei vibrate back and forth around the (equilibrium) bond length d_0. At d_0 the molecule has minimum potential energy (and maximum kinetic energy). At any shorter or longer internuclear distance, the potential energy of the molecule increases (and its kinetic energy decreases).

12.14 HEAT, *q*

The quantity of heat energy absorbed or liberated by the system when it is subjected to change. Note that if heat is *absorbed* by the system, q will have a positive sign; if heat is *liberated* by the system, q will have a negative sign. When the heat transfer occurs under conditions of constant pressure, it is labeled q_p; for constant volume heat transfer, it is q_v. When heat transfer occurs during a reversible process, it is labeled q_{rev}; if the process is irreversible, the label is q_{irrev}.

12.15 INTERNAL ENERGY *U* (OR *E*)

A thermodynamic state function that describes the sum of the kinetic and potential energy of a system. Changes in internal energy (ΔU) for closed systems result from a exchange of work and/or heat between the system and its surroundings:

$$\Delta U_{\mathrm{system}} = q + w \tag{12.15}$$

where q is heat energy absorbed *by* the system and w is work done *on* the system. A Greek uppercase delta Δ represents the difference between the final value of U and the initial value, that is, $\Delta U = U_{\mathrm{final}} - U_{\mathrm{initial}}$. When ΔU is positive, the internal energy of the system has increased by virtue of the heat absorbed and/or the work done on the system. Since the most common type of work done on a chemical system is pressure/volume work, a process carried out at a constant volume will involve no work, and therefore, $\Delta U = q_v$.

12.16 ENTHALPY (HEAT CONTENT) *H*

A thermodynamic state function defined by

$$H = U + PV \tag{12.16}$$

where U is the internal energy, P the pressure, and V the volume of the system. The standard enthalpy H^0 is the enthalpy of a system in its standard state.

12.17 ENTHALPY OF REACTION ΔH_r

The change in enthalpy when a system undergoes a chemical reaction:

$$\Delta H_r = H_{\text{final}} - H_{\text{initial}} = \Delta U + \Delta(PV) = \Delta U + V\Delta P + P\Delta V \qquad (12.17a)$$

where U is the internal energy, P the pressure, and V the volume. For reactions occurring at a constant pressure ($\Delta P = 0$, as with most reactions in solution), the change in enthalpy is the heat of reaction q_p:

$$\Delta H_r = q_p = \Delta U + P\Delta V \qquad (12.17b)$$

If the volume change is also negligible ($\Delta V = 0$, when no gases are involved), $\Delta H_r = \Delta U = q_v$.

12.18 EXOTHERMIC REACTION

A chemical reaction in which heat is liberated by the system to the surroundings. ΔH_r and q have negative values for an exothermic reaction.

12.19 ENDOTHERMIC REACTION

A chemical reaction in which heat is absorbed by the system from the surroundings. ΔH_r and q have positive values for an endothermic reaction.

12.20 STANDARD ENTHALPY (OR STANDARD HEAT) OF REACTION ΔH_r^0

The change in enthalpy ($H^0_{\text{final}} - H^0_{\text{initial}}$) when a system in its standard state undergoes a transformation and the final system also is in its standard state. Since pressure is constant for standard states (1 atm), the enthalpy change is equal to q_p. A positive value of ΔH_r^0 indicates that heat is absorbed by the system, and a negative value indicates that heat is given off by the system.

Example. The overall equation for photosynthesis:

$$6\,CO_2\,(g) + 6\,H_2O\,(l) \longrightarrow C_6H_{12}O_6\,(aq) + 6\,O_2\,(g)$$

$\Delta H^0_{298} = 2{,}900\,\text{kJ mol}^{-1}$ ($693\,\text{kcal mol}^{-1}$) indicates a large increase in the heat content of the system.

12.21 STANDARD ENTHALPY (OR STANDARD HEAT) OF FORMATION ΔH^0_f

The change in enthalpy when 1 mole of a substance in its standard state is formed from its constituent elements in their respective standard states. For an element in its standard state, ΔH^0_f is defined as zero.

Example. At 25°C (298 K) and 1 atm,

$$\text{C (s)} + 2\,\text{Cl}_2\text{ (g)} \longrightarrow \text{CCl}_4\text{ (l)}$$

The heat evolved in this reaction is $139.3\,\text{kJ mol}^{-1}$ ($33.3\,\text{kcal mol}^{-1}$). Therefore, the heat of formation of CCl_4 [$\Delta H^0_f(CCl_4, l)$]is $-139.3\,\text{kJ mol}^{-1}$.

12.22 HESS'S LAW OF CONSTANT HEAT SUMMATION

Because enthalpy changes are independent of the pathway taken (enthalpy is a state function), it is possible to calculate the enthalpy change of a reaction from related reactions whose enthalpy changes are known. This is Hess's law (after Henri Hess, 1802–1850).

Example. Suppose the heats of formation (ΔH^0_f) for CO (−109.4 kJ/mol) and CO_2 (−393.7 kJ/mol) are known and it is desired to calculate ΔH^0_{298} for the oxidation of CO to CO_2.

The heat liberated by the desired reaction, $CO + 1/2\ O_2 \rightarrow CO_2$, can be calculated from the algebraic sum of the two heat of formation reactions:

Reaction	ΔH_r
$\text{CO (g)} \longrightarrow \text{C (s)} + 1/2\ \text{O}_2\text{ (g)}$	109.4 kJ (26.2 kcal)
$\text{C (s)} + \text{O}_2\text{ (g)} \longrightarrow \text{CO}_2\text{ (g)}$	-393.7 kJ (-94.1 kcal)
$\text{CO (g)} + 1/2\ \text{O}_2\text{ (g)} \longrightarrow \text{CO}_2\text{ (g)}$	-284.3 kJ (-68.0 kcal)

If we use this approach, the heats of a wide variety of reactions can be predicted from the heats of formation of relatively few compounds.

12.23 HEAT CAPACITY *C* AND SPECIFIC HEAT CAPACITY *c* (SPECIFIC HEAT)

The ratio of heat absorbed to the increase in temperature of a system or substance:

$$C = q/\Delta T \tag{12.23a}$$

At constant volume, heat capacity is the (partial) derivative of internal energy with respect to temperature:

$$C_v = (\partial U / \partial T)_v \tag{12.23b}$$

At constant pressure, heat capacity is the (partial) derivative of enthalpy with respect to temperature:

$$C_p = (\partial H / \partial T)_p \tag{12.23c}$$

Both these are usually expressed on a "per mole" basis.

The specific heat capacity *c*, also known as the specific heat, is the heat capacity of a sample divided by its mass: $c = C/m$, that is, the heat capacity per gram.

Example.

Compound	C_p (J K^{-1} mol^{-1})	C_v (J K^{-1} mol^{-1})
CS_2	75.7	47.1
$CHCl_3$	116.3	72.8
CCl_4	131.7	89.5
C_6H_6	134.3	91.6

12.24 HEAT OF HYDROGENATION ΔH_{hydro}

The heat of reaction (enthalpy of reaction at a constant pressure) for the addition of molecular hydrogen to 1 mole of a reactant to form a particular hydrogenation product. Hydrogenation is always exothermic (i.e., ΔH_r is negative).

Example.

$$H_3CCH_2CH{=}CH_2 + H_2 \xrightarrow{\text{catalyst}} CH_3CH_2CH_2CH_3$$

ΔH_r = -127 kJ mol^{-1} (-30.3 kcal mol^{-1})

ΔH_r = -120 kJ mol^{-1} (-28.6 kcal mol^{-1})

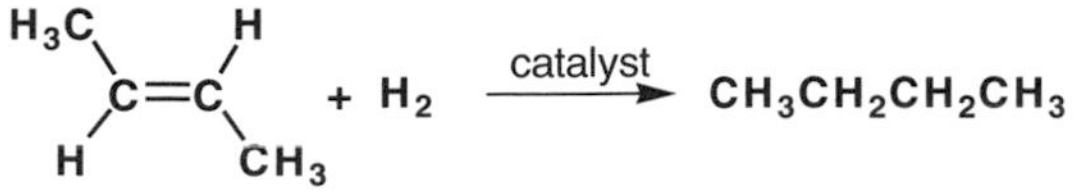

ΔH_r = -115 kJ mol^{-1} (-27.6 kcal mol^{-1})

12.25 HEAT OF COMBUSTION

The heat of reaction (enthalpy of reaction at a constant pressure) when 1 mole of a substance reacts with an excess of gaseous oxygen to convert all carbon atoms to carbon dioxide and all hydrogen atoms to water. Other elements are formally converted to specific oxidation products. Thermochemical adjustments are made on the experimental data to reflect the formal production of these specific compounds.

Example.

$$CH_4\,(g) + 2\,O_2\,(g) \longrightarrow CO_2\,(g) + 2\,H_2O\,(l)$$
$$\Delta H_r = -890.4\ \text{kJ mol}^{-1}\ (-212.8\ \text{kcal mol}^{-1})$$

$$CH_3I\,(l) + 1.75\,O_2\,(g) \longrightarrow CO_2\,(g) + 1.5\,H_2O\,(l) + 0.5\,I_2\,(s)$$
$$\Delta H_r = -807.9\ \text{kJ mol}^{-1}\ (-193.1\ \text{kcal mol}^{-1})$$

12.26 HEAT OF FUSION (OR MELTING) ΔH_m

The enthalpy change (heat absorbed) when 1 mole (*molar heat of fusion*) or 1 g (*specific heat of fusion*) of a substance is converted from the solid to the liquid phase at a specified pressure (usually, 1 atm) and a specified temperature (usually, the melting point of the substance). The process takes place isothermally.

Example. See the Table 12.27 below.

12.27 HEAT OF VAPORIZATION ΔH_{vap}

The enthalpy change (heat absorbed) when 1 mole (*molar heat of vaporization*) or 1 g (*specific heat of vaporization*) of a substance is converted from the liquid to the vapor phase at a specified pressure (usually, 1 atm) and a specified temperature (usually, the normal boiling point of the substance). The process takes place isothermally.

Example. Table 12.27.

TABLE 12.27. Melting Point, Heat of Fusion, Boiling Point, and Heat of Vaporization for Four Common Compounds

Substance	mp (°C)	ΔH_m (kJ mol^{-1})[a]	bp (°C)	ΔH_{vap} (kJ mol^{-1})[a]
C_6H_6	5.5	10.58	80.1	30.8
CH_3NH_2	−93.5	6.12	−6.2	25.8
CH_3OH	−97.9	3.17	64.7	35.3
H_2O	0	6.01	100	40.67

[a] At 1 atm.

12.28 HEAT OF SUBLIMATION ΔH_s

The enthalpy change (heat absorbed) when 1 mole (*molar heat of sublimation*) or 1 g (*specific heat of sublimation*) of the substance is converted directly from the solid to the vapor phase at a specified pressure (usually, 1 atm) and a specified temperature. The process takes place isothermally (see Sect. 10.18).

Example. The two isomeric hydrocarbons, $C_{10}H_8$, behave quite differently:

Substance	ΔH_s (kJ mol^{-1})[a]
Naphthalene	72.8
Azulene	95.4

[a] At 1 atm and 278 K.

12.29 BOND DISSOCIATION ENERGY (BOND STRENGTH, BOND ENTHALPY) $D^0_{(AB)}$

The enthalpy change that accompanies the homolytic cleavage of a bond in the gas phase at 298 K (25°C):

$$\text{A—B} \longrightarrow \text{A}\cdot + \text{B}\cdot \qquad (12.29a)$$

Example.

$$H_3C—CH_3 \longrightarrow 2\ CH_3\cdot \qquad (12.29b)$$

$$D° = 368 \text{ kJ mol}^{-1}\ (88 \text{ kcal mol}^{-1})$$

$$CH_3COO\text{-}H \longrightarrow CH_3COO\cdot + H\cdot \qquad (12.29c)$$

$$D° = 469 \text{ kJ mol}^{-1}\ (112 \text{ kcal mol}^{-1})$$

$$H_2C{=}O \longrightarrow H_2C{:} + {:}\ddot{\underset{\cdot}{O}}\cdot \qquad (12.29d)$$

$$D° = 732 \text{ kJ mol}^{-1}\ (175 \text{ kcal mol}^{-1})$$

12.30 AVERAGE BOND DISSOCIATION ENERGY D^a

The average of enthalpy changes for the homolytic cleavage of n equivalent bonds of a molecule. For example,

$$AB_n \longrightarrow \text{A (with n electrons)} + n\ B\cdot \qquad (12.30a)$$

$$D^a\ (\text{A-B}) = \Delta H_r / n$$

Example.

$$CH_4 \longrightarrow \cdot\ddot{C}\cdot + 4\,H\cdot \qquad (12.30b)$$

$$\Delta H_r = 1662 \text{ kJ mol}^{-1} \ (397 \text{ kcal mol}^{-1})$$

Four carbon–hydrogen bonds are broken:

$$D^a(\text{C–H}) = 1{,}662/4 \text{ kJ mol}^{-1} = 416 \text{ kJ mol}^{-1} \quad (99.4 \text{ kcal mol}^{-1})$$

although the actual dissociation energy of each bond is different:

$$CH_4 \longrightarrow \cdot CH_3 + H\cdot \qquad \Delta H_r = 435 \text{ kJ mol}^{-1} \qquad (12.30d)$$

$$\cdot CH_3 \longrightarrow \cdot\dot{C}H_2 + H\cdot \qquad \Delta H_r = 444 \text{ kJ mol}^{-1} \qquad (12.30e)$$

$$\cdot\dot{C}H_2 \longrightarrow \cdot\ddot{C}H + H\cdot \qquad \Delta H_r = 444 \text{ kJ mol}^{-1} \qquad (12.30f)$$

$$\cdot\ddot{C}H \longrightarrow \cdot\ddot{C}\cdot + H\cdot \qquad \Delta H_r = 339 \text{ kJ mol}^{-1} \qquad (12.30g)$$

The term bond energy is loosely used for either the average bond dissociation energy or an average of specific bond dissociation energies obtained from different molecules of a given type. For example,

	Average Bond Dissociation Energy[a]	
Bond	kJ mol^{-1}	kcal mol^{-1}
C–H	416	99
C–C	347	83
C–N	305	73
C–O	356	85
C–F	485	116
C=C	611	146
C=N	615	147
C=O	749	179

[a] 1 cal = 4.184 J.

12.31 IONIZATION ENERGY (IONIZATION POTENTIAL) (IE)

The energy required to remove a specific electron from a molecule (or atom) in its gaseous state:

$$M \xrightarrow{-\,e^-} M^{+\cdot}$$

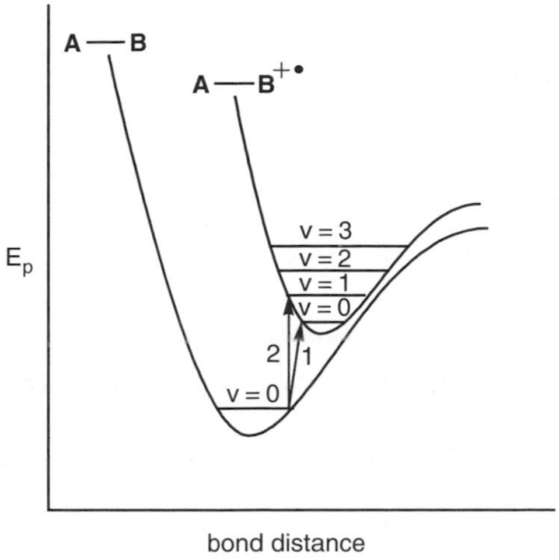

Figure 12.31. Morse curves for A−B and $A{-}B^{+\bullet}$ showing adiabatic (1) and vertical (2) ionization processes.

If the electron is removed from the molecule when it is in its lowest vibrational and rotational state to produce a radical-cation in its lowest vibrational and rotational state, the process is referred to as the *adiabatic ionization energy* (Fig. 12.31, process 1). The *vertical ionization energy* is the energy associated with an ionization process that occurs in accordance with the *Franck–Condon principle* (named after James Franck, 1882–1946, and Edward U. Condon, 1902–1974), that is, without a change in nuclear geometry or momentum (Fig. 12.31, process 2). This process usually requires more energy than an adiabatic ionization owing to the excess vibrational and rotational energy imparted to the ionized molecule (radical cation). Although ionization energy is most frequently used to indicate the minimum energy required to remove an electron from the highest occupied orbital, electrons in different orbitals each have a characteristic ionization energy.

12.32 ELECTRON AFFINITY (EA)

The energy liberated during the capture of an electron by an atom, ion, or molecule in the gas phase:

$$\mathrm{M} \xrightarrow{+\ e^-} \mathrm{M}^{-\bullet} + \text{energy}$$

The electron affinity is given a positive value in a convention that runs counter to usual thermodynamic conventions, where a negative value normally indicates that energy is released.

Example.

Atom	Electron Affinity (eV)[a]
C	1.25 + 0.03
O	1.465 + 0.005
F	3.448 + 0.005

[a] $1\,eV = 96.49\,kJ\,mol^{-1} = 23.06\,kcal\,mol^{-1}$.

12.33 HETEROLYTIC BOND DISSOCIATION ENERGY, $D(A^+B^-)$

The enthalpy change that accompanies the heterolytic cleavage of a bond. For example,

$$A{-}B \longrightarrow A^+ + B^-$$

$$D(A^+B^-) = \Delta H_r = \Delta H_f(A^+) + \Delta H_f(B^-) - \Delta H_f(A{-}B)$$

Example.

$$(CH_3)_3C\text{-}Br\ (g) \longrightarrow (CH_3)_3C^+\ (g) + Br^-\ (g)$$

$$\Delta H_f\,[(CH_3)_3C^+] = 736\,kJ\,mol^{-1} \quad (176\,kcal\,mol^{-1})$$

$$\Delta H_f\,[Br^-] = -234\,kJ\,mol^{-1} \quad (-56\,kcal\,mol^{-1})$$

$$-\Delta H_f\,[(CH_3)_3CBr] = -(-117\,kJ\,mol^{-1}) \quad -(-28\,kcal\,mol^{-1})$$

$$D\,[(CH_3)_3C^+\,Br^-] = 619\,kJ\,mol^{-1} \quad (148\,kcal\,mol^{-1})$$

By Hess's law (Sect. 12.22), it also is possible to calculate heterolytic bond dissociation energies from *bond dissociation energies* (D^0), adiabatic *ionization energies* (IE), and *electron affinities* (EA). Consider the following reactions:

(1)	$A{-}B$	$\longrightarrow A\bullet + B\bullet$	(D°)
(2)	$A\bullet$	$\longrightarrow A^+ + e^-$	(IE)
(3)	$B\bullet + e^-$	$\longrightarrow B^-$	(EA)
(1 + 2 + 3)	$A{-}B$	$\longrightarrow A^+ + B^-$	(D)

$$D(A^+B^-) = D^0(A{-}B) + IP(A^-) - EA(B^-)$$

For example,

$$H_3CCl\ (g) \longrightarrow CH_3^+\ (g) + Cl^-(g)$$

$$D^0(H_3C{-}Cl) = 351\ \text{kJ mol}^{-1} \quad (84\ \text{kcal mol}^{-1})$$
$$\text{IP}\ (CH_3^-) = 950\ \text{kJ mol}^{-1} \quad (227\ \text{kcal mol}^{-1})$$
$$-\text{EA}\ (Cl^-) = -368\ \text{kJ mol}^{-1} \quad (88\ \text{kcal mol}^{-1})$$
$$D(CH_3^+Cl^-) = 933\ \text{kJ mol}^{-1} \quad (223\ \text{kcal mol}^{-1})$$

12.34 ENTROPY S

A thermodynamic state function that is a measure of the degree of disorder or randomness of a system. For a change, from one state to another ΔS, at absolute temperature T,

$$\Delta S_{\text{system}} = q_{\text{rev}}/T \tag{12.34a}$$

where q_{rev} is the heat absorbed under reversible conditions (Sect. 12.4).

For an irreversible pathway (Sect. 12.5),

$$\Delta S_{\text{system}} > q_{\text{irrev}}/T \tag{12.34b}$$

An increase in disorder (decrease in order) results in an increase in entropy (positive ΔS); a decrease in disorder (increase in order) results in a decrease in entropy (negative ΔS).

Example.

$$C_2H_4\ (g) + H_2O\ (g) \longrightarrow C_2H_5OH\ (g)$$

less ordered system	more ordered system
2 particles	1 particle

At 298 K (25°C),

$$\Delta S = -125.6\ \text{J mol}^{-1}\text{K}^{-1} \quad (-30.02\ \text{cal mol}^{-1}\text{K}^{-1})$$

12.35 STANDARD ENTROPY S^0

The entropy of a substance at 298 K and 1 atm, based on S^0 being equal to zero for a perfect crystal of that substance at 0 K. S^0 can be calculated from heat capacities and enthalpies of phase changes by the following equation:

$$S^0 = \int_0^{T_{\text{mp}}} C_s\, dT + \frac{\Delta H_f^0}{T_{\text{mp}}} + \int_{T_{\text{mp}}}^{T_{\text{bp}}} C_1\, dT + \frac{\Delta H_v^0}{T_{\text{bp}}} + \int_{T_{\text{bp}}}^{T} C_v\, dT$$

Example.

$$C_2H_5OH\ (l)\ S_{298}^0 = 160.7\ \text{J mol}^{-1}\text{K}^{-1} \quad (38.4\ \text{cal mol}^{-1}\text{K}^{-1})$$
$$C_2H_5OH\ (g)\ S_{298}^0 = 282.0\ \text{J mol}^{-1}\text{K}^{-1} \quad (67.4\ \text{cal mol}^{-1}\text{K}^{-1})$$

These two values differ by the positive value (more disordered) of entropy of vaporization (ΔS^0_{vap}, 121.3 J mol$^-$K^{-1}), which can be calculated from ΔH^0_{vap} (36.3 kJ mol^{-1} at 298 K) divided by 298 K.

12.36 FIRST LAW OF THERMODYNAMICS

The energy content of the *universe* is fixed, that is, energy can neither be created or destroyed, only transferred. This can be expressed mathematically by

$$\Delta U_{\text{universe}} = \Delta U_{\text{system}} + \Delta U_{\text{surroundings}} = 0 \tag{12.36a}$$

$$\Delta U_{\text{system}} = q + w = -\Delta U_{\text{surroundings}} \tag{12.36b}$$

where ΔU is a change in internal energy, q is heat absorbed *by* the system, and w is work done *on* the system. Thus, any change in the energy content of a system must be accounted for in terms of heat and work flowing in and out of the system. For an isolated system where q and w must equal zero, no net change in the energy content of the system can take place; thus, the energy of the system is said to be conserved ($\Delta U_{\text{system}} = 0$).

12.37 SECOND LAW OF THERMODYNAMICS

Any *spontaneous* physical or chemical process must be accompanied by an increase in the entropy of the universe. This can be expressed mathematically by

$$\Delta S_{\text{universe}} = \Delta S_{\text{system}} + \Delta S_{\text{surroundings}} > 0 \tag{12.37a}$$

If the process is carried out *reversibly*, we get

$$\Delta S_{\text{universe}} = \Delta S_{\text{system}} + \Delta S_{\text{surroundings}} = 0 \tag{12.37b}$$

Thus, the sum of all (spontaneous) processes that occur in our universe results in an increase of the entropy (disorder) of the universe. Individual systems that are not isolated (e.g., chemical reactions) may either increase or decrease in entropy when they undergo a change, but the accompanying $\Delta S_{\text{surroundings}}$ will always assure that $\Delta S_{\text{universe}} > 0$.

12.38 THIRD LAW OF THERMODYNAMICS

A perfect (defect-free) crystal of a pure substance has perfect order and therefore an entropy of zero at absolute zero. This implies that 0 K cannot be attained because an infinite amount of work would be required to order the system perfectly.

Thus, the three laws can be summarized as follows:

First Law: You cannot win; the best you can do is break even.

Second Law: You can only break even at absolute zero.

Third Law: You cannot reach absolute zero.

As an aside, it should be pointed out that none of the three laws of thermodynamics can be proven.

12.39 GIBBS FREE ENERGY G

Named after Josiah W. Gibbs (1839–1903), a thermodynamic state function defined by the equation

$$G = H - TS \tag{12.39a}$$

where H is the enthalpy, T the absolute temperature, and S the entropy. For a change between two states at a constant temperature,

$$\Delta G = \Delta H - T\Delta S \tag{12.39b}$$

The free energy (change) is related to the amount of useful work the system (or process) can deliver. See also Sect. 12.46 (chemical potential).

12.40 STANDARD FREE ENERGY CHANGE ΔG^0

The free energy change between two systems in their standard states. At constant temperature,

$$\Delta G^0 = \Delta H^0 - T\Delta S^0 \tag{12.40}$$

where ΔH^0 is the standard enthalpy of reaction, T the absolute temperature, and ΔS^0 the standard entropy of reaction.

12.41 EXERGONIC (EXOERGIC) PROCESS

A process in which ΔG is negative, that is, free energy is transferred from the system to the surroundings. This is slightly different from exothermic, which involves only the ΔH for the reaction (see Sect. 12.18) and neglects ΔS.

12.42 ENDERGONIC (ENDOERGIC) PROCESS

A process in which ΔG is positive, that is, free energy must be supplied *by* the surroundings *to* the system to drive the reaction. This is slightly different from endothermic, which involves only the ΔH for the reaction (see Sect. 12.19) and neglects ΔS.

12.43 STANDARD FREE ENERGY OF FORMATION ΔG_f^0

The change in free energy when 1 mole of a substance in its standard state is formed from its constituent elements, each in their respective standard states, at 298 K. ΔG_f^0 of all *elements* in their standard states is by definition equal to zero.

Example. At 298 K,

$$\text{C (graphite)} \longrightarrow \text{C (gas)}$$

$$\Delta G_f^0(\text{C, gas}) = 697.9\,\text{kJ mol}^{-1} \quad (166.8\,\text{kcal mol}^{-1})$$

$$\text{C (graphite)} + 2\,\text{Cl}_2\,\text{(gas)} \longrightarrow \text{CCl}_4\,\text{(liquid)}$$

$$\Delta G_f^0(\text{CCl}_4, \text{liquid}) = -68.6\,\text{kJ mol}^{-1} \quad (-16.4\,\text{kcal mol}^{-1})$$

12.44 HELMHOLTZ FREE ENERGY *A*

Named after Herman L. von Helmoltz (1821–1894), a thermodynamic state function defined by the equation

$$A = U - TS \tag{12.44a}$$

where U is the internal energy, T the absolute temperature, and S the entropy. The difference between Gibbs free energy G and Helmholtz free energy A is

$$G - A = PV \tag{12.44b}$$

More important, the Gibbs and Helmholtz thermodynamic relationships are used in different situations. Equations involving G are generally more applicable to constant pressure processes, whereas those involving A are generally more applicable to constant volume processes.

12.45 ACTIVITY *a* AND ACTIVITY COEFFICIENT γ

Activity is a measure of the "effective" concentration of a condensed phase species, adjusted for any nonideal behavior. The activity, which is a dimensionless parameter, is equal to the concentration of a given species $[X_i]$ times its activity coefficient γ_i:

$$a_i = \gamma_i [X_i] \tag{12.45}$$

Because activities are dimentionless parameters, activity coefficients must have the units of inverse concentration.

12.46 CHEMICAL POTENTIAL

The free energy G_i of a component of a mixture is given by the equation

$$G_i = G_i^0 + RT \ln a_i \tag{12.46}$$

where G^0 is the standard-state free energy (at unit activity) of component i, a_i the actual activity (Sect. 12.45), R the gas constant, and T the absolute temperature. For most purposes in organic chemistry, concentrations (mol/L) are substituted for activities.

12.47 MASS ACTION EXPRESSION Q

The ratio of the product concentration (or actually, activities), each raised to the power of their respective stoichiometric coefficient, to the reactant concentrations (activities), each raised to the power of their respective stoichiometric coefficient. Thus, for the reaction

$$\mathrm{aA + bB \rightleftharpoons cC + dD}$$

$$Q = [C]^c\ [D]^d/[A]^a\ [B]^b$$

The value of Q will be dimensionless if activities are used; when mol/L concentrations are used, Q is treated as being dimensionless.

12.48 EQUILIBRIUM

In the general reaction,

$$\mathrm{aA + bB \rightleftharpoons cC + dD}$$

equilibrium is achieved when there are no longer any observable changes with time in the parameters that describe the system. At this point, the rate of the forward reaction equals the rate of the reverse reaction. Two arrows pointing in opposite directions are used to represent an equilibrium. Each concentration (activity) reaches a constant value, as do the mass action expression (Sect. 12.47), temperature, and pressure.

12.49 THERMODYNAMIC EQUILIBRIUM CONSTANT K

The value of Q (the mass action expression, Sect. 12.47) at equilibrium and given the symbol K or K_{eq}. The value of K is dimensionless because Q is dimensionless.

At equilibrium, the free energy of the system is at a minimum. Therefore,

$$\Delta G = 0 \tag{12.49a}$$

Since

$$\Delta G = (G_C + G_D) - (G_A + G_B) \tag{12.49b}$$

we can write

$$\begin{aligned} \Delta G &= G^0_C + RT\ln[C]^c + G^0_D + RT\ln[D]^d \\ &\quad - (G^0_A + RT\ln[A]^a + G^0_B + RT\ln[B]^b) \\ &= \Delta G^0 + RT\ln([C]^c\,[D]^d/[A]^a\,[B]^b) \\ &= \Delta G^0 + RT\ln K = 0 \end{aligned} \tag{12.49c}$$

Therefore,

$$\Delta G^0 = -RT\ln K$$

or

$$K = \exp(-\Delta G^0/RT) \tag{12.49d}$$

The equilibrium constant is related to the *Boltzmann distribution*, which states that, *at equilibrium*, the ratio between the population of any two energy states (P_i/P_j) is determined by the difference in energy between them and the absolute temperature:

$$P_i/P_j = \exp[-(E_i - E_j)/RT] = \exp(-\Delta E/RT) \tag{12.49e}$$

12.50 BRÖNSTED ACID AND BRÖNSTED BASE

A substance that can act as an H^+ (proton) donor in a chemical reaction is a Brönsted acid (after Johannes N. Brönsted, 1879–1947), and the species remaining after the loss of the proton is the *conjugate base* of that acid. A substance that accepts a proton is a Brönsted base and upon protonation becomes the *conjugate acid* of that base. Thus, the product(s) of a Brönsted acid–base reaction are themselves acids and bases. The concept of conjugate acids and bases was developed by T. M. Lowry (1878–1936), so the proton donor-acceptor concept is frequently referred to as the *Brönsted–Lowry* theory of acids and bases.

Example. In the equilibrium mixture

$$CH_3COOH + H_2O \rightleftharpoons CH_3COO^- + H_3O^+$$

TABLE 12.50. Conjugate Relationships

Brönsted Acid (Conjugate Acid)	Conjugate Base (Brönsted Base)
H_2O	HO^-
H_3O^+	H_2O
CH_5^+	CH_4
$CH_3CH_2^+$	$CH_2{=}CH_2$
CH_3COCH_3	$CH_3COCH_2^-$
$(CH_3)_2C{=}OH^+$	$(CH_3)_2C{=}O$

CH_3COOH and H_3O^+ are Brönsted acids (proton donors) and H_2O and CH_3COO^- are Bronsted bases (proton acceptors). The conjugate relationships are:

CH_3COOH, Brönsted acid (or conjugate acid)
CH_3COO^-, conjugate base (or Brönsted base)
H_2O, Brönsted base (or conjugate base)
H_3O^+, conjugate acid (or Brönsted acid)

In the equilibrium mixture

$$CH_3COOH + CF_3COOH \rightleftharpoons CH_3COOH_2^+ + CF_3COO^-$$

CF_3COOH and $CH_3COOH_2^+$ are Brönsted acids, and their conjugate bases are CF_3COO^- and CH_3COOH, respectively. Here, CH_3COOH is acting as a Brönsted base and not an acid, whereas, in the $CH_3COOH{-}H_2O$ mixture, CH_3COOH is acting as a Brönsted acid. Nitric acid is a very strong acid, but in the presence of concentrated sulfuric acid, it acts as a base. Compounds that can act either as acids or bases are described as *amphoteric*. Table 12.50 lists some additional conjugate relationships.

12.51 LEWIS ACID AND LEWIS BASE

Named after Gilbert N. Lewis (1875–1946), a Lewis acid is a substance (molecule or ion) that can act as an electron *pair* acceptor, and a Lewis base is a substance (molecule or ion) that can act as an electron *pair* donor. The species formed by the reaction of a Lewis acid with a Lewis base is a *Lewis adduct* (or complex).

Example.

Lewis Acid		Lewis Base		Lewis Adduct
$(CH_3)_3C^+$	+	$:\ddot{\underset{..}{Cl}}:^-$	$\rightleftharpoons$	$(CH_3)_3CCl$
$(CH_3)_3B$	+	$:NH_3$	$\rightleftharpoons$	$(CH_3)_3\overset{-}{B}\overset{+}{N}H_3$

The Brönsted–Lowry theory of acids and bases focuses on proton transfers, whereas the Lewis theory focuses on the fate of electron pairs (i.e., bonds).

12.52 STABILITY CONSTANT

The value of the equilibrium constant for formation of a Lewis acid–base complex.

12.53 HARD/SOFT ACIDS AND BASES (HSAB)

A qualitative classification of acidity and basicity of Lewis acids and bases proposed by R. Pearson, *J. Am. Chem. Soc. 85*, 3533 (1963). Lewis acids and bases may be classified as being hard, soft, or borderline. The hard and soft adjectives do not connote strong acids or bases.

Hard acids (electron pair acceptors) generally are small and do not possess unshared pairs of electrons in their valence shells. In addition, they afford Lewis adducts with highly localized positive charge density. A hard acid is characterized by high electronegativity and low polarizability. They are frequently cations or molecules with high-energy empty orbitals.

Hard bases (electron pair donors) are generally small, difficult to oxidize, and have no low-energy empty orbitals. They are frequently anions or molecules with low-energy filled orbitals. In addition, they afford Lewis adducts with a highly localized electron pair. A hard base is characterized by a highly electronegative donor atom and low polarizability.

Soft acids generally are large and highly polarizable, and have an empty valence orbital of low energy. Soft bases generally are large and highly polarizable, and have electrons that are easily removed by oxidizing agents because they have a filled valence orbital of high energy. Soft acids behave as electrophiles, and soft bases as nucleophiles (see Sects. 14.5 and 14.9).

Hard acids prefer to combine with hard bases, and soft acids prefer to bond to soft bases.

Example. Table 12.53 lists some hard and soft acids and bases. Ni(0) is a soft acid, and when surrounded by the soft base CO, it forms the stable liquid $Ni(CO)_4$ with a boiling point of 42°C. It is instructive to note that BF_3 is a hard acid and BH_3 is a soft acid. BH_3 forms more stable complexes with soft bases such as carbon monoxide and olefins than with hard bases. On the other hand, the hard acid–hard base complex $F_3B \bullet OR_2$ is more stable than $F_3B \bullet SR_2$, a hard acid–soft base complex. In BF_3 boron is largely B^{3+} because of the strongly electronegative fluorides; hence, BF_3 is hard. In BH_3 boron is largely neutral; hence, BH_3 is soft. It should also be noted that H_2O and HO^- are both hard bases although HO^- is a much stronger base than H_2O.

TABLE 12.53. Hard-Soft Classification of Acids and Bases

Acids		Bases	
Hard	Soft	Hard	Soft
H^+, Li^+, Na^+	Cu^+, Ag^+, Au^+, Ni(0)	H_2O	R_2S, R_3P
Mg^{2+}, Mn^{2+}	Pd^{2+}, Pt^{2+}	HO^-	RS^-
Al^{3+}, Sc^{3+}	Tl^{3+},Tl $(CH_3)_3$	F^-	CN^-
Cr^{3+}, Co^{3+}	RS^+	AcO^-	C_2H_4
Si^{4+}, Ti^{4+}	I^+, Br^+	ROH	C_6H_6
BF_3, $B(OR)_3$	BH_3	R_2O	H^-
$RSO_2{}^+$, SO_3	I_2, Br_2	RO^-	R^-
RCO^+, CO_2	R_2C:	RNH_2	CO

12.54 SUPERACID

An acidic medium that has a proton-donating ability greater than 100% sulfuric acid. Superacids are generally mixtures of fluorosulfonic acid (FSO_3H) and Lewis acids such as SO_3 and SbF_5. An equimolar mixture of FSO_3H and SbF_5 is known as *magic acid*. These acids have been used to study carbocations because such ions are long-lived in superacid solutions at low temperature.

12.55 IONIZATION CONSTANT

An equilibrium constant that is a measure of the degree of dissociation of a compound into its component ions.

Example. For water at 298 K,

$$H_2O\ (l) \rightleftharpoons H^+\ (aq) + OH^-(aq)$$

$$\begin{aligned} K_{ion} &= [H^+]\,[OH^-]/[H_2O] \\ &= [1.004 \times 10^{-7}]\,[1.004 \times 10^{-7}]/[55.5] \\ &= 1.86 \times 10^{-16}\,\text{mol L}^{-1} \end{aligned} \quad (12.55a)$$

In most aqueous solutions the concentration of water is not significantly changed by the small quantity that dissociates, and its value is combined with the equilibrium constant to form a new constant. For pure water, the new constant is given the symbol K_w. At 298 K,

$$K_w = [H_2O]\,K_{ion} = [H^+]\,[OH^-] = 1.008 \times 10^{-14}\,\text{mol}^2\,\text{L}^{-2} \quad (12.55b)$$

Note the different units for K_{ion} and K_w.

12.56 ACID DISSOCIATION CONSTANT K_a AND pK_a

An equilibrium constant that is a measure of the ability of a Brönsted acid to donate a proton to a specific reference base. The greater the value of K_a is, the stronger the acid. In dilute aqueous solution, water is the reference base:

$$\underset{\text{(Bronsted acid)}}{HA} + \underset{\text{(reference base)}}{H_2O} \rightleftharpoons H_3O^+ + A^-$$

$$K_a\,(HA) \text{ in mol L}^{-1} = [H_2O]\,K = [A^-]\,[H_3O^+]/[HA] \qquad (12.56a)$$

where K is the equilibrium constant for the system, and the concentration of water is a constant equal to 55.56 mol L^{-1}. At 50% ionization, K_a is equal to the hydrogen ion concentration. In other solvents,

$$HA + solv \rightleftharpoons solv\text{-}H^+ + A^-$$

$$K_a\,(HA)_{solv} = [solv]\,K = [A^-]\,[solv - H^+]/[HA] \qquad (12.56b)$$

Example. At 25°C for a dilute solution of acetic acid in water,

$$CH_3COOH + H_2O \rightleftharpoons CH_3COO^- + H_3O^+$$

$$K_a\,(CH_3COOH) = 1.76 \times 10^{-5}\,\text{mol L}^{-1} \qquad (12.56c)$$

For pure water at 25°C,

$$H_2O + H_2O \rightleftharpoons H_3O^+ + OH^-$$

$$K_a\,(H_2O) = [H_2O]\,K = [H_3O^+]\,[OH^-]/[H_2O] = 1.8 \times 10^{-16}\,\text{mol L}^{-1} \qquad (12.56d)$$

Thus, H_2O is a weaker acid than CH_3COOH.

The pK_a is defined by the equation

$$pK_a = -\log K_a \qquad (12.56e)$$

Note that as the strength of the acid increases, K_a increases, but pK_a *decreases* (or becomes more negative).

12.57 ACIDITY FUNCTION H AND HAMMETT ACIDITY FUNCTION H_0

A parameter that measures the proton-donating ability of a solvent system, as reflected by the ratio of indicator base B to its conjugate acid BH^+. The acidity function H is defined by

$$H = pK_a\,(BH^+) + \log[B]/[BH^+] \qquad (12.57)$$

The pK_a of indicator acid BH^+ is known and the relative concentrations of B and BH^+ are measured experimentally.

Example. The *Hammett acidity function* H_0 measures the acidity of solvent systems using substituted anilines as indicator bases B. The concentrations of BH^+ and B are measured spectroscopically, and the pK_a of each BH^+ is known. The solvent system consisting of 60 wt. % of sulfuric acid in water has an H_0 value of -4.32. The indicator base 2,4-dinitroaniline [pK_a (BH^+) $= -4.38$] was used to determine this value. Once an H_0 value is known for a solvent system, that system may be used to experimentally determine an unknown pK_a.

12.58 BASE DISSOCIATION CONSTANT K_b AND pK_b

An equilibrium constant that characterizes the ability of a Brönsted base to accept a proton from a reference acid. The larger the value of K_b is, the stronger the base. In dilute aqueous solution, water is the reference acid:

$$\underset{\text{(Bronsted base)}}{B} + \underset{\text{(reference acid)}}{H_2O} \rightleftharpoons BH^+ + OH^-$$

$$K_b\,(B) \text{ in mol L}^{-1} = [H_2O]\,K = [BH^+]\,[^-OH]/[B] \tag{12.58a}$$

where K is the equilibrium constant for the system (see Sect. 12.49). The larger the value of K_b is, the stronger the base. The product of the aqueous acid and base dissociation constants of conjugate pairs is equal to K_w, the ionization constant for water:

$$K_a\,(BH^+) \times K_b\,(B) = K_w \tag{12.58b}$$

Thus, if K_a (BH^+) is large (a strong acid), then K_b (B) is small (a weak base). The pK_b is defined by the equation

$$pK_b = -\log K_b \tag{12.58c}$$

Example.

Base	K_b	pK_b	Conjugate Acid	K_a	pK_a
H_2O	$10^{-15.74}$	15.74	H_3O^+	$10^{+1.74}$	1.74
$(CH_3)_2NH$	$10^{-3.27}$	3.23	$(CH_3)_2NH_2^+$	$10^{-10.73}$	10.73
CH_3O^-	$10^{+1.5}$	-1.5	CH_3OH	$10^{-15.5}$	15.5
HO^-	$10^{+1.74}$	-1.74	H_2O	$10^{-15.74}$	15.74

Note that as the strength of the base strength increases, K_b increases, but pK_b *decreases*.

12.59 PROTON AFFINITY (PA)

A measure of the basicity of a species as determined by the standard enthalpy of that species losing a proton in the gas phase. For the reaction the proton affinity is given by

$$BH^+ \longrightarrow B + H^+$$

$$\text{PA (B)} = \Delta H_f^0(\text{B}) + \Delta H_f^0(\text{H}^+) - \Delta H_f^0(\text{BH}^+) \qquad (12.59)$$

The proton affinity is a quantitative measure of the gas phase basicity of both Lewis and Brönsted bases. The higher the proton affinity is, the stronger the base.

Example.

	Proton affinity	
Species	kcal mol^{-1}	kJ mol^{-1}
CH_3NH_2	211.3	884.0
$CH_3COOCH_2CH_3$	198.0	828.4
CH_3COCH_3	194.6	814.2
CH_3CN	187.4	784.1
CH_3CH_2OH	186.8	781.6
Benzene	183.4	767.3
$H_2C{=}CHCH_3$	181.0	757.3
H_2O	168.9	706.7

12.60 HYDRIDE AFFINITY

A measure of the tendency of a species to accept a hydride as measured by the standard enthalpy of its losing a hydride ion in the gas phase. For the reaction

$$HA \rightleftharpoons H^- + A^+$$

$$\text{Hydride affinity } (\text{A}^+) = \Delta H_f^0(\text{A}^+) + \Delta H_f^0(\text{H}^-) - \Delta H_f^0(\text{HA}) \qquad (12.60)$$

Hydride affinity, equivalent to the heterolytic bond dissociation energy $D(\text{H}^-\text{A}^+)$, is one quantitative measure of the gas phase acidity of Lewis acids such as carbocations. The higher the hydride affinity is, the stronger the Lewis acid. Conversely, the lower the hydride affinity is, the greater the stability of the Lewis acid.

Example. The gas phase hydride affinities of cyclopentyl cation (Fig. 12.60*a*) and the norbornyl cation (Fig. 12.60*b*) are 245 and 234 kcal mol^{-1} (1,025 and 979 kJ mol^{-1}), respectively. Cyclopentyl cation is thus a stronger Lewis acid than norbornyl cation; conversely, the norbornyl cation in which the positive charge is delocalized is more stable than the cyclopentyl cation.

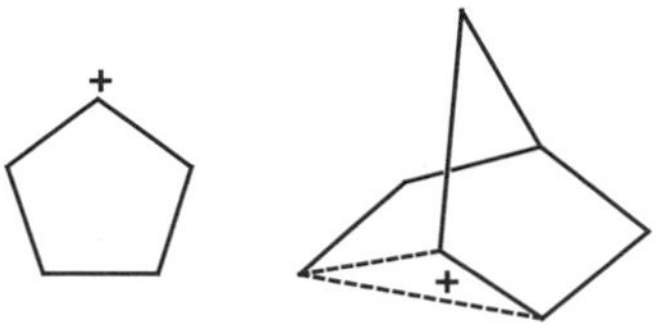

(a) (b)

Figure 12.60. (*a*) Cyclopentyl cation and (*b*) norbornyl cation; the hydride affinities are 245 and 234, respectively.

12.61 AUTOPROTOLYSIS

The proton transfer reaction between two identical molecules, one acting as a Brönsted acid and the other as a Brönsted base.

Example.

$$CH_3OH + CH_3OH \rightleftharpoons CH_3OH_2^+ + CH_3O^-$$

$$H_2O + H_2O \rightleftharpoons H_3O^+ + HO^-$$

12.62 APROTIC SOLVENT

A solvent that cannot function as a Brönsted acid (proton donor) under a specified set of reaction conditions.

Example. A solvent that does not contain an N–H, O–H, or S–H bond; for example, acetonitrile (H_3C–CN), dimethylsulfoxide [DMSO, $(H_3C)_2SO$], and hexamethyl phosphoramide [HMPA, $((H_3C)_2N)_3P{=}O$]. These solvents, all of which are highly polar with large dielectric constants, do not act as proton donors.

12.63 LEVELING EFFECT OF SOLVENT

A solvent that cannot distinguish the relative strengths of the acids in a series of strong acids, if all of them are stronger than the conjugate acid of the solvent (called the *lyonium* ion). Similarly, a solvent that cannot distinguish the relative strengths of the bases in a series of strong bases, if all of them are stronger than the conjugate base of the solvent (called the *lyate* ion). The solvent is said to have leveled, that is, made equal, the acidity or basicity of the acids or bases.

Example. If 1 mole each of HCl, HBr, and HI is placed separately into water, the three resulting solutions have the same concentration of H_3O^+ even though the acid strength is HI > HBr > HCl. All three hydrogen halide acids are much stronger than

H_3O^+ (the conjugate acid of H_2O), and thus all available protons of each of these acids protonate H_2O:

$$HX + H_2O \longrightarrow H_3O^+ + X^-$$

12.64 TRANSITION STATE

The potential energy maximum of the lowest potential energy barrier of all pathways lying on the path from reactants to products in a chemical reaction. This state thus corresponds to a local maximum in a plot of potential energy versus reaction path (see Sect. 12.69) and is represented by the abbreviation TS or a superscripted double dagger ‡. Some of the older literature refers to the hypothetical species at a transition state as the *activated complex*; however, more commonly today, the species also is referred to as the transition state.

Because transition states have a lifetime of approximately one molecular vibration (about 10^{-12} sec), they cannot be isolated. By convention, the bonds being made and broken in the transition state are represented by dotted lines to convey the idea of partial bond character, and the whole structure frequently is shown in brackets. Although reactions usually are considered to proceed in a specific direction, a transition state has an equal probability of forming the reactant or product, that is, the reactant and products are in equilibrium with the transition-state complex. (More accurately, this statement only holds if the transition state is momentumless.)

12.65 ELEMENTARY REACTION STEPS, MOLECULARITY

A transformation proceeding through a single transition state. The number of species involved in this single transformation is called the *molecularity* of that step.

Example. In the two examples below

$$A \longrightarrow B$$

$$A + B \longrightarrow C$$

the molecularity of the first reaction is 1 (*unimolecular*), and that of the second reaction is 2 (*bimolecular*).

12.66 CONCERTED REACTION

A reaction that proceeds through only one elementary reaction step and involves only one transition state, in which all bond making and bond breaking take place simultaneously.

Example. A typical concerted reaction is the S_N2 reaction:

$$Y^- + H_3CX \longrightarrow H_3CY + X^-$$

12.67 MULTISTEP REACTIONS

A sequence of two or more elementary reaction steps. Common to both reaction steps is a transient species called a *reactive intermediate* or just an *intermediate*, which is characterized by a lower potential energy than those states, lying on the reaction path, that are immediately adjacent to it. The intermediate corresponds to a local minimum in a plot of potential energy versus reaction path (see Sect. 12.69). An intermediate has a finite lifetime that is appreciably greater than the period of a molecular vibration ($>10^{-12}$ sec) and may even be on the order of hours.

Example. A one-step reaction involves one transition state and no intermediates, whereas a two-step reaction has two transition states and one intermediate, and an n-step reaction has n transition states and $(n - 1)$ intermediates. In the typical two-step reaction shown below, R_3C^+ is an intermediate:

$$Y^- + R_3CX \xrightarrow{\text{step 1}} Y^- + R_3C^+ + X^- \xrightarrow{\text{step 2}} R_3CY + X^-$$

12.68 MECHANISM OF A REACTION

A complete description of the elementary steps in a reaction, that is, the sequence of bond making and bond breaking in a chemical reaction. A complete description of a reaction includes characterization of all intermediates and transition states including their geometries and relative free energies. The mechanism must be consistent with all experimental data. Studies of reaction rates (kinetics, see Sect. 12.72) and stereochemistry are two of the powerful tools used to acquire these data. Experimental evidence can easily disprove a particular mechanism, but it can never prove one. Formulas including detailed geometries, electron arrow-pushing to represent bond making and breaking processes, and energy diagrams are the main tools used by the organic chemist to depict reaction mechanisms.

12.69 REACTION COORDINATE (PROFILE) DIAGRAM

A plot of potential energy versus changes in atomic bond distances in one set of reacting molecules as they proceed from reactants through transition states and/or intermediates to products. In the two-dimensional plot called a *potential energy profile*, the potential energy of the most favorable pathway is plotted as the ordinate, and the abscissa is a parameter called the reaction coordinate and is related to changes in bond lengths and bond angles. The reaction coordinate represents changes in the location of the reacting atoms, and these changes can be taken as a measure of the progress of the conversion of one molecule from reactant to product.

Example. The displacement of chloride ion in methyl chloride by iodide ion in the reaction

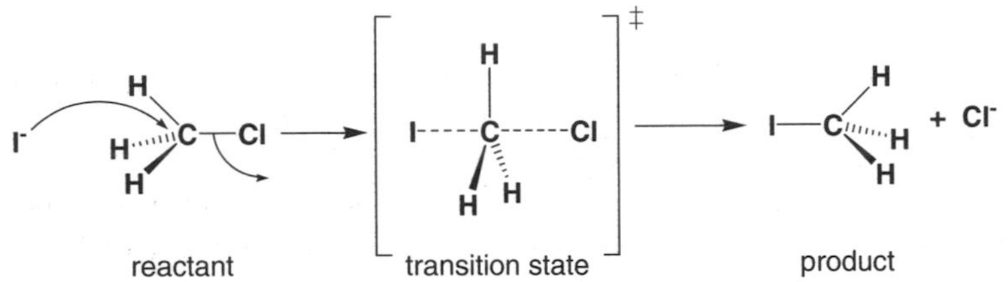

can be represented by the potential energy profile shown in Fig. 12.69*a*. The maximum in the curve is the *transition state*; it is represented by the bracketed structure with a superscripted double dagger in the reaction equation. Reaction coordinate diagrams represent *microscopic phenomena*. The difference in potential energy between the starting system and the transition state ε_a represents the energy barrier for the conversion of reactants to products, a microscopic quantity. If the average value for ε_a is multiplied by Avogadro's number N_0

$$\varepsilon_a \times N_0 = E_a \tag{12.69}$$

then a macroscopic quantity, the energy of activation per mole E_a, is obtained (see Sect. 12.81). Macroscopic quantities such as free energies, enthalpies, and Arrhenius energies of activation are usually presented (incorrectly) on plots where the reaction coordinate, a microscopic representation, is shown as the abscissa. Figure 12.69*b* shows a three-dimensional plot of two reaction coordinates against potential energy, giving rise to a *potential energy surface (hypersurface)*. The heavy line illustrates the lowest-energy pathway and is a representation of what is plotted in the two-dimensional energy profile (Fig. 12.69*a*). Figure 12.69*c* is the reaction coordinate diagram for the two-step reaction given in Sect. 12.67. The figure shows the relationship among the energies of the reactant, the product, the intermediate, and the two transition states.

12.70 FREE ENERGY DIAGRAM

A diagram that shows the relative Gibbs free energies of reactants, transition states, intermediates, and products which occur during the course of a reaction.

Example. For the two-step reaction

$$\mathrm{A} \longrightarrow \mathrm{B} \longrightarrow \mathrm{C}$$

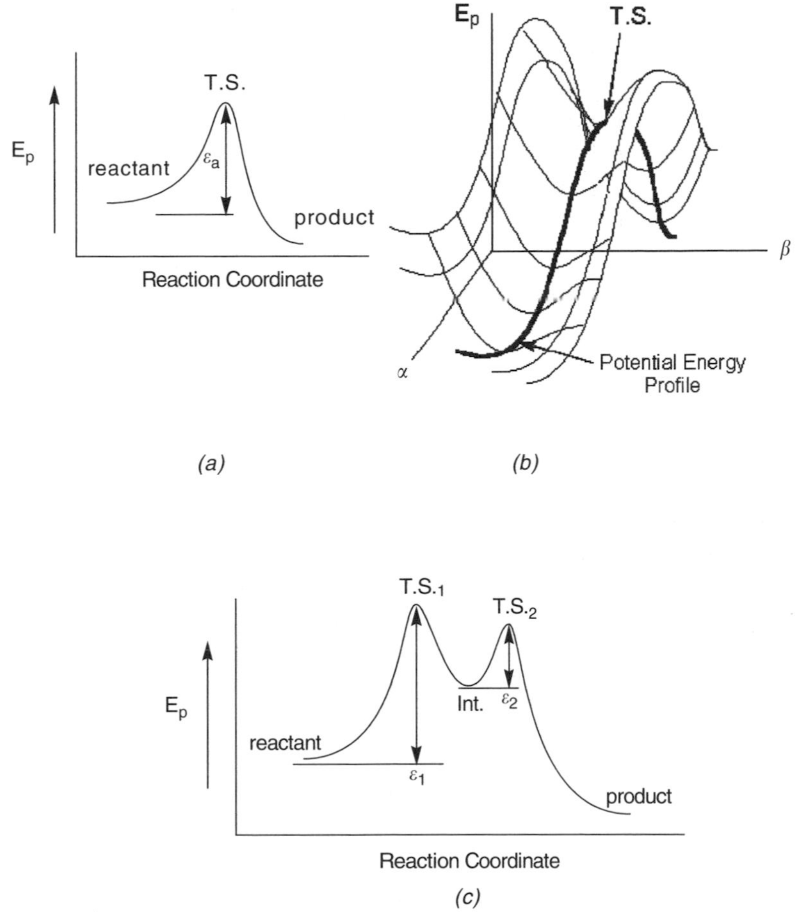

Figure 12.69. (*a*) A two-dimensional plot of the potential energy E_p versus the reaction coordinate; (*b*) a three-dimensional plot of the potential energy E_p versus reaction coordinates α and β; (*c*) a two-dimensional plot of potential energy E_p versus the reaction coordinate for a two-step reaction.

the free energy diagram (Fig. 12.70) shows the relative free energies of five states involved in the transformation of *A* to *C*: those that correspond to the energies of the reactant *A*, the intermediate *B*, and the product *C* which are minima on the potential energy profile, and the two transition states which are energy maxima on potential energy profile. The points are separated for clarity and generally placed in the same sequential order as the states they represent in the order of their occurence during the reaction. The points should not be connected by a continuous curve and the abscissa is not defined. This type of diagram is used to represent macroscopic properties and not the microscopic properties of individual reacting molecules, which are represented in reaction coordinate diagrams (see Sect. 12.69).

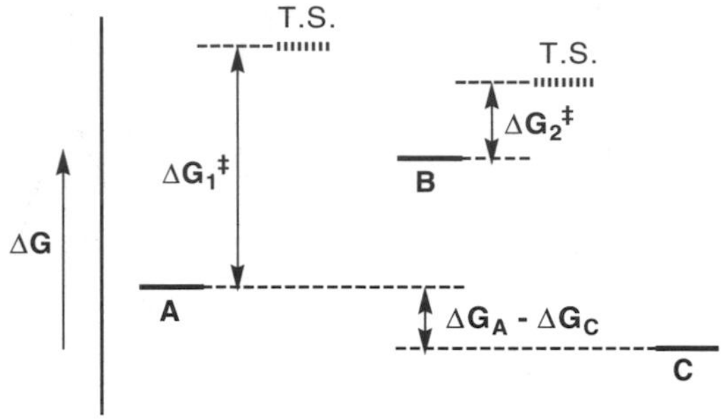

Figure 12.70. A free energy diagram of a two-step reaction.

12.71 HAMMOND'S POSTULATE

According to this concept, named after G. S. Hammond (1921–), in any elementary step in a reaction, the geometry and structure of the transition state of that step more closely resemble either the reactant or product *of that step*, whichever is higher in energy. (In this context, "resembles" means similar in structure, energy, and response to reaction variables such as temperature and solvent.) This implies that in an endothermic step the transition state resembles the products and that in an exothermic step the transition state resembles the reactants.

Example. Figure 12.71 shows the potential energy profile of the two-step reaction described by the equation

$$\underset{\text{reactants}}{Y^- + R_3CX} \xrightarrow[\text{step 1}]{\text{slow}} \underset{\text{intermediate}}{Y^- + R_3C^+ + X^-} \xrightarrow[\text{step 2}]{\text{fast}} \underset{\text{products}}{R_3CY + X^-}$$

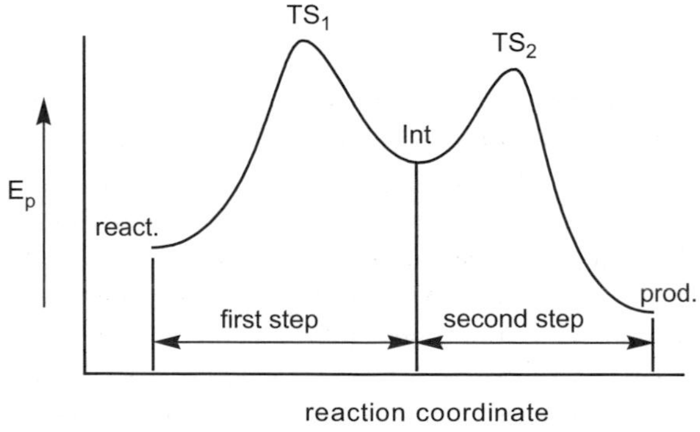

Figure 12.71. A potential energy profile illustrating the Hammond postulate.

The first step is *endothermic*, and therefore, TS_1 resembles the *product of that step*, which is the intermediate (Int) for the overall reaction, more than it resembles the reactant. The second step is *exothermic*, and therefore, TS_2 *resembles the reactant of that step*, Int, more than it resembles the reaction product.

12.72 KINETICS

The scientific discipline that deals with rates of reactions and the factors that influence those rates.

12.73 RATE OF REACTION ν

A quantitative measure of how fast a reaction occurs, usually expressed in mol/L · sec. For the elementary (one-step) reaction

$$aA + bB \longrightarrow cC$$

the rate of the reaction ν is related to the derivative of the concentration of the reactant or product with respect to time:

$$\nu = -(1/a)\, d[A]/dt = -(1/b)\, d[B]/dt = (1/c)\, d[C]/dt \qquad (12.73)$$

where a, b, and c are the stoichiometric coefficients, and $[A]$, $[B]$, and $[C]$ represent concentrations of the respective substances. In a one-step reaction, the *rate of disappearance*, $-d[A]/dt$ or $-d[B]/dt$, or the *rate of appearance*, $d[C]/dt$, is equal to the rate of reaction only when the stoichiometric coefficients of all species are the same, usually 1.

12.74 DIFFERENTIAL RATE EQUATION (RATE LAW), RATE CONSTANT

A differential equation that describes the dependence of the reaction rate on the concentrations of species present in the reaction mixture. The rate equation is frequently used to substantiate or negate a proposed reaction mechanism.

Example. For a one-step (concerted) reaction

$$A \xrightarrow{k} B$$

$$\text{Rate} = \nu = -d[A]/dt = d[B]/dt = k\,[A] \qquad (12.74a)$$

The symbols placed in brackets represent concentrations in mol/L and the k is the *rate constant*, which is a proportionality constant.

The rate constant is independent of concentration, but is temperature-dependent and characteristic for a specific reaction. In this example, the rate constant k must have the units of sec^{-1}, in order for the rate to have units of mol/L · sec. For the one-step reaction

$$\mathrm{A + B \xrightarrow{k} C}$$

the rate equation will match the stoichiometry:

$$\text{Rate} = \nu = -d[A]/dt = -d[B]/dt = d[C]/dt = k\,[A]\,[B] \qquad (12.74b)$$

Here the rate constant k must have the units of L/mol·sec, in order for the rate to have units of mol/L · sec.

12.75 INTEGRATED RATE EXPRESSION

An equation that gives the concentration of the reactants in a chemical reaction as a function of time. The equation is obtained by solving (integrating) the differential rate equation.

Example. For the first reaction shown in Sect. 12.74, the integrated rate expression is

$$\ln[A]_0 - \ln[A] = \ln([A]_0/[A]) = kt \qquad (12.75a)$$

and for the second reaction shown in Sect. 12.74, the integrated rate expression is

$$kt = \frac{1}{[B]_0 - [A]_0} \ln\left\{\frac{[B]_0\,[A]}{[B]\,[A]_0}\right\} \qquad (12.75b)$$

where $[A]$ and $[B]$ are the concentrations in moles/L of A and B at time t, $[A]_0$ and $[B]_0$ are the initial concentrations of A and B, and k is the rate constant.

12.76 RATE-DETERMINING (OR RATE-LIMITING) STEP

In a multiple-step reaction, the step that has the higher (highest) transition-state energy.

Example. Suppose the reaction

$$\mathrm{A + B \longrightarrow D}$$

follows the two-step mechanism:

$$\mathrm{A + B \xrightarrow{k_1} C} \quad \text{Step 1}$$

$$\mathrm{C \xrightarrow{k_2} D} \quad \text{Step 2}$$

Let us assume that step 1 is slow compared to step 2 ($k_1 << k_2$), that is, step 1 has a transition state of higher energy than step 2. The differential rate equations for steps 1 and 2 are

$$-d[A]/dt = d[C]/dt = k_1[A][B] \qquad (12.76a)$$

and

$$d[D]/dt = -d[C]/dt = k_2[C] \qquad (12.76b)$$

respectively. For step 2 to be much faster than step 1, C must be reacting as fast as it is formed. C is therefore said to reach a *steady-state* concentration. In general, this occurs when the (small) concentration of a particular (high-energy) intermediate does not change during the course of a reaction because the rate of material being formed is balanced by the rate of material being consumed. Therefore,

$$d[C]/dt = -d[C]/dt = 0 = k_1[A][B] - k_2[C] \qquad (12.76c)$$

and

$$k_1[A][B] = k_2[C] \qquad (12.76d)$$

and thus,

$$d[D]/dt = k_1[A][B] \qquad (12.76e)$$

This rate equation shows that the rate of reaction is not dependent on the concentration of C because C is involved after the rate-determining step of the reaction.

For a two-step reaction in which the first step is reversible and C is a *steady-state intermediate*, such as

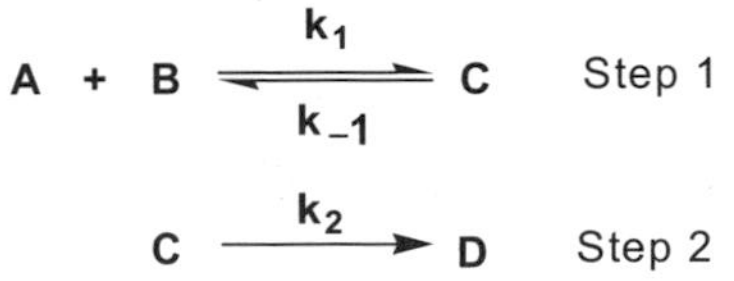

$$d[C]/dt = k_1[A][B] - k_{-1}[C] - k_2[C] = 0 \qquad (12.76f)$$

and

$$[C] = (k_1/(k_{-1} + k_2))[A][B] \qquad (12.76g)$$

and therefore,

$$d[D]/dt = k_2[C] = (k_1k_2/(k_{-1} + k_2))[A][B] \qquad (12.76h)$$

This whole equation must be used when $k_{-1} \approx k_2$. However, if one of the steps is much faster than the other, this equation can be modified. For example, if the second step is much faster than the return of C to the starting material ($k_2 >> k_{-1}$), then this equation simplifies to

$$d[D]/dt = k_1[A][B] \tag{12.76i}$$

This is the same result as seen above when the first step is irreversible.

On the other hand, if the return of C to the starting material is much faster than its conversion to D ($k_{-1} >> k_2$), then this equation simplifies to

$$d[D]/dt = (k_1k_2/k_{-1})[A][B] \tag{12.76j}$$

Another possibility for this two-step reaction sequence is that C is not a steady-state intermediate, and thus, step 1 is a *preequilibrium (prior equilibrium)*, that is, an equilibrium is established prior to the rate-determining step.

As above, the overall reaction rate is

$$d[D]/dt = k_2[C] \tag{12.76k}$$

but

$$[C] = k_1[A][B]/k_{-1} = K[A][B] \tag{12.76l}$$

where

$$K = k_1/k_{-1} \tag{12.76m}$$

Thus,

$$d[D]/dt = k_2K[A][B] \tag{12.76n}$$

12.77 MICHAELIS–MENTON KINETICS

This is a method for handling the kinetics of any catalyzed reaction, but usually used with enzyme-catalyzed reactions; it was proposed by Lenor Michaelis and Maud Menton, *Biochem. Z. 49*, 33 (1913).

Example. Consider the following two-step reaction:

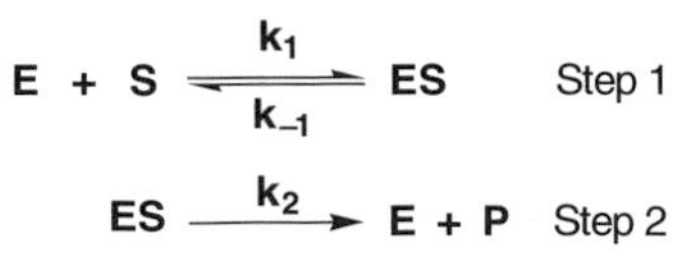

in which E is an enzyme, S is a substrate for that enzyme, ES is an enzyme-substrate complex that is a steady-state intermediate for this reaction, and P is the reaction product. Thus, just as in Sect. 12.76,

$$d[ES]/dt = k_1[E][S] - k_{-1}[ES] - k_2[ES] = 0 \tag{12.77a}$$

and therefore,

$$[E][S] = ((k_{-1} + k_2)/k_1)[ES] = K_M[ES] \tag{12.77b}$$

where K_M is the Michaelis–Menton constant.

12.78 PSEUDO-RATE LAW

This is the simplified rate law observed when the concentration of a species appearing in the rate equation is present in large excess, so its concentration is effectively a constant, and thus the observed rate equation does not reflect its involvement. This type of behavior commonly is observed when the solvent is involved in a reaction.

Example. Suppose the elementary reaction

$$\text{A} + \text{B} \longrightarrow \text{C}$$

follows the rate equation

$$\text{Rate} = v = -d[A]/dt = -d[B]/dt = d[C]/dt = k[A][B] \tag{12.78a}$$

The reaction is second-order. If the concentration of B is very large compared to that of A, then the rate equation would be

$$\text{Rate} = -d[A]/dt = k'[A] \tag{12.78b}$$

where $k' = k[B]$. The reaction is then said to follow pseudo-first-order kinetics, and k' is the pseudo-first-order rate constant.

12.79 HALF-LIFE $t_{1/2}$

The time required for the concentration of a reactant to fall to one-half its initial value.

Example. For the unimolecular integrated rate equation given in Sect. 12.75, substituting $t = t_{1/2}$, that is, the half-life, and $[A] = ([A]_0/2)$, and solving give

$$t_{1/2} = \ln([A]_0/([A]_0/2))/k = \ln 2/k = 0.0693/k \tag{12.79}$$

The half-life is independent of the initial concentration for a first-order reaction. This is not true for reactions of higher order.

12.80 LIFETIME τ

For a first-order reaction only, the inverse of the rate constant:

$$\tau = 1/k \tag{12.80a}$$

Here, k is the sum of the rate constants for all the paths by which the starting material reacts, and thus, τ is the time required for the concentration of the starting material to decrease by a factor of $1/e = 0.37$. (This situation appears in many photochemical processes; see Chap. 18.)

Example. For the unimolecular integrated rate equation given in Sect. 12.75, the rate expression can be rewritten as

$$\ln([A]_0/[A]) = kt = t/\tau \tag{12.80b}$$

12.81 ARRHENIUS ENERGY OF ACTIVATION E_a

Named after Svante Arrhenius (1859–1927), an experimentally determined quantity that relates rate constants to temperature by the equation

$$k = Ae^{-E_a RT} \tag{12.81a}$$

where k is a rate constant, A is a constant known as the preexponential or frequency factor, R is the gas constant (1.986×10^{-3} kcal.mol•K, 8.314×10^{-3} kJ/mol•K), and T is the absolute temperature (K). When we take the logarithm of both sides, the Arrhenius equation becomes

$$\ln k = \ln A - E_a/RT \tag{12.81b}$$

Thus, $-E_a$ is the slope of a plot of ln k versus $1/T$.

12.82 TRANSITION-STATE THEORY

For an elementary reaction step, the transition state can be viewed as being in equilibrium with the reactants and the products. This assumption allows the calculation of thermodynamic like parameters dealing with the rate of the reaction.

FREE ENERGY OF ACTIVATION $\Delta G^\ddagger$. The difference in Gibbs free energy between a transition state and the state of the reactants. The free energy of activation $\Delta G^\ddagger$ should not be confused with experimental energy of activation E_a in the Arrhenius equation. Whereas the free energy of activation ($\Delta G^\ddagger$) has both an enthalpy component and an entropy component (Eq. 12.82a), E_a has only an enthalpy component (Eq. 12.82b) and the entropy is included in the A term (Eq. 12.82c).

ENTHALPY (OR HEAT) OF ACTIVATION $\Delta H^{\ddagger}$. The difference in enthalpy between a transition state and the state of the reactants.

ENTROPY OF ACTIVATION $\Delta S^{\ddagger}$. The difference in entropy between a transition state and the state of the reactants.

These three parameters are related by the equation

$$\Delta G^{\ddagger} - \Delta H^{\ddagger} \quad T\Delta S^{\dagger} \tag{12.82a}$$

For a reaction in solution at a constant pressure, the enthalpy of activation is related to the Arrhenius energy of activation (Sect. 12.81) by

$$E_a = \Delta H^{\ddagger} + RT \tag{12.82b}$$

where R is the gas constant. At 25°C, the difference between E_a and $\Delta H^{\ddagger}$ is approximately 0.6 kcal mol^{-1}. The entropy of activation is related to the Arrhenius A factor (Sect. 12.81) by

$$\ln A = \Delta S^{\ddagger}/R + \ln(\kappa T/h) + 1 \tag{12.82c}$$

where R is the gas constant, T the absolute temperature, κ Boltzmann's constant, h Planck's constant.

The free energy of activation for any elementary reaction step may be calculated from the absolute rate equation

$$\Delta G^{\ddagger} = RT \ln(\kappa T/hk) \tag{12.82d}$$

where R is the gas constant, T the absolute temperature, κ Boltzmann's constant, h Planck's constant, and k the rate constant. This equation is commonly referred to as the *Eyring equation* (after Henry Eyring, 1901–1981); it is often written in the form

$$k = (\kappa T/h) \exp(-\Delta G^{\ddagger}/RT) \tag{12.82e}$$

12.83 THERMODYNAMIC STABILITY

A measure of the magnitude of either the heat of formation (ΔH_f^0) of a substance or the standard Gibbs free energy (ΔG^0) of a reaction. The smaller (or more negative) ΔH_f^0 or ΔG^0 is, the more stable the substance or reaction products, or the less stable the starting materials. *Stability* is synonymous with thermodynamic stability.

Example. The heats of hydrogenation (see Sect. 12.24) of 1-butene, *cis*-2-butene, and *trans*-2-butene, to give in each case butane, are −30.3, −28.6, and −27.6 kcal mol^{-1}, respectively. If we assume that ΔG^0 is approximately equal to ΔH^0, then we can determine that 1-butene is less stable than *cis*-2-butene, which in turn is less stable than *trans*-2-butene. This same conclusion about stabilities can be drawn from looking at the heats of formation: *trans*-2-butene with a $\Delta H_f^0 = -12.1$ kJ mol^{-1} = -2.9 kcal mol^{-1} is more stable than *cis*-2-butene with a $\Delta H_f^0 = -7.1$kJ mol^{-1} = -1.7 kcal mol^{-1}.

At 0°C, the following equilibrium might occur:

$$C_2H_4\ (g) + H_2O\ (g) \rightleftharpoons C_2H_5OH\ (g)$$

The standard Gibbs free energy difference for this reaction is -8.12 kJ mol^{-1} (-1.94 kcal mol^{-1}); hence, C_2H_5OH (*g*) is more stable than its component parts C_2H_4 (*g*) and H_2O (*g*).

12.84 KINETIC STABILITY

The lack of reactivity of a substance in a specified reaction.

Example. As shown in Table 12.93, $(CH_3)_3CBr$ is hydrolyzed about a million times faster than CH_3Br. To cite another example, ethane does not react with bromine in carbon tetrachloride, whereas ethylene, under the same conditions, is very reactive. The kinetic stability of a species should not be confused with stability or *thermodynamic stability* (see Sect. 12.83). Kinetic stability is determined by the energy of activation (E_a, see Sect. 12.81) for that specific reaction.

12.85 THERMODYNAMIC (OR EQUILIBRIUM) CONTROL OF A REACTION

At equilibrium (under *reversible* reaction conditions), the ratio of products from competing reactions is determined by the relative stability of each product, measured by its standard free energy of formation (ΔG_f^0). The composition of the equilibrium mixture does not depend on how fast each product is formed ($\Delta G^‡$) in the reaction.

Example. The addition of HCl to 1,3-butadiene can lead to two major products, both derived from a common intermediate, as shown in Fig. 12.85*a*. Figure 12.85*b* shows the potential energy profile for the second step in the addition. The more stable 1,4-addition product (*thermodynamic product*), even though it is formed more slowly than the 1,2-addition product, predominates at higher temperatures where equilibrium is attained. At lower temperatures, where the equilibrium is not yet attained, the 1,2-addition product (*kinetic product*) predominates because it is formed more rapidly. In some cases, the thermodynamic product may also be the kinetic product.

12.86 KINETIC CONTROL

The relative quantity of each product from competing reactions, under nonequilibrium conditions, is determined by how fast ($\Delta G^‡$) that product is formed and is not necessarily a function of its relative stability (ΔG_f^0).

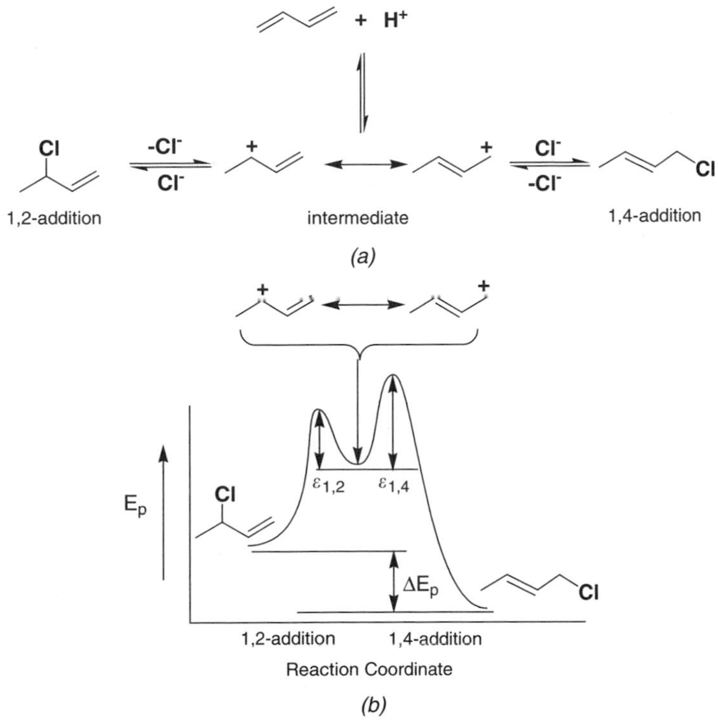

Figure 12.85. (*a*) The 1,2- and 1,4-addition of HCl to 1,3-butadiene and (*b*) potential profile for the second step in the addition of HCl to 1,3-butadiene.

Example. See thermodynamic control, Sect. 12.85.

12.87 CURTIN–HAMMETT PRINCIPLE

In a reaction where one product is generated by one conformer of the starting material while a different product is generated by a second conformer of the starting material, where both conformers rapidly interconvert and the products are kinetically controlled (Sect. 12.86), the product distribution is independent of the relative amount of each starting conformers. The product distribution is determined solely by the difference in free energy of activation ($\Delta\Delta G^{\ddagger}$) for the conversion of each of the conformations to its respective product, Fig. 12.87.

Example. Consider the reaction of a base with 1,2-diphenyl-1-chloroethane to generate *cis*- and *trans*-stilbene:

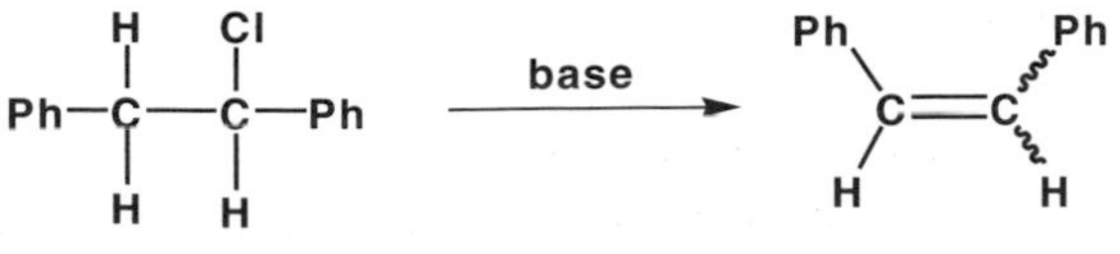

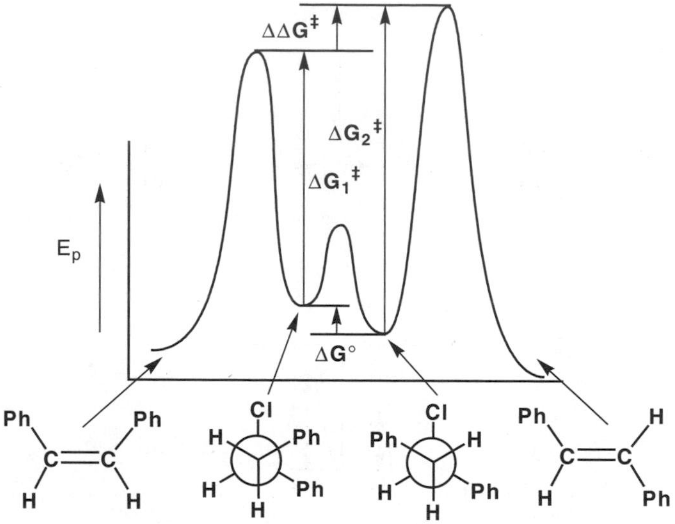

Figure 12.87. The product ratio, according to the Curtin–Hammett principle, is determined by the difference in free energy of activation ($\Delta\Delta G^{\ddagger}$) for the conversion of each of the conformations to its respective product.

The different *anti* conformations of the starting material are easily convertible, but the products are stable to the reaction conditions. The Curtin–Hammett principle states that the product ratio is not proportional to the ratio of the two *anti* conformations of the starting materials, but rather to the difference in activation barriers ($\Delta\Delta G^{\ddagger}$) for the conversion of each of the conformations to its respective product. (The activation barrier for the formation of the *trans* isomer is $\Delta G_2^{\ddagger}$ and for the *cis* isomer it is $\Delta G_1^{\ddagger} + \Delta G^0$.)

12.88 DIFFUSION- (OR ENCOUNTER-) CONTROLLED RATE

This is a bimolecular reaction that occurs with 100% efficiency every time the reaction partners undergo collision; hence, it is a bimolecular reaction that proceeds with the maximum rate constant possible under a given set of reaction conditions k_{diff}:

$$k_{\text{diff}} = 8RT/3{,}000\eta \qquad (12.88)$$

where R is the gas constant, T the absolute temperature, and η the viscosity of the medium in poise. Diffusion-controlled reactions have negligible or nonexistent free energies of activation and are exceedingly fast, typically, $k_{\text{diff}} = 10^{11}\,\text{L mol}^{-1}\,\text{sec}^{-1}$.

12.89 CATALYSIS

The phenomenon of altering the rate of a reaction by addition, in relatively small quantities of a substance that participates in the reaction, but that is continuously recycled and not consumed during the reaction. The principal function of the catalyst is to lower the activation barrier. The added substance need not be the actual catalyst under the reaction conditions but may be the precursor to the active catalyst, which then becomes part of the reaction cycle and is continuously used and regenerated. Catalysts are usually employed to enhance the rate of reaction by changing the mechanism, thereby lowering the energy of activation. The catalysis may be either *homogeneous* or *heterogeneous*. In the former case, the catalyst is soluble in the substrate or its solutions and is distributed uniformly in the reaction mixture. Reactions catalyzed by an acid or a base (see Sects. 12.90 and 12.91) are usually of this type. In heterogeneous catalysis the catalyst is a separate insoluble phase, which facilitates recovery and recycling. Most petroleum refining processes involve heterogeneous catalysis. Because the rate constant for a reaction is an exponential function of $\Delta G^{\ddagger}$, the free energy of activation (Sect. 12.82), a small decrease in this value results in a large increase in rate.

Example. At room temperature, the rate constant for the hydrolysis of urea, $(H_2N)_2C{=}O$, to NH_3 and CO_2, is $3 \times 10^{-10}\,\text{sec}^{-1}$ with $\Delta G^{\ddagger} = 30\,\text{kcal mol}^{-1}$. When the hydrolysis is carried out under the same conditions in the presence of the enzyme urease, a specific homogeneous catalyst for this reaction, the $\Delta G^{\ddagger}$ for the reaction is lowered to $11\,\text{kcal mol}^{-1}$ and the rate of hydrolysis is enhanced by a factor of 10^{14} or 100,000 billion times the rate in the absence of catalyst. It is probably not an exaggeration to state that every important biological and commercial organic reaction involves catalysis of some sort or other.

12.90 GENERAL ACID OR BASE CATALYSIS

This is the acceleration of the rate of a reaction due to the presence of any Brönsted acid (or base); each acid (base) present enhances the rate in an additive fashion. The acid (or base) is unchanged by the overall reaction.

Example. If an aqueous reaction is catalyzed by any acid HA_i, the reaction is subject to general acid catalysis. For a reaction involving the two-step sequence,

$$\mathrm{Y} + \mathrm{HA_i} \underset{}{\overset{\text{slow}}{\rightleftharpoons}} \mathrm{YH^+} + \mathrm{A_i^-}$$

$$\mathrm{YH^+} \xrightarrow{\text{fast}} \text{products} + \mathrm{H^+}$$

where HA_i represents any acid, HA_1, HA_2, and so forth, the rate equation is

$$\text{Rate} = \{k_1[\mathrm{HA_1}] + k_2[\mathrm{HA_2}] + \cdots\}\,[Y] \qquad (12.90)$$

Each acid independently affects the overall rate. General base catalysis involves Brönsted bases in a similar fashion.

12.91 SPECIFIC ACID OR BASE CATALYSIS

The rate of a reaction is accelerated by one specific acid or base (usually, the conjugate acid or base of the solvent) rather than by all acids (or bases) present. The acid or base is unchanged by the overall reaction.

Example. In the two-step, acid-catalyzed, aqueous reaction represented by

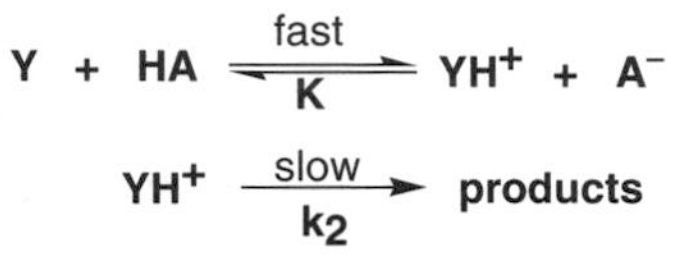

the rate of the reaction is

$$\text{Rate} = k_2K[Y][\text{HA}]/[\text{A}^-] \qquad (12.91\text{a})$$

Using the acid dissociation equation of HA (see Sect. 12.56), we can easily show that

$$[\text{HA}]/[\text{A}^-] = [\text{H}_3\text{O}^+]/(K_a[\text{H}_2\text{O}]) \qquad (12.91\text{b})$$

and thus,

$$\text{Rate} = k_2K[Y][\text{H}_3\text{O}^+]/K_a[\text{H}_2\text{O}] \qquad (12.91\text{c})$$

If additional acids are present in the medium, the strongest acid present will always be H_3O^+ (see Sect. 12.63), and only the effect these acids have on the concentration of H_3O^+ will appear in the above rate equation. Specific base catalysis involves a Brönsted base in a similar fashion. General acid catalysis can be experimentally distinguished from specific acid catalysis by observing the reaction rate at constant pH as a function of buffer concentration. If the reaction rate increases with an increase in buffer concentration, at constant pH, the reaction exhibits general acid catalysis.

12.92 REACTIVITY

A kinetic term that involves a comparison of rate constants for similar reactions of different substances. This is the opposite of kinetic stability (Sect. 12.84).

Example. The rate constant for the reduction of cyclopentanone to cyclopentanol by sodium borohydride in isopropyl alcohol at 0°C is 7×10^{-4} L mol^{-1} sec^{-1}, compared to a rate constant of 161×10^{-4} L mol^{-1} sec^{-1} for the analogous reaction of cyclohexanone. Cyclohexanone is said to be more reactive than cyclopentanone in this reaction by a factor of $(161 \times 10^{-4})/(7 \times 10^{-4}) = 23$.

12.93 RELATIVE RATE

The ratio of the rate constant for a particular reaction to the rate constant for a related reference reaction.

Example. Table 12.93 shows the relative rates of hydrolysis of alkyl bromides in water at 50°C. The hydrolysis of ethyl bromide is used as the reference reaction.

TABLE 12.93. Relative Rates of Hydrolysis of Alkyl Bromides in Water at 50°C

Alkyl Bromide	Relative Rate
CH_3Br	1.05
C_2H_5Br	(1)
$(CH_3)_2CHBr$	11.6
$(CH_3)_3CBr$	1.2×10^6

12.94 PARTIAL RATE FACTOR f

The rate of a substitution reaction at one particular position in a substituted benzene derivative relative to the rate of the same substitution reaction in benzene.

Example. For monosubstituted benzenes ϕ_z, the partial rate factors f for *para*, *meta*, and *ortho* positions are given by

$$f_p^{\phi z} = [(k_{\phi zp}/1)/(k_{\phi H}/6)] \cdot (\%\ para/100) \tag{12.94a}$$

$$f_m^{\phi z} = [(k_{\phi zm}/2)/(k_{\phi H}/6)] \cdot (\%\ meta/100) \tag{12.94b}$$

$$f_o^{\phi z} = [(k_{\phi zo}/2)/(k_{\phi H}/6)] \cdot (\%\ ortho/100) \tag{12.94c}$$

where, for example, $k_{\phi zp}$ is the rate constant for substitution of the hydrogen *para* to the Z-group. For *para* substitution, only one possible position is available. Since there are six possible substitution sites in benzene, $k_{\phi H}$ is divided by 6 in the $f^{\phi z}{}_p$ equation. For *meta* and *ortho* substitution, two equivalent sites are available, compared to 6 in benzene. Therefore, $k_{\phi z}$ is divided by 2 and $k_{\phi H}$ by 6 for these cases. Bromination of toluene proceeds 605 times faster than bromination of benzene. Three products are obtained: 66.8% *p*-bromotoluene; 0.3% *m*-bromotoluene; and 32.9% *o*-bromotoluene. The partial rate factors are

$$f_p^{\phi z} = 2425$$
$$f_m^{\phi z} = 5.4$$
$$f_o^{\phi z} = 597$$

If $f > 1$ for a given position, that position is activated compared to benzene; if $f < 1$, that position is deactivated.

12.95 SUBSTITUENT EFFECT

A change in the rate constant or equilibrium constant of a reaction caused by the replacement of a hydrogen atom by another atom or group of atoms. Such substituent effects result from the size (steric effects) and/or the electronic properties (electronic effects) of the substituent on the reaction site. The electronic effect of the substituent may be either *electron-releasing* or *electron-withdrawing*. Such electronic effects can be subdivided into inductive and resonance (mesomeric) effects. *Inductive effects* depend on a substituent's ability to supply or withdraw bonding electrons (i.e., its electronegativity) and are transmitted through σ bonds or through space. Inductive effects weaken rapidly as the distance between the substituent and reactive center increases. *Resonance effects* involve delocalization of electrons through the π system (see Sect. 3.4).

Example. An electrophilic aromatic substitution reaction (see Sect. 14.55) on anisole (methoxybenzene) can lead to *ortho*, *meta*, and *para* monosubstitution (Fig. 12.95). The presence of the methoxyl group influences the relative energies of the three possible transition states leading to the intermediates. In practice, such energies are assessed by examining the structure of the intermediates. The methoxyl substituent exerts an electron-releasing resonance effect ($+R$ effect), which causes delocalization of the positive charge. In the *ortho*- and *para*-substituted intermediates, the positive charge can be delocalized into the substituent, but this is not possible to any appreciable extent in the *meta*-substituted intermediates (Fig. 12.95). The methoxyl group, however, also exerts an electron-withdrawing inductive ($-I$) effect in all three intermediates. In the *ortho* and *para* cases, the result of the $+R$ and $-I$ effects of methoxyl group is a net electron-releasing effect (resonance effects dominate inductive effects), hence stabilization of these intermediates. In the *meta* case, the electron-withdrawing $-I$ effect dominates and the intermediate has a considerably higher energy than the *ortho* and *para* intermediates. *Ortho* substitution leads to the sterically less favorable *ortho* intermediate. Therefore, the distribution of final products in the reaction is *para* > *ortho* >> *meta* ($k_p > k_o >> k_m$). Anisole undergoes electrophilic aromatic substitution 1,000 times faster than benzene. The rate constants for reaction at the three different positions in anisole can be expressed as partial rate factors.

12.96 KINETIC ISOTOPE EFFECT

A change of rate that occurs upon isotopic substitution from a lighter isotope to a heavier isotope, generally expressed as a ratio of rate constants, k_{light}/k_{heavy}. A *normal isotope effect* is one in which the ratio k_{light}/k_{heavy} is greater than 1, and an *inverse*

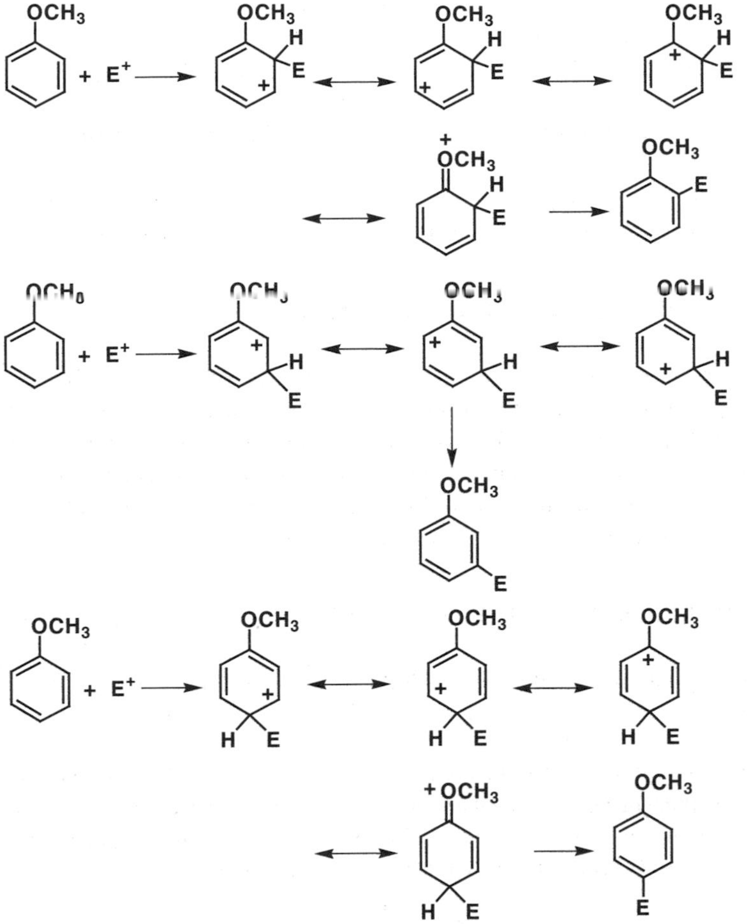

Figure 12.95. The electrophilic aromatic substitution of anisole showing the resonance stabilization of the *ortho*, *meta*, and *para* intermediates; the *meta* cannot involve delocalization involving the substituent.

isotope effect is one in which the ratio is less than 1. A *primary isotope effect* is one that measures the effect of isotopic substitution on a reaction that breaks the bond directly to the isotopically substituted atom. A *secondary isotope effect* measures the effect of isotopic substitution on an atom not directly involved in bond making or breaking.

Example. In the free radical bromination of toluene,

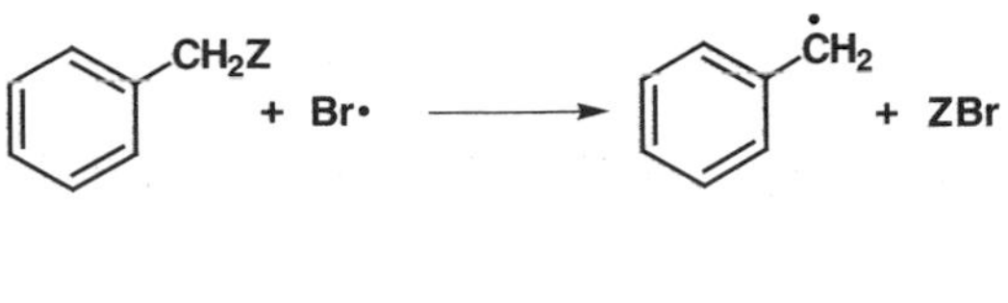

where $Z = H$ or D, k_H/k_D is 4.6. Primary isotope effects for k_H/k_D are generally between 2 and 7 if bond breakage occurs in the rate-determining step. In the solvolysis of isopropyl bromide,

$$(CZ_3)_2CHBr + H_2O \longrightarrow (CZ_3)_2CHOH + HBr$$

where $Z = H$ or D, k_H/k_D is 1.34. Secondary isotope effects for k_H/k_D are generally between 0.6 and 2.0 if isotopic substitution affects the rate-determining step.

12.97 DOUBLE LABELING EXPERIMENT

An experiment in which 2 parts of one molecule are isotopically labeled and then this compound is mixed with an unlabeled compound before they are allowed to undergo a chemical reaction. If during the reaction the 2 parts of the original molecule remain bonded, the product will either contain both labels or no label. In contrast, if the original starting material dissociates during the chemical reaction, the product may contain two, one, or no isotopic labels.

12.98 COMMON ION EFFECT

The rate suppression produced by addition of an ion that is being formed in the reaction.

Example. The addition of chloride ion decreases the rate of solvolysis of an alkyl halide. The presence of Cl^- from an independent source has a mass-law effect on the equilibrium, driving it back to the starting materials, and thus suppresses the overall rate of reaction. Consider the S_N1 reaction shown below:

$$RCl \underset{k_{-1}}{\overset{k_1}{\rightleftharpoons}} R^+ + Cl^-$$

$$R^+ + H_2O \xrightarrow[-H^+]{k_2} ROH$$

in which R^+ is a steady-state intermediate. Solving this in the usual manner, with $[H_2O]$ in large excess so that it is constant and, thus, can be part of the rate constant, gives

$$d[R^+]/dt = k_1[\mathrm{RCl}] - k_{-1}[R^+][\mathrm{Cl}^-] - k_2[R^+] = 0 \qquad (12.98a)$$

and

$$[R^+] = (k_1/(k_{-1}[\mathrm{Cl}^-] + k_2))[\mathrm{RCl}] \qquad (12.98b)$$

and therefore,

$$d[\mathrm{ROH}]/dt = k_2[R^+] = (k_1k_2/(k_{-1}\,[\mathrm{Cl}^-] + k_2))[\mathrm{RCl}] \qquad (12.98c)$$

Because [Cl^-] is in the denominator, the larger this concentration is, the slower the reaction rate.

12.99 ION PAIR

Two oppositely charged ions held in close proximity by solvation and Coulombic forces. An ion pair is treated as a single unit when considering colligative properties or kinetics. There are different types of ion pairs.

TIGHT (CONTACT) ION PAIR. The two ions are in direct contact with each other and are not separated by any solvent. This type of pairing is usually symbolized by A^+B^-.

SOLVENT SEPARATED (LOOSE) ION PAIR. The two ions are separated by one or more solvent molecules. This type of pairing is usually symbolized by $A^+\|B^-$.

12.100 SALT EFFECT

A change in the rate of a reaction caused by the addition of a salt (electrolyte) that generally does not have an ion in common with reactants or products. The addition of a salt changes the ionic strength of the solution, which in turn affects the free energies of ions that either are being formed or destroyed.

Example. The addition of $LiClO_4$ generally increases the rate of reaction. Ions are being formed

$$\mathrm{RX} \rightleftharpoons \mathrm{R}^+ + \mathrm{X}^-$$

and an increase in the ionic strength (polarity) of the solution favors their formation. $LiClO_4$ is used because the ClO_4^- is a very poor nucleophile and reacts very slowly with the free carbocation R^+.

In a reaction involving a solvent-separated ion pair, the ClO_4^- can react with this ion pair, which decreases that rate of the reaction of the ion pair returning to the starting material and, thus, increases the rate of formation of the highly reactive carbocation. This is referred to as the *special salt effect.*

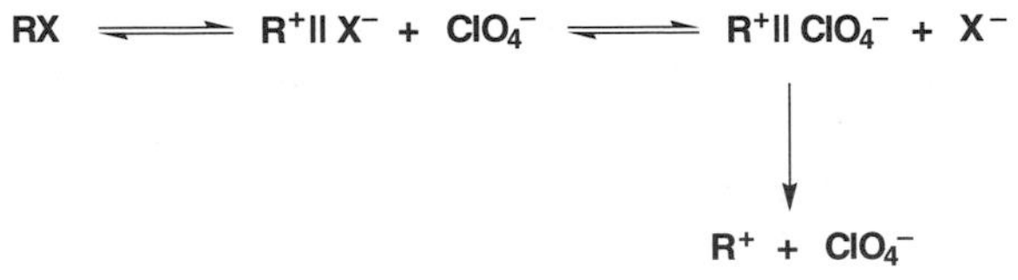

12.101 MIGRATORY APTITUDE

A term used to compare the relative rates at which different atoms or groups migrate to another atom during the course of a reaction. Migratory aptitudes are a function of the type of reaction and the conditions of the reaction.

Example. In a pinacol rearrangement, relative migratory aptitudes for the R groups, are *p*-methoxyphenyl, 500; *p*-tolyl, 15.7; *m*-tolyl, 1.95; *m*-methoxyphenyl, 1.6; phenyl, 1; *p*-chlorophenyl, 0.66. Electron-releasing groups have a favorable influence on the migratory aptitude in this reaction, because the migrating group is moving to a center of low electron density.

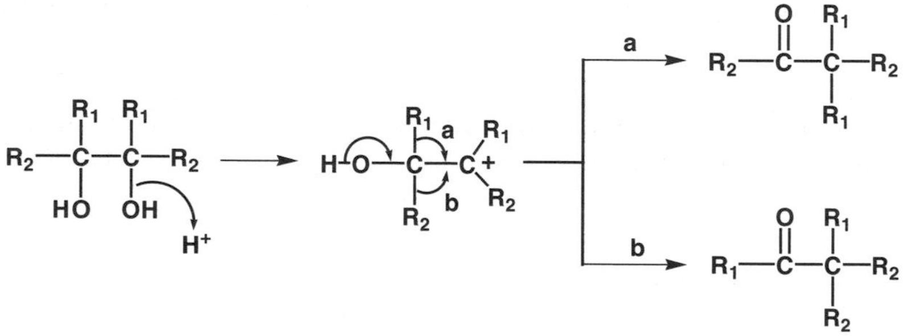

Figure 12.101. The pinacol rearrangement.

12.102 LINEAR FREE ENERGY RELATIONSHIPS

A mathematical expression in which the reactivity of a substituted species is compared to the reactivity of an unsubstituted species.

Example. The Hammett (Sect. 12.103), Taft (Sect. 12.104), Brönsted (Sect. 12.105), and Winstein–Grunwald (Sect. 12.106) equations.

12.103 HAMMETT EQUATION

Named after Louis P. Hammett (1894–1987), an equation of the form

$$\log (K_z/K_0) = \rho\sigma \qquad (12.103a)$$

where k_z is the rate constant of a reaction for a species carrying a substituent z, k_0 is the rate constant for the reaction of the unsubstituted species, σ (sigma) is a constant expressing the (electronic) effects of the substituent, and ρ (rho) is the reaction parameter that characterizes the sensitivity of the reaction to the substituent change. It is set at one for the standard reaction of the ionization of the benzoic acids. The values of σ are evaluated by comparing the acid dissociation constants of benzoic acid K_0 with substituted benzoic acids K_z, using the following relationship:

$$\sigma = \log(K_z/K_0) \qquad (12.103b)$$

TABLE 12.103*a*. Hammett Substituent Constants σ

Substituent	*meta*	*para*
–OH	+0.12	−0.37
$-OCH_3$	+0.12	−0.27
$-NH_2$	−0.16	−0.66
$-CH_3$	−0.07	−0.17
–H	0.00	0.00
–Cl	+0.37	+0.23
$-CF_3$	+0.43	+0.54
$-NO_2$	+0.71	+0.78

TABLE 12.103*b*. Rho (ρ) Values for *meta*- and *para*-Substituted Benzene Derivatives

Equilibria	ρ
$RC_6H_4COOH + H_2O \rightleftharpoons RC_6H_4COO^- + H_3O^+$	1.00
$RC_6H_4NH_3^+ + H_2O \rightleftharpoons RC_6H_4NH_2 + H_3O^+$	2.77
$RC_6H_4CHO + HCN \xrightleftharpoons{EtOH} RC_6H_4CH(OH)CN$	−1.49
Rates	
$RC_6H_4COOC_2H_5 + OH^- \xrightarrow[30°C]{EtOH} RC_6H_4COO^- + H_2O$	2.43
$RC_6H_4COOH + CH_3OH \xrightarrow[25°C]{H^+} RC_6H_4COOCH_3 + H_2O$	−0.23

A positive value for σ indicates that the substituent is electron-withdrawing ($K_z > K_0$), and a negative value indicates that it is electron-releasing ($K_z < K_0$). When log k_z/k_0 is plotted against σ, according to Eq. 12.103a, the result should be a straight line with a slope ρ. Reactions that are assisted by donation of electron density to the reaction site have negative ρ values, whereas reactions that are favored by withdrawal of electron density from the reaction site have positive ρ values. The ρ values for many types of equilibria can be ascertained by substituting equilibrium constants, K_0 and K_z, for rate constants in Eq. 12.103*a*. The Hammett equation is subject to certain limitations and has been improved by various modifications of the choice of σ values.

Example. Table 12.103*a* gives some values of Hammett constants σ, and Table 12.103b values of ρ for several equilibria and reaction rates.

12.104 TAFT EQUATION

Named after Robert Taft (1922–), the linear free energy relationship

$$\log (K_z/K_0) = \sigma_I \rho_I \qquad (12.104a)$$

σ_I is designed to assess inductive effects of substituents, and its values are obtained from the equation

$$\sigma_I = 0.262 \log(K_z/K_0) \quad (12.104b)$$

where K_z and K_0 are acid dissociation constants of substituted acetic acids and acetic acid, respectively.

12.105 BRÖNSTED EQUATION

Linear free energy relationships that relate the strength of an acid or base to the rate constant of an acid- or base-catalyzed reaction. For acids,

$$\log k = \alpha \log K_a + \text{constant} \quad (12.105a)$$

For bases,

$$\log k = \beta \log K_a + \text{constant} \quad (12.105b)$$

where k is the rate constant of the catalyzed reaction, K_a is the acid dissociation constant of the catalyst or its conjugate acid, and α/β are reaction constants similar to the Hammett ρ. If α or $\beta = 1$, the proton transfer essentially is complete in the transition state, and if they equal zero, essentially no proton transfer has occurred in the transition state.

12.106 WINSTEIN–GRUNWALD EQUATION

Named after Saul Winstein (1912–1969) and Ernest Grunwald (1923–), a linear free energy relationship used to determine the effect of the solvent on a chemical reaction involving ionization in the rate-determining step.

$$\log(k/k_0) = mY \quad (12.106)$$

where m is the reaction's sensitivity to solvent ionization power analogous to the Hammett ρ and Y is the ionizing power of the solvent analogous to the Hammett σ. The solvolysis of *tert*-butyl chloride is the standard reaction for which m is set equal to 1 in order to establish the Y scale, and 80% aqueous ethanol is the solvent for which $Y = 0$.

12.107 TUNNELING

Due to the uncertainty principle (see Sect. 1.4), quantum mechanically there is a finite probability that a molecule undergoing a chemical reaction will not follow the normal path shown in the reaction coordinate diagram (see Sect. 12.69), but rather, the molecule will follow a path in which it tunnels through the barrier. This phenomenon is only observed with light atoms (usually, H) and when the barrier is very narrow, that is, hydrogen atoms can penetrate barriers for short distances.

12.108 MICROSCOPIC REVERSIBILITY

For an elementary reaction step, the reaction path followed by the back reaction is the exact reverse of the reaction path followed by the forward reaction. For a multistep mechanism, the back reaction follows the identical mechanism to the forward reaction but in reverse. Microscopic reversibility is also known as the *principle of detailed balancing*.

Example. In the reaction,

$$\mathrm{A} \rightleftharpoons [\mathrm{B}]^{\ddagger} \rightleftharpoons \mathrm{C}$$

if $[B]^{\ddagger}$ is the transition state in the conversion of A to C, then under the same conditions, $[B]^{\ddagger}$ is the transition state in the conversion of C to A.

12.109 ISOKINETIC TEMPERATURE

For a series of similar reactions, it often is found that the enthalpy of activation $\Delta H^{\ddagger}$ and the entropy of activation $\Delta S^{\ddagger}$ are proportional to each other:

$$\Delta H^{\ddagger} = \beta\, \Delta S^{\ddagger} + \text{constant} \qquad (12.109)$$

where β has units of temperature and is the isokinetic temperature.

12.110 MARCUS THEORY

Named after Rudolph Marcus (1923–) who received the Nobel Prize in 1992, a theory that originally dealt with the rates of electron transfer reactions, that is, reactions in which a single electron is transferred from one compound to another. It now has been extended to include the transfer of a proton, a hydride, a hydrogen atom, and even a methyl group.

Example. Consider the methyl transfer reaction

$$\mathrm{Y^- + H_3CX \longrightarrow H_3CY + X^-}$$

The Marcus equation for the free energy of activation for this reaction is

$$\begin{aligned}\Delta G^{\ddagger} = \Delta G^{\ddagger}_{\text{int}} &+ (1/2)(\Delta G^{0} - w^{R} - w^{P}) \\ &+ [(\Delta G^{0} - w^{R} - w^{P})^{2}/16(\Delta G^{\ddagger}_{\text{int}} - w^{R})]\end{aligned} \qquad (12.110a)$$

$\Delta G^{\ddagger}_{\text{int}}$ is the intrinsic barrier, that is, the barrier that would exist even if the reactants and products had the same ΔG^0. This barrier is obtained by averaging the $\Delta G^{\ddagger}$ for the two energetically symmetrical reactions:

$$X^- + H_3CX \longrightarrow H_3CX + X^-$$

$$Y^- + H_3CY \longrightarrow H_3CY + Y^-$$

$$\Delta G^{\ddagger}_{\text{int}} = \Delta G^{\ddagger}_{x,y} = (1/2)(\Delta G^{\ddagger}_{x,x} + \Delta G^{\ddagger}_{y,y}) \qquad (12.110b)$$

ΔG^0 is the difference in free energy between the products and reactants.
w^R and w^P are work terms: the free energy needed to bring the reactants together and to separate the products. These terms are usually very small except when dealing with two ions.

If the free energy difference between the reactant and the product is much smaller than the intrinsic barrier for interconversion, then the $(\Delta G^0/\Delta G^{\ddagger}_{\text{int}})$ term can be neglected. This says that the free energy of activation is

$$\Delta G^{\ddagger} = \Delta G^{\ddagger}_{\text{int}} + (1/2)(\Delta G^0) = (1/2)(\Delta G^{\ddagger}_{x,x} + \Delta G^{\ddagger}_{y,y} + \Delta G^0_{x,y}) \qquad (12.110c)$$

In contrast, if the absolute value of the free energy difference, that is, $|\Delta G^0|$, and the intrinsic barrier are approximately the same, then a strange effect is predicted by Marcus theory. Namely, as the reaction become thermodynamically more favorable, that is, more exothermic, ΔG^0 becomes more negative, but $\Delta G^{\ddagger}$ increases, and, thus, the reaction slows down. This inverted phenomenon is due to the $(\Delta G^0)^2$ term increasing as ΔG^0 becomes more negative (see Sect. 18.172).

SUGGESTED READING

Carpenter, B. K. *Determination of Organic Reaction Mechanisms*. Wiley-Interscience: New York, 1984.

Carroll, F. A. *Perspectives on Structure and Mechanism in Organic Chemistry*. Brooks/Cole: Pacific Grove, CA, 1998.

"Hard and Soft Acids and Bases." *Chem. Eng. News*, February 17, 2002.

Isaac, N. *Physical Organic Chemistry*, 2nd ed., Longman: Essex, UK, 1995.

Lowry, T. H. and Richardson, K. S. *Mechanism and Theory in Organic Chemistry*, 3rd ed., Harper and Row: Cambridge, MA, 1987.

13 Reactive Intermediates (Ions, Radicals, Radical Ions, Electron-Deficient Species, Arynes)

13.1	Reactive Intermediate	507
13.2	Species	508
13.3	Charged Species	508
13.4	Ion	508
13.5	Cation	508
13.6	Anion	509
13.7	Uncharged Species	509
13.8	Neutral Species	509
13.9	Electron-Deficient Species	510
13.10	Free Radical	510
13.11	Radical	511
13.12	Biradical (Diradical)	511
13.13	Divalent (or Dicovalent) Carbon	511
13.14	Carbene, R_2C:	512
13.15	Methylene	512
13.16	Electron Spin Multiplicity	512
13.17	Singlet Carbene	513
13.18	Triplet Carbene	513
13.19	Carbenoid Species	513
13.20	α-Ketocarbene, $RC(O)-\ddot{C}-R$	514
13.21	Acylcarbene	514
13.22	Nitrene, $R-\ddot{N}$:	514
13.23	Acyl Nitrene, $RC(O)-\ddot{N}$:	515
13.24	α-Ketonitrene	515
13.25	Charge-Localized (or Localized) Ion	515
13.26	Charge-Delocalized (or Delocalized) Ion	515
13.27	Carbocation	516
13.28	Classical Cation	517
13.29	Nonclassical Cation	517
13.30	Bridged Cation	518
13.31	Carbonium Ion	519
13.32	*-Enium* Ion	520
13.33	Carbenium Ion, R_3C^+	520

The Vocabulary and Concepts of Organic Chemistry, Second Edition, by Milton Orchin, Roger S. Macomber, Allan Pinhas, and R. Marshall Wilson

13.34 Other 2nd Row *-Enium* Ions: Nitrenium (R_2N^+), Oxenium (RO^+), and Fluorenium (F^+) Ions 521
13.35 3rd Row *-Enium* Ions: Silicenium (R_3Si^+), Phosphenium(R_2P^+), Sulfenium (RS^+), and Chlorenium (Cl^+) Ions 521
13.36 *-Onium* Ions 522
13.37 2nd Row *-Onium* Ions: Ammonium (R_4N^+), Oxonium (R_3O^+), and Fluoronium ($R_2\,F^+$) Ions 522
13.38 3rd Row (and Higher) *-Onium* Ions: Phosphonium (R_4P^+), Sulfonium (R_3S^+), and Chloronium (R_2Cl^+) Ions 523
13.39 Diazonium Ion, $R{-}N^+{\equiv}N$ 523
13.40 Acylium Ion, $RC{\equiv}O^+$ 523
13.41 Siliconium Ion, R_3Si^+ 524
13.42 Aminium Ion, $R_3\dot{N}^+$ 524
13.43 Iminium Ion, $R_2C{=}N^+RR'$ 524
13.44 Carbanion 524
13.45 Nitrogen Anion (Nitranion) 525
13.46 Oxygen Anion (Oxyanion) 525
13.47 Alkoxides, M–OR 526
13.48 Enolate Anion 526
13.49 Radical Ion 526
13.50 Radical Cation 527
13.51 Radical Anion 527
13.52 Ion Pair, Radical Pair 528
13.53 Tight Ion Pair, Tight Radical Pair [Geminate Ion (Radical) Pair, Intimate Ion (Radical) Pair, Contact Ion (Radical) Pair] 528
13.54 Solvent-Separated (or Loose) Ion Pair 528
13.55 Counterion 529
13.56 Gegenion 529
13.57 Zwitterion (Dipolar Ion) 529
13.58 Betaine 529
13.59 Ylide 530
13.60 Aryne 530
13.61 Strained Ring Systems 531
13.62 Tetrahedral Intermediates from Carbonyls 532
13.63 Summary Table of Intermediates 532

Many chemical reactions occur by a mechanism with two or more elementary steps. In the first step of such a process, the reactants must overcome an energy barrier to form an intermediate that, in a second step, proceeds over a second energy barrier to go on to product:

$$A+B \xrightarrow{k_1} \text{Int} \xrightarrow{k_2} C$$

In many cases, the first energy barrier in going from reactants to the intermediate via transition state 1, with a rate constant k_1, is greater than the second energy barrier in going from the intermediate, via transition state 2, to the product with a rate constant k_2, resulting in $k_2 > k_1$. This situation is frequently expressed in terms of a two-dimensional plot of the reaction coordinate (progress of the reaction) as a function of the potential energies of the following entities: the reactants; the intermediate; the product (all at energy minima); and the two transition states (at energy maxima). (See Chap. 12 and Fig. 12.69*c*.) The intermediate is represented by a minimum between two maxima. Such a plot, for example, describes the solvolysis reaction of *t*-butyl chloride. The slow reaction involving the ionization to give the intermediate *t*-butyl cation is followed by the relatively fast reaction of the cation with H_2O to give *t*-butyl alcohol. In some fewer cases (e.g., 1,4- vs. 1,2-addition to a conjugated diene), the energy barrier in going from intermediate back to reactants is smaller than that in going from intermediate to product (see Fig. 12.85*b*). Such cases involve an equilibrium reaction leading to either kinetic or thermodynamic control, depending on the temperature and/or reaction time. In any case, the shape of the well (relatively deep or shallow) representing the intermediate determines the relative reactivity of the intermediate. A deep well implies that a high vibrational state is required for the intermediate to pass on to product or to revert to reactants and, therefore, the possibility of trapping such an intermediate is enhanced. This is sometimes the situation where carbenes are intermediates. Indeed, the work of Jack Hine and co-workers in the early 1950s, which launched the modern era of carbene chemistry, involved the hydrolysis of chloroform to give an intermediate dichlorocarbene that was trapped by water.

Reactive intermediates can be divided into two large classes: charged and neutral species. The charged species include cations, anions, radical cations, and radical anions. Neutral species include electron-deficient entities such as radicals, biradicals, carbenes, and nitrenes. As an intermediate, *atomic carbon* is in a class by itself. Developments in the preparation of carbon fibers, carbon nanotubes, and fullerenes, all of which probably are generated from atomic carbon, are testimony to its importance. These substances are discussed in other chapters and will not be further discussed here. Arynes such as benzyne and many other strained systems such as *trans* cyclic olefins are also neutral intermediates. Intermediates play an important role in the mechanisms of most organic reactions, with the exception of truly concerted reactions. The role of intermediates is exemplified by the mechanisms of the many reactions discussed in Chap. 14. Intermediates are important also in excited-state chemistry (Chap. 18), in addition to practically all catalytic reactions, and thus a complete discussion of all intermediates as a separate topic would be repetitive as well as impractical. Accordingly, we have largely focused on classes of intermediates rather than intermediates in specific reactions. Some intermediates have been discussed in earlier chapters, especially the chapter dealing with the processing of petroleum. However, we think some repetition is necessary to present an integrated treatment of reaction intermediates.

13.1 REACTIVE INTERMEDIATE

An intermediate is a species, which is not the final product formed during one specific reaction. Because this intermediate, by definition, goes on to react further, it is called (perhaps redundantly) a *reactive* intermediate. In plots showing the progress of the reaction (the reaction coordinate) as a function of potential energy (see Fig. 12.69*c*), such intermediates are represented by minima betweeen transition-state maxima.

13.2 SPECIES

Although this noun has a precise meaning in some disciplines, such as biology, in chemistry-speak it is used as an imprecise, but convenient term, to denote either atoms, ions, radicals, intermediates, fragments of molecules, or other chemical entities sometimes of uncertain or imprecisely defined structure.

13.3 CHARGED SPECIES

A species in which the number of electrons and the number of protons are unequal. Most charged species can be written with the charge localized on a particular atom, in which case the number of valence electrons surrounding that atom will be fewer than (positive charge) or greater than (negative charge) the number of valence electrons normally associated with that atom in its elementary state. Charged species are reactive because of electrostatic attractions.

Example. Since the element carbon has four valence electrons, methyl cation, CH_3^+ (Fig. 13.3*a*), with only three carbon valence electrons and an empty orbital is positive charged, and methyl anion, CH_3^-, with five valence electrons contributed by carbon (Fig. 13.3*b*) is negative charged. In this chapter only, all charges in the figures will be circled in order to avoid ambiguity.

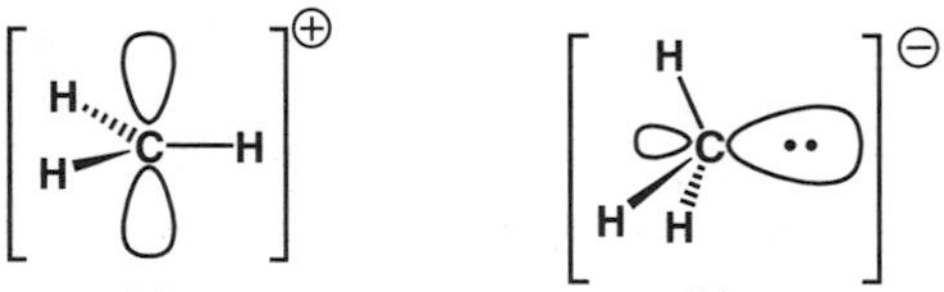

(*a*) (*b*)

Figure 13.3. (*a*) Methyl cation and (*b*) methyl anion.

13.4 ION

Synonymous with charged species.

13.5 CATION

A positively charged atom or group of atoms.

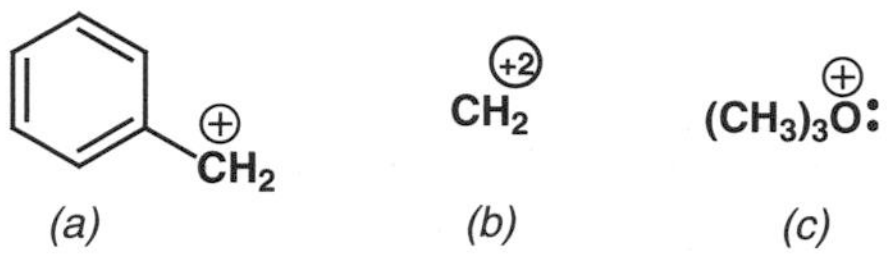

Figure 13.5. (*a*) Benzyl cation; (*b*) methylene dication; (*c*) trimethyloxonium ion.

Example. Benzyl cation (Fig. 13.5*a*); methylene dication (Fig. 13.5*b*); trimethyloxonium ion (Fig. 13.5*c*).

13.6 ANION

A negatively charged atom or group of atoms.

Example. 2-Naphthyl anion (Fig. 13.6*a*) and phenylmethylene dianion (Fig. 13.6*b*).

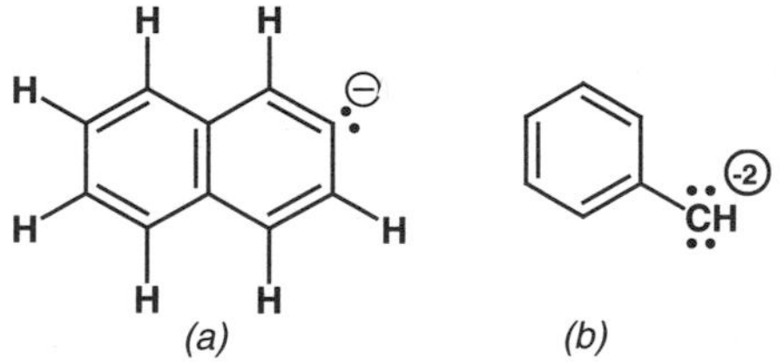

Figure 13.6. (*a*) 2-Naphthyl anion and (*b*) phenylmethylene dianion.

13.7 UNCHARGED SPECIES

A species that has neither a net positive nor a net negative charge.

Example. The species shown in Figs. 13.9*a* and *b*.

13.8 NEUTRAL SPECIES

This species is synonymous with uncharged species; it is not to be confused with the word "neutral" when it is used in the description of acid–base properties.

Example. Free radicals (Sect. 13.10); benzyne (Sect. 13.60).

13.9 ELECTRON-DEFICIENT SPECIES

A molecular species that possesses fewer than the maximum possible number of electrons in the valence shells of its atoms.

Example. Methyl radical (Fig. 13.9*a*); methylene (Fig. 13.9*b*); and methyl cation (Fig. 13.9*c*). The full complement of valence electrons surrounding carbon and all other 2nd row elements in their respective compounds is eight, whereas in the first three examples shown, the species possess seven, six, and six valence electrons, respectively. Under this definition, Lewis acids (Sect. 12.51), such as boron trifluoride (BF_3), and compounds containing multicenter bonds, for example, diborane (B_2H_6) (Fig. 13.9*d*) that possesses two, two-electron, three-center bonds (B–H–B), are electron-deficient. There are 8 bonds normally requiring 16 electrons, but only a total of 12 electrons are available: 6 from the 2 boron atoms and 6 from the 6 hydrogen atoms.

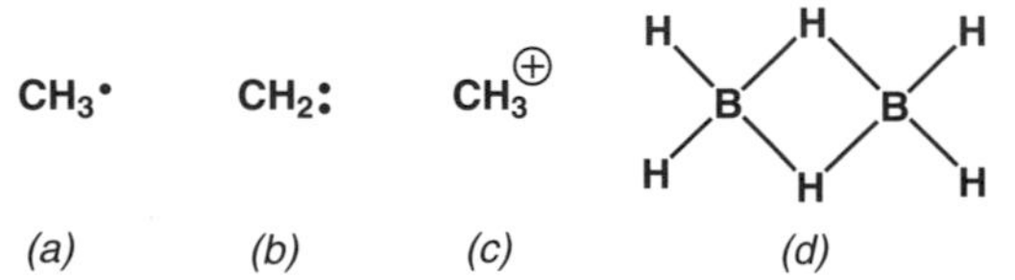

Figure 13.9. (*a*) Methyl radical; (*b*) methylene; (*c*) methyl cation; (*d*) diborane.

13.10 FREE RADICAL

An uncharged species possessing one or more unpaired electrons. Radicals are essentially small magnets and owe their reactivity to this fact. If the radical species bears a charge, it is known as a radical ion. Transition metal complexes that have one or more unpaired electrons associated with the metal are not generally referred to as free radicals. If a radical is associated with another radical counterpart in a solvent cage, the system is referred to as a *radical pair*, or *geminate radical pair* (Sect. 13.53).

Example. Benzyl radical (Fig. 13.10*a*) and methoxyl radical (Fig. 13.10*b*). The name of a monovalent radical always ends in *-yl*. In the standard convention, a dot (•) or single arrow (↑) represents an unpaired electron. Two dots (:) or two arrows facing in opposite directions (↑↓) represent two paired electrons. Two arrows facing

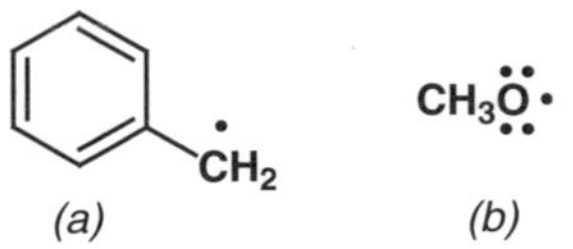

Figure 13.10. (*a*) Benzyl radical and (*b*) methoxyl radical.

in the same direction (↑↑) represent two unpaired electrons (biradical), but these cannot, of course, occupy the same orbital and so some higher-energy (usually, antibonding) orbital must be involved.

13.11 RADICAL

Synonymous with and used interchangeably with the term *free radical*. Unfortunately, common group names such as methyl, isopropyl, and so on, when referred to in naming compounds, as in methyl chloride (IUPAC chloromethane) or isopropyl alcohol (IUPAC 2-propanol), are frequently called radicals, but these species are now referred to as groups even though the naming system using these groups is still called radicofunctional nomenclature (Sect. 6.7). The radical functional name is constructed from the (misnamed) radical (methyl, isopropyl) followed by the functional group (chloride, alcohol, etc.).

13.12 BIRADICAL (DIRADICAL)

An uncharged species possessing two unpaired electrons, thus theoretically capable of existing in two electronic states with different electron spin multiplicity, that is, singlet (paired) and triplet (unpaired) states.

Example. Cyclopentanediyl (Fig. 13.12*a*) and dimethylenediaminyl biradical (Fig. 13.12*b*). Compounds such as triplet methylene (Fig. 13.12*c*) have been called biradicals, but are preferably referred to as triplet carbenes. (See Sect. 13.18.) The *biradical 1,4-dehydrobenzene* (Fig. 13.12*c*), an isomer of benzyne (Sect. 13.60), is thought to be an intermediate in the cyclization of 1,5-hexadiyne-3-ene.

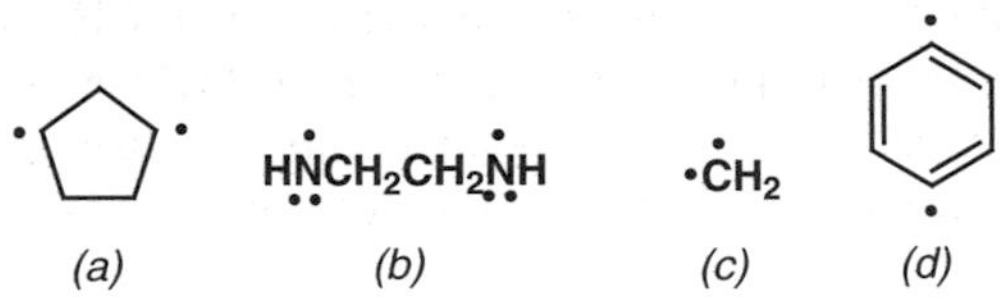

Figure 13.12. (*a*) 1,3-cyclopentanediyl; (*b*) dimethylenediaminyl biradical; (*c*) triplet methylene; (*d*) 1,4-dehydrobenzene biradical.

13.13 DIVALENT (OR DICOVALENT) CARBON

This is a carbon atom having only two covalent bonds attached to it; it may be charged or neutral.

Example. Methylene (Fig. 13.13*a*); mcthylmethylene radical cation (Fig. 13.13*b*); and phenylmethylene radical anion (Fig. 13.13*c*). The italicized carbons are divalent.

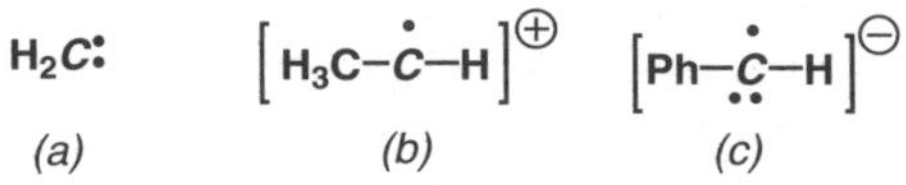

Figure 13.13. (*a*) Methylene; (*b*) methylmethylene radical cation; (*c*) phenylmethylene radical anion.

13.14 CARBENE, R_2C:

A generic name for an uncharged species containing a dicovalent carbon atom. Because this carbon atom is surrounded by a sextet rather than an octet of valence electrons, it is electron-deficient and, hence, is electrophilic.

Example. Singlet dichlorocarbene (Fig. 13.14*a*); triplet dichlorocarbene (Fig. 13.14*b*); and methylenecarbene (Fig. 13.14*c*). Organometal carbenes (Sect. 9.44) are a class of organometal complexes involving a divalent carbon atom doubly bonded to a transition metal and are called carbenes, although the carbon is no longer bonded to only two other atoms. Carbenes are involved in many types of reactions, for example, metathesis reactions and the Wolff rearrangement.

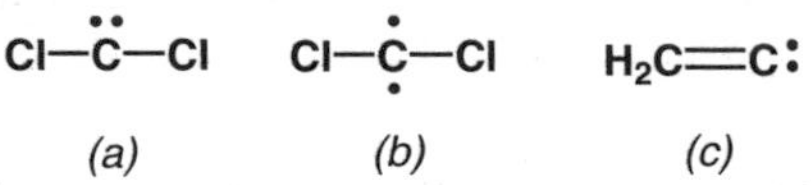

Figure 13.14. (*a*) Singlet dichlorocarbene; (*b*) triplet dichlorocarbene; (*c*) singlet methylene carbene.

13.15 METHYLENE

The one specific carbene that has the molecular formula :CH_2. The carbene, methylene, should be distinguished from the dicovalent methylene *group* $-CH_2-$ involving tetravalent carbon, which occurs in an infinite number of real and possible compounds as well as the carbon atom of a C-terminal double bond, $=CH_2$.

13.16 ELECTRON SPIN MULTIPLICITY

The number of spin orientations observed when the species in question is placed in a magnetic field. The multiplicity is determined by the number of unpaired electron spins n and the spin angular momentum quantum number $s = (\pm\frac{1}{2})$ and $|ns| = S$.

$$\text{multiplicity} = 2\,|ns| + 1 \qquad (13.16)$$

Example. CH_4, S = 0, multiplicity = 1 (singlet); $CH_3\uparrow$, S = 1/2, multiplicity 2 (doublet); $CH_2\uparrow\uparrow$, S = 1, multiplicity = 3 (triplet).

13.17 SINGLET CARBENE

A carbene that has an electron spin multiplicity of 1 (no unpaired electrons).

Example. Singlet methylene (Fig. 13.17*a*) and singlet phenylcarbene (Fig. 13.17*b*). The pair of nonbonding electrons occupies a single sp^2 orbital. The empty p orbital is responsible for the singlet carbene's electrophilic character as illustrated in the reaction with olefins, Fig. 13.7*c*.

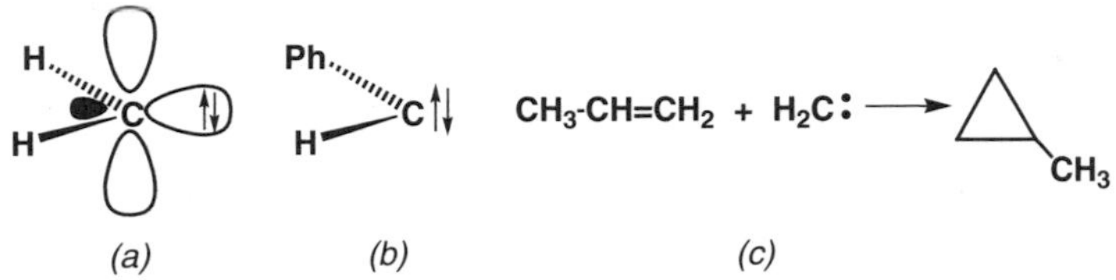

Figure 13.17. (*a*) Singlet methylene (H–C–H angle = 102°); (*b*) singlet phenylcarbene; (*c*) the reaction of singlet methylene with propene.

13.18 TRIPLET CARBENE

A carbene that has an electron spin multiplicity of 3 (two unpaired electrons).

Example. Triplet methylene (Fig. 13.18*a*) and triplet methylcarbene (Fig. 13.18*b*) (ground-state triplets). In the ground-state triplets, the unpaired electrons occupy separate sp^3 orbitals, The H–C–H angle is approximately 136°.

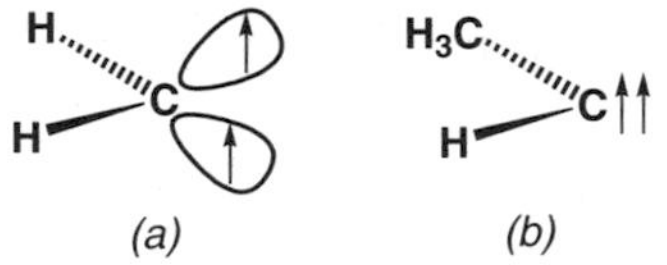

Figure 13.18. (*a*) Triplet methylene (H–C–H angle = 136°) and (*b*) triplet methylcarbene.

13.19 CARBENOID SPECIES

A species that, when undergoing reaction, has the characteristics of a carbene, but in fact is not a free carbene.

Example. The carbenoid species, iodomethylzinc iodide (Fig. 13.19), reacts with ethylene to yield the same product, cyclopropane, as the reaction of ethylene with methylene, a free carbene.

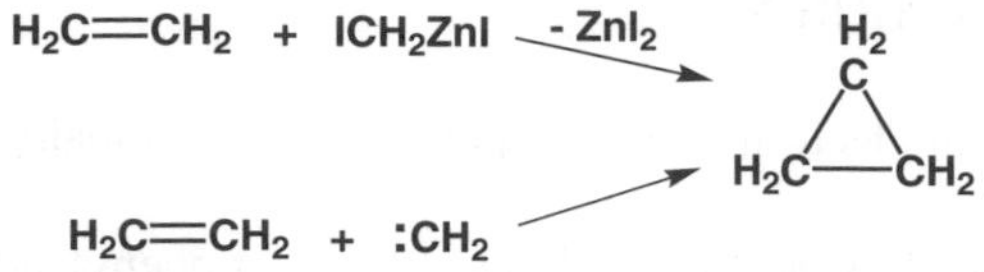

Figure 13.19. The reaction of iodomethylzinc iodide with ethylene.

13.20 α-KETOCARBENE, $RC(O)-\ddot{C}-R$

A carbene in which at least one of the groups attached to the dicovalent carbon atom is an acyl group.

Example. Acetylcarbene (Fig. 13.20).

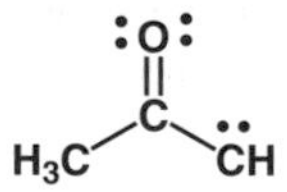

Figure 13.20. Acetylcarbene.

13.21 ACYLCARBENE

Synonymous with α-ketocarbene.

13.22 NITRENE $R-\ddot{N}:$

This is an uncharged species containing a monocovalent nitrogen atom; this nitrogen is surrounded by a sextet of valence electrons and hence is electron-deficient and, like carbene, acts as an electrophile. Nitrenes, like carbenes, can exist in two spin states, singlet and triplet, with the latter as the lower-energy ground state.

Example. Phenylnitrene (Fig. 13.22). Low-temperature (77 K) matrix studies of the ultraviolet spectrum of triplet phenylnitrene show absorption bands between 300 and 400 nm, and the ESR spectrum suggests delocalization of a single electron into the ring.

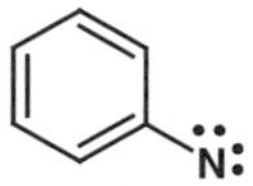

Figure 13.22. Phenylnitrene (singlet).

13.23 ACYL NITRENE, RC(O)−N̈:

This is a nitrene in which an acyl group is attached to the monocovalent nitrogen atom; it is analogous to an α-ketocarbene. Nitrenes are intermediates in a variety of reactions, for example, the Curtius reaction, Hofmann degradation, Schmidt degradation, and Lossen reaction. All these reactions lead to the formation of isocyanates and are thought to proceed through reactive intermediate acyl nitrenes.

Example. Benzoylnitrene and its rearrangement (Fig. 13.23).

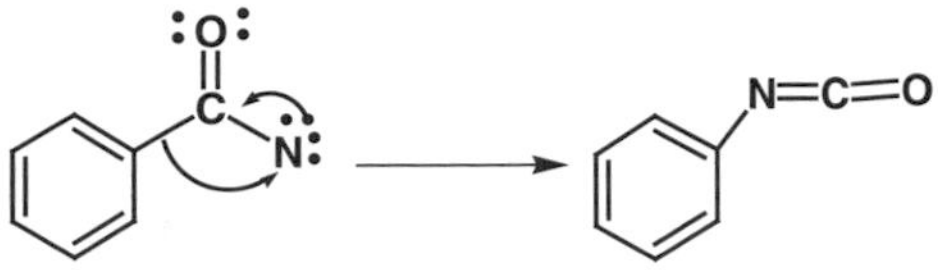

Figure 13.23. Benzoylnitrene rearrangement to phenyl isocyanate.

13.24 α-KETONITRENE

Synonymous with acyl nitrene.

13.25 CHARGE-LOCALIZED (OR LOCALIZED) ION

An ion in which the charge resides principally on a single atom.

Example. *t*-Butyl cation (Fig. 13.25*a*) and the cyclohexyl anion (Fig. 13.25*b*); the electrical charges on these ions are represented as being on the italicized carbon atoms.

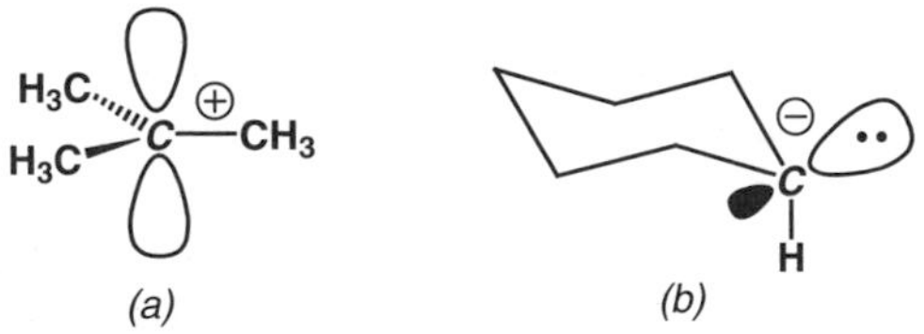

Figure 13.25. (*a*) *t*-Butyl cation and (*b*) cyclohexyl anion.

13.26 CHARGE-DELOCALIZED (OR DELOCALIZED) ION

An ion in which the charge is distributed over more than one atom. These ions are frequently represented by resonance structures to reflect the delocalization.

Example. Allyl cation (Fig. 13.26*a*), benzyl anion (Fig. 13.26*b*), and acetyl cation (Fig. 13.26*c*).

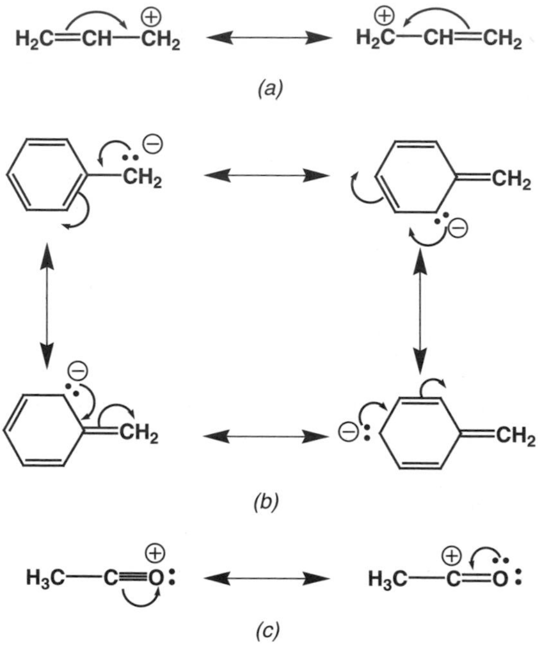

Figure 13.26. Resonance structures of (*a*) allyl cation; (*b*) benzyl anion; (*c*) acetyl cation.

13.27 CARBOCATION

A cation in which the positive charge may be formally localized on one or more carbons.

Example. In the most frequently encountered examples, the carbon atom is trivalent and thus surrounded by a sextet of valence electrons, as in the *t*-butyl cation (Fig. 13.27*a*). In many cases, the charge on carbon may be delocalized on an adjacent hetero atom, as in one of the resonance structures of protonated acetone (Fig. 13.27*b*). These examples demonstrate some ambiguity in the definition since, as in the acetone case, they may be called oxonium (Sect. 13.37) ions as well. It is probably best to refer to the resonance structure with the charge on carbon atom as the *carbocation form* and the resonance structure with the charge on oxygen as the *oxonium ion form*, bearing in mind, of course, that resonance structures are not separate species. All carbenium ions, such as Fig. 13.27*a* (Sect. 13.33), and all carbonium ions (Sect. 13.31), such as Fig. 13.27*c* (Sect. 13.31), the protonated methane or methonium ion, are carbocations.

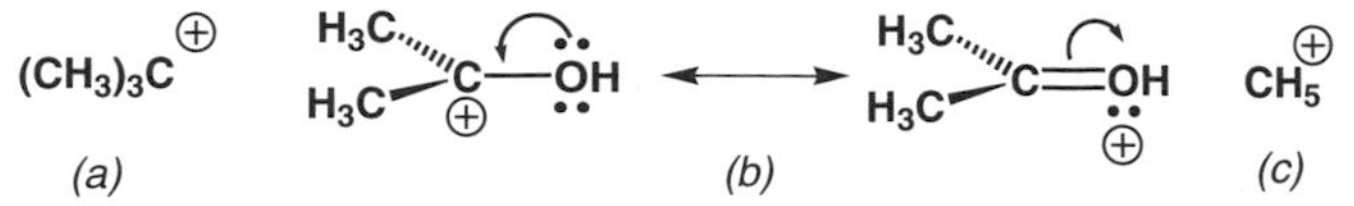

Figure 13.27. (*a*) *t*-Butyl cation; (*b*) resonance structures of protonated acetone; and (*c*) protonated methane or methonium ion. First prepared by George Olah and co-workers.

13.28 CLASSICAL CATION

A cation in which the charge is either localized or delocalized, but is not distributed by means of a closed (bridging) multicenter bond (Sect. 3.46).

Example. Methyl cation (Fig. 13.28*a*) and 1-hydroxyethyl cation (Fig. 13.28*b*). All carbenium ions (Sect. 13.33) are classical cations. Species in which the charge is on carbon, such as methyl cation, are frequently called carbonium ions, but according to IUPAC guidelines, because they are electron-deficient, they should be called carbenium ions (see Sect. 13.33).

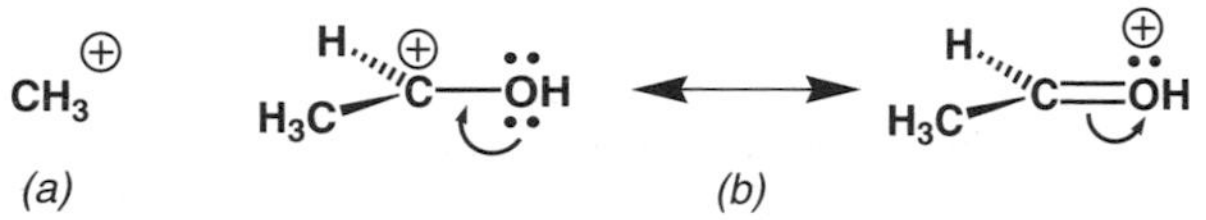

Figure 13.28. (*a*) Methyl cation and (*b*) 1-hydroxyethyl cation (protonated acetaldehyde).

13.29 NONCLASSICAL CATION

A delocalized bridging cation in which the charge is distributed by means of closed multicenter bonding (Sect. 3.46).

Example. The most controversial example involves the norbornyl cation (Fig. 13.29*c*) generated during acetolysis of 2-norbornyl tosylate. Here, the positive charge is distributed between carbons 1 and 2 via a two-electron, three-center bond (carbons 1, 2, and 6), giving rise to what appears to be a five-coordinate carbon atom 6 (Fig. 13.29*c*). Carbon 6 may also be considered to be tetravalent with three normal and two one-half bonds, the latter shown as the broken lines in Fig. 13.29*c*. This situation results from the involvement of the two electrons in the σ bond between carbons 1 and 6 in the shifting of the positive charge from carbon atom 2 (Fig. 13.29*a*) to carbon atom 1 (Fig. 13.29*b*).

The controversy in the norbornyl polemic involves the question of whether the two cations (Figs. 13.29*a* and *b*) are a rapidly interconverting pair of classical carbenium ions (the so-called windshield-wiper effect in which the single bond, pivoting on carbon 6, swings between carbons 1 and 2) or the nonclassical bridging cation of Fig. 13.19*c*. It may be more satisfactory to represent the nonclassical norbornyl

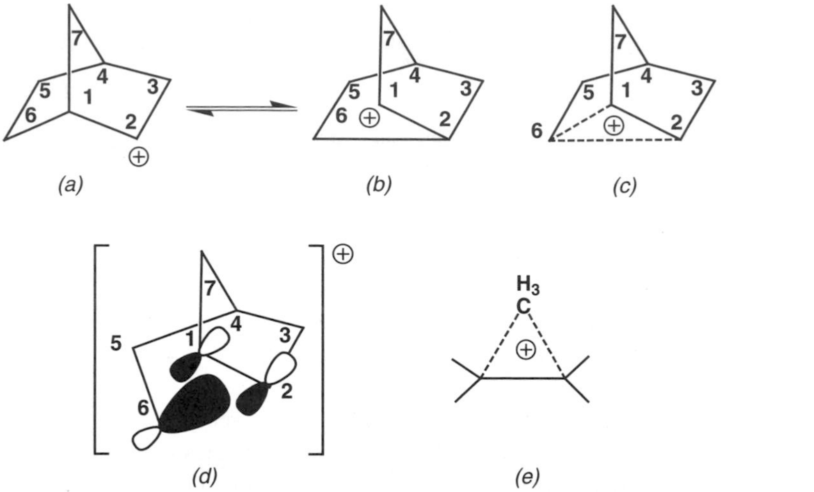

Figure 13.29. (*a*) and (*b*) The two classical ion structures of the norbornyl cation; (*c*) the nonclassical norbornyl ion single structure; (*d*) a representation of the orbital overlap in the norbornyl cation; (*e*) a representation of the transition state in a 1,2-shift of a neighboring methyl group onto a carbocation.

structure from a molecular orbital point of view. In this model, a carbon sp^3 orbital on C-6 combines with the two carbon *p* orbitals on C-1 and C-2 (Fig. 13.29*d*) to generate three molecular orbitals. In order of increasing energy, these molecular orbitals are bonding, weakly antibonding, and strongly antibonding orbitals. Because only two electrons are involved, these occupy the bonding orbital. Essentially, the same type of two-electron, three-center bonding is involved at the transition state in a typical 1,2-shift of a methyl group onto an adjacent carbocation (Fig. 13.29*e*). The important distinction between these two kinds of nonclassical cations is that the norbornyl (Fig. 13.29*c*) has a finite existence, that is, it is an intermediate corresponding to a minimum on a potential energy surface, whereas the nonclassical cation of Fig. 13.29*e* is a postulated transition state corresponding to a maximum on a potential energy surface. Indeed, it is claimed that the norbornyl cation has been observed by ^{13}C-NMR spectra in superacid media, in which case it would be a true intermediate.

13.30 BRIDGED CATION

A closed three-center, four-electron cation.

Example. The bromonium ion shown in Fig. 13.30*a*, and the phenonium ion in Fig. 13.30*b*, are examples of classical ions (Sect. 13.28), but they are sometimes (ambiguously) referred to as bridged ions. Both involve three-center bonding, but since two covalent bonds are involved in the bridge, four electrons are required in the bonding schemes. In the bromonium ion, there are as many valence electrons

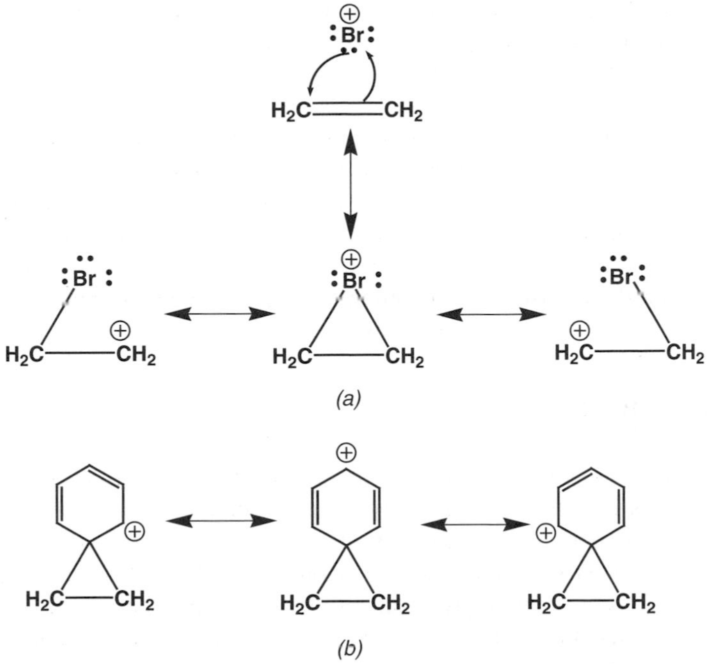

Figure 13.30. (*a*) A bromonium ion and (*b*) a phenonium ion.

available as are required for separate localized bonds; the bridging involves the participation of a lone pair on the bromine atom of the open form. From the molecular orbital point of view, it may be considered that the three-centered ion is generated by carbon–carbon $p\pi$ orbitals overlapping a bromine p orbital to form three molecular orbitals: one strongly bonding, one essentially nonbonding, and one antibonding. However, in contrast to the nonclassical case of the norbonyl cation in which two electrons of the σ bond are involved, there are now four electrons involved: two from the bromine and two from the carbon–carbon π bond. These four electrons would occupy the bonding and nonbonding orbitals. It is likely that the nonbonding orbital is, in fact, slightly bonding. Note that the bromonium ion involves participation of lone pair nonbonding electrons on the bromine atom in bonding to the charged carbon. In the case of a β-phenethyl cation, participation of the π electrons in the delocalization of the positive charge leads to the bridging phenonium ion (Fig. 13.30*b*). Bridged cation forms are of special interest in connection with neighboring group effects (Sect. 14.25) and anchimeric assistance (Sect. 14.26).

13.31 CARBONIUM ION

A hypervalent carbocation, that is, a positively charged carbon having a valence (coordination number) greater than 3. Considerable confusion continues to exist with respect to this definition because in some literature and in some older texts, the term carbonium ion is used frequently, but incorrectly, as a generic term for a species that should be called a carbocation.

Example. In the nonclassical norbornyl cation, written in the delocalized form of Fig. 13.29*c*. C-1, C-2, and C-6 each have a coordination number greater than 3 (C-1 and C-2 a coordination number of 4, and C-6 a coordination number of 5) and hence this structure may be called a carbonium ion. However in the localized forms (Figs. 13.29*a* and *b*), the carbons bearing the charge have a coordination number of 3 and therefore these forms are carbenium ions (Sect. 13.33). Fortunately, the confusion can be avoided by referring to this nonclassical ion as a carbocation.

13.32 -*ENIUM* ION

A generic suffix (pronounced en-ee-um) used in the name for the cationic portion of an ionic species in which the positively charged nonmetallic atom has two fewer electrons than the normal complement in its valence shell. Thus, the charged atom has one less covalent bond than the corresponding uncharged species and the *ene* part of the name is meant to imply unsaturation. In a formal sense, such ions may be regarded as generated from their parent species by the removal of hydride. *Enium* is usually attached as a suffix to a root name.

Example. See carbenium (Fig. 13.33) and nitrenium (Fig. 13.34).

13.33 CARBENIUM ION, R_3C^+

An -*enium* ion in which the (unsaturated) atom carrying the charge is carbon.

Example. Methylcarbenium ion (ethyl cation, Fig. 13.33*a*); trimethylcarbenium ion (*t*-butyl cation, Fig. 13.33*b*). *Vinyl carbenium ions* are assumed to be intermediates in the addition of electrophiles to terminal acetylenes such as the carbenium ion (Fig. 13.33*c*), formed from the addition of a proton to phenylacetylene. Each of these ions has a total of six electrons in the valence shell surrounding the positively charged carbon atom and they are therefore electron-deficient species. There are some rare carbenium ions (e.g., Fig. 13.33*d*) that are so stable they can be bottled along with their counterion and are available commercially.

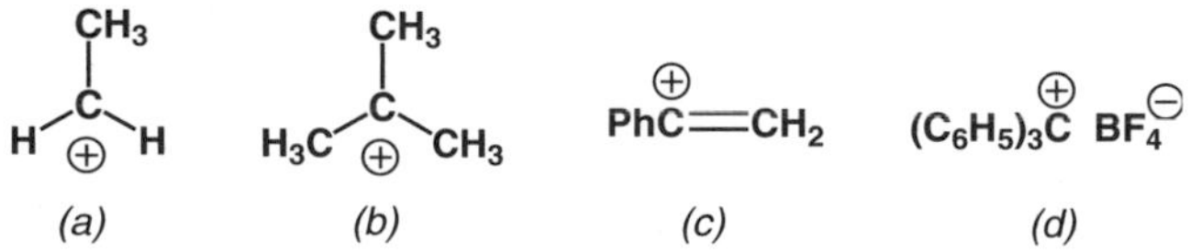

Figure 13.33. (*a*) Ethyl cation (methylcarbenium ion); (*b*) *t*-butyl cation (trimethylcarbenium ion); (*c*) phenylvinyl cation; (*d*) triphenylcarbenium tetrafluoroborate.

The effect of structure on the relative thermodynamic stabilities of carbenium ions can be demonstrated by comparing standard heat of formation data. Thus, $\Delta H_f^0(CH_3CH_2^+) - \Delta H_f^0\,(CH_3CH_3) = 240 \pm 4\,kcal\,mol^{-1}$ ($1{,}004 \pm 17\,kJ\,mol^{-1}$); ΔH_f^0 $[(CH_3)_3C^+) - \Delta H_f^0\,[(CH_3)_3\,CH = 202 \pm 4\,kcal\,mol^{-1}$ ($845 \pm 17\,kJ\,mol^{1}$). These data are in accord with the accepted generalization that tertiary carbenium ions are easier to form and are more stable than primary carbenium ions.

13.34 OTHER 2ND ROW -*ENIUM* IONS: NITRENIUM (R_2N^+), OXENIUM (RO^+), AND FLUORENIUM (F^+) IONS

-*Enium* ions in which the charged atoms are nitrogen, oxygen, and halogen, respectively.

Example. Diphenylnitrenium ion (Fig. 13.34*a*); methyloxenium ion (Fig. 13.34*b*); fluorenium ion (Fig. 13.34*c*). A set of nonbonding electrons in each of these three species may be either paired (singlet) or unpaired (triplet) and it is likely that the ground states, at least in *a* and *b*, are triplets. However, when such species are generated thermally, they are initially formed in the singlet states.

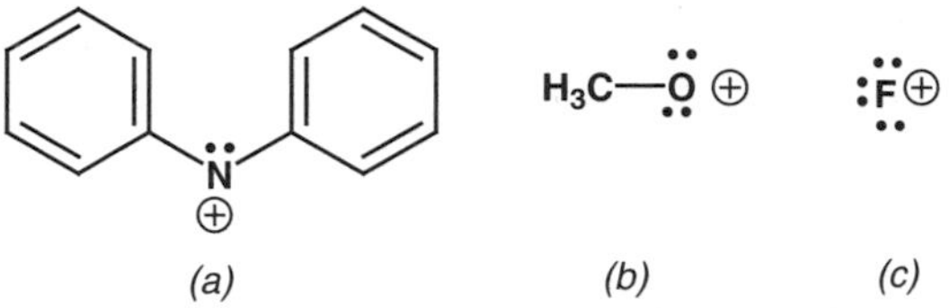

Figure 13.34. (*a*) Diphenylnitrenium ion; (*b*) methyloxenium ion; (*c*) fluorenium ion.

13.35 3RD ROW -*ENIUM* IONS: SILICENIUM (R_3Si^+), PHOSPHENIUM(R_2P^+), SULFENIUM (RS^+), AND CHLORENIUM (Cl^+) IONS

-*Enium* ions of the 3rd row elements silicon, phosphorus, sulfur, and chlorine, respectively, all of which are positively charged and electron-deficient.

Example. Methylsilicenium ion (Fig. 13.35*a*); dimethylphosphenium ion (dimethylphosphino cation, Fig. 13.35*b*); methylsulfenium ion (methylsulfanyl cation, Fig. 13.35*c*).

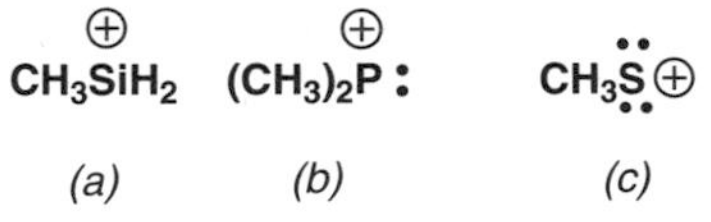

Figure 13.35. (*a*) Methysilicenium ion; (*b*) dimethylphosphenium ion (dimethylphosphino cation); (*c*) methylsulfenium ion (methylsulfanyl cation).

13.36 -*ONIUM* IONS

The *-onium* suffix (as distinguished from the *-enium* suffix denoting a positive charge and unsaturation) denotes a positively charged ion (*other than carbon or silicon*) in which the atom bearing the charge possesses a closed shell (rare gas) electronic configuration. In a few cases, the "on" in *-onium* is omitted, giving the suffix *-ium* that has the same meaning. The root name to which the suffix is attached reflects the parent from which the ions were generated.

Example. 1-Methylpyridinium ion (Fig. 13.36*a*); diphenylammonium ion (Fig. 13.36*b*); diethyloxonium ion (protonated diethyl ether, Fig. 13.36*c*). The methyl carbocation (Fig.13.28*a*) does not conform to this definition and hence should not be called methyl carbonium ion.

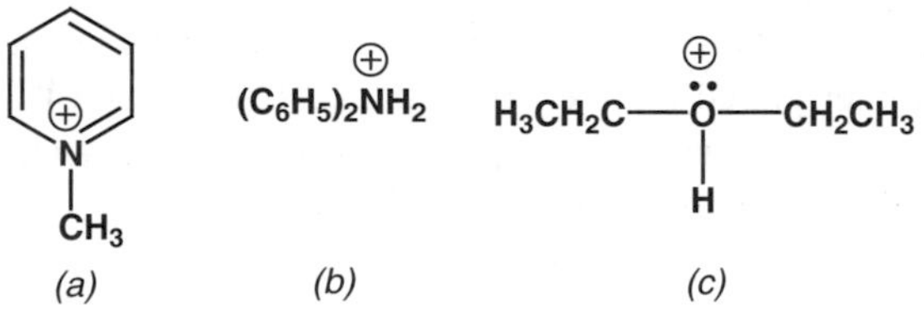

Figure 13.36. (*a*) 1-Methylpyridinium ion; (*b*) diphenylammonium ion; (*c*) diethyloxonium ion (protonated diethyl ether).

13.37 2ND ROW -*ONIUM* IONS: AMMONIUM (R_4N^+), OXONIUM (R_3O^+), AND FLUORONIUM (R_2 F^+) IONS

An *-onium* ion named so as to emphasize the identity of the charged heteroatom (nitrogen, oxygen, fluorine, etc.) and the groups attached to it. The *-onium* is usually attached as a suffix to a root name. R_4N^+ species as a class are called *quaternary ammonium* ions; such ions are frequently used as carriers to transport anions like HO^- from an aqueous phase to a hydrocarbon phase.

Example. Tetraethylammonium ion (Fig. 13.37*a*); triethyloxonium ion (Fig. 13.37*b*); and diphenylfluoronium ion (Fig. 13.37*c*). These ions all have eight electrons in the

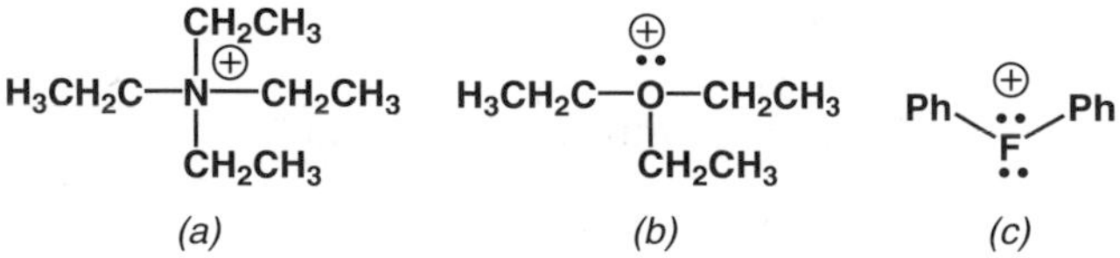

Figure 13.37. (*a*) Tetraethylammonium ion; (*b*) triethyloxonium ion; (*c*) diphenylfluoronium ion.

valence shell of the charged atom. Ph-$^+NH_3$ may be called either an anilinium ion or phenylammonium ion.

13.38 3RD ROW (AND HIGHER) -*ONIUM* IONS: PHOSPHONIUM (R_4P^+), SULFONIUM (R_3S^+), AND CHLORONIUM (R_2Cl^+) IONS

Onium ions of phosphorus, sulfur, and chlorine, respectively. Although positively charged, these 3rd row hetero atoms all are surrounded by an octet of valence electrons. The same nomenclature applies to appropriate higher-row elements as well.

Example. Trimethylphenylphosphonium ion (Fig. 13.38*a*); trimethylsulfonium ion (Fig. 13.38*b*); diphenylchloronium ion (Fig. 13.38*c*).

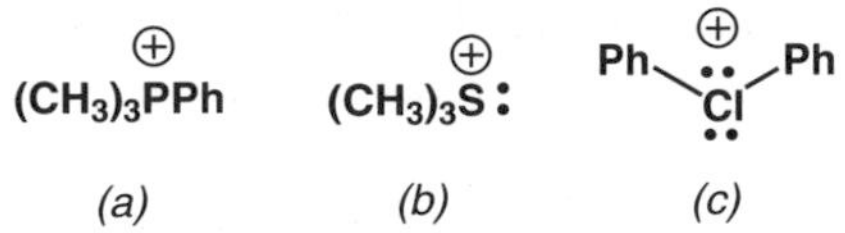

Figure 13.38. (*a*) Trimethylphenylphosphonium ion; (*b*) trimethylsulfonium ion; (*c*) diphenylchloronium ion.

13.39 DIAZONIUM ION, R−N$^+$≡N

An *-onium* ion in which the positive charge resides primarily on a tetracovalent nitrogen atom that is bonded to a carbon atom and also is part of a nitrogen–nitrogen triple bond.

Example. Phenyldiazonium ion (Fig. 13.39). It should be appreciated that the positive charge is partially delocalized over the second nitrogen atom as well, illustrated by the resonance structure shown in Fig. 13.39*b*. However, Fig. 13.39*a* represents a much more important contributing structure than Fig. 13.39*b*, because in the former both nitrogens are surrounded by an octet of electrons.

Figure 13.39. (*a, b*) Resonance forms of phenyldiazonium ion.

13.40 ACYLIUM ION, RC≡O$^+$

A generic name for an acyl group that bears a positive charge.

Example. Acetyl cation (Fig. 13.40).

Figure 13.40. Resonance forms of acetyl cation.

13.41 SILICONIUM ION, R_3Si^+

Commonly used synonymously with what should be called a silicenium ion (Sect. 13.35), analogous to the misuse of carbonium for carbenium.

13.42 AMINIUM ION, $R_3\dot{N}^+$

The radical cation of an amine produced by the loss of a single electron (oxidation) from a parent primary, secondary, or tertiary amine.

Example. These species are usually unstable, reactive intermediates. However, in the case of some of them derived from triarylamines, the radical cations are very stable, albeit reactive, and they can be stored indefinitely, see Fig. 13.50*b* and *c*.

13.43 IMINIUM ION, $R_2C{=}N^+RR'$

An *-ium* ion in which a positive charge resides primarily on a tetracovalent nitrogen atom connected to a carbon atom by a double bond. Iminium ions are intermediates in the Mannich reaction. Nucleophillic attack on the carbon atom of this intermediate can be rationalized on the basis of the resonance structure that places the positive charge on that carbon atom:

Example. Cyclohexaniminium ion (Fig. 13.43).

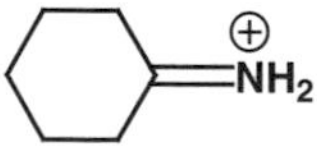

Figure 13.43. Cyclohexaniminium ion.

13.44 CARBANION

A generic name for a negatively charged ion with an even number of electrons in which the charge formally resides on one or more carbon atoms.

Example. Isopropyl anion (Fig. 13.44*a*) and allyl anion (Fig. 13.44*b*).

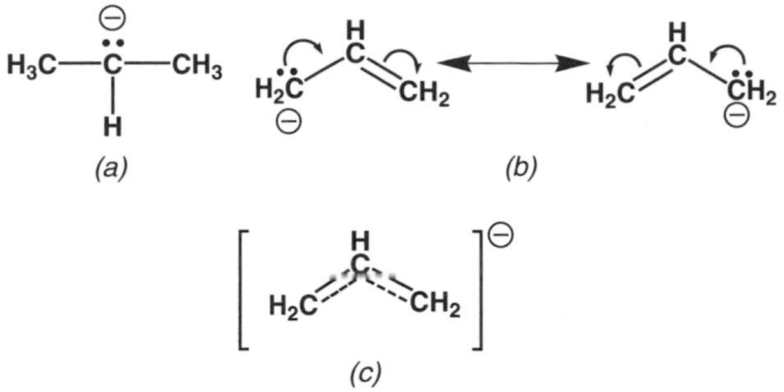

Figure 13.44. (*a*) Isopropyl anion; (*b*) allyl anion; (*c*) a single structure representing (*b*).

13.45 NITROGEN ANION (NITRANION)

A negatively charged ion with an even number of electrons in which the charge formally resides on a nitrogen atom. This anion may be regarded as being the result of removing a proton from an amine. Nitranion is a term that is not presently in use, but such a name would be convenient and unambiguous.

Example. Dimethylaminyl anion (Fig. 13.45). In a commonly used nomenclature certain compounds consisting of nitrogen anions are called amides, for example, sodium amide, $Na^+ \, {}^-NH_2$, and lithium dimethylamide, $Li^+ \, {}^-N(CH_3)_2$. Naming such compounds as amides is confusing in view of the fact that compounds having the structure $RC(O)NH_2$, as well as their derivatives, are called amides.

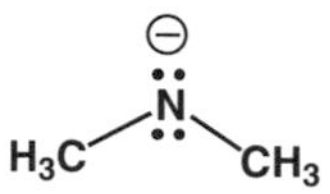

Figure 13.45. Dimethylaminyl anion.

13.46 OXYGEN ANION (OXYANION)

A negatively charged ion with an even number of electrons in which the charge formally resides on an oxygen atom. Oxyanion, like nitranion, is a term that is not presently in use, but such a name would be systematic and convenient.

Example. Methoxyl anion (Fig. 13.46*a*) and acetate anion (Fig. 13.46*b*).

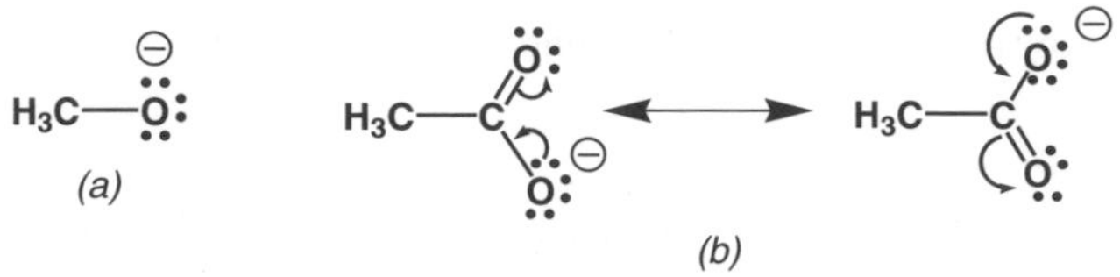

Figure 13.46. (*a*) Methoxyl anion and (*b*) acetate anion.

13.47 ALKOXIDES, M–OR

Salts in which the negatively charged oxygen in the anion is bonded to an alkyl group.

Example. Sodium methoxide, $NaOCH_3$, and aluminum isopropoxide, $Al[OCH–(CH_3)_2]_3$.

13.48 ENOLATE ANION

The delocalized anion remaining after removal of a proton from an enol or from the carbonyl compound in equilibrium with the enol.

Example. Acetone enolate anion (Fig. 13.48). It can be generated either by the loss of a proton from the methyl group adjacent to the carbonyl group or by loss of a proton from the –OH group of the tautomeric enol in equilibrium with the ketone.

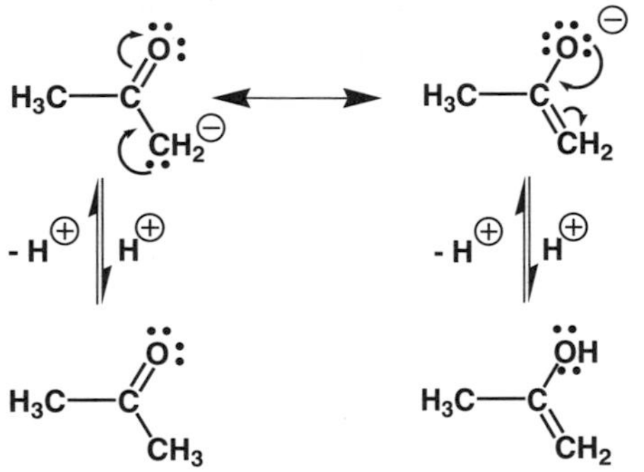

Figure 13.48. Keto-enol equilibrium via the acetone enolate anion.

13.49 RADICAL ION

An ion with an unpaired electron. A dot (•) with a charge (+ or −) on the appropriate atom or adjacent to the molecular formula or outside a bracket represents an ion with an odd electron.

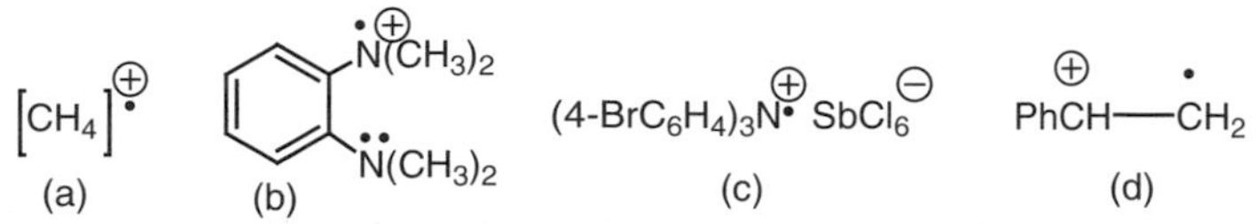

Figure 13.50. (*a*) Methane radical cation; (*b*) *N,N,N′N′*-tetramethyl-*o*-phenylenediamine radical cation; (*c*) tris(4-bromophenyl)aminium hexachloroantimonate (BAHA) is a commercially available compound (this and other radical cations derived from amines are commonly, but incorrectly according to Chemical Abstract usage, called aminium ions, e.g.); (*d*) styrene radical cation.

13.50 RADICAL CATION

A cation with an unpaired electron. A few radical cations (along with their counterions of course) are so stable that they can be bottled and sold commercially (e.g., Fig. 13.50*c*).

Example. Radical cations may be visualized as being formed by removing an electron from the neutral parent compound. Indeed, this is exactly what occurs in the mass spectrometer where a high-energy electron beam on collision with a neutral compound ejects an electron from it, giving a radical cation called the molecular ion, M^+, for example methane radical cation (Fig. 13.50*a*). *N,N,N′N′*-Tetramethyl-*o*-phenylenediamine radical cation (Fig. 13.50*b*) belongs to a class of deeply colored compounds of similar structure called *Wurster's salts*; tris(4-bromophenyl)aminium hexachloroantimonate (BAHA, Fig. 13.50*c*). The loss of one electron (oxidation) from an olefin gives rise to a radical cation where the plus charge is placed (arbitrarily) on one carbon of the double bond and the unpaired electron is placed on the other (Fig. 13.50*d*). Such radical cations serve as powerful dienophiles in the Diels–Alder reaction (**RCDA**) because they are electron-deficient.

13.51 RADICAL ANION

An anion with an unpaired electron. These anions may be visualized as being formed by adding an electron to an antibonding orbital of the parent neutral species. The donor species, usually Na or K, becomes the countercation.

Example. Radical anions (>C$^{\bullet}$–O$^-$) formed from the addition of an electron donated by Na or K to a carbonyl group are called *ketyls*. Such ketyl intermediates readily dimerize to pinacols. Radical anions formed by a similar reaction with esters are involved in the acyloin condensation. Aromatic hydrocarbons react with electron donors such as Na or K to give a radical anion (Fig. 13.51).

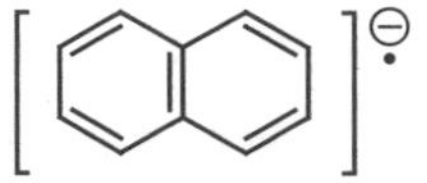

Figure 13.51. Naphthalene radical anion. The electron transferred from the donor (e.g., Na) occupies the lowest antibonding orbital of naphthalene.

13.52 ION PAIR, RADICAL PAIR

An ion pair is a general term used to describe a closely associated cation and anion that behave as a single unit. A radical pair is a general term used to describe two closely associated radicals that behave as a single unit.

13.53 TIGHT ION PAIR, TIGHT RADICAL PAIR [GEMINATE ION (RADICAL) PAIR, INTIMATE ION (RADICAL) PAIR, CONTACT ION (RADICAL) PAIR]

Alternate names for ion and radical pairs in which individual ions or radicals retain their stereochemical configuration. No solvent molecules separate the cation from the anion or the radicals from each other.

Example. An asymmetric carbenium ion with its counteranion (Fig. 13.53*a*) and a tight radical pair (Fig. 13.53*b*).

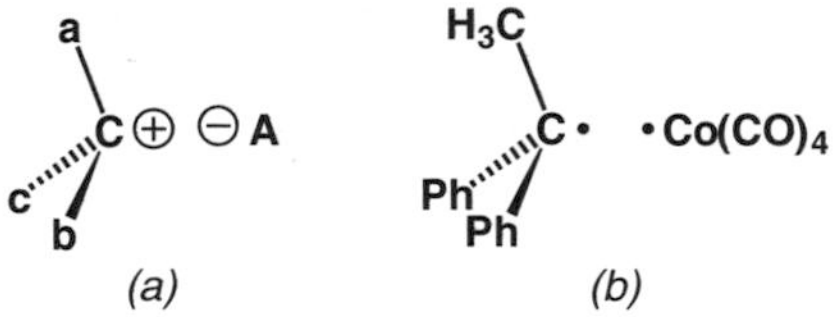

Figure 13.53. (*a*) A tight (geminate) ion pair and (*b*) a tight (geminate) radical pair. Geminate radical pairs are detected by chemically induced dynamic nuclear polarization (CIDNP) in the NMR spectrum.

13.54 SOLVENT-SEPARATED (OR LOOSE) ION PAIR

An ion pair in which the individual ions are separated by one or more solvent molecules. The ions may or may not retain their individual stereochemistry, depending on reaction conditions. The symbol generally used to represent this type of ion pair consists of two parallel vertical lines separating the two ions: $R^+||X^-$.

13.55 COUNTERION

An ion associated with another ion of opposite charge.

Example. In sodium acetate (Fig. 13.55), the sodium cation is the counterion of the carboxylate anion and *vice versa*.

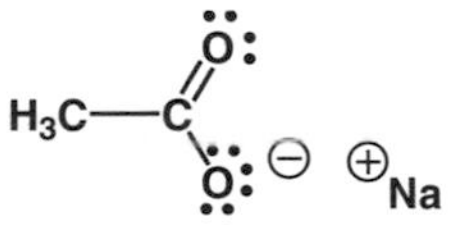

Figure 13.55. Sodium acetate.

13.56 GEGENION

From the German word meaning "counter"; synonymous with counterion.

13.57 ZWITTERION (DIPOLAR ION)

A net uncharged molecule that has separate cationic and anionic sites. In the case of amino acids, the dipolar ion can be visualized as arising from the transfer of a proton from the carboxyl group to the amino group.

Example. The zwitterion (dimethylammonio)acetate (Fig. 13.57*a*) in an aqueous solution is in equilibrium with the corresponding α-amino acid. The position of this equilibrium varies with the particular amino acid and is of biological significance.

(a) (b)

Figure 13.57. (*a*) The zwitterion (dimethylammonio)acetate and (*b*) *N,N*-dimethylaminoacetic acid.

13.58 BETAINE

Pronounced bayta-ene, a special class of zwitterions in which the positive site has no hydrogen atoms attached to it.

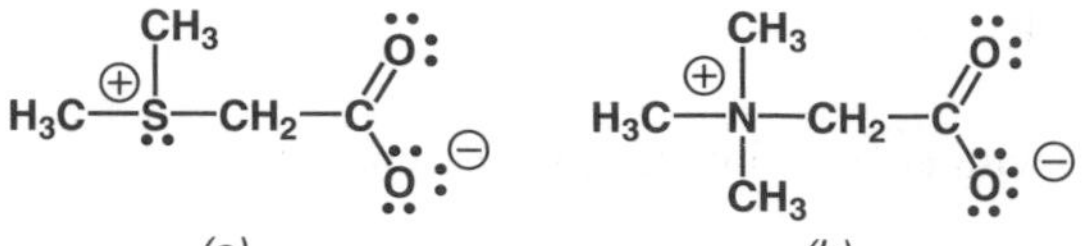

Figure 13.58. (*a*) (Dimethylsulfonio)acetate and (*b*) betaine [(trimethylammonio)acetate].

Example. (Dimethylsulfonio)acetate (Fig. 13.58*a*). (Trimethylammonio)acetate (Fig. 13.58*b*) has the common name betaine, which has now taken on a generic meaning.

13.59 YLIDE

A neutral species whose structure may be represented as a 1,2-dipole with a negatively charged carbon or other atom having a lone pair of electrons, bonded to a positively charged heteroatom. The term is usually preceded by the name of the hetero-atom.

Example. The phosphorus ylide methylenetriphenylphosphorane (Fig. 13.59*a*) and nitrogen ylide (trimethylammonio)methylide (Fig. 13.59*b*).

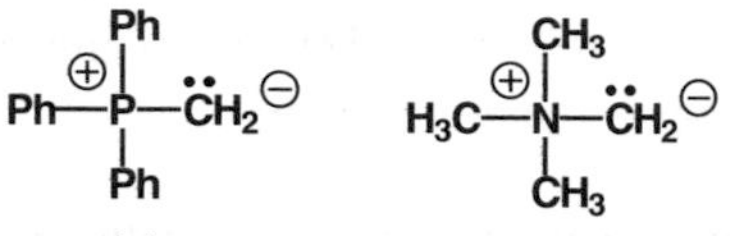

Figure 13.59. (*a*) The phosphorus ylide methylenetriphenylphosphorane and (*b*) the nitrogen ylide (trimethylammonio)methylide.

13.60 ARYNE

An uncharged species in which two adjacent atoms of an aromatic ring lack substituents, thus leaving two atomic orbitals (perpendicular to the aromatic π system) to form a weakly bonding molecular orbital occupied by two electrons.

Example. Singlet benzyne (Fig. 13.60*a*) and triplet 2,3-pyridyne (Fig. 13.60*b*). Although benzyne appears to have a triple bond, it is not an acetylene-type triple bond because the "third bond" is generated from σ sp^2 orbitals and not from $p\pi$ orbitals. Benzene may be considered as having six C–H σ bonds accounting for 12 electrons. If two hydrogen atoms, each with its one electron, are removed from adjacent carbons, there remain 10 σ electrons, and thus, the aryne is an electron-deficient species. Accordingly, arynes are electrophilic in character and react readily with nucleophiles and bases. In the preparation of arynes from aryl halide precursors,

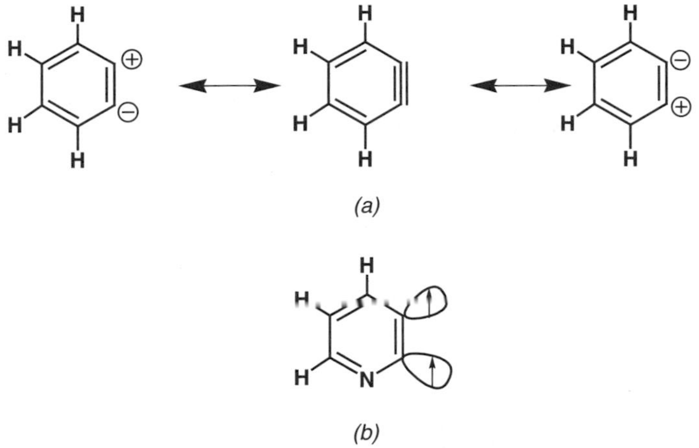

Figure 13.60. (*a*) Singlet benzyne and (*b*) triplet 2,3-pyridyne.

bases must therefore be absent. The first cycloaddition reaction of benzyne (by Georg Wittig in 1955) involved the following sequence:

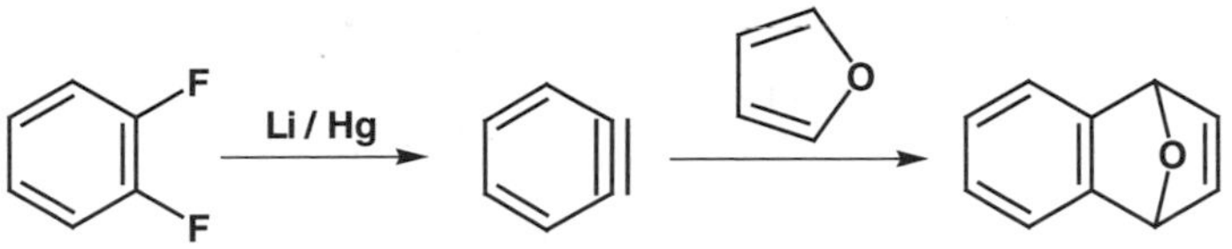

13.61 STRAINED RING SYSTEMS

This category includes a variety of ring system intermediates that owe their reactivity to steric constraints which are relieved by further reaction.

Example. Cyclobutadiene and cyclopentadienone are antiaromatic compounds (Sect. 3.27) as well as being strained ring systems, and they and other relatively small ring compounds having a *trans* double bond, such as *trans*-cycloheptene and *trans*-cyclooctene (Fig. 13.61), are very reactive. Cyclopropanones are intermediates in the Favorsky rearrangement.

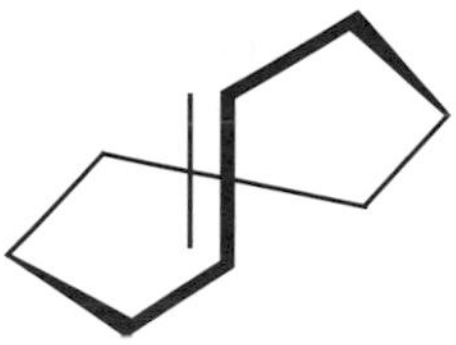

Figure 13.61. *trans*-cyclooctene.

13.62 TETRAHEDRAL INTERMEDIATES FROM CARBONYLS

The reaction of carbonyl groups at the sp^2 carbon, leading to the eventual substitution of one of the groups on the carbonyl by an incoming group, proceeds through a sp^3 intermediate formed by nucleophilic addition in the initial step.

Example. The hydrolysis of esters, amides, and acyl halides, ester exchange, and many related reactions. Most of these reactions can be catalyzed by either acid or base. The hydrolysis of an ester by hydroxide (Fig. 13.62) is typical.

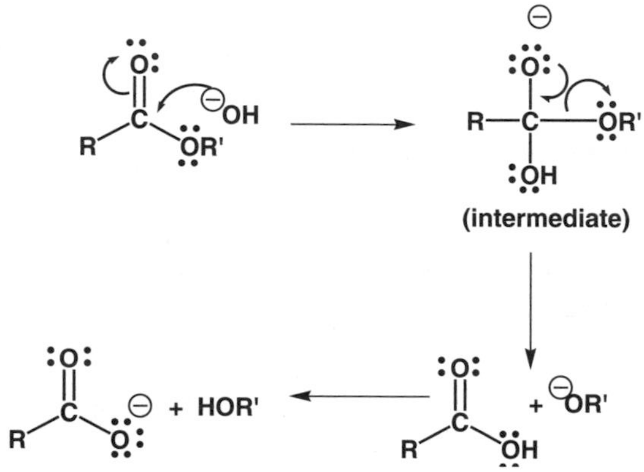

Figure 13.62. The hydrolysis of an ester with base.

13.63 SUMMARY TABLE OF INTERMEDIATES

Some of the information on the structure and names of intermediates described in this chapter is summarized in Table 13.63.

TABLE 13.63. Some Examples of Organic Intermediates of 2nd Row Elements

	Structures (Names)		
Species	C	N	O
-Onium ion	$CH_5^{\oplus}$ (carbonium ion or methonium)	$(CH_3)_4N^{\oplus}$ (tetramethyl-ammonium)	$(CH_3)_3\ddot{O}^{\oplus}$ (trimethyloxonium)
		$Ph{-}N^{\oplus}{\equiv}N:$ (phenyldiazonium)	

TABLE 13.63. (Contd.)

	Structures (Names)		
Species	C	N	O
-Enium ion	$(CH_3)_3\overset{\oplus}{C}$ (trimethylcarbenium)	$(CH_3)_2\overset{\oplus}{N}$ (dimethylnitrenium)	$CH_3O^{\oplus}$ (methyloxenium)
-Ene	$CH_2:$ (methylene carbene)	$CH_3\ddot{N}:$ (methylnitrene)	$:O:$ (oxene)
Anion	$Ph_3C:^{\ominus}$ (triphenylmethyl carbanion)	$(CH_3)_2\ddot{N}:^{\ominus}$ (dimethylaminyl anion)	$CH_3\ddot{\underset{..}{O}}:^{\ominus}$ (methoxyl anion)
Cation	$(CH_3)_3\overset{\oplus}{C}$ (trimethylcarbenium)	$(CH_3)_4\overset{\oplus}{N}$ (tetramethyl-ammonium)	$(CH_3)_3\overset{\oplus}{O}:$ (trimethyloxoniumm)
Radical cation	$Ph—\overset{\oplus}{C}H—\overset{\bullet}{C}H_2$ (styrene radical cation)	$Ph\overset{\bullet}{N}H_2^{\oplus}$ (aniline radical cation)	
Radical anion	$Ph—\overset{\ominus}{C}H—\overset{\bullet}{C}H_2$ (styrene radical anion)		
Ylide	$Ph_3\overset{\oplus}{P}—\overset{\ominus}{\ddot{C}}H_2$ methylene triphenylphosphorane ylide	$(H_3C)_3\overset{\oplus}{N}—\overset{\ominus}{\ddot{C}}H_2$ trimethylammonio ylide	
Aryne	 (benzyne)	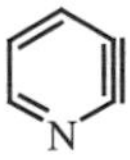 (2,3-pyridyne)	

SUGGESTED READING

Abramovitch, R. A. ed. *Reactive Intermediates*, Vols. 1–3. Plenum Press: New York, 1980, 1982, 1983.

Bellville, D. J., Wirth, D. N., and Bauld, N. L. *J. Am. Chem. Soc. 103*, 718 (1981); Bellville, J. and Bauld, N. L. *J. Am. Chem. Soc. 104*, 2665 (1982). On radical cations in the Diels–Alder reaction.

Leffler, J. E. *The Reactive Intermediates of Organic Chemistry*. Wiley-Interscience: New York, 1956. In this book, the author illustrates the benefits one can achieve by speculating on the role of intermediates. As Mark Twain once said (in *Life on the Mississippi*, Chap. XVII): "There is something fascinating about science. One gets such wholesale returns of conjecture out of such a trifling investment of fact."

Moody, C. J. and Whitham, G. H. *Reactive Intermediates*. Oxford University Press: New York, 1992.

Moss, R. A., Platz, M. S., and Jones, M. *Reactive Intermediate Chemistry*. Wiley-Interscience: New York, 2004.

Olah, G., Klopman, G., and Schlosberg, R. H. *J. Am. Chem. Soc. 91*, 3261 (1969). On methonium cation.

14 Types of Organic Reaction Mechanisms

14.1	Electron-Pushing (Arrow Pushing)	538
14.2	Homolytic Cleavage	538
14.3	Heterolytic Cleavage	538
14.4	IUPAC Symbols for Reaction Mechanisms	539
14.5	Nucleophile (Literally, Nucleus-Loving)	540
14.6	Nucleophilicity	540
14.7	Leaving Group	540
14.8	Nucleofuge	540
14.9	Electrophile (Literally, Electron-Loving)	541
14.10	Electrophilicity	541
14.11	Electrofuge	541
14.12	Reagent	541
14.13	Substrate	541
14.14	Substitution	542
14.15	Substitution Nucleophilic Bimolecular, S_N2 (A_ND_N)	542
14.16	Ambident Ion or Radical (Literally, Bothsides of the Tooth)	542
14.17	Substitution Nucleophilic Bimolecular with π Bond Rearrangement, S_N2'(3/1/ A_ND_N)	543
14.18	Substitution Bimolecular Nucleophilic via Addition-Elimination ($A_N + D_N$)	543
14.19	Substitution Nucleophilic Unimolecular, S_N1 (D_N * A_N)	544
14.20	Substitution Nucleophilic Unimolecular with π Bond Rearrangement, S_N1' (1/$D_N^{\neq}$ * 3/A_N)	544
14.21	Solvolysis (Literally, A Cleavage (Lysis) by Solvent)	545
14.22	Hydrolysis, Alcoholysis, and Aminolysis	545
14.23	Substitution Nucleophilic Internal, S_Ni [D_N (+D) + A_N]	545
14.24	Substitution Nucleophilic Internal with π Bond Rearrangement, S_Ni' [1/D_N (+D) + 3/A_N]	546
14.25	Neighboring Group Participation	546
14.26	Anchimeric Assistance	546
14.27	Substitution Electrophilic Unimolecular, S_E1 ($D_E + A_E$)	547
14.28	Substitution Electrophilic Bimolecular, S_E2 (D_EA_E)	548
14.29	Substitution Electrophilic Bimolecular with π Bond Rearrangement, S_E2' (3/1/A_ED_E)	548
14.30	Substitution Electrophilic Internal, S_Ei (cyclo-$D_EA_ED_NA_N$)	549
14.31	Substitution Electrophilic Internal with π Bond Rearrangement, S_Ei' (cyclo-1/3/ $D_EA_ED_NA_N$)	550

The Vocabulary and Concepts of Organic Chemistry, Second Edition, by Milton Orchin, Roger S. Macomber, Allan Pinhas, and R. Marshall Wilson

14.32	Elimination	550
14.33	α (or 1,1)-Elimination ($A_nD_E + D_N$)	550
14.34	β (or 1,2)-Elimination ($1/D_E2/D_N$)	550
14.35	Saytzeff (Zaitsev) Elimination	551
14.36	Hofmann Elimination	552
14.37	Bredt's Rule	552
14.38	*trans* (or *anti*) Elimination	552
14.39	*cis* (or *syn*) Elimination	553
14.40	Elimination Unimolecular, E1 ($D_N + D_E$)	553
14.41	Elimination Bimolecular, E2 ($A_nD_ED_N$)	554
14.42	Elimination Unimolecular Conjugate Base, El_{cb} ($A_nD_E + D_N$)	555
14.43	γ (or 1,3)-Elimination	555
14.44	Concerted Unimolecular Elimination (cyclo-DD)	556
14.45	Addition [A (+) A]	556
14.46	Nucleophilic Addition to Carbon–Carbon Multiple Bonds ($A_N + A_E$)	556
14.47	Conjugate Nucleophilic (or Michael) Addition	557
14.48	Electrophilic Addition to Carbon–Carbon Multiple Bonds ($A_E + A_N$)	558
14.49	Markovnikov's Rule	558
14.50	π Complex	559
14.51	σ Complex	559
14.52	*cis* Addition	560
14.53	*trans* Addition	560
14.54	Addition Electrophilic Bimolecular, Ad_E2 ($A_E + A_N$)	560
14.55	Substitution Electrophilic Aromatic, S_EAr ($A_E + D_E$)	561
14.56	*ortho* and *para* Director	562
14.57	*meta* Director	562
14.58	Substitution Nucleophilic Aromatic, S_NAr ($A_N + D_N$)	563
14.59	Benzyne (or Aryne) Mechanism	563
14.60	Chain Reaction	564
14.61	Chain-Initiating Step	565
14.62	Initiator	565
14.63	Chain-Propagating Steps	565
14.64	Chain Carriers	566
14.65	Chain-Terminating Step	566
14.66	Inhibitor	566
14.67	Substitution Homolytic Bimolecular, S_H2 ($A_rD_R + A_RD_r$)	566
14.68	Substitution Radical Nucleophilic Unimolecular, $S_{RN}1$ ($T + D_N + A_N$)	567
14.69	Free Radical Addition	567
14.70	Anti-Markovnikov Addition	567
14.71	Addition to the Carbonyl Group	568
14.72	Cram's Rule of Asymmetric Induction	570
14.73	Ester Hydrolysis	570
14.74	Saponification	570
14.75	Esterification	570
14.76	Acid-Catalyzed (A) Bimolecular (2) Hydrolysis with Acyl-Oxygen (AC) Cleavage, $A_{AC}2$ ($A_h + A_N + A_hD_h + D_N + D_h$)	571
14.77	Base-Catalyzed (B) Bimolecular (2) Hydrolysis with Acyl-Oxygen (AC) Cleavage, $B_{AC}2$ ($A_N + D_N$)	571

14.78 Acid-Catalyzed (A) Unimolecular (1) Hydrolysis with Alkyl-Oxygen (AL) Cleavage, $A_{AL}1$ ($A_h + A_N + D_N + D_h$) 572
14.79 Base-Promoted (B) Bimolecular (2) Ester Hydrolysis with Alkyl-Oxygen (AL) Cleavage, $B_{AL}2$ (A_ND_N) 572
14.80 Oxidation Number 573
14.81 Oxidation State 573
14.82 Oxidation and Reduction 573
14.83 Redox Reaction 574
14.84 Intramolecular and Intermolecular 574
14.85 Isomerization 574
14.86 Rearrangement 575
14.87 Degenerate Rearrangement 575
14.88 Pericyclic Reaction 575
14.89 Electrocyclic Reaction 575
14.90 Conrotatory and Disrotatory 576
14.91 Orbital Correlation Diagram 576
14.92 Principle of Conservation of Orbital Symmetry 578
14.93 State Correlation Diagram 578
14.94 Allowed Reaction 580
14.95 Forbidden Reaction 581
14.96 Cycloaddition 581
14.97 Cycloreversion 582
14.98 Retrograde Cycloaddition 582
14.99 Alder Rules 582
14.100 1,3-Dipole 583
14.101 1,3-Dipolar Addition 583
14.102 Sigmatropic Shifts 584
14.103 [1,*n*]-Sigmatropic Shift 584
14.104 [*m*,*n*]-Sigmatropic Shift ($m \neq 1$) 584
14.105 Cheletropic Reaction 586
14.106 Group Transfer Reaction 586
14.107 Suprafacial *s* and Antarafacial *a* 586
14.108 Woodward–Hoffmann Rules 586
14.109 Frontier-Orbital Approach 588
14.110 Hückel–Möbius (H-M) Approach 589
14.111 Baldwin's Rules 590

In Chapter 12, the elements of a reaction mechanism were described. Essentially, the mechanism of a reaction is a description of the sequence and energetics of bond-making and bond-breaking steps that leads from reactant(s) to product(s), and a description of all intermediates involved. A great advance in modern organic chemistry is the realization that all the thousands of known organic transformations can be partitioned among a relatively limited number of types of mechanisms. In this chapter, the most common types of organic reaction mechanisms are described.

14.1 ELECTRON-PUSHING (ARROW-PUSHING)

A formalism used to depict the making and breaking of bonds, that is, the movement of electrons. Movement of a single electron is shown by a curved arrow with a single barb (a "fishhook" arrow); movement of an electron pair is shown by a curved arrow with a double barb. Electron-pushing is used extensively in this and succeeding chapters to rationalize electronic reorganizations associated with chemical reactions, and is also useful in showing the electronic interconversions between resonance structures (Sects. 3.4 and 3.5). However, care must be taken to avoid attaching too much literal significance to the formalism. The uncertainty principle (Sect. 1.4) prevents one from knowing the exact location and identity of individual electrons. When depicting the movement of electrons, it is important to remember that the number of electrons around an atom can never exceed two (one pair) for H or eight (four pairs) for B, C, N, O, or F.

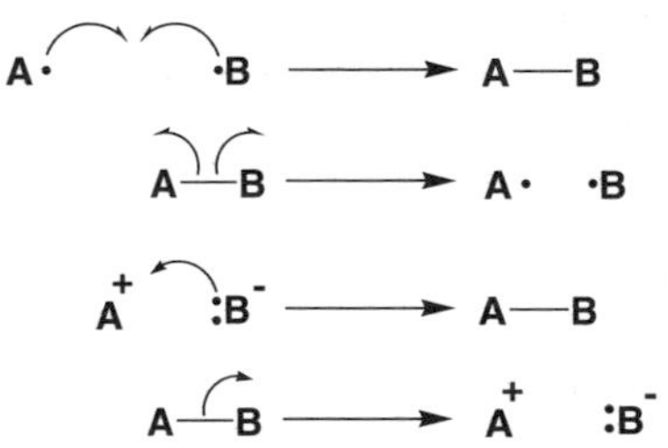

Figure 14.1. Bond-making and bond-breaking reactions.

14.2 HOMOLYTIC CLEAVAGE

The breakage of a single (two-electron) bond, leaving one of the electrons on each of the resulting fragments (Fig. 14.2). Thus, to show this cleavage requires the use of two single-barb arrows. *Homogenic bond formation* is the reverse of homolytic cleavage. The formal charge (Sect. 2.11) of an atom is unaffected by either of these processes.

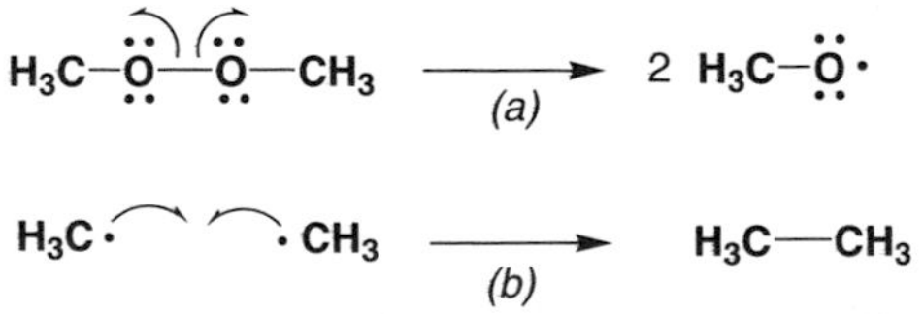

Figure 14.2. (*a*) Homolytic cleavage and (*b*) homogenic bond formation.

14.3 HETEROLYTIC CLEAVAGE

The breakage of a single (two-electron) bond, leaving both the electrons in a filled nonbonding orbital on one of the resulting fragments and an empty orbital on the other

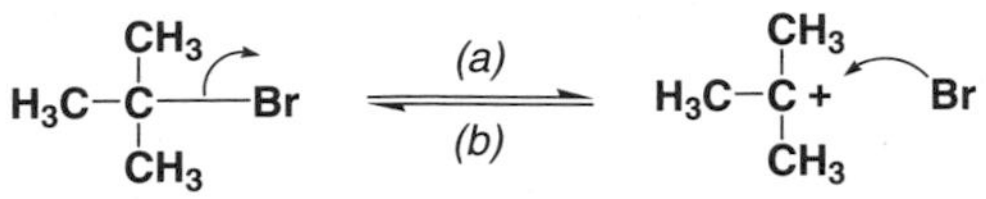

Figure 14.3. (*a*) Heterolytic cleavage and (*b*) heterogenic bond formation.

(Fig. 14.3). Thus, to show this cleavage requires the use of one double-barb arrow. The formal charge on the atom that receives the electrons becomes more negative by 1, while the formal charge of the atom losing the electrons becomes more positive by 1. *Heterogenic bond formation* is the reverse of heterolytic cleavage.

14.4 IUPAC SYMBOLS FOR REACTION MECHANISMS

The International Union of Pure and Applied Chemistry (IUPAC) has developed a system of symbols for labeling the steps in a reaction mechanism. The symbols used to abbreviate each step are taken from the list provided in Table 14.4.

TABLE 14.4. Symbols Used in the IUPAC System for Reaction Mechanisms

Symbol	Placement	Meaning
A	On line	Bond making (association)
D	On line	Bond breaking (dissociation)
+	On line	Stepwise process
*	On line	Same as +, except that the reaction intermediate is short-lived
E	Subscript	Electrophilic or electrofugic at core atom
N	Subscript	Nucleophilic or nucleofugic at core atom
R	Subscript	Homolytic at core atom
e	Subscript	Same as E, but at a peripheral atom
n	Subscript	Same as N, but at a peripheral atom
r	Subscript	Same as R, but at a peripheral atom
H	Subscript	Same as E, with H^+ as electrophile or electrofuge
h	Subscript	Same as H, but at a peripheral atom
xh	Subscript	Bond making or breaking between an H^+ and a base atom
C	On line	Diffusional combination
P	On line	Diffusional separation
int	Subscript	Molecules or ions that are weakly complexed
ss	Subscript	Solvent-separated ion pair or molecular equivalent
$\neq$	Superscript	Preceding step is rate-limiting
{ }	On line	Repeated sequence

(A core atom is directly at the reaction site; a peripheral atom is not directly at the reaction site.) For example, A_ND_N represents a one-step (concerted) mechanism consisting of one nucleophile making a bond to the reaction center at the same time a nucleofuge is departing the reaction center; see S_N2, Sect. 14.15. By contrast, $D_N * A_N$ represents a two-step (i.e., two transition states) mechanism that accomplishes the same net result; see S_N1, Sect. 14.19. See also Guthrie, R. D. and Jencks, W. P., *Accounts of Chemical Research* 22, 343 (1989).

14.5 NUCLEOPHILE (LITERALLY, NUCLEUS-LOVING)

An electron pair donor, that is, a Lewis base. Examples include any species with an unshared pair of electrons or a π bond.

Example. The hydroxide ion Figures 14.13 and 14.19.

14.6 NUCLEOPHILICITY

The relative (kinetic) reactivity of a nucleophile. In general, nucleophilicity is a function of the polarizability of the pair of electrons and, as such, increases with atomic number as one goes down a group in the Periodic Table. For nucleophiles with the same attacking atom, nucleophilicity increases with basicity and decreases with increasing solvation. Although any relative order of nucleophilicity depends on conditions and substrate, a typical order is $RS^- > R_3P > I^- > CN^- > R_3N > RO^- > Br^- > PhO^- > Cl^- > RCO_2^- > F^- > CH_3OH > H_2O$. (Also see "Hard/Soft Acid–Base Theory," Sect. 12.53.)

14.7 LEAVING GROUP

A generic term for the atom or group of atoms that departs during a substitution or elimination reaction. The departing group may be a nucleofuge (see Sect. 14.8) or an electrofuge (see Sect. 14.11).

14.8 NUCLEOFUGE

The leaving group in a nucleophilic substitution or elimination reaction that departs *with* the electron pair that originally bonded it to the remainder of the substrate molecule. The formal charge of the core atom of the nucleofuge decreases (becomes more negative) by 1 as it departs.

Example. The Cl^- in Figs. 14.13, 14.17, and 14.18, or the $PhSO_3^-$ in Fig. 14.19.

14.9 ELECTROPHILE (LITERALLY, ELECTRON-LOVING)

An electron pair acceptor, that is, a Lewis acid. Examples include any species with an empty valence orbital or with a highly polarizable σ bond or π bond, such as an alkyl halide or an α,β-unsaturated ketone.

Example. The carbocation in Figure 14.19.

14.10 ELECTROPHILICITY

The relative (kinetic) reactivity of an electrophile. The electrophilicity of a series of electrophiles generally increases as the Lewis acidity increases and as the solvation (or coordination to other molecules) decreases.

14.11 ELECTROFUGE

The leaving group in an electrophilic substitution or elimination reaction that departs *without* the electron pair which originally bonded it to the remainder of the substrate molecule. The formal charge of the core atom of the electrofuge increases (becomes more positive) by 1 as it departs.

Example. The $HgBr^+$ in Figs. 14.29 and 14.31, and the H^+ in Fig. 14.55.

14.12 REAGENT

The compound that serves as the source of the molecule, ion, or free radical which is arbitrarily regarded as the attacking species, that is, the species that causes the chemical change to occur in a reaction.

Example. Figure 14.13.

14.13 SUBSTRATE

The compound that arbitrarily is regarded as being attacked by the reagent. Usually, it is the transformation of the substrate that is the important part of the reaction because the substrate usually provides the majority of the carbon atoms of the product.

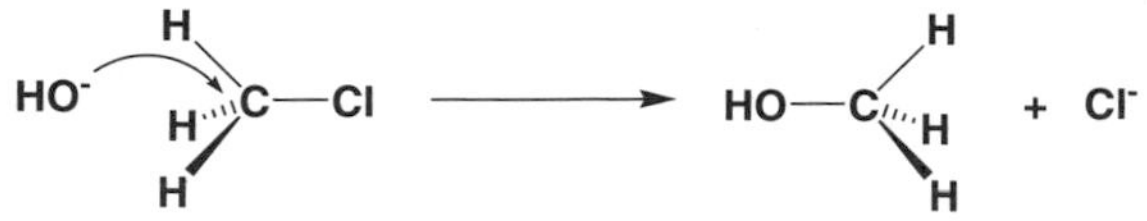

Figure 14.13. The reagent HO^- attacking the substrate H_3C–Cl.

14.14 SUBSTITUTION

A reaction in which an attacking species of one type (nucleophile, electrophile, or free radical) replaces another group of the same type, as, for example, the reaction shown in Fig. 14.13.

14.15 SUBSTITUTION NUCLEOPHILIC BIMOLECULAR, S_N2 (A_ND_N)

The concerted displacement of one nucleophile by another. The site of substitution is usually an sp^3 hybridized carbon. This mechanism involves stereospecific backside approach by the attacking nucleophile relative to the nucleofuge, causing inversion of configuration at the reaction site. The reaction is kinetically first-order in both attacking nucleophile and substrate, and thus second-order overall. The S_N2 reaction is highly sensitive to steric effects around the reaction center, and thus the order of substrate reactivity is methyl > primary > secondary >> tertiary.

Example. The reaction shown in Figure 14.13.

14.16 AMBIDENT ION OR RADICAL (LITERALLY, BOTH SIDES OF THE TOOTH)

A species with two or more possible attacking sites (teeth), or a species with two or more possible sites of attack.

Example. Figure 14.16; the arrows indicate the possible reactive sites.

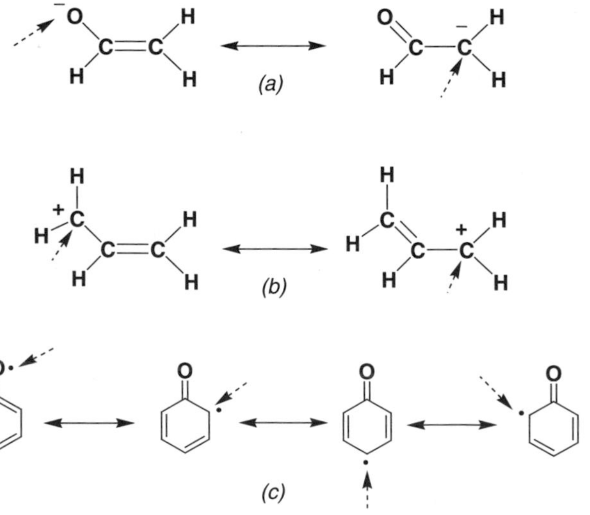

Figure 14.16. Ambident species: (*a*) nucleophile; (*b*) electrophile; (*c*) radical.

14.17 SUBSTITUTION NUCLEOPHILIC BIMOLECULAR WITH π BOND REARRANGEMENT, S_N2' ($3/1/A_ND_N$)

A concerted nucleophilic displacement in which the site of nucleophilic attack is at an atom other than the original point of attachment of the nucleofuge (usually, one multiple bond separated from the original point of attachment). This mechanism is kinetically indistinguishable from the S_N2 reaction, but the attack/departure stereochemistry is *cis*.

Example. Figure 14.17.

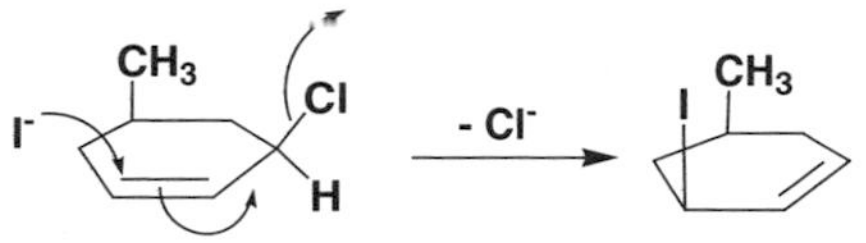

Figure 14.17. An S_N2' reaction.

14.18 SUBSTITUTION BIMOLECULAR NUCLEOPHILIC VIA ADDITION-ELIMINATION ($A_N + D_N$)

A multistep nucleophilic interchange normally involving a carbanion intermediate in which bond making precedes bond breaking. This mechanism occurs at unsaturated carbon or saturated 3rd row elements (Si, P, S), but it is unknown at saturated carbon. For this mechanism to occur at an olefinic carbon, the carbanion intermediate must be stabilized by electron-withdrawing (ewg) substituents. This mechanism is kinetically indistinguishable from the S_N2 mechanism under conditions where the intermediate is highly reactive. The stereochemistry of the reaction depends on the details of how the nucleofuge departs from the intermediate.

Example. Substitution at carbon is shown in Fig. 14.18*a* and substitution at silicon in Fig. 14.18*b*. The trigonal bipyramidal intermediate in the silicon substitution reaction has a geometry which is similar to that for the transition state of an S_N2 substitution reaction at carbon.

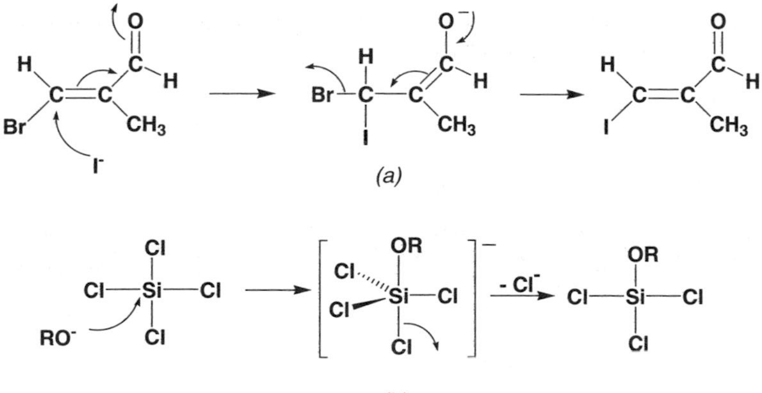

Figure 14.18. (*a*) Substitution at carbon; and (*b*) substitution at silicon.

14.19 SUBSTITUTION NUCLEOPHILIC UNIMOLECULAR, S_N1 (D_N * A_N)

A multistep nucleophilic interchange with bond breaking (the rate-limiting step) preceding bond making. This mechanism normally involves a carbocation intermediate and therefore requires a polar solvent. The reaction is first-order in substrate, but zero-order in attacking nucleophile. Attack by the nucleophile on the carbocation can usually occur from either side, leading to racemic product. Only substrates that can lead to relatively stable carbocations react by this mechanism, the relative reactivity being tertiary (or resonance-stabilized ions such as benzyl or allyl) >> secondary >> primary > methyl.

Example. A typical reaction is shown in Fig. 14.19.

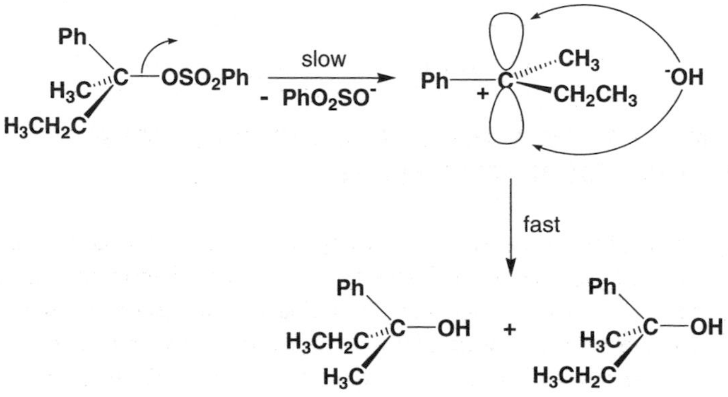

Figure 14.19. An S_N1 reaction.

14.20 SUBSTITUTION NUCLEOPHILIC UNIMOLECULAR WITH π BOND REARRANGEMENT, S_N1' ($1/D_N^{\neq}$ * $3/A_N$)

Kinetically and stereochemically similar to the S_N1 mechanism, except that the attacking nucleophile becomes attached to a different atom than the one originally attached to the nucleofuge. Either a carbenium ion rearrangement or an allylic ion rearrangement is often involved in this mechanism.

Example. Figure 14.20.

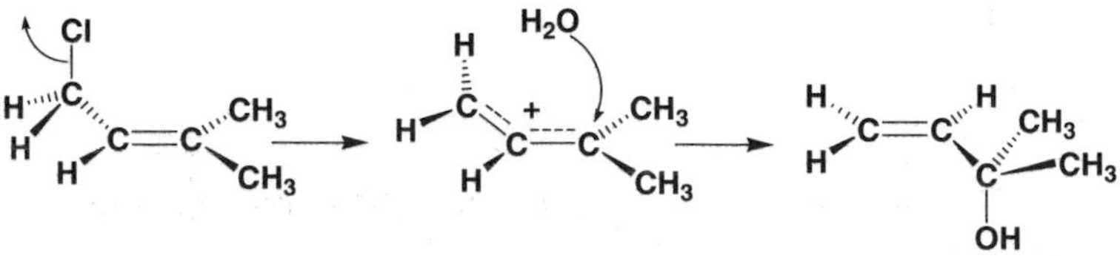

Figure 14.20. An S_N1' reaction.

14.21 SOLVOLYSIS [LITERALLY, A CLEAVAGE (LYSIS) BY SOLVENT]

A nucleophilic substitution in which the solvent serves as the attacking nucleophile. Even if the reaction is bimolecular (e.g., S_N2), its kinetic order in the solvent is indistinguishable from zero because the solvent is present in great excess (see Sect. 12.78).

Example. Figure 14.21.

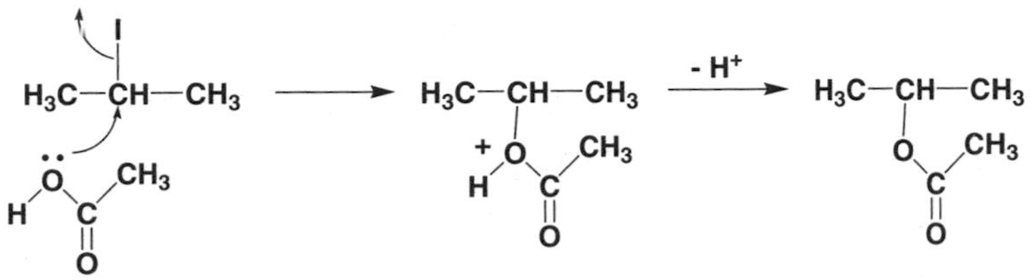

Figure 14.21. An S_N2 solvolysis with acetic acid (acetolysis).

14.22 HYDROLYSIS, ALCOHOLYSIS, AND AMINOLYSIS

A solvolysis in which water, an alcohol, or an amine is the solvent, respectively.

Example. Figure 14.22.

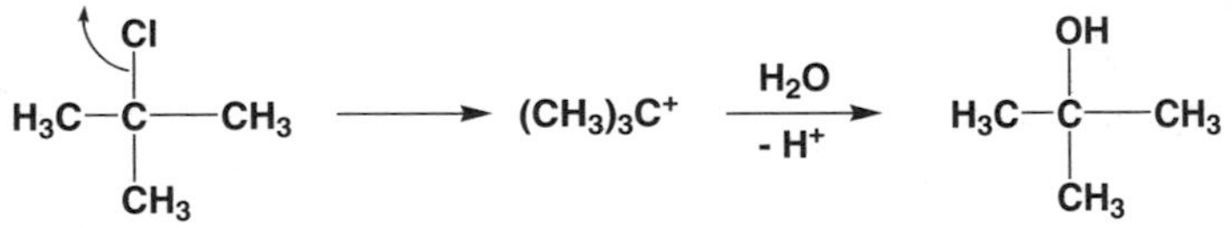

Figure 14.22. Hydrolysis by an S_N1 mechanism.

14.23 SUBSTITUTION NUCLEOPHILIC INTERNAL, S_Ni [D_N (+D) + A_N]

A stepwise intramolecular (unimolecular) nucleophilic interchange in which the attacking nucleophile is part of the original substrate. The reaction resembles an S_N1 process, except that introduction of the nucleophile occurs with retention of configuration at the reaction site. Ion pairs are often implicated in this type of mechanism.

Example. Figure 14.23.

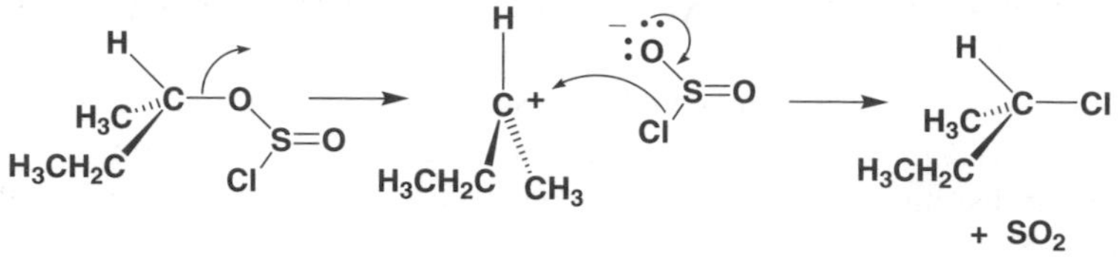

Figure 14.23. An S_Ni reaction.

14.24 SUBSTITUTION NUCLEOPHILIC INTERNAL WITH π BOND REARRANGEMENT, S_Ni' [1/D$_N$ (+D) + 3/A$_N$]

Similar to the S_Ni mechanism, except that the internal nucleophile becomes attached to an atom different from (e.g., allylic) the original point of attachment. The stereochemical path involves a *cis* departure/attack relationship.

Example. Figure 14.24.

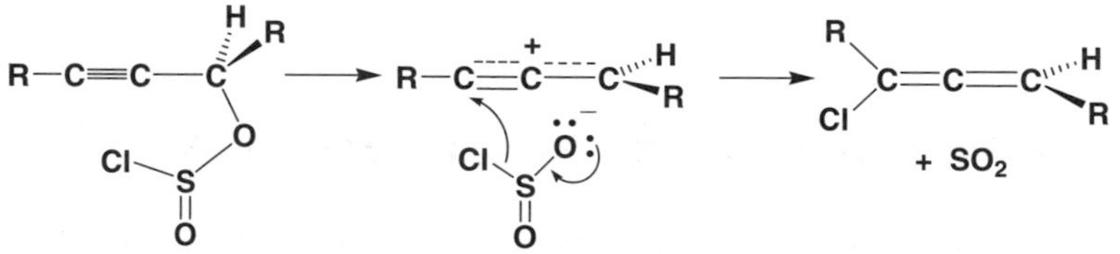

Figure 14.24. An S_Ni' reaction.

14.25 NEIGHBORING GROUP PARTICIPATION

The intramolecular involvement of one functional group (via its *n*, π, or σ electrons) in the reaction at another functional group. The term is most commonly encountered in reactions involving carbocations, where the neighboring group shares a pair of electrons with the positively charged carbon. Such interactions can take place before, during, or after the rate-determining step, and can involve retention of configuration (due to a double inversion) in the unrearranged products. Rearranged products, if formed as a result of participation, can involve inversion at both the carbenium carbon and the neighboring-group-bearing atom.

Example. Figures 14.25*a*, *b*, and *c* show *n*, π, and σ electron participation, respectively.

14.26 ANCHIMERIC ASSISTANCE

From the Greek words *anchi* and *meros*, meaning "Neighboring Parts," neighboring group participation in the rate-determining step of a reaction. This term is most often encountered in reactions involving carbocation intermediates, where *neighboring group participation* occurs in the rate-limiting (ionization) step that generates this

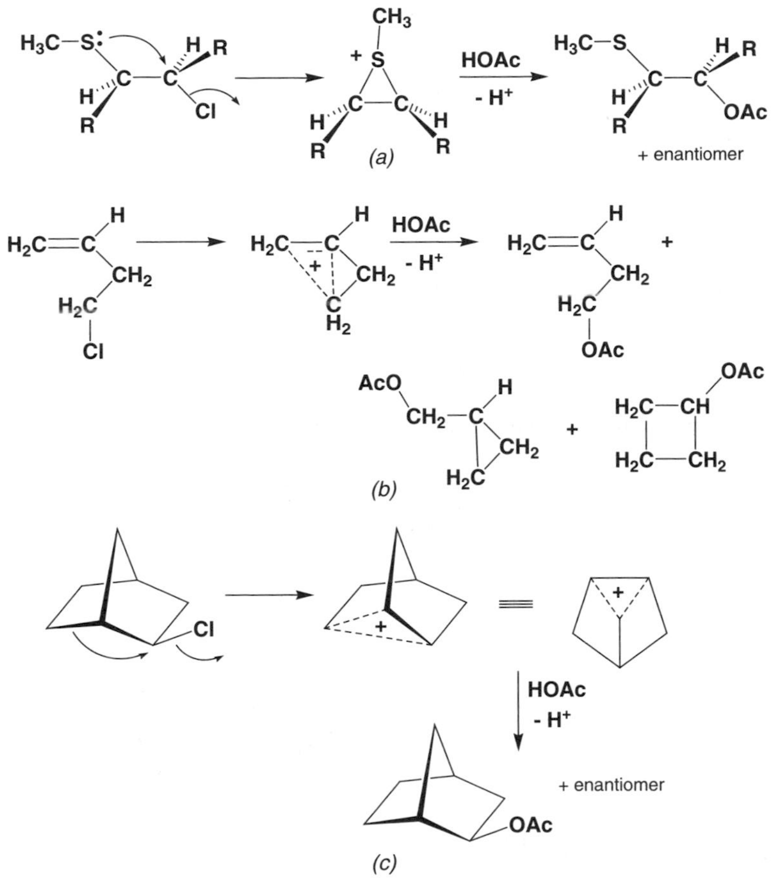

Figure 14.25. Neighboring group participation: (*a*) lone pair; (*b*) π electron; (*c*) σ electron participation.

ion. Such assistance causes an increase in the reaction rate compared to a model reaction in which participation is absent.

Example. Figure 14.26.

14.27 SUBSTITUTION ELECTROPHILIC UNIMOLECULAR, S_E1 ($D_E + A_E$)

A multistep electrophile interchange with bond breaking (the rate-limiting step) preceding bond making. The reaction is first-order in substrate and zero-order in attacking electrophile. The intermediate is a carbanion, which generally inverts more rapidly than it is attacked by the electrophile, leading to racemization (or epimerization) if the reaction site is a chiral center. The most common examples of the S_E1 mechanism

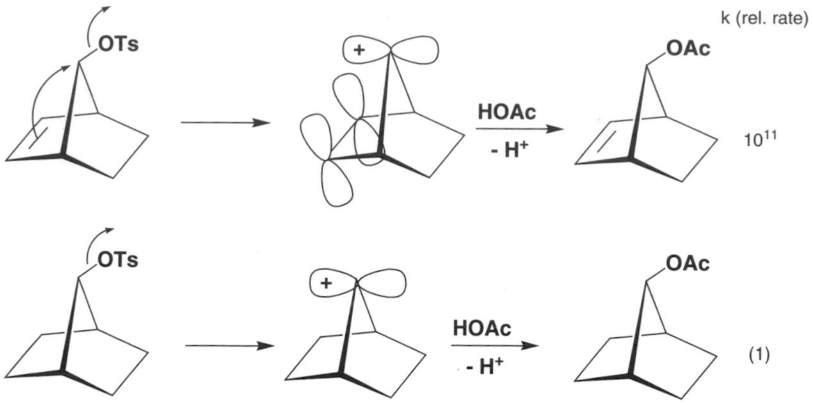

Figure 14.26. Anchimeric assistance and its effect on reaction rate.

involve removal of H^+ attached to carbon by reaction with a strong base ($A_{xh}D_h + A_E$). S_E1 is the electrophile analog of the S_N1 mechanism.

Example. Figure 14.27.

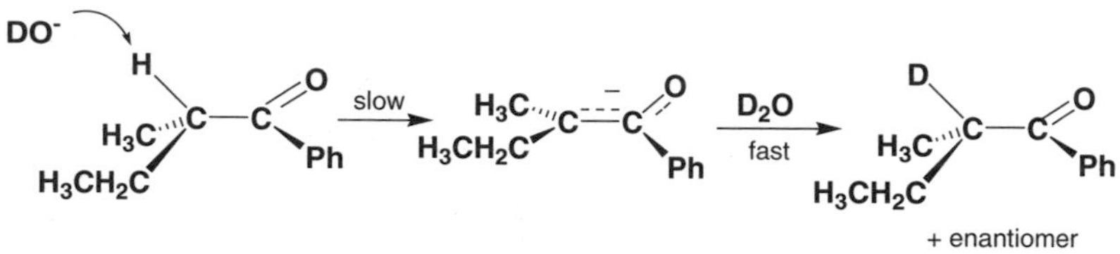

Figure 14.27. An S_E1 reaction (racemization).

14.28 SUBSTITUTION ELECTROPHILIC BIMOLECULAR, S_E2 (D_EA_E)

The concerted substitution of one electrophile by another. This mechanism, which is relatively uncommon, involves stereospecific frontside approach by the attacking electrophile relative to the electrofuge, causing retention of configuration at the reaction center. The reaction is kinetically first-order in both substrate and attacking electrophile and, thus, second-order overall. S_E2 is the electrophile analog of the S_N2 mechanism.

Example. Figure 14.28.

14.29 SUBSTITUTION ELECTROPHILIC BIMOLECULAR WITH π BOND REARRANGEMENT, S_E2' ($3/1/A_ED_E$)

A mechanism kinetically identical to the S_E2 process, but involving formation of a rearranged product.

Example. Figure 14.29.

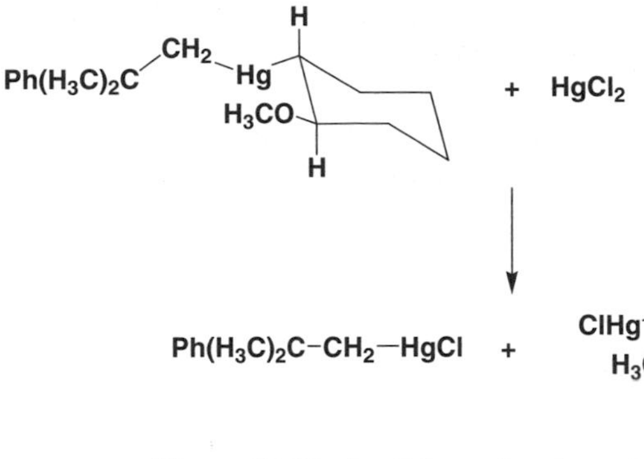

Figure 14.28. An S_E2 reaction (retention).

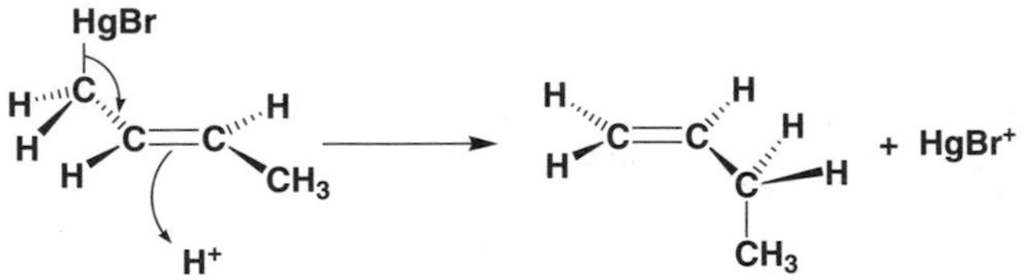

Figure 14.29. An S_E2' reaction (rearrangement).

14.30 SUBSTITUTION ELECTROPHILIC INTERNAL, S_Ei (CYCLO-$D_EA_ED_NA_N$)

A concerted four-center electrophilic interchange, kinetically and stereochemically indistinguishable from the S_E2, except that at the transition state the nucleophile associated with the attacking electrophile becomes coordinated to the electrofuge in a cyclic transition state.

Example. Figure 14.30.

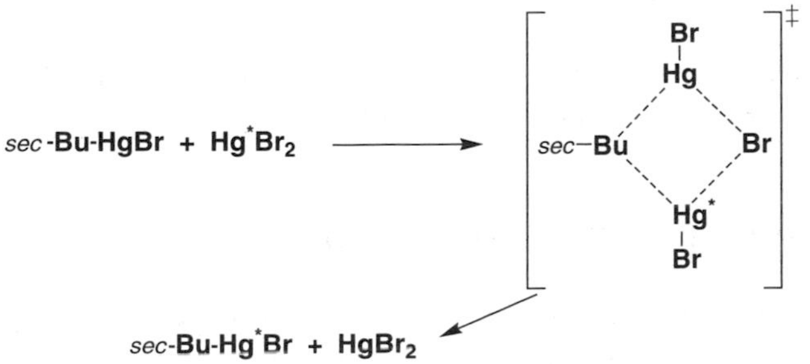

Figure 14.30. An S_Ei reaction (internal, no rearrangement).

14.31 SUBSTITUTION ELECTROPHILIC INTERNAL WITH π BOND REARRANGEMENT, S_Ei' (CYCLO-1/3/$D_EA_ED_NA_N$)

A concerted electrophile interchange, kinetically identical to the S_Ei, but involving a multicenter transition state and formation of rearranged product.

Example. Figure 14.31.

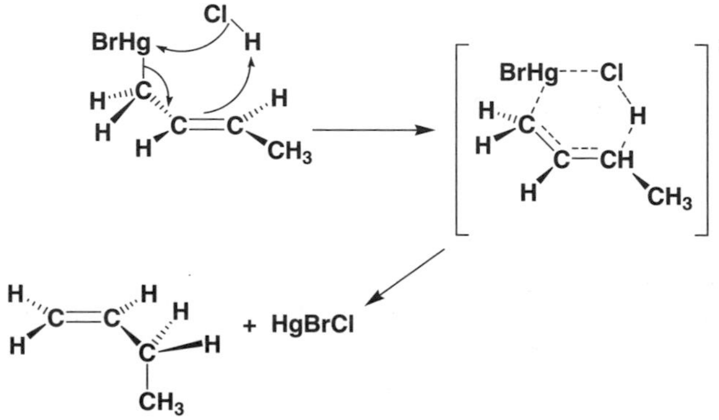

Figure 14.31. An S_Ei' reaction (rearrangement).

14.32 ELIMINATION

A reaction that involves a loss of two groups or atoms from the substrate and, therefore, an increase in the degree of unsaturation of the substrate. See the examples below.

14.33 α (OR 1,1)-ELIMINATION ($A_nD_E + D_N$)

Loss of two atoms or groups from the same atom leading to a hypovalent neutral species (e.g., carbene, nitrene). The reaction generally follows an $E1_{cb}$ mechanism (Sect. 14.42), except when N_2 is the leaving group.

Example. Figure 14.33*a* and *b*. There are certain rare cases in which loss of the nucleofuge precedes loss of the electrofuge (Fig. 14.33).

14.34 β (OR 1,2)-ELIMINATION (1/D_E2/D_N)

An elimination reaction in which the two leaving groups (usually, an electrofuge and a nucleofuge) are originally attached to adjacent atoms in the substrate.

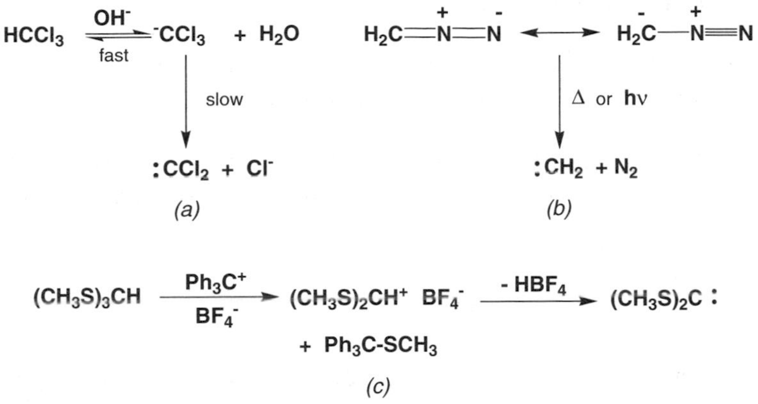

Figure 14.33. α-Elimination to form carbenes: (*a*) via carbanion (*b*) by direct loss of nitrogen and (*c*) via carbenium ion.

Their departure causes formation of a new π bond between the two adjacent atoms.

Example. Figure 14.34 shows a β-elimination that involves dehydrohalogenation.

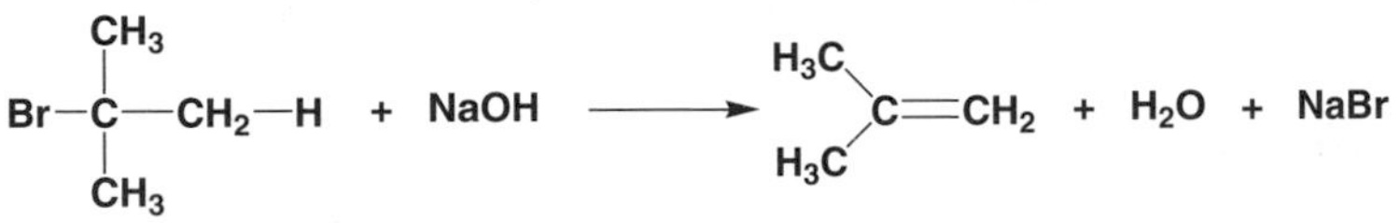

Figure 14.34. A β-elimination to dehydrohalogenated product.

14.35 SAYTZEFF (ZAITSEV) ELIMINATION

Named after Alexander M. Saytzeff, (1841–1910), a 1,2-elimination that results in the most substituted (and hence, most stable) alkene product.

Example. Figure 14.35 shows a dehydrohalogenation that characteristically leads to predominant Saytzeff product *(a)*.

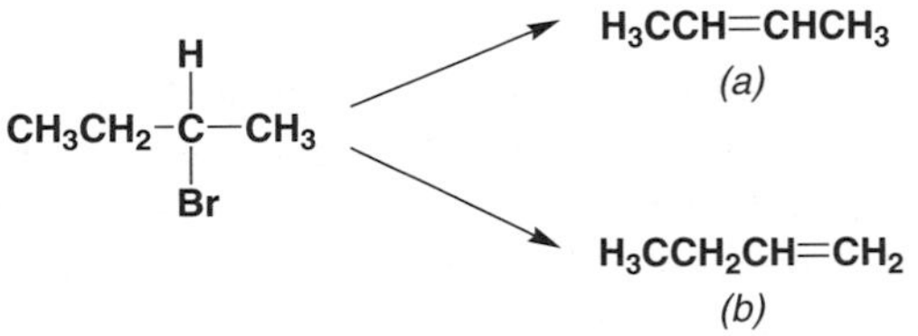

Figure 14.35. A β-elimination to: *(a)* Saytzeff product and *(b)* Hofmann product.

14.36 HOFMANN ELIMINATION

Named after A. W. Hofmann (1818–1892), a kinetically controlled (Sect. 12.86) 1,2-elimination that results in the least substituted alkene product.

Example. The β-elimination achieved by pyrolysis of quaternary ammonium hydroxides (Figs. 14.36*a* and *b*) characteristically leads to almost exclusive Hofmann regiochemistry.

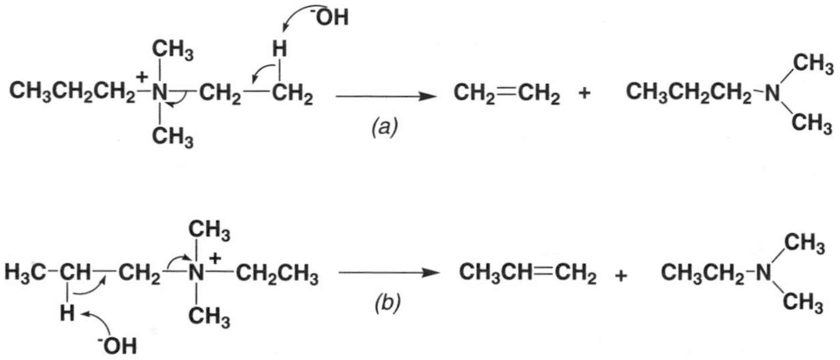

Figure 14.36. A Hofmann elimination: (*a*) less substituted olefin (Hofmann regiochemistry); (*b*) the more substituted olefin (Saytzeff regiochemistry) is not formed.

14.37 BREDT'S RULE

Named for Julius Bredt (1855–1937), this rule specifies that when a 1,2-elimination in a bicyclic system with $x \geq 1$ occurs, the bridgehead atoms are not involved unless the ring bearing the incipient *trans* double bond has at least eight atoms, that is, when $n + m > 4$ (Fig. 14.37).

Example. Figure 14.37. When $n + m > 4$, the bridgehead carbon can be involved; however, bridgehead double bonds in molecules with four or fewer carbon atoms between the bridgehead atoms ($n + m \leq 4$) are too strained and reactive to be isolated.

14.38 *TRANS* (OR *ANTI*) ELIMINATION

A description of the stereochemistry of a 1,2-elimination in which the two leaving groups depart from opposite sides of the incipient π bond. Other than the stereochemical result, no other mechanistic information is necessarily implied. Note that *trans* elimination may give *cis* or *trans* (*Z* or *E*) product, depending on the conformer (or stereoisomer) of the starting material.

Example. Figure 14.38.

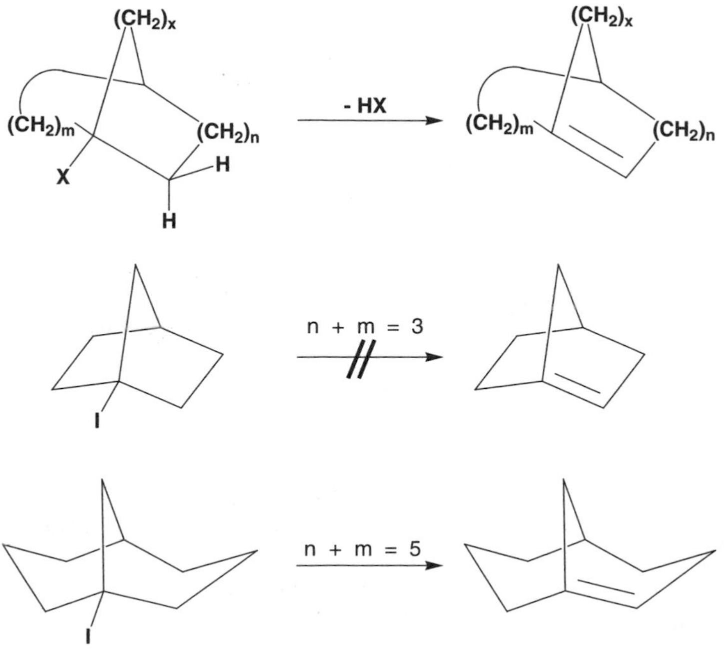

Figure 14.37. Application of Bredt's rule.

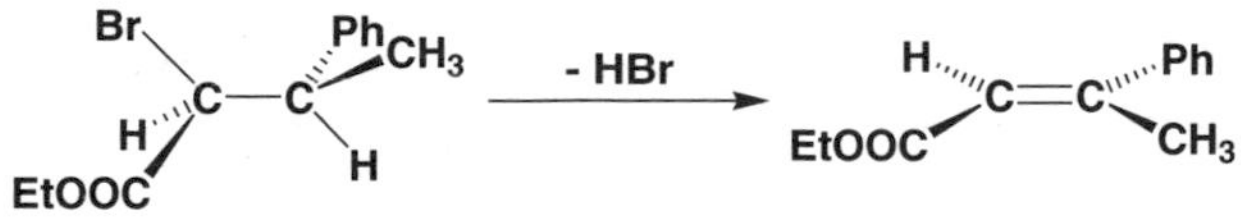

Figure 14.38. A *trans* elimination.

14.39 *CIS* (OR *SYN*) ELIMINATION

The stereochemistry of a 1,2-elimination in which the two leaving groups depart from the same side of the incipient double bond. No other mechanistic information necessarily is implied. Note that *cis* elimination may give *cis* or *trans* (*Z* or *E*) product, depending on the conformer (or stereoisomer) of the starting material.

Example. Figures 10.39*a* and *b*; part *b* shows acetate pyrolysis leading to the *Z* isomer. (In this reaction, some of the *E* isomer is formed as well.)

14.40 ELIMINATION UNIMOLECULAR, E1 ($D_N + D_E$)

A multistep 1,2-elimination mechanism in which the nucleofuge is lost in the first (rate-limiting) step and the electrofuge is lost in the second step. The intermediate is

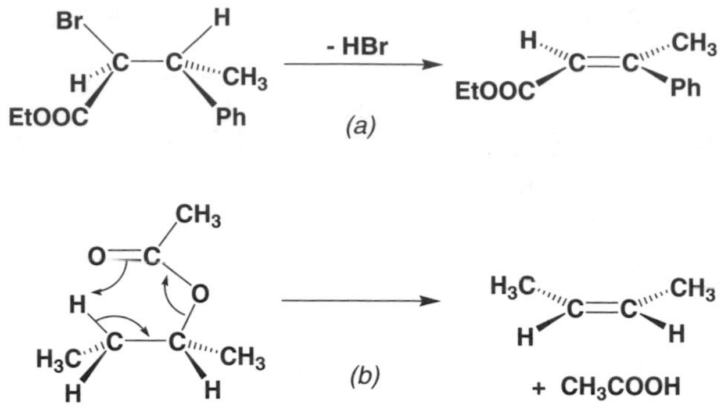

Figure 14.39. A *cis* elimination: (*a*) dehydrohalogenation and (*b*) acetate pyrolysis.

a carbocation, and this mechanism often competes with S_N1 processes. Saytzeff regiochemistry is generally observed. The stereochemistry of the elimination depends on the exact reaction conditions; often, competitive *cis* and *trans* elimination is observed. Since the first step is rate-limiting, the elimination/substitution ratio (E1/S_N1) is independent of the nature of the leaving group.

Example. Figures 14.40*a*, *b*, and *c* show three possible products from a common intermediate.

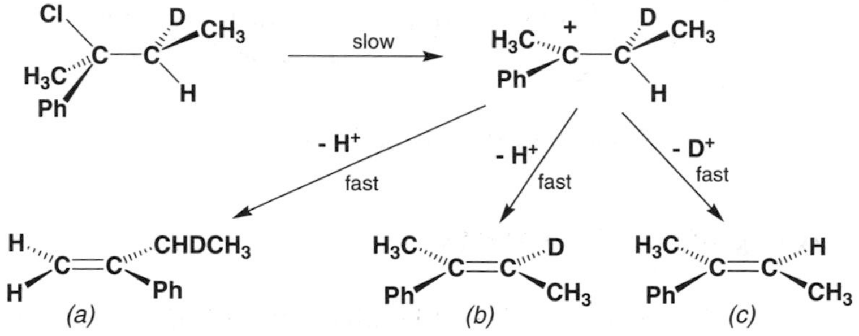

Figure 14.40. An E1 elimination: (*a*) Hofmann product; (*b*) Saytzeff elimination (*trans*); (*c*) Saytzeff elimination (*cis*).

14.41 ELIMINATION BIMOLECULAR, E2 ($A_nD_ED_N$)

A concerted 1,2-elimination that is first-order in both substrate and base. *trans* Elimination is generally favored. Because the attacking base is sensitive to the steric environment around the electrophile, Hofmann orientation is generally preferred. The E2 mechanism often competes with the S_N2 reactions.

Example. Figure 14.41.

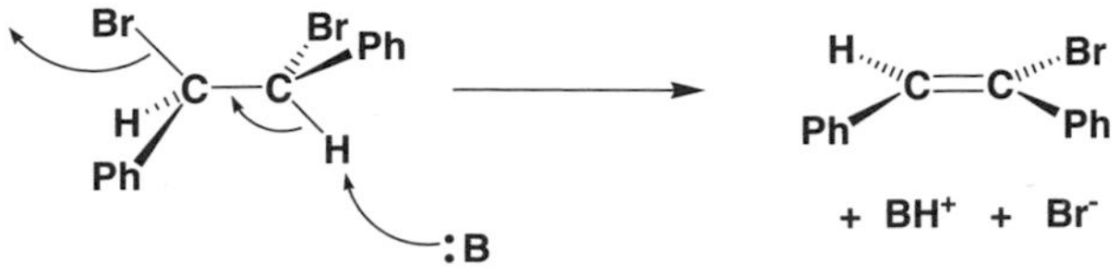

Figure 14.41. An E2 reaction.

14.42 ELIMINATION UNIMOLECULAR CONJUGATE BASE, E1$_{CB}$ (A$_n$D$_E$ + D$_N$)

A multistep elimination mechanism in which loss of the electrofuge (usually, H^+) occurs before loss of the nucleofuge. This mechanism involves a reversibly formed carbanion intermediate, the formation of which requires the presence of an electron-withdrawing substituent on the carbanion carbon. Usually, the reaction is first-order in both substrate and base, hence second-order overall. Hofmann regiochemistry is generally preferred; thc elimination stereochemistry may be *cis*, *trans*, or a combination.

Example. Figure 14.42.

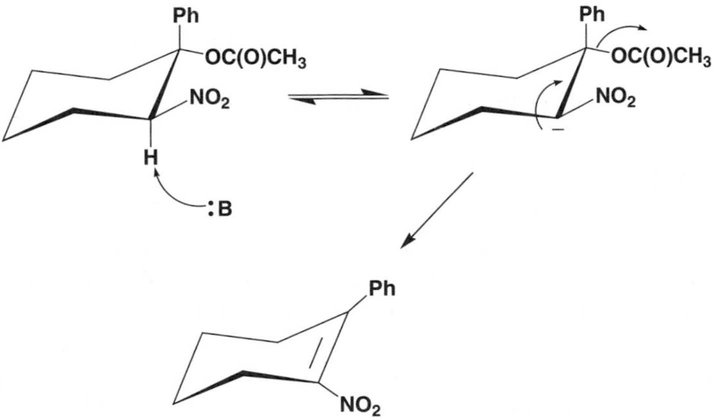

Figure 14.42. An E1$_{cb}$ reaction.

14.43 γ (OR 1,3)-ELIMINATION

An elimination reaction in which the two leaving groups are originally on atoms separated by another atom. The loss of the two groups leads to the formation of a new bond between the two atoms originally connected to the leaving groups, forming a three-membered ring. The E1$_{cb}$ mechanism is generally required.

Example. Figure 14.43.

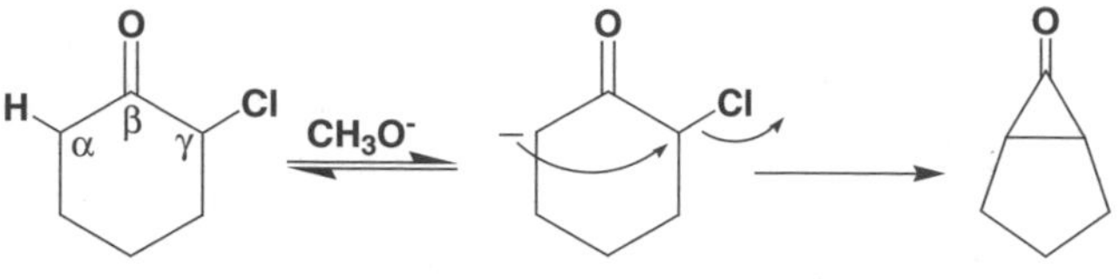

Figure 14.43. A γ-elimination such as occurs in a Favorsky rearrangement.

14.44 CONCERTED UNIMOLECULAR ELIMINATION (CYCLO-DD)

A one-step elimination, generally following a *cis* stereochemical course. These reactions are discussed more fully in the context of pericyclic cycloreversions (see Sect. 14.97).

Example. Xanthate ester pyrolysis (Fig. 14.44) and acetate ester pyrolysis (Fig. 14.39*b*).

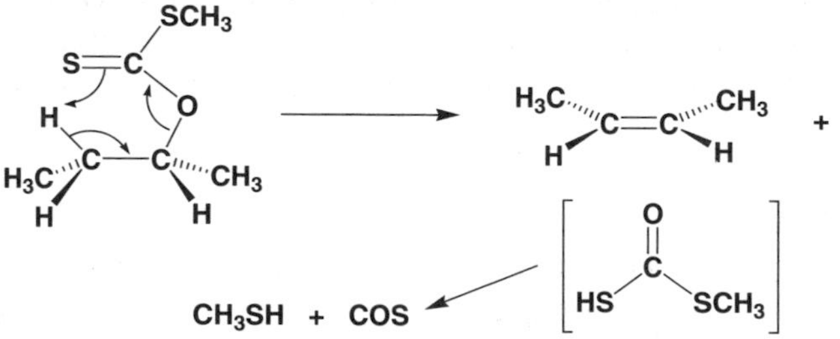

Figure 14.44. Xanthate ester pyrolysis.

14.45 ADDITION [A (+) A]

A reaction that involves an increase in the number of groups attached to the substrate and, therefore, a decrease in the degree of unsaturation of the substrate. Addition reactions are the reverse of elimination reactions. Most commonly, an addition involves the gain of two groups or atoms (one electrophile and one nucleophile) at each end of a π bond (1,2-addition) or ends of a conjugated π system (e.g., 1,4- or 1,6-addition). There are, however, examples of addition to certain highly reactive σ bonds (e.g., cyclopropane addition).

Example. Figure 14.45.

14.46 NUCLEOPHILIC ADDITION TO CARBON–CARBON MULTIPLE BONDS ($A_N + A_E$)

An addition reaction in which attachment of the nucleophile precedes attachment of the electrophile. Such reactions (usually occurring under basic conditions) involve carbanion intermediates, and thus are important only when the nucleophile is extremely reactive (e.g., is itself a carbanion) and/or the incipient carbanion is

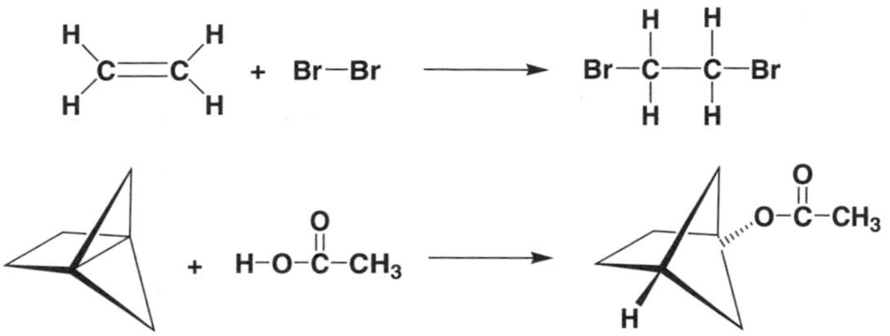

Figure 14.45. 1,2-Additions.

stabilized by electron-withdrawing substituents (i.e., the π bond is polarized). Carbon–carbon triple bonds are somewhat more reactive toward nucleophilic addition than comparably substituted double bonds.

Example. Figures 14.46*a* and *b*.

14.47 CONJUGATE NUCLEOPHILIC (OR MICHAEL) ADDITION

Named for Arthur Michael (1853–1942), nucleophilic addition to a carbon–carbon π bond that is conjugated with an electron-withdrawing group such as a carbonyl. The reaction involves first a 1,4-addition to the ends of the conjugated system, followed by a migration of the electrophile (usually, H^+) to the 2 position (ketonization of the enol); the reaction results in net 1,2-addition to the π bond.

Example. Figure 14.47.

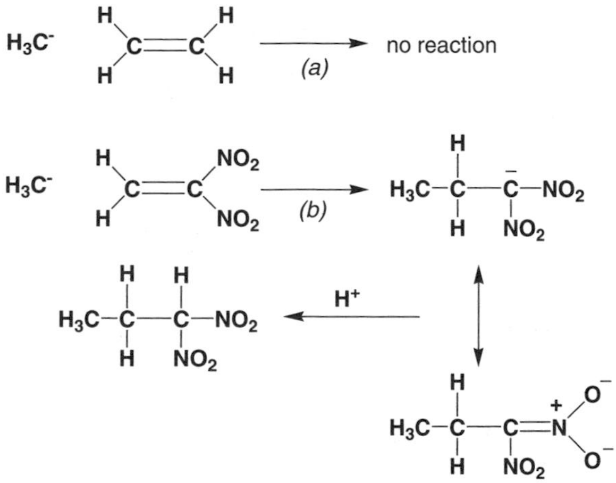

Figure 14.46. Nucleophilic addition to a double bond: (*a*) alkenes do not react and (*b*) reaction via a stabilized anion. (The methide ion can be generated e.g., from CH_3Br; the overall reaction is the addition of CH_4 across the double bond and may be considered as a Michael addition, Sect. 14.47)

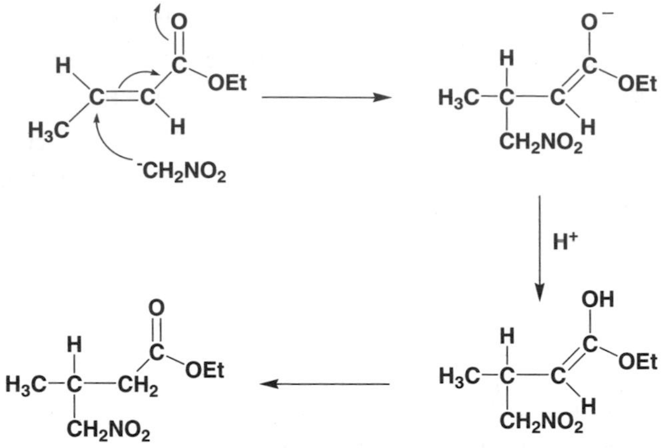

Figure 14.47. Michael addition.

14.48 ELECTROPHILIC ADDITION TO CARBON–CARBON MULTIPLE BONDS ($A_E + A_N$)

An addition reaction (usually, a 1,2-addition to a π bond) in which electrophilic attachment precedes nucleophilic attachment; hence, a reaction involving a carbocation intermediate. Such reactions generally occur under acidic or neutral conditions and can involve open or cyclic cationic intermediates. Olefins are generally more reactive than acetylenes in electrophilic additions.

Example. Figure 14.48. Also see Fig. 14.49.

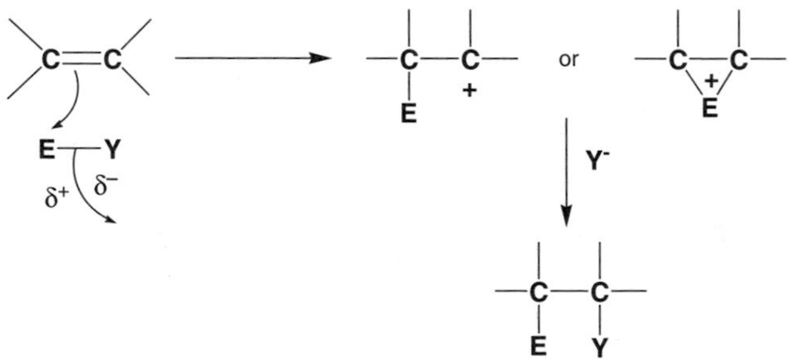

Figure 14.48. Stepwise electrophilic addition.

14.49 MARKOVNIKOV'S RULE

Named for Vladimir W. Markovnikov (1838–1904), this rule originally stated that the addition of HX (X = halogen) to an unsymmetrically substituted alkene occurs with attachment of the hydrogen to the carbon atom of the double bond having more hydrogens. More generally, the rule states that in the addition of E^+–Nu^- to a carbon–carbon

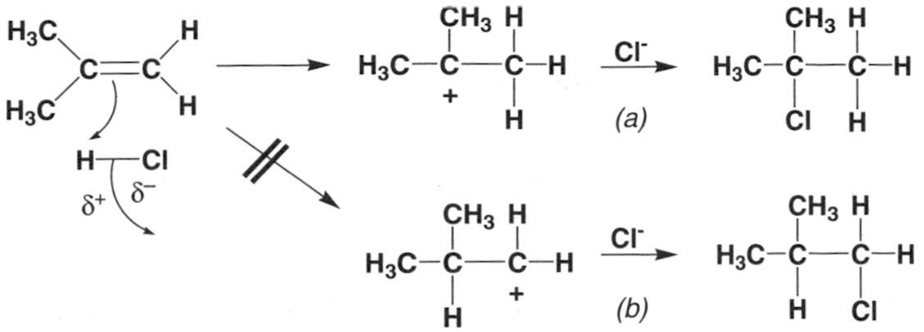

Figure 14.49. Markovnikov addition: (*a*) Markovnikov product via a tertiary carbenium ion and (*b*) anti-Markovnikov product (minor product) via a much less stable primary carbenium ion.

multiple bond, the electrophilic group E adds to the less substituted carbon. It is now recognized that this preference is a result of formation of the more stable carbocation intermediate. (See Sect. 14.70, "Anti-Markovnikov Addition.")

Example. Figures 14.49*a* and *b*.

14.50 π COMPLEX

The species that results from the interaction of an electrophile with the π electrons of the substrate. In the resulting complex, the electrophile is not bonded to a particular atom. An empty orbital on the electrophile overlaps a filled π-type orbital of the donor. A π complex may be regarded as a *charge-transfer complex*. Formation of the π complex is often difficult to establish, but it is believed to be rapid and reversible. Such complexes may also be involved in some electrophilic substitution and addition reactions. The π complex is usually a precursor to the σ complex, Sect. 14.51

Example. Figures 14.50*a*, *b*, and *c*.

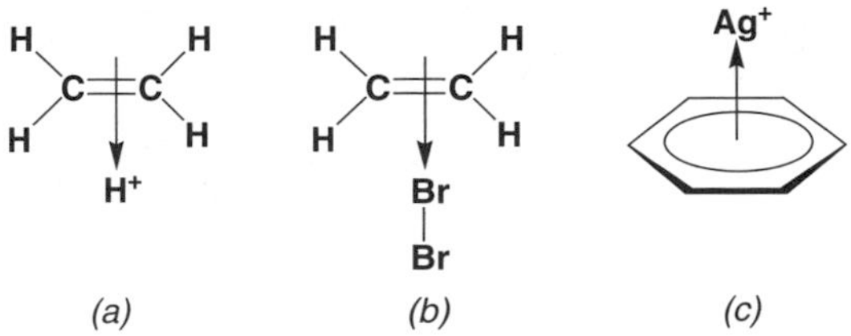

Figure 14.50. π Complexes: (*a*) with a proton; (*b*) with Br_2; (*c*) with Ag^+. The arrow is used to indicate the interaction of the filled π system with an empty orbital on the electrophile.

14.51 σ COMPLEX

The species that results from the interaction of an electrophile with a σ bond. An empty orbital on the electrophile overlaps a filled σ orbital on the nucleophile to give a three-center, two-electron interaction similar to that of borane (Sect. 3.45).

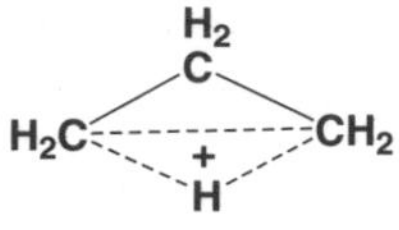

Figure 14.51. σ Complexe: edge-protonated cyclopropane

Example. Figure 14.51*a* shows edge-protonated cyclopropane. The complexes shown in Figs. 14.51*b*, *c*, and *d*, resulting, respectively, from the electronic rearrangement of the π complexes shown in Figs. 14.50*a*, *b*, and *c*, also are called σ complexes to distinguish them from their precursor π complexes.

14.52 *CIS* ADDITION

The stereochemical result of a 1,2-addition reaction involving formal attachment of both groups (or atoms) to the same side of the π bond.

Example. Figure 14.53*a*.

14.53 *TRANS* ADDITION

The stereochemical result of a 1,2-addition reaction involving formal attachment of both groups (or atoms) to opposite sides of the π bond.

Example. Figure 14.53*b*.

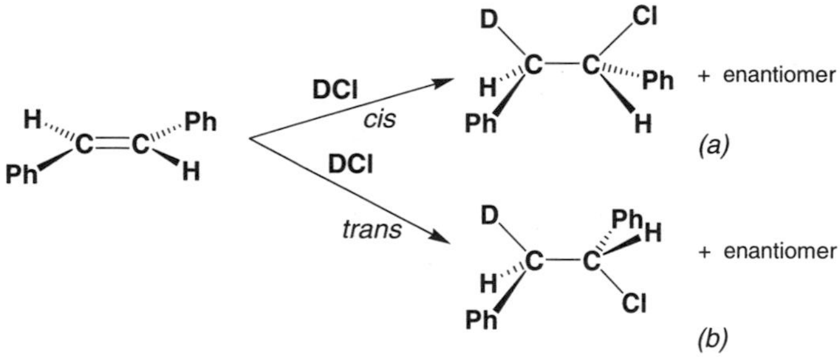

Figure 14.53. Addition reaction to an *E* (*trans*) isomer: (*a*) *cis* addition giving the *threo* product and (*b*) *trans* addition giving the *erythro* product.

14.54 ADDITION ELECTROPHILIC BIMOLECULAR, Ad_E2 ($A_E + A_N$)

A two-step addition in which attachment of the electrophile occurs first (the rate-determining step) and is followed by a rapid nucleophilic attack on the cationic intermediate. The reaction is first-order in substrate and first-order in electrophile. The stereochemistry of the addition depends on the structure of the intermediate and the degree of association of the nucleophile to the electrophile. Certain electrophiles such as Br^+ (from Br_2), HgX^+ (from HgX_2), and OH^+ (from RCO_3H) give cyclic intermediates (unless the open ion is sufficiently stabilized by electron-donating substitutes), followed by nucleophilic attack that leads to net *trans* addition.

Example. Figures 14.54*a*, *b*, and *c*.

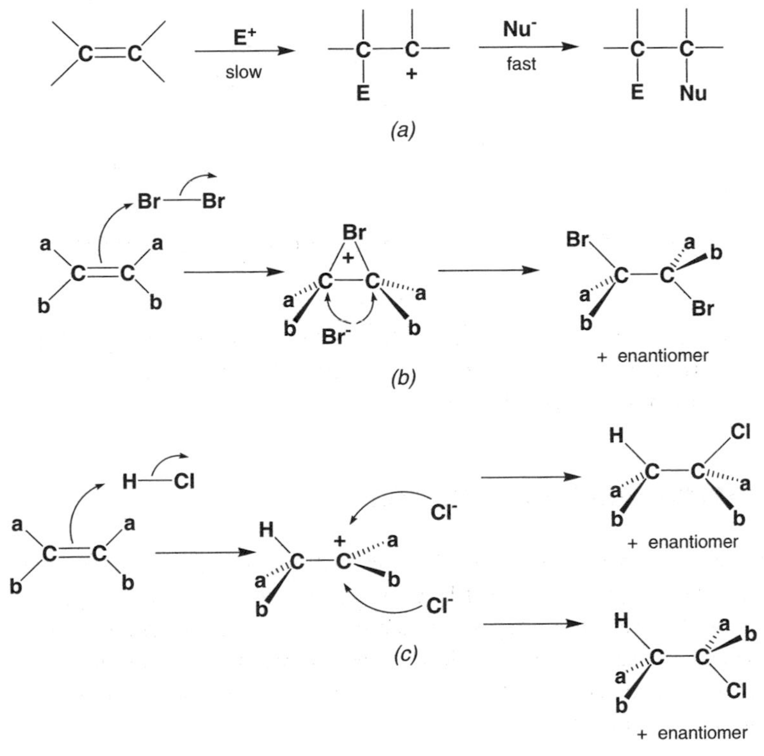

Figure 14.54. Ad_E2 reactions: (*a*) general reaction; (*b*) addition of bromine; (*c*) addition of HCl.

14.55 SUBSTITUTION ELECTROPHILIC AROMATIC, S_EAr ($A_E + D_E$)

The stepwise replacement of an electrophile attached to an aromatic ring by an attacking electrophile. The most common electrofuge is H^+. The mechanism involves at least two steps (more if π complexes are involved): attachment of the attacking electrophile to give a cyclohexadienyl cation (sometimes referred to as a σ complex or benzenonium ion), which then loses an electrofuge *from the same atom that was attacked* to regenerate the aromatic π system. Nucleophilic attachment to the intermediate (i.e., net addition to the aromatic π system) occurs only very rarely because of the cost of losing the resonance energy associated with an aromatic system.

Example. Figure 14.55.

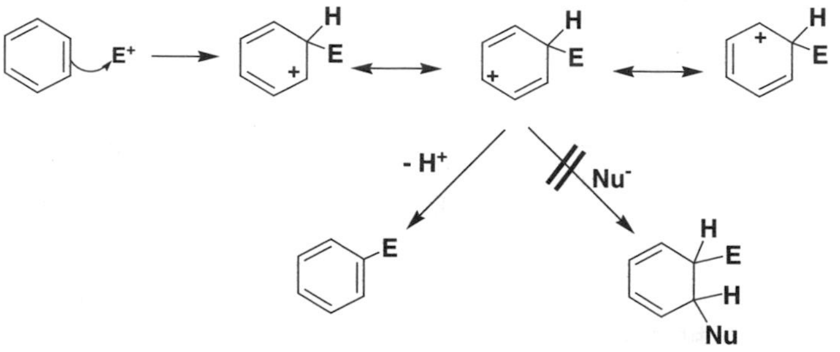

Figure 14.55. Aromatic electrophilic substitution, S_EAr, of H by E.

Because the aromatic ring is a nucleophile, electron-withdrawing substituents on the aromatic ring inhibit attack on the electrophile, electron-donating groups activate the ring toward attack on the electrophile.

14.56 *ORTHO* AND *PARA* DIRECTOR

A substituent on an aromatic ring that facilitates electrophilic substitution at the positions *ortho* and *para* to itself. Electron-donating groups (alkyl,–OR, $-NR_2$, etc.) and halogens fall into this category; attachment of an electrophile *ortho* or *para* to them leads to a more stable σ complex because of resonance involving the electron-donating substituent. On the other hand, *meta* attack does not permit any resonance stabilization or interaction.

Example. Figure 12.95.

14.57 *META* DIRECTOR

A substituent on an aromatic ring that directs electrophilic substitution *meta* to itself. This occurs because the σ complex resulting from *meta* attack is less destabilized by like-charge repulsion than are the σ complexes from *ortho* or *para* attack. Typically, *meta* directing groups are electron-withdrawing, such as $-NO_2$, –CN, and –C(O)R.

Example. Figure 14.57.

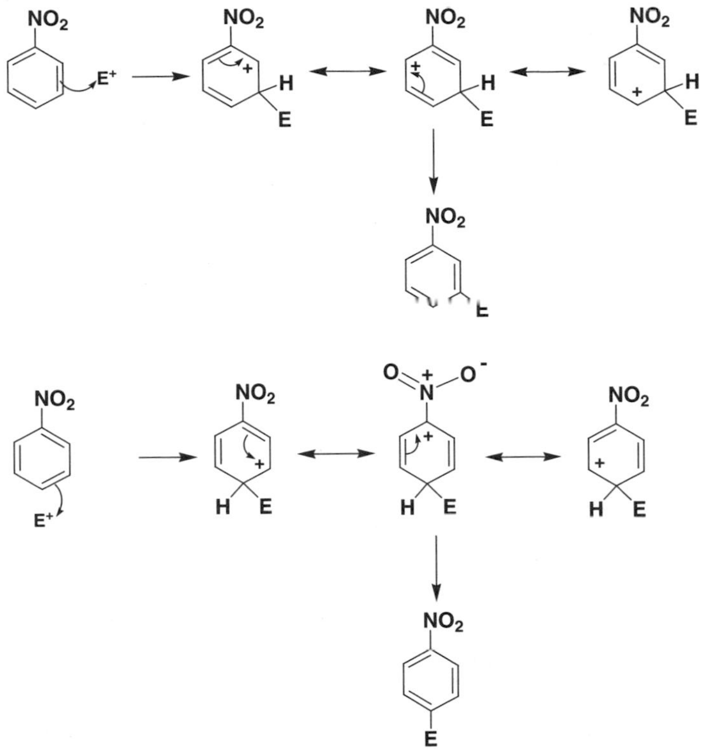

Figure 14.57. *Meta* is the preferred substitution with an electron-withdrawing substituent. *Para* (and *ortho*) substitution would involve high-energy resonance structures owing to adjacent positive charges.

14.58 SUBSTITUTION NUCLEOPHILIC AROMATIC, S_NAr ($A_N + D_N$)

Substitution of one nucleophile for another at an aromatic carbon. Although typical aromatic compounds are resistant toward nucleophilic attack, the presence of one or more strong electron-withdrawing groups on the ring can stabilize the anionic intermediate (this is called a *Meisenheimer complex* after J. Meisenheimer, 1879–1934), which subsequently loses a nucleofuge to reform the aromatic π system.

Example. Figure 14.58. This type of reaction frequently is used for determining which amino acid is the *N*-terminal residue in a polypeptide.

14.59 BENZYNE (OR ARYNE) MECHANISM

A nucleophilic substitution at an aromatic carbon atom under strongly basic conditions involving an elimination–addition mechanism with a benzyne intermediate.

Example. Figure 14.59.

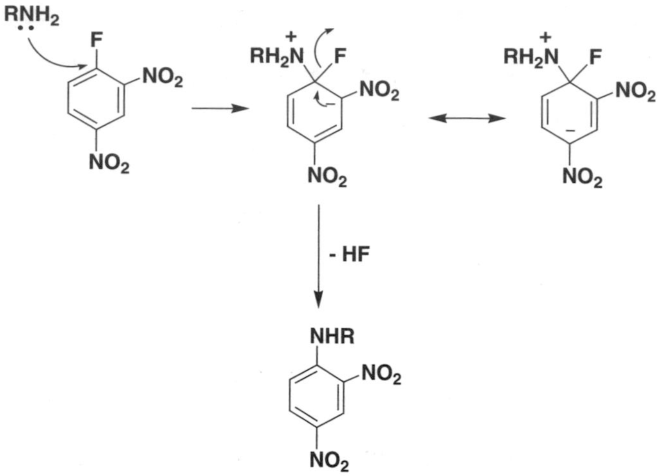

Figure 14.58. Nucleophilic aromatic substitution, S_NAr.

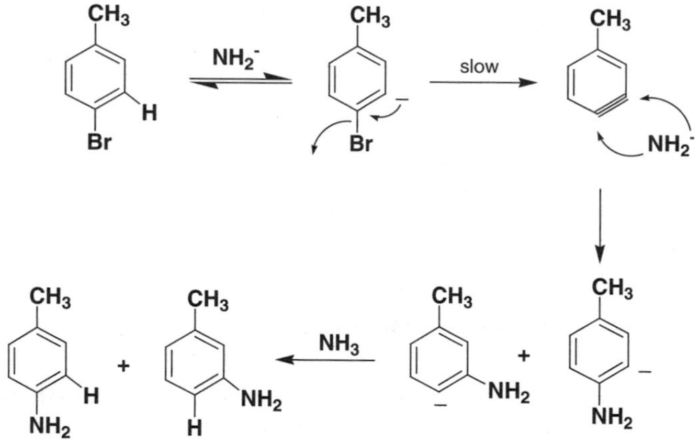

Figure 14.59. Nucleophilic substitution involving a benzyne intermediate.

14.60 CHAIN REACTION

A multistep mechanism, most commonly (but not always) involving free radicals, in which one of the products from the last step is one of the starting materials for the first step. Free radical reactions related to polymer formation have been discussed in Chap. 8.

Example. The chain reaction for the chlorination of ethane, in which the ethyl radical is the product from step 2 and the starting material for step 1 in Fig. 14.60.

$$H_3CH_2C\cdot + Cl_2 \longrightarrow H_3CH_2C\text{-}Cl + Cl\cdot \quad (1)$$

$$Cl\cdot + H_3CCH_3 \longrightarrow H\text{-}Cl + H_3CH_2C\cdot \quad (2)$$

$$H_3CCH_3 + Cl_2 \longrightarrow H_3CH_2C\text{-}Cl + HCl \quad (3)$$

Figure 14.60. A chain reaction. See also Sect. 11.6.

14.61 CHAIN-INITIATING STEP

The nonrepeating preliminary step or steps in which the chain carrier is first formed.

Example. For the free radical chain reaction in Fig. 14.60, two possible chain initiating sequences are shown in Fig. 14.61.

$$Cl_2 \xrightarrow{h\nu} 2\ Cl\cdot$$

or

$$(CH_3)_3CO\text{-}OC(CH_3)_3 \xrightarrow{\Delta} 2\ (CH_3)_3CO\cdot$$

$$(CH_3)_3CO\cdot + H_3CCH_3 \longrightarrow (CH_3)_3COH + H_3CH_2C\cdot$$

Figure 14.61. Chain initiation.

14.62 INITIATOR

A compound that brings about the chain-initiating step. Free radical initiators, such as peroxides (RO–OR) and aliphatic azo compounds (R–N=N–R), which when heated or irradiated, readily undergo homolytic bond dissociation to generate free radicals.

Example. The di-*t*-butylperoxide in Fig. 14.61 and the azobisisobutyronitrile in Fig. 8.11.

14.63 CHAIN-PROPAGATING STEPS

The repeating steps of a chain reaction. For example, steps 1 and 2 in Fig. 14.60 occur many times, in a single-chain sequence, until a chain termination step (Sect. 14.65) occurs. The sum of steps 1 and 2 gives the overall net reaction 3.

14.64 CHAIN CARRIERS

The intermediates that are alternately formed and then destroyed in a chain reaction. For example, in the free radical reaction in Fig. 14.60, both the CH_3CH_2• and Cl• are chain carriers.

14.65 CHAIN-TERMINATING STEP

A reaction in which two chain carriers react together without regenerating a new chain carrier. The chain reaction is therefore stopped.

$$2\ Cl\bullet \longrightarrow Cl_2$$

$$2\ H_3CH_2C\bullet \longrightarrow H_3CCH_2CH_2CH_3$$

$$H_3CH_2C\bullet + Cl\bullet \longrightarrow H_3CH_2C\text{-}Cl$$

$$2\ H_3CH_2C\bullet \longrightarrow H_3CCH_3 + H_2C{=}CH_2$$

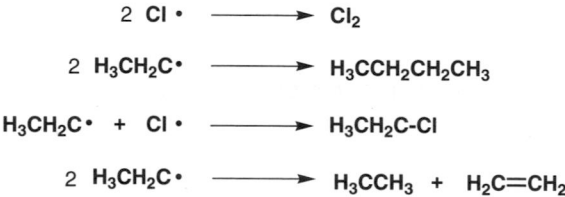

Figure 14.65. Radical chain termination reactions.

Example. Possible termination reactions for the chain reaction shown in Fig. 14.65. The last reaction shows termination by *disproportionation.*

14.66 INHIBITOR

A reagent that suppresses a chemical reaction. In free radical chain reactions, compounds such as molecular oxygen and butylated hydroxytoluene (BHT) intercept the chain carrier(s), generate very stable radicals, and thus divert the chain carriers from the chain reaction.

Example. Figure 14.66.

14.67 SUBSTITUTION HOMOLYTIC BIMOLECULAR, S_H2 ($A_rD_R + A_RD_r$)

The attack of a free radical (or atom) on a terminal atom (an atom bonded to only one other atom, such as H) of another molecule. Such reactions are usually chain-propagating steps in a free radical chain reaction mechanism.

Example. Reactions 1 and 2 in Fig. 14.60 are examples of S_H2 reactions.

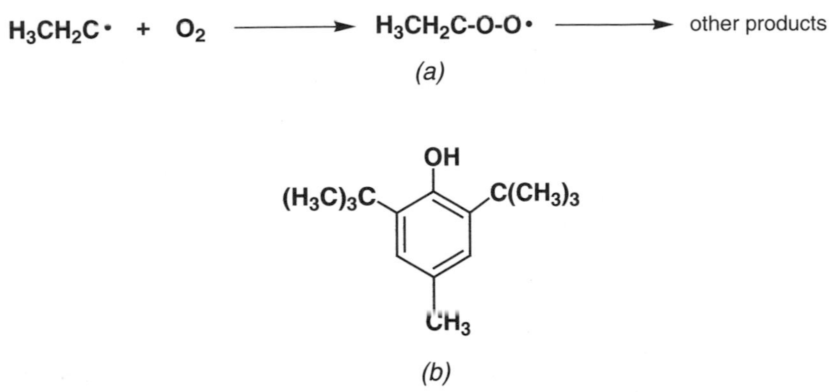

Figure 14.66. *(a)* Molecular oxygen as an inhibitor and *(b)* 2,5-di-*tert*-butyl-4-methylbenzene, known in the trade as butylated hydroxytoluene (BHT), a common food preservative.

14.68 SUBSTITUTION RADICAL NUCLEOPHILIC UNIMOLECULAR, $S_{RN}1$ ($T + D_N + A_N$)

A nucleophilic substitution reaction that proceeds by a radical anion chain mechanism.

Example. The substitution of NH_2 for I in an aryl iodide can occur by the benzyne mechanism (see Sect. 14.59) or the $S_{RN}1$ mechanism (Fig. 14.68).

14.69 FREE RADICAL ADDITION

The addition of a free radical to an unsaturated center. The radical prefers to add to the least substituted carbon atom of the unsaturated center.

Example. Figure 14.70.

14.70 ANTI-MARKOVNIKOV ADDITION

Addition of H–Z to an unsymmetrically substituted π bond occurs with attachment of the hydrogen to the carbon atom of the double bond having the fewer hydrogens. Z may be halogen, BR_2, and so on.

Example. Under free radical chain conditions, the addition of HBr to an unsymmetrical olefin proceeds via the more stable radical intermediate (tertiary > secondary > primary) to give the opposite (anti-Markovnikov) product from that produced

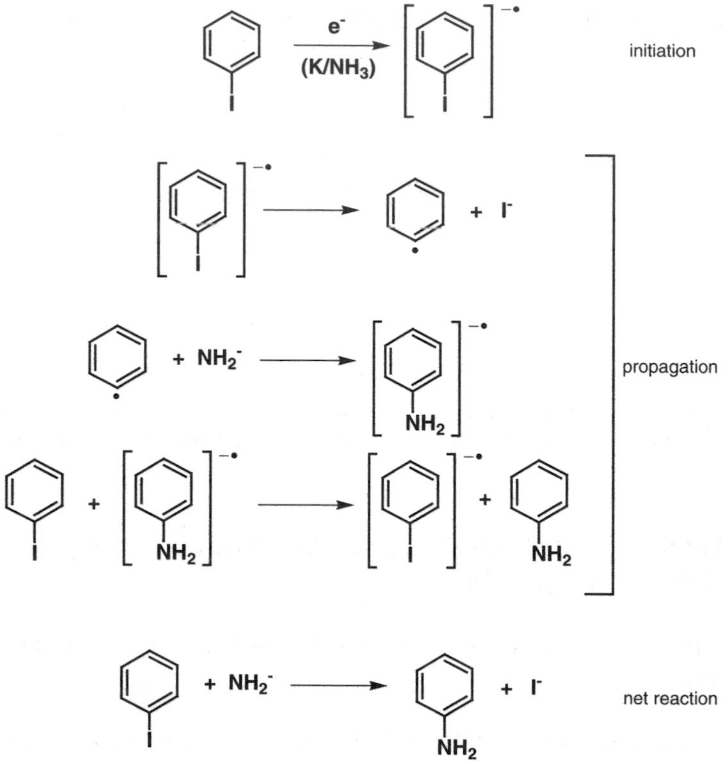

Figure 14.68. Mechanism of radical substitution of iodide by amide, $S_{RN}1$. The electron in the first step is supplied by a reducing agent

under ionic conditions, which leads to Markovnikov addition (Sect. 14.49). Boron hydrides also generate the anti-Markovnikov adduct because the BH_2 is the electrophile and the H is the nucleophile.

Notice that in Fig. 14.49 and Fig. 14.70 the group that becomes attached to the least substituted carbon atom is the first group that is attacked by the double bond.

14.71 ADDITION TO THE CARBONYL GROUP

Addition to the carbonyl group occurs with attachment of the nucleophile to the carbonyl carbon and the electrophile to the carbonyl oxygen.

Example. As can be seen in Fig. 14.71*a*, under acid conditions the electrophile (H^+) adds first, whereas under basic conditions the nucleophile adds first. Addition is irreversible if Nu = H^- or R^-, but generally is reversible (favoring the carbonyl) if

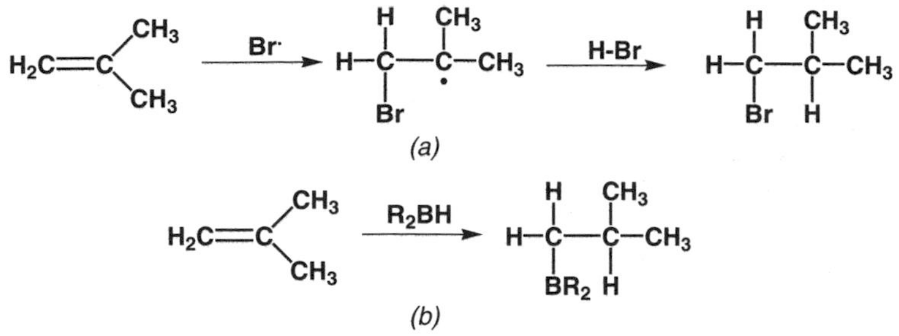

Figure 14.70. Anti Markovnikov addition: (*a*) free radical addition of HBr and (*b*) addition of boranes.

Nu = RO^-, $:NHR_2$, or a halide. Both $-C=N$ and $-C\equiv N$ multiple bonds behave similarly to the carbonyl function. Because the coordination number of the carbonyl carbon increases from 3 in the carbonyl to 4 in the product, the reaction is sensitive to steric factors.

Figure 14.71*b* shows the π and π* orbitals of a carbonyl. Note that in the π orbital, the greater electron density resides on the oxygen, and in the π* orbital, the major contribution is on the carbon. This is a molecular orbital representation of the valence bond resonance structure of the carbonyl group with a plus charge on carbon and a negative charge on oxygen.

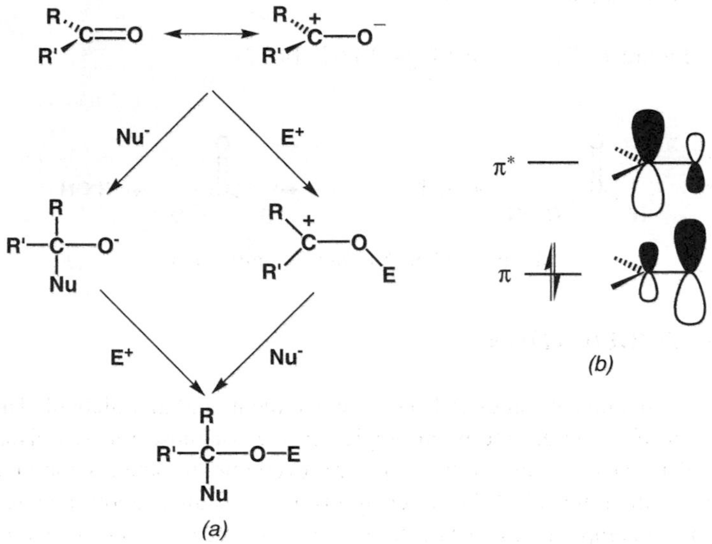

Figure 14.71. (*a*) Carbonyl addition reactions and (*b*) π and π* orbitals of a carbonyl.

14.72 CRAM'S RULE OF ASYMMETRIC INDUCTION

A rule named after D. Cram for predicting the epimer that preferentially is formed by attack at a carbonyl group with a neighboring chiral center. For a discussion of Cram's rule, see Sect. 7.95.

14.73 ESTER HYDROLYSIS

The conversion of an ester, by reaction with water, into its acid and alcohol constituents. The hydrolysis may involve cleavage at either the acyl oxygen or alkyl oxygen bond; the details of the mechanism are strongly dependent on the nature of R′, as well as the conditions of the reaction. This reaction may be reversible, see Sect. 14.75.

Example. Figure 14.73. (see also Figs. 13.62, 14.76, 14.77)

Figure 14.73. An ester hydrolysis.

14.74 SAPONIFICATION

Irreversible base-induced ester hydrolysis to form the salt of an acid (which has soaplike properties) and an alcohol.

Example. Figure 14.74. (see also Figs. 13.62, 14.77)

Figure 14.74. An ester saponification.

14.75 ESTERIFICATION

The reaction that creates an ester from its constituent acid and alcohol. The mechanisms of esterification are the microscopic reverses of those for hydrolyses (Sects. 14.76–14.79). Because many of the steps are reversible, to obtain good yields of the ester requires the removal of the water by-product (usually, azeotropically) to force the reaction to completion. Esterification is an example of a *condensation reaction*; such reactions are characterized by either intra- or intermolecular loss of a small molecule such as H_2O, ROH, and so on, between two reacting sites, thereby condensing two species into one (intermolecular) or condensing the size of a molecule by ring formation (intramolecular).

Example. The reaction going from right to left in Fig. 14.73.

14.76 ACID-CATALYZED (A) BIMOLECULAR (2) HYDROLYSIS WITH ACYL-OXYGEN (AC) CLEAVAGE, $A_{AC}2$ ($A_h + A_N + A_hD_h + D_N + D_h$)

Nucleophilic attack by water on a protonated ester to give a tetrahedral intermediate that rapidly collapses to the alcohol and protonated acid. This is the most common acid-catalyzed hydrolysis mechanism.

Example. Figure 14.76.

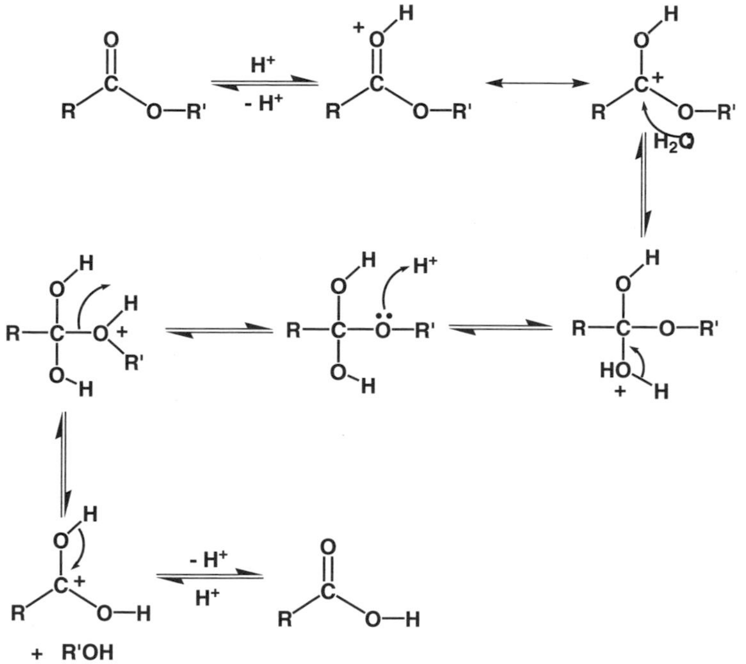

Figure 14.76. Acid-catalyzed hydrolysis, $A_{AC}2$.

14.77 BASE-CATALYZED (B) BIMOLECULAR (2) HYDROLYSIS WITH ACYL-OXYGEN (AC) CLEAVAGE, $B_{AC}2$ ($A_N + D_N$)

Nucleophilic attack by base on the ester carbonyl carbon to give a tetrahedral ion, followed by rapid collapse to the alkoxide and acid. The ultimate products are the alcohol and conjugate base of the acid formed with consumption of one mole

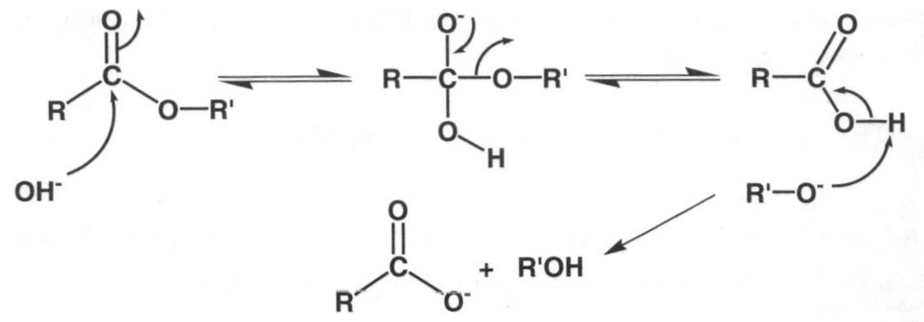

Figure 14.77. Base-catalyzed hydrolysis, $B_{AC}2$.

equivalent of base. This is the most common mechanism for "base-catalyzed" ester hydrolysis.

Example. Figure 14.77. If $R' = Ph$, the phenol also is converted to its conjugate base with overall consumption of two equivalents of base.

14.78 ACID-CATALYZED (A) UNIMOLECULAR (1) HYDROLYSIS WITH ALKYL-OXYGEN (AL) CLEAVAGE, $A_{AL}1$ ($A_h + A_N + D_N + D_h$)

Fragmentation of a protonated ester to give the carboxylic acid and a carbocation, which is rapidly trapped by water or loses a proton to form an alkene. This mechanism only occurs when R'^+ is a relatively stable carbocation.

Example. When R'^+ is a stable carbocation (e.g., *t*-butyl) and $RCOO^-$ is a very stable/ nonnucleophilic anion (e.g., *p*-nitrobenzoate), the hydrolysis can occur without acid catalysis, but otherwise proceeds through the acid-catalyzed mechanism shown in Fig. 14.78.

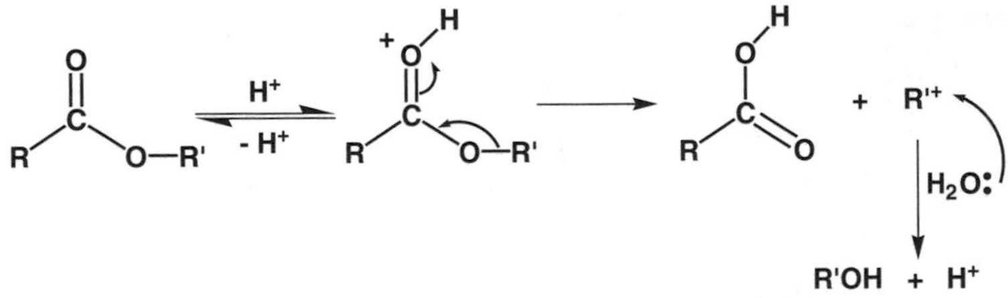

Figure 14.78. Acid-catalyzed alkyl oxygen hydrolysis, $A_{AL}1$.

14.79 BASE-PROMOTED (B) BIMOLECULAR (2) ESTER HYDROLYSIS WITH ALKYL-OXYGEN (AL) CLEAVAGE, $B_{AL}2$ (A_ND_N)

Although not strictly an ester hydrolysis, it is an ester cleavage involving nucleophilic displacement at the alcohol carbon.

Example. Figure 14.79. Because this is an S_N2 reaction, the R′ alcohol must be methyl, primary, or secondary, but not tertiary.

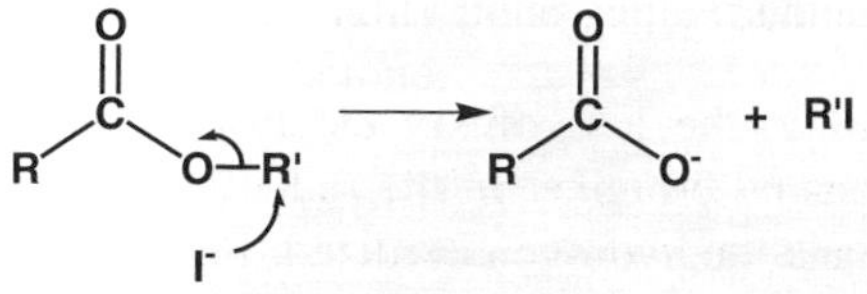

Figure 14.79. Base-promoted alkyl oxygen ester cleavage, $B_{AL}2$.

14.80 OXIDATION NUMBER

A number assigned to an atom in a molecule, representing the atom's formal ownership of its the valence electrons. All the electrons in a given bond are formally assigned to the more electronegative partner. Thus, divalent oxygen is usually assigned an oxidation number of -2 (exceptions: $OF_2 = +2$; $O_2 = 0$; $H_2O_2 = -1$), hydrogen is usually $+1$ (exceptions: metal hydrides $= -1$; $H_2 = 0$), halogens are generally -1 (except in the elemental state). The algebraic sum of the oxidation numbers of all the atoms in a molecule must equal the charge on the molecule or ion. (See Sect. 2.10 and 9.11)

Example. Table 14.80. Notice that the oxidation number of carbon can range from -4 to $+4$.

TABLE 14.80. Oxidation Numbers for the Underlined Carbon Atom (R is not an H).

-4	-3	-2	-1	0	1	2	3	4
$\underline{C}H_4$	$H_3\underline{C}R$	$H_2\underline{C}R_2$	$H\underline{C}R_3$	$\underline{C}R_4$	$R_3\underline{C}OH$	$R_2\underline{C}(O)$	$R\underline{C}(O)OH$	$\underline{C}O_2$
	$\cdot\underline{C}H_3$	$Cl\underline{C}H_3$	$R\underline{C}H_2Cl$	$\underline{C}H_2Cl_2$	$R\underline{C}HO$	$H\underline{C}(O)OH$		$\underline{C}F_4$
		$H_3\underline{C}OH$		$\underline{C}$ (solid)		$\underline{C}O$		
				$H_2\underline{C}O$				

14.81 OXIDATION STATE

Synonymous with oxidation number.

14.82 OXIDATION AND REDUCTION

An oxidation is the increase of an oxidation number to a more positive value; a reduction is the decrease of an oxidation number to a more negative value. An oxidation must always be accompanied by a reduction somewhere in the system; that is, the total net change in oxidation numbers must be zero. Stated alternatively, the loss of electrons (oxidation) by one atom or compound must be matched by the gain of electrons (reduction) by another.

Example. In Fig. 14.82*a*, the loss of six electrons from carbon is matched by the gain of six electrons by bromine. In the reduction of a carbonyl (Fig. 14.82*b*), the carbon is reduced and the hydrogen oxidized. Figure 14.82*c* shows an intramolecular reduction–oxidation in which one carbon is reduced and another carbon is oxidized.

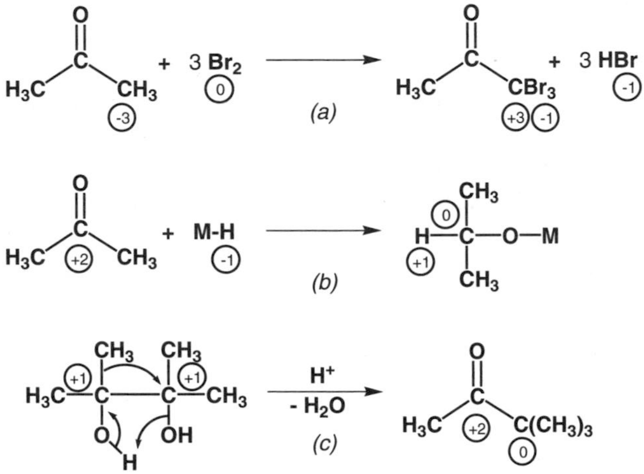

Figure 14.82. (*a*) Six-electron oxidation; (*b*) carbonyl reduction; (*c*) intramolecular reduction–oxidation, the pinacol rearrangement.

14.83 REDOX REACTION

A chemical reaction involving reduction and oxidation.

14.84 INTRAMOLECULAR AND INTERMOLECULAR

Intramolecular means within a single molecule or species, and intermolecular means involving two or more (either identical or different) molecules or species.

14.85 ISOMERIZATION

A reaction in which a molecule or species is transformed into a different molecule or species with the same molecular formula (i.e., an isomer, see Sect. 7.7).

14.86 REARRANGEMENT

A reaction involving a change in the bonding sequence, that is, which atoms are bonded to which other atoms, within a molecule. Most organic chemists use the term to denote reactions involving a change in the connectivity of the carbon skeleton of a molecule; however, reactions such as double bond migration are sometimes referred to as rearrangements as well.

14.87 DEGENERATE REARRANGEMENT

A rearrangement that leads to a product which is indistinguishable from the starting material. All fluxional molecules (Sect. 7.146) undergo degenerate rearrangements.

Example. Figure 7.146.

14.88 PERICYCLIC REACTION

Intra- or intermolecular processes involving *concerted* reorganization of electrons within a closed loop of interacting orbitals. Pericyclic reactions are subdivided into five classes: electrocyclic; cycloaddition; sigmatropic shift; cheletropic; and group transfer reactions, all of which are discussed below.

14.89 ELECTROCYCLIC REACTION

An intramolecular pericyclic reaction involving opening (or closing) of a ring by conversion of σ to π bonds (or the reverse).

Example. Figure 14.89.

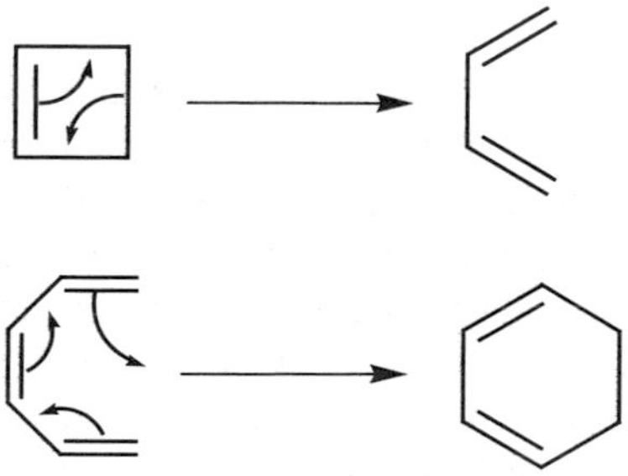

Figure 14.89. *(a)* Electrocyclic ring opening of cyclobutene to butadiene and *(b)* electrocyclic ring closure of hexatriene to cyclohexadiene.

14.90 CONROTATORY AND DISROTATORY

During an electrocyclic reaction the terminal atoms (last carbon atoms and their substituents) of the open isomer must each rotate approximately 90° around the bond to which they are attached to form the σ bond of the product (or vice versa). If both termini rotate clockwise or both counterclockwise, the ring closing (or opening) is said to be *conrotatory*. If the two termini rotate in opposite directions, one clockwise and the other counterclockwise, the cyclization (or ring opening) is said to be *disrotatory*.

Example. Figure 14.90. The electrocyclic ring opening of cyclobutene (ring closure of butadiene), in principle, can occur by either of two conrotatory paths or two disrotatory paths.

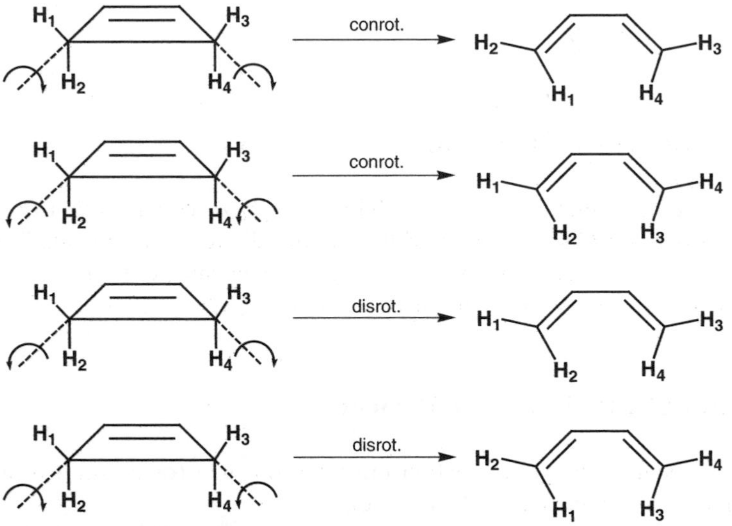

Figure 14.90. Conrotatory and disrotatory ring opening of cyclobutene.

14.91 ORBITAL CORRELATION DIAGRAM

A diagram that shows the correspondence in energy and symmetry between relevant reactant and product orbitals. This method can be used for any pericyclic reaction, but is usually utilized only for electrocyclic reactions. The relevant orbitals are those that undergo change during the reaction. The diagram is constructed by listing the relevant orbitals of reactant and product in separate columns according to their relative energies. Each orbital is classified on the basis of the symmetry elements retained at all points along the reaction coordinate. Correlation lines then are drawn connecting the lowest-energy orbital of the reactant with the lowest-energy orbital of the product with the same symmetry.

Example. The orbital correlation diagram for the conrotatory ring opening of cyclobutene to 1,3-butadiene requires an examination of the relevant molecular orbitals. The four relevant molecular orbitals of cyclobutene are the σ and π molecular orbitals and their corresponding σ* and π* shown in Fig. 14.91*a*. In the ground or normal state, the σ and π molecular orbitals, which are the two lowest-energy molecular orbitals, are each occupied by two electrons. The four relevant molecular orbitals of 1,3-butadiene are the two π orbitals formed in the product and their corresponding π* orbitals, shown in Fig. 14.91*a*. These are designated ψ_1, ψ_2, ψ_3, and ψ_4 in order of increasing energy. Both cyclobutene and butadiene possess a C_2 axis (the C_2 operation is a 180° rotation around the two-fold symmetry axis), and this symmetry element is preserved in going from one compound to the other. The eight molecular orbitals are now classified

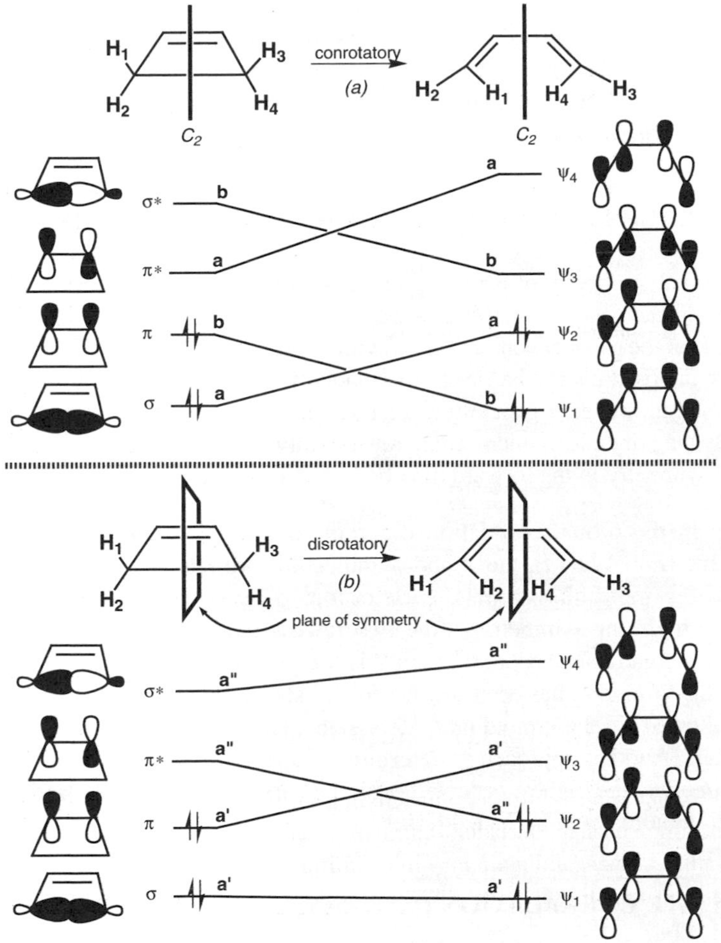

Figure 14.91. Orbital correlation diagrams: *(a)* conrotatory and *(b)* disrotatory ring opening of cyclobutene.

according to their behavior with respect to the C_2 operation. If the orbital does not change sign under this operation (symmetric behavior), it is denoted as belonging to a symmetry species designated as *a*; if the orbital does change sign (antisymmetric behavior), it is designated as belonging to a symmetry species, designated as *b* (see the character table for C_{2v}, Sect. 4.26). With this information, the orbital correlation diagram for the reaction can be drawn as shown in Fig. 14.91*a*.

As a further illustration, but without elaboration, the orbital correlation diagram for the disrotatory reaction is shown in Fig. 14.91*b*. In the disrotatory mode, a plane of symmetry is preserved during the transformations and the orbitals are classified as a' for symmetric behavior with respect to this symmetry operation and a'' for antisymmetric behavior (the a' and a'' notation comes from the character table for molecules possessing only a plane of symmetry).

Notice for the conrotatory opening the occupied orbitals of cyclobutene correlate with the occupied orbitals of butadiene, and thus, the reaction is thermally allowed (see Sect. 14.93). However, for the disrotatory opening, the occupied orbitals of cyclobutene do not correlate with the occupied orbitals of butadiene, and thus, this reaction is thermally forbidden (see Sect. 14.95).

14.92 PRINCIPLE OF CONSERVATION OF ORBITAL SYMMETRY

Overall orbital symmetry of a pericyclic reaction is said to be conserved (i.e., maintained) when each occupied orbital of the reactant transforms smoothly into an occupied orbital of the product with the same symmetry. Under these conditions, there is no symmetry-imposed energy barrier to the reaction, although other factors (e.g., steric, ring strain, entropy, etc.) generally give rise to a finite *activation energy* (see Sect. 12.81). By the principle of microscopic reversibility, if a reaction involves conservation of orbital symmetry in the forward direction, it also does in the reverse direction.

Example. In the orbital correlation diagram for the conrotatory ring opening of cyclobutene (Fig. 14.91*a*), the σ and π molecular orbitals of cyclobutene, the two lowest-energy molecular orbitals, each occupied by two electrons in the ground state, have the same symmetry as the two lowest-energy molecular orbitals of the product, 1,3-butadiene. Accordingly, in this transformation, as well as in the reverse reaction, the symmetry has been conserved and such a process is energetically favorable (is allowed) in the ground or normal state. This is not the case for the disrotatory process. Another approach to determine if a reaction is allowed or forbidden is to look simply at the highest occupied orbital of the reactant and the product. This is discussed in more detail in Sect. 14.109.

14.93 STATE CORRELATION DIAGRAM

A correlation diagram for a pericyclic reaction in which the energies and symmetry of the electronic states of the reactants and the products are determined from the energies and symmetry of the individual relevant orbitals. Each electron in a relevant molecular orbital must be considered in determining the symmetry of the state. The electrons are paired in the molecular orbitals of practically all molecules in their ground or normal state (a notorious exception is dioxygen, O_2), and hence, these molecules are in their singlet state. When molecules are exposed to ultraviolet light, they may undergo electronic transitions from their ground state to states of higher energy (excited states) that are usually characterized by having only single electrons in molecular orbitals. The energy and symmetry of the excited states must be specified to construct the state correlation diagram. As with the orbital correlation diagram, the states are listed in the increasing order of their energies and finally correlations are made analogous to the orbital correlations.

Example. In order to construct the state correlation diagram for the conrotatory ring opening of cyclobutene to 1,3-butadiene, it is useful to refer to the information developed in Fig. 14.91 for the orbital correlation diagram for this interconversion. To obtain the state correlation diagram corresponding to the orbital correlation diagram, attention must be paid to the symmetry of each occupied molecular orbital. To help develop state symmetry properties, the characters of +1 for electrons in the *a* symmetry species and −1 for electrons in the *b* symmetry species are assigned. The symmetry of the state can then be determined by multiplying the character (+1 or −1) assigned to each electron, depending on whether it occupies an *a* or *b* orbital. In the ground state of cyclobutene, the electronic configuration is $\sigma^2\pi^2$ corresponding to $a^2 \bullet b^2 = (+1)^2 \bullet (-1)^2 = +1 = A$, where *A* refers to a state that is totally symmetric. The electronic configuration of the ground state of 1,3-butadiene is $\psi_1^2\psi_2^2$ and the character is $b^2a^2 = (-1)^2 \bullet (+1)^2 = +1 = A$. Thus, the ground or lowest-energy state of each partner in the conrotatory interconversion has the symmetry *A*; hence, the interconversion is allowed in the ground state.

Now consider the conversion of the first excited state of cyclobutene to the first excited state of butadiene. The electronic configuration of cyclobutene in the first excited state is $\sigma^2\pi\pi^* = a^2ba = 1^2(-1)(1) = -1 = B$. The orbital correlation diagram (Fig. 14.91*a*) shows that σ in cyclobutene correlates with ψ_2 of butadiene, π with ψ_1, and π^* with ψ_4; hence, the $\sigma^2\pi\pi^*$ excited state of cyclobutene correlates with the $\psi_2^2\psi_1\psi_4 = a^2ba = B$ in butadiene. However, the butadiene configuration $\psi_2^2\psi_1\psi_4$ is not the lowest excited state of butadiene, which is $\psi_1^2\psi_2\psi_3 = b^2ab = B$. Accordingly, the probability of cyclobutene in the lowest-energy excited state being transformed to butadiene in the correlated excited state is quite low because of the uphill energy requirement. A higher excited state of cyclobutene has the configuration $\sigma\pi^2\sigma^* = B$ corresponding to $\psi_1^2\psi_2\psi_3 = B$ of butadiene, its lowest-energy excited state, and these two states are correlated. This information can be placed in a correlation diagram that constitutes a state correlation diagram (Fig. 14.93*a*) for the conrotatory process.

By exactly the same type of reasoning, the correlation diagram for the disrotatory process, using +1 for a' and −1 for a'', can be constructed and is shown in Fig. 14.93*b*. The straight lines (either solid or broken) connect the symmetry states that correlate with each other. However, because of the *noncrossing rule* [orbital mixing (see Sects. 2.33 and 2.42) prevents states of the same symmetry from crossing], the solid lines and curves represent the true correlations. In contrast to the conrotatory process, the state correlation diagram in Fig. 14.93*b* shows that the ground-state (thermal) disrotatory conversion of cyclobutene to butadiene has an energy barrier (the hump in the solid correlation). On the other hand, the controtatory photochemical process from the first excited state of cyclobutene to the first excited state of butadiene has an energy barrier and is forbidden, while the disrotatory mode is allowed since there is no barrier.

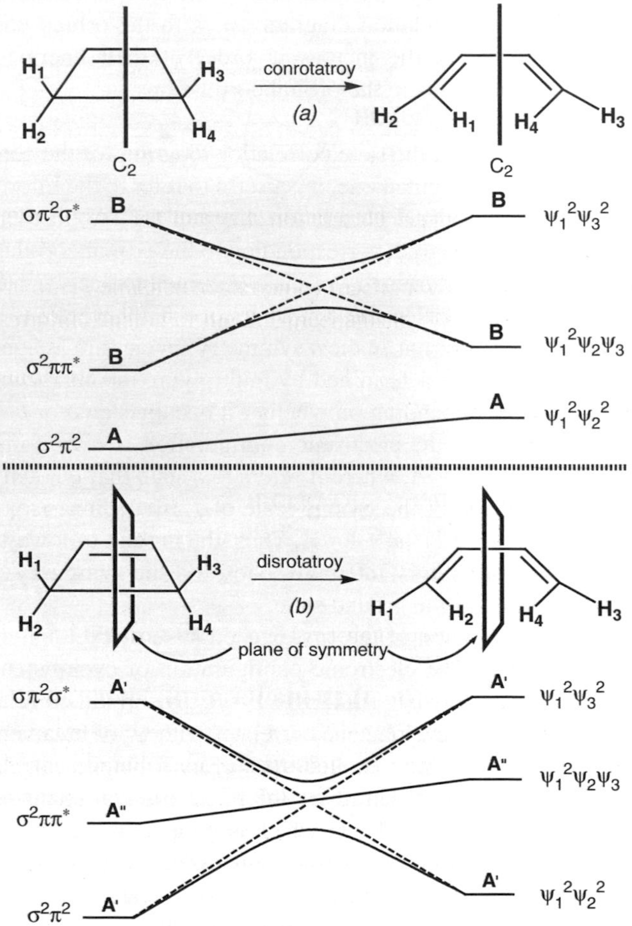

Figure 14.93. State correlation diagrams: *(a)* conrotatory and *(b)* disrotatory ring opening of cyclobutene.

14.94 ALLOWED REACTION

A pericyclic reaction for which orbital symmetry is conserved. Such reactions must involve conversion of the ground electronic state of reactant into the ground state of product. Another possibility is conversion of the first excited state of reactant into the first excited state of the product. The former type of conversion is referred to as being thermally allowed, and the latter as photochemically allowed.

Example. In the state correlation diagram for the cyclobutene-butadiene reaction (Fig. 14.93*a*), the conrotatory mode is an allowed reaction in the ground state because this is an A to A conversion from ground state to ground state. The orbital correlation diagram (Fig. 14.91*a*) also shows the allowed nature of the conrotatory ring-opening reaction in the ground state because filled orbitals correlate only with filled orbitals.

14.95 FORBIDDEN REACTION

A pericyclic reaction in which orbital symmetry is *not* conserved, that is, in which occupied orbitals of the reactant transform into unoccupied orbitals of the product with the same symmetry. If a reaction is forbidden, it may still occur in a concerted mode at a high temperature, or it may occur via some nonconcerted pathway, such as through a diradical or by metal catalysis.

Example. For the conrotatory conversion of cyclobutene to butadiene, the first excited state of cyclobutene correlates with the second excited state of butadiene. Although both these states have the same B symmetry, this interconversion is not energetically favorable. The reaction is, therefore, photochemically forbidden.

For the disrotatory conversion of cyclobutene to butadiene, the ground state of cyclobutene correlates with a higher excited state of butadiene. Again, although both these states have the same symmetry, this interconversion is not energetically favorable, and thus, the disrotatory conversion is thermally forbidden. The orbital correlation diagram (Fig. 14.91*b*) also shows the forbidden nature of the disrotatory ring-opening reaction in the ground state because a filled orbital correlates with an empty orbital of butadiene.

14.96 CYCLOADDITION

An addition of the two termini of one π system to the termini of a second π system. Because the two termini of each π system remain connected, the reaction generates a product that has a new ring. Cycloadditions frequently are called $[m + n]$-cycloadditions, in which m is the number of electrons in one π system and n is the number of electrons in the other. The mechanisms of cycloaddition are often, but not always, concerted.

Example. Different types of cycloadditions are shown in Figs. 14.96*a*, *b*, and *c*, including the [2 + 2]-cycloaddition, Diels–Alder or [4 + 2]-cycloaddition, and [4 + 4]-cycloaddition.

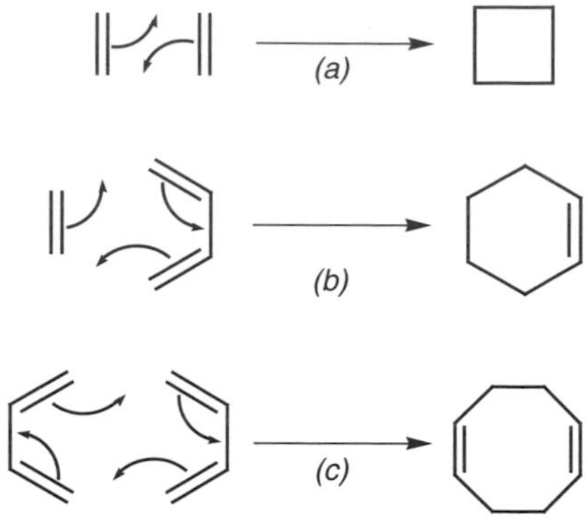

Figure 14.96. Cycloadditions: (*a*) [2 + 2]-cycloaddition; (*b*) Diels–Alder or [4 + 2]-cycloaddition; (*c*) [4 + 4]-cycloaddition.

14.97 CYCLOREVERSION

The microscopic reverse of a cycloaddition reaction.

Example. The reverse of the reactions shown in Fig. 14.96.

14.98 RETROGRADE CYCLOADDITION

Synonymous with cycloreversion.

14.99 ALDER RULES

Named for Kurt Alder (1902–1958, who received the Nobel Prize in 1950 jointly with Otto Diels), these also are known as the Alder–Stein rules. These rules predict the stereochemistry and regiochemistry of the kinetic product expected from a Diels–Alder reaction:

1. The diene must adopt an *s-cis* geometry, and addition to both the diene and dienophile must occur in a *cis* manner, Fig. 14.99*a*.
2. The dienophile adds in an *endo* manner to the diene, Fig. 14.99*b*.
3. When starting with a monosubstituted diene and a monosubstituted dienophile, the 1,2- or 1,4-disubstituted product is more likely to occur than the 1,3-disubstituted product, Fig. 14.99*c*.

Example. Figures 14.99*a*, *b*, and *c* correspond respectively to the three Alder rules given above.

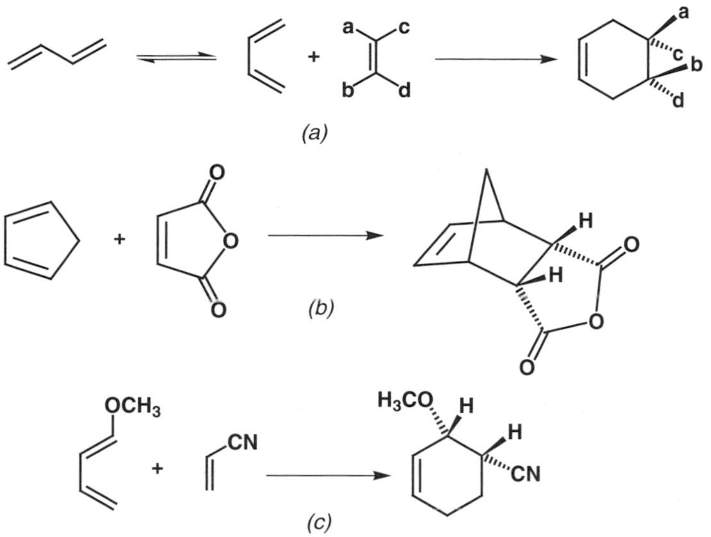

Figure 14.99. Examples of the products predicted by the Alder rules.

14.100 1,3-DIPOLE

A three-atom linkage, one resonance form of which has a negative charge at one end and a positive charge (hence an empty orbital) at the other. Such species have 4π electrons. (Notice that the electrons in the orthogonal π system are not included because they do not undergo a change during the reaction.)

Example. Figure 14.100.

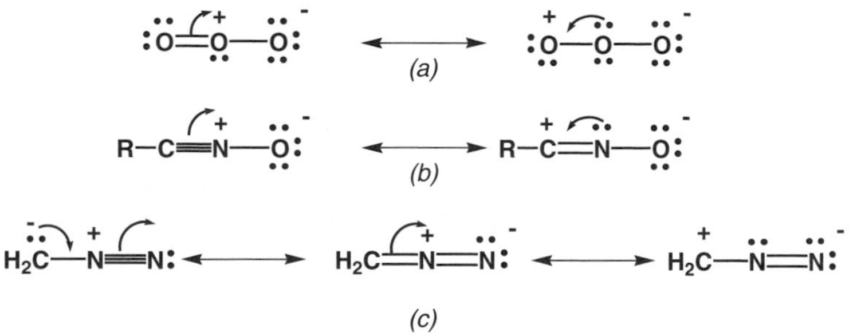

Figure 14.100. 1,3-Dipoles: (*a*) ozone; (*b*) a nitrile oxide; (*c*) diazomethane.

14.101 1,3-DIPOLAR ADDITION

A cycloaddition involving a 1,3-dipolar molecule or fragment and another (usually, π) fragment. If the second fragment were a single π bond, such a cycloaddition would be described as [4 + 2]-addition, and would follow the same symmetry selection rules *(vide infra)* as any other [4 + 2]-cycloaddition.

Example. Figure 14.101.

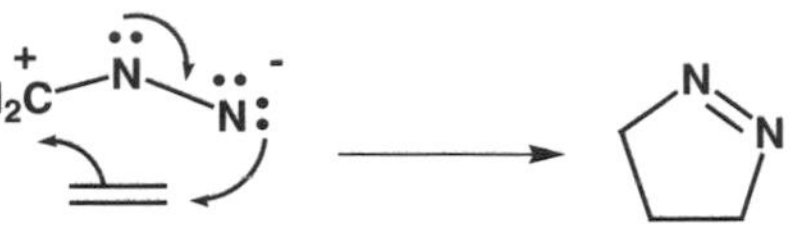

Figure 14.101. 1,3-Dipolar addition.

14.102 SIGMATROPIC SHIFTS

Rearrangements that consist formally of the migration of a σ bond (actually, the σ electrons) and the group attached to this bond from one position in a chain or ring to a new position in the chain or ring. The migrating σ bond often is at an allylic position, and during the rearrangement the σ electrons are accepted into a *p* orbital at the other terminus of the allylic system.

14.103 [1,*n*]-SIGMATROPIC SHIFT

A sigmatropic shift in which there is no rearrangement within the migrating group. The number 1 is assigned to the atom of the chain or ring attached to the migrating σ bond. The number *n* is assigned to the new point of attachment of the σ bond. Thus, a group has migrated from the first to the *n*th position along the chain or ring. This group may migrate with inversion or with retention of its stereochemistry.

Example. Figures 14.103*a*, *b*, and *c* show a [1,2]-shift, [1,3]-shift, and [1,5]-shift, respectively, all with retention of stereochemistry at the migrating center. Figure 14.103*d* shows a [1,3]-shift with inversion of stereochemistry at the migrating center.

14.104 [*m*,*n*]-SIGMATROPIC SHIFT ($m \neq 1$)

In its most general sense, a sigmatropic shift may be regarded as the rearrangement of two molecular fragments, one fragment containing *m* atomic orbitals and the other *n* atomic orbitals. All $m + n$ atomic orbitals are involved in the shift process. The [1,*n*]-sigmatropic shift (Sect. 14.103) is, therefore, a special case in which only a single orbital on one of these fragments is involved in the shift process. The sigmatropic shift itself involves the breaking of the bond linking the two fragments at the 1 and 1′ positions and the formation of a new bond at the *m*-position of one fragment and the *n*-position of the other.

Example. The Cope rearrangment (Fig. 14.104*a*) and Claisen rearrangement (Fig. 14.104*b*) are examples of [3,3]-sigmatropic shifts. The Cope rearrangement of the parent 1,5-hexadine best illustrates this process. Two allyl fragments are involved in the shift, which is best seen in the transition state for the shift. These allyl fragments

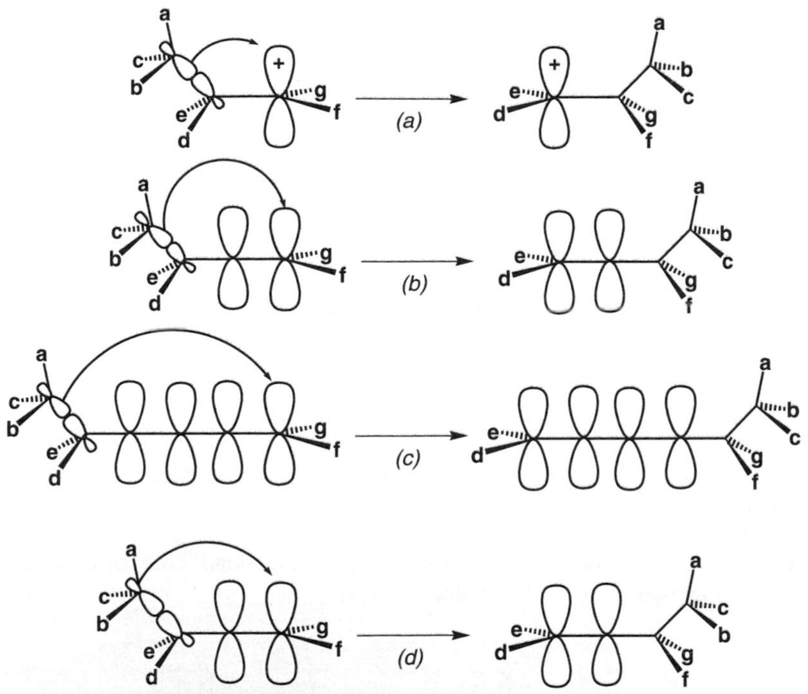

Figure 14.103. [1,*n*]-Sigmatropic shifts: (*a*) [1,2]-shift with retention; (*b*) [1,3]-shift with retention; (*c*) [1,5]-shift with retention; (*d*) [1,3]-shift with inversion.

are originally connected through a bond between the 1 and 1′ carbon atoms and become linked through a new bond between their 3 and 3′ carbon atom, hence a [3,3]-sigmatropic shift. In many cases, and this one in particular, the distinction of which group "migrates" is artificial, since at least here the two groups are identical. In face the reaction in Fig. 14. 104a is an example of a degenerate rearrangement (see Sect. 14.87), since, without the appropriate labeling, the product is indistinguishable from the starting material.

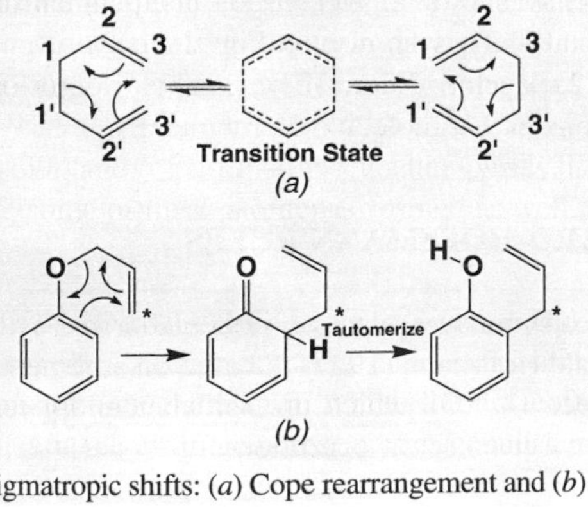

Figure 14.104. [3,3]-Sigmatropic shifts: (*a*) Cope rearrangement and (*b*) Claisen rearrangement.

14.105 CHELETROPIC REACTION

A cycloaddition in which one of the reacting species acts through a single atom possessing both a filled and empty orbital.

Example. The addition of a carbene to an alkene (Fig. 14.105).

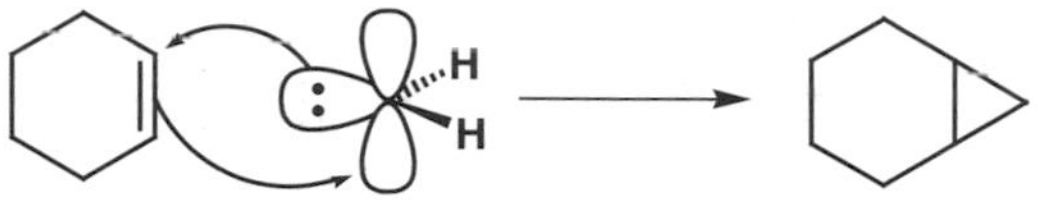

Figure 14.105. A cheletropic reaction.

14.106 GROUP TRANSFER REACTION

The one-step *cis* addition of two atoms or groups to a π (or σ) bond via a cyclic transition state.

Example. Figure 14.106. The reduction of a double bond with diimide (N_2H_2) involving the formal transfer of a molecule of H_2.

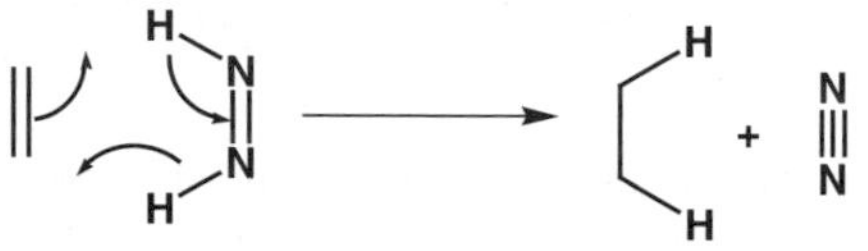

Figure 14.106. Group transfer reaction of hydrogen to a double bond.

14.107 SUPRAFACIAL *s* AND ANTARAFACIAL *a*

Terms that describe the stereochemical relationship between the reacting termini of a molecule involved in a pericyclic reaction. For suprafacial, the attachments occur on the same face (side) of the molecule, whereas for antarafacial, the attachments occur on opposite sides of the molecule.

Example. Figure 14.98*b* shows a suprafacial [1,3]-sigmatropic shifts. The corresponding antarafacial shift is shown in Fig. 14.107a. Figure 14.107b shows a [2*s* + 2*s*]- and [2*s* + 2*a*]-cycloaddition. (The number preceding the *s* or *a* indicates the number of electrons in that molecule or fragment.)

14.108 WOODWARD–HOFFMANN RULES

A series of generalized symmetry selection rules elaborated by Robert B. Woodward (1917–1979) and Roald Hoffmann (1937–), based on arguments such as orbital correlation diagrams (Sect. 14.105), which predict whether a given pericyclic reaction will be allowed under a given set of conditions.

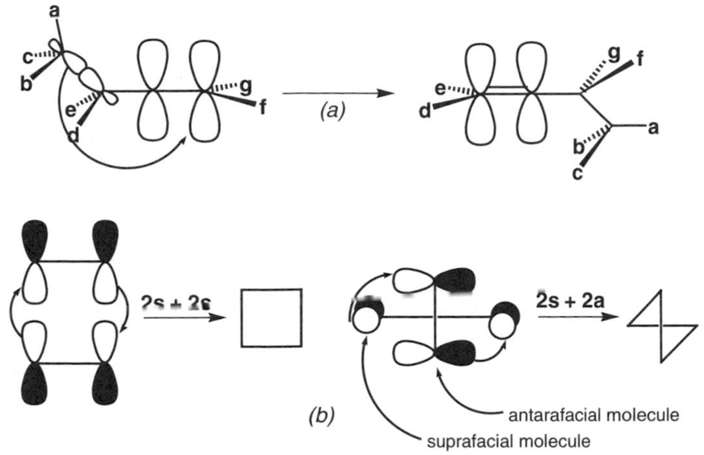

Figure 14.107. (*a*) Antarafacial [1,3]-sigmatropic shift and (*b*) [2*s* + 2*s*]- and [2*s* + 2*a*]-cycloadditions.

In its most general form, the rules state that a pericyclic reaction is thermally (ground state) allowed if the total number of (4*m*+ 2) suprafacial and (4*n*) antarafacial reacting components (electronic fragments) is odd, and is photochemically (first excited state) allowed if the total number of (4*m*+ 2) suprafacial and (4*n*) antarafacial reacting components is even, where *m* and *n* are any integer including zero. Note that the (4*m*+ 2) antarafacial and (4*n*) suprafacial components are not counted. The rules can be restated more clearly for individual subclasses of pericyclic reactions.

ELECTROCYCLIC. In such reactions, there is only one reacting component to be considered. If it contains (4*m*+ 2) π electrons (2, 6, 10, . . .), then the process is disrotatory (suprafacial) in the ground state and conrotatory (antarafacial) in the excited state. If it contains (4*n*) π electrons (4, 8, 12, . . .), then the process is conrotatory (antarafacial) in the ground state and disrotatory (suprafacial) in the excited state. For example, the cylobutene/butadiene conversion in Fig. 14.89 is a four-electron process and will occur in a conrotatory manner. In contrast, the cyclohexadiene/hexatriene conversion is a six-electron process and will occur in a disrotatory manner.

CYCLOADDITIONS. In such reactions, the number of electrons and the stereochemical relationship between the reacting termini for each component must be considered. For example, in the [2*s* + 2*s*] reaction in Fig. 14.107*b*, the number of (4*m*+ 2)s and (4*n*)*a* components is two, and thus, this is a forbidden reaction. In contrast, for the [2*s* + 2*a*] reaction, the number is one and, thus, is the allowed reaction pathway.

The Diels–Alder reaction shown in Fig. 14.99*b* is an example of an allowed [4*s* + 2*s*] reaction because the number of (4*m*+ 2)*s* and (4*n*)*a* components is one.

SIGMATROPIC SHIFTS. In the application of the Woodward–Hoffmann rules to such reactions, the "migrating bond" is treated as though it were heterolytically cleaved and the electrons assigned to the migrating group. For example, consider the [1,3]-shifts shown in Figs. 14.103*b*, 14.103*d*, and 14.107*a* that are a suprafacial shift with retention, a suprafacial shift with inversion, and an antarafacial shift with retention, respectively. The first is an example of a [2*s* + 2*s*] reaction and is forbidden, the second is a [2*s* + 2*a*] reaction and is allowed, and the third is a [2*a* + 2*s*] reaction and also is allowed.

14.109 FRONTIER-ORBITAL APPROACH

A methodology developed by Kenichi Fukui (1918– , who shared the Nobel prize with Roald Hoffmann in 1981) for quickly predicting whether a given pericyclic reaction is allowed or forbidden by examining the symmetry of the highest occupied molecular orbital (HOMO) and, if the reaction is bimolecular, the lowest unoccupied molecular orbital (LUMO) of the second partner.

Example. The HOMO of butadiene is ψ_2, which is shown in Fig. 14.91. For this orbital to close to form a σ bond and not the σ^* bond, the orbitals must rotate in a conrotatory manner; therefore, conrotatory is the allowed reaction pathway.

Cycloadditions are handled by examining the HOMO of one fragment and the LUMO of the other. As shown in Fig. 14.109*a*, the [2*s* + 2*s*] reaction is forbidden and the [2*s* + 2*a*] reaction is allowed. Figure 14.109*b* shows that the [4*s* + 2*s*] Diels–Alder reaction is allowed, regardless of which HOMO/LUMO pair is picked. This is true for any cycloaddition reaction.

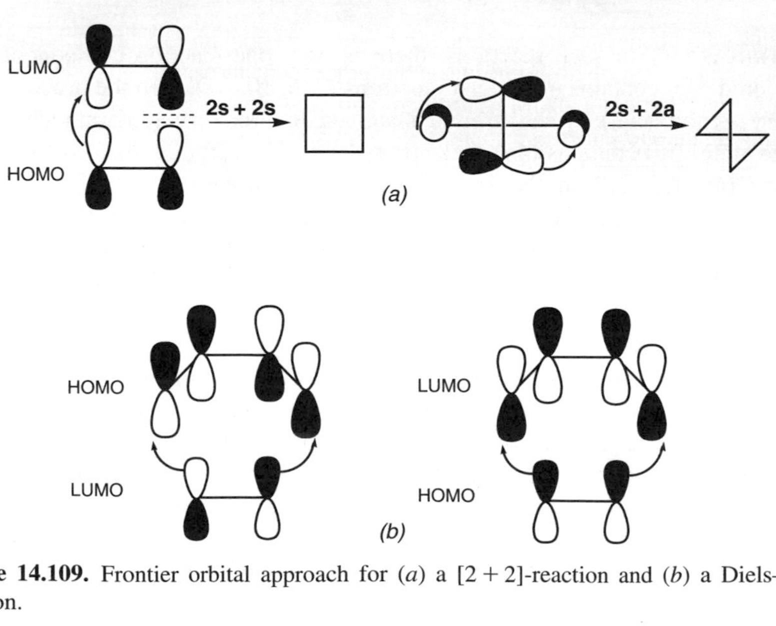

Figure 14.109. Frontier orbital approach for (*a*) a [2 + 2]-reaction and (*b*) a Diels–Alder reaction.

14.110 HÜCKEL–MÖBIUS (H–M) APPROACH

Devised by Erich Hückel (1896–1980) and August Möbius (1790–1868), this method, popularized by H. E. Zimmerman [*Accts. Chem. Res. 4*, 272 (1971)] and M. J. S. Dewar [*Angew. Chem., Int. Ed. 10*, 761 (1971)], for quickly assessing whether a given pericyclic process is allowed or forbidden examines the cyclic array of orbitals at the transition state of the pericyclic reaction. If this array has no nodes or an even number of nodes between atoms (called a Hückel array) and if it contains $(4q + 2)$ electrons, then it is stabilized and the reaction is allowed. If, on the other hand, it has an odd number of nodes between atoms (called a Möbius array), then it must possess $4r$ electrons to be stabilized. In assessing whether the array is Hückel or Möbius, nodes through atoms are not counted.

Example. In Fig. 14.110*a*, the conrotatory ring closure of butadiene is a four-electron process with one node so it is allowed, whereas the disrotatory ring closure is a four-electron process with no nodes so it is forbidden. In Fig. 14.110*b*, the $[2s + 2s]$-reaction is a four-electron process with no nodes so it is forbidden, whereas the $[2s + 2a]$-reaction is a four-electron process with one node so it is allowed. In Fig. 14.110*c*, the Diels–Alder reaction is a six-electron process with no nodes so it is allowed.

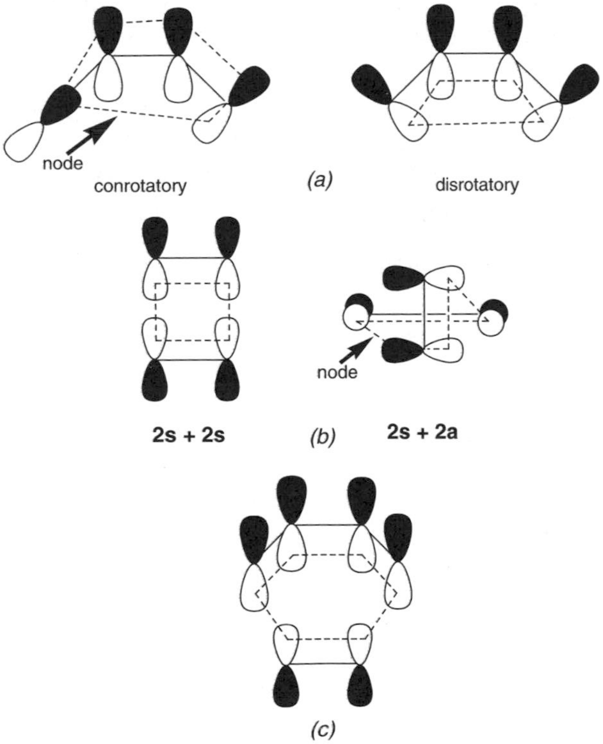

Figure 14.110. Application of the Hückel–Möbius rules: (*a*) electrocyclic reaction; (*b*) [2 + 2]-reaction; (*c*) Diels–Alder reaction.

14.111 BALDWIN'S RULES

Empirical rules, named after J. Badwin, for the mode of ring closure in a nucleophilic or radical reaction. See Sect. 7.148. In these rules, favored and disfavored indicate how likely a type of reaction is to occur; the number indicates the size of the ring formed in the product; the *tet*, *trig*, and *dig* refer to attack at an sp^3, sp^2, or *sp* hybridized carbon, respectively; and *endo* and *exo* refer to whether the carbon that was attacked becomes *endo* or *exo* in the newly formed ring system.

	Favored	Disfavored
tet	3–7 exo	5–6 endo
trig	3–7 exo and 6–7 endo	3–5 endo
dig	5–7 exo and 3–7 endo	3–4 exo

Example. Figure 14.111 compares a 5-exo-trig favored reaction and a 5-endo-trig disfavored reaction.

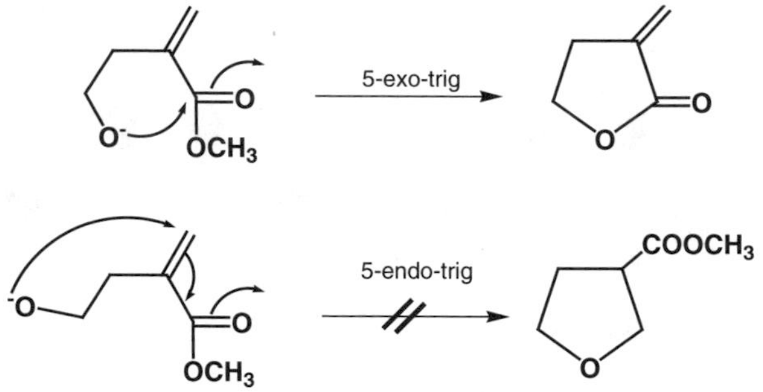

Figure 14.111. Baldwin's rules: the favored 5-exo-trig reaction and the disfavored 5-endo-trig reaction.

SUGGESTED READING

Carpenter, B. K. *Determination of Organic Reaction Mechanisms*. Wiley-Interscience: New York, 1984.

Carroll, F. A. *Perspectives on Structure and Mechanism in Organic Chemistry*. Brooks/Cole: Pacific Grove, CA, 1998.

Isaac, N. *Physical Organic Chemistry*, 2nd ed. Longman Scientific and Technical: Essex, UK, 1995.

Lowry, T. H. and Richardson, K. S. *Mechanism and Theory in Organic Chemistry*, 3rd ed. Harper and Row: Cambridge, MA, 1987.

March, J. *Advanced Organic Chemistry*, 4th ed. Wiley-Interscience: New York, 1992.

15 Nuclear Magnetic Resonance Spectroscopy

15.1	Nuclear Spin Quantum Number *I*	594
15.2	Magnetic Moment μ	594
15.3	Nuclear Quadrupole	594
15.4	Nuclear Spin State	594
15.5	Multiplicity (of Nuclear Spin States)	595
15.6	Nuclear Spin Orientation Quantum Number *m*	595
15.7	Nuclear Zeeman Effect	595
15.8	Precession	595
15.9	Larmor Frequency ω	596
15.10	Magnetogyric Ratio γ	596
15.11	Energy *E* of a Nuclear Spin State	597
15.12	Population *P* of Nuclear Spin States	597
15.13	Saturation	597
15.14	Spin Flip	598
15.15	High Field	598
15.16	Polarized Spin-State Population	598
15.17	Resonance Condition	598
15.18	Relaxation	598
15.19	Spin-Lattice (Longitudinal) Relaxation (T_1, Spin-Lattice Relaxation Time)	599
15.20	Phase-Coherent Spins	599
15.21	Spin-Spin (Transverse) Relaxation	599
15.22	Correlation Time τ_c	599
15.23	Laboratory Frame (of Reference)	600
15.24	Rotating Frame (of Reference)	600
15.25	Net Magnetization Vector **M**	601
15.26	NMR Spectrometer	602
15.27	Locking (the External Magnetic Field)	602
15.28	Shimming (the External Magnetic Field)	604
15.29	NMR Signal Generation	604
15.30	Free Induction Decay (FID) Signal	604
15.31	Fourier Transformation	605
15.32	The (Frequency Domain) NMR Spectrum	606
15.33	(Chemically) Equivalent Nuclei	607
15.34	Integral (of a Signal)	607
15.35	Homonuclear	608

The Vocabulary and Concepts of Organic Chemistry, Second Edition, by Milton Orchin, Roger S. Macomber, Allan Pinhas, and R. Marshall Wilson

15.36 Reference Compound/Signal 608
15.37 Signal Position $\delta\nu$ in Hz 609
15.38 Chemical Shift δ 609
15.39 Data Acquisition Parameters 609
15.40 Spectral (or Sweep) Width 610
15.41 Pulse Width 610
15.42 Signal-to-Noise Ratio (S/N) 610
15.43 Computer-Averaged Transients (CAT) 610
15.44 Dwell Time 610
15.45 Acquisition Time 611
15.46 Resolution 611
15.47 Delay Time 611
15.48 Line Width 611
15.49 Diamagnetic Shielding (Lenz's Law) 612
15.50 Paramagnetic Deshielding 614
15.51 Magnetic Anisotropy of Different Types of Chemical Bonds 614
15.52 Ring Current 617
15.53 Upfield (Shift) 618
15.54 Downfield (Shift) 618
15.55 Chemical Shift Ranges for ^{1}H and ^{13}C 618
15.56 Shoolery's Rules 620
15.57 Spin Coupling (Splitting) 620
15.58 Multiplicity of an NMR Signal 620
15.59 The $n + 1$ Rule 621
15.60 Relative Intensities of Lines Within a Signal 624
15.61 Coupling Constant J (Coupling Constant Magnitude and Sign) 624
15.62 Coupling Diagrams 625
15.63 The Karplus Equation 627
15.64 Weak (First-Order) Coupling 628
15.65 Strong (Second-Order) Coupling 628
15.66 Long-Range Coupling 629
15.67 Magnetic Equivalence 629
15.68 Pople Spin System Notation 630
15.69 *AB* Quartet 631
15.70 Subspectrum Analysis 631
15.71 Deceptively Simple Spectra 632
15.72 Virtual Coupling 632
15.73 Dynamic NMR Spectroscopy 633
15.74 (Site) Exchange 633
15.75 Slow Exchange Limit 634
15.76 Fast Exchange Limit 634
15.77 Coalescence 635
15.78 Reversible Complexation 635
15.79 Chemical Shift Reagents 637
15.80 Chiral Shift Reagents 637
15.81 Solvent-Induced Shift 638
15.82 Double Resonance Experiments 638
15.83 Spin Decoupling 638
15.84 Pulse Sequence 641

15.85	Nuclear Overhauser Effect (NOE)	641
15.86	Gated Decoupling	642
15.87	Inverse Gated Decoupling	643
15.88	Attached Proton Test (APT)	644
15.89	Distortionless Enhancement by Polarization Transfer (DEPT)	644
15.90	Off-Resonance Decoupling	646
15.91	Two-Dimensional NMR	646
15.92	Homonuclear Correlation (HOMCOR), H,H-COSY	646
15.93	Heteronuclear Correlation (HETCOR), C,H-COSY	648
15.94	Nuclear Overhauser Enhancement and Exchange Spectroscopy (NOESY)	650
15.95	Two-Dimensional Incredible Natural Abundance Double Quantum Transfer Experiment (2D INADEQUATE)	651
15.96	Magnetic Resonance Imaging (MRI)	653
15.97	Chemically Induced Dynamic Nuclear Polarization (CIDNP)	654
15.98	The Net Effect	654
15.99	The Multiplet Effect	654
15.100	The Radical Pair Theory of CIDNP Effects	654

Nuclear magnetic resonance (NMR) spectroscopy involves the interactions of atomic nuclei simultaneously with two different magnetic fields: a strong fixed magnetic field (the *external field* or *applied field* B_0) and a much weaker field (B_1) oscillating perpendicular to B_0 at radio frequencies. Fig. 15.0 shows the configuration of these component parts.

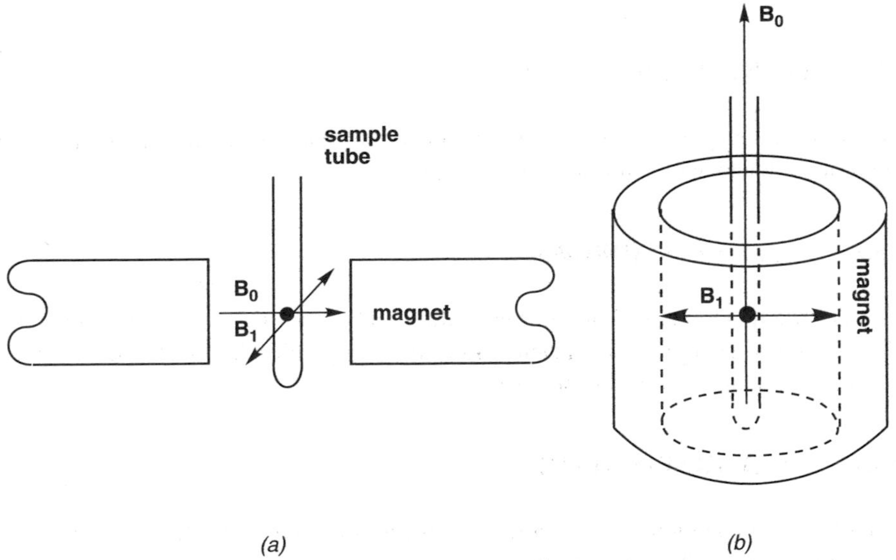

Figure 15.0. The configuration of fixed magnetic field B_0 and oscillating magnetic field B_1 in a conventional or electromagnetic (*a*) and a superconducting magnetic (*b*).

The spectroscopic signals generated by the nuclear–magnetic interactions give incredibly detailed information about the molecular environment of each atom, which is equivalent to molecular structure information including connectivity, geometry, proximity, dynamics, and so on. Indeed, it can be argued that NMR is the single most powerful spectroscopic technique for the elucidation of molecular structure and dynamics.

15.1 NUCLEAR SPIN QUANTUM NUMBER I

A specific number for each isotope of every element, always a multiple of 1/2, that describes the behavior of that isotope when immersed in an external magnetic field. Each proton and each neutron have a nuclear spin of 1/2. Therefore, the nuclear spin of any isotope is determined by the parity (odd or even) of the numbers of protons and the number of neutrons in its nucleus.

Example. Isotopes whose nuclei consist of even numbers of both protons and neutrons (e.g., ^{12}C, ^{16}O, ^{18}O, and ^{32}S) exhibit a nuclear spin of zero (0 times 1/2); nuclei with $I = 0$ are nonmagnetic and cannot be detected by NMR methods. A nucleus with odd numbers of both protons and neutrons exhibits a nuclear spin that is an even multiple of 1/2 (i.e., an integer); examples include ^{2}H ($I = 1$), ^{10}B ($I = 3$), and ^{14}N ($I = 1$). A nucleus with an odd number of protons and an even number of neutrons, or vice versa, exhibits a nuclear spin that is an odd multiple of 1/2; examples include ^{1}H ($I = 1/2$), ^{13}C ($I = 1/2$), ^{17}O ($I = 5/2$), ^{19}F ($I = 1/2$), and ^{31}P ($I = 1/2$). Any isotope with a nonzero nuclear spin has its own magnetic field and can be studied by NMR methods.

15.2 MAGNETIC MOMENT μ

A number that expresses the strength (in units of nuclear magnetons) of the magnetic field of a nucleus with a nonzero nuclear spin.

15.3 NUCLEAR QUADRUPOLE

The electric quadrupole of any nucleus with $I > 1/2$ resulting from the nonspherical distribution of charge in the nucleus. Such nuclei exhibit additional complications when being studied by NMR methods.

15.4 NUCLEAR SPIN STATE

A particular orientation of the magnetic moment of a nucleus, relative to an external magnetic field. Each such state has its own specific energy (Sect. 15.11).

15.5 MULTIPLICITY (OF NUCLEAR SPIN STATES)

The number (also called spin multiplicity) of different nuclear spin states that a given nucleus can adopt in the presence of an external magnetic field. This multiplicity is equal to $2I + 1$, where I is the nuclear spin of the nucleus (Sect. 15.1).

Example. Table 15.5 shows the relationship between the value of I, the spin multiplicity of the nucleus, and the allowed magnetic spin quantum numbers (Sect. 15.6).

TABLE 15.5. Correlation of Spin Quantum Number and Multiplicity

I	Multiplicity	m Value
0	1	0
1/2	2	+1/2, −1/2
1	3	+1, 0, −1

15.6 NUCLEAR SPIN ORIENTATION QUANTUM NUMBER *m*

The label of each nuclear spin state allowed for a nucleus with nuclear spin quantum number I. This label is related to the energy of the corresponding spin state (Sect. 15.11). A positive value of m indicates that the magnetic moment makes an acute angle with the external field (alignment "with" the field), a negative value indicates an obtuse angle (alignment "against" the field), and a value of zero indicates that the magnetic moment is perpendicular to the external field. The $2I + 1$ allowed values of m for a nucleus with nuclear spin I range from $+I$ to $-I$.

15.7 NUCLEAR ZEEMAN EFFECT

The term given to the observation that when a collection of nuclei of an isotope with $I > 0$ is immersed in an external magnetic field, the nuclei partition themselves among $2I + 1$ different nuclear spin states.

15.8 PRECESSION

The wobbling motion (similar to that of a spinning top) of the magnetic moment of an individual nucleus around the axis of the external magnetic field.

Example. See Fig. 15.8.

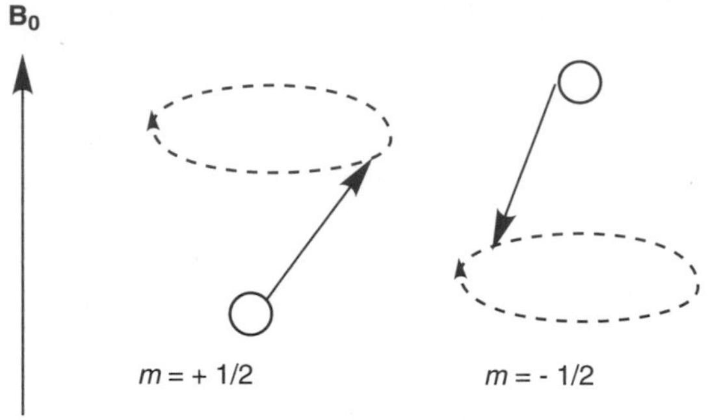

Figure 15.8. Precession of the magnetic moment of each of the two possible spin states of an $I = 1/2$ nucleus in external magnetic field B_0. The slanting arrows represent the precessing magnetic moment vectors.

15.9 LARMOR FREQUENCY ω

The frequency of nuclear precessional motion (Sect. 15.8), expressed either as angular frequency (ω, radians per second) or linear frequency (ν, Hz). The relationship between the two frequencies is given by $\omega = 2\pi\nu$. The Larmor frequency for a given nucleus is directly proportional to the strength of the external magnetic field B_0. It is the Larmor frequency of a nucleus that is the critical parameter in NMR spectroscopy.

15.10 MAGNETOGYRIC RATIO γ

The proportionality constant for a given isotope that relates Larmor frequency to external magnetic field strength, as in the equation $\omega = \gamma B$ or $\nu = \gamma B/2\pi$.

Example. Table 15.10 gives the value of γ (in units of 10^6 rad T^{-1} sec^{-1}) and the corresponding Larmor frequency (in units of MHz T^{-1}) for several common isotopes. A proton has the largest magnetogyric ratio of any isotope.

TABLE 15.10. Magnetic Characteristics of Several Important Isotopes in a Magnetic Field of 1T

Isotope	$\gamma \times 10^{-6}$ rad T^{-1} sec^{-1}	$\nu \times 10^{-6}$ Hz
^1H	267.5	42.6
^{13}C	67.3	10.7
^{19}F	251.7	40.0
^{31}P	108.3	17.2

Note: T represents Tesla, a unit of magnetic field strength; 1 T = 10^4 Gauss.

15.11 ENERGY *E* OF A NUCLEAR SPIN STATE

The energy E_i of nuclear spin state *i* is directly proportional to three variables: the magnetogyric ratio of the nucleus (γ, Sect. 15.10), the nuclear spin quantum number of the *i*th spin state (m_i, Sect. 15.6), and the strength of the external magnetic field B_0 according to the equation: $E_i = -m_i \gamma h B_0/2\pi$, where *h* is Planck's constant (Sect. 1.3). See Fig. 15.11. Note that the more positive m_i is, the lower the energy (and the greater the stability) of the corresponding spin state.

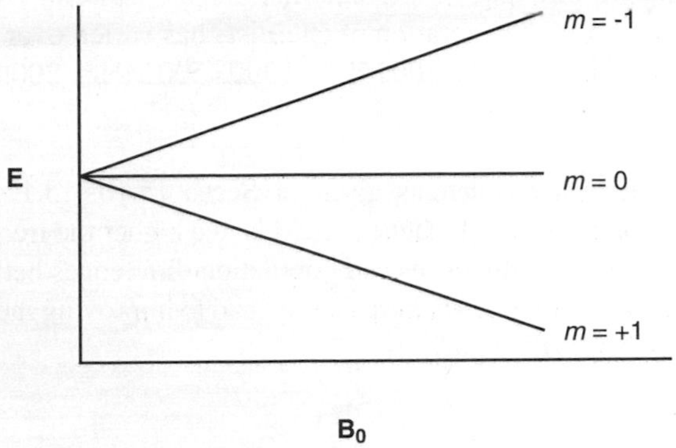

Figure 15.11. The energies of the three possible nuclear spin states ($m_i = +1, 0, -1$) of an isotope with $I = 1$.

15.12 POPULATION *P* OF NUCLEAR SPIN STATES

The fraction of nuclei in each nuclear spin state (Sect. 15.4) under a given set of conditions.

Example. At thermal equilibrium, the population of each spin state is determined by the relative energy of that state (Sect. 15.11) and the absolute temperature *T*, according to a *Boltzmann distribution*: $P_i \propto \exp(-E_i/kT)$, where *k* is the *Boltzmann constant*. Because the difference in energy between nuclear spin states is exceeding small at room temperature (e.g., on the order of 10^{-25} J for ^1H at 5.87 T), the population ratios are near unity, with only a few parts per million more in the lower state(s) compared to the higher state(s). Unfortunately, the strength of an NMR signal is proportional to this difference in populations between the states.

15.13 SATURATION

The condition where the populations (Sect. 15.12) of all spin states of a given isotope are equal, that is, the population differences are zero. At this point, the collection of nuclei cannot generate an NMR signal.

15.14 SPIN FLIP

The conversion of a magnetic nucleus from one spin state (i.e., $m = +1/2$) into another spin state (i.e., $m = -1/2$). A conversion that decreases the m value increases the energy of the nucleus (Sect. 15.11) and, thus, requires the absorption of energy.

15.15 HIGH FIELD

Refers to the relative strength of the external magnetic field. The highest field instrument available to each generation of chemists has varied over the years: 1960, 60 MHz; 1970, 100 MHz; 1980, 300 MHz; 1990, 500 MHz; 2000, 800 MHz and 2004, 900 MHz.

Example. According to the equations given in Sects. 15.10, 15.11, and 15.12, the greater the strength of the external magnetic field is, the higher the frequency of precession, the greater the energy differences and population differences between spin states, and the less chance of saturation. All these factors lead to improving the quality and usefulness of the resulting NMR signal.

15.16 POLARIZED SPIN-STATE POPULATION

A nonequilibrium population of a nuclear spin state (Sect. 15.12).

Example. Whenever a collection of magnetic nuclei in an external magnetic field, initially at thermal equilibrium, absorbs energy in such a way that some nuclei in the lower-energy spin state flip their spins, a polarized distribution results. A saturated spin-state distribution is an example of a polarized population.

15.17 RESONANCE CONDITION

The irradiation of a collection of magnetic nuclei, at thermal equilibrium in an external magnetic field (B_0), with a weaker oscillating magnetic field (B_1) whose frequency exactly matches the precessional frequency of the nuclei. Under these conditions, some of the nuclei can absorb energy from the oscillating field and undergo spin flips. This interaction is essential to the generation of an NMR signal.

15.18 RELAXATION

The conversion of a polarized spin-state distribution back to thermal equilibrium. There are several ways by which relaxation can occur, but regardless of the exact relaxation mechanism, the process is governed by an exponential decay of the form

$\exp(-t/T)$, where t is the time since the beginning of relaxation and T is the time constant for the particular type of relaxation.

15.19 SPIN-LATTICE (LONGITUDINAL) RELAXATION (T_1, SPIN-LATTICE RELAXATION TIME)

The mechanism for relaxation by which the excess energy of a polarized set of nuclear spins is dissipated as thermal energy (heat) to the bulk medium (the "lattice"), returning the system to thermal equilibrium. The time constants for spin-lattice relaxation, labeled T_1 (the spin-lattice relaxation time $= 1/k_1$), range from hundreds of seconds (inefficient spin-lattice relaxation) in crystalline solids to milliseconds (efficient spin-lattice relaxation) in lattices containing quadrupolar nuclei or paramagnetic molecules.

15.20 PHASE-COHERENT SPINS

The condition in which the magnetic moments of nuclei in a set precess in phase. This condition results from irradiation of the set of nuclei with oscillating magnetic field B_1.

Example. When a set of, for example, $I = 1/2$ nuclei is at equilibrium in an external magnetic field (B_0), the magnetic moments of the individual nuclei precess (Sect. 15.8) randomly (out of phase) around the axis of B_0, as shown in Fig. 15.20*a*. When the set of nuclei is irradiated by B_1, the magnetic moments "bundle up," precessing in phase with the oscillating magnetic field (Fig. 15.20*b*).

15.21 SPIN-SPIN (TRANSVERSE) RELAXATION (T_2, SPIN-SPIN RELAXATION TIME)

The mechanism for relaxation by which a phase-coherent bundle of nuclear magnetic moments becomes randomly dephased (Fig. 15.20*c*) after B_1 is turned off. The time constant for spin-spin relaxation, labeled T_2 (the spin-spin relaxation time), is usually shorter than T_1, that is, spin-spin relaxation is generally much faster than spin-lattice relaxation. Note that the relaxation time T_2 is inversely related to the rate constant for spin-spin relaxation. In most cases, spin-spin relaxation is hastened by inhomogeneities in B_0 (Sect. 15.28), giving rise to a shorter effective spin-spin relaxation time T_2^*. The magnitude of T_2^* determines the line width of a frequency domain NMR signal.

15.22 CORRELATION TIME τ_c

The time constant, labeled τ_c, for certain types of molecular motion (e.g., rotation and translation) that affect the rate of spin-lattice relaxation. In effect, τ_c is the average

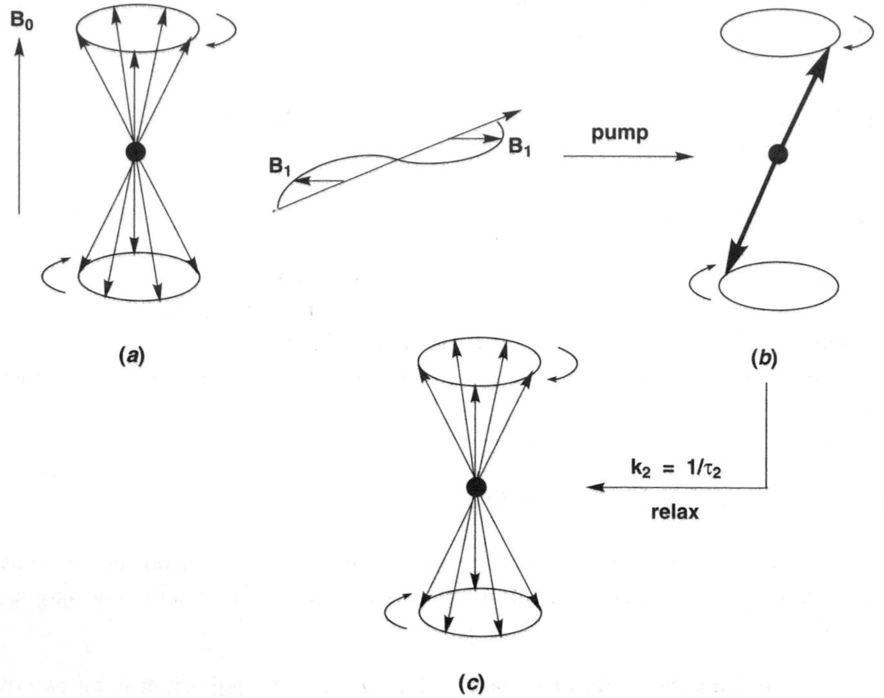

Figure 15.20. (*a*) Random precession of a set of nuclei around B_0. (*b*) Phase coherent precession of the set of nuclei after B_1 is turned on. (*c*) Random precession of a set of nuclei after relaxation.

length of time that two neighboring magnetic dipoles remain in the appropriate relative orientation to interact.

15.23 LABORATORY FRAME (OF REFERENCE)

A Cartesian coordinate system with axes labeled x, y, and z, where the z-axis is defined as parallel to the external magnetic field B_0. In the laboratory frame, nuclear magnetic moments (μ, Sect. 15.2) precess around the z-axis (Fig. 15.23).

15.24 ROTATING FRAME (OF REFERENCE)

A Cartesian coordinate system with axes labeled x', y', and z, where the z-axis is parallel to the external magnetic field B_0, while the x' and y' axes rotate around the z-axis at exactly the same frequency as a given nuclear magnetic moment precesses.

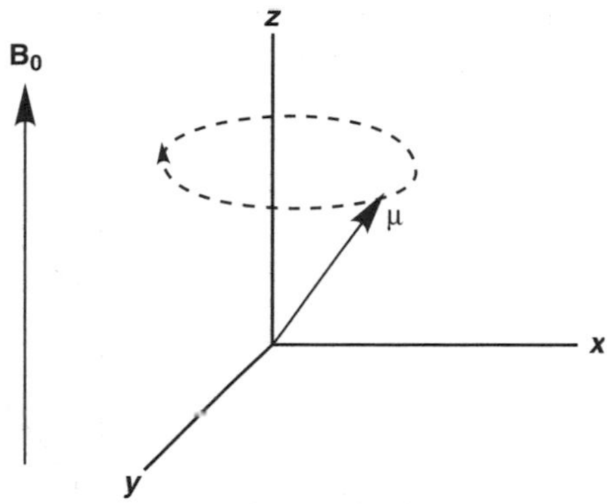

Figure 15.23. A single nuclear magnetic moment precessing around the z-axis in the laboratory frame.

Because the rotating frame and nuclear magnetic moment precess together, the magnetic moment appears static (unmoving) in the rotating frame. The rotating frame allows us to simplify pictorial representations of the behavior of precessing nuclear magnetic moments.

Example. Figure 15.24 shows the same nuclear magnetic moment depicted in Fig. 15.23, as it would appear in the rotating frame.

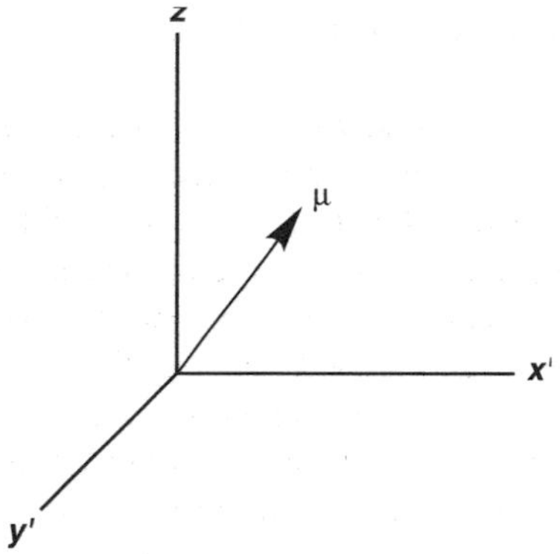

Figure 15.24. A precessing nuclear magnetic moment as it appears in the rotating frame. Note that the primes on the x and y-axis labels indicate that these axes are precessing in sync with the nuclear magnetic moment. Compare with Fig. 15.23.

15.25 NET MAGNETIZATION VECTOR M

The vector sum of all nuclear magnetic moments precessing in a given system.

Example. Figure 15.25 shows a rotating frame view of the net magnetization vector for the precessing nuclear magnetic moments in Figs. 15.20*a* and *b*.

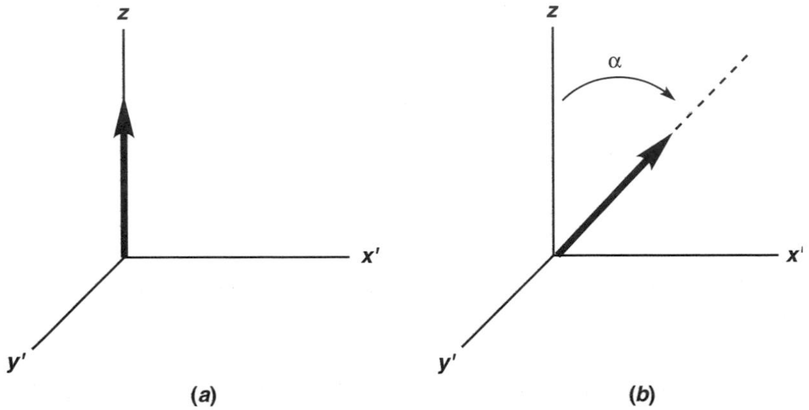

Figure 15.25. Pictorial rotating frame depiction of the net magnetization vector **M** for the nuclear magnetic moments in (*a*) Fig. 15.20*a* and (*b*) Fig. 15.20*b*.

15.26 NMR SPECTROMETER

An instrument designed to generate, collect, and display the signals of magnetic nuclei precessing in a strong magnetic field. NMR spectrometers are characterized by their ^{1}H operating frequency, for example, the precessional frequency of ^{1}H nuclei at the magnetic field strength of the instrument.

Example. A cutaway view of the key elements of a modern NMR spectrometer is shown in Fig. 15.26. The external magnetic field (B_0) is generated in the superconducting solonoid coils cooled to 4 K (−269°C) by liquid helium. This is necessary since the niobium-tin magnet core is only superconducting at a very low temperature. The *probe region*, at the center of the field, is not within the liquid helium, so its temperature can be set to any desired value from −100°C to +150°C. The sample to be investigated (usually, a solution) is contained in a glass tube, which is then inserted into the middle of the probe and spun by an air turbine. The oscillating magnetic field (B_1, Sect. 15.17) is generated by passing radio frequency current through the transmitter/receiver coil. If the magnet is designed to provide a ^{1}H precessional frequency of 400 MHz, the instrument is described as a 400-MHz NMR spectrometer.

15.27 LOCKING (THE EXTERNAL MAGNETIC FIELD)

An electronic feedback circuit that accurately maintains (stabilizes) the external magnetic field (B_0) at the nominal fixed value. In most modern NMR spectrometers, a deuterium (^{2}H) lock system is used. A substance (usually, the solvent in the sample tube)

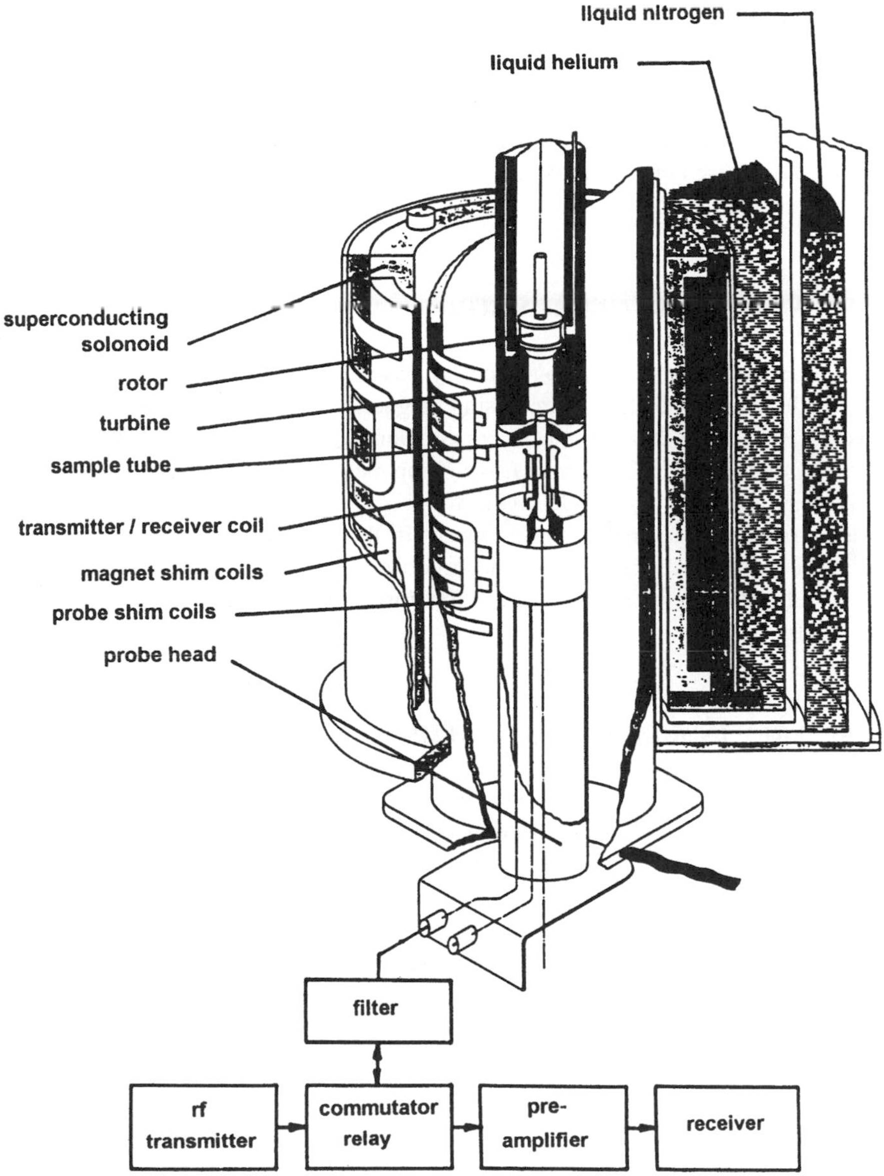

Figure 15.26. Cutaway diagram of a modern NMR spectrometer.

contains deuterium, and the resonance (precession) frequency of the deuterium nuclei is constantly compared to a reference oscillator frequency. A small change in the field strength of B_0 causes a corresponding change in the deuterium frequency (Sect. 15.10). The locking circuit detects this frequency drift and sends a signal to coils in the superconducting magnet that adjusts the field strength of B_0, bringing back the deuterium frequency to consonance with the reference oscillator.

15.28 SHIMMING (THE EXTERNAL MAGNETIC FIELD)

Electronic adjustments made to maximize the *homogeneity* (uniformity throughout the probe region) of the external magnetic field B_0. Very small electric currents are sent through a series of shim coils surrounding the probe (see Fig. 15.26). These currents induce small magnetic fields that compensate for any inhomogeneities in B_0.

15.29 NMR SIGNAL GENERATION

A strong but short pulse of radio frequency current sent through the transmitter coil (Fig. 15.26) causes the already precessing magnetic nuclei to bundle up, as shown in Figs. 15.20*b* and 15.25*b*. The net magnetization vector **M**, which has been rotated by *tip angle* α (Fig. 15.25*b*), now has a component rotating in the *x*,*y*-plane at the Larmor frequency of the nuclei. This rotating magnetic component induces (by Faraday induction) a small electric current in the receiver coil (Fig. 15.26). This induced current oscillates at the Larmor frequency (or frequencies, if several different sets of nuclei are precessing) and is the primary NMR signal of the sample (see Fig. 15.30*b*).

15.30 FREE INDUCTION DECAY (FID) SIGNAL

The primary NMR signal described in Sect. 15.29, collected digitally as a function of time. This signal is the composite of all contributing Larmor (precessional) frequencies in the sample. The FID signal is also known as the *time domain spectrum*. See Sect. 15.39.

Example. Figure 15.30*a* shows the ^{1}H FID signal for toluene. The observed signal is a composite of damped harmonic oscillations associated with each type of proton in the molecule. The corresponding FID for a single type of proton is shown in Fig. 15.30*b*. When the receiver coil is oriented with its axis perpendicular to that of the applied field, it will detect the rotating component of magnetic moment spinning at the Larmor frequency. This process produces a current in the receiver coil that decays as a damped harmonic oscillation. The envelope of the signal decays with a time constant of T_2^* (Sect. 15.21). Thus, the intensity is given by the formula

$$I = I_0 \exp(-t/T_2^*) \qquad (15.30)$$

The first step in the analysis of complex FID data (Fig. 15.30*a*) is the deconvolution (separation) into the individual damped harmonic components (Fig. 15.30*b*) associated with each type of nuclei present in the molecule under observation.

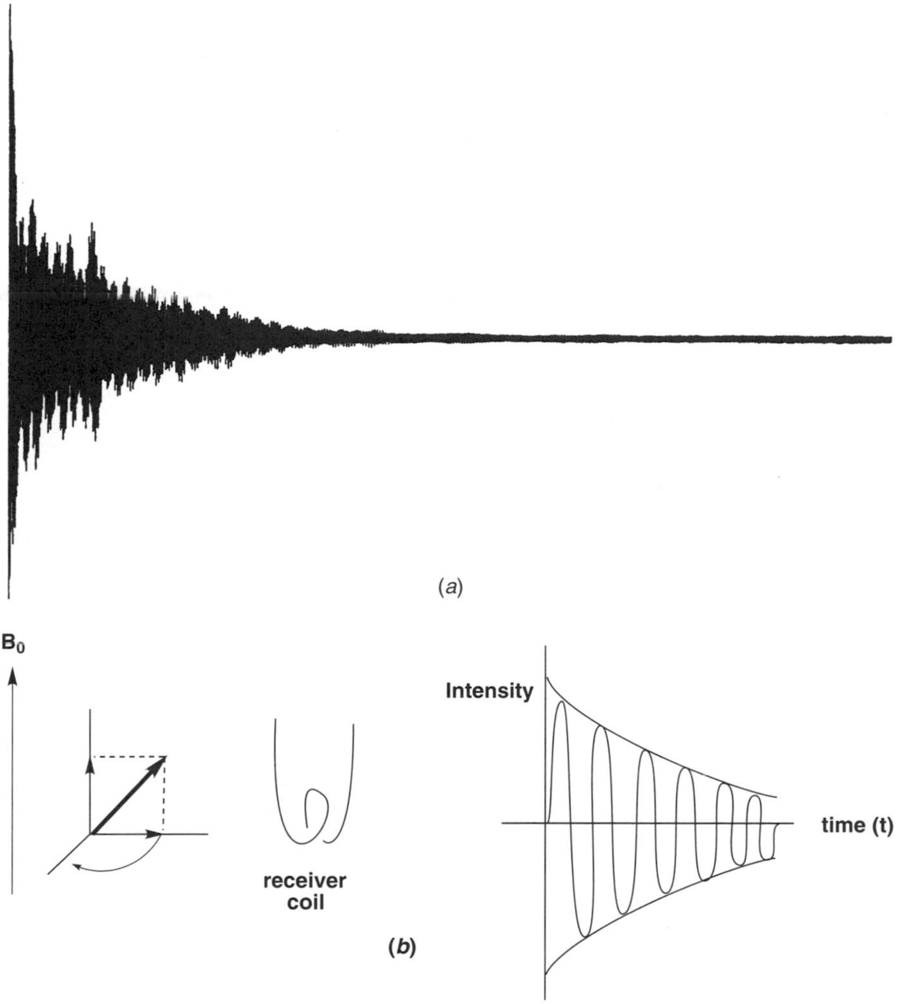

Figure 15.30. The FID ^{1}H signals: (*a*) for toluene and (*b*) for a single type of proton.

15.31 FOURIER TRANSFORMATION

The mathematical process by which time domain data is converted to frequency domain data, and *vice versa*. A plot of the transformed data will reveal one signal for each contributing Larmor frequency that made up the original FID signal. The intensity of each signal is related to, among other factors, the number of nuclei giving rise to that signal.

Example. Once the component damped harmonic oscillations have been deconvoluted from the total FID as outlined in Sect. 15.30, the periods of oscillation t can be

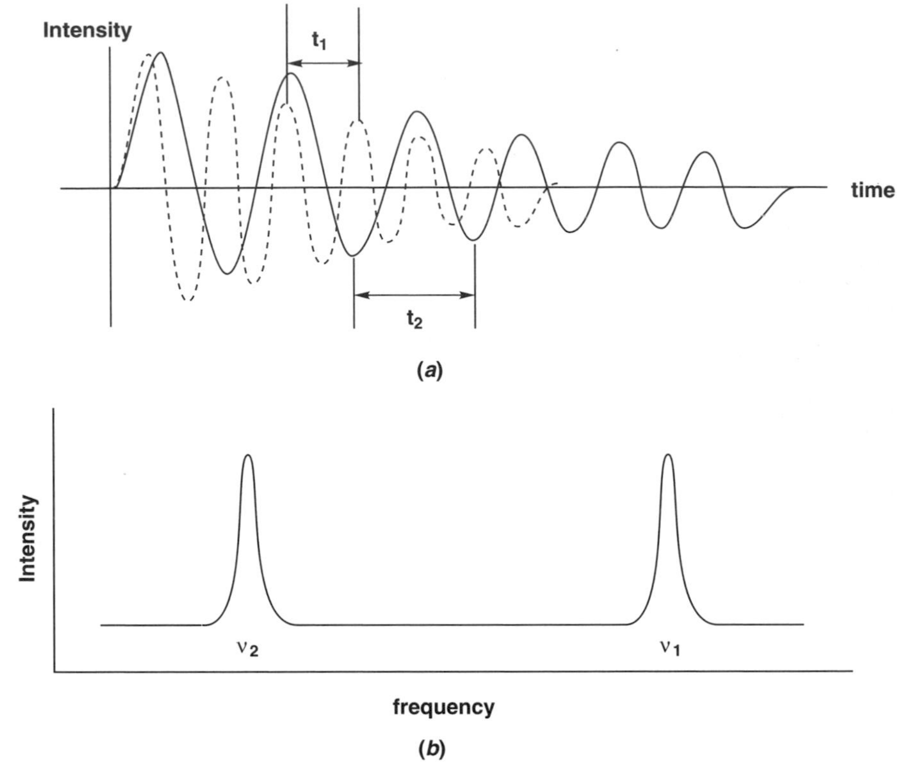

Figure 15.31. (*a*) Two resolved FID components and (*b*) FID components transformed into a frequency distribution.

determined for each component (Fig. 15.31*a*). The reciprocals of these periods are the frequencies of the oscillation components. It is the distribution of these frequencies that constitutes the conventional NMR spectrum.

15.32 THE (FREQUENCY DOMAIN) NMR SPECTRUM

A plot of the frequency domain signal resulting from Fourier transformation of the FID signal. This plot has frequency (or a parameter related to frequency), increasing to the left, as the x-axis and signal intensity as the y-axis.

Example. Figure 15.32 is the ^{1}H NMR spectrum of toluene, the Fourier transform of the FID signal in Fig. 15.30. This spectrum consists of one sharp signal at 2.34 ppm (Sect. 15.38) and a complex signal around 7.2 ppm. The signal at 0.0 ppm (Sect. 15.38) is due to TMS, the reference compound.

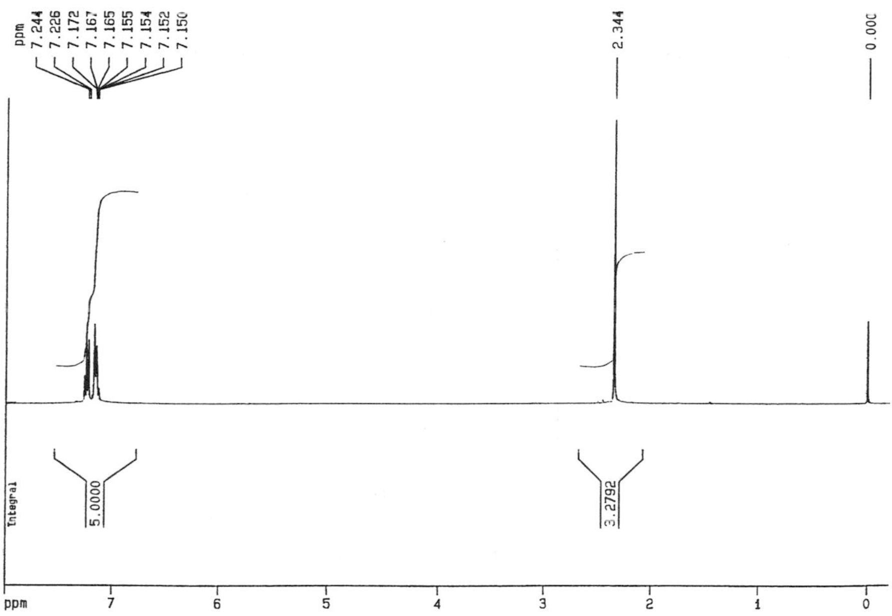

Figure 15.32. The 400-MHz ^{1}H spectrum of toluene.

15.33 (CHEMICALLY) EQUIVALENT NUCLEI

Two or more nuclei in a structure that are equivalent by symmetry (C_n or S_n axes, see Sects. 4.3 and 4.7) and, therefore, have identical precessional frequencies. Two nuclei that are not equivalent by symmetry must have different environments within a molecule and will therefore exhibit slightly different precessional frequencies. The number of NMR signals that a compound exhibits is equal to the number of nonequivalent sets of equivalent magnetic nuclei present in the structure.

15.34 INTEGRAL (OF A SIGNAL)

The relative intensity of a signal in the NMR spectrum, measured electronically as the area under the signal curve. The integral of a signal can, at the option of the spectroscopist, be plotted on the spectrum.

Example. In Fig. 15.32, the integral of the two signals is (left to right) 5.00 to 3.28. In the structure of toluene, the three methyl hydrogens are equivalent, the two *ortho* aromatic hydrogens are equivalent, and the two *meta* aromatic hydrogens are equivalent; the *para* aromatic hydrogen is not equivalent to any of the others. So the ^{1}H NMR spectrum should exhibit four signals with relative intensity 3:2:2:1. Figure 15.32

shows the methyl signal at 2.344 ppm; the three aromatic signals overlap to give the complex pattern around 7.2 ppm. Similarly, toluene should exhibit five carbon signals: methyl, ipso, *ortho*, *meta*, and *para*:

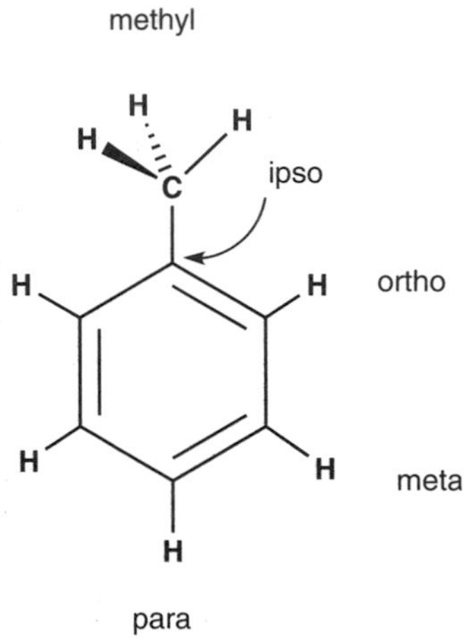

15.35 HOMONUCLEAR

Restricted to one isotope.

Example. Each NMR spectrum of a compound is homonuclear in the sense that the signals of only one isotope can be displayed in a given spectrum. Thus, for example, a sample of toluene can be examined under conditions of field strength and operating frequency that provide its ^{1}H NMR spectrum (Fig. 15.32). Then the operating frequency can be changed to that value appropriate for ^{13}C, and the ^{13}C NMR spectrum can be obtained. Each NMR spectrum shows only the signals for nuclei of the target isotope.

15.36 REFERENCE COMPOUND/SIGNAL

A sharp NMR signal used to define "zero" of the frequency axis of an NMR spectrum (Sect. 15.32).

Example. In order to determine the positions (precessional frequencies) of NMR signals, it is necessary to have a reference signal (i.e., not from the compound being analyzed) of the isotope under investigation (i.e., a ^{1}H signal for a ^{1}H spectrum, a ^{13}C signal for a ^{13}C spectrum, etc.) that serves as a "zero" frequency marker. This signal is due to a small amount of a *reference compound* added to the sample, a compound that exhibits a sharp signal for the isotope under investigation. This signal determines 0.00 Hz (Hertz, cycles per second) of the frequency scale. The compound *tetramethylsilane* [TMS, $(H_3C)_4Si$], which exhibits one sharp ^1H signal (for the 12 equivalent hydrogens), one sharp ^{13}C signal (for the four equivalent carbons), and one sharp silicon signal, is used as the reference compound for ^1H (Fig. 15.32, 0.00 ppm signal), ^{13}C, and ^{29}Si NMR spectra. NMR spectra of other isotopes (e.g., ^{19}F, ^{31}P) require reference compounds with the appropriate nuclei. Internal standards, such as TMS, are added directly to the sample, and external standards are measured in a separate sample tube.

15.37 SIGNAL POSITION $\delta\nu$ IN HZ

The position of each signal in an NMR spectrum is determined electronically by measuring its frequency (in Hz) *relative* to the reference signal frequency.

Example. A signal in an ^1H spectrum occurring at a frequency 937.6 Hz higher than (to the left of) the TMS signal (Sect. 15.36) is said to have a position $\delta\nu$ of 937.6 Hz (relative to TMS). A signal occurring at a frequency 937.6 Hz lower than (to the right of) the TMS signal is said to have a position $\delta\nu$ of -937.6 Hz (relative to TMS). Note that the position of an NMR signal, expressed in Hz, is directly proportional to the operating frequency of the instrument, which in turn is directly proportional to the magnetic field strength of the instrument.

15.38 CHEMICAL SHIFT δ

The position of an NMR signal δ, expressed in parts per million (ppm) relative to the operating frequency of the instrument and reference signal. The chemical shift of a given nucleus is sensitive to its exact magnetic environment.

Example. The chemical shift δ of an NMR signal is its single most important characteristic. It is related to it position $\delta\nu$ in Hz by the equation $\delta = \delta\nu/OF$, where OF is the operating frequency in MHz of the instrument. Thus, a 400-MHz ^1H NMR signal at 937.6 Hz (to the left of TMS) has a chemical shift *d* of $937.6/400 = 2.344$ ppm. A signal at 937.6 Hz to the right of TMS has a chemical shift of -2.344 ppm.

15.39 DATA ACQUISITION PARAMETERS

The set of adjustable instrument parameters that result in generation and collection of the FID signal (Sect. 15.30). These include spectral width, pulse width, number of scans, dwell time, acquisition time, and delay time.

15.40 SPECTRAL (OR SWEEP) WIDTH

The range of frequencies (in Hz) to be acquired in an NMR spectrum.

Example. In ^{1}H NMR most signals occur over a span of approximately 15 ppm (Sect. 15.38), or 6,000 Hz at an operating frequency of 400 MHz. By contrast, in ^{13}C NMR the spectrum may cover a span of 300 ppm, or 30,000 Hz at an operating frequency of 100 MHz.

15.41 PULSE WIDTH

The length of time, in μsec, that the oscillating magnetic field B_1 is turned on, that is, the width in μsec of the B_1 pulse. See Sect. 15.29. This pulse must bring into coherence (Sect. 15.20) all target nuclei. The pulse width is inversely related to the spectral width (Sect. 15.40): spectral width $\leq$ (pulse width)$^{-1}$.

Example. In typical NMR experiments, the pulse width is set to approximately one-fourth of the inverse of the sweep width. Thus, a sweep width of 6,000 Hz requires a pulse width of $(1/4)(6{,}000\ \text{Hz})^{-1} = 4.167 \times 10^{-5}\ \text{sec} = 41.67\ \mu\text{sec}$.

15.42 SIGNAL-TO-NOISE RATIO (S/N)

The ratio of the height of a frequency domain NMR signal to the average height of the random baseline noise in the frequency domain spectrum (Sect. 15.32). This noise is an artifact of the electronic circuitry in an NMR instrument. The greater the S/N is, the more precisely an NMR signal can be identified and integrated. The S/N can be increased by signal averaging techniques.

15.43 COMPUTER-AVERAGED TRANSIENTS (CAT)

A technique by which a set of digital spectroscopic data (e.g., an FID signal, Sect. 15.30) can be averaged over several measurements to provide an increased signal-to-noise ratio (S/N). This procedure is also known as *signal averaging*.

Example. By repeating the acquisition of a spectroscopic data set a number of times n (where n is number of *scans*, also called "shots" or "reps"), then adding all the data together, signal height increases directly with n, while noise increases with $(n)^{1/2}$. Therefore, the S/N increases by $n/(n)^{1/2} = (n)^{1/2}$. Thus, by repeating the acquisition of data 16 times, the S/N improves by a factor of 4.

15.44 DWELL TIME

The time interval between successive samplings (points) of the digital FID signal.

Example. The *Nyquist theorem* (from information theory) states that in order to determine the frequency of each contributor to an FID signal (see Fig. 15.31), one must acquire at least two points along each cycle of the highest-frequency contributor. Since, for a sweep width of 6,000 Hz, the highest frequency to be measured is 6,000 Hz (166 μsec per cycle = $1/\tau_1$ in Fig. 15.31*a*), one must measure the signal no less than once every 83 μsec (166 μsec/2).

15.45 ACQUISITION TIME

The length of time over which the FID signal is monitored. The acquisition time is inversely related to the desired resolution.

15.46 RESOLUTION

The smallest separation between two frequency domain signals that still allows the signals to be distinguished. The resolution provided in an FID data set is the inverse of the acquisition time.

Example. If the desired level of resolution in a given NMR experiment is to be no more than ±0.5 Hz, the acquisition time must be at least 2 sec. If the dwell time (Sect. 15.44) is, for example, 83 μsec, an acquisition time of 2 sec will generate 24,096 data points. The number of data points generated in an experiment must not exceed the RAM capacity of the computer being used to acquire the data.

15.47 DELAY TIME

The length of time between the end of one data acquisition sequence and the beginning of the next pulse.

Example. After one pulse/acquisition sequence (one scan, also called a "shot" or "rep"), the nuclei must be given the opportunity to undergo spin-lattice relaxation before the next B_1 pulse is initiated. For hydrogen nuclei, 1 or 2 sec is usually sufficient; for some nuclei that relax more slowly, longer delays are necessary.

15.48 LINE WIDTH

The width of a frequency domain signal, measured (in Hz) at half-height, that is, the half-width. The half-width of a Lorentzian frequency domain NMR signal is the inverse of the fastest relaxation time for the nuclei giving rise to the signal. Usually, this is the effective spin-spin relaxation time T_2^*.

15.49 DIAMAGNETIC SHIELDING (LENZ'S LAW)

The reduction in effective magnetic field strength that occurs when an induced magnetic field opposes the applied or external field at the nucleus under observation.

Example. Lenz's law [Heinrich Lenz (1804–1865)] states that an induced field will always oppose an applied field (the inducing field) at the center of the induced field. This situation is represented in Fig. 15.49*a*. In the case of a typical atom (a nucleus surrounded by its associated electrons), the applied field B_0 will induce a magnetic field B_I in the electron cloud surrounding the nucleus. According to Lenz's law, this induced field will oppose the applied field at the central nucleus. Therefore, the magnitude of the magnetic field at the nucleus, the effective field B_E, will be the applied field strength B_0 reduced by the strength of the induced field B_I (Fig. 15.49*b*):

$$B_E = B_0 - B_I \tag{15.49a}$$

As a result, the nucleus will be exposed to a lower field strength than that of the applied field, and consequently, will precess more slowly than would be predicted by the magnitude of the applied field (see Sect. 15.10). Since an induced field will not interact with the applied field to the same degree or in the same sense in all regions of space, the space associated with an induced field can be divided into two regions as shown in Fig. 15.49*c*. In the inverted biconical- or hourglass-shaped region of space, the induced field will always oppose the applied field. In this region, *diamagnetic shielding* will occur to varying degrees and be strongest at the central nucleus. In all other regions of space, the induced field will reinforce the applied field, and *paramagnetic deshielding* will occur to varying degrees (see Sect. 15.51).

Of course in the complex electronic environment of a molecule, induced field components will be generated by the electron surrounding all the atoms and in all the bonds of the molecule. At any given nucleus, the net induced field will be composed of a strong diamagnetic component produced by the electrons immediately surrounding that nucleus, as well as lesser paramagnetic components produced by electrons surrounding immediately adjacent atoms. This situation is represented in Fig. 15.49*d*, where it can be seen that the effective field at any given nucleus generally will be slightly different from that at another nucleus in the molecule. Consequently, at a fixed frequency, while each ^{13}C nucleus in a molecule will achieve resonance at the same effective field strength B_E, the induced field strength B_I at each ^{13}C nucleus will be slightly different. Thus, a slightly different applied or external field strength B_0 will be required to achieve the resonance condition at each nucleus. Since an NMR spectrum is a plot of nuclear absorption versus applied field strength, the ^{13}C nuclei of a molecule will appear to resonate at different magnetic field strengths. Alternatively, at a fixed applied or external magnetic field strength B_0, each nucleus will resonate at a slightly different frequency. This situation may be represented by a modification of the equation in Sect. 15.10, where σ is the *shielding constant* for a particular nucleus in the molecule. In this representation,

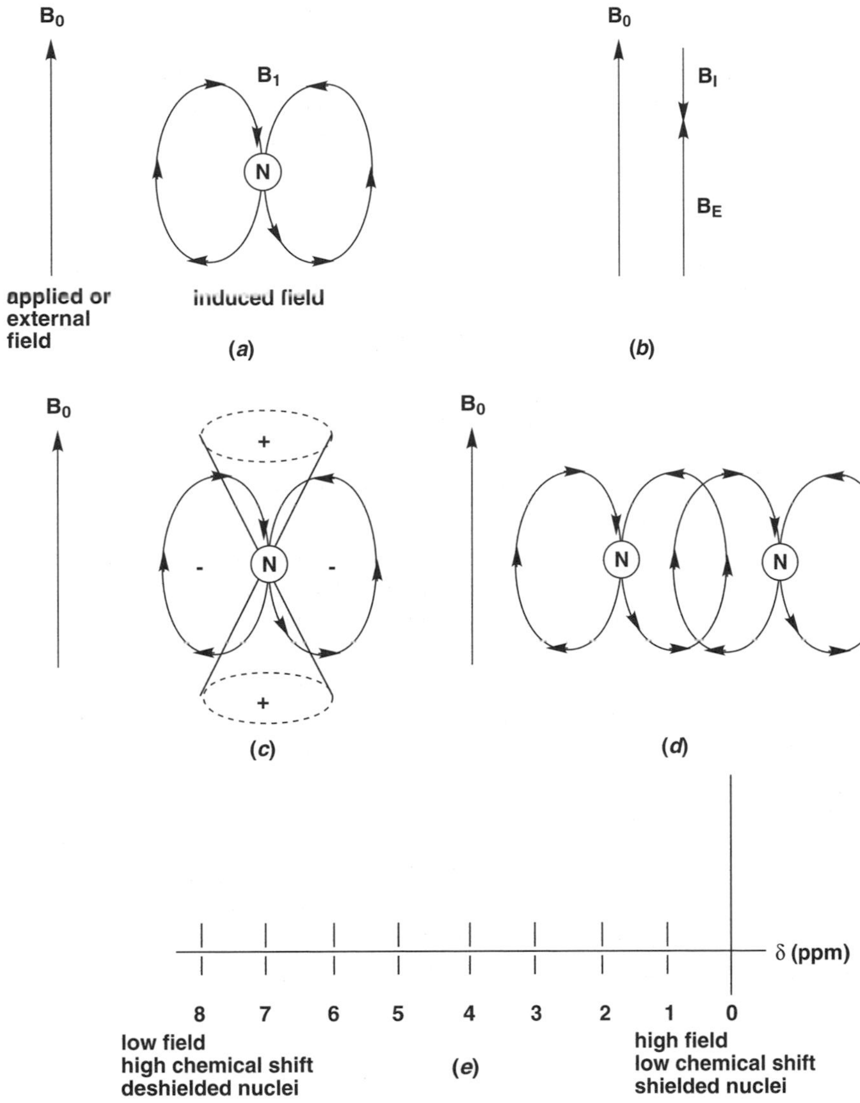

Figure 15.49. (*a*) Lenz's law relationship between applied and induced fields for an atom. (*b*) Relationship of applied, induced, and effective fields. (*c*) Diamagnetic and paramagnetic regions of space produced by induced field. (*d*) Relationship between induced field components. (*e*) Terminology used in describing regions of NMR spectra.

each different type of nucleus in a molecule will have its own characteristic shielding constant, and thus, at fixed B_0, will absorb microwave radiation at a slightly different frequency (ν):

$$\nu = (\gamma/2\pi)\, B_E = (\gamma/2\pi)(B_0 - B_I) = (\gamma/2\pi)\, B_0\,(1 - \sigma) \qquad (15.49b)$$

This concept of more or less additive induced magnetic field components provides an extremely powerful tool for the qualitative prediction of chemical shifts of ^{1}H and ^{13}C nuclei. As a foundation for the application of this tool, it is most important to understand the following general situations: (1) Electron-rich molecules or functional groups having regions of high electron densities will produce relatively large induced fields. As a result, higher applied fields will be necessary to overcome the shielding effects of these large induced fields, and the nuclei in electron-rich molecules will display NMR signals in the high-field (low chemical shift) regions of the spectrum (Fig. 15.49e). (2) Exactly the opposite situation applies to electron-deficient molecules or groups. The lack of shielding electron density leads to small induced fields, and NMR signals in the low-field (high chemical shift) regions of the spectrum.

15.50 PARAMAGNETIC DESHIELDING

The increase in magnetic field strength that occurs when an induced magnetic field enhances the applied or external field at the nucleus under observation (see Fig. 15.49*c*). This is the opposite situation from that which occurs in diamagnetic shielding. In general, paramagnetic deshielding will be produced by fields induced in regions of electron density adjacent to, but not centered at, the nucleus under observation (Fig. 15.49*d*). Furthermore, the term paramagnetic as used in this context should not be confused with its application to molecules with unpaired electrons. In such cases, the spins of the unpaired electrons become aligned with the applied or external field. This creates very powerful deshielding paramagnetic fields that can extend much longer distances than the relatively weak induced field components discussed above.

Example. Paramagnetic molecules (those with one or more unpaired electrons, such as O_2 and certain metal ion complexes) cause nearby nuclei to experience a net magnetic field that is significantly stronger than the external field. This causes the nearby nuclei to precess faster than in the absence of the paramagnetic species.

15.51 MAGNETIC ANISOTROPY OF DIFFERENT TYPES OF CHEMICAL BONDS

Each type of chemical bond has its own characteristic isotropic induced field. These induced fields seem to arise from the circulation of electron density within the bond or array of bonds, and are always polarized in accordance with Lenz's law (Sect. 15.49).

Example. Magnetic field anisotropy of single bonds seems to arise from the circulation of electron density within the single bond, as shown in Fig. 15.51*a*. Consequently, the hydrogens attached to carbon–carbon bonds are deshielded by the induced field of the single bond. The hydrogens of a methyl group are deshielded by a single such induced field, while the hydrogens of a methylene group are deshielded by two such induced fields (Fig. 15.51*b*), and that of a methine group by three such induced fields.

As a result, the protons of a methyl group (δ = 0.8–0.9 ppm) are deshielded less than those of a methylene group (δ = 1.2 ppm), and these, in turn, are deshielded less than that of a methine group (δ = 1.5–1.6 ppm).

Magnetic field anisotropy of double bonds seems to arise from circulation of electron density within the π bond, as shown in Fig. 15.51*c*. Two factors influence the chemical shifts of protons attached to doubly bonded carbons. The first of these is the induced magnetic field associated with the double bond, which deshields these protons, as

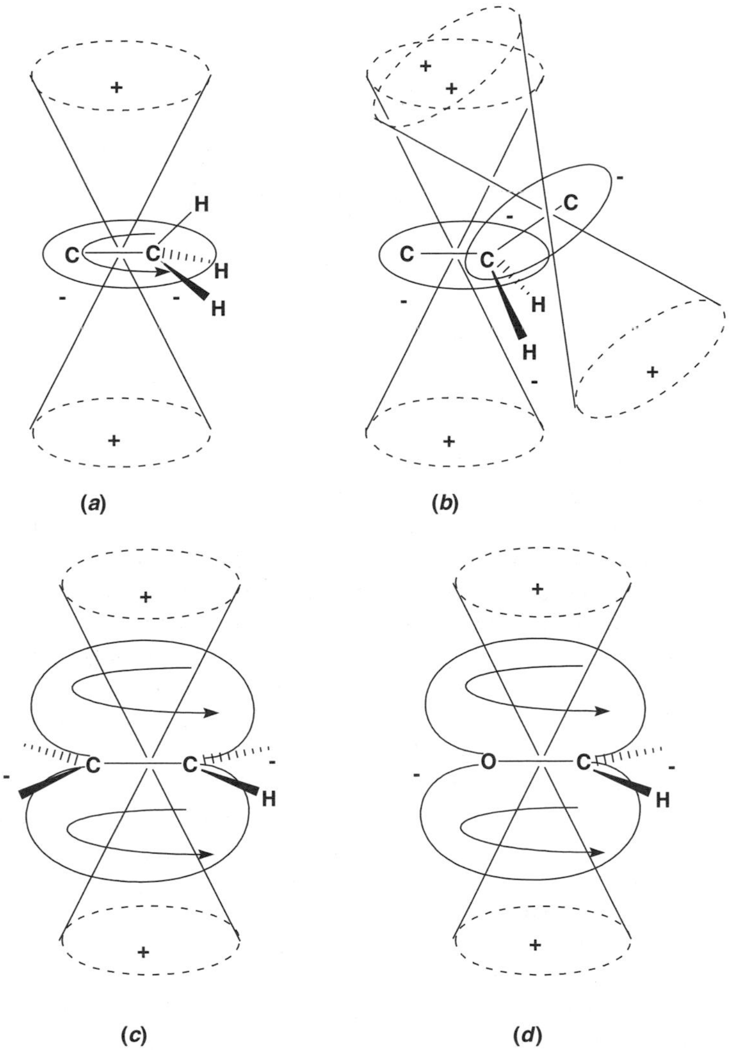

Figure 15.51. Magnetic anisotropy of different types of chemical bonds. (*a*) Carbon–carbon single bond; (*b*) carbon–carbon single bond; (*c*) carbon–carbon double bond; (*d*) carbon–oxygen double bond; (*e*) triple bonds.

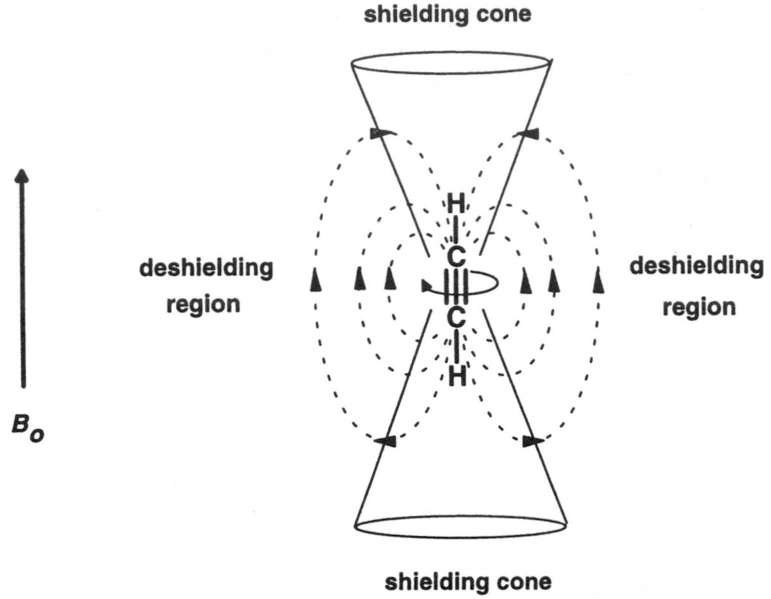

(*e*)

Figure 15.51. (Contd.)

shown in Fig. 15.51*c*. The second factor is the sp^2 hybridized carbons of the double bonds, which causes them to be more electronegative then sp^3 hybridized carbons. Thus, more electron density is withdrawn from protons attached to sp^2 hybridized carbons of double bonds than from protons attached to sp^3 hybridized carbons in saturated systems. This greater electron withdrawal further deshields such protons on doubly bonded carbons. Since both effects reinforce each other, protons attached to doubly bonded carbons would be expected to have signals at much lower fields than those on saturated carbons. This is indeed the case, since protons on saturated, sp^3 hybridized carbons typically have chemical shifts in the range of $\delta = 0.5$ to 2.5 ppm, while protons on doubly bonded, sp^2 hybridized carbons have chemical shifts in the range of $\delta = 5.5$ to 8 ppm.

A particularly instructive example of this interplay between the effects of induced fields and electron-withdrawal deshielding is found in carboxylic acids and aldehydes. In carboxylic acids, the strong electron withdrawal from the acidic proton leads to extreme deshielding of that proton ($\delta = 10$–13 ppm). Since this proton is not attached directly to the carbonyl carbon, the short-range, deshielding induced field of the carbonyl group plays a relatively minor role in the deshielding of carboxylic acid protons. On the other hand, the aldehydic proton is deshielded effectively by both the electron-withdrawing effect of the carbonyl oxygen, and since it is directly attached to the carbonyl carbon, by the induced field of the carbonyl group as well (Fig. 15.51*d*). Thus, aldehydic protons typically have chemical shifts in the same range as carboxylic acid protons ($\delta = 9$–10 ppm). Students sometimes equate this deshielding of aldehydic protons to electron withdrawal alone and, thus, regard aldehydic protons as acidic, which is not the case. In aldehydes, the induced magnetic field makes a major contribution to the deshielding of the aldehydic proton, and aldehydic protons are not appreciably acidic. Exactly the opposite effect occurs with metal hydrides. In $HCo(CO)_4$, the induced magnetic fields produced by the electron surrounding the metal atom strongly shield the "hydride" to the extent that it affords an NMR signal at −11.6 ppm (11.6 ppm above TMS)! On the basis of this chemical shift value alone, one might assume that this hydrogen has a full complement of electrons and significant hydride character. However, $HCo(CO)_4$ is a very strong acid with a pK_a of about −3!

Magnetic field anisotropy of triple bonds seems to arise from the circulation of electron density around the triple bond axis, as shown in Fig. 15.51*e*. In this case, the *sp* hybridized carbon effectively withdraws electron density from terminal acetylenic protons, causing the deshielding of this type of proton. However, the field induced by the triple bond is strongly shielding and counteracts this effect. The net result is that terminal acetylenic protons tend to have chemical shifts between those of protons attached to carbon–carbon single and double bonds ($\delta = 2.0$–3.0 ppm).

15.52 RING CURRENT

The term used to describe the apparent circulation of π electrons around the periphery of an aromatic ring, which gives rise to shielding and deshielding regions. See

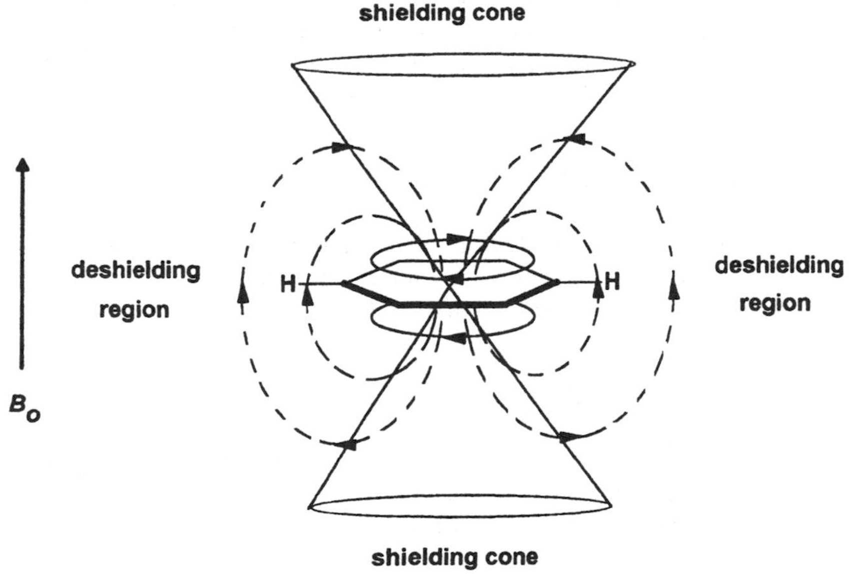

(*a*)

Figure 15.52. (*a*) Apparent circulation of π electrons around the periphery of an aromatic ring and (*b*) [18]annulene.

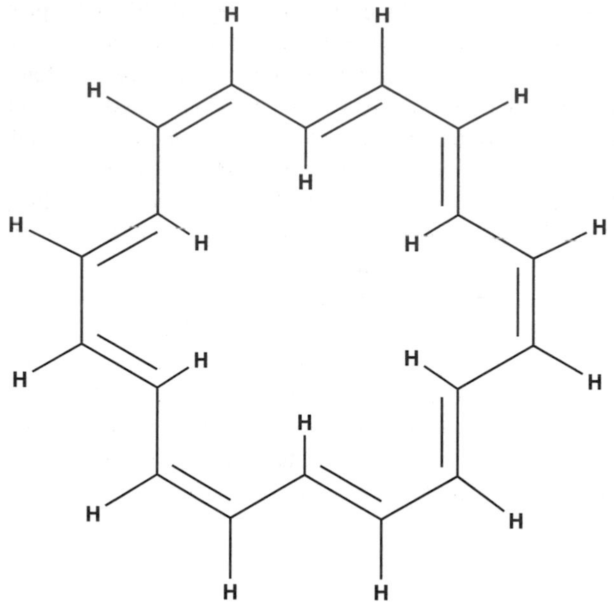

(*b*)

Figure 15.52. (Contd.)

Fig. 15.51*a*. In sufficiently large aromatic rings, protons may project into the shielding region within the ring, as well as the deshielding region around the periphery of the ring. Thus, [18]annulene (Fig. 15.52*b*) displays signals at $\delta = 9.3$ ppm for the external protons and $\delta = -3.0$ ppm for the internal protons.

15.53 UPFIELD (SHIFT)

The movement of an NMR signal toward the right of the spectrum, that is, to lower frequency and lower chemical shift. Such a movement is usually caused by an increase in shielding of the nuclei, giving rise to the signal.

15.54 DOWNFIELD (SHIFT)

The movement of an NMR signal toward the left of the spectrum, that is, to higher frequency and higher chemical shift. Such a movement is usually caused by an decrease in shielding of the nuclei, giving rise to the signal.

15.55 CHEMICAL SHIFT RANGES FOR ^{1}H AND ^{13}C

Figures 15.55*a* and *b* depict the span of chemical shifts for hydrogens and carbons in various molecular environments. Notice that the span of chemical shifts (in ppm)

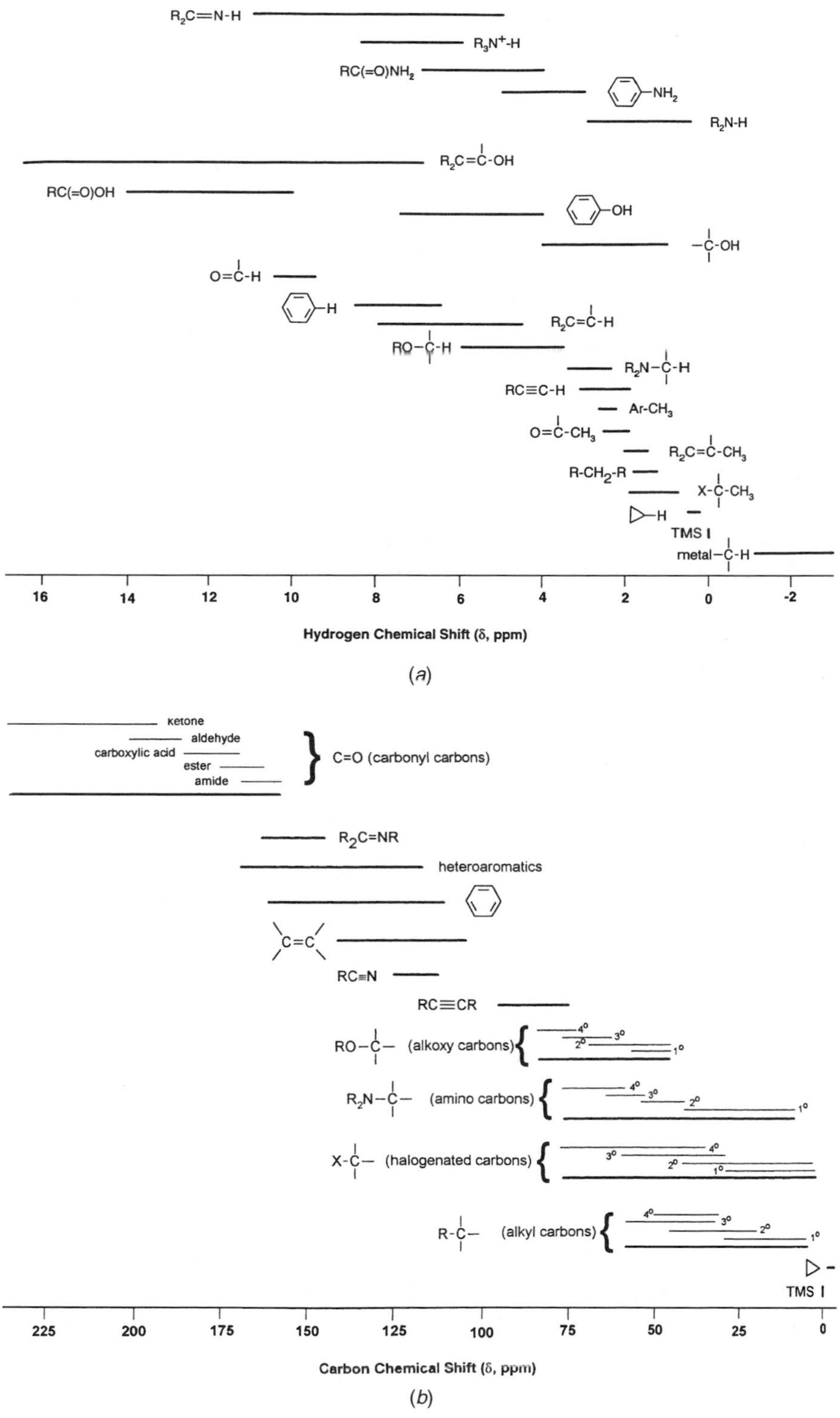

Figure 15.55. The chemical shift ranges for (*a*) ^{1}H and (*b*) ^{13}C nuclei.

for carbon is over 10 times greater than that for hydrogen. Notice also how the relative chemical shifts of hydrogens tend to parallel the relative chemical shifts of the carbons to which they are attached.

15.56 SHOOLERY'S RULES

A set of empirical rules, developed by Jim Shoolery of Varian, Inc., for determining the chemical shift of a proton in a molecule of the general formula A-CH_2-B. To the base value for CH_2 is added the substituent constants for groups A and B. Tables of these substituent constants may be found in standard NMR textbooks (e.g., Silverstein and Webster, Chap. 4, Appendix B, p. 203, and Macomber, Table 6.2, p. 72). In addition to the original set of "rules," base values and substituent constants have been expanded to include vinyl hydrogens and carbon spectra.

Example. Consider the molecule phenylacetone, Ph–CH_2–C(O)CH_3. The base value is 0.23 δ. To this is added the substituent constant of 1.83 δ for the phenyl (Ph) and 1.50 δ for the acyl group (COR) for a total of $0.23 + 1.83 + 1.50 = 3.56$ δ. The observed value is 3.67 δ. In general, the calculated value is within 0.1 or 0.2 δ of the observed value.

15.57 SPIN COUPLING (SPLITTING)

The interaction between the nuclear magnetic fields of two or more magnetic nuclei that causes each of their NMR signals to be split into more than one line. Spin coupling can occur between nonequivalent nuclei of the same isotope (*homonuclear coupling*) or between nuclei of different isotopes (*heteronuclear coupling*).

15.58 MULTIPLICITY OF AN NMR SIGNAL

The number of lines into which the signal of a single nucleus or set of equivalent nuclei is split by nonequivalent adjacent nuclei. The multiplicity equation

$$\text{Multiplicity} = 2nI + 1 \qquad (15.58)$$

indicates the number of lines or peaks to be expected, where n is the number of equivalent adjacent nuclei and I the spin quantum number of those nuclei (Sect. 15.10). The magnetic environment of each nucleus is affected by the spin state of adjacent, nonequivalent nuclei. The spin states of the adjacent nuclei affect the spins of the surrounding electrons and these electrons and, in turn, relay this spin information through the bonds to the electronic environment of the nucleus under observation. The number of possible different magnetic environments is equal to the multiplicity.

15.59 THE $n + 1$ RULE

A simplified rule for predicting the multiplicity of an NMR signal, where every nucleus involved has a spin of 1/2. This is the situation for ^{1}H and ^{13}C, two of the most commonly encountered nuclei in NMR spectroscopy. When applied to these nuclei, the multiplicity equation reduces to

$$\text{Multiplicity} = n + 1 \tag{15.59a}$$

where n is the number of adjacent nuclei that are not equivalent to the nucleus under observation.

Example. Several commonly encountered situations in ^{1}H NMR spectroscopy are shown in Fig. 15.59*a*. In this figure, the trivial situation, where there are no adjacent protons, for example, $CHCl_3$, has been left out. In this situation, all the observed protons of this type will be in a single magnetic environment, and will give rise to a single NMR peak, a singlet. When a single nonequivalent proton (H_2) is adjacent to the proton under observation (H_1), about half the molecules will have an H_2 spin state aligned with the external field, and the other half with an H_2 spin state aligned against the external field. These two different spin configurations will produce two different magnetic environments for the proton under observation (H_1), and this, in turn, will lead to two peaks in the NMR signal for H_1, a *doublet*. An adjacent, nonequivalent set of methylene protons will give rise to *triplet* signal and a set of methyl protons to a *quartet* signal.

It should be noted that this process works in both directions, as shown in Fig. 15.59*b*. Thus, H_1 will appear as a triplet, since n_2 is 2, also equal to the integration of the H_2 signal, and H_2 as a doublet, since n_1 is 1, also equal to the integration of the H_1 signal.

More complex situations occur in which the proton under observation is in close proximity to several different types of nonequivalent protons that are themselves nonequivalent (Fig. 15.59*c*). In this situation, the multiplicity equation expands to

$$\text{Multiplicity} = (n_1 + 1)\,(n_2 + 1)\,(n_3 + 1) \cdots \tag{15.59b}$$

where n_1, n_2, and n_3 refer to the number of different types of adjacent protons. In the case shown in Fig. 15.59*c*, H_2 and H_3 both give rise to doublet signals, since each is adjacent to a single nonequivalent proton, H_1. However, H_1 is adjacent to two different

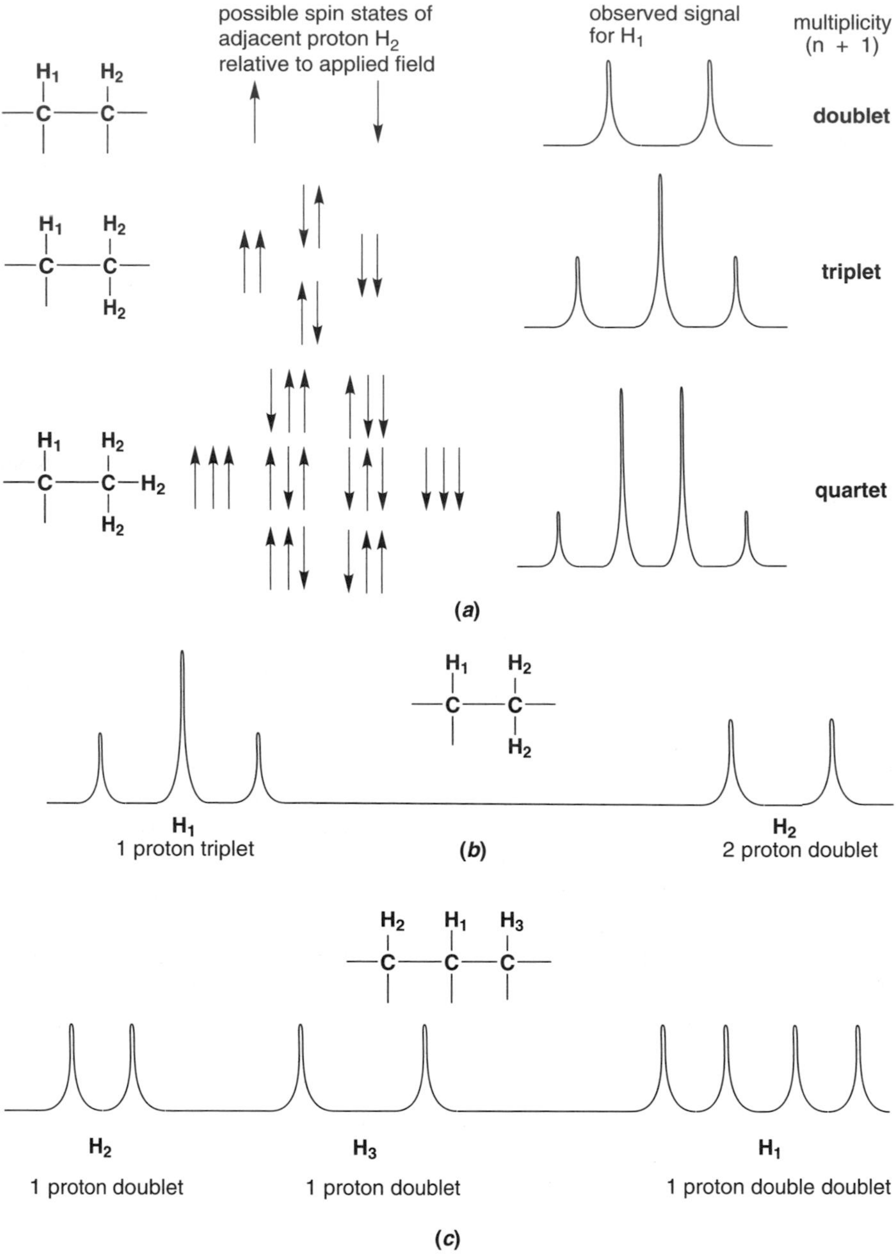

Figure 15.59. (*a*) Effect of nonequivalent adjacent nuclei on the magnetic environment of a proton under observation; H_1 is the proton being observed and H_2 are the nonequivalent adjacent protons that contribute to the magnetic environment of H_1. (*b*) Relationship between multiplicity and number of coupled protons. (*c*) Complex (multiple term) coupling relationship between more than one type of nonequivalent adjacent proton. (*d*) The 250-MHz ^{1}H spectrum of pure isobutyl alcohol.

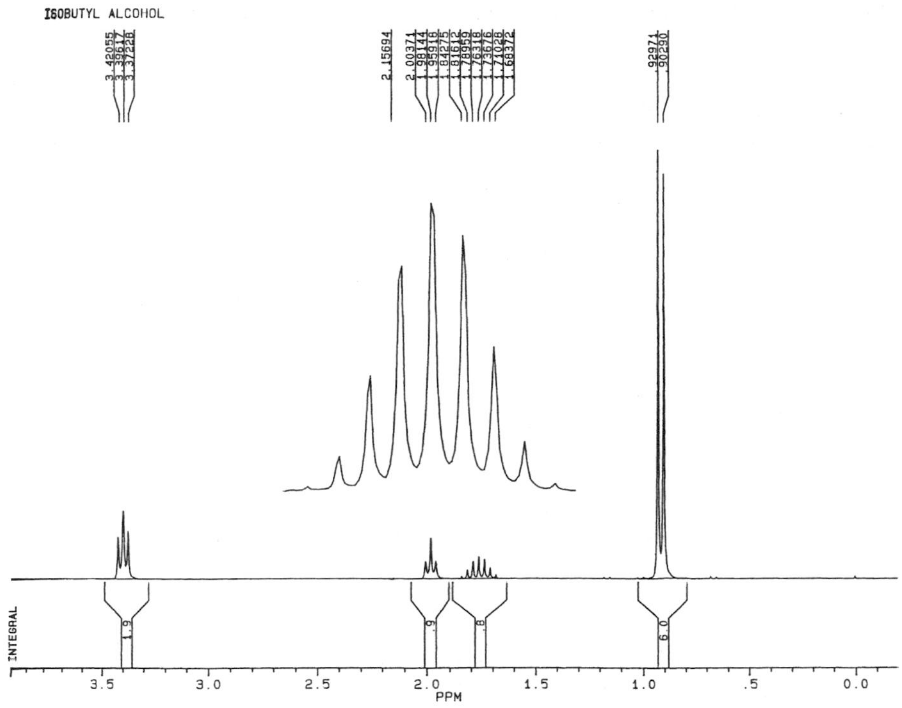

(*d*)

Figure 15.59. (Contd.)

types of nonequivalent protons (H_2, $n_2 = 1$, and H_3, $n_3 = 1$), which according to the expanded multiplicity equation should give rise to a maximum of four peaks (Sect. 15.62). Although such a signal might be described as a quartet, since it has four lines, it is more accurately described as a double doublet, since this terminology reflects the terms in the expanded multiplicity equation. It should be mentioned that the presence of two or more different types of adjacent protons can give rise to complex signals that are not readily resolved into their multiplicity components. Such signals are commonly referred to as multiplets.

Finally, an important variation of this multiplicity rule is encountered in the recognition of solvent signals in NMR spectroscopy. Since deuterated solvents are most frequently used in the preparation of NMR samples, and since deuterium has a spin quantum number *I* of 1, deuterated solvents give rise to a signal with higher-order multiplicities. For instance, dimethysulfoxide-D_6 will always contain a small amount of dimethylsulfoxide-H_1D_5. This material will give rise to a five-line signal in the ^{1}H NMR spectrum at $\delta = 3.35$ $(2 \times 2 \times 1 + 1)$. Also deuterochloroform will give rise to a three-line signal in ^{13}C spectra at $\delta = 77.0$ $(2 \times 1 \times 1 + 1)$.

15.60 RELATIVE INTENSITIES OF LINES WITHIN A SIGNAL

The relative intensity of the lines within a signal depends on the nuclear spin of the nuclei that are coupled. If all the nuclei have a nuclear spin of 1/2, the relative intensities can be determined from the coefficients of the binomial distribution (i.e., *Pascal's triangle*, Blaise Pascal 1623–1662):

n	Multiplicity	Intensity Ratio
0	1	1
1	2	1 1
2	3	1 2 1
3	4	1 3 3 1
4	5	1 4 6 4 1

Example. For the signals in the example of Sect. 15.60, the quartet would exhibit lines with relative intensity 1:3:3:1, while the doublet would exhibit lines with relative intensity 1:1. See Fig. 15.59.

15.61 COUPLING CONSTANT *J* (COUPLING CONSTANT MAGNITUDE AND SIGN)

For a simple doublet, triplet, or quartet, the separation between adjacent lines or peaks in Hz. For more complicated coupling patterns such as a doublet of doublets, there are two coupling constants. The magnitude of one constant J is the ΔHz between lines 1 and 2, or between lines 3 and 4. The magnitude of the second constant J' is the ΔHz between lines 1 and 3, or between lines 2 and 4. For a more complete explanation about how to obtain coupling constants from more complex patterns, see Sect. 15.62.

If two nuclei are coupled to each other, their respective coupling constants will be of equal magnitude. The *magnitude of a coupling constant* between two nuclei is determined by several factors, including the magnetogyric ratios (Sect. 15.10) of the two nuclei, the number of bonds between the coupled nuclei, the fraction of *s*-character of the orbitals involved in these bonds, and a geometric parameter that describes the three-dimensional relationship between the nuclei in a molecule. It is dependent only on the structure of the molecule containing the coupled nuclei; it is independent of the external magnetic field and instrument operating frequency. In order to observe the effects of coupling between two nuclei, they must be nonequivalent.

Coupling constants can have either positive or negative signs. The *coupling constant sign* indicates the relative stability of the parallel spin state (magnetic moments in the same direction) versus the antiparallel spin state (magnetic moments in the opposite directions). A positive J indicates that the antiparallel state is more stable than the parallel state, whereas a negative J indicates the converse.

However, the coupling constant sign normally has no affect on the appearance of an NMR spectrum.

Example. For the signals in the example in Sect. 15.59, the ΔHz between adjacent lines of the quartet will all be the same, and will be equal to the separation between the two lines of the doublet. Coupling constants are often designated with a leading superscript that indicates the number of bonds separating two coupled nuclei within a molecule. Thus, ^1J indicates a coupling constant between two nuclei directly bonded to each other, for example, C–H. ^2J indicates a two-bond (geminal) coupling constant, for example, H–X–H or H–X–F, and ^3J indicates a three bond (vicinal) coupling constant, for example, H–X–Y–H or H–X–Y–F, and so on. In general, the magnitude of the coupling constant J decreases as the number of intervening bonds increases. For example, $J_{\text{C-H}}$ values range from ca. 125 to 250 Hz, depending on the hybridization at carbon, while $J_{\text{H-C-H}}$ and $J_{\text{H-C-C-H}}$ values range from 0 to 18 Hz. Furthermore, one-bond J values are positive, two-bond J values are negative, and three-bond J values are again positive.

15.62 COUPLING DIAGRAMS

A diagram that correlates the coupling constants of an NMR signal with the coupling constants of other signals in the spectrum.

Example. Coupling diagrams are most useful in the analysis of complex coupling patterns. The multiplicity of each signal may be resolved into its component, where each component is associated with a single coupling constant J. For instance, a proton that is coupled to two different types of protons may display four lines, as shown in Figs. 15.62*a* and *b*. According to the multiplicity relationship, $m = (1 + 1)(1 + 1) = 4$. Even though the signal consists of four lines, it is not a quartet, since these four lines have intensity ratios of 1:1:1:1, not 1:3:3:1, which is the intensity ratio characteristic of a true quartet. This signal is more precisely termed a double doublet in accordance with the terms in the multiplicity expression. If the smallest of these coupling constants J were removed from the pattern, as shown in Fig. 15.62*a*, two lines would remain separated by the larger coupling constant J'. If J' were removed from this residual pattern, a single line would remain, and this line corresponds to the chemical shift of the entire double doublet pattern. Notice that it does not matter whether this double doublet signal is deconvoluted as shown above, or whether the larger coupling constant is removed first, followed by the smaller coupling constant. Compare Figs. 15.62*a* and *b*. Notice, also, that this analysis can be applied in the opposite sense. Thus, a single line located at the chemical shift of the signal can be split into a doublet of magnitude J, and each of the resulting branches further split into doublets of magnitude J' to reproduce the observed NMR signal.

Coupling diagrams are very useful in the interpretation of an NMR spectrum. Consider the three-line pattern shown in Figs. 15.62*c* and *d*. These two signals look

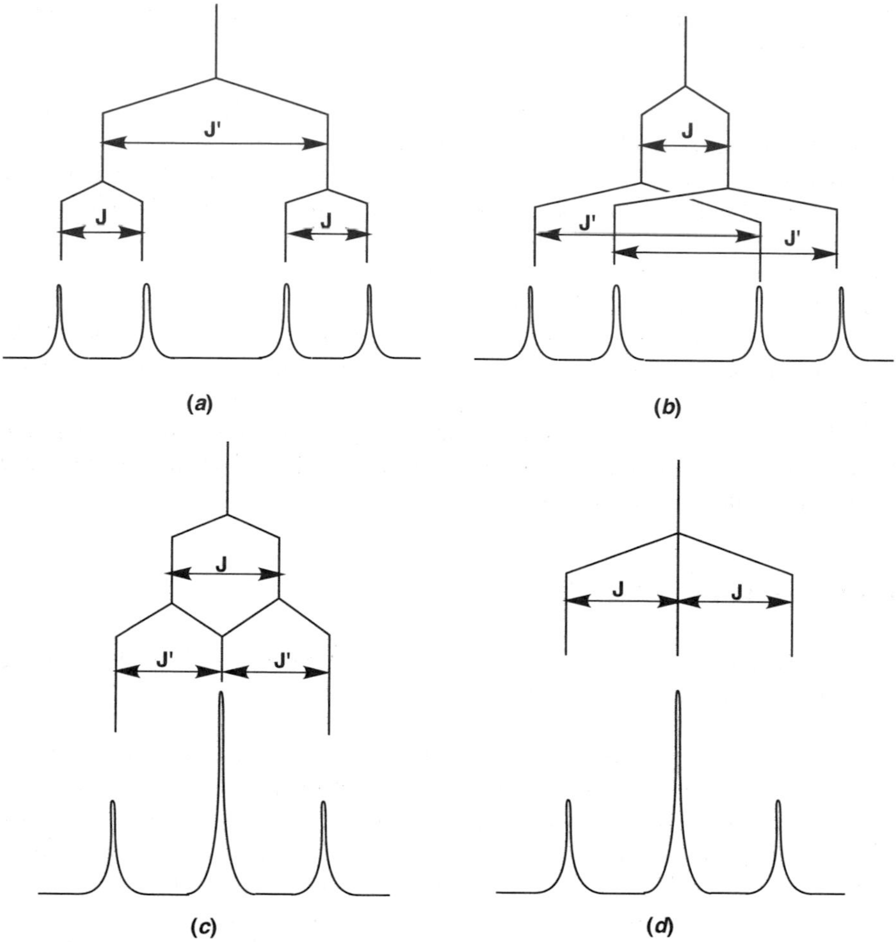

Figure 15.62. Coupling diagrams (*a*) and (*b*) show alternative deconvolution sequences for a double doublet ($J\,J'$), (*c*) coupling diagram for a double doublet ($J = J'$), (*d*) coupling diagram for a triplet.

identical, but can arise from a very different atomic configuration within the molecule. Such a signal might be interpreted either as a double doublet (Fig. 15.62*c*) or triplet (Fig. 15.62*d*). If it were the former, then J must equal J', and the proton that gives rise to the signal must be adjacent to two different types of protons, as H_1 is in Fig. 15.59*c*. This is a commonly encountered situation, and similar J's often give rise to pseudo-triplets in NMR spectra. Alternatively, if this signal were a true triplet, a single coupling constant J would be involved (Fig. 15.62*d*), and the proton that gives rise to the signal would be adjacent to two equivalent protons. In order to distinguish between these two superficially similar situations, decoupling experiments would have to be conducted in order to determine the number of coupling constant associated with the signal.

15.63 THE KARPLUS EQUATION

The relationship between the magnitude of the vicinal (three-bond) coupling constant J and the dihedral (torsional, ϕ) angle between the coupled nuclei first proposed by Martin Karplus (1930–):

$$J = A \cos 2\phi + C \quad \text{for } \phi \ 0^\circ \text{ to } 90^\circ \tag{15.63a}$$

$$J = A' \cos 2\phi + C \quad \text{for } \phi \ 90^\circ \text{ to } 180^\circ \tag{15.63b}$$

$$A' > A$$

where A, A', and C are scaling constants such that $J_{0^\circ} = A + C$, $J_{90^\circ} = C$, and $J_{180^\circ} = A' + C$ (Fig. 15.63).

Example. The magnitude of ^3JH–C–C–H will vary as shown in Fig. 15.63. In general, the minimum coupling constant will occur at a dihedral angle of 90°, and the maxima at angles of 0° and 180° with the coupling constant at 180° being slightly larger than that at 0°.

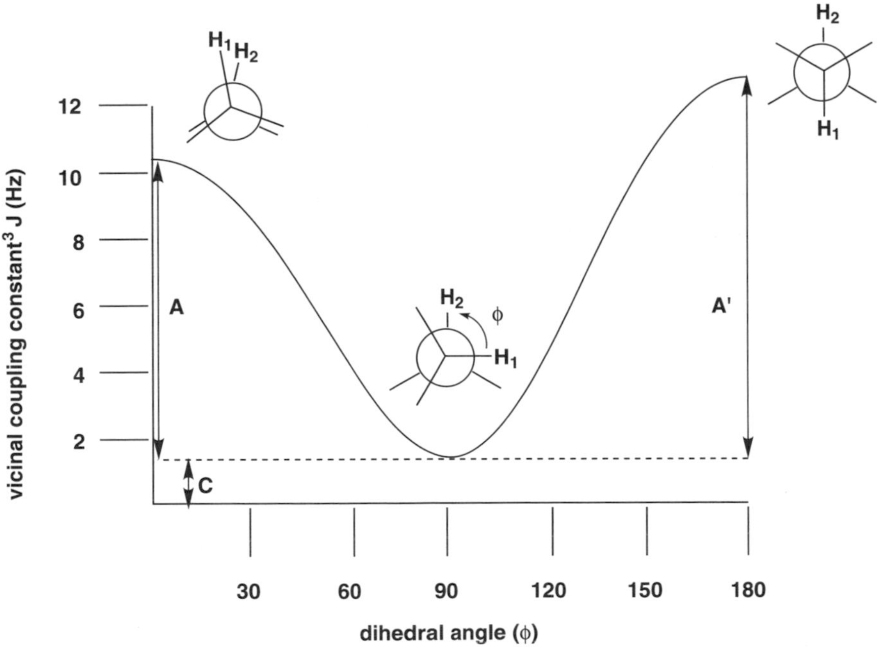

Figure 15.63. Karplus curve.

15.64 WEAK (FIRST-ORDER) COUPLING

Coupling between nuclei where the ratio of the chemical shift difference (in ΔHz) to the coupling constant (in ΔHz) is greater than 10. Spin systems that are weakly coupled follow the $n + 1$ rule and the relative multiplet intensities described in Sect. 15.60.

Example. Suppose the NMR spectrum in Fig. 15.62 was obtained at 250 MHz, the chemical shifts are 4.60 ppm (δ_1) and 1.23 ppm (δ_2), and the coupling constant they share is 10 Hz. The difference in chemical shift ($\Delta\delta$) is $4.60 - 1.23 = 3.37$ ppm, which equals $3.37 \times 250 = 842.5$ Hz at 250 MHz. The ratio of $\Delta\delta$ to J is 842.5 Hz/10 Hz = 84, so this spin system is first-order (weakly coupled) and will obey the $n + 1$ rule. Note that all heteronuclear coupling is weak by virtue of the large differences in precessional frequencies between different isotopes.

15.65 STRONG (SECOND-ORDER) COUPLING

Coupling between nuclei where the ratio of the chemical shift difference (in ΔHz) to the coupling constant (in ΔHz) is significantly less than 10 (see Sect. 15.64). The NMR spectra of strongly coupled spin systems exhibit multiplets that depart significantly from expectations based on first-order rules. These "second-order" effects include *multiplet slanting* where the inner lines of a multiplet increase in relative intensity, while the outer lines decrease (Fig. 15.65). Other second-order perturbations include deceptive simplicity and virtual coupling.

Example. Suppose the NMR spectrum in Fig. 15.62 were obtained at 100 MHz, the chemical shifts are 1.60 ppm (δ_1) and 1.23 ppm (δ_2), and the coupling constant they share is 10 Hz. The difference in chemical shift ($\Delta\delta$) is $1.60 - 1.23 = 0.37$ ppm, which equals $0.37 \times 100 = 37$ Hz at 100 MHz. The ratio of $\Delta\delta$ to J is 37 Hz/10 Hz = 3.7, so this spin system is strongly coupled. In such a case, the spectrum would actually appear as shown in Fig. 15.65. One advantage of using the highest possible operat-

Figure 15.65. The NMR spectrum of a strongly coupled spin system. Note how the inner lines have increased in intensity, and the outer lines have decreased, compared with Fig. 15.62.

ing frequency is that the value of $\Delta\delta$ (in Hz) increases with operating frequency, while the magnitude of J is fixed, so the ratio increases and the spectrum become more first-order in appearance.

15.66 LONG-RANGE COUPLING

Coupling between two nuclei separated by more than three bonds.

Example. Coupling over more than three bonds is exceptional, but it can be exhibited by nuclei separated by one or more π bonds, or by nuclei held in a W-shaped spatial relationship. Molecular examples are shown in Fig. 15.66 (normal $^4J_{HCCH} < 0.1$ Hz):

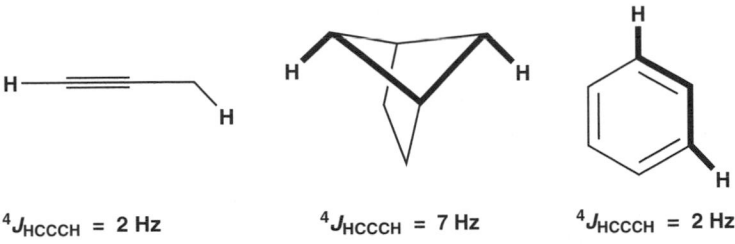

$^4J_{HCCCH}$ = 2 Hz $^4J_{HCCCH}$ = 7 Hz $^4J_{HCCCH}$ = 2 Hz

Figure 15.66. Examples of long-range coupling

15.67 MAGNETIC EQUIVALENCE

Two (or more) nuclei that are both chemically equivalent (Sect. 15.33) and equally coupled to any other selected nucleus.

Example. We saw in Sect. 15.33 that toluene has four sets of hydrogens: three equivalent methyl hydrogens, two equivalent *ortho* hydrogens, two equivalent *meta* hydrogens, and one *para* hydrogen. The methyl hydrogens are not coupled to any other hydrogens (because all other hydrogens are at least four bonds away; Sect. 15.61). Thus, they are equally coupled to any one of these other hydrogens, with a J value of zero! Therefore, the three methyl hydrogens are magnetically equivalent to each other. Each *ortho* hydrogen is coupled to its closer *meta* neighbor with a J value of 10 Hz, to the *meta* hydrogen across the ring with a J value of 1 Hz, and with the *para* hydrogen with a J value of 2 Hz (Fig. 15.67). So, although the two *ortho* hydrogens are chemically equivalent, they are not magnetically equivalent because their coupling to a given *meta* hydrogen is different. By the same reasoning, the two *meta* hydrogens are chemically equivalent, but not magnetically equivalent. Because protons, which are separated by four or fewer bonds, and which are not magnetically equivalent, couple to each other, the NMR spectrum may appear more complicated than if these protons were magnetically equivalent:

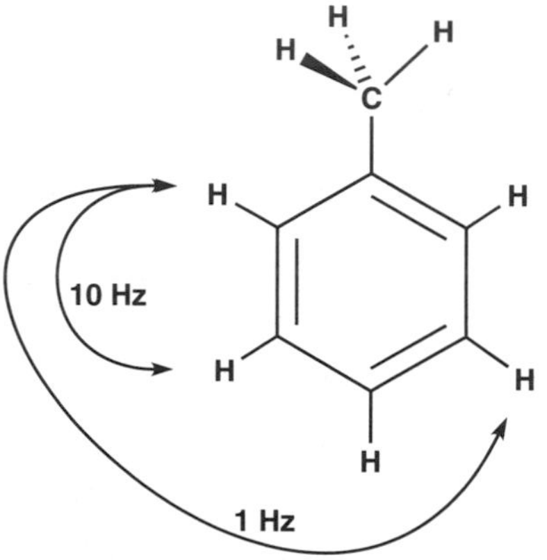

Figure 15.67. Proton coupling in toluene

15.68 POPLE SPIN SYSTEM NOTATION

A method to label sets of coupled nuclei so that their NMR spectra can be more readily recognized or predicted. Each nucleus in the system is assigned a capital letter; the closer the letters are in the alphabet, the more strongly coupled (Sect. 15.65) the nuclei are. Nuclei that are magnetically equivalent are assigned the same letter. Nuclei that are chemically, but not magnetically, equivalent are indicated with primes (e.g., A and A').

Example. In an ethyl group ($H_3C{-}CH_2{-}$), there are two sets of coupled hydrogens. The three methyl hydrogens are magnetically equivalent, as are the two methylene hydrogens. If these two sets share a coupling constant that is small compared to their difference in chemical shift, the spin system is weakly coupled (Sect. 15.64). Therefore, the Pople notation for this system is A_3X_2, with the methyl hydrogens corresponding to the three A spins, and the methylene hydrogens corresponding to the two X spins. The NMR spectrum for a spin system consists of a quartet for the X nuclei (coupled to three magnetically equivalent A nuclei) and a triplet for the A nuclei (coupled to two magnetically equivalent X nuclei, Sect. 15.59). If the difference in chemical shifts between the two sets of hydrogens was smaller (i.e., the system was strongly coupled, Sect. 15.65), the designation would be A_3B_2.

The phenyl group of toluene constitutes a strongly coupled $AA'BB'C$ system, with A and A' indicating the chemically but not magnetically equivalent *ortho* hydrogens, B and B' the *meta* hydrogens, and C the *para* hydrogens. The *ortho* hydrogen signal will appear as a doublet (due to coupling with the closer *meta* hydrogen) of doublets (due to coupling with the *para* hydrogen) of doublets (due to coupling with the other *meta* hydrogen; this last coupling may be too small to observe). The *meta* hydrogen signal will also appear as a doublet of doublets of doublets for the same reason. The *para* hydrogen signal will appear as a triplet (due to coupling with the two *meta* hydrogens that, as far as the *para* hydrogen is concerned, act as if they are magnetically equivalent) of triplets (due to the *ortho* hydrogens). The phenyl group therefore gives rise to a very complex pattern, as seen in Fig. 15.32.

15.69 *AB* QUARTET

A four-line spectroscopic pattern that resembles a quartet (Sect. 15.59), but instead of resulting from a set of nuclei coupled to three other equivalent nuclei, it arises from two strongly coupled (Sect. 15.65) nuclei. The reason that the pattern is not simply two doublets (i.e., an *AX* pattern) is because the line intensities are slanted by the effects of the strong coupling, Fig. 15.69.

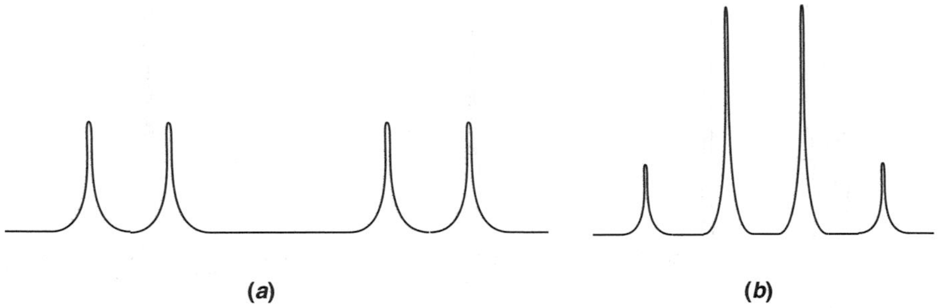

Figure 15.69. Comparison of (*a*) an *AX* spectrum with (*b*) an *AB* quartet.

15.70 SUBSPECTRUM ANALYSIS

Recognition of the Pople spin systems (Sect. 15.68) that give rise to specific signal patterns (subspectra) within an NMR spectrum.

Example. Consider the 250-MHz ^{1}H NMR spectrum in Fig. 15.70. A practiced NMR spectroscopist would immediately recognize the triplet at 1.34 δ and quartet at 4.26 δ (which share a coupling constant of 7.0 Hz), as characteristic of an A_3X_2 spin system such as CH_3CH_2O-. Furthermore, the doublets at 4.43 and 7.69 δ (sharing a *J* value of 16 Hz) are characteristic of an *AX* spin system such as XCH=CHY. The complex pattern from 7.36 δ to 7.54 δ is typical for the *AA'BB'C* system of a phenyl group. These three structural features are well accommodated by ethyl cinnamate, the structure of which is shown below.

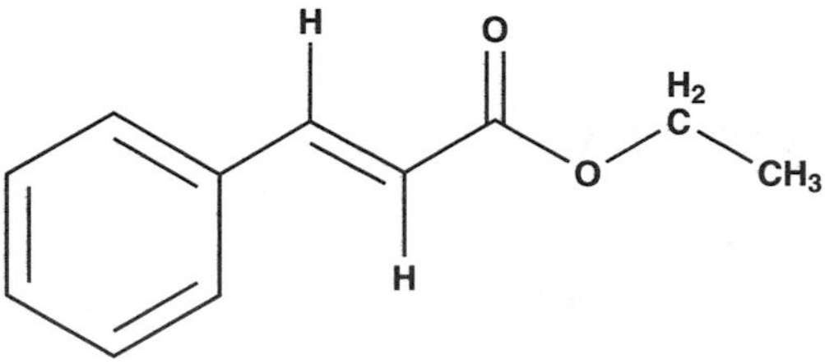

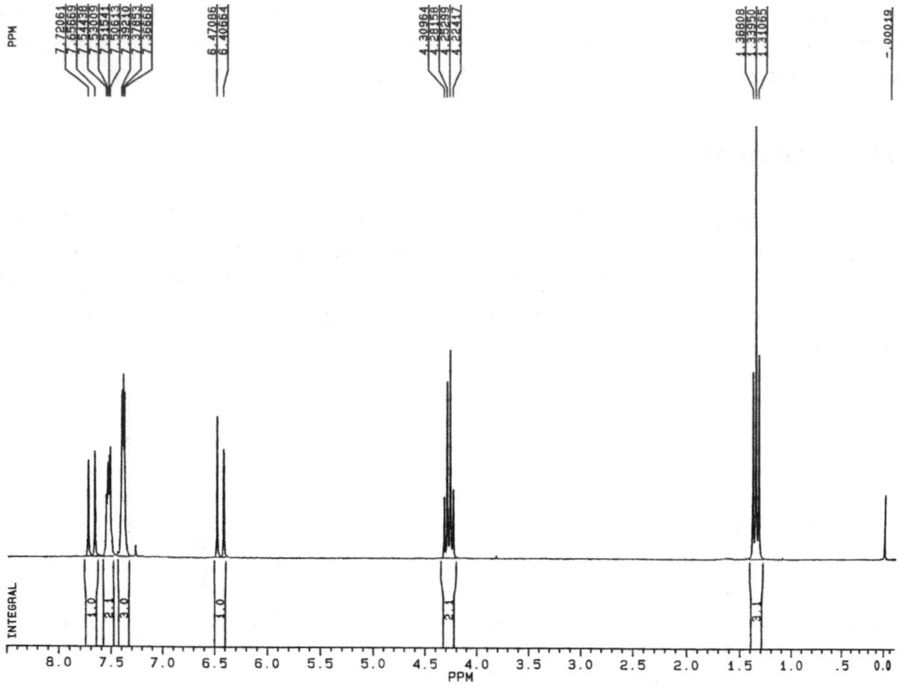

Figure 15.70. The 250-MHz ^{1}H NMR spectrum of ethyl cinnamate.

15.71 DECEPTIVELY SIMPLE SPECTRA

An NMR spectrum that exhibits fewer lines than the number expected on the basis of applying the $n + 1$ rule to the Pople spin system involved.

Example. The NMR spectrum of an $AA'X$ spin system depends on two chemical shifts, $\delta A = \delta A'$ and δX, and two coupling constants, J_{AX} and $J_{A'X}$. The $n + 1$ rule predicts that the A and A' nuclei should appear as two cocentered doublets and the X signal as a doublet of doublets. But, in fact, the signal for X appears to be a triplet, and the spacings between the lines are neither J_{AX} nor $J_{A'X}$, but rather the average of the two. This deceptive simplicity is a quantum mechanically predictable manifestation of second-order coupling effects.

15.72 VIRTUAL COUPLING

The appearance of extra lines in the multiplets of an NMR spectrum, suggesting there is coupling between two nuclei that are, in fact, not coupled. These extra lines are a quantum mechanically predictable consequence of second-order coupling perturbations (Sect. 15.65).

15.73 DYNAMIC NMR SPECTROSCOPY

An NMR spectroscopic study of a chemical system undergoing some type of reversible process such as a chemical reaction (e.g., a rearrangement) or a physical process (e.g., rotation around bonds). Thus, dynamic processes involve the reversible interconversion of two (or more) distinct chemical structures, and each structure gives rise to a different pattern of NMR signals.

Example. Fig. 15.73: The "ring flip" of a cyclohexane derivative that shifts the substituent from an equatorial position to the less stable (and therefore less populated) axial position is another example of a reversible dynamic physical process. An enol-keto tautomerization is an example of a reversible chemical reaction; the keto form is generally more stable (and more populated).

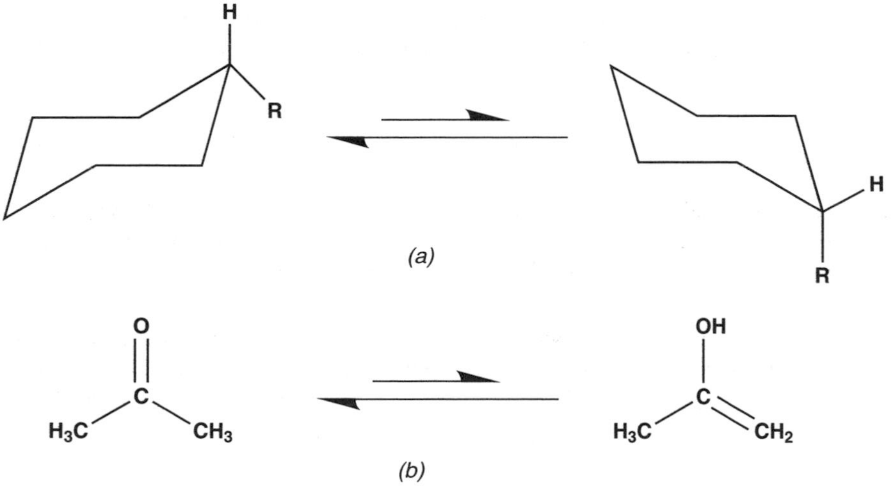

Figure 15.73. (*a*) "Ring flip" of a cyclohexane derivative and (*b*) enol-keto tautomerization.

15.74 (SITE) EXCHANGE

When a reversible chemical or physical process causes two atoms (or groups of atoms) to interchange their molecular environments.

Example. Dimethylformamide (DMF), Fig. 15.74, exists predominantly in a planar conformation with the two methyl groups occupying nonequivalent sites. One methyl group (shown in italics in the left-hand structure in Fig. 15.74 below) is *syn* to the oxygen, while the other methyl group is *anti* to the oxygen. Rotation around the C–N bond exchanges the sites occupied by the two methyl groups.

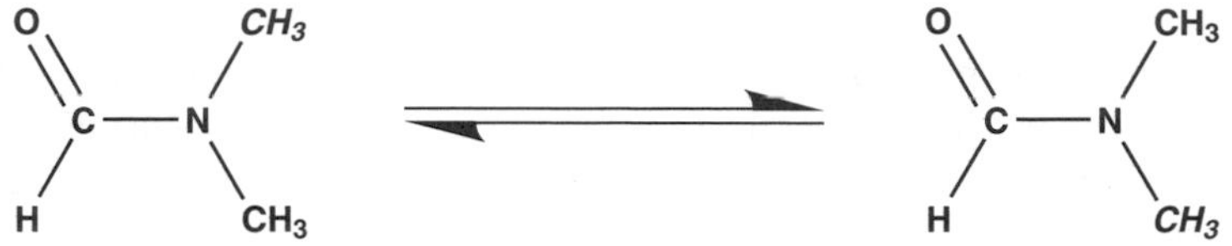

Figure 15.74. Site exchange of the two methyl groups in dimethylformamide.

15.75 SLOW EXCHANGE LIMIT

The temperature at which the rate constant k for site exchange (Sect. 15.74), expressed in sec^{-1}, is much smaller that the difference in chemical shift between the two sites, expressed in Hz, that is, $k << \nu_1 - \nu_2$. (*Note*: The dimension of Hz is sec^{-1}.) Under conditions of slow exchange, separate signals for the groups in each site are observed in the NMR spectrum. The rate constant is given by the following expression:

$$k = \pi(w - w_0) \tag{15.75}$$

where w is the width at half-height of the broadened peak and w_0 is the width at half-height of the unbroadened peak. The rate for a process can be increased or decreased by raising or lowering, respectively, the temperature of the sample.

Example. In the 250-MHz ^1H NMR spectrum of DMF (Sect. 15.74), the two methyl group signals appear at 2.74 and 2.91 δ, separated by 42.5 Hz. Thus, if the rate constant for exchange (i.e., for C–N bond rotation) is much slower than 42.5 sec^{-1}, both signals will be observed in the spectrum. See also Fig. 15.77.

15.76 FAST EXCHANGE LIMIT

The temperature at which the rate constant k for site exchange, expressed in sec^{-1}, is much larger than the difference in chemical shift between the two sites, expressed in Hz, that is, $k >> \nu_1 - \nu_2$. Under conditions of fast exchange, the signals for the groups in each site are averaged to give a single sharp signal in the NMR spectrum. The rate constant is given by the following expression:

$$k = \pi(\nu_1 - \nu_2)^2/2(w - w_f) \tag{15.76}$$

where ν_1 and ν_2 are the chemical shifts in Hz at the slow exchange limit, w is the width at half-height of the broadened peak, and w_f is the width at half-height of the unbroadened peak at the fast exchange limit. The rate for a process can be increased or decreased by raising or lowering, respectively, the temperature of the sample.

Example. In the 250-MHz ^{1}H NMR spectrum of DMF (Sect. 15.74), the two methyl group signals appear at 2.74 and 2.91 δ, separated by 42.5 Hz. Thus, if the rate constant for exchange (i.e., for C–N bond rotation) is much faster than 42.5 sec^{-1}, only one signal will be observed for the methyl groups, and it will occur at $\delta = (2.74 + 2.91)/2 = 2.83$.

15.77 COALESCENCE

The temperature at which the rate constant k for site exchange (Sect. 15.74), expressed in sec^{-1}, is equal to 4.4 times the difference in chemical shift between the two sites, expressed in Hz, that is, $k = 4.4\ (\nu_1 - \nu_2)$. Under conditions of coalescence, the signals for the groups in each site merge to give a single very broad signal in the NMR spectrum. The rate constant is given by the following expression:

$$k = \pi(\nu_1 - \nu_2)/\sqrt{2} \qquad (15.77)$$

Example. In the 250-MHz ^{1}H NMR spectrum of DMF, the two methyl group signals appear at 2.74 and 2.91 δ, separated by 42.5 Hz. Thus, if the rate constant for exchange (i.e., for C–N bond rotation) equals 4.4 (42.5 sec^{-1}) = 187 sec^{-1}, only one extremely broad signal will be observed for the methyl groups, and it will be centered at δ (2.74 + 2.91)/2 = 2.83. Figure 15.77 shows the 250-MHz ^{1}H NMR spectrum of DMF as a function of the exchange rate.

15.78 REVERSIBLE COMPLEXATION

A type of reversible chemical reaction where two dissimilar molecules form a weakly bonded bimolecular aggregate. The larger of the original molecules is termed the *host*, the smaller is the *guest*, and the aggregate is the *host-guest complex* or *supramolecule*. The complex has a specific structure and a bonding energy as measured by its equilibrium constant for formation ($K_{\text{formation}}$). Since the NMR signals of uncomplexed guest and host molecules are different from the signals of their complexed counterparts, both the structure and value of $K_{\text{formation}}$ can be probed by dynamic NMR methods.

Example. Such complexation is the basis of what is currently called *molecular recognition*. One biochemical example of guest-host complexation is the formation of an

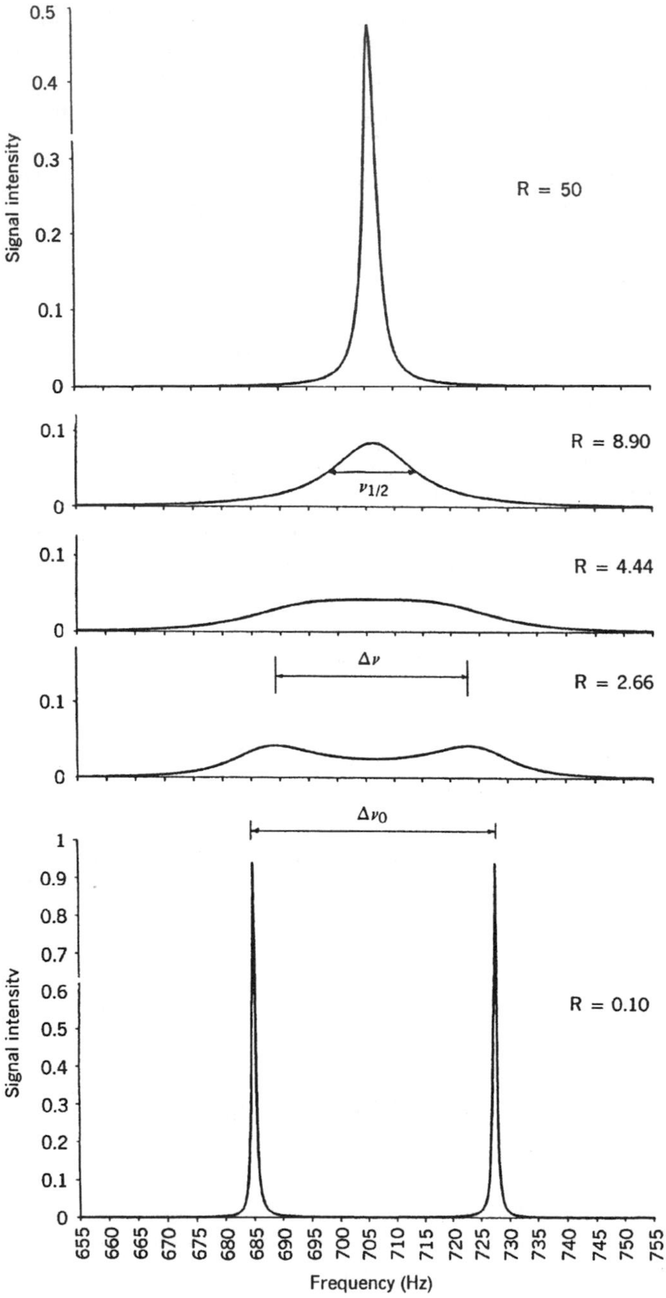

Figure 15.77. The 250-MHz ^{1}H NMR spectrum of DMF as a function of exchange rate. The symbol R represents the ratio of the exchange rate constant (in sec^{-1}) to the chemical shift difference (42.5 Hz in this case). The slow exchange limit appears at the bottom ($R = 0.10$), the fast exchange limit at the top ($R = 50$), and coalescence in the middle ($R = 4.44$).

enzyme–substrate complex prior to conversion of the substrate to its ultimate product. The weak attraction between the guest and the host is predominantly hydrogen-bonding in nature.

15.79 CHEMICAL SHIFT REAGENTS

Molecules, such as europium and other lanthanide ions, that cause substantial deshielding of nuclei in nearby solute molecules. This deshielding is caused by the magnetic properties of the unpaired electrons in the lanthanide ion and is manifested when the shift reagent molecules form transient guest-host complexes (Sect. 15.78) with the solute molecules. The nearer in the complex a solute molecule nucleus is to the lanthanide ion, the larger the deshielding effect will be. The result is that the various nuclei in the solute are deshielded to different degrees, causing an expansion of the spectrum.

Example. One commercially available chemical shift reagent is $Eu(fod)_3$, the structure of which is shown in Fig. 15.79.

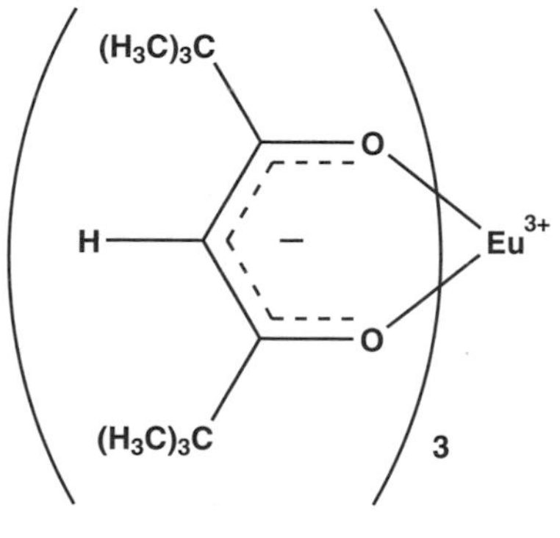

Figure 15.79. $Eu(fod)_3$.

15.80 CHIRAL SHIFT REAGENTS

Chemical shift reagents that possess chiral ligands, such that the entire reagent molecule is chiral. When such chiral shift reagents form guest-host complexes with chiral solute molecules, they cause different levels of deshielding in the two enantiomers of the solute. This is because the two enantiomers of the shift reagent can form two diastereomerically related guest host complexes with the two enantiomers of the solute. Thus, the two enantiomers (or enantiotopic nuclei) of the solute will exhibit separate NMR signals.

Example. One commercially available chiral shift reagent is $Pr(hfc)_3$, whose structure is shown in Fig.15.80.

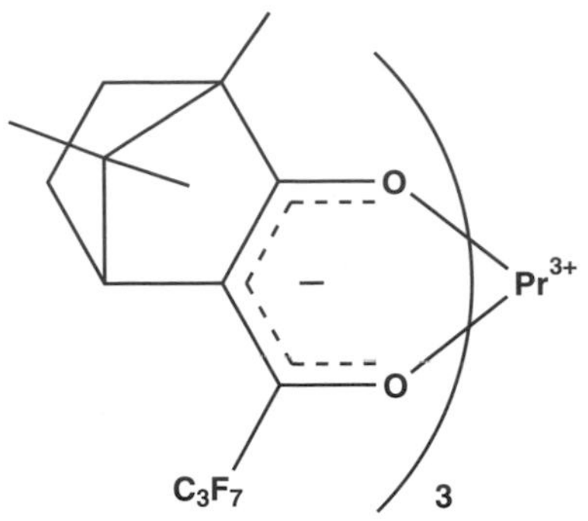

Figure 15.80. Pr(hfc)$_3$.

15.81 SOLVENT-INDUCED SHIFT

A change in the chemical shifts of various nuclei of a solute molecule when the solute is dissolved in a different solvent. The magnitude of the solvent-induced shift depends on the nature of the guest-host complex (Sect. 15.78) formed between the solvent and solute molecules. Aromatic solvents, such as benzene or pyridine, cause the largest solvent-induced shifts.

15.82 DOUBLE RESONANCE EXPERIMENTS

An NMR technique in which a collection of magnetic nuclei at equilibrium in a strong static applied field (B_0) is irradiated by two different oscillating magnetic fields: the observing field (B_1) and the irradiating field (B_2). Nuclei whose precessional frequency matches the frequency of the observed nuclei (B_1) undergo resonance to generate the signals in the NMR spectrum; nuclei whose precessional frequency matches the frequency of the irradiated nuclei (B_2) undergo resonance in such a way as to influence the intensity or multiplicity of the signals of the observed nuclei. The notation $X\{Y\}$ indicates an NMR double resonance experiment where nuclei X are observed while nuclei Y are irradiated. Double resonance experiments can be either homonuclear (X and Y are the same isotope) or heteronuclear (X and Y are different isotopes).

15.83 SPIN DECOUPLING

An $H_A\{H_B\}$ double resonance experiment where the frequency and intensity of the irradiating or decoupling field $h\nu_2$ and its associated magnetic field B_2 are adjusted so as to remove the effects of coupling of H_2 to H_1. The mechanism by which this occurs is outlined in Fig. 15.83*a*. The two branches of the doublet signal for H_1 arise from the existence of two different magnetic environments for H_1. These two magnetic environments each have three components. The applied magnetic field B_0 and the magnetic field induced by the electronic environment of H_1, B_I, are of the same magnitude for each branch. However, the magnetic field component derived from the nuclear spin of H_2, B_2, can be either additive or subtractive, as shown in Fig. 15.83*a*. The number of different values of B_2 is governed by the spin multiplicity rule (Sect. 15.59) and in the simple case shown in Fig. 15.83*a* is two. When H_2 is irradiated with an intense oscillating magnetic field B_2 ($h\nu_2$), net absorption occurs initially, since the population of the ground spin state is slightly higher than that of the excited state. However, a condition of saturation is rapidly achieved in which the populations of the two states become equal, and there is no net absorption.

It is important to note that under saturation conditions, a given H_2 nucleus does not have a long residence time in any spin state. Thus, according to Einstein's relationships for absorption (Sect. 15.118), given an equal population for ground and excited states, a two-level system such as the one under consideration here will have equal probability for absorption and stimulated emission. Consequently, at saturation, the spin states of H_2 are rapidly interchanging. In fact, this interchange is so rapid that a form of electronic hysteresis causes the magnetic field component from H_2 at H_1 to average to zero.

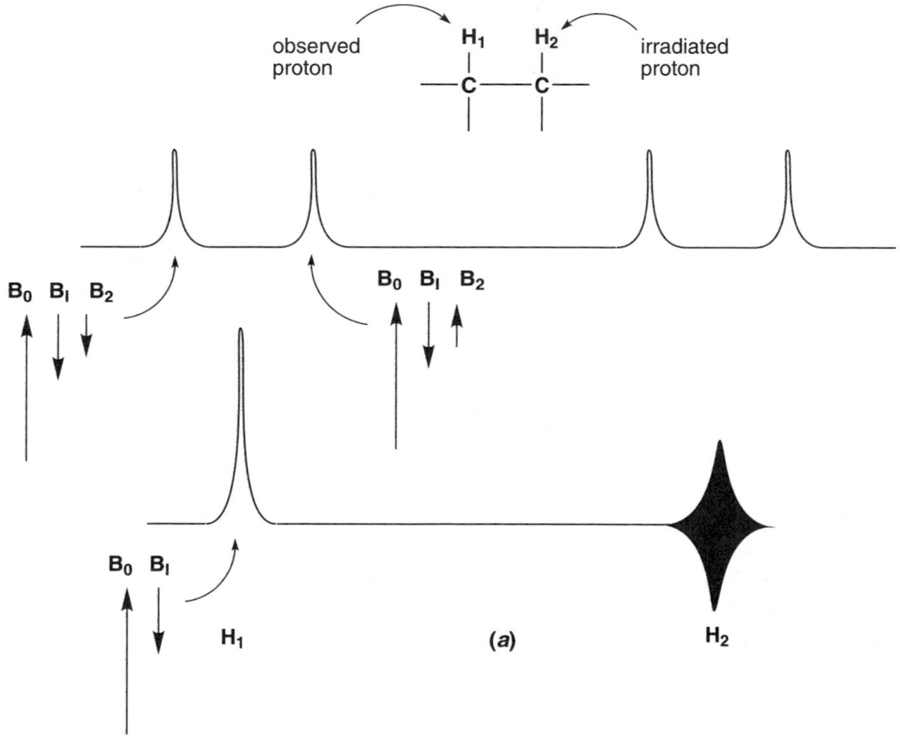

Figure 15.83. (*a*) Field components associated with the two branches of the signal for proton H_1 that is coupled to H_2, and the change in these field components generated by irradiation of H_2. (*b*) Three possible decoupling experiments that might be conducted on the three-spin system of H_1, H_2, and H_3.

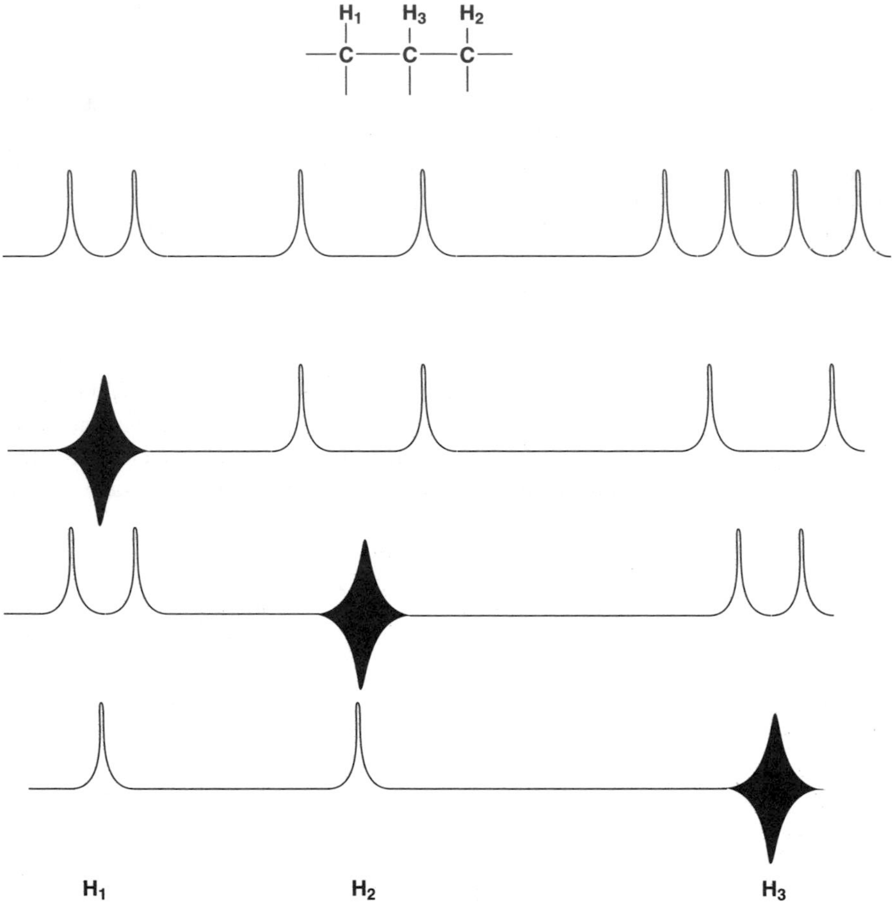

(*b*)

Figure 15.83. (Contd.)

The result is that the signal for H_1 no longer exhibits the coupling constant due to H_2, and it collapses to a singlet, as shown in Fig. 15.83*a*.

Spin decoupling experiment can be either heteronuclear (e.g., $^{13}C\{^1H\}$) or homonuclear (e.g., $^1H_1\{^1H_2\}$). Spin decoupling can also be used to establish unambiguously which signals are coupled to which other signals. To affect decoupling in a pulsed mode NMR experiment, the decoupling field must be on during acquisition of the FID signal.

Example. In the ^1H NMR spectrum shown in Fig. 15.70, irradiation of the quartet at 4.26 δ would cause the triplet at 1.34 δ to collapse to a singlet. Similarly, irradiation of the doublet at 6.43 δ would cause the doublet at 7.69 δ to collapse to a singlet. Only those coupling constants associated with a particular type of proton are lost when that type of proton is irradiated. This behavior is illustrated in Fig. 15.83*b*.

Thus, irradiation of H_1 and H_2 leads to the loss of the small and large coupling constants in H_3, respectively, whereas irradiation of H_3 leads to the loss of all coupling constants in both H_1 and H_2.

Most ^{13}C NMR spectra are generated with the aid of *broadband ^1H decoupling* ($^{13}C\{^1H\}$), that is, irradiation of *all* ^1H nuclei, rather than just certain signals. This removes all coupling between ^1H nuclei and the observed ^{13}C nuclei, so that (unless there are other magnetic isotopes present) each ^{13}C signal will appear as a singlet.

15.84 PULSE SEQUENCE

A detailed description of the timing, duration, and power of the operation of the B_1 and B_2 magnetic fields before and during the acquisition of the FID signal in an NMR experiment. Variations in the pulse sequence give rise to very useful changes in the nature of NMR signals, and provide additional and more direct information about the molecule structure of the sample.

15.85 NUCLEAR OVERHAUSER EFFECT (NOE)

The effect of altering the population difference (ΔP) of nuclear spin states by magnetic dipole-dipole relaxation. In general, the NOE depletes the population of the upper nuclear spin state (X_α in Fig. 15.85) and enhances the population of the lower spin state (X_β). This nuclear polarization usually results in an increase in intensity of the NMR signal associated with the polarized nuclear population.

Example. Figure 15.85 provides a stepwise accounting of the processes by which nuclear polarization occurs. If two magnetically active nuclei (X and Y both with $I = 1/2$) are in close proximity to one another, four group spin states will exist, as shown in Fig. 15.85. Irradiation of nuclei Y with an intense field of B_2 will cause the population of that nuclei to approach saturation ($\Delta P_y = 0$) (Fig. 15.85*a*). Relaxation of this high-energy spin state Y_α can be induced by magnetic dipoles in close proximity to Y—in this case, the magnetic dipole of X. This induced relaxation process can occur only if an equal, but opposite, change occurs in the angular momentum of the two nuclei Y and X; both spin states must change simultaneously: $Y_\alpha \rightarrow Y_\beta$ and $X_\alpha \rightarrow X_\beta$ (Fig. 15.85*b*). This process tends to lead to a depletion of the high-energy spin state of X and an enhancement of the population of its lower-energy spin state. The net result is that the population difference of the spin states of X (ΔP_x) is increased. Thus, when X is irradiated with a second, lower-intensity frequency B_1, the observing frequency, it will absorb more intensely.

The magnitude of the intensity change is dependent, among other factors, on the distance between the X and Y nuclei, increasing as the distance R decreases ($1/R^6$). Thus, the magnitude of the effect can be used as a gauge of proximity of two nuclei within a molecule. Because the magnitude of the effect also requires a finite amount of irradiation time to evolve, the irradiating field in an NOE experiment must be *on* for a few seconds prior to the B_1 pulse and FID acquisition (Sects. 15.29 and 15.30).

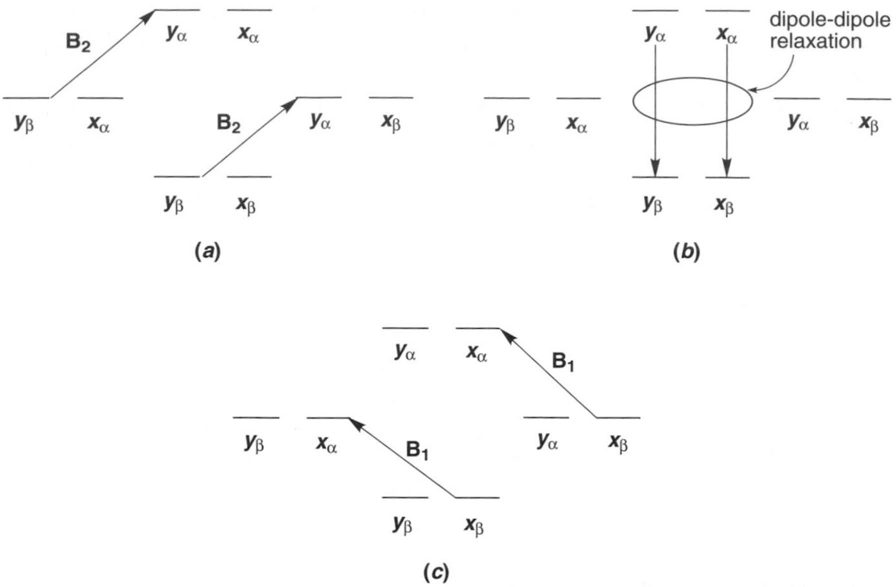

Figure 15.85. Nuclear overhauser effect. (*a*) Saturation of the spin-polarizing nucleus *Y* with an intense irradiation field B_2. (*b*) Generation of spin polarization in the population of nucleus *X* via dipole-dipole relaxation. (*c*) Observation of spin polarization of nucleus *X* using a second low-intensity irradiation field B_1.

A *difference NOE* is an NMR spectrum where the plotted signals represent the computed difference between the spectrum acquired with NOE and the spectrum acquired without NOE. In such an experiment, nuclei corresponding to one particular signal are irradiated with B_2. Positive signals in the difference NOE spectrum correspond to those nuclei that are nearby the irradiated nuclei in the structure of the sample molecule.

In the slow exchange ^{1}H NMR spectrum of DMF (Sect. 15.75), irradiation of the methyl signal at 2.91 δ causes an 18% increase in intensity of the formyl hydrogen (HC=O) signal, whereas irradiation of the methyl signal at 2.74 δ has essentially no effect on the formyl signal. These results indicates that the signal at 2.91 δ belongs to the methyl group, which is closer (*cis*) to the formyl hydrogen. The broadband ^{1}H decoupling used in $^{13}C\{^1H\}$ experiments (Sect. 15.83) also causes an NOE enhancement of the carbon signals. The magnitude of this enhancement tends to increase with the number of hydrogens attached to the carbon.

15.86 GATED DECOUPLING

An $X\{Y\}$ double resonance experiment (Sect. 15.82) where the irradiating field (B_2) is on before the initiating B_1 pulse, but is then turned off during FID acquisition. The result is to preserve the NOE, Fig. 15.86, but avoid any decoupling.

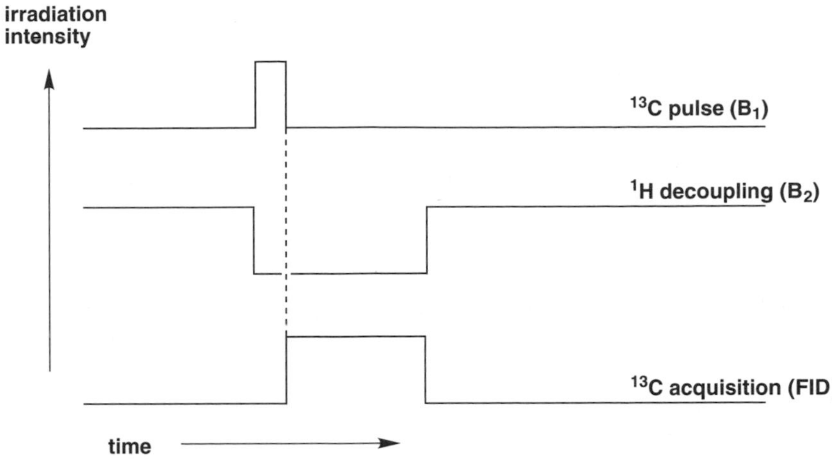

Figure 15.86. Pulse sequence used in gated decoupling for the acquisition of a ^{1}H-^{13}C NOE NMR spectrum.

15.87 INVERSE GATED DECOUPLING

An $X\{Y\}$ double resonance experiment where the irradiating field (B_2) is off before the initiating B_1 pulse, but is then turned on during FID acquisition. The result is to bring about the desired decoupling, but avoid any NOE enhancements, Fig. 18.57. (Sect. 15.85).

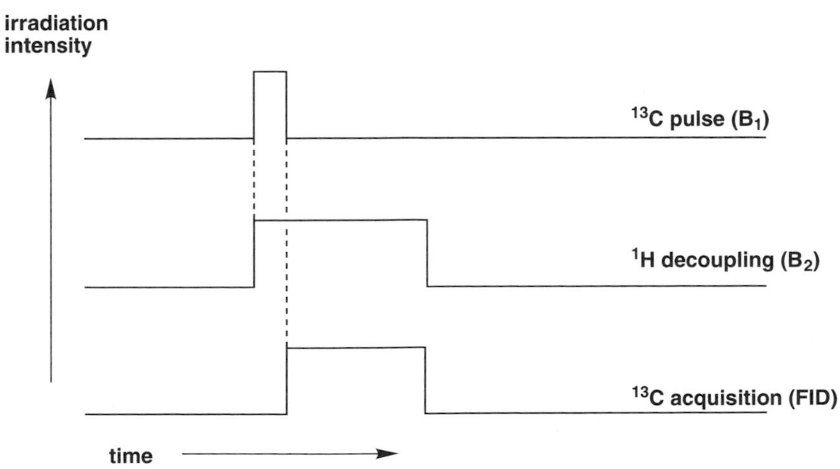

Figure 15.87. Pulse sequence used in inverse gated decoupling for the acquisition of a ^{1}H decoupled ^{13}C NMR spectrum.

15.88 ATTACHED PROTON TEST (APT)

A pulse sequence (Sect. 15.84), most commonly applied in ^{13}C spectroscopy, which discriminates carbon signals based on the number of hydrogens attached directly to the carbon, giving rise to the signal. By careful manipulation of the timing and duration of B_1 pulses and B_2 activation, the intensity of a carbon signal can be made to vary from +100% of its normal intensity to zero or even −100%, depending on the number of hydrogens attached to the carbon and the values of the one-bond C–H coupling constants ($^1J_{C\text{-}H}$). A series of spectra are collected digitally, each one differing from the previous one by a small change in the interpulse delay time between B_1 pulses. If all the $^1J_{C\text{-}H}$ values are approximately equal, an interpulse delay of $[2(^1J_{C\text{-}H})]^{-1}$ will cause all carbon signals to disappear except those due to quaternary (i.e., non-hydrogen-bearing) carbons. Similarly, an interpulse delay of $(^1J_{C\text{-}H})^{-1}$ will cause all quaternary and CH_2 carbon signals to exhibit normal (100%) intensity, while all CH and CH_3 carbons exhibit −100% (negative) intensity. Applying this technique to the ^{13}C spectrum of an unknown compound allows the number of hydrogens attached to each carbon to be deduced.

Example. Figure 15.88 shows the simulated APT ^{13}C spectrum of 2-chlorobutane, using an interpulse delay of $(^1J_{C\text{-}H})^{-1}$.

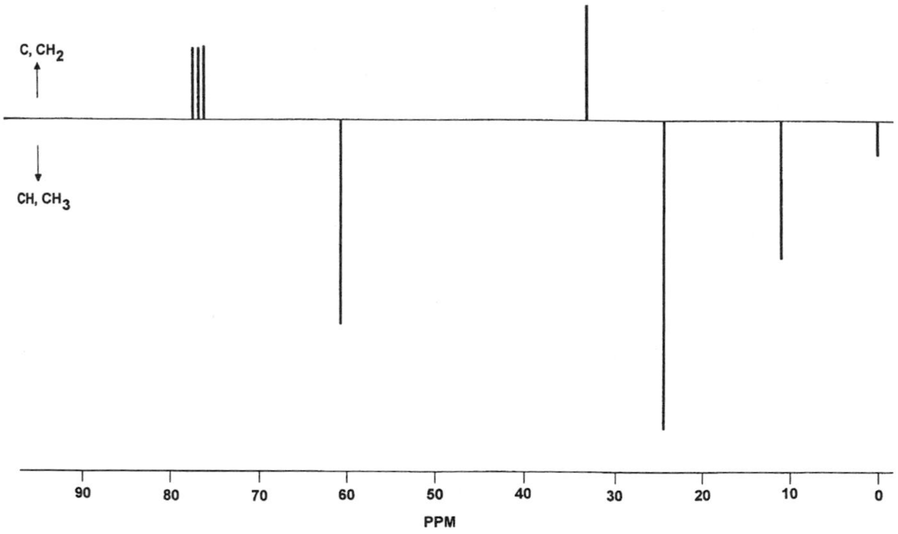

Figure 15.88. The simulated APT ^{13}C spectrum of 2-chlorobutane. The three-line signal at 75 δ is due to the solvent $CDCl_3$.

15.89 DISTORTIONLESS ENHANCEMENT BY POLARIZATION TRANSFER (DEPT)

A more sophisticated software-driven version of the APT pulse sequence (Sect. 15.88). A DEPT experiment first determines how each ^{13}C signal varies with interpulse delay time and then automatically edits (separates) the spectroscopic data into separate spectra: one for the CH_3 signals, one for the CH_2 signals, and one for the CH signals. Signals that appear in the normal ^{13}C spectrum, but not in one of the edited spectra, are due to quaternary carbons.

Example. Figure 15.89 shows the edited DEPT ^{13}C spectrum of the molecule shown below (artemisinin):

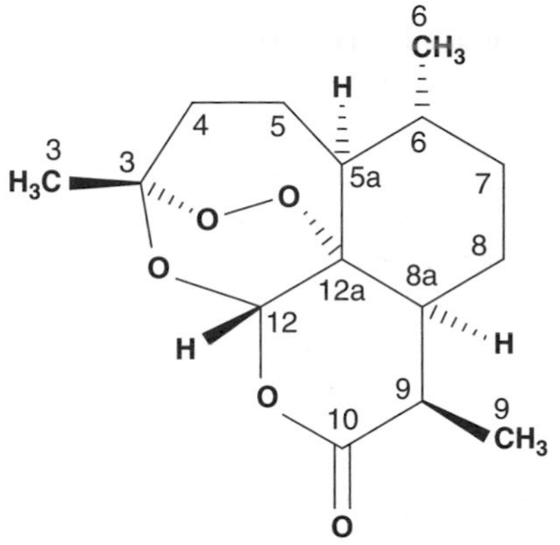

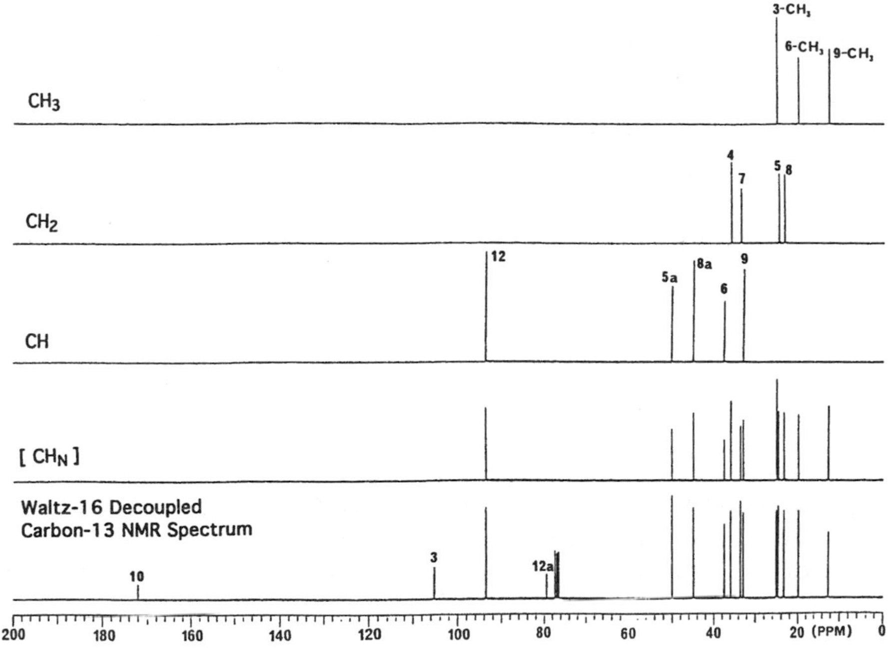

Figure 15.89. The edited DEPT ^{13}C spectrum of artemisinin.

15.90 OFF-RESONANCE DECOUPLING

An $X\{Y\}$ double resonance experiment where the frequency and intensity of the irradiating (or decoupling) field B_2 are adjusted to reduce, but not eliminate, the magnitude of spin coupling constants between nuclei X and nuclei Y. The result is that only the coupling interactions with the largest coupling constants are observed in the multiplets of the X nuclei. Note that the observed magnitude of the coupling constant in such an experiment is reduced to a fraction of its original value.

Example. In an off-resonance $^{13}C\{^{1}H\}$ experiment, the two- and three-bond C–H couplings are reduced to the point that they can no longer be resolved. The one-bond C–H coupling constants, although reduced from their original values (ca. 125–250 Hz), are still large enough to produce multiplets. Thus, signals for methyl carbons appear as quartets, methylene carbons appear as triplets, methine carbons as doublets, but nonhydrogenated carbons remain singlets. In this way the assignment of specific signals in the spectrum to specific carbons in the structure becomes easier.

15.91 TWO-DIMENSIONAL NMR

A two-dimensional (2D) NMR spectrum is a plot of NMR signal intensity as a function of two frequency variables, rather than just one as in a normal (or one-dimensional) spectrum. The frequency variables involved, which are determined by the nature of the pulse sequence, determine the specific type of two-dimensional spectrum produced. A two-dimensional spectrum is plotted either in a stacked plot (Fig. 15.91*a*) or, more commonly, a contour plot (Fig. 15.91*b*). In this example, the first frequency parameter (F_1) is the pulse width parameter $\Theta y'$, and the second frequency parameter (F_2) is the ^{13}C chemical shift scale.

Example. The most common types of two-dimensional NMR spectra include HETCOR (C,H-COSY), HOMCOR (H,H-COSY), NOESY, and 2D INADEQUATE (Sects. 15.92–15.95, respectively).

15.92 HOMONUCLEAR CORRELATION (HOMCOR), H, H-COSY

An H,H-COSY (H,H-correlation spectroscopy) spectrum, an example of homonuclear shift-correlated spectra (HOMCOR), is a two-dimensional NMR technique that correlates pairs of ^{1}H signals of a given compound. Each off-diagonal H,H-COSY signal constitutes a cross-correlation between two specific ^{1}H signals. Thus, the hydrogens, giving rise to the COSY signal, are spin-coupled to each other in the sample molecule.

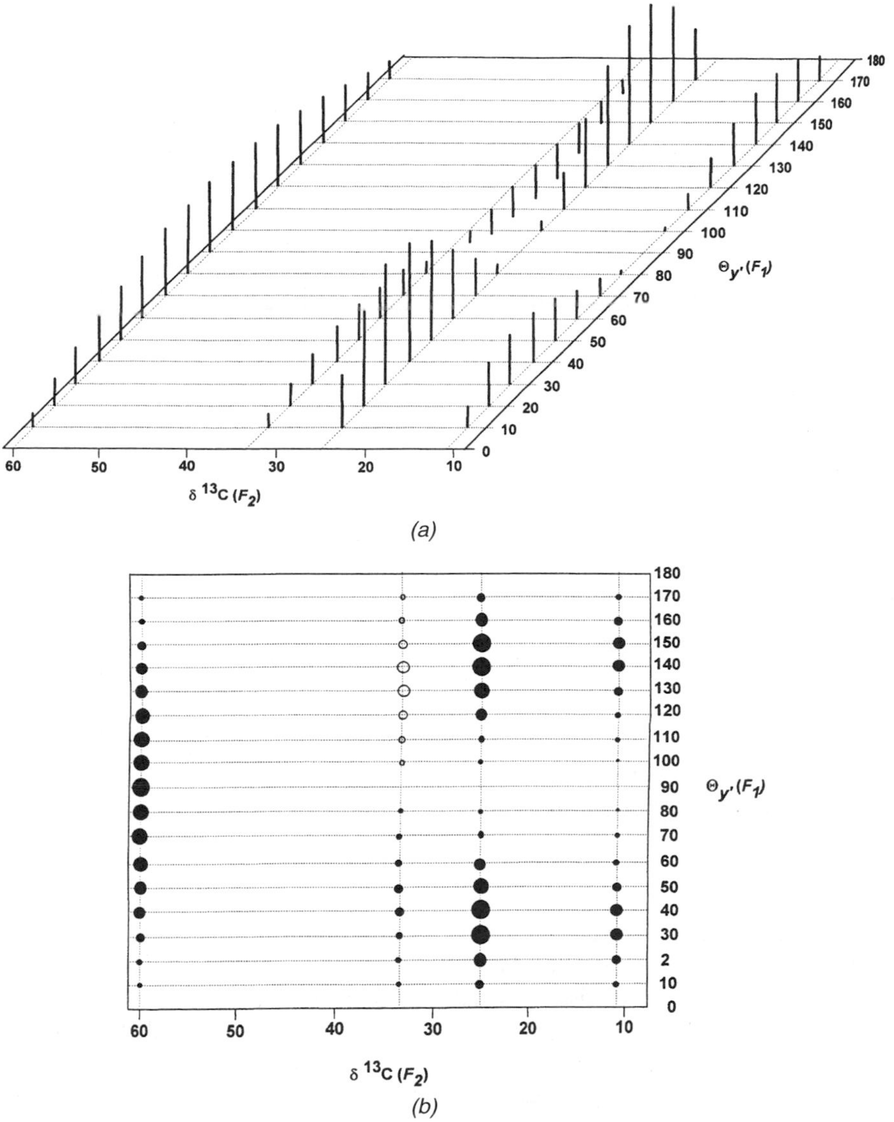

Figure 15.91. (*a*) Stacked plot format of the ^{13}C DEPT spectrum (Sect. 15.93) of 2-chlorobutane as a function of pulse width $\Theta_{y'}$. (*b*) A contour plot of the same data set.

Example. Figure 15.92 shows the H,H-COSY spectrum of 2-chlorobutane, with the same ^1H chemical shift scale (and spectrum) along both axes. The signals of the spectra appearing along each axis are numbered from 1 (lowest) to 4 (highest) chemical shift. Each off-diagonal signal indicates the correlation between the pairs of axis signals at the ends of the dashed lines. (Note that the off-diagonal signals above the

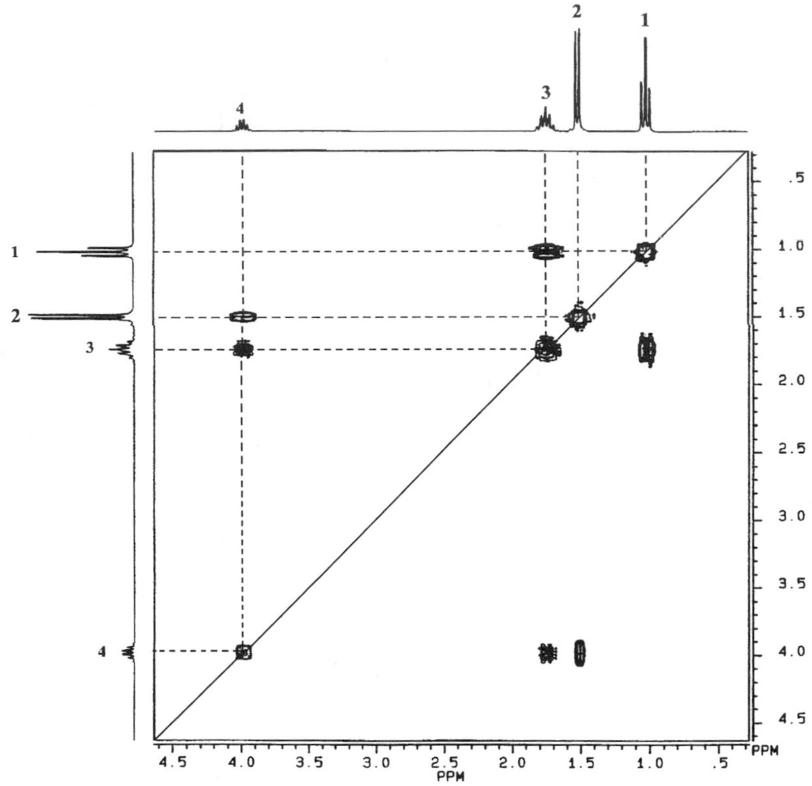

Figure 15.92. The H,H-COSY spectrum of 2-chlorobutane, with correlations indicated by dashed lines.

diagonal are the mirror image of those below the diagonal and give the same information.) The correlations indicated are H1/H3, H2/H4, and H3/H4. This shows that the hydrogen which gives rise to signal 1 is coupled to (and, therefore, not more than three bonds away from) the hydrogen that gives rise to signal 3, that 2 is coupled to 4, and that 3 is coupled to 4. A combination of the H,H–COSY and C,H–COSY data allows the chemist to put together the carbon backbone of the molecular structure by applying the following logic: If H1 is correlated with C1, H2 is correlated with C2, and if H1 is correlated with H2, the molecular connectivity H1–C1–C2–H2 is established.

15.93 HETERONUCLEAR CORRELATION (HETCOR), C,H-COSY

A C,H-COSY (C,H-correlation spectroscopy) spectrum, an example of heteronuclear shift-correlated spectra (HETCOR), is a two-dimensional NMR technique that correlates the ^{1}H and ^{13}C NMR signals of a given compound. Each C,H-COSY signal constitutes a cross-correlation between a specific ^{13}C signal and one or more specific ^{1}H signals. Thus, the carbon and hydrogen(s), giving rise to the COSY signal, are spin-coupled to each other by a one-bond coupling constant (i.e., are directly bonded to each other in the sample molecule).

Example. Figure 15.93 shows the C,H-COSY spectrum of 2-chlorobutane, with the ^{1}H chemical shift scale (and spectrum) along the *y*-axis, and the ^{13}C chemical shift scale (and spectrum) along the *x*-axis. The signals of the spectra appearing along each axis are numbered from 1 (lowest) to 4 (highest) chemical shift. Each elongated contour signal indicates the correlation between the pairs of axis signals at the ends of the dashed lines. The correlations indicated are C1/H1, C2/H2, C3/H3, and C4/H4. This shows that the carbon which gives rise to ^{13}C signal 1 is directly bonded to the hydrogen that gives rise to ^{1}H signal 1, and so on. The fact that the correlation is roughly diagonal (i.e., the most shielded hydrogen is attached to the most shielded carbon) is often, although not always, observed. See also Sect. 15.92.

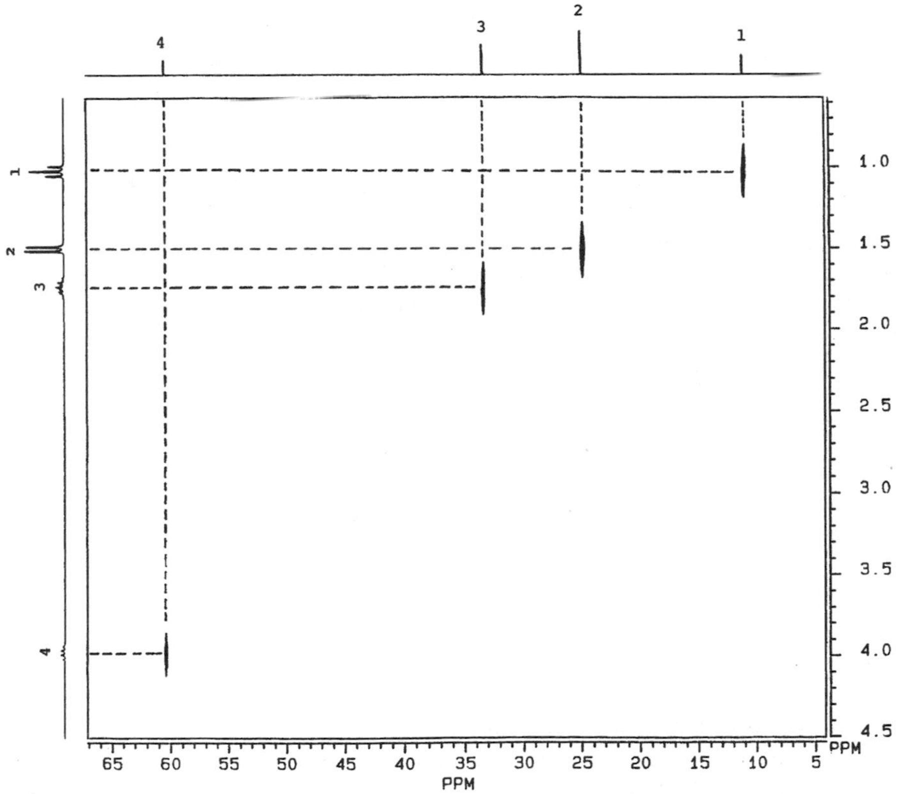

Figure 15.93. The C,H-COSY spectrum of 2-chlorobutane, with correlations indicated by dashed lines.

15.94 NUCLEAR OVERHAUSER ENHANCEMENT AND EXCHANGE SPECTROSCOPY (NOESY)

A type of two-dimensional COSY (correlation spectroscopy) spectrum (such as in Sect. 15.91) where each off-diagonal signal indicates that the associated nuclei are involved in an NOE polarization transfer. This indicates that the nuclei giving rise to the NOESY signal are either spatially proximate in the molecular structure in the sample molecule, or they are spin-coupled as in a normal H,H-COSY.

Example. Consider Figure 15.94 shows the H,H-NOESY spectrum of the molecule below:

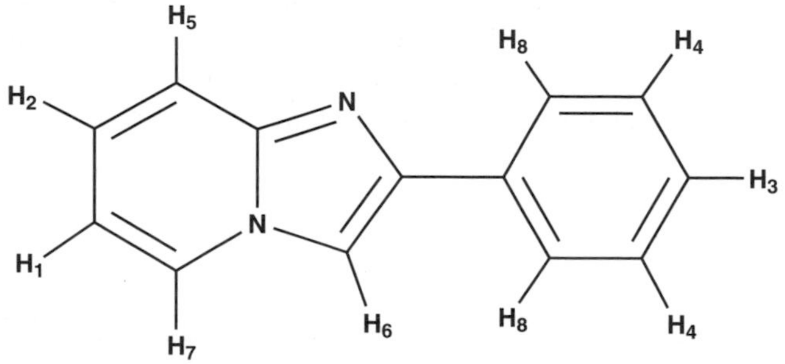

Both axes of the two-dimensional spectrum are ^{1}H chemical shifts, although the ^{1}H signals of the one-dimensional spectrum only appear along the top axis. These are numbered from 1 (lowest) to 8 (highest) chemical shift. Each off-diagonal NOESY signal indicates the correlation between the corresponding pair of axis signals. (Note that the off-diagonal signals above the diagonal are the mirror image of those below the diagonal and give the same information.) The indicated correlations are tabulated below (w = weak).

Hydrogen	Exhibits NOE with Hydrogen
1	2, 7
2	1, 5
3	4
4	3, 8
5	2, **6(*w*)**
6	**5(*w*), 7, 8**
7	1, **6, 8(*w*)**
8	4, **6, 7(*w*)**

The NOESY signals in boldface do *not* appear in the H,H-COSY, so they are solely the result of direct through-space NOEs, not spin-coupling. Note especially the correlations of proton 6, which is not spin-coupled to any of the other hydrogens (giving rise to a singlet in the one-dimensional spectrum), but is rigidly fixed in a position near hydrogens 7 and 8.

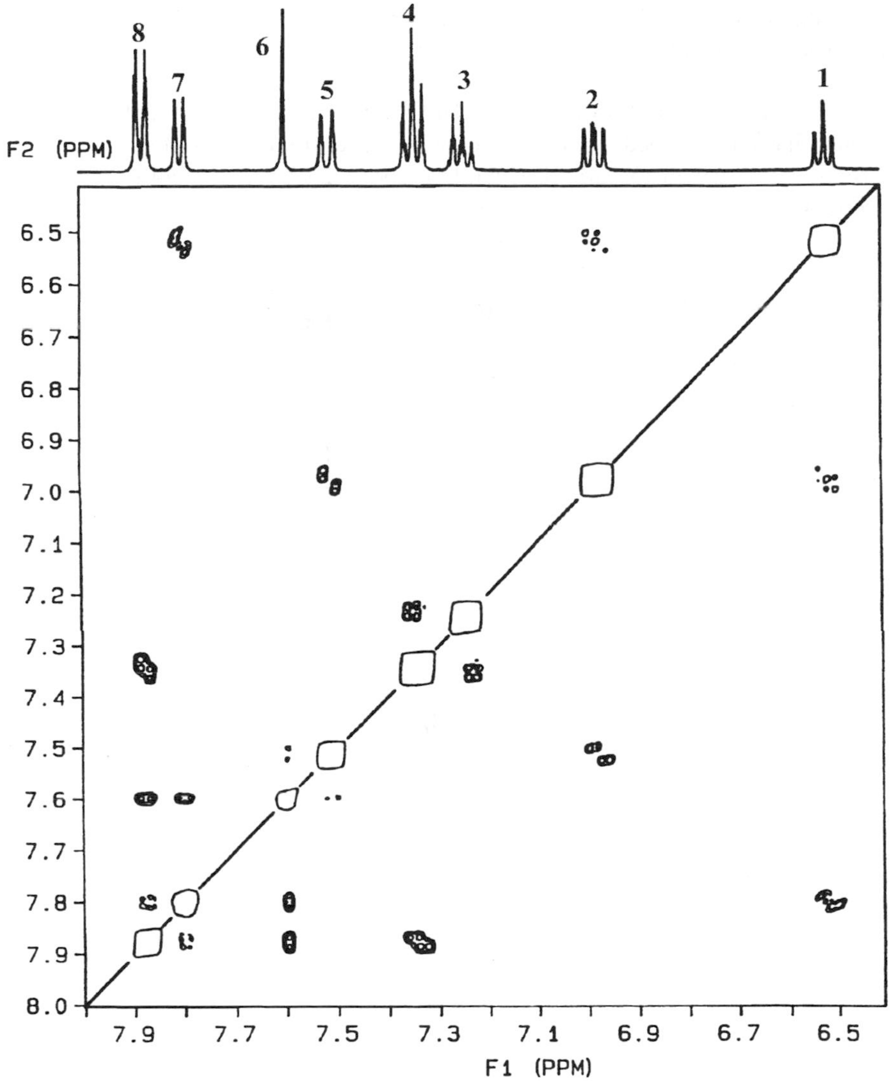

Figure 15.94. The H,H-NOESY spectrum of the tricyclic molecule whose structure is shown in Sect. 15.94.

15.95 TWO-DIMENSIONAL INCREDIBLE NATURAL ABUNDANCE DOUBLE QUANTUM TRANSFER EXPERIMENT (2D INADEQUATE)

This is a two-dimensional NMR technique (Sect. 15.91) that could be called C,C-COSY (see Sects. 15.92 and 15.93) because both axes are the ^{13}C chemical shift scale, and each pair of INADEQUATE footprint signals indicates a pair of correlated ^{13}C signals. Each such correlation results from direct one-bond C–C coupling. The two-dimensional INADEQUATE spectrum directly provides a complete map of all carbon–carbon connectivities, that is, the carbon skeleton of the sample molecule.

Example. Figure 15.95 shows the 2D INADEQUATE ^{13}C spectrum of 2-chlorobutane. Dashed lines are added to show the pairs of correlated "footprint" signals: Carbon 1 is attached to carbon 3, while carbons 2 and 3 are both attached to carbon 4.

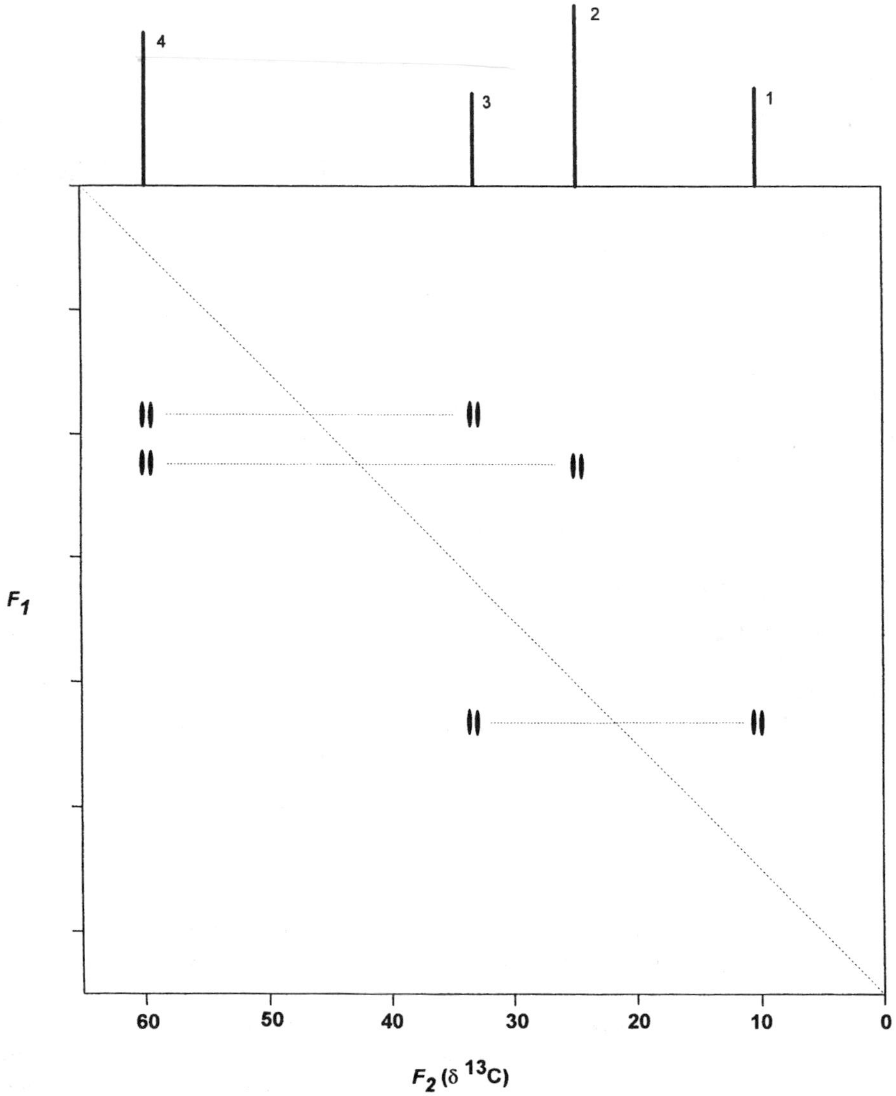

Figure 15.95. The 2D INADEQUATE ^{13}C spectrum of 2-chlorobutane.

15.96 MAGNETIC RESONANCE IMAGING (MRI)

The process by which the spatially dependent NMR properties of an object are recorded and present a three-dimensional image. In most medical applications, the data are presented as a cross section. Because a nuclear spin's Larmor (precession) frequency (Sect. 15.9) is proportional to the magnetic field strength experienced by the nucleus, a magnetic field that varies in intensity as a function of position within the object can be used to set up a correspondence between position in space with position in the NMR spectrum.

Example. A linear magnetic field gradient (a field that changes linearly with position along a single spatial dimension), say, along the x-axis, of a compound with a simple one-line spectrum (e.g., the proton spectrum of water) creates a one-to-one correspondence between spins located in a plane at each x position with position in the frequency spectrum. The NMR frequency axis, which normally represents the chemical shift, now corresponds to the x position. In general, the application of multiple magnetic field gradients along different spatial directions can be used to encode spatial information in two or three dimensions in the NMR signal. The decoding of this spatially encoded NMR signal (in other words, the reconstruction of the image) is performed by any of several mathematical algorithms (filtered back projection, Fourier transformation, algebraic reconstruction, etc.).

No matter how the spatial encoding and image reconstruction are accomplished, the result is essentially a multidimensional NMR spectrum whose frequency axes represent positions along spatial dimensions. Rather than representing these NMR "spectra" conventionally as line graphs or contour plots showing signal intensity versus frequency, NMR image data are displayed with signal intensity represented as brightness, making these displays appear very much like images (pictures). However, it is important to remember that they are not photographic images, but rather spatial maps of NMR signal intensity. Just as in conventional NMR spectroscopy, the signal intensity is determined by many factors, including NMR instrument parameters (radiofrequency and magnetic field gradient pulse amplitudes, durations, and interpulse timings, etc.) and the physical and chemical properties of the imaged nuclear spins (concentrations, relaxation parameters, chemical shifts, spin-spin couplings, molecular diffusion coefficients, flow rates, etc.).

In the case of biological objects (the human head, e.g.), the proton image contrast (light-dark variation) is due primarily to variations in water proton relaxation parameters (spin-lattice and spin-spin relaxation times), and only to a minor degree to variations in the concentration of water (the compound that contributes almost all the signal in conventional clinical MRI). It is possible to produce multidimensional data sets in which some dimensions are spatial, and some correspond to the conventional NMR spectrum; this process is most generally known as *chemical shift imaging*. Finally, it is possible to use the principles of MRI to produce one- or two-dimensional NMR spectra from small selectable regions within an object; these techniques are known collectively as *localized NMR spectroscopy*.

15.97 CHEMICALLY INDUCED DYNAMIC NUCLEAR POLARIZATION (CIDNP)

Anomalies in the NMR signals of products being formed from certain chemical reactions involving free radicals. These anomalies are of two types: (1) the *net effect* (Sect. 15.98), which involves large positive or negative changes in the signal intensities, compared to normal expectations, or (2) the *multiplet effect*, (Sect. 15.99), the appearance of multiplets with relative line intensities that are radically different from normal expectations based on Sect. 15.60. These effects may be observed separately or in tandem, depending on the type of reaction and nature of the product.

15.98 THE NET EFFECT

Anomalously large positive or negative signals in a CIDNP-NMR spectrum. The positive signals are the result of enhanced absorption (an *A* net effect), whereas the negative signals result from emission of radiation (an *E* net effect). The enhanced absorption or emission net effects are caused by *polarized* (i.e., nonequilibrium) nuclear spin-state populations (Sect. 15.12). If the lower-energy spin state is overpopulated, an *A* effect is observed, whereas overpopulation of the higher-energy spin state leads to an *E* effect. This polarization arises when a pair of free radicals, each with a highly magnetic unpaired electron, interacts to form a product molecule in the strong magnetic field of an NMR spectrometer.

15.99 THE MULTIPLET EFFECT

A symmetrical disturbance of the relative intensities of the lines in a multiplet arising from a compound formed under CIDNP conditions. The individual lines exhibit either *E* or *A* effects (Sect. 15.98). When the *E* effect (*e.g.*, negative) line is to the left of the *A* (enhanced positive) line in the spectrum, the multiplet is described as exhibiting an *E/A* multiplet effect. An *A/E* multiplet effect has the positive line to the left of the negative line.

15.100 THE RADICAL PAIR THEORY OF CIDNP EFFECTS

The currently accepted theory that correlates the observed CIDNP effects with specific properties of product molecular structures and mode of product formation. Thus, as developed by Robert Kaptein [*J. Chem. Soc. D*, 732 (1971)], the type of net effect (*E* or *A*) is directly correlated with the sign (positive for *A*, negative for *E*) of the product of four parameters: (1) the spin multiplicity of the original caged radical pair at its "birth" (+ for triplet, − for singlet), (2) the type of reaction leading to the product (+ for recombination within the cage, − for escape from the cage), (3) the difference in *g* factor between the radical containing the nucleus giving rise to the signal, and the other radical, and (4) the sign of the hyperfine coupling constant in the radical possessing the observed nucleus. Similarly, the type of multiplet effect in the signals of two coupled nuclei is correlated with the sign (positive for *E/A*, negative for *A/E*) of the product of six parameters: (1) and (2) are the same as above, (3) and (4) are the hyperfine coupling constants of the two coupled nuclei in their respective radicals, (5) the sign of the internuclear coupling constant (Sect. 15.61) shared by the nuclei giving rise to the multiplet, and (6) + if the coupled nuclei are in the same radical, − if they are not.

Example. Consider the reaction of diphenylethylene ($Ph_2C{=}CH_2$) with $HCo(CO)_4$, whose mechanism is shown in Fig. 15.100*a*. The NMR spectrum of the methyl group of Ph_2CHCH_3 immediately after mixing the two reagents is shown in Fig. 15.100*b*

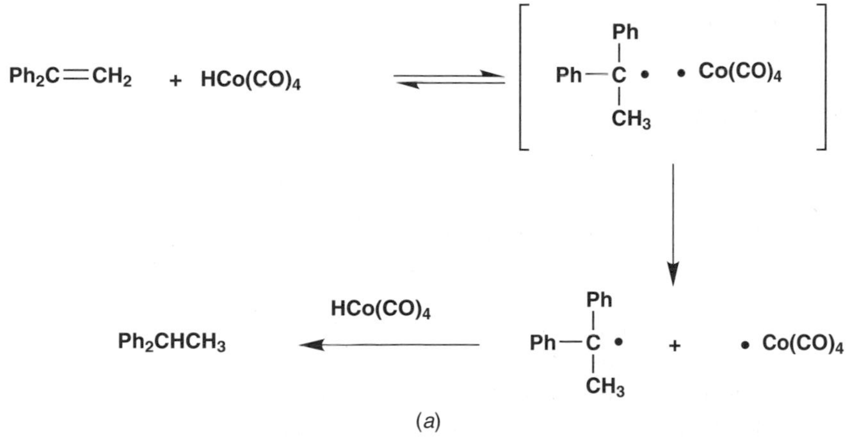

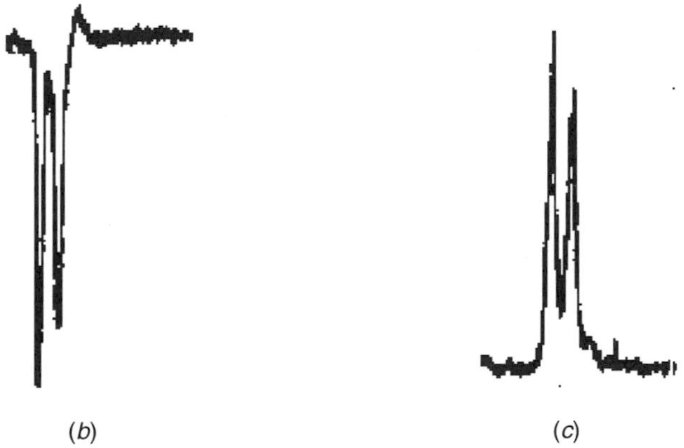

Figure 15.100. (*a*) Mechanism of the reaction of diphenyl ethylene ($Ph_2C{=}CH_2$) with $HCo(CO)_4$; (*b*) NMR spectrum of the methyl group of Ph_2CHCH_3 immediately after mixing the two reagents; (*c*) NMR spectrum 20 sec later (T. E. Nalesnik and M. Orchin, *Organometallics,* **1**, 222 (1982)).

(an emission spectrum) and the spectrum 20 sec later is shown in Fig. 15.100*c* (a normal NMR spectrum).

The emission spectrum can be predicted using the four parameters of the Kaptein equation:

1.	Singlet precursor	-1
2.	Escape from the cage	-1
3.	*g* Factor for $Ph_2(CH_3)C\bullet$ is smaller than for $(CO)_4Co\bullet$	-1
4.	HC–C• has a positive hyperfine coupling constant	$+1$

The product of $(-1)(-1)(-1)(+1) = -1$, which is consistent with the observed emission spectrum T. Nalesnik.

Acknowledgment. The authors thank Professor Jerry Ackerman of Harvard for his help with the definition of MRI.

SUGGESTED READING

Derome, A. E., *Modern NMR Techniques for Chemical Research,* Pergamon Press Oxford, 1987.

Lambert, J. B., and Mazzola, E. P. *Nuclear Magnetic Resonance Spectroscopy, An Introduction to Principles, Applications, and Experimental Methods,* Pearson Prentice Hall: Upper Saddle River, NJ, 2004.

Macomber, R. S. *A Complete Introduction to Modern NMR Spectroscopy.* Wiley-Interscience: New York, 1998. (This reference was the source of many figures in this chapter).

Seco, J. M., Quinoa, E., and Riguera, R. *Chem. Rev.*, *104*, 17 (2004).

Silverstein, R. M. and Webster, F. X. *Spectrometric Identification of Organic Compounds*. John Wiley & Sons: New York, 1998.

16 Vibrational and Rotational Spectroscopy: Infrared, Microwave, and Raman Spectra

16.1	Wavenumber $\tilde{\nu}$	661
16.2	The Spectroscopically Important Regions of the Electromagnetic Spectrum	661
16.3	Microwave Spectroscopy	662
16.4	Hooke's Law	662
16.5	Force Constant k	663
16.6	Potential Energy Function	663
16.7	Harmonic Oscillator	665
16.8	Rotational (Microwave) Spectra; Diatomic Molecules	666
16.9	The Rigid Rotor Model	666
16.10	Reduced Mass μ	667
16.11	The Principal Axes	667
16.12	Moments of Inertia I (Principal Moments of Inertia)	667
16.13	Angular Velocity ω	668
16.14	Kinetic Energy (KE) of the Rigid Rotor	668
16.15	Angular Momentum Quantum Number J	669
16.16	Energies of Rotational States E	669
16.17	Selection Rules for Rotational Spectra of Rigid Linear Molecules	669
16.18	The Nonrigid Rotor	669
16.19	Infrared Absorption Spectrum	669
16.20	Vibrational Spectrum	671
16.21	Vibrational Spectra of Diatomic Molecules	671
16.22	Vibrational Levels of the Harmonic Oscillator	672
16.23	Energies Associated with Infrared Absorptions	672
16.24	Allowed Energy Levels	673
16.25	Vibrational State	674
16.26	Ground Vibrational State	674
16.27	Excited Vibrational States	674
16.28	Selection Rules for the Harmonic Oscillator	674
16.29	Morse Curve	674
16.30	Anharmonic Oscillator	675
16.31	Fundamental Vibration	676
16.32	Spectroscopic Dissociation Energy	676

The Vocabulary and Concepts of Organic Chemistry, Second Edition, by Milton Orchin, Roger S. Macomber, Allan Pinhas, and R. Marshall Wilson

16.33 Dissociation Energy 676
16.34 Zero-Point Energy 676
16.35 Number of Degrees of Freedom 677
16.36 Number of Fundamental Vibrations 677
16.37 Genuine and Nongenuine Vibrations 677
16.38 Normal Vibrations; Normal Modes 678
16.39 Stretching Vibrations 678
16.40 Symmetric (Stretching) Vibration 679
16.41 Antisymmetric (Stretching) Vibration 680
16.42 Bending Vibrations 681
16.43 Degenerate Vibrations 682
16.44 In-Plane Bending 682
16.45 Out-of-Plane Bending 682
16.46 Group Frequencies 683
16.47 Forbidden and Allowed Transitions 687
16.48 Infrared Active Vibrations, Dipole Moment Considerations 687
16.49 Overtones 689
16.50 Combination Bands 689
16.51 Difference Bands 690
16.52 Accidental Degeneracy 690
16.53 Normal Coordinate 690
16.54 Fermi Resonance 691
16.55 Asymmetric Top Molecules 691
16.56 Symmetric Top Molecules 692
16.57 Spherical Top Molecules 692
16.58 Vibrational-Rotational Spectra 692
16.59 *R* and *P* Branches 693
16.60 *Q* Branch 693
16.61 Rayleigh Scattering 694
16.62 Raman Spectra, Raman Shifts 694
16.63 Stokes and Anti-Stokes Line Shifts 695
16.64 Polarizability 697
16.65 Symmetry Selection Rules: Intensities of Infrared Bands 698
16.66 Symmetry Selection Rules: Raman Spectra 699
16.67 The Exclusion Rule for Molecules with a Center of Symmetry 699
16.68 Fourier Transform (FT) 700
16.69 Fourier Transform Infrared Spectroscopy (FTIR) 701

In 1800, the English musician, turned astronomer, Sir Friedrich Wilhelm (William) Herschel (1738–1822) conducted an experiment to determine which color of the spectrum was responsible for the heat that apparently adversely affected the vision of some experienced solar astronomers. He wrapped a black cloth around a thermometer and then placed it in the various colors obtained when white light was passed through a glass prism. He found that the full red fell just short of the maximum of heat. Hence, the radiation responsible for the heat was, in what Herschel called, the infrared (beyond the red) region—actually, what is now known as the near infrared (Sect. 16.2) since glass is opaque to the middle infrared region. The heat associated with the absorption of infrared radiation by a molecule results from the transfer of energy from its vibrational excited states to neighboring molecules whose increased kinetic energy is expressed in the form of heat.

Molecules are in constant motion. Each individual atom within a molecule is free to move randomly in any direction, but such motion can be resolved into x, y, and z components. To specify completely the position of all atoms of a molecule containing n atoms will thus require $3n$ coordinates, corresponding to $3n$ degrees of freedom. There are three different types of motion that a molecule may undergo. Two of these, translational and rotational motion, involve movements of the entire molecule as an intact unit. The translational motions result in the movement of the center of mass and are of little interest to chemists. The rotational motion, that is, changes in angular orientation, can give rise to rotational spectra, and although these provide valuable information on small molecules, generally, they are of limited interest to organic chemists and require gas phase techniques. If a molecule is placed in an imaginary coordinate system, its random translational motion T can be resolved into component translations, T_x, T_y, T_z along the x-, y-, and z-axes. Similarly, the rotational motion R can be resolved into R_x, R_y, R_z rotational components. These six component motions of the entire molecule are sometimes called *nongenuine* vibrations since such motions do not change the relative positions of the atoms. In addition to such motions, all molecules, including those in the solid phase, are constantly undergoing *vibrational* motions that *do* change the position of the constituent atoms of a molecule relative to each other. These motions are called *genuine* vibrations; they are responsible for infrared spectra and are of great interest, especially to organic chemists.

To calculate the number of such genuine vibrations in a molecule containing n atoms, the (nongenuine) three-component translational and three-component rotational motions must be subtracted from the total $3n$ degrees of freedom, giving $3n - (3T + 3R) = 3n - 6$ genuine vibrations (for linear molecules, this number will be $3n - 5$ because rotation around the internuclear axis is not meaningful). These may be stretching vibrations (altering interatomic bond distances), bending vibrations (changing bond angles between atoms), or combinations of these motions. Bending vibrations may be further classified as being either in-plane or out-of-plane. If the vibrating molecule is exposed to electromagnetic radiation in the infrared region, the molecule can absorb the energy of radiation and produce a transition from the ground vibrational energy level to a higher vibrational energy level. In order for such a transition to occur, the frequency (energy) of the light source and the frequency (energy) of the vibration must exactly match, that is, be in resonance. Transitions involving rotational levels require energy in the range of 10 cal/mol (microwave spectroscopy), whereas energy of 10 kcal/mol (a factor of 1,000) is required for vibrational transitions (infrared or IR spectroscopy). Molecular vibrations occur very rapidly (10^{-12} sec) and such motions are even faster than conformational changes, for example, that which occurs between axial and equatorial substituted cyclohexanes (Sect. 7.122). Thus, IR spectroscopy can be used to discriminate between such conformational isomers. Infrared spectra provide structural information of great value in many subdisciplines of chemistry and are of particular interest to organic chemists. The *position* (frequency) of an absorption band in the IR spectrum of a compound depends on the energy difference between a vibrational ground state and a vibrational excited state and can be calculated from the Bohr–Einstein equation: $\Delta E = h\nu$ (Eq. 1.2). The *intensity* (absorbance or transmittance) of an absorption band depends on the probability that the transition will occur and is controlled by various selection rules. Such rules are derived, in part, from symmetry and dipole moment considerations. In order for an absorption to occur, that is, be allowed, in the IR region a change in the dipole moment of the molecule during the vibration is required. Even though a molecule may not have a permanent dipole moment in the ground state (e.g., O=C=O), if it develops a temporary one during the vibration (for CO_2, the antisymmetric stretch), it will be IR active, that is, give rise to an IR absorption.

Each functional group in a molecule gives rise to characteristic IR bands in a narrow range of the higher-frequency region of the spectrum of the molecule. Fortunately, these bands are almost independent of the rest of the molecule to which the functional group is attached. In addition, every organic compound shows a unique series of bands in the lower-frequency part of its IR spectrum, the fingerprint region. Accordingly, IR spectra can be a powerful tool in the identification of unknown structures and confirmation of suspected structures.

When a compound is exposed to light, some of the light passing through the solution is scattered at all angles with respect to the incident light (isotropic scattering). Some of this scattered light will emerge with no change of frequency (elastic scattering). However, the light resulting from inelastic scattering (usually collected at right angles to the incident light) may have frequencies slightly different from that of the incident light, and these small differences (Raman shifts) correspond to vibrational transitions in the absorbing molecule. Such Raman shifts constitute a Raman spectrum. The intensities of Raman bands (like those of IR bands) are determined by symmetry-based selection rules. Whereas there must be a change in the equilibrium *dipole moment* during a particular vibration in order for a transition to be IR active, there must be a change in the *polarizability* of the molecule in order for the transition to be Raman active.

IR and Raman spectra, two of several optical spectroscopies, are particularly valuable tools in determining the structures in the vast and important field of organometal coordination compounds. Thus, the fact that the IR spectrum of $Ni(CO)_4$ shows only one C–O stretching frequency is consistent with its tetrahedral structure and the equivalence of the four CO ligands. The C–O stretching frequency of uncomplexed or free CO is observed at 2,150 cm^{-1}, while the single CO band in the $Ni(CO)_4$ complex occurs at 2,057 cm^{-1}. The 93 cm^{-1} shift to lower frequency on the complexation of CO to nickel is ascribed to the transfer of electron density (back-bonding) from the C–O bond in the complex into the Ni–C bond of the complex, thereby reducing the C–O bond order (and thereby the vibration frequency) and increasing the Ni–C bond order.

Finally, it should not go unnoticed that some of the vocabulary of vibrational spectroscopy of interest to chemists has some commonality with that of music. Thus, for example, many of the acoustical features of music, such as frequency, fundamentals, harmonics, anharmonicity, overtones, and the octave (the interval between a fundamental and its first overtone), have their counterparts in vibrational spectroscopy. Perhaps it was no accident that Herschel was a musician before he became an astronomer and discovered the effect of infrared radiation.

16.1 WAVENUMBER $\tilde{\nu}$

The number of light wave cycles, expressed in units of reciprocal centimeters (cm^{-1}), which occur within a distance of 1 centimeter (cycles/cm). This number equals the reciprocal of the light wavelength λ (cm) and can be related to frequency since $\lambda = c/\nu$, where c equals the speed of light (3×10^{10} cm sec^{-1}) and ν is the light frequency (sec^{-1}) (Sect. 1.9):

$$\tilde{\nu} = \frac{1}{\lambda\ (\text{cm})} = \frac{\nu\ (\text{sec}^{-1})}{c\ (\text{cm sec}^{-1})} \tag{16.1a}$$

Example. Infrared radiation of a wavelength λ and of $2.5\,\mu$ [$1\,\mu$ (micron) $= 10^{-6}$ m $= 10^{-4}$ cm] has a frequency of

$$\nu = \frac{c\ (\text{cm-sec}^{-1})}{\lambda\ (\text{cm})} = \frac{3 \times 10^{10}}{2.5 \times 10^{-4}} = 1.2 \times 10^{14}\ \text{sec}^{-1} \tag{16.1b}$$

which when expressed in reciprocal centimeters equals

$$\tilde{\nu} = \frac{1}{\lambda} = \frac{1}{2.5 \times 10^{-4}} = 4{,}000\ \text{cm}^{-1} \quad \text{or} \quad 4{,}000\ \text{wavenumbers}$$

Infrared radiation is usually described in frequency units $\tilde{\nu}$ (cm^{-1}) but it is also, particularly in the older literature, expressed in wavelength units λ, microns μ, or the synonymous units, micrometers mμ. Frequency and wavelength units may be interconverted by use of the following equation:

$$\tilde{\nu} = \frac{10^4}{\lambda(\mu)} \tag{16.1c}$$

Frequency units rather than wavelength units are preferred because frequency units are linear with energy ($E = h\nu$), whereas wavelength units are not ($E = hc/\lambda$). Although wavelength, wavenumber, and frequency are each expressed in different units, they all are a quantity of energy.

16.2 THE SPECTROSCOPICALLY IMPORTANT REGIONS OF THE ELECTROMAGNETIC SPECTRUM

The portion of the electromagnetic spectrum of greatest interest to most chemists is shown in Fig. 16.2 and extends from the far ultraviolet region (highest frequency, shortest wavelength) to the microwave region (lowest frequency, longest wavelength).

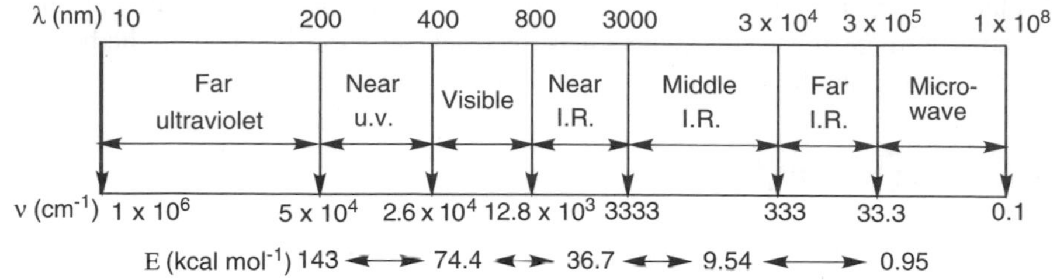

Figure 16.2. The electromagnetic spectrum.

The adjectives near and far describe the position of the spectrum with respect to the visible region as the reference region. Thus, the near ultraviolet region is the ultraviolet region closest to the visible and is the longer-wavelength (smallest-frequency) portion of the ultraviolet region (usually expressed in nanometers, nm, $1\,\text{nm} = 10^{-9}\,\text{m}$), while the near IR region is the IR region closest to the visible and is the shorter-wavelength (largest-frequency) portion of the IR region.

16.3 MICROWAVE SPECTROSCOPY

Absorption spectra due to rotational changes of molecules; microwave spectra occur in the frequency range of 0.01 to $33.3\,\text{cm}^{-1}$. In this range, the frequency is more commonly expressed in terms of cycles per second: (1 cycle/sec equals 1 Hz and 10^6 cycles/sec is equal to 1 megahertz, MHz). In order for a microwave spectrum to be observed, the molecule must have a permanent dipole moment. Accordingly, the simple homonuclear diatomic molecules such as dihydrogen and dinitrogen do not exhibit microwave spectra. Microwave spectroscopy gives information on the structure of a single molecule (bond distances and angles) that would otherwise be very difficult to obtain, but thus far such information has been obtained only on relatively small molecules.

Example. If a molecule absorbs radiation in the microwave region at $\tilde{\nu} = 10\,\text{cm}^{-1}$, the corresponding frequency (Eq. 16.1b) $\nu = \tilde{\nu} \times c = 10 \times 3 \times 10^{10} = 3.0 \times 10^{11}\,\text{Hz} = 3.0 \times 10^5\,\text{MHz} = 0.30\,\text{GHz}$, corresponding to 28.5 cal (Eq. 16.23b). Kitchen microwave ovens operate at 2.45 GHz, corresponding to the rotational frequency of H_2O.

16.4 HOOKE'S LAW

Named after Robert Hooke (1635–1703), the relationship that describes the restoring force f acting on a particle when the particle is attached to a fixed spring that is stretched from its equilibrium position. The restoring force f is proportional to the displacement Δx from the resting position:

$$f \alpha (-\Delta x) \tag{16.4}$$

The minus sign signifies that the restoring force of the spring f is acting in the direction opposite to the displacement caused by the stretching force.

16.5 FORCE CONSTANT k

The proportionality factor k in Hooke's law that converts Eq. 16.4 into an equality:

$$f = -k(\Delta x) \tag{16.5}$$

It is a measure of the stiffness of the spring to which the particle is attached and is equal to the restoring force per unit displacement. The units of k are usually given in dynes cm^{-1} and the sign of the force constant is normally positive. It is useful to recall the distinction between mass and weight. Mass m is expressed in grams, but weight, which is a force, is expressed in dynes. On the earth's surface the *weight* of a mass of 1 g is 980 *dynes*, but on the moon the *weight* of the same mass of 1 g is 163 *dynes*. It is only because the effect of gravity (force) on the surface of the earth is practically the same everywhere that the distinction between mass and weight loses its significance and objects are compared by measuring their mass. *Newton's law* (after Isaac Newton, 1643–1726) states that $f = m \bullet a$, where f = force, m = mass, and a = acceleration. In order to obtain units of force, it is necessary to specify units of mass, distance, and time. The force required to accelerate 1 kg of mass 1 m/sec^2 (meter-kilogram-second, the *mks system*) is called the Newton (N):

$$1\,\mathrm{N} = 1\,\mathrm{kg\,m\,sec^{-2}}$$

In the *cgs system* (centimeters-gram-second), the unit of force is the dyne:

$$1\ \mathrm{dyne} = 1\,\mathrm{g\,cm\,sec^{-2}}$$

Example. For the series of hydrogen halides, HF, HCl, HBr, HI, the force constants in dynes/(cm $\times 10^5$) are 8.8, 4.8, 3.8, and 2.9, respectively (Sect. 16.21); for the carbon–carbon double bond the force constant is 10×10^5 dynes/cm.

16.6 POTENTIAL ENERGY FUNCTION

In the single particle-spring system, the potential energy U is a function of the displacement x of the particle from its equilibrium position. The work that must be done to displace the particle a distance dx is $f_{\mathrm{applied}}\,dx$ and this work is stored as potential energy U. Hence,

$$dU = f_{\mathrm{applied}}\,dx = \text{work} \tag{16.6a}$$

The applied force acts against the force f that the spring exerts on the particle, that is, $f = -f_{\text{applied}}$, and hence,

$$dU = (-f)\,dx \quad \text{or} \quad \frac{dU}{dx} = -f \tag{16.6b}$$

Equation 16.6b shows an important and general relationship between potential energy and force. In the case of Hooke's law (Eq. 16.5), the potential energy derivative is $dU/dx = -kx$ that on integration, if we assume the equilibrium position is taken as zero potential energy, gives

$$U = \frac{1}{2}kx^2 \tag{16.6c}$$

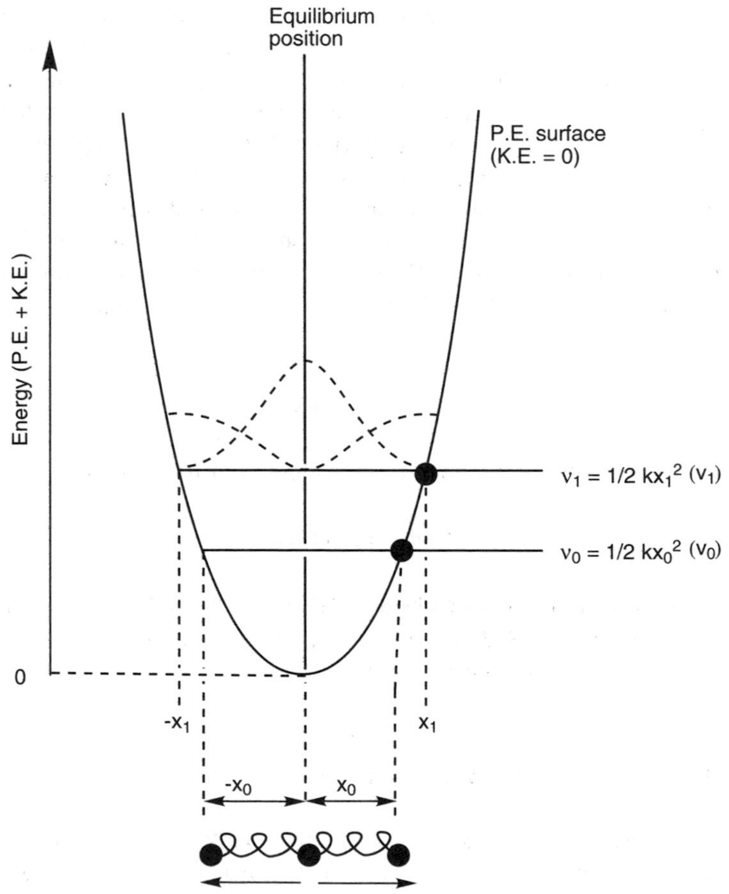

Figure 16.6. A plot of Eq. 16.6c showing the parabolic relationship according to Hooke's law between the potential energy and the extent of displacement in the particle and spring system. The velocity distribution and the probability distribution of the particle as a function of the displacement are shown, respectively, as the upper and lower broken lines inside the parabola.

Equations 16.6c and 16.5 are equivalent and both correspond to Hooke's law. The plot of Eq. 16.6c is a parabola (Fig. 16.6). The force constant k that measures the stiffness of the spring is represented by the curvature of the parabola; the larger the k, the steeper the slope of the parabola. An understanding of Fig. 16.6 is essential to ascertaining the probability (and hence the intensity) that a particular transition will occur (whether it is vibrational or electronic, the Franck–Condon factors). In a given energy level (u_0 or u_1, Fig. 16.6) the total energy of the particle is constant and equal to PE + KE. At a given displacement from the equilibrium position x, the PE is defined by the parabola (potential energy surface) at position x. Thus, if the parabola minimum is defined to have PE = 0, then at the equilibrium position of any energy level, PE = 0 and the KE is maximum. Where the energy levels intersect the parabola (potential energy surface), PE is maximum and the KE = 0 (i.e., the particle velocity is zero). And the KE is the difference in energy between the energy level u and the parabola at position x. Thus, the particle will always be moving most rapidly as it passes through the equilibrium position and slow to zero velocity (i.e., reverse direction) as it intersects the PE surface. Alternatively, the particle will spend the least time in the vicinity of the equilibrium position and the most time in the vicinities of the boundaries defined by the PE surface. This gives rise to the probability distribution of particle positions shown in the figure, which indicates that the particle is least likely to be found at the equilibrium position, and most likely to be found in the vicinities of the two intersections with the PE surface.

16.7 HARMONIC OSCILLATOR

If an oscillating (vibrating) motion, established when a particle attached to a spring is stretched and released, obeys Hooke's law (Eq. 16.7), the system is a harmonic oscillator. The model system consists of a spring attached at one end to a fixed stationary object like a wall and a particle attached to the other end of the spring. A particle with mass m attached to a spring with a force constant k will, when the spring is stretched and then released will vibrate only with the frequency ν, determined by

$$\nu = \frac{1}{2\pi}\sqrt{\frac{k}{m}} \tag{16.7}$$

the equation of the true harmonic oscillator.

Example. When a mass of 3 g is attached to a spring with a force constant of 27 dynes cm^{-1}, the frequency of vibration can be calculated from Eq. 16.7:

$$\nu = \frac{1}{2\pi}\sqrt{\frac{27\text{ g-cm/sec}^2\text{-cm}}{3\text{ g}}} = \frac{1}{2\pi}\sqrt{9(\text{sec}^{-2})} = \frac{3}{2\pi}\text{ sec}^{-1} \quad \text{(Hertz, Hz)}$$

16.8 ROTATIONAL (MICROWAVE) SPECTRA; DIATOMIC MOLECULES

The starting point for consideration of the theory of rotational spectra is a model (the rigid rotor) representing a diatomic molecule.

Example. The transition from the lowest rotational level of a carbon monoxide molecule to the next higher rotational level requires the absorption of a photon of $7.7\,\text{cm}^{-1}$ wave numbers corresponding to a frequency of 2.3×10^5 MHz (Eq. 16.1a). The energy associated with $7.7\,\text{cm}^{-1}$ is $0.022\,\text{kcal mol}^{-1}$ (Eq. 16.23b).

16.9 THE RIGID ROTOR MODEL

Such a model consists of two atoms of masses m_1 and m_2 (considered to be point masses) located at the ends a weightless rigid rod and separated by a distance r. The center of mass in this system is at the point where $m_1r_1 = m_2r_2$, r_1 and r_2 being the distances from the points m_1 and m_2 to the center of mass (Fig. 16.9). When the rod and its attached atoms are rotating around the center of mass, the moment of inertia I (Sect. 16.12) is expressed by Eq. 16.9a:

$$I = m_1r_1^2 + m_2r_2^2 \tag{16.9a}$$

Since $r = r_1 + r_2$ and $m_1r_1 = m_2r_2$, it follows that

$$r_1 = \frac{m_2}{m_1 + m_2} \times r \quad \text{and} \quad r_2 = \frac{m_1}{m_1 + m_2} \times r \tag{16.9b}$$

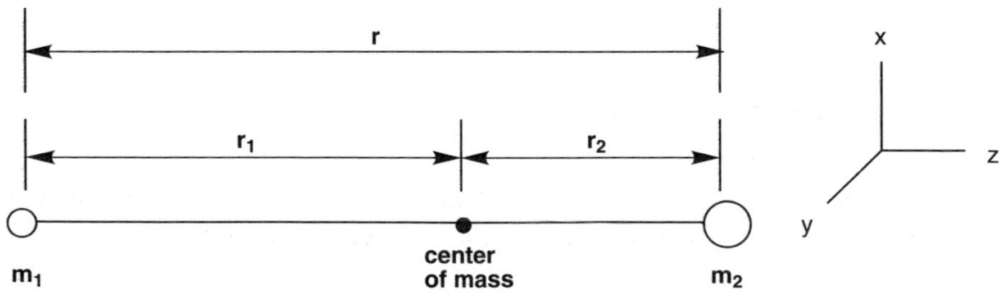

Figure 16.9. The rigid rotor model ($m_2 > m_1$).

Substituting the values of r_1 and r_2 in Eq. 16.9b into Eq. 16.9a gives

$$I = \frac{m_1 m_2}{m_1 + m_2} \times r^2 \tag{16.9c}$$

The quantity $\frac{m_1 m_2}{m_1 + m_2}$ is called the reduced mass (Sect. 16.10) and is usually designated by μ. The use of the reduced mass permits the two masses m_1 and m_2 to be treated as a single mass and hence Eq. 16.9c becomes

$$I = \mu r^2 \tag{16.9d}$$

This equation shows that in the system described by Eq. 16.9a, the moment of inertia generated by the two rotating atoms may be treated as though a single mass were rotating about a point at a distance *r* from the axis of rotation. Such a system is called a simple rigid rotor.

16.10 REDUCED MASS μ

The quantity required to deal with the oscillations of *two* particles connected by a spring. It is the mechanical analog of the diatomic molecule and is calculated from the masses of the two atoms m_1 and m_2 by dividing their product by their sum:

$$\mu = \frac{m_1 m_2}{m_1 + m_2} \tag{16.10}$$

16.11 THE PRINCIPAL AXES

A set of three orthogonal (mutually perpendicular) axes that are used to describe the rotational motion of a molecule. The principal axes have their origin at the center of mass of the molecule and are fixed in the molecular framework and considered to rotate with the molecule. They are oriented in such a way as to simplify the mathematical calculations of spectra.

Example. Figures 16.56 and 16.57.

16.12 MOMENTS OF INERTIA *I* (PRINCIPAL MOMENTS OF INERTIA)

The moments of inertia *I* around the principal axes are called the principal moments of inertia and are defined as follows:

$$I = \sum I_i = \sum m_i d_i^2 \tag{16.12}$$

where I_i is the principal moment of inertia of an atom i within the molecule, m_i is the mass of the atom i, and d_i is the perpendicular distance from atom i to the chosen principal axis. The total moment of inertia I is the sum of the moments of inertia around a principal axis of the individual atoms in a molecule.

Example. Figure 16.12 shows the three principal axes (in this case, the three coordinate axes) for a (unspecified) molecule with the center of mass at the origin. The moment of inertia around the z-axis for a particular atom i in this molecule depends on the distance d_i of this atom from the z-axis and the mass m_i of the atom.

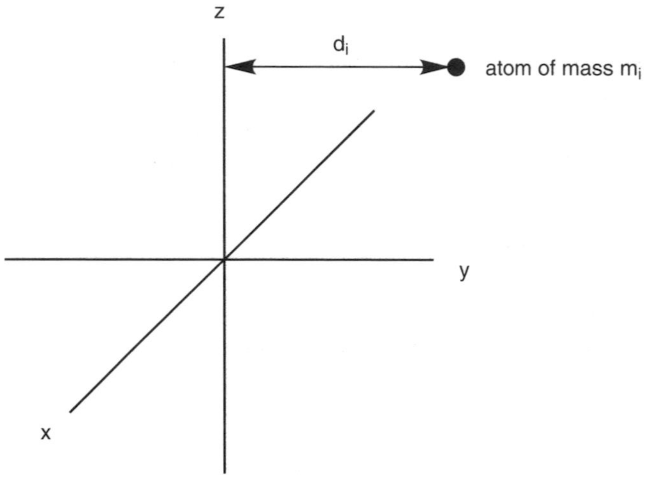

Figure 16.12. The moment of inertia of an atom i around the principal axis z in a molecule whose center of mass is placed at the origin of a coordinate system can be calculated from d_i and m_i.

16.13 ANGULAR VELOCITY ω

The number of rotations that a rigid body makes in space per unit time. Since polar coordinates are usually employed in rotational spectroscopy, ω is the number of radians of angle swept out in unit time by the rotating system.

16.14 KINETIC ENERGY (KE) OF THE RIGID ROTOR

The kinetic energy depends on the moment of inertia and the angular velocity:

$$\mathrm{KE} = 1/2\, I\omega^2 \tag{16.14}$$

In accordance with Eq. 16.14, which is based on classical mechanics, there are no restrictions on the allowed rotational energy, whereas in quantum mechanics there are such restrictions (Sect. 16.15).

16.15 ANGULAR MOMENTUM QUANTUM NUMBER *J*

In the quantum mechanical description of a rotating molecular system, a quantum number J must be defined that normally has integral units. Only rotational transitions where $\Delta J = 0, \pm 1$ are permitted (see, however, Sect. 16.60).

16.16 ENERGIES OF ROTATIONAL STATES *E*

The allowed energies of rotational states are governed by quantum mechanical restrictions. In the case of linear molecules, these can be calculated from the equation

$$E = \frac{J(J+1)h^2}{8\pi^2 I} \tag{16.16}$$

where J is the angular momentum quantum number, h Planck's constant, and I the moment of inertia, 16.12.

16.17 SELECTION RULES FOR ROTATIONAL SPECTRA OF RIGID LINEAR MOLECULES

In addition to the selection rule of $\Delta J = 0, \pm 1$, if rotational (microwave) absorptions are to be observed, the linear molecule must have a permanent dipole. The spectrum of such molecules will usually exhibit strong $\Delta J = \pm 1$ transitions.

16.18 THE NONRIGID ROTOR

The rigid rotator model assumes no change of bond lengths of a molecule during rotation. However, in real molecules, the rotational motion has a stretching effect due to centrifugal force. Hence, the moment of inertia must be corrected for the actual distorted bond length and angles, and this model is called the nonrigid rotor.

16.19 INFRARED ABSORPTION SPECTRUM

The spectrum that results when molecules of a compound absorb light in the infrared region, resulting in transitions between different vibrational energy levels. The frequency of the photon of light must exactly match the frequency of the periodic vibration of the molecule in order for absorption to occur. Infrared spectra are presented as a plot of the intensity of the absorption (as the ordinate) as a function of the wavenumber in cm^{-1} (or another measure of associated energy as the abscissa). The plot of the spectrum of a compound in the condensed phase, either neat or in solution, appears as bands rather than as lines. Each vibrational energy level has associated with it many rotational levels. Also each vibrational change of the molecule of interest is affected by the presence of surrounding molecules, giving rise to inhomogeneous broadening. A series of very closely spaced lines gives the appearance of a band (envelope). The *position* of an absorption band is regarded as the energy (usually, the wavenumber) where the intensity is at its maximum. The *intensity* of an absorption band at this position is expressed either as transmittance T, where $T = I/I_0$, the ratio of the radiant power transmitted I to the radiant power of the incident light I_0, or as absorbance (A_{max}), where $A = \log_{10}(1/T) = \log_{10} I_0/I$. When the ordinate is plotted as Transmittance, the absorption "peaks" appear as minima (pointing down), which is the usual way infrared spectra are generated and reported, whereas if the ordinate is plotted as absorbance, the peaks appear as maxima (pointing up), which is the customary way electronic spectra are plotted. When quantitative measurements based on infrared spectra are desired, A_{max} is divided by the molar concentration c (in mol/L) of the compound and the length l (in cm) of the path the radiant light traverses. This quantity is called the *molar extinction coefficient*, ε: $\varepsilon = A/(c \times l)$, and has units of L mol^{-1} cm^{-1}. The intensity of a band is determined by the area under the band equal to $\int \varepsilon d\nu$. This quantity is rather cumbersome to determine and an approximation is achieved by multiplying ε_{max} by $\nu_{1/2}$, where $\nu_{1/2}$ is the width at half height, that is, the width of the band in cm^{-1} at $\varepsilon = 1/2\ \varepsilon_{max}$. Figure 16.19 illustrates an ideal infrared band and the half-width concept.

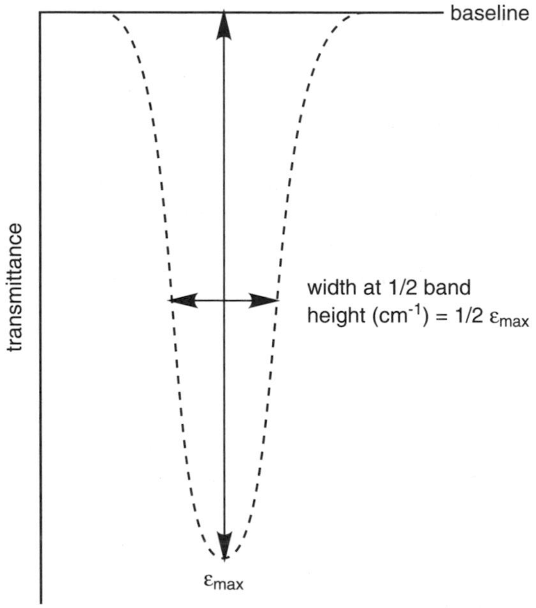

Figure 16.19. An ideal infrared band and the width at 1/2 ε_{max} whose use in the following equation: Intensity = $\varepsilon_{max} \cdot \nu_{1/2}$ approximates the area under the band in units of L mol^{-1} cm^{-2}.

16.20 VIBRATIONAL SPECTRUM

More inclusive than infrared spectrum since it is used to include Raman spectra as well (Sect. 16.62).

16.21 VIBRATIONAL SPECTRA OF DIATOMIC MOLECULES

The frequencies at which diatomic molecules absorb infrared radiation can be calculated from the formula used in the case of the single particle model, that of the harmonic oscillator (Sect. 16.7), but expanded to include the mass of the two particles or atoms. The result is the fundamental infrared equation:

$$\nu = \frac{1}{2\pi}\sqrt{\frac{k}{\mu}} \tag{16.21a}$$

where ν is the frequency, k the force constant (see Sect. 16.5), and μ the reduced mass (see Sect. 16.10). By using the reduced mass μ, the vibrations of the two atoms are reduced to the vibrations of a single point of mass m that was used in the equation of the harmonic oscillator (Eq. 16.7, where $\mu = m$).

Example. The stretching frequency for the diatomic molecule, carbon monoxide C≡O, occurs at 2,150 cm^{-1}. If this value is placed in Eq. 16.21a and the appropriate units are employed in the calculation, the force constant k calculates to 16.8×10^5 dynes cm^{-1}. The energy associated with absorption at 2,150 cm^{-1} calculated from Eq. 16.23b is 6.1 kcal mol^{-1} (25.6 kJ). The stretching frequency for a carbonyl group >C=O in a simple ketone is at lower frequency, 1,715 cm^{-1}, as expected, since it takes much less energy to stretch a double bond (4.9 kcal mol^{-1} or 20.5 kJ) than it does to stretch a triple bond, as in C≡O. In the series of hydrogen halides where the reduced mass is approximately equal for all halides, Eq. 16.21a shows that in such a case there should be a direct correlation between the observed frequency and the square root of k. Table 16.21 gives the relevant data for the series of hydrogen halides. It should be noted that, in general, if $m_1 = m_2$, then $\mu = 1/2m_1$ or $1/2m_2$. The greater the difference between the two masses, the more closely μ approaches the mass of the lighter atom. In the case of the hydrogen halides, the reduced masses are all close to 1.0, and that of HI approaches 1.0 more closely than that of HF.

TABLE 16.21. Infrared Spectral Data for the Hydrogen Halides

HX	μ	k(dynes/cm) $\times 10^5$	$\sqrt{k}$	ν(cm^{-1})	D_e(kcal/mol)[a]
HF	0.95	8.8	9.38	3,958	134
HCI	0.97	4.8	6.93	2,885	102
HBr	0.99	3.8	6.16	2,559	87
HI	0.99	2.9	5.39	2,230	71

[a]Dissociation energy (Sect. 16.33).

The plot of the calculated $\sqrt{k}$ against the observed frequency gives an almost perfect straight line. Since the reduced mass is essentially constant, it can be neglected, and the force constant can be calculated from the observed frequency alone. In polyatomic molecules, if only small changes are made from one compound to another, the force constant k (in contrast to the reduced mass) may be considered to be relatively constant. This is the situation when a hydrogen atom on a particular atom, usually carbon, in a molecule R-H is substituted by a *deuterium atom*, where the group R of mass M_R is relatively large, and hence the force constants k (R-H) $\cong k$ (R-D). In such a case, according to Eq. 16.21a, the frequency should be inversely proportional to μ, the reduced mass:

$$\frac{\nu(\text{C-D})}{\nu(\text{C-H})} = \frac{\sqrt{\mu(\text{R-H})}}{\sqrt{\mu(\text{R-D})}} = \sqrt{\frac{1 \times M_R/(1 + M_R)}{2 \times M_R/(2 + M_R)}} \tag{16.21b}$$

If M_R is relatively large compared to either the integers 1 or 2, the last expression is approximately equal to

$$\sqrt{\frac{1 \times M_R/M_R}{2 \times M_R/M_R}} = \sqrt{\frac{1}{2}} = 0.707 \tag{16.21c}$$

Thus, if a particular C–H bond in a large molecule is replaced by a C–D bond, a shift of the associated C–H stretching frequency to a lower frequency equal to 0.707 times the C–H stretch should be observed. However, in a relatively small molecule such as chloroform, Cl_3C–H, the C–H stretch occurs at 3,016 cm^{-1}, while the C–D stretch in Cl_3C–D absorbs at 2,253 cm^{-1}, a factor of 0.747. Here M_R is not sufficiently large to ignore the 2 in the quantity $(2 + M_R)$ in the denominator of Eq. 16.21b. Anharmonic effects also cause significant deviations from Eq. 16.21b.

16.22 VIBRATIONAL LEVELS OF THE HARMONIC OSCILLATOR

The various energy levels of the harmonic oscillator. These levels are represented by the lines that connect equal values on the potential energy curve at the two positions corresponding to maximum displacement from the equilibrium position of the vibrating particles.

Example. The two horizontal lines in the parabola (Fig. 16.6) represent two vibrational levels, v_0 and v_1.

16.23 ENERGIES ASSOCIATED WITH INFRARED ABSORPTIONS

The energy gap between two vibration levels. In order to calculate the energy associated with the absorption of light at any particular wavelength, use is always made of the Bohr–Einstein equation $\Delta E = h\nu$ (Sect. 1.2), where h is Planck's constant (Sect. 2.20) and ν the light frequency (Sect. 1.9). For use with electromagnetic radiation in the infrared region, the Bohr–Einstein equation may be expressed as follows:

$$E = hc\tilde{\nu} \tag{16.23a}$$

where $\tilde{\nu}$ is in wavenumbers. The unit of energy is the erg $= 10^{-4}$ kJ (4.184 kJ = 1 kcal). In dealing with absorption in the infrared region, organic chemists are accustomed to expressing energy in kcal (or kJ) and frequency as wavenumbers. When the constants in Eq. 16.23a are evaluated and collected and appropriate units used, this equation can be expressed for convenience in the following form:

$$E(\text{kcal/mol}) = 2.8635 \times 10^{-3}\tilde{\nu} \tag{16.23b}$$

Example. For the energy associated with infrared absorption at 2,000 cm^{-1},

$$E(\text{kcal/mol}) = 2.8635 \times 10^{-3} \times 2{,}000 = 5.73\,\text{kcal mol}^{-1} \quad (33.94\,\text{kJ})$$

Many physical chemists use electron volts (eV) for units of energy. The *electron volt* is the kinetic energy acquired by an electron upon acceleration through a potential of 1 V: 1 eV $= 1.602 \times 10^{-19}$ J. The corresponding molar energy is obtained through multiplication by *Avogadro's number* (6.022×10^{23}); thus, 1 eV $= 23.06\,\text{kcal mol}^{-1} = 96.46\,\text{kJ mol}^{-1}$.

16.24 ALLOWED ENERGY LEVELS

In the classical harmonic oscillator any value for the energy is allowed because the total energy depends only on the force constant and the magnitude of the displacement. However, application of quantum mechanical restrictions leads to only certain allowed energy levels characterized by a vibrational quantum number v:

$$E_v = (v + 1/2)\frac{h}{2\pi}\sqrt{\frac{k}{\mu}} \tag{16.24a}$$

where v is the vibrational quantum number, an integer ≥ 0. If the term equal to ν in Eq. 16.21a is substituted into Eq. 16.24a, the latter becomes

$$E_v = (v + 1/2)\,h\nu \tag{16.24b}$$

Equation 16.24b gives the following information:

1. The energy of the harmonic oscillator can only have values of positive half-integer multiples of $h\nu$.
2. The energy levels are evenly spaced in the harmonic oscillator.

3. In the lowest possible energy state where $v = 0$, a half-quantum of vibrational energy is still present, the *zero-point energy* (Sect. 16.34):

$$E(v_0) = 1/2\, h\nu \tag{16.24c}$$

16.25 VIBRATIONAL STATE

The characterization of a molecule possessing a particular amount of vibrational energy as defined by a specific vibrational quantum number v.

16.26 GROUND VIBRATIONAL STATE

The vibrational state having a vibrational quantum number of zero ($v = 0$). Since each electronic state has a number of vibrational states associated with it, there is a ground (or lowest-energy) vibrational state for each electronic state.

16.27 EXCITED VIBRATIONAL STATES

Vibrational states having energies greater than the ground vibrational state, that is, $v > 0$.

16.28 SELECTION RULES FOR THE HARMONIC OSCILLATOR

Only transitions between adjacent vibrational levels are allowed, that is, $\Delta v = \pm 1$. Ideally, energies involved in transitions between any two adjacent vibrational levels should be equal; such a system is a perfect harmonic oscillator. However, real molecules do not behave as perfect harmonic oscillators. Actually, the only transition that is observed at low light levels is the *fundamental vibration* $v_0 \rightarrow v_1$.

16.29 MORSE CURVE

Named after Phillip M. Morse (1903–1985), a plot showing the relationship between the potential energy of a chemical bond between two atoms as a function of the internuclear distance.

Example. The solid curve in Fig. 16.29 (as well as Fig. 2.18 where the potential energy is given in kJ) shows the Morse curve for dihydrogen, H_2. On stretching the bond (increasing r), the potential energy asymptotically approaches a final value, which is usually defined as zero on the potential energy scale, and represents dissociation into constituent atoms. Compressing the bond to a distance shorter than its equilibrium position is strongly resisted and the potential energy rises abruptly as the

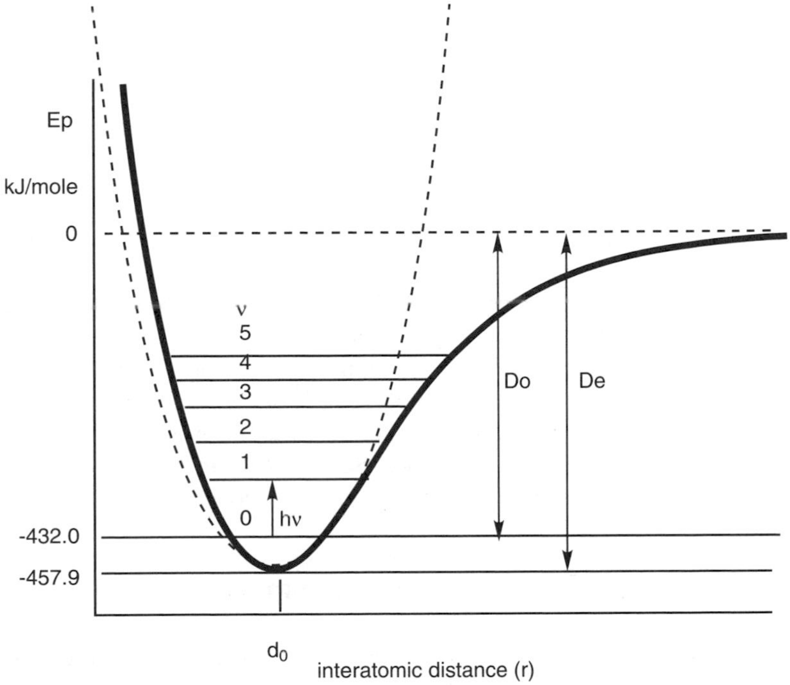

Figure 16.29. The potential energy curve of the dihydrogen molecule showing the vibrational levels. The dashed curve shows the parabola of the harmonic oscillator. The lower the vibrational level, the more closely the molecule behaves like a harmonic oscillator. The spacing between the levels decreases as their energy increases. D_0 is the spectroscopic dissociation energy (Sect. 16.32) and D_e is the dissociation energy (Sect. 16.33).

distance is shortened. The broken line in Fig. 16.29 represents the potential energy curve of a harmonic oscillator. Note that the Morse curve, especially close to the minimum energy, is similar to, but not exactly like, the parabola that characterizes the harmonic oscillator. The horizontal lines in the figure represent the first six vibrational levels of the molecule.

16.30 ANHARMONIC OSCILLATOR

The oscillating system that is distorted from the strictly parabolic relationship between potential energy and bond length or particle separation that characterizes the harmonic oscillator.

Example. The potential energy of the anharmonic oscillator is given by a modification of the energy equation (16.24b) of the harmonic oscillator and is described by a series equation:

$$E_v = (v + 1/2)h\nu - (v + 1/2)^2 h\nu x_e + (v + 1/2)^3 h\nu y_e - (v + 1/2)^4 h\nu y_e + \cdots \quad (16.30)$$

where x_e, y_e, ... are anharmonicity constants. Anharmonicity results in unequal spacing of the energy levels, with higher levels being more closely spaced than lower levels.

16.31 FUNDAMENTAL VIBRATION

The vibration associated with the transition from the ground-state (lowest) vibrational level v_0 to the first vibrational level v_1.

Example. The vibration associated with the transition $v_0 \rightarrow v_1$ in Fig. 16.29. The fundamental vibration frequency for this transition in the hydrogen molecule occurs at 4,395 cm^{-1}, corresponding to an energy of 6.4 kcal/mole. Because real molecules are anharmonic oscillators, the selection rule $\Delta v = \pm 1$ (item 2 in Sect. 16.24) does not rigorously apply and the resulting overtones (Sect. 16.49) can be detected and do not occur at exact multiples of the fundamental vibration frequency.

16.32 SPECTROSCOPIC DISSOCIATION ENERGY

The energy required to separate (in the gas phase) the atoms in a diatomic molecule in its lowest vibrational level v_0 from its equilibrium interatomic distance to an infinite distance. This energy can be determined experimentally.

Example. The energy represented by D_0 in Figs. 2.18 and 16.29 and which, for the dihydrogen molecule, corresponds to about 36,097 cm^{-1} (431.1 kJ or 103 kcal/mol).

16.33 DISSOCIATION ENERGY

In a diatomic molecule the energy required to separate (in the gas phase) the atoms to an infinite distance when the molecule is initially in an imaginary state corresponding to the lowest point on the potential energy curve where the potential energy is zero.

Example. The energy represented by D_e in Figs. 2.18 and 16.29 and which, for the dihydrogen molecule, corresponds to about 38,300 cm^{-1} (458.3 kJ or 109.5 kcal/mol) and is greater than the spectroscopic dissociation energy by the zero-point energy of about 2,203 cm^{-1} (27.2 kJ or 6.5 kcal/mol).

16.34 ZERO-POINT ENERGY

The vibrational energy that a molecule possesses when it is in its lowest vibrational energy level, $v = 0$. The molecule still has this energy at 0 K.

Example. The kinetic energy corresponding to ν_0 at the equilibrium position in Fig. 16.29. This energy corresponds to 1/2 $h\nu_0$ above the minimum (the unrealizable zero) in the potential energy curve. For dideuterium, D_2, the zero-point energy is almost 2 kcal mol^{-1} less than for H_2. This difference in the zero-point energy between the isotopes is responsible for the deuterium isotope effect on reaction rates observed in numerous reactions and frequently gives clues as to the mechanism of a reaction.

16.35 NUMBER OF DEGREES OF FREEDOM

In order to completely specify the location of all atoms in a molecule possessing n atoms, three coordinates for each atom or a total of $3n$ coordinates are required; these correspond to $3n$ degrees of freedom. However, from these $3n$ degrees of freedom, six must be subtracted (five for linear molecules) since the *position* of the molecule in space (actually, its center of mass) is determined by three degrees of freedom or three coordinates. And the *orientation* of the molecule in space is determined by three additional degrees of freedom or coordinates, two angles to locate the principal axis and a third angle to define the rotational position about this axis (the three Euler angles) [Leonard Euler (1707–1783)]. If the molecule is linear, this last angle is unnecessary since different rotational positions about the internuclear axis are equivalent and one degree of freedom is thereby lost. Thus, in a nonlinear molecule there are $3n - 6$ (and in a linear molecule $3n - 5$) degrees of freedom that define the positions of the atoms relative to one another, and hence the bond distances and angles and their changes, that is, the vibrations of a molecule.

16.36 NUMBER OF FUNDAMENTAL VIBRATIONS

The (genuine) vibrations that correspond to the $3n - 6$ (or for linear molecules the $3n - 5$) degree of freedom, that is, the number of changes in bond distances and bond angles that are theoretically possible in molecules when they absorb infrared radiation. Each of these degrees of freedom theoretically should give rise to a fundamental vibration, and hence there should be $3n - 6$ fundamental vibrations for nonlinear, and $3n - 5$ for linear, molecules.

Example. Ethylene, $H_2C{=}CH_2$, consists of six atoms and should display $3(6) - 6 = 12$ fundamental vibrations in its infrared spectrum.

16.37 GENUINE AND NONGENUINE VIBRATIONS

The $3n - 6$ vibrations that involve changes in relative positions of the atoms in a molecule are genuine vibrations. Nongenuine vibrations are those movements of a molecule that do not result in a change in bond distances or angles between the constituent atoms; they are the three translations and the three molecular rotations (for nonlinear molecules) that constitute the minus six in $3n - 6$.

Example. The translation T of a molecule in any direction in space can be broken into component translations along the three Cartesian coordinates T_x, T_y, and T_z. The three translations are nongenuine vibrations since they do not affect the relationship between the constituent atoms of the molecule. Similarly, rotations R may be broken up into component rotations around the three axes, R_x, R_y, and R_z, and the set of these three rotations are also nongenuine vibrations, giving a total of six nongenuine vibrations. For all nonlinear molecules there are $3n - [3 \text{ (translations)} + 3 \text{ (rotations)}] = 3n - 6$ genuine vibrations (5 and $3n - 5$, respectively, for linear molecules).

16.38 NORMAL VIBRATIONS; NORMAL MODES

The $3n - 6$ *fundamental vibrations* ($3n - 5$ for linear molecules) into which the random internal motions of a vibrating molecule can be resolved. The resolution is achieved in such a way that all the atoms move in phase with the same frequency in a periodic motion called the normal vibration (normal mode); all atoms reach their extreme positions as well as their equilibrium positions simultaneously. Also in a normal vibration, heavier atoms do not move as far as lighter atoms, that is, mass effects must be taken into account because the center of mass must remain fixed throughout the vibration cycle in order to avoid translations or rotations.

Example. The water molecule shows a total of $3n - 6 = (3 \times 3) - 6 = 3$ normal vibrations. Frequently, the motions of atoms in a vibration are depicted by arrows that show direction (but not magnitude) because such simplified depictions suffice to determine the symmetry involved. However, if normal vibrations are to be more accurately depicted, the mass effects should be included. The terms normal vibrations, normal modes, and fundamental vibrations are used interchangeably.

16.39 STRETCHING VIBRATIONS

Those motions that take place along the bond axis, also called valence vibrations. Of the $3n - 6$ fundamental vibrations of a nonlinear molecule, $n - 1$ must be stretching vibrations or stretching modes.

Example. In a common convention for indicating the direction of atomic displacements in a vibration, the atoms are shown in their equilibrium positions as circles and their in-plane motions during the vibration are shown as attached arrows or arrows over the atoms. In the triatomic molecule of water, there are $n - 1 = 2$ stretching vibrations. Figure 16.39*a* shows the symmetric stretch of the water molecule in which the hydrogen atoms are moving, in phase, away from the oxygen atom. In order to preserve the center of mass (and thus avoid translation, a nongenuine

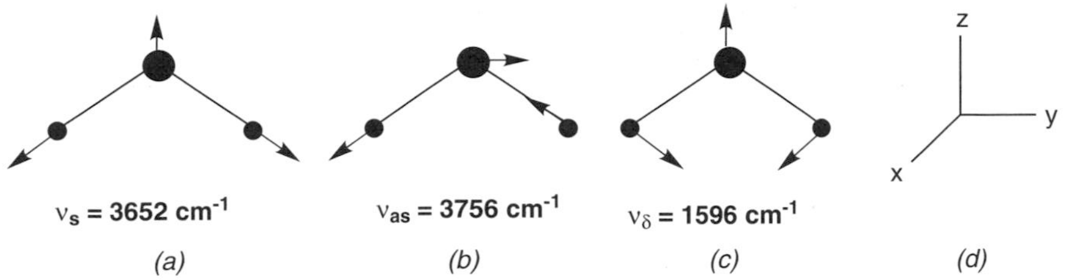

Figure 16.39. The (*a*) symmetric stretching vibration; (*b*) the antisymmetric stretching vibration; and (*c*) the scissoring (in-plane bending) vibrations of H_2O. (*d*) The coordinate system used for designation of symmetry species (Sect. 16.40).

vibration), the oxygen atom must also move, in phase, as shown in the figure. However, since the oxygen atom is much heavier than the hydrogen atoms, it does not move as far as the hydrogen atoms. After the stretching half of the vibrational cycle shown in the figure, all three atoms move in the opposite direction, again all in phase and the bonds undergo contraction back to their original geometries. Figure 16.39*b* shows the (antisymmetric) stretch that involves the simultaneous stretching of one and compression of the other O–H bond. The simultaneous in-phase movement of all atoms is shown by the arrows. This motion is followed by the simultaneous movement of all atoms in the opposite direction with all motion arrows being reversed. Stretching modes are usually abbreviated as ν_s and ν_{as} for the symmetric and antisymmetric (asymmetric) stretch, respectively.

16.40 SYMMETRIC (STRETCHING) VIBRATION

All vibrations (indeed all dynamic properties) of a molecule must be either symmetric or antisymmetric (Sect. 4.16) with respect to each symmetry operation that can be performed on the molecule. In order to test a vibration for this property under a particular operation, it is convenient to utilize the arrows showing the displacement of the atoms during a vibration. (The combination of arrows is called the *symmetry coordinate*.) If under *every* symmetry operation appropriate to the molecule, a particular stretching (or other) vibration is symmetric, the vibration is a totally symmetric stretch, or more simply, a symmetric stretch.

Example. The stretching vibration of H_2O shown in Fig. 16.39*a*. The normal modes of vibration are frequently designated by the symmetry species to which they belong. In order to determine such symmetry species, reference must be made to the character table that describes the symmetry properties of the molecule. The C_{2v} character table appropriate for the water molecule shown in Table 4.26 is repeated for convenience in Table 16.40.

The symmetric stretch (arrows) shown in Fig. 16.39*a* can now be analyzed with respect to each of the symmetry properties of the molecule. With respect to the two-fold rotation around the symmetry axis (C_2) in the plane of the paper, the behavior is

TABLE 16.40. The Character Table for C_{2v}

C_{2v}	E	C_2	σ_{xz}	σ_{yz}	
a_1	1	1	1	1	$\nu_s\,\nu_\delta$: z, α_{xx}, α_{yy}, α_{zz}
a_2	1	1	-1	-1	R_z, α_{xy}
b_1	1	-1	1	-1	x, R_y, α_{xz}
b_2	1	-1	-1	1	ν_{as}: y, R_x, α_{yz}

symmetric, $+1$ (i.e., all the arrows point in the same direction after the 180° rotation around this axis as before the rotation) and all the arrows remain pointing in the same direction as well after reflection in either of the two planes of symmetry (σ) that characterize the water molecule (or, for that matter, any nonlinear X_2Y molecule). Because this stretching vibration is symmetric ($+1$) under all symmetry operations in C_{2v}, the vibration is symmetric, (ν_s) and belongs, according to Table 16.40, to species a_1.

16.41 ANTISYMMETRIC (STRETCHING) VIBRATION

A stretching (or other) vibration that is antisymmetric with respect to *at least one* of the symmetry operations appropriate to the molecule.

Example. The stretching vibration of H_2O shown in Fig. 16.39*b* is an antisymmetric stretch. It is antisymmetric because after the C_2 operation (180° rotation, Fig. 16.41), all the arrows depicting this stretching motion point in the direction opposite to that shown before the rotation. Thus, the arrow on the left is pointing *away* from the oxygen before the rotation, but is pointing *toward* the oxygen after the rotation. Such behavior is called antisymmetric (Sect. 4.16) and is designated as -1 (Table 16.40). Under reflection in the xz-plane, which is perpendicular to the molecular plane yz, this stretching vibration is also antisymmetric. However, it is symmetric (unchanged) with respect to reflection in the molecular plane σ_{yz}, as well as being symmetric with respect to the identity operation E, two other operations in point group C_{2v}. Its symmetry

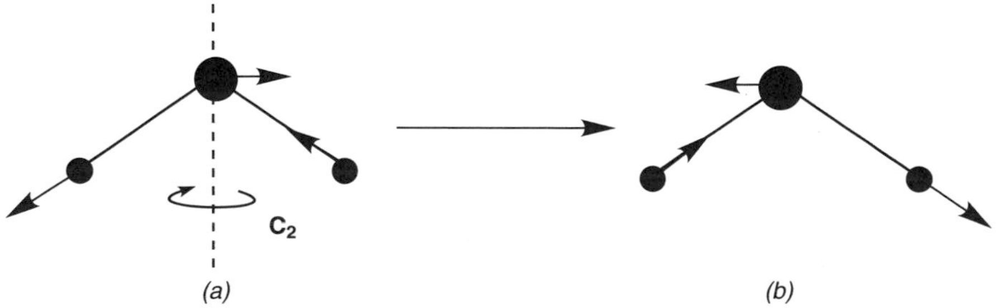

Figure 16.41. The antiymmetric stretch of the water molecule showing its antisymmetric behavior with respect to the C_2 operation.

behavior (+1, −1, −1, +1) classifies it as belonging to species b_2 in C_{2v}. Should this stretch be called antisymmetric, even though it is symmetric with respect to two operations in C_{2v}? Some authors designate it as an *asymmetric* rather than an antisymmetric stretch, but this is confusing as well since asymmetric means lacking any symmetry. Fortunately, the designation for it as ν_{as} is ambiguous and can be interpreted either way, although antisymmetric is probably preferable. In molecules with a center of symmetry, such as O=C=O, all normal vibrations that are symmetric are designated by the symmetry notation *g* (gerade, Sect. 4.17), and all that are antisymmetric are designated as *u* (ungerade, Fig. 4.17). The two stretches of CO_2 are shown in Fig. 16.42*a*. In most texts the symmetric and antisymmetric stretches are designated simply as ν_s and ν_{as}, respectively. CO_2 belongs to point group $D_{\infty h}$; the molecular axis (*z*) is an infinite-fold rotational axis and perpendicular to this axis are an infinite number of two-fold axes. A $D_{\infty h}$ molecule also has an infinite number of planes of symmetry that include the molecular axis as well as a plane of symmetry (σ_h) perpendicular to this axis. These symmetry properties require the molecule also to have a center of inversion *i*. The $3n - 5 = 4$ vibrations consist of two stretches and two bends. The notation in parentheses shown in Fig. 16.42 gives the symmetry species in $D_{\infty h}$ to which the stretches belong and it can be seen that ν_s is symmetric (+1) with respect to the operation *i*, whereas ν_{as} is antisymmetric (−1) with respect to *i*.

16.42 BENDING VIBRATIONS

Vibrations that result in angle deformation. The bending vibrations of a molecule occur at lower frequencies than the stretching vibrations since it requires much less energy to bend than to stretch the bonds of a molecule.

Example. Figure 16.39*c* shows the single bending vibration, which along with the two stretching vibrations, makes up the three ($3n - 6$) fundamental vibrations of the water molecule. This in-plane bending (scissoring) vibration is abbreviated as ν_δ and occurs at 1,596 cm^{-1} in the infrared spectrum, at considerably lower frequency than

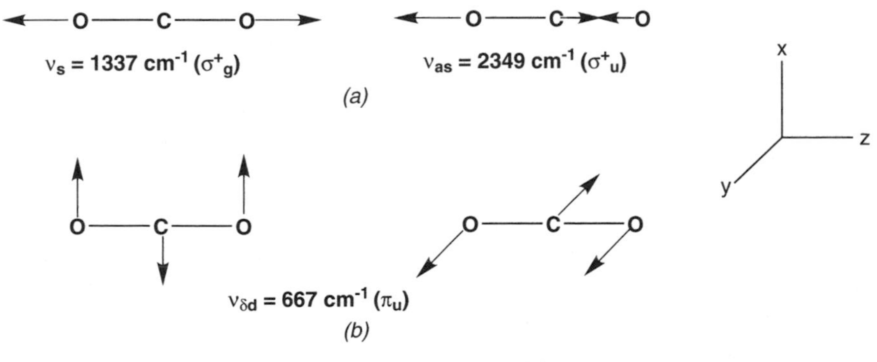

Figure 16.42. The four normal vibrational modes of CO_2: (*a*) the two stretching vibrations and (*b*) the two degenerate bending (scissoring) vibrations.

the stretching vibrations. This vibration is totally symmetric with respect to all the operations in C_{2v} and belongs to species a_1 (Table 16.40). In the case of carbon dioxide, the two bending motions of this linear molecule, one in the *xz*-plane and the other in the *yz*-plane, are shown in Fig. 16.42*b*. These two bending motions are interconvertible by a 90° rotation around the internuclear axis; they both constitute an identical molecular distortion and therefore require the same energy. There are, of course, an infinite number of vibrations of this type that differ only in their orientation with respect to the perpendicular molecular axis. Any of them, however, can be resolved into the two degenerate vibrations shown.

16.43 DEGENERATE VIBRATIONS

Vibrations that occur at precisely the same frequency (energy) and are therefore observed as a single absorption band.

Example. The two bending vibrations of CO_2 shown in Fig. 16.42*b* are degenerate. Together they give rise to one doubly degenerate vibrational band, at 667 cm^{-1}, abbreviated $\nu_{\delta d}$. A doubly degenerate vibration can occur only with molecules having an axis greater than two-fold ($>C_2$); CO_2 being a linear molecule has an infinite-fold axis, the internuclear axis. Triply degenerate vibrations can occur only with molecules having more than one threefold axis (C_3) such as the tetrahedral molecule, methane CH_4, which has four such axes (Sect. 4.7). On rare occasions two unrelated vibrations may fortuitously occur at nearly the same frequency and give rise to *accidental degeneracy*.

16.44 IN-PLANE BENDING

The bending vibration that occurs in the plane defined by any three atoms of interest in a polyatomic molecule. In the very common case of the structural unit RCH_2R', there are two in-plane bending modes of the $-CH_2-$ vibrations: one called a *rocking vibration* and the other a *scissoring vibration*.

Example. Figure 16.44*a* shows a $-CH_2-$ rocking vibration and Fig. 16.44*b* a $-CH_2-$ scissoring vibration that is analogous to the in-plane bending of H_2O (Fig. 16.39*c*).

16.45 OUT-OF-PLANE BENDING

The bending vibration that occurs out of the plane defined by the three atoms of interest in a polyatomic molecule, or out of the plane of a planar molecule. In the very common case of a RCH_2R' molecule, the out-of-plane bending mode of the $-CH_2-$ group can be either a *wagging vibration* or a *twisting vibration*.

Example. Figure 16.45*a* shows a $-CH_2-$ wagging vibration and Fig. 16.45*b* a $-CH_2-$ twisting vibration. Figure 16.45*c* shows an out-of-plane bending vibration for an AX_4

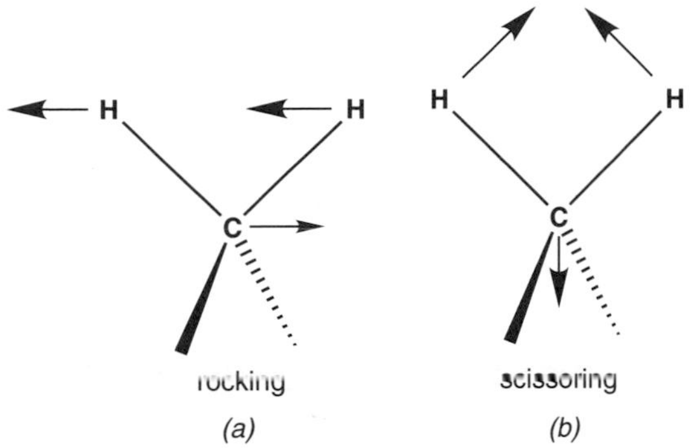

Figure 16.44. In-plane bending vibrations of $-CH_2-$: (*a*) rocking vibration and (*b*) scissoring vibration.

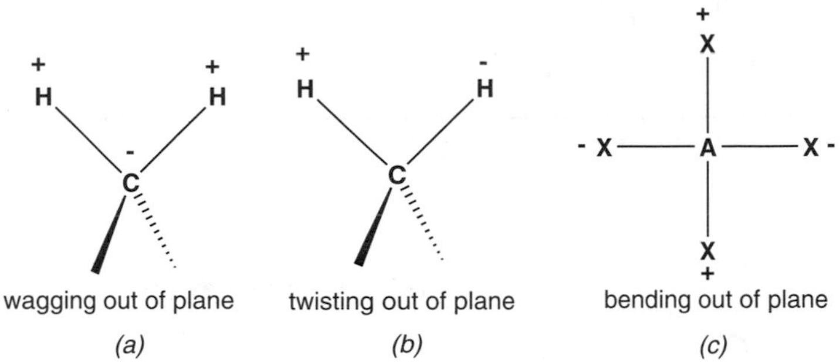

Figure 16.45. Out-of-plane bending vibrations: (*a*) the $-CH_2-$ wagging vibration; (*b*) the CH_2 twisting vibration; and (*c*) an out-of-plane vibration of a square planar molecule.

planar molecule. The − signs signify motion toward the observer and the + signs motion away from the observer.

16.46 GROUP FREQUENCIES

Absorption frequencies assigned empirically to a particular grouping of atoms whenever that grouping appears in a molecule. The frequencies of these absorptions are fortunately largely independent of the other atoms present in a molecule, and hence when they appear in a spectrum, the assignment to the particular grouping can be made with considerable confidence.

Example. It is, of course, impractical to discuss here the IR spectra of all functional groups in organic molecules so just a few examples are chosen to illustrate the group frequencies concept. Two common groupings in organic chemistry are the methyl (CH_3-) and methylene ($-CH_2-$) groups, and they occur together in many molecules. The presence of these groups gives rise to four stretching vibrations at relatively high frequencies. Two of these are methyl stretches at $2{,}959\,cm^{-1}$ (antisymmetric) and

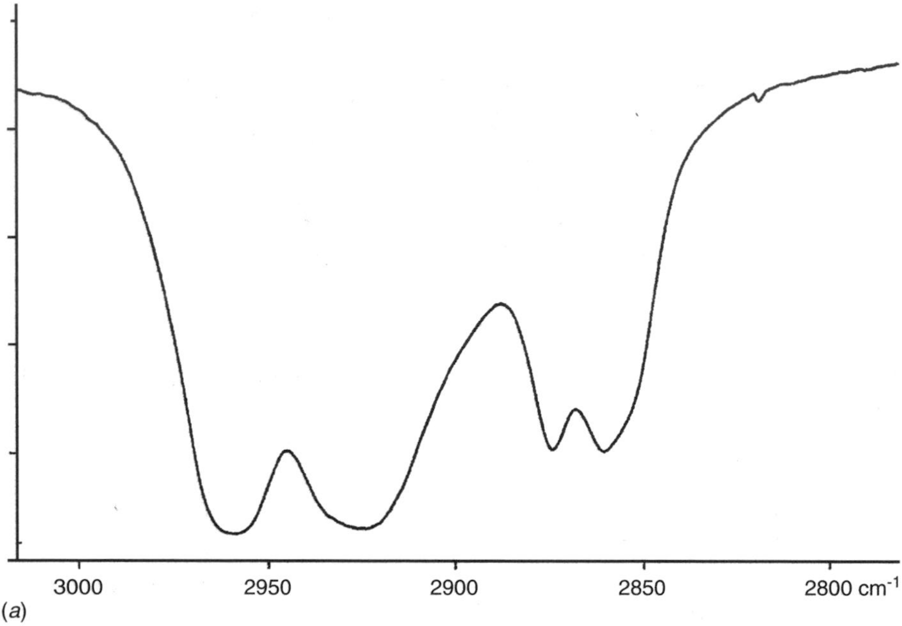

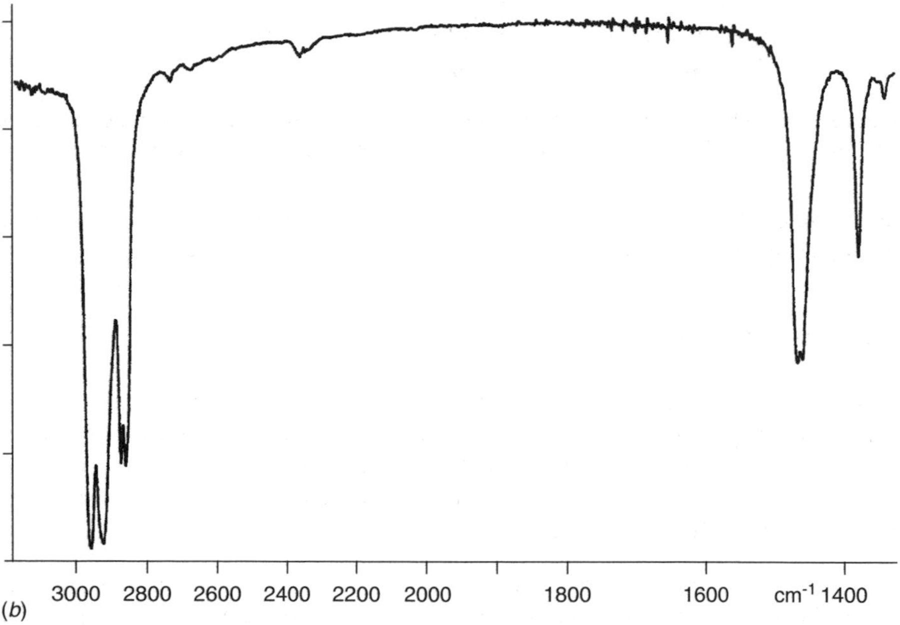

Figure 16.46. (*a*) The partial infrared spectrum (expanded) of hexane showing the four methyl and methylene stretching vibration frequencies and (*b*) the low-resolution IR spectrum (3,000–1,300 cm^{-1}) of hexane.

2,873 cm^{-1} (symmetric), and two are methylene stretches at 2,925 cm^{-1} (antisymmetric) and 2,860 cm^{-1} (symmetric). These four stretching vibrations are shown in the partial IR expanded spectrum of hexane in Fig. 16.46*a*. In addition to the stretching vibrations, the $-CH_2-$ group exhibits four bending vibrations: two in-plane (1,466 and 720 cm^{-1}) and two out-of-plane; these latter vibrations give rise to two weak bands, both of which occur in the 1,306 cm^{-1} range. The CH_3- group exhibits three bending vibrations at 1,739, 1,466, and 1,141 to 1,132 cm^{-1}. The full spectrum of hexane taken neat as a thin film pressed between NaCl plates is shown in Fig. 16.46*b*. The positions of the stretching and bending vibrations of the methyl and methylene groups in IR spectra are largely, but not entirely, independent of the structure of the rest of the molecule. Actually, only some of the bands observed in the full spectrum of hexane shown in Fig. 16.46*b* can be assigned to group frequencies. Many of the bands observed in high-resolution spectra, especially in the lower-frequency range, the "fingerprint" region, are due to normal modes of vibration that are characteristic of the molecule as a whole.

The carbonyl group, >C=O, is a functional group that occurs either alone or as a portion of other functional groups in a great variety of organic molecules, and the exact position of the carbonyl stretch in the infrared spectrum of a molecule is a powerful diagnostic tool. The exact position reflects the C–O bond order. The greater the double bond (order) character, the more difficult it is to stretch the C–O bond and the higher the frequency; the more like a single bond, the lower the frequency. A simple ketone, RC(O)R, typically shows a carbonyl stretch at about 1,715 cm^{-1}. Electron-donating groups attached to the carbonyl carbon such as multiple bonds or a nitrogen atom with lone-pair electrons reduce the C–O bond order (resonance effects) and lower the frequency while electron-withdrawing groups or atoms such as −OR increase the bond order (inductive effect) and increase the frequency. If the carbonyl group is part of a ring containing less than six atoms, the bond order and the carbonyl stretching frequency are lowered (bond angle effect). Fortuitously, but fortunately, the increments and decrements to the stretching frequency of the simple ketone are all approximately 30 cm^{-1}, and since the effects described above are additive, it is posssible to develop empirical rules that allow predictions to be made about where a carbonyl group in a particular environment will absorb. These empirical rules are given in Table 16.46*a*

TABLE 16.46*a*. Rules for Predicting the Effect of Substituents on the 1,715 cm^{-1} Band of the Carbonyl Group

Rule	Substituent on >C=O	Increase (+) or Decrease (−)
1,1a[a]	Electron-withdrawing group or atom (−OR)	+30
2	Electron-releasing group or atom $-(NR_2)$	−30
3	Conjugation with multiple bond (Ar; α,β-unsaturation)	−30
4	Carbonyl as part of five-membered ring	+30
5	Carbonyl as part of four-membered ring	+(2 × 30)

[a]For conjugation in the R group, add an additional increment of 30.

TABLE 16.46*b*. Effect of Structure on the 1,715 cm^{-1} Band of the Carbonyl Group

Structure	Rule No.	Effect	Calculated	Observed
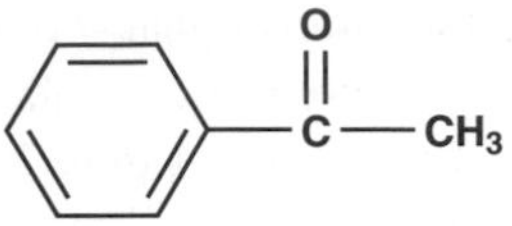	3	−30	1,685	1,685
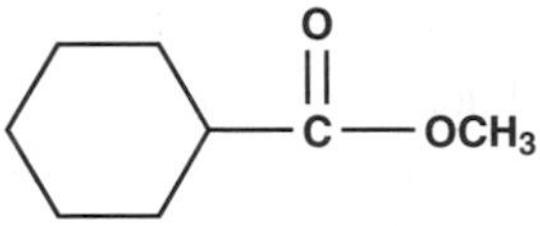	1	+30	1,745	1,735
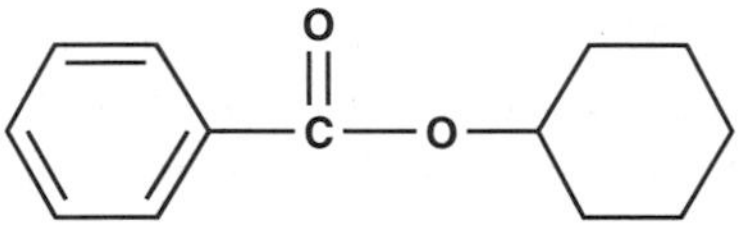	1,3	+30, −30	1,715	1,715
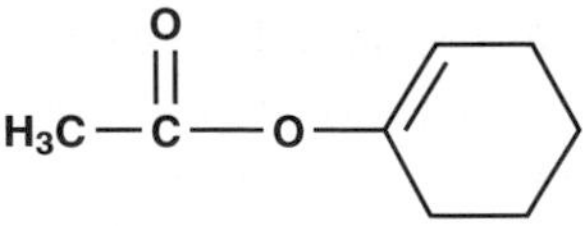	1,1a	+30, +30	1,775	1,766
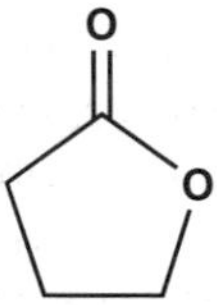	1,4	+30, +30	1,775	1,775
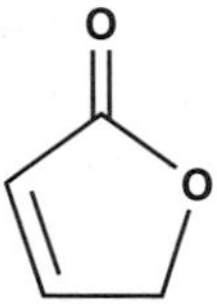	1,3,4	+30, −30, +30	1,745	1,750
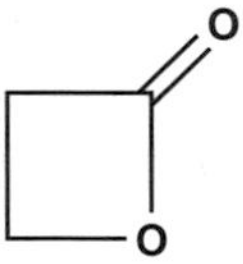	1,5	+30, +60	1,805	1,818

and applied in Table 16.46*b*. The infrared spectrum of methyl benzoate is shown in Fig. 16.46*c*.

Most of the group frequencies discussed above occur in the 4,000 to 1,000 cm^{-1} region that constitutes about two-thirds of the infrared region. The other third of the infrared region, 1,000 to 400 cm^{-1}, is the so-called *fingerprint region*, and unlike the functional group frequencies, which are largely independent of the structure of the surrounding molecule, the spectrum in this region reflects the structure of the entire molecule and shows a pattern that is unique for each compound.

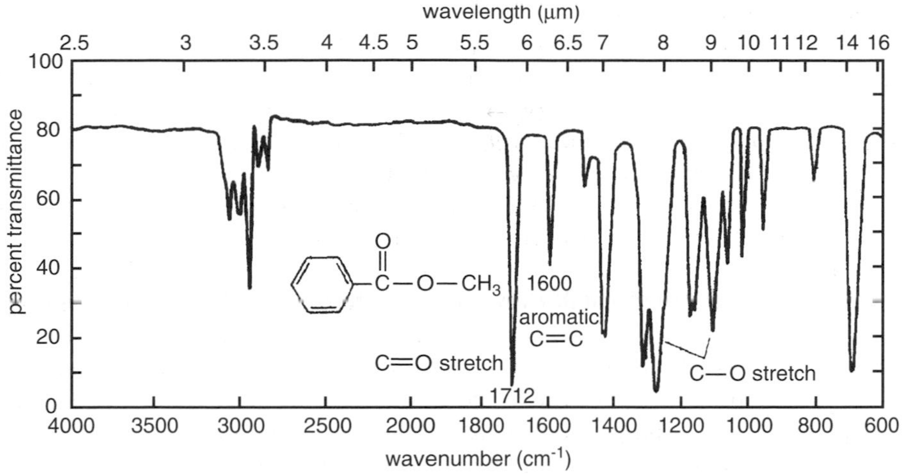

Figure 16.46*c*. The infrared spectrum of methyl benzoate (note that the effect of the phenyl substituent cancels that of the OCH_3 group and thus the carbonyl band occurs approximately at the same position as that of a simple ketone).

16.47 FORBIDDEN AND ALLOWED TRANSITIONS

Transitions that are predicted to have zero intensity are called *forbidden transitions* and those predicted to have nonzero intensity are called *allowed transitions*. Most forbidden transitions do occur to some extent, that is, the word "forbidden" should not be taken literally, and, as in life, the greater the penalty for violation, the more strictly forbidden is the event. Forbidden transitions generally give rise to weak bands, while allowed transitions generally give rise to strong bands. Whether transitions are forbidden or allowed depends on the operation of selection rules. The selection rule for the harmonic oscillator, that is, allowed transitions occur only between adjacent vibrational levels ($\Delta v = \pm 1$), is one such rule (Sect. 16.28). Other selection rules for vibrational spectroscopy are based on dipole moment considerations (Sect. 16.48) and on symmetry considerations involving the symmetry properties of an intensity-determining integral (Sects. 16.65 and 16.66). It is relatively easy, using group theory principles, to determine whether for a particular transition this integral vanishes (is zero) or is nonzero; if the former, it is forbidden, and if the latter, it is allowed. A low-intensity band may not necessarily reflect a forbidden band; it may be associated with a relatively small value of the intensity-determining integral.

16.48 INFRARED ACTIVE VIBRATIONS, DIPOLE MOMENT CONSIDERATIONS

Vibrations involving a change in the dipole moment (Sects. 2.13 and 4.31) are infrared active; they undergo allowed transitions, giving rise to observable bands in the infrared spectrum.

Example. Homonuclear diatomic molecules such as O_2, N_2, and so on, do not have a permanent dipole μ [$\mu = q \times d$, where q is the charge, either (+) or (−) of the dipole, and d is the distance separating the charges; see Sect. 2.13]. The vibrational transitions of such molecules ($v_0 \rightarrow v_1$, $v_1 \rightarrow v_2, \ldots$) are not associated with a change in dipole moment; hence, such transitions are not induced by infrared radiation and are infrared forbidden. Molecules that do not have a permanent dipole can still be infrared active if they develop a charge during a vibration. Consider carbon dioxide, O=C=O, that has a center of symmetry and therefore does not have a permanent dipole. During the symmetric stretch v_s (Fig. 16.42*a*), a dipole moment is not generated and this vibration is IR forbidden. (The frequency shown for the symmetric stretch is obtained from the Raman spectrum; see Sect. 16.62.) However, during the antisymmetric stretch v_{as} (Fig. 16.42*a*), a dipole moment clearly develops. While one bond is being stretched, the other C–O bond is being compressed. Thus, although the directions of the bond dipoles are opposed and tend to reduce one another, the magnitudes of the two bond dipoles differ, and hence there is a net instantaneous dipole moment produced during the vibration. Accordingly, the antisymmetric stretch is infrared active. The doubly degenerate bending vibration (Fig. 16.42*b*) is also infrared active. A vibration involving a change in the dipole moment of a molecule with a permanent dipole (no center of symmetry) is illustrated by the symmetric stretch of the water molecule (Fig. 16.39*a*). The equilibrium positions of the atoms (Fig. 16.48*a*) and the positions when the bonds are at their longest (Fig. 16.48*b*) and shortest (Fig. 16.48*c*) and the resulting dipole moment in each case (represented by the length of the arrows) are shown in the figure. (Of course, the dipole moment is changing continuously between these extremes but is never zero.) It is obvious that although in this case the direction of the dipole moment remains unchanged during the vibration, its magnitude changes and therefore the symmetric stretching mode is allowed and is infrared active.

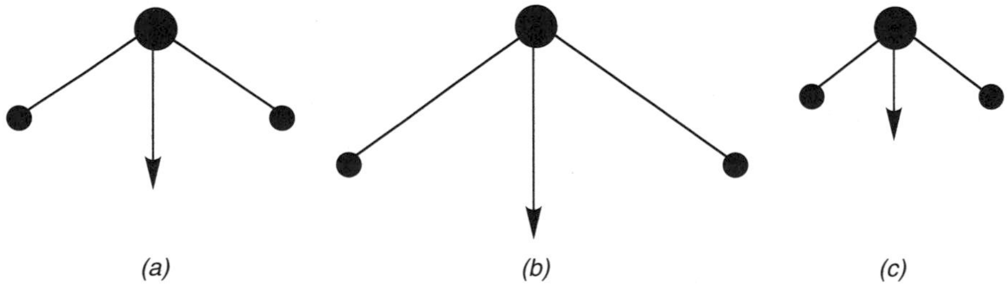

Figure 16.48. The change in dipole moment during the symmetric stretching vibration of the water molecule: (*a*) the equilibrium position of the atoms; (*b*) the position of the atoms at the maximum stretch; and (*c*) the same at the minimum stretch. The arrows represent the magnitude of the resulting dipole moment in each case.

16.49 OVERTONES

Absorptions that occur at a frequency which is approximately an integral multiple of the frequency of the fundamental vibration. The harmonic oscillator predicts equally spaced overtones, but real molecules do not behave as true harmonic oscillators. And because of anharmonicities, the overtones are not exact multiple integers of the fundamental vibrational transition. Overtones are forbidden transitions and the intensity of the overtones are therefore much weaker (a tenth to a hundredth) than the fundamental. The vibrational spacings ordinarily decrease with increasing values of the vibrational levels.

Example. The fundamental vibrational mode for the H–Cl stretch, that is, $v_0 \rightarrow v_1$, $\Delta v = +1$, occurs at 2,885.9 cm^{-1}, while the first overtone, $v_0 \rightarrow v_2$, $\Delta v = +2$, which is forbidden, occurs as a much weaker band at 5,668.0 cm^{-1} instead of 5,771.8 cm^{-1} (twice the fundamental frequency). However, if the correct anharmonicity constant is used in the anharmonicity equation (Sect. 16.30), the calculated overtone band agrees very well with the theoretical prediction.

16.50 COMBINATION BANDS

Absorption bands that appear at frequencies near the sum of two (or more) different frequencies associated with two fundamental transitions in the same molecule $(v_0 \rightarrow v_1) + (v'_0 \rightarrow v'_1)$. Combination bands result from a transition from the ground level to a new combination level that is not involved in the fundamental transitions. The frequency of a combination band is not the exact sum of the two fundamentals because of anharmonicity and thus a combination band is of low intensity.

Example. Figure 16.50 shows the infrared spectrum of HCN, a linear molecule. The two fundamental stretching frequencies occur at 3,270 cm^{-1} (C–H) and 2,085 cm^{-1}

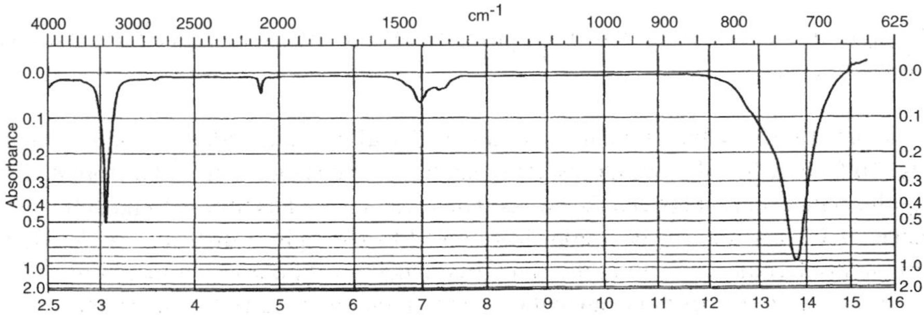

Figure 16.50. The infrared spectrum of HCN.

(C≡N). The doubly degenerate bending vibration is responsible for the band at $727\,cm^{-1}$. The weak band at $1{,}433\,cm^{-1}$ is the first overtone of the bending vibration at $727\,cm^{-1}$, which according to the harmonic oscillator model, should occur at $1{,}454\,cm^{-1}$. The very weak band at $2{,}800\,cm^{-1}$ is a combination band of the C≡N stretch plus the H–C–N bend ($2{,}085 + 722 = 2{,}812\,cm^{-1}$).

16.51 DIFFERENCE BANDS

Bands associated with transitions from an excited level of one vibration (usually of low frequency) to an excited level of another higher frequency vibration. Difference bands are not due to transitions from ground vibrational levels, and because no new levels are involved, the frequency should be exactly equal to the frequency difference of the the two fundamentals involved. Anharmonicity is irrelevant. Difference bands are not very common, are of low intensity, and can sometimes be eliminated by operating at low temperature.

Example. The spectrum of carbon dioxide exhibits a band due to a bending (doubly degenerate) vibration at $667\,cm^{-1}$ (Sect. 16.42) and two very weak bands on either side of this band: one at $721\,cm^{-1}$ and the other at $618\,cm^{-1}$. These are difference bands and result from the interaction (see "Fermi Resonance," Sect. 16.54) of the bending vibration with transitions associated with bands at 1,388 and $1{,}285\,cm^{-1}$: Thus, $1{,}388 - 667 = 721\,cm^{-1}$ and $1{,}285 - 667 = 618\,cm^{-1}$.

16.52 ACCIDENTAL DEGENERACY

When two (or more) vibrational levels belonging to different vibrational families have the same (or nearly the same) energy, the levels are said to be accidentally degenerate.

16.53 NORMAL COORDINATE

A single coordinate along which the progress of a single normal mode of vibration can be followed.

Example. Consider the stretching vibration of carbon monoxide, C≡O, shown in Fig. 16.53 in which the two atoms are separated by a distance r different from r_0, the equilibrium distance between the atoms. The carbon atom has moved to the left a distance $\Delta r(\mathrm{C})$, and the oxygen atom has moved to the right a distance $\Delta r(\mathrm{O})$. Both atomic displacements occur at the same frequency and in phase. Furthermore, since the center of mass does not change during the vibration, $m_C \Delta r(\mathrm{C}) = m_O \Delta r(\mathrm{O})$, where m_C and m_O are the atomic weights of carbon and oxygen. A normal vibration will occur in $^{12}C^{16}O$ if the carbon atom moves 16/12 as far as the oxygen atom The

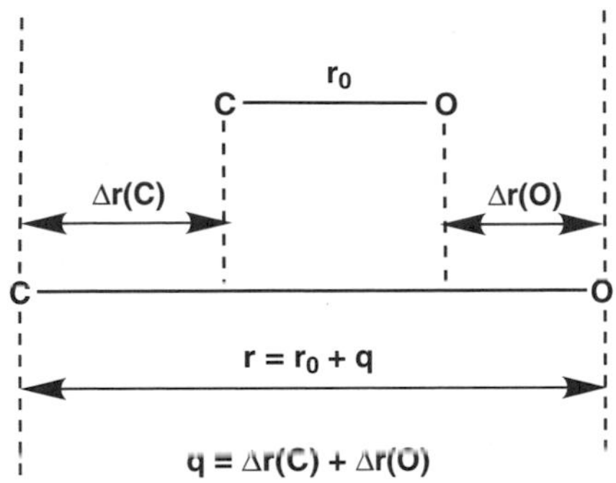

Figure 16.53. The normal coordinate q for the carbon monoxide stretching vibration.

normal coordinate q is equal to the sum of the displacements from the equilibrium atomic separation that is, $q = r - r_0 = \Delta r\,(\mathrm{C}) + \Delta r\,(\mathrm{O})$. Thus, the stretching vibration of CO is characterized by a single normal coordinate q that varies periodically.

16.54 FERMI RESONANCE

Named for Enrico Fermi (1901–1954, who received the Nobel Prize in 1938), this is the interaction (or "resonance") of accidentally degenerate or nearly degenerate vibrational levels; the interaction leads to a splitting that further separates the two levels so that they both are shifted from their normal or expected value. In order to interact, the two vibrations must have the same symmetry and be generated from atomic groupings that are located close to each other in the molecule.

Example. The symmetric stretching vibration of CO_2 (1,337 cm^{-1}, see Fig. 16.42*a*) and the first overtone of the doubly degenerate bending vibration ($2 \times 667 = 1{,}334\ cm^{-1}$) happen to be very close in energy, that is, accidentally degenerate. The overtone is actually split into two levels, owing to some rotational coupling, and one of these has the same symmetry as the symmetric stretch. As a result of Fermi resonance, the two extremely close levels of the same symmetry are split into a level at 1,388 cm^{-1} and one at 1,285 cm^{-1}. Frequently, the symmetric stretch of CO_2 is reported as the average of the two levels (1,337 cm^{-1}), but this is a calculated and not an observed value.

16.55 ASYMMETRIC TOP MOLECULES

Molecules in which all three principal moments of inertia are unequal, $I_a \neq I_b \neq I_c$. Such molecules lack any three-fold or higher-fold axis of symmetry. This requirement also eliminates molecules with a center of symmetry.

Example. CH_2Cl_2, H_2O, SO_2, CH_2O, CH_3OH are all asymmetric top molecules.

16.56 SYMMETRIC TOP MOLECULES

Molecules having two equal principal moments of inertia, and the third principal moment of inertia is unequal to the two equal ones: $I_a \neq I_b = I_c$. The one unequal axis may be greater than the two equal ones, in which case this axis is *oblate*, or the unequal axis may be smaller than the equal pair, in which case the axis is *prolate*.

Example. Molecules with one three-fold or higher-fold axis; (a) cyclobutane (Fig. 16.56*a*); the two axes with equal moments of inertia pass through carbon atoms at opposite corners of the square or through opposite sides of the square; (b) benzene (Fig. 16.56*b*). Other molecules in this category are $CHCl_3$ and BF_3.

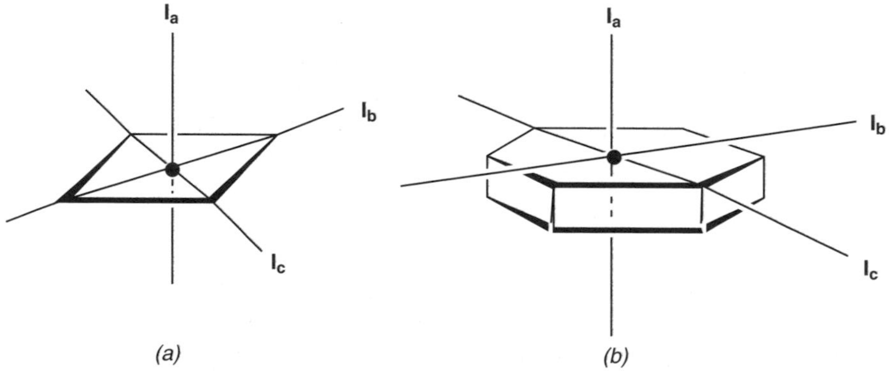

Figure 16.56. Symmetric top molecules where $I_a \neq I_b = I_c$: (*a*) square cyclobutane and (*b*) hexagonal benzene.

16.57 SPHERICAL TOP MOLECULES

Molecules having three equal principal moments of inertia, $I_a = I_b = I_c$.

Example. Tetrahedral and octahedral molecules belong in this category. In methane the three axes with equal moments of inertia are the three S_4 axes, easily recognized as the axes that bisect opposite faces of the cube in which a tetrahedron molecule can be inscribed (Fig. 16.57). All molecules that possess more than one rotational axis of greater than two-fold symmetry are spherical top molecules.

16.58 VIBRATIONAL-ROTATIONAL SPECTRA

Each vibrational level of a molecule has associated with it an infinite set of rotational levels so that the energy required for the vibration results in rotational transitions as well. A transition from the lowest or ground-state vibrational level, $v = 0$, having no rotational energy (ground rotational state, $J = 0$) can lead to vibrationally excited

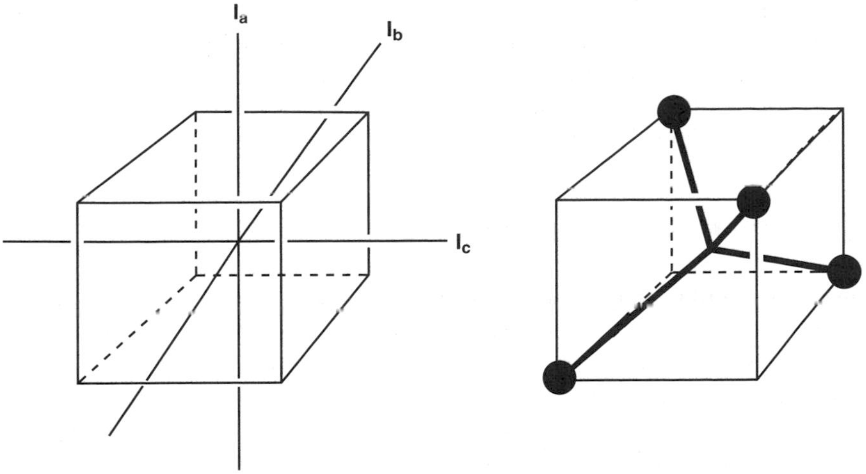

Figure 16.57. Spherical top molecules where $I_a = I_b = I_c$. (*a*) The three principal moments of inertia for the tetrahedral molecule methane pass through the centers of opposite faces of the cube that inscribes (*b*) the methane molecule.

molecules that exist in the first vibrationally excited state $v = 1$, but which are in different rotational levels, $J \neq 0$. In cases where there is no interaction between the vibration and rotation and if it is assumed that the molecule is performing simple harmonic motion and is rotating as a rigid rotor, then the energy for the transition is the sum of the energies of the harmonic oscillator and the rigid rotor.

16.59 *R* AND *P* BRANCHES

The allowed transitions for the diatomic harmonic oscillator are $\Delta v = \pm 1$ and for the rigid rotor they are $\Delta J = \pm 1$. Thus, in going from v_0 to v_1, there can be two changes in the quantum number for such a rigid rotor: $J = J' + 1$ and $J = J' - 1$, where J' is the rotational quantum number for $v = 0$. The simultaneous vibration-rotation transitions thus give rise to two series of lines: one ($\Delta J = +1$) on the high wavenumber (higher-energy) side of the pure vibration known as the positive or *R* branch, and one ($\Delta J = -1$) on the low wavenumber side (lower-energy) of the pure vibration known as the negative or *P* branch of the band. The spacings of these branches are not equal because real diatomic molecules do not behave as exact rigid rotors.

16.60 *Q* BRANCH

In addition to the *P* and *R* branches obtained in vibration-rotation spectra, it is possible to obtain a third branch in the case of linear molecules with more than two atoms. The third branch arises when $\Delta v = \pm 1$ and $\Delta J = 0$, for example, a transition from $v = 0$, $J = 0$ to $v = 1$, $J = 0$. The Q branch is usually missing in the spectra of diatomic molecules. For linear polyatomic molecules where the vibrational motion is perpendicular to the molecular axis, the transition $J = 0$ is allowed, whereas the motion (stretching mode) parallel to the molecular axis, $J = 0$, is forbidden.

16.61 RAYLEIGH SCATTERING

Named after Lord Rayleigh (John W. Strutt, 1842–1915, Nobel prize, Physics, 1904) the change in direction of electromagnetic radiation without a change in frequency when light is passed through a homogenous medium. When the incident photon collides with a molecule that is either in its ground or excited vibrational state and there is no exchange of energy, then the scattered photon has the identical frequency as the initial photon, and Rayleigh (or elastic) scattering has occurred. The intensity of light scattered by the solute in a solution is inversely proportional to the fourth power of the wavelength of light used.

16.62 RAMAN SPECTRA, RAMAN SHIFTS

Named for C. V. Raman (1888–1970, who received the Nobel Prize in 1930), the spectrum of scattered radiation observed when molecules collide with an incident photon. As a result of the collision, the vibrational or rotational energy of the molecule is changed (inelastic scattering), resulting in frequencies ν' of the electromagnetic radiation different from that of the incident light ν_0. The monochromatic exciting electromagnetic radiation (a laser source) is usually in the visible range. The energy difference (about 100–400 cm^{-1}) between the exciting light and the scattered light, $\Delta E = h|\nu_0 - \nu'|$ is called the *Raman shift*. Thus, if Raman lines (shifts) are observed shifted to either side of the Rayleigh line by $\pm 400\,cm^{-1}$, the molecules possesses a vibrational mode of this frequency. The series of shifts constitutes the *Raman spectrum* (see Fig. 16.63*b*) and corresponds to the vibrational spectrum produced by infrared light absorption. Thus by the use of light energy in the visible region (which is responsible for electronic transitions), vibrational spectra characteristic of infrared radiation are generated. Raman spectra are very similar to infrared spectra, but fewer transitions are forbidden in Raman spectra because of the different selection rules. Generally, but not always, the scattered light has a frequency lower than the exciting light because energy is lost to the absorbing molecule.

Example. The instrumentation used for the determination of infrared and Raman spectra are compared in the crude block diagram shown in Fig. 16.62. In infrared spectroscopy the incident light is usually generated by a Nernst filament or Globar (silicon carbide) heated to about 1000 to 1800°C. Such light contains all the different wavelengths in the infrared region. Passage through a reference cell compensates for solvent effects. In Raman spectroscopy the light source is usually a monochromatic

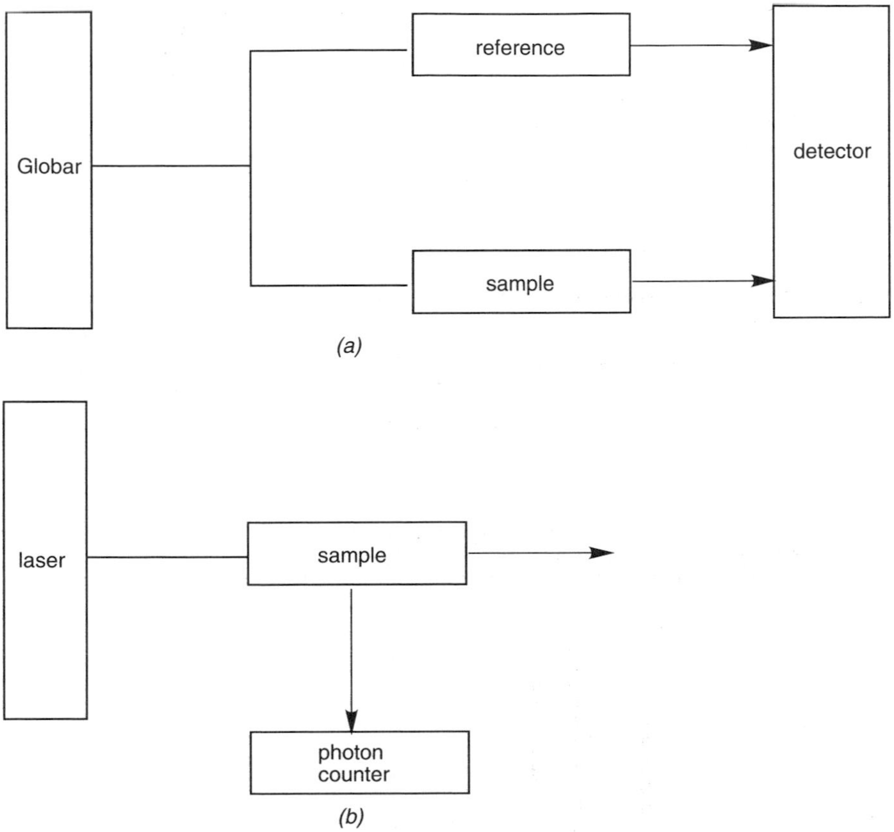

Figure 16.62. Simple block diagram of the (*a*) infrared and (*b*) Raman spectrometers.

(one wavelength) visible laser (commonly a YAG laser with emission at 1,064 nm). There is no reference cell. The solution of the sample is placed in a glass capillary tube and the incident light travels down the length of the tube. Most of the incident light passes through the sample, but about 0.1% is scattered. A small fraction of the scattered light is collected at right angles to the incident light and has slightly shifted frequencies relative to the incident light. These shifts correspond to the Raman spectrum.

16.63 STOKES AND ANTI-STOKES LINE SHIFTS

Named for George G. Stokes (1819–1903) the frequency difference between the exciting light and scattered light. The photon frequency difference is the same as the molecular vibrational frequency. When the scattered light has a lower frequency, the usual case, it is said to be Stokes shifted, while similar lines shifted to higher frequencies are said to be anti-Stokes shifted. The Stokes lines arise when the photon of the exciting

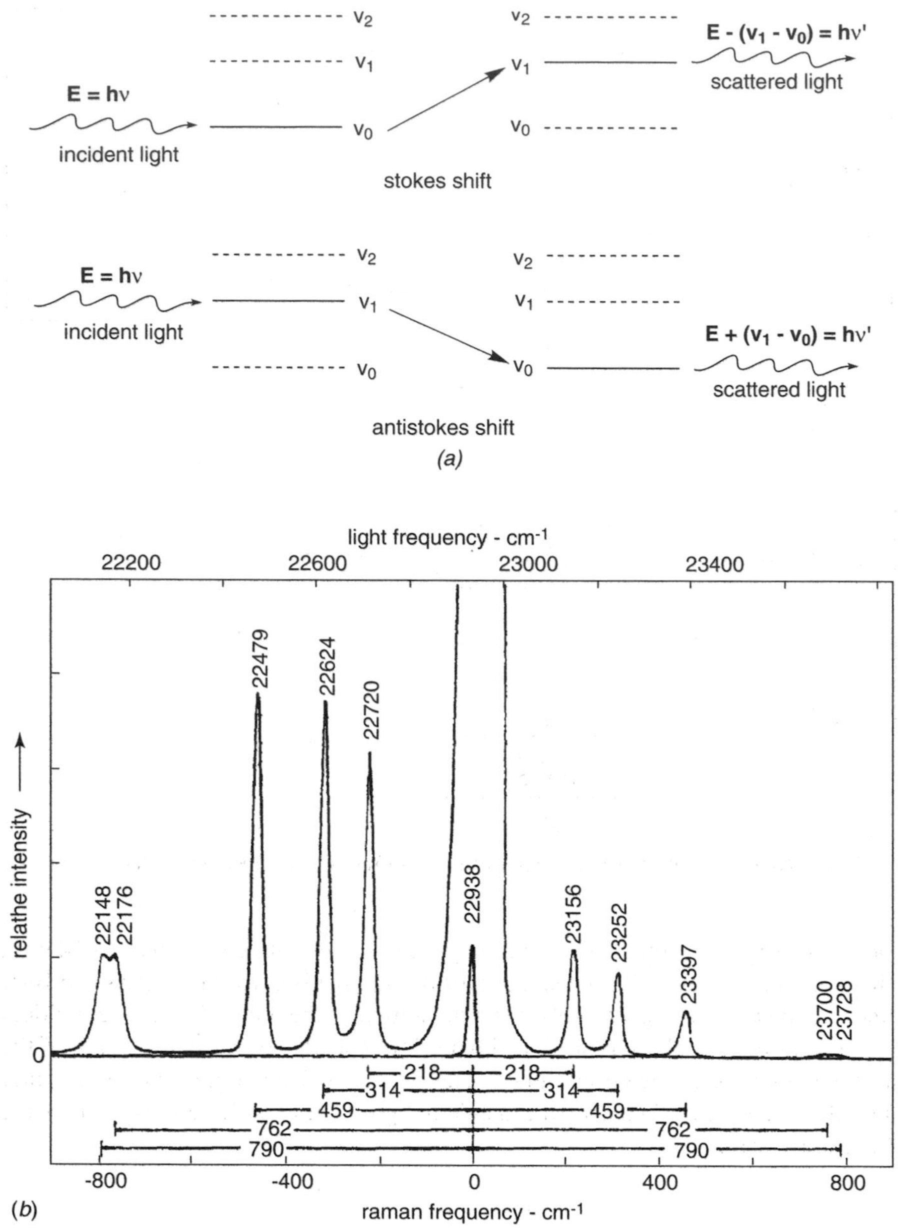

Figure 16.63. The scattering of light by a molecule giving rise to Stokes and anti-Stokes shifts: (*a*) the general scheme and (*b*) the Raman spectrum of CCl_4 showing both the Stokes and anti-Stokes portion of the spectrum.

light interacts with the vibrational *ground state* of the molecule, transfers some of its energy to it, and excites the molecule to a *higher* vibrational energy state. Thus, a photon of *lower* frequency than the incident light is produced in the scattered radiation. The anti-Stokes lines arise when the photon of the exciting light interacts with a vibrationally *excited* molecule and the molecule gives up energy to the photon so that a photon of *higher* frequency than the incident light is produced in the scattered radiation. The anti-Stokes lines are of lower intensity than the Stokes lines. Similar shifts to higher and lower frequencies occur in electronic spectroscopy and the same terminology is applied in these cases as well (see Sect. 18.136, 18.37).

Example. A general scheme illustrating the interactions leading to Stokes and anti-Stokes shifts is shown in Fig. 16.63*a* and an actual Raman spectrum showing both types of shifts is shown in Fig. 16.63*b*.

16.64 POLARIZABILITY

The deformability of the electron cloud of a molecule induced by the electric field component of electromagnetic radiation. (For polarizability of molecules by neighboring molecules in the absence of radiation, see Sect. 10.60.) When a molecule is subjected to electromagnetic radiation, the protons in the nuclei of the atoms making up the molecule attract the negative region of the electric field, while the electrons are attracted to, and move toward, the positive region of the electric field. This interaction gives rise to an induced dipole moment in the polarized molecule. The induced dipole moment μ, divided by the strength of the electric field, is the polarizability α:

$$\alpha = \mu / E \qquad (16.64)$$

The induced dipole moment μ, as well as the electric field E, are vectors, that is, they have both magnitude and directional character. For most molecules exposed to electromagnetic radiation, the polarizability α may be different in the x, y, and z directions. In such cases, the electric field component E_x will induce a dipole moment that has a component in the x direction, but that also may have a component in the y and z direction as well. The resulting polarizability can be expressed by the matrix multiplication of the x-, y-, and z-components of the polarization vector of the molecule by the x-, y-, and z-components of the reciprocal of the light electric field:

$$\begin{pmatrix} x \\ y \\ z \end{pmatrix}(x, y, z) = \begin{pmatrix} xx & xy & xz \\ yx & yy & yz \\ zx & zy & zz \end{pmatrix}$$

The resulting 3×3 matrix having nine elements is symmetrical around the diagonal from upper left to lower right, which means that $xy = yx$, $xz = zx$, and $yz = zy$. This leaves six distinct components of α, the polarizability, namely the three diagonal elements, xx, yy, zz, and three off-diagonal elements, xy, xz, and yz. The overall polarizability is the whole system of these α coefficients, and such a system of coefficients that establishes a linear relationship between vectors (in this case the two vectors, the dipole moment μ and and the electric field E) is called a *tensor* and, accordingly, the polarizability α is a tensor. Character tables usually show the symmetry species to which each of the six polarizability elements belongs (see, e.g., Table 16.40).

16.65 SYMMETRY SELECTION RULES: INTENSITIES OF INFRARED BANDS

The intensity I of a band due to a vibrational transition between two energy levels of a vibrational state is proportional to the integral below:

$$\sqrt{I} \propto \int \chi_i \mathbf{M} \chi_f \, d\tau \tag{16.65}$$

where χ_i and χ_f are vibrational wave functions of the initial and final vibrational states, respectively, $\mathbf{M}$ is the *dipole moment vector*, and the integration is taken over all space. $\mathbf{M}$ may be factored into three components, $\mathbf{M}_x + \mathbf{M}_y + \mathbf{M}_z$, where x, y, and z are related to the axes in a Cartesian coordinate system and represent translations (vectors) in the x, y, and z directions. A (nonmathematical) determination as to whether this integral is zero (corresponding to a forbidden transition) or nonzero (corresponding to an allowed transition) can be made as follows. The symmetry of the molecule under study is determined and reference is made to the character table that describes the symmetry of the molecule. The symmetry species to which the three translations belong is found on the right-hand side of every complete character table and is indicated there either by the letters x, y, and z or by the letters T_x, T_y, and T_z, the T meaning translations in the indicated direction. The symmetry of the particular molecular vibration under consideration is examined by looking at the arrows showing the motion of the atoms as they undergo a specific vibration, and this vibration is assigned to the corresponding symmetry species in the character table. If the vibration belongs to the same symmetry species as any one of the x, y, and z translations, the fundamental vibration will be allowed and give rise to an absorption band in the infrared region. Otherwise, that vibration will be forbidden and of low intensity. This recipe is, of course, an oversimplied description of the evaluation of the intensity determining Eq. 16.65. Functions such as $y = x^2$ (a description of a parabola) are even functions, that is, y is symmetric in x, and on integration they give positive values, whereas functions such as $y = x^3$, where y is antisymmetric with respect to x, are odd functions and their integration (the area under the curve) vanishes, that is, is zero. In order for the integrand of Eq. 16.65 to be symmetric, the product of the three terms under the integral sign must be symmetric, that is, belong to symmetry species A, A_1, A_g, or A' of the character tables ($+1$ under all symmetry operations of the character table). It is assumed that the initial vibrational wave function χ_i is symmetric, and it remains to be seen whether any x-, y-, z-component of $\mathbf{M}$ multiplied by χ_f is symmetric or antisymmetric. Now if either x, y, or z belongs to the same symmetry species as the vibration, the product **M** times χ_f will be +1 under all symmetry operations (the product of two identical signs always gives a plus sign) and, as a result, the total integrand will be plus or symmetric. Accordingly in this case, the integrand will be even and have a positive value and the transition will be allowed, giving rise to an observable band in the infrared spectrum of the molecule.

Example. The water molecule, H_2O, belongs to point group C_{2v} whose character table is given in Table 16.40. The three normal vibrational modes for the water molecule are described in Fig. 16.39. The symmetric stretch v_s and the bending vibration v_δ under all the symmetry operations of C_{2v} are +1, and hence these two normal modes belong to symmetry species a_1. Reference to Table 16.40 shows that the z-coordinate also belongs to a_1; hence, these two fundamental vibrations are infrared active and both are said to be polarized in the z direction. The stretching vibration v_{as} belongs to species b_2 and the character table shows that the y-coordinate also belongs to symmetry species b_2; hence, this fundametal vibration is also infrared active. Thus, we can expect that the infrared spectrum of H_2O will show three relatively strong bands in the infrared region, as indeed it does.

16.66 SYMMETRY SELECTION RULES: RAMAN SPECTRA

In considering the intensity of a Raman spectral band, an integral analogous to the integral used for the infrared case is employed, except that the dipole moment vector **M** in Eq. 16.65 is replaced by **P**, the *induced dipole moment vector:*

$$\int \chi_i \mathbf{P} \chi_f d\tau \qquad (16.66)$$

The six quadratic functions of the Cartesian coordinates that are the components of the polarizability tensor **P** (Sect. 16.64) are almost always listed on the right side of the character tables. A fundamental vibration will be Raman active (i.e., give rise to a Raman shift) if the normal mode involved belongs to the same symmetry species as one or more of the components of **P**, the polarizability tensor of the molecule.

Example. The character table for C_{2v} to which water belongs (Table 16.40) shows that each of the four symmetry species has an α (polarizability) component so all three fundamental vibrations of water are *Raman active* (as well as infrared active). No vibration of water belongs to species a_2, but if there were one, it would be Raman active. However, because no x-, y-, or z-component of **M** belongs to species a_2, such a vibration would be *infrared inactive*.

16.67 THE EXCLUSION RULE FOR MOLECULES WITH A CENTER OF SYMMETRY

If a molecule possesses a center of symmetry, all translations in either the x, y, and z directions (or Cartesian coordinates) are transformed to minus themselves on reflection through the center; hence, under this operation the x-, y-, and z-components of the transition moment **M** in Eq. 16.65 are all antisymmetric and ungerade (u). Reference to the character table to which a centrosymmetric molecule belongs will show the ungerade symmetry species to which x, y, and z belong and the vibrations that belong to such species will be *infrared active.* On the other hand, the binary product of two Cartesian coordinates, for example, α_{xy} or α_{zz}, does not change sign on inversion ($-1 \times -1 = +1$) and therefore all the components of the polarizabilty tensor **P** in Eq. 16.66 are gerade (g). Reference to the character table will show the gerade symmetry species to which each of the six α belong and the vibrations that belong to such species will be *Raman active.* These considerations lead to the exclusion rule, which states that in a molecule with a center of symmetry, no infrared active vibrations (all of which are u) can be Raman active and no Raman active vibrations (all of which are g) can be infrared active. This rule means that if the same band is observed in both the infrared and Raman spectrum of a compound, the compound does not possess a center of symmetry.

Example. Ethylene, $CH_2{=}CH_2$, has a center of symmetry and belongs to point group D_{2h}. Two of the $n - 1 = 5$ stretching modes are shown in Fig. 16.67. Figure 16.67*a* illustrates a stretching mode that is gerade (symmetry species A_g) and has polarizability tensors that also belong to the same symmetry species. This vibration is thus Raman active but infrared inactive. The stretching mode shown in Fig. 16.66*b* is ungerade and belongs to species B_{1u} in D_{2h}. Since the coordinate z also transforms as B_{1u}, this stretching vibration is infrared active but not Raman active.

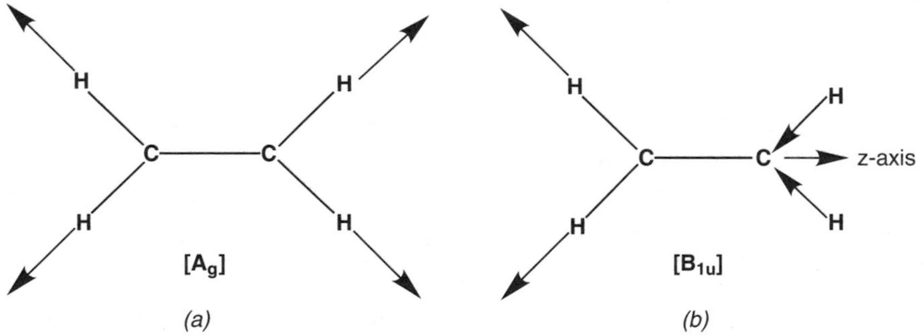

Figure 16.67. Two stretching modes of the ethylene molecule: (*a*) the stretching mode [A_g] that is Raman active but infrared inactive and (*b*) a stretching mode [B_{1u}], where translation in the z direction transforms as species B_{1u} and hence this mode is Raman inactive but infrared active.

16.68 FOURIER TRANSFORM (FT)

Named after Joseph Fourier (1768–1830), a mathematical methodology for transforming wave functions (e.g., vibrations or muscial notes) from functions of intensity versus time (the time domain spectrum) into functions of intensity versus frequency (the frequency domain spectrum).

Example. If a single string in a violin is bowed, the resulting sound, if recorded on an oscilloscope, would show an interference pattern or interferogram (a plot of total sound intensity vs. time), indicating that the fundamental frequency generated by the string is mixed with other harmonic frequencies to give the many overlapping waves that characterize the interference pattern. The resolution of the interferogram into the set of corresponding frequencies that produced it is the Fourier transform (FT) of the original wave function. Fourier transforms can be applied (by appropriate instrumentation) to time domain data from infrared, ultraviolet, visible, nuclear magnetic resonance, and mass spectrometers.

16.69 FOURIER TRANSFORM INFRARED SPECTROSCOPY (FTIR)

The indirect generation of an IR absorption spectrum accomplished by first generating the time domain (or to be exact, the distance domain) IR absorption spectrum interferogram, then Fourier-transforming the interferogram to generate the conventional spectrum.

Example. The interferogram is generated with an interferometer based on the design developed by Arthur Michaelson (1852–1931, the first American to win the Nobel prize in physics, in 1907) shown schematically in Fig. 16.69.

In this instrument, radiation from an infrared light source is passed through a beam splitter, a half-silvered mirror, that transmits half the beam to a movable mirror and reflects the other half to a fixed mirror. Both the transmitted beam and the reflected beam are recombined and passed to a detector. If the path length from the

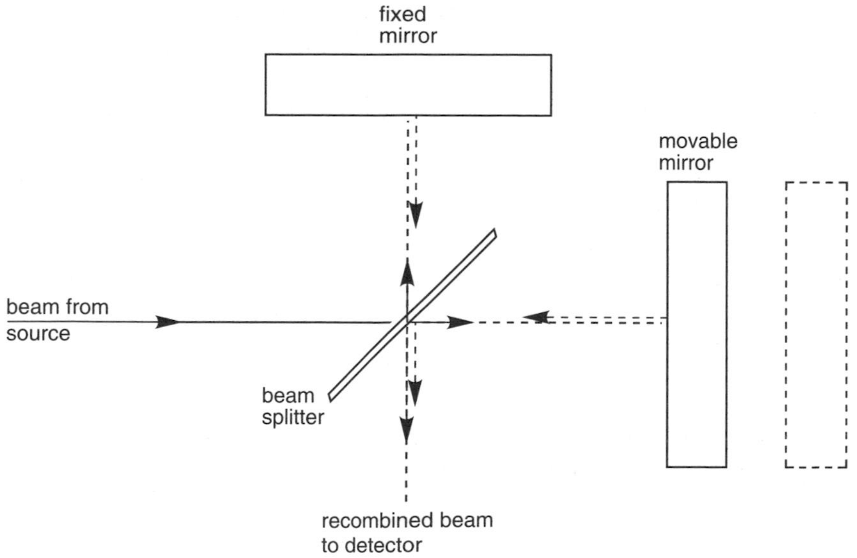

Figure 16.69. Schematic of the Michaelson interferometer.

beam splitter to the two mirrors differs by an integer multiple of wavelengths, the combination of the two beams leaving the mirrors will recombine with constructive interference and high-intensity radiation will reach the detector. The distance between the beam splitter and the movable mirror can be slowly changed, and when the two beams differ by one-half wavelength, the emergent beams recombine with destructive interference and low-intensity radiation reaches the detector. The minute movement (scanning) of the mirror (which is measured by a laser) allows light and dark images to reach the detector as a function of the distance, resulting in the interferogram. When infrared radiation passes through a scanning Michaelson interferometer, Fourier transformations of the resulting interferogram give a plot of beam intensity absorbed versus frequency. When a sample of a compound is placed in the path of the beam, the compound absorbs particular frequencies so that their intensities are reduced in the interferogram. The ensuing FT, when subtracted from that of the beam itself, is the infrared spectrum of the sample. The advantages of the FTIR are its increased speed and sensitivity compared to the simple IR. FTIR instruments can also perform other mathematical manipulations of the spectroscopic data, including signal averaging, comparisons of spectra, and kinetic analysis.

Acknowledgment. We wish to thank Prof. B. Ault for helpful discussion.

SUGGESTED READING

Coulthup, N. B., Daly, L. H., and Wiberley, S. E. *Infrared and Raman Spectroscopy*, 2nd ed. Academic Press: New York, 1978.

Conley, R. T. *Infrared Spectroscopy*. Allan and Bacon: Boston, MA, 1966.

Cotton, F. A. *Chemical Applications of Group Theory*. Wiley-Interscience: New York, 1963.

Harris, D. C. and Bertolucci, M. D. *Symmetry and Spectroscopy*. Oxford University Press: London, 1978.

Herzberg, F. H. *Infrared and Raman Spectra of Polyatomic Molecules*. Van Nostrand Co.: Princeton, NJ, 1956.

Jaffe, H. H. and Orchin, M. *Symmetry in Chemistry; Orbitals and Spectra*. John Wiley & Sons: New York, 1965.

Kemp, W. *Organic Spectroscopy*, 3rd ed. W. H. Freeman: New York, 1991.

Laszlo, P. and Stang, P. *Organic Spectroscopy*. Harper and Row: New York, 1971.

Nakamoto, K. *Infrared Spectra of Inorganic and Coordination Compounds*. John Wiley & Sons: New York, 1963.

Orchin, M. and Jaffe, H. H. *Symmetry, Orbitals and Spectra*. John Wiley & Sons: New York, 1971.

Silverstein, R. M., Bassler, G. C., and Morrill, T. C. *Spectrometric Identification of Organic Compounds*, 4th ed. John Wiley & Sons: New York, 1981.

17 Mass Spectrometry

17.1	Mass Spectrometer	705
17.2	Ionization Chamber	705
17.3	Mass Analyzer	706
17.4	Ionization	706
17.5	Electron Impact (EI)	706
17.6	Chemical Ionization (CI)	706
17.7	Field Desorption Ionization (FDI)	707
17.8	Fast Atom Bombardment (FAB)	707
17.9	Electrospray Ionization (ESI)	707
17.10	Matrix-Assisted Laser Desorption Ionization (MALDI)	707
17.11	Mass Number (Nominal Mass)	708
17.12	Mass-to-Charge Ratio (m/z)	708
17.13	Magnetic Analysis of m/z	709
17.14	Mass Spectrum	710
17.15	Resolution R	711
17.16	Double Focusing Mass Spectrometer	711
17.17	ICP-Mass Spectrometry (ICP/MS)	711
17.18	Ion Cyclotron Resonance (ICR) Spectroscopy	711
17.19	Fourier Transform Ion Cyclotron Resonance (FTICR) Spectroscopy	712
17.20	Time-of-Flight Mass Spectrometer (TOF)	712
17.21	Quadrupolar Mass (Filter) Spectrometer	712
17.22	Ion Trap Mass Spectrometer	713
17.23	Gas Chromatography/Mass Spectrometry (GC/MS)	714
17.24	Liquid Chromatography/Mass Spectrometry (LC/MS)	715
17.25	Isobaric Ions	715
17.26	Exact Mass	715
17.27	Odd Electron Cation (Radical Cation)	716
17.28	Even Electron Cation	716
17.29	The Nitrogen Rule	717
17.30	Molecular Ion ($M^{\bullet +}$)	717
17.31	Molecular Weight and Formula Determination by Exact Mass Spectrometry	717
17.32	Fragmentation	718
17.33	Precursor (Parent) Ion	718
17.34	Product (Daughter, Fragment) Ion	718
17.35	Fragmentation Pathways	718

The Vocabulary and Concepts of Organic Chemistry, Second Edition, by Milton Orchin, Roger S. Macomber, Allan Pinhas, and R. Marshall Wilson

17.36 Base Peak 719
17.37 Metastable Ion 719
17.38 Metastable Peak 719
17.39 Isotope Cluster 720
17.40 $M + 1$, $M + 2$ Peaks 721
17.41 Molecular Formula Determination from $M + 1/M + 2$ Data 722
17.42 Molecular Structure from the Mass Spectrum 722
17.43 Mass Spec/Mass Spec (MS/MS) 724
17.44 Proteomics 724

When identifying the structure of a molecule, the usual first step is to determine its molecular formula. Historically, this was accomplished by first establishing the empirical formula of the compound (by combustion analysis) and then determining the molecular weight of the compound by some colligative property (e.g., vapor pressure osmometry, melting point depression, etc.). All these techniques have inherent experimental errors that, for even moderately large molecules, introduce substantial uncertainty in the resulting molecular formula. Modern mass spectrometry allows a chemist, using only about a microgram of a pure compound, to determine its exact molecular weight and, thus, the exact molecular formula of a compound. This can be done for small molecules, as well as large biomolecules, such as proteins. Moreover, a substantial amount of structural information about the molecule can be obtained from the fragments detected in its mass spectrum.

In NMR (Chap. 15) or IR (Chap. 16) spectroscopy, where bulk samples containing a large ensemble of molecules are analyzed, both intra- and intermolecular effects are observed. In mass spectrometry, at high vacuum (10^{-6} torr), individual ions (charged molecules) are detected. Thus, with very few exceptions (see, e.g., Sect. 17.6), chemistry that occurs in a mass spectrometer is unimolecular. Because a mass spectrometer deals with individual ions and, thus, with extremely small masses, the mass unit is not the usual gram, but rather the Dalton, which until recently was called the atomic mass unit (amu). This unit is based on the arbitrary assignment of the mass of one atom of ^{12}C as 12.00000 Daltons, although the experimentally determined mass of a ^{12}C atom is slightly different. One mole of ^{12}C atoms has a mass of 12.00000 g. The experimentally measured mass of an ^{1}H atom is 1.6735×10^{-24} g, corresponding to 1.0078 Daltons. The factor for converting from grams to Daltons is the reciprocal of *Avogadro's number* (6.02214×10^{23} atoms per mole).

Notice that this chapter is entitled "Mass Spectrometry" and not "Mass Spectroscopy." The word "spectrometry" implies that some quantity, which is characteristic of a compound, is being measured, while the word "spectroscopy" implies that the quantity being measured is electromagnetic radiation. Thus, all forms of spectroscopy are forms of spectrometry, but not all forms of spectrometry are spectroscopy.

17.1 MASS SPECTROMETER

An instrument designed to first volatilize and then ionize uncharged molecules of the *analyte* (compound to be analyzed), and then to separate and analyze these ions on the basis of their mass-to-charge ratio (m/z).

Example. Figure 17.1 is a diagrammatic sketch of a mass spectrometer with a single focusing magnetic sector. The sample (1 µg or less) is introduced into the inlet system (1) as a solid, liquid, or gas. The sample is then vaporized into the sample bulb (2) by a combination of low pressure in the bulb, obtained by vacuum pump (3), and heat if necessary. The vaporized molecules are bled into the ion source (4), where they are ionized, usually by bombardment with high-energy electrons, and the resulting positive ions are accelerated between oppositely charged plates (5). The resulting ion beam passes through focusing slits (6) and into a magnetic field (7), which causes the ions to separate, each ion following a curved path that depends on its mass (measured in Daltons or atomic mass units) and charge (usually, +1). Finally, the flow of ions (*ion current*) is collected and measured as ion current in the detector (8), which is usually an electron multiplier. The bulk of the system is kept at high vacuum (ca. 10^{-6} torr) to prevent reactions of the ions with unionized neutral molecules.

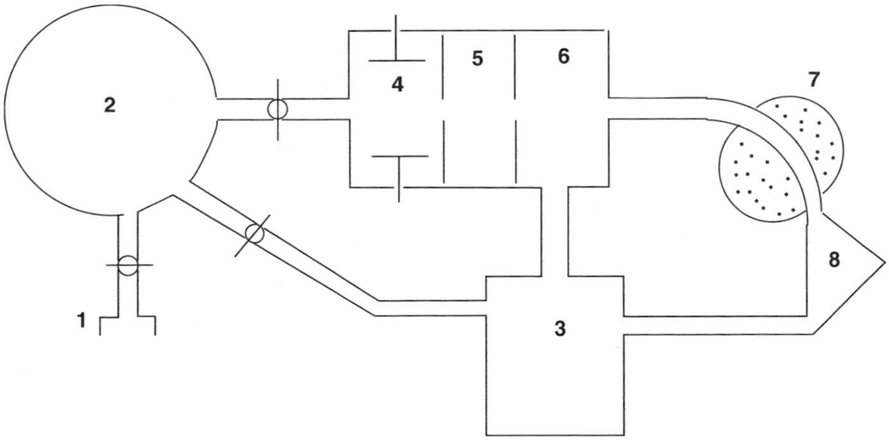

Figure 17.1. Schematic diagram of a mass spectrometer with a single focusing magnetic sector, including inlet system (1), sample bulb (2), vacuum pump (3), ion source (4), accelerating region (5), slits (6), magnetic field (7), dots represent lines of magnetic flux perpendicular to the page, and detector (8).

17.2 IONIZATION CHAMBER

The section of a mass spectrometer that includes the ion source, accelerating plates, and focusing slits, parts (4), (5), and (6) in Fig. 17.1.

17.3 MASS ANALYZER

The section of a mass spectrometer where the ions are separated according to their mass-to-charge ratio (m/z).

Example. The magnetic field depicted as 7 in Fig. 17.1 represents the mass analyzer section in this spectrometer design. Other types of mass analyzers utilize either an electric field, a quadrupolar field (Sect. 17.21), time-of-flight analysis (Sect. 17.20), or some combination of these.

17.4 IONIZATION

In the context of mass spectrometry, ionization is the creation of positive (or negative) ions from neutral molecules (M) by removing (or adding) one or more electrons, or adding one or more protons (H^+). Most mass spectrometry involves the detection of positive ions. The methods for accomplishing this ionization are discussed below.

17.5 ELECTRON IMPACT (EI)

A very common ionization technique in mass spectrometry, in which bombardment with a beam of high-energy electrons causes a molecule (M) to eject one or more of its electrons. To accomplish this, the electrons in the beam must possess very high energy (20–70 eV), which exceeds the ionization potential (see Sect. 12.31) of the analyte molecules by a considerable amount. In most cases, only singly charged ions are formed, but doubly and triply charged ions are sometimes formed. One characteristic of this technique is that the large excess energy of the beam electrons causes the formation of "hot" ions possessing large amounts of vibrational energy. These so-called hot ions subsequently break up into fragment ions of lower mass (see Sect. 17.32).

17.6 CHEMICAL IONIZATION (CI)

An ionization technique in which an ion of one type (a primary ion, usually CH_5^+ from added CH_4), formed by electron impact, creates a secondary ion by reacting with a neutral molecule (M), forming MH^+ by proton transfer.

Example. A mixture of methane (CH_4) and the compound of interest (M) in the ratio of 10^6:1 is bled into the ionization chamber. The methane, being in high abundance, is the source of virtually all the primary ions, including such species as CH_5^+ (protonated methane, a true carbonium ion, Sect. 13.31), $C_2H_5^+$, $C_3H_5^+$, etc. CH_5^+ is extremely reactive and transfers a proton to M:

$$CH_5^+ + M \longrightarrow M\text{–}H^+ + CH_4$$

Finally, the M–H^+ ion (with a mass 1 Dalton greater than the mass of M) is subjected to the usual sequence of acceleration, mass analysis, and detection. Fragmentation of M–H^+ ions is a relatively infrequent process, owing to the lower energy of these ions compared to those formed during electron impact.

17.7 FIELD DESORPTION IONIZATION (FDI)

An ionization technique in which ions are formed by the application of a strong electric field (i.e., a potential of +8 kV) through a wire, on which the solid analyte is deposited. Electrons from the analyte molecules are attracted to the wire, leaving behind positive ions, which desorb from the wire. These ions are lower in energy than those formed during electron impact and are, therefore, less prone to undergo fragmentation.

17.8 FAST ATOM BOMBARDMENT (FAB)

An ionization technique in which the analyte (M) is dissolved in a viscous polar solvent, such as glycerol, and instead of being bombarded with electrons as in EI (Sect. 17.5), M is bombarded with high-energy (fast-moving) xenon atoms. Typically, the adduct MH^+ is the dominant species generated by a proton transfer from the glycerol.

17.9 ELECTROSPRAY IONIZATION (ESI)

A ionization technique in which a dilute solution of the analyte is pumped through a capillary tube at a low flow rate (J. E. Fenn 1917– , Nobel Prize 2000). A high voltage, either positive or negative, is applied to the end of the capillary tube. This voltage causes charges to be generated at the surface of the liquid. When the Coulombic repulsion of the charges is greater than the surface tension, droplets are ejected from the tip of the capillary. These charged droplets then enter the mass spectrometer for analysis, where the solvent droplets rapidly evaporate, depositing their charge on the dissolved analyte molecules. Analytes that are acids or bases are easily charged by this method. For neutral analytes, an acid or a salt is added to the solution, and thus, charged species are generated by forming adducts with ions such as Cl^-, H^+, or Na^+. Multiply charged ions can be formed when the original neutral molecule bonding to more than one of these species. Instruments of this type have demonstrated sensitivity in the range of several hundred attomoles (about 2×10^{-16} mol).

17.10 MATRIX-ASSISTED LASER DESORPTION IONIZATION (MALDI)

An ionization technique in which a low concentration of the analyte, which must only weakly absorb light, is embedded in a solid matrix of a highly light-absorbing compound. Common matrices are nicotinic acid, dihydrobenzoic acid, and urea. This matrix is then irradiated with a short pulse of light from a laser. This irradiation causes the compound to ionize.

17.11 MASS NUMBER (NOMINAL MASS)

The mass of an ion, rounded to the nearest integer value in Daltons. This integer equals the sum of the number of protons and neutrons in all the atoms of the ion.

Example. Table 17.11 shows the mass number of several ions.

TABLE 17.11. Mass Number of Some Typical Ions

Ion	Mass Number
CH_3^+	15
$C_2H_5^+$	29
CO_2^+	44
N_2O^+	44
$C_2H_4O^+$	44

17.12 MASS-TO-CHARGE RATIO (m/z)

The ratio of the mass of an ion (m, in Daltons) to its charge (z). This ratio is sometimes given the abbreviation m/e. The charge of an ion equals the number of protons in the nuclei minus the number of electrons.

Example. Table 17.12 shows several ions and their m/z values expressed as the mass number divided by charge.

It is obviously possible for several different molecular formulae to have the same m/z. A mass spectrometer that works with only unit masses cannot distinguish an ion with a mass of 104 and a charge of +1 from an ion of mass 208 and charge of +2, because both have $m/z = 104$ (but see Sect. 17.26).

TABLE 17.12. Some Common Ions and their m/z Values

Ion	m/z
CH_3^+	15
$C_2H_5^+$	29
CO_2^+	44
CO_2^{2+}	22
N_2O^+	44
$C_2H_4O^+$	44

17.13 MAGNETIC ANALYSIS OF *m/z*

In a mass spectrometer, ions are separated on the basis of their m/z ratio. Two equations describe the motion of ions in a mass spectrometer using magnetic field analysis (separation). Once accelerated by the plates, each ion has kinetic energy (KE):

$$\mathrm{KE} = (1/2)\, mv^2 = zeV \tag{17.13a}$$

where m = mass of the ion, v = velocity of the ion, z = charge on the ion, e = the charge (in coulombs) of 1 electron, and V = accelerating potential through which the ion has passed (1–10 kV, not to be confused with the ionizing voltage used in electron impact, region 5 in Fig. 17.1).

As the ion enters the magnetic field, it follows a curved path of radius r given by

$$r = mv/ze\mathbf{B} \tag{17.13b}$$

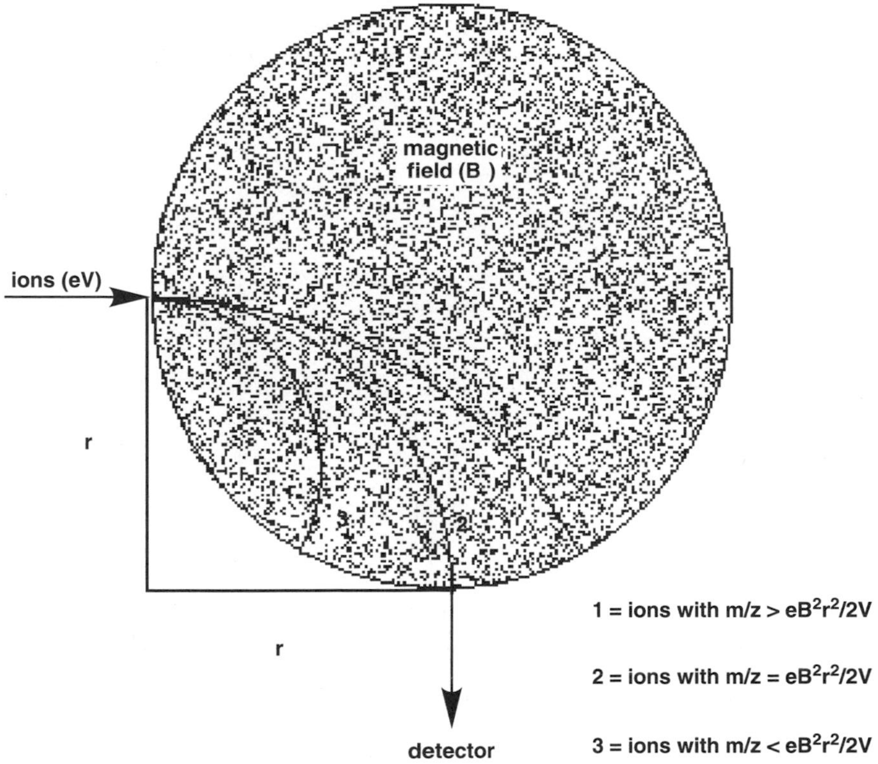

Figure 17.13. Illustration of the path of ions with differing m/z, as they pass through a magnetic analyzer.

where m, v, z, and e are as given above, and $\mathbf{B}$ = magnetic field strength. These two equations can be combined, eliminating v, to give

$$m/z = e\mathbf{B}^2r^2/2V \tag{17.13c}$$

If the instrument is designed with a set analyzer radius r, then ions with different m/z values can be brought into focus by varying either $\mathbf{B}$ or V. When $\mathbf{B}$, V, and r are all fixed, only ions with the corresponding m/z will reach the detector (Fig. 17.13). See also Sect. 17.20.

17.14 MASS SPECTRUM

A graphical display of the m/z values for a collection of ions versus their relative abundance. The abundance of an ion, which is related to its stability and ease of formation, is measured as a current of ions of each m/z reaching the detector. Such data can also be displayed in tabular form.

Example. Figure 17.14 shows the mass spectrum of *trans*-2-butene (C_4H_8). The ion at m/z 56 is the *molecular ion* ($M^{\bullet+}$, Sect. 17.30); the other ions are *fragment ions* (Sect. 17.32) with fewer carbons and hydrogens. The most intense peak, in this case

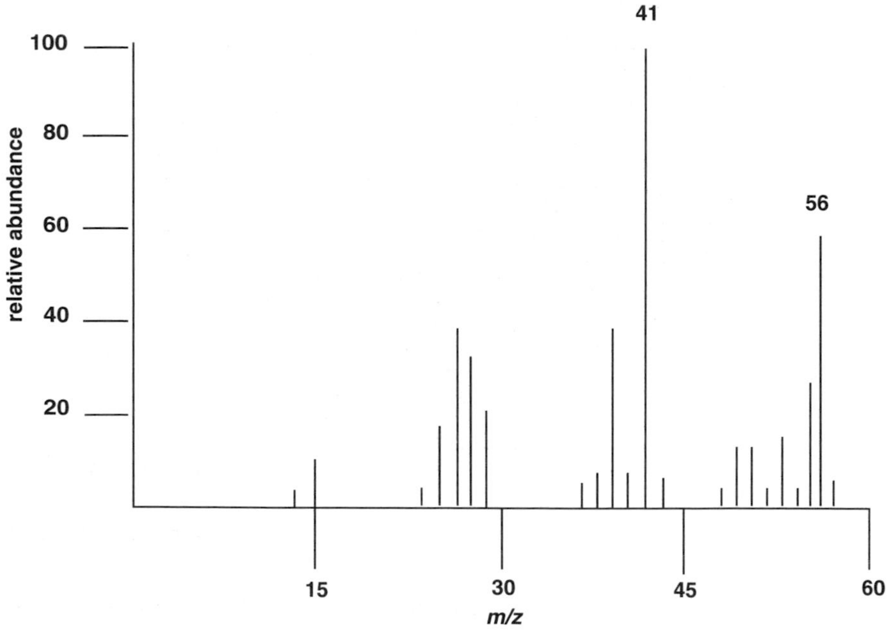

Figure 17.14. The mass spectrum of *trans*-2-butene: a graph of relative abundance (i.e., ion current) versus m/z.

at m/z 41, is called the *base peak* (Sect. 17.36) and is arbitrarily assigned an intensity of 100%.

17.15 RESOLUTION *R*

A parameter characteristic of a given mass spectrometer, which expresses how close two ions can be in terms of m/z and still be differentiated (with less than 10% overlap of their signals) on that instrument. Resolution R is defined as

$$R = \mathrm{M}/\Delta\mathrm{M} \tag{17.15}$$

where M is m/z of a given ion, ΔM is the difference in m/z between M and the next higher m/z signal that can be resolved from M. Thus, if an instrument can differentiate m/z 1,000 from m/z 1,002 (but not 1,001), $R = 1{,}000/2 = 500$. Low-resolution instruments are those with $R < 1{,}000$, whereas high-resolution instruments are those with $R > 10{,}000$. A resolution of 10^5 also allows the mass of an ion with m/z 100 to be determined with a precision better than ±0.001 Daltons.

17.16 DOUBLE FOCUSING MASS SPECTROMETER

A mass spectrometer with two analyzers in tandem, usually an electrostatic or quadrupolar analyzer (energy analyzer), followed by a magnetic analyzer (mass analyzer). Double focusing is required for the generation of high-resolution mass spectra.

17.17 ICP-MASS SPECTROMETRY (ICP/MS)

A inductively coupled plasma mass spectrometer for detecting trace elements. A solution of metal complexes is digested under acidic conditions ($\mathrm{pH} < 1$) to bring all metals as ions into solution. This solution is sprayed into a torch and inductively heated to about 10,000°C to generate a plasma. The positive ions in the plasma are then introduced into a mass spectrometer. Trace quantities of almost every metal and many nonmetals can be detected by this method.

17.18 ION CYCLOTRON RESONANCE (ICR) SPECTROSCOPY

A relatively high-pressure mass spectrometric technique by which relatively long-lived, low-kinetic-energy ions are forced to follow circular (cyclotron) paths within the confines of a magnetic field. The sample chamber also contains neutral gas phase molecules with which the ions can react. Ions with specific m/z are then excited by radiofrequency irradiation, while monitoring the intensity (abundance) of other ions.

When irradiation of ions with one specific m/z causes a change in abundance of ions at another m/z, these two ions are related by either a fragmentation process or an ion–molecule reaction.

17.19 FOURIER TRANSFORM ION CYCLOTRON RESONANCE (FTICR) SPECTROSCOPY

As with all forms of Fourier transform spectroscopy (see, e.g., Sect. 15.31 and 16.69), time domain data from an ICR (Sect. 17.18) are converted into frequency domain data. The frequency domain data are easily converted to a mass spectrum. FTICR is one of the best methods of mass spectrometry in terms of resolution and accuracy of mass determination. m/z values of greater than a million may readily be determined because an FTICR can analyze multiply charged ions generated by, for example, ESI (Sect. 17.9).

17.20 TIME-OF-FLIGHT MASS SPECTROMETER (TOF)

A mass spectrometer that determines the m/z of an ion by measuring the time required for each ion to travel a specific straight-line distance. No external magnetic filed is involved in this type of mass spectrometer (compare with magnetic analysis, Sect. 17.13).

Example. As shown in Sect. 17.13, any ion accelerated through a potential of V volts will acquire a velocity v given by

$$v = (2zeV/m)^{1/2} \tag{17.20a}$$

Since the time of flight t required for such an ion to traverse a distance d is simply d/v, the relation between m/z and t is given by

$$m/z = 2eVt^2/d^2 \tag{17.20b}$$

Therefore, since instrument parameters V and d are known, measurement of the time of flight leads directly to the m/z of the ion. Arrival times of successive ions at the detector can be less than a microsecond.

17.21 QUADRUPOLAR MASS (FILTER) SPECTROMETER

An instrument designed to select (i.e., separate or filter) only ions of a specific m/z. A linear stream of ions (with differing m/z values) passes through an electric quadrupole, generated by the application of a direct current voltage and a radiofrequency voltage to the rods shown in Fig. 17.21. This potential is rotated between the electrodes at a controlled frequency. As a result, the potential field in the interelectrode

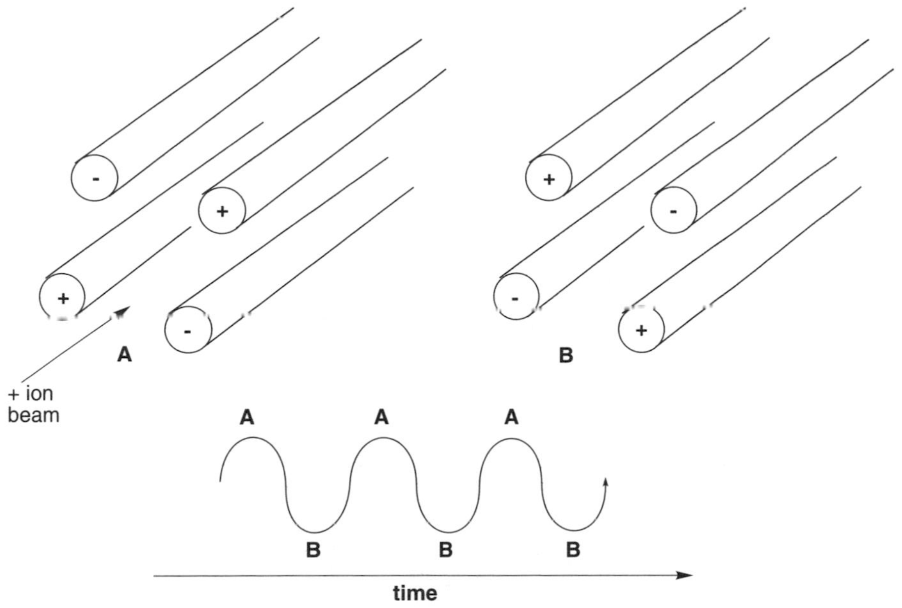

Figure 17.21. An electric quadrupole consisting of two sets of four parallel rods with the indicated charge. The ion stream with selected *m*/*z* passes through the center of the field, parallel to the rods. The charges on the rods oscillate between *A* and *B* as a function of time.

space resembles a "spinning saddle," and the positive ions are attracted alternately toward different sets of opposing electrodes. The result is that certain ions of selected masses will follow stable oscillating, helical trajectories through the interelectrode region and be detected. In contrast, other ions of different masses will be driven on "unstable" trajectories that will cause them to collide with the electrodes and be discharged. These ions will not be detected. Stable trajectories for ions of different masses can be selected by varying the electrode potentials and frequency of potential oscillation. Thus, scanning the entire range of stable trajectories will produce a mass spectrum of about unit resolution.

17.22 ION TRAP MASS SPECTROMETER

An instrument similar to a quadrupole mass (filter) spectrometer (Sect. 17.21), in which the ions are trapped in a three-dimensional quadrupolar field, as shown in Fig. 17.22. Ions are injected through one of the circular electrodes into the interelectrode space. Ions of different masses will accumulate in the trap, where they will trace out stable oscillating figure 8-like trajectories. Ions of selected masses can be irradiated with pulses of radiofrequency (rf) energy that destabilizes their orbits, causing them to exit the trap. By varying the frequency of this rf pulse, ion of different masses can be sequentially released from the trap and detected, affording a conventional mass spectrum.

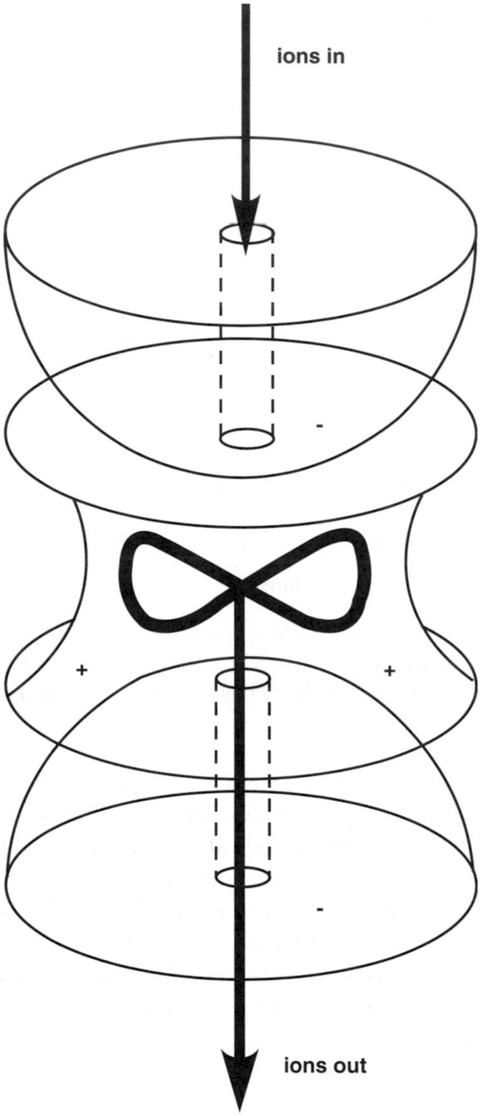

Figure 17.22. Electrode configuration and ion trajectory in an ion trap mass spectrometer.

17.23 GAS CHROMATOGRAPHY/MASS SPECTROMETRY (GC/MS)

The tandem attachment of a gas chromatograph (GC, see Sect. 10.30) to a mass spectrometer. The GC first separates the compounds of a vapor phase mixture, and each compound is passed separately into the ion source of the ms, where its mass spectrum is determined. Typical mass spectrometers for this type of instrument are a quadrupolar mass filter or an ion trap.

17.24 LIQUID CHROMATOGRAPHY/MASS SPECTROMETRY (LC/MS)

The tandem attachment of a liquid chromatograph (LC, see Sect. 10.31) to a mass spectrometer. The LC first separates the compound of a liquid phase mixture, and each compound is volatilized separately in the ion source of the ms, where its mass spectrum is determined.

17.25 ISOBARIC IONS

Ions with the same nominal mass number and charge, but different composition, for example, CO_2^{+} and N_2O^{+}, both $m/z = 44$.

17.26 EXACT MASS

The mass of a specific atom or ion in Daltons, expressed to a level of precision up to four or five decimal places. Such measurements require high-resolution instrumentation such as an FTICR (Sect. 17.19) or a TOF (Sect. 17.20).

Example. Table 17.26*a* shows the calculated exact mass of the stable isotopes of some of the most common elements found in organic compounds, based on the arbitrary assignment of ^{12}C as 12.00000 Daltons. It is important to distinguish the exact mass of a given isotope from the *atomic weight* of the element (in the Periodic Table), the latter being the *average* mass over all stable isotopes of that element weighted for the natural abundance of these isotopes.

Table 17.26*b* lists several isobaric ions, all having the *nominal* mass number 120, with small but significantly different *exact* masses calculated from the exact masses

TABLE 17.26*a*. The Exact Mass and Natural Abundance of Some Common Ions

Isotope	Exact Mass (Daltons)	Natural Abundance (%)
^{1}H	1.00783	99.98
^{2}H (D)	2.01410	0.02
^{12}C	12.00000	98.9
^{13}C	13.00336	1.1
^{14}N	14.0031	99.6
^{15}N	15.0001	0.4
^{16}O	15.9949	99.76
^{17}O	16.9991	0.04
^{18}O	17.9992	0.20
^{35}Cl	34.9689	75.5
^{37}Cl	36.9659	24.5
^{79}Br	78.9183	51
^{81}Br	80.9163	49

TABLE 17.26*b*. The Exact Mass of Ions of the Same Nominal Mass

Formula	Exact Mass
$C_2H_4N_2O_4$	120.0171
$C_6H_4N_2O$	120.0324
$C_4H_8O_4$	120.0423
$C_3H_8N_2O_3$	120.0535
C_8H_8O	120.0575
$C_7H_8N_2$	120.0687
$C_5H_{12}O_3$	120.0786
C_9H_{12}	120.0939

of the constituent atoms. Such ions can be readily distinguished using a high-resolution instrument mass spectrometer.

17.27 ODD ELECTRON CATION (RADICAL CATION)

A cation with an odd number of electrons (see Sect. 13.50). Such species are written $R^{\bullet +}$ to emphasize both their ionic character (by virtue of the net charge) and their free radical character (by virtue of the unpaired electron). (See Table. 12.27.)

Example.

TABLE 17.27. The *m*/*z* Values of Some Radical Cations

Ion	*m*/*z*
$CH_3OH^{\bullet +}$	32
$C_6H_6^{\bullet +}$	78
$CH_3NH_2^{\bullet +}$	31

17.28 EVEN ELECTRON CATION

A cation with an even number of electrons, all paired, and written Z^+ (See Table 17.28).

Example.

TABLE 17.28. The *m*/*z* Values of Some Common Cations

Ion	*m*/*z*
CH_3CO^+	43
$C_3H_3^+$	39
$H_2C{=}NH_2^+$	30
$C_6H_5N_2^+$	105

17.29 THE NITROGEN RULE

A molecule that contains an even number of nitrogens must also contain an even number of hydrogens (plus halogens) and must have an even molecular weight; a molecule that contains an odd number of nitrogens must also contain an odd number of hydrogens (plus halogens) and must have an odd molecular weight. This rule only holds for molecules containing C, H, N, O, and/or halogen atoms. The nitrogen rule can be generalized for any ion in the mass spectrum. For cations containing an even number (including zero) of nitrogen atoms (all nonionic forms of nitrogen), odd electron cations will have an even mass number, whereas even electron cations will have an odd mass number, For cations containing an odd number of nitrogen atoms, odd electron cations will have an odd mass number, whereas even electron cations (quaternary ammonium salts) will have an even mass number. This rule is of considerable value in suggesting a structure, or at least a molecular formula, for an ion in a mass spectrum.

Example. See the entries in Tables 17.27 and 17.28.

17.30 MOLECULAR ION ($M^{\bullet +}$)

The radical cation (odd electron cation) formed in the mass spectrometer after ejection of one electron from a neutral analyte molecule *M*. The molecular ion will be the highest mass ion, excluding isotope peaks (see Sect. 17.39), in a mass spectrum, and its exact mass will yield the exact molecular formula; this assumes there are no ion–molecule reactions.

Example. The vast majority of organic molecules are even electron, diamagnetic species; therefore, molecular ions are odd electron ions. The ions shown in Table 17.27 are the molecular ions from methanol, benzene, and methyl amine, respectively.

17.31 MOLECULAR WEIGHT AND FORMULA DETERMINATION BY EXACT MASS SPECTROMETRY

In an EI mass spectrum (Sect. 17.5), the exact mass of the molecular ion is numerically equal to the exact molecular weight of that substance. Thus, if one can identify which ion in a mass spectrum is the molecular ion, one knows the molecular weight of the substance. In general, the molecular ion will be the one with the largest *m*/*z* in the spectrum (but see also Sect. 17.39). However, in many cases the molecular ion is too weak to identify, because it undergoes fragmentation prior to analysis. In such cases, one must resort to lower EI voltages, or other ionization techniques. In CI (Sect. 17.6) or ESI (Sect. 17.9) mass spectrum, the molecular ion is formed by protonation of a neutral analyte molecule to give the species $(MH)^+$, an even electron ion, which has an exact mass 1.00783 Daltons greater than the exact mass of the analyte. As pointed out in Sect. 17.26, the exact mass of an ion also provides its molecular formula.

17.32 FRAGMENTATION

The unimolecular decomposition of one ion into two or more fragments, usually an ion and a neutral fragment. Provided fragmentation takes place before entry into the ion-accelerating region of the spectrometer, the fragment ions will appear at their correct m/z. See also Sect. 17.35.

17.33 PRECURSOR (PARENT) ION

Any ion that undergoes fragmentation and, therefore, can be regarded as the direct precursor of a given fragment ion. See also Sect. 17.35. This is a generic term and, thus, includes M, which is derived from the loss of an electron from the neutral analyte.

17.34 PRODUCT (DAUGHTER, FRAGMENT) ION

An ion resulting from fragmentation of its precursor ion. The product ion is generally more stable than the precursor ion.

17.35 FRAGMENTATION PATHWAYS

Each type of carbon skeleton and functional group undergoes characteristic fragmentations, and from these patterns the molecular structure can often be deduced. All fragmentation pathways fall into one of four classes, depending on the nature of the precursor and product ions. Class I is the most common, followed by class II; class III and IV are relatively rare.

Example. One example each of fragmentations I, II, and III is shown in Fig. 17.35. The exact mechanisms of many such fragmentations are not known with certainty.

TABLE 17.35.

Class	Precursor Ion	Product Ion	Other Fragment
I	Odd electron	Even electron	Free radical
II	Odd electron	Odd electron	Neutral molecule
III	Even electron	Even electron	Neutral molecule
IV	Even electron	Odd electron	Free radical

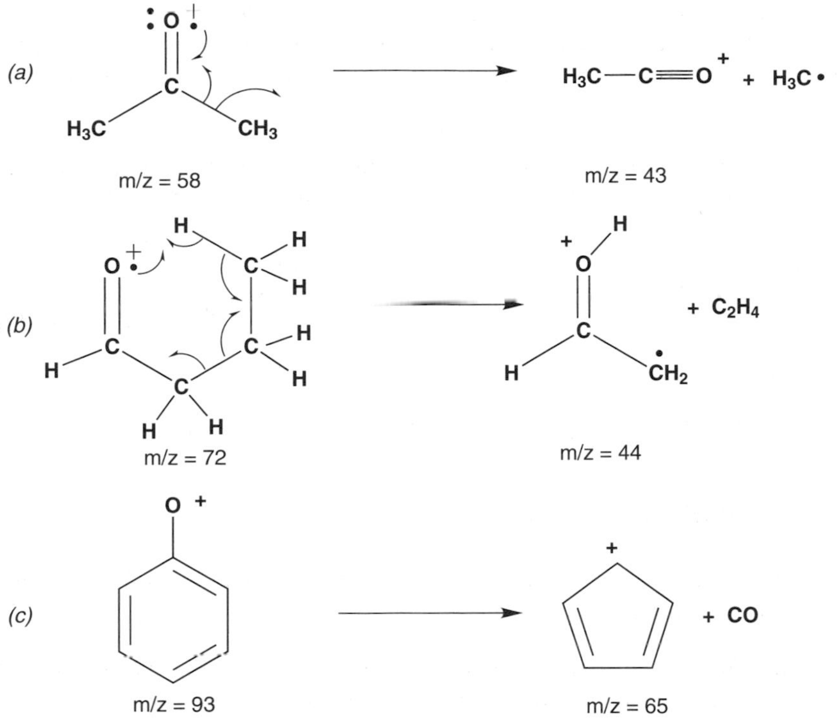

Figure 17.35. Three common fragmentation pathways: (*a*) type I (α cleavage); (*b*) type II (McLafferty rearrangement); (*c*) type III (decarbonylation).

17.36 BASE PEAK

The ion in a mass spectrum with the largest relative abundance. The base peak is generally assigned a relative abundance of 100%, with the intensities of other ions referred to it.

Example. The base peak in Fig. 17.14 occurs at *m*/*z* 41.

17.37 METASTABLE ION

An ion formed in the ionizing region of a mass spectrometer that is stable enough to survive acceleration, but undergoes fragmentation during passage through the magnetic mass analyzer. See also Sect. 17.38.

17.38 METASTABLE PEAK

A broad bell-shaped peak in a mass spectrum centered at a nonintegral *m*/*z* (labeled m*) that results from fragmentation of a metastable precursor ion of mass number

m_1 into a product ion of mass m_2. Under such conditions, $m^* = (m_2)^2/m_1$. The appearance of a metastable ion is a direct indication of which precursor and product ions are related. This type of peak cannot be observed in mass spectrometers with unit mass resolution.

Example. If the loss of methyl from *trans*-butene molecular ion occurred during mass analysis, a broad peak would be observed at $m^* = (41)^2/56 = 30.0$.

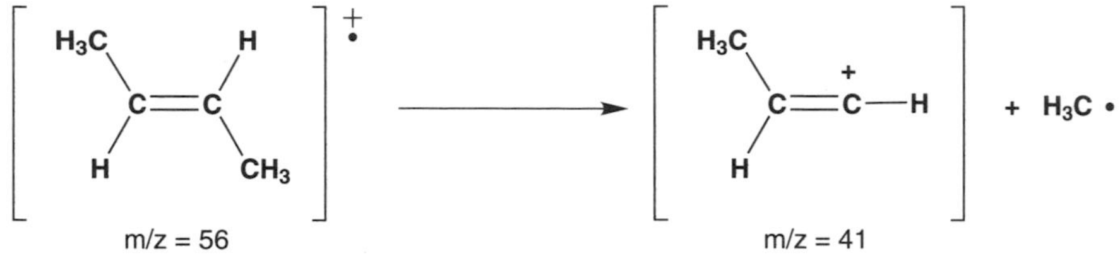

17.39 ISOTOPE CLUSTER

A group of peaks in a mass spectrum that share the same molecular formula, but differ in isotope distribution, and hence *m/z*.

Example. The mass spectrum of bromomethane is shown in Fig. 17.39. The molecular ion has composition $CH_3Br^{\bullet+}$, but there are 16 isotopic permutations of this composition. The most abundant of these are shown in Table 17.39; the relative abundance of each ion can be calculated from the data in Table 17.26*a*. (Ions with two deuterium atoms have a relative abundance of less than 0.001% and, thus, are not included in the table.)

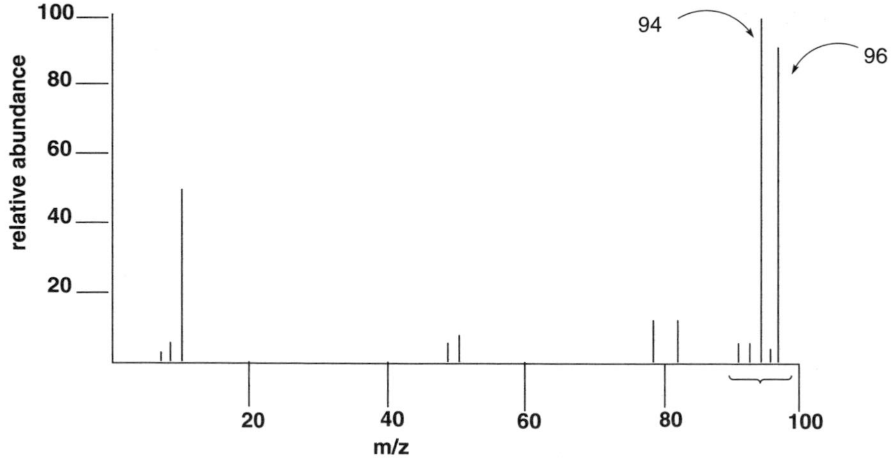

Figure 17.39. The mass spectrum of CH_3Br, showing the isotope cluster for the molecular ion (*m/z* 94–97).

TABLE 17.39. Isotope Clusters

Ion	Mass Number	Relative Abundance (%)
$^{1}H_3{}^{12}C^{79}Br$	94[a]	100
$^{1}H_2{}^{2}H_1{}^{12}C^{79}Br$	95	0.05
$^{1}H_3{}^{13}C^{79}Br$	95	1.1
$^{1}H_3{}^{12}C^{81}Br$	96	98
$^{1}H_2{}^{2}H_1{}^{12}C^{81}Br$	97	0.05
$^{1}H_3{}^{13}C^{81}Br$	97	1.1

[a]By definition, this is the molecular ion, because the molecular ion is the sum of the masses of the most abundant isotope of each atom.

Thus, the "molecular ion" will actually appear as a cluster of ions with *m/z* 94 to 97 (although some peaks may be too weak to observe). In principle, any ion will give rise to an isotope cluster whose appearance is predictable on the basis of the natural abundance of its isotopes. However, unless the elements in the compound have isotopes of comparable abundances (e.g., chlorine or bromine), the cluster will consist of one major peak and several much smaller peaks.

17.40 $M + 1, M + 2$ PEAKS

These are peaks in an isotope cluster at one and two units higher *m/z* than the nominal mass number; they are also referred to as $P + 1$ and $P + 2$ peaks. In the absence of exact mass data, the relative abundances of the $M + 1$ and $M + 2$ peaks can yield molecular formula information.

TABLE 17.40. The Abundance of + 120 Peaks

Formula	$\%M + 1$	$\%M + 2$
$C_2H_4N_2O_4$	3.15	0.84
$C_2H_6N_3O_3$	3.52	0.65
$C_3H_6NO_4$	3.88	0.86
$C_3H_8N_2O_3$	4.25	0.67
$C_4H_8O_4$	4.61	0.88
$C_3H_{10}N_3O_2$	4.62	0.49
$C_3H_{12}N_4O_2$	5.36	0.52
$C_5H_{12}O_3$	5.71	0.74
$C_5H_4N_4$	6.99	0.21
$C_6H_4N_2O$	7.35	0.43
$C_6H_6N_3$	7.72	0.26
$C_7H_8N_2$	8.46	0.32
C_8H_8O	8.81	0.54
$C_8H_{10}N$	9.19	0.37
C_9H_{12}	9.92	0.44

Example. The compound $C_cH_hO_oN_n$ will exhibit a molecular ion at $m/z = 12c + h + 16o + 14n$. The percent abundances of the $M + 1$ peak ($m/z = 12c + 1h + 16o + 14n + 1$) and $M + 2$ peak ($m/z = 12c + 1h + 16o + 14n + 2$) in the isotope cluster (with M set to 100%) can be estimated as follows:

$$\%\ M + 1 = (1.1)c + (0.016)h + (0.38)n$$

$$\%\ M + 2 = \{[(1.1)c]^2 + [(0.016)h]^2 + (40)o\}/200$$

Extensive tables are available that list $M + 1$ and $M + 2$ contributions for all possible combinations of C, H, O, and N up to m/z 300. A sample portion of such a table for $M = 120$ is shown in Table 17.40.

17.41 MOLECULAR FORMULA DETERMINATION FROM $M + 1/M + 2$ DATA

The molecular formula of a compound can, in principle, be established from either exact mass measurement or isotope cluster ratios once the molecular ion has been identified.

Example. Suppose an unknown compound gives an unequivocal molecular ion at a mass number m/z 120, and high-resolution exact mass data are unavailable. From Table 17.40, application of the nitrogen rule (Sect. 17.29) excludes the five molecular formulas with an odd number of nitrogens, since 120 is even, and therefore, these ions must be fragment ions. Even a low-resolution instrument can provide accurate $M + 1$ and $M + 2$ abundances. Suppose these came out as $\%\ M + 1 = 8.7$ and $\%\ M + 2 = 0.5$ (see Fig. 17.42*a*). The closest fit in the above table is C_8H_8O.

17.42 MOLECULAR STRUCTURE FROM THE MASS SPECTRUM

Once the molecular formula of an unknown compound has been determined, it is often possible to deduce the structure, or at least a significant portion of the structure, by a careful analysis of the fragmentation patterns. There are also numerous computer databases of mass spectra and software to compare them with the observed spectrum of a compound to be identified.

Example. The complete mass spectrum of the compound C_8H_8O, described in Sect. 17.41, is shown in Fig. 17.42*a*. The base peak occurs at m/z 77, which is most commonly due to $C_6H_5^+$. The peak at m/z 105 suggests the loss of CH_3 ($120 - 15 = 105$) to give $C_6H_5CO^+$. The peak at m/z 43 is either $C_2H_3O^+$ or $C_3H_7^+$, the former being more likely, considering hydrogen distribution. The composition of any of these could be confirmed by either exact mass determination or $M + 1/M + 2$ ratios. One structure that would be predicted to give exactly this spectrum is acetophenone, $PhC(O)CH_3$. The various possible fragmentations are shown in Fig. 17.42*b*. To confirm this assignment, one could compare the spectrum of the

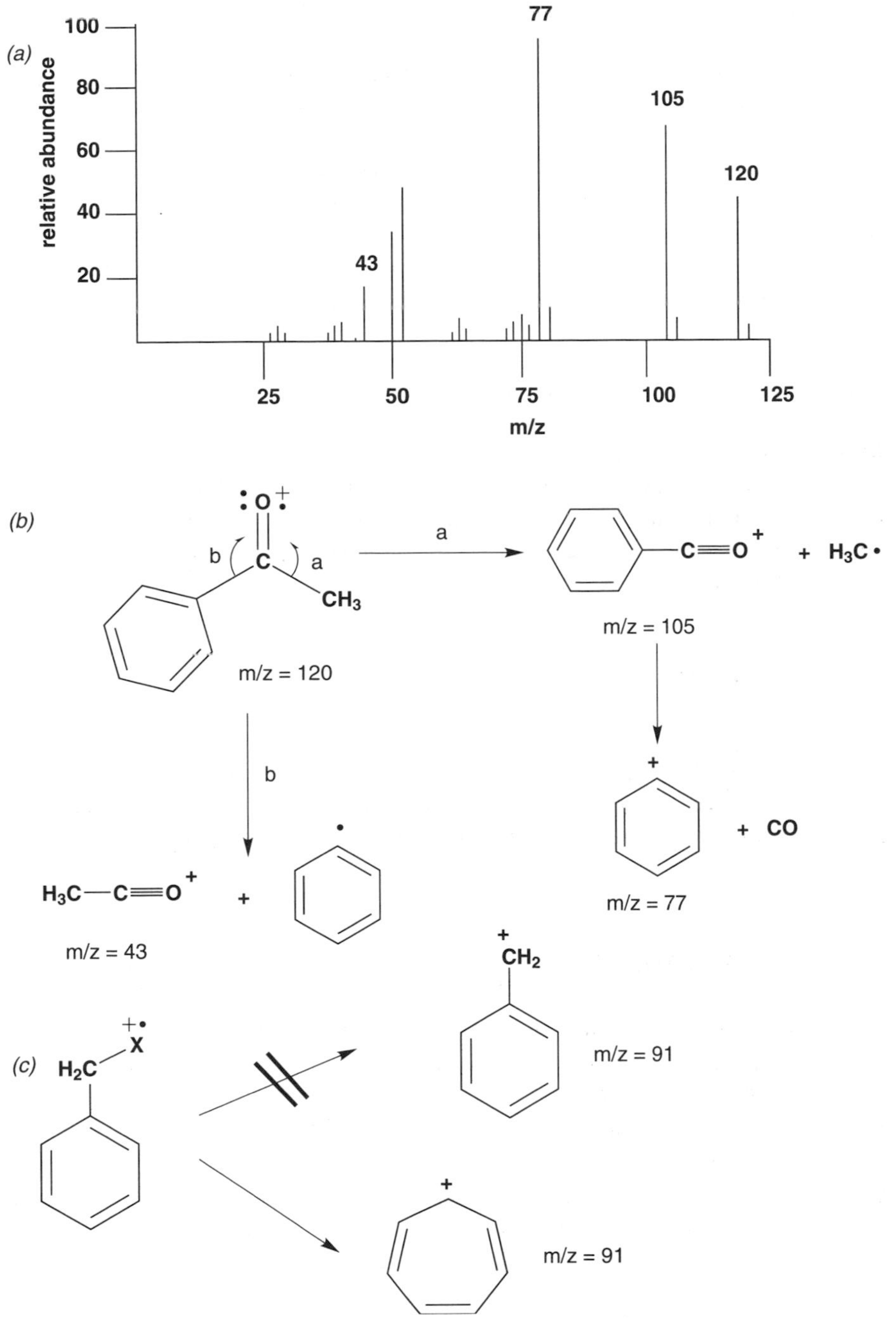

Figure 17.42. *(a)* The mass spectrum of C_8H_8O; *(b)* the fragmentation pathways leading to the ions with $m/z = 105$, 77, and 43; *(c)* the correct structure of the ion with $m/z = 91$.

unknown with the mass spectrum of authentic acetophenone taken under the same conditions.

It is important to realize, however, that *m*/*z* information really only gives composition information; structural information must be inferred. There are many cases where the "obvious" structure for an ion is incorrect, because substantial rearrangements can take place under the high-energy conditions associated with electron impact in the gas phase. For example, compounds possessing a benzyl group ($PhCH_2$–) often exhibit a large peak at *m*/*z* 91. This was once assumed to be the resonance-stabilized benzyl cation, but eventually it was shown that the correct structure is the tropylium ion (Fig. 17.42*c*).

17.43 MASS SPEC/MASS SPEC (MS/MS)

MS/MS, also known as MS^2, consists of a sequence of four steps: (1) ionization of the neutral molecule, (2) using the first mass spectrometer for selecting one particular *m*/*z*, (3) fragmentation of that ion via collision with inert gas atoms, and (4) finally mass analysis of the resulting product ions. Instruments that do MS^2 are presently available. These instruments can repeat the above procedure *n* times.

17.44 PROTEOMICS

The study of proteins in a mixture, cell, tissue, or an organism. Mass spectrometry plays a pivotal role in the emerging field since it provides the most effective means of determining the sequence of amino acids in the protein. Typically, due to the very high molecular weight of most proteins, this analysis is done by using multiply charged ions.

SUGGESTED READING

Cech, N. B. and Enke, C. G. *Mass Spectrom. Rev. 20*, 362 (2001).

DeHoffmann, E., Charette, J., and Stroobant, V. *Mass Spectrometry, Principles and Applications.* John Wiley & Sons: New York, 1996.

Hillenkamp, F. and Karas, M. *Anal. Chem. 63*, A1193 (1991).

Marshall, A. G., Hendrickson, C. L., and Jackson, G. S. *Mass Spectrom. Rev. 17*, 1 (1998).

Silverstein, R. M. and Webster, F. X. *Spectrometric Identification of Organic Compounds*, 6th ed., John Wiley & Sons: New York, 1998.

Smith, R. D., Shen, Y., and Tang, K. *Accts. Chem. Res. 37*, 269 (2004).

18 Electronic Spectroscopy and Photochemistry

18.1	Electromagnetic Radiation	729
18.2	Phase of Waves	729
18.3	Plane-Polarized Light	730
18.4	Circularly Polarized Light	731
18.5	Elliptically Polarized Light	732
18.6	Photon (Quantum)	734
18.7	Visible Region	734
18.8	Ultraviolet Region	734
18.9	Near Ultraviolet Region	734
18.10	Far Ultraviolet Region	734
18.11	Vacuum Ultraviolet Region	734
18.12	Ultraviolet A Region (UVA)	735
18.13	Ultraviolet B Region (UVB)	735
18.14	Ultraviolet C Region (UVC)	735
18.15	Light Intensity I	735
18.16	Einstein	735
18.17	Refractive Index n	735
18.18	Dispersion	737
18.19	Normal Dispersion	737
18.20	Anomalous Dispersion	737
18.21	Optical Rotatory Dispersion (ORD)	737
18.22	Cotton Effect	738
18.23	Octant Rule	739
18.24	Rotational Strength R_j	740
18.25	Brewster's Angle θ_B	741
18.26	Birefringence	742
18.27	Kerr Effect	743
18.28	Pöckels Effect	743
18.29	Optical Kerr Effect	743
18.30	Kerr Cell	744
18.31	Pöckels Cell	744
18.32	Self-Focusing of Optical Beams	745
18.33	Polarization Field P	746
18.34	Frequency Doubling (Second Harmonic Generation)	746
18.35	Harmonic Generation	747
18.36	Frequency Mixing	747

The Vocabulary and Concepts of Organic Chemistry, Second Edition, by Milton Orchin, Roger S. Macomber, Allan Pinhas, and R. Marshall Wilson

18.37	Refractive Index Matching	747
18.38	Absorption of Light	748
18.39	Electronic Absorption Spectrum	748
18.40	Beer–Lambert Law	748
18.41	Absorbance A	749
18.42	Optical Density	749
18.43	Molar Absorptivity (Molar Extinction Coefficient ε)	750
18.44	Transmittance T	750
18.45	Absorption Maximum λ_{max}	750
18.46	Absorption Shoulder	750
18.47	Absorption Band	750
18.48	Intensity of an Absorption Band	751
18.49	Half-Band Width $\Delta\nu_{1/2}$	751
18.50	Chromophore	751
18.51	Bathochromic Shift (Red Shift)	751
18.52	Hypsochromic Shift (Blue Shift)	752
18.53	Hyperchromic Shift	752
18.54	Hypochromic Shift	752
18.55	Isosbestic Point	753
18.56	Isoabsorptive Point	753
18.57	Ground State	753
18.58	Morse Curve	754
18.59	Vibrational State	754
18.60	Ground Vibrational State ($\nu = 0$)	755
18.61	Zero-Point Energy	755
18.62	Excited Vibrational State ($\nu > 0$)	755
18.63	Excited Electronic State	755
18.64	Bohr Condition	755
18.65	First Law of Photochemistry	756
18.66	0-0 Transition (0-0 Band)	756
18.67	Vibronic Transition (Vibronic Band)	756
18.68	Hot Transition (Hot Band)	756
18.69	Vibrational Fine Structure	756
18.70	Multiplicity	756
18.71	Singlet Electronic State	757
18.72	Triplet Electronic State	758
18.73	Doublet Electronic State	759
18.74	Manifold of States (Singlet Manifold, Triplet Manifold)	759
18.75	Jablonski Diagram	759
18.76	Allowed Transition	759
18.77	Forbidden Transition	760
18.78	Oscillator Strength f	761
18.79	Dipole Strength D_j	761
18.80	Born–Oppenheimer Approximation	762
18.81	Polarization of Electronic Transitions	762
18.82	Polarization of Absorption and Emission Bands	764
18.83	Selection Rule	764
18.84	Multiplicity (Spin) Selection Rule	764
18.85	Symmetry Selection Rule	765

18.86	Parity Selection Rule	765
18.87	La Porte Selection Rule	765
18.88	Franck–Condon Principle	765
18.89	Predissociation	767
18.90	Photoelectric Effect	767
18.91	π,π^* State	767
18.92	π,π^* Transition	768
18.93	n,π^* State	768
18.94	n,π^* Transition	768
18.95	n,σ^* State	769
18.96	n,σ^* Transition	770
18.97	σ,σ^* State	770
18.98	σ,σ^* Transition	770
18.99	*d,d* State	770
18.100	*d,d* Transition	771
18.101	Charge-Transfer (C-T) Excited States	771
18.102	Charge-Transfer (C-T) Transition	772
18.103	Rydberg Transition	773
18.104	Dimole Transition	773
18.105	Double-Photon Transition	774
18.106	Woodward–Fieser Factors (Rules) for Determining Polyene Absorption	776
18.107	Woodward–Fieser Factors (Rules) for Determining α,β-unsaturated Ketone and Aldehyde Absorption	777
18.108	*B* and *L* Excited States	778
18.109	Radiative Transition	780
18.110	Nonradiative Transition (Radiationless Transition)	780
18.111	Internal Conversion (IC)	780
18.112	Intersystem Crossing (ISC)	780
18.113	Spin-Orbit Coupling	781
18.114	Heavy Atom Effect	783
18.115	Emission	784
18.116	Spontaneous Emission	784
18.117	Stimulated Emission	784
18.118	Einstein's Relations for Adsorption and Emission	784
18.119	Quantum Yield Φ	785
18.120	Absolute Quantum Yield	786
18.121	Relative Quantum Yield	786
18.122	Primary Photochemical Process	786
18.123	Second Law of Photochemistry	787
18.124	Actinometry	787
18.125	Actinometer	787
18.126	Merry-Go-Round Reactor	787
18.127	Observed Radiative Lifetime τ_{obs}	787
18.128	Natural Radiative Lifetime τ	789
18.129	Fluorescence	790
18.130	Phosphorescence	790
18.131	Emission Spectrum	791
18.132	Excitation Spectrum	792
18.133	Delayed Fluorescence	792

18.134 Thermally Activated Delayed Fluorescence 793
18.135 Bimolecular Delayed Fluorescence 793
18.136 Stokes Band Shift 794
18.137 Anti-Stokes Band Shift 794
18.138 Förster Cycle 795
18.139 Laser (Light Amplification by Stimulated Emission of Radiation) 796
18.140 Coherence of Light 797
18.141 Spatial Coherence of Light 797
18.142 Temporal Coherence of Light 798
18.143 C-W Laser (Continuous-Wave Laser) 798
18.144 Pulsed Laser 798
18.145 Laser Mode 799
18.146 Normal Mode Laser 799
18.147 Q-Switched Laser 799
18.148 Mode-Locked Laser 800
18.149 Dye Laser 802
18.150 Flash Photolysis (Laser Flash Photolysis (LFP)) 802
18.151 Chemiluminescence (Biochemiluminescence) 803
18.152 Electrochemiluminescence 803
18.153 Photochromism 805
18.154 Transfer of Electronic Energy (Excitation Transfer) 806
18.155 Radiative Energy Transfer 806
18.156 Radiationless Energy Transfer (Coupled or Resonent Transfer) 806
18.157 Energy Transfer via Dipole–Dipole Interaction 808
18.158 The Förster Relationship 808
18.159 Energy Transfer via Exchange Interaction 809
18.160 Wigner's Spin Rule 810
18.161 Photosensitization 810
18.162 Quenching of Excited States 812
18.163 Stern–Volmer Relationship 813
18.164 Diffusion-Controlled Reaction 814
18.165 Vertical (Spectroscopic) Transition 814
18.166 Nonvertical Transition 814
18.167 Exciton 815
18.168 Excimer 816
18.169 Exciplex 816
18.170 Photoelectron Transfer (PET) 817
18.171 Rehm–Weller Equation 820
18.172 Inverted Marcus Region 820
18.173 Ellipticity θ 822
18.174 Circular Dichroism (CD) 822
18.175 Faraday Effect 824
18.176 Magnetic Optical Rotatory Dispersion (MORD) 825
18.177 Magnetic Circular Dichroism (MCD) 825
18.178 Ionization Energy (Potential) I_p 827
18.179 Vertical Ionization Energy (Potential) I_{pv} 828
18.180 Photoelectron Spectroscopy (PES) 828
18.181 Electron Spectroscopy for Chemical Analysis (ESCA) 829
18.182 Acousto-Optic Spectroscopy 830

Electronic spectroscopy has been one of the traditional methods for acquiring structural information. In recent years, methods that provide more detailed structural information have replaced electronic spectroscopy for many applications. However, electronic spectroscopy remains the primary tool for the study of rapid chemical processes and for acquiring information about the nature of excited states and other transient species. Thus, the absorption of light (electronic spectroscopy) not only provides entry to the excited state, from which all photochemistry originates, but it also provides information about the structure and behavior of excited states. Because all molecules have many excited states, and each of these has its own characteristic chemistry, photochemistry might be regarded as constituting more than half of all chemistry. Although this is surely a somewhat simplistic view of the chemical world, photochemistry has become a viable area of specialization over the past 40 years. For years, the absorption and emission of light have been used passively to observe molecules and their behavior. Today, the photochemical processes that arise from those absorption processes are being increasingly applied in the manipulation of molecules. The spatial and temporal control provided by the photoinitiation of reactions is uniquely suited for many modern applications that are conducted on a micro- or nanoscale. Consequently, in this chapter, the frequently interrelated vocabulary of electronic spectroscopy and that of photochemistry are developed in coordination with each other.

18.1 ELECTROMAGNETIC RADIATION

Time-dependent electric and magnetic fields oscillating sinusoidally in orthogonal planes and normal to the direction of wave propagation (see Sect. 7.37). Such an electromagnetic wave has a characteristic wavelength λ (Sect. 1.8) and frequency ν (Sect. 1.9), with both the magnetic and electric field components oscillating according to the following equations, where $\mathbf{B}_0$ is the maximum magnetic field strength, $\mathbf{E}_0$ is the maximum electric field strength, and ν is the frequency at which the electromagnetic wave oscillates:

$$\mathbf{B} = \mathbf{B}_0 \sin(2\pi\nu t) \quad \text{and} \quad \mathbf{E} = \mathbf{E}_0 \sin(2\pi\nu t) \tag{18.1}$$

Light is defined loosely as electromagnetic radiation with the appropriate energy for producing electronic absorption (Sect. 7.37). Other forms of electromagnetic radiation include cosmic rays, gamma rays, x-rays, infrared, microwaves, and radiowaves.

18.2 PHASE OF WAVES

The relative displacement of two waves or wave components. The phase of two waves of the same frequency is characterized by the *phase angle*, which is the number of radians or degrees by which one wave leads the other.

Example. The two waves shown in Fig. 18.2*a* are said to be in phase, since they have coincident nodes, phase angle = 0°. The waves in Fig. 18.2*b* are 90° out of phase,

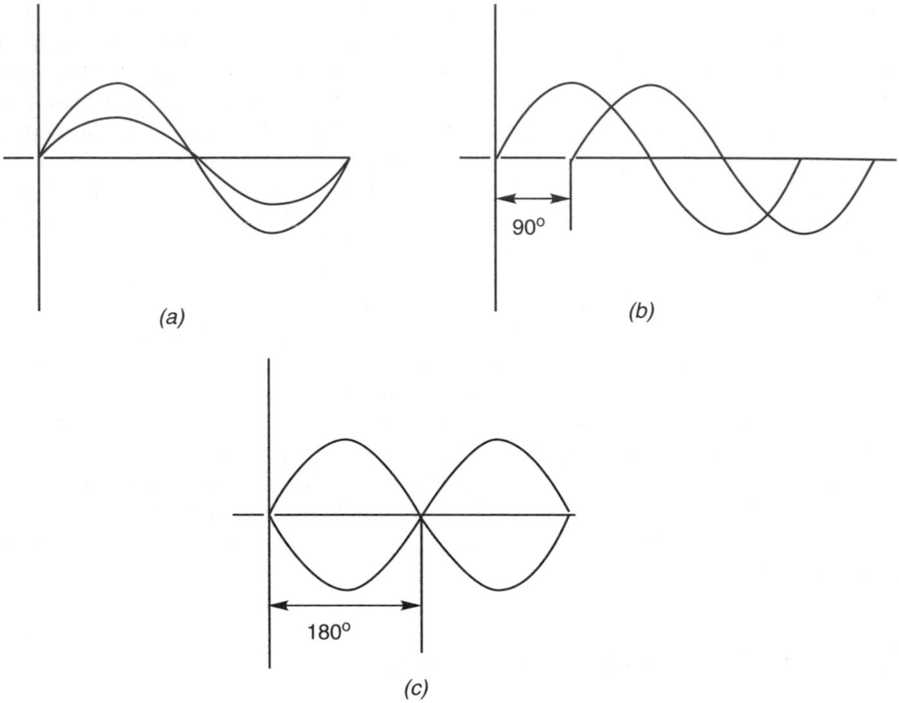

Figure 18.2. The phase of two waves that are (*a*) in phase with a phase angle of 0°, (*b*) 90° out of phase, and (*c*) 180° out of phase.

since the nodes on the positive slopes are displaced from each other by 90°, and those in Fig. 18.2*c* are 180° out of phase.

18.3 PLANE-POLARIZED LIGHT

The form of light in which all the electric field components of the beam are oscillating in parallel planes (see Sect. 7.38).

Example. Any plane-polarized wave is the resultant ($\mathbf{E}_R$) of either two in-phase (phase angle = 0°) plane-polarized components ($\mathbf{E}_{0°}$ and $\mathbf{E}_{90°}$ in Fig. 18.3*a*), or two equally intense circularly polarized components of opposite chirality ($\mathbf{E}_r$ and $\mathbf{E}_l$ in Sect. 18.4). Conversely, unpolarized light may be regarded as being a mixture of plane-polarized components of all possible orientations about the axis of propagation. Unpolarized light may be factored into two equally intense and orthogonal plane-polarized waves by passing through a (Paul) Glan prism (Fig. 18.3*b*).

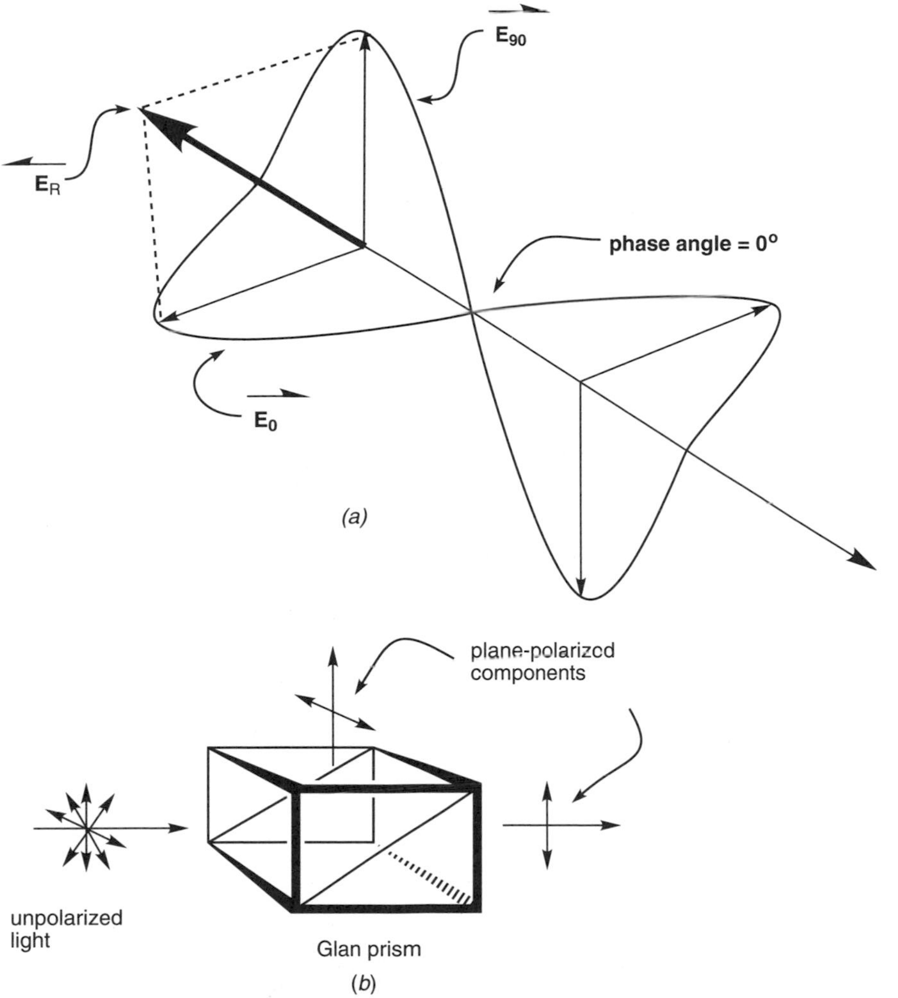

Figure 18.3. (*a*) Factoring a plane-polarized light wave ($\mathbf{E}_R$) into its two, orthogonal plane-polarized components ($\mathbf{E}_0$ and $\mathbf{E}_{90}$). (*b*) Resolution of unpolarized light into its two plane-polarized components with a Glan prism.

18.4 CIRCULARLY POLARIZED LIGHT

The form of light in which the electric field strength remains constant and rotates about the axis of light propagation. The chirality of the light depends on whether the electric field rotates in a clockwise or counterclockwise direction (Fig. 18.4*a*, see Sect. 7.41).

Example. A circularly polarized wave is the resultant of two equally intense plane-polarized waves (**E** and **E**′) polarized at 90° to each other and shifted 90° out of phase with each other (Fig. 18.4*b*). Circularly polarized light may be obtained by passing plane-polarized light through an oriented quartz crystal of the appropriate thickness so as to retard one of the plane-polarized components shown in Fig. 18.3*a* by 1/4 wavelength (90° phase shift). Quartz devices used for this purpose are called 1/4-wave plates.

Conversely, as indicated in Sect. 18.3, two coaxial and equally intense circularly polarized waves of opposite chirality combine to yield a resultant plane-polarized wave, as shown in Fig. 18.4*c*.

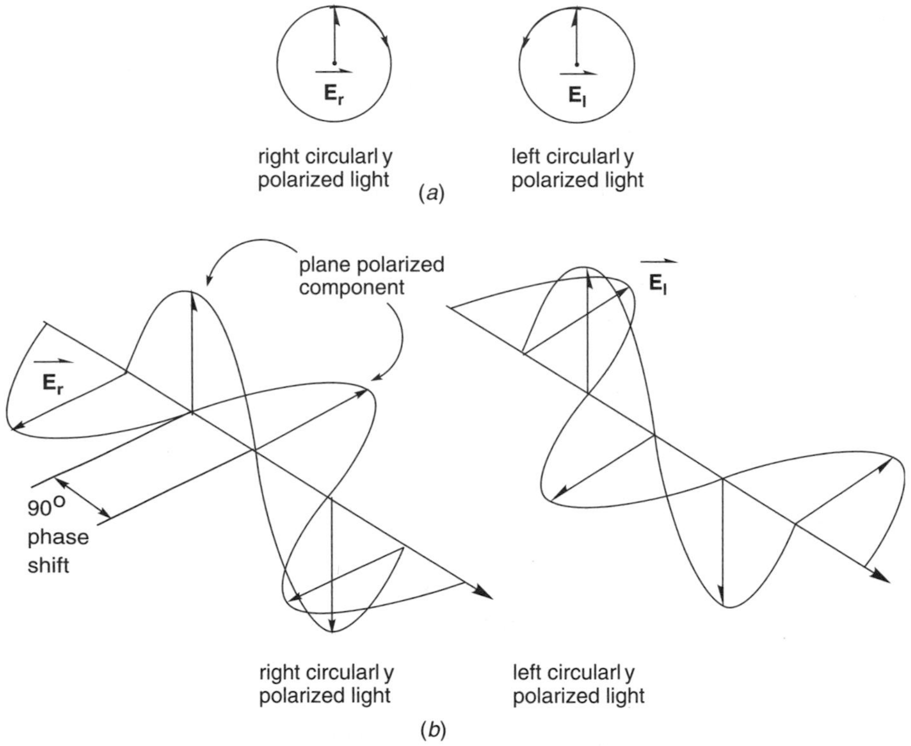

Figure 18.4. (*a*) Electric field rotation of the two forms of circularly polarized light as seen when viewed along the axis of propagation. (*b*) Relationships between the plane-polarized components and the chirality of right and left circularly polarized light. (*c*) [next page] Relationship between right and left circularly polarized components when viewed along the axes of propagation of the circularly polarized waves and the resultant plane-polarized wave.

18.5 ELLIPTICALLY POLARIZED LIGHT

Any form of light intermediate between plane- and circularly polarized light, that is, any light with a phase angle between the plane-polarized components of >0° and <90°. In an elliptically polarized wave, the electric field varies in intensity, as well as rotates about the axis of propagation (Fig. 18.5, see Sect. 7.43).

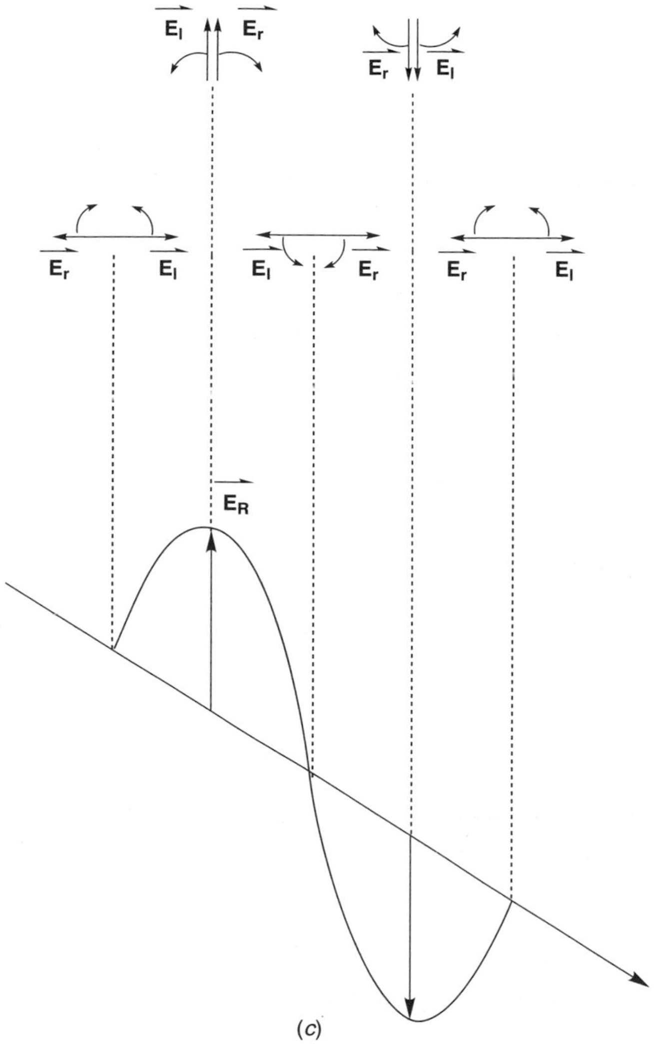

(c)

Figure 18.4. (Contd.)

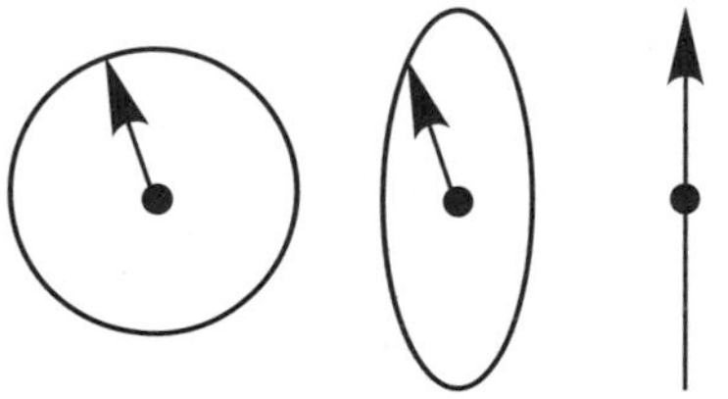

(a) (b) (c)

Figure 18.5. Electric field behavior of the various polarization forms of light: (*a*) circularly polarized light; (*b*) elliptically polarized light; (*c*) plane polarized light.

18.6 PHOTON (QUANTUM)

The most elementary unit of electromagnetic radiation. Associated with each photon is a discrete quantum of energy E and momentum p, which depend on the frequency ν of the particular photon (see Sect. 1.1):

$$E = h\nu \quad (18.6a)$$

$$E\ (\text{kcal/mol}) = 28.653 \times 10^3/\lambda\ (\text{nm}) = 2.786 \times 10^{-3}\nu\ (\text{cm}^{-1}) \quad (18.6b)$$

$$p = h\nu/c = h/\lambda \quad (18.6c)$$

where λ is the wavelength of light and c the speed of light in a vacuum, 3×10^8 m/sec. Planck's constant has the value of 6.6256×10^{-27} J-sec.

18.7 VISIBLE REGION

The region of the electromagnetic spectrum extending from about 400 to 800 nm (see Fig. 16.2).

18.8 ULTRAVIOLET REGION

The region of the electromagnetic spectrum extending from 10 to about 400 nm. This region of the spectrum is partitioned by photochemists into the three regions defined in Sect. 18.9 through 18.11, and by photobiologists into the three regions defined in Sect. 18.12 through 18.14 (see Fig. 16.2).

18.9 NEAR ULTRAVIOLET REGION

The region of the ultraviolet spectrum lying nearest to the visible region of the spectrum, and extending from about 200 to 400 nm (see Fig. 16.2).

18.10 FAR ULTRAVIOLET REGION

The region of the ultraviolet spectrum lying farthest from the visible region of the spectrum, and extending from 10 to 200 nm (see Fig. 16.2).

18.11 VACUUM ULTRAVIOLET REGION

This term is synonymous with the far ultraviolet region (Sect. 18.10). Because simple gases such as oxygen and nitrogen, as well as all organic compounds, absorb below about 200 nm, light below this wavelength must be manipulated in a vacuum (see Fig. 16.2).

18.12 ULTRAVIOLET A REGION (UVA)

The region of the ultraviolet spectrum with wavelengths below 290 nm. Wavelengths in this region are absorbed by proteins and nucleic acid, and thus, they are detrimental to life and described as abiotic.

18.13 ULTRAVIOLET B REGION (UVB)

The region of the ultraviolet spectrum with wavelengths between 290 and 330 nm. Wavelengths in this region cause pigmentation of the skin and produce vitamin D. These wavelengths are thus beneficial and described as biotic.

18.14 ULTRAVIOLET C REGION (UVC)

The region of the ultraviolet spectrum with wavelengths between 330 nm and the onset of the visible region at 400 nm (Sect. 18.7). Wavelengths in this region are absorbed principally by biological pigments such as carotinoids, flavins, and porphyrins.

18.15 LIGHT INTENSITY *I*

The brightness of light is referred to as its intensity I. In terms of the wave model, the intensity is proportional to the square of the maximum electric field strength $\mathbf{E}_0$ ($I \propto \mathbf{E}_0^2$). In the photon model, intensity is a measure of the photon flux density ($I = \text{photons/cm}^2\ \text{sec}$).

18.16 EINSTEIN

One mole of photons (6.03×10^{23} photons) is called an Einstein.

Example. The Einstein provides a photochemically convenient measure of light intensity. Einstein/cm^2-sec provides a measure of the number of photons that have impinged on a sample, and can readily be compared to the moles of starting material transformed into photochemical products in order to determine quantum yields of photochemical reactions (see Sect. 18.121).

18.17 REFRACTIVE INDEX *n*

A dimensionless number, which is a measure of the apparent speed of light passing through matter c' relative to the speed of light passing through a vacuum c (see Sect. 10.50):

$$n = c/c' \qquad (18.17a)$$

Example. The refractive index of a substance varies with the frequency of the observing light ν (Fig. 18.17) according to the following expression (also see Fig. 18.20):

$$n = 1 + h/(\nu_0^2 - \nu^2) \tag{18.17b}$$

Where ν_0 is a frequency of light absorbed by the substance and h Planck's constant. Thus, for visible light passing through glass, $\nu_0 >> \nu$, $n > 1$, the apparent speed of light in the glass will be less than that in a vacuum, $c > c'$. In general, refractive indices of organic substances have values >1, indicating that light travels more slowly through matter than through a vacuum. Furthermore, the more polarizable the electrons of the matter through which the light travels are (see Sect. 18.33), the greater the apparent retardation. Thus, molecules with more loosely bound π electrons and heavier elements tend to have larger refractive indices, as shown in Table 18.17.

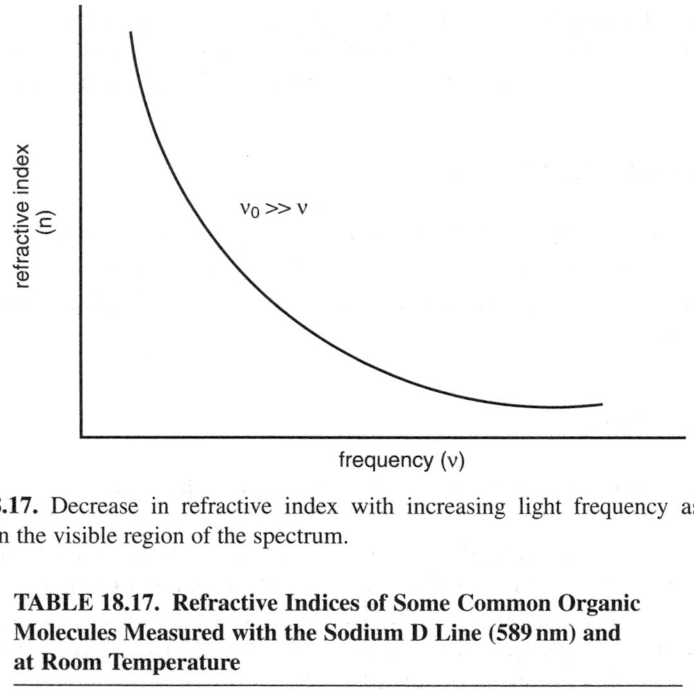

Figure 18.17. Decrease in refractive index with increasing light frequency as typically observed in the visible region of the spectrum.

TABLE 18.17. Refractive Indices of Some Common Organic Molecules Measured with the Sodium D Line (589 nm) and at Room Temperature

Substance (Formula)	Refractive Index
Hexane (C_6H_{14})	1.3757
Dichloromethane (CH_2Cl_2)	1.4237
Chloroform ($CHCl_3$)	1.4464
Carbon tetrachloride (CCl_4)	1.4631
Benzene (C_6H_6)	1.5011
Methyl iodide (CH_3I)	1.5293
Carbon disulfide (CS_2)	1.6295

However, for X-rays passing through graphite, $\nu_0 << \nu$, $n < 1$, the apparent speed of light in the graphite is faster than that in a vacuum, $c < c'$. These apparent violations of the universal constant c are easily reconciled when one realizes that the light that impinges on the matter is attenuated as it passes through the matter. Simultaneously, a new light wave of the same frequency is generated that has undergone a phase shift relative to the impinging light. If a delay in phase occurs, the light will appear to have traveled through the matter more slowly than through a vacuum, but if an advance in phase occurs, the light will appear to have traveled through the matter more rapidly than through a vacuum (see Sect. 18.33). This retardation or advancement in the phase of an observing light beam is referred to as its *phase velocity*.

18.18 DISPERSION

The change in refractive index of a substance as a function of the wavelength of light. This term originated with the observation of the dispersal of the colors of the visible spectrum when "white" light is passed through a glass prism.

18.19 NORMAL DISPERSION

The dispersion relationship observed when the frequency of the observing light is far removed from the frequencies of light characteristically absorbed by the substance through which the light is passing. In Eq. 18.17b, $\nu_0 >> \nu$ or $\nu_0 << \nu$ (Fig. 18.17).

18.20 ANOMALOUS DISPERSION

The dispersion relationship observed when the frequency of the observing light is scanned through an absorption band of the substance under examination (Fig. 18.20). Equation 18.17b is not valid in spectral regions where $\nu_0 \approx \nu$.

18.21 OPTICAL ROTATORY DISPERSION (ORD)

The relationship between the specific rotation $[\alpha]$ of an optically active substance and the wavelength of the observing plane-polarized light.

Example. Plane-polarized light is the resultant of two circularly polarized components of opposite chirality (Fig. 18.6*c*). The refractive indices for these two circularly polarized components (n_{RCP} and n_{LCP}) are not equal in optically active media. Thus, upon passing through an optically active medium, one circularly polarized form will be retarded relative to the other. As shown in Fig. 18.21, this retardation will produce a rotation in the plane of polarization of the plane-polarized resultant. Consequently, an ORD curve provides a measure of the differences in the dispersions of the right and left circularly polarized light.

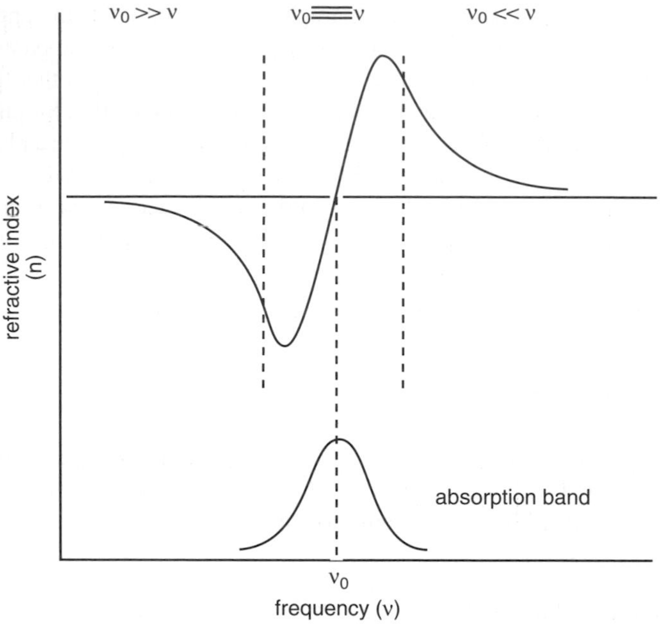

Figure 18.20. Anomalous dispersion occurs in the region of absorption bands. Outside of this region, normal dispersion is observed, as shown in Fig. 18.17.

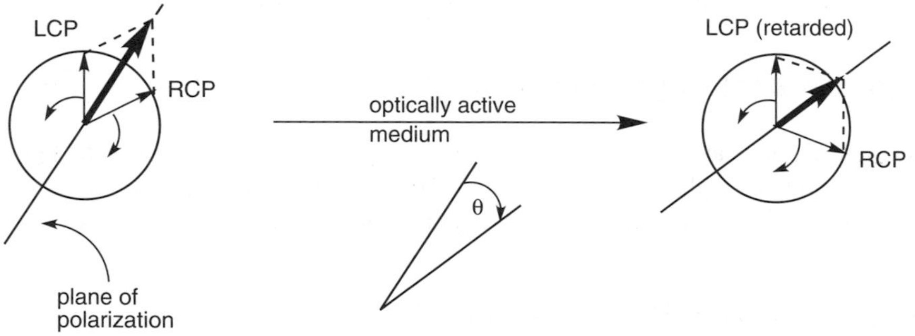

Figure 18.21. The rotation of plane-polarized light by the angle θ upon passing through an optically active medium results from the relative retardation of one of the circularly polarized components (retardation of the left circularly polarized component shown here).

18.22 COTTON EFFECT

The anomalous dispersion observed in the ORD curve of a substance when the wavelength of the plane-polarized light is scanned through an absorption band.

Example. Cotton effect curves have the same form as anomalous dispersion curves found in refractive index measurements with unpolarized light. A Cotton effect curve

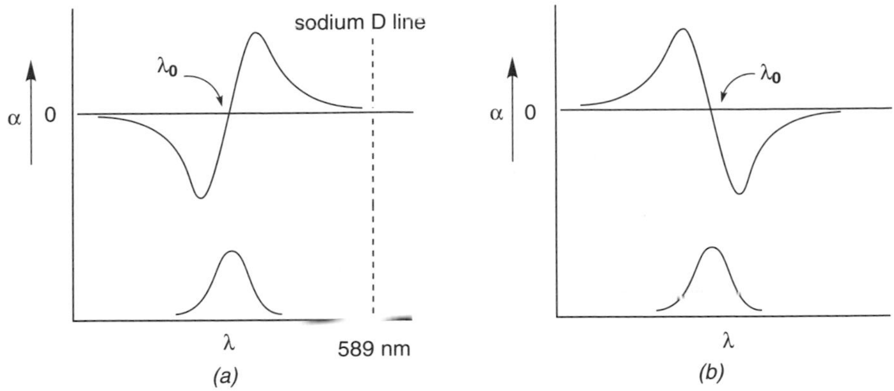

Figure 18.22. (*a*) Positive and (*b*) negative Cotton effects in optical rotatory dispersion showing the relationship between specific rotation, $[\alpha]_D$, and absorption band.

may have either a positive or negative sense (Fig. 18.22), and is characterized by λ_0, and the sign and magnitude of the long-wavelength maximum or minimum. As shown in Fig. 18.22*a* a positive Cotton effect exhibits a positive maximum to the long-wavelength side of λ_0, while a negative Cotton effect (Fig 18.22*b*) exhibits a negative minimum to the long-wavelength side of λ_0. It is interesting to note that the classical specific rotation at the sodium D line, $[\alpha]_D$, which has been used for years to characterize optically active organic substances, is but a single point on the ORD curve of the substance.

18.23 OCTANT RULE

An empirical model used to predict the sign and magnitude of the Cotton effect for optically active ketones.

Example. The space surrounding the carbonyl group is divided into eight regions, octants (Fig. 18.23*a*). With rare exception, the sign of the contribution of each substituent to the observed Cotton effect is determined by the region of space in which that substituent occurs. The algebraic sum of the contributions of the substituents determines the sign and magnitude of the Cotton effect curve. In Fig. 18.23*b*, the carbon skeleton of (+)-*trans*-3-*t*-butyl-4-methylcyclohexanone and the signs of the rear octants are projected onto a plane. The contributions of carbon atoms 3 and 5, and 2 and 6 cancel each other. The contributions of carbon atoms 1,4 and the methyl group are negligible, since they are on one of the boundary planes. Therefore, the contribution of the *t*-butyl group, which is positive, will determine the sign of the Cotton effect of the entire molecule. In fact, since carbonyl substituents rarely occur in the front octants, the signs of the rear octants almost always determine the signs of rotation of chiral carbonyl compounds.

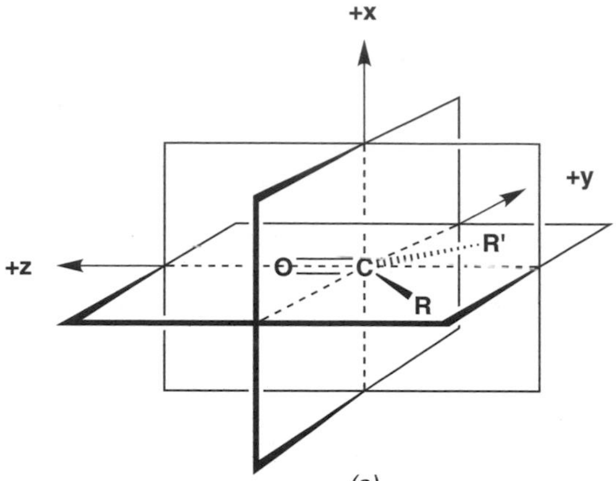

(a)

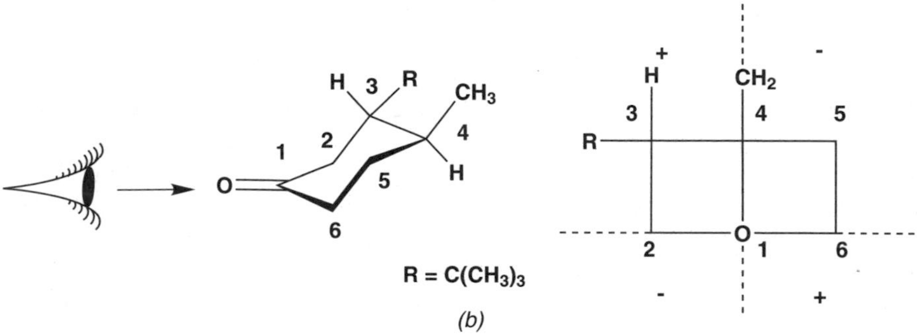

(b)

Figure 18.23. (*a*) The eight regions of space surrounding a carbonyl group. The predicted signs of rotational contributions of substituents in those regions of space are $(+x, +y, +z) = +$, $(-x, +y, +z) = -$, $(-x, -y, +z) = +$, $(+x, -y, +z) = -$, $(+x, -y, -z) = +$, $(+x, +y, -z) = -$, $(-x, +y, -z) = +$, $(-x, -y, -z) = -$. (*b*) Projection of the structure of (+)-*trans*-3-*t*-butyl-4-methylcyclohexanone on the four rear octants.

18.24 ROTATIONAL STRENGTH R_j

The parameter that relates the degree of optical rotation for a particular molecule to the quantum mechanical properties of that molecule (see Sect. 18.79).

Example. The rotational strength R_j may be calculated from the optical rotation of a substance (ϕ degrees/unit path length) with the following relationship:

$$\phi = (52.8N/hc) \sum_j [R_j \nu^2/(\nu_j^2 - \nu^2)]$$

where N = the number of molecules/unit volume

h = Planck's constant

c = the speed of light

ν = the frequency of the observing light

The rotational contributions of each of the jth electronic transitions at frequencies ν_j are then summed over the electronic absorption bands of the substance. For the theoretical determination of R_j, see Sect. 18.79.

18.25 BREWSTER'S ANGLE θ_B

The angle of incidence θ_B of an unpolarized beam of light upon an interface, which gives rise to the maximum intensity in the plane-polarized reflected beam (Fig. 18.25). If we assume that one of the interface constituents is a vacuum or gas, Brewster's angle will be

$$\theta_B = \tan^{-1} n$$

where n is the refractive index of the solid or liquid interface constituent named after Sir David Breuster (1781–1868).

Example. The reflected beam will always be exclusively plane-polarized in the plane normal to the plane of incidence (Fig. 18.25). Consequently, Brewster's angle

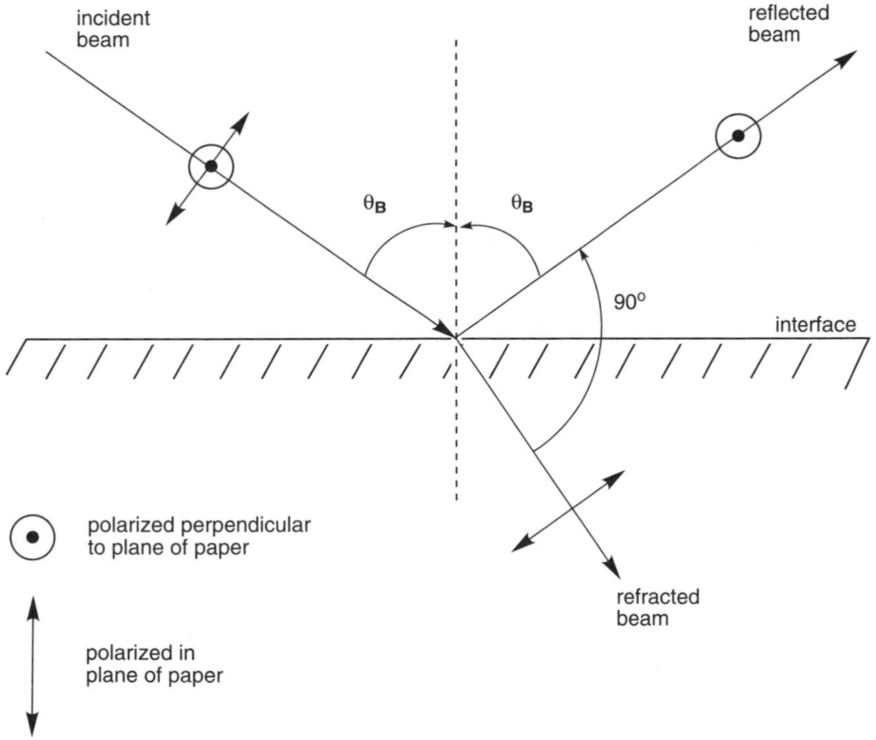

Figure 18.25. Reflection of a beam of unpolarized light from a surface: The reflected beam is polarized perpendicular to the plane of incidence and the refracted beam is polarized in the plane of incidence.

can be employed to minimize reflection losses when manipulating plane-polarized light or to produce plane-polarized light from unpolarized light.

18.26 BIREFRINGENCE

A property of all anisotropic forms of matter (those having some degree of molecular order) arising from the occurrence of two refractive indices, $n_{\parallel}$ and $n_{\perp}$ in Fig. 18.26, one for each of the two mutually perpendicular components of plane-polarized light. This property is manifested in a variety of ways, but is most frequently observed as an alteration in the forms of light polarization upon passing through such an anisotropic medium.

Example. Birefringence is responsible for the beautiful colors seen when most crystals or liquid crystals are viewed under a polarizing microscope. It is also the property on which many optical devices are based. For instance, Glan prisms, which are used to produce plane-polarized light from unpolarized light, as well as retardation plates and compensators, which are used to interchange all polarization forms of light, plane ⇆ elliptical ⇆ circular, all make use of birefringence. This latter application is described in greater detail in Sect. 18.4 for a 1/4-wave retardation plate. Thus, if one

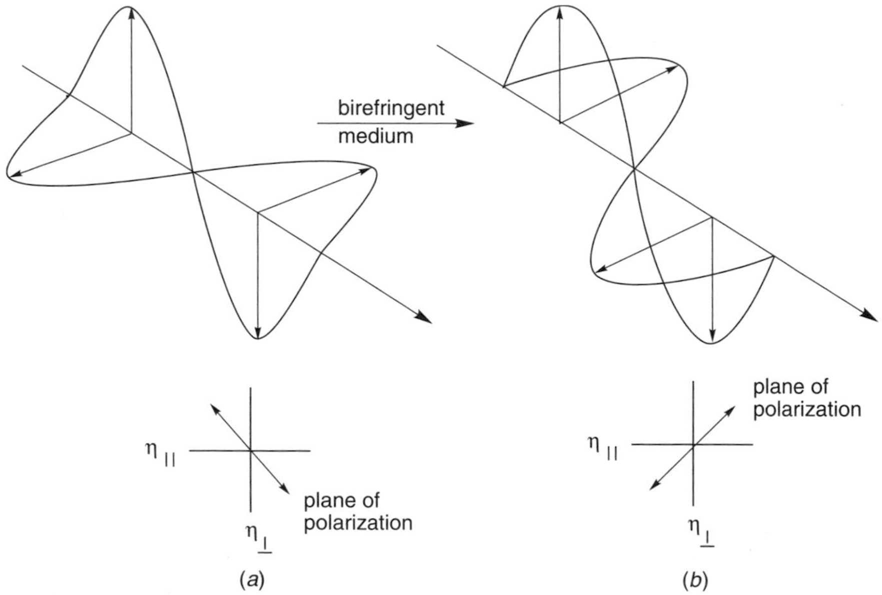

Figure 18.26. Rotation of plane-polarized light by 90° upon passing through a birefringent material: (*a*) impinging light beam and (*b*) transmitted light beam (180° phase shift, 90° rotation in the plane of polarization).

starts with plane-polarized light oriented at 45° with respect to the $n_{\parallel}$ and $n_{\perp}$ axes as shown in Fig. 18.26, one may alter the polarization form as follows:

Phase Shift	Form of Light Produced
90°	Right circularly polarized
180°	Plane-polarized (rotated 90°)
270°	Left circularly polarized
360°	Plane-polarized (rotated 180°)
All intermediate angles	Elliptically polarized

It should be noted that birefringence, while it can result in the rotation of plane-polarized light, differs from optical activity (Sect. 18.21) in that all possible polarization forms can be formed as emphasized above. In optical activity, on the other hand, only rotated forms of plane-polarized light are produced.

18.27 KERR EFFECT

The induction of birefringence in an isotropic medium such as a liquid, amorphous solid, or centrosymmetric crystal by the application of a strong electric field named after Reverend John Kern (1819–1902).

Example. In the Kerr effect, the change in the refractive index depends on the square of the electric field strength: $\Delta n \propto \mathbf{E}^2$. For liquids, the Kerr effect is associated with a very rapid ($<10^{-11}$ sec) electric-field-induced orientation of the molecules in the medium.

18.28 PÖCKELS EFFECT

Similar to the Kerr effect except that it occurs only with crystals lacking inversion symmetry, and that the change in refractive index displays a linear dependence on the electric field strength: $\Delta n \propto \mathbf{E}$ named after Friedrich Pöckels (1865–1913).

18.29 OPTICAL KERR EFFECT

The induction of birefringence in a liquid by the application of an intense plane-polarized laser light beam.

Example. This phenomenon gives rise to most of the unusual effects observed with high-intensity laser light (see Sect. 18.32). At light intensities greater than about 10^6 W/cm^2 (10^6 J/cm^2 sec), a direct electric field component is produced as the light passes through matter (see Sect. 18.34). This direct electric field gives rise to the requisite molecular orientation and the associated optical Kerr effect.

18.30 KERR CELL

A device that employs the Kerr effect to alter the polarization form of light.

Example. A polar medium, usually nitrobenzene, is placed between two electrodes. Upon application of an electric field, two refractive indices are generated: one for light polarized parallel to the electric field $n_{\parallel}$, and the other for light polarized perpendicular to the electric field $n_{\perp}$. If plane-polarized light is passed through the cell as shown in Fig. 18.26, any form of polarized light desired can be obtained by simply adjusting the electric field strength to achieve the appropriate phase shift. In this way, a Kerr cell can serve to manipulate a plane-polarized light beam as either an optical modulator or a very fast shutter (Fig. 18.30).

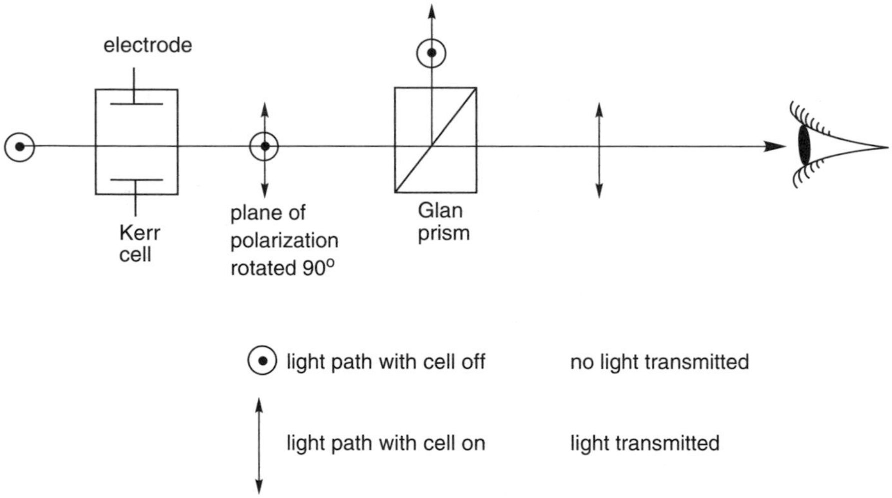

Figure 18.30. Use of a Kerr cell as an ultra-fast optical shutter.

18.31 PÖCKELS CELL

A device, which employs the Pöckels effect (Sect. 18.28), to alter the polarization form of light is called a Pöckels cell.

Example. A Pöckels cell functions in the same fashion as a Kerr cell (Sect. 18.30), except that a crystal, usually potassium dihydrogen phosphate (KH_2PO_4), is the active medium.

18.32 SELF-FOCUSING OF OPTICAL BEAMS

The collapse of an intense beam of laser light into small filaments of extremely high-intensity light upon passage through a solid or liquid medium.

Example. This phenomenon is produced by the optical Kerr effect (Sect. 18.29). The refractive index of a medium n is a function of light intensity $\mathbf{E}^2$:

$$n = n_0 + n_2\mathbf{E}^2 \tag{18.32a}$$

where n_0 is the normal refractive index and n_2 the second-order refractive index coefficient. At high light intensities (~10^6 W/cm^2), the second-order term becomes significant, and the refractive index within the light beam ($n_0 + n_2\mathbf{E}^2$) becomes appreciably greater than that of the surrounding medium n_0. This refractive index gradient across the beam cross section produces a lens effect, just as if a converging lens had been placed in the beams path, and the beam converges. In more quantitative terms, if the *beam divergence* θ_D is equal to or less than the *critical angle* θ_C, self-focusing becomes possible (Fig. 18.32):

$$\theta_D = 1.22\lambda/n_0D \tag{18.32b}$$

$$\theta_C = \cos^{-1}[n_0/(n_0 + n_2\mathbf{E}^2)] \tag{18.32c}$$

Condition for self-focusing: $\theta_D \leq \theta_C$

where θ_C is the maximum angle at which a beam can impinge on an interface and undergo total internal refraction, λ the wavelength of light, and D the beam diameter. It is the extremely high intensity of light produced by self-focusing that gives rise to many of the interesting effects of laser light.

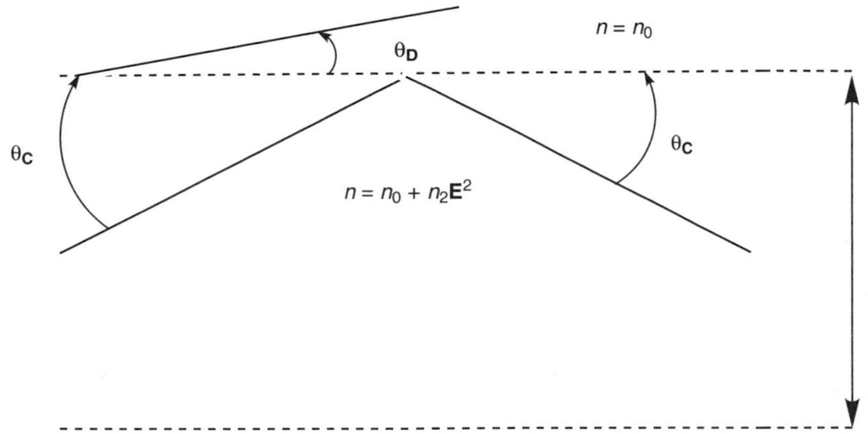

Figure 18.32. Relationship of the beam divergence and critical angle under conditions of self-focusing.

18.33 POLARIZATION FIELD *P*

The field induced in the electron clouds of an array of molecules by the oscillating electric field **E** of light. The first two terms of this field have the following form:

$$\mathbf{P} = \chi\mathbf{E}(1 + \alpha_2\mathbf{E} + \cdots)$$

where **E** is the electric field of the light, χ the polarizability of the molecules involved, and α_2 the second-order nonlinear coefficient (see Sect. 16.64 where alternate symbols are used).

Example. The electric field of light **E** polarizes the electron cloud of a molecule, inducing a dipole or field **P** by displacing the electrons to one side of the nucleus (Fig. 18.33). In the classical framework, this oscillating dipole **P** will give rise to a new light field **E**′. In this way, light propagates through matter, $\mathbf{E} \rightarrow \mathbf{P} \rightarrow \mathbf{E}'$. Thus, the wave forms of the impinging and transmitted light are coupled by the polarization field. It is through an understanding of the relationship of these fields that one can account for such phenomena as *phase velocity* (Sect. 18.17), *harmonic generation* (Sect. 18.35), and *frequency mixing* (Sect. 18.36).

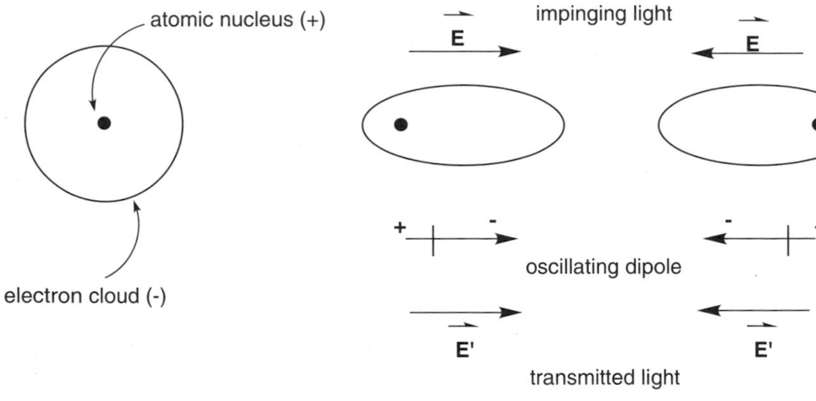

Figure 18.33. Polarization model for the transmission of light through matter.

18.34 FREQUENCY DOUBLING (SECOND HARMONIC GENERATION)

The process by which coherent light (Sects. 18.140–18.142) of one wavelength (λ_1) is passed through a mixing medium and converted into light (λ_2) of one-half the original wavelength ($\lambda_2 = \lambda_1/2$) or twice the original frequency ($\nu_2 = 2\nu_1$). Also known as *second harmonic generation*.

Example. At high light intensities, the second-order term in the polarization field expression (Sect. 18.33) becomes significant. Substitution of the wave form of the primary light ($\mathbf{E} = \mathbf{E}_0 \sin 2\pi\nu_1 t$) into this expression gives the form of the resulting polarization field and of the transmitted light as follows:

$$\mathbf{P} = \chi\mathbf{E}_0 \sin 2\pi\nu_1 t + 1/2\chi\alpha_2\mathbf{E}_0^2 - 1/2\chi\alpha_2\mathbf{E}_0{}^2 \cos 4\pi\nu_1 t \qquad (18.34)$$

The first term of this expression ($\chi\mathbf{E}_0 \sin \nu_1 t$) is associated with transmitted light of the original or primary frequency (ν_1). The second term ($1/2\chi\alpha_2\mathbf{E}_0^2$) is associated with a nonoscillating, direct electric field, which is the source of the optical Kerr effect (Sect. 18.29) caused by high light intensities. Finally, the third term ($1/2\chi\alpha_2\mathbf{E}_0^2 \cos 2\nu_1 t$) is associated with a new light wave of twice the frequency of the original light wave ($2\nu_1$). This new light wave is called the second harmonic. It is the application of frequency conversion that makes lasers one of the most versatile of light sources.

18.35 HARMONIC GENERATION

The general process of combining two or more photons of the same frequency in a mixing medium to form new photons of higher frequency.

Example.

$$n(h\nu_1) \rightarrow h\nu_2, \quad \text{where } \nu_2 = n\nu_1$$

$$n = 2, \quad \text{second harmonic generation}$$

$$n = 3, \quad \text{third harmonic generation}$$

The same considerations apply here as in frequency doubling (Sect. 18.34).

18.36 FREQUENCY MIXING

The most general situation in which two or more photons of any frequency interact with one another in a mixing medium to generate new photons of either higher or lower frequency:

$$h\nu_1 + h\nu_2 \rightarrow h\nu_3 + h\nu_4$$

The frequency up-conversion process ($\nu_4 = 0$, $\nu_3 = \nu_1 + \nu_2$) is most commonly employed and the same considerations apply here as in frequency doubling (Sect. 18.34).

18.37 REFRACTIVE INDEX MATCHING

One of the most important criterion for selecting a frequency mixing medium. Thus, only media that have the same refractive index for light of the primary frequency or frequencies and harmonic or mixed frequency will give rise to frequency doubling or mixing.

Example. Using frequency doubling as an example (Sect. 18.34), it can be seen that the frequency-doubled light wave components will be spawned all along the path of the primary light wave as it passes through the mixing medium. Under the normal circumstances, light of the doubled frequency will move through the medium more slowly than light of the primary frequency. If this difference in speed or phase velocity is too great, newly generated doubled light wave components will rapidly become 180° out of phase with previously generated light wave components, and the two component waves will annihilate each other. Thus, frequency doubling of light will only be observed in significant intensity if the newly generated, doubled wave components are in phase with those doubled wave components generated previously. In order for this to occur, the primary and doubled waves must be traveling through the medium with the same velocity.

18.38 ABSORPTION OF LIGHT

The irreversible transference of energy from a light beam to the matter through which it is passing. The production of *excited states* (Sect. 18.63) of the molecules involved in the absorption process is the usual consequence of light attenuation. However, certain scattering processes also lead to the attenuation of light and, therefore, have the same effect as absorption processes. For instance, *Raman scattering* leads to the conversion of photon energy into vibrational or thermal energy and *Brillouin scattering* leads to the conversion of photon energy to phonon or acoustic energy.

18.39 ELECTRONIC ABSORPTION SPECTRUM

This is the probability of light absorption by molecules or atoms to form excited electronic states as a function of light wavelength λ or frequency ν; a plot of ε versus λ or ν is given in Fig. 18.39.

18.40 BEER–LAMBERT LAW

The requirement that as a monochromatic light beam passes through a homogeneous, isotropic sample, the incremental attenuation of light ($-dI$) per increment of beam path length (dl) will be proportional to the light intensity and the concentrations of absorbing species (c):

$$-dI/dl \propto cI \quad (18.40a)$$

$$\log(I_0/I) = acl \quad (18.40b)$$

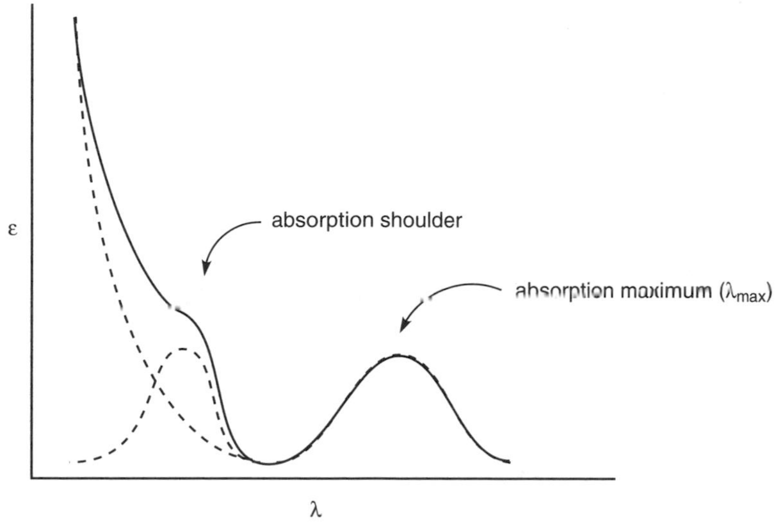

Figure 18.39. An electronic absorption spectrum displayed as a function of molar absorptivity ε (Sect. 18.43) and wavelength of light λ. The broken lines represent contributions to the observed solid line.

where I_0 is the initial or incident light intensity, I the final or transmitted light intensity, l the beam path length through the sample, and a the *absorptivity* or absorbance per unit concentration (mg/mL) and l, the path length named after Johann Lambert (1728–1777) and August Beer (1825–1863).

18.41 ABSORBANCE *A*

The logarithm of the ratio of the incident light intensity I_0 to the transmitted light intensity I in the Beer–Lambert expression (Sect. 18.40):

$$A = \log(I_0/I) = acl \qquad (18.41)$$

Example. For pure liquids, A is determined as indicated above. However, for solutions, the absorbance of the solute is the ratio of the light intensity transmitted through the pure solvent (I_0) to the light intensity transmitted through the solution (I), where both determinations are done with cells having the same dimensions.

18.42 OPTICAL DENSITY

This term is synonymous with absorbance (Sect. 18.41).

18.43 MOLAR ABSORPTIVITY (MOLAR EXTINCTION COEFFICIENT ε)

The proportionality constant ε in the Beer–Lambert relationship (Sect. 18.40) expressed in terms of mol/L and cm, that is, $A = \varepsilon cl$.

18.44 TRANSMITTANCE T

The ratio of the intensities of the light transmitted to the incident light (I/I_0), both measured as indicated in Sect. 18.41. Note that the transmittance is the reciprocal of absorbance. Accordingly, at each point in a spectrum where the absorbance passes through a maximum, the transmittance passes through a minimum.

18.45 ABSORPTION MAXIMUM λ_{max}

Any wavelength of light at which a maximum occurs in an electronic absorption spectrum (Fig. 18.39).

18.46 ABSORPTION SHOULDER

Any wavelength of light at which an inflection point occurs in an electronic absorption spectrum. This feature is indicative of a small absorption band (see broken line in Fig. 18.39) being partially masked by a larger absorption band.

18.47 ABSORPTION BAND

The region of the absorption spectrum in which the absorbance passes through a maximum or a shoulder is called an absorption band (Fig. 18.39).

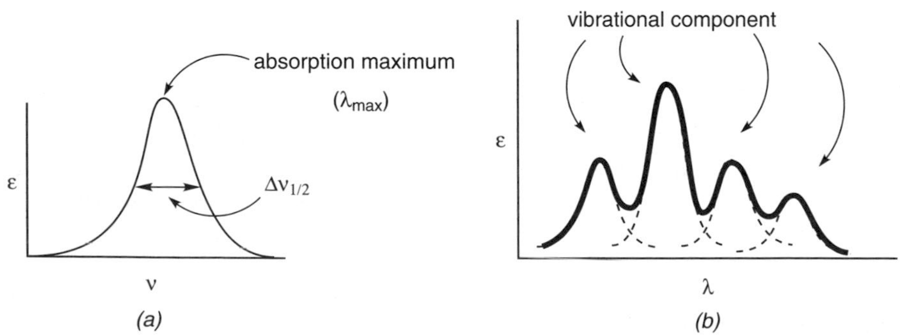

Figure 18.47. (*a*) A single electronic absorption band without vibrational structure. (*b*) A single electronic absorption envelope for a family of vibrational components.

Example. In the strictest sense, an absorption band should be taken to mean a single vibrational component of a single electronic transition (Fig. 18.47*a* and Sect. 18.67). However, this term is often applied to the envelope of vibrational components of a single electronic transition (Fig. 18.47*b*).

18.48 INTENSITY OF AN ABSORPTION BAND

A measure of the probability of an absorption event taking place, the *y*-axis in a plot of ε versus λ. In order to obtain a theoretically meaningful value for the absorption intensity, the area under the absorption curve $\int \varepsilon_v \, d\nu$ of the absorption band or envelope must be evaluated. However, this property is often characterized by the molar absorptivity ε at the absorption maximum λ_{max}.

Example. Absorption intensities range from as high as about $\varepsilon = 10^6$, unit probability, for absorption events that occur every time the molecule collides with a photon to about $\varepsilon = 10^{-3}$ for highly inefficient absorption events such as singlet-triplet absorption (Sect. 18.84).

18.49 HALF-BAND WIDTH $\Delta\nu_{1/2}$

The width of an absorption band ($\Delta\nu$) half-way between the baseline and its absorption maximum ($\Delta\nu$ at $\varepsilon_{max}/2$) (Fig. 18.47*a*) (see also Sect. 16.19).

Example. This parameter is particularly useful in the comparison of the intensities of similar bands in spectra of related compounds, since the area under the absorption curve $\int \varepsilon_v \, d\nu$ is approximately equal to $\varepsilon_{max}\Delta\nu_{1/2}$.

18.50 CHROMOPHORE

The functional group or portion of a molecule that gives rise to an absorption band in the absorption spectrum of that molecule.

Example. The carbonyl group of acetone (CH_3–C(O)–CH_3).

18.51 BATHOCHROMIC SHIFT (RED SHIFT)

The displacement of a particular absorption band to longer wavelength (lower energy). This behavior is usually associated with a change in chromophore substitution or solvent and medium effects (Fig. 18.51*a*).

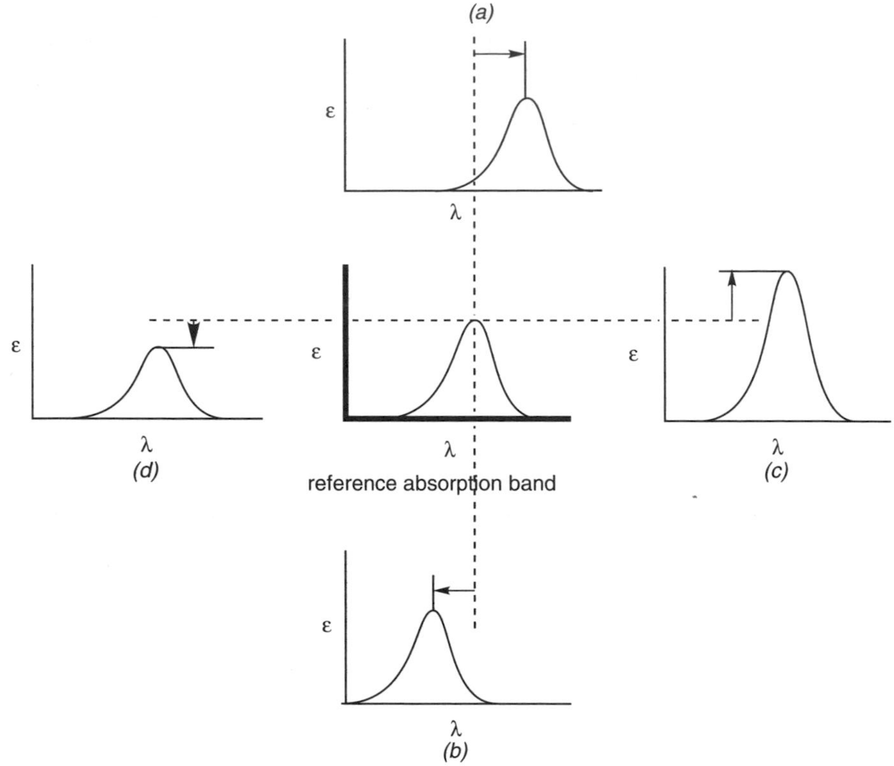

Figure 18.51. (*a*) Bathochromic shift to longer wavelength. (*b*) Hypsochromic shift to shorter wavelength. (*c*) Hyperchromic to higher ε. (*d*) Hypochromic shift to lower ε.

18.52 HYPSOCHROMIC SHIFT (BLUE SHIFT)

The displacement of a particular absorption band to shorter wavelength (higher energy). This behavior is usually associated with a change in chromophore substitution or solvent and medium effects (Fig. 18.51*b*).

18.53 HYPERCHROMIC SHIFT

The increase in intensity of a particular absorption band. This behavior is usually associated with a change in chromophore substitution or solvent and medium effects (Fig. 18.51*c*).

18.54 HYPOCHROMIC SHIFT

The decrease in intensity of a particular absorption band. This behavior is usually associated with a change in chromophore substitution or solvent and medium effects (Fig. 18.51*d*).

18.55 ISOSBESTIC POINT

The wavelength at which two interconvertible substances have equal absorptivities.

Example. Isosbestic behavior is illustrated in Fig. 18.55 for an absorbing acid at different pHs. The two absorbing species are AH and A^-

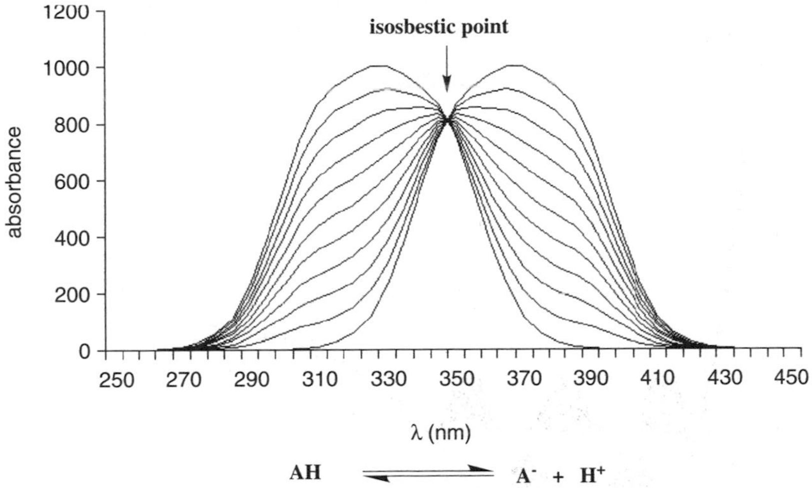

Figure 18.55. Isosbestic behavior in a reversible acid–base process.

18.56 ISOABSORPTIVE POINT

Similar to isosbestic point (Sect. 18.55), but applied to irreversible reactions. In a rigorous sense, the wavelength at which the absorbance of the starting material consumed in an irreversible reaction equals the sum of the absorbances of the products formed.

Example. Any reaction that exhibits an isoabsorptive point, upon approaching completion, generally will be an uncomplicated, high-yield reaction. For the simplest case, $A \rightarrow B$, the isoabsorptive point(s) will occur at the wavelength(s) where $\varepsilon_A = \varepsilon_B$, since $\varepsilon_A c_A$(consumed) $= \varepsilon_B c_B$(formed); $\varepsilon(c_A + c_B)$ remains constant throughout the reaction. However, isoabsorptive behavior also may be observed in more complex reactions such as $A \rightarrow B + C + \ldots$. Under these circumstances, the isoabsorptive point(s) will occur at the wavelength(s) where $\varepsilon_A c_A$(consumed) $= \varepsilon_B c_B$(produced) $+ \varepsilon_C c_C$(produced) $+ \ldots$, where c_A, c_B, and c_C are the concentrations of the respective species. In any case, isoabsorptive behavior will not be observed if one of the products undergoes a subsequent reaction, $A \rightarrow B \rightarrow D$.

18.57 GROUND STATE

The lowest-energy electronic and vibrational state of a particular molecule (Figs. 18.58 and 18.75).

18.58 MORSE CURVE

A curve that expresses the relationship between the potential energy of a diatomic or chemical bond and the bond distance r (Fig. 18.58, see Sects. 2.18 and 16.29).

Example. The potential energy increases at values smaller than d_0 in Fig. 18.58 due to the onset of nuclear repulsion, and increases again at values larger than d_0 due to the decrease in orbital overlap. As the internuclear distance r increases further, the potential energy asymptotically approaches a final value, which for the free atoms is usually taken as zero on the energy scale.

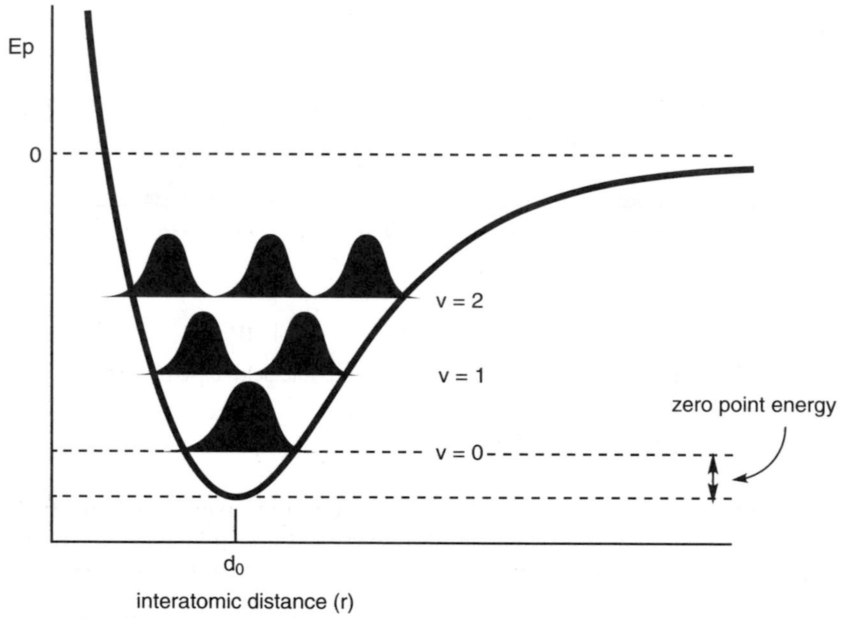

Figure 18.58. Morse curve for a chemical bond with superimposed vibrational energy levels and probability distributions. (For zero point energy, see Sect. 18.61.)

18.59 VIBRATIONAL STATE

A state that defines the energy and type of oscillation along and about the various chemical bonds of a molecule. The energy quantum numbers of these states are usually designated by v.

Example. The quantum numbers for the vibrational energy levels along with the probability of finding a molecule with a given bond length may be represented within the framework of the Morse curve as illustrated in Fig. 18.58. The shaded areas are the probability density distributions. See also Sect. 18.75.

18.60 GROUND VIBRATIONAL STATE ($\nu = 0$)

Lowest-energy vibrational state available for a particular bond or molecular in a given electronic state. This state is referred to as the 0 level or the $\nu = 0$ state (Figs. 18.58 and 18.75, see Sect. 16.26).

18.61 ZERO-POINT ENERGY

The energy associated with the ground vibrational state, $\nu = 0$, of a chemical bond (Fig. 18.58). This energy cannot be removed from the system even at absolute zero. For the hydrogen molecule, it amounts to 6.2 kcal/mol (see Sect. 16.34).

18.62 EXCITED VIBRATIONAL STATE ($\nu > 0$)

All vibrational states of higher energy than the 0 level in a particular electronic state (see Sect. 16.27).

18.63 EXCITED ELECTRONIC STATE

Any electronic state of higher energy than the ground electronic state. The energy necessary to form these states is usually provided by the absorption of a photon. In the first approximation, an excited state may be regarded as being produced by the promotion of a single electron from one of the filled or half-filled orbitals of the ground state into an empty or half-filled antibonding or nonbonding orbital.

Example. Excited electronic states are transient species, which usually return to the ground state within less than a second. A molecule in the excited state exhibits physical properties and chemistry that are distinct from those of the same molecule in its ground state. For instance, acidity may be either increased or diminished by a millionfold when promoted into its singlet excited state (Sect. 18.138). For specific examples of excited states, see Sects. 18.91, 18.93, 18.95, 18.97, 18.99, and 18.101.

18.64 BOHR CONDITION

The energy of an absorbed (or emitted) photon is equal to the difference in energy between the initial E_i and final E_f states of the molecule undergoing the transition:

$$\Delta E = E_f - E_i = h\nu$$

Named after Neils Bohr (1855–1962) Nobel prize in physics, 1922.

18.65 FIRST LAW OF PHOTOCHEMISTRY

Only the light that is absorbed by a molecule can be effective in producing photochemical change in the molecule.

18.66 0-0 TRANSITION (0-0 BAND)

A transition that occurs between the two lowest vibrational levels ($\nu = 0$) of any two electronic states. The absorption (or emission) band associated with that transition is the 0-0 band (see Fig. 18.75 process 6).

18.67 VIBRONIC TRANSITION (VIBRONIC BAND)

A transition that occurs between the lowest vibrational level ($\nu = 0$) of the initial electronic state and a higher vibration level ($\nu > 0$) of the final electronic state. The absorption (or emission) band associated with that transition is a vibronic band (see Fig. 18.75 process 7, 12, and 14). The term vibronic is a contraction for vibrational-electronic.

18.68 HOT TRANSITION (HOT BAND)

A transition that originates from one of the upper vibrational levels ($\nu > 0$) of the initial electronic state. The absorption (or emission) band associated with that transition is a hot band (see Fig. 18.75).

18.69 VIBRATIONAL FINE STRUCTURE

The band pattern frequently observed within the absorption (or emission) envelope of a single electronic transition. In order for fine structure (a band pattern) to be prominent in the spectrum of a compound, the 0-0 transition must be forbidden (Sect. 18.77). Only those vibronic transitions that are allowed give rise to fine structure (Sect. 18.76, Fig. 18.69).

18.70 MULTIPLICITY

The number of spin orientations observed when an electron, nucleus, or any other magnetic particle is placed in a magnetic field. The multiplicity is determined by the number of unpaired spins in the system n and the spin angular momentum quantum number s for the particle in question:

$$\text{Multiplicity} = 2|ns| + 1 \tag{18.70}$$

Example. The spin quantum number s of either an electron or a proton is $\frac{1}{2}$. Therefore, a single electron or proton is a doublet, $2|(1)\frac{1}{2}| + 1 = 2$. One of these spin orientations

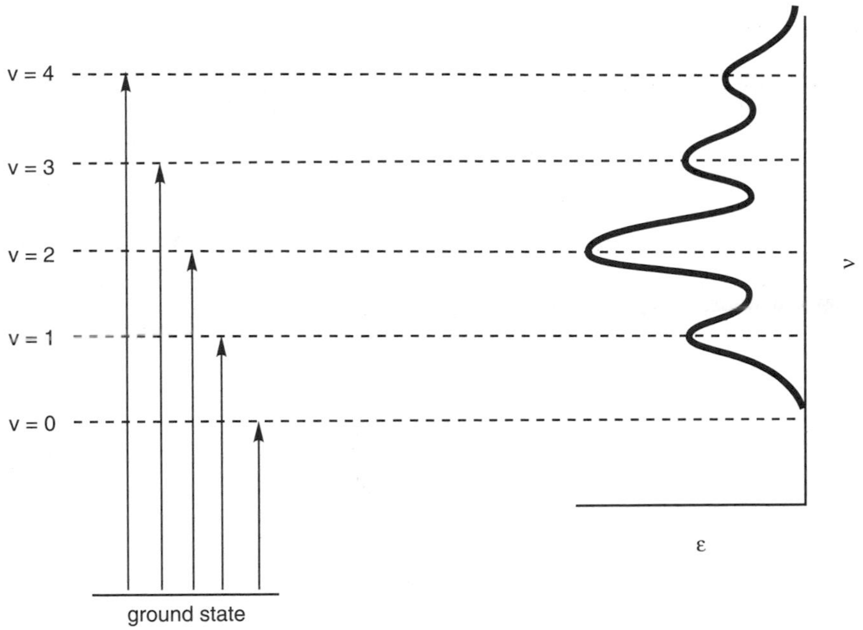

Figure 18.69. Relationship between vibrational fine structure in an absorption spectrum and vibronic transitions of a forbidden electronic transition.

will be aligned with the magnetic field and the other will be aligned against the magnetic field (see Sect. 15.59).

18.71 SINGLET ELECTRONIC STATE

The electronic configuration in which all electron spins are paired ($n = 0$) and the *total* spin angular momentum s equals zero. Such a state has a multiplicity of $2|(0 \times 1/2)| + 1 = 1$. Hence, it is called a singlet. When two paired electrons are involved, they may be in the same atomic or molecular orbital or in separate orbitals, and are frequently represented by two arrows pointing in opposite directions.

Example. In most singlet systems, the spin angular moments of two paired electrons will exactly cancel each other. The spin angular momentum is a vector quantity and its projection on the z-axis can be either $+1/2$ or $-1/2$. This can be envisaged by orienting two spins in an external magnetic field: one spin with the field and one spin against the field (Fig. 18.72*a*). Projections M_s on the z-axis of these two vectors cancel, with a net spin angular momentum of zero. However, if the two electrons occupy different half-filled molecular orbitals, but remain paired, their electron spin angular momenta may not exactly cancel each other. In such a case, the molecule will display a net magnetic moment even in the singlet state. Singlet oxygen is an example of such a situation.

18.72 TRIPLET ELECTRONIC STATE

The electronic configuration in which two unpaired electron spins ($n = 2$) exist within a system, and the *total* spin angular momentum s is 1. Such a state has a multiplicity of $2|(2 \times 1/2)| + 1 = 3$. Hence, it is called a triplet. The two unpaired electrons must of necessity ("Hund's Rule," Sect. 1.40) occupy separate atomic or molecular orbitals.

Example. It is easy for organic chemists to envisage two of the three *sublevels of the triplet state*, and these are often represented as, for example, in molecular orbital energy diagrams, as separate arrows in each of two orbitals with arrows both pointing

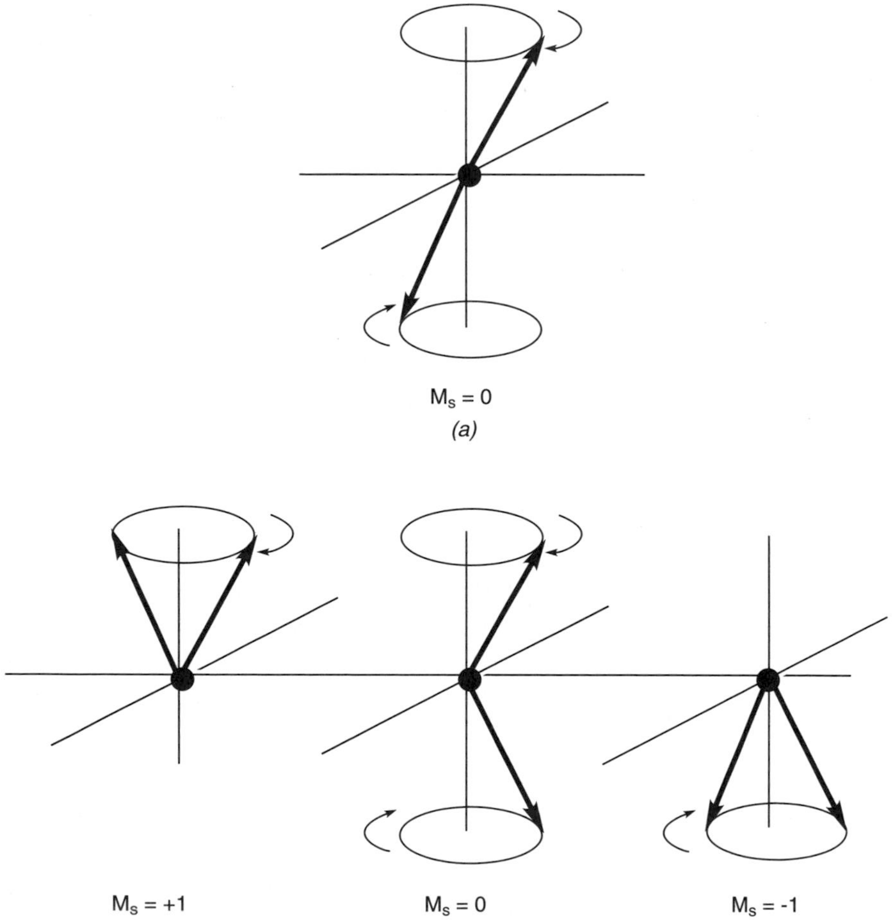

Figure 18.72. The z-component M_s of the possible spin configurations of a pair of electrons: *(a)* single configuration of the singlet state and *(b)* the three electronic spin configurations of the triplet state.

down, or with both arrows pointing up to indicate their parallel spins. However, the physical origins of the third sublevel are obscured by this simple notation and there is no analogous representation possible. However, all three sublevels may be better understood as illustrated in Fig. 18.72, in which the electron spins are oriented by an external magnetic field. The total spin will have the same magnitude in all three cases ($s = 1$). However, the three sublevels differ by their projections on the z-axis, M_s, which are $+1, 0$, and -1. The zero sublevel of the triplet state will differ from the singlet state only in the precession phase angle, which will be 0° instead of 180°, respectively. (See also Sect. 18.71 and 18.75.)

18.73 DOUBLET ELECTRONIC STATE

That electronic configuration containing an odd number of electrons in which only the odd electron is of necessity unpaired. Such a system will have a total spin angular momentum of $\frac{1}{2}$, $s = \frac{1}{2}$, and a multiplicity of $2|(1)\frac{1}{2}| + 1 = 2$. All free radical species containing a single unpaired electron exist in a doublet state.

18.74 MANIFOLD OF STATES (SINGLET MANIFOLD, TRIPLET MANIFOLD)

A family of electronic states with the same electron spin configurations.

Example. The ground singlet state and all the excited singlet states are referred to as the *singlet manifold.* Likewise, the lowest triplet state and all higher triplet states are referred to as the *triplet manifold.*

18.75 JABLONSKI DIAGRAM

A representation of the relative energies of the electronic, vibrational, and rotational states of a molecule into which arrows indicating the various photophysical processes that interconnect these states are incorporated (Fig. 18.75) named after Alexander Jablonski (1898–1980).

18.76 ALLOWED TRANSITION

Any interconversion of states that occurs with high probability (for absorption of light, see Sect. 18.47).

Example. This term is usually utilized to characterize absorption and emissions processes. Thus, allowed absorption or emission processes will exhibit intense absorption or emission bands, and forbidden transitions will exhibit low-intensity absorption (Sect. 18.77).

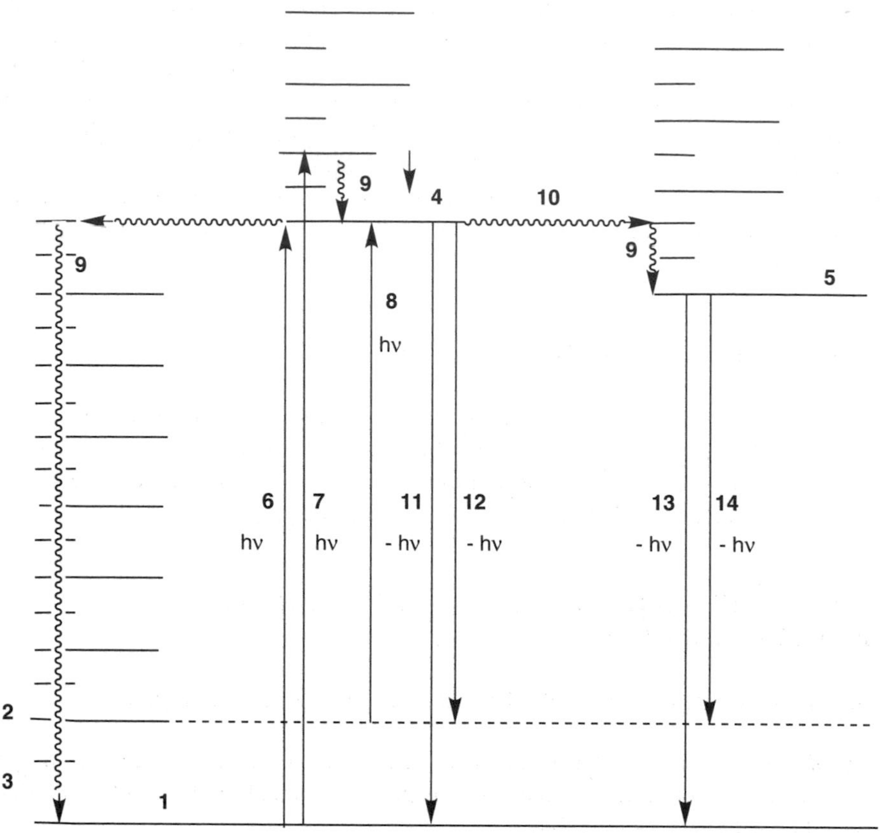

Figure 18.75. Jablonski diagram incorporating state energy levels and the most important photophysical processes: (1) Ground singlet state (Sect. 18.57). (2) Vibrational states (Sect. 18.59). (3) Rotational states. (4) Excited singlet state (Sect. 18.71). (5) Lowest triplet state (Sect. 18.72). (6) Absorption, 0-0 transition (Sect. 18.66). (7) Absorption, vibronic transition (Sect. 18.67). (8) Absorption, hot transition (Sect. 18.68). (9) Internal conversion, IC (Sect. 18.111). (10) Intersystem crossing, ISC (Sect. 18.112). (11) Fluorescence, 0-0 transition (Sect. 18.129). (12) Fluorescence, vibronic transition. (13) Phosphorescence, 0-0 transition (Sect. 18.130). (14) Phosphorescence, vibronic transtion. In general, straight arrows designate radiative process, those that require the absorption or emission of a photon. The wavy arrows designate nonradiative or radiationless processes (Sect. 18.110), those that involve release of energy as heat to the surrounding medium or are isoenergetic.

18.77 FORBIDDEN TRANSITION

Any interconversion of states that occurs with low probability (for absorption of light, see Sect. 18.47).

Example. This term is usually utilized to characterize absorption and emission processes. Thus, forbidden absorption or emission processes will exhibit weak absorption or emission bands. There are various degrees of forbiddeness (Sect. 18.83), some of which lead to weaker transition intensities than others. The boundary between forbidden and allowed is not well defined. Nevertheless, for absorption spectra, a band is usually considered to arise from a forbidden transition if it exhibits a molar extinction coefficient ε (Sect. 18.43) of less than about 1,000, and from an allowed transition if it exhibits an ε greater than about 1,000. Intersystem crossing (process 10 in Fig. 18.75 and Sect. 18.112) is an example of a forbidden transition that is not associated with an absorption or emission process, since the two coupled states are isoenergetic.

18.78 OSCILLATOR STRENGTH f

A theoretical measure of the intensity of an absorption or emission band. This parameter is derived from the quantum mechanical properties of the transition by the following relationship:

$$f = (8\pi 2mc/3h)\, G_j \nu D_j \tag{18.78a}$$

and is related to the absolute intensity of the absorption or emission band by

$$f = 0.102(mc2/N\pi e^2)\int \varepsilon\, d\nu = 4.315 \times 10^{-9}\int \varepsilon\, d\nu \tag{18.78b}$$

where m is the mass of an electron, c the speed of light, h Planck's constant, G_j the number of degenerate states to which absorption or emission can lead, ν the frequency of absorption or emission, D_j the dipole strength (Sect. 18.79), N Avogadro's number, and ε the molar extinction coefficient, which is integrated over the entire absorption or emission band.

18.79 DIPOLE STRENGTH D_j

The term D_j in the oscillator strength expression of transition intensity (Sect. 18.78) that provides the quantum mechanical characteristics of the transition:

$$\mathrm{I} \propto D_j = (i|\mu_e|j)^2 \tag{18.79a}$$

where i and j are the total wave functions for the initial and final states, respectively. The *electric dipole transition moment* μ_e equals

$$\mu_e = \sum e\mathbf{r} \tag{18.79b}$$

where **r** is the radius vector between the center of gravity of positive charge, the nuclei, and an electron. This property is evaluated for all electrons in the system. The expression for D_j is the purely electric analog of the expression for the rotational strength R_j (Sect. 18.24). The expression for R_j has a similar form to that for D_j, except that it is the imaginary scalar product of the electronic and magnetic transition moments of the *j*th electronic transition:

$$R_j = \mathrm{Im}\{(i\,|\boldsymbol{\mu}_e|\,j) \bullet (i\,|\boldsymbol{\mu}_m|\,j)\} \quad (18.79c)$$

where $\boldsymbol{\mu}_m$ is the magnetic transition moment.

Example. In order for light to be absorbed, D_j must not be zero. According to the expression for D_j, the displacement of charge produced by the removal of electron density from molecular orbital *i* and the injection of electron density into molecular orbital *j* is associated with an electric dipole transition moment $\boldsymbol{\mu}_e$. If this charge displacement has dipolar symmetry, for example, if the product of the symmetries of $i \times j$ has the same symmetry as that of an arrow (vector) oriented along any one of the three molecular axes (*x*, *y*, or *z*), then one of the possible light electric field components, $\mathbf{E}_x$, $\mathbf{E}_y$, or $\mathbf{E}_z$, will be suitable for driving the charge displacement. Under these conditions, the expression for D_j will be totally symmetric, for example, symmetry $(i \times j)$ = symmetry $\boldsymbol{\mu}_{ex}$, $\boldsymbol{\mu}_{ey}$, or $\boldsymbol{\mu}_{ez}$, = symmetry $\mathbf{E}_x$, $\mathbf{E}_y$, or $\mathbf{E}_z$ and, therefore, the expression for D_j and intensity *I* will be nonzero (see Sect. 18.81 for a specific example).

18.80 BORN–OPPENHEIMER APPROXIMATION

The assumption that the total wave function of a molecule can be factored into an electronic wave function Ψ, a vibrational wave function χ, and a spin wave function *s* named after Max Born (1882–1973) and J. Robert Oppenheimer (1904–1967).

Example. Application of the Born–Oppenheimer approximation to the dipole strength expression (Sect. 18.79) yields the integral

$$D_j = \{(\psi_i|\boldsymbol{\mu}_e|\psi_j)\,(\chi_i\,\chi_j)\}^2 \quad (18.80)$$

where the first integral evaluates the electronic wave functions over the electronic coordinates, and may be used to determine the transition intensity of the 0-0 transition. The second integral evaluates the vibrational wave functions over the nuclear coordinates, and may be used in combination with the electronic term to determine the intensities of all the vibronic transitions.

18.81 POLARIZATION OF ELECTRONIC TRANSITIONS

The orientation, within the framework of the molecular coordinate system, of the electric dipole transition moment $\boldsymbol{\mu}_e$. However, since $\boldsymbol{\mu}_e$ must coincide with the electric vector of the light **E**, in order for an absorption to occur, the polarization of a transition is defined more meaningfully as the orientation of the **E** vector of the light that is necessary to induce that transition.

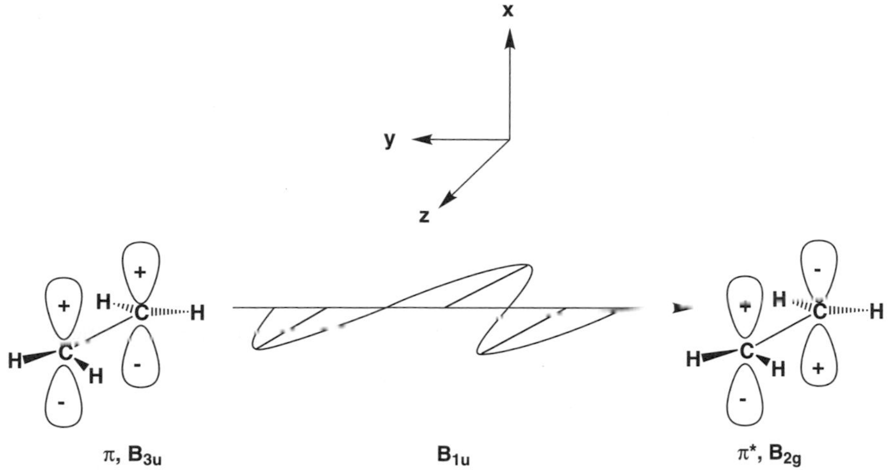

Figure 18.81. The absorption of z-polarized light by ethylene. The character table for the D_{2h} point group, of which ethylene is a member, is shown in Table 18.81.

TABLE 18.81. Character Table for D_{2h}

D_{2h}	E	$C_{2(z)}$	$C_{2(x)}$	$C_{2(y)}$	i	$\sigma_{(xz)}$	$\sigma_{(yz)}$	$\sigma_{(xy)}$	
A_g	+1	+1	+1	+1	+1	+1	+1	+1	
A_u	+1	+1	+1	+1	−1	−1	−1	−1	
B_{1g}	+1	+1	−1	−1	+1	−1	−1	+1	R_z
B_{1u}	+1	+1	−1	−1	−1	+1	+1	−1	z
B_{2g}	+1	−1	−1	+1	+1	+1	−1	−1	π^*, R_y
B_{2u}	+1	−1	−1	+1	−1	−1	+1	+1	y
B_{3g}	+1	−1	+1	−1	+1	−1	+1	−1	R_x
B_{3u}	+1	−1	+1	−1	−1	+1	−1	+1	π, x

Example. In the case of ethylene, the excitation of an electron from the π orbital into the π* orbital is allowed (Fig. 18.81). Ethylene belongs to the point group D_{2h}, and the transition in question will have the following dipole symmetry:

$$D_{\pi^*} = (\pi|\mu_{ez}|\pi^*)^2 = (B_{3u}|B_{1u}|B_{2g}) = A_g$$

$$\pi = + - + - - + - + = B_{3u}$$

$$\underline{\pi^* = + - - + + + - - = B_{2g}}$$

$$\pi \times \pi^* = + + - - - + + - = B_{1u}$$

$$\underline{\mu_z = + + - - - + + - = B_{1u}}$$

$$\pi \times \pi^* \times \mu_z = + + + + + + + + = A_g$$

The symmetry of the $\pi \times \pi^*$ product requires that μ_e be oriented along the z-molecular axis (z-polarized) in order for the expression to be totally symmetric (A_g) and, thus, allowed. Consequently, the $\pi \rightarrow \pi^*$ transition is said to be z-polarized, and will be observed only when the ethylsene molecule is exposed to light of the appropriate wavelength, which is polarized along the ethylene z-axis (the π bond axis).

18.82 POLARIZATION OF ABSORPTION AND EMISSION BANDS

The experimentally observed consequences of the polarization of electronic transitions (Sect. 18.81). The absorption or emission of only certain orientations of plane-polarized light relative to the coordinate system of the sample.

Example. This phenomenon is observed under a variety of experimental conditions. Polarized absorption and emission are both observed when dealing with samples in which some degree of molecular ordering or selection has been achieved. Such ordering will occur when the active molecule is in crystalline form, dispersed in an oriented liquid crystal medium, a stretched polymer film, or an isotropic solution that is exposed to a strong electric field. Polarized emission will be observed also in solution when certain molecular orientations are selectively excited by either unpolarized or plane-polarized light. In this case, polarized fluorescence may be observed in fluid media for a short period of time following excitation. This polarization decays rapidly as molecular tumbling quickly establishes a random orientation distribution in the excited-state population. In rigid media, polarized emission will be observed for a much longer time until a randomized excited-state population is established by energy transfer (Sect. 18.154).

18.83 SELECTION RULE

A criterion for determining whether a particular transition will be allowed or forbidden (see Sects. 18.84–18.87).

18.84 MULTIPLICITY (SPIN) SELECTION RULE

Transitions between states of the same multiplicity are allowed, whereas those between states of different multiplicities are forbidden.

Example. This selection rule arises from the fact that angular momentum is not conserved if a photon induces a simultaneous and uncoordinated change in both the spin and orbital angular momenta. Both singlet → singlet and triplet → triplet transitions are allowed by this selection rule, but singlet → triplet and triplet → singlet transitions are strongly forbidden, $\varepsilon << 1$.

18.85 SYMMETRY SELECTION RULE

The symmetry of the transition, for example, the symmetry of the initial state multiplied by the symmetry of the final state, must be dipolar, that is, have the symmetry of a vector oriented along any of the axes of the molecular coordinate system.

Example. See the example in Sect. 18.81. This rule arises from the fact that electric dipole transitions, as opposed to electric quadrupole and higher-order transitions, are most readily driven by the electric field of light. Transitions that are forbidden by this rule are moderately weak, $10 < \varepsilon < 1{,}000$, vibronic in nature, and do not display a 0-0 band, which is strongly forbidden.

18.86 PARITY SELECTION RULE

A special case of the symmetry selection rule (Sect. 18.85) for molecules that contain a center of symmetry *i*. The symmetry of an allowed transition, that is, the symmetry of the initial state multiplied by the symmetry of the final state, must be *ungerade* (dipolar). Alternatively, this rule is usually expressed as follows: Transitions between states of unlike parity, *ungerade* (u) $\rightarrow$ *gerade* (g) or $g \rightarrow u$, are allowed, while transitions between states of like parity, $g \rightarrow g$ or $u \rightarrow u$, are forbidden.

Example. See the example in Sect. 18.81. Due to the higher order of symmetry involved in the parity selection rule, transitions forbidden by this selection rule tend to be somewhat weaker, $1 < \varepsilon < 100$, than those forbidden by the more general symmetry selection rule (Sect. 18.85).

18.87 LA PORTE SELECTION RULE

Equivalent to the parity selection rule, but applied to the transitions of metal atoms. Thus, metal atom $d \rightarrow d$ transitions are said to be La Porte-allowed or -forbidden. Named after Otto La Porte (1902–1971).

18.88 FRANCK–CONDON PRINCIPLE

Neither the geometry of a molecule nor the momenta of its constituent nuclei are changed during the absorption event (see Sect. 12.31).

Example. This principle holds because absorption of a photon takes place much faster (~10^{-18} sec) than the time required for a single vibration (~10^{-12} sec). The Franck–Condon principle does not determine the intensity of an electronic absorption band system. However, it does determine the distribution of intensity among the constituent vibronic bands and, hence, the shape of the absorption band system.

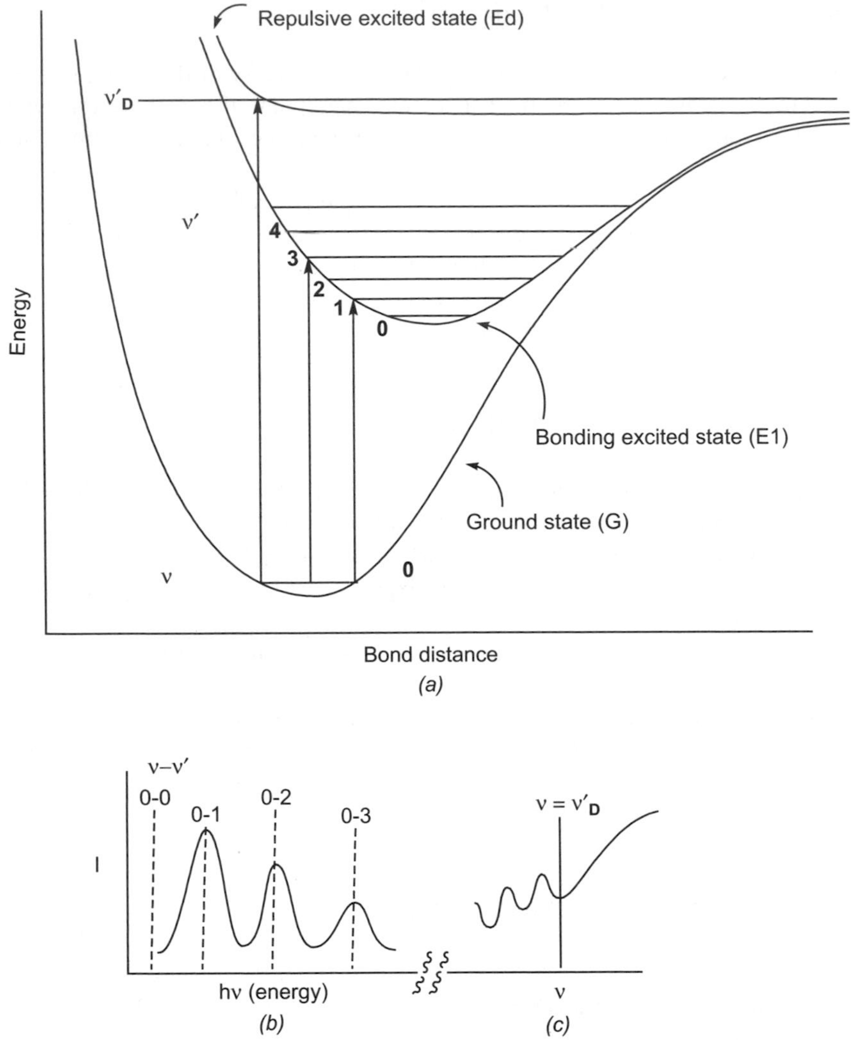

Figure 18.88. Franck–Condon model for vibronic transition intensity distribution. (*a*) Jablonski diagram for vibronic transitions to various excited states. (*b*) Intensity distribution for the corresponding vibronic transitions. (*c*) Intensity distribution for a predissociative excitation process.

Figure 18.88 illustrates this point for a simple molecule containing one bond. In general, an excited state E_1 will have a longer equilibrium bond length than the ground state G. In the model shown in Fig. 18.88*a*, all the possible ground-state bond lengths (G, $\nu = 0$) are shorter than the shortest bond length in E_1 ($\nu = 0$). Therefore, the transition G ($\nu = 0$) $\rightarrow E_1$ ($\nu' = 0$) is Franck–Condon-forbidden, since a change in molecular geometry would have to occur during the transition. By this same criterion, the transitions G ($\nu = 0$) $\rightarrow E_1$ ($\nu' = 1,2,3,\ldots$) are Franck–Condon-allowed. However, the momenta of the nuclei will tend toward zero near the walls of the potential energy surface, and be maximum in the center of the well. Thus, transitions G ($\nu = 0$) $\rightarrow E_1$ ($\nu' = 2$) can occur with little change in momenta and is more allowed than the G ($\nu = 0$) $\rightarrow E_1$ ($\nu' = 1$) transition, which will involve a significant change in momenta. These considerations would lead one to predict an absorption band fine structure similar to that shown in the insert (Fig. 18.88*b*). The 0-0 band should be absent, the 0-1 band intense, the 0-2 band weak, and the 0-3 band weak.

18.89 PREDISSOCIATION

A vibronic transition that results in bond dissociation within a single vibration.

Example. This phenomenon is observed when excitation leads to a higher vibrational level such as ν'_D [Fig. 18.88*a*, G ($\nu = 0$) $\rightarrow E_1$ ($\nu' = \nu'_D$)], which at some point crosses a repulsive excited state (E_d) potential energy surface. This process manifests itself by the disappearance of the vibrational fine structure, and the onset of an absorption continuum in the spectrum of the molecule (Fig. 18.88*c*).

18.90 PHOTOELECTRIC EFFECT

The production of free electrons when light of sufficient energy impinges on a metal surface.

Example. In order for this effect to occur, the light must exceed some threshold energy ($E_T = h\nu_T$). This threshold energy is equal to the work required to remove an electron from the surface of the metal. The operation of all photomultiplier tubes is based on this phenomenon.

18.91 π,π* STATE

The excited state produced by the transfer of electron density from a bonding π orbital into an antibonding π* orbital.

Example. The spectroscopic π,π* state of ethylene is shown in Fig. 2.38. However, since the π bond is net antibonding in this state, rotation will rapidly occur about the excited π bond to a minimum energy geometry, in which the two *p* orbitals are twisted at 90° to each other and the hydrogen atoms are as far apart as possible (Fig. 18.91, Sects. 18.165 and 18.166). It is this twisted geometry that is the source of much of the photochemistry observed from π,π* states.

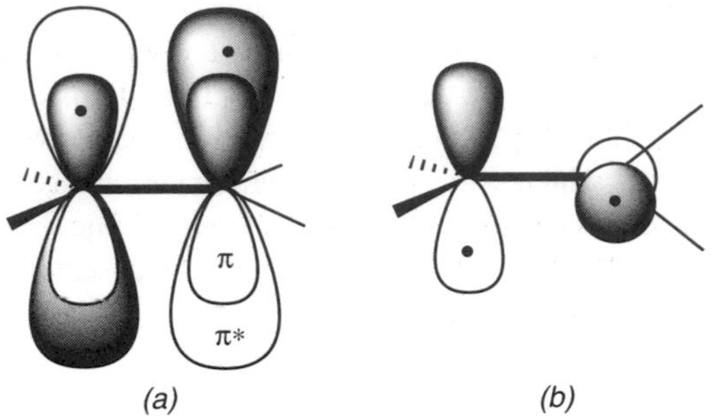

Figure 18.91. Electronic configuration and geometries of π,π* excited state: (a) planar (high-energy) configuration and (b) nonplanar (low-energy) configuration.

18.92 π,π* TRANSITION

The transition or absorption process leading to a π,π* excited state. It is usually characterized by an intense (allowed), structureless absorption band occurring anywhere in the near ultraviolet or visible region of the spectrum. The band will usually undergo a slight red shift in more polar solvents and a slight blue shift in less polar solvents. The transition is polarized along the double bond axis (Fig. 18.81).

18.93 *n*,π* STATE

The excited state produced by the transfer of electron density from a nonbonding orbital *n* into an antibonding π* orbital.

Example. The two constituent orbitals, each possessing one electron, of the carbonyl *n*,π* state, after the electron in the *n* orbital has been promoted to the π* orbital, are shown in Fig. 18.93*a*. Since electron density has been rotated 90° about the carbonyl bond axis in the production of this excited state, a quadrupolar distribution of charge exists. The two π quadrants contain a net negative charge, and the two nonbonding quadrants contain a net positive charge (Fig. 18.93*b*). There is, thus, a net shift of electron density away from the carbonyl oxygen toward the carbonyl carbon in the excited state, opposite to the polarization of the ground state. The chemistry of the *n*,π* carbonyl state consists of the reaction of electrophiles with the π system at carbon and the reaction of nucleophiles with the nonbonding orbitals on oxygen. This is exactly the opposite chemistry of the carbonyl in the ground state!

18.94 *n*,π* TRANSITION

The transition or absorption process leading to an *n*,π* excited state. It is usually characterized by a weak (forbidden), structured absorption band system since the

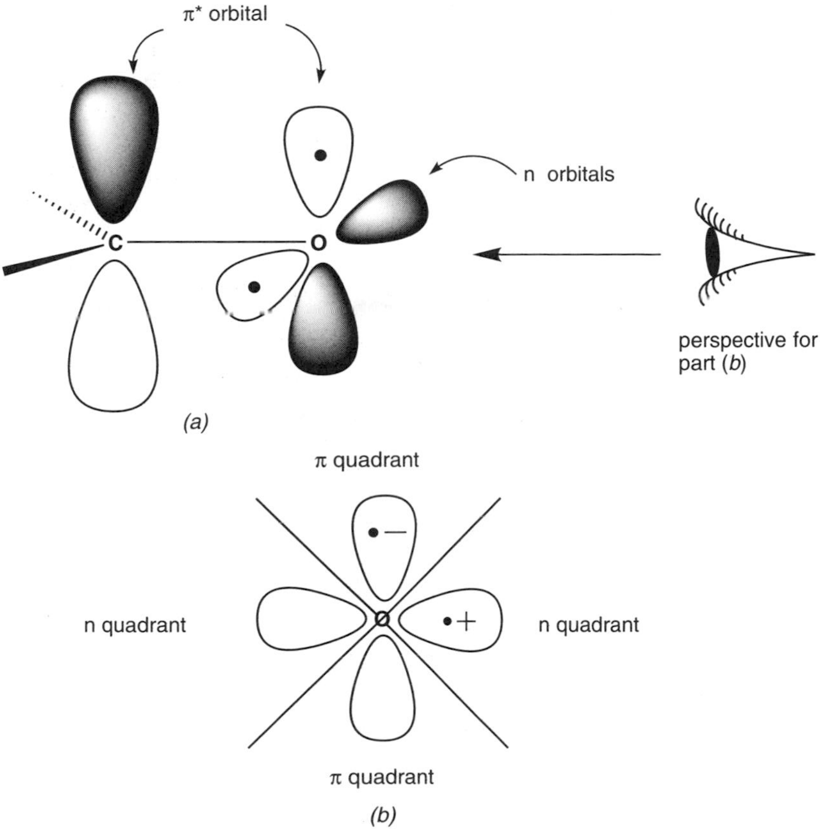

Figure 18.93. (*a*) Electronic configuration of the n,π^* carbonyl excited state with one electron in each of the n and π^* orbitals. (*b*) Radial distribution of charge and spin about the carbonyl in the n,π^* excited state.

two orbitals involved are orthogonal. Absorption occurs anywhere in the near ultraviolet or visible region of the spectrum. This band will usually undergo a slight blue shift with loss of fine structure in more polar solvents and a slight red shift with enhancement of fine structure in less polar solvents.

18.95 n,σ^* STATE

The excited state produced by the transfer of electron density from a nonbonding orbital into a σ^* orbital.

Example. The constituent orbitals of the n,σ^* state of a carbon–iodine bond are shown in Fig. 18.95. Because electron density has been introduced into the σ^* orbital, the carbon–iodine bond length will be increased and will tend to undergo dissociation. Accordingly, the photochemistry associated with n,σ^* states usually involves σ bond breakage.

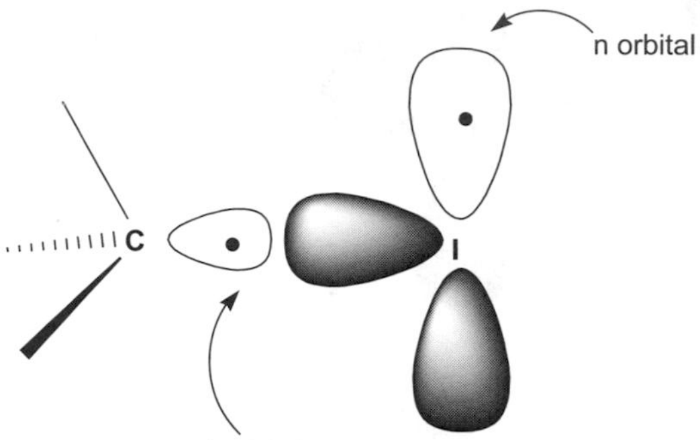

Figure 18.95. Electronic configuration of the n,σ^* excited state of the carbon–iodine bond showing one electron in each orbital.

18.96 n,σ^* TRANSITION

The transition or absorption process leading to an n,σ^* excited state. It is usually characterized by a strong (allowed) absorption band in the ultraviolet region of the spectrum.

18.97 σ,σ^* STATE

The excited state produced by the transfer of electron density from a bonding σ orbital into an antibonding σ* orbital.

Example. The σ,σ^* state of a carbon–carbon bond is net antibonding, and hence, the C–C bond will undergo rapid dissociation. The chemistry associated with σ,σ^* states thus involves σ bond breakage.

18.98 σ,σ^* TRANSITION

The transition or absorption process leading to a σ,σ^* excited state. It is usually characterized by a strong (allowed) absorption band occurring in the far or vacuum ultraviolet region.

18.99 d,d STATE

The excited state produced by the transfer of electron density from one *d* orbital usually on a metal atom to another *d* orbital on the same atom.

Example. Various *d,d* excited states may be formed in transition metal ion systems. The formation of this type of state is associated with an angular redistribution of electron density about the metal atom. Since this type of electronic redistribution tends to weaken metal–ligand bonds, the photochemistry associated with *d,d* excited states is usually that of ligand loss or exchanged (Fig. 18.99).

$$Cr(NH_3)_6^{3+} + H_2O \xrightarrow{h\nu} Cr(NH_3)_5(H_2O)^{3+} + NH_3$$

Figure 18.99. Ligand exchange associated with a *d,d* transition of a chromium complex.

18.100 *d,d* TRANSITION

This is the transition or absorption process leading to *d,d* excited states; each transition is characterized by weak (forbidden) absorption band in the visible and near ultraviolet regions of the spectrum of organometal complexes. The wavelengths and intensities of these bands are strongly influenced by the ligands attached to the metal.

18.101 CHARGE-TRANSFER (C-T) EXCITED STATES

The excited state produced by the partial or complete transfer of an electron from a bonding (occupied) orbital of a donor species into the antibonding (unoccupied) orbital of an acceptor species. Many excited states have some degree of C-T character. It is only when a significant, albeit arbitrary, fraction of the charge of an electron has been transferred between two relatively independent molecular subunits that an excited state is termed C-T.

Example. C-T excited states are encountered in transition metal ligand systems. Three types exist: C-T from the metal to the ligand (CTML), C-T from the ligand to the metal (CTLM), and C-T between the ligands (CTLL). The photochemistry associated with a CTLM state is loss of the ligand with reduction of the metal (Fig. 18.101*a*). In contrast, CTML states may lead to electron expulsion with oxidation of the metal (Fig. 18.101*b*).

$$(a)\quad Co(NH_3)_6^{3+} \xrightarrow[H^+]{h\nu} Co^{2+} + 5\,NH_4^+ + 1/2\,N_2$$

$$(b)\quad Fe(CN)_6^{4-} \xrightarrow{h\nu} Fe(CN)_6^{3-} + e^-$$

Figure 18.101. Charge transfer excitation processes: (*a*) CTLM with reduction of cobalt and (*b*) CTML with electron expulsion and oxidation of iron.

18.102 CHARGE-TRANSFER (C-T) TRANSITION

The transition or absorption process leading to a C-T state. C-T transitions may exhibit either intense or weak absorption bands, depending on the degree of overlap between the donor and acceptor orbitals. C-T bands may occur either in the visible or near ultraviolet region, and the positions of their absorption maxima are very sensitive to the polarity of the solvent. As a matter of fact, the shift in the C-T band can be used to evaluate the polarity of solvents.

Example. *N*-alkylpyridinium iodides exhibit C-T transitions due to electron transfer from the iodide ion to the pyridinium moiety (Fig. 18.102*a*). Since this transfer of charge produces an excited state that is less polar than the ground state, the transition requires less energy in nonpolar solvents. In fact, the solvent shifts observed in this system may be more than 100 nm into the red (lower frequency) upon going from a polar to a nonpolar solvent. If the ground state were nonpolar and the excited

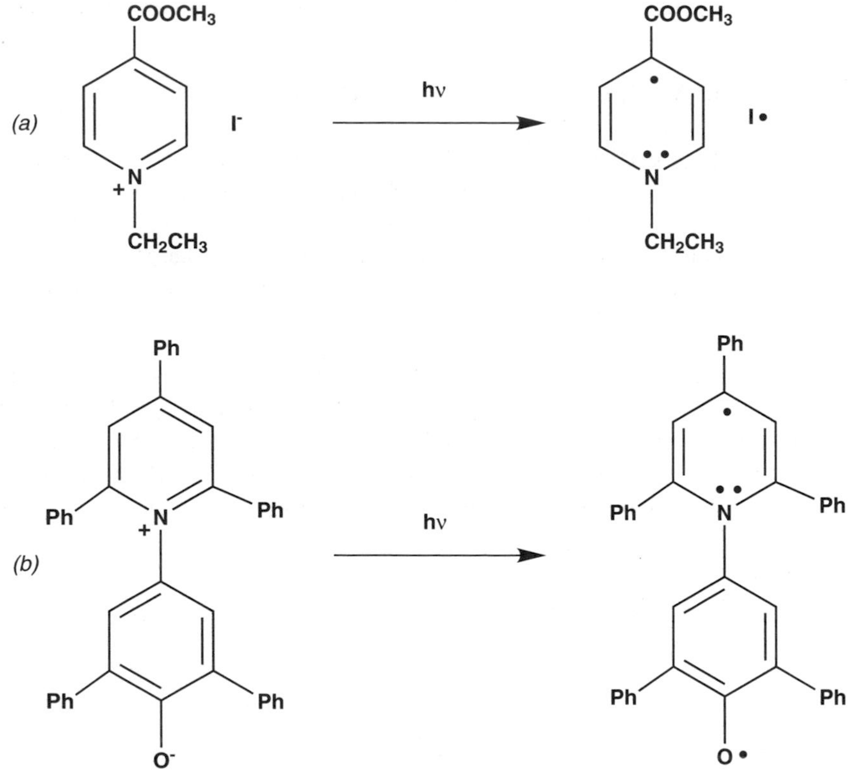

Figure 18.102. Charge-transfer transitions that produce a less polar species. (*a*) "Intermolecular" C-T process that forms the basis of Kosower Z scale of solvent polarity. (*b*) Intramolecular C-T process that forms the basis of the $E_T(30)$ scale of solvent polarity. Both processes involve increase red shifts with increasing polarity.

TABLE 18.102. Solvent Polarity as Measured on the Kosower Z Scale and the $E_T(30)$ Scale

Solvent	Kosower Z Scale	$E_T(30)$ Scale
Water	94.6	63.1
Methanol	83.6	55.4
Acetonitrile	71.3	45.6
Dimethylsulfoxide	71.1	45.1
N,N-dimethylformamide	68.5	43.8
Acetone	65.7	42.2
Hexamethylphosphoramide	62.8	40.9
Methylene chloride	64.2	40.7
Chloroform	63.2	39.1
Ethyl acetate	—	38.1
Tetrahydrofuran	—	37.4
Diethyl ether	—	34.5
Benzene	54.0	34.3
n-Hexane	—	31.0

state polar, a blue shift (higher frequency) would be observed upon going from a polar to a nonpolar solvent. The red-shifted pyridium salt systems shown in Fig. 18.102 have been used to establish two scales of solvent polarity. The "intermolecular" C-T band observed in the pyridium iodide shown in Fig. 18.102*a* is used to provide values for the Kosower Z scale, and the intramolecular C-T band observed in the highly twisted betaine shown in Fig. 18.102*b* is used to provide values for the $E_T(30)$ scale. Representative values for these two scales are shown in Table 18.102.

18.103 RYDBERG TRANSITION

Any transition or absorption process in which an electron is promoted to an orbital with a larger principal quantum number n than that of the initial orbital. In ethylene, the $2p\pi \rightarrow 3s$ transition is a Rydberg transition. Rydberg transitions are moderately intense transitions occurring in the short-wavelength (higher energy) of the near ultraviolet or far ultraviolet named after the physicist Janne Rydberg (1854–1919).

18.104 DIMOLE TRANSITION

A transition or absorption process involving the supramolecule formed by two molecules undergoing collision (*collision complex*). The energy of the absorbed photon may be localized in either of the collision partners or distributed between the two.

Example. The formation of singlet oxygen from the normal- or ground-state triplet oxygen by the absorption of light is not observed, since the transition is both highly

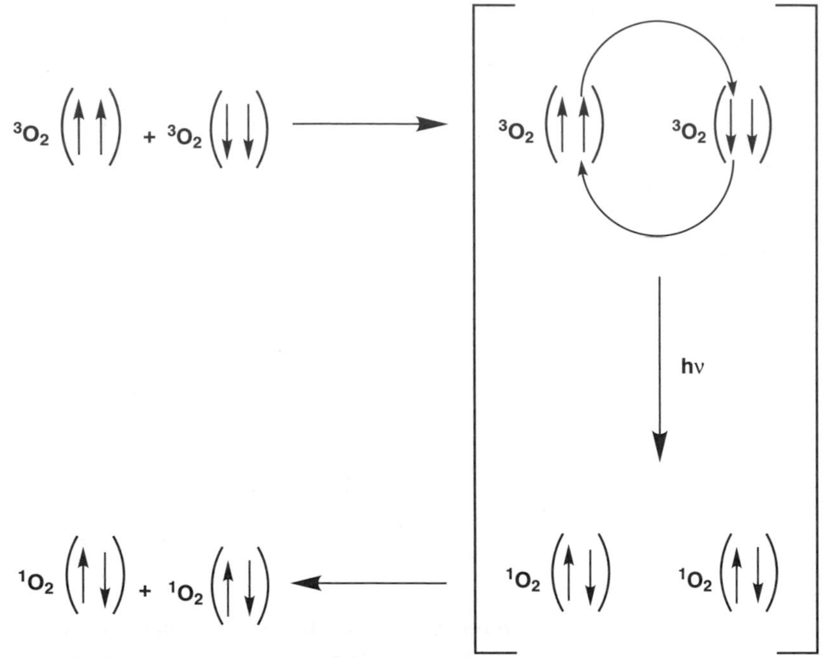

Figure 18.104. Dimole transition involving the collision complex of triplet molecular oxygen to form singlet molecular oxygen.

symmetry- and spin-forbidden. However, this process becomes allowed in the O_2–O_2 collision complex, since no net inversion of electron spin is required (Fig. 18.104). Dimole transitions are very weak due to the low probability of a three-body collision (two molecules and one photon). Nevertheless, dimole tranistions of molecular oxygen may be observed in the infrared, visible, and near ultraviolet regions of the spectrum at high oxygen pressure (150 atm) or in liquid oxygen.

18.105 DOUBLE-PHOTON TRANSITION

A transition or absorption process induced by the simultaneous or near simultaneous absorption of two photons, each one of which by itself does not have enough energy to produce the lowest-energy excited state available to the molecule.

Example. These transitions are produced only by extremely high intensity (~MW/cm^2, ~MJ/cm^2 sec) laser light. Given an energy separation of ΔE between the ground state and lowest excited state of a molecule (Fig. 18.105), the *uncertainty principle* indicates that the maximum lifetime Δt of a collision complex between a molecule and a photon of energy $\Delta E'$, where $\Delta E > \Delta E'$, will be

$$\Delta t \leq \hbar/(\Delta E - \Delta E') \tag{18.105}$$

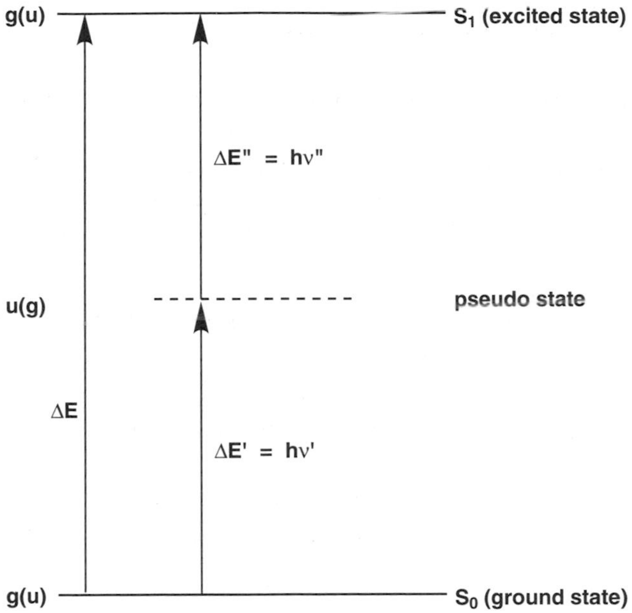

Figure 18.105. "Simultaneous" double-photon absorption via an uncertainty principle-governed pseudo-state.

where $\hbar$ is Planck's constant over 2π, that is, $h/2\pi$. If a second photon of energy $\Delta E''$, such that $\Delta E' + \Delta E'' \geq \Delta E$ collides with the initial collision complex within a time period Δt equal to about 10^{-15} sec following its formation, a conventional excited state may result. If a second collision does not occur within the time period Δt, the original photon is emitted, and net absorption does not take place. The intermediate collision complex is often called the *pseudo-state*, and the selection rule for this type of "simultaneous" *double-photon absorption* is opposite to that of the conventional single-photon absorption (Sect. 18.86). Thus, for double-photon absorption the parity-allowed transitions are

Ground state + $h\nu \rightarrow$ pseudo-state ($g \rightarrow u$ or $u \rightarrow g$ are both allowed)

Pseudo-state + $h\nu \rightarrow$ excited state ($u \rightarrow g$ or $g \rightarrow u$ are both allowed)

Ground state + $2h\nu \rightarrow$ excited state ($g \rightarrow g$ or $u \rightarrow u$ are both allowed)

It should be mentioned that double-photon absorption is sometimes used to refer to the process in which higher excited states are produced by the sequential (not simultaneous) absorption of two photons. In *sequential double-photon absorption*, the second photon is absorbed by a conventional intermediate, such as a relatively stable and long-lived excited state (not a pseudo-state). In this type of process, the uncertainty principle plays no role, and the usual single-photon selection rules apply (Sect. 18.86).

18.106 WOODWARD–FIESER FACTORS (RULES) FOR DETERMINING POLYENE ABSORPTION

A set of empirically determined wavelength increments that when added to the absorption maximum of a parent or reference diene system often provides good estimates for the position of the long-wavelength absorption maximum (λ_{max}) of complex polyenes. These factors (rules) are most reliably applied to polyenes incorporated into rigid ring systems, and must be used with caution when dealing with cross-conjugated systems (Sect. 5.33):

- Two types of dienes are used as the parent species to be incremented:
  1. 214 nm for heteroannular, open-chain, or *s-trans* dienes (Fig. 18.106*a*)
  2. 253 nm for homoannular or *s-cis* dienes (Fig. 18.106*b*)
- The following wavelength increments are added to the parent diene or base values:

1. 5nm	Exocyclic double bonds
2. 30 nm	Additional double bonds
3. Substituents	
5 nm	Alkyl, -R
6 nm	Alkoxy, -OR
0 nm	Acyloxy, -OCOR
30 nm	Mercapto, -SR
60 nm	Amino, $-NR_2$
5 nm	Halide

Example. For molecules such as the tetraene in Fig. 18.106*c*, which contains both hetero- and homoannular diene systems, the homoannular diene is taken as the parent

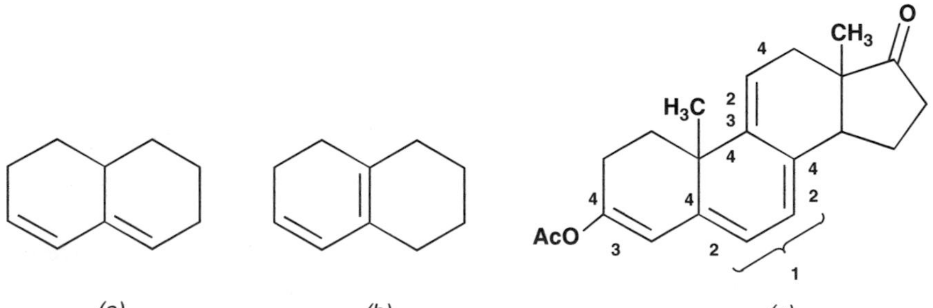

Figure 18.106. Woodward–Fieser rules for polyene absorption. (*a*) Parent heteroannular diene (*s-trans*). (*b*) Parent homoannular diene (*s-cis*). (*c*) Application of the Woodward–Fieser rules to a complex diene system.

diene. If one adds the substituent contributions to this parent diene as indicated in Fig. 18.106*c*, one would predict a λ_{max} of:

1. Homoannular diene	253 nm
2. Exocyclic double bonds (3 × 5 nm)	15 nm
3. Additional double bonds (2 × 30 nm)	60 nm
4. Alkyl groups (5 × 5 nm)	25 nm

λ_{max} (obs) = 355 nm, λ_{max} (calc) = 353 nm

18.107 WOODWARD–FIESER FACTORS (RULES) FOR DETERMINING α,β-UNSATURATED KETONE AND ALDEHYDE ABSORPTION

A set of empirically determining wavelength increments that when added to the absorption maximum of a parent or reference enone system often provides good estimates for the intense π,π*, long-wavelength absorption maximum (λ_{max}) of a complex α,β-unsaturated enone. These factors (rules) are applied in the same fashion as the polyene factors (Sect. 18.106), except that the influence of substituents varies depending on their proximity to the carbonyl moiety [Louis Fieser, (1899–1977)]:

- The parent enones base values are:
  1. 207 nm α,β-Unsaturated aldehyde
  2. 215 nm α,β-Unsaturated ketones
     Acyclic or cyclohexenone
  3. 202 nm Cyclopentenone
- The following wavelength increments are added to the parent enone base values (see Sect. 18.106):
  1. 39 nm Homoannular diene
  2. 5 nm Exocyclic double bond
  3. 30 nm Additional double bond
  4. Substituents as indicated in Table 18.107

Example. The trienone shown in Fig. 18.107*b* would be predicted to exhibit the following λ_{max}:

1. Cyclohexenone parent	215 nm
2. Homoannular diene	39 nm
3. Exocyclic double bonds	5 nm
4. Additional double bonds (2 × 30 nm)	60 nm
5. Alkyl substituents β	12 nm
>δ (3 × 18 nm)	54 nm

λ_{max} (obs) = 385 nm, λ_{max} (calc) = 385 nms

TABLE 18.107. Substituent Factors for α,β-Unsaturated Ketones and Aldehydes[a]

Substituent	α (nm)	β (nm)	γ (nm)	δ (nm) and >
Alkyl, -R	10	12	18	18
Hydroxy, -OH	35	30	50	—
Alkoxy, -OR	35	30	17	31
Acyloxy, -OCOR	6	6	6	—
Mercapto, -SR	—	85	—	—
Chloro, -Cl	15	12	—	—
Bromo, -Br	25	30	—	—
Amino, $-NR_2$	—	95	—	—

[a]See Fig. 18.107*a*.

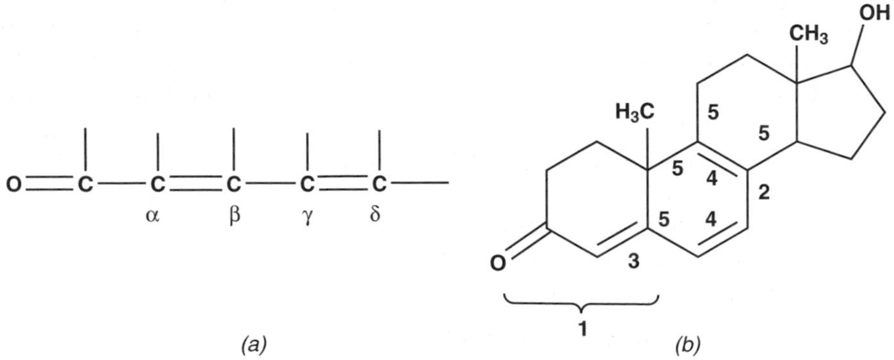

Figure 18.107. Woodward–Fieser rules for α,β-unsaturated enones. (*a*) Parent α,β-unsaturated enone with positions designated. (*b*) Application of Woodward–Fieser rules to a complex enone system.

18.108 *B* AND *L* EXCITED STATES

The excited states of catacondensed aromatics (Sect. 5.41) exhibit a total orbital angular momentum of $Q = \Sigma q = \pm 1$ for a *B* state or $2n + 1$ for a *L* state in a molecule with $2(2n + 1)$ π electrons, where $n = 0, 1, 2, \ldots$ and $q = 0, \pm 1, \pm 2, \pm 3, \ldots$. The transitions leading to these excited states from the ground state give rise to the *B and L absorption bands*, respectively. This nomenclature was designed by John R. Platt [*J. Chem. Phys*. 17, 484 (1949)].

Example. The free electron model (Sect. 3.11) predicts that there will be four low-energy excited states for catacondensed aromatics. These arise as shown in Fig. 18.108*a*. Each doubly degenerate set of orbitals ($q \neq 0$) may be regarded as having one orbital in which the electrons are circulating clockwise with angular momentum $+q$ and the other in which they are circulating counterclockwise with angular momentum $-q$. Using naphthalene ($n = 2$) as an example, we can see (Fig. 18.108*a*) that the ground state *A* has $Q = 0$. Upon excitation (Fig. 18.108*b*), two excited states will be

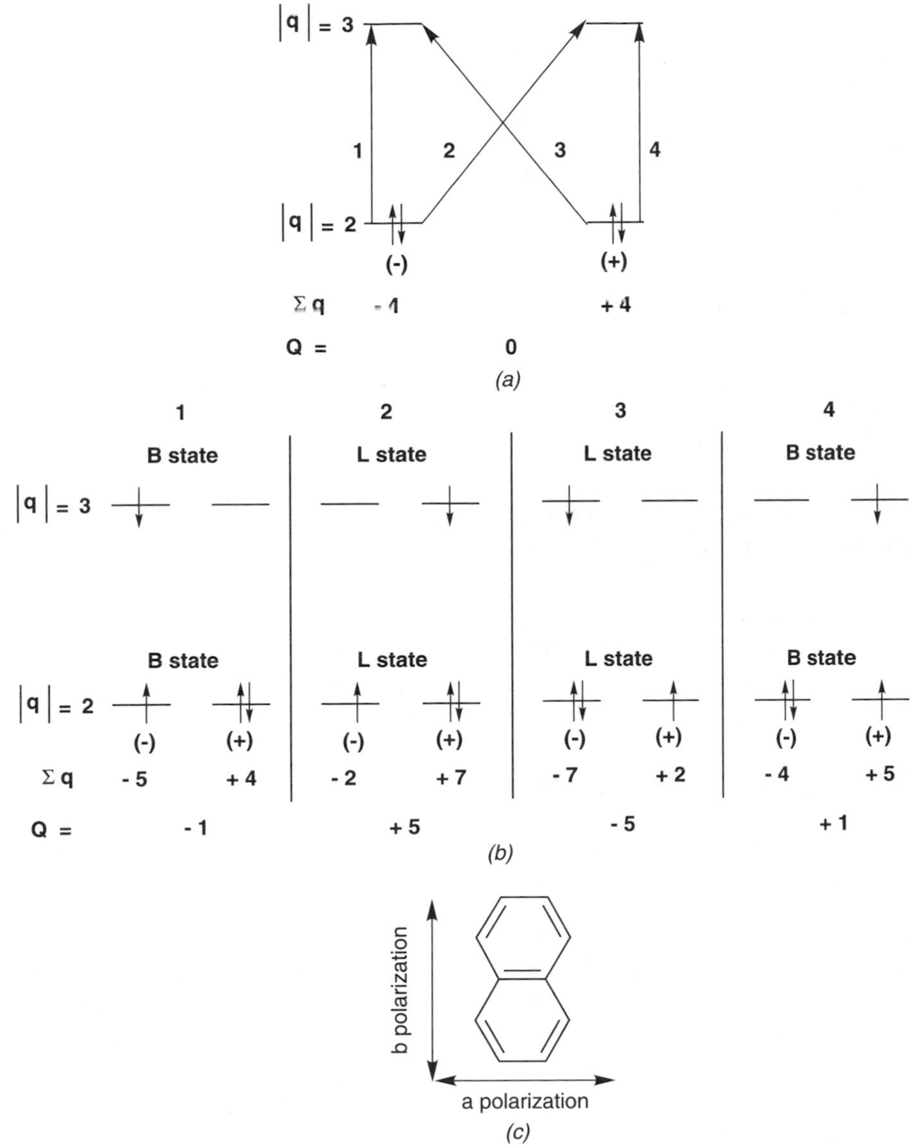

Figure 18.108. *B* and *L* excited-state model. (*a*) Angular momentum of naphthalene ground state and possible transitions to excited states. (*b*) Angular momentum of naphthalene excited states with *B* and *L* designations. (*c*) Alternative polarizations *a* and *b* of naphthalene transitions.

produced with low orbital angular momentum ($Q = \pm 1$). These are the *B* states. In addition, two excited states will be produced with high orbital angular momentum ($Q = \pm 5$). These are the *L* states. Naphthalene and other catacondensed aromatics are not circular because of the linkage between bridgehead carbons. Accordingly, they have potential energy wells that are distorted relative to the circular well of the free electron model, and hence the four excited states will not be degenerate as indicated by the model for benzene (Sect. 3.25). The L states will occur at lower energy than the B states. Both the B and L states will further split, depending on whether the transitions leading to these states are polarized along the long or the short molecular axis, as shown in Fig. 18.108*c*. These two polarizations are denoted by the subscripts b and a, respectively. The long-axis polarized b transitions occur at lower energy and the short-axis polarized a transitions at higher energy. The following transitions observed for naphthalene are assigned as follows: $A \rightarrow L_b$ (~300 nm), $A \rightarrow L_a$ (~270 nm), $A \rightarrow B_b$ (220 nm), $A \rightarrow B_a$ (~167 nm).

18.109 RADIATIVE TRANSITION

Any transition that involves the absorption or emission of a photon: the absorption of light (Sect. 18.38), the emission of light as fluorescence (Sect. 18.129), or phosphorescence (Sect. 18.130).

18.110 NONRADIATIVE TRANSITION (RADIATIONLESS TRANSITION)

Any transition that does not involve the absorption or emission of a photon: internal conversion (IC, Sect. 18.111), intersystem crossing (ISC, Sect. 18.112), and excited-state chemical reactions. Nonradiative transitions are usually designated by wavy arrows, as shown by processes 9 on the Jablonski diagram in Fig. 18.75.

18.111 INTERNAL CONVERSION (IC)

Any process by which nonradiative transitions between vibrational and rotational states of like multiplicity dissipate excited-state energy.

Example. Several examples of internal conversion processes are illustrated in the Jablonski diagram in Fig. 18.75 (processes labeled 9). Vibrational energy may be dissipated by transfer either intramolecularly to vibrational modes not associated with the chromophore, or intermolecularly to molecules in the immediately surrounding medium. This latter process has been called *external conversion*. Thus, excess excited singlet state vibrational energy is dissipated by decay down the vibrational ladder to the 0 vibrational level of that state. Further decay from the 0 level of the excited singlet state first requires isoenergetic transitions to a higher vibrational levels of either the ground singlet state or the triplet state. Once on the ground singlet or triplet state ladders, internal conversion down the ladders leads to the 0 levels of these states.

18.112 INTERSYSTEM CROSSING (ISC)

The process by which a nonradiative transition couples isoenergetic levels of two states with different multiplicities. This is normally a forbidden process, but becomes allowed because of spin-orbit coupling.

Example. In Fig. 18.75, the intersystem crossing from the excited singlet state to the triplet state is designated as process 10 (see also Sect. 18.113).

18.113 SPIN-ORBIT COUPLING

The interaction of the electron magnetic moment arising from the spinning charge within the electron, with the orbital magnetic moment arising from the circulation of charge about the nucleus. This interaction mixes pure states of different multiplicity and, thus, makes transitions between these mixed states more probable.

Example. In transitions between states of different multiplicity, the change in the electron magnetic moment or angular momentum must be balanced by the change in orbital angular momentum so that the total angular momentum of the system remains unchanged. In order for this to occur, the electron spin and orbital angular momenta must be coordinated or coupled, and one must consider them as a unit rather than independently, as in the Born–Oppenheimer approximation (Sect. 18.80).

The expression that defines the probability of intersystem crossing (ISC) from a singlet to a triplet state (Sect. 18.112) is the following:

$$({}^1\Psi_i S\,|H_{\mathrm{so}}|\,{}^3\psi_f T_f) \tag{18.113a}$$

where ${}^1\psi_i$ and ${}^3\psi_f$ are the initial and final electronic wave functions, respectively. The initial state usually is a singlet and is associated with a singlet electronic spin configuration S, and the final state usually is a triplet associated with one of the three triplet electronic spin configurations T_f. The spin-orbit Hamiltonian H_{so} and the singlet spin configuration S are both totally symmetric (a_1) for the carbonyl group (Fig. 18.113) the three possible triplet spin configurations T_f have the symmetries of rotations (T_x, T_y, T_z; b_2, b_1, a_2, respectively) about the three axes of the molecular coordinate system (Fig. 18.113*a*).

If only the n,π^* and π,π^* excited states are considered, and if the lowest-energy singlet state is n,π^*, then one finds that ISC cannot lead directly to the n,π^* triplet state, (Fig. 18.113*b*) since there is no component of the spin configuration T_f with a_1 symmetry. This transition is thus said to be spin-orbit-forbidden. The same situation applies for ISC between the singlet π,π^* and the triplet π,π^* state, which is also spin-orbit-forbidden:

$$[{}^1(n,\pi^*) \bullet S\,|H_{\mathrm{so}}|^3 \bullet (n,\pi^*) \bullet T_f] \tag{18.113b}$$

$$[a_2 \bullet (a_1\, a_1) \bullet a_2 \bullet T_f] = [a_1 \bullet T_f] = 0 \tag{18.113c}$$

In contrast, if ISC leads from the n,π^* singlet directly to a π,π^* triplet state, the transition is spin-orbit-allowed, since the required a_2 triplet spin configuration does exist. It should be noted that the reverse process is also spin-orbit-allowed; the electronic wave functions in the equation below can be interchanged without altering the overall symmetry of the equation (Fig. 18.113*c*):

$$[^1(n,\pi^*) \bullet S\,|H_{\rm so}|^3 \bullet (\pi,\pi^*) \bullet T_f] \quad (18.113\text{d})$$

$$[a_2 \bullet (a_1\, a_1) \bullet a_1 \bullet T_f] = [a_2 \bullet T_f] \quad (18.113\text{e})$$

$$T_f = T_z = a_2 \quad \text{and} \quad [a_2 \bullet T_z] = [a_1] \neq 0 \quad (18.113\text{f})$$

It is instructive to apply the information provided by these symmetry arguments in a graphical fashion more easily understood by organic chemists. This has been done in

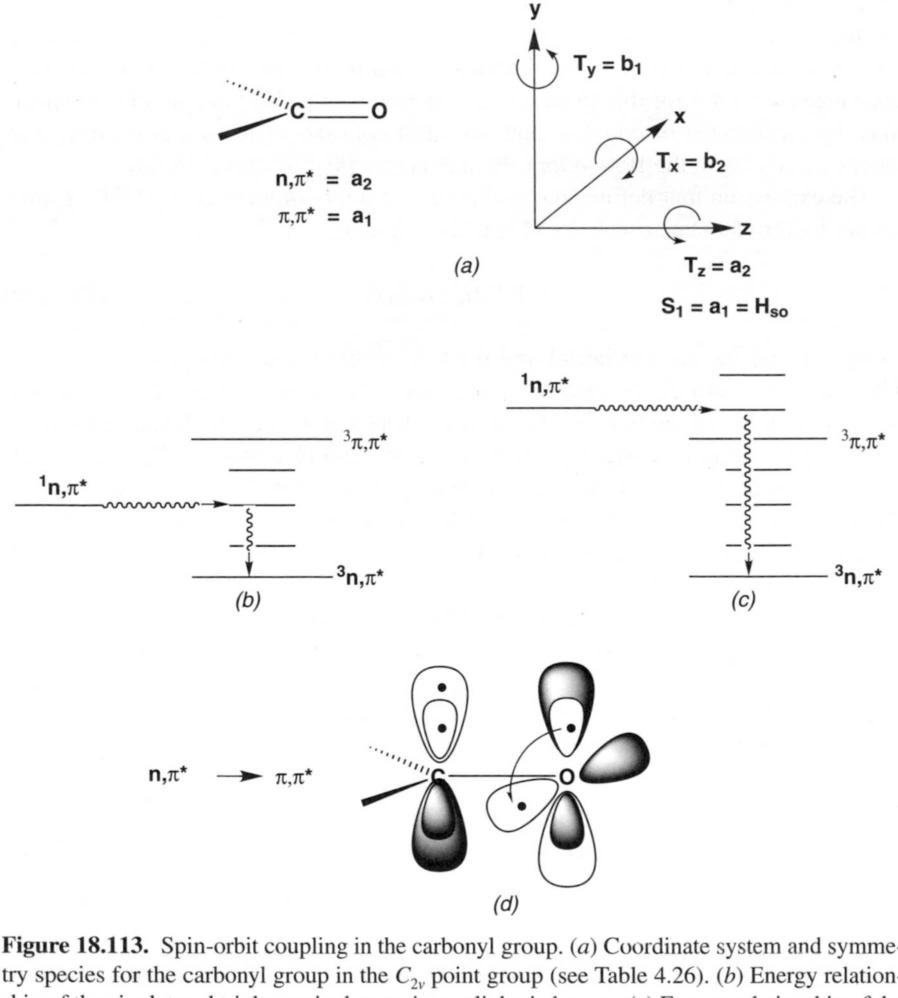

Figure 18.113. Spin-orbit coupling in the carbonyl group. (*a*) Coordinate system and symmetry species for the carbonyl group in the C_{2v} point group (see Table 4.26). (*b*) Energy relationship of the singlet and triplet excited states in an aliphatic ketone. (*c*) Energy relationship of the singlet and triplet excited states in an aryl ketone. (*d*) Relationship between spin and orbit angular momenta during ISC between the *n*,π* and π,π* excited states of a carbonyl group.

Fig. 18.113*d*. When the singlet to triplet state transition occurs, electron density rotates about the carbonyl axis (z-axis) as it shifts from the π orbital to the *n* orbital. The same rotation of charge will occur whether the transition is $n,\pi^* \rightarrow \pi,\pi^*$ or $\pi,\pi^* \rightarrow n,\pi^*$. This rotation of charge creates a magnetic moment $\mathbf{M}_1$ along the carbon–oxygen bond (z-axis). In order to balance this change in angular momentum, the electron spin must simultaneously become polarized so as to produce an equal, but opposite, magnetic moment $\mathbf{M}_2$. Under these circumstances, the net change in angular momentum is zero. The only way that this can occur is for the singlet state to couple with the T_z sublevel of the triplet state during the ISC process. This constitutes a simultaneous transition in both spin and electronic states, spin-orbit coupling.

Efficient ISC will occur in carbonyl groups only when a triplet π,π^* state exists at or below the energy of the singlet n,π^* state (Fig. 18.113*c*). This is not always the case. Aliphatic carbonyl groups can have high-energy π,π^* triplet states, as shown in Fig. 18.113*b*, and thus, they tend to undergo inefficient ISC. In aryl carbonyl groups, conjugation with the aromatic ring lowers the π,π^* triplet state so that it is now available for ISC, as shown in Fig. 18.113*c*, and ISC will be highly efficient.

The same factors increase the probability of ISC in heterocyclic aromatic molecules. Here again, the nonbonded and π orbitals are orthogonal to each other so electron transfer can occur from one to the other in the same fashion as shown in the above example. However, simple aromatic hydrocarbons are lacking the nonbonding orbitals necessary for ISC. As a result, ISC to the triplet states in aromatic hydrocarbons frequently will not occur at all. In cases such as this, triplet photosensitizers are required to produce the triplet states. Aryl ketones are excellent photosensitizers for the reasons noted above, and once their triplet states are formed, they can transfer energy to aromatic hydrocarbons to produce the otherwise inaccessible triplet states of these molecules (see Sect. 18.161).

18.114 HEAVY ATOM EFFECT

The increase in the probability of transitions between states of different multiplicity (ISC) induced by the presence of a heavy atom.

Example. This effect is due to the heavy-atom-induced increase in the efficiency of ISC. The role of the heavy atom is to facilitate spin-orbit coupling that occurs more readily as the electron approaches relativistic velocities near the heavy atom nucleus. This phenomenon is significant with phosphorus, sulfur, and the heavy halogens, bromine and iodine. It becomes more pronounced with the increasing atomic weight of the perturbing atom. It is manifested in several ways. One is the quenching of fluorescence and the associated enhancement of phosphorescence caused by more efficient ISC of a heavy atom. Another heavy-atom-induced process is the enhancement of singlet to triplet absorption. The heavy atom may exert its influence either intramolecularly when it is incorporated into the substrate molecule, the *internal heavy atom effect*, or intermolecularly when it is present in a solvent molecule, the *external heavy atom effect*.

18.115 EMISSION

The production of light in the radiative decay of excited states.

Example. See Fig. 18.75, processes 11 to 14.

18.116 SPONTANEOUS EMISSION

The production of light from the radiative decay of an excited state that occurs without the perturbation of an external radiation field (see Sect. 18.118).

Example. All low-intensity emission phenomena are of this type. Photons produced by spontaneous emission have a random phase; the light produced by spontaneous emission is not coherent (Sects. 18.140–18.142).

18.117 STIMULATED EMISSION

The production of light from the radiative decay of an excited state that is stimulated by the perturbation of an external radiation field (see Sect. 18.118).

Example. This is a high-intensity emission phenomenon that is usually only associated with laser sources. The photons produced by stimulated emission have a phase determined by the perturbing radiation field; the light produced by stimulated emission is coherent (Sects. 18.140–18.142).

18.118 EINSTEIN'S RELATIONS FOR ABSORPTION AND EMISSION

There are two such relationships:

1. The first relationship governs the intensities of absorption and corresponding stimulated emission. The probability B of a transition between two states is equal for both absorption and stimulated emission processes, provided that neither the initial state a nor the final state j is degenerate. When degeneracies exist, this expression has the form shown below, where g_a is the number of degenerate ground states and g_j the number of degenerate excited states:

$$g_a B_{aj} = g_j B_{ja} \tag{18.118a}$$

2. The second defines the relationship between the probabilities for spontaneous emission A_{ja} and stimulated emission B_{ja} as shown in the equation below:

$$B_{ja} = \frac{A_{ja}}{(8\pi h/c^3)\eta^3\nu^3} \tag{18.118b}$$

where h is Planck's constant, ν the frequency of the light emitted, η the refractive index of the medium, and c the speed of light. It can be seen from this equation that it will be much more difficult to produce stimulated emission at short wavelengths (high frequencies) of light.

Example. From the Einstein relationships, it is possible to determine whether a particular sample will absorb light or undergo stimulated emission. The familiar expression of the Beer–Lambert law (Sects. 18.40 and 18.43) has the form

$$I = I_0 e^{-\varepsilon c_a l} \qquad (18.118c)$$

where c_a is the concentration of the ground state of a substance at low light intensities, where the concentration of excited-state species c_j can be neglected. However, at high light intensities where c_j is no longer negligible, the Beer–Lambert law must be expanded to the following form, which applies for all light intensity levels:

$$I = I_0 e^{-\varepsilon\left(c_a - \frac{g_a}{g_j} c_j\right)l} = I_0 e^{\varepsilon N l} \qquad (18.118d)$$

$$N = \left(\frac{g_a}{g_j} c_j - c_a\right) \qquad (18.118e)$$

In these equations, N is the *population inversion* term. If we assume no degeneracies exist, that is, both g_a and g_j are 1, the exponential in the above equation will remain negative so long as $c_a > c_j$. Under these conditions, a light beam passing through the sample will be attenuated and normal absorption will take place. However, when a population inversion exists, $c_j > c_a$, the exponential becomes positive, and a light beam passing through the sample will be amplified rather than attenuated. This phenomenon is known as stimulated emission (Sect. 18.117) or less frequently as *negative absorption.*

18.119 QUANTUM YIELD Φ

A dimensionless number related to the probability of a particular photochemical event taking place.

Example. The quantum yield is expressed as the ratio of the number of times a particular photochemical event takes place to the number of photons absorbed. Alternatively, the quantum yield may be expressed as the ratio of the rate constant k for the process in question to the sum of all competing rate constants. For singlet state processes, the quantum yields or probabilities may be expressed as the ratio of the rate constant for the relevant photochemical process to the sum of the rate constants of all competing singlet processes. Quantum yields for triplet state processes are of the same form reduced by the probability of triplet state formation (the quantum yield for intersystem crossing, Φ_{isc}).

Quantum yields for several important photochemical processes are defined as follows:

1. Fluorescence Φ_f

   = number of fluorescence photons emitted/number of photons absorbed

   $$= k_f / (k_f + k_{\text{isc}} + k_{nr}) \qquad (18.119a)$$

2. Intersystem crossing Φ_{isc}

   = number of intersystem crossing events/number of photons absorbed

   $$= k_{\text{isc}}/(k_f + k_{\text{isc}} + k_{nr}) \qquad (18.119b)$$

3. Phosphorescence Φ_p

   = number of phosphorescence photons emitted/number of photons absorbed

   $$= \Phi_{\text{isc}} k_p/(k_p + k_{nr} + k_q[Q]) \qquad (18.119c)$$

   where the subscript designates the following processes: fluorescence (f), intersystem crossing (isc), nonradiative decay (nr), phosphorescence (p), and quenching (q), and $[Q]$ is the quencher concentration.

18.120 ABSOLUTE QUANTUM YIELD

A quantum yield determined by simultaneously measuring the number of photochemical events and the number of photons absorbed by the same sample.

18.121 RELATIVE QUANTUM YIELD

A quantum yield determined by a comparison of a photochemical property of the molecule under study with that of a standard substance for which the appropriate absolute quantum yield is known.

Example. A relative quantum yield of fluorescence (Φ_f') would be obtained by quantitatively measuring the fluorescence intensity of the molecule being studied (I_f'), and comparing this value with the fluorescence intensity of a standard substance (I_f) determined under the same conditions. Since the absolute quantum yield of the standard (Φ_f) is known:

$$\Phi_f' = (I_f'/I_f)\,\Phi_f \qquad (18.121)$$

18.122 PRIMARY PHOTOCHEMICAL PROCESS

In the words of William A. Noyes and co-workers, this is "the series of events beginning with the absorption of a photon by a molecule, and ending either with the disappearance of that molecule or with its conversion to a state such that its reactivity is statistically no greater that that of similar molecules in thermal equilibrium with their surroundings." [Noyes, W. A., Jr., Porter, G., and Jolley, J. G. *Chem. Rev. 56*, 49 (1956).]

18.123 SECOND LAW OF PHOTOCHEMISTRY

For single-photon absorption events, the sum of the quantum yields for primary processes must be unity:

$$\Phi_f + \Phi_p + \Phi_{nr} + \Phi_{pr} + \Phi_q = 1 \qquad (18.123)$$

where the subscripts designate the following processes: fluorescence (f), phosphorescence (p), nonradiative decay (nr), photochemical reactions (pr), and quenching (q).

18.124 ACTINOMETRY

The measurement of the number of photons absorbed by a sample (*dosage*) or the intensity of light (photons/cm^2-sec). Today, the term has become synonymous with the measurement of quantum yields.

18.125 ACTINOMETER

A physical device or chemical system used for the measurement of light dosage or intensity.

18.126 MERRY-GO-ROUND REACTOR

A mechanical device commonly used for the determination of relative quantum yields.

Example. A typical example of a merry-go-round reactor is shown in Fig. 18.126. Samples are mounted in the merry-go-round at an equal distance around a central light source. Generally, alternate positions in the merry-go-round contain samples of the photochemical systems to be studied. The remaining alternate positions contain chemical actinometers, which are used to determine the amount of light to which the samples are exposed. The central light source and light filters are fixed and the merry-go-round rotates slowly around them. In this way, variations in the light intensity around the source are averaged over all the samples uniformly.

18.127 OBSERVED RADIATIVE LIFETIME τ_{obs}

The experimentally determined time required for an excited-state population to decay to $1/e$ (36.79%) of its initial value after the exciting light has been turned off.

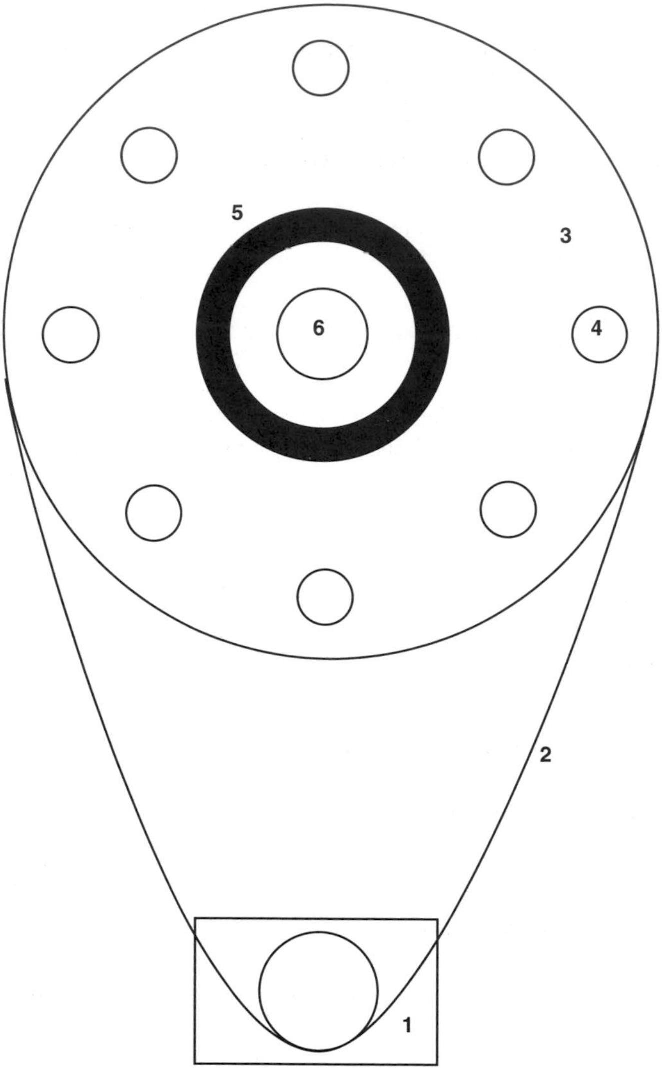

Figure 18.126. Merry-go-round photochemical apparatus for the determination of relative quantum yields (see Sect. 18.121). (1) Drive motor, (2) pulley or gear linkage, (3) merry-go-round, (4) sample mounts, (5) chemical or glass light filter, (6) light source in water-cooled jacket.

Example. For most low-intensity emission processes, the emission decay will be first-order:

$$I = I_0 \exp[-k_{\text{obs}}t] = I_0 \exp[-t/\tau_{\text{obs}}] \tag{18.127}$$

where I is the emission intensity at time t and I_0 the emission intensity at time zero. The observed first-order rate constant for emission decay k_{obs} is the sum of all first-order processes leading to excited-state decay, and τ_{obs} is the observed radiative lifetime

($\tau_{obs} = 1/k_{obs}$). The lifetime τ_{obs} is not an intrinsic property of the emitting molecule, but depends on experimental parameters such as temperature, solvent, and the presence of trace impurities (see Sect. 18.128). Note in the above equation that when $I = I_0 e^{-1}$, $t = \tau_{obs}$, a period of time equal to one lifetime has elapsed.

18.128 NATURAL RADIATIVE LIFETIME τ

An intrinsic property of an excited molecule that equals the reciprocal of the first-order rate constant for its decay. For fluorescence, $\tau_f = 1/k_f$, and for phosphorescence, $\tau_p = 1/k_p$, where k_f and k_p are the first-order rate constants for fluorescence and phosphorescence decay, respectively. The natural radiative lifetime τ would equal the observed lifetime τ_{obs} of an excited state if either fluorescence or phosphorescence were the only processes contributing to excited-state decay.

Example. The natural radiative lifetimes for fluorescence and phosphorescence are determined from the observed lifetimes in the following manner. Since the observed lifetimes τ_{obs} are equal to the reciprocal of the sum of all competing rate constants associated with the decay of the excited state (see Sect. 18.119 for a definition of terms and Sects. 18.129 and 18.130), for fluorescence, which usually originates from an excited singlet state:

$$\tau_{f(obs)} = 1/k_{f(obs)} = 1/(k_f + k_{nr} + k_{isc}) \qquad (18.128a)$$

For phosphorescence, which usually originates from the lowest triplet state:

$$\tau_{p(obs)} = 1/k_{p(obs)} = 1/(k_p + k_{nr} + k_q[Q]) \qquad (18.128b)$$

The natural radiative lifetimes of singlet excited states are equal to the observed lifetimes divided by the quantum yields Φ for the respective processes:

$$\tau_f = \tau_{f(obs)}/\Phi_f = 1/k_f = \{1/(k_f + k_{nr} + k_{isc})\}/\{k_f/(k_f + k_{isc} + k_{nr})\} \qquad (18.128c)$$

The natural radiative lifetime of triplet states must take into consideration the efficiency of triplet state formation, the quantum yield for intersystem crossing (Sect. 18.119):

$$\begin{aligned}\tau_p &= \tau_{p(obs)}(\Phi_{isc}/\Phi_p) = 1/k_p \\ &= \{1/(k_p + k_{nr} + k_q[Q]\} \\ &\quad \times\{k_{isc}/(k_f + k_{isc} + k_{nr})\}/\{k_{isc}/(k_f + k_{isc} + k_{nr})\} \\ &\quad \times\{k_p/(k_p + k_{nr} + k_q[Q])\} \qquad (18.128d)\end{aligned}$$

In essence, the natural radiative lifetime is the lifetime that would be observed if the singlet and triplet excited states were populated with unit efficiency, and if no other processes competed with the process being observed, fluorescence and phosphorescence in the examples above.

18.129 FLUORESCENCE

The emission from an excited state to produce a lower-energy state of the same multiplicity. In organic chemistry, it is the emission associated with the radiative decay of the lowest-energy excited single state back to the singlet ground state (Fig. 18.75, processes 11 and 12).

Example. Since fluorescence arises from a multiplicity-allowed transition, it is a rapid process that usually has a lifetime in the range of 10^{-13} to 10^{-8} sec. A fluorescence spectrum will usually appear on the low-energy side of the absorption spectrum with the only overlap in the two spectra occurring in the region of their respective 0-0 bands (Fig. 18.129). Furthermore, since the vibrational spacings in the ground and excited states will usually be about the same, there will be a rough mirror-image relation between the absorption and fluorescence spectra.

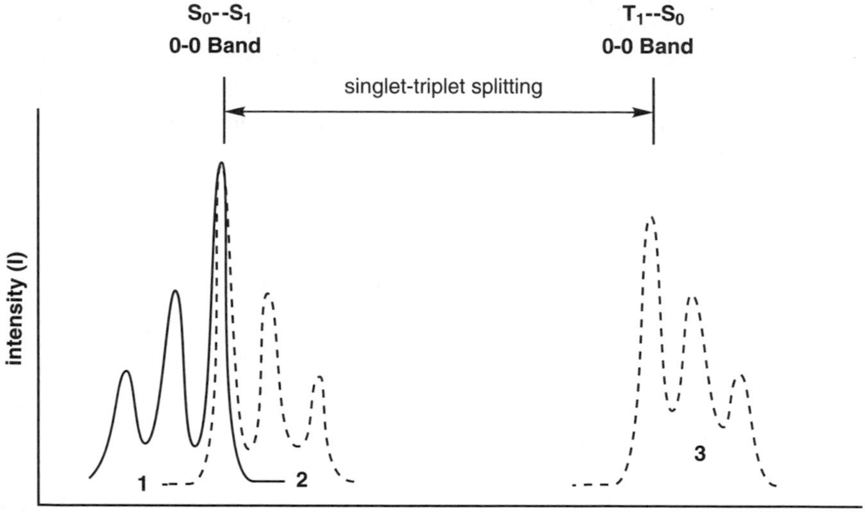

Figure 18.129. General relationship between an absorption spectrum (1, solid line), a fluorescence spectrum (2, broken line), and a phosphorescence spectrum (3).

18.130 PHOSPHORESCENCE

The emission from an excited state to produce a lower-energy state of a different multiplicity. In organic chemistry, phosphorescence is the emission associated with the radiative decay of the lowest triplet excited state back to the singlet ground state (Fig. 18.75, processes 13 and 14, and Fig. 18.129).

Example. Since phosphorescence arises from a multiplicity-forbidden transition, it is a slow process that usually has a lifetime in the range of about 10^{-8} sec to about 60 sec. Some cases of gas phase phosphorescence have lifetimes of up to 1 hour.

A phosphorescence spectrum will also appear on the low-energy side of the absorption spectrum, but there will be no overlap between the two spectra (Fig. 18.129). The 0-0 bands of the two spectra will be separated by an energy equal to or slightly greater than the *singlet–triplet splitting*, the energy between the lowest singlet and lowest triplet excited states. No mirror image relationship is found between absorption and phosphorescence spectra.

18.131 EMISSION SPECTRUM

The probability of light emission by molecules or atoms in their excited states as a function of the wavelength or frequency of the light *emitted*. A plot of emission light intensity (I) versus λ or ν (Fig. 18.129).

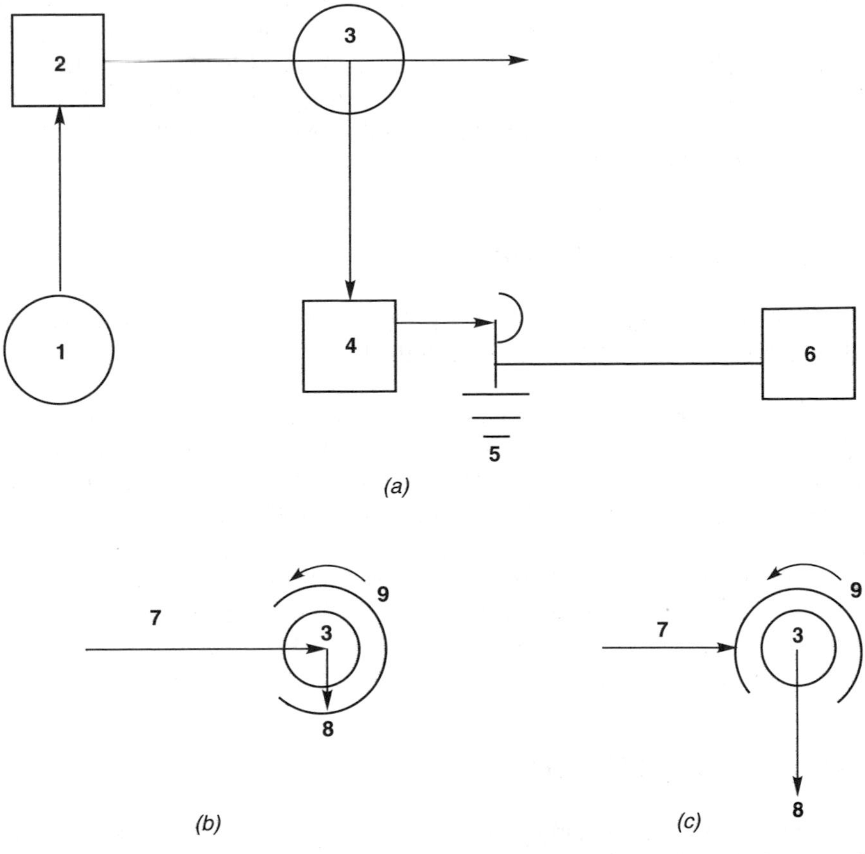

Figure 18.131. (*a*) General configuration of a simple apparatus for the acquisition on emission spectra: (1) excitation source, (2) excitation monochrometer, (3) sample, (4) emission monochrometer, (5) photodectector, (6) recorder. (*b*) and (*c*) Acquisition of long-lived phosphorescence spectra using a rotating shutter to coordinate excitation and data acquisition, (7) exciting light, (8) emitted light, and (9) rotating shutter.

Example. The configuration of the apparatus commonly used to determine emission spectra is illustrated in Fig. 18.131*a*. An appropriate wavelength of light is selected with the excitation monochromator, and this wavelength is used to irradiate the sample throughout the acquisition of the emission spectrum (see Sect. 18.132 for an alternative procedure). The emission from the sample usually is observed at 90° to the excitation light beam and analyzed by scanning the emission monochromator. The *total emission spectrum* is obtained by recording the spectrum of the light emitted at the same time as the sample is being irradiated. In order to obtain the *phosphorescence spectrum*, a shutter or shutters must be placed between the sample and both monochromators. In the configuration *a*, shown in Fig. 18.131*b*, the shutter first is opened to the excitation monochromator, but closed to the emission monochromator. The sample is excited and the shutter rotates until it becomes closed to the excitation monochromator and opened to the emission monochromator, configuration *b* in Fig. 18.131*c*. During the time that it takes for shutter rotation, any short-lived fluorescence emission has decayed and only the longer-lived phosphorescence emission remains. In modern instruments, the rotating shutter, shown in Fig. 18.131, is replaced by a pulsed light source and synchronized, electronically gated photodetector. Phosphorescence spectra are usually acquired at low temperatures (77 K). The *fluorescence spectrum* usually is taken to be the difference between the total emission spectrum and the phosphorescence spectrum.

18.132 EXCITATION SPECTRUM

The probability of light emission by molecules or atoms in their excited states as a function of the wavelength or frequency of the light *absorbed*.

Example. The apparatus configuration for the determination of an excitation spectrum is the same as that used in the determination of an emission spectrum (Fig. 18.131*a*). The procedure differs in that the excitation spectrum is determined by scanning the excitation monochromator and observing the emission at a fixed wavelength. The excitation spectrum of a substance will usually be very similar to the absorption spectrum of that substance. Excitation spectra are useful for the determination of the absorption spectra of very dilute fluorescent substances in either pure form or mixtures. In photochemical studies, they are useful when the purity of a sample is in question, or when the origin of a weak or unusual emission band has to be determined.

18.133 DELAYED FLUORESCENCE

Emission that exhibits the normal fluorescence spectrum, but which decays with a much longer lifetime. Lifetimes associated with delayed fluorescence are in the range expected for phosphorescence (triplet state) decay rather than normal fluorescence (singlet state) decay. There are two general mechanisms that give rise to delayed fluorescence; these are described in Sects. 18.134 and 18.135.

18.134 THERMALLY ACTIVATED DELAYED FLUORESCENCE (EOSIN- OR E-TYPE DELAYED FLUORESCENCE)

Delayed fluorescence that exhibits an increase in intensity with increasing temperature. This type of delayed fluorescence exhibits the same lifetime as triplet state decay and occurs with an intensity that is proportional to the exciting light intensity. It is the type of delayed fluorescence exhibited by the dye eosin and is accordingly named *E-type delayed fluorescence.*

Example. Delayed fluorescence of the E-type originates from the thermally activated intersystem crossing from the longer-lived triplet state back to the excited singlet state, followed by emission to the ground singlet state (Fig. 18.134). This process is only observed for molecules with small singlet–triplet splittings.

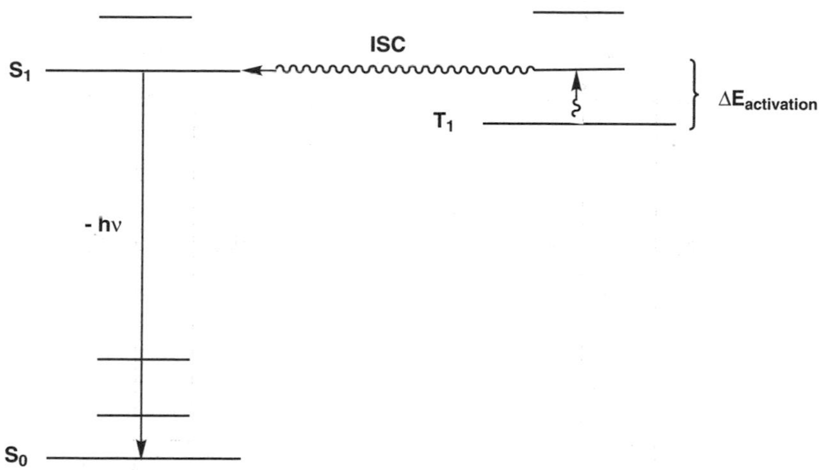

Figure 18.134. Mechanism for thermally activated or E-type delayed fluorescence. $\Delta E_{\text{activation}}$ = singlet–triplet splitting.

18.135 BIMOLECULAR DELAYED FLUORESCENCE (PYRENE- OR P-TYPE DELAYED FLUORESCENCE)

Delayed fluorescence that exhibits an increase in intensity that is proportional to the square of the exciting light intensity. This type of delayed fluorescence exhibits a lifetime equal to one-half that of the triplet state decay and with an intensity that does not follow an exponential temperature law. It is the type of delayed fluorescence exhibited by the aromatic hydrocarbon pyrene and is accordingly called *P-type delayed fluorescence.*

Example. Delayed fluorescence of the P-type originates from a triplet–triplet annihilation process (see Sect. 18.159, example 5) that produces an excited singlet species which can undergo fluorescence emission with return to the ground state.

18.136 STOKES BAND SHIFT

The displacement of a fluorescence emission band to lower energy (longer wavelength) relative to the absorption band (see Sect. 16.63).

Example. This is the usual situation observed in fluorescence spectroscopy. Excitation produces a vibronic singlet excited state, which first undergoes dissipation of vibrational energy via internal conversion to the 0 level of the excited singlet state, followed by emission in the 0-0 band (Fig. 18.136*a*).

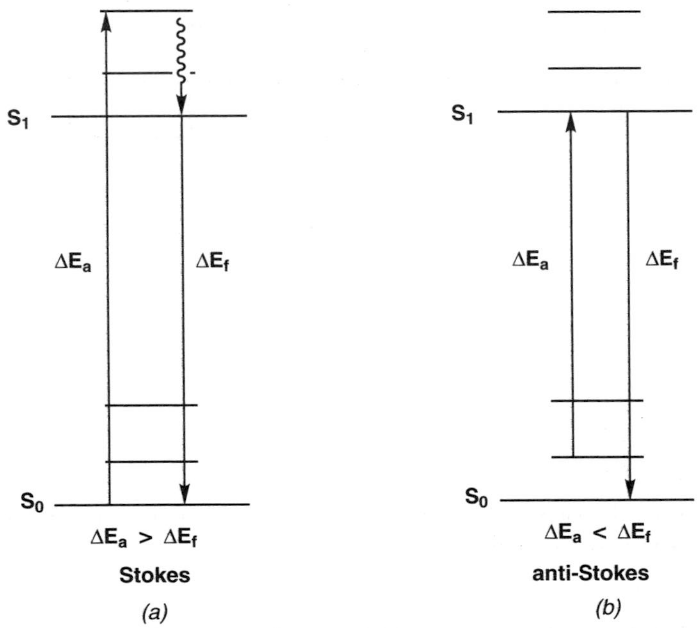

Figure 18.136. Energy shifts of bands in fluorescence spectrum relative to those in the absorption spectrum: (*a*) Stokes shift is to lower energy and (*b*) anti-Stokes is to higher energy.

18.137 ANTI-STOKES BAND SHIFT

The displacement of a fluorescence emission band to higher energy (shorter wavelength) relative to the absorption band (see Sect. 16.63).

Example. This is an uncommon phenomenon. It arises from the excitation of a hot transition (Sect. 18.68) to the 0 level of the excited singlet state followed by emission in the 0-0 band (Sect. 18.66, Fig. 18.136*b*).

18.138 FÖRSTER CYCLE

A thermodynamic cycle that relates the acidities of ground and excited states via the absorption and emission energies of acidic compounds and their conjugate bases. Named after Theador Förster.

Example. A Förster cycle for a molecule whose acidic form absorbs/emits at shorter wavelength than its conjugate base is shown in Fig. 18.138*a*. In this diagram, ΔE and $\Delta E'$ are the 0-0 band energies (Sect. 18.66) of the acid and its conjugate base, respectively, and ΔH and ΔH^* are the enthalpies of ionization in the ground and excited states, respectively. The following relationships can be derived from this cycle:

$$\Delta E + \Delta H^* = \Delta E' + \Delta H \tag{18.138a}$$

$$\Delta E - \Delta E' = \Delta H - \Delta H^* \tag{18.138b}$$

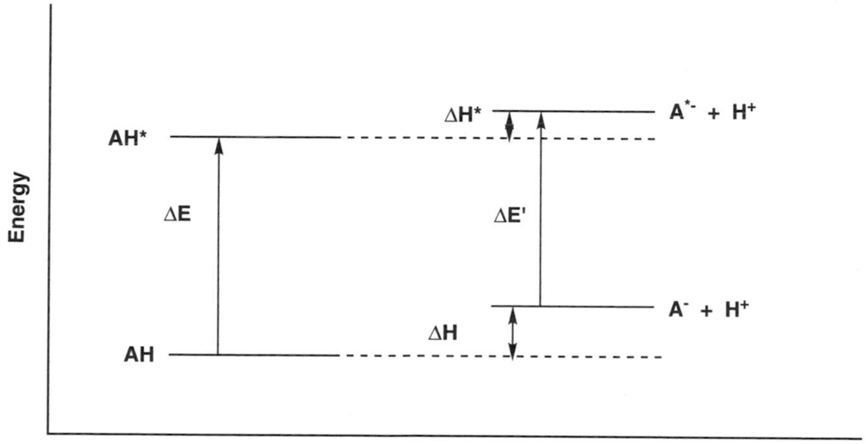

(a)

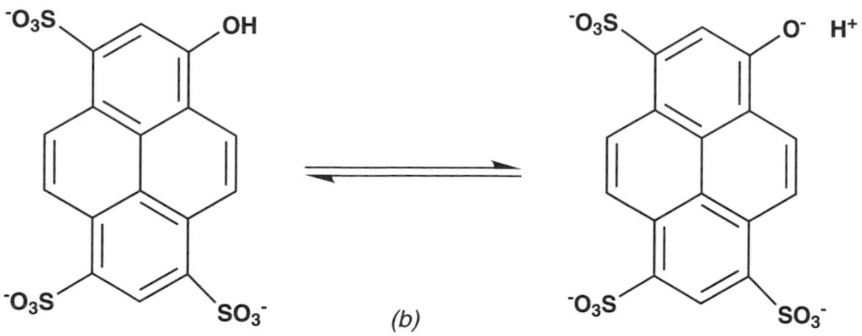

(b)

Figure 18.138. (*a*) Förster cycle for a molecule whose conjugate base absorbs/emits at longer wavelength than the acidic form. (*b*) A pyrene derivative that is greater than 10^6 times more acidic in its lowest singlet excited state than its ground state.

If one makes the assumption that $\Delta S = \Delta S^*$, then

$$\Delta G - \Delta G^* = \Delta H - \Delta H^* = \Delta E - \Delta E' = RT \ln(K_a^*/K_a) \qquad (18.138c)$$

and

$$(\Delta E - \Delta E')/RT = pK_a - pK_a^* \qquad (18.138d)$$

Since pK_a, ΔE, and $\Delta E'$ are readily determined experimentally, pK_a^* can be easily calculated. In the case of the pyrene derivative shown in Fig. 18.138*b*, the pK_a of the ground state is 7.3, whereas fluorescence data indicate that the pK_a^* of the singlet excited state is 1.0. In general, excited singlet states are found to be about a million times more or less acidic than their ground states. In contrast, triplet excited states have about the same acidities as their ground states.

18.139 LASER (LIGHT AMPLIFICATION BY STIMULATED EMISSION OF RADIATION)

A device with which an excited-state population inversion (see Sects. 18.117 and 18.118) can be produced and maintained long enough to give rise to a beam of coherent light (Sects. 18.140–18.142) by a stimulated emission process.

Example. A typical laser consists of an emitting medium in which an excited-state population inversion is produced either by an intense flash of light or by electron bombardment (Fig. 18.139). The medium usually is situated within a resonance cavity between two mirrors (*Fabry–Perot configuration*). Upon achievement of a population inversion, only those photons that are emitted spontaneously along the optical axis of the cavity will produce a significant avalanche of photons by stimulated emission. As this light is reflected (resonates) back and forth between the cavity mirrors, it will gain in intensity (be amplified) with each pass through the

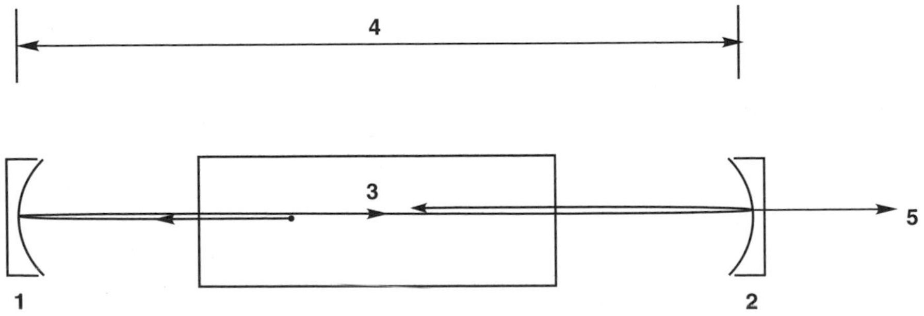

Figure 18.139. Fabry–Perot laser cavity configuration for amplification of light by stimulated emission: (1) back mirror, (2) front mirror, (3) emitting medium with inverted excited population, (4) resonance cavity, (5) laser output.

emitting medium. The cavity is designed so that the back mirror reflects 100% of the laser light and the front mirror reflects only about 95% of the laser light. The 5% of the light transmitted by the front mirror constitutes the laser output. Laser emission will be observed only if the amplification within the emitting medium exceeds the losses within the cavity due to scattering, spurious reflections, and absorption.

18.140 COHERENCE OF LIGHT

Those properties of a light beam that determine the degree to which the beam behaves as a single three-dimensional wave of electromagnetic radiation (Sects. 18.141 and 18.142).

18.141 SPATIAL COHERENCE OF LIGHT

The symmetrical variation of the light electric field strength across a beam cross section.

Example. For a circular beam, the electric field strength $\overline{\mathbf{E}}^2$ must vary in a symmetric fashion, as shown in Fig. 18.141. These electric field patterns correspond to

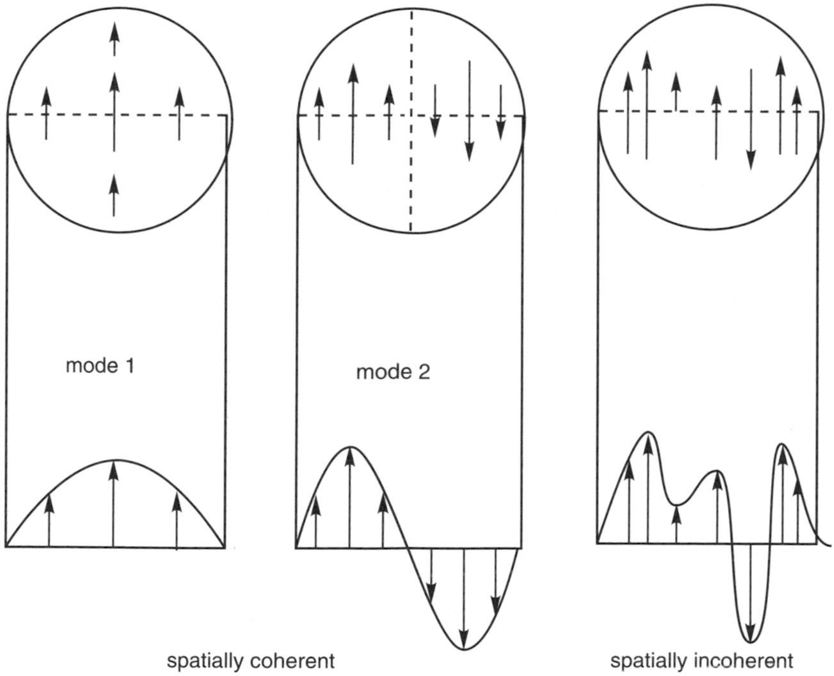

Figure 18.141. Light electric field patterns associated with various modes of laser resonance.

various modes of laser resonance (see also discussion of the transverse mode in Sect. 18.145).

18.142 TEMPORAL COHERENCE OF LIGHT

The generation of a light wave of a single frequency and polarization form for a significant period of time. The distance such a wave travels before being attenuated or undergoing a frequency shift is the *coherence length* of the beam (Fig. 18.142). The longer the coherence length is, the greater the temporal coherence of the beam.

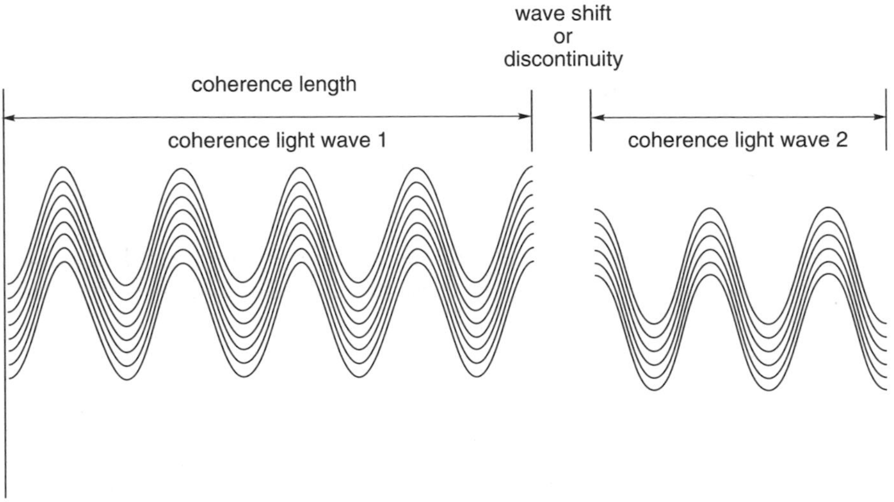

Figure 18.142. Temporal beam coherence as limited by (*a*) beam discontinuity or attenuation and (*b*) frequency shift in the light.

18.143 C-W LASER (CONTINUOUS-WAVE LASER)

Any laser with a continuous (not pulsed) output of light is a C-W laser.

Example. Commonly encountered continuous-wave (C-W) lasers are helium-neon (632.8 nm), argon ion in (488.9 and 514.5 nm), and carbon dioxide (10,632 nm) lasers.

18.144 PULSED LASER

Any laser with an intermittent (not continuous) output of light is a pulsed laser.

Example. Commonly encountered lasers of this type are ruby (693.4 nm) neodymium-YAG (1,064.8 nm) and titanium-doped sapphire (700–1000 nm) lasers.

18.145 LASER MODE

Any three-dimensional standing wave that exists within the resonance cavity of a laser constitutes a mode. The three-dimensional electromagnetic waves associated with these modes must differ by an integral multiple of 2π radians in order to be in resonance upon completing a round trip of the laser cavity.

Example. These electro-magnetic *oscillation modes* may be resolved into two-dimensional components: a *transverse mode*, which is perpendicular to the optical axis of the cavity (Fig. 18.141), and an *axial mode*, which is parallel to the optical axis of the cavity (Fig. 18.145, mode 2). In the axial mode, mode 1 will have no gain since the forward and reverse waves annihilate each other. In contrast, mode 2 will have high gain since the forward and reverse waves reinforce each other.

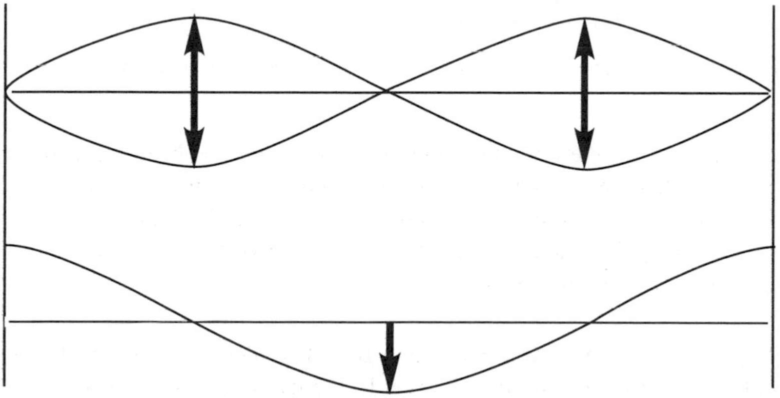

Figure 18.145. Two axial resonance modes for light within a Fabry–Perot resonance cavity: Mode 1 results in beam annihilation and mode 2 results in beam amplification.

18.146 NORMAL MODE LASER

This is the simplest type of pulsed laser consisting of nothing more than the basic Fabry–Perot cavity and active laser medium. Such a laser will emit randomly in a number of different modes and produce light as a series of closely spaced spikes, the envelope of which will be several microseconds (10^{-6} sec) in duration (Fig. 18.146).

18.147 Q-SWITCHED LASER

A normal mode, pulsed laser into the cavity of which has been placed an ultrafast light shutter (quality factor or *Q-switch*) (Fig. 18.147*a*). When the flashlamps fire, the shutter is closed so that resonance within the cavity cannot occur. Under these conditions, the excited-state population is increased beyond the levels achieved with the normal

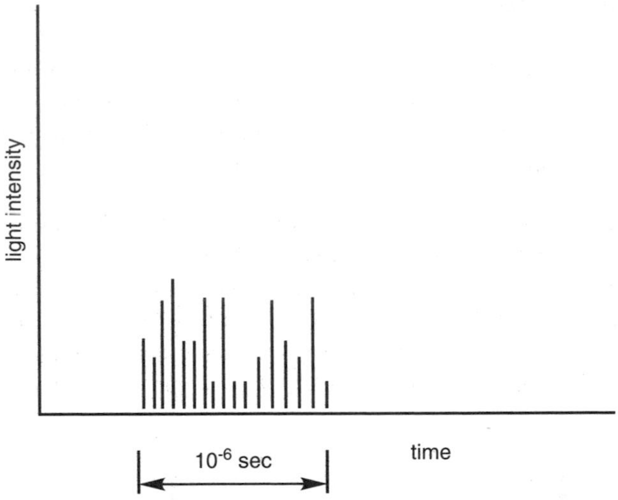

Figure 18.146. Characteristic light pulse pattern of a typical normal mode laser.

mode configuration. When the optimum population inversion has been reached, the shutter is opened suddenly, establishing a resonance condition within the cavity. At this point, all the available energy stored in the active laser medium is emitted in a single giant pulse. Such giant pulses have a width of a few nanoseconds (10^{-9} sec) and peak powers of up to several gigawatts (10^9 W).

Example. A Q-switch may be a Kerr cell (Sect. 18.30), a Pöckels cell (Sect. 18.31), or most simply, a solution of a bleachable dye. In the latter case, the concentration of the dye solution is adjusted so that at low light intensities, all the light along the optical axis of the cavity is absorbed. Since Q-switch dyes are selected specifically to have excited states that do not absorb the emitted light, as the population of the dye excited state increases, the dye solution becomes more transparent (bleaches) at the wavelength of the emitted light. Finally, the emitted light "burns" its way through the dye solution and laser action ensues. Since the dye excited states decay to the ground states in a few picoseconds (10^{-12} sec), the dye "shutter" (Q-switch) closes very rapidly. Consequently, only those modes that have the highest amplification (gain) will develop sufficient intensity to pass through the dye cell. The nanosecond (10^{-9} sec) laser pulses produced by a Q-switched laser are, in fact, envelopes of picosecond pulses derived from the modes of highest gain (Fig. 18.147*b*).

18.148 MODE-LOCKED LASER

A pulsed or C-W laser that has been adjusted (tuned) so that the output is derived from a single high-gain mode. In either case, this output consists of a train of gigawatt (10^9 W) single-mode pulses with pulse widths from about one-half to several hundred

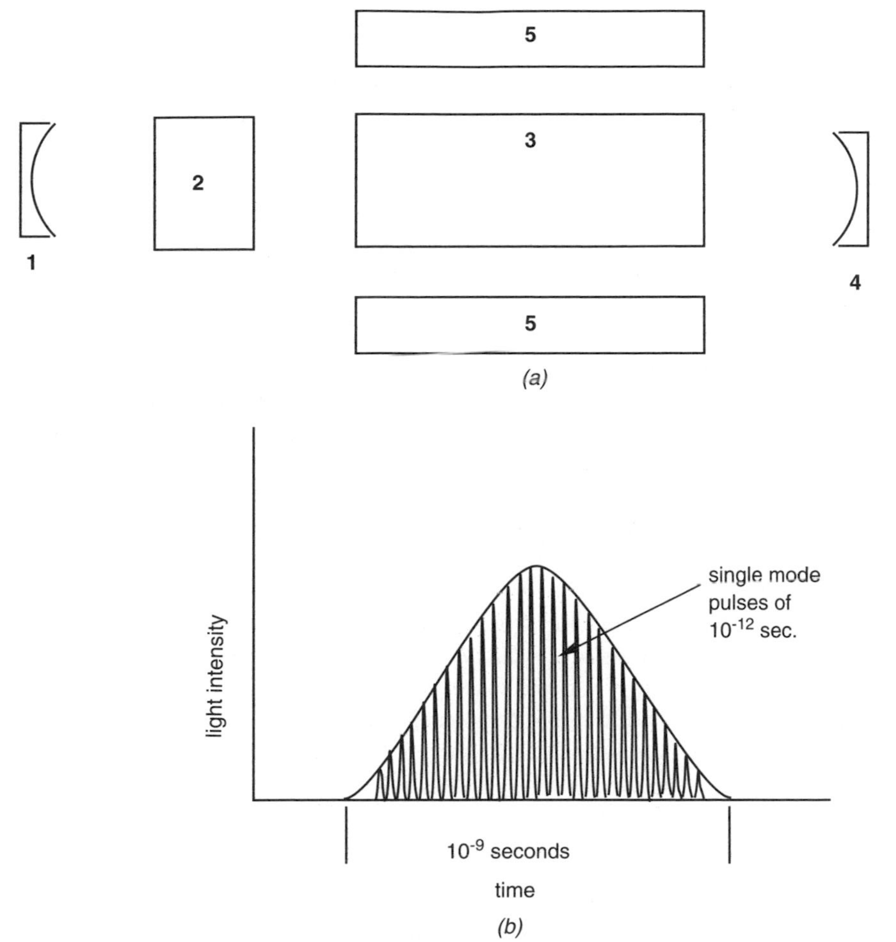

Figure 18.147. (*a*) Application of a Q-switch (ultra-fast light shutter) in the generation of giant laser pulses: (1) back mirror, (2) bleachable dye or electro-optic shutter, (3) laser material, (4) front mirror, (5) flash lamp. (*b*) Expanded view of a "single" Q-switched laser pulse.

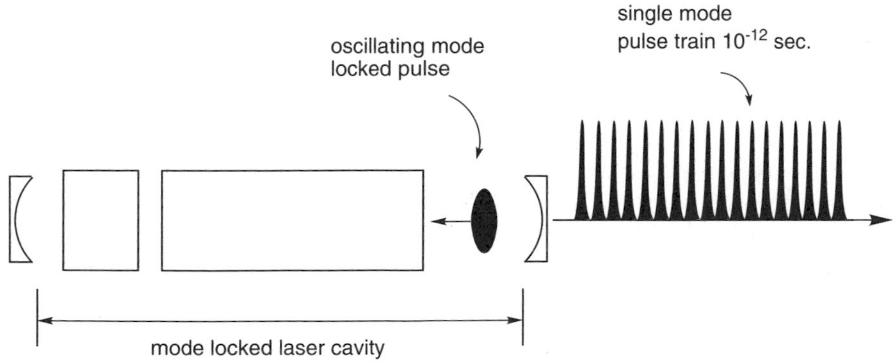

Figure 18.148. Optical characteristics of a mode-locked laser.

picoseconds (10^{-12} sec). Each pulse in the train is separated from the previous pulse by the time Δt that it takes a pulse to make a round trip up and down the laser cavity:

$$\Delta t = 2L/c \tag{18.148}$$

where L is the optical length of the cavity (the cavity length adjusted for the refractive indices of the various cavity components) and c is the speed of light in a vacuum (Figs. 18.147*b* and 18.148).

Example. Mode-locking a pulsed laser consists of simply adjusting the concentration of the Q-switch (Sect. 18.147) so that only the mode with the highest gain can bleach the dye cell and have enough residual intensity to be amplified. Mode-locked, pulsed lasers produce a train of picosecond pulses that last for several nanoseconds. A continuous train of mode-locked pulses can be obtained from a mode-locked, C-W laser by means of an intracavity acousto-optic modulator. This technique can be refined further by making both cavity mirrors 100% reflecting, allowing the mode-locked pulse to oscillate within the cavity, and be amplified over several round trips. At the appropriate time, the acousto-optic modulator switches and "dumps" the entire pulse out of the cavity. This technique is called *cavity-dumping* and has the advantage of providing more intense pulses with a longer variable spacing between pulses. Mode-locked lasers have revolutionized the study of very fast chemical phenomena, since they make possible the direct spectroscopic observation of extremely rapid events such as those that occur in photosynthesis, the photochemistry of vision, and many other very rapid processes.

18.149 DYE LASER

A laser based on stimulated emission in a fluorescence band of a dye.

Example. The advantage of this type of laser is that it may be tuned continuously over the more intense wavelengths of the dye fluorescence. A series of dyes that fluorescence at different wavelengths affords a dye laser system that may be tuned over the entire visible region of the spectrum (C-W dye laser) and even into the near ultraviolet region (pulsed dye laser).

18.150 FLASH PHOTOLYSIS (LASER FLASH PHOTOLYSIS (LFP))

The technique of exposing a photoreactive sample to an intense flash of light, which generates high concentrations of short-lived species such as excited states and transient reaction intermediates. These species then can be observed spectroscopically, and their structure and reaction characteristics determined. A typical flash photolysis apparatus is shown in Fig. 18.150. Strictly speaking, the term *photolysis* means to cleave with light (from the Greek word, *-lysis*, meaning "loosening" or "breaking"). Nevertheless, this term is generally applied even though not all photochemical reactions result in bond breakage.

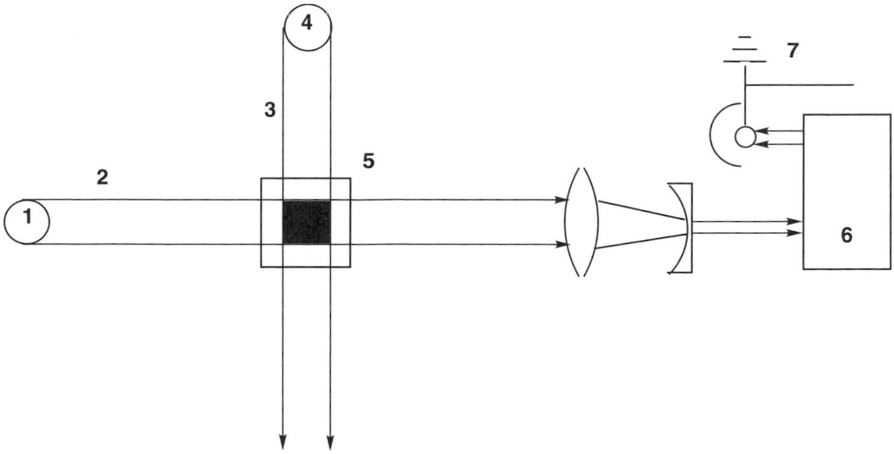

Figure 18.150. Flash photolysis apparatus for the generation and observation of transient species: (1) broad spectrum source, (2) analyzing beam, (3) excitation beam, (4) flash lamp or laser excitation source, (5) sample, (6) monochrometer, (7) photodetector output to oscilloscope or computer.

18.151 CHEMILUMINESCENCE (BIOCHEMILUMINENSCENCE)

The emission of light from excited-state molecules that are produced with the energy released in the thermal decomposition of high-energy molecules.

Example. In order for chemiluminescence to occur, the free energy change associated with the thermal decomposition plus a small contribution of Boltzmann energy must provide enough energy to populate the excited state of one of the product molecules. Consequently, chemiluminescence is always associated with the decomposition of highly strained molecules (Fig. 18.151*a*), or molecules that contain high-energy bonds, usually peroxide linkages (Fig. 18.151*b*). The most famous examples of this phenomenon involve molecules that contain both high-energy bonds and are also highly strained such the dioxetane shown in Fig. 18.151*b*. This latter process is the source of firefly light, and hence, it is an example of *biochemiluminenscence*.

18.152 ELECTROCHEMILUMINESCENCE

The emission of light observed when an alternating electric current is passed through a solution of a fluorescent substance.

Example. This phenomenon arises from the production of radical cations during the oxidizing phase of the voltage cycle and radical anions during the reducing phase. These two species then undergo a mutual annihilation via an electron transfer process

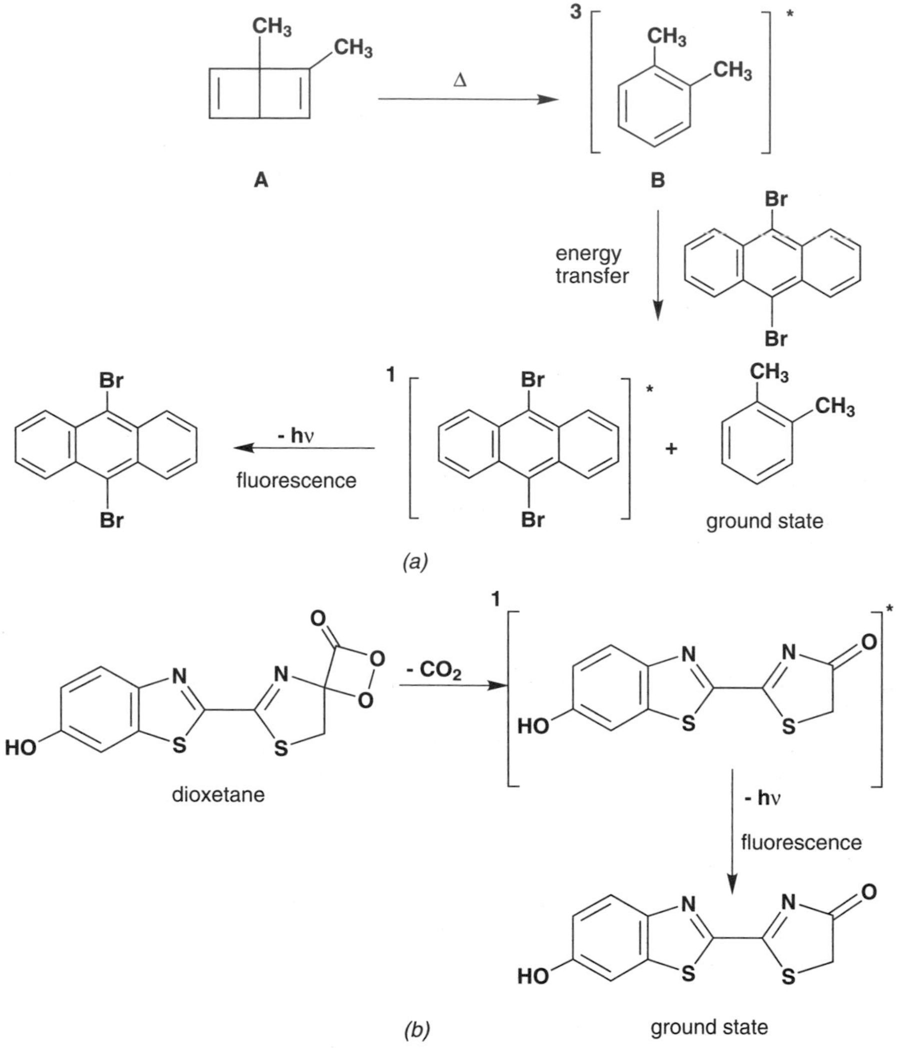

Figure 18.151. Chemiluminescent systems: (*a*) thermally induced triplet formation and chemiluminescence induced by a highly strained Dewar benzene. $A \rightarrow B$ $\Delta G = -88$ kcal/mol and $B \rightarrow {}^3B^*$ $E_T = 82$ kcal/mol, and thus, the net reaction is -6 kcal/mol exothermic. (*b*) Dioxetane decomposition is the source of firefly biochemiluminescence.

(Sect. 18.170) to produce a ground-state molecule and an excited-state molecule, as shown in Fig. 18.152. Excited-state formation might occur by either of two pathways: transfer of the unpaired radical anion electron to the radical cation (pathway 1 in Fig. 18.152*b*), or transfer of a paired electron to the radical cation (pathway 2 in Fig. 18.152*b*). The fluorescence observed results from the emission of the excited state species.

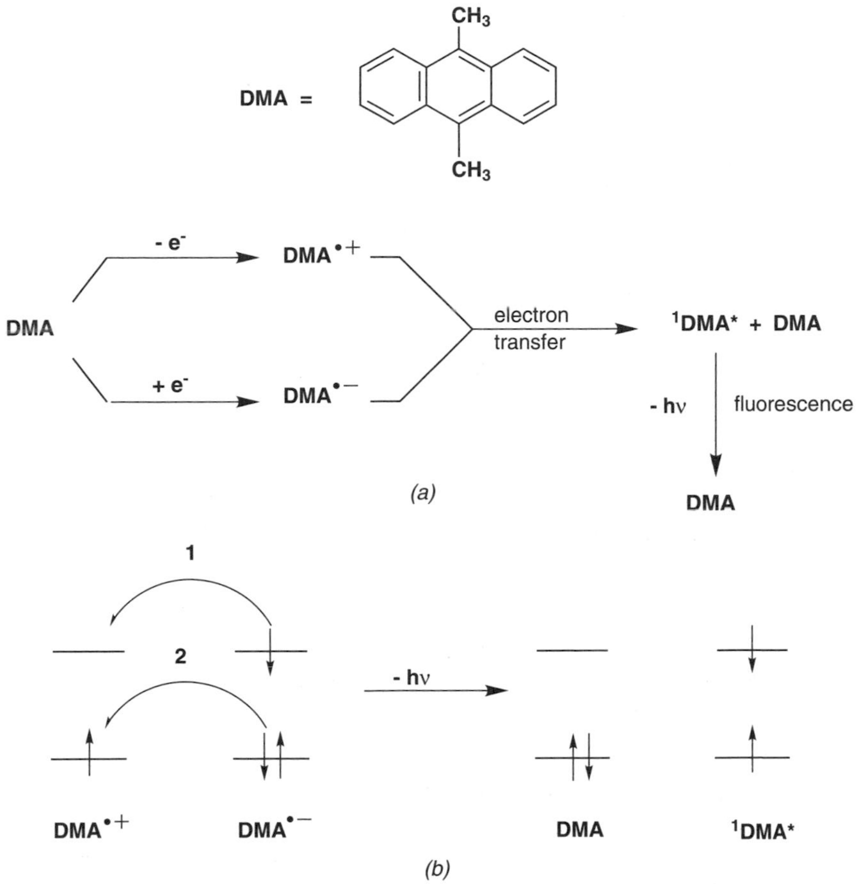

Figure 18.152. Mechanism of electrochemiluminescence. (*a*) Electrochemiluminescence of 9,10-dimethylanthracene. (*b*) Alternative mechanisms for radical ion annihilation and associated excited-state formation.

18.153 PHOTOCHROMISM

A reversible process by which a material changes color upon exposure to light and reverts to its original color upon storage in the dark.

Example. This phenomenon arises from a photochemical transformation of a usually colorless material into a new, brightly colored substance (Fig. 18.153). The colored substance then slowly reverts to the colorless starting material in a dark reaction that occurs over several seconds to several days.

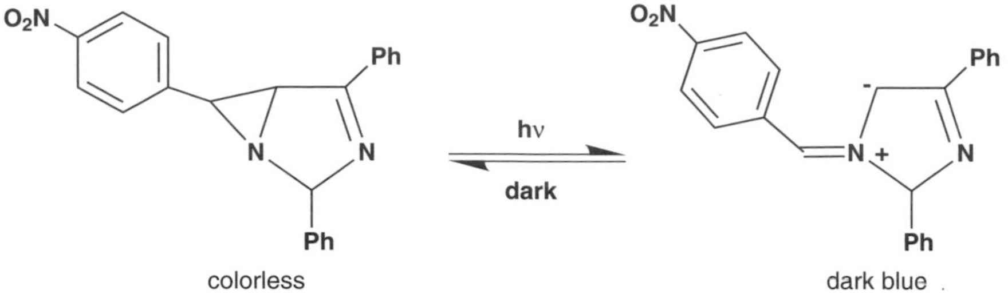

Figure 18.153. Photochromism of an aziridine.

18.154 TRANSFER OF ELECTRONIC ENERGY (EXCITATION TRANSFER)

The transfer of excited-state electronic energy from a donor molecule D^* to an acceptor molecule A. The electronic energy that is transferred may be converted into vibrational, rotational or translational energy, but most frequently it will promote the acceptor molecule into an electronic excited state A^*:

$$D^* + A \rightarrow D + A^*$$

This electronic energy transfer can occur by several mechanisms. These are described in Sects. 18.155 through 18.157 and 18.159.

18.155 RADIATIVE ENERGY TRANSFER

A straightforward, two-step energy transfer process in which a photon is emitted by the donor D^* and absorbed by the acceptor A:

$$D^* \rightarrow D + h\nu \qquad (18.155a)$$

$$h\nu + A \rightarrow A^* \qquad (18.155b)$$

Example. In order for this process to occur, the emission spectrum of the donor must overlap the absorption spectrum of the acceptor (Fig. 18.155). This type of energy transfer can occur over very great distances, as, for example, from the sun to the earth. It is further characterized by a reabsorption probability that varies as R^{-2}, and a donor lifetime that is not influenced by the presence of the acceptor. This is the core process in photosynthesis where Eq. 18.155*a* occurs on the sun and 18.155*b* occurs on earth.

18.156 RADIATIONLESS ENERGY TRANSFER (COUPLED OR RESONANT TRANSFER)

A single-step energy transfer process that does not involve a photon, but instead proceeds by the coupled transfer of energy or the synchronous demotion of the donor ($D^* \rightarrow D$) and promotion of the acceptor ($A \rightarrow A^*$) (Fig. 18.156).

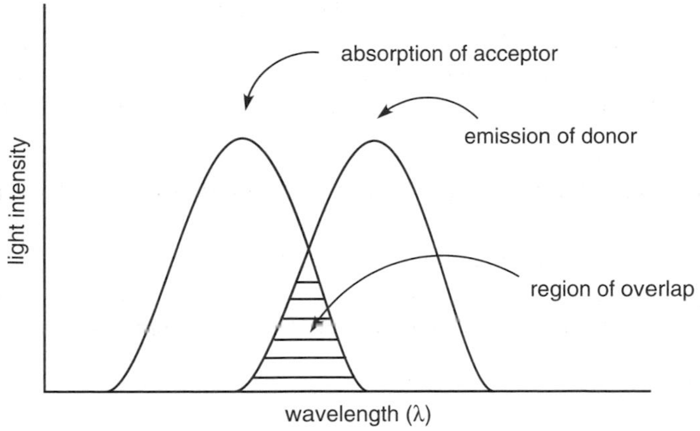

Figure 18.155. Spectral overlap condition for radiative energy transfer between two molecules. Only photons emitted in the region of spectral overlap will have a finite probability of absorption by the acceptor molecule.

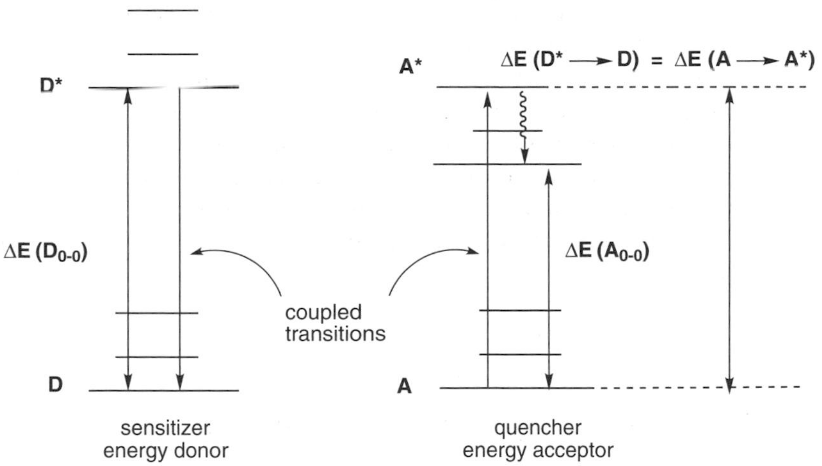

Figure 18.156. Energy relationships between donor and acceptor states in radiationless energy transfer.

Example. This type of energy transfer may proceed by either an electrostatic interaction between the donor–acceptor pair (Sect. 18.157), or an exchange of electrons between the donor–acceptor pair (Sect. 18.159). Both mechanisms require that the initial $(D^* + A)$ and final $(D + A^*)$ states be degenerate (Fig. 18.156). Energy transfer usually will take place between the zero-vibrational level of the excited donor and higher vibrational level of the acceptor. This requires that the excited-state energy (0-0 level) of the acceptor be equal to or lower than that of the donor $[\Delta E(D_{0\text{-}0}) \geq \Delta E(A_{0\text{-}0})]$. Under these circumstances, the transfer usually will become irreversible upon the loss of the energy associated with internal conversion in the acceptor. If both the demotion ($D^* \to D$) and promotion ($A \to A^*$) are vertical or spectroscopic transitions (Sect. 18.165), they will occur at the same energy and are said to be *coupled* or *resonant transitions*. This situation requires that the emission spectrum of the donor overlap with the absorption spectrum of the acceptor (Fig. 18.155).

18.157 ENERGY TRANSFER VIA DIPOLE–DIPOLE INTERACTION

Long-range (50–100 Å), electrostatic, radiationless energy transfer processes that proceed by an interaction between the dipole transition moment of the donor, $\mu_D(D^* \to D)$, and the dipole transition moment of the acceptor, $\mu_A(A \to A^*)$.

Example. The dipole–dipole interaction energy ΔU (cm^{-1}) is

$$\Delta U = (K|\mu_D||\mu_A|)/\eta^2 R^3 \qquad (18.157)$$

where the μ's are the donor and acceptor transition moments, K is an orientation factor for the alignment of these transition moments (↔ ↕, no interaction; ⇆, moderate interaction; ↔ ↔, strong interaction), η is the refractive index of the medium, and R is the distance between the donor and acceptor. For crystals and solids, a strong interaction is required (10^3–10^4 cm^{-1}), and the transfer can be very long-range with R^{-3} dependence and is called *exciton transfer*. For the more commonly encountered situation of energy transfer in solution, weaker interactions are involved (10–100 cm^{-1}). This type of transfer is long-range, as noted above, with R^{-6} dependence and is called *resonance transfer*. Dipole–dipole energy transfer selection rules are governed by the same selection rules that apply to their spectroscopic components ($D^* \to D$) and ($A \to A^*$). Thus, processes **1** and **2** are dipole–dipole-allowed and **3** and **4** are dipole-dipole-forbidden.

1. $^1D^* + {}^1A \to {}^1D + {}^1A^*$
2. $^1D^* + {}^3A \to {}^1D + {}^3A^*$
3. $^3D^* + {}^1A \to {}^1D + {}^1A^*$ (see Fig. 18.151*a*)
4. $^3D^* + {}^3A \to {}^1D + {}^3A^*$

Nevertheless, all these processes occur with about equal probability per excited species generated. This is because even though the rate constants k for process **3** and **4** are much smaller than those for **1** and **2**, the lifetime τ_D of $^3D^*$ is much greater than that of $^1D^*$, and the transfer probability per excited species generated is a function of the product $\tau_D k$.

18.158 THE FÖSTER RELATIONSHIP

A quantitative expression for the rate constant of dipole–dipole energy transfer (k_{D^*-A}) in terms of parameters that can be obtained experimentally:

$$k_{D^*-A} = (8.8 \times 10^{-25} K^2 \Phi_D/\eta^4 \tau_D R^6) \int F_D(\nu)\, \varepsilon_A(\nu)\, (d\nu/\nu^4) \qquad (18.158a)$$

where the spectral overlap integral (see Fig. 18.155) contains the frequency of light ν in wave numbers; $F_D(\nu)$, the donor emission spectrum in terms of quanta normalized to unity; and $\varepsilon_A(\nu)$, the molar extinction coefficients of the acceptor absorption spectrum. The remaining terms are K, an orientation factor that for a randomly oriented array of donor and acceptor molecules equals $(2/3)^{1/2}$; Φ_D, the quantum yield for donor emission in the absence of an acceptor; η, the refractive index of the medium; τ_D, the observed lifetime of the donor in the absence of acceptor (in sec) (see Sects. 18.127 and 18.128); and R, the distance separating the donor and acceptor molecules (in cm).

A useful variation of this equation provides the *critical radius for dipole–dipole energy transfer* R_0:

$$R_0^6 = (8.8 \times 10^{-25} K^2 \Phi_D/\eta^4) \int F_D(\nu)\, \varepsilon_A(\nu)\, (d\nu/\nu^4) \qquad (18.158b)$$

This distance R_0, usually given in Angstroms, is the separation between donor and acceptor molecules at which energy transfer proceeds with 50% efficiency.

Another variation provides the *critical acceptor concentration for dipole–dipole energy transfer* C_A^0:

$$C_A^0\,(\text{mol/L}) = (4.8 \times 10^{-10}\, \eta^4/K)\, [\Phi_D \int F_D(\nu)\varepsilon_A(\nu)(d\nu/\nu^4)]^{-1/2} \qquad (18.158c)$$

This is the acceptor concentration C_A^0 at which energy transfer proceeds with 50% efficiency.

18.159 ENERGY TRANSFER VIA EXCHANGE INTERACTION

Short-range (< 20 Å), radiationless energy transfer processes that proceed through donor–acceptor collision complexes and the exchange of electrons between the collision partners in these complexes (Fig. 18.161*a*).

Example. This type of energy transfer does not depend on overlap between the donor emission and the acceptor absorption spectra. Its rate is thought to be governed by an exponential dependence on the distance R separating the donor and acceptor molecules (e^{-CR}, where c is a constant depending on the orbital radius of the donor and acceptor molecules). In order for exchange energy transfer processes to be allowed, they must conserve spin, and therefore, they are governed by Wigner's spin rule (see Sect. 18.160 for application of Wigner's spin rule to the reaction of two triplet species). According to these criteria, the following energy transfer processes are allowed:

1. $^3D^* + {}^1A \rightarrow {}^1D + {}^3A^*$ (Sects. 18.161 and 18.162)
2. $^3D^* + {}^3A \rightarrow {}^1D + {}^5A^*$ (Sect. 18.160)
3. $^3D^* + {}^3A \rightarrow {}^1D + {}^3A^*$ (Sect. 18.160)
4. $^3D^* + {}^3A \rightarrow {}^1D + {}^1A^*$ (Fig. 18.161*a*)
5. $^3D^* + {}^3D^* \rightarrow {}^1D + {}^1D^*$ (Sect. 18.135)

Examples 1 and 4 are typical energy transfer processes utilized in sensitizing photochemical reactions (Sect. 18.161), and example 5 is the process usually referred to as *triplet–triplet annihilation.*

18.160 WIGNER'S SPIN RULE

The criterion that must be satisfied in order for spin S to be conserved in processes involving the interaction of two particles. For the general process,

$$S_1 + S_2 \rightarrow S_3 + S_4$$

When one of the possible initial spin states $(|S_1 + S_2|, |S_1 + S_2 - 1|, \ldots |S_1 - S_2|)$ matches one of the possible final spin states $(|S_3 + S_4|, |S_3 + S_4 - 1|, \ldots |S_3 - S_4|)$, the overall process will be spin-allowed named after Eugenewigner (1902–1995).

Example. For the interaction of two triplets, $s = 1$ (see Sect. 18.72), one would predict the following spin-allowed processes (Sect. 18.159):

Initial Spin State		Final Spin State
$^3D^* + {}^3A$	$\rightarrow$	$^1D + {}^5A^*$
$\|1 + 1\| = 2$ (quintet)		$\|+2\| = 2$ (quintet)
$^3D^* + {}^3A$	$\rightarrow$	$^1D + {}^3A^*$
$\|1 + 1 - 1\| = 1$ (triplet)		$\|+1\| = 1$ (triplet)
$^3D^* + {}^3A$	$\rightarrow$	$^1D + {}^1A^*$
$\|1 - 1\| = 0$ (singlet)		$0 = 0$ (singlet)

However, the statistics for collision complex formation indicate that these three processes do not occur with equal probability. Thus, nine possible collision complexes could arise from the collision of two triplets. Of these, five are quintet complexes and lead to $^5A^*$, three are triplet complexes and lead to $^3A^*$, and one is a singlet complex and leads to $^1A^*$. Therefore, if we assume a random orientation of collision partners, $^5A^*$ will be formed with a probability of 5/9, $^3A^*$ with a probability of 3/9, and $^1A^*$ with a probability of 1/9.

18.161 PHOTOSENSITIZATION

The process of initiating a photochemical reaction in a substrate by means of radiationless energy transfer from a donor molecule rather than by direct absorption of light by the substrate molecule.

Example. The photosensitized formation of singlet oxygen by energy transfer to triplet oxygen from the *photosensitizer* rose bengal (see Sect. 18.159, process 4, and Fig. 18.161*a*) illustrates an extremely useful photosensitization. The direct excitation of triplet oxygen to singlet oxygen is a spin forbidden transition; hence, the absorption coefficient for this transition is essentially zero. Photosensitization thus provides the only method of photochemically producing singlet oxygen in appreciable quantities. Singlet oxygen is an extremely reactive species; it reacts with a variety of substrates that are unreactive with triplet oxygen (Fig. 18.161*b*). Photosensitization also finds great utility in those systems where intersystem crossing from the excited singlet state to the triplet state proceeds with low efficiency or not at all. In such systems, photosensitization provides a means of bypassing the

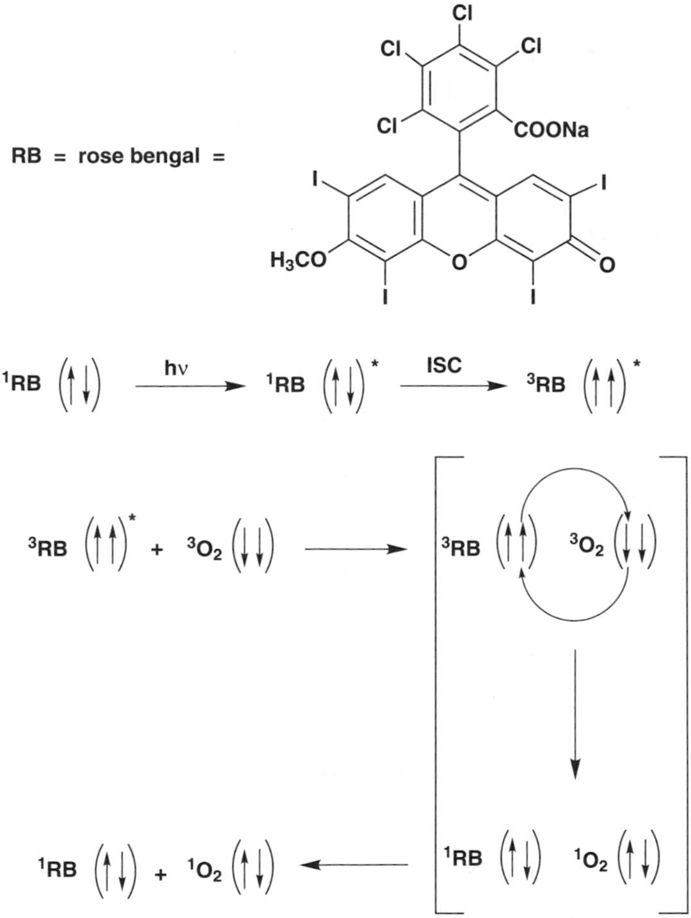

(*a*)

Figure 18.161. Photosensitized processes. (*a*) Sensitized formation of singlet oxygen via electron exchange in the collision complex with triplet rose bengal. (*b*) Peroxide formation in the Diels–Alder reaction of singlet oxygen with cyclopentadiene. (*c*) Direct formation of triplet diradicals from azoalkanes.

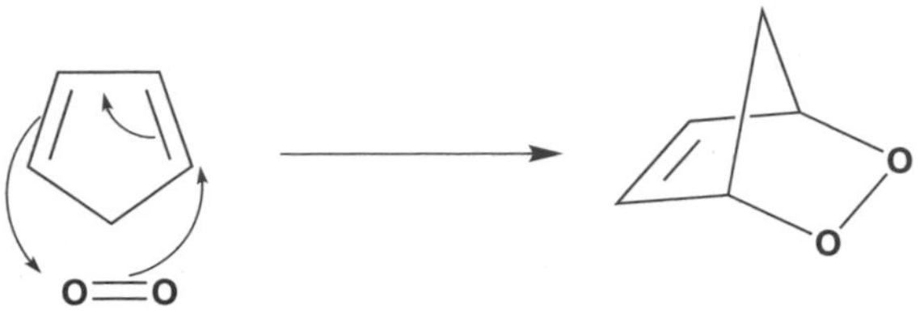
(b)

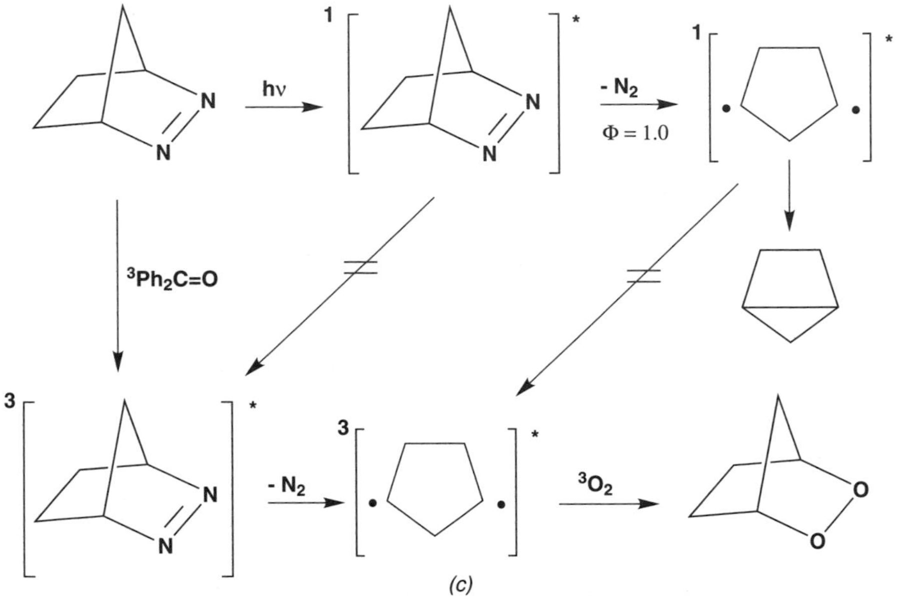

(c)

Figure 18.161. (Contd.)

singlet state and entering the triplet manifold directly (Sect. 18.159, process 1, and Fig. 18.161*c*). See also Sect. 18.162.

18.162 QUENCHING OF EXCITED STATES

The process of excited-state demotion back to the ground state by means of radiationless energy transfer processes.

Example. In every energy transfer process, the energy acceptor (*quencher*) is said to be sensitized by the energy donor (*sensitizer*), and the energy donor is said to be quenched by the energy acceptor (Fig. 18.156). One sometimes refers to an energy transfer process as a sensitization when the result is photochemically productive and as a quenching when the result is photochemically unproductive. In the latter case, for example, trace impurities frequently quench photochemically productive excited states.

18.163 STERN–VOLMER RELATIONSHIP

A correlation between the quantum yield of a photochemical reaction Φ and the concentration of added quencher $[Q]$:

$$\Phi_0/\Phi_q = 1 + k_q\,\tau\,[Q] \tag{18.163}$$

where Φ^0 and Φ_q are the quantum yields of the photochemical reaction in the absence and the presence of added quencher, respectively; k_q is the rate constant of the quenching process; τ is the lifetime of the excited state that is quenched in the absence of added quencher; and $[Q]$ is the concentration of quencher that has been added. Named after Otto Stern (1888–1969) Nobel prize in Physics (1943); Max Volmer (1885–1965).

Example. It has become standard procedure among photochemists to use Stern–Volmer relationships to test photochemical reaction mechanisms. Thus, if one experimentally determines a plot of Φ^0/Φ_q versus $[Q]$ for a particular photochemical reaction, one might observe a constant slope over all quencher concentrations (Fig. 18.163, case 1), or a larger slope at lower quencher concentrations than at higher concentrations (Fig. 18.163, case 2). The behavior displayed in case 1 is indicative of the involvement of a single excited state, usually the triplet state, in a photochemical reaction, whereas the behavior displayed in case 2 indicates the involvement of more than one excited state. Thus, at low quencher concentration only the longer-lived excited state, perhaps the triplet state, is being quenched, and at higher quencher concentrations the shorter-lived excited state, perhaps the singlet excited state, is being quenched as well. Finally, if the photochemical reaction under examination proceeds through a very short-lived excited state, usually a singlet excited state, the reaction may be unquenchable. In such cases, a horizontal line would be observed in the Stern–Volmer plot.

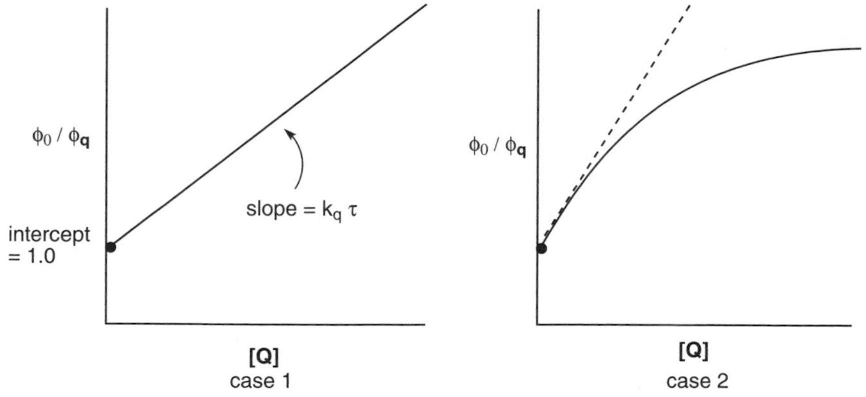

Figure 18.163. Case 1: form of the Stern–Volmer plot for the quenching of a single excited species. Case 2: form of the Stern-Volmer plot when more than one excited species on the reaction pathway is being quenched.

In the Stern-Volmer relationship, the plots should display an intercept of 1.0 and a slope that is equal to $k_q\tau$. It is frequently possible to estimate the lifetime of the excited species being quenched using this slope. In order to do this, one generally must make the assumption that the rate of the quenching process is diffusion-controlled (see Sect. 18.164).

18.164 DIFFUSION-CONTROLLED REACTION

A bimolecular reaction that occurs with 100% efficiency. Every time the reaction partners collide, they undergo reaction, and thus, a bimolecular diffusion-controlled reaction proceeds with the maximum possible rate constant k_d under a given set of reaction conditions:

$$k_d = 8\ RT/3{,}000\ \eta \qquad (18.164)$$

Where R is the gas constant, T the absolute temperature, and η the viscosity of the reaction medium in poise (see Sect. 12.88).

Example. Diffusion-controlled rate constants, k_d are frequently encountered in photochemical reactions or electron-exchange quenching processes (Sect. 18.162), for example, $k_q = k_d$ in the Stern–Volmer relationship (Sect. 18.163). Diffusion-controlled rate constants vary from about 10^{10} L/mol-sec in a nonviscous solvents such as hexane to 5×10^6 L/mol-sec in very viscous solvents such as glycerol.

18.165 VERTICAL (SPECTROSCOPIC) TRANSITION

Any excitation to a higher electronic state that proceeds in accordance with the Franck–Condon principle (Sect. 18.88, Fig. 18.165).

Example. Transitions of this type occur as the result of the absorption of a photon (*spectroscopic transition*) or an exothermic radiationless energy transfer process [e.g., the relationship of excited state energies is E (donor) > E (acceptor) Sect. 18.156–18.158]. Both these processes project a molecule into its excited state so rapidly that there is no time for its constituent atoms to change their positions or momenta, and accordingly, the excited state has exactly the same geometry as the ground state.

18.166 NONVERTICAL TRANSITION

Any transition to a higher electronic state that proceeds with an associated change in molecular geometry (Fig. 18.165).

Example. Transitions of this type are Franck–Condon-forbidden, and their occurrence has been established unequivocally only for certain endothermic triplet energy transfer

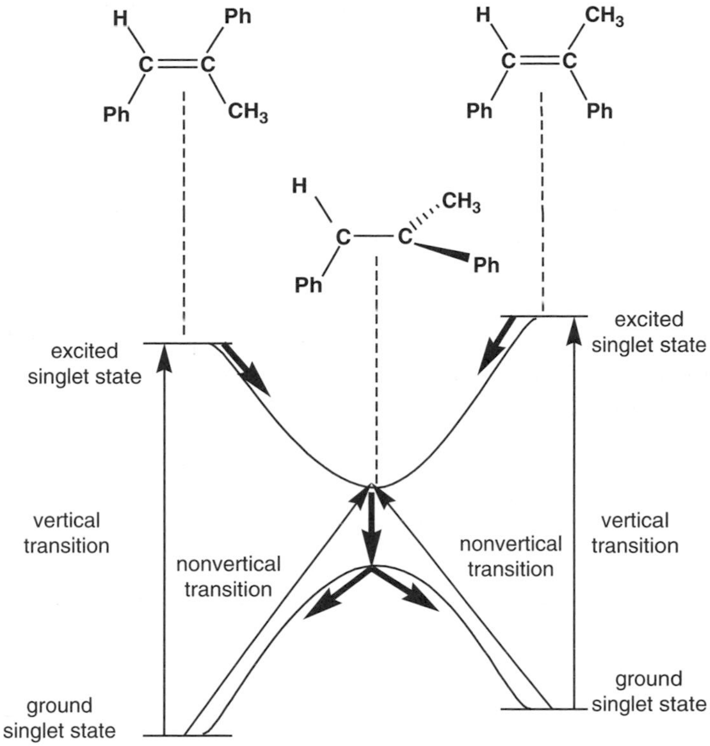

Figure 18.165. Relationship between the ground- and excited-state potential energy surfaces in this *cis-trans* photoisomerization of α-methylstilbene and possible transitions between those surfaces.

processes [e.g., the spectroscopic E (donor) $<$ E (acceptor)]. Apparently, the energy transfer step occurs in a collision complex between the donor and acceptor in which the geometry of the acceptor has been distorted significantly, as shown in Fig. 18.165.

18.167 EXCITON

An excited-state entity that encompasses a group of identical or very similar molecules or chromophores at any one time, and that moves through a uniform molecular environment from one molecule or site to another.

Example. Excitons occur in molecular crystals, polymers, and liquids. In these environments, an excited state is not localized on a single molecule or chromophore due to strong interactions with the neighboring molecules or chromophores. Thus, the exciton will move very rapidly through these media until it encounters an impurity or site with significantly lower excited-state energy (a trap). The exciton is trapped at this point and converted to a conventional localized excited state (see Sect. 18.168).

18.168 EXCIMER

A complex that is formed by the association of a ground-state atom or molecule with the excited state of the same atom or molecule (Fig. 18.168*a*). An excimer reverts into its constituents upon demotion to the ground state.

Example. Excimers are stabilized by means of exciton interactions that distribute the excited-state wave function equally between the two constituent molecules or atoms of the excimer (Fig. 18.168*b*).

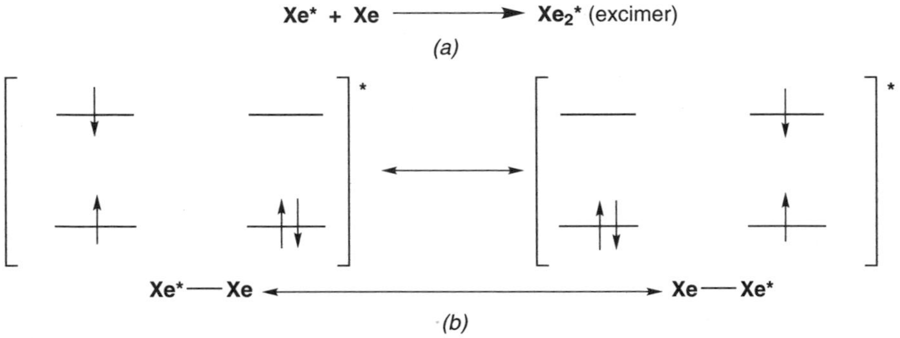

Figure 18.168. (*a*) Formation of a xenon excimer. (*b*) Delocalization of the exciton in a xenon excimer; note that for the resonance structure on the left, the atom on the left is in an excited state and the one on the right is in the ground state, and for the resonance structure on the right, the atom on the right is in an excited state and the one on the left is in the ground state.

18.169 EXCIPLEX

A complex that is formed by the association of a ground-state atom or molecule with the excited state of a different atom or molecule (Fig. 18.169*a*). The exciplex may be viewed as a heteroexcimer.

Example. Exciplexes are stabilized by a charge-transfer interaction between the constituent atoms or molecules (Fig. 18.169*b*). At present, there is some uncertainty as to whether or not exciplexes are anything more than the excited states of weak ground-state charge-transfer complexes that become more tightly bound in the excited state.

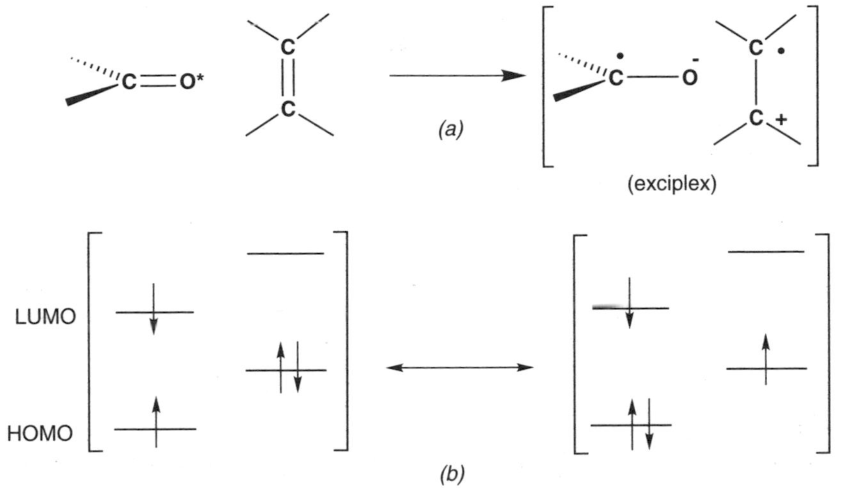

Figure 18.169. (*a*) Formation of an exciplex between a carbonyl excited state and an olefin ground state. (*b*) Charge transfer interaction in the formation of the carbonyl-olefin exciplex.

18.170 PHOTOELECTRON TRANSFER (PET)

The process by which a single electron is transferred (*single-electron transfer, SET*) between a molecule in its excited state and another molecule in its ground state.

Example. Electron transfer between molecules occurs in all redox reactions. If the electromotive driving force is sufficiently high, thermal electron transfer can occur between any two molecules in their ground states (Fig. 18.170*a*). However, in most cases, thermal electron transfer would be too endothermic to occur under ambient conditions. This energy barrier can often be overcome with the energy provided by a photon. Thus, although electron transfer may not be favorable between ground-state molecules, it can become quite favorable if one of the reaction partners is in its excited state. There are two possible routes by which PET might occur: (1) The excited-state species might serve as the electron acceptor (Fig. 18.170*b*), or (2) the excited species might serve as the electron donor (Fig. 18.170*c*). Which of these alternative pathways occurs will depend on the relative energies of the HOMO and LUMO of the acceptor and donor and their relative excited-state energies. In the reaction shown in Fig. 18.170*d*, the acceptor is the species excited, and so this reaction proceeds in accordance with the general scheme shown in Fig. 18.170*b*.

Reactions of this type constitute some of the most important reactions in all of biological chemistry. For instance, the reaction shown in Fig. 18.170*e* is a photochemical process by which blue light activates the repair of DNA damaged by ultraviolet light. This reaction is an example of the general process in which the excited state serves as the donor (Fig. 18.170*c*). Photosynthesis is a complex system of photoelectron transfer processes, which provides the energy that sustains most life on this planet.

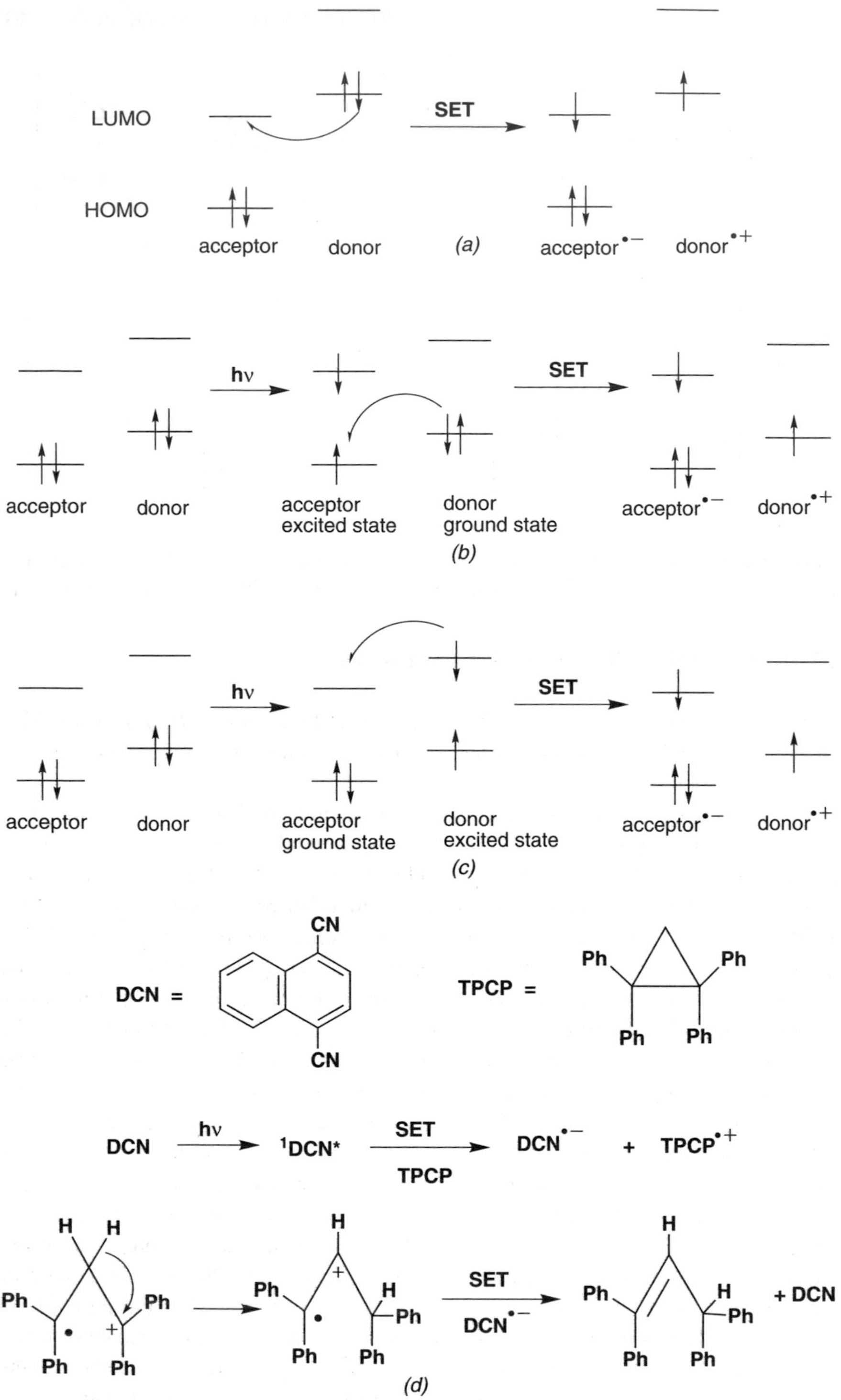

Figure 18.170. (Contd.)

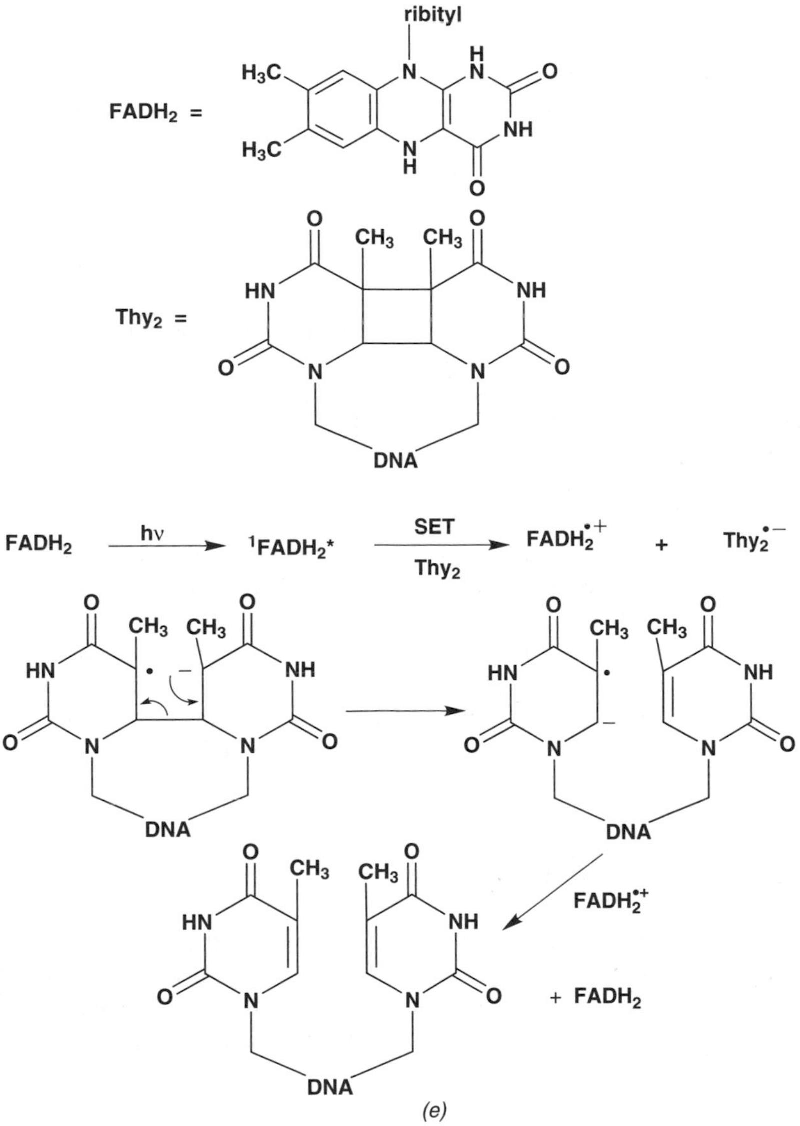

Figure 18.170. Relationships between HOMOs and LUMOs of acceptor and donor molecules in: (*a*) thermal electron transfer, (*b*) PET where the excited state serves as the acceptor, (*c*) PET where the excited state serves as the donor, (*d*) an excited-state acceptor (DCN) (case b) in the photocatalyzed isomerization of 1,1,2,2-tetraphenylcyclopropane (TPCP) to 1,1,3,3-tetraphenylpropane with 1,4-dicyanonapthtalene (DCN) as the catalytic electron transfer photosensitizer, and (*e*) an excited-state donor (case c) in the photochemical repair of thymine dimers (THY_2) in DNA by riboflavin ($FADH_2$).

18.171 REHM–WELLER EQUATION

A widely used equation (Dieter Rehm and Albert Weller, *Israel J. Chem.*, **8** 159 (1970)) used for evaluation the thermodynamics of photoelectron transfer processes:

$$\Delta G = 23.06\,(E^{1/2}_{\text{OX}} - E^{1/2}_{\text{RED}}) - \Delta E_{0\text{-}0} - (e^2/\varepsilon\,\alpha) \qquad (18.171)$$

where $E^{1/2}_{\text{OX}}$ is the oxidation potential of the donor, $E^{1/2}_{\text{RED}}$ is the reduction potential of the acceptor with both potentials for the ground state and referenced to a common electrode, $\Delta E_{0\text{-}0}$ is the energy of the singlet excited state as measured by the absorption 0-0 band (Sect. 18.66), and $(e^2/\varepsilon\alpha)$ is the coulombic attraction term, a measure of the capacity of the solvent to support the separation of charge. In order for a photoelectron transfer to proceed, ΔG must be ca. 5 kcal/mol or less. Reactions with ΔG's greater than 0 kcal/mol will require some degree of thermal activation, and those with ΔG's greater than 5 kcal/mol will not proceed without replacing the sensitizer with another sensitizer having a higher excited-state energy ($\Delta E_{0\text{-}0}$).

Example. The thermodynamic driving force for the photoelectron transfer reaction shown in Fig. 18.170*d* can be evaluated as follows. The oxidation potential ($E^{1/2}_{\text{OX}}$) of TPCP is +1.2 2V, and the reduction potential ($E^{1/2}_{\text{RED}}$) of DCN is −1.67 V, where both are measured for the ground-state molecules using a $Ag/AgNO_3$ electrode. The singlet excited-state energy of the photosensitizer DCN ($\Delta E_{0\text{-}0}$) is 86.4 kcal/mol. If the reaction was conducted in acetonitrile, the coulombic attraction term is +1.3 kcal/mol. Thus, ΔG for this reaction would be −21.1 kcal/mol, and it should proceed at a reasonable rate.

18.172 INVERTED MARCUS REGION

The region of negative slope in a plot of the logarithm of the rate constant for electron transfer versus ΔG_{et} for electron transfer (Fig. 18.172*a*). The contra-intuitive prediction of Marcus theory for electron transfer is that the rate of electron transfer will enter a region where it will decrease with increasing driving force ($-\Delta G_{et}$). Marcus theory predicts that as the vicinity of $\Delta G_{et} = 0$ is approached (Fig. 18.172*b*), a barrier to electron transfer will be encountered. This barrier λ is due to the energy required to reorganize the system to an optimum configuration for electron transfer and has two components: the energy λ_v required to align the donor and acceptor moieties and the energy λ_s required to reorganize the surrounding solvent molecules. The maximum in the rate constant profile (Fig. 18.172*c*) occurs at the point where the exothermicity of the reaction exactly matches the energy required for reorganization of the system (λ). At this point, electron transfer is a barrierless process. In the inverted Marcus region, electron transfer must occur from the ground vibrational state ν_0 of the reactants to a higher vibrational state ν_n of the products. As n increases, diminishing Franck–Condon overlap (see Sect. 18.88) factors make this process less efficient and k_{et} decreases (Fig. 18.172*d*) (see Sect. 12.110).

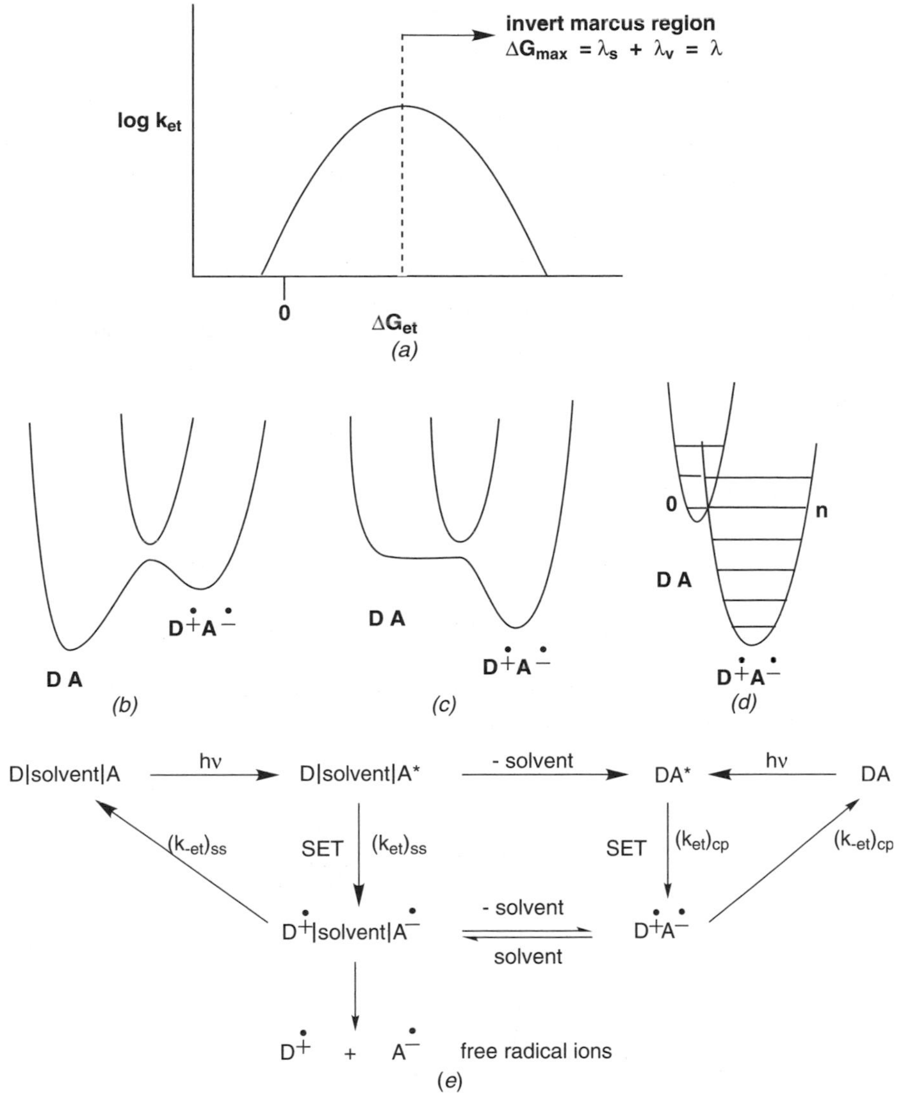

Figure 18.172. (*a*) Inverted Marcus region shown on the reaction rate constant profile plotted against the free energy for electron transfer. Schematic representation of the overlap of molecular coordinated of the reactants and products for (*b*) endothermic electron transfer, (*c*) barrierless electron transfer, and (*d*) exothermic electron transfer. (*e*) Formation of free radical ion pairs and back electron transfer in photochemical processes.

Example. Various possible scenarios can occur in photochemical electron transfer, as outline in Fig. 18.172*e*. In photochemical reactions, back electron transfer always leads to inefficiencies in the production of radical ions. Radical ion annihilation processes have been extensively studied using laser transient spectroscopy. Solvent-separated radical ion pairs (as in Fig. 18.172*e*) can be produced by irradiation of isolated donor or acceptor molecules in polar solvents, and contact radical ion pairs (cp) can be produced by irradiating weak ground-state charge transfer complexes of the same donor–acceptor pairs. Radical ion annihilation processes studied in these systems are the reverse of those described for radical ion formation (Figs. 18.172*b*–*d*). Nonetheless, the back electron transfer profile conforms to that predicted by Marcus theory (Fig. 18.172*a*). In addition, these studies have led to the surprising observation that in some reactions with high driving forces (ΔG_{et}), electron transfer in solvent-separated ion pairs can be faster then in contact ion pairs.

18.173 ELLIPTICITY θ

The parameter θ that characterizes elliptically polarized light (Sect. 18.5):

$$\theta = \tan^{-1}(\text{minor axis/major axis})$$

The major and minor axes refer to the axes of the ellipse circumscribed by the electric vector of the elliptically polarized light (Fig. 18.173).

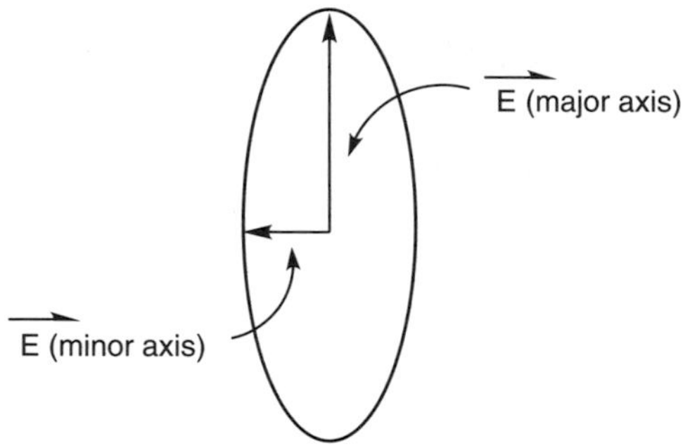

Figure 18.173. Major and minor axes of elliptically polarized light.

18.174 CIRCULAR DICHROISM (CD)

The difference in the absorbance ΔA of right and left circularly polarized light by asymmetric (Sect. 4.19) or dissymmetric (Sect. 4.18) molecules:

$$\Delta A = A_L - A_R \qquad (18.174a)$$

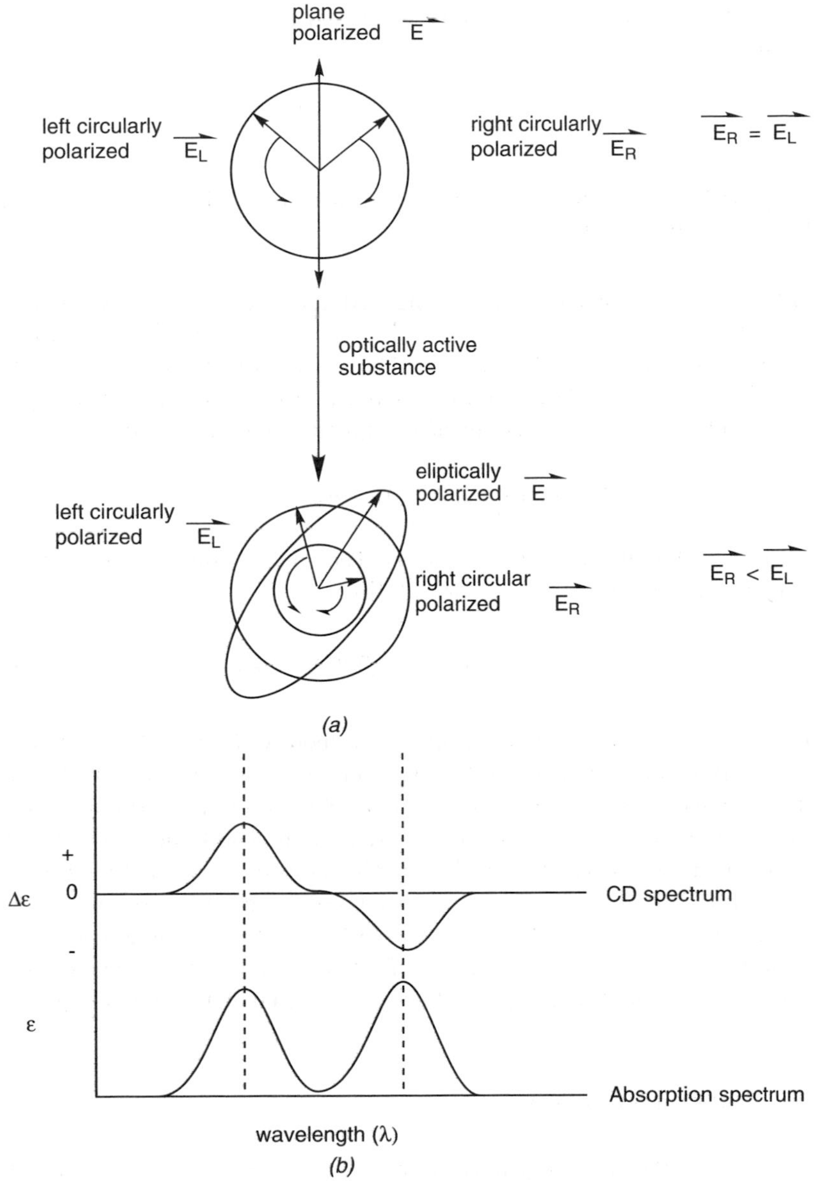

Figure 18.174. (*a*) Selective absorption of the right circularly-polarized component of plane-polarized light to form elliptically polarized light. (*b*) Relationship between absorption and circular dichroism spectra.

Alternatively, this phenomenon may be expressed in terms of the *molar circular dichroism* ($\Delta\varepsilon$):

$$\Delta\varepsilon = \varepsilon_L - \varepsilon_R \tag{18.174b}$$

where ε_L and ε_R are the molar extinction coefficients (Sect. 18.43) for left and right circular-polarized light, respectively.

Example. In CD measurements, plane-polarized light (Sect. 18.3) is passed through an optically active sample. The sample will selectively absorb one of the circularly polarized components of the plane-polarized light to afford elliptically polarized light (Sect. 18.5, Fig. 18.174*a*). The degree of ellipticity θ (Sect. 18.173) of this elliptically polarized light is determined experimentally and related to ΔA as follows:

$$\theta\ (\text{rad/cm}) = 2.303\ (A_L - A_R) / 4 \times 1 \tag{18.174c}$$

where l is the cell path length in centimeters. The *molar ellipticity* $[\theta]$ is referred to more commonly, and it is related to $\Delta\varepsilon$ as follows:

$$[\theta]\ (\text{rad-L/cm-mol}) = 3{,}298\ (\varepsilon_L - \varepsilon_R) \tag{18.174d}$$

Cotton effects (Sect. 18.22) are displayed in both CD and ORD spectra (Sect. 18.21). The principle difference in these two representations of the Cotton effect is the shape of the curves. The CD Cotton effect is either a simple maximum (+ Cotton effect) or minimum (− Cotton effect) centered on an absorption maximum of the compound being examined (Fig. 18.174*b*). This is to be contrasted with the more complex form of the Cotton effect curve displayed in an ORD spectrum (compare Figs. 18.22 and 18.174*b*). Although CD and ORD provide the same type of stereochemical information, CD is the method of choice in most stereochemical studies because CD curves are less complex, more easily deconvoluted, and interpreted.

18.175 FARADAY EFFECT

The induction of optical activity in an optically inactive substance when placed in a strong magnetic field Michael Faraday (1791–1867).

Example. In order to observe *magneto-optical activity* (the Faraday effect), the sample is placed in a strong magnetic field, and a beam of plane-polarized observing light is directed along the magnetic field (H) axis (Fig. 18.175) through a hole in both magnet pole faces. Both magnetic optical rotatory dispersion (MORD, Sect. 18.176) and magnetic circular dichroism (MCD, Sects. 18.174 and 18.177) are measured in this fashion.

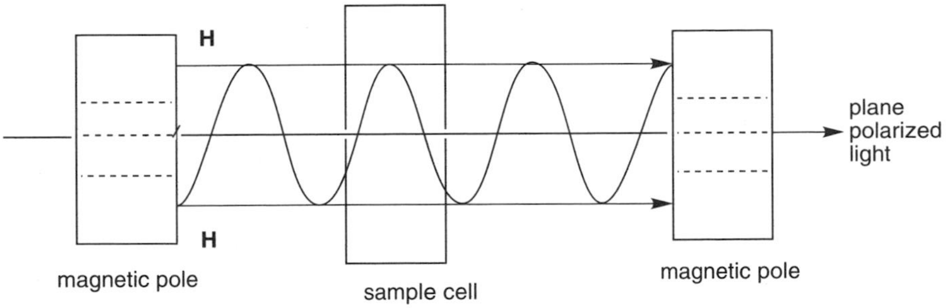

Figure 18.175. Experimental configuration for conducting Faraday effect experiments.

18.176 MAGNETIC OPTICAL ROTATORY DISPERSION (MORD)

The rotation of plane-polarized light observed when an optically inactive (achiral) substance is placed in a strong magnetic field.

Example. When the Faraday effect (Sect. 18.174) is applied in MORD, one observes the rotation of plane-polarized light by an angle of $\alpha°$, just as in the case of natural optical activity. However, MORD differs from natural optical activity in that reflection of the observing plane-polarized beam back through the sample along its original path leads to an observed rotation of $2\alpha°$ in the case of magnetic optical activity and 0° in the case of natural optical activity. Furthermore, the type of information that can be obtained from these magnetic measurements is quite different from that which is obtained from conventional ORD and CD studies (see Sects. 18.21 and 18.174, respectively).

18.177 MAGNETIC CIRCULAR DICHROISM (MCD)

The difference in the absorbance of right and left circularly polarized light, usually expressed as $\Delta\varepsilon = \varepsilon_L - \varepsilon_R$, observed when an optically inactive (achiral) substance is placed in a strong magnetic field.

Example. When the Faraday effect (Sect. 18.175) is applied to MCD, three factors can influence the appearance of the curve observed: (1) the splitting of paramagnetic ground and/or excited states into two components, which leads to two transitions of different energies: one driven by right, and the other by left, circularly polarized light (Fig. 18.177*a*). For the simple case in which only the ground state is paramagnetic, the two polarization forms will be absorbed at frequencies ν_R and ν_L:

$$\nu_R = \nu_0 - \mathbf{mgH} \tag{18.177a}$$

$$\nu_L = \nu_0 + \mathbf{mgH} \tag{18.177b}$$

where ν_0 is the frequency of the unperturbed transition, **mg** the magnetic moment of the ground state, and **H** the magnetic field strength. (2) The different probabilities of

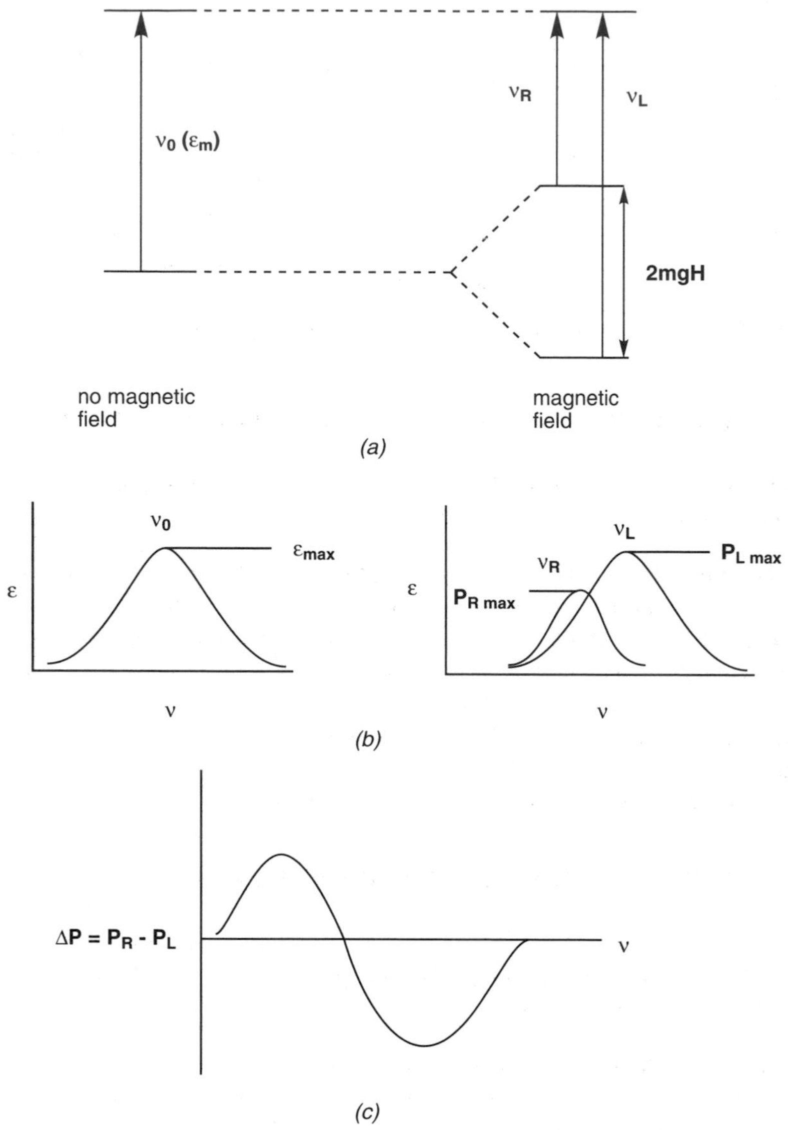

Figure 18.177. (*a*) Splitting of a paramagnetic ground state into two components by a strong magnetic field. (*b*) The different probabilities for absorption of right P_R and left P_L circularly polarized light to afford two absorption bands with their λ_{max} slightly displaced from one another. (*c*) The MCD spectrum obtained as a function ΔP.

absorption for right and left circularly polarized light, P_R and P_L, are shown in Fig. 18.177*b* and have the following relationships to each other:

$$P_R = \varepsilon_m (1 - a\mathbf{H}) \tag{18.177c}$$

$$P_L = \varepsilon_m (1 + a\mathbf{H}) \tag{18.177d}$$

where ε_m is the extinction coefficient for the unperturbed transition and u a linear coefficient. (3) These two factors, weighted by the different Boltzmann populations of the two sublevels of the ground state, combine to produce different extinction coefficients for right and left circularly polarized light (Fig. 18.177*c*). Even though the shape of MORD and MCD curves is influenced by the structure and stereochemistry of the molecule, no simple rules have yet been developed that establish MORD and MCD as important tools in structural chemistry. Instead, MORD and MCD are of the most value in the study of the magnetic properties of the ground and excited states.

18.178 IONIZATION ENERGY (POTENTIAL) I_p

The energy required to remove an electron from an atom or molecule to produce the corresponding radical cation (also see Sect. 12.31).

Example. An atom or molecule has a variety of ionization potentials, each of which corresponds to the energy necessary to remove an electron from one of the atomic or molecular orbitals present in that atom or molecule. In organic chemistry, the term ionization potential usually refers to the lowest energy of these processes (*adiabatic*

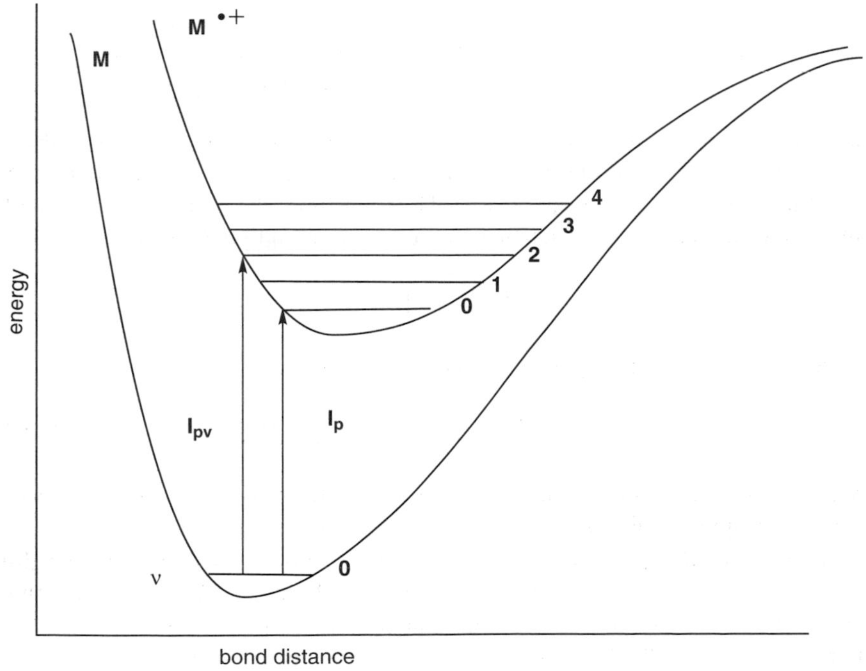

Figure 18.178. Relationships between potential energy surfaces in an un-ionized bond of a neutral molecule M, its singly ionized analog $M^{\bullet+}$, and various possible ionization pathways from M to $M^{\bullet+}$.

ionization energy). Specifically, the adiabatic ionization potential is the energy necessary to remove an electron from the HOMO of a molecule M in its ground vibrational state and produce the corresponding radical cation $M^{\bullet +}$ in its ground electronic and vibrational state (I_p in Fig. 18.178; see Sects. 12.31 and 16.6).

18.179 VERTICAL IONIZATION ENERGY (POTENTIAL) I_{pv}

The energy required to remove an electron from a molecule, in its ground electronic and vibrational state, to form the radical cation having exactly the same geometry as the starting molecule. This is the usual situation in photoionizations that is governed by the Franck–Condon principle (Sect. 18.88) in much the same way as are vertical electronic transitions (Sect. 18.165).

Example. Since the bond lengths and angles for the ground state of a radical cation will be different from those of the molecule from which it is derived, a vertical ionization process I_{pv} will usually result in a vibrationally excited ion, as shown in Fig. 18.178. Consequently, vertical ionization energies I_{pv} are always greater than the corresponding adiabatic ionization energies I_p by an amount equal to the vibrational energy of the resulting ion.

18.180 PHOTOELECTRON SPECTROSCOPY (PES)

A technique for the measurement of the ionization potentials $\mathbf{I_n}$ of a molecule associated with the removal of various valence shell electrons.

Example. This technique usually consists of irradiating a small molecule in the gas phase with light in the far ultraviolet region of the spectrum, usually the helium resonance line (HeI) at 58.4 nm (21.22 eV). This far ultraviolet radiation causes photoionization of the molecule, producing a radical cation and photoelectron:

$$\mathbf{M \xrightarrow{h\nu} \overset{+}{M^{\bullet}} + e^{\bar{\bullet}}}$$

The energy of the photon $\mathbf{E_{h\nu}}$ is partitioned as follows:

$$\mathbf{E_{h\nu} = I_n + \overset{+}{T}{}^{\bullet}_{n} + \overset{-}{T}{}^{\bullet}_{n} + \overset{+}{E}{}^{\bullet}_{vib} + \overset{+}{E}{}^{\bullet}_{rot}} \quad (18.180a)$$

where $\mathbf{I_n}$ is the ionization potential of a valence electron in the nth molecular orbital (highest occupied molecular orbital, $n = 1$; second highest occupied molecular orbital, $n = 2$, etc.); $\mathbf{\overset{+}{T}{}^{\bullet}_{n}}$ and $\mathbf{\overset{-}{T}{}^{\bullet}_{n}}$ are the kinetic (translational) energies of the resulting radical cation and photoelectron, respectively; and $\mathbf{\overset{+}{E}{}^{\bullet}_{vib}}$ and $\mathbf{\overset{+}{E}{}^{\bullet}_{rot}}$ are the vibrational and rotational energies of the resulting radical cation. Since the magnitudes of $\mathbf{\overset{+}{T}{}^{\bullet}_{n}}$ and $\mathbf{\overset{+}{E}{}^{\bullet}_{rot}}$ are very small, the above expression may be reduced to

$$\mathbf{E_{h\nu} = I_n + \overset{-}{T}{}^{\bullet}_{n} + \overset{+}{E}{}^{\bullet}_{vib}} \quad (18.180b)$$

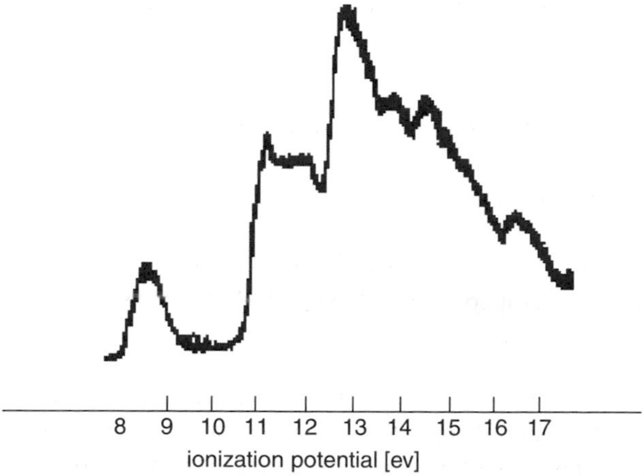

Figure 18.180. Photoelectron spectrum for 1,4-dimethyl-2,3-diazabicyclo[2.2.2]oct-2-ene and the approximate energies of the antisymmetric nonbonding and π orbitals (This figure taken from Houle, Chang, and Engel, *J. Am. Chem. Soc.*, **97,** 1824 (1975) with permission of the American Chemical Society).

The kinetic energy $\mathbf{T}^{\bullet-}_{\mathbf{n}}$ of the photoelectron may be determined by means of a magnetic spectrometer and recorded as a plot of photoelectron intensity (counting rate) versus $(\mathbf{I}_{\mathbf{n}} + \mathbf{E}^{\bullet+}_{\mathbf{vib}})$ (often referred to as ionization potential, eV) (Fig. 18.180). In low-resolution spectra, $\mathbf{I}_{\mathbf{n}}$ corresponds to the on-set of the band (e.g., $\mathbf{E}^{\bullet+}_{\mathbf{vib}} = 0$). Thus, in Fig. 18.180, the expulsion of an electron from the antisymmetric nonbonding azo orbital requires 8.06 eV, as compared to 10.48 eV for the expulsion of azo π electron. These experimentally determined ionization potentials are of great value in determining the energies $\mathbf{I}_{\mathbf{n}}$ of the various molecular orbitals in a molecule and, thus, for testing the accuracy of molecular orbital calculation procedures.

18.181 ELECTRON SPECTROSCOPY FOR CHEMICAL ANALYSIS (ESCA)

A technique for measuring ionization potentials that is very similar to PES (Sect. 18.180). It differs in that *X*-rays instead of ultraviolet light are used to initiate photoionization, resulting in the ejection of K-shell electrons instead of valence shell electrons.

Example. The ionizing radiation frequently employed in ESCA is Mg K_α *X*-rays (1253.6 eV). Just as in PES (Sect. 18.180), the kinetic energies of the photoelectrons are measured and these data used to determine the binding energies of the core electrons of gases, liquids, and solids. Since the binding energies of core electrons increase with an increasing oxidation state or the effective charge of the atom, ESCA can be used to determine not only the types of atoms involved, but also their oxidation states, in situations where other types of spectroscopy are not applicable. For instance, ESCA is finding widespread application in the determination of the oxidation state of metals on catalyst surfaces.

18.182 ACOUSTO-OPTIC SPECTROSCOPY

A form of spectroscopy in which the heating of the medium produced by the internal conversion of excited states (Sect. 18.111, process 9 in Fig. 18.75) is converted to acoustic waves by means of modulating the exciting light. These acoustic waves are detected with a microphone and the signal plotted versus the wavelength of the exciting light.

Example. This technique may be used with gases, liquids, or solids, and has several advantages over conventional optical absorption spectroscopy. The absorption spectrum of the surface layers of unwieldy solid samples such as flower petals, or inorganic solids, can be obtained using this method. Furthermore, acousto-optic spectroscopy also allows one to directly observe the occurrence of a photochemical reaction. Thus, the photodissociation of NO_2:

$$NO_2 \xrightarrow{h\nu} O(^3P_2) + NO(^2\Pi_{1/2})$$

occurs at wavelengths shorter than about 3890 Å. The optical–acoustic spectrum of NO_2 (Fig. 18.182) closely resembles the optical spectrum except at around 4100 Å where there is a distinct drop in the acoustic signal. This is due to the onset of photodissociation and appears as a drop in signal strength, since optical energy is no longer being completely dissipated as heat, but now is being partially stored as chemical energy in the dissociated species.

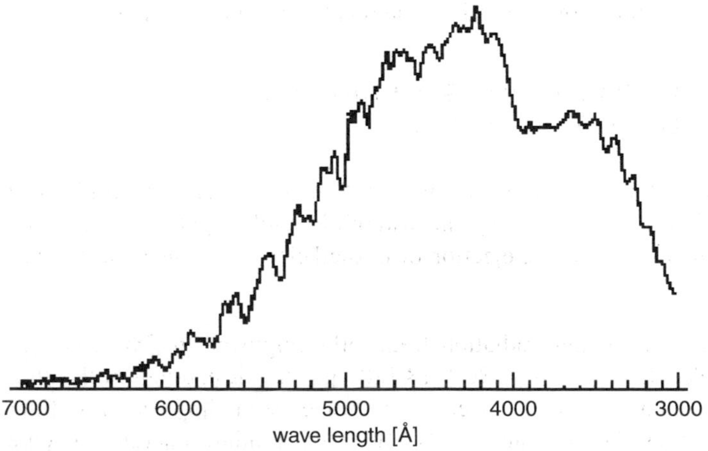

Figure 18.182. Acousto-optical spectrum for the dissociation of nitrogen dioxidical (This figure taken from Harshbarger and Robin, *Accts. Chem. Res.* **10**, 9 (1971) with permission of the American Chemical Society).

Acknowledgment. We wish to thank IUPAC for its useful Glossary of terms used in *Photochemistry, 3rd Ed.* July 2004.

SUGGESTED READING

Jaffe, H. H. and Orchin, M. O. *Theory and Applications of Ultraviolet Spectroscopy*. John Wiley & Sons: New York, 1962.

Kasha, M. and McGlynn, S. P. *Ann. Rev. Phys. Chem.* *7*, 403 (1956).

Michl, J. and Bonacic-Koutecky, V. *Electronic Aspects of Organic Photochemistry*. Wiley-Interscience: New York, 1990.

Pitts, J. N., Wilkinson, F., and Hammond, G. S. *Advances in Photochemistry*, Vol. 1. Wiley-Interscience: New York, 1963, p. 1.

Turro, N. J. *Modern Molecular Photochemistry*. Benjamin/Cummings: Menlo Park, CA, 1978.

NAME INDEX

Abramovitch, R. A., 533
Ackerman, J., 656
Adams, D. J., 418
Albright, T. A., 378
Alder, K., 582
Alexander, J. J., 24, 386
Allinger, N. L., 252, 281, 290
Angyal, S. J., 281, 290
Arduenco, A. J., 386
Arrhenius, S., 488
Atkins, P. W., 24
Ault, B. S., 702

Baekeland, L., 314
Baldwin, J., 288, 590
Barkla, C. G., 18
Barton, D. H. R., 225
Basolo, F., 386
Bassindale, A., 289
Bassler, G. C., 702
Bauld, N. L., 533
Beck, T., 24
Beer, A., 749
Bellville, D. J., 533
Bergius, F., 440
Berry, R. S., 377
Bertolucci, M. D., 53, 109, 702
Berzelius, J. J., 111, 293, 346
Billmeyer, Jr., F. W., 341
Bohr, N., 2, 3, 672
Bonacic-Koutecky, V., 831
Born, M., 762
Bowser, J. R., 53
Bredt, J., 138, 552
Brewster, J. H., 252
Brewster, Sir D., 741
Brintzinger, H. H., 386
Broche, H., 440
Brönsted, J. N., 470
Brookhart, M., 360
Bruice, P. Y., 341
Brunauer, S., 412
Burdett, J., 53, 82

Cahn, R. S., 236
Cantwell, W. J., 342
Capell, L. T., 138
Carothers, W., 294, 329
Carpenter, B. K., 504, 590
Carroll, F. A., 504, 590
Cech, N. B., 724
Charette, J., 724
Chatt, J., 346
Cleveland, J. M., 24
Collins, A. N., 289
Collman, J. P., 386
Condon, E. U., 463
Conley, R. T., 702
Corey. E. J., 270
Cotton, F. A., 84, 109, 353, 702
Coulson, C. A., 24, 53
Coulthup, N. B., 702
Cox, D. M., 137
Crabtree, R. H., 386
Cram, D., 269
Crosby, J., 289
Crutzen, P., 147
Curl, Jr., R. F., 136, 138

Daly, L. H., 702
de Broglie, L., 7
Debye, P., 33
DeHoffmann, E., 724
Denney, R. C., 418
Dermer, O. C., 138, 220
Derome, A. E., 656
Dewar, M. J. S., 346, 589
Douglas, B., 24, 386
Duncanson, L. A., 346
Dyson, P. J., 418

Ebbesen, T. W., 138
Edison, T., 138
Einstein, A., 3, 672
Eliel, E. L., 252, 281, 290
Emmett, P., 412
Enke, C. G., 724
Euler, L., 677
Eyring, H., 489

Faraday, M., 127, 824
Fenn, J. E., 707
Fenske, M., 394
Fermi, E., 691
Ferraro, J. R., 109
Fieser, L., 389, 777
Finke, R. G, 386
Fischer, E., 227
Fischer, E. O., 374, 376
Fischer, F., 445
Fletcher, J. H., 138, 220
Flory, P., 294
Förster, T., 795
Fourier, J.-B. J., 700
Fox, R. B., 138, 220
Franck, J., 463
Fried, J., 341
Fukui, K., 84, 588
Fuller, Richard Buckminster, 136

Gamow, G, 24
Gibbs, J. W., 467
Gil-Av, E., 390
Glan, P., 730
Goodyear, C., 337
Grable, B., 294
Gray, H. B., 53
Green, M. L. H., 360
Green, M. M., 449
Grignard, V., 348
Grubbs, R. H., 375
Grunwald, E., 502
Gubitz, G., 418
Guthrie, R. D., 540

Haber, F., 447
Hall, L. H., 109
Hamilton, Lord, 66
Hammett, L. P., 500
Hammond, G. S., 482, 831
Hampden-Smith, M., 386
Hantzsch, A., 193
Hargittai, I., 109
Hargittai, M., 109
Harris, D. C., 53, 109, 702
Hassel, O., 225
Heath, J. R., 137
Heathcock, C. H., 82
Heck, R. F., 384
Heegar, A., 321
Hegedus, L. S., 386
Heisenberg, W., 3
Heitler, W., 26
Helmholtz, H. L. von, 468
Henderson, C. L., 724
Herschel, Sir F. W., 658
Herzberg, F. H., 702
Herzberg, G., 84
Hess, H., 458
Hillenkamp, F., 724
Hine, J., 507
Hoffmann, R., 84, 378, 386, 586, 588
Hofmann, A. W., 552
Hofmann, P., 378
Hooke, R., 662
Hückel, E., 55, 67, 589
Hund, F., 17

Iijima, S., 137
Ingold, C., 236
Isaac, N., 504, 590

Jablonski, A., 759
Jackson, G. S., 724
Jaffe, H. H., 53, 82, 109, 386, 702, 831
Jean, Y., 53, 82
Jencks, W. P., 540
Jensen, W. B, 24
Johnson, R., 386
Jolley, J. G., 787
Jones, M., 386, 534
Juaristi, E., 290

Kaldor, A., 137
Kaptein, J., 654
Karas, M., 724

Karo, W., 220
Karplus, M., 627
Kasha, M., 831
Katritsky, A. R., 220
Kealy, T. J., 346
Kemp, W., 702
Kent, J. A., 449
Kerr, Rev. J. C., 743
Kikuchi, O., 24
Klopman, G., 534
Kodas, T. T., 386
Kosower, E. M., 82, 773
Kroto, H. W., 136, 137

La Porte, O., 765
Lambert, J., 749,
Lambert, J. E., 656
Laplace, P. S., 9
Laszlo, P., 252, 702
LeBel, J. A., 225
Leffler, J. E., 533
Lehn, J.-M., 154, 269
Lenz, H., 612
Lewis, G. N., 28, 471
Liberles, A., 82
Lieber, S., 386
London, F., 26, 410
Lowry, T. H., 504, 590
Lowry, T. M., 470

MacDiarmid, A., 321
Macomber, R. S., 620, 656
Mandelkern, L., 341
March, J., 590
Marcus, R., 503
Mark, H., 293
Markovnikov, V. W., 558
Marshall, A. G., 724
Martin, A. S. P., 390
Mazzola, E. P., 656
McDaniel, D. H., 24, 386
McGlynn, S. P., 831
Meisenheimer, J., 563
Menton, M., 486
Michael, A., 557
Michaelis, L., 486
Michaelson, A., 701
Michl, J., 831
Mislow, K., 109
Miyaura, N., 385
Möbius, A., 589
Molina, M., 147
Moody, C. J., 534
Morawetz, H., 342
Morrill, T. C., 702
Morrison, G. A., 281, 290
Morse, P. M., 35, 674
Morton, J., 342
Moss, R. A., 386, 534
Mulliken, R. S., 27, 30

Nakamoto, K., 702
Natta, G., 319
Negishi, E., 386
Newman, M. S., 89, 290
Newton, I., 663
Nobel, A., 185
Nogradi, M., 290
Noller, C. R., 138
Norton, J. R., 386
Novak, B. M., 342
Noyes, Jr., W. A., 787

O'Brien, S. C., 137
Ogston, A., 225, 261
Olah, G., 517, 534
Oppenheimer, J. R., 762
Orchin, M., 53, 82, 109, 383, 386, 449, 702, 831

Parshall, G. W., 369, 386
Pascal, B., 624
Pasteur, L., 224
Patai, S., 144, 220
Patterson, A. M., 138
Pauli, W., 17
Pauling, L., 24, 26, 29, 55
Pauson, P. J., 346
Pearson, R., 84, 472
Pederson, C., 269
Pines, H., 386
Pinhas, A. R., 378
Pirkle, W. H., 418
Pitts, J. N., 831
Pitzer, K., 272
Planck, M., 3
Platt, J. R., 778
Platz, M. S., 534
Pochapsky, T. C., 418
Pöckels, F., 743
Poiseuille, J. L. M., 407
Porter, G., 787
Potapov, V. M., 290
Pott, A., 440
Prelog, V., 236
Priestley, J., 334

Quin, L. D., 220
Quinoa, E., 290, 656

Raman, C. V., 694
Raoult, F. M., 391
Rappoport, Z., 144, 220
Rayleigh, Lord (J. W. Strutt), 694
Reddy, G. S., 369
Rehm, D., 820
Reynolds, O., 431
Richardson, K. S., 504, 590
Riguera, R., 290, 656
Roberts, J. D., 82
Robinson, R., 59
Roelen, O., 382
Rohlfing, E. A., 137
Rosenblum, M., 347
Rowland, F. S., 147
Rydberg, J., 773

Sandler, S. R., 220
Saytzeff, A. M., 551
Schlenk, W. J., 348
Schlosberg, R. H., 534
Schmitz, J., 138
Schrock, R. R., 374
Schroedinger, E., 10
Schurig, V., 390
Seco, J. M., 290, 656
Sheldrake, G. N., 289
Shen, Y., 724
Shirakawa, H., 321
Shoolery, J., 620
Silverstein, R. M., 620, 656, 702, 724
Slater, J., 26
Small, H., 418
Smalley, R. E., 136, 137, 138
Smith, R. D., 724
Solomons, T. W. G., 138
Stalcup, A., 418
Stang, P., 252, 702
Staudinger, H., 293
Stern, O., 813
Stille, J. K., 385
Stock, A., 349
Stokes, Sir, G. G., 695
Streitwieser, A., 53, 82
Stroobant, V., 724
Summerfeld, A., 3
Suzuki, A., 385
Suzuki, K., 24
Synge, R. L. M., 390
Szwarc, M., 305, 342

Taft, R., 501
Tang, K., 724
Tavener, S. J., 418
Taylor, E. C., 220
Tebbe, F. N., 369
Teller, E., 412
Thompson, D., 84
Thouet, H., 138
Trnka, T. M., 375
Tropsch, H., 445
Tsvet, M., 389
Turro, N. J., 831

van Beylen, M., 342
van der Waals, J. D., 39
van't Hoff, J. H., 225, 406
Volatron, F., 53, 82
Volmer, M., 813

Walker, D. F., 138
Weber, J. H., 193
Webster, F. X., 620, 656, 724
Weissberger, A., 220
Weller, A., 820
Wells, A. F., 109
Wender, I., 449
Werner, A., 345
Weyl, H., 84
Whitham, G. H., 534
Whiting, M. C., 347
Wiberley, S. E., 702
Widman, O., 193
Wigner, E., 84
Wilen, S. H., 290
Wilkinson, F., 831
Wilkinson, G., 345
Winstein, S., 502
Wirth, D. N., 533
Witcoff, H. A., 449
Wittig, G., 531
Woehler, F., 111, 421
Woodward, R. B., 84, 111, 347, 389, 586

Yanagi, T., 385

Zeise, W. C., 346, 352
Ziegler, K., 319
Zimmerman, H. E., 589
Ziomek, J. S., 109

COMPOUND INDEX

Numbers in **bold** indicate the page on which the term is a topic.

Acene, 130
Acetaldehyde, 161, 381
Acetal, **162**, 163, 263
Acetate, 168, 525, 526
Acetic acid, 167, 471
  acidity of, 474
  catalytic synthesis of, 382
  Monsanto synthesis, **382**
  sustituted, 502
Acetic anhydride, 170
Acetone, 162, 163, 164, 381
Acetonitrile, 183, 477
Acetophenone, 150, 163
Acetyl cation, 516
Acetylacetonato anion, 351
Acetylcarbene, 514
Acetylene, 321, 424
  geometry of, 42
Aci form of the nitro group, **184**, 555, 557
Acid
  azide, **188**
  halide, **168**
    inorganic acid, 218, **220**
Acidimide
  chloride, 172
  substituted, **172**
  ester, 172
Acrilan, 333
Acrylic acid, 178, 183
Acrylonitrile, 183, 336, 429, 447
Acyl group, 168
  azide, **188**
  halide, **168**, 210
    hydrolysis of, 532
  nitrene, **515**
  carbene, **514**
Acylium ion, 407, 516, **523**
Acyloin, 178
Adamantane, 124, 135
Adamantylidene adamantane, 199
Adipic acid, 75
Alanine, 338
Alcohol, **147**, 148, 153, 162, 163
  polyhydric, **149**
  primary, 147
  secondary, 147
  tertiary, 147
Aldehydes, 160–167, 179, 358, 382, 777
Aldohexose, D- and L-, 250
Alkali metal, 21
Alkaline earth metal, 21, 349
Alkane, 112–114, 431, 446
  branched-chain, 115
  cyclo-, 116, 119
  iso-, 117
  straight-chain, 114, 117, 435
Alkene, 112, 114, 120, 230
Alkoxide, 526
Alkoxyl group, 148
Alkyl group, 116, 144
  iso-, 118
  normal, 117
Alkyl halide, 145, 147, 357, 541
Alkylate, 432
Alkylbenzenesulfonic acid, 414
Alkyne, 112, 114, 127
Allene, 127
Alloy, 340

Allyl
anion, 74, 286, 525
cation, 74, 286, 516
radical, 74, 286
Aluminosilicate, 427
Aluminum isopropoxide, 526
Amide, **170**, 172, 186, 195
hydrolysis of, 532
ion, 268, 269, 303, 525, 568
secondary, **171**
substituted, **171**
tertiary, **171**
Amidine, 172
Amine, 163, **179–182**, 274, 525
hydroxyl-, **181**
oxidation, 524, 527
oxides, **181**
primary, 167, 180
secondary, 180, 186, 187
tertiary, 180–182, 328
Aminium ion, **524**
Amino acid, 529, 563
Aminoaldehyde, β-, 179
Aminocaproic acid, E-, 329
Aminocarboxylic acid, α-, 179
Aminoketones, β-, 179
Aminolysis, 172, **545**
Ammonia, 274
clock, 274
synthesis, **447**
geometry of, 42
Ammonium
chloride, 182
cyanate, 421
salts, 182, 403, 522
quaternary, **182**, 403, **415**
Aniline, 150
substituted, 475
ion, 522
Anisole, 150, 152, 496
Annulene, **77**, 233
[14], 233
[18], 618
Anthanthrene, 131
Anthracene, 75, 131
Antimony
oxide/uranium salt, 429
pentafluoride, 473
Arene, **75**
Aromatic hydrocarbon, **75**, 125, 783
Aroyl halide, **168**
Arsenic hydride, 380
Artemisinin, 645
Aryl group, **128**, 144
halide, **145**, 384, 385
Aryne, 530, 533, 563, 564
Ascorbic acid, 165
Ash, 437, **439**, 443
Asphalt, 423
Atomic carbon, 507
Azide, **188**
acid, **188**
acyl, **188**
Azinic acid, 184
Aziridine, 193, 274, 806
Azo compound, **189**, 190, 235, 565
Azoalkane, 812, 829
Azobenzene, 189
Azobisisobutronitrile, AIBN, 299, 565
Azomethane, 189
Azoxy compound, **190**
Azoxybenzene, 190
Azulene, 76

BAHA (tris (4-bromophenyl)aminium hexachloroantiminate), 527
Bakelite, 293, 314
Benz[a]anthracene, 75, 130, 132
Benzaldehyde, 150, 161
Benzene, 127
allyl-, 128
substitution of, 495
D_6, 638
Benzenecarbothioic O-acid, 207
Benzenediazonium chloride, 190
Benzenesulfonic acid, 203
Benzenethiol, 201
Benzenoid hydrocarbon, **127**
Benzenonium ion, 561
Benzil, 172
Benzo[a]pyrene, 129
Benzoic acid, 150
Benzoin, 178
Benzonitrile, 150
Benzoquinone, p-, 154, 155, 178
Benzoyl
cation, 722
peroxide, 308, 309
Benzyl, 76
anion, 516
cation, 509
tropylium cation, 723, 724
radical, 510
Benzyne, 507, 509, 530, 531, 563, 564
Betaine, **529**, 530, 772, 773
Bicyclo[2.2.1]heptene, 235
Bicyclo[5.5.5]heptadecane, 283
Biopolymers, 293
Biphenyl isomers, 282
Bis(cyclopentadienyl)dimethylzirconium, 322

Bis(diphenylphosphino)ethane, 1,2-, dppe, 351
Bismuth molybdate, 429
Bisulfate, **204**
Borabicyclo[3.3.1]nonane, 9- (9-BBN), 196
Borane, **196**, 472
  heterocyclic, **196**
Boric acid, 196
Boron
  hydride, 368
  trifluoride, 472, 510
Boronic acid, 385
Bromobenzene, 146, 396
Bromobutane, 2-, 285
Bromocyclopentadiene, 5-, 260, 261
Bromooctane, 1-, 415
Buckminsterfullerene, 136, 137
Bucky ball, 136, 137
Bullvalene, 287, 288
Buna rubber, 336
Butadiene, 1,3-, 576-581
  s-*cis*-, 277, 278
  s-*trans*-, 277, 278, 459
Butanedithioic acids, 207
Butanesulfenic acid, 205
Butanethiol, 1-(methylthio)-1-, 211
Butene
  1-, 459, 489
  cis-2- or Z, 230–237, 258, 259, 489
  trans-2- or E, 230–237, 258, 259, 489, 710, 720
Butraldehyde, n-, 383
Butyl
  bromide, t-, 464
  esters, 572
  hydroperoxide, 197
  peroxide, di-, 198, 565
  thiol, 442
  4-methylbenzene, 2,5-di-, 566, 567
  4-methylcyclohexanone, (+)-trans-3-, 739, 740
Butylated hydroxytoluene, BHT, 566, 567
Butyrolactone, γ-, 195, 310, 311

Caffeine, 398
Calcium
  carbonate, 340, 444
  oxide, 444
Calixarene, **157**, 158
Caprolactam, 195, 310, 311, 447
Carbaborane, 197
Carbamate, 180
Carbamic acid, 180
Carbene
  α-keto-, **514**
  Fischer, 374
  Schrock, 374
  singlet, **513**
  triplet, 511, **513**
Carbenium ion, 516, 517, **520**, 533, 544, 559
  metal stabilized, 374
Carbocation, 424–432, 498, 499, **515–520**, 544–562
  hypervalent, 519
  primary, 424
  secondary, 424
  tertiary, 425
Carbodiimide, **192**
Carbon
  dioxide
    infrared spectrum, 681–690
    super critical, 398
  disulfide, 339, 396
  fibers, **137**, 507
  monoxide, 346, 357, 382, 666, 671
  nanotubes, 112, **137**, 507
  tetrachloride, 301
    Raman spectrum, 696
Carbonate, 58, 59, 60
Carbonium ion, 516, 517, **519**, 532, 706
Carbonyl compound, **160–166**, 258, 768
  β-dicarbonyl, **174**
  derivatives of, 162, 163, 164, 167
  hydroxy-, **178**
  infrared spectra of, 685–687
  ketone, 162
  table of derivatives, 167
  unsaturated, **177**
    α, β-, **178**, 557
Carborane, 159, **197**, 368
Carboxylate anion, **168**, 184, 529
Carboxylic acid, **165–168**, 175, 179, 198, 403, 616
  anhydrides, **169**
  derivatives, **168–171**, 188
Carbyne, 347
Catenane, 154, **159**
Cationic surfactant, **415**
Cellobiose, 338
Celluloid, 293
Cellulose, 338, 339, 404
  acetate, 293
Chiral shift reagents, **637**
Chloral hydrate, 174
Chlorine radical, 565, 566
Chlorobutane, 2-, 644–652
Chloroethylene, 260
Chlorofluorocarbon, **146**
Chlorofluoromethane, 257
Chloroprene, 336
Chloropropane, 3-, 257
Cinnamaldehyde, trans, 178
Cinnamate, ethyl, 631

Citrate, 261
Clairain, 438
Cobalt
  hexafluoride, 370, 371
  hydride tetracarbonyl, 617, 655
  octacarbonyl, 348
Coffee, 398
Coronand, **155**
Cresian, 333
Crown ether, 155, 157, 351, 404
Crude oil, **422–426**, 431, 436, 442
Cryptand, 155, **156**
Cryptophane, **158**
Cubane, 124
Cumulene, **127**
Cyanamide, 192
  trimer, 314
Cyanate, **190**, 191
Cyanic acid, 191
Cyano group, 183
Cyanohydrins, **164**, 270
Cyclobutadiene, 72, 531
Cyclobutane, 271
Cyclobutene, 576, 577, 578, 579
Cycloheptatrienyl cation, 70, 71
Cyclohexadiene, 1,3-, 575
Cyclohexane, 279
Cyclohexaniminium ion, 524
Cyclohexanone, 164
Cyclohexyl anion, 515
Cyclonite, RDX, 186
Cyclooctatetraene, 363
Cyclooctene, *trans*-, 121, 531
Cyclopentadiene, 812
Cyclopentadienide anion, 287, 288
Cyclopentadienone, 531
Cyclopentadienyl
  anion, 70, 71
  complexes, 321, 322, 354
Cyclopentanediyl, 1,3-, 511, 812
Cyclopentane, 271, 285
Cyclopentene, 311, 312
Cyclopentenone, 777
Cyclopentyl cation, 477
Cyclophane, **133**
Cyclopropane, edge-protonated, 560
Cyclopropanone, 256, 271, 531,556
Cyclopropenyl cation, 70, 71
Cysteine, 201, 202, 337
Cystine, 201, 202

DABCO, 193, 328
DDT, 146
Decalin, *cis*- and *trans*-, 233
Dehydrobenzene, 1,4-, 511
Dendrimers, 329, 330
Deuterochloroform, 623
Dewar benzene, 58, 804
Diacetamide, 171
Diacetone alcohol, 178, 179
Dialkylimidazolium salts, *N, N'*-, 418
Diazabicyclo[2.2.1]heptene, 2,3-, 812
Diazabicyclo[2.2.2]octane, 1,4-, 193, 328
Diazene, 189
Diazo
  group, 187
  ketones, α-, **188**
Diazoacetophenone, α-, **188**
Diazoalkane, **187**
Diazomethane, 187, 583
Diazonium ion, **190**, **523**, 532
Diazonium salt, aromatic, 190
Dibenzoyl peroxide, 198, 199
Diborane, 196, 510
Dibromobutane, 2,3-, 239
Dicarbadodecaborane(12), 1,2-, 197
Dicarboxylic acids, **175**
Dichlorocarbene, 507, 512
Dichlorocyclohexane
  cis-1,3-, 233
  cis-1,2-, 257, 260
Dichlorocyclopropane, *trans*-1,2-, 96, 255, 260
Dichlorocyclopropanone, *trans*-2,3-, 256
Dichloroethane, 1,2-, 336
Dichloroethylene
  cis-1,2-, 87, 91, 256
  trans-, 103
Dichloroketene, 177
Dichloromethylsilane, 336
Dicobaltoctacarbonyl, 383
Dicyanonaphthalene, 1,4-, DCN, 818, 819, 820
Dicyclohexylcarbodiimide, DCC, 192
Diethyl ether, 152
Diethylaluminum chloride, 311
Diethylmagnesium, 348
Diethyloxonium ion, 522
Diglyme, 153
Dihydrobenzoic acid, 708
Diimine, 189, 586
Diiron nonacarbonyl, 348, 351, 352
Diironenneacarbonyl, 348, 351, 352
Diketone
  α-, 165, **172**
  β-, 174
Dimethyl sulfate, 204
  Dimethyl-1-ethylphosphoranylbenzene, 2,3-, 213
  Dimethyl-2,3-diazabicyclo[2.2.2]oct-2-ene,
    1,4-, 829

Dimethylallene, 1,3-, 230, 231, 238
Dimethylaminoacetic acid, *N,N*-, 529
Dimethylaminyl anion, 525
Dimethylcyclohexanone, 2,5-, 263
Dimethylenediaminyl, 511
Dimethylether, 428
Dimethylformamide, DMF, 633, 634, 635
Dimethylhydrazine, 1,1-, 189
Dimethylphosphenium ion, 521
Dimethylphosphine, 213
Dimethylphosphinothioic O-acid, 219
Dimethylphosphonomonothioic S-acid, 219
Dimethylsulfonio acetate, 530
Dimethylsulfoxide, DMSO, 477
  D_6, 623
Dinitroaniline, 2,4-, 475
Dinitrogen complex, **364**
Dinitrosopentamethylenetetraamine, 336
Dioctyl phthalate, 313
Diosphenol, 173
Dioxabicyclo[2,2,1]hept-5-ene, 199
Dioxabicyclo[2.2.1]heptane, 2,3-, 812
Dioxacyclobutadione, 1,2-, 199
Dioxane, 1,4-, 192
Dioxetane, **199**, 803, 804
Dioxin, 146
Dioxolane, 163
Dioxygen complex, **364**
Diphenyl-1-chloroethane, 1,2-, 491
Diphenylammonium ion, 522
Diphenylchloronium ion, 523
Diphenyldiazomethane, 187
Diphenylethylene, 655
Diphenylfluoronium ion, 522
Diphenylketene, 177
Diphenylnitrenium ion, 521
Disubstituted cyclohexanes
  cis, 279, 280
  trans, 280
Disulfides, **201**
Dithiohemiacetal, **211**
Dithiohemiketal, 211
Dithioic
  acid, **207**
  ester, **208**
Divinylbenzene, 307
  DNA, 154, 817, 819
Dodecahedrane, 124
Dynamite, 185

Enamine, 180
Endiol, 1,1- and 1,2-, **165**
Enium ion, **520**, **521**, 533
Enol ether, **166**
Enolate anion, **526**, 547, 557, 558
  acetone, 526
Enol, **164**–178, 286, 287, 526, 557
Enzyme, 261, 283, 404, 487
Eosin, 93
Epichlorohydrin, 331
Epoxide, 193, 310
Erythrose, 240, 260
Ester, 160, **169**, 195, 570
  dithioic acid, 208
  phosphoric acid, 214
  phosphorous acid, 217
  sulfenic acid, 206
  sulfinic acid, 205
  sulfonic acid, 204
  sulfuric acid, 204
  thioic O-acid, 207
  thioic S-acid, 206
Ethane, 87, 88, 273
Ethanesulfenamide, 210
Ethanethial, 208
Ethanethioate, S-methyl, 206
Ethanethiol, 200
Ethanethioyl chloride, 210
Ethanol, 445
Ether-alcohol, **153**
Ethers, **152**, 153, 166, 404
Ethyl
  cation, 520
  mercaptan, 200
  methyl sulfide, 201
  methyl sulfoxide, 202
  radical, 564, 565, 566
Ethylene, 273, 381, 424–427, 677, 767–773
  glycol, 149, 163, 395
  oxide, 310, 311
Ethylenediaminetetraacetic acid, 350
Ethylmagnesium bromide, 346
Ethylmethylamine, 274
Ethylphosphonous acid, 215
Ethylsulfinic acid, 205
Ethyltrimethylphosphonium bromide, 219, 220
Eu(fod)3, 637

Fatty acid salt, 414
Ferric
  ferrocyanide ion, 347
  hexacyanoferrate, $Fe_4[Fe(CN)_6]_3$, 354
Ferrocene, 346–363
Fluoranthene, 76, 130
Fluorocarbons, **146**
Fluorosulfonic acid, 473
Formaldehyde, 161, 314, 338
Formic acid, 166, 167

Freon, 147
Fuel
  oil, **423**
  Diesel, 423, **435**, 446
Fullerenes, 112, **135**, 137, 507
Fulminate, **191**
  n-propyl, 191
Fulminic acid, 191
  salt, 191
Fulvalene, 126
Fulvene, 76, 126
Fumaric acid, 176
Furan, 77, 78

Gallium arsenide, 380
Gasohol, 434, **445**
Gasoline, 422–435, 445, 446
  Glass, 246
Glucose
  α- and β-, 264
  D-, 228, 250, 263, 264, 338
Glutaric acid, 176
Glutarimide, 173
Glyceraldehyde, D- and L-, 250
Glyceric acid, 250
Glycerol, 149, 707
Glycine, 338
Glycol, 149
Glyme, 153
Graphite, 112, **135**, 137, 448, 737
Gutta-percha, **335**

Halophosphorus compound, **218**
Heme, 350, 351
Hemiacetal, **162**, 263
Hemiaminal, **163**
Hemiketal, 163
Hemoglobin, 351
Heptacene, 130
Heptalene, 133
Hexacene, 130
Hexadecane, 435
Hexadiyne-3-ene, 1,5-, 511
Hexahelicene, 230, 231
Hexamethylenediamine, 329
Hexamethylphosphoramide, HMPA, 477
Hexatriene, 1,3,5-, 575
Homocysteine, 377
Hydrazide, 189, 190
Hydrazine, 167, 189
Hydrazobenzene, 189
Hydrazoic acid, 188
Hydridocobalt tetracarbonyl, 382
Hydridorhodium tetracarbonyl, 382
Hydrocarbon, 111, 422, 431
  acyclic, 113
  alicyclic, **119**
  aliphatic, 113, 114, 119
  alternant, 64, 71, **73**, 76
  aromatic, **75**, 125, 783
  benzenoid, **127**
  branched-chain, 115
  bridged, **121**, **122**
  catacondensed aromatic, **130**, 778, 779
  conjugated, 73, 125
  cross-conjugated, **126**, 776
  cyclic, 77, 119, 120
  even alternant, 71, 73, 76
  fluorinated, 336, 448
  nonalternant, 73, **76**, 130
  odd alternant, **73**, 74
  polycyclic, 122, 125
  polycyclic aromatic, 132
  polynuclear aromatic, **129–131**
  saturated, **112**, 113, 119
  straight-chain, 114
  unsaturated, 112, 120
Hydroclone, **441**
Hydrogen, 397, 431, 674, 675, 676
  atom, 13, 52, 503
  chloride, 689
  cyanide, 689
  fluoride, 35
  molecule, 35, 52, 456, 755
  peroxide, 198
  primary, **116**
  secondary, **116**
  tertiary, **116**
Hydroperoxide, **197**
Hydroquinone, 154, 155
Hydroxamic acid, 171
Hydroxy carbonyl compound, **178**
Hydroxyethyl, 1-, cation, 517
Hydroxyl group, 147, 149
Hydroxylamine, 167, **181**
Hydroxyproline, 338
Hypophosphorous acid, **215**

Imidazol-2-ylidine, 373
Imide, **172**
Imine, 167, **180**, 182
Iminium ion, **524**
Iodobenzene, 568
Iodomethylzinc iodide, 513, 514
Iron pentacarbonyl, 346, 377
Isoalkanes, **117**
Isoalkyl group, **118**
Isobutane, 432

- Isobutyl alcohol, 623
- Isobutylene, 336, 432
- Isocyanate, **191**, 326–337, 515
- Isocyanide, **183**, 357
- Isonitrile, 183
- Isooctane, 432, 434, 435
- Isoprene, 335, 336
- Isopropanol, 151
- Isopropoxyl anion, 148
- Isopropyl
  - alcolhol, 151
  - anion, 525
  - methyl sulfone, 202
- Isothiocyanates, **209**
- Isothiocyanatobenzene, 209

- Kerosine, 423
  - Ketal, **163**
- Ketene, **177**, 191
- Keto
  - acid, **175**
  - form, 164, 174, 178, 286, 287, 526
- Ketocarbene, α-, **514**
- Ketol, α-, 165, 178
- Ketone, **160–167**, 175, 179, 252–269, 739–740
  - α, β-unsaturated, 541
  - derivatives, 163, 164, **166**, 212
- Ketonitrene, α-, **515**
- Kevlar, 326, 327, 333, 334
- Knockout drops, 174
- Kodel, 333

- Lactam, 171, **195**, 310
- Lactic acid, 341
- Lactone, 169, **195**, 203, 310
- Lanthanide ion, 637
- Latex, 311
- Lithium
  - dimethylamide, 525
  - iodide, 38
  - perchlorate, 499
  - stearate, 436

- Maleic acid, 176
- Malonic acid, 176
- Mannich base, 179
- Melamine, 314
- Mercaptans, **200**, 201
- Mesityl oxide, 178
- Mesylate, methyl, 204
- Metacyclophane, [2,3], 134
- Metal
  - carbene complexes, 311, 321, 322, 347, **360–375**, 512
  - carbonyl
    - Co, 358, 368, 369, 383
    - Fe, 352, 353, 367, 368
    - Ir, 356, 364, 367
    - Mn, 358
    - Ni, 356
    - Rh, 354, 368
    - Ru, 367
    - W, 375
  - cluster compounds, **366**, 367
  - hydrido, 359, 363
- Metallacycle, 320, 321, **368**, 375
- Metallocene, 312, **321**, 347, **363**
- Metallocyclobutane, 320
- Methane, 27, 88, 94, 113, 114, 682, 706
  - bromochloroiodo-, 238
  - protonated, CH_5^+, 516, 517, 706
  - Methane radical cation, 527
- Methanediimine, 192
- Methanesulfinate, ethyl, 205
- Methanoic acid, 167
- Methanol, 98, 381, 445, **446**
- Methine group, 101, **119**
- Methionine, 377
- Methonium ion, 516, 517, 706
- Methoxyl
  - anion, 148, 525, 526
  - radical, 510
- Methyl
  - anion, 508
  - benzoate, 687
  - bromide, 721
  - cation, 508, 510, 517
  - chloride, 464, 465
  - -2-decalone, (-) and (+) *trans*-10-, 253, 254
  - ester, 187
  - hydrogen sulfate, 204, 205
  - iodide, 381
  - lithium, 346, 348
  - radical, 510
  - t-butyl ether, MTBE, 435
  - vinyl ether, 166
  - vinyl ketone, 179
- Methylaluminoxane, MAO, 321, 363
- Methylamine, 716
- Methylcarbene, triplet, 513
- Methylcarbenium ion, 520
  - Methylcyclopentadienyltricarbonylmanganese, η^5, MMT, 435
- Methyldiazonium ion, 190
- Methylene, 512
  - carbene, 512
  - chloride, 145
  - group, 101, **118**, 145, 512

Methylene, (*Continued*)
singlet, 510, **512**, 513
triplet, 511, 513
Methylenetriphenylphosphorane, 530
Methylformamide
s-*cis*-*N*-, 277, 278
s-*trans*-*N*-, 277, 278
Methylide, (trimethylammonio)-, 530
Methylidene (methylene) group, 117
Methylnaphthalene, 1-, 435
Methyloxenium ion, 521
Methylphosphonic diamide, *P*-, 219
Methylpyridinium chloride, *N*-, 182
Methylpyridinium ion, *N*-, 522
Methylsilicenium ion, 521
Methylstilbene, α-, 815
Methylsulfenium ion, 521
Molecular oxygen, 364, 566, 567, 614, 734, 774
Molozonide, 200
Morpholine, 192
Myoglobin, 351

Nanhydrin, 173
Nanotube, carbon, 112, **137**
Naphtha, 423, 427, 444
Naphthacene, 130
Naphthalene, 75, 76, 778, 779
dicarboxylic acid/ethylene glycol, PEN, 326
radical anion, 304, 528
Naphthalenecarbodithioate, methyl 1-, 208
Naphthyl anion, 2-, 509
Naural rubber, 315
Neoprene, 336
Nickel
tetrakis(triphenylphosphine), 344, 345
tetracarbonyl, 345, 472, 660
Nicotinic acid, 708
Nitramine, **186**
Nitranion, **525**
Nitrate
anion, 364
ester, **185**
Nitrene, 507, **514**, 515, 533, 550
α-keto-, 515
Nitrenium ion, 520, **521**, 533
Nitrile, 164, **183**
oxide, 191, 583
Nitrite
ester, **186**
ethyl, 186
Nitro group, 184
Nitroalkane, **184**
Nitrobenzene, 563, 744
Nitrocellulose, 338
Nitrogen
anion, **525**
ylide, 530
Nitroglycerin, 185, 186
Nitromethane, 184
Nitrones, **182**
Nitronic acid, 184
Nitrosamide, **187**
Nitrosamine, **187**
Nitroso group, 185, 336
Nitrosoalkane, **185**
Nitrosoamine, *N, N*-dimethyl, 187
Nitrosobenzene, 185
Nitrosourea, *N*-methyl-*N*-, 187
Nitrous
acid, 187
oxide cation, 715
Noble gas, 22
Norbornene, 122
Norbornyl cation, **477**, 517, 518
Nylon, 195, 294, **329**
6-, 310, 327, 329, 333
6,6-, 315, 327, 329, 333

Olefins, **120**
Organoborane, **196**
Organoborate, **196**
Organomercury, 549
Organometallic compounds, 344, 345, **347**, 380
carbene complexes, 311–322, 347, **360–375**, 512
main group, 347, **348**
transition metal, 347, **349**
Organophosphorus, **212**
acid, 214
acid amides, **219**
thioacids, **219**
Orthophosphoric acid, 214
Oxalic acid, 176
Oxene, 533
Oxenium ion, **521**, 533
Oxetane, **193**
Oxime, 167, 185, 235
Oxirane, 193
Oxygen
anion, **525**
molecular, 364, 566, 567, 614, 734, 774
singlet, 757, 773, 810
triplet, 773, 810
Ozone, 147, 200, 583
Ozonide, **200**

Palladium
acetate, 384
tetrakis(triphenylphosphine), 348
Paracyclophane
[10], 134
[3,4], 134
Paraffin, **114**
Penicillin, 195
Pentaborane, 196
Pentacene, 130
Pentaerythritol, 152
Pentalene, 133
Pentane, 397
Peracid, 198
Perbenzoic acid, 198
Perester, **198**
Peroxide, 565, 645
diacyl, **198**
dialkyl, **198**
diaroyl, **198**
Peroxy ester, 198
Peroxybenzoic acid, 198
Perylene, 131
Petroleum, **421–423**
Phenanthrene, 75
Phenanthroline, 77, 78
Phenes, **131**
Phenethyl, β-, cation, 519
Phenols, **149**, 150, 164
Phenonium ion, 518, 519
Phenyl, 128, 149
azide, 188
cation, 722
cyanate, 191
isocyanate, 191
isocyanide, 183
mercaptan, 201
Phenylactone, 620
Phenylcarbene, 513
Phenyldiazonium ion, 523
Phenylenediamine radical cation, *N,N,N',N'*-tetramethyl-o-527
Phenylnitrene, 514
Phenylphosphinic acid, 215
Phosphane, 212
Phosphate, **214**
trimethyl, 214
Phosphenium ion, **521**
Phosphine, **212**
imide, **214**
oxide, **213**
Phosphinic acid, **214**, 215
alkyl or aryl, **215**
dialkyl or diaryl, **216**
ethylmethyl, 216
Phosphinous acid, **216**
dialkyl or diaryl, **216**
dimethyl, 216
Phosphite, **217**
Phosphonate, dimethyl methyl, 217
Phosphonic acid, **216**, 217
alkyl or aryl, **217**
Phosphonium
ions, **523**
salts, **219**
Phosphonous acid, 214, **215**
alkyl or aryl, **215**
Phosphoranes, **213**, 533
Phosphoric acid, 214
Phosphorous acid, 216, **217**
Phosphorus
pentachloride, 42
pentafluoride, 284
ylide, 530
Phthalic acid, 175, 176
ester, 313
Phthalimide, 172, 173
Pimilic acid, 176
Pinacol, 527
Piperazine, 192
Piperidine, 192
Plastics, 293, **313**
Platinacyclobutane, 369, 370
Podand, **153**
Poly-
(acrylonitrile), 138, 300
(dimethylsiloxane), 315
(ethylene glycol), PEG, 310, 415
(ethylene terphthalate), PET, 326, 327, 341
(ethylene), 294, 298, 299, 300, 324, 427
(glycolic acid), 340
(hexamethyleneadipamide), 329
(isoprene), *cis*-1,4-, 311
(lactic acid-co-glycolic acid), 340, 341
(methyl methacrylate), 300, 304
(phenylene sulfide), 313
(phosphazene), **340**
(propylene), 300, 316, 427
(tetrafluoroethylene), 300
(vinyl acetate), 300
(vinyl alcohol), 300
(vinyl chloride), PVC, 299, 300, 313
acetylene, 321
acrylamide, 404
amide, **326**, 328, 333
boron hydride, 197

Poly- (*Continued*)
carbonate, **326**
carboxylic acid, **175**
chlorobiphenyl, PCB, 146
ene, conjugated, 55, 59, **125**, 776
ester, **326**, 333
ether, 351
propylene, 317–319
high-density, HDPE, 300, **312**, 320, 341
low-density, LDPE, 300, **312**
hydric alcohol, **149**
imide, **326**
ol, **149**, 337
peptide, 563
styrene, 295, 296, 300
urethane, 193, **326–328**, 336, 337
Porphyrin, 239, 364
Potassium
dihydrogen phosphate, 744
ferrocyanide, 345
$Pr(hfc)_3$, 637, 638
Prismane, 124
Proline, 338
Propanesulfenate, methyl, 206
Propanesulfinamide, 2-, 210
Propanesultone, 1,3-, 203
Propanethioate
O-ethyl, 207
S-acid, 206
Propanol, 2-, 151
Propargylbenzene, 128
Propellane, 122
Propiolactone, β-, 195
Propiolic acid, 178
Propriolactam, β-, 195
Propylene, 381, 383, 424–427
Protein, 179, 337, 340
Prussian blue, 347
Pyrazine, 77, 78
Pyrene, 131, 793
Pyridine, 77, 78
D_5, 638
N-oxide, 4-methyl, 181
Pyridinium hydrochloride, 182
Pyridyne, 530, 531
Pyrimidine, 77, 78
Pyrrole, 77, 78
Pyrrolidine, 192

Quaternary ammonium salt, 182
Quinhydrone, 154, 155

Rayon, 293, **338**
Riboflavin, $FADH_2$, 819
Rocket fuel, 189
Rope, 137
Rose Bengal, 810
Rotaxane, 154, **159**
Rubber, **334–336**
foamed, **336**
natural, **335**
silicon, 315
Ruthenium carbene, di(tricyclohexyl)phosphine-, 375

Salicylaldehyde, 178
Saran, 300
Sarin, 217
Sasol, 446
Schiff base, 167, 180
Schwartz's reagent, 359
Sebacic acid, 436
Semicarazide, 167
Semicarbazone, 167
Sepharose, 404
Serine, 338
Silanol, 331
Silicon carbide, 694
Silicones, 331, 336
Siliconium ion, **524**
Silyl enol ether, 166
Soap, 413, 414
invert, **415**
Sodium
amide, 525
bisulfite, 167
fluoride, 38
methoxide, 526
naphthalide, 304
oleate, 413, 414
thiophenoxide, 415
Spherand, **157**
Spiran, **125**
Stilbene, *cis*- and *trans*-, 491
Styrene, 150, 295, 336
radical anion, 304
radical cation, 527
Suberic acid, 176
Succinic acid, 176
Succinic anhydride, 170
Sulfanilamide, 209
Sulfate, **204**
dimethyl, 204
Sulfenamide, **210**
Sulfenate, **206**
Sulfenic acid, **205**
Sulfenium ion, **521**
Sulfenyl halide, 205

Sulfide, 201
Sulfinamide, **210**
Sulfinate, **205**
Sulfinic acid, **205**
Sulfinyl halide, 205
Sulfonamide, **209**
Sulfonate, **204**
Sulfone, **202**
Sulfonic acid, **203**, 403
  salt, 414, 415
Sulfonium ion, **523**
Sulfonyl halide, 203
Sulforaphane, 209
Sulfoxide, **202**, 209
Sulfur, 336
  heterocyclic compound, **210**
  hexafluoride, 42, 285
  trioxide, 473
Sulfurane, **212**
Sulfuryl chloride, 203, 220
Sultone, **203**

Tartaric acid, meso, 97
Tebbe's reagent, 369
Terylene, 333
Tetracene, 130
Tetraethylammonium ion, 522
Tetraethyllead, TEL, 346, 435
Tetrahydrofuran, 192
Tetramethylammonium chloride, 182
Tetramethylsilane, TMS, 609
Tetraphenylcyclopropane, 1,1,2,2-, 818, 820
Tetraphenylsulfurane, 212
Thalidomide, 172, 173, 225
Thiacyclopentane, 211
Thioacetal, **211**
Thioalcohol, **200**
Thioaldehyde, **208**, 211
Thiobenzophenone, 208
Thiocarboxylic acid, **206**
Thiocyanate, **209**
  anion, 364
Thiocyanatopropane, 2-, 209
Thiohemiacetal, **211**
Thioic
  O-acid, **207**
  O-ester, **207**
  S-acid, **206**
  S-ester, **206**
Thioketal, 212
Thioketone, **208**
Thiokol, 336
Thiols, **200**, 201, 211
Thionyl chloride, 202, 220
Thiophene, 77, 78
Thiophenol, **201**
Thioyl halide, **210**
Threose, 240, 260
Thymine, 819
  dimer, 819
Titanium
  dioxide, 340
  ethoxide, 340
  hydroxide, 340
  tetrachloride, 319
Toluene, 150
Triacetamide, 171
Triammoniacobalt(III) chloride, 345
Tricyclo[2.1.1.0^{1,4}]hexane, 271
Triethyl borate, 196
Triethylaluminum, 319
Triethyloxonium ion, 522
Triethylphosphorane, 213
Triflate, methyl, 204
Trifluoroacetic acid, 471
Triketone, 1, 2, 3-, 172, 173
Trimethlyamine oxide, 181
Trimethylammonio acetate, 530
Trimethylammonium chloride, 182
Trimethylcarbenium ion, 520
  tetrafluoroborate, 520
Trimethylenemethane, 65
Trimethyloxonium salt, 509
Trimethylpentane, 2,2,4-, 432, 434, 435
Trimethylphenylphosphonium ion, 523
Trimethylsulfonium ion, 523
Trimethyl orthoformate, 169
Triphenylphosphine, 344, 351, 358
  imide, P,P,P-, 214
  oxide, 213
Tris(ethylenediamine)cobalt(III) chloride, 348
Tritium, 406
Tungstacyclopentane, 3,4-dimethyl-, 369, 370
Tungsten
  carbonyl carbene, diphenyl-, 375
  hexachloride, 311
Tygon, 313, 336

Uranocene, 363
  Urea, 170, 708
Urease, 493
Urethane, 180, 326, 327, 328

Vinyl, **120**
  chloride, 299
  isocyanate, 309
  polymer, 296-303

Vinylcarbenium ion, 520
Vinylcyclohexene, 4-, 120, 121
Vitamin
  B_{12}, 376
Vitamin (*Continued*)
  C, 165, 166
  D, 735
Vitrain, **437**, 438
Vitrinite, 438

Wacker catalyst, Pd+Cu, 381
Water
  critical point, 397
  infrared, 681, 682
  orbitals, 108
  gas shift, 422, **444**, 447
Wurster's salt, 527

Xanthate ester, 556
Xenon, 707
Xylene, 150

Ylide, **530**, 533
  nitrogen, 533
  phosphorus, 533

Zeise's
  salt, 346, 352, 353, 361, 381
  dimer, 351, 352
Zeolite, 424, **427**, 429, 430
  chabazite, 428
  erionite, 428
  faujasite, 428
  mordenite, 427, 428
  ZSM-5, 428
Zirconium, bis(cyclopentadienyldimethyl), 363

GENERAL INDEX

Numbers in **bold** indicate the page on which the term is a topic.

10 *Dq*, 370
1*s* Orbital, **12**
2*p* Orbital, 12, **15**
2*s* Orbital, **13**
A values, **281**
$A_{AC}2$, **571**
$A_{AL}1$, **572**
Abalone, 340
Abiotic, 735
Absorbance, *A*, 670, **749**, 750, 753, 824
Absorption, **412**
Absorption and stimulated emission, relative intensities of, 784
Absorption band shape, 765
Absorption band,
  intensity of an, **751**, 761
  shift of, 751, 752
  ultraviolet, 737, 739, **750**, 751, 772, 778, 779, 780
    allowed, 768, 770
    forbidden, 768, 771
    width of, 751
Absorption continuum, 767
Absorption maximum, λ_{max}, 749, **750**, 751, 776, 777, 823
Absorption of light, 672, 729, **748**, 759–763, 780
Absorption shoulder, 749, **750**
Absorption,
  negative, 785
  singlet-triplet, 751, 764, 783
Absorptivity,
  A, 749
  molar, ε, 670, 748-**750**, 751, 761, 809
Accelerating potential, 709
Acceptor,
  electron, 154, 349, 817-820
  electron pair, 471, 472
  energy, 806–808, 812
  photochemical, 771
  proton, 470
Accidental degeneracy, 682, **690**, 691
Acenes, 130
Acetals, **162**, 163, 263
Acetate, 168, 525, 526
Acetic acid, 167, 471
  acidity of, 474
  catalytic synthesis of, 382
  Monsanto synthesis of, **382**
Acetic acids, substituted, 502
Acetone, 162–164, 381
  fragmention, 719
  protonated, 517
Acetyl cation, 516
Acetylacetonato anion, 351
Acetylcarbene, 514
Acetylene, 321, 424
  geometry of, 42
Achiral, 97, **232**, 256
Aci form of the nitro group, **184**, 555, 557
Acid azides, **188**
Acid dissociation constant, K_a, and pK_a, **474**, 475, 494, 500, 502
Acid halides, **68**
Acid halides of inorganic acids, 218, **220**
Acid-catalyzed bimolecular hydrolysis with acyl-oxyen cleavage, $A_{AC}2$, **571**
Acid-catalyzed unimolecular hydrolysis with alkyl-oxygen cleavage, $A_{AL}1$, **572**
Acidimide, substituted, **172**

The Vocabulary and Concepts of Organic Chemistry, Second Edition, by Milton Orchin, Roger S. Macomber, Allan Pinhas, and R. Marshall Wilson

Acidimido ester, 172
Acidity function, *H*, **474**
Acidity of Halogen acids, HX, 477
Acoustic waves, 830
Acousto-optic spectroscopy, **830**
Acquisition time, 609, **611**
Acrilan, 333
Actinometer, **787**
Actinometry, **787**
Activated complex, 478
Activation barrier, 493, 503
Activity coefficient, γ, **468**
Activity, *a*, **468**, 469
Acyl azides, **188**
Acyl group, 168
Acyl halides, **168**, 210
Acyl halides, hydrolysis of, 532
Acyl nitrene, **515**
Acylcarbene, **514**
Acylium ion, 407, 516, **523**
Acyloin condensation, 527
Addition electrophilic to
  carbon-carbon multiple bonds, 557, **558**–561, 567
Addition of nucleophiles to
  carbonyl groups, 569
  cyano groups, 569
Addition
  polymerization, **298**
  polymers, 297
  reaction, **556**
Addition,
  1,2-, 556, 557, 560, 561
  1,4-, 556, 557
  1,6-, 556
  anti-Markovnikov, 559, **567**
  cis, 359, **560**, 582, 586
  Markovnikov, 558, 559, 568
  trans, 265, **560**, 561
Addition to
  carbon-carbon double bonds, 558, 567
  conjugated π-systems, 556
  cyclopropanes, 556
  imine groups, 569
  multiple bonds, 556, 557
Ad_E2, **560**
Adiabatic Process, **454**
Adsorption, **411**
  chemical, **411**
  physical, **411**
Aerosol, 147, **415**
Agarose gel, 404
Aggregation, 414
  molecular, 413, 414
  self-, 414
Agostic hydrogen, **360**
Alcohol,
  primary, 147
  secondary, 147
  tertiary, 147
Alcohols, **147**, 148, 153, 162, 163
  polyhydric, **149**
Alcoholysis, 172, **545**
Aldehyde derivatives, 162, **166**, 211
Aldehydes, 160, **161**, 167, 179, 358, 382, 777
  magnetic anisotropy of, 615, 616
Alder rules, **582**
Alder-Stein rules, 582
Aldol, 178, 179
Alkali metal, 21
Alkaline earth metal, 21, 349
Alkanes, 112, **113**, 114, 431, 446
  branched-chain, **115**
  cyclo-, 116, **119**
  iso-, **117**
  straight-chain, **114**, 117, 435
Alkene
  formation, 551, 552, 572
  reaction with carbenes, 586
Alkenes, 112, 114, **120**, 230
  coordination number of, 355
Alkoxides, **526**
Alkoxyl group, **148**
Alkyl group, **116**, 144
  iso-, **118**
  normal, **117**
Alkyl halide, **145**, 147, 357, 541
Alkylate, **432**
Alkylation, 433
Alkylidenation reaction of olefins, *trans*-, **375**
Alkylidene group, 117
Alkylidyne complexes, **375**
Alkynes, 112, 114, **127**
  coordination number, 355
Allowed
  energy levels, **673**
  photochemically in the excited state, 581, 587
  reaction, **580**
  thermally in the ground state, 578, 581, 587
  transition, 756, **759**, 764–770
Alloys, 340
Allyl
  anion, 74, 286, 525
  cation, 74, 286, 516
  radical, 74, 286
  system, charge density of, 64
Allylic isomers, **286**

Ambident
electrophile, 542
ion, **542**
ligands, **364**
nucleophile, 542
radical, **542**
American Petroleum Institute, 423
Amide Ion, 268, 269, 303, 525, 568
Amides, **170**, 172, 186, 195
hydrolysis of, 532
secondary, **171**
substituted, **171**, 277
tertiary, **171**
Amine oxides, **181**
Amine,
primary, 167, 180
secondary, 180, 186, 187
tertiary, 180–182, 328
Amines, 163, 179, **180**–182, 274, 525
Amines,
hydroxyl-, **181**
oxidation of, 524, 527
Aminium ion, **524**
Amino
acids, 529, 563
carbonyl compounds, **179**
Aminoaldehyde, β-, 179
Aminocarboxylic acid, α-, 179
Aminoketones, β-, 179
Aminolysis, 172, **545**
Ammonia, 274
clock, 274
geometry of, 42
synthesis, **447**
Ammonium salts, **182**, 403, **522**
quaternary 182, 403, 415
Amorphous, 246
Amphiphilic copolymers, 307
Amphoteric, 471,
Analyte, 401, 705
Analyzer, 249
Anancomeric ring, **282**
Anchimeric assistance, 519, **546**
Anderson-Schultz-Flory equation, 446
Andiron structure, **227**
Anesthetic, 152
Angle, bond, **37**
Angular
dependence of orbital, 15
groups, **123**
momentum, 11
orbital, 778
quantum number, *J*, **669**
velocity, **668**
Anharmonic
effects, 672
oscillator, **675**, 676
Anharmonicity, 689
constants, 675, 676
Anilines, substituted, 475
Anilinium ion, 522
Animal fibers, **337**
Anion, 288, **509**
complexed, 159
exchange, 403
metal complexes, 365
radical, 154, 304, 512, **527**, 533, 567, 803–805
Anion,
alkoxyl, 148, 525, 526, 533
amide, 525, 533
aminyl, 533
carboxylate, **168**, 184, 526, 529
cyclopentadienyl, 70, 71
cyclopropenyl, 72
enolate, **526**
naked, 156
radius, 38, 39
Anionic
initiation, 309
ligands, 364
polymerization, 298, **303**, 304, 324
surfactant, 415
Anions, electrophoretic migration of, 400
Aniostropic, **416**
forms of matter, 742
Annulenes, **77**, 233
Anode, 400, 406,
Anomer, **263**
Anomeric, 263
Anomerization, 263
Ansa
bridge, 321
ligands, 363
Antarafacial, *a*, **586**, 587
Anthracite, 436
Anti conformation, **277**, 279
Antiaromatic compounds, **72**, 531
Antibody, 404
Antibonding
molecular orbital, **43**, 511, 527, 528
σ molecular orbital, **45**
Anticlinal eclipsed conformation, **277**
Anticrowns, **159**
Antifreeze, 149
Antiknock index, 434
Anti-Markovnikov addition, 559, **567**
Anti-periplanar conformation, **277**
Anti-Stokes band or line shift, **695**, **794**

Antisymmetric
behavior, **94**
stretching vibrations, 679, **680**
transformation, 578
vibrations 679, **680**
vibrations of methyl and methylene groups 683, 684, 685
API gravity, **423**
Apical and equatorial positions
of a trigonal bipyramid, **284**, 377
Apicophilicity, **284**
Aprotic solvent, **477**
Aqua complexes, 345
Arachno clusters, **368**
Aramid, 326, 333, 340
Arenes, **75**
Aromatic hydrocarbons, **75**, 125, 783
Aromatic protons, chemical shifts of, 617
Aroyl
group, 168
halides, **168**
Arrhenius energy of activation, E_a, 480, **488**, 493, 578
Arrow pushing, **538**
Arrow,
double-headed, 58, 60, 538
looped (rearrangement), 200
single-headed (Fish hook), 59, 60, 538
Aryl
group, **128**, 144
halide, elimination of HX, 530, 531
halides, **145**, 384, 385
Aryne mechanism, **563**
Arynes, **530**, 533, 563, 564
Ash, 437, **439**, 443
Aspect ratio, 137
Asphalt, 423
Asymmetric, **97**, 107
center, **231**
induction **267**, 269, 569
intermolecular, 267, 268
intramolecular, 267, 268
synthesis, **267**, 268
top molecules, **691**
Asymmetry, 261
Atactic polymer, **319**
A-Tel, 333
Atomic
carbon, 507
core, **22**
mass unit, amu, 704
number, 18
orbital, **7**, 42, **45**
coefficients, **50**, 51, 63, 64, 67, 69
hybridization Index, **23**
nonequivalent hybrid, **23**
pπ, **45**
orbitals for many electron atoms, **17**
orbitals,
equivalent hybrid of, **23**
hybridization of, **22**, 50
linear combination of, LCAO, 27, **43**, 47, 56, 64, 109, 379
radius, **37**
Atoms,
pro-(*R*), 257
pro-(*S*), 257
Atropisomer, 274, **282**, 283
Attached proton test, APT, **644**
Attomole, 707
Attrital coal, 437, **438**
Aufbau principle, **17**, 43, 48, 50, 372
Autoprotolysis, **477**
Average bond dissociation energy, table of, 462
Avogadro's number, 673, 704
Axial
mode, 799
positions, **279**–282, 633
substituent, 279–281
Axial/equatorial group interchange, 279, 280
Axis of
rotation, C_n, **85**, 255, 256, 261, 279, 577, 607
symmetry, alternating or improper, S_n, **88**, 89, 232, 257, 258, 261, 607
Azeotrope, **394**
Azeotrope,
alcohol/water, 394
maximum boiling, **395**
minimum boiling, **394**
ternary, water/benzene/ethanol, 394
water/HCl, 395
Azides, **188**
acid, **188**
acyl **188**
Azimuthal quantum number, *l*, **11**
Aziridines, 193, 274, 806
Azo compounds, **189**, 190, 235, 565
cis-trans (*E/Z*) isomers of, 189
Azoalkane, 812, 829
Azoxy compounds, **190**
cis-trans (*E/Z*) isomers of, 190

B absorption bands, 779, 780
$B_{AC}2$, **571**, 572
Backbiting, 301
Back-bonding or -donation, 352, 362

Back-bonding,
  CO to metal, 660
  metal-to-ligands, **352**
Backside attack, 542
Baeyer strain, **271**
Bakelite, 293, 314
$B_{AL}2$, **572**
Baldwin's rules, **288**, 289, **590**
  dig, tri, tet, endo, exo, 289, 590
Band, 0-0, **756**, 760, 762, 765–767, 790, 794, 795, 807, 820
Banded coal, **437**
Barrel, 423
Base dissociation constant, K_b, and pK_b, **475**
Base-catalyzed bimolecular hydrolysis
  with acyl-oxygen cleavage, $B_{AC}2$, **571**, 572
Base-promoted bimolecular ester hydrolysis
  with alkyl-oxygen cleavage, $B_{AL}2$, **572**
Bathochromic shift, **751**, 752
Bay region, **132**
Beckwith's rules, **289**
Beer-Lambert law, **748**–750, 785
Bending vibrations, **681**
  in water, 699
  in plane, 659
  out-of-plane, 659
Bending,
  in-plane, 679, **682**
  out-of-plane, **682**
Benzene,
  molecular ion of, 716
  resonance energy of, 62
  substitution of, 495
Benzenoid hydrocarbons, **127**
Benzenonium ion, 561
Benzoyl
  cation, 722
  peroxide as radical initiator, 308, 309
Benzyl
  anion, 516
  cation, 509
  radical, 510
  system, 76
Benzyl/tropylium cation, 723, 724
Benzyne, 507, 509, 530, 531, 563, 564
  mechanism, **563**, 564, 567
Bergius process, **440**
Berry pseudo-rotation, **284**, 285, 377
BET equation, **412**
Betaine, **529**, 530, 772, 773
Bicyclic molecules, 283, 552
Bimolecular, 478
  nucleophilic substitution via
  addition-elimination, **543**
Binding energy of core electrons, 829
Biochemiluminenscence, **803**
Biodegradable polymers, **340**
Biodeisel Fuel, 435
Biopolymers, 293
Biotic, 735
Biphenyl isomers, 282
Biradical 507, **511**, 812
  triplet, 72, 812
Birefringence, **245**, 247, **742**, 743
Bis, 152
Bisulfates, **204**
Bite angle, 351
Bituminous, 436, 439
Block polymers, **306**
Blue shift, **752**, 768, 769, 773
Boat conformation, **280**
Boghead coal, 438
Bohr equation or condition, Bohr-Einstein equation
  or Planck-Einstein equation, **3**, 660, 672, 673, **755**
Boltzmann distribution, 470, 597, 827
Boltzmann's constant, 597
Bombyx mori (silk moth), 338
Bond
  angle strain, **271**
  cleavage,
    heterolytic, 36, 464, 476, **538**
    homolytic, 36, 356, 461, **538**
  dissociation, **36**, 767, 770, 830
    energy, average, D^a **461**
    energy, $D^o_{(A\text{-}B)}$ **37**, **461**, 464, 674, 675
    energy, heterolytic, **464**, 476
    energy, specific, 462
  length, **36**, 754
    in excited state, 766
    in ground state, 766
  order, ρ_{rs} **64**
  slippage, 361
  strength, **461**
  stretching strain, **271**
Bond,
  carbon-carbon, 770
  carbon-iodine, 769, 770
  chemical, **27**
  closed multicenter, 517
  coordinate covalent, **40**, 202
  covalent, **28**
  dative, **40**
  delocalized, **56**, 65
  electron pair, **28**
  hydrogen, **40**, 154
  ionic, **35**

Bond, (*Continued*)
  localized π, **46**
  nonpolar covalent, 28, **32**
  partially ionic, **34**
  polar covalent 28, **34**
  polarized, **410**
  σ, **44**, 47
  single, double, and triple, **35**
Bonding
  energy, 74
  molecular orbital, **43**
Bonding,
  closed three-center, two-electron, **80**
  closed three-center, four-electron, 518
  delocalized, **56**
  localized two-center, two-electron, **28**, 56
  open three-center two-electron, **80**
  σ multicenter, **79**
  three-center two-electron, **80**, 517, 518, 559
Boranes, **196**, 472
  heterocyclic, **196**
Born-Oppenheimer approximation, **762**, 781
Boron
  hydride addition to double bonds, 569
  hydrides, 368
Bottled gas, 442
Bowsprit, 280, **281**
Bow-stern interaction, 280, 281
Bredt's rule, **123**, **552**
Brewster's
  angle, Φ_B, **741**
  rules, **252**
Bricks, 340
Bridged cation, 517, **518**
Bridgehead
  carbon atoms, **121**-123
  substituents, 283
Bridging groups, names of, **366**
Brisänce, 186
British thermal units, BTU, 437
Bromonium ion, 518, 519, 561
Brönsted
  acid, 427, **470**, 474, 477, 493, 494
  base, **470**, 475–477, 493, 494
  equation, 500, **502**
Brönsted-Lowry Theory, 470, 472
Bulletproof vest, 26
Buna rubber, 336
Butadiene, 1,3-, 576–581
  addition of hydrogen chloride, 491
  polymers of, 298, 336
  molecular orbital of, 57
  secular determinant of, 68
Butterflies, 338
Butyl
  cation, *t*-, 515, 516, 517, 520
  chloride, *t*-, solvolysis of, 502, 507
  esters, *t*-, 572
Butyrolactone, γ-, 195, 310, 311
  polymerization of, 310, 311

C_{2v} point group, 92, 578
Cabenium ion, asymmetric, 528
Cage
  escape, 654
  recombination, 654
Calixarenes, **157**, 158
Cannels coal, 438
Caprolactam, 195, 310, 311, 447
  polymerization of, 310, 311
Carbaborane, 197
Carbamates, 180
Carbanion, 515, 516, **524**, 533, 547, 555, 556
Carbene, 118, 183, 379, 507, 511, **512**, 513, 533, 550
  complexes, organometallic, 311, 312, 320, 321, 347, 360, 369, **373**
  reaction with alkenes, 586
  trapping, 507
Carbene,
  α-keto-, **514**
  Fischer, 374
  Schrock, 374
  singlet, **513**
  triplet, 511, **513**
Carbenes, stable, 373
Carbenium ion, 516, 517, **520**, 533, 544, 559
  metal stabilized, 374
Carbenium ions, thermodynamic stability of, 521
Carbenoid species, **513**
Carbocation, 61, 424-426, 432, 473, 476, 498, 499, 515, **516**-520, 544-547, 554, 558-562, 571
Carbocation,
  Hypervalent, 519
  primary, 424
  secondary, 424
  tertiary, 425
Carbocations, relative stabilities of, 425
Carbocyclic, 14
Carbodiimides, **192**
Carbohydrate chemistry, 263
Carbon dioxide
  cation, 715
  bending modes of, 681, 682
  IR spectrum, 681, 682, 688, 690
  stretching modes of, 96, 681

super critical, 398
symmetry allowed or forbidden IR transition, 688
valence bond structure of, 26
Carbon
elemental crystalline forms, **135**
fibers, **137**, 507
monoxide, 346, 357, 382, 666, 671
stretching vibrations, 690, 691
nanotubes, 112, **137**, 507
primary, **115**, 116, 147
quaternary, 115
radicals, relative stabilities of, 425
secondary, **115**, 116, 147
skeleton, connectivity change in, 575
tertiary, **115**, 116, 147
tetrachloride, 301
raman spectrum of, 696
Carbon-13, C, CH, CH_2, and CH_3 signals, 644, 645
Carbonate, 58, 59, 60
resonance structure of, 58
Carbon-carbon double bond faces, 256, 259
Carbonium ion, 516–**519**, 532, 706
Carbonyl compounds, 164, 166, 258
dicarbonyl, β-, **174**
hydroxy-, **178**
unsaturated, **177**
Carbonyl group
bond order, 685
charge distribution, 569, 769
derivatives of, 162–164, 167
faces of, 269, 270
infrared spectra of, 685
irreversible addition to, 568
ketone, 162
π molecular orbitals, 569
reaction with electrophiles, 568, 569
reversible addition to, 568
table of derivatives, 167
Carbonyl groups, 144, **160**, 164, 193, 768, 782
addition to, 211, 269, 532, **568**, 569
Carbonyl rotational energy levels, 666
Carbonyl stretching frequencies
effect of conjugation, 685, 686
effect of electron-releasing groups, 685, 686
effect of electron-withdrawing groups, 685, 686
effect of small rings, 685, 686
of ketones, 685–687
tables of, 685, 686
Carbonyl, α, β-unsaturated, **178**, 557
Carboranes, 159, **197**, 368
Carboxylate anion, **168**, 184, 529
Carboxylated polystyrene, cross-linked, 403
Carboxylic acid, 165, **166**, **168**, 175, 179, 198, 403, 616
anhydrides, **169**
derivatives, **168**–171, 188
Carbyne, 347
complexes, **375**
Carcinogen, 129, 187
Carcinogenic properties, 132
Cartesian coordinates, 8
Casinghead gas, 442
Catalysis, **492**
acid, 502
base, 502
general acid, **493**
general base, **493**
heterogeneous, 380, 412, **429**, 430, 433, 434, 493
homogeneous, 380, **429**, 430, 433, 434, 493
metallocene, 312, **321**
specific acid, **494**
specific base, **494**
with organometal compounds, 380
Ziegler-Natta, 312, 317–**319**
Catalyst surface characterization, 830
Catalyst,
Co, 382, 382
cobalt-thoria, 446
Fe, 446, 447
Ni, 384
Pd, 384, 385
phase-transfer, **415**
Pt, 448
Rh, 382, 383
zinc oxide, 446
Catalysts, 428
dual function or bifunctional, **430**, 431
metal carbene, 375
polymerization, 363
single-site, 321, 363
Catalytic
activation, **362**
activity, 356
coupling reactions, **384**, 385
cracking, **424**, 431
hydrogenation, 112, 362
synthesis of acetic acid, 382
Catalyzed reactions, 486, 502
Catenanes, 154, **159**
Caterand, 160
Cathode, 400, 406
Cation, **508**, 516, 517
allyl, 74
bridged, **518**
classical, **517**, 518

Cation (*Continued*)
complex, 155–160
cycloheptatrienyl, 70, 71
cyclohexadienyl, 561
cyclopropenyl, 70, 71
even electron, **716**
exchange, 402
nonclassical, **517**, 518
odd electron, **716**
radical, 154, 463, 512, 524, **527**, 533, **716**, 803–805, 827, 828
radius, 38, 39
resonance stabilization of, 497, 562, 563
Cationic surfactant, **415**
Cations, electrophoretic migration of, 400
C-C double bond, magnetic anisotropy of, 615
C-C single bond, magnetic anisotropy of, 615
C-C triple bond, magnetic anisotropy of, 616
Cellobiose, 338
Celluloid, 293
Cellulose, 338, 339, 404
acetate, 293
regenerated, 338, 339
Center of mass, 678, 690
Center of symmetry, *i*, **87**, 91, 232, 239, 257, 258, 765
Ceramics, 40
Cetane number, **435**
cgs system, 663
Chain
branched, 115, 117
carriers, **566**
growth polymer, 310
growth polymerization, 297, **298**, 299, 305, 310
reaction, **564**, 565, 566
straight, 116
transfer, **301**
Chain-initiating step, **565**
Chain-propagating steps, **565**
Chair conformation, **279**
Chair-chair flip, 279, 280
Challenger, shuttle, 316
Character of transformation matrix, **105**
Character tables, 102, 105
C_{2v}, 680
D_{2h}, 763
of degenerate point groups, 101, **106**
of nondegenerate point groups, 101
of point groups, **101**
Charge,
density, **64**
distribution, **64**
formal, **32**, 144, 181, 538–541
Charged species, **508**
Charge-delocalized ion, 168, 286, **515**
Charge-localized ion, **515**
Charge-transfer excited state, **771**, 772
CTLL, 771
CTLM, 771
CTML, 771
Charge-transfer transition, **772**
Charge-transfer, CT, 154, 771, 772, 816
intermolecular of intramolecular, 772
Cheletropic reaction, 575, **586**
Chemical
adsorption, **411**
bond, **27**
formula of constitutional repeating Units, **296**
ionization, CI, **706**, 717
potential, **469**
Chemical shift
imaging, 653
ranges for ^{1}H and ^{13}C, **618**, 619
reagents, **637**
Chemical vapor deposition, CVD, **380**
Chemically equivalent atoms or groups, **256**, 629
Chemically induced dynamic nuclear polarization, CIDNP, 528, **654**
Chemiluminescence, 199, **803**
Chemisorption, **411**
Chemzymes, **270**
Chiral, 97, 224, **232**, 258
auxiliary, **268**
carbonyl groups, 739, 740
catalysts, 380
center, **231**, 267, 285, 316, 318
chromatography, 402, **404**
environment, 256
molecules, 237–239, 822
pool, **269**
selector, 404
selector,
cyclodextrin, 404
macrocyclic antibiotics, 404
peptide, 404
polysaccharide, 404
protein, 404
separations, 390
shift reagents, **637**
interaction of enantiomers with, 637
solvents, **269**
Chirality, 237, 247, 261
of amines, 274
Chlorenium, **521**
Chlorine radical, 565, 566
Chlorocarbons, **146**
Chlorofluorocarbons, **146**
Chloronium ions, **523**

Cholesteric mesophase, **417**
Chromatograph, gas-liquid, 401
Chromatography, **398**
affinity, 390, 402, **404**
chiral, 402, **404**
column, **398**
gas-liquid partition, GLPC, GLC, GC, 390, **401**, 405, 714
gel permeation, 402
high performance liquid, HPLC, 390, **402**, 404, 715
ion, 402, **403**, 404
ion exchange, **404**
molecular exclusion, **402**
paper, **399**, 400
reversed-phase, **402**
size-exclusion, **324**
thin-layer, **399**, 400
Chromophore, 189, **751**, 752
CIDNP,
multiplet effect in, **654**, 655
multiplet effect, *A/E,* 654, 655
multiplet effect, *E/A,* 654, 655
net effect, **654**
net effect, *A*, 654
net effect, *E*, 654
CIP (Cahn-Ingold-Prelog) priority, **236**, 237, 251, 259
Circular dichroism, CD, 252, **822**
magnetic, MCD, 824, **825**, 826
molar, $\Delta\varepsilon$, 824
spectrum, 824, 826
Circularly polarized light, **243**, 246–248, **731**, 733, 737, 822–824
right- and left-, 732, 826
Cis and *trans* configurations, **232**
Cis-addition, 359, **560**, 582, 586
Citrate, 261, 261
Clairain, 438
Claisen rearrangement, 584, 585
Classical cation, **517**, 518
Cleavage, β-, 424
Closo clusters, **368**
Cn axis, **85**, 255, 256, 261, 279, 577, 607
C-O double bond, magnetic aniostropy of, 615
CO_2-acceptor process, **444**
Coal, 421, **436**, 444
anthracite, 436
attrital, 37, **438**
banded, **437**
bed methane, 442
bituminous, 436, 439
boghead, 438
cannels, 438
clairain, 438
durain, 438
fixed carbon, 436
fusain, 437, **438**
gas, **442**
gasification, **442**
Lurgi, **443**
heating value of, **437**, 441
macerals, **438**
MAF, **439**
mineral matter in, **438**, 439
nonbanded, **438**
pyrolysis, 442
proximate analysis of, **436**, 440
rank of, **436**
solvent-refined, **440**
subbituminous, 436
vitrain, **437**, 438
Coalescence, **635**, 636
Cocoon, 337
Coefficient of viscosity, **407**, 492, 814
Coefficients of atomic orbitals, **50**, 52, 69
Coffee, 398
Cogeneration of gas and electricity, 444
Coherence length, 797
Coherence of light, **797**
Coherent light, 784, 796, **797**
Coinage metal, 349
Coke, **439**, 442
Coke oven gas, 442
Coking, delayed, **439**
Collagen, 338
Colligative properties, 322, 325, **408**, 499
Collision
Complex, 773, 809–811, 815
induced fragmentation of ions, 724
three-body, 774
Colloid, **412**
Colloidal
dispersion, 340
systems, 415
Column chromatography, **398**
capillary, 401
packed, 401
Combination bands, **689**
Combinations,
macroscopic, 340
microscopic, 340
Common ion effect, **498**
Compensator, 247, 742
Complex ion, 345, 354, 365, 366, 373
Complex,
cation, 155–160

Complex (*Continued*)
charge-transfer, 154, 155, 559, 816
endohedral, 136
enzyme-substrate, 487
excited state, 816
π, **352**, **559**, 561
σ, 73, **559–**562
Complexation, reversible, **635**
Complexed, anion, 159
Complexes,
aqua, 345
coordination, 22, 320, 321, 345, 350, 355, 356, 360, 361, 371, 771
dinitrogen, **364**
dioxygen, **364**
high-spin, **372**
homoleptic, 349
low-spin, **372**
olefin-metal, 346
sandwich-type organometallic, 347
transition metal, 285, 355–359, 510, 771
Composites, 138, **339**
Compression molding, **314**
Computer-averaged transients, CAT, **610**
Concerted
displacement, 542,
reactions, **478**
unimolecular elimination, **556**
Condensation
polymers, 297
reaction, 570
Condensed structure, 226
Conductivity
cell, 404
electrical, 344
thermal, 344
Cone angle, **362**
Configuration, 224, **226**, 227, 232, 252, 262, 273, 316, 317
Configuration of
atoms, electronic, **18**
disubstituted chair cyclohexanes, **279**, 280
molecules, electronic, **48**, 70, 74, 79, 347, 579, 580
valence electrons, 349
Configuration,
absolute, **238**, 249, 250
amphi, 234
cis-like, 237
D and *L*, 232
d^{10}, 349
d^9, 349
d^n, 349, **356**, 372
double inversion of, 546,
erythro, 232, **240**
exo and *endo*, 232, **234**
Fabry-Perot, 796, 799
inversion of, **250**, 251, 263, 265, 274, 285, 542
like *l* and unlike *u*, 232, **262**
meso, 232, **239**
noble gas, 22, 347
octahedral, **285**
re and *si*, 232, 259, 262
relative, **238**, 250, 251
retention of, **251**, 358, 545–549
spin, 622, 757–759, 781
syn and *anti*, 232, **234**
threo, 232, **240**
trans-like, 237
Configurational base unit, **316**, 318, 319
enantiomeric, 316, 318
Configurational isomer, **230**
Configurations, 230
(*R*) and (*S*), 232, **237**, 238, 239, 257, 262, 317
α and β, 232, **235**
cis and *trans*, **232–**234
E and *Z* 232 –234, **237**, 273
Conformation, 224, 227, 232, 239, 272, **273**, 283, 491
Conformation,
anti, **277**, 279, 492
anticlinal eclipsed, 276, **277**
anti-periplanar, 276, **277**
boat, **280**
chair, **279**
cis, **276**
crown, **282**
eclipsed, **275**
envelope, **282**, 285
folded, **282**
gauche, **276**, 279
Newman projection of, 89, 269, 279, 280
puckered, 282
puckered, 278
sickle, **277**
skew, **276**
staggered, 87, 89, **275**, 279, 293, 294
synclinal, **276**
syn-periplanar, **276**
trans, **276**
tub, **282**
W, **277**
zig-zag, 114, 227
Conformational equilibrium, rapid, 491
Conformational isomer, **273**
Conformer, **273**, 491
Conglomerate, **251**
Conical symmetry, **99**

Conjugate
acid, 470, 471, 477, 494
acid/base, table of, 471
base, 470, 471, 477, 494, 555
Conjugated
double bonds, 55, 59, 62, 75, 76, 125, 126, 132, 174, 178, 321, 557
polyenes, **125**
systems, 55, **59**–62, 67, 72, 557
Conjunctive names, 150, **175**, 176, 183
Connectivity, 224, **225**, 226, 283
Conrotatory, **576**, 578, 580, 581, 587–589
Conservation of orbital symmetry, 578, 580
Constitution, **225**
Constitutional
isomers, 225, 228, **229**, 260, 286, 287
repeating unit, bivalent, 308
repeating unit, CRU, **295**, 296, 316–319, 329
tetravalent, 309
Contact
ion pair, **499**, **528**
radical ion pair, 821, 822
time, **433**
Continuous-wave laser, **798**
Contour diagram of orbital, 15
Coordinate covalent bond, **40**, 181
Coordinates,
cartesian, 8
polar, 8
spherical, 8
Coordination
chemistry, 345
complexes, 22, 320, 321, 345, 350, 355, 356, 360, 361, 371, 771
compounds, 345, **350**, 355,
number, 345, **355**–357, 360, 519, 520
polymerization, **319**
sphere, 350
Coordinative unsaturation, **356**
Cope rearrangement, 59, 60, 584, 585
Copolyimerization, 306
radical, 335
Copolymer,
alternating, 305, 306
block, 304, 305, 338
random, 300, 305, 335
Copolymers, 301, **305**, 307
Coronands, **155**
Correlation diagram,
orbital, **576**–581
state, **578**, 581
Correlation time, τ_c, **599**
Cotton, 332, **338**
effect, 252, 253, **738**, 739, 823, 824
linters, 338
Coulomb integral, α_i, 64, **66**, 67, 70
Coulombic attraction term, 820
Counter ion, **529**
Coupled energy transfer, **806**–808
Coupling constant, *J*, **624**–630
magnitude, **624**
sign, **624**
hyperfine, 655
Coupling constants
in transition metal complexes, 360
^{1}H-^{13}C, 646
Coupling diagrams, **625**
Coupling reaction, catalytic, **384**, 385
Coupling rules, first-order, 628
Coupling,
AB quartet, **631**
deceptively simple, 628
first-order, **628**
heteronuclear, 620
homonuclear, 620, 647, 648
long range, **629**
rotational, 691
second-order, **628**, 632, 633
strong, **628**, 630
virtual, 628, **632**
W-, 629
weak, **628**, 630
Covalence, **30**
Covalent bond, **28**
Cracking,
catalytic, **424**, 431
steam, **424**–427
thermal, **424**
Cram's rule of asymmetric induction, **269**, **570**
Cresian, 333
Critical acceptor concentration for energy transfer, 809
Critical
angle, 745
point, 397
pressure, **397**
radius for dipole-dipole energy transfer, 809
state, 397
temperature, **397**
velocity, 431
Cross-linked polystyrene, sulfonated, 402, 403
Cross-linking, **307**, 329, 337
Cross-linking agent, 336
Crown ethers, **155**, 157, 351, 404

Crude oil, 422, **423**, 424, 426, 431, 436, 442
sour, **423**
sweet, 423
Cryoscopic constant, **406**
Cryptands, 155, **156**
Cryptophanes, **158**
Crystal field
splitting, **370**–373
stabilization energy, CFSE **371**
theory, CFT **369**
theory, modified, 373
Crystallinity, test for, 246
Crystallites, 315
Cumulenes, **127**
Cuprammonium rayon, 340
Curtin-Hammett principle, **491**
Curved-arrow notation, **59**, 538
C-W laser, **798**, 802
mode-locked, 800
Cyanamide, 192
trimerization of, 314
Cyanates, **190**, 191
Cyano group, 183
Cyanohydrins, **164**, 270
Cyclic compounds, 227
peroxides, 199, 200
Cyclization,
nucleophilic, 590
radical, 590
Cycloaddition reactions,
[m+n], 527, 531, 575, **581**–583, 586, **587**–589, 812
[2+2], 375
[4+2], 584
retrograde, **582**
Cycloalkanes, 117, **119**, 431
Cycloalkenes, **120**, 311
Cycloheptatrienyl cation, 70, 71
Cyclohexanone, 164
reduction of, 494
Cyclohexyl anion, 515
Cyclopentadienyl
anion, 70, 71, 287, 288
complexes, 321, 322, 354
Cyclopentanediyl, 1,3–511, 812
Cyclopentanone, reduction of, 494
Cyclopentene, 311, 312
polymerization of, 311, 312
Cyclopentyl cation, 477
Cyclophanes, **133**
Cyclopropane formation, 513, 514
Cyclopropane,
edge-protonated, 560
carbene synthesis of, 586
Cyclopropenyl cation, 70, 71
Cycloreversion, **582**
Cyclotron, 711
Cylindrical symmetry, 45, **98**

d orbtial, 11, 12, **16**, 202
splitting, **370**, 371
d orbitals, electrons in, 356
d,l pair, **251**, 265
D_{4h} point group of, 106
d^6 configuration, 347
Dacron, 333
Dalton, 704
Dalton's Law, 396
Damped harmonic oscillation, 604, 605
Data acquisition
parameters, **609**
sequence, 611
Dative bond, **40**
Daughter ion, 706, 710, **718**
de Broglie relationship, **7**, 10
Debye, 33
Decarboxylation, 180
Deceptively simple spectra, **632**
Deconvolution of coupling patterns, 626
Decoupling, 627
broad band ^{1}H, 641
broadband, 642
field/FID relationship, 640
gated, **642**
heteronuclear, 640
homonuclear, 640
Inverse gated, **643**
off-resonance **646**
spin, **638**
Degeneracy, accidental, 682, **690**, 691
Degenerate, 106
orbitals, **12**, 15, 17, 48, 71, 106, 370, 778
point groups, **94**, 101
rearrangements, 287, **575**
states, 784
number of, 761
vibrations, **682**, 689, 690
doubly, 682
triply, 682
Degrees of freedom (3n-6) and (3n-5), 659, **677**
Degrees of unsaturation, 123, **133**, 550, 556
Dehydrobenzene, 1,4-, 511
Dehydrogenation, 431
Dehydrohalogenation, 551, 554
Delay time, 609, **611**
Delayed
coking, **439**

fluorescence, bilmolecular, **793**
fluorescence, thermally activated **793**
Delocalization
energy, 62, **71**
of charge, 476
Delocalized bond, **56**
Delocalized bonding, **56**
Dendrimer, **329**, 330
generations, 330
Denier, **334**
Density, 154
Desalination of water, 409
Descriptor, **232**, 286
endo/exo, 199
Deshielding
by aromatic rings, 617
by shift reagents, 637
of double bond protons, 615, 616
of methinyl groups, 614, 615
of methyl groups, 614, 615
of methylene groups, 614, 615
Detection of trace metals or elements, 711
Detector,
electron capture, **406**
flame ionization, **405**
hydrogen flame, **405**
refractive index, 402
thermal conductivity, **405**
ultraviolet and visible, 402
Detergents, **414**, 415
Determination of polymer molecular weight **324**
Detonator, 191
Deuterium
in NMR, 623
lock, 602
zero point energy, 677
Dextrorotatory, 250
Dialkylimidazolium salts, *N*, *N'*-, 418
Diamagnetic shielding, **612**, 613, 614,
Diamond, 37, 112, **135**
Diastereocenter, **231**, 250
Diastereogenic, **231**
center, 230, **231**, 237–239, 262–264
Diastereoisomer, **230**
Diastereomer, **230–**232, 239, 259–262, 263
Diastereomers, giving rise to, 231
Diastereoselective reaction, **266**, 267, 270
Diastereotopic
atoms or groups, **259**–261, 268
faces, **260**, 261, 269
Diaxial interactions, 1,3-, 281
Diazo group, 187
Diazo ketones, α-, **188**
Diazoalkanes, **187**
Diazonium ion, **190**, **523**, 532
Diazonium salt, aromatic, 190
Diblock
copolymers, 307
polymers, 306
Dibromobutane, 2,3-, 239
Dichlorocarbene, 507, 512
Dicarboxylic acids, **175**
Dichloroethylene, *cis*-1,2-, 87, 256
symmetry of, 91
Dicovalent carbon, **511**
Dielectric constant, ε, **410**
Dielectric constants of common solvents, table of, 411
Diels-Alder reaction, 527, 531, 581, 582, 589, 812
of anthracene, 131, 132
endo geometry, 582, 583
regiochemistry of, 582, 583
Diene,
heteroannular, 776
homoannular, 776, 777
Dienophiles, 527
Diesel fuel, 423, **435**, 446
Difference bands, **690**
Differential rate equation, **483**
Diffusion, 408
Diffusion-controlled reaction and rate, **492**, **814**
Dihapto, 353
Dihedral angle, **272**, 275, 276, 627
Diimine, 189
reduction with, 586
Diisocyanates, 337
Diketone,
α-, 165, **172**
β-, 174
Dimole transition, **773**
Dinitrogen complexes, **364**
Dioxetane, **199,** 803, 804
Dioxin, 146
Dioxolane, 163
Dioxygen complexes, **364**
Dipolar
addition, 1,3-, **583**
ion, **529**
Dipole, 33
Dipole moment, **33**, 34, 107, 410, 687
of polyatomic molecules, **33**
requirement for infrared absorption, 662
vector, 698
induced, 699
change of, Δμ, 660, 687, 688
symmetry of, 107

Dipole strength, D_j, **761**, 762
Dipole,
1,2-, 530
1,3-, **583**
induced, 410, 697
transient, 410
Dipole-dipole
energy transfer, 808
interaction, **410**, 808
Diradicals, 507, **511**, 812
Director,
meta, **562**
ortho and *para*, **562**
Dismutation reaction of
acetylene, 376
olefins, **375**, 376
Dispersion, 407, **737**
Dispersion forces, London, **410**
Dispersion,
anomalous, **737**, 738
normal, **737**
optical rotatory, 249, 252, 253, **737**, 738
Disproportionation of radical, 566
Disrotatory, **576**, 578–581, 587–589
Dissociation, 473
Dissociation energy, D_e, 671, 674–**676**
of molecular hydrogen, 676
Dissociation,
bond, **36**, 767, 770, 830
heterolytic, 36, 464, 476, 538
homolytic, 565
Dissymmetric, **96**, 97, 256, 257
Distillation, **390**
plates, 393
extractive, **395**
fractional, 389, **392**, 394
steam, **395**, 424
vacuum, 436
Distortionless enhancement by
polarization transfer, DEPT, **644**
Disubstituted cyclohexanes,
cis, 279, 280
trans, 280
Disulfides, **201**
in proteins, 202
Dithiohemiacetals, **211**
Dithiohemiketals, 211
Dithioic acids, **207**
Dithioic esters, **208**
Divalent carbon, **511**, 512
DNA, 154
damage and repair, 817, 819
Domains,
amorphous, 315
crystalline, 315
glassy, 315
Donar-acceptor pairs, 155
Donor,
electron, 154, 349, 817–820
electron pair, 471, 472, 540
energy, 806-808, 812
photochemical, 771
proton, 470, 472
Dosage of light, 787
Double bond, **35**, 133, 232, 233
bridgehead, 552
carbon-carbon, 120, 123, 133, 237, 552
carbon-metal, 347
carbon-nitrogen, 524
cis-, 233
conjugated, 55, 59, 62, 75, 76, 125, 126, 132, 174, 178, 321, 557
cumulative, 177
diimide reduction of, 586
endocyclic, **126**, 776
essential, **61**, 65
exocyclic, **126**, 776, 777
faces, 256, 259
metal-metal, 347
migration, 575
radical cation, 527
strained *trans*-, 531
trans-, 233
Double focusing mass spectrometer, **711**
Double labeling experiment, **498**
Double resonance experiments, **638**, 642, 643
heteronuclear, 638
homonuclear, 638
Double-photon absorption,
sequential, 775
simultaneous, 775
Double-photon transition, **774**
Double-stranded polymers, 309
Doublet
electronic state, **759**
spin state, 512, 756
Duality of electronic behavior, **7**
Ductility, 19, 344
Durain, 438
Dwell time, 609, **610**, 611
Dye, 189
bleachable, 800, 801
fluorescence, 802
laser, **802**
Dynamite, 185
Dyne, 663
Dyneema, 332
Dynel, 333

E1, **553**, 554
$E1_{CB}$, 550, **555**
E2, **554**
Ebullioscopic constant, **395**
Eclipsed conformation, **275**, 276
Eclipsing strain, **272**, 280
Edge-bridging ligands, **367**
Effect,
  -I, 496
  +R, 496
Effective atomic number, EAN, **355**, 356, 374
Eigenfunction, **10**, 11, 66
Eigenvalue, 10, **11**, 66
Einstein, **735**
Einstein's relations for absorption/emission, 639, **784**
Elasticity, 334
  modulus of, 315, 334
Elastomeric fibers, 333, **337**
Elastomers, 306, **334**
Electric dipole
  transition, 765
  transition moment, μ_e, 761, 762, 808
Electric field
  chiral, 243
  of light, 240, 241, 729
  strength, 743, 744, 745, 746
  strength of light, 735
Electric quadrupole filter, 712, 713
Electrical insulators, 326
Electricity generation, 436
Electrochemiluminescence, **803**
Electrocyclic reaction, 575, **587**–589
Electrofuge, 540, **541**
Electromagnetic radiation, 240, **729**
Electromagnetic spectrum
  far infrared region, 662
  far ultraviolet region, 662, **734**, 770, 773, 828
  infrared region, 662, 774
  microwave region, 662
  near infrared region, 662
  near ultraviolet region, 662, **734**, 768, 771–774, 802
  spectroscopically important regions of the, **661**
  ultraviolet region, 662, **734**
  vacuum ultraviolet region, **734**, 770
  visible region, 662, **734**, 768, 771, 774, 802
Electron, 4
  acceptor, 154, 349, 817-820
  affinities of atoms, table of, 464
  affinity, EA, 30, **463**, 464
  bombardment, 705
  capture detector, **406**
  delocalization, 62, 70, 71, 75, 77, 184, 497, 514
  density, **64**, 78
  donating group, 472, 500, 562
  donor, 154, 349, 817–820
  ejection, K-shell 829
  exchange, 809, 811
  impact, **706**, 717
  pair bond, **28**, 36
  photoexpulsion, 771
  spin, 12
  spin multiplicity, 511, **512**, 513, 757, 758, 781
  spin quantum number, m_s, **12**, 512, 756
Electron transfer, 155, 304, 348, 527, 568, 771, 803 804, 805, 820
  barrier, λ, 820, 821
  photosensitizer, 818, 819
  rate constant, k_{et}, 820
  back or reverse, 821, 822
  photochemical, **817**–820
  thermal, 817, 818
Electron volt, 673
Electron,
  bonding, 80
  spectroscopy for chemical analysis, ESCA, **829**
Electron-deficient
  compounds, π-, **78**
  groups, 614, 616
  species, **510**, 512, 530
Electron-donating groups, 496, 501, 614
Electronegativity, **29**, 33, 34, 78, 472
  of sp^3, sp^2 and sp carbon atoms, 615–617
Electron-excessive compounds, π-**78**
Electronic absorption
  bands, 741
  spectroscopy, 729, 750, 830
  spectrum, **748**, 750, 756, 761, 766, 790, 792, 806-808, 823, 830
Electronic behavior, duality of, **7**
Electronic configuration, 17, **18**, 22, **48,** 74, 79, 379
  of atoms, **18**
  of metal complexes, 349, 356, 372
  of molecules, **48**, 70, 74, 79, 347, 579, 580
Electronic dipole transition moment, symmetry of, 763
Electronic effects, 496, 500
Electronic states, 674, 759, 760
  excited states, 577–581, 587, 729, 748, **755**, 767
  ground states, **753**, 755, 760
  doublet states, **759**
  singlet states, 511, 521, **757**, 759
  triplet states, 511, 513, 521, **758**–760
Electronic wavefunction, 762
Electron-pushing notation, **59**, 479, **538**,
Electron-rich groups, 614, 616

Electrons,
  lone, nonbonding or unshared pair, 40, 50, 77, 144, 202, 274, 519
  upaired, 144, 373, 511-513, 526, 527, 758
  valence, 19, 22, 28, 41, 347, 510, 512, 828
Electron-withdrawing group, 496, 501, 543, 555, 557, 562, 563, 616
Electrophile, 472, 520, 524, 527, 530, **541**
Electrophilic addition
  bimolecular, Ad_E2, **560**, 561
  to carbon-carbon multiple bonds, 557, **558**–561, 567
Electrophilic aromatic substitution, S_EAr 495, 496, **561**, 562
Electrophilic character, 513, 514
Electrophilic substitution, 73
  bimolecular with π-bond rearrangement, S_E2', **548**, 549
  bimolecular, S_E2, **548**, 549
  internal with π-bond rearrangement, S_Ei', **550**
  internal, S_Ei, **549**
  unimolecular, S_E1, **547**
Electrophilicity, **541**
Electrophoresis, **400**
  capillary, 401
Electrospray ionization, ESI, **707**, 712, 717
Elementary reaction steps, **478**, 479
Elements, B group, 349
Elimination reaction, 266, 540, **550**
Elimination
  unimolecular conjugate base, El_{cb} **555**
  unimolecular, El, **553**, 554,
  1,1-, **550**
  1,2-, **550–**554
  1,3-, 555
  α-, **550**
  anti, **552**, 554, 555
  β-, **550–**554
  β-hydrogen, **358**, 359, 369, 374, 384
  bimolecular, E2, **554**
  cis or *syn*, **553**, 556
  concerted unimolecular, **556**
  electrophilic, 541
  γ–, **555**
  Hofmann, **552**
  Saytzeff, **551–**554
  trans-, **552**, 554
Elimination/substitution ratio, 554
Elimination-addition mechanism, 563
Elliptically polarized light, **245**, **732**, 733, 743, 822, 823
Ellipticity, **822**
Emission decay, 784, 789
  rate, k_{obs}, 788
Emission of light, 729, **784**
Emission spectrum, 756, 759–761, 790, **791**, 792, 806–807
  total, 792
Emission,
  fluorescence, 760, 780, 790, 794
  laser, 796
  phosphorescenece, 760, 780, 790
  polarization of, 764
  spontaneous, **784**
  stimulated, 639, **784**, 796
Emulsifying agent, 311, **414**
Emulsion, 311, **412**–414
  polymerization, **311**
Enamine, 180
Enantiocenter, **231**, 250
Enantiogenic, 230, **231**, 237
  center, **231**, 237, 238, 261, 269
Enantiomer,
  (-)-, **249**
  (+)-, **249**
  D- and L-, **250**
  dextrorotatory, *d*-, **249**
  levorotatory, *l*-, **249**
Enantiomeric
  excess, **254**
  invertomer, 274
  purity, **254**
Enantiomers, 225, **230**, 231, 237, 248, **249**, 251, 252, 254–258, 262, 269, 283
Enantiomers,
  giving rise to, 231
  resolution of, 404
  separation of, 404
Enantiomorph, **230**
Enantioselective reaction, **266**, 267, 270
Enantiotopic
  atoms or groups, 256, **257**, 261, 268, 283, 637
  faces, 256, **258**, 261
Encounter-controlled reaction and rate, 492, 814
End groups, 295
Endergonic, 467
Endiol,
  1,1-, 165
  1,2-, 165,
Endo cyclization, 288, 289
Endocyclic double bond, **126**, 776
Endoergic, **467**
Endohedral complex, 136
Endoperoxides, **199**
Endothermic, 467, 483
  reaction, **457**
Enediols, **165**
Energies associated with infrared absorptions, **672**

Energy, 453, 466
barrier, 286, 478, 480, 578, 580
to rotation, 273
Energy levels,
allowed, **673**
rotational, 659, 670
vibrational, 659, 669, 670, 748, 754, 780
Energy of
activation, 361, 578
$\Delta\Delta G^{\ddagger}$, 491
E_a, 480, 488, 490, 492
Energy of
nuclear spin state, 594–**597**, 598
rotational states, **669**
Energy operator, 66, 67
Energy transfer dependence on distance, 783
R^{-2}, 806
R^{-6}, 808
R^{-3}, 808
Energy transfer,
donor-acceptor separation distance, 808, 809
electronic, 806
extremely long range, 806
long range, 808
radiationless, **806**, 808 810, 812
radiative, **806**
rate constant for, 808, 809
short range, 809
via dipole-dipole interaction, **808**, 809
via exchange interaction, **809**, 814
Energy trap, exciton, 815
Energy,
acoustic, 748
adiabatic ionization, 463, 827, 828
average bond dissociation, **461**
Boltzmann, 803
bond, 74, 462
dissociation, **37**, **461**
crystal field
splitting, 370, 371
stabilization, **371**
d orbitals, 370, 371
delocalization, 62, **71**–74
dissociation, 671, 674–**676**
electron, 10, 66, 67
excited state, 580
free, 470
heterolytic bond dissociation, **464**, 521,
hydrogen bond, 40, 41
internal, *U*, 455, **456**, 457, 466, 468
ionization, 30, **462**, 827
kinetic, 66, 454, **455**, 456, 665
localization, **72**
orbital, 11, 12, 17, 18, 43, 47, 51, 56, 57, 62, 63, 70, 71–74, 79, 109, 580,
pairing, 371, **372**
phonon, 748
photon, 3, 661, 673, 734, 755
potential, 62, 66, **455**, 456, 478, 479, 665, 754
resonance, 58, **62**, 562
rotational, 455
specific bond dissociation, 462
spectroscopic dissociation, 675, **676**
strain, 271, 272
transition state, 361, 478, 481, 484
translational, 455
vertical ionization, 463, **828**
vibrational, 455, 463, 674, 748, 754
zero-point, 37, 674–**676**, 754, **755**
Engine knock, 434
Engineering thermoplastic, 313
Enium ions, **520**, **521**, 533
Enol ethers, **166**
Enolate
anion, **526**, 547, 557, 558
of acetone, 526
Enolic hydrogen, 164
Enols, **164**, 165, 173, 174, 178, 286, 287, 526, 557
Entgegen, (*E*), 237
Enthalpy of
activation, $\Delta H^{\ddagger}$, **489**, 503
reaction, ΔH_r, **457**
formation, standard, ΔH°_f, **458**, 476
reaction, standard, ΔH°_r, **457**
Enthalpy, *H*, 455, **456**, 465, 467
bond, **461**
change in, ΔH, 467
standard, H°, 457
Entropy of activation, $\Delta S^{\ddagger}$, **489**, 503
Entropy, S, 455, **465–468**
change in, ΔS, 467
standard, S°, **465**, 467
Environmental chemistry, 146
Enzyme, 261, 404, 487
reactions, 283
Enzyme-substrate complex, 637
Eosin- or E-type delayed fluorescence, **793**
Epimerization, **263**, 547
reversible, 263
Epimers, **262**–265
Epoxides, 193, 310
Epoxy resin, 331
Equatorial
positions, **279**–282, 633
substituent, 279–282
Equilibrium, **469**, 470
conditions, 490, 491

Equilibrium constant,
  axial/equatorial, 281
  thermodynamic, *K*, **469**, 472–474, 496, 501
Equilibrium control of a reaction, **490**, 491
Equivalent atoms or groups, 87, 88, **100**, 136, 257, 285, 607
  chemically, **256**, 261, 629
  symmetry, **100**, **255**-261
Equivalent
  bonds, 68
  faces, **256**, 261
  hybrid atomic orbitals, **23**
  orientations, **90**, 92
  sides, 131
  structures, 58, 287
  symmetry operations, **91**
Equivalent nuclei,
  chemically, **607**, 630
  magnetically, 629, 630
  number of NMR, *n*, 620
  symmetry, 607
Erythro isomers, 560
Esters
  formation of, 570
  hydrolysis of, 532, **570**–572
Esterification, **570**
Esters, 160, **169**, 195, 570
  pyrolysis, of 553-556
Esters of
  dithioic acids, 208
  phosphoric acid, 214
  phosphorous acid, 217
  sulfenic acids, 206
  sulfinic acids, 205
  sulfonic acids, 204
  sulfuric acid, 204
  thioic *O*-acids, 207
  thioic *S*-acids, 206
ET(30) solvent polarity scale, 773
Ethane, 273
  geometry of, 87, 88
Ether-alcohols, **153**
Ethers, **152**, 153, 166, 404
Ethyl
  cation, 520
  radical, 564–566
Ethylene, 273, 381, 424-427, 677, 767, 768, 773
  geometry of, 42
  oxide, 310, 311
    polymerization of, 310, 311
  polymerization of, 321
  stretching modes of, 700
  twisted, 96, 767, 768
Ethylidene group, 117
Euler angles, 677
Europium ions, 637
Even electron cation, **716**
Exact mass, **715**, 717
Exact mass spectrum,
  molecular formula determination, **717**
  molecular weight determination, **717**
Exchange
  of heat or work, 453,
  of mass, 453,
  site, **633**–635
Excimer, **816**
  xenon, 816
Exciplex, **816**
  carbonyl-olefin, 817
Excitation
  spectrum, **792**
  transfer, **806**
Excited singlet state,
  acidity constant of, K^*_a and pK^*_a, 796
Excited state, 79, 577–581
  chemistry, 55
  complex, 816
  decay, 787, 789
  acidity, 795
Excited state,
  aliphatic carbonyl groups, 783
  aromatic hydrocarbons, 783
  aryl carbonyl groups, 783
  charge-transfer, **771**
  d,d, **770**, 771
  electronic, 577–581, 587, 729, 748, **755**, 767
  heterocyclic aromatic molecules, 783
  ionization of, 795
  lifetime, 813
  n,π^*, **768**, 782
  n,σ^*, **769**, 770
  π,π^*, 48, **767**, 768, 782
  quenching of, 786, 787, **812**, 813
  σ,σ^*, 770
  singlet, 760, 781, 790, 813
  thermal generation of, 803
  triplet, 781,796, 813
  absorption of, 775
  B and *L*, **778**
  spectroscopic detection of, 802
  acidity constant of, K^*_a and pK^*_a, 796
Excited vibrational state, 660, **674**, 697, **755**
Exciton, **815**
  interaction, 816
  transfer, 808, 815
Exclusion rule for molecules with a center of symmetry, **699**

Exergonic, **467**
Exinite, 438
Exo cyclization, 288, 289
Exocyclic double bonds, **126**, 776, 777
Exoergic, **467**
Exothermic, 467, 483
  reaction, **457**
Explosive limits, 396
Extensive properties, 454
External conversion, 780
Extinction coefficient, molar, ε, 670, 748–750, 761, 809
Extractive distillation, 395
Exxon donor solvent process, 441
Eyring equation, 489

Face-bridging ligands, 367
Faraday effect, 824, 825
  induction, 604
Fast atom bombardment, FAB, **707**
Fatty acid salts, 414
Favorsky rearrangement, 531, 556
Fenske equation, 394
Fermi resonance, 690, **691**
Fertilizers, 447
Fibers, 329, **332**, 333
  acrylic, 334
  animal, **337**
  carbon, **137**, 507
  elastomeric, **337**
  synthetic, **332**, 338
  vegetable, **338**
Fibroin, 338
Field desorption ionization, FDI, **707**
Fingerprint region, 660, 685, 686
Firefly glow, 803
First
  harmonic, **6**
  overtone, **6**, 660, 661, 689–691
Fischer
  assay, **447**
  projection, **227**, 228, 250
Fischer-Tropsch process, 382, 422, **445**
Fixed
  bed operation, **430**
  carbon, 436
Flagpole, 280, **281**
Flame ionization detector, FID, **405**
  retardant, 152
Flash
  lamp, 803
  photolysis, 802, 803
  point, **396**
    of gasoline, 396
    of lubricating oil, 396
  separator, 441
Flax, 332
Flow,
  laminar, **431**
  turbulent, **431**
Fluidized bed operation, 429, **430**, 441
Fluorenium ion, **521**
Fluorescence, 760, 780, 786, 787, **790**, 794, 795, 804
  spectrum, 790–794
  delayed, **792**
    bimolecular, pyrene- or P-type, **793**
    thermally activated, eosin- or E-type, **793**
  lifetime of, 789, 793
  polarized, 764
  quantum yield of, 786, 789
  quenching of, 783
Fluorocarbons, **146**
Fluoronium ion, **522**
Fluxional molecules, **287**, 288, 575
Fly ash, **439**
Foamed polyurethane, **328**
Foaming agent, nitrogen, 336
Food preservative, 567
Forbidden
  reaction, 580, **581**
  transition, 756, 759, **760**, 764, 765, 768, 771
  thermally, 578
Force constant, k, **663**, 665, 671
Formal charge, **32**, 144, 181, 538–541
Förster
  cycle, **795**
  relationship, **808**, 809
Fourier transform, FT, **605**, 606, **700**
  infrared spectroscopy, FTIR, **701**
  ion cyclotron resonance spectroscopy, FTICR, **712**, 715
  nuclear magnetic resonance, FTNMR, 60–607
Fractional distillation, 389, **392**, 394
Fractionation column, 392
Fragment ion, 706, 710, **718**
Fragmentation, 717, **718**
  classes of ions I-IV, 718
  pathways, **718**, 723
  patterns, 722
  process, 712
  delayed, 719, 720
Franck-Condon
  allowed or forbidden, 766, 767
  factors, 665
  overlap, 820
  principle, 463, **765**, 814, 828

Free electron method, FEM, 55, **62**, 63, 69
Free energy, 470
  diagram, **480**
  of activation, $\Delta G^{\ddagger}$, **488**, 503
  of change, standard, ΔG^{o}, 455, **467**, 469
  of electron transfer, ΔG_{et}, 820
  of formation, standard, ΔG^{o}_{f}, **468**, 490, 504
  of product separation, w^{p}, 504
  of reactant approach, w^{r}, 504
Free energy,
  change in, ΔG^{o}, 281, 467
  Gibbs, *G*, 455, **467**–469, 480, 482, 489
  Helmholtz, *A*, **468**
  linear relationships, **500**–502
Free induction decay
  of signal, FID, **604**–606, 609, 611, 641
  acquisition of, 642, 643
Free swelling index, FSI, **439**
Free valence index, Fr, **65**
Freon, 147
Frequencies of electromagnetic radiation, 661, 662
Frequencies, group, **683**
Frequency, **5**
  domain, 605, 606, 611
    Lorentzian, 611
  doubling, **746**, 747
  factor, *A*, 488
  fundamental, 701
  infrared, 661
  Larmor, **596**, 598, 600–605, 638, 653
  mixing, 746, **747**
  infrared absorption band, 660
  precessional, 596, 598, 600–605, 638, 653
  rotational, 662, 666
  upconversion, 747
Frontier orbitals **79**
Frontier-orbital approach, 79, 379, **588**
F strain, **272**
Fuel cells, 422, **448**
  proton exchange membrane, 448
Fuel
  oil, 423
  diesel, 423, **435**, 446
Fullerenes, 112, **135**, 137, 507
Fulminates, **191**
Fulminic acid, 191
  salts, 191
Functional group, 142, **144**
  nomenclature, **147**, 149, 150, 152, 161, 175, 176, 177
Fundamental
  frequency, 701,
  vibrations, 674, **676**–678, 689, 699
  number of, **677**
  wave, **6**
Fusain, 437, **438**
Fusinite, 438

Gas chromatography-mass spectrometry, GC/MS, 405, **714**
Gas-liquid partition chromatography, GPLC, GLC, GC, 390, **401**, 405, 714
Gasohol, 434, **445**
Gasoline, 422, 423, 428, 431, 434, 435, 445, 446
  straight-rum, 423
Gated decoupling, 642
Gauche conformation, **276**, 279
GC, 390, **401**, 405, 714
Gegenion, **529**
Gel, 340, 400
  permeation chromatography, 402
  agarose, 404
Gem, **274**
Geminal, **274**
  coupling, 625
Geminate radical pair, 510, **528**
General acid catalysis, **493**
General base catalysis, **493**
Geometrical isomers, **232**
Gerade, *g*, 47, 48, **95**, 681, 698, 700, 765
g factor, 654, 655
Gibbs free energy, *G*, 455, **467**–469, 480, 482, 489
Glan prism, 730, 742
Glass, 246
Glass transition temperature, T_g, **315**
GLC, 390, **401**, 405
Globar, 694
GLPC, 390, **401**, 405
Glycols, 149
Gossypium spp., 338
Graft polymer, **307**
Graphene, 137
Graphite, 112, **135**, 137, 448, 737
Graphitize, 334
Greases, **436**
Green chemistry, 418
Grignard
  reaction, 268
  reagent, 348
Ground state, 79, 577–581, 587
Ground state,
  electronic, **753**, 755, 760
  triplet, 514
Ground vibrational state, 660, **674**, 676, 697, **755**
Group
  frequencies, **683**
  molecular orbitals, 61

orbitals, **107**
transfer reaction, 575, **586**
Guest, 156, 635
Gutta-percha, **335**

Haber process, 447
Half-band width, $\Delta\nu_{½}$, **751**
Half-life, $t_{1/2}$, **487**
Halophosphorus compounds, **218**
Hamiltonian operator, **66**, 67
spin-orbit, 781
Hammett acidity function, H_o, **474**
Hammett
equation, **500**, 501
substituent constant, σ, 500, 502
substituent constant, table of, 501
Hammond's postulate, **482**
Hantzsch-Widman hetercyclic nomenclature, **193**
Hapto, **353**
Haptotropic conversion, 353
Hard/soft
acid-base theory, 540
acids and bases, HSAB, **472**
acids/bases, table of, 473
Harmonic
frequencies, 701
generation, 746, **747**
second, **746**
oscillator, **665**, 671–675, 689, 690, 693
selection rules for, **674**, 687
vibrational levels of, **672**
first, **6**
second, **6**, 747
H-coal process, **441**
He resonance line, 828,
Heat, 452,
absorbed under irreversible conditions, q_{irrev}, 465,
absorbed under reversible conditions, q_{rev}, 465,
Heat capacity
at constant pressure, C_p, 459,
at constant volume, C_v, 459,
C, 454, **458**, 465
specific, *c*, **458**
Heat content, *H*, **456**
Heat of
activation, $\Delta H^{\ddagger}$, **489**
combustion, **460**
formation, standard, $\Delta H^o{}_f$, **458**
fusion, ΔH_m, **460**
fusion, molar, 460
fusion, specific, 460
hydrogenation, ΔH_{hydro}, 62, **459**, 489, 490
melting, ΔH_{m}, **460**
reaction at constant pressure, q_p, 457
reaction at constant volume, q_v, 457
reaction, hydrogenation, 459
reaction, standard, $\Delta H^o{}_r$, 432, **457**
sublimation, ΔH_s, **461**
sublimation, molar, 461
sublimation, specific, 461
vaporization, ΔH_{vap}, **460**
vaporization, molar, 460
vaporization, specific, 460
Heat transfer, 430, 456
at constant pressure, 456
at constant volume, 456
irreversible, 456
reversible 456
Heat,
q, 453–456, 466, 658, 659
absorption of, 456
liberation of, 456
Heating value
of coal, **437**, 441
of synthesis gas, 443
Heavy atom effect, **783**
external, 783
internal, 783
Heck reaction, **384**
Height equivalent to a theoretical plate, HETP, **393**
Heisenberg uncertainty principle, **3**, 502, 538, 774, 775
Helicity (M, P), **286**
Helmholtz free energy, A, **468**
Heme, 350
proteins, 351
Hemiacetals, **162**, 263
Hemiaminals, **163**
Hemiketal, **163**
Hemp, 332
Hertz, Hz, 5
Hess's law of constant heat summation, **458**, 464
HETCOR, **648**
Heteroaromatic compounds, **77**, 78
Heteroatom, 77, **144**, 192, 193
Heterocyclic boranes, **196**
Heterocyclic
compounds, tables of, 193, 194
nitrogen compounds, table of, 194
phosphorus compounds, **218**
polymers, 310
sulfur compounds, 210
systems, **192**, 193, 196, 210, 218
Heterogeneous catalysis, 380, 412, **429**, 430, 433, 434, 493
Heterogenic bond formation, 539

Heterolytic bond cleavage, 36, 464, 476, **538**, 539
Heteronuclear corrleation spectroscopy, HETCOR, C,H-COSY, **648**, 649
Hexane, infrared spectrum of, 684, 685
High explosive, 185, 186, 191
High field, **598**
High performance liquid chromatography, HPLC, 390, **402**, 404, 715
Highest occupied molecular orbital, HOMO, **79**, 588, 817, 828
High-spin complexes, **372**
Hofmann
  degradation, 515
  elimination, **552**
  regiochemistry (less stable olefin), 552–555
Homeomorphic isomers, **283**
Homogeneous catalysis, 380, **429**, 430, 433, 434, 493
Homogenic bond formation, 538
Homoleptic complexes, 349
Homologs, **116**
Homolytic bond cleavage, 36, 356, 461, **538**
Homonuclear, **608**
Homonuclear correlation spectroscopy, HOMCOR, H,H-COSY, **646**–650
Homopolymer, **297**, 301
  single strand, 297
Homopolymerization, 306
Homotopic
  atoms or groups, **255**, 256, 260, 261
  faces, 258
Hooke's law, **662–**665
Host, 156, 635
Host-guest complex, 156, 635-638
Hot
  band, **756**
  transition, **756**, 760, 794
HPLC, 390, **402**, 404
Hückel
  array, 589
  inscribed polygon method, 71
  molecular orbital energy levels, (4*n* + 2), **70**
  molecular orbital theory, HMO, 56, 66, **67**–71, 81
    coefficient determination, 50, 52, **69**
Hückel-Möbius (H-M) approach, **589**
Hückel's rule (4*n*+2), **70**–72, 130
Hund's rule, **17**, 48, 372, 758
Hybridization
  index, **23**
  of atomic orbitals, **22**
  of carbons in cumulative double bonds, 177
Hydrated carbonyl, 173
Hydrazides ,189, 190
Hydride
  abstraction, 424
  affinity, **476**
  migration, 1,2-, 301, 302, 320, 424, 426
Hydrocarbons, 111, 422, 431
  acyclic, **113**
  alicyclic, **119**
  aliphatic, 113, **114**, 119
  alternant, 64, 71, **73**, 76
  aromatic, **75**, 125, 783
  benzenoid, **127**
  branched-chian, 115
  bridged, **121**, 122
  bridged bicyclic, **122**–125
  catacondensed aromatic, **130**, 778, 779
  conjugated, 73, 125
  cross-conjugated, **126**, 776
  cyclic, 77, 119, 120
  even alternant, 71, **73**, 76
  fluorinated, 336, 448
  nonalternant, 73, **76**, 130
  odd alternant, **73**, 74
  peri-condensed polynuclear aromatic, **131**
  polycyclic, 122, 125
    aromatic, 132
  polynuclear aromatic, **129**, 130
  saturated, **112**, 113, 119
  straight-chain, 114
  unsaturated, **112**, 120
Hydroclone, **441**
Hydrocracking, **431**
Hydroformylation reaction, **382**, 430
Hydrogen, 397, 431, 674–676
  abstraction, α-, 59, 60, **360**
    intermolecular and intramolecular, 301
  activation, 362
  agostic **360**
  atom, 13, 52, 503
    migration, 1,2-, 301, 302, 320, 425, 426
  bond, **40**, 154, 174, 179, 198, 329, 337
    ammonia-water, 40
  bonding, 637
  chloride, stretching vibrations of, 689
  cyanide, Infrared spectrum of, 689
  deficiency index, 123, **133**, 550, 556
  elimination, β-, **358**, 359, 369, 374, 384
  flame detector, **405**
  molecule, 35, 52, 456, 755
  primary, **116**
  secondary, **116**
  tertiary, **116**
Hydrogenation, 440
  catalyst, 357
  of olefins, 358

Hydrogen-based economy, 422
Hydroliquefaction of coal, 441
Hydrolysis, **545**
  of acyl halides, 532
  of alkyl bromides, 495
  of amides, 532
  of esters, 532, **570**–572
  of urea, 493
Hydroperoxides, **197**
Hydrophilic, 143, **413**
Hydrophobic, 143, **413**
Hydroxy carbonyl compounds, **178**
Hydroxyethyl, 1-, cation, 517
Hydroxyl group, 147, 149
Hydroxylamines, 167, **181**
Hydrozirconation, 359
Hyperchromic shift, **752**
Hyperconjugation, **61**, 425
Hypersurface, 480, 481
Hypochromic shift, **752**
Hypsochromic shift, **752**

i point, 87
I strain, **271**
Icosahedral symmetry, 136,
ICP-Mass spectrometry, ICP/MS, **711**
Ideal
  gas, 455
  solution, **391**
Identity operation, *E*, **91**, 94, 101
Ignition, 396
  temperature, 396
Imides, **172**
Imines, 167, **180**, 182
Iminium ion, **524**
Incredible natural abundance
  double quantum transfer experiment,
  two dimensional, 2D INADEQUATE, **651**, 652
Index of unsaturation, 123, **133**, 550, 556
Indistinguishable, 84
Induced dipole, 410, 697
Inductive effects, 496, 502
Inductively-coupled plasma mass spectrometer,
  ICP/MS, 711
Inelastic deformation, 334
Information theory, 611
Infrared
  absorption band intensity, 660, 670, 687
  absorption bands, 660, 669, 698–670
  absorption energies, **672**
  active vibrations, **687**, 699, 700
  active vibrations,
    dipole moment considerations, **687**
  allowed transition, **687**, 688, 698
  deuterium isotope effect, 672, 677
  dipole moment considerations, **687**, 688
  forbidden transition, **687–**689, 698
  inactive vibrations, 699, 700
  inhomogeneous line broading, 670
  middle region, 659
  near region, 659
  normal modes, **678**
  Q branch, **693**
  R and *P* branches, **693**
  radiation, 659, 661
  spectrometer, 694, 695
  spectroscopy, 659
    Fourier transform, FTIR, **701**
  spectrum
    absorption, 659, **669**, 671, 677, 694
    frequency domain, 700
    of water, 678, 688, 699
    time domain, 700, 701
  water stretching modes, 95, 679, 688
Inhibitor, **566**
Initiating groups, 295
Initiation steps of radical chain reactions, 299,
  425, 565
Initiator,
  catalytic, 304, 305
  cation, 303
  radical, 299, 307, 309, **565**
Injection molding, **315**
In-plane bending, 679, **682**
Insertion
  of carbon monoxide, 358
  of olefins, 358
  migratory, **357**
  reaction, intramolecular, **357**
Integrated rate expression, **484**
Intensities of absorption bands, relative, 762
Intensities
  of infrared bands, 698, 699
  of lines within an NMR signal, **624**, 654
Intensity of an absorption band, **751**, 761
Intensive properties, **454**
Interatomic distance (bond length), **36**,
  754
Interfacial tension, **412**, 414
Interferogram, 701
Interferometer, Michaelson scanning, 701
Intermediate, synthetic, 269
Intermolecular, **574**
  pericyclic reactions, 575
Internal
  conversion, IC, 760, **780**, 794, 807, 830
  energy, *U*, 455, **456**, 457, 466, 468
  refraction, total, 745

Intersystem crossing, ISC, 760, 761, **780**–783, 785, 793, 811
  quantum yield of, 785, 786, 789
Intimate
  ion pair, **528**
  radical ion pair, 821, 822
Intramolecular, **574**
  cyclizations, 288, 289
  pericyclic reactions, 575
Intrinsic barrier, $\Delta G^{\ddagger}_{int}$, 504
Inverse
  gated decoupling, **643**
  operation, **93**
Inversion
  of configuration, **250**, 251, 263, 265, 274, 285, 542
    migration with, 584, 585
  of population, 785, 796, 800
  symmetry, *i*, **87**, 91, 107, 257, 699, 700, 765
Inverted Marcus region, **820**
Invertomer, **274**
Ion, **508**
  chromatography, **403**
  current, 405, 406, 705, 710
Ion,
  acylium, **523**
  ambident, **542**
  aminium, **524**
  carbenium, **520**, 544, 559
  charge-delocalized, 286, **515**
  charge-localized, **515**
  closed three-centered four-electron, 518
  daughter, **718**
  diazonium, **523**
  dipolar, **529**
  enium, **520, 521**
  even alternate, 70
  fragment, 706, 710, **718**
  hot, 706, 828
  lyate, 477
  lyonium, 477
  molecular, 527, 710, 717
  nonintegral mass, 719
  odd alternate, 70, 76
  onium, **522**
  parent, **718**, 719
  precursor, **718**, 719
  primary, 706
  product, **718**, 719
  quaternary ammonium, 522
  secondary, 706
  silconium, **524**
Ion cyclotron resonance spectroscopy, ICR, **711**
  Fourier transform, FTICR, **712**, 715
Ion-dipole interaction, 369, **410**
Ion exchange
  chromatography, 402–**404**
  resin, 427
Ion-molecule reactions, 712
Ion pair, **499**, **528**, 545
  contact, **499**, **528**
  intimate, **528**
  loose, **499**, **528**
  solvent-separated, **499**, **528**
  tight, **499**, **528**
Ion radical, 527 820–822
Ion trajectory radius in magnetic field, 709
Ion trap mass spectrometer, **713**, 714
Ionic
  addition, 266
  bond, 34, **35**
  character, 35
  liquids, **418**
  radius, **38**
  resonance energy, Δ, 30
  strength, 499
  valence, **30**
Ionization, **706**
  chamber, **705**
  constant, 473
    of water, K_w, 473
  energy, 30, **462**, **827**
    adiabatic 463, 464, 827, 828,
    vertical 463, **828**
  potential, 66, 454, **462**, 706, **827**, 828
  chemical, **706**
  field desorption, **707**
  Matrix-Assisted Laser Desorption, **707**
Ionizing
  radiation, 308
  voltage, 709
Ionomer, 306
Ionophore, 156
Ions,
  collision induced fragmentation of, 724
  isobaric, **715**
  oxonium, 516
  vibrationally excited, 828
Irreducible representations, **102**
Irreversible
  process, **453**, 570, 753
  reaction, 453
Isoabsorptive point, **753**
Isoalkanes, **117**
Isoalkyl group, **118**
Isobaric ions, **715**
Isocyanate polymerization, 309
Isocyanates, **191**, 326–328, 337, 515

Isokinetic temperature, **503**
Isolobal
 analogies, **378**
 fragment, 378
Isomer,
 configurational, 230
 positional, **229**
 skeletal, **229**
Isomerization, 431, **574**
Isomers, 100, 228, 574
 allylic, **286**
 cis/*trans*, 233, 237
 conformational, **273**, 283
 constitutional, 225, 228, **229**, 260, 286, 287
 endo/*exo*, **234**
 erythro, 560
 functional group, **229**
 geometrical, **232**
 homeomorphic, **283**
 in(side) and out(side), **283**
 inversion, **274**
 keto-enol, **164**, 165
 optical, **248**
 positional, **229**, 260, 266, 267
 skeletal, **229**
 structural, **229**
 syn/*anti*, 234, 235
 threo, 232, 560
 valence, **287**
Isonitrile, 183
Isosbestic point, **753**
Isotactic, 311
 polymer, **316**, 320
Isothermal process, **454**, 460, 461
Isothiocyanates, **209**
Isotope, 594, 715
 cluster, **720**–722
 distribution, 720
 spin quantum number, 594
Isotope effect,
 inverse, 496, 497
 kinetic, **496**
 normal, 496
 primary, 497
 secondary, 497
Isotopes,
 magnetogyric ratio of, 596
 natural abundance of, 715, 720, 722
 table of natural abundance, 715
Isotopic substitution of deuterium for hydrogen, $CHCl_3/CDCl_3$, 672
Isotopomer, **283**
Isotropic, **416**, 417
 media, 743

IUPAC, 21, 116, 539
 system of nomenclature, **116**, 127, 147, 150, 151, 161, 162, 167, 175–178, 183, 200, 202, 318, 319, 329, 350, 517
  for homopolymer, **297**
  metal complexes, **365**
chemical formula of
  constitutional repeating units, 296
 symbols for reaction mechanisms, table of, **539**

Jablonski diagram, **759**, 780
Jute, 332

K region, **132**
Kappa convention, κ, 351,
Kapton, 326, 327
Karplus
 curve, 627
 equation, **627**
Kekulé structures, **75**, 226
Kelvin, 455
Kernel, **22**
Kerogen, **447**
Kerosine, 423
Kerr
 cell, **744**, 800
 effect, **743**, 744
Ketals, **163**
Ketenes, **177**, 191
Keto acids, **175**
Keto form, 164, 174, 178, 286, 287, 526
Ketocarbene, α–, **514**
Keto-enol tautomerism, 526, 633
Ketol, α-, 165, 178
Ketone derivatives, 163, 164, **166**, 212
Ketone, α, β-unsaturated, 541
Ketones, 160, **162**, 167, 175, 179, 252, 258, 269, 739, 740
Ketonitrene, α, **515**
Ketyl radical anion, 527
Kevlar, 326, 327, 333, 334
Kinetic
 control of a reaction, **490**, 491
 energy, 66, 454, **455**, 456, 665
  of ions, 709
 isotope effect, **496**
 product, 490, 491, 582
 stability, 355, **490**, 494
Kinetics, 452, 479, **483**
 1st order, 488
 enzyme, 486, 487
 Michaelis-Menton, **486**
 of reactions involving solvent, 487
 pseudo first-order, 487

Knockout drops, 174
Kodel, 333
Koppers-Totzek gasifier, **443**
Kosower Z-scale, 772, 773

L absorption bands, 779, 780
L region, **132**
La Porte selection rule, **765**
Laboratory frame of reference, **600**
Lactams, 171, **195**, 310
Lactones, 169, **195**, 203, 310
Ladder polymers, **308**
Lanthanide ions, 637
Laplacian operator **9**
Larmor frequency, ω, **596**, 598, 600–603, 605, 638, 653
Laser
  beam divergence, 745
  desorption ionization, matrix-assisted, **707**
  flash photolysis, LFP, **802**, 803
Laser light, high-intensity, 743
Laser mode, **799**
Laser,
  light amplification by
    stimulated emission of radiation, 694, 695, 745, 774, 784, **796**
  argon, 798
  carbon dioxide, 798
  cavity-dumped, 802
  continuous wave, CW, **798**
  dye, **802**
  helium-neon, 798
  neodymium-YAG, **695**, 798
  normal mode, **799**
  pulsed, **798–**802
  Q-switched, **799**
  ruby, 798
  Ti-sapphire, 798
  wavelength tuning of, 802
Latex, 311, 702
Lattice, 599,
Law of thermodynamics,
  first, **466**
  second, **466**
  third, **466**
Leaded gasoline, 434, **435**
Leather, **338**
Leaving group, 187, 204, **540**, 541
Length, bond, **36**
Lenz's law, **612**, 614
Leveling effect of solvent, 477
Levorotatory, 250
Lewis
  acid/base complex, 471, 472
  acids, 159, 270, **471**, 472, 510, 541
  adduct, 471, 472
  base, **471**, 472, 476, 540
  complex, 29, 471, 472
    of boron, 472
  electron (dot) structures, **28**
Lexan, 326, 327
Lifetime of, 478, **488**, 787, 789
  fluorescence, 789, 793
  phosphorescence, 789
  triplet state, 808
Lifetime,
  excited state, 813
  natural radiative, τ, **789**
  observed radiative, τ_{obs}, **787**
Ligand
  exchange of chromium ammonium complex, 771
  slippage, 353
Ligand field theory, LFT, **373**
  field geometries, *d* orbital energies, 371
  field splitting strength of ligands, 370
Ligand substitution, **360**
  associative, 353, 360
  dissociative, 360
  intimate mechanism, 361
  stoichiometric mechanism, 361
Ligands, **155**, 344, **349**
  ambidentate, **364**
  ansa, 363
  bidentate, 350
  bridging, **351**, 364
  carbene, 374
  chelating, **350**
  dianionic donor, 374
  edge-bridging, **367**
  edge-sharing, 351
  face-bridging, **367**
  face-sharing, 351
  hexadentate, 351
  monodentate, 351
  pentadentate, 351
  polydentate, 351
  σ-bonded, 357
  tetradentate, 351
  tridentate, 350
  unsaturated, 357
Light, **240**
  intensity, **735**, 787
  modulation, 830
  scattering, 323–**325**, 695, 696, 697
  stick, 199
Light, circularly polarized, **243**, 246–248, **731**, 733, 737, 822–824

Light,
  coherent, 784, 796, **797**
  electric field of, 240, 241, 729
  elliptically-polarized, **245**, **732**, 733, 743, 822, 823
  frequency of, 240
  generation of plane-polarized, **242**
  high intensity, 774
  magnetic field of, 240, 241, 729
  plane-polarized, **240**–242, 246–249, **730**, 732, 733, 737, 741, 743, 824, 825
  relationship between plane- and circularly-polarized, **244**
  spatial coherence of, **797**
  temporal coherence of, **798**
  wavelength, 240, 737
Lignite, 436
Line formulae, 366
Linear free energy relationships, **500**–502
Linear molecules, rotation of, 669
Liquefied
  natural gas, LNG, **442**
  petroleum gas, LPG, **442**
Liquid chromatography-mass spectrometry, LC/MS, **715**
Liquid crystals, 154, 240, **415**, 416
  lyotropic, **416**
  thermotropic, **416**
Living polymers, **304**, 306
Local symmetry, **98**
Localized
  π bond, **46**
  two-center, two-electron bond, **28**, 56
Locking the external magnetic field, **602**
London dispersion forces, **410**
Lone pair electrons, **28**
  localized, 472
Longitudinal relaxation, **599**
Loose ion pair, **499**, **528**
Lossen reaction, 515
Lowest unoccupied molecular orbital, LUMO, **79**, 588, 817
Low-spin complexes, **372**
Lubricant, 135
Lubricating oils, **436**
Lucite, 300
Lurgi
  coal gasification, **443**
  slagging gasifier, **443**
Lyate ion, 477,
Lyonium ion, 477,
Lyotropic liquid crystal, **416**
Macromolecule, **294**, 295
Macroscopic properties/quantities, 480, 481
MAF coal, **439**
Magic acid, 473
Magnetic
  analysis of *m/z*, **709**
  circular dichroism, MCD, 824, **825**, 826
  equivalence, **629**
  optical rotatory dispersion, MORD, 824, **825**
  resonance imaging, MRI, **653**
  transition moment, μ_m, 762
Magnetic field anisotropy of different types of chemical bonds, **614**
  double bonds, **615**
  single bonds, **614**
  triple bonds, 617
Magnetic field
  of light, 240, 241, 729
  strength, 709, 825
  effective, B_E, 612, 613, 614
  applied or external, B_o, 593, 595–597, 600, 609, 612–614, 638, 639
  homogeneity of B_o, 604
  induced, B_I, 612–614, 639
  adjacent nuclear spin component, B_2, 639
  oscillating, B_1, 593, 598–610
  strong, 824, 825
Magnetic moment of, 757, 825
  electron orbital motion, 781
  electron spin, 781
Magnetically equivalent nuclei, 629, 630
Magnetogyric
  ratio, γ, **596**, 597, 624
  ratios of isotopes, table of, 596
Magneto-optical activity, 824
Main group elements, 19
Malleability, 19, 344
Manifold of states, **759**
Mannich
  base, 179
  reaction, 179, 524
Marcus theory, **503**, 820–822
Markovnikov addition (Markovnikov's Rule), **558**, 559, 568
Mass, 453, 663
  action expression, *Q*, **469**
  analyzer, **706**
  average, 715
  center of, 667
  exact, 715, **717**
  nominal, **708**, 715, 721
  number, **708**
  reduced, 666, **667**
Mass spectrometer, 324, 405, **705**
  ion trap, **713**
  time-of-flight, **712**, 715

Mass spectrum, **710**
  base peak, 710, **719**
  high resolution, 711
  isotope peaks, *M*+1, *M*+2, **721**, 722
  low resolution, 711
  molecular structure, **722**
Mass spectrum/mass spectrum, *MS/MS*, **724**
Mass-to-charge ratio, *m/z*, 705, 706, **708**
Matrix-assisted laser desorption ionization, MALDI, **707**
McLafferty rearrangement, 719
Mechanism of a reaction, 59, **479**, 539
Meisenheimer complex, 563
Melmac, 314
Melt transition temperature, T_m, **315**
Mercaptans, **200**, 201
Merry-go-round reactor, **787**
Meso
  carbon atoms, **131**, 136
  compound, 317
  configuration, **239**, 240
  isomers, 239, 261, 264
Mesophase, 240, **416**
  cholesteric, **417**
  nematic, **417**
  smectic, **416**, 417
Meta director, **562**
Metal carbene catalysts, 375
Metal-carbene complexes, 311, 321, 322, 347, 360, 369, **373**, 375, 512
Metal
  cluster compounds, **366**, 367
  complexes, nomenclature of, **365**
  hydrides, 359, 363
  hydrides, chemical shifts of, 617
Metallacycles, 320, 321, **368**, 375
Metallocene catalysis, 312, **321**
Metallocenes, 321, 347, **363**
Metallocyclobutane, 320
Metalloid, 19
Metal-olefin bond, 352
Metals, 19, 344
Metals, main group, 345
Metal-to-ligand back-bonding, **352**
Metastable
  ion, **719**
  peak, **719**
Metathesis, ring-closure, 375
Methanation, 444, **445**
Methane, 113, 114, 682, 706
  geometry of, 88
  protonated, CH_5^+, 516, 517, 706
  radical cation, 527
  structure of, 27
  symmetry axes in, 94
Methanol, 381, 445
  molecular ion of, 716
  symmetry of, 98
  synthesis of, **446**
Methanol-to-gasoline process, MTG, 428
Methine group, 101, **119**
Methonium ion, 516, 517, 706
Methoxyl
  anion, 148, 525, 526
  radical, 510
Methyl
  anion, 508
  benzoate, infrared spectrum of, 687
  carbene, triplet, 513
  carbenium ion, 520
  cation, 508, 510, 517
  diazonium ion, 190
  esters, synthesis of, 187
  group, stretching frequencies of, 683
  radical, 510
  shift, 1,2-, 518
Methylamine, molecular ion of, 716
Methylene, 512
  carbene, 512
  group, 101, **118**, 145, 512
  group,
    in-plane bending modes, 682
    out-of-plane bending modes, 682, 683
    stretching frequencies of, 683
  singlet, 510, **512**, 513
  triplet, 511, 513
Methylformamide,
  s-cis-N-, 277, 278
  s-trans-N-, 277, 278
Methylidene (methylene) group, 117
Methyloxenium ion, 521
Methylpyridinium ion, *N-*, 522
Methylsilicenium ion, 521
Methylsulfenium ion, 521
Micelle, 154, **413**
Michael addition, **557**
Michaelis-Menton
  constant, 487
  kinetics, **486**
Micrinite, 438,
Microscopic
  properties/quantities, 480, 481
  reversibility, **503**, 578, 582
Microwave
  spectra of diatomic molecules, **666**
  spectroscopy, 659, **662**, 669

Migratory
aptitude, **500**
insertion, **357**
Mineral matter in coal, **438**, 439
Mixed functional molecules, 143
mks system, 663
Mobile phase, 398–401
polar, 402
Möbius array, 589
Mode,
axial, 799
laser normal, 799
oscillation, 799
transverse, 798, 799
Mode-locked laser, **800**
Modes, infrared normal, **678**
Modulus
of elasticity, 315, 334
stiffness, 332
Molal
boiling point elevation constant, K_b, 323, **395**, 408
freezing point depression constant, K_f, **406**, 408
Molar
absorptivity, ε, 670, 748–**750**, 751, 761, 809
ellipticity, [Θ], 824
extinction coefficient, ε, 670, 748–**750**, 761, 809
Molding, **314**
Molecular
connectivity, 648, 652
exclusion chromatography, **402**
faces, 256
nitrogen, 734
recognition, 155, 635
refractivity, 408
sieves, **429**
structure, 224, **226**
wires, 242
Molecular ion, 527, 710, **717**, 720
formula determination from $(M+1)/(M+2)$ data, **722**
isotope peaks, 721
Molecular orbital, **42**, 43, 55, 56, 67, 107, 426
energy diagram (MOED), **47**, 71, 72, 161
2nd row heteroatomic molecules, **50**
2nd row homodiatomic molecules, **48**
for N_2, 49
for O_2, 49
theory, 26, 27, **43**
Hückel, HMO, 56, 63, 66, **67–70**
extended Hückel, **81**
antibonding, **43**
antibonding π*, **46**, 56, 57
antibonding σ*, 43, **45**
bonding, **43**
bonding, π, **45**, 56, 57
bonding, σ, **44**
delocalized, π, **56**
highest occupied, HOMO, **79**, 588, 817, 828
lowest unoccupied, LUMO, **79**, 588, 817
nonbonding, 43, 50, **74**
energy, π, **70**
Molecular orbitals of
carboxylate anion, 168
nitro group, 184
carbonyl group, 161
Molecular orbitals, group, 61, **107**
Molecular oxygen, 364, 566, 567, 614, 734, 774
biradical character of, 49
Molecular structure, representations of
configuration, **227**
connectivity, **226**
Molecular weight
of polymers, **322**
number average, **322**, 325
weight-average, **323**, 325
Molecularity, **478**
Molecules, organized, 245, 246
Moments of inertia, *I*, 666, **667**, 668, 691, 692
oblate axis, 692
principal, **667**
prolate axis, 692
Momentum, **p**, **6**, 11
angular, 11
of a photon, 734
Monohapto, 353
Monomeric units, 194, **295**
Morse curve, **35**, 463, **674**, **754**
for molecular hydrogen, 675
Mother of pearl, 340
Moths, 337
Motions, translational, vibrational, and rotational, 659
Multicenter σ bonding, **79**
Multiple-step reaction, 484, 485
Multiplet slanting, 628, 631
Multiplicity, **756**
allowed transition, 790
forbidden transition, 790
of nuclear spin states, **595**
of unpaired electron spins, 512
Multistep reactions, **479**
Mutarotation, **263**
Mylar, 333

Nafion, 448
Names of bridging groups, **366**

Nanotubes, carbon, 112, **137**
Naphtha, 427, 444
heavy, 423
light 423
Naphthalene, 75, 76, 778, 779
dicarboxylic acid/ethylene glycol, PEN, 326
radical anion, 304, 528
Naphthyl anion, 2-, 509
Natural gas, 421, **442**
liquefied, **442**
substitute, **444**
Naural rubber, 315
Neighboring group participation, 519, **546**
Nematic mesophase, **417**
Neoprene, 336
Nernst filament, 694
Net magnetization vector, **M**, **601**, 604
Neutral species, **509**
Newman projection, **89**, 96, **227**, 228, 269, 279, 280
Newton, 663
Newton's law, 663
Nickel tetracarbonyl, 345, 472
infrared spectrum of, 660
Nido clusters, **368**
Nitramines, **186**
Nitranion, **525**
Nitrate
anion, 364
esters, **185**
Nitrene, 507, **514**, 515, 533, 550
α-keto-, **515**
singlet, 514
triplet, 514
Nitrenium ions, 520, **521**, 533
Nitrile, 164, **183**
oxide, 191, 583
Nitrite esters, **186**
ethyl, 186
Nitro group, 184
Nitroalkanes, **184**
Nitrobenzene, 744
electrophilic aromatic substitution of, 563
Nitrogen
anion, **525**
containing radical cations, mass of, 717
fixation, 365
generators, 336
rule, **717**, 722
ylides, 530
Nitrones, **182**
cis-trans (*E-Z*) isomers of, 182
Nitrosamides, **187**
Nitrosamine, **187**
Nitroso group, 185, 336
Nitrosoalkanes, **185**
Nitrous oxide cation, 715
NMR, nuclear magnetic resonance, 593
^{13}C spectrum, 608, 610
^{1}H spectrum, 608, 610
chemical shift
reagent, **637**
δ, **609**, 613, 615, 616, 618, 620, 628
contour plot, 646, 647, 648, 649
coupling relationships between nonequivalent adjacent protons, 622
decoupling field, 638, 639
deshielded region, 613
deshielding, 614–618
downfield shift, **618**
dynamic spectroscopy, **633**
fast exchange limit, **634**, 636
frequency domain spectrum, **606**
high field, 613
irradiating field, B_2, 638, 639
line width, **611**
localized spectroscopy, 653
low field, 613
$n + 1$ rule, **621**
observing field, B_1, 638, 639
of keto-enol tautomers, 633
operating frequency, 609
pulse
delay, 644
sequence, **641**
width, **610**
reference compound or signal, **608**, 609
reversible process, 633
shielded region, 613
shielding, 614–618
by aromatic rings, 617
by triple bond, 616, 617
constant, σ, 613
signal
averaging, 610
generation, **604**
integration, **607**, 621
multiplicity, **620**, 621
double doublet, 622–626
doublet, 621, 622
multiplets, 623
pseudo-triplet, 626
quartet, 621, 622
singlet, 621
triplet, 621, 622, 626
position δν in Hz, **609**
resolution, **611**
slow exchange limit, **634**, 636

solvent signals, 623
solvent-induced shift, **638**
spectral width, **610**
spectrometer, **602**, 603
spin coupling, **620**
splitting, **620**
stacked plot, 646, 647
sweep width, **610**
time domain spectrum, 604, 605
two-dimensional, 2D, **646**
upfield shift, **618**
Nobel Prize, 185
Noble gas, 22
Nodal
plane or surface, 5, **14**, 57, 71
point, 4, **5**
Node, 4
Nomenclature, American ABA, 348
Nomenclature, IUPAC system, **116**, 127, 147, 150, 151, 161, 162, 167, 175–178, 183, 200, 202, 318, 319, 329, 350, 517
Nomenclature
alkyl groups, 117
bridging groups, 366
carbinol, 148
Hantzsch-Widman heterocyclic, **193**
metal complex ligands, 365
metal complexes, 351, 353, **365**
oxa-aza, **193**, 196, 210
priority of functional group suffixes, table of, 150
radicofunctional, **147**, 151, 163, 183, 200, 511
ligands, 365
replacement, prefixes and suffixes, **193**, 196, 210
substitutive, **149**, 197, 200
von Baeyer system, 122
Nominal mass, **708**, 715, 721
Nonbonding molecular orbital, 43, 50, **74**
Nonbonding orbitals, antisymmetric, 829
Nonclassical cation, **517**, 518
Noncrossing rule, 579
Nondegenerate point groups, 94
Nonequilibrium conditions, 490, 491
Nonequivalent hybrid atomic orbitals, **23**
Nonmetals, 19
Nonpolar covalent bond, **32**
Nonradiative decay, 787
Nonradiative or radiationless transition, 760, **780**
Nonrigid rotor, **669**
Nonsuperimposible mirror image, 96, 97
Nonvertical transition, **814**
Norbornyl cation, 477, 517, 518
orbital overlap, 518
Normal
coordinate, *q*, **690**
mode laser, **799**
vibrations, **678**
Normalization, **52**, 62, 69
Normalized orbitals, **51**
Nuclear
charge, nonspherical distribution of, 594
magnetic moment, μ, **594**, 595, 599–601
magnetic resonance, NMR, 593
Overhauser effect, NOE, **641**
difference spectrum ,642
enhancement and exchange spectroscopy, NOESY, **650**, 651
spectrum of dimethylformamide, 642
spectrum, ^{1}H-^{13}C, 642
quadrupole, **594**
spin quantum number, *I*, **594**–597, 620, 621
spin state, **594**, 595, 598, 642
energy of, 595, **597**, 598
multiplicity of, **595**
population of, **597**
Zeeman effect, **595**
Nucleofuge, **540**
Nucleophile, 184, 472, **540**
Nucleophilic
addition to carbon-carbon multiple bonds, 269, **556**
conjugate addition, **557**
displacement, 543
substitution
aromtic, S_NAr, **563**
internal, S_Ni, **545**
π-bond rearrangement, S_Ni', **546**
bimolecular, S_N2, 285, **542**
addition-elimination, **543**
π-bond rearrangement, S_N2', **543**
Nucleophilicity, **540**
Nucleotides, separation of, 400
Number of degrees of freedom, **677**
Number of fundamental vibrations, **677**
Nyquist Theorem, 611

Obital angular dependence, 15
Observed rotation, 248, 249
Octahedra configuration, **285**
Octane number, 431–**434**
of ethanol, 445
blending, 434, 445
motor, 434, 435
research, 434, 435, 445
Octant rule, **252**, 253, **739**
Octet rule, **29**
Odd electron cation, **716**

Oder, 200
Off-resonance decoupling, **646**
Ogston's principle, **261**
Oil
  bunker, 423
  crude, 422, **423**–426, 431, 436
  fuel, 423
  gas, 423
  heating, 423
  lubricating, **436**
  shale, 421, **447**
Olefiant, 120
Olefin, **120**
  activation, 353
  formation of more stable, 551
  isomers, *cis-trans*, 815
  metathesis, 311, **375**, 376, 512
Oligo-, 153
Oligoethers, **153**, 156–158
Oligomer, **295**
Onium ions, 516, **522**, **523**, 532
Open crowns, 153
Operation, inverse, **93**
Operator, Laplacian, **9**
Optical activity, **247**, 248, 317, 737, 743
  induced, 824, 825
Optical
  antipodes, 248
  beams, self-focusing of, **745**
  density, **749**
  isomers, **248**
  Kerr effect, **743**, 745, 747
  modulator, 744
  purity, **254**
  rotation, **248**, 249, 252
    rules for correlation of configuration, **252**
  stability, **255**
  yield, **254**
Optical rotatory dispersion, ORD, 249, 252, 253, **737**
  magnetic, MORD, 824, **825**
Orbital
  1*s*, **12**
  2*p*, 12, **15**
  2*s*, **13**
  atomic, **7**, 42, **45**
  contour diagram, 15
  d, 11, 12, **16**, 202
  f, 11, **16**
  energy, **11**, 12, 17, 18, 43, **47**, 51, 56, 57, 62, 63, 70–74, 79, 109, 580
  frontier, 588
  hybridized, 22, 23
  molecular, **42**, 56, 67
  orthonormal, **52**
  p, 11, **14**
  s, 11, **12**
  symmetry, 109
  valence, **21**
Orbital symmetry, 576
  conservation of, 578, 580
  correlation diagram, **576**–579, 581
Orbitals
  basis set of, **44**, 108
  degenerate, **12**, 15, 17, 48, 70, 106, 370, 778
  high-energy filled, 472
  low-energy empty, 472
  normalized, **51**
  orthogonal, **52**
Order of classes, symmetry, **92**
Organoboranes, **196**
Organoborates, **196**
Organomercury chemistry, 549
Organometal compounds as catalysts, **380**
Organometallic carbene complexes, 311, 321, 322, 347, 360, 369, **373**, 375, 512
Organometallic compounds, 344–**347**
  main group, 347, **348**
  transition metals, 347, **349**
    σ, 347
    delocalized π, 347
    π, 347
Organophosphorus
  acid amides, **219**
  acids, **214**
  thioacids, **219**
O-rings, 316,
Orlon, 300, 333
Ortho
  director, **562**
  esters, **169**
ortho, meta, and *para*
  substitution, partial rate factor for, 495
  substitution/substituent, **129**, 496
Orthogonal orbitals, **52**
Oscillator
  strength, *f*, **761**
  anharmonic, **675**
Osmosis, **408**, 409
  reverse, **409**
Osmotic pressure, 323–**325**, **408**
Out-of-plane bending, **682**
  of planar AX_4 molecules, 682, 683
Overlap integral, 52, **66**, 67
Overtone, first, **6**, 660, 661, 689–691
Overtones, 676, **689**
Oxene, 533
Oxenium ions, **521**, 533

Oxetanes, 193
Oxidation **573**, 574
  number or state, **31**, 32, 345, **354**–357, 365, 374, **573**, 574
  CO and H in $HCo(CO)_4$, 354
  of coordination complexes, 354
  of iron in Ferric hexacyanoferrate, $Fe_4[Fe(CN)_6]_3$, 354
  of amines, 524, 527
  states of metals, 829, 830
Oxidative addition, **357**, 362, 382, 384
Oxime, 167, 185, 235
Oxiranes, 193
Oxo reaction, **382**
Oxonium ions, 516, **522**, 533
Oxyanion, **525**
Oxygen
  anion, **525**
  molecular, 364, 566, 567, 614, 734, 774
  singlet, 757, 773, 810
  triplet, 773, 810
Ozonides, **200**
π bond
  order, ρ_{rs} **64**
  bonding molecular orbital, **45**, 56, 57
  complex, **559**
    d-π* bonding, 352
    σ-bonding, 352
  orbital, 11, **14**
  antibonding molecular orbital, **46**, 56, 57
π,π* State, 48, **767**, 768

Pairing energy, 371, **372**
Paper chromatography, **399**, 400
Para director, **562**
Paraffins, **114**
Paramagnetic
  deshielding, 612, 613, **614**
  molecules, 373, 599, 614, 825, 826
Parent ion, **718**
Parity selection rule, **765**, 775
Partial rate factor, *f*, **495**
Partially ionic bond, **34**
Participation of lone pairs, 546
  π-bonds 546
  σ-bonds 546
Particle, β-, 406
Partition coefficient, **398**, 399
Pascal's Triangle, 624
Pauli exclusion principle, **17**, 43
π-bond formation, 551
Peak width at half height, 670
Peat, 436
Peptides
  synthesis of, 180
  separation of, 400
Peracids, **198**
Percent conversion, **433**
Perester, 198
Pericyclic reaction, **575**–578, 580, 581, 586, 588, 589
Periodic table, American ABA scheme, **19**, 20, 349
Peroxides, 565, 645
  diacyl, **198**
  dialkyl, **198**
  diaroyl, **198**
Peroxy ester, 198
Petrochemicals, 422, **426**
Petrographic composition, **438**
Petroleum, 421, **422**, 423
Phase
  angle, 244, 729
    precession, 759
  diagram, 397
  of waves, **729**
  velocity, **246**, 737, 746
Phase-coherent spins, **599**
Phase-transfer catalyst, **415**
Phenes, **131**
Phenethyl cation, β-, 519
Phenols, **149**, 150, 164
Phenonium ion, 518, 519
Phenyl
  carbene, 513
  cation, 722
  diazonium ion, 523
  group, 128, 149
  nitrene, 514
Phenylenediamine radical cation, *N,N,N',N'*-tetramethyl-*o*-, 527
Phosphanes, 212
Phosphates, **214**
Phosphenium ion, **521**
Phosphine, **212**
  imides, **214**
  oxides, **213**
Phosphinic acid, **214**, 215
  alkyl or aryl, **215**
  dialkyl or diaryl, **216**
Phosphinous acid, **216**
  dialkyl or diaryl, **216**
Phosphites, **217**
Phosphonic acid, **216**, 217
  alkyl or aryl, **217**
Phosphonium
  ions, **523**
  salts, **219**

Phosphonous acid, 214, **215**
alkyl or aryl, **215**
Phosphoranes, **213**, 533
Phosphorescence, 760, 780, 786, 787, **790**
spectrum, 790–792
enhancement of, 783
lifetime of, 789
quantum yield of, 786, 789
Phosphorous
acids, 216, **217**
pentachloride, geometry of, 42
ylides, 530
Photochemical process, primary, **786**, 787
Photochemistry, 729
of chlorine, 565
first law of, **756**
second law of, **787**
Photochromism, **805**
Photodetector, 791
Photodissociation, 830
Photoelectric effect, **767**
Photoelectron, 828
kinetic energy, 828
spectroscopy, PES, **828**, 829
transfer, PET, **817**–820
Photoionization, 829
Photoisomerization, cis-trans, 815
Photolysis, 802
laser flash, **802**, 803
Photomultiplier tube, 767
Photon, **3**, **734**, 735
flux density, 735
Photooxidation
of iron complex, 771
of metals, 771
Photophysical processes, 759
Photoreduction of cobalt complex, 771
Photosensitization, 783, **810**, 812
Photosensitizer, 807, 810, 812
electron transfer, 818, 819
Photosynthesis, 422, 457, 458, 802, 806, 817
Phthalic acid, 175, 176
esters of, 313
Pinacols, 527
Pitzer strain, **272**
Planck-Einstein equation, **3**
Planck's constant, *h*, **3**
Plane of symmetry, σ, **86**, 91, 232, 239, 240, 257, 258, 264, 578
Plane-polarized light, **240**–242, 246-249, **730**, 732, 733, 737, 741, 743, 824, 825
absorption of, 764
components of, 730
emission of, 764
Plastic flow, 34
Plasticizer, **313**
Plastics, 293, **313**
recycling of, **341**
Plate,
1/4-wave, 246, 732, 742
retardation, 742
Pöckels
cell, **744**, 800
effect, **743**, 744
Podands, **153**
Point group, **91**1
nondegenerate and degenerate, **94**
Point symmetry operations, 92
Polar
coordinates, 8
covalent bond, **34**
Polarimeter, **249**
Polarizability
of electron pair, 540
of metals, 370
Polarizability, χ or α, 102, 242, **410**, 472, 660, **697**, 746
tensor, 699, 700
Polarization field, **746**
second-order nonlinear coeficient, α_2, 746
Polarization
forms of light, 742, 743
of absorption and emission bands, **764**
of electronic transitions, **762**
of light, 245, 246, 744
Polarized
bond, **410**
spin-state population, **598**
Polarizer, 242, 249
Polmers, oriented, 312
Poly-, 153
(phosphazene), **340**
amide, **326**, 328, 333
boron hydrides, 197
carbonates, **326**
carboxylic acids, **175**
chlorobiphenyls (PCBs), 146
Polydispersity index, **324**, 325
Polyenes, conjugated, 55, 59, **125**, 776
Polyesters, **326**, 333
Polyethers, 351
Polyethylene,
high-density, HDPE 300, **312**, 320, 341
low-density, LDPE, 300, **312**
Polyfunctional compounds, 143
symmetrical, **152**
Polyhydric alcohols, **149**
Polyimide, **326**

Polymer
  amorphous, 315
  architecture, **310**
  atactic, **319**
  block, 305, **306**
  branched, **301**
  elastomeric, 315
  graft, **307**
  isotactic, **316**
  linear, **299**
  molecular weight, determination of, **324**
  single strand, **299**
  stereoregular, **318**
  stretched, 242
  syndiotactic, **318**
  tactic, **316**–319
Polymeric, 293
  polyols, 337
Polymerization, 293, **297**
  catalysts, 319–321, 363
  initiation with
    anions, 298
    cations, 298
    radicals, 298, 299
    transition metal complexes, 298
  of electron-deficient vinyl compounds, 304
  of ethylene, 363
  of heterocycles, ring-opening chain-growth, 310
  of propylene, 363
Polymerization,
  addition, **298**
  anionic, 298, **303**, 304, 324
  average degree of, 296
  cationic, **303**
  chain-growth, **298**
  condensation, **325**
  coordination, **319**
  dianionic, **304**
  emulsion, **311**
  radical, **299**
  regiospecific, 299
  ring-opening metathesis, ROMP, **311**, 324, 375
  step-addition, 326
  step-growth, **325**
  transition metal-catalyzed, 311
Polymers, 143, 293, **295**
  biodegradable, **341**
  branched polyethylene, 312
  conducting, 321
  ladder, **308**
  linear polyethylene, 312
  living, **304**
  molecular weight, **322**
  recycle of, **341**
  thermoplastic, **313**
  thermosetting, **313**
Polyols, **149**
Polypeptide, 563
Polypropylene,
  atactic, 319
  isotactic, 317, 318
  syndiotactic, 318
Polystyrene, 295, 296
  resin, 403
Polyurethanes, 193, **326**-328, 336, 337
  foamed, **328**
Pople spin system notation, **630**
Population
  inversion, 785, 796, 800
  of polarized nuclear spin states, 641, 654
  of states, 470
Porphyrins, 239
  transition metal, 364
Positional isomers, **229**, 260, 267
Potential energy, 62, 66, **455**, 456, 478, 479, 665
  diagram, profile or surface, 479–482, 490, 491, 507
  function, **663**
Potential,
  oxidation, 820
  reduction, 820
Pott-Broche process, **440**
Power gas, **443**
Precession, **595**
  in phase, 599
Precursor ion, **718**, 719
  even-electron, 718
  odd-electron, 717, 718
Predissociation, **767**
Pre-exponential factor, A, 488
Prelog strain, **272**
Principal
  axes, **667**, 677
  quantum number, *n*, **11**, 18, 19, 349, 773
Principle of
  conservation of orbital symmetry, **578**
  detailed balancing, 503
Prism, calcite, 242
Probability distribution within energy well, 664
Probe region for NMR, 603
Processes, radiative, 760
Prochiral, 261
  atoms or groups, **257**, 283
  faces, **258**
Producer gas, **443**
Product ion, **718**, 719
  even-electron, 718
  odd-electron, 718

Propagation steps of radical chain reactions, 299, 305, 425, 565
Proper axis, 85
Propylene, ammoxidation of, 429
Protecting group, 180
Protein amino acid sequence, 724
Proteins, 179, 337, 340
  separation of, 400
Proteomics, **724**
Proton
  affinity, **476**
    table of, 476
  tautomers, 286, 287, 526
  transfer, 471
Prototropic tautomers, 286
Prototropy, 165
Proximate analysis of coal, **436**, 440
Pseudo-
  aromatic compound, **132**
  asymmetric center, **264**
  axial, **280**
  equatorial, **280**
  noble gas, 22
  rate law, **487**
  rotation, 285, **377**
  state, 775
Puckered conformation, 278
Pulse sequence, **641**
Pulsed laser, **798**–802
  mode-locked, 800–802
Pyramidal inversion, 274
Pyrene, P-type delayed fluorescence, **793**
Pyridinium salts as solvent polarity indicators, 772
Pyridyne, 530, 531

Q branch for linear polyatomic molecules, 694
Qiana, 333
Q-switch, 799
Q-switched laser, **799**, 800
Quadrupolar
  field, 712
    three dimensional, 713
  nuclei, 599
Quantum, **3**, **734**
  mechanics, 2, **4**
Quantum number,
  angular momentum, *J*, **669**
  azimuthal (angular momentum), *l*, **11**
  electron spin, *s*, **12**, 512, 756
  magnetic spin, 595
  magnetic, m_l, **12**
  nuclear spin orientation, m_i, **595**
  principal, *n*, **11**, 18, 19, 349, 773
Quantum yield, 735, **785**, 787, 789, 813
  absolute, **786**
  relative, **786**, 787
Quarter-wave plate, 246, 732, 742
Quaternary ammonium salt, 182
Quencher, 807, 812
Quenching
  of excited states, 786, 787, **812**, 813
  rate constant, k_q, 813
Quinhydrone, 154, 155

R group, definition of, 143
Racemate, **252**
Racemic
  mixture, 248, **251**
  modification, **251**, 252
Racemization, 544, 547
Racemize, **252**
Radiation, electromagnetic, **240**, 729
Radiationless energy transfer, **806**
Radiative
  decay, 790
  energy transfer, **806**
  transition, **780**
Radical, **511**
  addition, **567**
  allyl, 74
  ambident, **542**
  chain reaction, 299, 305, 424, 425, 564–566
    initiating steps, 299, 425, 565
    propagating steps, 299, 305, 425, 565, 566
    terminating steps 299, 425, 565, **566**
  free 299, 305, 509, **510**, 511, 716, 718, 759
  group, R 143, 147
  intermediates, 356
  nature of $[Co(CN)_5]^{-3}$, 356
  pair, 510, **528**, 654
    theory of CIDNP effects, 654
    geminate, 510, **528**
    singlet, 654, 655
    triplet, 654, 655
  polymerization, **299**, 302
  species, 143
  stabilities, 567
Radical anion, 154, 304, 511, 512, **527**, 533, 803, 804, 805
  chain reaction, 567, 568
  phenylmethylene, 511, 512
Radical cation, 154, 463, 511, 512, 524, **527**, 533, **716**, 717, 803–805, 827, 828
  methylmethylene, 511, 512
  containing nitrogen, mass, of 717
Radical ion, 510, 511, **526**
  annihilation, 803–805, 822

free, 821
pair, 348
contact, 821, 822
solvent separated, 821, 822
Radical-cation Diels-Alder reaction, RCDA, 527
Radicals in
NMR, 654
radicofunctional nomenclature, 511
Radicofunctional names, **147**
Radio frequency signal, 604, 712, 713
Radius, atomic, **37**
Raman
active vibrations, 660, 699, 700
band intensity, 699
inactive vibrations, 700
shifts, 660, **694**, 699
spectrometer, 694, 695
spectroscopy, 688
spectrum, 660, 671, **694**, 696
Rank of coal **436**
Raoult's law, **391**, 394, 395
Rate
constant, **483**, 488, 496, 500
equation,
1st order, 484, 488
2nd order, 484
differential, **483**
expression,
general, 483
integrated, **484**
law, **483**
of diffusion- or encounter-controlled reaction, **492**, 814
of reaction, 469, **483**–487, 498, 499, 504
Rate-determining or rate-limiting step, **484**, 546
Rates, relative, **495**
Rayleigh scattering, **694**
Rayon, 293, **338**
Re face, **259**, 262, 270
Reaction
barrier, 493, 503
coordinate, 576, 577
diagram or profile, **479**, 502, 503, 508
mechanisms, 59, **479**, 537, 539
Hammett parameter, ρ, 500
table of, 501
Reaction rate, 469, **483**–487, 498, 499, 504
catalysis, 492
Reaction,
allowed, **580**
cheletropic, 575, **586**
concerted, **478**, 540
diastereoselective, **266**, 267
diffusion- or encounter-controlled, **492**, 814
electrocyclic, **575**, 588, 589
enantioselective, **266**, 267
endothermic, **457**
enthalpy of, ΔH_r, **457**, 459
exothermic, **457**
forbidden, 580, **581**
group-replacement, 542
group-transfer, **586**
hydride-transfer, 503
mechanism of, 59, **479**, 537, 539
methyl-transfer, 503
multistep, **479**
pericyclic, **575**
proton-transfer, 503
redox, **574**
regioselective, **266**
regiospecific, **267**
ring-closure, 576
ring-opening, 576
stereoselective, **265**
stereospecific, **264**
Reactive intermediates, 479–481, 506, **507**, 508, 532, 533
allylic species, 286, 491, 516, 544
anion, 507, 508, 515, 533, 563
aryne, 507, 530–533
atomic carbon, 507
biradical or diradical, 507, 511
carbene, 118, 507, 511, 512, 533, 544
carbocation, 482, 491, 498, 507, 508, 515, 517, 518, 520, 532, 533, 544
enium ion, 520, 521, 533
excited state, 507
metallacyclic, 375
nitrene, 507, 515, 533
radical, 507, 511
radical ion, 348, 507, 524, 527, 528, 533
spectroscopic detection of, 802
steady-state, 485, 486, 487, 498, 524, 527,
strained multiple bond, 507, 530–533
tetrahedral, 532, 543, 571, 572
trigonal bipyramidal, 543
summary table of, **532**, 533
Reactivity, 490, **494**
ratios, 306
relative, 540, 541
Reagent, **541**
Rearrangements, 59, 60, 543, **575**, 584
1,3- or allylic 286, 544, 546, 549, 550
cis allylic, 546
Cope 59, 60
degenerate, 287, **575**
ozonide, 200

Rearrangements (*Continued*)
pinacol, 500, 574
transition metal complexes, 357
Receptor protein, 404,
Rectus (*R*), 237
Recycle of polymers, **341**
Red shift, **751**, 752, 768, 769, 772
Redox reaction, **574**
Reduced mass, m, 666, **667**, 671
Reduction, 261, **573**, 574
of aromatic hydrocarbons, 527
of carbonyl groups, 527
of esters, 527
Reductive elimination, **357**, 377, 384
Reductones, 166
Refection, **86**
Reference bridge, 234
Reference, *r*, **234**
Refineries, 423
Reflux ratio, **394**
Refluxing, **390**
Reformate, 431
Reforming reactions, 430
table of, 432
Reforming, high quality gasoline, **431**
Refractive index, 252, **407**, **735**, 736, 741–744, 745, 808, 809
matching, **747**
dependence on electric field strength, 743
second-order coefficient, 745
Refractive indices
of circularly-polarized light forms, 737
of common organic molecules, table of, 736
Refrigerant, 147
Regiochemistry, of
ring closure, **288**, 289
double bond cyclizations, 288, 289
electrophile addition to double bonds, 558, 559
Regioselective reaction, **266**
Regiospecific reaction, **267**
Rehm-Weller equation, **820**
Reinforcing agents, 337
Relationships between K_a and K_b, table of, 475
Relative rates, **495**
of hydrolysis of alkyl bromides, table of, 495
Relative reactivity, 540, 541
Relative volatility, α, **391**
Relaxation, **598**, 599
magnetic dipole-dipole, 641
Repeating unit, bivalent, 299
Resid, 423, 424
Resin, 313
Resin,
anion exchange, **403**
cation exchange, **402**
epoxy, **331**
ion exchange, 427
Resinite, 438
Resolution, **252**, **711**
Resolution of invertomers, 274
Resonance
condition, **598**
effects, 496
energy, **62**, 70
of benzene, 62
integral, β, **67**, 70
structures, **58–60**, 78, 129 168, 183, 184, 187, 188, 202, 203, 374, 516, 517, 523–526
of carbonate, 58
transfer of energy, 808
Resonant energy transfer, **806**
Restoring force, 662
Retention of configuration, **251**, 358
migration with, 584, 585
Retrograde cycloaddition, **582**
Reversed-phase chromatography, **402**
Reversibility, 466
Reversible
process, **453**, 485, 753
reaction, 453
Reynolds number, **431**
R_f value, **400**
Rigid rotor
model, **666**, 668, 669, 693
simple, 667
Ring closure
regiochemistry, **288**, 289
stereochemistry, **288**, 289
metathesis, 375
Ring
current, **617**
flipping, **279**–282, 633
puckering, **277**
strain, **272**
strained systems, **531**
Ring-opening
chain-growth polymerization of heterocycles, **310**
metathesis polymerization, ROMP, **311**, 324, 375
Rocket fuel, 189
Rope, 137
Rotating frame of reference, **600**
Rotation of linear molecules, 669
Rotational
axis, fold of, **85**
isomerism, 377
spectra of diatomic molecules, **666**

state, 463, 759, 760
strength, R_j, **740**, 761, 762 740
Rotaxanes, 154, **159**
Rotomer, **273**
Rubber, **334**
bands, 337
tree (Hevea brasiliensis), 335
foamed, **336**
natural, **335**
styrene-butadiene rubber, SBR, 336
Rubbers, synthetic, **335**, 336
Rule of 18, **355**, 356, 374
Rydberg transition, **773**
σ
bond, **44**, 47
complex, 73, **559**, 562
donor-π acceptor concept, 352
orbital, 11, **12**
plane, 86
skeleton, **47**
σ* orbital, **45**

Salt effect, **499**
special, 499
Saponification, **570**
Saran, 300
Saturation of NMR signal, **597**, 639
Sawhorse structure, **227**, 228
Saytazeff regiochemistry (more stable olefin), 552
Scalemic mixture, **254**
Scattering,
Brillouin, 748
elastic, 660, 694
inelastic, 660, 694–697
isotropic, 660
Raman, 748
Rayleigh, **694**
Schiff base, 167, 180
Schlenk equilibrium, 348
Schmidt degradation, 515
Schroedinger equation, **10**, 62, 66, 67
angular, 11
hydrogen atom, **11**
spatial, 11
Schultz-Flory α, 446
Schwartz reagent, 359
s-cis, **276**, 582, 583, 776, 777
Scission, β, 424
S_E1, **547**
S_E2, **548**, 549
S_E2', **548**, 549
S_EAr, 496, **561**, 562
Second harmonic, **6**, 747
Secular determinant, 67, **68**, 69
S_Ei, **549**
S_Ei', **550**
Selection rules, **764**
multiplicity, **764**
spin, **764**
for energy transfer, 808
for harmonic oscillators, **674**
for infrared absorption, **698**
for Raman spectroscopy, **699**
for rotational spectra of rigid linear molecules, **669**
for ultraviolet and visible absorption, **765**
orbital symmetry, 584, 586
symmetry, 584, 586, **698**, 699, **765**
Selectivity, **33**
Selector, 404
Semicarazide, 167
Semicarbazone, 167
Semiconductor, 321
Semifusinite, 438
Semimetals, 344
Semipermeable membrane, **408**
S_H2, **566**
Shale oil, **447**
Sheep, 337
Shell designation, **18**
Shift, sigmatropic, 575, **584**, 585, **588**
Shimming the external magnetic field, **604**
Shoolery's rules, **620**
Shutter,
electro-optic, 801
fast, 744
Si face, **259**, 262, 270
Sickle conformation, **277**
Sigmatropic shift, 575, **584**, 585, **588**
[1,2]-, 584, 585
[1,3]-, 584–587
[1,5]-, 584, 585
[1,n]-, **584**
[3,3]-, 584, 585
[m,n]-, **584**, 585
Signal-to-noise ratio, S/N, **610**
Silanols, 331
Silica gel, 398
Silicenium ion, **521**, 524
Silicon rubber, 315
Silicones, **331**, 336
Siliconium ion, **524**
Silk, 332, **337**
Silyl enol ether, 166
Single bond, **35**
essential, **60**
Single electron transfer, SET, 348, 817–819
Single-site catalysis, SSC, 321

Singlet
  carbene, **513**
  manifold, **759**
  oxygen, 199
  spin state, 512
  state, 511, 521, **757**, 759
Singlet-triplet splitting, 791, 793
Sinister (*S*), 237
Skeletal
  isomers, **229**
  structure, 226
Skew
  boat, **281**
  conformation, **276**
Skin pigmentation, 735
Smectic mesophase, **416**
Sn axis, **88**, 89, 232, 257, 258, 261, 607
S_N1, 360, 498, **544**, 545
  order of reactivity of alkyl groups in 544
S_N1', **544**
S_N2, 187, **542**, 573
  order of reactivity of alkyl groups in, 542
S_N2', **543**
S_NAr, **563**
S_Ni, **545**
S_Ni', **546**
Soap, 413, 414
  invert, **415**
Sol, 340
Sol-gel method, 340
Solid waste, 341
Solvation of ions, 410,
Solvent
  aprotic, **477**
  ionizing power, 502
  polarity, 772, 773
    table of Kosower Z and ET(30) scales, 773
  separated ion pair, **499**, **528**
Solvent-refined coal, SRC, **440**
Solvent-separated radical ion pair, 821,
Solvolysis, 498, 502, **545**
SP^3 hybridized carbon, 542
Space
  time yield, **433**
  velocity, **432**
Spandex, 33
Spatial coherence of light, **797**
Species, **508**
  attacked, 541
  attacking, 541
  electron-deficient, **510**, 530
  symmetry, **102**
Specific
  acid catalysis, **494**
  base catalysis, **494**
  heat capacity, *c*, **458**
  rotation, [α], **248**, 254, 737, 739
Spectral overlap integral, 809
Spectrometry, 704
Spectroscopic
  dissociation energy, *D*0, 675, **676**
  transition, 463, **814**, 815, 828
Spectroscopy, 704
Speed of light, apparent, 735
Spherands, **157**
Spherical
  coordinates, 8
  symmetry, 13, **98**
  top molecules, **692**
  top, methanol, 692
Spiders, 337
Spin
  angular momentum, total, 757, 758
  configuration,
    doublet, 512
    singlet, 512
    triplet, 512
  decoupling, **638**
  flip of nuclear spin, **598**
  forbidden, 773
  multiplicity, 595
  statistics, 810
Spin state,
  antiparallel, 624
  doublet, 756
  parallel, 624
Spin-allowed processes, 810
Spin-lattice
  relaxation, **599**, 611, 653
    time, T_1, **599**, 653
Spinneret, 332, 339
Spinning, **332**
  dry, 332
  wet, 332
  melt, 332
Spin-orbit coupling, 373, 780, **781**, 783
Spin-spin
  coupling, 377
Spin-spin relaxation, **599**
  time, effective, T_2*, 599, 604, 611
  time, T_2, **599**
Spirans, **125**
Spiro atom, 125
Sponge, 336
Spontaneous and stimulated emission, probabilities of, 784

Spontaneous
  emission, **784**
  ignition temperature, **396**
  process, 466
$S_{RN}1$, **567**
Stability, 489, 490
  constant, **472**
  kinetic, **490**
  thermodynamic, **489**
Staggered conformation, 87, 89, **275**, 279, 293, 294
Standard
  entropy, S^o, **465**
  free energy of formation, $\Delta G^o{}_f$, **468**, 490, 504
  mass element, ^{12}C, 704
  state, **455**–457, 468
Standing wave, **4**
  three-dimensional, 799
Star procedure, alternating, 76
State, **454**
  correlation diagram, **578**, 581
  function, **454**, 458, 465–468
  variable, **454**
  B, 778
  d,d, **770**, 771
  L, 778
  n,π*, **768**, 782
  n,σ*, **769**, 770
  π,π*, 48, **767**, 768, 782
  σ,σ*, **770**
  singlet, 511, 521, **757**, 759
  standard, **455**–457, 468
  transition, 274, 360, 361, **478**–486, 502, 503, 506, 507, 518
  triplet, 511, 513, 521, **758**, 759, 760
Stationary phase, 398–401
  chiral, 404
  nonpolar, 402
Stationary wave, **4**
Steady-state approximation, 485, 487
Steady-state intermediate, 485, 487, 498
Steam
  cracking, **424**–427
  distillation, **395**
  reforming, **445**, 447
    to produce synthesis gas, **445**
Steel, 439
Step growth polymerization, 297
Stereochemical isomer, **230**
Stereochemically nonrigid molecules, **377**
Stereochemistry, **225**, 226, 479
  of ring closure, **288**, 289
  cis-attack/departure, 543
Stereoelectronic effect, **288**
Stereogenic, **230**
  center, **230**, 232, 238, 240, 268
Stereoisomers, 228, **230**, 265, 283
  giving rise to, 230
Stereoregular polymer, 318, 320
Stereoselective reaction, **265**
Stereoselectivity, 267, 268
Stereospecific reaction, **264**
Steric
  bulk of phosphines, 362
  effects, 496
  strain, 271, **272**
Stern-Volmer relationship, **813**, 814
Steroids, 123, 235, 236
Stiffness, modulus of, 332
Stille reaction, **385**
Stimulated emission, 639, **784**, 796
Stoichiometric coefficients, 469, 483
Stokes band or line shift, **695**, **794**
Strain, **270**
  bond angle, **271**
  bond-stretching, **271**
  FBI, 271
  ring, **272**
  steric, 271, **272**
  torsional, 271, **272**, 275, 276
Strained ring systems, **531**
s-trans, **277**, 776
Strength, bond, **461**
Stretching frequencies
  in transition metal complexes, 360
  of ketone carbonyl groups, 671
Stretching frequency, 671
  of carbon monoxide, 671, 690
  of hydrogen halides, table of, 671
Stretching vibrations, 95, 659, **678**–680
  antisymmetric, 679, **680**
Structural isomers, **229**
Styrene, 150, 295, 336
  radical anion, 304
  radical cation, 527
Subbituminous, 436
Sublimation, **396**
Subspectrum analysis, **631**
Substituent effects of, **496**
  electron-donating groups, 496, 501
  electron-withdrawing groups, 496, 501
Substituents, 149, 151
Substitute natural gas, SNG, **444**
Substitution, **542**
Substitution electrophilic
  aromatic, S_EAr, 496, **561**, 562
  bimolecular
    S_E2, **548**, 549

Substitution electrophilic (*Continued*)
with π bond rearrangement, S_E2', **548**, 549
internal
S_Ei, **549**
with π bond rearrangement, S_Ei', **550**
unimolecular, S_E1, **547**
Substitution homolytic bimolecular, S_H2, **566**
Substitution nucleophilic
aromatic, S_NAr, **563**
bimolecular
via addition-elimination, **543**
S_N2, 360, **542**, 573
with π bond rearrangement, S_N2', **543**
unimolecular
S_N1, 360, 498, **544**
with π-bond rearrangement, S_N1', **544**
Substitution radical nucleophilic
unimolecular, $S_{RN}1$, **567**
Substitution reaction, 540, 542
associative ligands, 353
Substitution,
aromatic, 73, 496, 497
ligand, **360**
Substitutive nomenclature, **149**
Substrate, 261, 487, **541**
Sulfates, **204**
Sulfenamides, **210**
Sulfenates, **206**
Sulfenic acids, **205**
Sulfenium ion, **521**
Sulfenyl halide, 205
Sulfides, **201**
Sulfinamides, **210**
Sulfinates, **205**
Sulfinic acids, **205**
Sulfinyl halide, 205
Sulfonamides, **209**
Sulfonates, **204**
Sulfones, **202**
Sulfonic acids, **203, 403**
salts of, 414, 415
Sulfonium ions, **523**
Sulfonyl halide, 203
Sulfoxides, **202**, 209
chirality of, 202, 209
Sulfur, 336
heterocyclic compounds, **210**
hexafluoride geometry of, 42
Sulfuranes, **212**
Sultones, **203**
Summary table of reactive intermediates, **532**, 533
Super critical, 397
Superacid, **473**, 518
Superimposible, 84
Suppressor column, 404
Suprafacial, *s*, **586**, 587
Supramolecules, 143, **154**, 635, 773
Surface
active agent, **414**
area, **412**
of solids, 412
tension, **411**, 412
Surfactant, 413, **414**
anionic, 415
cationic, **415**
nonionic, **415**
Surroundings, 453, 466
Suzuki reaction, **385**
Symmetric
behavior, **94**
stretching vibrations, **679**
of water, 699
top molecules, **692**
benzene, 692
cyclobutane, 692
transformation, 577, 578
vibrations, **679**
of methyl and methylene groups, 683–685
Symmetrical polyfunctional compounds, **152**
Symmetry and dipole moments, **107**
Symmetry
classification of atoms, groups, and faces, **261**
coordinate, 679
element, **84**
classes, order of, **92**
classes of, **92**, **93**
equivalent atoms or groups, **100**, **255**–259
forbidden transition, 774
number of, **99**
ammonia, 99
ethylen, 99
water, 99
operation, **84**, 91
orbital, **107**
selection rules, 584, 586, **698**, **765**
species, **102**
Symmetry,
alternating or improper axis of, S_n, **88**, 89, 232, 257, 258, 261, 607
axis of, C_n, **85**, 255, 256, 261, 279, 577, 607
center of, *i*, **87**, 91, 232, 258
conical, **99**
cylindrical, 45, **98**
local, **98**
plane of, σ, **86**, 91, 232, 239, 240, 257, 258
rotation, 85

rotation-reflection, **88**, 89, 232, 257, 258, 261, 607
spherical, 13, **98**
Synclinal, **276**
Syndiotactic, 317
polymer, **318**
Syn-periplanar conformation, **276**
Synthesis gas, 382, 443, **444**, **445**, 447
production of, 443
Synthesis, asymmetric, 267
Synthetic fibers, **332**
table of, 333
Synthetic rubbers, table of, 336
System, **453**, 454
closed, 453, 456
isolated, 453
macroscopic, 453
microscopic, 453

Tactic polymer, **316**–319
Taft equation, 500, **501**
Tanning, 338
Tar
acids, 421
bases, 421
sand, 421, **448**
Athabasca, 448
Tautomerism,
imine-enamine, 180
keto-enol, 526, 633
Tautomers, **286**, 287
proton, 165, 184, 185, 286, 287
valence, **287**
T_d point group, 94
Tebbe's reagent, 369
Tedlar, 300
Teflon, 298, 300
Telechelic polymers, 307
Telomer, 301
Temperature,
absolute, K, 455
isokinetic, **503**
melt transition, T, 315
Template effect, 160
Temporal coherence of light, **798**
Tensile strength, 332
Tensor, 698
Teratogen, 225
Teratogenic activity, 172
Terminating groups, 295
Termination steps of radical chain reactions, 299, 425, **566**
Tetrahedral intermediates of carbonyl groups, **532**, 571, 572
Tetrakis, 152
Texaco gasifier, **443**
Theoretical plate, 392, **393**
height equivalent to, HETP, **393**
Thermal
conductivity detector, **405**
cracking, **424**
Thermistor, 405
Thermochemistry, **453**
Thermodynamic
control of a reaction, **490**, 491
cycle, 795
function, 454
product, 490, 491
stability, 70, 355, **489**
Thermodynamics, 452, **453**
first law of, **466**
second law of, **466**
third law of, **466**
Thermoplastic polymers, **313**
Thermoplastics, 315, 326
Thermoset, 314
Thermosetting polymers, **313**, 315
Thermotropic liquid crystals ,**416**
Thin film deposition, 380
Thin-layer chromatography, TLC, **399**, 400
Thioacetals, **211**
Thioalcohols, **200**
Thioaldehydes, **208**, 211
Thiocarboxylic acids, **206**
Thiocyanate, **209**
anion, 364
Thiohemiacetals, **211**
Thioic
O-acids, **207**
O-esters, **207**
S-acids, **206**
S-esters, **206**
Thioketals, 212
Thioketones, **208**
Thiokol, 336
Thiols, **200**, 201, 211
oxidation of, 201
Thiophenols, **201**
Thioyl halide, **210**
Third harmonic generation, 747
Thread, 334
Threo isomers, 232, 240, 560
Thymine deimer, 819
Tight
ion pair, **499**, **528**
radical pair, **528**
Time domain, 605
Time-of-flight mass spectrometer, TOF, **712**, 715

Tin alkyl complexes, 385
Tip angle, α, 602, 604
Tires, 337
TMS reference signal, 606
Toluene, 150
  bromination of, 495, 497
Torsional
  angle, **272**, 275–277, 627
  strain, 271, **272**, 275, 276
Total reflux, 393
Town gas, **442**
Trace of transformation matrix, **105**
Trans
  conformation, **277**
  effect, **361**
  kinetic, 362
  influence, 361
  trans-addition, 265, **560**, 561
Transfer of
  β-hydride, intramolecular, **358**
  electronic energy, **806**
Transformation
  matrices, **104**
  matrix, trace or character, **105**
Transient
  dipole-dipole interaction, 410
  intermediates, spectroscopic detection of, 802
Transistion metal
  catalysts, 363
  complexes, 285, 355, 356–359, 510, 771
Transition metals, 19, 22, 346, 349, 384, 771
  inner, 19, 349
  outer, 19, 349
  triads of group VIIIB, 21
Transition moment interaction,
  orientation factor for, 808, 809
Transition state, 274, 360, 361, **478**–486, 502, 503, 506, 507, 518
  between cyclohexane conformations, 280
  cyclic, 549, 550, 586
  four-center, 359
  three-center, two-electron, 426
Transition,
  0-0, **756**, 760
  allowed, 756, **759**, 764, 765, 768, 770
  charge-transfer, **772**
  coupled, 806–808
  d,*d*, 765, **771**
  dimole, **773**
  dipolar, 765
  double-photon, **774**
  forbidden, 756, 759, **760**, 764, 765, 768, 771
  n,π*, **768**
  n,σ*, **770**
  nonradiative or radiationless, 760, **780**
  nonvertical, **814**
  π,π*, **768**, 777
  radiative, **780**
  rotational, 669
  Rydberg, **773**
  σ,σ*, **770**
  singlet-singlet, 764
  singlet-triplet, 751, 764, 783
  spectroscopic, 463, **814**, 815, 828
  triplet-triplet, 764
  ungerade, 765
  vertical, 463, **814**, 815, 828
  vibronic, **756**, 760, 762, 765
Transition-state theory, **488**
trans-metallation, 384
Transmethylation, **376**
Transmittance, 670, **750**
Transportation fuels, 422
Transverse
  mode, 798, 799
  relaxation, **599**
Triblock copolymers, 307
Tricyclic and polycyclic compounds, **124**
Triethyloxonium ion, 522
Trigonal bipyramid, 42, 101, 284, 285, 377, 543
  apical position, 284
  apical/equatorial position exchange, 285
  equatorial position, 284
Trihapto, 353
Triketone, 1, 2, 3-, 172, 173
Trimethylcarbenium ion, 520
Trimethylenemethane, 65
Trimethyloxonium salt, 509
Trimethylphenylphosphonium ion, 523
Trimethylsulfonium ion, 523
Triple
  bond, **35**
  point, 397
Triplet
  carbene, **513**
  manifold, **759**, 812
  sensitization, 809
  spin state, 512
  state, 511, 513, 521, **758**–760
    lifetime, 808
    sublevels, 758
Triplet-triplet
  annihilation, 793, 810
  interactions, 809
Triply degenerate orbitals, 106
Tris, 152
Truncated icosahedron, 136
Tunneling, **502**

Turnover
number, **434**
rate, **434**
Turnstyle rotation, 378
Twist boat, **281**
Tygon, 313, 336

Ultraviolet region,
UVA, **735**
UVB, **735**
UVC, **735**
Uncertainty principle, Heisenberg, **3**, 502, 538, 774, 775
Uncharged species, **509**
Ungerade, *u*, 46, 48, **95**, 681, 698, 700, 765
Unimolecular, 478
Unsaturated carbonyl, α,β-, **178**, 557
Urethane, 180, 326–328

Valence, **30**
dsp orbitals, 355
isomers, **287**
orbital, **21**
shell, expansion of, 202
tautomers, **287**
vibrations, 678
Valence,
ionic, **30**
primary, 345
secondary, 345
Valence bond (VB) theory, 26–**28**, 55, 58
wave functions, **52**
Valence Shell Electron-Pair Repulsion, VSEPR, **41**
van der Waals
attraction, 39
forces, 410, 411
interactions, 154
radius, **39**, 135, 272
repulsion, 39
van't Hoff factor, *i*, **406**
Vapor
deposition, chemical, **380**
pressure lowering, 323
Vapor-liquid equilibrium, 390, **391–**394
diagram, 391
Vectorial addition of dipole moments, **33**
Vegetable fibers, **338**
Velocity
distribution within energy well, 664
angular, **668**
Vertical
ionization energy or potential, **828**
transition, 463, **814**, 815, 828

Vesicles, 154
Vibrational
bands, 750
energy levels, 673
0-, 780,
spacing of, 675, 676
excited state, 660, **674**, 697, **755**
fine structure, **756**
frequency, 665, 671
ground state, 660, **674**, 676, 697, **755**
ladder, 780
quantum number, *n*, 673, 674
spectra, **671**
of diatomic molecules, **671**
state, 463, **674**, **754**, 759, 760, 766, 767
transition, 660, 669, 674, 676
wavefunction, 762
Vibrational-rotational
spectra, **692**
transition, 693, 694
allowed, 693
selection rules for, 693, 694
Vibrations,
antisymmetric, 679, **680**
bending, **681**
degenerate, **682**, 689, 690
genuine, 659, **677**
nongenuine, 659, **677**
normal, **678**
rocking, 682, 683
scissoring, 682, 683
stretching, 95, 659, **678–**680
symmetric, **679**
symmetric stretching, **679**
twisting, 682, 683
wagging, 682, 683
Vibronic
band, **756**
transition, **756**, 760, 762, 765–767, 794
Vicinal
bonds, 275
coupling, 625, 627
vic, **275**, 279
Vinyl
group, **120**
isocyanate polymers, 309
polymers, 296
head-to-head (tail-to-tail) orientation, **303**
head-to-tail orientation, **302**
table of, 300
Vinylcarbenium ion, 520
Virtual coupling, 628, **632**
Viscometer, **407**
Viscose rayon, 338

Viscosity, **407**, 814
  coefficient of, **407**, 492
  index, **407**
  measurements, **325**
  of polymers, 324, **325**
Vision, 802
Visualization techniques, 399
Vital force, 111, 421
Vitrain, **437**, 438
Vitrinite, 438
von Baeyer system of nomenclature, 122
Vulcanization, **336**
Vulcanizing agents, 337

W conformation, **277**
Wacker
  catalyst, Pd+Cu, 381
  reaction, **381**
Walden inversion, **284**, 285
Water,
  bending modes of, 681
  critical point of, 397
  in-plane bending modes of, 682
  orbitals of, 108
  super critical, 397
Water-gas
  shift converter, 444
  shift reaction, 422, **444**, 447
Wave equation in
  one dimension, **9**
  three dimensions, **9**
Wave function, **8**, 62, 63, 66–69, 94
  probability interpretation of the, **9**
  in valence bond theory, **52**
Wave mechanics, **4**, 10,
Wavelength, **5**, 7
  of light, 240, 737
Wavenumber, **661**
Weight, 663
Werner's coordination theory, 345
Wetting agents, **414**
Wheatstone bridge, 405
Wigner's spin rule, 809, **810**
Windshield-wiper effect, 517, 518
Winstein-Grunwald equation, 500, **502**
Wire array, 242
Wolff rearrangement, 512
Woodward-Fieser factors
  for α,β-unsaturated ketone and aldehyde absorption, 777
  for polyene absorption, **776**
  tables of, 776, 777, 778
Woodward-Hoffmann rules, **586**
Wool, 332, **337**
Work, w, 452, 453, 456, 466
Wurster's salt, 527

Xanthate ester, pyrolysis of, 556
X-rays, 829

Yarn, 332
Yield, **433**
Ylide, **530**, 533
  nitrogen, 533
  phosphorus, 533
-ylidene, 145

Zeeman effect, nuclear, **595**
Zeigler-Natta catalysts, 363, 364
Zero-point energy, 37, 674–**676**, 754, **755**
Ziegler-Natta catalysis, 312, 317-**319**
Zigzag structure, **227**, 228
Zusammen (*Z*), 237
Zwitterion, **529**